Soils: Genesis and Geomorphology

Second Edition

In its first edition, *Soils: Genesis and Geomorphology* established itself as the leading textbook in the fields of pedology and soil geomorphology. Expanded and fully updated, this second edition maintains its highly organized and readable style. This edition is no small upgrade; it is a major revision. Suitable as a textbook and a research-grade reference, the book's introductory chapters in soil morphology, mineralogy, chemistry, physics, and organisms prepare the reader for the more advanced treatment that follows. This textbook devotes considerable space to discussions of soil parent materials and soil mixing (pedoturbation), along with dating and paleoenvironmental reconstruction techniques applicable to soils. Although introductions to U.S., Canadian, and international soil classification systems are included, theory and processes of soil genesis and geomorphology form the backbone of the book.

This thoroughly updated second edition has been revised, rewritten, and restructured to make topics flow more smoothly and read more easily, without sacrificing depth. Every figure has been revised, sharpened, and improved, and the glossary has been updated. The book now boasts an introductory chapter on soil chemistry, and new sections have been developed on soil descriptions, geoarchaeology, digital soil mapping, soil production functions, bombturbation, and the World Reference Base for Soil Resources. Replete with more than 550 high-quality figures and photos and a detailed glossary, this book will be invaluable for anyone studying soils, landforms, and landscape change anywhere on the globe.

RANDALL J. SCHAETZL is a professor of geography and geosciences at Michigan State University. His research has been published in all the leading soils, geomorphology, and geography journals. He is the editor of the Soils section for the *International Encyclopedia of Geography* (Association of American Geographers) and a Fellow of the Geological Society of America. He is an expert in the soils and landforms of the Great Lakes region, and he is editor in chief of *Michigan Geography and Geology* (2012). His expertise on podzolization and pedoturbation has led him to publish numerous papers on these widespread soil processes.

MICHAEL L. THOMPSON is a professor of soil science at Iowa State University. For more than twenty years, he taught pedology and clay mineralogy courses, and for the past ten years, he has taught courses in soil chemistry. Thompson's research has dealt with paleosols, soil organic matter, and the fate of actinides, heavy metals, and organic contaminants in waste-amended soils. His research seeks to identify the chemical and physical conditions that favor stability, transformations, and movement of soil organic matter, anthropogenic contaminants, and clay minerals in soils. He is a Fellow of the Soil Science Society of America, the American Society of Agronomy, and the American Association for the Advancement of Science.

Praise for the Second Edition:

'Schaetzl and Thompson have done it. They took a book that was already unique among soil science books and made it better. This comprehensive textbook is a perfect read for soil scientists ... The book is the perfect teaching tool because the instructor who reads and uses it becomes so much more informed about pedology.... I enjoyed reading the first version so much I will buy a copy of the second edition for myself for professional and pleasure reading.'

– John M. Galbraith, Virginia Tech

'...the definitive textbook for those wishing to understand the complexity and beauty of soil genesis. The revised edition is a fantastic update, including rich additions of subject material and many new and/or improved figures. The well-organized text clearly conveys the importance of interdisciplinary research in understanding soil genesis through time, and it can be used for introductory and advanced courses ... well done authors!'

– Patrick Drohan, The Pennsylvania State University

'...a remarkably well-measured fresh cut across the disciplines of soil science, geology, and ecology. Neglected paths and boundaries between these disciplines finally get the light they deserve. This book is a rare success in the difficult act of balancing descriptive versus process-oriented perspectives on soils ... an approachable writing style and delightful figures and photos ... a deep treatment of the theory and history of studying soils, and it is encyclopedic in its geographical and temporal scopes.'

– Kyungsoo Yoo, University of Minnesota

'...a depiction of soils as a global system that links to – and interacts with – all other natural systems critical to human endeavors. The book is an asset to geomorphologists and pedologists alike.'

– Martha Cary (Missy) Eppes, University of North Carolina

'... this now classic textbook ... has been significantly updated, even including much about the on-going and important discussions about soils as human-natural systems. The book should be on the shelves of all environmental scientists and managers.'

– Daniel Richter, Duke University, author of *Understanding Soil Change*

'This book comprises the most important up-to-date theories and concepts of soil genesis ... very well written; you can feel that the authors are really fascinated by the world of soils, and their enthusiasm is infectious. And, what I think is extremely important ... this book imparts not only comprehensive knowledge on soils but also profound understanding of the geomorphological processes shaping landscapes and their interaction with soil formation.'

– Daniela Sauer, Dresden University of Technology

Praise for the First Edition:

'The writing is clear and concise, and the authors' enthusiasm for their subject material is obvious.... an excellent textbook for upper-level undergraduate and graduate level courses in soil geography, pedology, and geomorphology.'

– *The Canadian Geographer*

'... a big book in size, concept, and ideas.... lavishly illustrated ... Each section is often approached in quite fresh and new ways.... wonderful.'

– *Environmental Conservation*

'... a rare textbook: well-written, comprehensive, up-to-date, thought-provoking, and refreshingly opinionated.... well-suited for any course in pedology or soil geomorphology, whether it is taught in a geography, geology, or soil science program.'

– Joseph A. Mason, Department of Geography, University of Wisconsin-Madison

'... an excellent summary of pedogenic theory and should occupy the shelves of all pedologists and students of soil science.'

– *Vadose Zone Journal*

Soils

Genesis and Geomorphology

Second Edition

Randall J. Schaetzl
Michigan State University

and

Michael L. Thompson
Iowa State University

CAMBRIDGE
UNIVERSITY PRESS

University Printing House, Cambridge CB2 8BS, United Kingdom

One Liberty Plaza, 20th Floor, New York, NY 10006, USA

477 Williamstown Road, Port Melbourne, VIC 3207, Australia

314-321, 3rd Floor, Plot 3, Splendor Forum, Jasola District Centre, New Delhi-110025, India

79 Anson Road, #06-04/06, Singapore 079906

Cambridge University Press is part of the University of Cambridge.

It furthers the University's mission by disseminating knowledge in the pursuit of education, learning and research at the highest international levels of excellence.

www.cambridge.org
Information on this title: www.cambridge.org/9781107016934

First edition published 2005
Second edition published 2015
Reprinted 2016

A catalogue record for this publication is available from the British Library

Library of Congress Cataloging in Publication data
Schaetzl, Randall J., 1957– author.
Soils : genesis and geomorphology / Randall J. Schaetzl, Michael L. Thompson. – Second edition.
   pages   cm
Includes bibliographical references and index.
ISBN 978-1-107-01693-4 (hardback)
1. Soils.   2. Soil formation.   3. Soil structure.   4. Geomorphology.
I. Thompson, Michael L., author.   II. Title.
S591.S287   2015
631.4–dc23   2014033154

ISBN 978-1-107-01693-4 Hardback

Dedication

We dedicate this volume to those who have inspired us to write it –
through their lifelong scholarship, insatiable curiosity about the world around them,
and their willingness to share it …
innovative thinkers who have made many, including us,
stop and think about the world through different intellectual "filters"…
scholars who make soils fun and exciting, but mostly,
genuinely good people who inspired everyone around them to be better.

Francis D. Hole in 1978
(1932–2002)

Photo courtesy Univ. of Wisconsin,
Photo Media Center

Donald L. Johnson in 2009
(1934–2013)

Photo by RJS

Peter W. Birkeland in 1997
(1934–)

Photo by D Muhs

A New Psalm (inspired by Psalm 19)
The earth beneath the feet of all runners and walkers
Declares the glory of God, our cherisher!
The roots of trees and grasses, the mole
And all organisms in the rich realm of darkness …
These are God's handiwork.
Our life in the realm of sunlight
Is upheld by the vital earth. God made it so.
All creatures that live on the land depend on the soil,
Which is like a strong parent,
Providing for all peoples and
All creatures that live above the waters.
Praise be to the holy ground that is softly under our feet;
Praise be to God who has blessed the living carpet
That He has spread for our walking,
In the days of our living in the flesh,
And into which our rich residues will return.

Francis D. Hole, TNS

(For years, Dr. Hole had appended the honorary title, "TNS – temporarily not soil," to his name, a sign of a bond that ties us all together, and of his love for the soil. We are proud to carry that notion forward.)

Contents

Preface

This book is about pedogenesis and soil geography. To some, this is a difficult and challenging area of study, because a geographic approach to soils requires synthesis of the many physical, chemical, and biological systems that have interacted to form the soils that blanket most landscapes. Plus, these landscapes change through time. Our purpose in writing this book is to assert that only through a study of the spatial interactions of soils *on landscapes* can soil and landscape evolution be resolved and understood.

This book can be used in courses on soil geography, soil genesis, pedology, and soil geomorphology. Our assumption is that the readers have had some background in the natural sciences and are eager to learn more about soils. We do not assume, nor does the reader need, a substantial background in soils to read and comprehend this book. Difficult as the task may seem, our goal was to write a soils text that could serve both as an initial soils text and as a cutting-edge resource book of research grade.

Our emphasis, beyond that of pedogenesis and soil geography, is deliberately intended to be broad and comprehensive. Other similar books (Daniels and Hammer 1992, Gerrard 1992, Birkeland 1999) focus more on geomorphology and the initial geologic setting as a guiding framework for the understanding of soil landscape evolution. We emphasize these issues in later chapters. Fanning and Fanning (1989) and Buol *et al.* (2011) focus on soil genesis while emphasizing classification. Our book tries to walk the middle ground amid these approaches to the study of soils.

This book relies heavily on concepts and imagery – mental and graphical – to convey ideas. We have compiled a suite of figures, images, and graphics that, in and of themselves, convey messages that cannot be put into words. Throughout the text we include brief "out-takes" on soil landscapes from around the world. We call these excursions "Landscapes," and hope that they convey, with pictures and graphics, what would otherwise take many hundreds of words to explain.

We believe in the necessity and importance of soil classification; we define, explain, and use its terminology in the book, but we do not focus on it. Because we feel that one of the best ways to "learn" and use taxonomy is to examine it in the context of landscapes, we include taxonomic information within many of the "Landscapes" excursions.

Soils research is exciting. It is enlightening to see and read about cutting-edge research. Therefore, we have spent considerable effort linking this book and its concepts to the literature. We are proud of the extensive literature listed herein. We have tried to cite many of the major works in the field, both the classic ones and the recent state-of-the-art papers. If we have missed something, we urge our readers to call it to our attention; we will be receptive. Where possible, we have tried to cite mainly papers and studies that are readily accessible in major academic libraries. That is, we have steered clear of obscure papers or those in the gray literature, as well as theses and dissertations, unless we felt that they were essential reading. The end result is a book that relies heavily on work published in national and international scholarly journals and books. If you wish to have a digital copy of our References section entries, just e-mail us (soils@msu.edu, mlthomps@iastate.edu) and ask.

The Glossary is rich in terms, many of which are only marginally touched upon in the text. Our philosophy with regard to the Glossary was simple: If the reader might need to know a term to understand something in the book, include it in the Glossary and define it clearly. The Glossary adds length to the book but makes it more "readable."

Our goal for this second edition was to update the book – with respect to concepts, theoretical advances, and the literature – without necessarily making it longer. We have added a chapter on soil chemistry as well as new sections on digital soil mapping, soils and archaeology, and the World Reference Base for Soil Resources, among others.

We are constantly striving to make the book more readable and comprehensive. We encourage you, the reader, to help us. For example, if you wish any topics to be added to the Glossary or the body of the book, contact us with your request. More importantly, alert us to your papers, send reprints and citations, e-mail or write to inform us of new findings or breakthroughs; we will include them as best we can. Contact us with your perceptions of the book, positive or otherwise. Help us make this book better and we promise to continue to work hard toward this goal.

Acknowledgments

It is impossible to thank adequately the many, many people who have made this book possible by inspiring us, assisting us, and, yes, funding us. All we can say is THANK YOU again and again. This is your book too. Too many to name, those of particular note are:

- Dr. Sharon Anderson wrote two chapters in edition 1 and, because of a heavy teaching and administrative load at her home institution, was unable to participate in edition 2. We thank her for her important and well-written contributions to that edition, and for her good nature and patience through that tough writing task. We have missed her positive energy, not to mention her expertise and contributions in mineralogy and soil chemistry in this edition. This will always be partly *her* book.
- My (RJS) family: my wife, Julie Brixie, has been a steadfast supporter of me, my work, my career, and this book, not to mention a solid proofreader and chapter/figure organizer. I could not have done it without her support and encouragement. My (RJS) children (Madeline, Annika, and Heidi) have helped with innumerable small and large tasks associated with it and have put up with their dad being at the office far too much; I will be home more now. My parents, John and Florence Schaetzl, instilled within me, through example rather than spoken word, the importance and payoffs of hard work.
- I (RJS) also want to thank the many graduate and undergraduate students who have worked with me over the years on soils projects, in the field, on the computer, and in the lab. They make each day at work a delight and a place where I love to be. Alhough they are too numerous to name, I nonetheless want to single out Beth Weisenborn, Joe Hupy, Trevor Hobbs, Paul Rindfleisch, Mike Luehmann, Brad Miller, Kevin Kincare, and Kristy Gruley for the effort and insight they have brought to the table and the accomplishments they have left behind. You have helped me as much as, or more than, I have you. It is for you and people like you that this book is meant. Always remember: Work hard and keep your mouth shut, and good things will happen.
- My (MLT) wife, Jan, forester par excellence, who will never let me win the argument about which came first – the tree or the soil; my colleagues in soil science at Iowa State University, Jon Sandor, David Laird, Bob Horton, Lee Burras, and Ali Tabatabai; my mentors in pedology from long ago, Neil Smeck, Jerry Bigham, and George Hall; my students, who have always challenged me to dig deeper; and the Agronomy Department at Iowa State for financial support.
- Matt Lloyd, our editor at Cambridge University Press, has been an unbelievably strong and unfailing supporter of the book – right from its inception. He deserves special mention and thanks. Matt has been everything to us that an editor could possibly be, so easy to work for and with, we could not have asked for a better colleague at Cambridge. We are fortunate to have had him, behind the scenes, as the rudder and engine of this ship. Our goal was to make Matt happy with and proud of this second edition; we hope that we have succeeded.
- No amount of thanks or praise can suffice for the incredible work done on the figures of this book by Ha-Jin Kim. She edited or created every single figure in edition 2, with her tremendous skill, diligence, and attention to detail. She single-handedly improved every figure in this edition, and did it with care, professionalism, and class. If the figures in edition 2 seem better than the ones in edition 1 – and we believe they are – it is because of her talent, work ethic, and dedication. What would we have done without her? Thank you very much, Ha-Jin!
- Despite her own busy schedule, including a Ph.D. dissertation that needed to be written, Kristy Gruley helped us out by proofreading nearly the entire second edition text. She did an outstanding job. She has a careful eye for detail and love of science, and it shows in her work. But she gave more than proofreading skills to the job. Her insight and ability to think outside the box added a lot of extra zest to the final product. Thank you, Kristy!
- Don Johnson, a true academic free spirit, genuinely inspired thinker, and tireless academic who was not afraid to look at the world through different glasses, and who never grew weary of the field or the library: Don, a giant of soil geomorphology and a wonderful person, passed away in 2013, but his memory and ideas permeate this book.
- Francis Hole, a one-of-a-kind scholar who will always hold a special place in my (RJS) heart and mind. I (RJS), like so many, would not have found the disciplines of soil science and soil geography were it not for Francis Hole.
- François Courchesne, a driving force behind the development of the first edition of this book.
- Scott Isard, my (RJS) academic conscience and motivator, always willing to discuss academics and scholarship.
- Curt Sorenson, who taught me (RJS) simply to love soils, instilled within me a passion to excel, and gave me the confidence to do it.
- Leon Follmer, who taught me (RJS) to look closely at soils and made me realize that soils and paleosols are truly remarkable things.
- Duke Winters, who was able to convince his young colleague (RJS), infatuated with soil science, why it is important to view soils through geographic lenses.

Those who otherwise assisted in the production, editing, or compilation of the book deserve special mention:

- Matt Mitroka, Ellen White, Beth Weisenborn, Peter Dimitriou, and Beth Kaupa assisted Paul Delamater in the production of the figures for edition 1, arguably the strength of the book. Paul was the consummate QA/QC person for the edition 1 graphics. Thank you, Paul! Then, Mike Luehmann also helped edit and craft some of the edition 2 figures.
- Ron Amundson, Alan Arbogast, Jim Arndt, Szandra Baklanov, Brenda Buck, Dave Cremeens, Ron Dorn, Muhsin Eren, Chris Evans, Leon Follmer, Jacquelyn Gill, Alexandra Golyeva, John Hunter, Joe Hupy, Diana Johnson, Don Johnson, Mark Johnson, Norbert Juergens, Mikhail Kanevskiy, Tim Kemmis, Tanzhou Liu, Warren Lynn, Tim Messner, Xiaodong Miao, Leslie Mikesell, Brad Miller, Wes Miller, Curtis Monger, Cristine Morgan, Héctor Morrás, Jeff Munroe, Fritz Nelson, Jenny Olson, Ãkos Pető, Marty Rabenhorst, Paul Reich, C. Mario Rostagno, Pavel Samonil, Phil Schoenenberger, Kent Syverson, John Tandarich, Charles Tarnocai, Judy Turk, Michael Velbel, Pat Webber, Beth Weisenborn, Antoinette Winkler Prins, Barbara Woronko, and Leo Zulu provided images, graphs, charts, and figures of soils and landscapes that have been reproduced within the book. Without these images, the book would have been so much weaker.
- Exceptional editing and reviews of individual chapters (and parts thereof) were provided by Bob Ahrens, Alan Arbogast, Linda Barrett, Art Bettis, Janis Boettinger, Julie Brixie, Alan Busacca, Joe Chiaretti, François Courchesne, David Cremeens, Missy Eppes, Kathryn Fitzsimmons, Leon Follmer, Ryan Haag, Jon Hempel, Robert Horton, Vance Holliday, Geoff Humphreys, Christina Hupy, Joe Hupy, Don Johnson, Christina Kulas, Zamir Libohova, Johan Liebens, Wes Miller, Piotr Owczarek, Jonathan Phillips, Marcus Phillips, Greg Pope, Paul Rindfleisch, Mark Stolt, Julieann van Nest, Natasa Vidic, Beth Weisenborn, Gary Weissmann, Indrek Wichman, Kathy Woida, and Catherine Yansa.
- Patricia Brixie proofread the first edition of the book in its entirety. It was then reviewed by Art Bettis, Vance Holliday, Donald Johnson, and Dan Muhs.
- Bill Dollarhide, John Gagnon, Charles Gordon, Bill Johnson, Mike Risinger, Richard Schlepp, Bruce Thompson, and Cleveland Watts (of the USDA–NRCS), and Dave Cremeens provided us with data and block diagrams from various NRCS soil surveys.
- Many others were quick to help or offer advice when a question or issue arose: Bob Ahrens, Alan Arbogast, Art Bettis, Steve Bozarth, George Brixie, Brenda Buck, Joe Chiaretti, David Cremeens, Bob Engel, Leon Follmer, Bill Frederick, Kristy Gruley, John Hempel, Vance Holliday, Don Johnson, Bruce Knapp, Zamir Libohova, Mike Luehmann, Rolfe Mandel, Brad Miller, Curtis Monger, Bruce Pigozzi, John Tandarich, Greg Thoen, and Dan Yaalon.

Funding for the many costs associated with the development of a book of this type was provided by various agencies of Michigan State University: the Agricultural Experiment Station, the Office of the Vice President for Research and Graduate Studies, and the Department of Geography. Some of the data, findings, and insights into the soils and landforms of the Great Lakes region was developed in conjunction with NSF grants made to RJS and colleagues (NSF awards BCS-9819148, SBR-9319967, BCS-0422108, and BCS-0851108); any opinions, findings, and conclusions or recommendations expressed in this material are, however, those of the authors and do not necessarily reflect the views of the National Science Foundation or Michigan State University.

In addition to Matt Lloyd, we thank the professional staff at Cambridge University Press and their affiliates for their help on, and support of, this book project. The staff at Cambridge, especially Susan Francis, Sarika Narula, Jayne Aldhouse, Sally Thomas, and Anna Hodson, made the typescript and figures into a book. We also thank the many behind-the-scenes people, e.g., copy editor Susan Thornton, as well as Jayashree and her staff, whose tireless work often goes unseen and unappreciated; we did it!

Last, we acknowledge that we have approached this book from the perspective of St. John Vianney, the Curé of Ars, when he said, "I have been privileged to give great gifts from my empty hands."

Part I

The Building Blocks of Soil

Chapter 1

Introduction

Soil landscapes and their spatial variation are exciting and complex. But to understand soils fully, they must be studied in space *and* time. Indeed, we embrace Daniels and Hammer's (1992) statement that soils are four-dimensional systems, not simply the one-dimensional profile. In this book we incorporate these ideas by synthesizing complex, overlapping topics and use this knowledge to help answer the questions: How do soil landscapes form? How and why do they change through time?

Soil genesis and geomorphology, the essence of this book, cannot be studied without a firm grasp on the processes that shape the *distributions* of soils – their complex patterns. Unfortunately, we will never fully understand the complex patterns of the Earth's soils. And even if we do aim to understand them, we must be mindful that the pattern is ever-changing. Again we quote Daniels and Hammer (1992: xvi): "One cannot hope to interpret soil systems accurately without an understanding of how the *landscape and soils have coevolved* over time" (emphasis ours). Every percolation event translocates some material (however minute) within a soil, while every runoff event moves material across its surface, changing the soil landscape ever so slightly. The worms, termites, and badgers that continually burrow, mix, and churn soils make them different than they were yesterday. Chemical and biochemical reactions within soils weather minerals and enable microbes to decompose organic matter, perpetuating the cycle from living matter to humus to chemical elements and back again. Like landscapes, soils evolve; changing patterns of soils over time are a reflection of a multitude of interactions, processes, and factors, replete with feedbacks, inertia, and flows of energy and mass. Yes, soils are a challenge! For that reason, we provide information, tools, resources, and background data to draw the reader closer to deciphering this most complicated – and important – of natural systems.

Whitehead (1925) wrote, "It takes a genius to undertake the analysis of the obvious." All people who walk the Earth's surface depend on the soil, yet the soil is not obvious to all. It is seemingly everywhere, and yet comparatively few study it. Additionally, soils are usually hidden from view and require excavation to be revealed. Neither are soils discrete entities like trees, insects, or lakes, which have clearly defined outer boundaries. Instead, soils grade continuously, one into another, until they end at the ocean, a sheer rock face, or a lake. When broken into discrete entities, in the way a geologist might break apart a rock, soils appear to lose their identity. This soil science – it's not easy. But therein lies the challenge!

We believe that a geographic approach is one of the most fruitful avenues to study soils (Boulaine 1975). Like most of the components of Earth science, soils are spatial things, varying systematically across space. To study soils completely, we must grasp not only *what* they are, but also how they relate to their adjoining counterparts. Soil geography focuses upon the geographic distributions of soils. It emphasizes their character and genesis, their interrelationships with the environment and humans, and their history and likely future changes. It is operationalized at many scales, from global to local. Soil geography *encompasses* soil genesis. Soil patterns cannot be explained without knowing the genesis of the soils that compose that pattern. Likewise, soil patterns cannot be fully explained or understood without knowledge of the *geomorphic evolution* of the landforms and rocks of which they form the skin. An understanding of how the Earth's surface may change over time, as a result of erosion, deposition, or weathering, is also necessary if we are to predict future changes in the soil landscape.

As the title says, this book is about soil genesis and soil geomorphology, and all that these disciplines encompass (Table 1.1). Tandarich *et al.* (1988) used the term "geopedology" to refer to the intersection of the disciplines of geology, geography, and soil science. We embrace that term and view it as the central motif of this book.

Pioneers of Soil Science, Soil Survey, and Soil Geography

Pedology (Russian *pedologiya*, soil speech) is the science of soil genesis, classification, and distribution. Many

Table 1.1 | Some of the academic domains of pedology

Distribution of soils and soil taxa across the landscape

Soil survey and mapping

Soil genesis, both within and among pedons

Human impacts on soils: Anthropedology

Paleopedology and the study of landscapes of the past

Soil geomorphology

Soil-slope and soil catena studies

Soil landscape analysis and explanation of soil patterns

Pedometrics

Spatial representation of soils and the use of spatial soil data

Evolution of soils and landscapes

Note: Not an exhaustive list; in no particular order.
Source: Modified from Hole and Campbell (1985).

discourses have been written that focus on the nature and future of the field of pedology (Arnold 1992, Dudal 1987, Daniels 1988, Churchward 1989, Jacob and Nordt 1991, Clayden 1992, Brasher 1997, Bockheim *et al.* 2005, Richter 2007), although to many, it is roughly synonymous with *soil science*, and to most, it has a clear and strong *field* component. The application of pedology is often best manifested in soil mapping and survey. In recent years the field has evolved (and is evolving) into one that has more direct societal consequences and practical applications (Baveye 2006, Baveye *et al.* 2006), particularly as it relates to anthropogenic impacts on soils (White 1997, Tugel *et al.* 2005, Richter and Markewitz 2001).

Because soils have sustained human life since its inception, you may think that pedology has a long history. Not so. In fact, soil science was a late arrival among the natural sciences (Hole and Campbell 1985). Many attribute its founding to Vasili Dokuchaev (1846–1903), a Russian scholar and teacher. Others acknowledge the work of Charles Darwin (1809–1882), perhaps the world's most underappreciated soil scientist (Johnson 2002, Johnson *et al.* 2005, Johnson and Schaetzl 2014). Regardless of who gets the credit for jump-starting this discipline, pedology is unquestionably little more than a century and a half old! Our brief overview of the founders of soil science (later) should underscore that they were multifaceted thinkers who understood that the soil landscape was a complex system, requiring that it be studied using a geographic approach. More detailed accounts of the personalities involved in the development of the field are presented elsewhere (Kellogg 1974, Cline 1977, Tandarich and Sprecher 1994).

Dokuchaev is often called the father of soil science, although he acknowledged the influence of several others (particularly in the field of agricultural chemistry) in the development of his ideas (Tandarich and Sprecher 1994; Fig. 1.1A). Trained in Russia, he wrote his most reputed works on the soils of the Russian steppes, primarily Chernozems. In his work, he developed and used concepts on the nature and genesis of soil profiles, as well as soil landscapes. Dokuchaev and his students produced the first scientific classification of soils and developed soil mapping methods, laying the foundation for the modern fields of pedology and soil geography (Buol *et al.* 1997). He is known

Fig. 1.1 Three influential scholars in the field of soil science: (A) Vasili V. Dokuchaev (1846–1903), Russian agriculturalist, geographer, and pedologist. Image courtesy of J. Tandarich. (B) Curtis F. Marbut (1863–1935), American agriculturalist, soil scientist, and early developer of the U.S. soil classification system. Image courtesy of J. Tandarich. (C) Hans Jenny (1899–1992), Swiss pedologist and agricultural chemist; professor at the University of California. Image by R. Amundson.

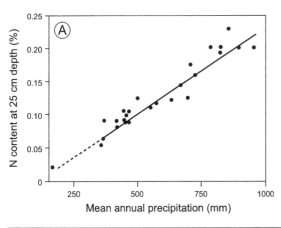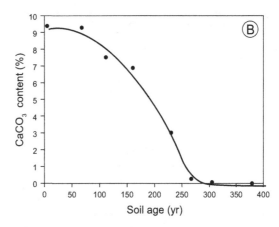

Fig. 1.2 Examples of two functional relationships that Hans Jenny produced for his 1941 book, *Factors of Soil Formation.*

for developing the basic A–B–C horizon nomenclature, and a factorial model of soil development, in which soils and soil patterns were seen as a function of independently varying state factors of the environment. Although not universal, this model remains, in various revised forms, the primary explanatory model for soils worldwide (see Chapter 12). Dokuchaev's model led to the development of the concept of the *zonal soil*, which characterized vast tracts of land and was thought to represent the end point of soil development for that region. Zonal soil concepts, although obsolete today, essentially jump-started soil survey and mapping worldwide and made the complex world of soils more understandable to the masses. Dokuchaev's teachings, carried across the Atlantic by E. W. Hilgard (1833–1903), were highly influential on many prominent soil scientists.

Darwin was a contemporary of Dokuchaev. Unfortunately, by omitting Darwin's ideas from his writings, Dokuchaev would inadvertently bury them. Darwin focused on local-scale biological origins of many soil properties and on biomechanical processes in soils, such as mixing by worms (Darwin 1881, Johnson 1999). The lack of soil terminology in his works, his general lack of students to spread his approach, coupled with the growing acceptance of Dokuchaev's factorial model for soil development doomed Darwin's *biomechanical soil processes* to the theoretical back seat, until resurrected decades later (see Chapter 11).

In 1899, the United States started its soil survey program, under the direction of Milton Whitney (1860–1927), primarily using geological concepts of soils, e.g., granite soils and alluvial soils (Shaler 1890). This approach continued for a little more than a decade (e.g., Marbut *et al.* 1913). A major sea change later occurred when Curtis F. Marbut (1863–1935), who earned his Ph.D. in geology at Harvard under the eminent geographer William Morris Davis (1850–1932), was appointed soil scientist in charge of the U.S. Bureau of Soils (Tandarich *et al.* 1988; Fig. 1.1B).

While at Harvard, Marbut had been influenced by the writings of Konstantin Glinka (1897–1927), a student of Dokuchaev's, and the soils-related work of Nathaniel Shaler (1841–1906). Marbut had translated Glinka's book *Die Typen der Bodenbildung* from German into English and applied many of the ideas within to the budding soil survey program (Cline 1977, Tandarich and Sprecher 1994). Marbut's impact on soil science in the United States proved to be strong and long-lasting. Indirectly but strongly influenced by the ideas of Dokuchaev, he changed the way soils were viewed, emphasizing that they should be classified and mapped on the basis of horizon and profile characteristics, thereby reducing the influence of geology. Marbut eventually developed a multicategorical soil classification system that stood as the U.S. standard for decades (Marbut 1927a, b, 1935; see Chapter 8).

In 1941, Hans Jenny (1899–1992), professor of soil science at the University of California-Berkeley (Fig. 1.1C), published a landmark treatise entitled *Factors of Soil Formation.* Much of this book is devoted to his functional-factorial model of soil formation, following on the work of Dokuchaev. In this model, soils are seen as being influenced by five interacting factors: climate, organisms, relief, parent material, and time (see Chapter 12). Jenny developed many numerical soil functions, each an equation showing how soils change as four of the factors are held constant and one is allowed to vary (Fig. 1.2). Jenny was both a soil geographer and a soil scientist. He noted (1941a: 262) that "the goal of the soil geographer is the assemblage of soil knowledge in the form of a map. In contrast, the goal of the functionalist is the assemblage of soil knowledge in the form of a curve or an equation." He commented that soil maps display areal arrangement but give no insight into causal relationships, and that mathematical curves reveal dependency of soil properties on state factors, but the conversion of such knowledge to the field is impossible without a soil map (Arnold 1994). Thus, Jenny proposed that the union of geographic

Table 1.2 | Some of the major advances in the field of soil science, from its inception to the present day

Date(s)	Conceptual/theoretical/methodological advances
Pre-1800	Soils classified on the basis of relative productivity
1800–1880	Concepts of soil as (1) a medium for plant growth and (2) a weathered rock layer; soil classifications based on geologic/physical soil properties of surface horizon; A-B-C horizon designations introduced
1880–1920	Appearance of fundamental soil geography concepts: (1) environmental correlations and (2) soil as a natural body; introduction of zonal classification based on climate-vegetation relationships; links between soil horizons, profiles, and factors introduced; U.S. Soil Survey established; development of soil series concept
1920–1940	Widespread appearance and adoption of fundamental pedology concepts: (1) soil as a natural body and (2) soil-forming factors; development of first regional soil classification systems, often based on zonal soil concepts; A-B-C horizon designations and solum concept become widely accepted; focus on collection of physical and chemical soils data; organization of soil series into regional soil classification systems; development of catena and soil cover pattern concepts
1940–1960	Factors of soil formation refined and clarified; development of global soil taxonomic systems; intensified soil mapping facilitated by development of functional relationships for quantitative study; aerial photography enhances soil mapping
1960–1990	Introduction of pedon and polypedon concepts; development of quantifiable, properties-based taxonomic systems; development of new models of soil formation, including first "process" models; recognition of coevolution of soils and landforms; recognition of regressive-progressive nature of pedogenic processes, and of polygenesis; importance of pedoturbation as a soil-forming process increasingly recognized; expansion of paleopedology; introduction of M-S-W horizon designations for tropical soils; methodological advances in soil micromorphology
1990–2005	Increased understanding and modeling of pedogenic and soil-geomorphic processes; refinement of global soil models and global soil taxonomic systems; development of statistical and computer-based soil information systems, and the rise of the discipline of pedometrics; beginnings of digital soil mapping; enhanced recognition of biomantles in soils; concepts of soils as a key component (1) in interrelated Earth physical systems and (2) as complex, nonlinear systems; expansion of absolute dating techniques applicable to soils; soil science and ecosystem sustainability concerns surface; increased attention to soil C cycles and stores
2005–present	Expansion and increased availability of soil geographic data in digital form, and GIS utilization/applications thereof; increased recognition of the importance of humans as a soil-forming factor; recognition of soils as a key component of Earth's critical zone; efforts to create first worldwide digital soil map

Source: Modified from Bockheim *et al.* (2005), with contributions from Mermut and Eswaran (2001).

and functional methods provided an effective pedological research motif. Arnold (1994:105) restated this idea as follows – spatial soil patterns need to be understood through functional relationships of the soil-forming factors in *space and time*. Jenny's (1941a) model stands today as one of the most geographic of the several soil models, because it is used subliminally or overtly by almost every soil mapper. More recent models, which refine and elaborate on Jenny's, as well as those that propose very different ways of looking at the soil landscape (Johnson and Hole 1994), are discussed in Chapter 12. Table 1.2 provides a summary of the major conceptual advances that have occurred in pedology, from its beginnings to the present.

Things We Hold Self-Evident

Following the lead of Buol *et al.* (1997) and Hole and Campbell (1985), we provide in the following listing some concepts or truisms in soil science and soil geography, slightly modified from their original sources.

- Soil complexity is more common than simplicity.
- Because soils lie at the interface of the atmosphere, biosphere, hydrosphere, and lithosphere, a thorough grasp of the workings and nuances of soils requires some understanding of meteorology, climatology, ecology, biology, hydrology, geomorphology, geology, and many other Earth sciences.

- The state factor model of soil formation (climate, organisms, relief, parent material, and time) is a useful conceptual approach to understand the spatial variation in pedogenic processes and soils.
- The characteristics of soils and soil landscapes include the number, sizes, shapes, and arrangements of soil bodies, each of which can be characterized on the basis of horizons, degree of internal homogeneity, slope, landscape position, age, and other properties and relationships.
- Distinctive bioclimatic regimes or combinations of pedogenic processes produce distinctive soils. Thus, morphological features, e.g., illuvial clay accumulation in B horizons, are produced by combinations of pedogenic processes operating over time.
- Pedogenic processes act both to create and to destroy order (anisotropy) within soils, and these opposing sets of processes can and do proceed simultaneously. Soil profiles reflect the balance of these processes, present and past.
- Contemporary soils carry imprints of pedogenic processes that were active in the past, even if they are difficult to observe or quantify. A succession of different soils may have developed, eroded, and/or regressed at any particular site, and during that time, pedogenic and site factors, e.g., vegetation, sedimentation, geomorphology, have changed. Thus, an understanding of paleoecology, paleogeography, glacial geology, and paleoclimatology is important to studies of soil genesis. These studies constitute a basis for predicting future soil changes and for interpreting *paleosols* – soils of past environments.
- The geologic principle of *uniformitarianism* applies to soils, i.e., pedogenic processes active in soils in the past are similar to those that are active today. These processes, however, may vary in expression and intensity over space and time.
- There are relatively few old soils (in a geological sense). Little of the soil continuum dates back beyond the Tertiary Period, and most soils and land surfaces are no older than the Pleistocene Epoch. Why? Over time, soils are eroded or buried by geological events, or they are modified by shifts in pedogenic processes. In short, soils exist at a vulnerable location – the skin of the Earth.
- Knowledge of pedogenesis and geomorphology is critical to effective soil classification and mapping. Nonetheless, soil classification systems cannot be based entirely on *perceptions* of soil genesis, because genetic processes are seldom observed and are difficult to measure directly. Classification systems must be based on observable and measurable soil characteristics, as informed by an understanding of pedogenesis.

- Soils are natural clay factories. Shales worldwide are, often, simply soils that have been eroded and deposited in the ocean basins, to become lithified at a later date.
- Humans can and do alter soils, inadvertently and purposefully. It follows that an understanding of pedogenesis is basic to wise land use and management practices, and knowledge of how humans affect soils is essential to interpreting their current morphology and chemistry.

The Framework for This Book

This book has three major parts. We introduce the building blocks of soil in Part I. We continue adding to the basic knowledge base in Part II (Chapters 9–13), but add a great deal more material on theory and soil genesis/processes. In Chapter 12, for example, we introduce a large dose of pedogenic and geomorphic *theory*, which in combination with the previous chapters allows us to discuss soil genesis and pedogenic *processes* at length in Chapter 13. An understanding of soil genesis provides important information to scientists who classify and map them. Finally, we pay considerable attention in Part III (Chapters 14–16) to examining soil landscapes over time and how soils can be used as dating tools and as keys to understanding past environments. Part III is the synthesis section, for within it we pull together concepts introduced previously and apply them to problems of dating landscapes and understanding their evolution. Lateral flows of materials and energy link soil bodies to adjoining ones on the landscape, helping to reinforce the all-important three-dimensional component – an emphasis of Part III. Woven into the book are studies and examples of soil landscapes in three dimensions, often through the use of traditional block diagrams. We hope that the reader will gain from these applications and discussions a *holistic* perspective on soils, and begin to appreciate that they are integrated across and within landscapes, and that they have a history and a future.

We introduce, throughout the book, many classic studies and examples of how the evolution of soils has been effectively worked out, in order to tie certain concepts together and expose the reader to some of the classic literature. To be sure, our book has a North American focus – we live there, and it is the focus of a large proportion of the soil literature. However, we have gone to great lengths to include the global soils community in this book.

In sum, we think this book will be of use to "land lookers" worldwide (Hole 1980). We hope it is enjoyable, intellectually stimulating, and, most importantly, useful to you, the reader. We thank you for choosing *Soils: Genesis and Geomorphology*.

Chapter 2

Basic Concepts: Soil Morphology

What Is Soil?

Soil means different things to different people. To a farmer or horticulturalist, it is a medium for plant growth. To an engineer, it is something to build on, or remove, before construction can occur, or it may actually be a type of medium used for road building, house foundations, or septic drain fields. To a hydrologist, soil functions as a source of water purification and supply. To some geologists, it is the post-Pliocene overburden that is covering up the rocks!

We use a slightly modified statement from the one offered by Johnson (1998a) as the best and most widely applicable definition of soil: Soil is organic or lithic material, normally at the surface of planets and similar bodies, that has been altered by biological, chemical, and/or physical agents. This definition and the one proposed by Richter and Markewitz (1995) – "Soil is the biologically excited layer of the Earth's crust" – give equal rank to biological processes and agents, which were ignored by some previous definitions that focused on the physical and chemical processes of soil formation. Pedologists and soil geographers often use a similar definition: A soil is a natural, three-dimensional body that has formed at the Earth's surface through the interactions of at least five soil-forming factors (climate, biota, relief or topography, parent materials, and time). This definition emphasizes that soils are "naturally occurring" bodies and introduces the five soil-forming factors into the equation, front and center. All of these definitions have merit and fit with the ways in which soils are discussed in this book.

Soils are composed of clastic particles (mineral matter), organic materials in various stages of decay, living organisms, water (or ice), and gases within pores of various sizes (Fig. 2.1). The absolute amounts of each, and their arrangement into a particular fabric, are the sum of soil morphology. Every soil has a distinct morphology, defined as its structure or form. Soil morphology is all that can be seen and felt about a soil. It includes not only "what is there" but also how it is "put together" – its architecture. To many, the main components of soil morphology include *horizonation, texture, color, redoximorphic features, porosity, structure,* and *consistence,* i.e., the look and feel of the soil. In this chapter, we will discuss the main features associated with soil morphology, many of which are normally included in a standard soil profile description.

Soil Profile Descriptions

Soil scientists often start their study of soils by excavating a pit, using an exposure, such as a road cut, or by extracting an undisturbed soil core using a hydraulic sampling probe, and then *describing* the soil they see. Soil descriptions are a standard way of communicating information about soils, as they occur in the field. They represent the most fundamental data of soil genesis. The current best guide for making soil profile descriptions is a *Field Book* published by the USDA-Natural Resources Conservation Service (Schoeneberger *et al.* 2012). It contains instructions, definitions, and concepts for making or reading soil descriptions and for sampling soils, as presently practiced in the United States. It draws heavily from the *Soil Survey Manual* (Soil Survey Division Staff 1993).

There is no one best way to describe a soil, just as there is no one best list of features to describe. Soil profile descriptions are tailored to the investigation. Nonetheless, most descriptions include at least the following components: (1) date, (2) location, (3) slope gradient and aspect, (4) landscape position and likely geomorphic origin, and (5) soil horizonation. Then, for each horizon, the scientist describes the following morphological characteristics: (1) depths of the top and bottom of the horizon; (2) color(s); (3) texture, including an estimate of coarse fragment content and characteristics; (4) structure; (5) consistence; (6) degree of effervescence (if calcareous); and (7) redoximorphic features that are indicators of wetness. Presence of a water table or other notable features, e.g., ped or rock coatings, roots, pores, animal burrows, concretions or nodules, forms of disturbance, or discontinuities, are also noted. The shape (topography) and distinctness of the soil

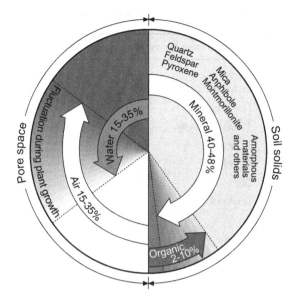

Fig. 2.1 Volumetric composition of soil pores and solids. The broken line between water and air indicates that these proportions fluctuate as the soil wets and dries. Similarly, organic matter contents of soils vary from zero to nearly 100%, although 2–8% is a common range for most soils.

horizon boundaries are also included in the description. Taken together, the data of a soil profile description constitute a powerful tool for the interpretation of soil and landscape genesis (see Landscapes box).

Texture

Mineral (clastic) soil particles are usually first divided into the fine earth fraction (<2 mm diameter) and coarser fractions. Geologists commonly use the *phi* scale when referring to the sizes of individual particles, whereas pedologists usually refer to particle diameters in millimeters or micrometers, following the USDA system (Fig. 2.2). *Texture* is a term that refers to the relative proportions of differently sized particles in a soil. Textural class names usually include descriptors for the fine earth fraction only, e.g., silt loam, sandy clay, unless the amount of coarse fraction is large enough to warrant inclusion, e.g., gravelly loamy sand. First, let us discuss the coarse fraction – the "gravelly" in the previous texture class example.

The names given to coarse fragments vary among naming systems, depending on size, shape, and lithology (Alexander 1986, Poesen and Lavee 1994; Table 2.1, Fig. 2.2). Except for the larger ones, e.g., cobbles, stones, and

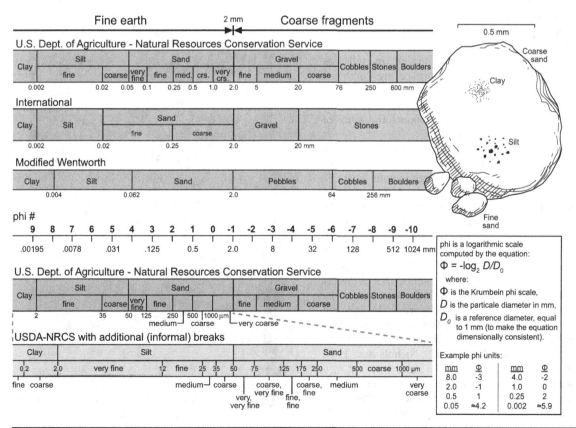

Fig. 2.2 Particle size limits and names in the major systems in use today. The graphical portrayal of clay, silt, and sand grains is drawn to the USDA system.

Table 2.1	Names given to rock fragments of various sizes and shapes	
Shape and size	Name	Adjectival term for soil texture class
Round (spherical, cubelike, or equiaxial)		
2–75 mm diameter	Pebbles	Gravelly
2–5 mm diameter	Fine pebbles	Fine gravelly
5–20 mm diameter	Medium pebbles	Medium gravelly
20–75 mm diameter	Coarse pebbles	Coarse gravelly
75–250 mm diameter	Cobbles	Cobbly
250–600 mm diameter	Stones	Stony
>600 mm diameter	Boulders	Bouldery
Flat		
1–150 mm long	Channers	Channery
150–380 mm long	Flagstones, flags	Flaggy
380–600 mm long	Stones	Stony
>600 mm long	Boulders	Bouldery

Source: Soil Survey Division Staff (1993).

boulders, most coarse fragments in soils are gravel-sized; in the USDA system that includes all clasts between 2 and 76 mm diameter (Fig. 2.2). In all cases, coarse fragments must be strongly cemented or resistant to rupture. Aggregates of fine earth particles are not coarse fragments; they should be disaggregated to determine their true textural composition.

Coarse fragments are very important in soils, as they affect percolation rates and surface area and can greatly impact root growth and tillage operations. Geomorphologists can often infer the genetic history of a sediment or soil by knowing the amounts and kinds of coarse fragments that it comprises.

Another way in which coarse fragments affect soils is through potential void space. Rocks and other coarse fragments can take up considerable volume in soils (Fig. 2.3). Thus, soil processes are compressed into less space than if the same soil had no coarse fragments (Schaetzl 1991b). Rock fragments also help soils resist compaction and erosion and retain good structure (Poesen *et al.* 1990, van Wesemael *et al.* 1995). Indeed, soils with high amounts of coarse fragments tend to have lower bulk densities, probably because the fine earth fraction cannot pack as closely to the large particles as it can to itself (Stewart *et al.* 1970). Many coarse fragments are not impermeable and can retain some soil water, thereby affecting soil water characteristics in ways beyond just their impact on void space (Coile 1953, Hanson and Blevins 1979, Nichols *et al.* 1984, Ugolini *et al.* 1996).

Coarse fragment modifiers are only added to the textural class name, e.g., gravelly loamy sand, when the fragments are present in sufficient amounts. In the USDA scheme, this lower limit is usually set at 15% coarse fragments by volume (Soil Survey Division Staff 1993; Table 2.2). For example, a

Fig. 2.3 A Typic Torriorthent from Imperial County, southern California, with large amounts of coarse fragments within the profile. This extremely gravelly, weakly developed soil is formed in coarse-textured alluvium.

Table 2.2 | Texture class modifiers, based on coarse fragment content

Coarse fragment content (volumetric%)	Texture class modifier
<15%	No texture adjective is used, e.g., loam, loamy sand
15–<35%	Use adjective, e.g., gravelly, cobbly, stony
35–<60%	Use "very" plus adjective, e.g., very gravelly, very channery
60–<90%	Use "extremely" plus adjective, e.g., extremely gravelly, extremely flaggy
≥90%	No adjective modifier; use the appropriate noun of the dominant coarse fragment to describe the texture, e.g., gravel, cobbles, channers

Source: Schoeneberger *et al.* (2012).

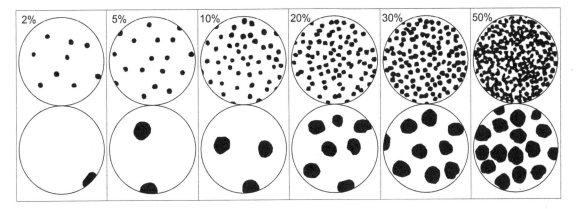

Fig. 2.4 Charts typically used to estimate the percentage of any isolated feature, e.g., coarse fragments, mottles, burrows, or roots, in soils.

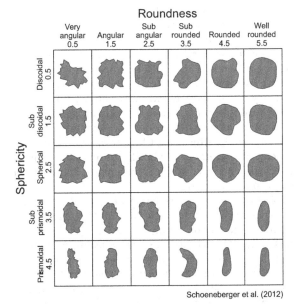

Schoeneberger et al. (2012)

Fig. 2.5 Standard chart used to estimate the sphericity and roundness of clasts.

sandy loam with 17% gravel would be considered a gravelly sandy loam and a clay loam with 40% cobbles and 19% gravel would be a very cobbly, gravelly clay loam. However, the amount of the coarse fragments is a volumetric estimate (Fig. 2.4), because it is too difficult to obtain data on coarse fragment content; very large samples are required for complete accuracy (Alexander 1982).

It is often important to note not only the volume of the soil occupied by coarse fragments (Fig. 2.4, Table 2.1), but also their shape and lithology. Shape is captured by two variables: *sphericity* and *roundness* (Fig. 2.5). Sphericity relates to the overall shape of a feature, irrespective of the sharpness of its edges; it is a measure of the degree to which it resembles a sphere. Roundness is a measure of how much a particle's corners and edges are rounded; it is a measure of smoothness. Coarse clasts often become more rounded and spherical as they become more weathered, and thus information of this kind may be a useful as a weathering or comminution index. Alternatively, fragments may be deposited in different stages of roundness or sphericity, and so data of this type can provide discriminating information about the soil's parent material history. Information about lithology is also important, because it

Table 2.3 | Some general properties of sand, silt, and clay

Property	Sand	Silt	Clay
Size range (mm)	2.0–0.05	0.05–0.002	<0.002
Means of observation	Naked eye or hand lens	Light microscope or hand lens	Electron microscope
Dominant mineral types	Primary	Primary and secondary	Mostly secondary
Attraction of particles for water	Low	Medium	High
Cohesiveness	Low	Medium	High
Surface area	Very low	Low–medium	High–very high
Water-holding capacity	Low	Medium–high	High
Plant-available water capacity	Low	High	Medium
Aeration	Good	Medium	Poor
Compaction potential	Low	Medium	High
Resistant to pH change	Low	Medium	High
Ability to retain chemicals and nutrients	Very low	Low	Medium–high
Susceptibility to wind erosion	Moderate (especially fine sand)	High	Usually low, depends on water content
Susceptibility to water erosion	Low (unless fine sand)	High	Depends on degree of aggregation
Wet consistence	Loose, gritty	Smooth	Sticky, malleable
Dry consistence	Very loose, gritty	Powdery, some clods	Hard clods

Source: Brady and Weil (1999).

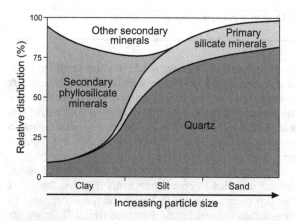

Fig. 2.6 | General relationship between particle size and particle mineralogy in the fine earth (< 2 mm) fraction.

can be used to discriminate among parent materials. It can also provide an estimate of the potential release of cations to the soil, as the coarse fragments weather. Perhaps because they are so important, coarse fragments should be given more consideration in soil characterization analyses (Corti *et al.* 1998).

Within the fine earth fraction, particles are divided, on the basis of size, into sand, silt, and clay (Soil Survey Division Staff 1993) (Fig. 2.2). Sand, silt, and clay are each referred to as soil separates. Each of these three components imparts its own character to the soil and often has a distinct mineralogy (Fig. 2.6, Table 2.3). Sand (and most of the silt fraction) is composed of primary minerals, mainly quartz, whereas many clay-sized particles are secondary minerals, formed from the weathering of primary minerals. This raises an important point – *clay is a size fraction irrespective of mineralogy*, whereas the term *clay mineral* is generally associated with a family of phyllosilicate minerals, e.g., kaolinite, smectite, and vermiculite, along with oxide clay minerals including hematite and goethite (see Chapter 4). Not all clay-size particles, however, are phyllosilicate minerals; many of the coarser clay-sized particles are quartz, feldspar, and/or amorphous materials.

Soil texture refers to the relative proportions of sand, silt, and clay within the fine earth fraction (Figs. 2.2, 2.7). It is commonly thought of, simply, as the "feel" of the sample, e.g., sandy, silty, sticky (rich in clay), or loamy (no one dominant "feel"). In the field, soil texture can be approximated by rubbing a moist sample between the thumb and forefinger. Clayey soils form a ribbon. Sandy soils feel gritty. Silt imparts a smooth feel. After a bit of practice, one can become quite proficient at determining soil texture by its feel alone. Soils dominated by larger particles, e.g., sands,

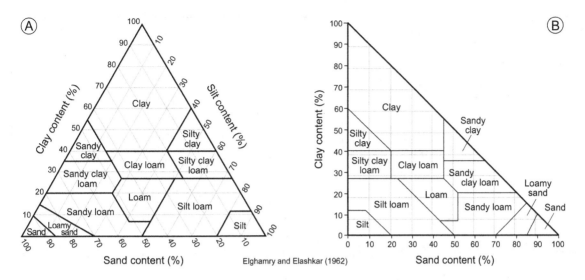

Fig. 2.7 Two types of textural triangles. (A) The traditional USDA triangle. (B) An alternative triangle, much like an *X-Y* plot in a traditional Cartesian coordinate system. Boundaries between the classes are the same in both presentations.

are referred to as *coarse-textured*. *Fine-textured* soils have larger amounts of silt and clay.

Before the amounts of sand, silt, and clay can be quantitatively determined, the fine earth fraction must first be dispersed, so the soil particles behave as independent units, and not as aggregated clumps. Usually this is accomplished in one of two ways: (1) using settling principles as described by Stokes' law or (2) using laser diffraction (Konert and Vandenberghe 1997, Beuselinck *et al.* 1998, Dinis and Castilho 2012). In the former, using either a hydrometer or pipette method, a soil sample is shaken in water containing a dispersing agent and then allowed to settle. By making assumptions about the average density and shape of the particles, one can calculate the rates at which the various soil separates settle. In the pipette method, subsamples are removed from the suspension at predetermined times, and their oven-dry weights are determined (Ulmer *et al.* 1994). These weights are corrected for subsampling and then converted to proportions of sand, silt, and clay by dividing them by the initial sample weight. In laser diffraction, a soil-water suspension is forced to flow between two glass panes. Lasers interact with the particles as they pass between the panes. Diffraction refers to the relationship between the particles and the angle and intensity of scattered laser light. Light scatters more intensely, and at smaller angles, off larger particles, i.e., the analyzer itself does not measure particle size, but rather, the angle and intensity of light scattered from the particles. Particle size distributions are then determined by mathematically analyzing the laser scattering pattern, using an optical model (Loizeau *et al.* 1994, Eshel 2004). For many decades, pipette analysis was the "gold standard" for particle size analysis, but it is gradually being replaced by data from laser diffraction

units. Fortunately, after some small conversion, laser data have been found to provide comparable results to previously accepted methods of particle size characterization (Beuselinck *et al.* 1998, Eshel *et al.* 2004, Arriaga *et al.* 2006, Miller and Schaetzl 2012).

Landscapes: The Exciting Additional Detail, Insight, and Interpretation Provided by Laser Diffraction Data

It is not something that many people would get excited about – additional and better particle size data! But for many soil scientists, geomorphologists, and sedimentologists, the added detail that laser diffractometry provides about soils and sediments is truly exciting.

Using the time-consuming, traditional (pipette) methods of particle size analysis (psa), researchers were happy to have accurate data for clay, silt, and sand contents of soils. Additional lab work allowed for five or more "sand splits," e.g., coarse sand or fine sand, to be determined. These data were the best available, and a tremendous amount of high quality research was accomplished using these traditional psa methods. For example, part A of the figure shows a detailed depth plot of sand, silt, and clay data for the Farm Creek stratigraphic section of loess (wind-blown silt) units and tills in central Illinois (Follmer *et al.* 1979). At the time, such data were indeed impressive.

But look what can be accomplished today, in a fraction of the time. The data in part B of the figure are from some samples that we have recovered from a deep (>5 m) core in Wisconsin, where thick loess overlies pre-Wisconsin till, which in turn rests on sandstone residuum. In that respect, the site is not unlike the one at Farm Creek. But look – instead of three (or seven) particle size breaks along the 0–2 μm spectrum, most laser diffraction units provide a hundred or more! These data can then be plotted as continuous curves. These types of data provide a great amount of detail that can be used to

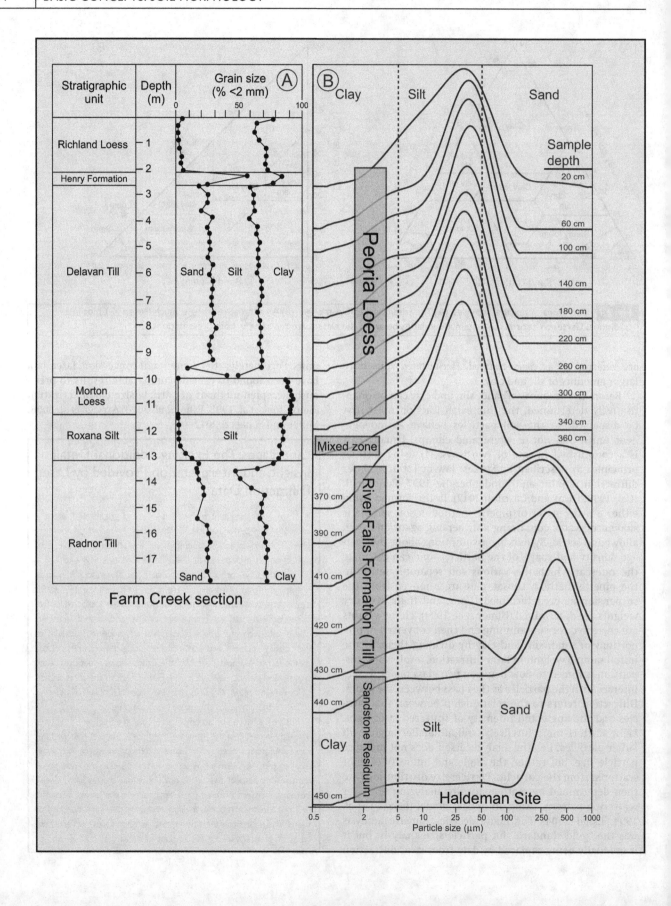

Farm Creek section

Haldeman Site

interpret sedimentological and pedological history (Mason and Jacobs 1999, Hobbs *et al.* 2011, Schaetzl and Luehmann 2013). Note how the increase in clay (the Bt horizon) at 20 cm is shown in the laser data – as a broad, uplifted "shoulder" between 2 and ~10 μm. The detail is tremendous, and the potential interpretive gains – exciting!

Nonetheless, the data from the laser diffraction unit are more difficult to plot along equal depth increments, because of line overlap. An additional disadvantage of the laser diffraction data is that clay is viewed differently from that measured in the pipette method. When compared to the pipette method, laser diffractometry commonly underestimates the amount of <2 μm clay (Loizeau *et al.* 1994, Beuselinck *et al.* 1998, Mason *et al.* 2007), although there is good correlation between the two methods. For this reason, Konert and Vandenberghe (1997) suggested that a clay–silt break of 8 μm be utilized in laser diffractometry, in order to facilitate comparisons with traditional particle size analysis data. In our lab, we use a clay–silt break at 6 μm, instead of 2 μm, as defined in most particle size schemes (Fig. 2.2).

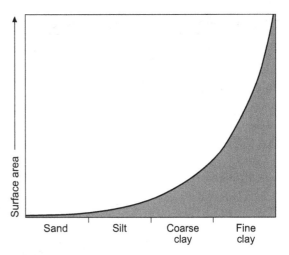

Fig. 2.8 General relationship between particle size and surface area in soils.

Sand, silt, and clay content data, when plotted on a ternary diagram called a *textural triangle*, are used to determine the textural class of a soil. The standard textural triangle (Fig. 2.7A) has been in use for decades for this purpose. Elghamry and Elashkar (1962) developed a textural triangle that looks quite different, but functions similarly (Fig. 2.7B), much like an *X-Y* plot in a traditional Cartesian coordinate system.

Particle size classes that are totally dominated by one size fraction are simply named for that fraction, e.g., *sand*. Alternatively, *loamy* textures are not dominated by any one size fraction, but have the feel of more than one. Note that a sample with equal proportions of clay, silt, and sand has a *clay loam* texture, because the larger surface area of the clay particles dominates the "feel" of the sample. Names of most texture classes have a modifier, e.g., loamy sand or silty clay. In these cases, the dominant feel of the sample is given by the last name, while the modifier (first name) implies that its texture grades toward another texture class, loam or silt.

Like coarse fragment content, texture is important for a number of reasons (Jury and Bellantouni 1976, Poesen and Lavee 1994, Ugolini *et al.* 1996). Most importantly, texture affects pores sizes, and thus, the way water moves through and is retained within the soil (Salter and Williams 1965, Harden 1988b, Bennett and Entz 1989, Lin *et al.* 1999). In saturated flow, water moves rapidly through coarse-textured soils such as sands and those with large amounts of coarse fragments, because they have larger pores and little surface area to attract the water with matric (suction) forces (Brakensiek and Rawls 1994). In clays and fine-textured soils, pore spaces are usually smaller and often not well interconnected, leading to low permeabilities and rates of saturated flow.

Texture and coarse fragment content greatly affect soil surface area. Surface area is important to soils because particle surfaces attract/retain water, cations, anions, and nutrients. Surfaces also acquire coatings, which impart color to the soil. Surface area increases as texture class becomes finer and is especially high in fine clay (Fig. 2.8). The high surface area of clayey soils means that they can retain much more water than coarse-textured soils. Nonetheless, much of this water is held so tightly that plants cannot extract it from the clay surfaces.

The clay fraction of a soil is the most chemically active fraction. It includes both minerals, such as layer silicates and oxides, and organic matter. The more clay that exists in a soil, the more chemical reactions such as ion exchange and adsorption will influence the soil's behavior. Besides regulating plant growth and the fate of soil amendments, these chemical reactions have enormous impact on the pathways of pedogenesis in soil.

Color

Color is perhaps the first thing we notice about soil. We see a soil's color long before we touch, smell, or taste it.[1] Color in soils is a function of the types of pigments (Table 2.4), as well as their abundance and distribution within the soil matrix. Pigments can occur as coatings (cutans) on the

[1] Yes, some of us do smell and taste soil! Taste is a good way to differentiate silt from sand and clay. Perhaps "taste" is the wrong word, because most soil "tasters" do little more than gently pass the soil sample between their teeth, to determine its grittiness. Taste is, nonetheless, a convenient way to distinguish (in the field) salic horizons from those that lack soluble salts.

And while you're at it, smell a newly exposed soil sometime. It smells great. Much of that "earthy" smell is the aroma of actinomycetes that are active in the soil, especially after a warm rain.

Table 2.4 | Soil colors and the primary pigmenting agents that create them

Color of coating[a]	Typical form of coating
Black or brown	Humus or magnetite
Black or bluish black	Reduced manganese (Mn^{2+}) oxides, pyrite, and iron monosulfides
White	Sodium salts, carbonate salts, gibbsite, or uncoated quartz grains
Amber yellow or brown (2.5YR)	Jarosite [$KFe_3(SO_4)_2(OH)_6$]
Light gray	Uncoated quartz grains; salts
Brown and yellowish brown (7.5YR to 2.5Y hues; 10YR most common)[b]	Goethite [α-FeOOH]
Reds, browns, and oranges[b]	Iron oxide minerals or amorphous iron oxides —Deep red (5YR–2.5YR or redder): hematite [α-Fe_2O_3] —Reddish brown (5YR [to 7.5YR]): ferrihydrite [$Fe_2O_3 \cdot 0.5(H_2O)$] —Reddish brown: maghemite [γ-Fe_2O_3] —Orange (7.5YR 5/8, 6/8, 7/8) and bright orange: lepidocrocite [γ-FeOOH]

[a] White or light-colored soil horizons are often devoid of coatings, allowing the color of the quartz particles that dominate their mineralogy to show through.

[b] Based on data provided from various sources, including Davey et al. (1975), Hurst (1977), Soileau and McCracken (1967), Torrent et al. (1983), and Vepraskas (1999); these characteristics should be used as guidelines only; many exceptions can and do occur; hues are not considered reliable indicators when color values and chromas are <3.5, because of masking by (usually) organic matter.

larger, more visible grains, e.g., sand, or as small independent particles between these larger grains. The late Francis D. Hole, a soil scientist at the University of Wisconsin (see the Dedication section of this book), made frequent reference to "soil paint" when discussing the coatings on soil particles. If no coatings exist, we mainly see the clean soil particles, which usually impart a light or white color, if the main minerals are quartz. Occasionally pigments exist in soils as small, clay-sized particles, isolated between larger soil separates, e.g., sand grains. In this case, much more pigment will be needed to create a deep soil color, as compared to thin coatings of a pigment on the grains per se.

The Munsell Color System

Qualitative verbal descriptions of color imply different things to different people. What is reddish brown to one person is brownish red or dusky red to another. For this reason, soil scientists have objectively reported color by comparing samples to standardized color chips in Munsell© Color Charts. These books are standard equipment for anyone needing to describe accurately the color of a soil or rock sample.

The Munsell system takes advantage of the fact that color is composed of three elements: *hue*, *value*, and *chroma*. Hue refers to the chromatic composition of the light (wavelengths) that emanates from the object; think of it as the actual spectral color, like red or yellow. Each page of the Munsell charts contains several color chips, all with the same hue. *Hue* is most commonly represented by the abbreviations R for red, Y for yellow, and YR for yellow-red. Hue ranges from 2.5 to 10, e.g., going from red to yellow the hues are 10R, 2.5YR, 5YR, 7.5YR, 10YR, 2.5Y, 5Y, etc. Most

well-drained, midlatitude soils have 10YR or 7.5YR hues. The factor that most influences hue is clay mineralogy; red soils are hematite-rich, brown soils are goethite-rich, and white soils can be rich in salts or carbonate minerals (Table 2.4). *Value* describes the darkness or lightness of the color. Some refer to value as the intensity of color. It ranges from 0 to 10 and is displayed along the vertical axis of each Munsell page. As the color value changes, the amount of white or black pigment that is added to the color changes: 0 is black and 10 is pure white. High color values are very light, as if the chip has been faded by exposure to sunlight. Low color values imply dark colors, which generally correspond to high amounts of organic carbon (humus) and/or wetness (Fig. 2.9). For this reason, soil colors should always be measured at consistently "field moist" conditions, unless a dry color is required (Soil Survey Staff 1999). *Chroma* refers to the purity, strength, or grayness of the color; it is also ranked on a scale of 0 to 10. At a chroma of 0, all hues converge to a single scale of neutral grays, referred to as N0 (N for neutral). In essence, chroma is changed (reduced) as more and more gray is introduced into the color.

When referring to a soil color, the three components are listed in this order: hue value/chroma. For example, 5YR 5/3 describes a reddish brown soil, while a 2.5Y 6/8 soil is olive yellow. Each color chip in the Munsell book has been assigned a color descriptor or adjective that can be used to facilitate communication of the color. Humans think in color names, not color chip numbers! Although these descriptors are provided to help us visualize a soil color, some of them are difficult to imagine without looking at the book, e.g., light greenish gray.

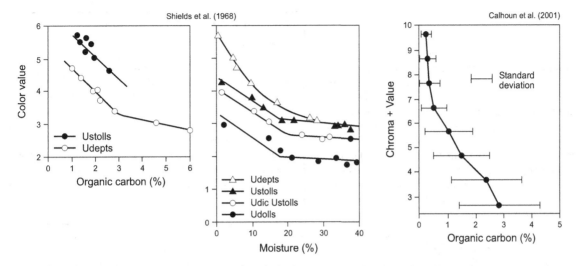

Fig. 2.9 Relationships between Munsell color and various soil properties. (A) Color value versus organic carbon content. (B) Color value versus soil moisture. (C) Color chroma + value versus organic carbon content.

Soil color is often a difficult soil parameter to describe in a repeatable manner, because people see colors slightly differently and because different light conditions affect soil color considerably (Anton and Ince 1986, Reheis *et al.* 1989). For this reason, determining the color of a soil sample should be done in uniform, bright light, preferably sunlight. Although it is impossible to match every soil sample exactly to a color chip, one should report the chip that is the closest match, or perhaps the two closest matches. Beware, however, that all Munsell books are not the same; the color chips can become soiled and faded with age, especially if they have been exposed to direct sunlight too often. Use of older and used Munsell books adds another level of error and uncertainty into soil color descriptions and should be avoided.

Origins of Soil Color

The color of a soil particle, soil sample, or soil horizon is a function of the type and degree of its various coatings (Table 2.4), and almost all soil particles have some degree of this "paint" on them. Dark colors usually imply organic matter coatings, although manganese concretions are also usually black-blue. Red colors result from various iron-bearing minerals. Only in E horizons (and some strongly gleyed horizons) are soil materials stripped so "clean" of coatings that the color of the primary minerals shows through. Because most soils are dominated by quartz, at least in the sand fraction, and quartz is dominantly clear, white, or light pink in color, E horizons may be white, light brown, or pale-colored. Color provides a great deal of information about the particle and ped coatings in the soil, which immediately impart genetic clues. For example, dark colors suggest wetness and high organic matter contents (Plice 1942). Prolonged anaerobic conditions, typical of wet soils, produce muted gray colors and low chromas, a condition referred to as *gleyed*. On the other hand, bright (high-chroma) reds and browns are generally associated with well-drained, upland soils (see Chapters 13 and 14).

As any painter knows, the less area to be covered, the more coats that can be applied with the same amount of paint, and the deeper and richer the color will be. Thus, soils with low surface areas, i.e., the coarse-textured ones, are most easily and quickly colored, or stripped of their coatings/color. This is why sands can change color rather quickly, in response to pedogenic processes. As a result of this characteristic, Soil Taxonomy (Soil Survey Staff 1999) does not allow cambic horizons (subsurface horizons with minimal evidence of soil development) to be sandy – the coloration associated with cambic-type development simply develops too quickly in sandy soils (see Chapter 8). Similarly, because sandy horizons are more quickly and completely stripped of their soil "paint," not only can E horizons form much more rapidly in sandy than in loamy soils, but sandy soils also usually have higher Munsell values.

While many soils are a potpourri of color, others are quite uniform. Intra- and interhorizon differences in soil color are usually due to different coatings or variations in degree (completeness) of the coating; they are worthy of description because they can provide important clues about genesis. "Spots" of one color set amid a matrix of another, e.g., mottles, might imply some sort of accumulation process, e.g., when sodium salts precipitate into a darker soil matrix. For mottled soils, the colors of the matrix *and* the mottles should be described, and the abundance of the latter noted, often by using charts such as those in Fig. 2.4.

Redoximorphic Features

Wet soils, especially those that undergo periodic saturation and aeration cycles, commonly acquire a patchy network of color patterns that result from changes in the oxidation-reduction status of the soil solution. Previously called mottles, *redoximorphic features* provide important information about the long- and short-term wetness status of a soil, as described further in Chapter 13. The description of redox features is a standard component of soil profile descriptions. Normally, the soil scientist will describe four components of redox features: (1) their quantity, as a percentage; (2) their size; (3) their color; and (4) their color contrast (Vepraskas 1999, Schoeneberger *et al.* 2012).

Landscapes: A Typical Soil Profile Description

We present in the following a profile drawing and description of the Antigo silt loam, the state soil of Wisconsin, in the United States. Note the repeated sequence of descriptors for each horizon: horizon name, depth to the top and bottom of the horizon, color, texture, structure, consistence, roots, other distinguishing characteristics, coarse fragment content, pH, and boundary. The Antigo soil is formed in a stacked sequence of two different parent materials – loess over glacial outwash, with a mixed zone between. The horizon designations show this distinction by the use of numbers as prefixes on the horizon names (see Chapter 3).

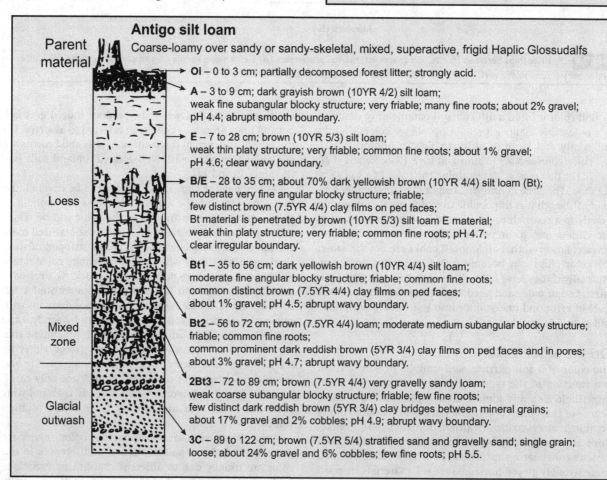

Antigo silt loam

Coarse-loamy over sandy or sandy-skeletal, mixed, superactive, frigid Haplic Glossudalfs

Parent material

Oi – 0 to 3 cm; partially decomposed forest litter; strongly acid.

A – 3 to 9 cm; dark grayish brown (10YR 4/2) silt loam; weak fine subangular blocky structure; very friable; many fine roots; about 2% gravel; pH 4.4; abrupt smooth boundary.

E – 7 to 28 cm; brown (10YR 5/3) silt loam; weak thin platy structure; very friable; common fine roots; about 1% gravel; pH 4.6; clear wavy boundary.

Loess

B/E – 28 to 35 cm; about 70% dark yellowish brown (10YR 4/4) silt loam (Bt); moderate very fine angular blocky structure; friable; few distinct brown (7.5YR 4/4) clay films on ped faces; Bt material is penetrated by brown (10YR 5/3) silt loam E material; weak thin platy structure; very friable; common fine roots; pH 4.7; clear irregular boundary.

Bt1 – 35 to 56 cm; dark yellowish brown (10YR 4/4) silt loam; moderate fine angular blocky structure; friable; common fine roots; common distinct brown (7.5YR 4/4) clay films on ped faces; about 1% gravel; pH 4.5; abrupt wavy boundary.

Mixed zone

Bt2 – 56 to 72 cm; brown (7.5YR 4/4) loam; moderate medium subangular blocky structure; friable; common fine roots; common prominent dark reddish brown (5YR 3/4) clay films on ped faces and in pores; about 3% gravel; pH 4.7; abrupt wavy boundary.

2Bt3 – 72 to 89 cm; brown (7.5YR 4/4) very gravelly sandy loam; weak coarse subangular blocky structure; friable; few fine roots; few distinct dark reddish brown (5YR 3/4) clay bridges between mineral grains; about 17% gravel and 2% cobbles; pH 4.9; abrupt wavy boundary.

Glacial outwash

3C – 89 to 122 cm; brown (7.5YR 5/4) stratified sand and gravelly sand; single grain; loose; about 24% gravel and 6% cobbles; few fine roots; pH 5.5.

Pores, Voids, and Bulk Density

About half of the volume of many soils is simply voids that are filled with air and water (Fig. 2.1). A well-aerated soil may have up to two-thirds void space. In compacted soils, voids are reduced in size and number. The amount of void space in a soil is termed *porosity* and is usually expressed on a percent-by-volume basis, e.g., 45% porosity. Soils with few voids or pores are said to be densely packed. Porosity should not be confused with sorting, which refers to the variety of grain sizes within a sample or soil. Well-sorted soils are dominated by a narrow range of grain sizes, while poorly sorted soils have a wide variety of grain sizes. In general, however, poorly sorted soils tend to be closely packed, as they have many small particles that can fill large pore spaces between the larger particles.

One of the main factors that maintain or create high porosities in soils is soil biota, especially macroscopic soil fauna. Worms, termites, and many other forms of fauna move through the soil and, in so doing, leave behind preferred pathways as biopores (Dexter 1978). High amounts of aggregation also help to maintain high porosity, as large pores can be easily maintained between soil aggregates (or peds). Other processes that help form and maintain porosity included freeze-thaw and wet-dry activity. Pore space tends to decrease with depth, where soil biota and aggregation, as well as the frequency of the cycles mentioned previously, are increasingly less influential.

Soil scientists often group pores into two main categories: *macropores* and *micropores*; pores of intermediate size are sometimes called *mesopores*. There is no firmly agreed-on, predetermined size cutoff between the two pore types, but a common standard is that macropores are larger than 0.075 mm (75 μm) while micropores are smaller than 30 μm (Kay and Angers 2000). Gases and liquids move through large pores faster and more efficiently than through small pores, although connectivity and tortuosity of pores are also important properties that affect flow rates. Macropores are commonly formed by roots and biota, e.g., worms, and can readily accommodate most roots. They can occur between sand grains in a sandy soil and between peds (aggregates) in medium- and fine-textured soils, where they are called *packing pores* and take on a crudely planar shape. When formed by roots or burrowing animals, the resulting *biopores* tend to be tubular in shape. Macropores conduct water and air rapidly, and they drain quickly after being filled with water (Edwards *et al.* 1988). Micropores, on the other hand, retain water for long periods, as a result of surface tension and matric forces. They tend to be the main type of pore in fine-textured soils, with the exception of interped pores. For a soil description, the quantity and size of cylindrical pores that reflect insect or worm burrows or abandoned root channels are assessed by observing a vertical soil face.

The soil solution, consisting of liquids and solutes, is stored within pores and passes through them. Pores that are large and well connected to each other allow fluids like water and air to pass more readily than do small, isolated voids. Many clayey soils have high amounts of pore space but do not conduct water and gases at rapid rates because the pores are so small. In contrast, large and interconnected pores in sandy soils facilitate rapid movement. The property of *permeability*, representing the ability of a soil horizon to transmit fluids, is determined by pore size and continuity. Permeability is one factor that determines *flux density*, i.e., the rate at which a fluid will move through a soil horizon. Flux density is measured in terms of the fluid volume passing through a unit area per unit of time, e.g., cm^3 cm^{-2} h^{-1}, which reduces to length per unit time, e.g., cm per hour. The rate of movement generally varies depending on the direction of flow. For example, vertical fluxes may be faster than horizontal fluxes. Flux also varies as a function of the hydraulic gradient, whether the soil is saturated with water or unsaturated, and with temperature and viscosity of the soil solution (see Chapter 6).

Contrasting with porosity is the volume occupied by clastic and other solid materials. The mass of these materials per unit volume of soil is referred to as *bulk density*. A soil with no voids or pores whatsoever would have a bulk density of ~2.65 g cm^{-3}, because the particle density of most soil-forming minerals is about 2.65–2.70 g cm^{-3}. Thus, 2.65 forms a theoretical upper limit of soil bulk density, although in reality only a few very dense, compacted soils achieve bulk densities >2.3 g cm^{-3}. Most organic soils and O horizons have bulk densities well below 1.0; in theory these soils would float on water, which has a bulk density of 1.0 g cm^{-3}. Most mineral soils, in their natural state, have bulk densities of 1.1 to 1.8 g cm^{-3} (Manrique and Jones 1991). Bulk density determined on moist samples usually best reflects the field condition. It is measured by dividing the oven-dry (100 °C) mass of a soil sample by the moist volume of the sample; the weights and volumes of any coarse fragments are determined and excluded, so that a few large rocks within the sample do not skew the data (Reinhart 1961, Curtis and Post 1964, Vincent and Chadwick 1994).

Any of several methods can be used to obtain bulk density information for soils (Shipp and Matelski 1965, Blake and Hartage 1986, Laundré 1989, Vincent and Chadwick 1994). In theory, a sample of known volume is removed from the soil, and after drying, the coarse fragments are removed. The mass of the remaining fine earth is divided by the initial sample volume (minus the volume of the coarse fragments) to arrive at the bulk density value.

Factors that affect bulk density and porosity have been the subject of some study. Bulk densities tend to increase with depth in soils, just as porosities decrease. Soils and horizons high in organic matter tend to have lower bulk densities, probably because many of these samples are from near-surface horizons where biological activity is high and because organic matter tends to attract soil fauna, which create pores (Adams 1973, Alexander 1980; Fig. 2.10). Somewhat surprisingly, clayey and silty soils tend to have low bulk densities while soils with sandy loam and sandy clay loam textures are more prone to compaction. High bulk densities are often associated with loamy soils, where clay and silt can fill in the large voids between sand grains, optimizing the packing of the soil matrix. Clayey soils have many micropores, which are difficult to eliminate, helping to explain their relatively low bulk densities (Heinonen 1960, Bernoux *et al.* 1998).

Structure

Structure refers to the arrangement of primary soil particles, e.g., sand, silt, and clay, into natural aggregates called *peds*. Aggregates formed by tillage or other human-induced practices – involving compaction – are called *clods*. Aggregates are naturally occurring clusters or groups of soil particles in which the forces holding the particles together are stronger than the forces between them (Chenu and Cosentino

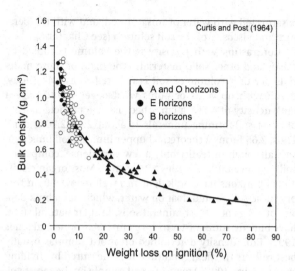

Fig. 2.10 Relationship between bulk density and loss on ignition in the O, A, and B horizons of some soils of the Green Mountains, Vermont. Loss on ignition is a surrogate for organic matter content.

2011). Aggregation implies the coexistence of fine pores (within the aggregates) and coarse pores (between them), leading to longer-term structural stability and favorable water-flow dynamics. Large amounts of coarse fragments tend to inhibit the formation of stable structural aggregates and render structure less important to properties like infiltration capacity or porosity. Conversely, fine-textured soils tend to have greater structural stability.

Soil horizons have their own unique structures, depending on the amount of organic matter and biotic activity, texture, numbers of freeze-thaw and wet-dry cycles, and other factors. Soils that lack pedogenic structure are termed massive (in fine-textured materials that possess no preferred planes of weakness) or single-grained (in sands where each grain behaves independently). How structure forms and is maintained, and what the particular structure type implies about pedogenesis, are our focus in the sections that follow.

Peds, or soil aggregates, must have strong internal cohesion that prevents their destruction by the disruptive forces that continually occur in soils. What kind of binding agents perform this function in soils? Humus (highly decomposed organic matter) is a common "glue" in A horizons, as are fungal hyphae and the various fluids secreted by soil fauna, e.g., saliva and urine (Degens et al. 1996, Guggenberger et al. 1999). No one questions that soil microorganisms are important agents of structure formation, stabilization, and destruction, and, as such, they influence soil structural dynamics (Andrade et al. 1998, Baldock 2001, Chenu and Cosentino 2011); their role in aggregate stabilization was recognized years ago (Allison 1968, Martin and Waksman 1940). To that end, many peds are simply fecal pellets of macrofauna, held together by body fluids and prior compaction (Sutton and Sheppard

1976, Martin and Marinissen 1993, Zhang and Schrader 1993, Tisdall et al. 1997). In clayey soils, peds may be stable because of the attraction that clays have for each other, i.e., cohesion. Salts and oxides of iron and aluminum may also act as binding agents (Giovannini and Sequi 1976).

As with porosity, wet-dry cycles, freeze-thaw cycles, root activity, and fauna all play a role in the formation of soil structure (Materechera et al. 1992). Shrinking and swelling are common to certain types of clays when they undergo alternating desiccation and wetting. An index of the degree to which a soil will expand upon wetting and shrink upon drying is the coefficient of linear extensibility (COLE). Soils with higher COLE values often have smaller peds, other characteristics being equal. Finally, we note that structures formed in fine-textured soils tend to be more persistent and less likely to be destroyed by tillage, compaction, and so on, because their cohesive forces are stronger.

The shape and size of peds are characteristic for the most stable aggregate that the multitude of processes that operate within a soil can maintain, and when enough pedogenic glue is available, a particular ped type (shape and size) will become dominant. Opposing this natural aggregation are the forces that break peds apart, many of which are the same ones that create porosity; these all decrease in intensity with depth. For that reason, ped sizes generally increase with depth, and in weathered parent material, soils can be structureless or massive, or in some cases they may display inherited, geologic structure.

Three important attributes of soil structure are recorded in soil descriptions: ped grade, size, and shape. *Grade* describes the distinctness of structural units, i.e., their ease of separation into discrete peds and how the peds hold together when the soil is handled. The classes of structural grade are weak, moderate, and strong. These class names are often confused with soil consistence (soil strength), so it is often helpful to think of the grade classes as weakly expressed, moderately expressed, and strongly expressed, The *size* of peds, e.g., 2–5 cm in diameter, is determined by comparison with standard charts (Schoeneberger et al. 2012).

Perhaps the most readily identifiable aspect of structure is its *shape*. Different structural shapes can be used to infer specific kinds of pedogenesis. Peds in near-surface A horizons are usually *granular* in shape and are typically only about 1–5 mm in diameter. The granular (in older literature, *crumb*) structure common to A horizons owes much of its existence to soil biota, as many of these more or less spherical peds are worm casts, or they have been reworked many times by soil fauna or have been formed by the cement of organic exudates from small roots (Blanchart 1992, Zhang and Schrader 1993). Buntley and Papendick (1960) described soils that were so intensively worked by worms that they had granular and small subangular blocky structure down to several tens of cms below the surface. Dense networks of fine roots help maintain this kind of structure by proliferating on ped faces and only

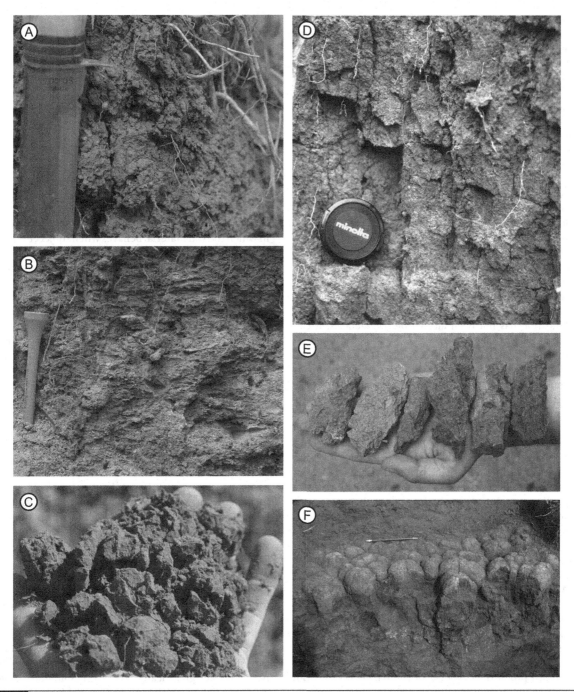

Fig. 2.11 Shapes of common soil structures. Photos by RJS unless otherwise indicated. (A) Granular soil structure, from an Ustoll formed under native grassland vegetation. Knife blade for scale. (B) Platy soil structure. Golf tee for scale. (C) Angular blocky structure. Lens cap for scale. (D) Prismatic structure from an Entisol in Kansas. (E) Prismatic peds removed from a buried paleosol formed in loess. (F) Columnar structure in a Typic Natrustoll from Grant County, North Dakota. Pencil for scale. Photo by J. Arndt.

infrequently penetrating the ped itself (Fig. 2.11A). Their continued growth literally squeezes the peds and helps to maintain ped integrity. High amounts of organic matter (humus), typical of A horizons, contribute to the strong intrapedal cohesive forces. Granular peds tend to be porous, and because they do not fit together tightly, horizons with granular structure have comparatively high porosities and permeabilities (Buntley and Papendick 1960). *Platy* structure consists of thin (<4 mm thick) platelike peds that are aligned parallel to the soil surface (Fig. 2.11B). This type of structure is common in E horizons and if well developed can impede infiltration. Platy structure is usually due to wet-dry and/or freeze-thaw cycles, which are transmitted into the soil along planes that parallel the surface (Fedorova and Yarilova 1972, Howitt and Pawluk 1985b). High amounts of organic matter, and the large numbers of soil fauna that go with it, tend to disrupt platy structure, which may help explain why it is most common in humus-poor E horizons. One question must always be asked when describing platy structure: "Is it pedogenic or geologic?" Many surficial sediments are initially stratified, and if this stratification has not been altered by pedogenesis, it can appear as plates or laminae.

In the lower part of the profile, peds become larger and more blocklike. *Angular* and *subangular blocky* structure is common in B horizons (Fig. 2.11C). In angular blocky structure, the edges of the blocks are sharp and the rectangular facets between them are distinct, whereas the more common subangular blocky structure has more rounded edges. Commonly, blocky peds are held together by coatings (cutans) of material that has been translocated into this horizon from above. This material has moved from the surface A horizon into the subsurface B horizon in suspension or solution, in the large pores between peds. Eventually, some of the percolating water soaks into the peds and in so doing any material being carried along is plastered on the ped face (see Chapter 13). As with granular structure, root proliferation between blocky peds helps to maintain them.

At still greater depths, the forces that act to break up the soil matrix into distinct peds not only become less frequent, but tend to be more vertically oriented (Fig. 2.11D, E). Blocks that are taller than they are wide, prisms, will form under this scenario. *Prismatic structure* is, therefore, typical of the lower B horizon and C horizons. Prisms can reach mammoth proportions; widths of the flat prism tops can be up to 150 cm, and lengths of 200 cm are not uncommon. A particular type of B horizon, a fragipan, consists of large, distinct prisms that many believe harken back to times when permafrost and freezing processes extended to several meters depth (see Chapter 13).

In areas where soils have high amounts of exchangeable sodium, the tops of the prisms become rounded and sometimes appear white from an abundance of clean quartz grains; this type of structure is called *columnar* (Fig. 2.11F). The whitish caps on the columns are said to resemble "biscuit tops." The tops of the prisms become rounded as soil water is forced to run off the tops and flow down between them, because pore size limits the flow of water within the peds. Thus, the edges are worn away and take on a rounded shape, but the bases may still be flat and angular, as in prismatic structure.

Consistence

Consistence refers to soil strength. The Soil Survey Division Staff (1993) provided a more comprehensive definition of consistence, including (1) resistance of soil material to rupture; (2) resistance to penetration; (3) plasticity, toughness, and stickiness of puddled soil material; and (4) the manner in which the soil material behaves when subject to compression. Although consistence can be evaluated at wet, moist, and dry conditions, it is highly dependent upon moisture content, so the moisture content at the time of description must be noted. Examples of moist consistence classes are loose, very friable, friable, firm, very firm, and extremely firm. *Friable* is a common type of moist consistence, referring to the ease with which a soil crumbles when pressure is applied; friable soils crumble easily between the fingers. Determining soil consistence is somewhat subjective.

Presentation of Soil Profile Data

In soil genesis studies, soils are usually described on a horizon basis, and samples are collected to represent the entire horizon (Schoeneberger *et al.* 2012). Less commonly, soils are sampled at different depths along predetermined increments, regardless of horizons. In order to evaluate and compare these data, they are commonly presented either in a table or as a *depth plot* or function. Depth plots are the standard within the field, providing a graphical way of visualizing changes in soil properties throughout the profile.

Graphically, the user has several options for presenting soil data in depth plots. In a traditional depth plot, the data for each horizon are shown as a single point at the depth centroid of the horizon (Fig. 2.12). However, many different methods exist for plotting data (Ponce-Hernandez *et al.* 1986, Bishop *et al.* 1999, McBratney *et al.* 2000). Therefore, care should be taken when choosing the best way to analyze soil data graphically (Fig. 2.12). Some also advocate acquiring data from the soil surface and from the very bottom of the pit/profile, and using these data as "anchors" for the depth function.

Soil Micromorphology

Pedologists and soil geographers routinely use microscopic examinations of intact soil material to determine the arrangement of the various components of the soil matrix.

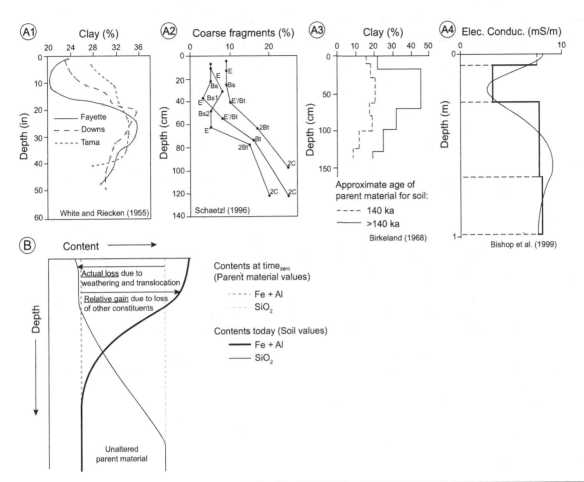

Fig. 2.12 Examples of the variety and utility of depth plots of soils data. (A) Examples of different ways to portray soil data acquired at various depths graphically, i.e., different types of depth plots. (A1) Smooth lines, drawn freehand, connecting symbols placed at the center of each horizon. (A2) Straight lines connecting symbols placed at the center of each horizon. (A3) Bars drawn and connected together as in a step function, allowing the reader to visualize horizon thicknesses better. (A4) Equal-area quadratic splines. (B) Examples of how depth plot data can be used to evaluate and determine morphologic changes – and assess pedogenic processes – in soils over time.

One application of this approach is to determine whether a sediment sample, taken out of context, is from a soil, i.e., it has pedogenic fabric – or from a geologic deposit. This application is particularly useful in paleopedology – the study of buried soils (see Chapter 16). This aspect of soil science – *soil micromorphology* – involves the study of the materials and fabric of individual peds, or parts thereof. Usually, microscopic examinations of soil fabric are performed to help understand the processes that have interacted to form the soil matrix, e.g., clay translocation, accumulation of salts, burrowing of fauna, and weathering. Many processes of soil formation involve additions or removals of material from the soil as well as transformations of materials, the manifestations of which are visible in the soil fabric.

Why study soil micromorphology? Perhaps the most ingenious and straightforward answer to this question was provided by Kubiëna (1970), who imagined someone who wanted to learn about watches and how they work, yet had never seen one before. How would/should he or she begin to understand them? There would be any of several ways:

1. Put each watch in a mortar, grind it to a powder, and determine the chemical composition of the whole.[2] This method would provide information about bulk composition, but would not provide any insight into the working parts, such as chains, levers, gears, and so forth. One would have learned only that watches are made of different metals and nonmetals. This analysis is similar to whole soil, elemental, and oxide analyses done by soil scientists in the early and mid-twentieth century.

[2] For example, see Sherman and Alexander (1959).

Table 2.5	Names traditionally given the various types of voids and pores in soils
Name	Characteristics
Packing void	Voids formed between larger particles that, because of their size and shape, do not adequately pack together
Vugh	A small cavity or unconnected void, usually with an irregular shape; most common in fine-textured soils
Vesicle	Unconnected void with smooth walls and (usually) rounded shapes
Chambers and channels	Connecting passageways
Planes	Voids elongated along one plane or axis, commonly formed at ped faces

Source: After Brewer (1976).

2. Take each watch apart, sort the myriad pieces into groups, and determine the sizes and ratios of these groups. In this way, one could classify the watches into logical groups; however, this information will still not help answer questions about how they work. Nor might one know even which watch parts interacted with others. This analysis is similar to fractionation procedures in soil science, where amounts of certain soil components, e.g., sand, clay, organic matter, feldspar, or fecal pellets, are determined and possibly ratioed to one another.

3. Each watch could be carefully inspected, while intact, such that the placement of each part would be determined, relative to each other and to the whole. Then the watches could carefully be dismantled and again, each part examined with respect to its interactions with other parts. This analysis is akin to traditional soil micromorphology, in which the parts of the soil are examined each in relation to the other.

4. Last, one could inspect the watches as they operate, tick, and chime, i.e., *while they work*. Stopping them for the sake of analysis (as was done in 1–3) deprives the investigator of important information about function. This fourth method of analysis is synonymous with the measurement of soil processes in situ. Examples include removal and analysis of soil solutions percolating through the soil, observation and measurement of the mixing activities of ants or mobile tracers, or the continuous monitoring of soil water tables or temperature.

All soils are composed of voids and solids. The solids can be divided into immobile skeletal grains, known as *skeleton*, and those solids that are also mobile, i.e., *plasma* (Corti *et al.* 1998, Egli *et al.* 2001). The skeleton is that part of the soil that is generally not mobile or readily translocated within the soil, except perhaps by large biota or freeze-thaw processes (see Chapter 11). Plasma materials include translocated clay, organic matter, and other soluble materials that are subject to mobilization or have been mobilized in the past. Features that are distinctly pedological in origin are usually identified as such because of perceived origin, plasma concentrations, mineralogy, or some

type of rearrangement of constituents (Cady *et al.* 1986). The linings of voids (Table 2.5) are preferred locations for plasma to be deposited as coatings or *cutans*. Generally, the cutan is named for its composition, e.g., calcan for calcite (Table 2.6), and location, e.g., channel cutan. These reference works provide further detail on soil micromorphological terms and methods: Kubiëna (1970), Brewer (1976), Bullock and Murphy (1983), Douglas and Thompson (1985), Courty *et al.* (1989), Nahon (1991), FitzPatrick (1993), Kapur *et al.* (2008), Stoops (2003), and Stoops *et al.* (2010).

Tools Used to Study Soil Micromorphology

Soil micromorphological analyses are most powerful when combined with soil data of other kinds and from various scales, such as a profile description and chemical data (Eswaran *et al.* 1979; Fig. 2.13). Various tools exist with which to view and study soils at different *scales* of observation. At the broadest scale, one might view the entire landscape using digital soil data in a geographic information system (GIS) or with an aerial photograph. To examine soil fabric, however, one must get close enough to the soil to actually see individual components. This can be done by examining a hand sample with the naked eye or a hand lens. At increasingly finer scales of analysis, one might view (1) whole soil samples and natural aggregates using visible light in a dissecting microscope, (2) thin sections of soils using plane-polarized light in a petrographic microscope, (3) very small but otherwise intact fragments of soils with an electron microscope. In all cases, the goal is to observe various parts of the soil in as natural a state as possible, and to determine how the parts are related to one another. For example, the micromorphologist will look for arrangement of sand and silt grains, coatings on voids and ped faces, and evidence of biotic materials.

A traditional but still useful form of micromorphological research involves low-power magnification of intact soil aggregates and fragments, often using a binocular light

Table 2.6 | Types of cutans (ped coatings) and cemented zones in soil horizons

Component	Cutan name[a]	Cemented name
Clay	Argillan, clay skin, clay film, Tonhautchen	
Silt	Silan, siltan	
Humus	Organ, organan	
Manganese	Mangan	
Crystalline salts, e.g., carbonates, chlorides, and sulfates	Soluan	
Skeleton grains, e.g., silt, very fine sand	Skeletan, neoskeletan	
Sesquioxides of iron and aluminum	Sesquan	Ortstein, placic
Ca and Mg carbonates	Calcan, calcitan	Caliche, calcrete, nari, kankar, croute calcaire
Iron	Ferran	Iron pan, laterite, plinthite, ferricrete, ironstone
Goethite	Goethan	
Allophane	Allan	
Sodium salts, especially halite	Halan	Salcrete
Gypsum	Gypsan	Gypcrete
Silica		Silcrete, duripan
Various combinations of materials	E.g., ferri-argillan, organoargillan	

[a] Cutan names are also occasionally used when referring to pore linings. Prefix modifiers such as "neo-" and "quasi-" are, respectively, used when the cutans are minimally developed or "almost like" the root cutan name, for example, neo-argillan or quasi-sesquan.

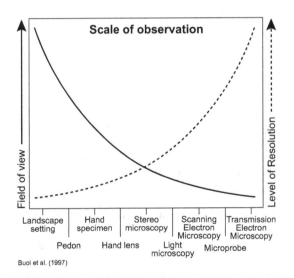

Buol et al. (1997)

Fig. 2.13 Diagram illustrating the various scales at which soils and soil (or landscape) fabric can be examined, and the tools that are used at those scales.

microscope. Little pretreatment, other than drying (which is not even necessary), is required. At this scale, typically at magnifications of 10–100× and under reflected light, it is possible to examine features that are too large for more traditional micromorphological techniques, but too small to see with the naked eye. This form of investigation provides information about the arrangement of skeletal grains and plasma and has a great deal of merit because (1) it requires little preparatory time and thus offers a good return on (time) investment and (2) it spans the scale between examination of soils in the field (macromorphology) and by the higher magnification tools of traditional micromorphological methods.

Similar types of soil materials, i.e., peds and fragments of peds, can be examined at much higher magnifications with the scanning electron microscope (SEM). Electron microscopy uses a beam of electrons to form images of a sample's surface, unlike light microscopy, which utilizes visible light. By using electrons, the method provides resolving power thousands of times greater than that of a light microscope (Smart and Tovey 1982). In addition, SEM provides a larger depth of field and greater resolution than do many optical methods.

In SEM, a high-energy focused electron beam is scanned across a sample in a raster (or pattern of parallel lines). Prior to imaging, the sample is usually dried and gold- or carbon-coated under a vacuum; the coating makes the sample surface more conductive. When the electron beam hits the sample, it interacts in various ways, and secondary or backscattered electrons are emitted. The secondary electrons are captured by a detector and used to produce a detailed map of the sample surface. The intensity of emission of both secondary and backscattered electrons is very

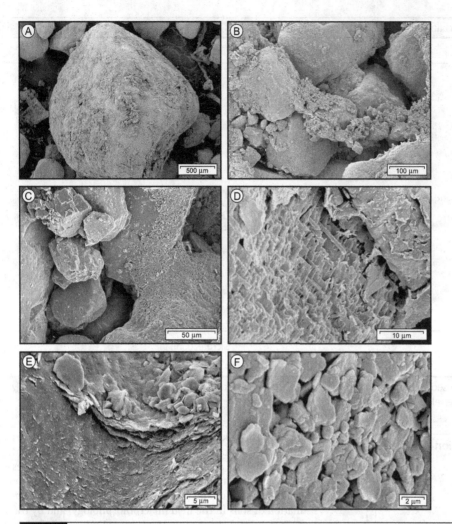

Fig. 2.14 SEM photomicrographs of material removed from Spodosols and Alfisols in Michigan, illustrating some of the various scales and resolutions that are possible in electron microscopy. Images by B. Weisenborn. (A) Coatings of amorphous and organic materials, including fecal pellets, on a large sand grain. (B) Randomly packed fine sand grains with bridges of clay and fine silt. (C) Bridge of fine silt and clay between two sand grains. (D) Pore (probably a channel) lining of amorphous material filling in and binding clay platelets along the weathered edge of a soil mineral. (E) and (F) Close-up views of phyllosilicate clays in an argillan (oriented clay coating).

sensitive to the angle at which the electron beam strikes the surface, i.e., to topographical features on the specimen. Parts of the sample that are higher or stand above others are usually brighter because they give off more secondary electrons. Because secondary electrons that are produced deeper than about 5 nm into the sample are absorbed by it, the detectors capture primarily those electrons that interact with the very surface of the sample. Thus, the image created by the detector is essentially a relief map of the surface (Cremeens *et al.* 1988; Figs. 2.14, 2.15). SEM can provide very high resolution images of particles down to fine clay size (<0.2 μm), but SEM alone cannot identify the minerals. Therefore, although SEM images have excellent detail and resolution, they are limited in some respects.

There is another very robust and useful tool can often be used in conjunction with an SEM: energy dispersive spectroscopy (EDS) or energy dispersive X-ray analysis (EDXRA). SEM/EDS analysis is very useful for the identification of the locations and amounts of certain atomic elements within a sample (Hill and Sawhney 1971, Bisdom *et al.* 1983, Pye and Croft 2007, Mujinya *et al.* 2011). This technique, performed within a modified electron microscope, takes advantage of the fact that the interaction of the electron beam with the sample also generates X-rays characteristic of the elements present. It permits nondestructive, quantitative analysis of elements with atomic numbers >11 (Na) and semiquantitative analysis of elements down to atomic number 4 (Be) (Sawhney 1986). These data can be displayed as element maps, showing

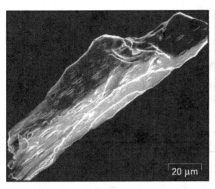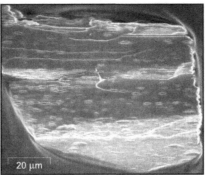

Fig. 2.15 Etch pits in amphibole grains, imaged using an SEM. Images by D. Cremeens.

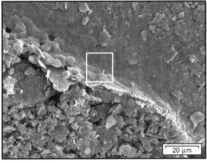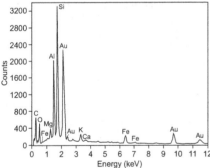

Fig. 2.16 Example of an energy-dispersive spectroscopic (EDS) spectrum collected from a soil sample with an amorphous cutan overlying a matrix of fine silt and clay particles. SEM image and associated spectrum by B. Weisenborn.

where particular elements in the sample are more dominant than elsewhere (EDXRA), or they can be displayed as an EDS spectrum, illustrating the relative amounts of various elements within a given part of the sample (Norton *et al.* 1983; Fig. 2.16). The elemental composition of materials (minerals, in this case) can be obtained with great spatial resolution. Thus, detection and mapping of the locations where various types of X-rays were produced on the sample help identify where specific minerals and amorphous materials exist at the surface of the sample (Fig. 2.16). EDS has

many applications in soil analyses of various kinds (Perry and Adams 1978, Sawhney 1986, Schaefer *et al.* 2004).

Another form of electron microscopy involves the transmission electron microscope (TEM). Very useful in studies of mineralogy, the TEM provides higher magnifications and greater detail than does the SEM (Chartres 1987). Although its main application has been in weathering-related studies of minerals (Calvert *et al.* 1980b, Ahn and Peacor 1987, Dong *et al.* 1998, Vacca *et al.* 2003), the TEM is being increasingly used to investigate undisturbed soil morphology as well (Foster 1994; Laird and Thompson 2009).

A third type of micromorphological analysis involves the examination of soil and rock thin sections with a petrographic microscope, e.g., FitzPatrick (1993). Although this method requires a significant amount of sample preparation, the rewards can be great. In this method, bulk samples a few centimeters across are removed, intact and as little disturbed as possible, from the soil profile. They are dried and then impregnated with a clear polyester or epoxy resin (Innis and Pluth 1970, Middleton and Kraus 1980). Impregnation of the sample without significantly changing its morphology is always tricky. Usually it is accomplished by immersing the dry sample in the resin while it is under vacuum to ensure that the resin fills all the pores. After the resin is allowed to harden, a thick section of about 1 cm is cut from it, polished, mounted on a glass slide, and then ground and polished to 25–30 μm thickness. Depending on the application, thin sections can be oriented in any direction. For example, if most of the pores are oriented vertically and the investigator wants to examine pore linings, horizontally oriented thin sections would be useful. If, however, the fabric of ped interiors were of most interest, thin sections might be cut vertically. Thin sections can also be examined with the SEM if the resin is hard enough and the section has been finely polished.

When examined under the petrographic microscope, soil thin sections provide a plethora of information.

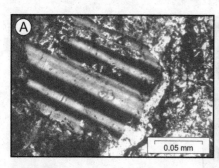 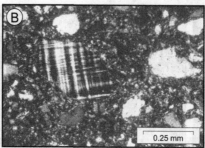

Fig. 2.17 Photomicrographs of twinned feldspar minerals in a soil thin section. The black-white stripes, or twinning pattern, alternate as the microscope stage is rotated. Cross-polarized light. Images by D. Cremeens. (A) A plagioclase feldspar from an Endoaqualf in Illinois. Note the preferential weathering of one twin versus the other. (B) A microcline feldspar from an Endoaqualf in Illinois. Note the apparent lack of weathering.

Fig. 2.18 Photomicrograph of a vugh (100 μm in width) or pore infilled with illuvial clay from the BC horizon of a Hapludalf in Pennsylvania. Plane polarized light. Image by D. Cremeens.

Minerals that are sand size or larger can often be identified in polarized light (Fig. 2.17). Pores and pore linings are clearly visible and they can be easily measured (Fig. 2.18). If desired, the thin section can be stained, bringing out the locations of certain minerals such as feldspars (Reeder and McAllister 1957, Norman 1974, Houghton 1980, Morris 1985, Ruzyla and Jezek 1987). Laminated coatings of clay (argillans) usually appear speckled with various colors as the stage is rotated in cross-polarized light. Edges and coatings of mineral grains provide information about weathering.

Disadvantages of this technique include the long preparatory times, the potential for failure during impregnation, and the inability of the analyst to view the soil in three dimensions. Advantages include the ability to see inside mineral grains and a different way to identify minerals than either of the two previous methods provided. The primary advantage of using thin sections to investigate soil morphology is that the scale of observation (ranging from 10s of μm to ~10 mm) is appropriate for evaluating many important pedogenic features such as coatings, pores, and mineral accumulations.

Chapter 3

Basic Concepts: Soil Horizonation ... the Alphabet of Soils

Soil Horizons

Most of the Earth's surface contains soils, and most soils have discernible, genetic horizons, that is, genetic layering. Horizons are what distinguishes soil material from nonsoil sediment. Soil horizons may be absent in areas where erosion rates exceed the rates of soil development, as on steep slopes. In other places, horizons may not occur because they have not yet formed in young sediment, or because the sediments are essentially inert material devoid of nutrients that would support vegetation. Regardless, the presence of one or more genetic horizons indicates the existence of a soil, and vice versa.

Early soil scientists used terms like *soil layer*, *vegetable mould*, *stratum*, and *level* to describe these genetic layers (Tandarich *et al.* 2002). For almost a century (Shaw 1927), a layer formed by pedogenic processes that is more or less parallel to the soil surface has been called a *soil horizon*. Moreover, because its origin is pedogenic, referring to it as a *genetic horizon* is also appropriate. Anderson (1987, 56) described soil horizons as the "working aggregates of the whole (soil) system, and, like the organs of an organism, ... generally adapted for the performance of specific functions." In this chapter, we describe the basic characteristics of the major soil horizons. These types of horizons are different from and not to be confused with *diagnostic horizons*, which are defined for the purposes of soil classification and will be discussed in Chapter 8.

Types of Soil Horizons

Soil horizons generally form within unconsolidated materials on geomorphically stable surfaces that have been subaerially exposed for a sufficient length of time. In fact, pedologists often use the presence of well-developed horizons as an indication that the surface below which the soil is forming has been relatively stable for a considerable period (see Chapter 14). Horizons form as material is translocated (upward, downward, or laterally) within the profile or as it is transformed (chemically or physically) in place (Simonson 1959; see Chapter 12). Pedogenic processes tend to form distinct horizons within the upper mantle of unconsolidated materials, i.e., certain types of horizons are often associated with a certain suite of pedogenic processes (see Chapter 13).

Ever since Vasili Dokuchaev and Nikolai Sibirtsev introduced them at the 1893 World's Columbian Exposition in Chicago, soil horizons have been divided into a few types of *master horizons*. Today, these are the O, A, V, E, B, C, and R horizons (Tandarich *et al.* 1988; Table 3.1, Fig. 3.1). We also recognize the D and K horizons, but acknowledge that their use is informal (see later discussion). O horizons are organic horizons that form above mineral soil materials, or bedrock. A horizons (Ah in the Canadian classification system) are traditionally assigned to "topsoil" layers rich in decomposed organic matter (humus). V horizons are highly vesicular horizons that typically form near the surface in desert soils. B horizons are zones of accumulation or alteration in the subsoil, and C horizons are seen as unaltered (or weakly altered) parent material. The letter E (Ae in Canada) is assigned to light-colored, *eluvial* horizons from which clay, organic matter, and other mobile substances have been removed (Guthrie and Witty 1982). R horizons are hard bedrock.

In soil descriptions, the master horizons are usually given suffixes that provide additional information about their characteristics, such as the presence of an *illuviated* ("washed in") substance, its degree of organic matter decomposition, or its density, among many others (Table 3.2). Capital letters are used to denote master horizons and lowercase letters are used as suffixes, e.g., Ap, Bx, and Cd horizons. Note: The suffixes are not subscripts. Horizons may have as many suffixes as are deemed appropriate, e.g., Btx, Bhsm, and BCtg (Fig. 3.2), and O and B horizons *always* have a suffix. The application of suffixes is at the discretion of the describer and is based on his or her interpretation of the horizon in question.

Horizons that are transitional between master horizons can take two forms (Table 3.3). *Gradational* horizons, where one horizon smoothly grades into another, are designated

Table 3.1 | The Master soil horizons

Master horizon	Characteristics
O	Layers dominated by organic material (litter and humus) in various stages of decomposition
A	Mineral horizons that formed at the surface or below an O horizon and (1) are characterized by an accumulation of humified organic matter intimately mixed with the mineral fraction or (2) have properties resulting from cultivation, pasturing, or similar kinds of disturbance
V	Mineral horizons that formed at the soil surface, or below a layer of rock fragments, e.g., desert pavement, or a physical or biological crust, and which are characterized by the predominance of vesicular pores
E	Light-colored mineral horizons in which the main feature is loss of weatherable minerals, silicate clay, iron, aluminum, and/or humus, resulting in a concentration of mostly uncoated quartz grains and other resistant materials (as a suffix "e" in Canadian system)
B	Subsurface mineral horizons dominated by (1) illuvial accumulations of clay, iron, aluminum, humus, and so on; (2) removal, addition, or transformation of primary carbonates or gypsum; (3) residual concentrations of sesquioxides; (4) loss of all or most nongeologic structure; (5) brittleness; and/or (6) gleying
C	Mineral horizons, excluding hard bedrock, that have been less affected by pedogenic processes and lack the properties of O, A, E, or B horizons and, therefore, likely retain some rock structure (if developed in residuum) or sedimentary structure (if developed in transported regolith); includes deeply weathered, soft saprolite (see Chapter 10)
D	Deep horizons that show virtually no evidence of pedogenic alteration, leaching of carbonates, or oxidation; retain geologic structure and are often dense and slowly permeable; are formed in unconsolidated sediments (*informal*)
R	Hard, continuous bedrock that is sufficiently coherent to make digging by hand impractical
M	Root-limiting subsoil layers consisting of nearly continuous, human-manufactured materials
W	Water/hydric layers in organic soils – often as segregated ice lenses (*Canada only*)
L	Limnic horizons or layers, i.e., materials originally deposited in water by aquatic organisms or derived from aquatic plants and subsequently modified by aquatic animals

Source: Modified from Guthrie and Witty (1982) and the Soil Classification Working Group (1998).

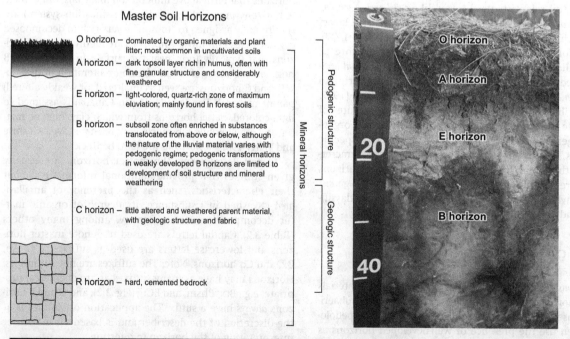

Fig. 3.1 The master soil horizons, shown (A) diagrammatically and (B) in a soil profile. Not all master horizons occur in every soil. The soil in (B) is in northern Michigan and shows an O horizon overlying a thin and somewhat wavy A horizon and a light-colored E horizon. The B horizon is specifically a Bs horizon, and the C horizon is below the depth shown. The tape scale is in centimeters. Photo by RJS.

Table 3.2 | Suffix symbols used with master soil horizons

Suffix	Used with which master horizons?	Characteristics/comments	Approximate Canadian System equivalent
O horizon suffixes			
a	O	Highly decomposed (sapric) organic material	Oh in Organic soils, H in forest litter
e	O	Organic material of intermediate (hemic) decomposition	Om in Organic soils, F in forest litter
i	O	Slightly decomposed (fibric) organic material	Of in Organic soils, L in forest litter
A horizon suffixes			
p	O, A	Plowed, tilled, or otherwise disturbed surface layer; disturbed A, E, B, and C horizons are designated Ap	p
B horizon suffixes associated with illuviation			
c	B, C	Presence of concretions or hard nonconcretionary nodules, usually of Fe, Al, Mn, or Ti	cc
h	B	Accumulations of amorphous and dispersable organic materials (humus); moist Munsell value and chroma ≤3	h
j	B, C	Accumulation of jarosite (iron hydroxyl sulfate), as either ped coatings or nodules	
k	B, C	Accumulation of secondary, pedogenic carbonates, commonly $CaCO_3$, as ped coatings, filaments, masses, or nodules, but <50% by volume	ca
kk	B, C	Engulfment of horizon by secondary, pedogenic carbonates, implying that the soil fabric is plugged (>50% by volume) and massive; corresponds to stage III (or greater) development (see Chapter 13)	
m	B	Horizon that is >90% cemented or indurated (fracturing is allowed); implies physical root restrictions	c
n	B, C	Accumulation of pedogenic, exchangeable sodium (Na), commonly as sodium salts	n
q	B, C	Accumulation of pedogenic, secondary silica (quartz), exclusive of fragipans	–
s	B	Accumulation of illuvial, amorphous, dispersible complexes of Fe and Al sesquioxides	f
t	B, C	Accumulation of silicate clay, as argillans on ped faces or lamellae (clay bands)	t
y	B	Accumulation of pedogenic gypsum	s or sa
yy	B	Horizon that is so dominated by secondary, pedogenic gypsum that pedogenic or lithologic features are obscured or disrupted by gypsum crystals	–
z	B	Accumulations of salts more soluble than gypsum	s
Other B horizon suffixes			
w	B	Development of color or structure in a horizon but with little evidence of illuvial materials	m
x	E, B	Horizon with fragipan characteristics, i.e., genetically developed firmness, brittleness, and/or high bulk density	x
o	B, C	Residual accumulation of sesquioxides	–

(cont.)

Table 3.2 | *(cont.)*

Suffix	Used with which master horizons?	Characteristics/comments	Approximate Canadian System equivalent
Suffixes associated with limnic layers			
co	L	Coprogenous earth	co
ma	L	Marl	–
di	L	Diatomaceous earth	–
Suffixes associated with cold soils			
f	Any except uppermost	Frozen subsoil horizon containing permanent, continuous ice (permafrost); not used for seasonally frozen layers	z
ff	Any except uppermost	Frozen subsoil horizon, not containing enough ice to be cemented by it; not used for seasonally frozen layers	z
Suffixes indicating pedoturbation			
ss	A, B, C	Presence of slickensides	ss
jj	Any	Evidence of cryoturbation	y
–	Any	Evidence of mixing by processes other than cryoturbation or argilliturbation, e.g., tree uprooting, animal burrowing, or mass movement	u
Other suffixes			
g	A, E, B, C	Strong gleying in which Fe has been reduced and removed, or in which Fe has been preserved in a reduced state, usually due to saturation with stagnant water; commonly have a moist Munsell chroma ≤2	g
v	B	Presence of plinthite – a firm, iron-rich, humus-poor, reddish material that often hardens irreversibly when exposed, dried, and wetted	–
r	C	Weathered or soft bedrock, including saprolite or dense till, that roots can easily penetrate; sufficiently incoherent to permit hand digging with a spade	–
d	A, B, C	Noncemented but nonetheless dense, root-restrictive layers with high bulk density and low numbers of connected pores, e.g., dense basal till (Cd) or plow pans (Ad)	–
m	B	Continuous or nearly continuous cementation or induration of the soil matrix; examples: carbonates (km), silica (qm), sesquioxides (sm), or carbonates and silica (kqm)	c
u	Any	Presence of human artifacts, e.g., asphalt, bricks, cardboard, plastic, and garbage	–
b	A, E, B	A buried, genetic, mineral soil horizon, i.e., horizon that is part of a paleosol; always used at the end of a suffix string	b
–	A, B, C	Presence of native carbonate, i.e., not illuvial accumulations	k
–	A, B	Weak expression; failure to meet the specified limits of the suffix it modifies; always used at the end of a suffix string	j
Special case			
^	Any	Used as a prefix to master horizon designations, to indicate human-transported material, moved there from a source area outside the pedon	

Source: Modified from: Guthrie and Witty (1982), the Soil Classification Working Group (1998), and the Soil Survey Staff (2010).

Table 3.3 | Types of transitional soil horizons

Type of transitional horizon	Characteristics	Examples	Possible types
Gradational	Horizons dominated by the properties of one, but having subordinate properties of another; characteristically more like the horizon designated first	AE: Transitional between A and E, but more like the A horizon BC: Transitional between the B and C, but more like the B horizon	AB, AE, AC EA, EB, EC BA, BE, BC CA, CB, CA
Mixed/interrupted	Used where discrete, intermingled bodies (individual parts) of two horizons exist within one horizon; intermingling so "congested" that separation of the bodies into individual horizons is not justified	E/B: composed of individual parts of E and B horizon components in which the E component is dominant and surrounds the B materials	A/E, A/C, A/B E/A, E/B B/A, B/E, B/C C/A, C/B

Source: After Guthrie and Witty (1982).

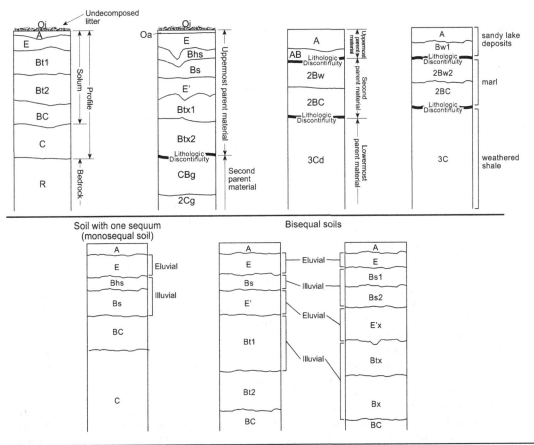

Fig. 3.2 | Illustrations showing how soil horizon terminology functions in different types of soils and lithologic situations.

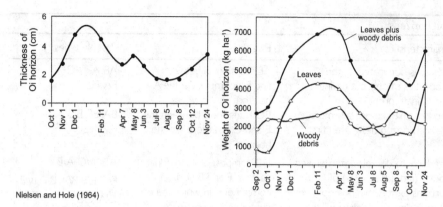

Nielsen and Hole (1964)

Fig. 3.3 Thickness and mass data for Oi horizons in a silty Hapludalf, at various times throughout the year.

with two capital letters, and their properties are more like those of the horizon designated first, e.g., AB or BC, (Fig. 3.2). Another type of transitional horizon, *mixed*, occurs where discrete parts of two horizons exist intermingled within one layer. When intermingled bodies of the two horizons are so "congested" that separation of each of the bodies into an individual horizon is not justified, the capital letters of the two master horizons are separated by a slash, e.g., E/B or A/E.

Arabic numerals (1, 2, 3, etc.) are used as suffixes to indicate vertical subdivisions within a horizon, due to, for example, changes in morphology or chemistry, e.g., Bt1, Bt2 or Eg1, Eg2. Arabic numerals (Roman numerals in Canada) are also used as prefixes to differentiate parent materials, e.g., 2C, 3C. *Lithologic discontinuities* are boundaries between different (stacked) parent materials that are distinguished from one another by properties such as mineralogy or texture, e.g., when loess overlies glacial till (Fig. 3.2; see Chapter 10). In such circumstances, horizons formed in the uppermost parent material are *not* preceded by a 1 (I in Canada), but lower horizons are preceded by a 1, 2, et cetera. The number 2 is used for the second parent material, i.e., the one below the uppermost parent material. If a third parent material is also present, horizons formed in it are designated with a 3, and so on. An example of this is shown in the third profile of Figure 3.2, where the A and AB horizons are formed in one parent material, the 2Bw and 2BC horizons are formed in a second parent material, and the 3Cd horizon is formed in a third parent material.

A prime (') symbol is used to indicate the second occurrence of a horizon within the same profile, but only in cases where an intervening horizon of a different type exists (Fig. 3.2). This type of designation can be tricky. Recall that if the second occurrence of the horizon is immediately below the first, an Arabic numeral suffix is used (e.g., A1, A2; E1, E2; Bs1, Bs2). But if the two horizons are separated by a different type of horizon(s), the prime is used to distinguish them (e.g., A, E, Bs, E', Bs'). The prime is used for horizons that have genetic links or are forming in the same profile; if the lower E, for example, is part of a buried soil

from an earlier period of soil formation, the prime would not be used. Instead, it would be designated an Eb horizon (the *b* is for buried; see Table 3.2).

O Horizons

O horizons are dominated by organic, not mineral, material (Table 3.1; Fig. 3.1). They include organic materials and decomposing debris on the mineral soil surface. They are usually located at the top of the profile as the uppermost master horizon. O horizons provide a buffer between the mineral soil and the atmosphere. They insulate the soil below from extremes of temperatures and moisture, and they provide mechanical protection from raindrop impact, runoff, and other erosional forces. When present, they facilitate infiltration of water.

In many soils O horizons are absent. For example, in plowed soils, O horizons are missing because they have been incorporated into the plow (Ap) horizon. In areas where decomposition rates are high, O horizons may not be present at all times of the year because they have decomposed. In dry or cold climates, production of litter may be so low that a continuous O horizon does not form. Finally, on steep slopes O horizons may erode so fast as to be absent.

O horizons form as organic (plant) debris accumulates on the mineral soil surface. The amount of litter in the O horizon at any one time is always in dynamic equilibrium between inputs and losses via decomposition and erosion (Sharpe *et al.* 1980). Litter inputs may be continuous throughout the year; however, in most ecosystems, the main influx of dead organic debris to the soil surface precedes a dry or cold season (Fig. 3.3). For this reason, O horizons usually attain their greatest thickness immediately after leaf fall. Although leaf fall is temporally predictable, additions of twigs and other woody debris occur randomly throughout the year, e.g., by wind and ice storms. Inputs of litter to O horizons are balanced by decomposition, or turnover, which also has seasonal patterns. Litter inputs, which are closely related to net primary productivity, are most rapid in warm and wet climates (Fig. 3.4), but because decomposition rates are also especially high in these

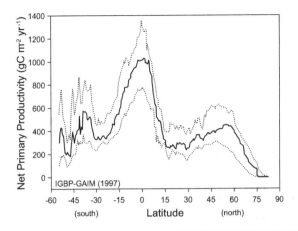

Fig. 3.4 Latitudinal distribution of annual net primary productivity (NPP). Median (solid line) values are enveloped by the 10th and 90th percentiles (dotted lines) of the mean NPP values. NPP is an excellent surrogate for litterfall.

climates, O horizons tend to be thin there. Instead, they are thickest in cool climates or locally where slowed decomposition rates, occasional standing water, and acidic conditions inhibit decomposition. This situation is best epitomized in bogs of mid- and high latitudes, where O horizons are so thick that the soils are Histosols (see Chapter 10).

O horizons form quickly. They can be lost just as quickly. It is not unusual for O horizons on fertile forested sites with abundant moisture and organisms (especially worms; see Kevan 1968) to be completely decomposed and incorporated into the mineral soil within months. On a longer timescale, Albert and Barnes (1987) found that the O horizon of a Michigan forest had essentially reformed 50 years after the forest had been clear-cut. O horizons in the virgin stand were actually thinner and weighed less per unit area than did the O horizons in the stand that had been cut 50 years previous. Schaetzl (1994) studied a chronosequence of sites in the Great Lakes region, where thick O horizons develop on sandy Spodosols. The study design centered on a series of plots that had been burned; after the burn, the forest and the O horizon regenerated. His data indicated that O horizons reform very quickly after fire, and that some sort of steady state was achieved in little more than a century (Fig. 3.5). Thus, at least for mid-latitude sites, O horizons appear to be able to recover from disturbance relatively quickly. Similar data are available for O horizons in different ecological settings (Peet 1971, Fox et al. 1979, Jacobson and Birks 1980, O'Connell 1987). Nonetheless, few data exist on the end point of O horizon development, i.e., the potential steady state thickness that an O horizon will attain, given complete freedom from disturbance. O horizons in cooler climates form more quickly and are thicker, but probably take longer to reach a steady state and attain greater thicknesses than in tropical or subtropical areas (cf. Raison et al. 1986, O'Connell 1987).

Types and Subdivisions of O Horizons

All O horizons must have a suffix, which refers to the degree of decomposition of the organic material within the horizon (Table 3.2). Raw or nearly raw organic material is called *fibric*, for the dominance of plant fibers. Thus, Oi horizons are composed of predominantly fibric material (*i* is the second letter in fibric). They normally have the lowest bulk densities of the three kinds of O horizons (Lee *et al.* 1988a). Highly decomposed O horizon material is given the adjective *sapric*, and horizons dominated by sapric materials are designated Oa. Note again that the second letter in sapric is *a*. Oe horizons have intermediate (*hemic*) levels of decomposition. In some older U.S. literature and in Canada, Oi, Oe, and Oa horizons are referred to as L, F, and H layers, respectively, referring to Litter, Fermentation (or *Formultningsskiktet*), and Humus (or *Humusamneskiktet*), or A_O and A_{OO} horizons.

The terminology used to describe O horizons has a long history (Chertov 1966, Green *et al.* 1993). Raw organic material on the soil surface is still referred to as *litter*. And the O horizon is still, collectively, referred to as the *forest floor* by foresters and biologists (Pritchett 1979, Fox *et al.* 1987, Melillo *et al.* 1989). This classification has its problems, however, since some foresters include the A horizon with the forest floor and the humus layer. Terms such as *raw humus*, *peat*, *acid humus*, *leaf mold*, *duff*, *mor*, *moder*, and *mull* are but a few of the terms that have been used to describe organic material on the surface of mineral soils (Romell and Heiberg 1931, Heiberg and Chandler 1941, Wilde 1950, 1966, Duchaufour 1976), beginning as far back as the late nineteenth century (Muller 1879). Obscure as they may seem to the uninitiated, these terms have great merit; forest humus forms have been used for the assessment of site quality, indicate different degrees of microbial activity, and correlate to geology, climate, tree growth, vegetation types, and soil faunal communities (Ponge *et al.* 2002, 2011). Litter quality and type have also been quantified, as in the *Humus Index* of Ponge *et al.* (2002), thereby offering an inexpensive and reliable method of comparing the properties of O horizons.

The earliest litter classification, still in use in some parts of the world, divides litter layers into mull (Danish *muld*) and mor (Danish *mor* or *maar*) types. *Mull* humus is most often associated with soils in broadleaf forests. It is dominated by bacterial decomposition at the micrometer scale and by worm activity at the centimeter scale. It has a pH >5.0, with a crumblike structure and diffuse lower boundary (Koshel'kov 1961). Soil macrofauna help form the character of the mull O horizon by mixing the raw organic materials with mineral material from the A horizon below (Bernier 1998). In particular, earthworms and arthropods are seen as vital to the development of mull humus from Ca-rich leaf litter. Thus, mull humus has significant incorporations of mineral matter, while mor humus (see later discussion) is essentially all organic matter, with much less mixing between the O horizon and the mineral soil.

The more acidic, slower to decompose, *mor* humus is associated with decomposition dominated by fungi, and a

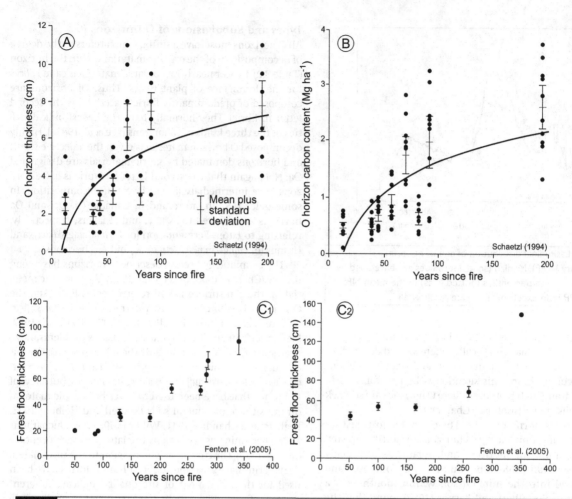

Fig. 3.5 Changes in O horizon character on some previously burned, mixed forest plots in northern Michigan. Data for the 10 samples per plot are shown as dots. (A) Thickness of the "Oi + Oe" horizons versus plot age. (B) Carbon contents of the O horizons versus plot age. (C) Thickening and development of O horizon with time in black spruce forests on fine-textured soils in western Quebec, Canada, under (1) high and (2) low severity fire frequencies. Each point represents the mean of 20 values for each site.

general lack of macrofauna. Mor usually forms in coniferous litter, with its matted roots, needles, and litter fragments (Heiberg and Chandler 1941). Some mors have pH values <4.0. Phenolic substances, common in vegetation of low base cation saturation, may form protective coatings on cellulosic plant materials, further retarding decomposition (Kevan 1968). The large amounts of phenolic substances also render the litter unpalatable to earthworms, limiting their role in the decomposition process and minimizing mixing within the litter layer. In fact, thick mats of mor humus are often stratified into layers according to varying degrees of decomposition, rather than being intimately mixed (Pritchett 1979). Here, fungi, protozoa, collembola, and mites are responsible for much of the (slower) breakdown of acidic, needle-leaf litter into mor humus. Litter with characteristics intermediate between mull and mor is termed *moder* or *duff mull* (Hoover and Lunt 1952).

Biological Degradation of Raw Litter in O Horizons
The process whereby fresh litter is changed, fragmented, and decomposed to form humus, i.e., *humification*, is continuously ongoing in O horizons, unless they are extremely dry or frozen (see Chapter 13). Breakdown and decomposition are performed by soil macro- and microorganisms, whose activity is in turn a function of temperature, moisture, and the characteristics of the litter (von Lützow *et al.* 2006, Kleber *et al.* 2007, Schmidt *et al.* 2011). Warm, moist environments favor litter decomposition and humus formation, and they also facilitate high levels of productivity of the terrestrial plants that produce the litter. A useful way to envision this breakdown/decomposition process is to realize that both macro- and microorganisms are feeding on it simultaneously, with the most palatable and easily decomposed materials going first, while some of the larger and more resistant materials must first be fragmented,

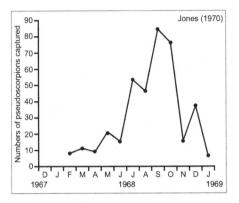

Fig. 3.6 Monthly numbers of *Chelonethi* pseudoscorpions captured from an O horizon in an ash-oak forest in Great Britain. Pseudoscorpions are small carnivorous soil fauna that feed primarily on insects, spiders, springtails, and mites.

Fig. 3.7 Thick O horizons, composed of leaves, woody debris, and moss, on top of a large limestone boulder in northern Michigan. Isolation of this litter from many soil macrofauna, coupled with the high pH of the rock, impeded its decomposition. Photo by RJS.

perhaps several times, by macrofauna. With time, the mass of organic material becomes more and more decomposed, until it is completely oxidized to CO_2 or humified.

After the raw organic debris (fibric material) falls to the surface, it begins to be consumed by organisms of all sorts and sizes as a food (energy) source (Burges 1968). Soil macrofauna also ingest the litter and increasingly fragment it (Schulmann and Tiunov 1999). The fragmentation process conditions the litter for further decomposition by microbial populations in that it increases its surface area and usually mixes it with soil minerals and microorganisms (Gunnarsson *et al.* 1988).

Because litter is a primary source of food for many soil organisms, which are in turn a food source for higher organisms, the numbers and mass of litter-dwelling organisms vary systematically throughout the year (Fig. 3.6). Often, litter dwellers are in greatest number in the warm season, such that by autumn (in the mildly acidic and well-drained soils of midlatitudes) the previous year's litter has been mostly consumed. Most litter feeders will therefore migrate into the mineral soil to estivate, or to lay eggs and die, before the onset of the dry/cold season.

Earthworms are particularly good at fragmentation; litter decomposes much faster in their presence. The worms comminute the organic matter they ingest and mix it with mineral soil; their casts are fertile ground for microorganisms and are well known as sites of enhanced microbial activity within the upper soil profile (Kevan 1968, Fiuza *et al.* 2011). Actinomycetes and bacteria live in the digestive tracts of earthworms, and their activity continues in the excrement (Kevan 1968). Earthworm casts are also important because, if excreted at the surface or within the O horizons, they bury other litter and create a more favorable microenvironment for its further decomposition.

Equally important in the fragmentation and mixing processes are millipedes, collembola, mites, and isopods (Cárcamo *et al.* 2000; see Chapter 7). Collembola are important

in this regard, as they readily consume the fecal pellets of larger soil arthropods and break this material down further; *coprogenic* materials make up a large part of Oa horizons. In tropical areas where termites are abundant (see Chapter 11), soils have low organic matter contents. One reason for this is that termites are so efficient at digesting cellulose that little is left for the soil itself. This situation contrasts with midlatitude environments where macrofauna like millipedes, mites, and earthworms pass most of the organic materials through their digestive tracts, with most of the material remaining in organic form. Also importantly, the organic matter gets mixed with clay into microaggregates, where it is physically protected from further decomposition – another reason why A horizons are well stocked with humus.

Over time, fecal pellets of soil meso- and macrofauna accumulate in the soil. According to Kevan (1968), the pellets are again fed upon by thecamoebae and enchytraeid worms, as well as by collembola (again), producing even finer microaggregates (~50–80 μm diameter) that are held together by humified organic matter (Thompson *et al.* 1990). Thus, multiple cycles of ingestion, fragmentation, mixing, and excretion must occur for all the litter to be in a finely divided and well-decomposed, i.e., sapric, state (Schnitzer and Monreal 2011).

In situations where the litter is isolated from most soil macrofauna, as happens where the soil is extremely shallow to bedrock, the decomposition process is much slower and is dominated by microorganisms. Here, litter may accumulate and thicken to the point that Folists (organic soils over bedrock or gravel) will develop (Fox and Tarnocai 2011; Fig. 3.7).

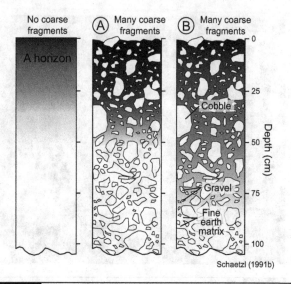

Schaetzl (1991b)

Fig. 3.8 Theoretical influence of the amount of coarse clasts on the formation of A horizons. In gravelly and rocky soils, the fine earth fraction of the A horizon occupies less volume, causing the organic materials within it (A) to become more concentrated in the limited volume, and hence, the A horizon is darker and more humus-rich throughout, than it otherwise would be, or (B) to accumulate in the A horizon to greater depths than it otherwise would.

Like any organic residue, litter under favorable conditions will be utilized by soil organisms as a source of carbon and energy (Kevan 1968, Martin and Haider 1971). For example, Knollenberg et al. (1985) found that, for every gram of live weight of earthworms in a Michigan soil, 8–27 mg of leaves (dry weight) were consumed per day. At this rate, they hypothesized that the worm population at their research site could consume almost 95% of the leaves that fell within four weeks. Decomposition is hastened by periodic additions of moisture, as soil fauna ingest the litter and excrete it in a partially or wholly digested form, and by warmer temperatures (Raw 1962). Regardless of how much physical fragmentation occurs, however, eventually litter is decomposed or biochemically transformed (Hayes 2009).

A Horizons

The A horizon is the dark, uppermost, *mineral* soil horizon (Fig. 3.1; Table 3.1). After being tilled or cultivated, this layer is usually referred to as topsoil. Mineral particles in the A horizon have dark coatings of humus and organic matter in advanced stages of decay. Isolated fragments of organic matter, fragments of dead plants and animals, fecal pellets, seeds, and pollen grains, as well as opal phytoliths, are also commonly found within A horizons. In uncultivated soils, A horizons often form beneath an O horizon, which contributes some of the partially decomposed organic matter to the A horizon. Mixing of partially decomposed organic materials from the O horizon into the mineral soil can occur as soil animals drag plant matter into their burrows (Nielsen and Hole 1964) or as small plant fragments wash down into pores and cracks and along root channels (Graham and Southard 1983, Marin-Spiotta et al. 2011). By depositing casts of mineral soil on and in the O horizon, burrowing animals such as ants and worms incorporate parts of the O horizon into the A and mix the two; ultimately, this mixed zone will revert to an A horizon as the organic materials decompose further. Organic materials mixed into the A horizon then further decompose into humus, which is a brown to black, amorphous, carbon-rich material. Most particle coatings in the A horizon are humus. Darkening of the mineral material by additions of organic matter and humus is called *melanization* (Schaetzl 1991b, Freppaz et al. 2010; see Chapter 13). Many A horizons form in soils that lack an O horizon, such as in grasslands and most dry regions. Here, most of the humus is derived from below-ground contributions from roots (Fig. 13.1).

In general, the amount of organic matter in the A horizon is dependent on the balance between long-term additions and losses. Additions occur from both above the ground, as litter fall, and below ground, as roots and soil organisms die and decay in place. Losses include translocation of organic matter to deeper horizons, erosion of organic-rich materials, and oxidation/mineralization of organic matter to CO_2. Because additions and losses vary considerably in space and time, A horizons also vary in characteristics, both spatially and temporally. They tend to be darkest where organic matter additions exceed losses, such as cold or wet sites, i.e., where oxidation of organics is slow. Grassland soils also have thick dark A horizons, because of the large, ongoing additions of organic matter by roots (Ponomareva 1974). In some grasslands, accumulation is assisted by the sometimes dry conditions, which inhibit decomposition by limiting microbial activity. Some forest soils, however, can also exhibit thick, dark A horizons, as, for example where the forest is thin and grasses occur in the understory (Afanas'yeva 1966, Nimlos and Tomer 1982), where forest has recently replaced grassland vegetation (Geis et al. 1970), or where the forest soil is very rocky (Gaikawad and Hole 1965, Small et al. 1990). High amounts of coarse fragments cause soils to have thicker and/or darker A horizons, as the large volume of coarse clasts in the soil limits the volume of soil in which organic matter can be distributed (Fig. 3.8).

In an interesting study of a forest soil, Nielsen and Hole (1964) removed litter from a forested research plot for 5 years. Organic matter in the upper 7.5 cm of the A horizon decreased, and the number of earthworm middens decreased from 64,000 to 22,000 ha^{-1}. On another research plot, leaf litter was imported, doubling the O horizon thickness. On this plot, soil organic matter increased, as did midden densities (from 64,000 to 125,000 ha^{-1}). These experiments illustrate the comparatively ephemeral nature of organic matter in the O horizon and upper zones of A horizons.

Plowed A horizons are designated Ap (Table 3.2). Ap horizons usually have an abrupt or clear lower boundary, occurring at the base of the plow zone. In areas where the

A horizon is thin, the Ap horizon will have incorporated parts of some of the lower horizons within it.

V horizons are highly vesicular horizons formed in desert soils (Fig. 13.76). The porosity is typically not well interconnected, leading to low permabilities. These horizons are normally found in the uppermost part of the profile, often taking the place of an A horizon. More information about V horizon genesis is found in Chapter 13.

E Horizons

E horizons are light-colored mineral horizons that show evidence that organic matter, clay, and/or iron or aluminum oxides have been lost, usually due to translocation of these substances out of the horizon, by infiltrating water. The loss of the above constituents, *eluviation*, usually implies that grains coated with these substances have been stripped clean, exposing the light-colored quartz grains that dominate most soils. Additionally, many of the dark, weatherable minerals, e.g. biotite, pyroxene, amphibole, have decomposed by weathering (see Chapters 9 and 13), adding to the light color of the horizon. Thus, E horizons are usually dominated by uncoated, sand- and silt-sized quartz and feldspar grains. In older texts, E horizons are referred to as A2 horizons, and in Canada, Ae horizons. Because the E horizon displays evidence of eluviation better than perhaps any other horizon, its symbol was changed from A2 to E in 1982 (Guthrie and Witty 1982).

E horizons form as vertically and/or laterally moving water strips sand and silt grains of their coatings and weathers many of the dark minerals. Thus, E horizons form primarily in humid climates, on freely draining sites (no high water tables, no water-impeding layers), or where water perches and moves laterally above a zone with low permeability. In soils where water perches above a slowly permeable layer, chemical reduction of Fe and Mn occurs (Kemp *et al.* 1998), and the reduced forms of these two elements may be translocated in laterally flowing water (see Chapter 14). The result is a horizon depleted of Fe- and Mn with many of the morphological characteristics of a typical, clay-impoverished E horizon.

A and E horizons both experience eluviation, where most coatings on larger grains have been stripped and translocated (in suspension or solution) to lower horizons. The difference between the two horizons centers on the balance between additions and losses of organic matter. Humus coatings remain intact in the A horizons, but even these are translocated out of E horizons. Thus, E horizons can be conceptually considered the lower portion of the eluvial zone, where eluviation is maximal.

Where A horizons are thick, as in grassland soils, E horizons are often absent. Even though eluviation of various types of compounds proceeds in such A horizons, additions of organic materials outweigh the losses. In contrast, E horizons are commonplace in uncultivated forested soils in humid climates, where translocation is frequent and intense and where A horizons are thin. In some soils with minimal pedoturbation, such as acidic sandy Spodosols, E horizons immediately underlie O horizons. Here, decaying organic matter is only very slowly (if at all) mixed into the mineral soil. As the litter decomposes within the O horizon, small organic molecules wash into the mineral soil and completely bypass the upper solum; deposition occurs in the B horizon below (see Chapter 13).

With time, an E horizon usually grows downward at the expense of the B horizon below. The boundary between the E and B horizon can be abrupt or gradual. The latter form of E horizon is essentially a degraded B, and it can be taken as clear evidence of continued profile deepening and development (Bullock *et al.* 1974; see Chapter 13). Degraded B horizons often have the outward appearance of an E, but they can contain fragments of B material within peds (Payton 1993b).

B Horizons

B horizons are sometimes referred to as subsoil materials. In contrast to the A and E horizons, which are indicative of eluviation, B horizons often show evidence of *illuviation* (Latin *il*, in, and *luv*, washed), i.e., translocation of materials *into* the horizon (Fig. 3.1). Often, illuvial substances or weathering by-products are carried downward by percolating water and deposited in the B horizon as the water soaks into peds, or as the wetting front stops. In this way, illuvial substances are plastered onto individual particles, pore walls, and ped faces as coatings (Table 2.6). Illuvial materials may also be deposited in B horizons if translocated ions precipitate into a solid, although not all precipitation results in obvious ped or pore coatings. The many types of B horizons and how they form will be explained in later chapters.

In the naming of illuvial B horizons, suffixes are used to convey information about the nature of the illuvial materials, e.g., clay, sesquioxides of iron and aluminum, carbonates, sodium, humus, gypsum, sulfur, and silica, alone or in combination (Table 3.2). Placing a suffix on a B horizon name, or indeed on any horizon name, as an indicator of illuvial materials is a judgment call. If one sees evidence for illuvial coatings, one adds the B horizon suffix, but there is no quantitative rule that must be met or some minimum that must be exceeded. Learning to recognize and describe illuvial coatings in B horizons is an important skill for the soil scientist, and it is one that develops with experience.

Weak B horizons (termed Bw horizons) are primarily expressed not by their illuvial products but by changes in color that are associated with mineral weathering and accumulation of weathering products, especially Fe oxides. The loss of rock structure and the concomitant development of soil structure are also associated with Bw horizons. In soils that have formed in carbonate-rich parent materials, e.g., some glacial sediments, the lower limit of the B horizon is taken as the depth of carbonate leaching.

Because B horizons take longer to form and to be destroyed, and by virtue of their location in the subsoil, they are less likely to be eroded than are A and E horizons.

For this reason, many soil classification schemes emphasize B horizon characteristics (Expert Committee on Soil Science 1987, Soil Survey Staff 1999). Similarly, paleopedologists often observe B horizons that have been buried by younger sediments with no intervening A horizons. Such instances are taken as evidence of significant erosion of the upper horizons immediately before burial – leaving only the B horizon as evidence of the prior soil (Sorenson 1977, Olson 1997, Olson and Nettleton 1998).

The Sequum Concept

A sequum is a genetic couplet of an eluvial horizon above an illuvial horizon, usually an E and an underlying B horizon. Many soil profiles in humid regions have an E-B sequum. Those that have two sequa are termed *bisequal soils* (Schaetzl 1996; Fig. 3.2). Often, bisequal soils form where two different periods of eluvial-illuvial processes have occurred in the same pedon, e.g., clay translocation at depth, forming a E-Bt sequum, and iron/aluminum translocation above, forming a E-Bs sequum. The second, or lower, E horizon in a bisequal soil is usually denoted with the prime, e.g., E'. Although it is the second E horizon in the profile, an intervening horizon occurs between it and the E of the upper sequum, necessitating the use of the prime.

The Solum Concept

The term *solum* (plural, *sola*) refers to the soil horizons above the C horizon, i.e., the O, A, E, and B horizons, if they are present (Fig. 3.2). The Soil Survey Division Staff (1993) defined the solum as a set of horizons that are related to one another through the same cycle of pedogenic processes. A modern solum includes all horizons now forming near the soil surface, but it excludes buried sola. Thus the solum can be thought of as an entire soil formed by pedogenesis. If the C horizon truly represents unaltered parent material (see later discussion), then the solum consists of all horizons that have been altered by pedogenesis.

Determining the lower limit of the solum often requires judgment and experience. In soils developed on calcareous parent materials, the solum is usually taken to be the zone from which $CaCO_3$ has been dissolved and removed by weathering and leaching. The C horizon, below the solum, remains calcareous and unleached. This boundary can sometimes be quite abrupt. In residual soils developed on saprolite, however, the lower boundary of the solum is diffuse. Unmistakable, clay-rich Bt horizons may grade into saprolite-like material with clay coatings and some degree of pedogenic structure, gradually yielding to unaltered saprolite below. Many times, these boundaries are difficult to define because illuvial clay commonly extends well below the zone of pedogenic structure, along vertical fissures in the saprolite, and because the change from rock structure to soil structure occurs gradually. Douglas *et al.* (1967) found that the lower boundary of the solum was often placed too shallowly in some Illinois soils. They noted that plant roots and illuvial clay extended well into what would otherwise

be considered the C horizon. Buol *et al.* (1997) suggested that the maximum lower limit of the solum must be the maximum depth of perennial roots, implying that to be defined as soil, the solum must have been influenced by biotic activity. Perhaps the best way to delineate the lower boundary of the solum is to examine the layer just below; if it shows little or no sign of pedogenic structure or color change, it is not considered part of the solum. In sum, the solum is a useful theoretical construct that focuses attention on the boundaries of pedogenic processes in a soil.

C Horizons

Defined by Charles Kellogg (1930) as "parent material that has been unaltered by soil-building forces," C horizons are viewed as the *parent material* from which the soil horizons above have formed (Fig. 3.1). Although shallow profiles that rest directly on bedrock may not have a C horizon (Willimott and Shirlaw 1960), where mineral materials are thick, a C horizon can usually be observed at depth. In soils derived from bedrock, the zone between the solum and fully lithified material is by definition the C horizon. However, these seemingly straightforward definitions are not always simple to apply in practice (Tandarich *et al.* 2002), because most soil materials described as C horizons are weathered or altered by pedogenic processes to some extent (Kellogg 1936). Many exhibit thin, patchy illuvial coatings of clay or other secondary minerals along joints and major ped faces. Others show evidence of oxidation, gleying, and jointing, and some can be within the depth of rooting. These types of minor pedogenic alterations are common in C horizons because water that enters is not without solids and solutes, regardless of where it originated within the profile. Oh and Richter (2005) documented losses of Si from soils and saprolites to depths exceeding 6 m (Fig. 3.9). As noted, the B-C (and hence, lower solum) boundary may be diffuse and somewhat arbitrarily placed (Simonson and Gardner 1960), and criteria used to choose the boundary must be weighted carefully (Fig. 3.9). Most studies that have examined solum depth have concluded that the upper boundary of the C horizon can be placed deeper than it might be on the basis of field criteria alone (e.g., Douglas *et al.* 1967, Richter and Markewitz 1995), especially in sandy soils (Harris *et al.* 2005).

Tandarich *et al.* (2002) reviewed the various concepts used to define C horizons, noting that various zones have also been differentiated within C horizons. For example, many soil and Quaternary scientists who work in landscapes dominated by glacial drift recognize three zones within thick C horizons – an upper zone in which Fe has been oxidized and from which $CaCO_3$ has been leached, a middle zone that is oxidized but unleached, and a lower zone that is both unoxidixed and unleached (Leighton and MacClintock 1930, Kay and Graham 1941, Birkeland 1974, Hallberg *et al.* 1978a). In calcareous parent materials, the upper boundary of the C horizon corresponds to the lower limit of carbonate leaching. However, even in these circumstances, the calcareous C horizon may show some

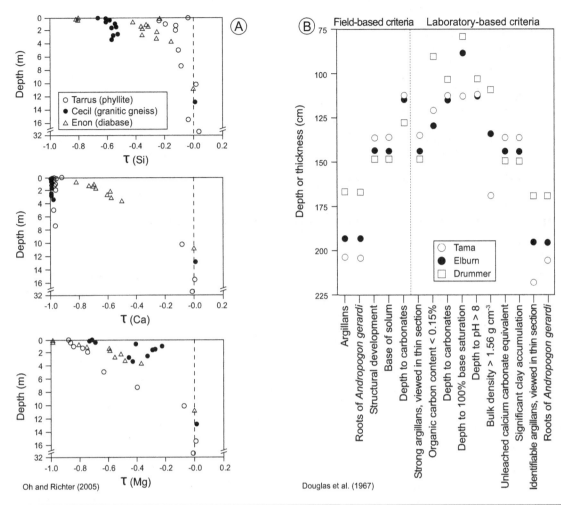

Fig. 3.9 Data that illustrate the weathered nature and the variable depth to the top of some C horizons. (A) Losses versus gains of Si, Ca, and Mg from three soils formed on saprolite in North Carolina, United States, illustrating the great depths to which weathering and pedogenesis can operate. (B) Solum thickness in three different soils, based on differing criteria. Note that criteria used to determine the depth to the bottom of the solum (i.e., the B-C boundary) are seldom in agreement.

evidence of alteration, especially with regard to illuviation, eluviation, acidification, or weathering (Godfrey and Riecken 1954, Douglas *et al.* 1967, Richter and Markewitz 1995). For example, even if most of a C horizon remains calcareous, areas along vertical fractures may be leached of carbonates. Still other C horizons that are calcareous may show evidence of illuviation of carbonates from above (Schaetzl *et al.* 1996). Indeed, Douglas *et al.* (1967) stressed that the depth to detectable carbonates may not be a suitable criterion for identifying the C horizon in some soils. Nonetheless, depth to carbonates remains a common criterion used to identify the C horizon (solum) boundary in many soils formed in calcareous materials.

The recognition that many soil processes continue well into the C horizon, even if they are not easily observable in the soil's field morphology, has led to the development of the field of science called *whole-regolith pedology*. This field of study spans the interface between pedogenesis and weathering. It is based on the observation that while pedogenic processes may originate or flourish near the surface, they can also continue deep into the subsurface (Cremeens *et al.* 1994, Harris *et al.* 2005, Schoeneberger and Wysocki 2005).

Many soils have formed in more than one parent material (Schaetzl 1998). As noted, the contact between two dissimilar parent materials in a soil is called a *lithologic discontinuity*, and successive parent materials in one soil are noted by placing an Arabic number before the horizon designation in the soil description (Fig. 3.2; see Chapter 10). Stacked parent materials with different origins may be designated as C horizons if they display minimal evidence of pedogenesis, even if it appears that no A or B horizons have developed within them.

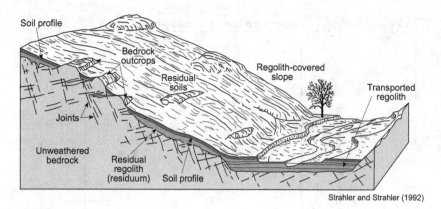

Strahler and Strahler (1992)

Fig. 3.10 Diagram of a regolith-covered slope, illustrating the differences among bedrock, residual regolith (residuum), and transported regolith.

Buried Soils

Soils that are or have been buried are referred to as *paleosols*, *geosols*, or fossil soils (see Chapter 16). They are introduced here because, once recognized, they may also be described using many aspects of the horizon nomenclature used to describe modern soils, except that a *b* follows the horizon designation. For example, a buried soil might have Ab, BAb, Btb, and C horizons. Normally, the paleosol C horizon does not receive the *b* suffix. If a buried paleosol is later exhumed, the *b* suffix is no longer applied.

Whether or not the *b* suffix is used depends on whether the soil is actually, by definition, buried. The Soil Survey Staff (1999) defines a buried soil as one that is covered with a surface mantle of new soil material that either is ≥50 cm thick or is 30–50 cm thick and is at least half as thick as the buried soil. Thus, the 50-cm minimum limit has priority in the definition. It is assumed that the surface mantle has been minimally altered by pedogenesis. Nonetheless, Schaetzl and Sorenson (1987) noted that any covering of sediment, if it is clearly identifiable as such, regardless of thickness, is enough to "bury" a soil. In situations where the covering is thin, pedogenic processes will quickly act to incorporate that new mantle into the buried soil, in effect "unburying" it. This process is called *soil welding* (Ruhe and Olson 1980; see Chapter 16).

Regolith, Residuum, Saprolite, and the Weathering Profile

Soil is different from sediment. Soils develop, via pedogenic processes, *from* sediment and *in* sediment. Knowing how to distinguish soil from sediment is critical to understanding how soils form (Mandel and Bettis 2001). Soils develop within weathered, unconsolidated (parent) materials at the Earth's surface, under the influence of

biota and climate. Soils cannot form within a solid rock, but they can form within the weathered by-products of that rock.

All unconsolidated material at the Earth's surface, overlying bedrock, is called *regolith* (Fig. 3.10). This term is used even if the material has been reconsolidated or cemented subsequently by pedogenic processes (Ollier and Pain 1996). Regolith can develop either by (1) forming in place as bedrock weathers or (2) deposition of preweathered material that has been transported by gravity, water, wind, ice, or another vector (Graham *et al.* 1990a, Stromsoe and Paasche 2011; see Chapter 10). Unconsolidated material that is directly derived from underlying lithified bedrock is referred to as residual regolith, or simply, *residuum*, and is most common on stable uplands in unglaciated parts of the world. In glaciated regions, residuum either is deeply buried or has been eroded. The second kind of regolith, transported regolith, can take many forms, such as alluvium, glacial sediment, or eolian sand. The base of the regolith is the bottom of Earth's "critical zone" (Brantley and Lebedeva 2011).

To quote Ollier and Pain (1996), regolith is "nobody's baby," existing in a kind of limbo between disciplinary regimes. For example, geologists may consider it debris that obscures a clear view of the rock below – a type of "post-Pliocene overburden." Pedologists view most regolith as parent material that may be transformed to something different – soil. Engineers are most concerned with regolith's measureable physical properties; it is a material with which to work. And geomorphologists view regolith as a key to explaining landform genesis. In short, regolith is something to everyone, but it is everything to almost no one (Cremeens *et al.* 1994).

In residual settings where bedrock has been weathering in situ for long periods, regolith often overlies a thick, weathered zone of soft rock that retains some geologic structure; this material is called *saprolite* (Figs. 3.11, 3.12). In such cases, soil material that has lost its geologic

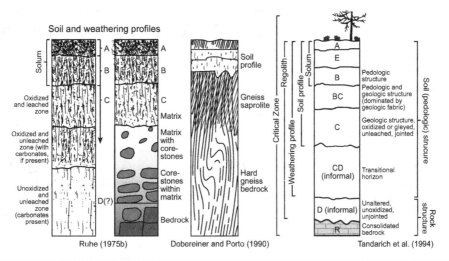

Fig. 3.11 Examples of soil profile and weathering profile terminology. In two of the profiles, unaltered (D horizon) material exists at depth.

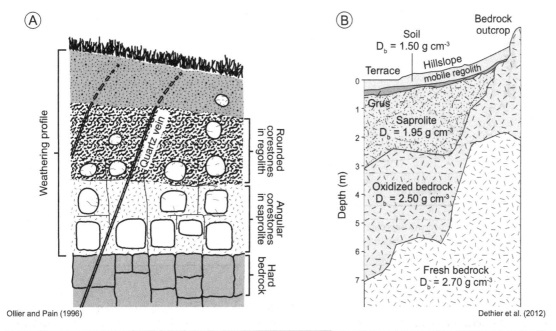

Fig. 3.12 Features of weathering profiles. (A) A typical, generalized weathering profile. (B) A weathering profile as described for the Front Range, Colorado.

structure through mixing and other pedogenic processes usually grades imperceptibly downward, through the saprolite into hard bedrock (Graham *et al.* 1990a, Stolt *et al.* 1991; Fig. 3.13). An intermittent or gradational interface with the bedrock is common, for example, in sandstone, where unconsolidated sandstone residuum grades downward to rock fragments of increasing size and, eventually,

to solid rock. Chemical weathering processes dominate in saprolite, whereas pedogenic processes, including biological activity and physical reorganization of parent material particles, dominate in the soil profile (Rice *et al.* 1985).

In some places, the rock-saprolite interface may be abrupt, as, for example, where glaciers have scoured it clean. A sharp contact between bedrock and regolith often

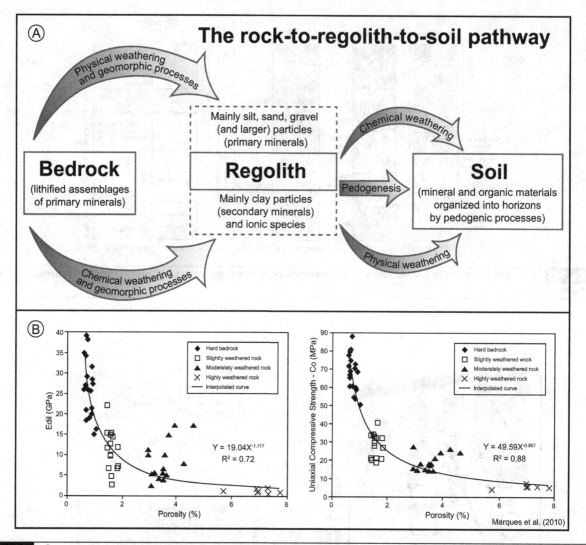

Fig. 3.13 Conceptualization of the bedrock-saprolite-residuum-soil transition and examples of how rock characteristics change through this transition. Part B shows porosity data for a gneiss bedrock-saprolite-soil transition in Brazil, compared to data on compressive rock strength and deformability.

suggests that the regolith was transported there. A transportational vector, e.g., glaciers or wind, would have eroded any soft bedrock or residuum that may have existed there. In the process, the uppermost bedrock contact became sharper and more abrupt. In transported regolith, one might observe zones immediately below the soil material that are only slightly modified by leaching and/or oxidation. At still greater depths, the regolith may be essentially unmodified by weathering, where minerals have not been dissolved, oxidized, or neoformed.

In both residual settings and transportational/depositional settings, the entire sequence of soil, weathered zones, and unweathered zones in the regolith constitutes a *weathering profile* (Tandarich *et al.* 2002; Figs. 3.11, 3.12).

The term "weathering profile" was defined and developed by geologists (Kay and Pierce 1920, Kay and Apfel 1929, Leighton 1958). For example, in their work with glacial sediments in Iowa, Hallberg *et al.* (1978a) subdivided the weathering profile of till or loess into three zones. From the top down, they are (1) an oxidized and leached zone, containing the soil profile; (2) an oxidized and unleached zone; and (3) an unoxidized and unleached zone (Bettis 1998). Weathering profiles grade downward into unweathered sediment, saprolite, or bedrock; they grade upward into soil. In general, within the weathering profile, zones closer to the surface are more weathered and depleted of mobile weathering products than are deeper zones (Phillips 2001b, Oh and Richter 2005).

D Horizons

The C horizon concept was originally devised by soil scientists who focused primarily on near-surface soil horizons that supported plant growth, not on deep subsurface weathering zones. Because materials that have been designated as C horizons are often chemically and physically altered, i.e., leached of carbonates, jointed, or oxidized, the idealized C horizon concept is difficult to apply consistently. To address the confusion, Tandarich *et al.* (1994) reintroduced the *D horizon* concept to refer to that part of the traditional C horizon that is *unaltered* by pedogenic processes, but that is not hard bedrock. They defined the D horizon as geogenic or nonpedogenic layers of sediment, excluding consolidated bedrock, characterized by original rock or sedimentary fabric, lack of tension joints, and lack of alteration features of biooxidation origin. We use the term "reintroduce," because the D horizon concept was originally introduced by Dokuchaev in 1900 for a horizon below the C (Tandarich *et al.* 2002). Earlier use of the term "D horizon" had applied it to bedrock. Then, 94 years later, Tandarich *et al.* chose to limit the C horizon to the modified part of the traditional parent material. Others, too, have noted that the 1930s definition of the C horizon was insufficient to connote its modified character, and they have variously referred to this material as Cu or C4, e.g., Follmer (1979b). In areas that are comparatively shallow to bedrock, the D horizon may never have existed, or it may have been quickly converted to C or solum material. On old landscapes that have been deeply weathered, D horizons may not occur. In short, D horizon material is unaltered, whereas the C horizon of Tandarich *et al. is* slightly chemically and/or physically altered. A, E, and B horizons are all biologically, chemically, *and* physically altered.

The D horizon can be considered part of the *pedoweathering* profile. Tandarich *et al.* (1994) embraced the concept of the pedoweathering profile as one that merges traditional soil profile concepts (see later discussion) with those of the geologic weathering profile (see Chapter 2). The latter has traditionally been used by Quaternary geologists (e.g., Kay and Pierce 1920, Frye *et al.* 1960b, Willman *et al.* 1966). According to Tandarich *et al.* (1994), an unwritten agreement between Curtis Marbut, then head of the soil survey program in the U.S. Bureau of Soils, and Morris Leighton, a Quaternary geologist in Illinois, led to the distinction between studies of the soil profile, which were to remain in the realm of soil scientists, and studies of the weathering profile, which were given over to those more geologically oriented. This dichotomy may have made sense at the time (1950s), because Quaternary geologists were focused on larger, regional-scale processes rather than horizon-scale processes. Similarly, many pedologists were seeking to differentiate the horizon-scale pedogenic processes that led to the remarkable variety of soil morphology at the Earth's surface. To this end, the pedoweathering profile concept of Tandarich *et al.* (1994) has remerged the geologic and pedologic views of the weathered portion of the Earth's crust into a coherent theme. The pedoweathering profile retains the concept of unaltered parent material, but that material is labeled as D horizon.

Although the D horizon concept has not yet been fully accepted, its usage seems to be a logical extension of horizon nomenclature into the deep subsurface, reflective of a greater interest in processes and materials that exist below the solum. As of this writing, no national-level soils agency, such as the Natural Resources Conservation Service, has adopted the D horizon concept. Nonetheless, we support the concept.

Soil Profiles and Soil Individuals

Soil scientists study soils at all scales, from the microscopic to the global (Fig. 3.14). The genetic soil horizons described previously are somewhere in the middle of that spatial range, and they are often the most accessible starting point for documenting and exploring soil properties and processes. The vertical sequence of genetic horizons at any point on the landscape is called a *soil profile* (Tandarich *et al.* 2002); it can be thought of as the upper part of the pedoweathering profile. The term "soil profile" is used in several ways. In some contexts, it is used simply to refer to a body of material in the uppermost few meters of the Earth's surface that may be exposed in house excavations, roadcuts, mines, or open graves. The term may also be used to describe the one-dimensional sequence of genetic horizons, e.g., A-Bt-BC-C. In this sense, it is a conceptual model of the real soil body. It can also be thought of as a plane, a two-dimensional entity that is perpendicular to the soil surface. Because a plane has no volume, it can be viewed but not sampled (Jones 1959).

But the soil we seek to understand is not just a plane. It is a real, natural body in which a characteristic sequence of horizons has developed from parent materials, over time. For that body, we can define three spatial dimensions and one temporal dimension. The sequence that we observe reflects the patterns and impacts of climate, of biological activity, and of topographic position, as they have acted together over time to form a soil. The complexity of these interactions usually leads to enormous complexity in the properties of horizons and thus, in the degree of horizon expression. Properties that we use to distinguish horizons from one another, e.g., thickness, color, texture, structure, and organic matter concentration, usually vary continuously (sometimes discontinuously) across the landscape, making the very definition of a "soil individual" a philosophical exercise. To help determine what is an actual soil, we must appreciate and document the range of soil characteristics in a given landscape and arbitrarily assign limiting values, quantitative or qualitative, that help us communicate with one another about what properties might occur in that soil that lies beneath our feet. Starting with a philosophical perspective, we explore some of these basic concepts in the next section.

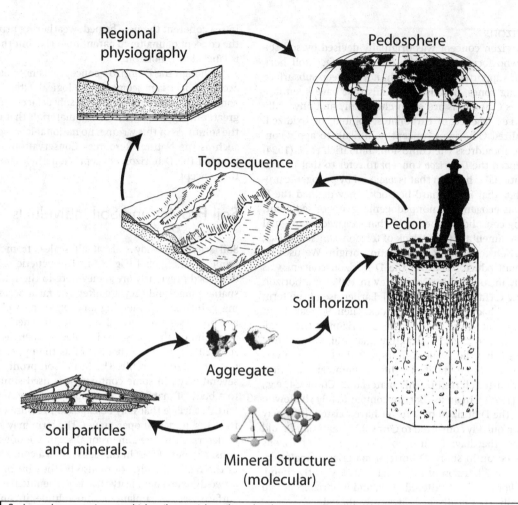

Fig. 3.14 Scales and perspectives at which soil materials, soils, and soilscapes can be studied.

The Material and the Conceptual

Humans perceive and interact with the material universe, i.e., that which is composed of mass or energy, by means of our senses, but also by use of ideas, words, and models. Both the material universe and the conceptual universe are real, and they are linked to one another through our consciousness (Hayakawa and Hayakawa 1990). We create ideas and words that reflect the material universe so that we can (1) communicate about it, (2) predict its future, and (3) better understand how its components are related to one another. In science, we constantly strive to refine our conceptual understanding of the material world, while recognizing that the words we use are always an approximation of what we mean or know. Although the material universe and the conceptual universe can be thought of as components of the great continuum of reality, in general it is often useful to make a distinction between imperfect words and the material things to which they point. In the discussion that follows, pay particular attention to the distinctions that we make between the *material* soil and the *conceptual* terms we use to communicate about it. We will see that our concepts and terms are inherently imperfect. How useful they are depends greatly on how they are defined and interpreted.

To begin, let us acknowledge the obvious fact that *soil individuals*, bodies with discrete, easily recognizable boundaries, are exceptionally rare, i.e., a soil is not analogous to an individual organism (Knox 1965). With the exception of recently constructed, remediated soils and horticultural growing media, the natural bodies we call soils are parts of a continuum across the landscape. The factors and processes that regulate soil formation vary continuously in time and across space to produce the soils we see. Therefore, the physical, chemical, and biological properties of soil vary continuously as well. Earlier, we considered the challenges associated with defining boundaries between weathered and unweathered parent material. Discerning the boundaries between soils

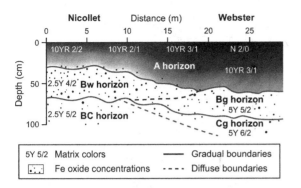

Fig. 3.15 Schematic variation in soil properties along a transect in a typical central Iowa landscape. The two soils, Nicollet and Webster, while distinct in concept, are not expressed as discrete individuals on the landscape.

that are adjacent to one another on the landscape is even more problematic. There are more characteristics, more horizons, to account for. Moreover, the properties that we use for differentiation change gradually in both space and time, and often at different rates. So to communicate, predict, and understand this pattern, we design models to help us create conceptual order from the chaos (see Chapter 12). Depending on how precise our communications or predictions need to be in time or space, we may define the components of our models loosely or tightly. But ultimately, the characteristics that we choose to set boundaries to distinguish one soil from another are most often arbitrary and chosen for convenience, economy, or purpose, and not for their independent existence in nature. That is, the boundaries are *ours*, not necessarily those of the soil landscape.

Consider, for example, two soils, Nicollet and Webster, that occur on the same landscape in central Iowa (Fig. 3.15). Both soils are derived from glacial till, and they have both formed over a period of several thousand years under the influence of tall grass prairie vegetation on gently sloping uplands. Most often, Nicollet occurs in slightly higher landscape positions. Although they have many traits in common, these soils may be distinguished from one another by the color of the upper part of the B horizon, e.g., dark grayish brown in Nicollet and olive gray in Webster, and by the presence or absence of Fe oxide concentrations in that zone. Nicollet soils have thinner A horizons with lower concentrations of organic carbon than do Webster soils. These subtle differences can be interpreted to reflect the length of time that water has saturated the soil, during the period of pedogenesis. If we were to map a field where both soils occur, we could use a soil probe to collect samples across a systematic transect of 25 m, describe the properties of the samples, and plot them on a graph (Fig. 3.15). Because the differentiating characteristics are likely to vary independently of one another at this scale, our choice about where to draw the boundary between these two soils will

depend on which properties are most important to us, how well we can measure them, and how much uncertainty we are willing to tolerate in the distinction. We will return to these ideas when we describe mapping units in Chapter 8. In sum, the soil boundary we choose to place on the map is dependent on our predefined perceptions of what each soil *is*; the boundary is a concept and does not preexist on the landscape.

This is an important point, especially when we need to predict soil properties or the outcomes of soil processes at particular locations or particular times. Although we can identify the general parameters that influence soil variation, making specific predictions about how soil-forming processes have been realized at a specific location is challenging. This is the challenge of soil scientists everywhere – how to predict soil properties and processes efficiently and accurately across space and time. The accuracy of our predictions depends on our models of the processes, on our knowledge of the properties, and, particularly, on the scale in which we choose to work.

Concepts in Soil Classification

One of the most important models that we use in soil science is the soil classification system. We will describe soil classification in greater detail in Chapter 8. But here, it is appropriate to refer to some basic elements of the U.S. soil classification system – *Soil Taxonomy* (Soil Survey Staff 1999). It is a widely used system, and it has also served as a model for soil classification systems developed in other countries. At the highest level of abstraction, soil concepts are arranged into twelve major groups, or orders. The orders are divided into suborders that are composed of great groups, which are in turn collections of subgroups. The lowest category of classification in *Soil Taxonomy* is the soil series, a concept intended to gather soils that have a similar sequence of horizons and homogeneous properties. Like all words, *a soil series is a concept*; it is not a material entity.

When Soil Taxonomy was developed in the mid-twentieth century, the term *polypedon* was coined for a concept that would link the properties associated with a soil series to a specific location on the landscape, i.e., a soil body (Soil Survey Staff 1975; Johnson 1963). The polypedon was intended to refer to a three-dimensional body on the landscape in which all soil properties were sufficiently homogeneous, so as to be consistent with the concept of a single soil series (Fig. 3.16). A polypedon could be considered a hypothetical soil individual. Polypedons are bounded either by *miscellaneous areas*, i.e., nonsoil bodies, or soils of dissimilar pedons (soils of a different taxonomic class). The polypedon concept has not been widely used, however (Ditzler 2005), because it is difficult to identify material soil bodies that exemplify the idealized concept of a polypedon. In short, the great variability and continuity of soil properties in nature lead to uncertainty in identifying appropriate boundaries, as described earlier.

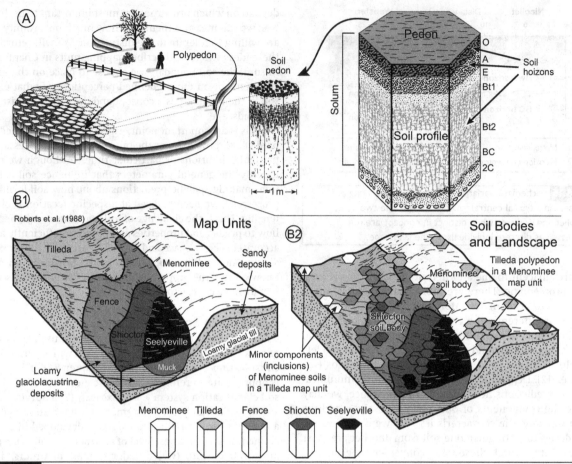

Fig. 3.16 Some of the basic soil and soil-landscape concepts. (A) Horizons, solum, profile, pedon, and polypedon. (B) Depiction of soils as they occur on a landscape, illustrating the concepts of pedon, soil body, polypedon, and map unit. 1. Map units on a hillslope. 2. Similar diagram, but illustrating the natural variability that occurs in soil landscapes. Map units attempt to outline soil bodies

A second concept developed in Soil Taxonomy is the *pedon*. The term refers to a body of soil with lateral dimensions large enough to permit the study of horizon sequences and boundaries (Johnson 1963, Campbell and Edmonds 1984). The dimensions of a pedon vary with the degree of lateral and vertical variation in soil properties, but in general the pedon is expected to consist of a volume of soil ranging from ~2 to ~20 m³ (Fig. 3.16). The pedon includes all of the soil's genetic horizons, as well as the rooting zone of most plants. For practical purposes, a lower limit of the pedon is often set at a depth of about 2 m, or bedrock, whichever is shallower. The fundamental idea of a pedon is that it must exhibit all the properties that would allow the three-dimensional body to be classified as a soil. Therefore, the pedon is a theoretical construct corresponding to a *sampling unit* used to represent a soil.

Like the polypedon concept, the original concept of the pedon is not widely used in the practice of soil science. For practical reasons, most soils are sampled from relatively small pits, with hand probes, or with hydraulic probes,

none of which provides the opportunity to document the lateral variation in soil properties that must be known before collecting a representative sample. And in many contexts, the 2-m limit is too shallow to capture all of the deep subsurface properties of the soil profile or the weathering profile (Cremeens *et al.* 1994). In practice, the pedon is the volume of soil from which a soil scientist chooses to collect physical samples. The descriptions and characterization data of those soil samples are then used to refine the predictions of soil maps and to expand our understanding of soil processes and genesis.

The pedon concept has been linked to other concepts. The pedon and the biota that exist within and above it were called a *tessera* by Jenny (1958). Also, a set of adjoining and genetically related pedons on the landscape, irrespective of their taxonomic classification, has been called a *soil body* (Hole and Campbell 1985; Fig. 3.16B). The next logical step to take is from the soil body to the soil *map (or mapping) unit*. We will discuss map units and their meaning in more detail in Chapter 8, but here we note that a map unit is a cartographic concept, designed to delineate soil bodies and miscellaneous areas on a soil map. Such a delineation on a soil map rarely represents a single soil body. Instead, it is assumed that inclusions of soil bodies with disparate properties will occur in map units. Nonetheless, they have been grouped together so that they can be effectively portrayed at the scale of the map (Fig. 3.16B; see Chapter 8).

Putting the Letters Together

In Chapter 2, we presented the morphological properties that soil scientists use to describe soils. At the beginning of this chapter, we used the "alphabet soup" analogy to introduce the names of genetic soil horizons. Now we can extend the analogy further. In the same way that letters are used to form words, it is the arrangement of soil horizons with different properties that distinguishes soils from each other. Just as a particular sequence of letters in a word conveys meaning to us if we recognize the language, a sequence of soil horizons and their properties can tell us about a soil if we recognize its pattern and comprehend that pattern in an appropriate context. Finally, just as words only tell a meaningful story when they are arranged in a recognizable pattern called a sentence, a soil's complete story can only be appreciated when it is told in the context of other soils on the landscape. To understand those patterns and contexts is a theme of immense importance and represents the motivation to which this book is dedicated.

Chapter 4

Basic Concepts: Soil Mineralogy

Minerals are naturally occurring, inorganic compounds that have a characteristic chemical composition and a regular, repeating three-dimensional array of atoms in a crystal structure. Minerals can be classified according to their chemical composition and crystal structure, or according to whether they are *primary* (having formed directly from magma) or *secondary* (having formed by chemical weathering of other, preexisting minerals, either primary or secondary). Primary minerals tend to dominate in coarser size fractions of soils, whereas secondary minerals are most abundant in the clay and fine silt fractions (Fig. 2.6). Knowing the structure, properties, and origins of soil minerals is essential for understanding the processes of mineral transformation and transport, all of which are important in soil genesis. This chapter emphasizes mineral structure, classification, properties, occurrence, and identification, whereas later chapters (especially Chapters 13 and 15) focus on the role of minerals in soil genesis and geomorphology, e.g., as indicators of soil genesis, age, or paleoenvironments.

Bonding and Crystal Structures

Understanding the chemical and structural classification of minerals requires knowledge of bonding and crystal structure in inorganic solids, particularly solids in which oxygen is an anion.

Oxygen (in the form of the anion O^{2-}) makes up about 47% (by mass) of the Earth's crust (Fig. 4.1). It is by far the most abundant element and anion in the crust. Consequently, nearly all major groups of soil minerals, including silicates, oxides, phosphates, carbonates, and sulfates, are ionic solids in which O^{2-} is the primary anion. The most common cations in soil minerals are Si^{4+}, Al^{3+}, and Fe^{3+}, reflecting their crustal abundance as well (Fig. 4.1).

The bonds between these cations and O^{2-} are predominantly ionic, and mineral structures commonly consist of large O^{2-} anions in close packing around a small polyvalent cation such as Si^{4+} or Al^{3+} (Fig. 4.2). The number of O^{2-} anions that can fit around the cation in a close-packed arrangement depends on the radii of the ions involved (Fig. 4.2). Table 4.1 lists the ionic radii, the radius ratio ($r_{cation}/r_{O^{2-}}$),

and the resulting number of O^{2-} anions that can surround (coordinate with) the cation. Because of its small size, Si^{4+} is coordinated with four O^{2-} anions to form a silica *tetrahedron* (SiO_4^{4-}), whereas the larger Al^{3+}, Fe^{3+}, Fe^{2+}, and Mg^{2+} cations are usually coordinated with six O^{2-} anions to make an *octahedral* structure (Fig. 4.2).

Various models have been devised to represent mineral structures. The sphere-packing or space-filling models of mineral structures (far left of Fig. 4.2) are the most accurate representations of mineral structures, but other schematic representations provide additional information about the three-dimensional arrangements of ions in minerals. Ball-and-stick diagrams (far right of Fig. 4.2) use small balls to represent ions and lines to depict the bonds between atoms. Polyhedral models (middle of Fig. 4.2) show lines that connect O^{2-} ions at the corners of tetrahedra and octahedra but do not show ions or bonds. Polyhedral models are most useful for showing the sharing of corners, edges, or faces (i.e., one, two, or three O^{2-} ions) between adjacent polyhedra.

Oxides

Metal oxides, which are ubiquitous in soils, have relatively simple chemical formulas and crystal structures (Tables 4.2, 4.3). Most metal oxides in soils are secondary minerals, formed during the weathering of primary minerals that contain iron or aluminum; only a few oxides in soils are inherited as primary minerals from igneous rocks. Secondary iron and aluminum oxides are major components of the clay fraction of highly weathered soils such as Oxisols and Ultisols (see Chapter 13). In younger soils, these same oxides do not constitute a significant fraction of the soil mass and are often present as grain coatings, where they can influence aggregation and retention of both cations and anions.

Oxide minerals are composed of aluminum, iron, or other cations that are octahedrally coordinated with oxygen anions (Fig. 4.3). They are classified according to the type of metal cation and the manner in which adjacent octahedra are arranged. Octahedra can share corners, edges, or faces; edge sharing is most common. In some cases, two or more

Table 4.1	Typical dimensions of the oxygen anion, common cations, their radii, and coordination numbers with O^{2-} in common minerals		
Ion	Ionic radius (nm)	Radius ratio with oxygen[a]	Coordination number in minerals[b]
O^{2-}	0.140	-	
Si^{4+}	0.042	0.34	4
Al^{3+}	0.051	0.41	4,6
Fe^{3+}	0.064	0.51	6
Fe^{2+}	0.074	0.63	6
Mg^{2+}	0.066	0.47	6

[a] Ratio of cation radius to O^{2-} radius.
[b] The number of cations that surround a single metal atom in the closest-packed arrangement.

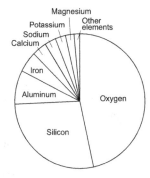

Fig. 4.1 Relative abundance (*mass%*) of the most common elements in the Earth's crust. When examined on a *volumetric* basis, oxygen composes about 94% of the crust, because of its comparatively larger size.

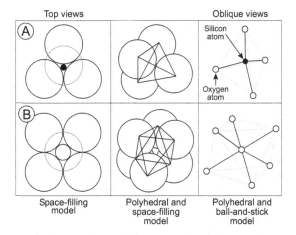

Fig. 4.2 Different views and models of tetrahedra and octahedra. (A) In the Si-O tetrahedron, the small silicon cation is surrounded by four O^{2-} anions. The corners of the tetrahedra represent the centers of O^{2-} anions surrounding the tetrahedral cation. (B) In an octahedron, the central cation, e.g., Fe^{3+}, Al^{3+}, or Mg^{2+}, is surrounded by six O^{2-} anions. In some minerals, some O^{2-} ions are replaced by OH^- ions.

minerals have identical crystal structures but differ in the arrangement of atoms in space, i.e., the manner in which adjacent octahedra share corners or edges.

Aluminum Oxides

The most abundant aluminum oxide in soils is *gibbsite*, $Al(OH)_3$, sometimes referred to as *hydrargillite* (Huang *et al.* 2002). When gibbsite forms discrete particles rather than coatings on other mineral grains, it typically occurs as hexagonally shaped plates (Huang *et al.* 2002). Gibbsite is composed of sheets of $Al(OH)_6$ octahedra that share edges (Fig. 4.3). In the octahedral sheets, one-third of the cation sites in the center of the octahedra are vacant, and two-thirds are occupied by Al^{3+} (Fig. 4.3). Gibbsite neoformation is promoted by high rainfall in freely drained settings, and by warm temperatures that promote weathering of primary minerals (Furian *et al.* 2002; see Chapter 13). Thus, crystalline gibbsite is most common in soils of warm, tropical climates; poorly crystalline $Al(OH)_3$ is more common in

less weathered soils. An abundance of gibbsite in midlatitude settings may suggest that the soil is old, having developed through weathering regimes more intense than that of today (Wang *et al.* 1981, Ogg and Baker 1999).

Other forms of $Al(OH)_3$ differ in their stacking of adjacent Al octahedral sheets. Bayerite, found in bauxite deposits, is rare in soils. Nordstrandite rarely occurs in soils (Table 4.2). In addition to $Al(OH)_3$, oxyhydroxides of aluminum can occur in two forms in highly weathered soils: diaspore (α-AlOOH) and boehmite (γ-AlOOH) (Gilkes *et al.* 1973, Sadleir and Gilkes 1976). These two polymorphs of AlOOH differ in the arrangement of double chains of octahedra (Fig. 4.3).

Table 4.2 | Characteristics of aluminum oxides in soils

Mineral	Frequency of occurrence in soils	Usual crystal shape	Intense X-ray diffraction peaks (nm)	Thermal analysis peaks (°C)
Gibbsite γ-Al(OH)$_3$	Most common Al oxide in soils and bauxite deposits	Pseudohexagonal plates (or rods)	0.485 0.437	Endotherm 300–330
Al(OH)$_3$(am)	Common in soils; intermediate in the formation of gibbsite	Amorphous	None	Endotherm 150–200
Bayerite α-Al(OH)$_3$	Uncommon in soils; common in bauxite deposits	Triangular pyramids	0.222 0.471 0.435	Endotherm 300–330
Boehmite γ-AlOOH	Not common in soils; occurs in bauxite deposits	Tiny particles	0.611 0.316	Endotherm 450–580
Diaspore α-AlOOH	Not common in soils; common in bauxite deposits	Tiny particles	0.399 0.232	Endotherm 540

Source: Hsu (1989).

Iron Oxides

Iron oxides are more abundant in soils than are aluminum oxides. Several Fe oxide minerals can be found in soils, depending on mineral inheritance and the soil's weathering environment (Table 4.3). Iron oxides form when Fe^{2+} in primary minerals is oxidized, released from the mineral structure, and hydrolyzed to form hydroxy iron polymers that crystallize into iron oxides. Iron oxides can occur both as coatings on mineral grains in mild and moderate weathering environments and as discrete mineral particles in stronger weathering environments (Bigham *et al.* 2002).

Because iron oxide particles are often small and have a high surface area, even a small amount can greatly enhance aggregation and affect soil color (Table 2.4). Iron oxides also promote aggregation and flocculation in soils. They have positively charged surface groups that attract negatively charged sites on clays and organic matter, and they also possess negatively charged sites that adsorb cations (Borgaard 1983).

Most iron oxides contain Fe^{3+} that is octahedrally coordinated to six O^{2-}, OH^-, or H_2O groups. Structurally, the various forms of iron oxide differ in the number of O atoms shared between octahedra and the distribution of vacant octahedra, i.e., lacking a central Fe^{3+} cation. In most Fe oxides, two-thirds of the octahedral sites are occupied and one-third are vacant. In goethite and hematite, Al^{3+} can substitute for Fe^{3+} in the crystal structure (Bigham *et al.* 2002; Table 4.3, Fig. 4.3). Aluminum substitution is more important in iron oxides of soils derived from aluminosilicate-rich parent materials (Allen and Hajek 1989).

Goethite (α-FeOOH), the most widely distributed iron oxide in soils, can be found in temperate to tropical, and in semiarid to humid climates. It imparts a brown to yellowish brown color to soils, although when hematite is also present the goethite hue may be masked (Table 2.4). Goethite particles are small (<0.1 μm), and in soils they have very irregular shapes (Bigham *et al.* 2002). Goethite consists of double chains of Fe-O octahedra. The octahedra form double chains (Fig. 4.3) that share corners or edges; the double chains are staggered, with H bonds between O and OH of neighboring octahedra that do not share O atoms. Goethite can have up to one-third of the Fe^{3+} ions substituted by Al^{3+} (Schulze 1984).

Lepidocrocite (γ-FeOOH) is a *polymorph* of goethite, i.e., it has the same chemical formula but a different crystalline structure. In lepidocrocite, double chains of Fe octahedra form zigzag sheets, with hydrogen bonding between the sheets (Fig. 4.3). Lepidocrocite is orange and is typically found in concretions or mottles, i.e., as redoximorphic features in soils that are subject to alternating oxidizing and reducing conditions (Bigham *et al.* 2002; see Chapter 13).

Hematite (α-Fe$_2$O$_3$) forms in strongly weathered soils, although it can also be inherited from the parent material. The name "hematite" is loosely translated as *blood rock* – because even small amounts of hematite will give soils a strong, almost blood red coloration (Table 2.4). Hematite crystals in rocks display a large variety of forms and sizes, including hexagonally shaped plates up to 50 nm in diameter (Schwertmann and Taylor 1989). In soils, hematite particles tend to be very irregular in shape (Bigham *et al.* 2002). In the hematite structure, two-thirds of the octahedral sites are occupied by Fe^{3+}, and the Fe-O octahedra are arranged so that each Fe-containing octahedron shares one face with another (Fig. 4.3). Hematite can have up to one-sixth of the Fe^{3+} ions substituted by Al^{3+} (Bigham *et al.* 1978).

Table 4.3 | Characteristics of iron oxides in soils

Mineral	Soil environment	Crystal shape	Density (g cm⁻³)	Maximum Al substitution	Intense X-ray diffraction peaks (nm)	Thermal analysis peaks (°C)
Goethite α-FeOOH	Most common Fe oxide in soils	Irregular in soils; needles, laths when synthesized	4.26	One-third of Fe^{3+} may be substituted	0.418[a] 0.245 0.269	Endotherm 280–400
Lepidocrocite γ-FeOOH	Poorly drained, noncalcareous soils	Laths	4.09	Little	0.626 0.329 0.247 0.194	Endotherm 300–350 Exotherm 370–500
Ferrihydrite $5Fe_2O_3 \cdot 9H_2O$ (6-line)[b]	Precursor to hematite; forms when Fe^{2+} is rapidly oxidized in presence of organic matter or silicates	Spherical or aggregated	3.96	Not known	0.254 0.224 0.197 0.173 0.147	Endotherm 150
Hematite α-Fe_2O_3	Highly weathered soils	Irregular in soils	5.26	Up to one-sixth of Fe^{3+} may be substituted	0.270 0.368 0.252	None
Magnetite Fe_3O_4	Primary mineral persists in coarse grains; contains Fe^{2+} and Fe^{3+}	Cubes	5.18	Possible	0.253 0.297	None
Maghemite γ-Fe_2O_3	Formed from magnetite by oxidation of Fe^{2+} to Fe^{3+}; may form under burning vegetation	Cubes	4.87	Possible	0.253 0.297	Exotherm 600–800

[a] Goethite's 0.418 nm peak may be obscured by the 0.426 nm peak of quartz; size fractionation to remove quartz is essential.

[b] In soils, ferrihydrite more commonly occurs in a less crystalline morphology that shows only two broad X-ray diffraction peaks near 0.254 nm and 0.151 nm; it is called 2-line ferrihydrite.

Source: Schwertmann and Taylor (1989). See also Table 2.4.

Ferrihydrite is a poorly crystalline, fine-grained Fe oxide with the approximate formula $5Fe_2O_3 \cdot 9H_2O$. Its reddish brown color is intermediate between the red of hematite and the yellow-brown of goethite (Table 2.4). Ferrihydrite crystals can form aggregates, with particle sizes between 2 and 5 nm. The small size, high surface area, and poor crystallinity (with only short-range ordering that produces very broad X-ray peaks) of ferrihydrite cause it to be very reactive and susceptible to dissolution. Thus, it dissolves in selective dissolution treatments that do not affect goethite and other crystalline Fe oxides. Consequently, many "amorphous Fe oxides" described in older literature are now presumed to have been ferrihydrite. Ferrihydrite is believed to be structurally similar to hematite, with more vacant Fe sites (fewer Fe^{3+} ions) and some of the O^{2-} anions being replaced by OH^- and H_2O. It is an intermediate in the formation of hematite (and of goethite) and forms in environments where Fe^{2+} is oxidized to Fe^{3+} in the presence of organic matter or silicate. Ferrihydrite has been reported in subtropical to cold climates, and it is also found in bogs and spodic horizons (Schwertmann and Taylor 1989).

Magnetite (γ-Fe_3O_4) is a primary mineral inherited from igneous rocks; it can be found as black, magnetic particles in the coarse fraction of many soils. Magnetite contains one-third Fe^{2+} and two-thirds Fe^{3+}. Magnetite and its weathering product maghemite (γ-Fe_2O_3) are found mainly in soils derived from basalt or other mafic igneous rocks. It can also form where other iron oxides, such as goethite, in organic-matter-rich surface horizons, have been heated by fire. Because magnetite in soils can become aligned with

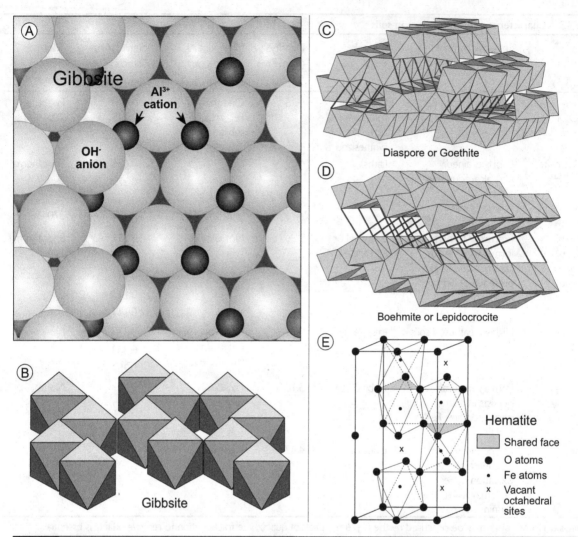

Fig. 4.3 Schematic diagrams of various oxide clay minerals. (A) Representation of the gibbsite $(Al(OH)_3)$ structure, showing small Al^{3+} cations surrounded by six larger OH^- ions – three below the cation and three above. The Al^{3+} ions occupy two-thirds of the octahedral positions. (B) Representation of the gibbsite structure, showing octahedra with shared edges. An Al^{3+} ion occurs in the center of each octahedron, and each corner represents OH^-. (C) Representation of either diaspore $(\alpha\text{-}AlOOH)$ or goethite $(\alpha\text{-}FeOOH)$, depending on whether Al^{3+} or Fe^{3+} is the octahedral cation. The figure shows staggered double chains of octahedra, with corners of octahedra shared between adjacent double chains. (D) Representation of either boehmite $(\gamma\text{-}AlOOH)$ or lepidocrocite $(\gamma\text{-}FeOOH)$, depending on whether Al^{3+} or Fe^{3+} is the octahedral cation. The figure shows zigzagging double chains of octahedra, with edges shared between adjacent double chains. (E) Representation of hematite $(\alpha\text{-}Fe_2O_3)$, showing shared crystal faces and vacant parts of the crystal structure.

the Earth's magnetic field, its presence in soils, especially in buried soils, can be valuable as a relative dating tool (see Chapter 14).

Manganese Oxides

Manganese oxides are more complex and less well characterized than are Al or Fe oxides. The study of pedogenic Mn oxides is hindered by their poor crystallinity, which makes distinguishing one Mn oxide from another, especially in mixtures, difficult.

Manganese oxides can contain Mn^{2+}, Mn^{3+}, and Mn^{4+} ions that are octahedrally coordinated to O^{2-}. They can form either tunnel or layer structures of Mn-O octahedra, with varying degrees of hydration. The Mn octahedra share edges to form single, double, or triple chains, or flat layers. Tunnel structures form when the chains share corners to form the edges of a tunnel – as in the tunnels of goethite (Fig. 4.3). The tunnels differ in height and width depending on whether a single, double, or triple chain forms the tunnel boundary. Both the tunnels and the interlayer spaces

Table 4.4 | Structural classification of primary silicates

Silicate class	General formula	Number of shared oxygens in each tetrahedron	Examples of minerals
Nesosilicates (orthosilicates) Isolated tetrahedra	SiO_4^{4-}	None	Olivine, garnet, zircon
Sorosilicates Double tetrahedra	$Si_2O_7^{6-}$	1	Epidote
Cyclosilicates Hexagonal ring formed by six tetrahedra, each sharing two corners	$Si_6O_{18}^{12-}$	2	Beryl, tourmaline
Inosilicates			
Single chain	SiO_3^{2-}	2	Pyroxenes: augite, diopside, enstatite, hypersthene
Double chain	$Si_4O_{11}^{6-}$	3	Amphiboles: hornblende, actinolite, tremolite
Phyllosilicates Layer (sheet) silicates	$Si_2O_5^{2-}$	3	Kaolin group: serpentines, talc, micas, vermiculite, smectites, chlorite
Tectosilicates (structural silicates) All four oxygens of each tetrahedron shared	SiO_2	4	Feldspars: orthoclase, albite, and anorthite Quartz Zeolites

can hold water or adsorbed cations. Of the Mn oxides, birnessite and vernadite (both layer structures) seem to be the most common, although todorokite (tunnel) and lithiophorite (layer) have also been reported in soils (White and Dixon 2002). In wet soils that undergo alternating periods of oxidation and reduction, manganese oxides are commonly found as discrete nodules, resembling fragments of black pepper (see Chapter 13). These features, commonly called "manganese shot," are a telltale indictor of wet soils.

Chlorides, Carbonates, Sulfates, Sulfides, and Phosphates

Chloride minerals (mainly halite, NaCl) have a simple structure, with each Na^+ ion octahedrally coordinated to six Cl^- anions, and each Cl^- anion surrounded by six Na^+ cations. Chlorides are extremely soluble and occur mainly as salt crusts on the surfaces of arid soils, particularly soils derived from saline parent material or influenced by saline waters or aerosols (Doner and Grossl 2002; see Chapter 13).

Carbonates such as calcite ($CaCO_3$) or dolomite [(Ca, Mg)(CO_3)$_2$] are less soluble than chlorides and are typically found in dry soils, in young soils, or in soils forming in calcareous parent materials (see Chapter 13). Calcite is the most common soil carbonate; it can be either pedogenic (forming in root zones where CO_2 concentrations are

high) or inherited from the parent material. Soil dolomite is believed to be inherited from calcareous sediments or eolian dust. Structurally, each Ca^{2+} or Mg^{2+} is octahedrally coordinated to six O^{2-} ions, one from each of six CO_3^{2-} anions. The octahedra are distorted because of ion sizes, so calcite and dolomite have rhombohedral crystal forms.

Sulfate minerals that contain Ca^{2+}, Na^+, or Mg^{2+} are relatively soluble and occur predominantly in dryland soils. Gypsum ($CaSO_4 \cdot 2H_2O$) is the most common sulfate mineral and can be either inherited or pedogenic (see Chapter 13). In contrast, jarosite ($KFe_3(SO_4)_2(OH)_6$) forms in acidic environments where FeS_2 from pyrite, coastal sediments, or mine spoils is exposed to air and oxidized (see Chapter 13). FeS_2 is the most common sulfide mineral in soils and is rapidly oxidized when exposed to air or oxygen-rich water (Fanning et al. 2002).

Phosphate minerals are not abundant in soils, though apatite ($Ca_5(F,Cl,OH)(PO_4)_3$) has been identified in young soils. In general, phosphate minerals are less soluble than either carbonates or sulfates (Harris 2002).

Silicates

Silicates are minerals composed of silica tetrahedra (Fig. 4.2), in which Si^{4+} cations are surrounded by four oxygen anions. Silicates differ in their crystal structures, chemical formulas (Table 4.4, Figs. 4.4, 4.5), and temperatures of

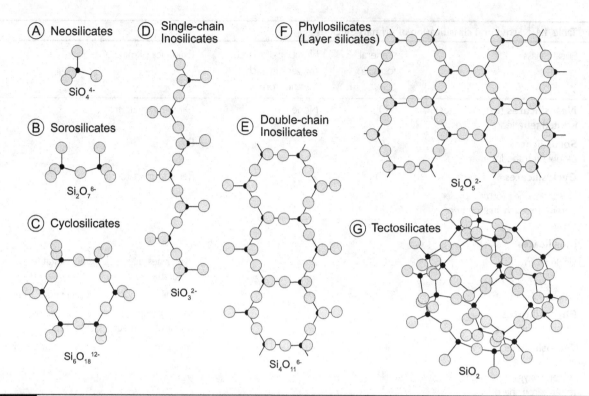

Fig. 4.4 Schematic (ball-and-stick) diagrams of silicate structures, arranged according to increasing sharing of O^{2-} at the corners of SiO_4^{4-} tetrahedron. (A) Nesosilicates or island silicates. (B) Sorosilicates, which share one corner of each tetrahedron. (C) Cyclosilicates, which share two corners of each tetrahedron. (D) Single-chain inosilicates, which share two corners of each tetrahedron. (E) Double-chain inosilicates, which share two or three corners of each tetrahedron. (F) Phyllosilicates or layer silicates, which share three corners of each tetrahedron, all in the same plane. (G) Tectosilicates, such as quartz, which share all four corners of each tetrahedron in a three-dimensional framework.

formation, all of which affect their stability in soils, as well as their cation exchange capacities (CECs). Many important primary minerals are silicates, including quartz, feldspars, micas, pyroxenes, and amphiboles. Primary silicate minerals are most common in the sand fraction of soils (Fig. 2.3), with the relative abundance of each depending upon the composition of the parent material and the extent of weathering. Secondary silicates, such as kaolinite and smectite, form by the weathering of primary silicates and are most abundant in the clay fraction.

Silicate minerals are classified according to how many corner O^{2-} anions each silica tetrahedron shares with other tetrahedra, as well as the geometric arrangement of neighboring tetrahedra (Fig. 4.5). Additional classification criteria will be addressed for tectosilicates and phyllosilicates, because they are so common in soils.

Nesosilicates, Sorosilicates, and Cyclosilicates

Silicates that consist of independent SiO_4^{4-} tetrahedra (not bonded to other, similar tetrahedra) are known as *nesosilicates* or orthosilicates (Fig. 4.4). Olivine (Mg, Fe)SiO_4 is a common nesosilicate in which neighboring silica tetrahedra are held together by electrostatic attraction between the SiO_4^{4-} tetrahedra and interstitial Mg^{2+} or Fe^{2+} cations. Olivine is a primary mineral that forms at high temperatures when magma is cooling to igneous rock, and it weathers readily in soils (Allen and Hajek 1989).

Sorosilicates such as epidote contain pairs of tetrahedra that share one corner O atom; they have the formula $Si_2O_7^{6-}$ (Fig. 4.4). *Cyclosilicates* (Fig. 4.4) typically have six tetrahedra arranged in a ring, each sharing two corner O atoms, to give the general formula $Si_6O_{18}^{12-}$. Sorosilicates and cyclosilicates such as beryl and tourmaline are not common in soils (Allen and Hajek 1989), but because of its resistance to weathering, the abundance of tourmaline has been widely used as an indicator of soil age or weathering status (see Chapter 15).

Inosilicates (Chain Silicates)

The *inosilicate* (or chain silicate) group (Fig. 4.4) includes pyroxenes and amphiboles, which originate in rocks that are intermediate between mafic and felsic classes. Single-

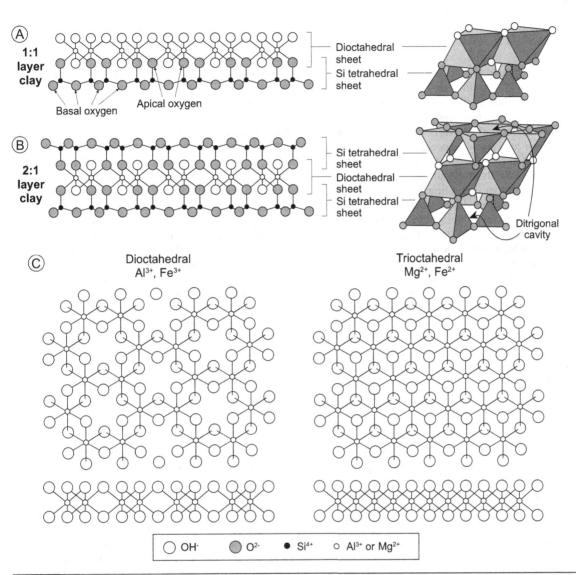

Fig. 4.5 Schematic diagrams of layer silicate (phyllosilicate) structures. (A) Ball-and-stick diagram of a 1:1 layer phyllosilicate, composed of one silica tetrahedral sheet composed of basal plane of O atoms, Si^{4+} cations, apical oxygens shared with an octahedral sheet, OH groups in the octahedral sheet (not shared), a trivalent octahedral cation occupying two-thirds of octahedral sites, and an octahedral, outer OH plane. A polyhedral model of the same phyllosilicate structure is located to the right. In this truncated polyhedral structure, not all the connections between tetrahedra and octahedra can be shown. (B) Ball-and-stick diagram of a 2:1 layer phyllosilicate, showing upper and lower silica tetrahedral sheets sharing apical oxygens with a dioctahedral sheet within the 2:1 layer. The octahedral OH groups in the 2:1 layer silicate are inside the crystal structure. A polyhedral model of the same phyllosilicate structure is located to the right. In this truncated polyhedral structure, not all the connections between tetrahedra and octahedra can be shown. (C) Comparison of a dioctahedral sheet (two-thirds of cation sites occupied by trivalent cations) and a trioctahedral sheet (all cation sites occupied by divalent cations), as seen from above. The large, open circles represent either O^{2-} or OH^- ions.

chain silicates such as pyroxenes have long chains of silica tetrahedra that share two corner O^{2-} ions and have the general formula SiO_3^{2-}. Double-chain silicates such as amphiboles are composed of parallel chains of silica tetrahedra in which half the tetrahedra are cross-linked to an adjacent chain. These linkages result in the general double-chain formula of $Si_4O_{11}^6$. Double-chain structures provide for greater resistance to weathering than do single chains. For example, amphiboles such as hornblende are more stable than pyroxenes, e.g., augite or diopside, but they are less stable than tectosilicates such as feldspars or quartz (Fig. 9.1). The most common cations that balance the negative charge of the silicate chains in amphiboles and pyroxenes are Ca^{2+}, Mg^{2+}, and Fe^{2+} (Allen and Hajek 1989).

Tectosilicates (Framework Silicates)

Quartz is the best known and most common *tectosilicate* (framework silicate), having the general formula SiO_2. The silica tetrahedra in quartz and all other framework silicates share corners with four other tetrahedra. All O^{2-} anions in each tetrahedron are shared between two tetrahedra to give a cross-linked, three-dimensional network of silica tetrahedra (Figs. 4.4, 4.5). This three-dimensional network of bonded tetrahedra (rather than a linear or planar structure as in other silicates) has no planes of weakness and, thus, is comparatively resistant to both physical and chemical weathering (Figs. 9.1, 9.2, Table 10.2). There are other polymorphs of SiO_2, including biogenic opals or phytoliths, and polymorphs that form at high temperature, but quartz is the most stable at the Earth's surface and the most common mineral in soils. It is also the second most abundant mineral in the Earth's crust, second only to another group of tectosilicates – feldspars (Allen and Hajek 1989).

In feldspars, one to two out of every four Si^{4+} ions are substituted by Al^{3+}, giving the framework structure a net negative charge and a general formula of MSi_3AlO_8, where M represents a metal cation (Ca^{2+}, Na^+, or K^+) that resides between silica tetrahedra and balances the negative charge of the silicate structure. Feldspars in which Ca^{2+} or Na^+ balances the negative structural charge are known as *plagioclases*. The plagioclase series or family includes two end members – one in which 100% of the charge is satisfied by Ca^{2+} cations (anorthite) and one in which it is satisfied by Na^+ (albite). Plagioclase feldspars can have any composition, but those that are nearer the albite end of the continuum are more weathering-resistant (Fig. 9.1). Because of Al substitution and interstitial cations, plagioclase feldspars are considerably more susceptible to weathering than is quartz. This characteristic is often used to assess the degree of weathering in soils by comparing the abundance of feldspar to that of quartz (Table 15.3, Fig. 14.64; see Chapter 15). Feldspars weather when their interstitial cations are removed, a state that promotes subsequent dissolution of the silicate structure. Potassium feldspars ($KAlSi_3O_8$) cannot form a solid solution with plagioclases because K^+ is a much larger cation than Na^+ or Ca^{2+}. Potassium feldspars (K-spars) are much more resistant to weathering than are plagioclases (Fig. 9.1). Microcline, orthoclase, and sanidine are the most common forms of K-spar.

Zeolites are framework silicates in which the silica tetrahedra form a more open and less uniform structure, with more interstitial space than is observed in quartz or feldspars. In zeolites, tetrahedra are linked to form open, cagelike structures. Zeolites differ in the number of linked tetrahedra and the size of the interstitial cavities. Like feldspars, zeolites have Al^{3+} substitution for some of the Si^{4+}, which produces a negative structural charge that is balanced by Na^+, K^+, Ca^{2+}, and Mg^{2+}, and sometimes by Ba^{2+} or Sr^{2+}. Although zeolites can form in basalts and other mafic igneous rocks, they also form as secondary minerals in sedimentary rocks, particularly those derived from volcanic rocks. Zeolites can be found in a variety of soil environments (Boettinger and Ming 2002).

Phyllosilicates (Layer Silicates)

Phyllosilicates (layer silicates) contain silica tetrahedra in which the three O^{2-} ions at the base of each silica tetrahedron are shared with adjacent tetrahedra, and the linked tetrahedra are arranged to form a sheet of pseudohexagonal rings, with the general formula $Si_2O_5^{2-}$ (Fig. 4.4). Oxygen ions at the base of the tetrahedra are called the basal plane, or siloxane surface. The apical O^{2-} ions of each tetrahedron are not shared with other silica tetrahedra, but they are in octahedral coordination with cations such as Al or Mg that occur in an immediately superjacent plane (Fig. 4.5). The combination of an $Si_2O_5^{2-}$ silica tetrahedral sheet with an aluminum octahedral sheet creates a mineral with the overall *unit cell*[1] of $Si_2Al_2O_5(OH)_4$ (kaolinite).

Phyllosilicates in which one silica tetrahedral sheet is bonded to one octahedral sheet are called 1:1 phyllosilicates, signifying that for each tetrahedral sheet there is one octahedral sheet. Kaolinite and serpentine are examples of 1:1 layer silicates (Table 4.5). The term *layer* should not be confused with *sheet*. *Layer* refers to the *combined* tetrahedral and octahedral sheets.

In 2:1 layer silicates, two silica tetrahedral sheets surround (or sandwich) one octahedral sheet (Fig. 4.5). The basal plane of oxygen atoms, i.e., the siloxane surface, of each tetrahedral sheet is on the outside of each 2:1 layer, creating an upper and lower siloxane surface. Some apical oxygens are shared with the octahedral sheet on the inside of the layer (Fig. 4.5). A 2:1 layer that contains two $Si_2O_5^{2-}$ tetrahedral sheets sandwiching an aluminum octahedral sheet has a unit cell formula of $Si_4Al_2O_{10}(OH)_2$. The formulas of 2:1 phyllosilicates have twice as many Si tetrahedra and only half as many OH groups per unit cell compared with the formula of 1:1 layer silicates, because oxygens of the octahedral sheet are shared with the adjacent tetrahedral sheets. Note that mineral formulas are written with the tetrahedral cations (Si^{4+}) listed before the octahedral cations (Al^{3+} in these examples); the O^{2-} and OH^- groups that balance the cation charge are listed last. Examples of phyllosilicates with 2:1 layers include micas, smectites, and vermiculite (Table 4.6).

In addition to being classified as 1:1 or 2:1 layers, phyllosilicates are classified according to the *type* of octahedral sheet they contain. As shown previously for gibbsite $Al(OH)_3$ (Fig. 4.3), trivalent octahedral cations only occupy

---

[1] The "unit cell" is the smallest set of atoms in the crystal structure that contains a complete sample of the crystal pattern that is repeated in space to form the mineral.

| Table 4.5 | Classification of 1:1 phyllosilicates according to the type of octahedral sheet (dioctahedral or trioctahedral) and octahedral cation, layer charge, formula, and interlayer forces |

Octahedral sheet and cation	Layer charge (moles/ formula unit)	Mineral or group name	Ideal formula
Kaolin group			
Dioctahedral	0	Kaolinite	$Si_2Al_2O_5(OH)_4$
Al^{3+}	0	Halloysite	$Si_2Al_2O_5(OH)_4 \cdot 2H_2O$
	0	Dickite	$Si_2Al_2O_5(OH)_4$
	0	Nacrite	$Si_2Al_2O_5(OH)_4$
Serpentinite group			
Trioctahedral	0	Chrysotile	$Si_2Mg_3O_5(OH)_4$
Mg^{2+} and some Fe^{2+}	0	Antigorite	$Si_2Mg_{3-3/m}O_5(OH)_{4-6/m}$
	0	Lizardite	$[Si_{2-x}Al_x][(Mg,Fe)_{3-x}Al_x]O_5(OH)_4$

two-thirds of the cation sites in the octahedral sheet; one-third of the sites are vacant (Fig. 4.5). Octahedral sheets that contain predominantly trivalent cations such as Al^{3+} or Fe^{3+} are called *di*octahedral sheets, because only *two* out of every three of the octahedral sites are filled (Fig. 4.5), giving a 1:1 layer formula of $Si_2Al_2O_5(OH)_4$. In contrast, divalent cations like Mg^{2+} or Fe^{2+} may occupy all three octahedral sites (Fig. 4.5). For example, a 1:1 layer composed of a silica tetrahedral sheet and a Mg octahedral sheet has the formula $Si_2Mg_3O_5(OH)_4$ (serpentine). Thus, the term *trioctahedral* can be remembered as signifying that there are *three* divalent octahedral cations in the chemical formula, whereas *dioctahedral* sheets have *two* trivalent octahedral cations per formula unit (Tables 4.5, 4.6).

Additional criteria for phyllosilicate classification include the amount of isomorphous substitution in the mineral structure and whether the substitution occurs in the tetrahedral or octahedral sheet. *Isomorphous substitution* is the replacement of one ion in the crystal structure by another ion of similar charge and radius (Table 4.1), but without altering the crystal form. Isomorphous substitution is only important in 2:1 clays; 1:1 layers have negligible isomorphous substitution. In clay minerals, isomorphous substitution typically involves substitution of lower-valence cations for higher-valence cations in the crystal structure, e.g., Al^{3+} for Si^{4+}. Each substitution by a lower-charge cation creates a net *negative charge* in the layer structure. The amount of charge per unit cell or per formula unit[2] is called the *layer charge*. The layer charge is neutralized by cations (or cationic polymers) that are adsorbed between two adjacent 2:1 phyllosilicate layers (Fig. 4.6). Isomorphous substitution can occur in either the

tetrahedral sheet (Al^{3+} for Si^{4+}) or the octahedral sheet, e.g., Mg^{2+} for Al^{3+}. Clay minerals can be categorized on the basis of the amount of tetrahedral charge or octahedral charge, i.e., isomorphous substitution in the tetrahedral or octahedral sheet, as well as on the basis of the total layer charge. Components in the interlayer region are also used as a basis of phyllosilicate classification.

Each of these classification criteria will be described more fully later, using specific examples. As shown in Tables 4.5 and 4.6, the type of layer (1:1 or 2:1) is the primary criterion for classifying phyllosilicates, with the next most important criteria being (1) total layer charge, (2) whether the layer charge is in the tetrahedral or octahedral sheet, and (3) whether the octahedral sheet is dioctahedral or trioctahedral. The type of interlayer cation or material is also important in some cases.

1:1 Layer Silicates: Kaolin Group

The kaolin group of minerals consists of 1:1 dioctahedral minerals in which Al^{3+} is the octahedral cation (Table 4.5). Kaolinite, dickite, and nacrite all have the formula $Si_2Al_2O_5(OH)_4$. Minerals in the kaolin group have little or no isomorphous substitution and hence a negligible layer charge. Instead, adjacent 1:1 layers are held together by hydrogen bonds between the OHs of the octahedral sheet and basal Os of the next tetrahedral sheet (Fig. 4.7). Kaolinite, dickite, and nacrite differ in the stacking arrangement of the neighboring 1:1 layers (Bailey 1980). Dickite and nacrite form in hydrothermal environments and are not usually found in soils. Kaolinite, which is very common in soils, can either have a hydrothermal or pedogenic origin. Pedogenic kaolinite particles are small (<1 μm) and typically are pseudohexagonal. Kaolinite can form by weathering of other aluminosilicates in slightly acidic, well-drained, and highly weathered soils in which silica and relatively mobile cations such as Ca^{2+} and Mg^{2+} have been removed by leaching (White and Dixon 2002).

---

[2] Unit cell charge and formula unit charge do not have the same value in 2:1 minerals because the unit cell is based on a 20-oxygen unit and the chemical formula is based on a 10-oxygen unit. So the formula unit layer charge is one-half the unit-cell layer charge.

Table 4.6 Classification of 2:1 phyllosilicates according to octahedral cation and sheet type (dioctahedral or trioctahedral), layer charge, charge location, and typical chemical formula

Octahedral sheet and cation	Layer charge (moles/formula unit)	Mineral or group name	Ideal formula	Location of isomorphous substitution
		Pyrophyllite-talc group		
Dioctahedral				
Al^{3+}	0.01	Pyrophyllite	$Si_4Al_2O_{10}(OH)_2$	
Fe^{3+}		Ferripyrophyllite	$Si_4Fe_2O_{10}(OH)_2$	
Trioctahedral				
Mg^{3+}	0.02	Talc	$Si_4Mg_3O_{10}(OH)_2$	
Fe^{2+}	0.01	Minnesotaite	$Si_4Fe(II)_3O_{10}(OH)_2$	
Ni^{3+}	0.04	Willemseite	$Si_4Ni_3O_{10}(OH)_2$	
		Mica group		
Dioctahedral				
Al^{3+}	1.0	Muscovite	$K(Si_3Al)Al_2O_{10}(OH)_2$	Tet sheet
Al^{3+} and Fe^{3+}	1.0	Celadonite	$K\,Si_4$ $(Al_{0.2}Fe^{3+}_{0.8}M^{2+}_{1.0})$ $O_{10}(OH)_2$	Oct sheet
Al^{3+} and Fe^{3+}	0.6–0.85	Illite (clay mica)	$K_{0.8}(Si_{3.4}Al_{0.6})$ $(Al_{1.4}Fe^{3+}_{0.4}M^{2+}_{0.2})$ $O_{10}(OH)_2$	Tet (oct) sheet
Al^{3+} and Fe^{3+}	0.6–0.8	Glauconite	$K_{0.8}(Si_{3.8}Al_{0.2})$ $(Al_{0.5}Fe^{3+}_{0.9}M^{2+}_{0.6})$ $O_{10}(OH)_2$	Oct (tet) sheet
Trioctahedral				
Fe^{2+} and Mg^{2+}	1.0	Biotite	$K(Si_3Al)\,(Mg, Fe)_3$ $O_{10}(OH)_2$	Tet sheet
Mg^{2+}		Phlogopite	$K(Si_3Al)\,(Mg)_3$ $O_{10}(OH)_2$	Tet sheet
Al^{3+} and Li^+		Lepidolite	$K(Si_3Al)\,(Li, Al)_{2.5-3}$ $O_{10}(OH)_2$	Tet sheet
		Vermiculites		
Mainly trioctahedral; some dioctahedral	0.6–0.9		$Mg_{0.4}(Si_{3.2}Al_{0.8})$ $(Mg_{2.6}Fe(II)_{0.4})$ $O_{10}(OH)_2$	Either
		Smectite group		
Dioctahedral				
	0.2–0.6	Beidellite	$Ca_{0.25}\,n(H_2O)$ $(Si_{3.5}Al_{0.5})Al_2$ $O_{10}(OH)_2$	Tet sheet
	0.2–0.6	Nontronite	$Na_{0.33}(Si_{3.67}Al_{0.33})Al_2$ $O_{10}\,(OH)_2$	Tet sheet
	0.2–0.6	Montmorillonite	$Ca_{0.25}\,n(H_2O)$ $Si_4(Al_{1.5}Mg_{0.5})$ $O_{10}(OH)_2$	Oct sheet

Table 4.6	(cont.)			
Octahedral sheet and cation	Layer charge (moles/formula unit)	Mineral or group name	Ideal formula	Location of isomorphous substitution
Trioctahedral				
	0.2–0.6	Saponite	$Na_{0.33}(Si_{3.67}Al_{0.33})Mg_3$ $O_{10}(OH)_2$	Tet sheet
	0.2–0.6	Hectorite	$Na_{0.33}Si_4(Mg_{2.67}Li_{0.33})$ $O_{10}(OH)_2$	Oct sheet
		Chlorite group		
Di, trioctahedral sheets +di, trioctahedral interlayers	variable		$(Si_3,Al)(Mg_3)$ $O_{10}(OH)_2 +$ $[(Mg_3)(OH)_6]$	Mainly tet sheet

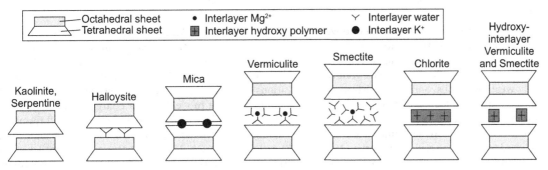

Fig. 4.6 Schematic diagrams of phyllosilicates common to soils, showing octahedral and tetrahedral sheets and the type of interlayer material. From left: 1:1 clays (kaolinite and serpentine); halloysite with interlayer water; mica group with interlayer K residing in the ditrigonal cavity in its basal plane of O atoms, yielding collapsed 2:1 layers; vermiculite (2:1) with hydrated interlayer cations such as Mg^{2+}; smectite group (2:1) with expanded interlayer regions and water molecules around each interlayer cation; chlorite (2:1:1) with a continuous metal hydroxide octahedral sheet in the interlayer position; hydroxy-interlayered vermiculite or smectite with discontinuous clusters of octahedral metal hydroxides.

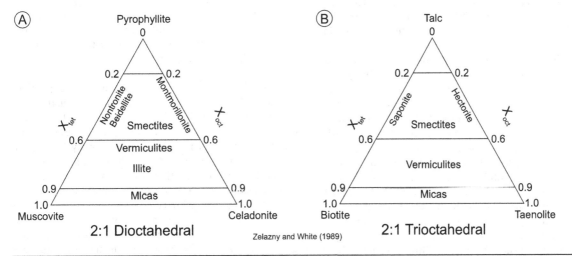

Zelazny and White (1989)

Fig. 4.7 Compositional diagrams of 2:1 clay minerals, showing their layer charge, X (moles of charge per 10-oxygen formula unit), on the vertical axis and location of layer charge on the horizontal axis. (A) Dioctahedral 2:1 minerals. (B) Trioctahedral 2:1 minerals.

Halloysite ($Si_2Al_2O_5(OH)_4 \cdot 2H_2O$) has the same general formula as kaolinite except that there are two interlayer water molecules per unit cell. Halloysite minerals can be tubular, lath-shaped, or even spheroidal. Halloysite that has been completely dehydrated is platy. It is poorly ordered, probably because of the effect of drying on the interlayer bonding. Halloysite forms by weathering in acidic, volcanic sediment and soils (White and Dixon 2002, Joussein *et al.* 2005).

1:1 Layer Silicates: Serpentine Group

Serpentine group minerals are 1:1 trioctahedral minerals that usually have Mg^{2+} in the octahedral sheet, although Co, Cr, Ni, and Al also can substitute (Table 4.5). The serpentine minerals are found in soils derived from ultramafic rocks (serpentinites) that originate in the ocean crust (see Chapter 8). Chrysotile ($Si_2Mg_3O_5(OH)_4$), which produces asbestos fibers, is composed of 1:1 layers that curl into cylindrical or spiral fibers; the curling occurs to maximize hydrogen bonding between layers. Antigorite and lizardite (Table 4.5) both have overall platy morphology (Bailey 1980).

Serpentines form by hydrothermal alteration of olivine and pyroxene and are also found in ultramafic igneous and metamorphic rocks. Serpentines weather easily, rendering them uncommon in the clay fraction of soils, except in very young, minimally weathered soils. Soils derived from serpentinites tend to be high in pedogenic chlorite, which is normally an unstable clay mineral, and smectite, especially iron-rich smectite, as well as more uncommon minerals such as talc and lizardite (Wildman *et al.* 1968, Parisio 1981). High Fe contents lead to the formation of iron oxides, especially goethite, and cause reddish-brown soil colors. Some authors report that the initial clay mineral suites of serpentinites are preferentially altered to vermiculite and smectite clays (Wildman *et al.* 1968, Ducloux *et al.* 1976).

Overview of 2:1 Layer Silicates

All 2:1 phyllosilicates contain two tetrahedral sheets that share apical O atoms with either a dioctahedral sheet (containing Al^{3+} or Fe^{3+}) or a trioctahedral sheet (containing Mg^{2+} or Fe^{2+}) (Fig. 4.6). There are many more types of 2:1 phyllosilicates than 1:1 phyllosilicates, because 2:1 layer silicates have differing amounts of isomorphous substitution. Thus, 2:1 clays are classified according to their layer charge (the net charge on the 2:1 layer caused by isomorphous substitution) *and* according to whether the isomorphous substitution occurs in the tetrahedral sheet (Al^{3+} substituting for Si^{4+}) or the octahedral sheet. The groupings shown in Table 4.6 reflect both the overall layer charge as well as its location. The classification of some common 2:1 phyllosilicates, based on total layer charge and the source of the layer charge, is also shown schematically in Figure 4.7. Another criterion used to classify 2:1 phyllosilicates is the type of interlayer cation or material. Interlayer cations (or cationic hydroxide sheets or polymers) balance the net negative charge in the 2:1 layer; the interlayer material is part of the mineral's overall chemical formula.

2:1 Layer Silicates: Talc-Pyrophyllite Group

Talc and pyrophyllite are the simplest 2:1 clay minerals in terms of chemical composition, consisting of an octahedral sheet that is sandwiched between two tetrahedral sheets. These prototype 2:1 minerals have no isomorphous substitution in either the tetrahedral sheet or the octahedral sheet and therefore, relatively simple formulas. Pyrophyllite ($Si_4Al_2O_{10}(OH)_2$) has an Al dioctahedral sheet; talc is trioctahedral and has an Mg octahedral sheet ($Si_4Mg_3O_{10}(OH)_2$). Because there is no isomorphous substitution, the 2:1 layers are uncharged and adjacent 2:1 layers are held together only by weak dispersion (van der Waals) forces. The weakness of the interlayer forces is responsible for the slippery or greasy feel of talc and for its lubricating properties; the layers readily slide past each other. Talc forms from ultramafic parent material in metamorphic and hydrothermal environments; it also forms in soils from weathering of pyroxenes and amphiboles. Both minerals are rare in soils (Zelazny *et al.* 2002).

2:1 Layer Silicates: Mica Group

Although there are many types of micas (Table 4.6), only muscovite and biotite are common in soils. In micas, the 2:1 layers have isomorphous substitution of lower-valence cations for higher-valence cations in the tetrahedral and/or octahedral sheet to yield a layer charge of about 1.0 mole per formula unit (which is one-half of a unit cell formula). The high negative charge on the 2:1 layers is balanced by K^+ cations that are held tightly between adjacent layers by strong electrostatic forces (Fig. 4.6). Micas are platy minerals with pronounced cleavage parallel to the plane of interlayer K^+ ions.

Muscovite, which is the most common soil mica, has an Al dioctahedral sheet. In addition to Al^{3+} in the octahedral sheet, Al^{3+} substitutes for approximately one-quarter of the Si^{4+} ions in the tetrahedral sheets. Because Al^{3+} has a smaller positive charge than does Si^{4+}, each Al^{3+} that substitutes for Si^{4+} creates a net negative charge. Comparison of the formula of muscovite, $K(Si_3Al)Al_2O_{10}(OH)_2$, with that of pyrophyllite, $Si_4Al_2O_{10}(OH)_2$, reveals that Al^{3+} substitutes for one out of every four Si^{4+} cations in the tetrahedral sheets and that the negative layer charge is balanced by interlayer K^+.

Biotite is the second most common soil mica; it can be locally important in young soils derived from mafic parent material. Like muscovite, biotite has a layer charge of 1.0, caused by Al^{3+} substituting for Si^{4+} in the tetrahedral sheet. Biotite differs from muscovite, however, by having a trioctahedral sheet in which Fe^{2+} is the dominant octahedral cation. The Fe^{2+} gives biotite a darker color than muscovite and allows biotite to weather more rapidly.

Soil micas are primary minerals that are normally inherited via physical weathering of the parent material; they do not readily form in situ. Muscovite is found in granitic igneous rocks, as well as in metamorphic rocks. It is also found in sericite, as a retrograde metamorphic alteration product of feldspar and other aluminosilicates. Biotite has a similar origin, but it is also found in rocks that are more mafic, consistent with biotite's high Fe and Mg content. Biotite tends to weather up to 100 times more rapidly than muscovite (Fig. 9.1) and is less common in soils, especially weathered soils. The difference in weathering rates can be attributed to different weathering mechanisms. In muscovite, the primary weathering mechanism is loss of interlayer potassium, which begins at particle edges. Thus, small muscovite particles weather faster than larger particles. In biotite, oxidation of Fe^{2+} to Fe^{3+} in the trioctahedral sheet enhances the weathering rate (Harris et al. 1985, Thompson and Ukrainczyk 2002).

Other micas (Table 4.6) are uncommon in soils, except locally where they are abundant in the parent material. Celadonite has isomorphous substitution mainly in its dioctahedral sheet; taenolite has isomorphous substitution in its trioctahedral sheet (Table 4.6, Fig. 4.7). Illite and glauconite, both of which form in marine environments, are dioctahedral micas that have isomorphous substitution in both the octahedral and dioctahedral sheets (Table 4.6, Fig. 4.7). In illite, there is more substitution in the tetrahedral than in the octahedral sheet, whereas glauconite has more octahedral than tetrahedral substitution. In older soil literature, the term "illite" is sometimes used to refer to fine-grained, partially weathered soil muscovite (also called hydrous mica or clay mica), but there have been efforts by mineralogists to reserve the term "illite" for micas formed via marine diagenesis, not pedogenesis. Although most micas have a layer charge close to 1.0, illite and glauconite, which form in marine environments, have a lower layer charge (0.6 to 0.8), which overlaps somewhat with that of vermiculite, as described later.

Most micas have K^+ as the interlayer cation. Interlayer K^+ is held tightly, or *fixed*, between adjacent mica layers, because the charge properties of the layers and of the K^+ ion provide ideal conditions for K^+ fixation. Because muscovite and biotite have high layer charges in the tetrahedral sheet, the electrostatic forces between the 2:1 layers and interlayer K^+ are very strong. And because K^+ is a large, low-charge cation, it is attracted strongly to the highly charged mica surfaces. The size of the K^+ ion provides a nearly perfect fit in the pseudohexagonal or ditrigonal ring formed by the basal-plane O atoms (Figs. 4.5, 4.6). Interlayer K^+ in micas rests inside the ditrigonal cavities, allowing the mica layers to collapse together around the K^+ ions (Fig. 4.6). Thus, the separation between adjacent 2:1 layers of micas is too small to allow entry of water molecules or other cations into the interlayer region, and as a result, the interlayer K^+ is fixed or nonexchangeable. This characteristic causes micas to have a relatively low capacity for retention or exchange of cations, because cation exchange is restricted to external surfaces of mica domains. This characteristic also plays an important role in identification of micas by X-ray diffraction, as described later.

2:1 Layer Silicates: Vermiculite

Vermiculite has a layer charge of 0.6 to 0.9 mole per unit cell, and it can be trioctahedral or dioctahedral (Table 4.6). The layer charge of vermiculite is similar to that of the low-charge micas illite and glauconite; vermiculite is distinguished from those micas by having Mg^{2+} or other readily exchangeable cations in the interlayer region to balance the layer charge, whereas illite and glauconite have interlayer K^+. The interlayer cations in vermiculite are typically divalent (Ca^{2+} or Mg^{2+}) and strongly hydrated, meaning that water molecules are strongly attracted to, and retained by, the cations. For example, the attraction between Mg^{2+} and H_2O is greater than the attraction between Mg^{2+} and the negatively charged basal surface of vermiculite, so adsorbed Mg^{2+} cations are surrounded by one or two shells of water molecules when in the interlayer region of vermiculite. Because the negative charge of the layers is lower than that of mica, divalent cations and their hydration shells can enter and leave the interlayer region easily. As long as the replacing cations are small and strongly hydrated, the interlayer region of vermiculite will remain open and accessible to other cations (Mallah 2002). Because vermiculite has a moderately high layer charge, it has a high cation exchange capacity (CEC) (160–200 $cmol_c$ kg^{-1}) (Barshad and Kishk 1969). However, if cations such as K^+ or NH_4^+ enter between the layers, these relatively large, low-charge cations cause vermiculite to collapse like mica, and the K^+ and NH_4^+ can be trapped in the interlayer region. This behavior is used to help identify vermiculite by X-ray diffraction.

Vermiculites typically obtain their negative charge from isomorphous substitution in the tetrahedral sheet. In some instances, particularly with trioctahedral vermiculites, isomorphous substitution in the octahedral sheet can actually produce a net positive charge. Vermiculites are reported to be unstable intermediates in the mica weathering process (Kittrick 1973). Trioctahedral vermiculite can form from biotite if the rate of interlayer K^+ loss exceeds the rate of iron release from the octahedral sheet. This transformation is favored at pH values >7.5, because a lower pH promotes dissolution of the 2:1 layers rather than simple transformation via loss of interlayer K^+ (Fanning et al. 1989).

2:1 Layer Silicates: Smectite Group

Smectites are a group of 2:1 phyllosilicates characterized by a low layer charge of 0.2 to 0.6 mole per formula unit. They can be either trioctahedral or dioctahedral, and they can have either tetrahedral or octahedral charge (Table 4.6). The type of octahedral sheet and the source of the layer charge are related to their mode of formation.

Soil smectites tend to occur in soils rich in Si^{4+}, Mg^{2+}, and Ca^{2+} (Reid-Soukup and Ulery 2002). Factors that favor smectite formation in soils include low-lying topography and/or poor drainage, i.e., areas with minimal leaching (including dry climates), and base-rich parent materials or alkaline soil-forming environments (Borchardt 1989,

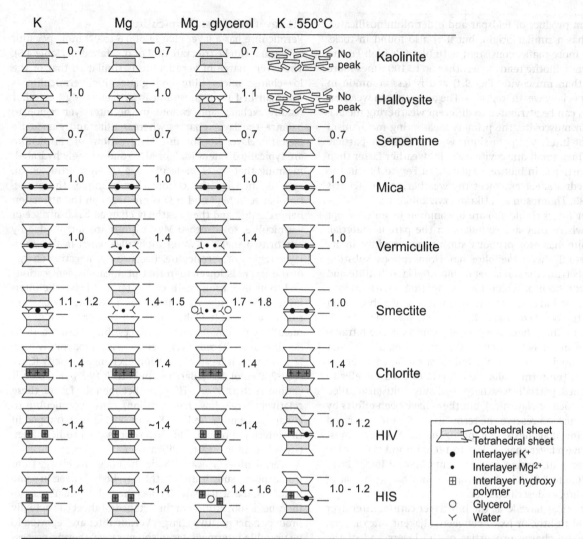

K	Mg	Mg - glycerol	K - 550°C	
0.7	0.7	0.7	No peak	Kaolinite
1.0	1.0	1.1	No peak	Halloysite
0.7	0.7	0.7	0.7	Serpentine
1.0	1.0	1.0	1.0	Mica
1.0	1.4	1.4	1.0	Vermiculite
1.1 - 1.2	1.4 - 1.5	1.7 - 1.8	1.0	Smectite
1.4	1.4	1.4	1.4	Chlorite
~1.4	~1.4	~1.4	1.0 - 1.2	HIV
~1.4	~1.4	1.4 - 1.6	1.0 - 1.2	HIS

Legend:
- Octahedral sheet
- Tetrahedral sheet
- ● Interlayer K$^+$
- · Interlayer Mg^{2+}
- ⊞ Interlayer hydroxy polymer
- ○ Glycerol
- Y Water

Fig. 4.8 Schematic diagram showing how clay mineral (001) d-spacings (in nm) change in response to four standard X-ray diffraction pretreatments – saturation by K$^+$, Mg^{2+}, Mg^{2+}, and glycerol, and heat treatment to 550°C. Collapse or expansion of smectite and vermiculite crystals can be detected by X-ray diffraction, depending on layer charge as well as the type of interlayer material in the clay.

Folkoff and Meentemeyer 1985, Kloprogge *et al.* 1999, Thiry 2000, Walsh and Humphreys 2010). Particularly influential in the formation of smectites is the presence of a long dry season. In soils, smectite is most abundant in the very fine clay fraction. Smectite minerals may be inherited from the soil parent material; they can also form by weathering of other layer silicates, or by neoformation. Dioctahedral smectite either can form by partial weathering of other 2:1 clays such as mica, vermiculite, and chlorite or can be neoformed after dissolution of other minerals with subsequent precipitation/recrystallization of the dissolved ions. Precipitation from solution is favored by near-neutral or mildly alkaline conditions and low concentrations of organic ligands. Trioctahedral smectites either are inherited from the sedimentary material or form by simple transformation (partial weathering without dissolution of 2:1 layers) of other trioctahedral 2:1 clays. Dioctahedral smectites are more stable and more abundant in soils than are trioctahedral smectites.

Montmorillonite, with its Al dioctahedral sheet and isomorphous substitution of Mg^{2+} (and other divalent cations) for octahedral Al^{3+}, is common in soils derived from Mg-rich mafic rocks. It also dominates in soils with parent materials derived from some marine shales, such as the Cretaceous Pierre Shale in western North America. This rock is a likely source of at least some of the montmorillonite found in

Pleistocene tills and loess in the midwestern United States, because Pleistocene glaciers crossed this shale as they moved to the southeast. Beidellite, which also has an Al dioctahedral sheet but has isomorphous substitution of Al^{3+} for Si^{4+} in the tetrahedral sheet, can form by direct transformation of other 2:1 clays, i.e., the 2:1 layers remain intact during weathering. Nontronite is an Fe^{3+}-rich smectite with tetrahedral-sheet charge that forms by simple transformation of iron-rich chlorite and vermiculite or glauconite (Robert 1973). Although the ideal formulas for smectites (Table 4.6) are relatively simple, smectites that form by precipitation in poorly drained environments contain a much wider range of octahedral cations and exchangeable cations, incorporating whatever ions are abundant in the soil solution at the time of formation.

When Na^+ is the interlayer cation, the low layer charge of smectite and the low charge density of Na^+ produce only weak attractive forces between Na^+ and the smectite surface. Such weak forces permit water to enter the interlayer region between the 2:1 layers and force the layers apart (Fig. 4.6). Interlayer separations in Na^+ montmorillonite in soil can be tens of times greater than when K^+ or Mg^{2+} is the interlayer cation. The extent of smectite swelling also depends on the total layer charge and on whether the charge is in the octahedral or tetrahedral sheet. Montmorillonite has octahedral layer charge and swells more in water and other polar solvents than does beidellite, which has tetrahedral charge. When isomorphous substitution occurs in the octahedral sheet, the charge is farther away from the interlayer cations than when the charge is in the tetrahedral sheet. The greater separation between the negative layer charge and the cations produces weaker electrostatic attraction, allowing more water molecules to enter the interlayer space, a process that in turn enables greater swelling between clay layers.

Natural shrink-swell processes in clayey, montmorillonite-rich soils during seasonal wet-dry cycles play an important role in soil genesis. Shrink-swell cycles drive a form of soil mixing called argilliturbation (Jayawardane and Greacen 1987; Table 11.1). In many montmorillonite-rich soils of wet-and-dry climates, frequent and strong argilliturbation processes produce morphological characteristics that define the soil order Vertisols, e.g., cracks and slickensides (ped faces smoothed by the swelling pressure of adjacent aggregates) (Wilding and Tessier 1988, Coulombe *et al.* 1996; see Chapter 11). It is important to keep in mind that seasonal shrink-swell processes do not occur at the scale of individual montmorillonite layers. Shrinking and swelling occurs when water moves out of and into pores that are much larger than the spaces between the 2:1 layers. The small size, flexibility, and low layer charge of smectite minerals all make it easy for water to move in and out of the micropores that occur between substacks of smectite layers. Soils with small amounts of smectite do not swell

extensively, yet their physical properties can be strongly influenced by smectite, particularly if soil pore waters are rich in Na^+.

Na weakens the interparticle forces between montmorillonite particles and promotes deflocculation (dispersion) of fine-grained smectite particles. The dispersed montmorillonite particles, many of which are <0.2 μm in diameter, can be translocated through the soil profile but eventually clog soil pores and may inhibit drainage. Dispersion of montmorillonite at the soil surface can produce surface crusts that dramatically decrease infiltration rates. Thus, the small size of montmorillonite particles, coupled with the presence of Na^+ in shallow water tables or irrigation water, often create conditions of poor tilth in dry climates, where this setting is most likely to occur.

2:1 Layer Silicates: Chlorite and Hydroxy-Interlayered Clays

True chlorites are primary trioctahedral 2:1 clay minerals with a layer charge similar to that of mica, but with a positively charged octahedral sheet of metal hydroxide between adjacent 2:1 layers, to balance the layer charge (Table 4.6, Fig. 4.6). The octahedral sheet within each 2:1 layer of chlorite is usually trioctahedral, dominated by either Mg^{2+} or Fe^{2+} (Barnhisel and Bertsch 1989; Table 4.6). Chlorite is sometimes called a 2:1:1 phyllosilicate because of the interlayer octahedral sheet. In chlorite, the interlayer hydroxide sheet is continuous and blocks the entire interlayer region; other ions cannot enter the interlayer region unless the interlayer hydroxide sheet dissolves.

Chlorites form in mafic igneous and metamorphic environments; when present in soils, they are inherited from the parent material. Chlorites are generally unstable in soils, usually limited to areas with weakly developed soils that are derived from chlorite-bearing parent materials, or to soils in cold climates where weathering is inhibited (Gao and Chen 1983, Yemane *et al.* 1996).

Hydroxy-interlayered vermiculite (HIV) and hydroxy-interlayered smectite (HIS) consist of vermiculite and smectite with positively charged Al^{3+} or Fe^{3+} hydroxide polymers in the interlayer regions. The most common interlayer material is aluminum hydroxide, although ferruginous parent materials (particularly chlorite) can produce iron hydroxide polymers in the interlayers (Barnhisel and Bertsch 1989). In some older literature, the terms "pedogenic chlorite," "chlorite-vermiculite intergrade," and "chlorite-smectite intergrade" have been used to refer to HIV and HIS. The interlayer hydroxide polymers in HIV and HIS can range from nearly complete interlayers that completely block ion adsorption sites and prevent smectite swelling or collapse, to discrete islands that only slightly modify the ion adsorption and swelling behavior of vermiculite and smectite. Although other ions can enter the interlayer region of HIV and HIS when the hydroxy interlayers are not extensive, the metal hydroxide polymers are strongly adsorbed and not readily replaced. Thus, HIV and HIS have lower CECs than do vermiculite or smectite. Because hydroxide interlayers tend to minimize the natural swelling tendencies of smectite, such interlayers may have a favorable effect on soil physical properties in smectite-rich soils.

HIV and HIS can form either by weathering of chlorite or by precipitation of hydroxide interlayers during weathering of mica and vermiculite; hydroxide interlayers form as silica is leached. The presence of hydroxide interlayers can stabilize vermiculite and smectite and slow the rate of silica loss. However, with pH >7.2, the interlayers tend to dissolve (Barnhisel and Bertsch 1989).

Identification of Phyllosilicates in Soils by X-Ray Diffraction

Soil clay fractions usually contain mixtures of phyllosilicates, typically aggregated by oxides, organic matter, and occasionally carbonates. Phyllosilicates in these mixtures can be identified using X-ray diffraction (XRD), after pretreatment to remove the aggregating agents (Kunze and Dixon 1986, Moore and Reynolds 1989, Poppe et al. n.d.).

XRD Theory

X-ray diffraction methods generally allow for the differentiation among different clay minerals in a sample, based on the expansion or contraction of the interlayer space – the region between adjacent 2:1 or 1:1 layers. This task is accomplished using different cations and solvents (Brown and Brindley 1980, Harris and White 2008). XRD not only allows specific clay minerals to be identified from complex mixtures, but can also permit semiquantitative estimates of their abundance (Brindley 1980, Brown and Brindley 1980, Moore and Reynolds 1989, Środoń et al. 2001, Kahle et al. 2002).

XRD relies on the principle that minerals are crystalline materials with periodic, repeating planes of atoms, and that these planes have uniform distances between them (Fig. 4.8). The distance between a plane of atoms in a 2:1 or 1:1 layer (for example, the plane of O^{2-} ions that form the siloxane surface) and that same plane of atoms in an adjacent layer is called the (001) d-spacing. The prefix (001) refers to those planes of atoms that intersect only the c-axis of the mineral's unit cell. Each layer silicate mineral has a characteristic (001) d-spacing. Thus, when using XRD to identify phyllosilicates, the (001) planes are the focus. In XRD, phyllosilicate clay samples are usually prepared so that they are oriented with their layers parallel to one another. Then, after being treated with different cations, solvents, and heat treatments, the d-spacings of the various clays in the samples can be determined and used to facilitate their identification.

Landscapes: Geophagy

Geophagy (or *geophagia*), the purposeful consumption of soil, is a transcultural phenomenon, which is particularly widespread in tropical nations, especially in Africa (Abrahams and Parsons 1996, 1997, Wilson 2003). Most geophagy involves eating clay, whether dried to a powder or shaped into disks, lozenges, or other forms. The source of the clay may be weathered B or C horizon material, although some people mine clay from insect mounds, such as those of termites or wasps (Hunter 1984a, 1993; see the figure "Landscapes geophagy"). After a good source of geophagic materials is located, most cultures return to the same site again and again.

Good heavens, why would people purposefully eat dirt? Most commonly, geophagy is practiced because it serves a purpose in people's lives – legitimately or as a placebo. Physiologically, geophagy is most common among pregnant and lactating women, suggesting that the clay is ingested as a supplemental source of inorganic nutrients. Persons who have intestinal parasites such as hookworm continually suffer gastric irritation and often resort to geophagy to assuage this condition (Hunter 1973). Lambert et al. (2013) argued that the ingestion of clay may have reinforced digestive barriers against alkaloids and toxins, conferring a selective advantage on people who practiced geophagy. Ingesting some kinds of clay may soothe heartburn and diarrhea, alleviate hunger pangs, or simply satisfy a craving. These reasons, along with curiosity, may be reasons that children eat clay (Vermeer 1966). When adults, these children are likely also to give clay to their children, perpetuating the culture. Hooda et al. (2004), however, found evidence that ingestion of soil, inadvertent or deliberate, can potentially reduce the absorption of already bioavailable nutrients. The jury is still out!

By far the most common physiological association of geophagy is with pregnancy, but it also is common during menstruation and lactation (Hunter 1984a, 1984b, 1993, Abrahams and Parsons 1997, Henry and Kwong 2003, Njiru et al. 2011). The cravings of pregnant women may be associated with a nutritional need, and in many Third World cultures this craving is satiated by eating clay. Hunter (1984a) analyzed clay tablets eaten by the people of Central America and found that they were rich in Fe, Ca, and Mn and contained not insignificant amounts of Cu, K, and Zn (see also Hunter 1984b, 1993). In 1966, Vermeer hypothesized that because the Tiv people in northern Nigeria do not drink milk, consumption of Ca-rich clay by pregnant Tiv women may have been vital to the health of their fetuses. Women in Guatemala overwhelmingly stated that they stop eating clay after giving birth. Thus, geophagy can be seen in the context of a nutritional supplement during a period of increased need, and, if so, it also fulfills a physiological need (Abrahams and Parsons 1997). Hunter (1984b) described a 21-year-old woman in Sierra Leone who was raising four children in a mud hut. She was pregnant again and eating an exceptionally large amount of mud (about 290 g daily) from a wasp nest. But in so doing she was providing herself with 160% of the recommended pregnancy supplementation of Zn, 62% of Ca, 56% of Mn, 47% of Fe and 30% of P. This practice may explain how she had given birth to four healthy children.

The roots of geophagy, however, may lie in spiritual and religious beliefs (Hunter 1984a). Religious geophagy almost always involves the ingestion of clay tablets, formed by craftspersons and later blessed by monks, priests, or holy persons. Clay tablets are made in many forms, shapes, and sizes; they are eaten in many parts of the world, fulfilling spiritual as well as physiological needs (Hunter 1984a, Hunter et al. 1989). They are commonly sold and traded as commodities (see figure).

With these two main motivations in mind, Hunter (1973) developed a culture-nutrition hypothesis to explain geophagy. He suggested that, over long periods, the cultural practice of geophagy subconsciously responds to the physiological needs of the body under stress. Pregnancy, for example, puts the woman in physiological stress, demanding a behavioral response. Long ago, the responses varied from person to person and place to place. Eventually, trial and error led to a subconscious knowledge that eating clay is beneficial in these circumstances. This knowledge is then passed on as a cultural behavior, which may become institutionalized and eventually develop into a cottage industry in which clay tablets are manufactured and sold (Hunter et al. 1989).

Geophagy is not limited to humans (French 1945, Setz et al. 1999, Krishnamani and Mahaney 2000). Many animals eat minerals

(Slamova et al. 2011), especially as a source of salt, as at salt licks. In the Tambopata-Candamo Reserve Zone of southeastern Peru, macaws lick and eat exposed clays in cliffs almost daily. As with human consumption, it is thought that the clay acts as a type of jungle antacid. Because the macaws eat a variety of seeds and fruit, and many of these items they eat are high in various types of toxins, the minerals in the clay may help detoxify the seeds and fruit.

The medicinal potential of clay has been exploited in modern Western cultures, too. Until the 1980s, the antidiarrheal drug product kaopectate was made from kaolinite and pectin (Pray 2005). Then, until the 1990s, it was made of attapulgite and pectin. Today's product, however, is made from bismuth subsalicylate – minerals are no longer involved!

Figure: (A) Two areas where soil is being mined for geophagical clays. Photos by J Hunter. (1) Termite (*Macrotermes*) mound in Zambia. (2) Pit in clay-rich soils in Guatemala. (B) Tablets made from geophagical clays. (a) Tablets drying in the sun, Ghana. (b) Marketplace display of more than a dozen different geophagical clays in various shapes and sizes.

X-ray diffraction requires the use of an X-ray *diffractometer*, which consists of an X-ray generator and tube, slits to collimate the X-ray beam emitted from the tube, a sample chamber (which holds the oriented clay sample), a goniometer that rotates the sample (and/or the detector) and measures the angle between the two, and a detector to measure the intensity of diffracted X-rays at different angles (Cullity 1978, Wilson 1987, Harris and White 2008; Fig. 4.9A). The wavelength or energy of X-rays employed is chosen by using a filtering system, a monochromator, or an energy-sensitive detector. The output from this process is an X-ray diffractogram (Fig. 4.10) – a quantitative

recording of the intensity of diffracted X-rays as a function of diffraction angle (Θ).

When X-rays impact a crystal, they pass through the solid until they strike atoms in the crystal structure, from which they may be reflected, i.e., diffracted. Because X-rays are diffracted from many atoms at different locations in the crystal, they are scattered in all directions, and most of the diffracted X-ray waves are out of phase with one another. However, at certain angles that depend upon the distance between parallel planes of atoms in the sample crystals, diffracted X-rays are *in phase* and reinforce one another. The condition for constructive interference

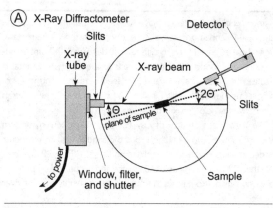

(A) X-Ray Diffractometer

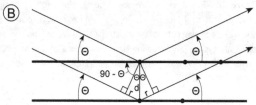

(B)

Fig. 4.9 Schematic diagrams designed to show the theory and workings of an X-ray diffractometer. (A) The X-ray source generates X-rays from a metal target, the shutter opens during analysis and closes for sample insertion, and slits collimate the beam of radiation. Also shown are the sample holder and detector. A metal filter may be used to select all but the desired wavelength of X-rays. Some diffractometers use a monochromator or a solid-state detector for this purpose. The angle θ is the angle between the incident radiation and the plane of the sample. The angle 2θ is the angle between the incident radiation and the diffracted radiation that reaches the detector. (B) Schematic diagram showing how incident X-radiation is diffracted by repeating planes of atoms (represented by heavy parallel lines) in a crystal. Diffracted X-rays interfere constructively when the path difference (2r) between X-rays hitting one plane of atoms and those hitting a parallel plane of atoms is 2r = nλ, where λ = the wavelength of the radiation.

(when the X-ray waves are in phase) is described by *Bragg's law*, which relates the wavelength λ of the X-rays to the distance d between parallel planes of atoms, i.e., the d-spacing of the clay minerals, and the sine of the angle of incidence Θ between the X-ray beam and the plane of atoms (Fig. 4.9B):

$$n\lambda = 2d \sin(\Theta)$$

where n refers to an integer.

Identification of clay minerals involves determining the angle at which the diffracted X-rays have interfered constructively to produce a peak in a diffractogram. Because the wavelength of the X-rays is controlled by the metal target in the X-ray tube, it is a simple matter to use Bragg's law

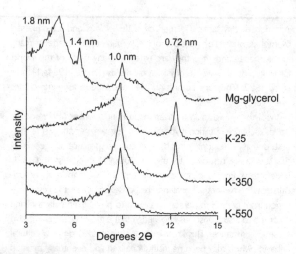

Fig. 4.10 Typical set of X-ray diffractograms used to identify layer silicate clay minerals. When saturated with Mg and glycerol, four clay minerals can be distinguished in this sample: smectite (1.8 nm), vermiculite (1.4 nm), mica (1.0 nm), and kaolinite (which accounts for most of the peak at 0.72 nm). Upon K saturation and drying at 25°C, the expandable peaks of smectite and vermiculite collapse toward 1.0 nm. Remaining water molecules are removed from the interlayer region when the K-saturated sample is then dried at 350°C, and both expandable minerals collapse to 1.0 nm. After heating to 550°C, kaolinite is destroyed by dehydroxylation, and the 0.72 nm peak disappears. In this sample, the dominant clay mineral (by mass) is smectite (~65%), with vermiculite (~5%), clay mica (~10%), and kaolinite (~20%) constituting the remainder.

to calculate the d-spacing that corresponds to each peak. Then the d-spacings in the diffractogram are compared to known d-spacings of various clay minerals (Fig. 4.8).

Different peak *orders* for the same mineral are commonly observed in a diffractogram. *First-order* diffraction peaks correspond to $n = 1$ in Bragg's law, and they give a true d-spacing of the mineral. *Second-order* diffraction peaks occur at angles where $n = 2$ in Bragg's law, giving an apparent d-spacing equal to half the true d-spacing; third-order peaks give apparent d-spacings of one-third the true d-spacing, etc. Thus, a mineral with a first-order or true d-spacing of 1.42 nm will give a second-order peak at 0.71 nm, a third-order peak at 0.474 nm. First-order peaks are normally used to identify layer silicate minerals, but occasionally the second-order peak from one mineral can overlap with the first-order peak of another mineral, complicating matters.

Clay Mineral Identification Strategies Using XRD

Before layer silicate clays can be identified by XRD, the soil must be pretreated to remove aggregating agents, e.g., carbonates, soluble salts, iron oxides, and organic matter

(Kunze and Dixon 1986, Moore and Reynolds 1997, Poppe *et al.* n.d.). The clay fraction is first separated from the rest of the soil sample, and then subjected to a series of cation, solvent, and heat treatments, each designed to *change the d-spacing* of certain layer silicates and facilitate their identification.

Specific strategies have long been used to identify important soil phyllosilicates (Figs. 4.8, 4.10). A portion of the sample is exposed to a solution containing small, strongly hydrated cations such as Ca^+ (Brindley and Brown 1980) or Mg^{2+} (Harris and White 2008) to saturate the negatively charged interlayer sites with these cations. Another subsample is exposed to a KCl solution to saturate the interlayer sites with K^+. These cation-saturated clay slurries are then either pipetted onto a glass slide or filtered through a porcelain plate to orient the clay layers and make them lie parallel to one another. The samples are then X-rayed to determine the various d-spacings. The Mg-saturated sample is subsequently equilibrated with either glycerol or ethylene glycol, whereas the K-saturated sample is heated to 550°C. Both the heat-treated and Mg-glycerol-treated samples are then X-rayed again. Treatments with K^+, Mg^{2+} (or Na^+), Mg-glycerol, and heat cause predictable changes in the d-spacings of different phyllosilicates, corresponding to each mineral's layer type, layer charge, type of interlayer material, and swelling characteristics (Fig. 4.8).

Mica is identified by the presence of a 1.0-nm peak in Mg-glycerol-treated samples. Micas give a 1.0-nm peak in every treatment because the layers are collapsed around interlayer K^+; only micas give a 1.0-nm peak in Mg-glycerol samples. In some cases, biotite can be distinguished from muscovite by comparing the ratios of the first-order peak at 1.0 nm with the intensity of the second-order peak at 0.5 nm. In muscovite, the 0.5-nm peak is about 50% as intense as the 1.0-nm peak; biotite's high iron content causes the second-order peak to be only 20% as intense as the 1.0-nm first-order peak (Brindley 1980, Fanning *et al.* 1989). Another means of distinguishing muscovite from biotite is based on the difference in unit-cell dimensions between dioctahedral and trioctahedral micas. These dimensions can be interpreted from XRD data, too, but the procedure requires a randomly oriented powder sample instead of the intentionally oriented sample described previously (Brindley and Brown 1980).

Smectite can be distinguished from other 2:1 phyllosilicates by the presence of a peak at 1.6 to 1.8 nm when Mg-saturated samples are treated with ethylene glycol or glycerol. In contrast, K^+ saturation causes smectite to collapse to ~1.0 nm. Although HIS may also expand in the presence of glycerol, it does not collapse with K-treatment. Vermiculite is characterized by 1.4 nm peaks with Mg and Mg-glycerol treatments and a 1.0-nm peak after K treatment. Vermiculite can be difficult to identify in samples that also contain large amounts of smectite because the broad smectite peak may overlap with that of vermiculite. Chlorite has a 1.4-nm d-spacing that persists in all treatments. It is the only mineral that retains a 1.4-nm peak when heated to 550°C.

HIV and HIS exhibit d-spacings of 1.2–1.4 nm with K and Mg treatments. When treated with glycerol, HIS may expand slightly to 1.6 nm; HIV does not expand. When K-saturated, both HIV and HIS collapse to smaller d-spacings when heated to 550°C. The less extensive the hydroxy interlayers, the greater the collapse. In samples that also contain chlorite and vermiculite, HIS can be identified by expansion with glycerol. HIV can be inferred from the ratios of the 1.4-nm and 1.0-nm peaks in different treatments – if HIV is present, the 1.4 nm/1.0 nm peak ratio will be greater in the room-temperature K-saturated sample than in the sample that was heated to 550°C.

As shown in Figures 4.8 and 4.10, kaolinite has a d-spacing of about 0.7 nm in all cation and solvent treatments; the d-spacing is unaffected by cation and solvent treatment because 1:1 clays have no isomorphous substitution and no interlayer cations to affect the distance between repeating planes of atoms. However, heating to 550°C causes kaolinite to decompose by dehydroxylation, and its 0.7-nm peak disappears. Serpentines, also 1:1 minerals, do not decompose when heated and can be more difficult to identify, but their sharper X-ray peaks (due to geologic, not pedogenic origins) facilitate their identification. Halloysite is an 1:1 clay with interlayer water. Replacement of that interlayer water with glycerol allows halloysite to expand, with characteristic d-spacings from 1.0 to 1.1 nm.

Kaolinite or serpentine can be identified easily in the absence of chlorite and HIV by the presence of a 0.7-nm peak in K-saturated samples. Although no other minerals have d-spacings of 0.7 nm, kaolinite's first-order peak at 0.7 nm coincides with the second-order peak for chlorite, making it difficult to identify kaolinite when chlorite is present. Because kaolinite decomposes at 550°C but chlorite retains its 1.4-nm d-spacing (and 0.7-nm second-order peak) at 550°C, kaolinite can be positively identified if the 0.7 nm/1.4 nm ratio decreases when the K-treated sample is heated to 550°C. In some cases, kaolinite can be identified in the presence of chlorite on the basis of peak width. Chlorite has very sharp XRD peaks, whereas pedogenic kaolinite may give a broad peak at 0.7 nm. Thus, a sharp 1.4-nm peak and a broad 0.7-nm peak may provide evidence that both chlorite and pedogenic kaolinite are present.

Identification of Iron and Aluminum Oxides in Soils

Iron and aluminum oxides in highly weathered soils, where they are present in sufficiently high concentrations, can usually be readily identified by XRD. But even in highly weathered soils, XRD identification of Fe and Al oxides usually requires particle-size separation to isolate the clay fraction, where concentrations of pedogenic oxides are greatest (Shaw 2001). Some iron oxides can also be separated magnetically. Another approach is differential XRD, in which diffractograms are obtained both for a sample that contains oxides and for a sample that has been chemically treated to remove oxides; the difference between the treated and untreated diffractograms yields a diffraction pattern attributable to the oxide minerals. Yet another approach involves treating the sample with concentrated NaOH to dissolve silicates, and then X-raying the remaining solid phase (Schwertmann and Taylor 1989). The most intense X-ray peaks of common iron and aluminum oxides are reported in Tables 4.2 and 4.3, respectively.

Mössbauer spectroscopy, which gives information about the chemical environment of Fe^{2+} and Fe^{3+} in iron oxides, can also be used to identify iron oxides (Bowen and Weed 1981, Parfitt and Childs 1988). The method is based on measurement of magnetic fields at different temperatures. The magnetic field of Fe in different oxides responds differently to changing temperature, depending on the specific chemical environment of Fe in the mineral.

Iron and aluminum oxides can also be identified by differential thermal analysis (DTA) (Karathanasis 2008). DTA involves measuring the difference in temperature between the sample and a reference material, as the sample is heated. When a mineral undergoes an exothermic reaction (energy is given off), the temperature of the sample is greater than the temperature of the reference material. An endothermic transition causes the sample to have a lower temperature than the reference material. Dehydroxylation (loss of H_2O when OH^- and H_3O^+ ions are lost from the mineral structure during heating) gives endotherms; recrystallization can produce exotherms. DTA is often used for quantitative analysis of gibbsite, which has a characteristic endotherm between 250°C and 350°C.

Selective chemical dissolution of Fe oxides allows for the distinction between ferrihydrite and the other secondary Fe oxide minerals, e.g., goethite, lepidocrocite, and hematite (Jackson *et al.* 1986). Typically the Fe dissolved by extracting a soil sample with Na-dithionite and Na-citrate is attributed to all Fe oxides (excluding magnetite), whereas Fe that can be dissolved in an extraction that employs acidified ammonium oxalate is attributed only to ferrihydrite (or to poorly crystalline iron oxides) (see Chapter 13). Because dissolution depends on factors such as crystallinity and surface area, selective dissolution approaches cannot be used to differentiate specific oxide minerals further. Still, by comparing selectively dissolved iron to total iron in a soil sample, the technique can be used to characterize the degree of weathering a soil has undergone. More highly weathered materials have a greater proportion of dithionite-extractable iron, and weakly or moderately weathered materials have a higher proportion of oxalate-extractable iron.

Chapter 5

Basic Concepts: Soil Chemistry

Soils are multiphase systems in which solids, liquids, gases, and colloids collectively interact. The solids in soils are crystalline or poorly crystalline minerals as well as organic matter. The liquid is water. Gases include nitrogen (N_2), oxygen (O_2), carbon dioxide (CO_2), methane (CH_4), and water vapor (H_2O). And colloids are minerals, humified organic matter, and many kinds of microorganisms that are generally <1 μm in diameter. Although some chemical and biochemical reactions in soils occur in a single, homogeneous phase, e.g., the liquid or solid phase, most reactions occur at the boundary between two phases, e.g., at the gas–liquid interface or at the liquid–colloid interface. Alternatively, they could be coupled with reactions that occur in more than one phase. Because both the architecture and the biological activity of soils vary by horizon and landscape position, the speed and direction of chemical and biochemical reactions also vary in complex and fascinating ways.

The Liquid Phase in Soils

Most chemical reactions in soils occur in the liquid phase, i.e., the *soil solution*, or at the interface between the liquid and solid phases. The soil solution consists of water in which cations, anions, ion pairs, small organic molecules, and gas molecules are dissolved and in which colloids are suspended. The chemical composition of the soil solution usually varies seasonally and depends on how much water is in the soil, on the minerals present, and on plant nutrient uptake. In many soils, Ca^{2+}, Mg^{2+}, Na^+, and K^+ (referred to as *base cations)* are the most abundant cations, whereas the dominant anions include HCO_3^-, Cl^-, and SO_4^{2-} (Table 5.1). Other cations and anions are present at relatively low concentrations. Most trace metals in solution, such as Fe or Al, occur in a variety of species that are determined by the pH of the solution and by the abundance of complexing anions or ligands. Soluble complexes of ions may be charged, e.g., $Fe(OH)_2^+$ or uncharged, e.g., $CaCO_3^\circ$.

The likelihood that any soluble species will react with another (i.e., its chemical *activity*) depends on its charge, size, and the concentrations of all other species that are in the solution (indexed by the solution's *ionic strength*). Ions shield one another from participating in chemical reactions, so the activity of any particular ion decreases as the ionic strength of the solution increases. In this text, we assume that the ionic strength of most soil solutions is sufficiently low that we can approximate the chemical activity of a soluble species by referring only to its concentration. The assumption certainly fails in salt-affected soils and in recently fertilized soils, but it is a reasonable simplification for understanding many solution-phase chemical reactions in the context of soil genesis. Concentrations of ions, complexes, and molecules in the soil solution are measured in mass units (mg L^{-1}) or molar units (moles L^{-1} or M). Typical concentrations of the major ions range from 10^{-5} to 10^{-3} M.

Simple cations and anions are *hydrophilic* (water-loving), and, hence, water molecules are clustered around them in solution. For example, in solution, Ca^{2+} ions are usually surrounded by a hydration shell of 9–10 water molecules that are oriented so the negatively charged dipole of each water molecule faces the positively charged cation (Fig. 5.1).

Acids and Bases in Soils

Acids and Bases

In water solutions, chemical species that can donate protons (H^+) to water molecules are *acids* and those that can accept protons are *bases*. When an acid molecule donates a proton, the chemical reaction is called *dissociation*, and the proton released is accepted by a water molecule. For a generic acid, HA (H represents a proton and A represents the *conjugate base* of the acid, usually an anion), the dissociation reaction can be written

$$HA + H_2O \rightleftarrows H_3O^+ + A^-.$$

For convenience, we often leave the water molecule out of our notation and write the hydronium ion (H_3O^+) just as H^+ and refer to it as if it were only a proton. Protons do

Table 5.1	Common soluble species found in the soil solution
Element	
	Typical cations or uncharged species
Ca	Ca^{2+}, $CaSO_4^{\circ}[aq]$, $CaHCO_3^{+}$
Mg	Mg^{2+}, $MgSO_4^{\circ}[aq]$, $MgCO_3^{\circ}[aq]$
Na	Na^{+}, $NaHCO_3^{\circ}[aq]$
K	K^{+}
N	NH_4^{+}, NH_3°, NO_3^{-}, $N_2[aq]$
Fe [III]	$FeOH^{2+}$, Org-Fe[a]
Mn	Mn^{2+}, $MnSO_4^{\circ}$, Org-Mn[2]
Zn	Zn^{2+}, $ZnSO_4^{\circ}[aq]$, Org-Zn
Cu	Org-Cu
Al	Al^{3+}, $AlOH^{2+}$, $Al[OH]_2^{+}$, $Al[OH]_3^{\circ}$, $Al[OH]_4^{-}$, Org-Al
	Typical anions or uncharged species
S	SO_4^{2-}, $CaSO_4^{\circ}[aq]$, $MgSO_4^{\circ}[aq]$
Cl	Cl^{-}
P	$H_2PO_4^{-}$, HPO_4^{2-}, $MgHPO_4^{\circ}[aq]$, $CaHPO_4^{\circ}[aq]$, $CaH_2PO_4^{+}$
Si	$H_4SiO_4^{\circ}[aq]$, $H_3SiO_4^{-}$
B	$H_3BO_3^{\circ}[aq]$, $B[OH]_4^{-}$
C	$CO_2[aq]$, H_2CO_3, HCO_3^{-}, CO_3^{2-}, organic C

Adapted from a compilation by R.J. Zasoski, University of California, Davis.

Note: The relative importance of the species depends on the pH of the soil solution.

[a] Org-: soluble organic complexes.

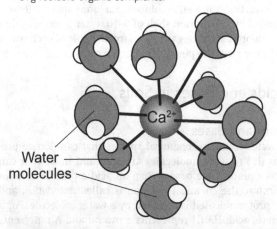

Fig. 5.1 "Exploded" view of a hydrated Ca^{2+} cation in bulk solution. The linkages between Ca^{2+} and H_2O molecules shown in this illustration represent ion–dipole bonds. Not drawn to scale.

not exist in aqueous solution by themselves; they are constantly being passed from one water molecule to another, forming H_3O^{+} cations.

Soil pH

The hydronium ion (simplified to H^{+}) in soil solution is a critical participant in many chemical reactions, so it is important to know its concentration. The greater the concentration of H^{+} ions in a system, the more active they can be in chemical reactions. Because the concentrations of H^{+} ions can vary over several orders of magnitude, we use a shorthand notation to refer to their concentration, not in moles per liter, but as the *negative logarithm of the concentration in moles per liter*. This value is called the *pH*, where *p* refers to the negative logarithm (base 10). Therefore, a solution that is 10^{-3} molar (moles L^{-1} or simply M) in H^{+} concentration is said to have a pH of 3. And a solution that is 3.16×10^{-6} M in H^{+} ($10^{-5.5}$ M) has a pH of 5.5. The pH of the soil solution is often measured with a glass electrode, after equilibrating a soil sample with water or a dilute salt solution, e.g., 0.01 M $CaCl_2$. Soils are classified as acid, neutral, or alkaline on the basis of their pH values (Fig. 5.2). The pH of the soil solution regulates so many chemical reactions that it is often called the "master variable" in soil.

Degree of Dissociation

The strength of an acid in water depends on how likely it is to give up, or dissociate, a proton to water molecules. *Strong acids* completely dissociate in water, leaving their corresponding anions, e.g., Cl^{-} or NO_3^{-}, in solution. Many acids in soil solutions are *weak acids* that only partially dissociate. Some common examples of weak acids, arranged in decreasing order of acid strength, are acetic acid (CH_3COOH), carbonic acid (H_2CO_3), phosphoric acid ($H_2PO_4^{-}$), silicic acid (H_4SiO_4), and, yes, even water (H_2O).

Consider the general case of the dissociation of a weak acid, HA:

$$HA \rightleftharpoons H^{+} + A^{-}.$$

The equilibrium constant for this reaction is given the name *acid dissociation constant*, or K_a.

$$K_a = \frac{[H^{+}][A^{-}]}{[HA]},$$

where A^{-} is the ion that has lost a proton. We can assess the willingness of acids to donate protons by comparing their K_a values. As with pH, it is often convenient to refer to the value of a dissociation constant of a given acid by the shorthand version, pK_a, i.e., the negative logarithm of the acid dissociation constant.

If we take the logarithm of both sides of the preceding K_a expression, multiply both sides of the resulting equation by −1, substitute for pH and pKa, and rearrange, we arrive at the *Henderson–Hasselbalch equation*, which is

$$pH - pK_a = \log \frac{[A^{-}]}{[HA]}.$$

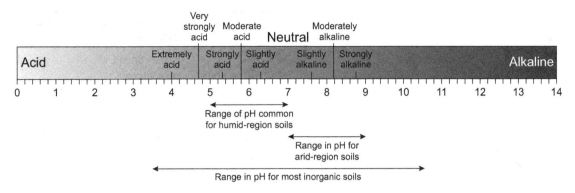

Fig. 5.2 The pH range scale, showing typical pH values for soils and the terminology used to refer to various soil pH values.

This useful relationship provides a pH benchmark to make predictions about how dissociated a given weak acid might be in the soil solution. When the system's pH is equal to the acid's pK_a, then

$$\log \frac{[A^-]}{[HA]} = 0 \text{ and } [A^-] = [HA],$$

which implies that half of the acid molecules are dissociated. If the pH of the solution is greater than the acid's pK_a, then $[A^-] > [HA]$, i.e., the dissociated species dominates. And when the pH of a system is less than the pK_a, $[HA] > [A^-]$, i.e., the undissociated acid dominates.

Carbonic Acid

Carbonic acid is a common and important weak acid in soil water, found wherever CO_2 gas and water coexist. In photosynthesis, plants use CO_2 in the air to fix carbon. But as they respire, both roots and soil microorganisms *release* CO_2(g) to the soil atmosphere. As a result, soil atmosphere contains considerably more CO_2 than does the atmosphere above the soil. The CO_2 in soil air dissolves in soil water and carbonic acid is formed, making it the most abundant weak acid in the soil solution. Protons dissociated from carbonic acid contribute to the weathering of primary minerals such as feldspar and secondary minerals such as calcite (see Chapter 9). CO_2 gas is also produced in some soils during methane fermentation, sulfate reduction, denitrification, and ammonification.

The average partial pressure of CO_2 gas in Earth's atmosphere (currently about 4×10^{-4} atm) is increasing, yet it is much smaller than the atmospheric concentrations of N_2 and O_2. Nevertheless, some very important reactions in soil water involve CO_2 directly or indirectly. CO_2(g) dissolves in H_2O in an amount that is proportional to its concentration in the atmosphere. The dissolution of CO_2 molecules in water is written

$$CO_2(g) + H_2O \rightleftarrows H_2CO_3^*(aq),$$

where $H_2CO_3^*$ refers to both dissolved CO_2 molecules as well as undissociated H_2CO_3 molecules, because it is difficult to distinguish H_2CO_3 from $CO_2 \cdot H_2O$ analytically. The equilibrium constant for this reaction, called the Henry's law constant for CO_2(g), is

$$K_H = \frac{[H_2CO_3^*]}{P_{CO_2(g)}[H_2O]}$$

In concentration-based equilibrium calculations, the concentration of pure water is assumed not to change significantly, and it is assigned a value of 1. Thus, the empirically determined value of K_H is the ratio of the concentration of dissolved CO_2 in the water to the partial pressure of CO_2 gas in the air above the water.

The Henry's law constant for CO_2 has been found to be 3.34×10^{-2} mole L^{-1} atm^{-1} at 25°C (National Institute of Standards and Technology 2011). So if we know the partial pressure of CO_2(g) in the air (see earlier discussion), we can use K_H to predict the activity of $H_2CO_3^*$ in a solution that is in equilibrium with the air:

$$[H_2CO_3^*] = K_H [P_{CO_2}].$$

Carbonic acid is a weak acid, and the degree of dissociation of protons from H_2CO_3 depends on the pH of the solution. We can determine the fraction of the acid that has dissociated by calculating the difference between the pH and the pK_a of each step in the dissociation process, as shown in the following relationships (cf. the Henderson–Hasselbalch equation earlier).

$$H_2CO_3^- \rightleftarrows H^+ + HCO_3^-$$

$$K_{a1} = \frac{[H^+][HCO_3^-]}{[H_2CO_3]}$$

$$pH - pK_{a1} = log \frac{[HCO_3^-]}{[H_2CO_3]}$$

$$HCO_3^- \rightleftarrows H^+ + CO_3^{2-}$$

$$K_{a1} = \frac{[H^+][CO_3^{2-}]}{[HCO_3^-]}$$

$$pH - pK_{a2} = log \frac{[CO_3^{2-}]}{[HCO_3^-]}$$

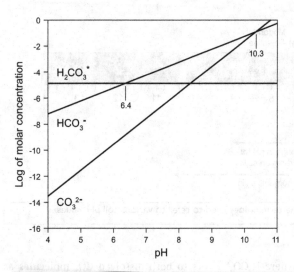

Fig. 5.3 Concentrations of aqueous carbonate species in equilibrium with CO_2 gas at atmospheric concentration, as a function of pH. $H_2CO_3^*$ represents the composite of both $CO_2 \cdot H_2O$ and $H_2CO_3(aq)$ species. Although the concentration of $H_2CO_3^*$ does not depend on pH, the concentrations of bicarbonate and carbonate ions do. Moreover, at most environmentally relevant pH values, bicarbonate ions are present in significantly higher concentrations than carbonate ions.

Table 5.2	Examples of typical, low-molecular-mass organic acids found in the soil solution	
Name	Typical concentration range (mmol L^{-1})	pK_a values
Formic	0.25–4.35	3.75
Acetic	2.7–5.7	4.75
Citric	0.014–0.21	3.14, 4.75, 6.4
Malic	1–4	3.40, 5.20
Tartaric	1–4	3.22, 4.85
Malonic	1–4	2.83, 5.69

Source: After Wolt (1994).

These relationships tell us that, at pH less than 6.4 (that is, the K_{a1} of carbonic acid), the undissociated form of carbonic acid will be the most abundant of the possible species. Between pH 6.4 and pH 10.3 (the K_{a2}), the bicarbonate species, HCO_3^-, will dominate, whereas only at pHs >10.3, CO_3^{2-} will dominate (Fig. 5.3).

Soluble Organic Acids

The liquid phase of soils includes dissolved *inorganic* acids such as carbonic acid, but also *organic* acids of low molecular mass. Organic acids are products of microbial metabolism and root exudates. They include both extracellular exudates and decomposition products, e.g., compounds left over from the degradation of plant components such as cellulose (Table 5.2). All are weak acids, and their willingness to give up protons depends on the pH of the soil solution. The most common functional group in these acids is the carboxylic acid group (R–COOH), where R represents any unspecified organic components to which the carboxylic acid C is bonded. When dissociated, the carboxylic acid group becomes a *carboxylate* group (R–COO⁻) that carries a single negative charge.

Hydrolysis of Metals

Unlike the base cations Ca^{2+}, Mg^{2+}, Na^+, and K^+, Fe^{3+} and Al^{3+} are metal cations with a sufficiently high ionic potential (charge·radius ratio) that water molecules in their hydration spheres are highly polarized and protons can be readily dissociated. The hydroxyl ions that remain bind strongly with the metal cation. Such hydrolysis reactions affect soil pH in ways that are similar to acid dissociation reactions. Take the hydrolysis of Al^{3+}, for example:

$$Al^{3+} + H_2O \rightleftharpoons AlOH^{2+} + H^+ \qquad \begin{array}{cc} K & pK \\ 10^{-5} & 5 \end{array}$$

Applying the Henderson–Hasselbalch model for weak acids, we can predict the extent of the hydrolysis reaction at a given pH. For example, if the pH of the soil solution is 6.5,

$$pH - pK = \log \frac{[AlOH^{2+}]}{[Al^{3+}]} = 1.5 \rightarrow \frac{[AlOH^{2+}]}{[Al^{3+}]} = 10^{1.5} = 31.6$$

This expression allows us to predict that at pH 6.5, the concentration of $AlOH^{2+}$ ions will be 31.6 times the concentration of Al^{3+} ions. Alternatively, at pH 4.5, $[AlOH^{2+}]/[Al^{3+}] = 0.316$, so there would be only 3.16 times as many Al^{3+} ions as $AlOH^{2+}$ ions in solution. Free Al^{3+} ions are fairly toxic to most crop plants, whereas $AlOH^{2+}$ ions are less so. Limiting uptake of Al^{3+} by crops is one reason that agronomists attempt to maintain soil pHs above 6 (Havlin *et al.* 2004).

Bases and Alkalinity

A solution's capacity to prevent, i.e., buffer, pH changes by neutralizing added acids is called its *alkalinity*. Soluble components that can accept protons are called bases. They will prevent the system's pH from declining, even when hydronium ions are formed by the dissociation of weak acids, or by the hydrolysis of Fe or Al. A number of soluble species can contribute to alkalinity in natural waters, e.g., bicarbonate, carbonate, hydroxide, phosphate, silicate, dissolved ammonia, and organic acids from which protons have been dissociated. All are proton-accepting species, although not all are anions. The total alkalinity, A_T, of a soil solution near pH 7 could be expressed as

$$A_T = [HCO_3^-] + 2[CO_3^{2-}] + [OH^-] + [HPO_4^-] + 2[HPO_4^{2-}] + [H_3SiO_4^-] + [NH_3] + [R-COO^-] - [H^+],$$

where $R-COO^-$ represents all organic acid anions. Usually bicarbonate (HCO_3^-) and carbonate (CO_3^{2-}) ions are the most abundant bases in the soil solution, and thus they contribute the most to alkalinity. In addition to soluble species, soil water may contain colloidal particulates that could buffer pH, e.g., organic colloids, poorly crystalline silica, e.g., biogenic opal, and microcrystalline $CaCO_3$.

Complexation Reactions

Metal–Ligand Complexes

Earlier we noted that Al hydrolyzes in aqueous solution to form soluble hydroxide complexes. Fe^{3+} behaves similarly to Al^{3+}, as indicated in the following expressions:

	pK
$Fe^{3+} + H_2O \rightleftarrows FeOH^{2+} + H^+$	2.2
$Fe^{3+} + 2H_2O \rightleftarrows Fe(OH)_2^+ + 2H^+$	5.7
$Fe^{3+} + 3H_2O \rightleftarrows Fe(OH)_3^0 + 3H^+$	13.1

In both cases the positively charged metal cation is associated with one or more hydroxide anions that coordinate with it. The hydroxide ions are called *ligands*, i.e., molecules or ions that match their negative charge or negative dipole with a positively charged, central metal ion. Water molecules occupy the remaining coordination positions around the metal ion and can also be considered ligands, so the first Fe hydroxide complex here could be written as $Fe(H_2O)_5OH^{2+}$.

In soil water, there are many potential ligands for metal ions in solution. Inorganic ligands include OH^-, F^-, Cl^-, HCO_3^-, CO_3^{2-}, and SO_4^{2-}. Organic ligands are primarily dissociated organic acids, e.g., acetate (CH_3COO^-) or oxalate ($^-OOC\text{-}COO^-$), although some N atoms in organic molecules may also act as ligands. Ligand complexes are not restricted to Fe and Al ions. Divalent transition-series metals such as Fe^{2+} or Mn^{2+} do not form hydroxide complexes as readily as Fe^{3+} or Al^{3+}, but they do form complexes with other inorganic or organic ligands. In soil solutions, the dominant aqueous species of a metal ion may be a ligand complex rather than the "free" metal ion (Table 5.1).

Chelation

When a metal is coordinated by more than one ligand from the same molecule, the complex is called a *chelate* (Fig. 5.4). Chelate complexes are very stable; they keep

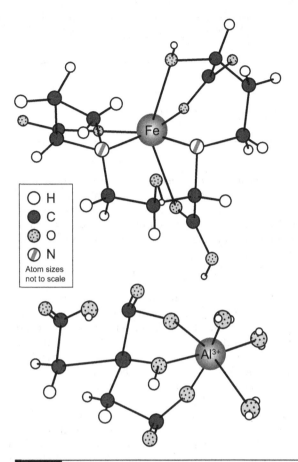

Fig. 5.4 Examples of chelated forms of Fe and Al cations. (A) A metallophore complex (Fe mugineic acid), and (B) an Al citrate.

metal ions in the solution phase and prevent the precipitation of metal oxides. Soluble organic compounds commonly contain more than one potential ligand, e.g., carboxylate or hydroxyl moieties, and readily form chelates with multivalent cations. Metal chelates are more mobile in the soil solution than are free metal cations, in part because they have lower positive charge and are less likely to be retained by the negative charges on mineral surfaces.

Some chelating molecules are essential to life in the soil, because they facilitate the transfer of otherwise insoluble but essential metals across cell membranes. Both plants and soil microorganisms produce *metallophores*, i.e., low-molecular-mass organic compounds that chelate micronutrient metal ions such as Fe or Zn. Once chelated, the metal ions become soluble and can be carried by water to a cell wall. Special transporter proteins then shuttle the chelated complex across the cell membrane and into the cell, where the metals participate in a number of physiological processes.

Table 5.3 | Permanent, pH-independent charges in layer silicate minerals, as estimated by cation exchange capacity

Mineral class	Cation exchange capacity (cmoles $(-)$ kg^{-1})
Micas (e.g., biotite, muscovite, illite)	20–40
Vermiculite	100–150
Smectites (e.g., montmorillonite, beidellite)	80–120
Kaolinite	1–10

The significance of metal complexes in pedogenesis is twofold. First, by reducing the concentration of free metal ions in solution, complexation promotes disequilibrium and increases the likelihood of mineral dissolution. For example, the dissolution of goethite may be portrayed in the expression

$$\alpha\text{-FeOOH} + 3H^+ \rightleftharpoons Fe^{3+} + 2H_2O$$

If the concentration of the product of this reaction, Fe^{3+}, is reduced by its subsequent reaction with oxalate ions and pH remains constant, the disequilibrium will promote dissolution and the reaction will shift to the right, releasing more Fe^{3+} to the solution. In other words, the formation of soluble metal chelates promotes the weathering of minerals in which those metal atoms occur. The second way in which metal complexation influences soil genesis is related to solubility and mobility. Complexation of metal ions by negatively charged, soluble ligands increases the potential for metal translocation in percolating water. Hence, complexation is a key part of podzolization, in which metal cations are translocated to B horizons via soluble complexes that may be chelated (see Chapter 13). Had they not been chelated, their solubility and hence, translocatability, would have been greatly reduced.

Sorption of Solution Species

Solution species interact with soil minerals and organic matter in many ways. The term *sorption* is used to describe any molecular-scale mechanism that results in the accumulation of a solute at a solid-phase surface. In this chapter, we classify sorption reactions into three categories: (1) ion exchange, (2) adsorption, and (3) precipitation. Before describing these concepts in more detail, we will first look at the nature of the charged mineral and organic surfaces that solution species interact with.

Charged Solid Surfaces

Major plant nutrients, such as Ca^{2+}, K^+, Mg^{2+}, Na^+, NH_4^+, NO_3^-, and phosphate ions, are retained in the soil by the charged surfaces of minerals and organic matter. Cations are attracted by negative charges on solid-phase surfaces, just as anions are attracted by positive charges. As described in Chapter 4, the basal surfaces of most layer silicate minerals exhibit a constant negative charge that originates within each mineral's crystal structure. Such charges are always present and are independent of the system's pH, so they are called constant or *permanent charges*. Negative charges also occur where carboxyl groups and phenols in organic matter have been deprotonated. In contrast, positive charges may occur on the surfaces of oxide minerals and at the edges of layer silicate minerals. The sign and concentration of charges on organic matter, oxides, and layer silicate edges depend on the nature and concentration of ions – especially H_3O^+ – in the solution. Hence, they are called pH-dependent or variable charges. Both kinds of charges are important to the movement or retention of solutes.

Constant, pH-Independent Charges

Layer silicates, e.g., kaolinite or montmorillonite, occur in the clay fraction of most soils (see Chapter 4). Because of their small size, large surface area, and charge characteristics, these minerals have a profound impact on chemical reactions in soil. As explained in Chapter 4, isomorphic substitution of cations in the tetrahedral or octahedral sheets of 2:1 layer silicates results in an excess of negative charge that is expressed at the basal surfaces of the minerals. This negative charge arises from the structure of the mineral and is not influenced by the pH of the soil solution. Therefore, the surfaces of layer silicate minerals have a constant, negative, pH-independent charge. The *cation exchange capacity* (CEC) of a layer silicate mineral reflects this structural negative charge on a *mass* basis (Table 5.3).

Structural charge and, hence, CEC vary widely among the layer silicates. For example, kaolinite is a mineral with essentially no isomorphic substitution, and therefore, almost no structural charge. (The small CEC of kaolinite particles arises from desorption of H^+, as described later.) Sufficient isomorphic substitution exists in the 2:1 minerals smectite and vermiculite, however, to produce a layer charge of ~0.2– ~0.6 and ~0.6– ~0.9 mole of charge per formula unit, respectively (see Chapter 4). At the high end of the scale, micas have a significant amount of isomorphic substitution, giving them the highest structural charge among the layer silicates. Isomorphic substitution in micas is so high that the negatively charged layers are collapsed around interlayer cations such as K$^+$ (Fig. 4.6). Cations in solution do not have ready access to these internal charges, however, because the layers are held together by their electrostatic attraction for K$^+$ ions. Therefore, the CEC of micas is low compared to those of vermiculites or smectites. Only when chemical weathering of mica lowers the structural charge do the layers open to allow the K$^+$ to escape or other cations to enter.

The CEC of vermiculite and smectite minerals in extremely acid and very acid soils (pH < ~5.5) can be considerably less than that in soils with higher pH. In such instances, weathering releases Al ions from the minerals, which then precipitate with hydroxide ions in the interlayer regions of the minerals. The precipitates hold the layers together, blocking access of base cations to charged sites and consequently diminish the CEC of the mineral by 20–80% (Inoue and Satoh 1992). Such minerals are called hydroxy-interlayered vermiculite and smectite (Chapter 4).

Variable, pH-Dependent Charges

pH-Dependent Charges at Silicate Edges

Hydronium (H$_3$O$^+$) ions are attracted to, and may be bound at, the edges of layer silicate sheets where broken bonds and unsatisfied negative charges occur (Fig. 5.5). The degree to which Al-O groups at the edge of a layer will accept an extra proton depends on the concentration of hydronium ions in the surrounding solution, making it a pH-dependent charge. The adsorption and desorption of protons may be illustrated as

$$\equiv\!AlOH_2^{+\frac{1}{2}} + H_2O \rightleftarrows \ \equiv\!AlOH^{-\frac{1}{2}} + H_3O^+,$$

where $\equiv$AlO represents an oxygen ion that is at the edge of the layer silicate crystal but that is not shared between the tetrahedral and octahedral sheets (Fig. 5.5). At pH values less than 6–6.5, the net charge at the edge of the layer is positive (Tombácz and Szekeres 2004, 2006). As pH increases above 6.5, the charge at the edge becomes increasingly negative.

On a mass basis, the positive charges at the edges of smectites and vermiculites are usually negligible when compared to the negative charges that arise from isomorphic substitution and that are expressed on the minerals' planar surfaces. But kaolinite and other 1:1 layer silicates have very little isomorphic substitution, so there is little permanent charge. Therefore, in 1:1 layer silicates at pHs common in highly weathered soils (e.g., 3.5–5.5), positive charges at layer edges contribute measurably to the ability of the mineral to retain anions, i.e., its *anion exchange capacity*.

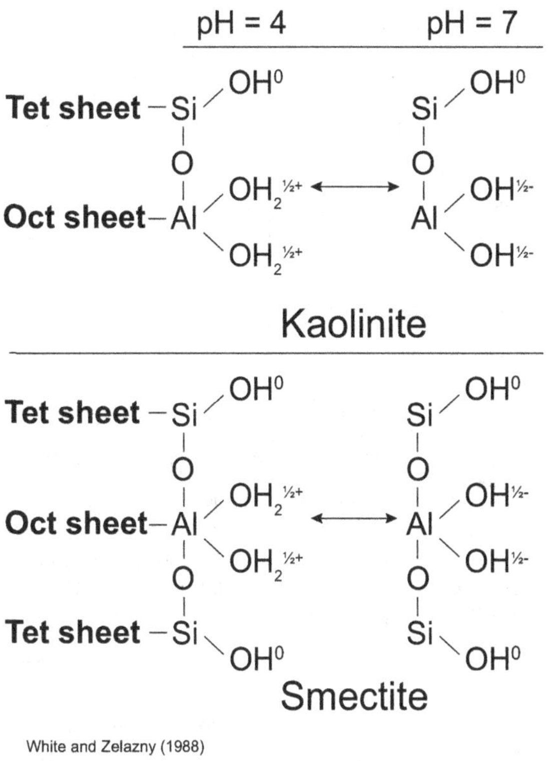

pH-Dependent Charges on Oxide Surfaces

Reactions in which H$^+$ is adsorbed or desorbed are also common at the surfaces of Al, Fe, and Mn oxide minerals,

Fig. 5.5 Diagram showing how positive charge develops at the edges of layer silicate minerals, illustrated for kaolinite and smectite at pH 4 and pH 7. The net charge at the edges depends on the pH of the solution.

e.g., gibbsite (Al(OH)$_3$) or goethite (FeOOH). For most oxides, permanent charges like those in layer silicates are negligible. Instead, the charge on oxide mineral surfaces results from interactions with ions from the solution. Both positively and negatively charged sites simultaneously occur on the mineral surface, and their density depends on pH. For example, at the surface of a ferric oxide such as goethite, we might find the reactions shown in Figure 5.6.

The Fe^{3+} ions shown in Figure 5.6 are bound to unseen oxygens or hydroxyls within the crystal (at left). At the crystal surface, the anions' negative charges do not perfectly balance the Fe cations' positive charges. Thus, the imbalance of charge at points on the surface allows a proton to be either adsorbed or desorbed, increasing or decreasing the net charge, respectively.

Because H$_3$O$^+$ and/or OH$^-$ ions are participants in these reactions, the extent of the reaction at equilibrium depends on the pH of the solution. At some pH value, a value that varies by mineral, the number of moles of positively charged sites on the mineral surface will equal the number of moles of negatively charged sites, i.e., [$\equiv$Fe-OH]$^{-\frac{1}{2}}$ = [$\equiv$Fe-OH$_2$]$^{+\frac{1}{2}}$, and the *net charge*

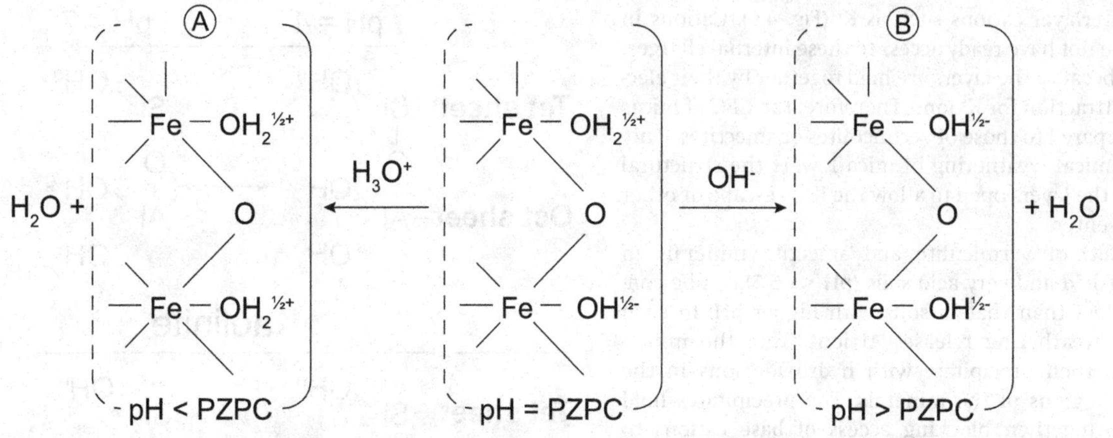

Fig. 5.6 Schematic illustration of the net charge on the surfaces of variable-charge minerals like Fe oxides. Such charges depend on the concentrations of charged species in the surrounding solution that have the potential to be adsorbed on the surface. In the simplest case, where no ions other than H^+ are likely to be adsorbed, the surface charge depends on the pH of the solution. The point of zero proton charge (PZPC) is the pH at which the positively charged sites at the surface are equal in number to the negatively charged sites. When the system pH is less than the PZPC, the net surface charge is positive (A). When the system pH is greater than the PZPC, the net surface charge is negative (B). The greater the difference between pH and PZC, the more positive or negative the net charge will be.

will be zero. If there are no other competing cations and anions in the system, i.e., H^+ and OH^- are the only ions likely to be adsorbed, then this pH is called the *point of zero proton charge* (PZPC). When pH < PZPC, $[\equiv Fe-OH]^{-\frac{1}{2}} < [\equiv Fe-OH_2]^{+\frac{1}{2}}$, the net charge is positive, and at pH > PZPC, $[\equiv Fe-OH_2]^{+\frac{1}{2}} < [\equiv Fe-OH]^{-\frac{1}{2}}$, the net charge is negative. (Bracketed terms refer to concentration of charge sites per unit area of the crystal.)

Predicting the actual net charge of a mineral surface in a soil is complicated, because there usually are other ions, e.g., Al^{3+} or $H_2PO_4^-$, that compete with H^+ and OH^- for sorption sites. Adsorption of these ions at specific sites on the mineral surface also changes the net charge of the surface. Nonetheless, we often start with the approximation that, at pH > PZPC, a sesquioxide surface will have a net negative charge and at pH < PZPC the surface will have a net positive charge (Table 5.4).

pH-Dependent Charges in Soil Organic Matter
Soil organic matter (SOM) consists largely of the residues of plant materials that either retain the fundamental chemical characteristics of polysaccharides, peptides, lignin, and lipids or have been incorporated into neoformed humic substances (Stevenson 1994). Both carboxyl groups (R–COOH) and phenols (aromatic OH groups) in soil organic matter have the potential to donate protons to water molecules in the same way that carboxyls in soluble weak acids dissociate.

$$R\text{-COOH} + H_2O \rightleftarrows R\text{-COO}^- + H_3O^+$$

Table 5.4	Point of zero proton charge for some common sesquioxide and silicate minerals
Mineral	PZPC
α-Al(OH)$_3$ (gibbsite)	5.0
α-FeOOH (goethite)	7.8
Fe$_2$O$_3$ (hematite)	8.5
δ-MnO$_2$ (pyrolusite)	2.8
SiO$_2$ (quartz)	2.0
Feldspars	2–2.4
Kaolinite	6.0–6.5
Montmorillonite	~6.5

Source: Stumm (1992); Tombácz and Szekeres (2004, 2006).

$$Ph\text{-OH} + H_2O \rightleftarrows Ph\text{-O}^- + H_3O^+,$$

where Ph–OH refers to any phenol and Ph–O$^-$ refers to phenolates in general.

Similarly, the degree of their dissociation depends on the pH of the system. The higher the pH, the more the carboxyl groups and aromatic OH groups are likely to be dissociated. In other words, as pH rises, the negative charge associated with SOM increases. Thus, like the oxide minerals, SOM is a source of pH-dependent charge, and part of the CEC of a soil can be attributed to soil organic

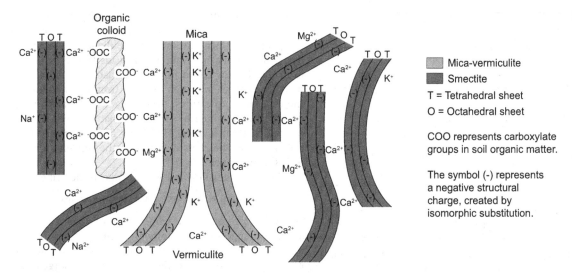

Fig. 5.7 Schematic diagram of a colloidal cation exchange complex (clay minerals and soil organic matter) in a pH-neutral soil, where negative charges arise from isomorphic substitution in clay minerals and from dissociation of carboxyl groups in soil organic matter. Each of the ions shown, except K^+ in the interior of the mica particle, is assumed to be hydrated. In addition, free water molecules (not shown) occur in the volume between solid-phase particles.

matter. A useful rule of thumb for predicting the CEC due to SOM is

$$CEC_{SOM} \text{ (cmol(-) kg}^{-1} \text{ of SOM)} = 50 \text{ pH} - 60.$$

This empirically derived equation (McBride 1994) indicates that at pH 7 SOM may contribute a negative charge to the soil of approximately 290 cmol(-) kg^{-1} of SOM. But in a pH 5 soil, SOM is likely to contribute a negative charge of approximately 190 cmol(-) kg^{-1}, a decline of one-third. The CEC of SOM is hard to measure directly because there is a close physical and chemical association between organic matter and clay particles. Some SOM sites are prevented from participating in exchange reactions because the organic matter is closely aggregated with high-surface-area clay minerals. In highly weathered soils that are dominated by low-surface-area and low-CEC kaolinite, SOM plays a much larger role in retaining nutrients because it provides a substantial portion of the soil's CEC.

Cation Exchange Capacity of Soil

The negative charge expressed by layer silicates and organic matter in soils is always balanced by the positive charges of base cations such as Ca, Mg, Na, K, and in acid soils by Al and H (Fig. 5.7). Base cations are attracted to negatively charged surfaces, but they are not bound there by covalent bonds. Instead, the attraction is electrostatic, allowing for a constant exchange of ions between the surface of the mineral and the soil solution. The cations retain a surrounding sphere of water molecules at all times (Fig. 5.1). Exchange reactions are called *outer-sphere reactions*, *nonspecific sorption*, or *physisorption*. They are reversible and rapid,

limited only by how fast ions can diffuse to and from the surface.

A soil's *cation exchange capacity* is a composite of various negative charges. They are derived from isomorphic substitution in layer silicates, adsorption or desorption of protons at oxide surfaces and the edges of layer silicates, and the dissociation of carboxyls and phenols in SOM. The CEC of a soil, reported in units of cmol(-) kg^{-1}, is often measured in a laboratory procedure by displacing the base cations, Al, and H with a flood of NH_4^+ or Ba^{2+} ions. By adding together the positive charges attributable to the displaced cations, a value is obtained for the total negative charge that retained them in the soil sample. The concentrations of Ca, Mg, Na, and K that are exchangeable with NH_4^+ are often assumed to be indices of plant-available nutrients, and high CEC values are usually associated with fertile soils (Havlin *et al.* 2004).

Knowledge of soil CEC is valuable for several reasons (Burt 2004). For example, one index of mineral weathering in a soil sample is the ratio of charges due to exchangeable base cations to the total CEC measured at pH 8.2. Converted to a percentage, this value is called *the base cation saturation* (BCS).

$$BCS(\%) = \frac{\text{Sum of charges (referred to soil mass),}}{CEC \, at \, pH \, 8.2} = \frac{\sum[Ca^{2+}] + [Mg^{2+}] + [K^+] + [Na^+]}{CEC \, at \, pH \, 8.2} \times 100,$$

where the terms in brackets have units of cmoles of *charge* (for each cation) per kg of soil. BCS values are also used in soil classification. For example, Ultisols are soils that typically formed under native forest vegetation and that have weathered B horizons. Low BCS values in the lower profile are an indication of intense or long weathering

Table 5.5	Index of clay mineralogy based on the CEC7/clay ratio
CEC7/clay ratio	Indicates clay mineral suites dominated by
>0.7	Smectite
0.5–0.7	Smectite or layer silicate mixtures
0.3–0.5	Layer silicate mixtures or illite
0.2–0.3	Kaolinite or layer silicate mixtures
<0.2	Kaolinite and / or oxides

during pedogenesis. Therefore, BCS values <35% are a distinguishing criterion for Ultisols (Soil Survey Staff 2014). In contrast, most forest-derived soils with BCS values >35% at comparable depths reflect less weathered parent material, and many of these are classified as Alfisols (see Chapter 8). Another example of the use of BCS values is in the recognition of umbric and mollic epipedons. Umbric epipedons are thick, organic matter–rich (>1% SOM) surface horizons that have low BCS values (<50%, when the CEC is measured at pH 7). In contrast, mollic epipedons are also SOM-rich surface horizons, but they are less weathered and have pH 7 BCS values that are >50%. A third example of the use of CEC values in soil classification is the natric horizon, a subsurface zone of clay accumulation in which levels of exchangeable Na exceed 15% of the CEC measured at pH 8.2.

Where SOM is low, the CEC of soil can also be used as an index of the kinds of clay minerals present. The CEC (cmol(−) kg^{-1} of soil, measured at pH 7), normalized by the soil's clay content (as a percentage), is referred to as the CEC7/clay ratio. This ratio reflects the minerals most likely to dominate chemical reactions in the soil (Table 5.5).

In sum, soil CEC reflects the soil's clay and organic matter concentrations, as well as its mineral composition and pH. Table 5.6 presents the chemical characteristics of A and B horizons of four soils that range from relatively unweathered to moderately and highly weathered. The CEC of the Iowa Alfisol is less than that of the Iowa Mollisol because its horizons have less clay, less organic matter, and a lower pH. The CEC of the Ohio Ultisol is less than that of the Iowa Alfisol because it has less clay and the clay fraction is dominated by hydroxy-interlayered vermiculite and not an expanding 2:1 clay like smectite. The Ultisol from Nigeria has the lowest CEC of the group mainly because its clay fraction is dominated by kaolinite and Fe oxides. As the intensity of weathering increases in the sequence of the four soils, the BCS percentage declines dramatically.

Cation Exchange Reactions
Cation exchange reactions regulate the concentrations of essential nutrients (Ca, Mg, Na, K, and NH_4^+) in solution and therefore influence their availability for uptake by plant roots. For example, as Ca^{2+} in solution is taken up by the plant, its concentration in the bulk solution declines (Fig. 5.8). But that disequilibrium leads to the release of exchangeable Ca^{2+} from mineral surfaces and its replacement by another ion, such as Mg^{2+}, from the solution. As the root takes up cations, overall electroneutrality in the system is maintained by H_3O^+ ions that the root releases in exuded solutes or by H_3O^+ ions formed when $CO_2(g)$ released from the roots dissolves in the soil solution and forms carbonic acid.

The exchange of Ca and Mg ions may be described as

$$Ca^{2+}(aq) + MgX_2(ex) \rightleftarrows Mg^{2+}(aq) + CaX_2(ex),$$

where X(ex) represents a single mole of negative charge at an exchange site on a mineral such as montmorillonite. Because Ca^{2+} and Mg^{2+} have similar ionic potentials (the ratio of charge to ionic radius) and access to charged sites on the montmorillonite surface is relatively unrestricted, montmorillonite does not strongly prefer Ca^{2+} to Mg^{2+}, or vice versa.

Ion exchange can also occur between cations with different charges. The exchange of Ca^{2+} and Na^+ ions, for example, regulates plant growth in arid and semiarid region soils that have excess Na. The exchange reaction can be written

$$0.5Ca^{2+}(aq) + NaX(ex) \rightleftarrows Na^+(aq) + Ca_{0.5}X(ex),$$

where X(ex) again represents a single mole of charge on the exchanging surface. Because the divalent Ca^{2+} ion has the higher ionic potential, its retention at accessible mineral surfaces is favored over that of the monovalent Na^+ ion. Still, when soils are irrigated for long periods with groundwater enriched in Na, $Na^+(aq)$ concentrations can be high enough to drive the exchange reaction shown earlier toward the left. Excess Na has deleterious effects on both soil structure and plant growth, and a decline in agricultural productivity normally follows (see Chapter 13). This kind of *salinization* or *sodification* is an example of soil change that is accelerated by anthropogenic activity.

In contrast to Ca-Na exchange, the exchange of Ca^{2+} for K^+ ions depends strongly on the nature of the exchanging mineral. Some layer silicates such as vermiculite prefer to retain K^+ over Ca^{2+} at interlayer exchange sites. The K^+ ion has a lower hydration energy than Ca^{2+} and is more likely to shed its hydration sphere once in the interlayer space. Without the water molecules, K^+ is the perfect size to nestle into the plane of oxygens that make up the basal surface of the tetrahedral sheets in vermiculite, and therefore to get closer to the sites of isomorphic substitution in the mineral structure (see Chapter 4). These examples illustrate that the preference of one cation over another in exchange reactions at silicate surfaces depends on both characteristics of the ions (such as ionic potential) and structural characteristics of the mineral.

Table 5.6 | Physical and chemical characteristics of A and B horizons in four contrasting soils; these examples illustrate impact of clay and organic matter contents, pH, and mineralogy on cation exchange properties of soil

Soil and Horizon	Depth (cm)	Sand	Silt	Clay	pH (H₂O)	Organic C (%)	CBD Fe[a] (mg kg⁻¹)	Cation exchange capacity					Total CEC[b]	BCS[c] (%)
		(————— % —————)						Ca (————— cmol(+) kg⁻¹ —————)	Mg	Na	K	H + Al	cmol(−) kg⁻¹	
^{d}Mollisol (Iowa, USA) (Dominant layer silicate: smectite)														
Ap	0 – 18	3	64	33	6.1	3.46	9000	20.4	7.1	0.1	0.4	8.9	36.9	76
B	46 – 58	3	60	36	5.6	1.04	10 000	18.1	6.9	0.1	0.5	8.6	34.1	75
^{d}Alfisol (Iowa, USA) (Dominant layer silicate: smectite)														
Ap	0 – 15	5	67	28	5.4	1.37	7000	10.0	2.8	–	0.5	10.2	23.5	57
Bt	15 – 32	5	66	29	5.8	0.96	10 000	10.8	3.9	–	0.5	9.2	24.4	62
^{e}Ultisol (Ohio, USA) (Dominant layer silicate: hydroxy-interlayered vermiculite)														
Ap	0 – 23	24	64	12	5.9	0.83	8710	3.0	1.6	–	0.3	3.7	8.6	57
Bt	33 – 41	20	58	22	4.7	0.21	15 300	2.2	1.3	–	0.2	7.4	11.1	33
^{f}Ultisol (Nsukka, Nigeria) (Dominant layer silicate: kaolinite)														
Ap	0 – 10	80	4	16	4.9	1.22	19 000	0.63	0.32	0.21	0.03	6.1	7.3	16
Bt	40 – 55	76	2	22	4.6	0.69	26 000	0.20	0.05	0.03	0.09	6.4	6.8	5

[a] Citrate-dithionite-extractable Fe, an index of secondary Fe oxides and oxyhydroxides.
[b] Sum of cations.
[c] Base cation saturation.

Sources: [d] Natural Resources Conservation Service (2013), [e] Thompson *et al.* (1981), [f] Mbila *et al.* (2001).

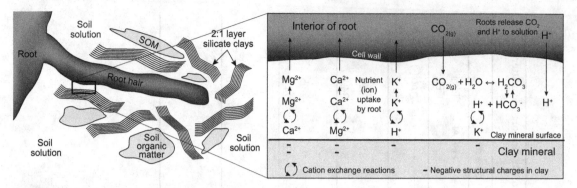

Fig. 5.8 Schematic diagram of ion exchange and uptake near a root.

Anion Exchange Reactions

Soils in which minerals are not highly weathered do not have significant anion exchange capacities. However, oxide-rich and kaolinite-dominated soils of tropical regions, e.g., Oxisols, often do have measurable anion exchange capacities (Garciamiragaya and Herreramarcano 1993, Beinroth et al. 2011). The pH of these soils is usually lower than the PZPC of the oxide surfaces, leading to a surplus of positive charges that attract anions, e.g., HCO_3^-, Cl^-, NO_3^-, and, in some contexts, SO_4^{2-}. Other anions of interest in soils, e.g., $H_2PO_4^-$, are not likely to be readily exchangeable with other anions. Instead, they are adsorbed with more energetic bonds (see later discussion).

Adsorption Reactions

In contrast to the rapid, reversible electrostatic reactions of ion exchange, chemical *adsorption* of ions and nonionic molecules (*sorbates*) at solid surfaces (*sorbents*) involves bonding mechanisms with higher total energy and at specific sites on the solid surface. The stronger bonds formed by the adsorption of *ions* can have both ionic and polar covalent characteristics, e.g., the bonding of oxyanions such as orthophosphate at the surfaces of Fe oxide minerals. Adsorption of *nonionic organic molecules* by soil clay and organic matter involves van der Waals forces, which are individually very weak but cumulatively strong when many atoms are involved. In either case, sorbates are retained by the solid phase, reducing their mobility in the soil and their bioavailability for uptake by roots, mesofauna, or microorganisms. Transition series metals such as Cu and Zn, oxyanions such as $H_2PO_4^-$ and HPO_4^{2-}, and nonionic organic molecules such as those in the waxes on leaves and plant stems are all examples of species that can be adsorbed by soil minerals and organic matter.

Adsorption reactions can be described as

$$\equiv S + [A] \rightarrow \equiv SA,$$

where $\equiv S$ represents an adsorption site on the solid-phase surface (e.g., "whole" soil or an individual mineral or organic phase) and A represents the adsorbate species. At equilibrium, the distribution of A between the solid phase and the solution phase is described by the *distribution coefficient*

$$K_d = \frac{\equiv SA}{[A]},$$

where $\equiv SA$ has units of mg kg^{-1}, [A] has units of milligrams per liter, and K_d has units of mg kg^{-1} (mg L^{-1})$^{-1}$ or liters per kilogram. The larger the value of K_d, the larger is the ratio of adsorbed ions or molecules to those remaining in solution. At constant pH and ionic strength, adsorption reactions are irreversible or only slowly reversible, and they are often referred to as *chemisorption* or *specific sorption* reactions. When the adsorbing species is an ion, no water molecules intervene between the ion and sorbing surface site, so it is termed an *inner-sphere complex*.

The potential for a soil material to adsorb ions and molecules at specific sites can be assessed experimentally. Soil (or adsorbent) samples of constant mass are mixed with a series of aqueous solutions in which the initial concentration of adsorbate is varied. At the end of an equilibration period, the final concentration of the adsorbate that remains in the solution is determined. Then the adsorbate's concentration on the solid phase is determined as the difference between the initial and final solution compositions divided by the sample mass. Plotted against one another, the final concentrations in the solution phase and the corresponding concentrations on the solid phase constitute an *adsorption isotherm*. At low concentrations of an adsorbate, the plot of an adsorption isotherm is often linear because the number of adsorption sites on the solid exceeds the maximum number of molecules or ions that were added to the samples. Linear isotherms may also indicate that there is little variation in bonding energy among the adsorbing sites. The slope of a linear isotherm is the K_d value noted earlier. K_d values are useful because they allow comparison of the adsorptive power of one soil material with that of another, or prediction of the buffering capacity in an unknown soil.

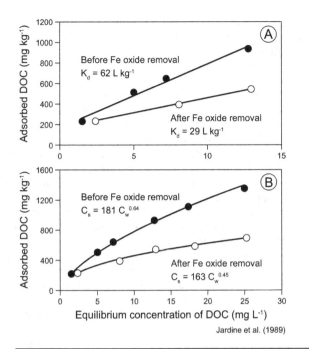

Jardine et al. (1989)

Fig. 5.9 Adsorption isotherms of dissolved organic C by the Bt horizon of a Paleudult before and after removal of Fe oxides and oxyhydroxides. (A) At low equilibrium concentrations of DOC, where the isotherm is linear and the K_d model is applicable. (B) Over the entire range of concentrations, where the isotherm is nonlinear and fits the Freundlich model.

As the concentration of adsorbate species in the soil solution increases and potential sites for adsorption are filled, or as binding mechanisms become less energetic, the soil's binding capacity commonly decreases. At this point, the adsorption isotherm becomes nonlinear. Many models have been developed to describe nonlinear isotherms quantitatively; one of the most useful is the Freundlich model. The Freundlich equation can be written

$$C_s = K_F C_w^n$$

where C_s is the concentration of adsorbate on the sorbent (mg kg^{-1}), C_w is the concentration of adsorbate in the solution (mg L^{-1}), K_F is an index of the sorption capacity of the sorbent (mg kg^{-1} (mg L^{-1})$^{-n}$), and the Freundlich exponent (n) is a unitless index of the heterogeneity of sorption sites or sorption mechanisms. The experimental values C_s and C_w are fit to the model, and values for K_F and n are derived. Although K_F values of different isotherms cannot be compared directly with one another, because they are dependent on the n values, K_F values are indicative of the amount of adsorbate retained by the sorbent. The n value also provides additional useful information. The lower the n value, the more heterogeneous is the range of adsorption mechanisms and energies involved in adsorption. On the other

hand, as n approaches 1, the more uniform the binding sites must be, and the Freundlich model becomes increasingly similar to the linear K_d model.

Adsorption reactions play an important role in attenuating the movement of dissolved organic matter and in promoting the accumulation of carbon in some subsurface horizons. For example, as noted previously, soluble organic compounds released during decomposition of plant residues contain deprotonated carboxylic and phenolic groups, so they carry a net negative charge as they move with percolating water. When they encounter the positively charged surfaces of Fe or Al oxides and oxyhydroxides, they are likely to be adsorbed and further movement is stopped. The chemical process can be described as

$$\equiv FeOH_2^+ + R-COO^- \rightarrow \equiv FeOOC-R + H_2O$$

where $\equiv FeOH_2^+$ represents a positively charged site on the solid surface and $R-COO^-$ represents the carboxylate portion of an organic molecule.

Jardine et al. (1989) used adsorption isotherms to demonstrate quantitatively the importance of Fe oxides to adsorption of dissolved organic matter in the Bt horizon of an Ultisol in Tennessee, in the United States (Fig. 5.9). After removing solid-phase organic matter from the soil samples, they measured the potential for adsorption of soluble organic matter both before (Fig. 5.9A) and after (Fig. 5.9B) treating the samples to remove Fe oxides. At low concentrations of dissolved organic C (DOC), the isotherms were linear, and there was more adsorption (i.e., the K_d value was greater) where Fe oxides were present (Fig. 5.9A). However, considered over the entire range of DOC concentrations, the isotherms were nonlinear. Application of the Freundlich model confirmed that the capacity for adsorption of organic matter declined when Fe oxides were not present (Fig. 5.9B). In addition, the lower n value after oxide removal (0.45 < 0.64) suggested that, even though the adsorptive capacity was smaller, additional mechanisms of adsorption attracted the organic compounds to the solid surfaces. Jardine et al. (1989) concluded that 50–70% of the total adsorbed organic matter could be attributed to adsorption by Fe oxides in the Bt horizon.

Another application of adsorption reactions is in the classification of soils derived from volcanic ash, i.e., Andisols (see Chapter 8). Weathering of ash deposits results in materials that are often enriched in ferrihydrite, allophane, or imogolite. These poorly crystalline minerals have a strong affinity for phosphate (McDaniel et al. 2011). Soil Taxonomy (Soil Survey Staff 1999) uses this affinity as one of several criteria to identify ash-derived zones in Andisols. To be classified with andic soil properties, a soil material must have a P retention value ≥25%.[1] Worldwide,

[1] P retention is determined by measuring the P adsorbed when a 5-g soil sample is equilibrated with 25 mL of a 1,000 mg kg^{-1} solution of P at pH 4.6 for 24 h (Burt 2004).

P retention values of soils composed of volcanic ash average about 75% (Batjes 2011).

Mineral Precipitation Reactions

A third way in which soluble ions may be retained in a soil is by precipitation, i.e., the neoformation of a solid phase, which can range widely in its crystallinity. Neoformation of secondary minerals by precipitation of ions is a common and important process during weathering of primary minerals and the genesis of soils over millennial-scale periods (Chapter 13). In principle, the likelihood of precipitation of an ionic solid can be predicted from the solution-phase concentrations of the mineral's constituent ions and the solubility product constant. For example, the solubility of $CaCO_3$ (calcite) is described by the reaction

$$CaCO_3(s) \rightleftharpoons Ca^{2+}(aq) + CO_3^{2-}(aq).$$

The idealized solubility product constant for this reaction can be written

$$K_{sp} = [Ca^{2+}][CO_3^{2-}] = 10^{-8.3},$$

where the ion concentrations are in molar units. When solution concentrations of Ca^{2+} and CO_3^{2-} ions are multiplied by one another, the resulting value is called an *ion activity product* (IAP). If the IAP = K_{sp}, the system is said to be in equilibrium with crystalline calcite. If IAP > K_{sp}, then solid-phase calcite is thermodynamically stable and may precipitate from the solution. If IAP < K_{sp}, then not only is calcite not predicted to precipitate, but any calcite that might be present in the soil is expected to dissolve, according to the principle of chemical equilibrium.

Mineral solubility or precipitation reactions, however, are not always consistent with predictions made for idealized conditions. For example, idealized predictions are based on thermodynamic models that predict *possible* chemical reactions but not their *rates*. In other words, minerals may precipitate or dissolve, but at rates that are too slow to measure under the real-world constraints of fluctuating temperature, ionic strength, and pH. Soil solutions may be supersaturated with respect to one or more mineral phases (i.e., IAP >> K_{sp}), but precipitation may still not happen because there is no appropriate surface available to "seed" crystallization. Coatings of organic matter or poorly crystalline phases on a mineral surface may also prevent access by ions, inhibiting further crystal growth. In addition, precipitation can be prevented or slowed because high concentrations of *other* ions in solution shield those that might precipitate from one another. Finally, a mineral may not precipitate because of competition – a second mineral competes with it for ions that are common to both.

Some minerals go through repeated cycles of precipitation and dissolution as the chemical conditions of the soil solution fluctuate. Calcite, ferrihydrite, and lepidocrocite are some common examples of minerals that may dissolve or precipitate at annual or decadal scales. In contrast, other minerals, e.g., kaolinite, are very stable once they are formed, and they are unlikely to dissolve. In many highly weathered soils, the most stable minerals are usually kaolinite, gibbsite, hematite, and goethite.

Chapter 6

Basic Concepts: Soil Physics

Soil physics is the branch of soil science that deals with the physical properties and processes of soils. Soil physics is generally concerned with the state and movement of matter and energy in soils; hence, most soil physicists study the movement of water in soils, and the changes in soil temperature, over time and space. Soil physical properties such as water content, texture, and structure, as well as soil physical processes such as water retention and transport, soil temperature and heat flow, and the composition of the soil atmosphere (mainly O_2 and CO_2), all affect weathering and soil genesis. Therefore, in this chapter we discuss the basic concepts of soil physics that are necessary as background to the discussions of soil genesis and geomorphology in more depth that follow.

Soil Water Retention and Energy

Many pedogenic processes begin and end with the flow of water in soils (see Chapters 13 and 14). Water is the main agent by which solids and ions are transported within soils. Knowledge of the forces acting on water flow is, therefore, important for understanding soil genesis, not to mention soil use and management.

Water is retained in soils in two ways. *Adsorbed* water molecules are retained at or very close to the *surfaces* of soil particles. They are held there by attractive forces acting between the water molecules and the surface, or between water molecules and ions near the particle surface. Water that is *absorbed* is taken *into* the pores of a solid (soil, mineral, rock particle, or organic substance) and is retained by the surface tension of the water molecules interacting with one another in small pores. Water retention in soils, in both adsorbed and absorbed forms, increases with increasing contents of clay and organic matter because of the affinity of those solids for water.

Liquid water has high surface tension because the H atoms in each H_2O molecule form strong hydrogen bonds to the O atoms of neighboring water molecules. The polar, hydrogen-bonding nature of water also leads to a strong, adhesive attraction between water molecules and most soil particles. For example, water molecules can form H-bonds with OH groups on oxide minerals and at the edges of clay minerals, as well as with NH and OH groups on soil organic matter. In addition, water molecules strongly solvate cations that are adsorbed by soil minerals. Because the H-bonds and ion-dipole bonds between water and soil particles are relatively strong, the first few layers of water molecules that are closest to particle surfaces are most tightly adsorbed (Fig. 6.1). Because of the strong attraction between water and charged and polar sites in clays and organic matter, water retention increases as the surface area of a soil increases and as the amounts of clay and organic matter increase. Adsorbed water can range from one or two layers of water molecules (a few Å thick) to tens of water molecules thick (ca. 80 Å) in smectite-rich soils.

Soil water is retained not only by adhesive forces between water and soil solids, but also by H-bonds between adjacent water molecules, i.e., cohesive forces. Water held in small pores above the saturated zone, sometimes called capillary water, is retained there by a combination of adhesive and cohesive forces. More water is held in fine-grained soils, therefore, because of their greater surface area and the greater number of smaller pores. Swelling soils that are rich in smectite contain many small pores and retain large amounts of water (see Chapter 4).

Soil water content can be expressed on a gravimetric basis, w_g (mass of water per mass of dry soil), or on a volumetric basis, θ (volume of water per unit volume of soil). Gravimetric water content is readily measured by weighing a sample of moist soil, oven-drying it at 105°C for 24 h (or until the mass stops decreasing), and then reweighing the oven-dry soil (Gardner 1986). Gravimetric water content is then calculated as

$$w_g = \frac{(\text{mass moist soil} - \text{mass oven dry soil})}{\text{mass oven dry soil}}.$$

Oven drying at 105°C removes most, but not all water; some water may be retained as tightly adsorbed molecules that

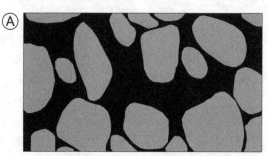

Soil is **saturated**; all pores are filled with water

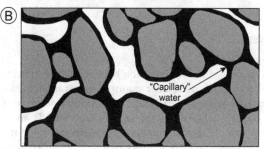

Soil has drained to **field capacity** (1/3 bar); water that remains is called capillary water

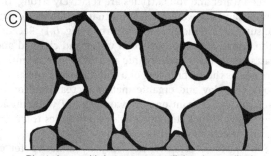

Plants have withdrawn water until they have wilted; this is the **permanent wilting point** (15 bar); water that remains is called difficult to remove, and some is in such close contact that it is permanently affixed (adsorbed) to the soil particles

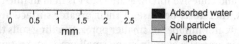

Fig. 6.1 Schematic diagram of soil pores showing water retained in small pores and as surface films; large pores remain air-filled.

surround cations in swelling layer silicates. Nevertheless, oven drying is a widely used method to determine w_g, and it requires little in the way of equipment other than an oven and a balance.

Volumetric water content (θ) is usually calculated from measurements of w_g and the soil's dry bulk density (ρ_b, mass of soil per unit volume of soil, e.g., g cm^{-3}), with the following equation:

$$\theta = \rho_b w_g / \rho_w$$

where ρ_w, the density of water, is equal to 1.0 g cm^{-3} at 25°C.

Volumetric water content can be determined nondestructively with specialized equipment that measures moisture-dependent properties of soils (Topp and Davis 1985, Gardner 1986). One of the most widely used nondestructive methods is time domain reflectometry (TDR), in which soil water content is estimated by measuring the dielectric constant (ϵ) of the soil, based on the travel time of a magnetic pulse between two probes. In TDR, multiple probes are inserted into the soil, typically vertically from the soil surface (horizontal probes can also be inserted into the wall of a soil pit). An electromagnetic pulse is then emitted and reflected back along each probe to the source, at which the travel time and velocity are determined. The velocity of the electromagnetic pulse depends on the soil's dielectric constant, which depends in turn on soil minerals, solute concentration, and – most importantly – water content. Calibration equations appropriate for the type of soil are then used to predict θ from ϵ (Topp *et al.* 1980, Roth *et al.* 1990). TDR is best suited for long-term monitoring of soil moisture in mineral soils at one site, because site-specific calibration curves must be developed for each soil. Other nondestructive methods of measuring water content are ground penetrating radar, neutron attenuation, and gamma-ray attenuation (Gardner 1986).

Although water content is an important soil property, the potential energy of soil water governs its retention and movement. Like other fluids, water flows from regions where it has greater potential energy to regions where it has less. The potential energy of soil water is expressed relative to the potential energy of pure liquid water in a reservoir where interactions of water with the edges of the reservoir are negligible. The total potential energy of soil water (ψ_t) is the sum of all of the forces acting on it. In unsaturated soils that do not shrink and swell significantly, these forces are usually separated into three major components: (1) gravity (ψ_g, gravitational potential), (2) the attraction between the soil particles and water (ψ_m, matric potential), and (3) the attraction between dissolved ions and water (ϕ_o, osmotic or solute potential) (Fig. 6.2).

In rigid, unsaturated soils, the total soil water potential (ψ_t) is approximated by the sum of potential energy attributable to gravity, matric, and osmotic forces.

$$\psi_t = \psi_g + \psi_m + \psi_o.$$

These components of soil water potential can be expressed in a variety of units, depending on the measurement technique and how the values might be used in combination with related parameters. For example, matric potential is often expressed in units of pressure (energy per unit volume of water, in bars, atmospheres, or pascals), because water potential is usually measured by applying pressure (or suction) to the soil and measuring the pressure (or tension) necessary to extract water from the soil. To combine

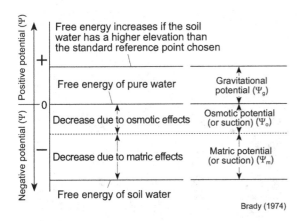

Brady (1974)

Fig. 6.2 Representation of water potential, with free, pure water having a potential of zero. Water ponded above the soil surface yields a positive potential, whereas the attraction of water for dissolved ions (osmotic forces) and the interactions of water with gas and particle surfaces (matric forces) decrease the total potential energy of soil water.

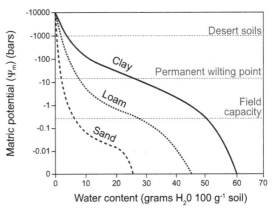

Brady and Weil (2001)

Fig. 6.3 Water characteristic curve (matric potential vs. water content) for soils of sand, loam, and clay textures. Field capacity and permanent wilting point are shown at −0.3 and −15 bars, respectively.

measurements of individual components of soil water potential with one another, they are commonly expressed in "head units," i.e., energy per unit weight of water. Head units at the scale most useful for soils can be expressed as a length, e.g., in centimeters or meters.

Although the osmotic potential in saline soils can affect the availability of water to plants, it is usually a much less important component of ψ_t in unsaturated soils than is the matric potential (ψ_m). In fact, in unsaturated, nonsaline soils, ψ_m is typically about 95% of ψ_t (Donahue *et al.* 1983). In situations where the soil air pressure is higher than the atmospheric pressure, or where the overlying soil exerts pressure on the soil water, additional potential terms would be needed to reflect the impacts of soil air pressure and overburden pressure on the total energy potential of soil water.

In saturated soils, the components of total potential change from those of unsaturated soils. At a water table, where all pore spaces are filled with water, the matric potential, ψ_m, is equal to zero, because there is no air-water boundary where surface tension can be exhibited. At points below the water table, the total potential, ψ_t, is positive (Fig. 6.2), controlled mainly by the overlying water that exerts a hydrostatic pressure, ψ_p, on the water at any point.

Water that drains freely from a soil under the influence of gravity is referred to as *gravitational water*. In gravitational water, the cohesive forces between neighboring water molecules exceed the adhesive, matric forces that attract water to soil particles, allowing it to drain freely within the soil (Fig. 6.1). As water drains from a soil and it becomes unsaturated, the matric potential (ψ_m) decreases and becomes negative. The negative values of ψ_m reflect

the fact that water is held in the soil by matric forces that exceed the force of gravity. Removing this water from the soil requires energy, i.e., oven-drying or suction. The water content of a soil after gravitational drainage ceases is sometimes called *field capacity* and it is often correlated with the water content when ψ_m = −0.3 bar. As shown in Figure 6.3, the water content at field capacity is greatest in clayey soils, because of their larger surface area and larger capillary porosity, and least in sands. As the soil dries further, weakly held water in larger pores (greater ψ_m) is removed first, but water adjacent to particle surfaces and held by stronger matric forces (more negative ψ_m) remains. The layer of adsorbed water closest to particle surfaces is estimated to have a water potential of −8,000 bars, whereas the outer layers of adsorbed water films have ψ_m ranging from −0.3 to −0.1 bar, depending on soil texture. Another important water potential benchmark is the *permanent wilting point*, which is the water content at the potential below which plants are unable to extract soil water, and hence, wilt. The permanent wilting point varies with plant type. Many crop plants have a permanent wilting point of about −15 bars, although desert plants can extract soil water at potentials as low as −100 bars.

The relationship between soil water potential and water content, called the *moisture characteristic function*, depends largely on soil texture. Moisture characteristic curves, which are plots of ψ_m versus 0, differ greatly between sandy soils and clay-rich soils (Fig. 6.3). At any given matric potential, clays have higher water contents than do coarser soils because of their greater surface area and greater cumulative porosity. Conversely, at a particular water content, clayey soils have a more negative ψ_m than do coarser-textured soils, because of the greater surface area

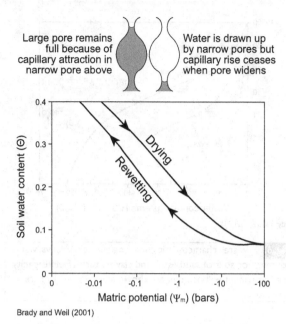

Brady and Weil (2001)

Fig. 6.4 Example of the hysteresis effect in soils. The water content of the soil, at the same matric potential, varies depending on whether the soil is being dried or wetted.

Fig. 6.5 A sandy soil a few hours after a 2-cm rainfall event. Note the clear, wavy wetting front, which is typical for infiltration into sandy soils.

of the clays and their much smaller pore sizes. Laboratory methods for determining the relationship between water content and matric potential are described by Dane and Hopmans (2002); field methods are detailed by Young and Sisson (2002).

The moisture characteristic curve (plot of ψ_m vs. θ; Fig. 6.3) for a given soil is not a unique function, as it depends on whether the soil is being wetted or dried. Figure 6.4 illustrates this *hysteresis effect* in soils. Hysteresis refers to the dependence of a system not only on its current environment but also on its past environment, i.e., a soil that is wetting up behaves differently than the same soil that is drying down. Hysteresis in the soil moisture characteristic curve is caused by pores emptying in a different order than they fill, and by entrapment of air during wetting (Fig. 6.4). During wetting, large pores may fill first, and trap air in small pores or in small cavities between pores. If wetting/drying were to occur slowly enough for equilibrium to occur, then the smallest pores would pull the water into them as air escaped from them, and the wetting and drying curves would be more comparable.

Different components of soil water potential can be measured with different methods. Tensiometers measure soil water potentials by allowing an airtight, porous ceramic tube (connected to a monometer) to equilibrate with soil. They can measure matric potentials of 0 to −1 bar (Young and Sisson 2002). Pressure plates can also be used to extract water from soil samples in a laboratory at

potentials down to −15 bars. Piezometer tubes can be used to measure hydrostatic and overburden pressure in saturated soils (Young 2002).

Soil Water Movement

Water flow in soils is vitally important in soil genesis, because percolating water dissolves minerals and transports ions, colloids, and metal-organic complexes downward in the soil profile. In addition, upward movement of water in response to evapotranspiration is vital to plant growth. In some soils upward water movement leads to salt accumulations at the surface (see Chapter 13).

The first step in soil water transport is *infiltration*, the process by which water moves into the soil from above the surface (Fig. 6.5). The infiltration rate of a soil, measured in units of volume per unit area per unit of time, e.g., 4.2 cm^{-3} cm^{-2} h^{-1}, depends upon factors such as soil texture and pore size, which are often affected by the presence of swelling clays (which can plug soil pores as they expand), as well as organic matter contents. Hydrophobic organic matter can actually repel water and therefore limit infiltration until water is ponded on the surface, unless large pores also exist at the surface. Infiltration rates are also affected by the initial water content in the soil below the leading edge of the wetted soil. In addition, soil temperature affects infiltration; decreasing temperatures cause a decrease in infiltration due to the greater viscosity of water at low temperatures. Of course, infiltration is negligible in most frozen soils.

Infiltration rates normally diminish with time after initial wetting. The reasons for this include the following: (1) the hydraulic gradient decreases as the soil wets (see later discussion); (2) clay minerals and other colloids may become suspended in the water, plugging pores as they infiltrate; (3) surface crusts of low permeability can form, especially in bare soils; and (4) clays swell as they wet, further decreasing pore space, size, and connectivity. Entrapment of air in soil pores during infiltration can also inhibit infiltration.

Once water has entered the profile, *percolation*, or redistribution of water in the soil, occurs. The leading edge of the wetted zone in a dry soil is called the *wetting front* (Fig. 6.5). Percolation can occur under *saturated* or *unsaturated* conditions. In saturated flow, essentially all soil pores are filled with water. Then the steady-state, vertical flux of water can be described by *Darcy's law*:

$$J_{sat} = -K_{sat}\,(\Delta H/\Delta z)$$

where J_{sat} refers to volume transported per unit area per unit time (flux density, e.g., cm³ of water per cm² of soil per day, which reduces to cm day⁻¹); K_{sat} is the saturated hydraulic conductivity of the soil; Δz is the distance between the top of the saturated soil, z_2, and a lower reference point, z_1, where water freely drains and is defined to have zero hydraulic potential; and ΔH is the hydraulic head difference, which is the distance between the upper surface of the water table and depth z_1 in the soil (Fig. 6.6). In a perfectly saturated soil column (with no water ponded on its surface), ΔH is the difference in gravitational potential, ψ_g, which is $\Delta z = z_2 - z_1$. However, if water is ponded on the surface, ΔH must also include the potential due to hydrostatic pressure.

Water flows from areas of greater total water potential to areas of lesser total water potential. In saturated soils, the hydraulic head or gravitational potential is the primary driving force for downward flow, and by convention, the negative sign is used to indicate downward flow, i.e., drainage. Saturated hydraulic conductivity is an empirical proportionality factor that relates water flux, J_{sat}, to a gradient ($\Delta H / \Delta z$). K_{sat} is an index of all the properties of the soil and water that regulate flow. Under saturated flow conditions, hydraulic conductivity of the soil depends on porosity and pore size distribution; K_{sat} is greater in sandy soils than in fine-textured soils. Although K_{sat} can be measured

Flux density = volume / unit area / unit time

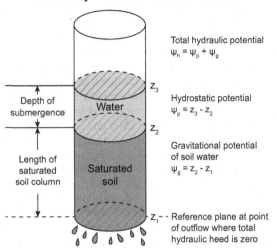

Total hydraulic potential
$\psi_h = \psi_p + \psi_g$

Hydrostatic potential
$\psi_p = z_3 - z_2$

Gravitational potential of soil water
$\psi_g = z_2 - z_1$

z_1 - - Reference plane at point of outflow where total hydraulic heed is zero

Fig. 6.6 Illustration of Darcy's law, which states that the flux density of water through a porous soil is proportional to the hydraulic gradient. The proportionality constant, K_{sat} (hydraulic conductivity), allows us to compare one soil to another and to make predictions about water flow. K_{sat} in the vertical direction is determined by measuring the flux density, J_{sat}, of water moving through a soil volume of known dimensions and where the hydraulic gradient is controlled. The hydraulic gradient is the ratio of the total hydraulic potential to the gravitational potential. For vertical water flow in saturated conditions, the total hydraulic potential is the potential due to gravitational forces plus any hydrostatic potential due to submergence. In this figure, we show an idealized column of saturated soil with water freely exiting at a basal reference plane. The length of the column ($z_2 - z_1$) represents the gravitational potential in head units, and the thickness of the ponded water at the surface ($z_3 - z_2$) represents the hydrostatic pressure in head units. Since, under saturated conditions, ψ_m is zero, $\Delta H = \psi_g + \psi_p$, and the hydraulic gradient is $\Delta H / \Delta z$. Therefore, $= -K_{sat} = J_{sat}/ (\Delta H / \Delta z)$, where the negative sign indicates downward flow by convention.

using a gradient of 1 (in which case, it has units of distance per unit of time), it is not a velocity term. Field conditions rarely if ever conform to the assumptions of Darcy's law because of the heterogeneity of porosity in different soil horizons, but this simple equation for one-dimensional saturated flow can still provide comparative predictions of water movement in soils that have different physical properties (Reynolds *et al.* 2002).

Water flow in *unsaturated* conditions – in which the soil pores are *not* all filled with water – depends on the soil's unsaturated hydraulic conductivity, K_{unsat}, value, but differences in water potential provide the primary driving forces for unsaturated water flow. When the soil is not saturated, some continuous, water-filled pores may still be accessible for water flow, but K_{unsat} decreases as the water content and matric potential decrease. In addition, because the larger pores drain and fill with air first, the hydraulic conductivity initially declines very rapidly, as water content decreases. Also, the air-filled pore spaces act as barriers to water flow, causing percolating water to follow more tortuous paths along thin water films on soil particles, and through small pores. Thus, water flux declines rapidly with decreasing water content. Methods for measuring unsaturated flow properties are described by Clothier and Scotter (2002).

In unsaturated conditions, where ψ_m dominates the total water potential, water flows from regions of higher water content (less negative ψ_m) to regions of lower (more negative) matric potential. Gravitational (downward) drainage can occur under unsaturated conditions only when the matric potential is greater than about −0.3 bar (field capacity). However, as noted previously, at matric potentials less than −0.3 bar, the attraction of water for particle surfaces exceeds the pull of gravity, so flow is governed by differences in matric potential, not by gravity. Consequently, unsaturated flow can be upward in response to evaporation at the surface, or downward as water is redistributed after a wetting event, or laterally.

In soils in which there are abrupt changes in soil texture at horizon boundaries, such as at *lithologic discontinuities* (see Chapters 3 and 10), rates of percolation can change markedly. When coarse-textured soil horizons overlie fine-grained layers, the lower hydraulic conductivity of the smaller pores in the fine-grained soil can cause water moving as saturated flow to stop (perch) at the boundary. However, the larger matric suction that is typical of the finer-textured soil below the discontinuity may eventually facilitate water flow into it, much as a sponge does. In contrast, when fine-grained horizons overlie coarse-textured layers, the flux of water through the upper layer to the discontinuity is insufficient to displace air from the larger pores in the coarse material until the hydraulic head over the layer is large enough. In short, water often will not flow into the coarser material below until the finer soil material above is saturated, or nearly so. After flow channels are established in the coarse layer below, preferential flow occurs through flow paths where larger pores in the sandy underlying horizon have been filled.

The term *preferential flow* is used to recognize that water movement is faster in large, continuous pores than in small, discontinuous pores. A number of terms have been used to refer to different aspects of preferential flow: Macropore flow, short-circuiting, fingered flow, and nonmatrix flow are a few (Bouma and Dekker 1978, Beven and Germann 1982, Bouma *et al.* 1982, Jury and Horton 2004). Preferential flow is important in soil genesis because it is through large, continuous planar pores or channels that colloidal suspensions carrying clay or organic matter can move downward most efficiently and rapidly. An idealized example of preferential flow is shown schematically in

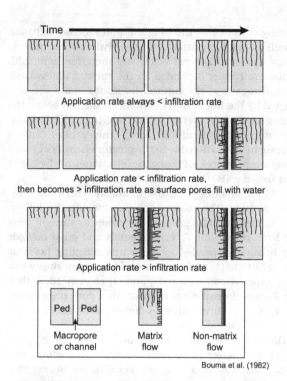

Bouma et al. (1982)

Fig. 6.7 Illustration of water flow in soils, as affected by pedality and infiltration rates. (A) Preferential, nonmatrix flow may occur when water is applied (by precipitation or irrigation) to dry peds at a rate that exceeds the infiltration rate. After Bouma *et al.* (1982).

Figure 6.7. If the rate of precipitation added to a soil surface is always less than the minimum infiltration rate, all water flow occurs only through the uniform soil matrix (first tier in Figure 6.7). But over time, infiltration rates decline as soil pores are filled, so at higher application rates, applied water spills over and travels mainly via continuous pores (between peds), i.e., as nonmatrix flow (second tier in Figure 6.7). In the third case, the water application rate is always greater than the infiltration rate, and nonmatrix flow occurs immediately, along with lateral infiltration into unsaturated peds. The morphological features associated with nonmatrix flow of colloidal suspensions are manifested as clay coatings occur alongside planar pores and channels.

Soil Temperature

Soil temperature affects many aspects of soils, especially plant growth, microbial activity, and water movement (Post and Dreibelbis 1942, Baker 1971, Berry and Radke 1995, Sharratt *et al.* 1995). Temperature data are important for estimating evaporation rates, mineral weathering rates, freeze-thaw processes, and the overall biological environment of a soil. In general, a 10°C increase in temperature causes a doubling in the rates of many biological and biochemical processes. Because they reflect longer-term trends than do air temperatures, soil temperatures might be better used to gauge trends in global climate (Gilichinsky et al. 1998, Isard et al. 2007). Temperatures also influence the movement of moisture through the soil matrix, e.g., in cold regions frozen soils can lower permeabilities and lead to runoff and erosion (Pierce et al. 1958, Zuzel and Pikul 1987, Todhunter 2001).

Soil temperatures are largely determined by the rate at which heat is exchanged between the soil and the soil-atmosphere interface, i.e., soil heat flux (Oliver et al. 1987). Soil heat flux is influenced by many internal factors, e.g., heat capacity, thermal conductivity, and soil moisture content, and their effect on latent heat capacity. External factors that affect heat flux include air temperature and the presence of insulating materials at the air-soil boundary, e.g., crop and vegetative cover, snowpack, and forest leaf litter (Pierce et al. 1958, Johnsson and Lundin 1991, Schaetzl and Isard 1996).

Important determinants of soil temperature are the amount of solar and long-wave (infrared) radiation that reaches the soil surface, as well as the many soil properties that govern its absorption and reflection. The amount of solar energy absorbed by atmospheric gases and particulate matter depends on altitude, latitude, aspect, and time of year, as well as time of day. At higher altitudes, the atmosphere is less dense, so incoming radiation interacts with fewer molecules before reaching the Earth's surface, and a higher proportion of the energy reaches the Earth's surface. The effects of latitude and season of the year are related to the angle of incidence of sunlight, which controls not only the distance the energy travels through the atmosphere before reaching the surface, but also the amount of surface reflection versus absorption. At high latitudes and in winter months when the sun is low in the sky, the sun's energy hits the Earth at low angles of incidence, and as much as half of the sun's energy can be reflected.

After solar radiation passes through the atmosphere and reaches the Earth, it is either reflected or absorbed. The fraction of incident energy reflected by the surface is known as albedo. Albedo depends not only on the angle at which the sun's energy strikes the Earth, but also on the angle and direction of any ground slope, and on ground cover, soil color, and soil moisture. At the poles, much of the radiation that reaches the Earth is diffuse radiation that has been scattered, so slope direction and angle have less effect on albedo. Snow both insulates the soil surface and reflects most of the sun's energy. Light-colored sand also has a very high albedo. The albedo of vegetation ranges from 5% to 30% because of variations in leaf color and ground cover. In contrast with the high albedo of snow, water reflects less than 10% of the incident radiation and is an excellent absorber of the sun's energy, except at very low solar incidence angles. Increasing soil moisture increases energy absorption and decreases reflection because wet soils are darker than dry soils.

Heat Flow in Soil

Soil temperature is controlled by energy inputs to the soil (described earlier), as well as by the thermal properties of the soil (described in this section). Heat energy in the soil can be (1) used to warm the soil, (2) radiated back to the atmosphere, (3) conducted to some depth below the soil surface, or (4) used to evaporate water or melt ice. Because soil is a heterogeneous, porous medium, heat flow through it depends on the relative proportions of solid, liquid, and gas in the soil; on the size of soil particles; on soil bulk density; and on the specific heat capacity and thermal conductivity of each soil constituent. The *specific heat capacity* of a substance is the heat energy required to raise the temperature of 1 g of that substance by 1°C. Water has a high heat capacity (4.18 kJ kg^{-1} °C^{-1}), which gives it the ability to absorb or lose a large amount of energy with relatively little change in temperature. In contrast, air has an extremely low heat capacity – it warms up rapidly in the day and cools down rapidly at night. The specific heat capacity of most soil minerals is about 0.7 kJ kg^{-1} °C^{-1}, or about one-sixth that of water, whereas organic matter has a heat capacity intermediate between that of minerals and water. Because the heat capacity of water is high, wet soils heat up more slowly and cool down more slowly than do dry soils. Also, because higher soil water contents promote absorption rather than reflection of the sun's energy, variation in soil water content has a much greater effect on soil temperature than does variation in either mineral content, bulk density, or organic matter content. Clearly, a soil's heat capacity and hence its temperature depend largely on the relative proportions of water, air, and solids within it at any one time.

Heat transport also plays a key role in determining soil temperature, because soils are sandwiched between two "boundary layers" that may have different temperatures – the atmosphere and the deep subsurface. Heat flow in soils can occur by (1) *radiation* of heat from soil solids and water to the atmosphere, (2) *conduction* of heat through solids and water, and (3) *advection* of heat-filled liquids or *diffusion* of heat-filled gases in soil pores. An example of heat transport involving advection is soil temperature that increases as a warm rain infiltrates an otherwise cool soil. Evaporation of soil water, diffusion of the water vapor, and its subsequent condensation into liquid water are mechanisms of *latent heat transport*. Water must absorb energy (2,260 kJ kg^{-1}) to evaporate, and when water vapor moves from one region of the soil to the other, or leaves the soil to return to the atmosphere, the vapor is transporting energy equal to the heat of vaporization. When the water condenses at

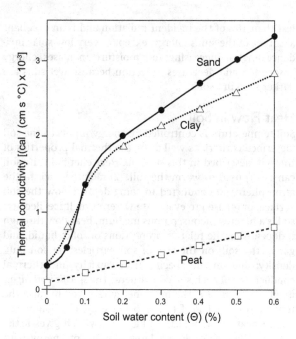

Fig. 6.8 Relationship between thermal conductivity and water content for soils with different textures, showing that sandy soils are better thermal conductors than are finer-textured soils. Adapted from the data of van Duin (1963).

another location, the heat is released. Thus, evapotranspiration by plants as well as direct evaporation of water from the soil surface are two ways that soils become cooler and lose heat.

Even so, heat transfer in soils occurs primarily via conduction through solids and water. Conduction is governed by the thermal conductivity of the soil (reported in units of mcal cm^{-1} s^{-1} °C^{-1} or W m^{-1} °C^{-1}). Quartz is a ubiquitous soil mineral, and its thermal conductivity is about 15 times greater than that of water and about 300 times greater than that of air. Because minerals are the best heat conductors in soil and air is a poor heat conductor, the thermal conductivity of soil increases as its bulk density increases, as a result of increased particle-to-particle contact. Similarly, thermal conductivity is greater in coarse-textured soils than in fine-textured soils (Fig. 6.8), because of better solid-phase continuity and the abundance of quartz grains. Because heat conduction is much greater in water films than in air-filled pores, increasing the water content of a dry soil can cause a tremendous increase in heat conduction – water fills small pores first and creates intergrain pathways for conduction.

Models and mathematical equations to describe heat flow near the surface are complex because radiation, conduction, advection, evapotranspiration, and diffusion of water vapor, as well as incident radiation, must all be taken into consideration, and these are rarely at steady state. Daily variations in soil temperature can be dramatic, especially near the surface (Fig. 6.9A). However, soil temperature measurements at different depths throughout a year provide a great deal of information about heat flow in soils. For example, Figure 6.9B shows idealized soil temperature variations in a soil, at depths down to 10 m. The graph is based on average daily soil temperature measurements over many months, so only long-term trends are shown. Note that the mean temperature is the same at all depths, that temperature variation becomes increasingly damped with depth, and that annual temperature changes can be fit with a sine function that has a period of 1 year. The amplitude of the sine wave, i.e., the difference between the maximum or minimum temperature and the average temperature, increases nearer the surface, because temperature variations are larger in the upper profile (Fig. 6.9C). At depth, soil temperatures vary less and less. Annual soil temperature variations are observable even to 10 meters, although the temperature only varies about the average by 0.2°C at this depth.

It is important to note that there is a lag between heating or cooling at the surface and heat transfer to or from the deeper soil. This is best shown in annual temperature profiles of soils where summer and winter extremes occur. In the example shown in Figure 6.9B, the maximum soil temperature is shown at the soil surface in July, at 200 cm in September, and at 500 cm in December. The increasing lag time with increasing depth reflects the time required for heat energy obtained at the surface to be transferred downward. Because of this lag time, heat flow is downward in the summer, when the soil surface is warmer than at depth. In winter, heat flow is upward, because the soil surface is colder than the deep soil, which experiences its maximum temperature in the winter (Fig. 6.9B).

In contrast with the annual temperature cycle, diurnal variations in soil temperature typically penetrate only the upper few centimeters of soil (Fig. 6.9C). Daily oscillations in temperature are nearly completely damped at 50 cm; that is one reason why mean annual soil temperatures, used to classify soil temperature regimes (see Chapter 8), are based on data taken at 50 cm depth (Smith *et al.* 1964, Carter and Ciolkosz 1980, Schmidlin *et al.* 1983, Smith 1986, Isard and Schaetzl 1995).

Soil Gas Composition and Transport

Gases and/or water are retained within all soil pores. The main source of gases in soil air is the atmosphere above the soil. Gases diffuse from the atmosphere into the soil, and vice versa; few gases are actually formed within the soil itself. The composition of gases in the soil atmosphere is,

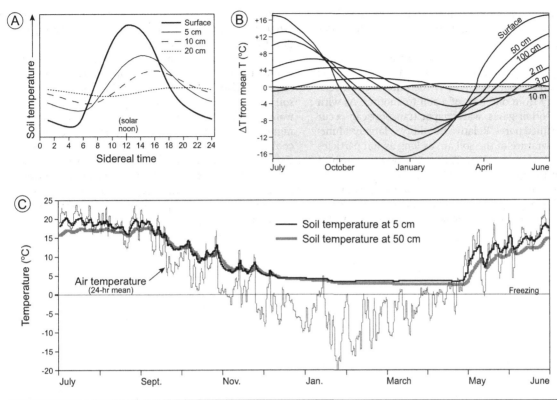

Fig. 6.9 Soil temperature patterns, through time and with depth, for well-drained soils. (A) Changes in daily soil temperature, theoretical and generalized, as a function of depth and time of day. (B) Annual temperature curves, theoretical and generalized, for various depths. (C) Actual air and soil temperatures recorded at an upland, forested site in the Upper Peninsula of Michigan. Data from RJS.

therefore, generally similar to the composition of air above the soil surface. The main difference between the two is in CO_2 and O_2 concentrations. Root and microbial respiration cause soil CO_2 concentrations to be much higher, and O_2 concentrations to be typically much lower, in soil than in the overlying atmosphere.

Gases move into and through soils mainly via diffusion; convection is often, but not always, negligible except in the uppermost few millimeters. Oxygen and CO_2 can dissolve in water and diffuse through water films and water-filled pores, yet diffusion in water is much slower than is gas diffusion, so the primary mode of transport is gas-phase diffusion. Diffusion is governed by *Fick's law of diffusion*, which specifies that it depends on the concentration gradient and on the *diffusion coefficient* of the gas. Diffusion in many soils can be quite slow, given that gas molecules have to travel through narrow pores and along tortuous pathways. Thus, the effective diffusion coefficient of a gas in soil is retarded by a factor known as the *gas tortuosity factor*, which is determined experimentally but depends

on soil texture and pore size distribution in the gas-filled pores.

Fine-textured soils with small pores have the most limited gas exchange with the air, especially when moist and filled (partially or completely) by water. Thus, in warm, clayey, moist soils, CO_2 concentrations may exceed 5% by volume, in contrast to about 0.035% CO_2 in the atmosphere. Conversely, CO_2 is produced as oxygen is consumed, so O_2 decreases under conditions where CO_2 increases. In clayey soils, under anaerobic conditions, it is possible for O_2 concentrations to fall to near-zero. Conversely, in dry, sandy soils, the soil atmosphere is closer to equilibrium with the atmosphere, so oxygen in soil pores is rapidly replaced via diffusion from the atmosphere, as it is consumed by roots and microbes.

The high CO_2 concentrations in soils create alkalinity in the soil solution, promote precipitation of carbonate minerals, and assist in the development of calcic horizons (see Chapter 13). Oxygen depletion below the soil surface can lead to chemically reduced zones in the soil, whereas oxidizing conditions promote dissolution of Fe^{2+} and Mn^{2+} from minerals and the precipitation of Fe and Mn oxides (see Chapters 13 and 14).

Although water vapor is a gas, water vapor movement in soils is different from flow of CO_2 and O_2. The driving force for water vapor transport is the difference in relative humidity from one part of a soil to another. As with transport of other gases, water vapor transport can occur through air-filled pores. Relative humidity is largely a function of temperature of the soil air, as long as soil particles are covered by a water film. Thus, water vapor transport depends on the gradients in soil air temperature as well as the temperature gradient of the soil solids. In addition, water-filled pores are accessible to the flow of water vapor because water vapor can condense on one side of a water "plug" and other water molecules can evaporate on the other side, with the net effect that water vapor appears to flow through water-filled pores.

In general, water vapor flows from areas of higher temperature to areas of lower temperature, where the relative humidity is lower. Thus, during the day, when the soil surface is hot and water evaporates from the surface, water vapor flows upward toward the drier surface. At night, vapor either can flow downward (if the soil surface cools sufficiently), or it may continue to flow upward (if the relative humidity is still lower at the surface). This process is important in cold soils, where water vapor can flow toward a freezing front, developing thick ice lenses and segregations there, and promote mixing by cryoturbation (see Chapter 11).

Chapter 7

Basic Concepts: Soil Organisms

Life is not only an integral part of the soil; it is the *essential* part – the part that distinguishes Earth's soil from biotically sterile and comparatively and stunningly simple soils of other planets. Organisms literally make the soil what it is (Verboom and Pate 2013). Ecologists have traditionally portrayed soil organisms as wholly contained within a black box labeled *decomposers*, i.e., a single trophic level through which all aboveground material, with its multiple trophic levels, is ultimately recycled (Sugden *et al.* 2004). Current research, however, has repeatedly shown that the soil food web is just as complex as its aboveground counterpart, or even more so.

Soils teem with life of all kinds, and soil organisms can be classified according to a variety of different criteria. Functional classification schemes group organisms according to their role in the food web (Table 7.1; Fig. 7.1), whereas phylogenetic classifications (the basis for the binomial system of scientific names) are based upon organisms' morphology, physiology, habitat, and genetic relationships. With respect to soils, organisms (particularly soil fauna) can be classified on the basis of *where* they spend most of their life cycle, e.g., wholly within the soil, partially within the soil, above the soil, and which parts of their life cycle are spent in each domain (Fig. 7.2). Hole (1981) also classified soil biota on the basis of how often and where they inhabited the soil, using slightly different criteria (Table 7.2).

Living things were originally divided between the plant and animal kingdoms, but organisms are now classified into three major domains: eucaryota, bacteria, and archaea (Woese *et al.* 1990). Eukaryotes – the higher order and hence, more complex, organisms – have their genetic material organized inside a nuclear membrane. They can be single-celled (algae, yeasts, most protozoa) or multicellular (most fungi and all the various plants and animals). Eukaryotes range from primary producers at the bottom of the food web to predatory animals at the top. In contrast, the cells of prokaryotes (Bacteria and Archaea) lack a nuclear membrane and an organized nucleus. Clearly important in swamps and in the initial colonization of extreme environments, Archaea also play a role in the

nitrogen cycle. Details of the role of Archaea in soil genesis are poorly understood at present but fast developing (Leininger *et al.* 2006, Hatzenpichler 2012, Zhalnina *et al.* 2012).

In this chapter, we examine some of the major groups of soil biota, i.e., plants, microorganisms, and *soil fauna* (animals). Our emphasis is on their role in the soil ecosystem. Soil organisms play vital roles in the soil system (Verboom and Pate 2013), and their potential utility in mankind's future is indeed very high. For example, in 1952 the Nobel laureate Selman Waksman (1888–1973) isolated the antibiotic streptomycin from soil bacteria, and the preservation of pedodiversity and biodiversity may assist similar research in the future (Yaalon 2000). The importance of organisms in soil genesis is discussed more fully elsewhere, especially in Chapters 11 and 13.

Primary Producers

Primary producers are photosynthetic organisms, placed generally at the bottom of the food web. Through the process of photosynthesis, they use the sun's energy to convert CO_2 into organic compounds and molecular oxygen, both of which are essential for the survival of other organisms that depend on primary producers. Algae, cyanobacteria, and plants are all primary producers (Bold and Wynne 1979). Cyanobacteria, formerly called "blue green algae," are single-celled photosynthetic bacteria that are also capable of N fixation. Algae are single-celled eukaryotes. Given ample water and warmth, both cyanobacteria and algae can grow on the surfaces of rocks and soils and can contribute considerable organic matter to them (Donahue *et al.* 1983). Cyanobacteria and algae can also form mutualistic or symbiotic associations with fungi to make lichen (see Chapter 15). In lichen, the algae or cyanobacteria photosynthesize to provide organic C for the fungus; the fungus provides nutrients to the photosynthetic partner and can also regulate light (Ahmadjian 1993). Cyanobacteria can also fix N for the fungus. Lichens play a role in colonizing

Table 7.1 | Classification of organisms based on their functionality in the soil system

Types of organisms	Main functional role	Trophic level	Details
Bacteria, fungi	Decomposers	Second	
	Mutualists and symbionts	Second	Rhizobia form symbiotic relationship with legume roots and fix N_2 for plants; fungi form mycorrhizal associations with roots and provide nutrients and water to plants; can dissolve P-containing minerals
Bacteria, fungi, nematodes, small arthropods	Pathogens	Second and third	Pathogens cause disease
	Parasites	Second and third	Parasites consume host organism
Nematodes, microarthropods, herbivorous burrowing mammals	Root feeders	Second	Feed on roots, sometimes killing the plant
Protozoa (eat bacteria), nematodes (eat fungi and bacteria), microarthropods (eat fungi)	Bacterial feeders and fungal feeders (grazers)	Third	Consume bacteria, typically releasing excess NH^{4+} and other nutrients into soil; consume beneficial and harmful bacteria
Earthworms, macroarthropods	Shredders	Third	Shred plant litter and ingest bacteria and fungi, which live in their digestive tract; increase soil aggregation and aeration
Large nematodes (eat small nematodes), large arthropods, mice, shrews, birds, voles, etc.	Higher-level predators		Control populations of lower-level predators; burrowing animals transport soil and organic matter in burrows; transport small organisms

rocks and other harsh environments by attaching to rocks and facilitating the early stages of weathering (Fig. 7.3; see Chapters 10 and 15). In coastal environments with frequent fog, lichens that hang from trees, e.g., "Spanish moss," cause the fog to condense and drip onto the soil. This condensed moisture can greatly increase soil water contents, affect the distribution of plants, and accelerate weathering.

Most primary producers are members of the plant kingdom (Plantae). In this chapter, we emphasize the relationship between plants and other soil organisms, in particular the effects of roots and litter. Most biological activity in soils is concentrated near the soil surface – in the rooting zone. O horizons and A horizons are sources of organic carbon for soil organisms and for soil development (see Chapter 3). Dead roots are important sources of organic matter and provide channels that influence aeration and infiltration.

Roots can constitute 30–50% of plant biomass (Donahue et al. 1983). Some plants have shallow, fibrous root systems, whereas others have deep taproots. Root cell walls are composed of lignin, cellulose, hemicellulose, and pectin. Cells at the root surface are coated with a protective layer of suberin and a mucilage of polysaccharides. The thin zone of soil immediately adjacent to and influenced by living roots is known as the *rhizosphere*. It is sometimes divided between the ectorhizosphere (around the outside of the root), the

rhizoplane (the root surface), and the endorhizosphere (the root cells that can be colonized by bacteria or fungi). The rhizosphere is rich with biological activity, containing bacteria that feed on sloughed-off plant cells, as well as the proteins and sugars released by roots. The boundary of the rhizosphere cannot be defined precisely because its extent depends on the relative rates of root exudation of soluble compounds and microbial utilization of those compounds. However, roots typically influence about 1 to 2 mm of surrounding soil (Pinton et al. 2001).

Living roots release organic compounds (exudates) as well as sloughed cells and root hairs (exfoliates); both can dramatically stimulate microbial activity. The amount of C entering the soil in this manner is not insignificant (Kuzyakov and Domanski 2000). By growing plants in radiocarbon labeled CO_2 and tracing the release of ^{14}C, Shamoot et al. (1968) showed that the roots of several perennial and annual crop plants released 20–50% more C to the soil in exudates and exfoliates than was present in the roots themselves at the end of a growing season. Root exudates can be divided into two categories: low-molecular-weight, water-soluble compounds that are easily metabolized by bacteria, e.g., sugars, amino acid, organic acids, hormones, and vitamins, and secondary metabolites, such as antibiotics and complex compounds, that are insoluble in water and need to be hydrolyzed by extracellular enzymes before they can be assimilated by bacteria (Cheng et al. 1993, Brimecombe

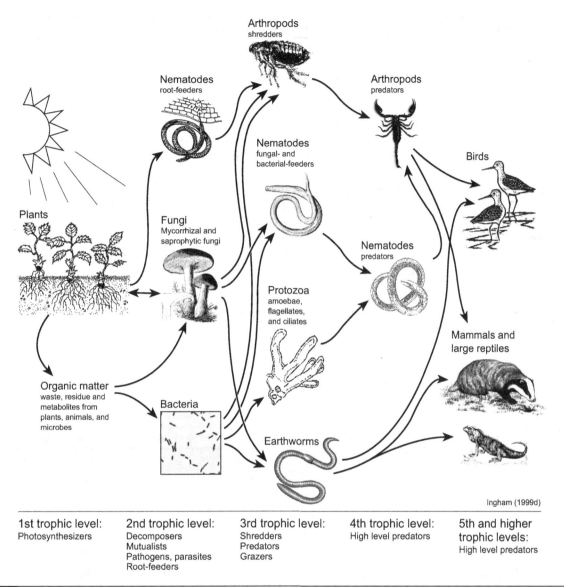

1st trophic level:
Photosynthesizers

2nd trophic level:
Decomposers
Mutualists
Pathogens, parasites
Root-feeders

3rd trophic level:
Shredders
Predators
Grazers

4th trophic level:
High level predators

**5th and higher
trophic levels:**
High level predators

Fig. 7.1 The soil food web and the functional relationships among soil organisms.

et al. 2001). About 80% of the C lost from living roots is in the form of polysaccharides known as mucigel (Hale *et al.* 1978). Soluble root exudates stimulate microbial growth, which explains why the numbers of bacteria in the rhizosphere are up to 1,000 times greater than in bulk soil, ranging from 10^9 to 10^{12} per gram of soil (Foster 1988).

Bacteria

Although the rhizosphere usually contains 10^8 to 10^{12} bacterial cells per gram, bacteria are so small that they usually occupy only a tiny fraction of the soil volume (Ingham 1999d). More than 10^4 bacterial species are known;

however, it is estimated that less than half of the bacteria in soils have been identified. They tend to be most abundant in nonacid soils.

Bacteria can be categorized according to phylogenetic relationships, according to their energy and nutritional requirements, or according to their functional role in soils. In the context of soils, it is most relevant to focus on the carbon and energy requirements of bacteria and their functional roles in the soil ecosystem. Bacteria that require *organic* carbon compounds for cell growth are known as *heterotrophs*, whereas *autotrophs* utilize CO_2 as their C source. *Photoautotrophs* are primary producers such as cyanobacteria that use the sun's energy to convert CO_2 into cellular organic C. The majority of soil bacteria are *heterotrophs* and

Table 7.2 | Classification of soil organisms based on their "incidence" in the soil

Category of soil organism	Explanation/definition	Examples
Permanent	All life stages reside in the soil	Most microorganisms, collembola, most earthworms
Temporary	One active stage lives in the soil; another active stage does not or is periodic	Many insects (larvae reside in the soil, adults do not)
Periodic	The animal moves in and out of the soil frequently	Many insects, some snakes and reptiles, prairie dogs
Alternating	One or more generations of the animal live in the soil; the other generation(s) live(s) above it	Potato aphid (*Rhopalosiphoninus*); oak-apple gall wasp (Biorhiza)
Transient	Inactive stages (eggs, pupae; hibernating stages) reside in the soil; active stages do not	Most insects
Accidental	The animal accidentally falls or is blown or washed into the soil	Insects in forest environments; many different kinds of animals that fall into cracks of Vertisols

Source: After Hole (1981).

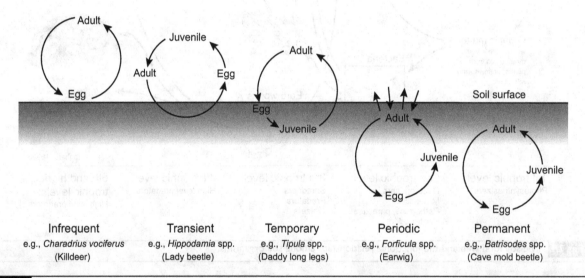

Infrequent
e.g., *Charadrius vociferus*
(Killdeer)

Transient
e.g., *Hippodamia* spp.
(Lady beetle)

Temporary
e.g., *Tipula* spp.
(Daddy long legs)

Periodic
e.g., *Forficula* spp.
(Earwig)

Permanent
e.g., *Batrisodes* spp.
(Cave mold beetle)

Fig. 7.2 A classification of fauna that interact or live within the soil, based on where (relative to the soil) they spend key stages of their life cycle. Most of the examples are from the class Insecta.

require a source of organic C. They can obtain energy by using either organic compounds (*organotrophs*) or inorganic compounds (*lithotrophs*) as electron donors. Lithotrophs get their energy by reducing inorganic compounds other than oxygen, e.g., nitrate, sulfate, ferrous iron, and manganese, and are active in waterlogged soils where oxygen is unavailable. Essentially all oxidation and reduction (redox) reactions in soil are biochemical reactions mediated by bacteria (see Chapter 14).

Heterotrophic bacteria can be divided into three functional groups based on the *source* of organic C they use for cell growth. *Pathogens* derive their C from a (living) host plant or animal and cause disease in their host. *Mutualists* form a mutualistic or symbiotic relationship with another organism, such as the association between cyanobacteria and fungi to create lichen. Another example is the symbiotic association of N-fixing bacteria (Rhizobia) with legume roots, in which the bacteria get organic C from the plant and the bacteria provide fixed N to plants (Fig. 7.4).

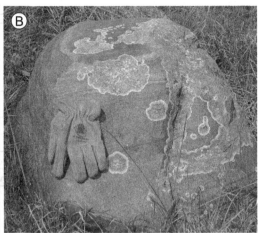

Fig. 7.3 (A) The lichen (*Xanthoparmelia lineola*) thalli on a rock on a basaltic ridge at 2,200 m elevation, near Ranchos De Taos, New Mexico. *X. lineola* grows slowly here because of the dry climate. This lichen species has been studied by Benedict (1967) in the Colorado Front Range. The knife is 8.5 cm in length. Photo by P Webber. (B) Thalli from an unidentified species of lichen on Sioux Quartzite boulder on a moraine in eastern Kansas, illustrating that lichens are also commonly found in warm, humid climates. Photo by RJS.

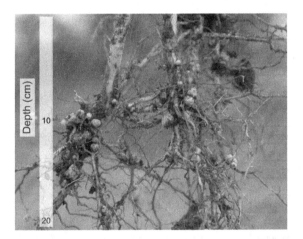

Fig. 7.4 Nodules on the roots of a soybean plant. Nodules such as these are common in plants in the legume (Fabaceae) family. The N-fixing bacteria – termed Rhizobia – within the nodules help the plant fix N_2 gas from the soil air, converting it into ammonia (NH_3). When the plant dies, the fixed N is released to the soil, making it available to plants. Photo by RJS.

organic C in the litter layer, root zone, and rhizosphere. In turn, they can promote plant growth by the release of hormones. They also release polysaccharides that promote aggregation and aeration, and assimilate and cycle nutrients in plant litter.

Actinomycetes, which are multicellular, filamentous bacteria, are also a type of decomposer. They produce the characteristic musty odor of wet soil. About 90% of the actinomycetes in soil are streptomycetes. They live in soil and plant litter and can degrade complex substances such as lignin, chitin, humin, aromatic compounds, keratin, and cellulose (Paul and Clark 1996).

Fungi

Fungi are aerobic, eukaryotic, heterotrophic organisms that obtain their C and energy from organic compounds in either organic litter or living plant and animal tissue. They include one-celled yeasts as well as a range of multicellular filamentous organisms such as molds, mildews, smuts, rusts, and mushrooms that range in size from a few micrometers to several meters in length (Ingham 1999b; Fig. 7.1). Most fungal cells grow as hyphae – tubular, elongated, and threadlike (filamentous) structures that can contain multiple nuclei and extend at their tips. Mats of hyphae are called *mycelia* or mycelia mats. Fungi can also be divided into the same three functional groups described for bacteria (pathogens, mutualists, and decomposers). Within each functional group, however, fungi perform slightly different roles than do bacteria. Fungi are the most common microbial-size decomposers in acidic soils (Millar 1974, Tate 1991).

Decomposers are the most abundant, and arguably most important, functional group of soil bacteria. They live primarily on plant litter, dead roots, and root exudates. Some decomposers utilize only simple, water-soluble organic compounds, whereas other decomposer bacteria can either metabolize large, complex compounds or release extracellular enzymes that hydrolyze complex molecules into simple compounds that can then be more readily assimilated. Decomposers depend upon plant-derived

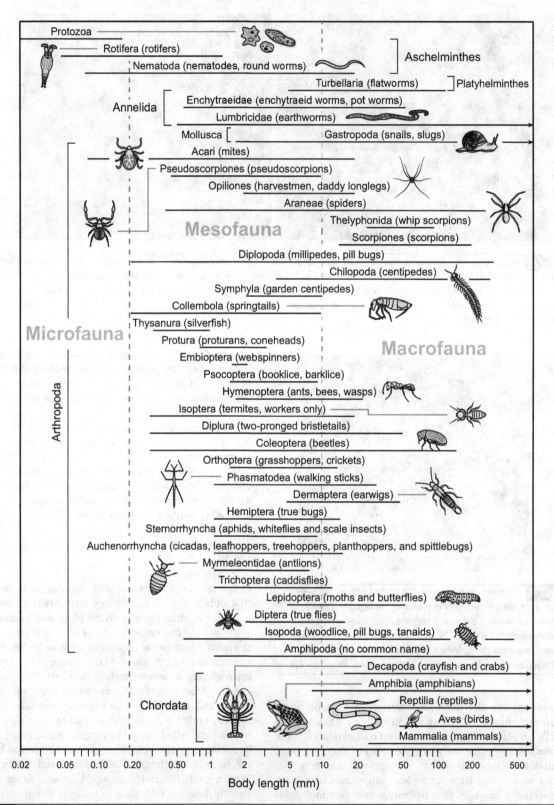

Fig. 7.5 Dimensions and sketches of selected soil organisms. Compiled from various sources.

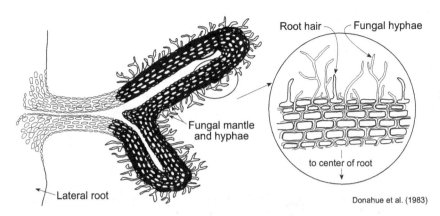

Donahue et al. (1983)

Fig. 7.6 Diagram showing a mantle of ectomycorrhizae around a root, with hyphae penetrating between epidermal cells.

Fungal decomposers, also known as saprophytic fungi, are the most numerous and visible fungi in soils. Decomposer fungi convert complex organic substances into simpler molecules, e.g., organic acids, that bacteria can use. Fungi are similar to actinomycetes in their ability to degrade compounds like cellulose, chitin, and aromatic rings, but they are more active at lower pH values than are actinomycetes, and are more prevalent in litter derived from wood. Because fungi need much less N than do bacteria, they can obtain energy from organic compounds with much wider C:N ratios than those that are favored by bacteria (Ingham 1999a). Fungi also play key roles in the humification of soil organic matter.

Fungi that colonize plant roots in a mutualistic or symbiotic relationship are known as *mycorrhizae* (literally, "fungus root"). Nearly all plants have mycorrhizal fungi associated with their roots, although plants with fibrous roots rely less on mycorrhizae for successful growth than do those with taproots. Most mycorrhizae obtain their C from organic compounds produced by the host plant. In exchange, the fungi provide the plant with nutrients by accessing parts of the soil solution that are inaccessible to the roots and by releasing extracellular enzymes that solubilize nutrients such as phosphorous. Ecologists and physiologists are discovering that mycorrhizae perform important functions in the soil, many of which were only recently uncovered. In addition to enhancing a plant's nutritional status, mycorrhizae also hinder water loss and erosion, protect against pathogens and minimize harm from toxic wastes, and regulate the flow of carbon (Pennisi 2004). In addition, new studies are showing that they link plants together, facilitating the transfer of nutrients not only among fungi but from plant to plant.

Seven different types of mycorrhizal associations are known, but about 90% of mycorrhizae fall into three categories: ectomycorrhizae, arbuscular mycorhizae, and ericoid mycorrhizae. Ectomycorrhizae, which include edible wild mushrooms such as truffles, infect trees and shrubs, mainly in temperate regions. They form a mantle or sheath around the root surface, and penetrate between, rather than through, root cortex cells (Fig. 7.6). Many different fungal species can form ectomycorrhizae, and sometimes a single tree can host several different species of mycorrhizae simultaneously. Fungal mycelia can produce enzymes that promote the release of phosphate and inorganic N for the plant. A second type of mycorrhizal association, arbuscular mycorrhizae (previously known as vesicular arbuscular mycorrhizae), can infect nearly all agricultural crops, many grasses, and numerous trees and shrubs, including legumes. They occur in soils that have near-neutral pH and typically are P-limited but not limited by available N (Paul and Clark 1996). Arbuscular mycorrhizaae penetrate the root interiors, even to the cortex, where they form arbuscules composed of finely branched hyphae that fill the cytoplasmic space of root cortical cells (Fig. 7.6) and transfer nutrients to the roots (Paul and Clark 1996). Some, though not all, also have hyphal extensions that form internal storage vesicles between cortical cells. In addition, some hyphae extend out from the root into the soil, where they can take up nutrients that would not be accessible to an uninfected root. Ericoid mycorrhizae form coils inside the root cells, with fungal hyphae extending a few millimeters beyond the roots. They are found mainly found on acidic, nutrient-poor soils.

Fungal hyphae are also important in soil genesis because they can penetrate cracks in rocks and even mineral grains, promoting physical and chemical weathering. In addition to their role in weathering, hyphae also play a critical role in the aggregation of mineral particles in surface horizons. By increasing aggregation, they increase both the infiltration and water-holding capacities of soils.

Soil Fauna

For purposes of communication, soil fauna are commonly divided into groups based on size. They range in size from tiny nematodes to earthworms to large mammals and reptiles that burrow in the soil and tread upon it (Fig. 7.5). Microfauna (<200 μm long) include protozoa, nematodes, and rotifers. Mesofauna are organisms that are ~200 μm to 10 mm in length and include large nematodes, springtails, many species of mites, and collenbolans. Macrofauna are >10 mm in length, e.g., millipedes, and most ants and termites. And finally, some scientists use the term megafauna to refer to large fauna that approach a kg or more in weight and are at least 20 mm long, e.g., badgers and pocket gophers. We also offer a term for those soil animals that live mainly in the soil – *soil infauna*. In the sections below, we provide some discussion of the roles of these fauna in the soil system.

Protozoa

Protozoa are single-celled, eukaryotic organisms in the Protista kingdom. They are the largest soil organisms outside the animal kingdom and are capable of movement. Most are grazers that feed mainly on bacteria, though some can also consume fungi; large predatory protozoa even prey on small protozoa.

Protozoa are grouped by size, morphology, and method of locomotion. The relationships among different protozoa, their prey and their predators are shown schematically in Figure 7.1. Ciliates are the largest protozoa (up to 500 μm) and can be either grazers or predators. Juvenile ciliates move by coordinated motions of tiny hair-like cilia, whereas adults lose their cilia and use tentacles to reach their food (Paul and Clark 1996). Because of their relatively large size, ciliates are most abundant in coarse-textured soils.

Amoebae are intermediate in size and move about by the use of a pseudopod – a temporary foot that is also used to trap bacteria and fungi. Some amoebae (testate amoebae) have a protective covering; naked amoebae do not. Most amoebae are grazers, though some are disease-causing parasites (Ingham 1999c).

Zooflagellates are the smallest protozoa and move by using a whip-like flagellum. Although some zooflagellates are free-living in the soil, most form commensal, symbiotic or parasitic associations with insects and vertebrate animals. Termites contain symbiotic zooflagellates in their guts and rely on them to digest wood (Paul and Clark 1996; see Chapter 11).

Protozoa are more abundant in the rhizosphere, where they feed on bacteria, than in the bulk soil. Protozoa play an important role in the soil N cycle by digesting bacteria, which have a higher N content than other soil biota. When protozoa digest bacteria, excess N is released as NH_4^+. Protozoa also play a key role in the soil ecosystem by regulating bacterial populations (Ingham 1999c).

Nematodes

Nematodes are non-segmented worms that range in size from ~0.1 to over 20 cm in length (Fig. 7.5). Most are less than 2.5 mm long. Because of their comparatively large size, they are more common in coarse-grained, porous soils, moving through the soil within water films on particle surfaces. Different nematodes operate at different trophic levels; some are plant parasites that feed on roots, whereas others are grazers that feed on bacteria or fungi (Paul and Clark 1996). Some larger nematodes are predators that prey other nematodes as well as other small organisms, whereas others

Fig. 7.7 Termite mounds in the Jalapao region of east-central Brazil. Photo by A Arbogast.

are omnivores that prey on a range of organisms or on different organisms at different stages of life. Predatory nematodes attach to their prey and suck out the contents of their prey, leaving only the exoskeleton (Paul and Clark 1996).

Nematodes that graze on bacteria and fungi fulfill the same ecological role as do protozoa: cycling nutrients and regulating microbial populations. They also disperse bacteria and fungi to different parts of the soil, transporting them either in their digestive system or attached to their exterior (Ingham 1999d).

Arthropods

The word "arthropod" means "jointed foot." These animals include insects (springtails, beetles, ants, and termites), crustaceans (sow bugs), arachnids (mites and spiders), and myriapods (centipedes, millipedes, and scorpions). Arthropods are both diverse and numerous in soils; several thousand different species of arthropods can exist in a hectare. The range of sizes for many soil arthropods is shown in Figure 7.5. Springtails and mites are dominant in many soils, although ants and termites prevail at certain sites (Fig. 7.7). Arthropods occupy a range of niches in the soil food web, including root feeders (herbivores), grazers, shredders, and predators. Herbivores such as cicadas, root maggots, and root worms eat roots and can damage plants. Springtails, mites, and silverfish are grazers that consume bacteria and fungi that are on or near roots. A related and very important functional group of soil fauna is the shredders, which live near the soil surface and shred plant litter while eating bacteria and fungi that adhere to the litter (see Chapter 3). Shredders will also eat roots and the microbes associated with roots if no litter is available. Shredders include millipedes, sow bugs, termites, roaches, and some types of mites (Fig. 7.8).

Larger arthropods such as spiders, ground beetles, scorpions, centipedes, pseudoscorpions, ants, and some mites are best viewed as predators. They are beneficial when they eat pests such as some nematodes, but some also prey on beneficial organisms. For example, centipedes will prey on earthworms. Most predators and shredders live near the soil surface; those that live deeper than about 5 cm are usually less than 2 mm long, and are blind (Moldenke 1999).

Fig. 7.8 A common millipede of midlatitude forests (*Narceus americanus*), with leaves for scale. The majority of the 1,400 known species of North American millipedes eat decaying vegetable matter and fungi and play a crucial role in decomposition of leaf litter and nutrient cycling. Photo by RJS.

All arthropods mix and aerate the soil as they feed (Lyford 1963). Soil particles get cycled repeatedly through the digestive tract of many types of soil organisms, where the soil is mixed with mucus and organic matter to make fecal pellets – an important type of aggregate in many soils (see Chapter 3). Arthropods that feed on bacteria and fungi help cycle and release nutrients by mixing them throughout the soil. Burrowing arthropods also dramatically affect porosity and infiltration (Moldenke 1999).

Arthropod fauna such as ants (Hymenoptera) and termites (Isoptera) are found worldwide and move vast amounts of soil material (Mandel and Sorenson 1982, Levan and Stone 1983). Termites are particularly abundant in Australia, Africa, and throughout the tropics, whereas earthworms are the dominant burrower in more humid locations like Europe and parts of North America (Lee and Wood 1971a, Lobry de Bruyn and Conacher 1990; see Chapter 11).

Termites are, arguably, the most widespread and important of the soil arthropods. The major differences between termites and many other soil fauna are: (1) in concert with protozoa and bacteria that live in the termite gut, they can digest cellulose, which most other infauna cannot, (2) they use their excreta, which is highly resistant to further decomposition, to line the mound rather than allowing it to immediately biocycle, leading to (3) longer periods of organic matter immobilization within the mounds, which themselves can take decades to fully deteriorate (Lee and Wood 1968, Grube 2001). In short, termites extract large quantities of organic matter and store it in intractable form in their mounds, dramatically affecting the carbon cycle (Lee and Wood 1968). Other soil animals, especially worms and millipedes, are most effective at comminuting soil organic matter, passing it through their guts and mixing

it with soil. These actions increase the surface areas of the organic materials and exposes cytoplasm, thereby *facilitating* the rapid breakdown and biocycling of litter (Nielsen and Hole 1964, Haimi and Huhta 1990, Scheu and Wolters 1991, Alban and Berry 1994, Bohlen *et al.* 2004).

Landscapes: Termite-Generated Fairy Circles

Many of us have heard of crop circles. Are they formed by aliens or mischievous kids? We'll leave that for you to decide.

But what of *fairy circles*? They have an equally mysterious and controversial history. First, what are they? Fairy circles are large (2–15 m dia.), slightly concave, circular patches that, oddly, lack vegetation (see figure). The circles only occur on very sandy/pebbly soils. They commonly are ringed with perennial grasses – hence, the name. Fairy circles occur in large numbers in the desert margin grasslands of southwestern Africa, particularly in Namibia. Locals agree that every now and then, new circles form.

But, how are they formed? In the oral myths of the local Himba people, the rings are said to be footprints of the gods. Theron (1979) hypothesized that an allelopathic compound released by dead *Euphorbia damarana* plants could be responsible for the lack of grasses. Others (Moll 1994, Becker and Getzin 2000) suggested that they were formed simply by termite herbivory; termites consumed all the grass seeds in the immediate vicinity of their nests. Ants have been similarly implicated (Picker *et al.* 2012). Albrecht *et al.* (2001) and Jankowitz *et al.* (2008) suggested that they were caused by an unknown, semi-volatile gas or chemical substance associated with termite nests. Measurements of carbon monoxide and hydrocarbons in the soil led to the proposal of similar geochemical origins (Naudé *et al.* 2011). Nonetheless, many of the early theories that focused on the actions of poisonous plants, ants, or termites as causal factors have been systematically tested and rejected, leaving the origin of fairy rings a mystery (van Rooyen *et al.* 2004), and thus, their role in local ecology, unclear.

However, recently a new hypothesis has been promoted. Norbert Juergens (2013) reported that the circles are generated by the actions of the sand termite (*Psammotermes allocerus*), which does not build mounds. It removes vegetation in ring-like patterns. The removal of vegetation in the first year often takes place on irregularly shaped patches a few square meters in area. The circular shape is later formed after the first rain percolates into the sandy soil and promotes growth of vegetation in the moist soil. This shape gets more circular with time because the surrounding dry soil is pulling water from the wetter area, equally in all directions, and because the termite colony is arranging its activity pattern around the main body of moist soil. Once the circle is generated, the sandy soil beneath the bare circle continues to collect water. Because of rapid percolation and lack of evapotranspiration from the bare patches, water is preferentially retained beneath them. Roots of the grasses adjacent to the circles extend 20–30 cm into them, taking advantage of the extra water. The result is a perennial belt (ring) of taller grass around a bare, circular patch. Vegetation growth within the circles proper is hindered by subterranean root-foraging of the sand termite. The fairy circles widen as termites continue to forage into the surrounding grasses, leaving behind dead grass stumps. As a result, these rings of perennial vegetation facilitate termite survival, locally increase biodiversity, and trap sediment blowing in the wind, sometimes forming a slightly raised rim. This termite-generated system can persist through prolonged droughts because, by the removal of all the plants, more water can accumulate underneath the circle after rain events, i.e., it becomes a water trap. Thus, the termites have found a way to develop a perennial water supply, facilitating their survival in this hostile desert. Plus, generation of a ring of perennial plant biomass also facilitates their survival during droughts.

Termites have changed this desert from one that would have been dominated by low-productivity, droughty plants, to a patchy one with increased biodiversity and areas of perennial grass. Fairy circles provide another example of a soil-related mystery solved by good science, with an eye to the impact of soil fauna.

Figure: Appearance and structure of African fairy circles. (A1) A fairy circle as viewed from the ground, showing thicker than normal vegetative growth, during a wet year. (A2) Oblique aerial view of the landscape, during a similar year. (A3) View of fairly circles in a dry-normal year; note the larger extent of barren surface. Photos by N Juergens. (B) Sketch of the structural elements within a fairy circle. The perennial belt is formed by a ring of large plants located around the bare patch. Grasses outside the circle die and disappear during drought, leaving the belt as the only surviving vegetation biomass during drought periods.

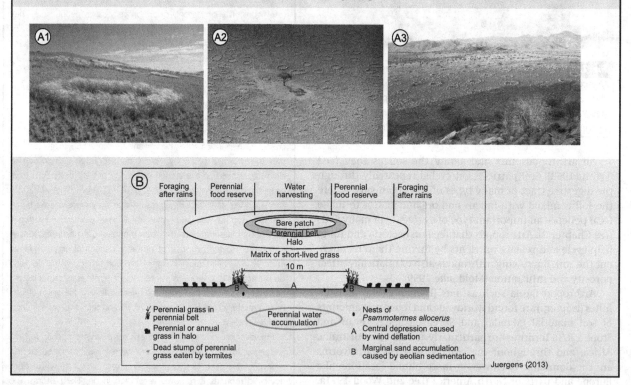

Juergens (2013)

Earthworms

Earthworms (annelid worms) are among the largest soil invertebrates and typically contribute more to soil biomass than do other soil invertebrates (Lee 1985, Edwards and Bohlen 1996). They are most abundant in medium-textured soils that are moist and not too hot. Earthworms also tend to avoid sandy soils. Over 700 genera and 7000 species of earthworms are documented. They burrow through soil or litter and ingest soil, organic matter, and attached microorganisms. As this material moves through their guts, the ingested organic matter is broken into increasingly smaller pieces, inoculated with microorganisms, and mixed with mineral particles (Scheu and Wolters 1991). For this reason, they are considered *shredders* in the soil ecosystem.

Litter-feeding earthworms and beetles burrow in surface litter and mix it into the upper mineral soil, thereby assisting in the formation of A horizons (Nielsen and Hole 1964). Along with slugs, woodlice, snails, and millipedes, they also assist in the decomposition and compaction of Oa horizon material from Oi material (see Chapter 3). By ingesting raw or slightly decomposed organic matter from the O horizon, earthworms facilitate its incorporation into the mineral soil in two ways: (1) directly, by depositing humus-rich casts in the subsurface, and (2) indirectly, by increasing the amount of macropores, especially vertically oriented macropores, thereby facilitating translocation of small particles of organic matter into the mineral soil (Stout and Goh 1980, Frelich *et al.* 2006). Through these actions, they facilitate *melanization* (Table 13.1), and are a major reason why the A horizons in many soils are thicker than expected; soils with abnormally thin A horizons are often worm-free.

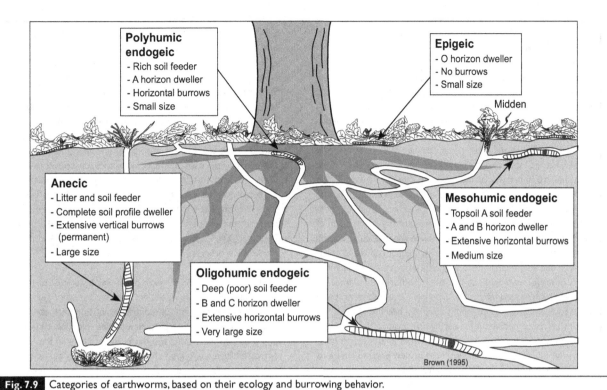

Polyhumic endogeic
- Rich soil feeder
- A horizon dweller
- Horizontal burrows
- Small size

Epigeic
- O horizon dweller
- No burrows
- Small size

Midden

Anecic
- Litter and soil feeder
- Complete soil profile dweller
- Extensive vertical burrows (permanent)
- Large size

Mesohumic endogeic
- Topsoil A soil feeder
- A and B horizon dweller
- Extensive horizontal burrows
- Medium size

Oligohumic endogeic
- Deep (poor) soil feeder
- B and C horizon dweller
- Extensive horizontal burrows
- Very large size

Brown (1995)

Fig. 7.9 | Categories of earthworms, based on their ecology and burrowing behavior.

Earthworms excrete small soil aggregates (casts) that are enriched in microorganisms, organic matter, and nutrients. Casts improve soil structure and enhance aeration, infiltration, and water-holding capacity (Donahue *et al.* 1983). By promoting aggregation, they also play an important role in the physical protection (storage) of soil organic matter. Aggregation in earthworm casts may actually increase C storage in mineral soils by physically protecting the carbon from microbes (Six *et al.* 2004). Also, some types of earthworms excrete Ca with their casts, which may further help to bind and protect soil organic matter from microbial degradation (Crang *et al.* 1968, Canti and Piearce 2003, Bossuyt *et al.* 2004, Lee *et al.* 2013).

Earthworms are categorized according to where they live in the soil and the types of burrows they form. *Epigeic* species, called "red wigglers" by fishermen, are small earthworms that live at the soil surface or in the litter layer. They require high organic matter concentrations but are adapted to a broad range of moisture conditions. *Endogeic* earthworms, in contrast, ingest organic matter in the upper layers of the mineral soil. As they burrow, they fill their burrows with casts. Subtypes of endogenic earthworms are shown in Figure 7.9. *Anecic* species primarily burrow vertically, deep in the soil. They help to translocate organic matter as they move soil between lower and upper horizons. Nightcrawlers are the best-known anecic

earthworms. They create permanent burrows that can be several meters deep. Anecic earthworms feed on organic matter found at the soil surface, piling it up as middens at the soil surface and eventually bringing it deep into their burrows and enhancing its decomposition (Edwards 1999; Fig. 7.10). If the openings of their burrows are not blocked with casts or plugs of organic matter, the burrows can facilitate the downward migration of water, pesticides, and nutrients; however, the high organic matter content of burrow walls may also adsorb dissolved ions and pesticides (Edwards 1999).

Landscapes: Invading versus Migrating Worms

Anglers, take note: that nightcrawler on your hook is an invader. Most bait worms in North America are actually European species. Indeed, Canada, New England, some mid-Atlantic states, and much of the upper Midwest are entirely populated by non-native earthworms. The rest of the continent contains a mixture of non-native and native earthworms. While the ancestry of your local earthworm might not seem like a big deal, scientists say otherwise: the encroachment of the non-native species has a profound effect on soil structure, soil genesis, and the soil carbon balance.

James has concluded that the dividing line between native and non-native species roughly coincides with the furthest advance of the Wisconsin ice sheet, about 22 000 years ago. North of that limit, few native earthworm species remain. Why? During glaciation, earthworms were obliterated from the landscapes that were covered with ice, and permafrost pushed many others even farther south. In the 180–220 centuries that followed, they simply have not retaken their original territory; migration of earthworms is a very slow process! Support for James' hypothesis comes from an unpublished Dutch study that measured the rate at which earthworm populations expand their range. Dutch researchers placed earthworm colonies in a field free of worms. After 1 year, colonies of European worms, species that are especially good colonizers, had expanded by ~10 m. If you take 10 m per year as the maximum rate of range expansion and multiply that by 22 000 years (see also Costello et al. 2011), that yields only about 220 km (137 miles). In other words, native North American earthworms have expanded their territory to the north of the glacial border by the length of a small state since the ice sheets left. Many earthworms to the north are, therefore, not native and could not have migrated there. Instead they are European and Asian immigrants, brought inadvertently by humans in the horticultural trade, or discarded by frustrated fishermen. As people moved plants, probably in the colonial period, worms were inadvertent hitchhikers.

Why be concerned about the earthworm invasion? In a wooded area with worms, soils have thinner O horizons and thicker A horizons, while in worm-free zones, like those in New England, northern Michigan or Minnesota, leaf litter is much thicker and lower horizon boundaries are more diffuse (Stout and Goh 1980, Suarez et al. 2006b, Lyttle et al. 2011, Holdsworth et al. 2012). A horizons thicken at the expense of the O and E horizons, and organic C contents in the A horizons increase. Lyttle et al. (2011) also found that, within less than five years after arrival of the earthworms, A horizons change from single grain to a strong medium granular structure. Nutrient cycling is also more prominent when worms are present. Resner et al. (2011) showed that biocycled Ca and P peak near the soil surface prior to earthworm invasion, but diminish or become diluted after invasion. Frelich et al. (2006) found that earthworm invasion can lead to reduced availability and increased leaching of N and P in some soil horizons. Seed banks and vegetation are affected, and generally, pH rises, after earthworms invade a forest (Hale et al. 2006, Hopfensperger et al. 2011). For example, their invasive presence can even be detected in the growth rates of the overstory trees (Larson et al. 2010). Yes, invasive earthworms change not only soils, but the ecology of their newfound ecosystem (Karberg and Lilleskov 2008, Corio et al. 2009, Nuzzo et al. 2009). And the invasion is still in progress, with worms moving into areas that have long been worm-free.

Hawaii provides another example. It was historically earthworm-free, until it started drawing settlers from all parts of the globe. It now has a vast diversity of introduced tropical and European earthworms. Humans also introduced pigs to Hawaii; these pigs now run wild. The pigs preceded the worms and made an easy living in the Hawaiian woods, finding enough food on the ground that they did little rooting below the surface. Once the worms arrived, though, the pigs developed a taste for them and are now tearing up the remaining natural forest in search of the non-native worms.

Yes, earthworms, they do make a difference.

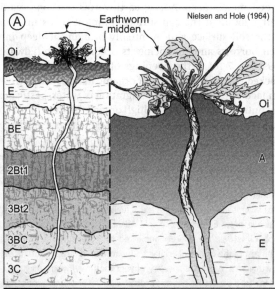

Fig. 7.10 Earthworm (*Lumbricus terrestris* – a "nightcrawler") middens. (A) Drawing of a midden and associated soil horizons. (B) Photo of a midden on the forest floor. Note how the litter around the midden has been largely cleared and dragged into a pile at the entrance to the earthworm's burrow. Nickel (5 cent) coin for scale. Photo by RJS.

Fig. 7.11 A shallow surface disturbance created by bison – a buffalo wallow – on the Konza Prairie, eastern Kansas. Photo by J Gill.

Mammals, Birds, Reptiles, and Amphibians

Most soil biology books or chapters do not devote a great deal of space to the roles of the largest fauna – the vertebrates – to soils. Perhaps this is because many vertebrates, e.g., birds, reptiles, and larger mammals, primarily live outside of the soil and their influence is not obvious. Nonetheless, these fauna can compact or loosen the soil, they transport and move it by scratching, scraping, digging, wallowing, and nesting (Fig. 7.11), and they all contribute feces and

thus function within various nutrient and sediment cycles (Pal and Bhattacharya 1986, Gutterman 1997, McMillan *et al.* 2000, Richards *et al.* 2011). Indirectly, these actions can denude vegetation and facilitate soil erosion/transport (Eldridge and Rath 2002). Soil compaction and denudation are particularly noteworthy around waterholes in grasslands and deserts where, for example, ungulates congregate in the dry season.

Mammals primarily influence the soil by burrowing and living within it. Many rodent and mole species are particularly adept at burrowing; their influence is primarily seen as associated with soil mixing (see Chapter 11). Künelt (1961) noted that, in California, gophers may turn over as much soil in 5 months as earthworms do in 5 years. Many amphibians also perform similar functions, although they are more geographically restricted to wet soils. Many snakes and skinks (*Scincus* spp.) are also proficient burrowers.

Although no bird species is considered a true soil animal, many build their nests on or within the soil and feed there. Many feed upon soil animals, such as worms and beetles, and in certain areas they can decimate the ground cover. In Antarctica, penguins and sea birds have profound influence on soils, primarily through their excrement. Recent work has focused on the effects of penguin guano as an important source of organic acids, driving podzolization in soils of penguin rookeries that otherwise would have little soil development (Beyer *et al.* 1997, 2000). Bird guano-influenced soils are referred to as *ornithogenic soils* (Tatur 1989, Myrcha and Tatur 1991, Beyer *et al.* 1997, Michel *et al.* 2006).

Part II

Soil Genesis: From Parent Material to Soil

Chapter 8

Soil Classification and Mapping

Soil classification and mapping are as old as farming (Michéli and Spaargaren 2011). By 8000 BC, farmers in Europe had already determined where the better agricultural soils were and had preferentially settled on them. By 2000 BC, evidence of farming communities exists for the fertile black soils formed on the Deccan Plateau of India, but not in areas where these soils are absent (Shchetenko 1968). Clearly, early peoples were capable of assessing soil character – the essence of soil classification – and were able to use this information to differentiate the productivity of soils across the landscape. Creating physical maps of this soils knowledge would occur later. Even then, a system whereby soils could be differentiated from each other, based on their characteristics, was needed, just as it is today. Captured in this statement is the essence of soil classification.

The essence of any classification system is to place a name on a "definable entity," in this case, a basic soil unit. Once this is done, it is possible to arrange it in an orderly system and establish its interrelationships with related entities and establish its value (Beckmann 1984). Most importantly, though, it allows for communication about the entities being classified.

Users of classification schemes agree that a name should convey the unit's unique and defined range of characteristics, whether it be a plant, e.g., *Pinus sylvestris*; rock, e.g., arkosic sandstone; or soil, e.g., an Aeric Fragiaquept. For soils, that range of characteristics may involve physical properties such as clay content of the B horizon, thickness of the A horizon, or color, among others. Over time, these ranges are altered and adjusted as new and more complete information about the universe of soils (or plants or rocks) accrues. Thus, most classification systems are open-ended, and periodic updates are assumed.

Before we try to classify a soil, we must first define it. We agree with Johnson (1998a), who defined *soil* as organic or lithic material at the surface of planets and similar bodies that has been altered by biological, chemical, and/or physical agents. But to classify soils, it is essential that we also agree on some sort of singular or *basic unit* (Johnson 1963). In many disciplines, it is relatively easy to determine the basic unit that is being classified: a plant, a rock, a virus. In soils, however, the elementary or basic unit of classification is not always clearly differentiable. How deep should it go? How areally extensive should it be? How much internal variability should be allowed? For example, Ollier *et al.* (1971) found that the Baruya people in New Guinea had many names for *soil materials*, but not for *profiles*, i.e., depth sequences of horizons. On the basis of their use of soil, the classification scheme of the Baruya worked well, illustrating the point that all classification schemes must fill a specified need.

The basic soil unit, like the basic units of all classification schemes, must be a small, arbitrary volume that must not be able to be further subdivided and is of a convenient size for study and measurement, such as a hand rock specimen in geology (Van Wambeke 1966, Arnold 1983, Beckmann 1984). Ideally, it should also be able to be sampled, i.e., it should have volume. It should have clear and definable boundaries and include the entire vertical thickness of the soil.

Some early classification systems used as the basic soil unit the soil profile when defined as a two-dimensional entity that, theoretically, could be sampled. Later, the *pedon* was chosen as the basic unit of sampling (Soil Survey Division Staff 1993), largely because it *could* be sampled. A pedon is defined as the smallest body of one kind of soil that is large enough to represent the nature and arrangement of horizons and natural soil variability (Soil Survey Division Staff 1993; Fig. 3.16). We stress, however, that a pedon is not a soil per se. Rather, it is a concept developed to represent a soil that exists on a larger scale, i.e., a soil landscape.

Soil Geography, Mapping, and Classification

Soils vary both gradually *and* abruptly across the landscape, as dictated by the continuously changing character

Fig. 8.1 An excavation showing an abrupt boundary between two soils. The pedon at the left has formed in sandy glaciofluvial sediment, while the one at the right has formed in clayey glaciolacustrine sediment. This abrupt contact formed as meltwater traversing the lake plain ripped gullies into it, and later filled them with sand. Photo by RJS.

Soils exist on most landscapes in patterns. Sometimes, the pattern is easily interpretable and mappable, because it corresponds nicely to one of the soil-forming factors, e.g., relief. In other instances, the soil pattern is extremely complex and the soil mapper is given few clues by the landscape as to where specific soil properties might occur. A soil mapper who tries to delineate a pure, contiguous unit of pedons, i.e., the concept of a polypedon (see Chapter 3; Fig. 3.16), is likely to be frustrated and ineffective. Taxonomic limits do not necessarily conform to natural breaks or features of the landscape. Nor does the landscape always give adequate clues about exactly where changes in soil properties might occur, even changes that are significant enough to place the soil in another taxonomic class. In essence, the polypedon concept is constrained by the limits of the taxonomic system, while the landscape obeys different rules. This is why some have called for the abandonment or minimization of the polypedon concept (Soil Survey Division Staff 1993, Ditzler 2005). As discussed in Chapter 3, there are also problems with the concept of a pedon. The pedon is defined as being a sampling unit that is fully representative of a polypedon. But just as a single tree cannot possibly exhibit all the possible characteristics of the species, neither can a pedon exhibit all the possible characteristics of a taxonomic class, e.g., one pedon cannot exhibit all possible ranges of horizonation.

Soil map units – those polygons you see on a soil map – are the mapper's best attempt to represent the complexity of the soil landscape by units with some degree of internal consistency. But soil map units do not represent polypedons; it is unrealistic to map pure polypedons (Ditzler 2005). Indeed, if one were to take weeks to observe every soil on a small landscape and truly delineate taxonomically pure map units, the map would likely resemble Swiss cheese! Such a complex map would have very limited utility for the intended use of traditional soil surveys and would be far too expensive to produce. Therefore, soil mappers attempt to distinguish between *soil bodies* (see Chapter 3) that are understood to include soils with properties outside the range of a single series but that are likely to be used and managed in a similar way. Map units may be named for a particular soil series, but they are assumed to include some components of similar or dissimilar soils. Although typical map units are not pure from a taxonomic standpoint, they are designed for their practical value in land management and other applications.

The System of Soil Taxonomy

In this book, we focus on one of many systems used worldwide to classify soils. Michéli and Spaargaren (2011) provided a splendid review of the many other systems, such as those of New Zealand, France, Russia, and Canada. Soil Taxonomy – the name for the soil classification system

of the five soil-forming factors (climate, organisms, relief, parent material, and time) across space (Johnson 1963; see Chapter 14). At places where one of the factors changes abruptly, such as at a bedrock or parent material contact, the soil may change character abruptly as well (Fig. 8.1). Most often, pedologic gradations are gradual. When we map these soils – this soil landscape – we must fit what is there into a taxonomic class or construct, and this task is not always easy.

A taxonomic class or *taxon* is a concept designed to represent a segment of the soil continuum. The challenge of soil classification is to set taxonomic limits, lines, and boundaries in the most appropriate way, so that they can actually be *mapped* and so that the map units can be *interpreted* uniformly (Cline 1963). Regardless of the class limits chosen, they will not fit all landscapes equally well. A class limit might work well on one landscape, where it nicely delimits soils formed from Pleistocene deposits from those formed in Holocene alluvium. On other landscapes, similar limits may not work, perhaps because the soils there have properties so close to the class limit that they cannot be easily differentiated. The challenge of the soil scientist is to define class limits that make taxonomic sense *and* are mappable.

used in the United States and other countries – (Soil Survey Staff 1999) grew out of the work of Guy D. Smith, a pioneer in the early soil survey of the United States (Smith 1979, 1983, 1986). Ahrens and Arnold (2011) provided an excellent review of this system and its details.

A Historical Overview

In order to understand the rationale for Soil Taxonomy best, it is useful to examine it in light of the historical events that led to its development (Ditzler and Ahrens 2006). In the United States, the earliest mapping was done under the jurisdiction of Milton Whitney, head of the U.S. Soil Survey (Simonson 1986). From its inception in 1899, the Soil Survey program has existed to map the soil resources of the nation and provide land-use interpretations, especially as related to crop yield.

Early concepts of pedogenesis in the United States were based primarily on soils as products of rock weathering (Simonson 1986). It made sense, therefore, to key the soil classification system to geology. The basic cartographic unit was the *soil type*, which was the precursor to the current concept of the *soil series* (Simonson 1952, Smith 1983). A soil type had similar kinds of horizons and was separated from others on the basis of the general texture of the profile; this type of classification facilitated the differentiation of different kinds of soil parent materials, e.g., loess, till, or residuum (Ahrens and Arnold 2011).

The soil series concept of the early 1900s was very different from today's. The series was viewed as spanning large physiographic regions. The Miami series, for example, was mapped on glacial deposits of all kinds. The Cecil series was widespread on the old, weathered landscapes of the Piedmont (Gibbs and Perkins 1966). Each series included a variety of textural classes; Miami silt loam, Miami loam, Miami clay loam, and others, were all *phases* of the Miami series. Because soils were regarded at this time as essentially weathered rock, parent material considerations far outweighed all others, and texture was a very important variable. As mapping progressed, more soils were being described and more series were created; the list was growing and becoming difficult to manage.

Earlier, in Russia, Valisy Dokuchaev had been espousing a different approach (see Chapter 1). His work, done largely between 1870 and 1900, viewed soil as a natural body with properties that resulted from the influence of climate and living organisms acting on parent material over time, as conditioned by relief (Smith 1983). These concepts were translated from Russian to German by Konstantin Glinka (1914). The then-head of the U.S. Soil Survey, Curtis Marbut, was so influenced by these ideas that he changed the way soils were classified in the United States, emphasizing that classifications should be based on characteristics of their *horizons*. Marbut considered the following horizon characteristics to be important for distinguishing one series from another: number, arrangement,

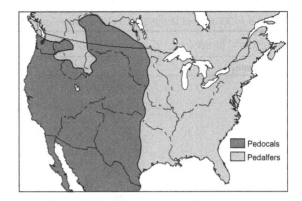

Fig. 8.2 Distribution of Pedocals and Pedalfers (old terms) in the United States and adjacent regions.

thickness, texture, color, chemical composition, and structure, as well as the geology of the parent material. He and his staff eventually developed a multicategorical classification system (Marbut 1927a, 1927b, 1935), one of the more important advances of which was the concept of the *normal soil*: a freely draining, mature soil having clearly expressed horizons (see Chapter 13). The idea of a mature soil, probably based on the ideas of the geographer William Morris Davis and his geographical cycle, in which landscapes are seen to move from youth through maturity and old age (Davis 1899), was a major part of the system. Soils that did not fit the normal soil concept, e.g., organic soils, alkali soils, and wet soils, were not included in the higher categories of the system, again emphasizing the importance of the upland, well-developed soil; all others were aberrations, of sorts. Normal soils were placed within two categories at the highest level: *Pedocals* and *Pedalfers* (Fig. 8.2). Pedocals are soils of dry regions that accumulate *calcium* in the profile. *Pedalfers* are soils of humid regions, enriched in *Al* and *Fe* in their subsurface horizons. These names continue to be used loosely and do retain some descriptive value (Table 8.1).

On the heels of the 1935 Marbut classification came the classification of 1938 (Baldwin *et al.* 1938). This system deemphasized the Pedocal/Pedalfer scheme, and in its place soils were classified into one of three major categories, or *orders*: Zonal, Azonal (soils with weakly expressed or incomplete profiles, i.e., young soils), and Intrazonal (soils whose profiles are dominated by local or particular factors of soil formation, such as volcanic ash, salts, or high water tables). The three categories were an outgrowth of Dokuchaev's normal, transitional, and abnormal soil orders. Again, zonal soils were considered typical of broad geographic regions, or belts. Soil orders were broken into suborders, and then into great soil groups (Tables 8.1, 8.2).

This first serious attempt at a U.S. soil classification scheme was not without problems. Many soils did not fit into a category; some fit into several. Most great soil groups

Table 8.1 Outline of the 1938 soil classification system

Category VI (order)	Category V (suborder)	Category IV (great soil groups)
Zonal soils		
Pedocals	Soils of the cold zone	Tundra soils
	Light-colored soils of arid regions	Desert soils
		Red Desert soils
		Sierozem
		Brown soils
		Reddish-Brown soils
	Dark-colored soils of the semiarid, subhumid, and humid grasslands	Chestnut soils
		Reddish Chestnut soils
		Chernozem soils
		Prairie soils
		Reddish Prairie soils
Zonal soils		
Pedalfers	Soils of the forest–grassland transition	Degraded Chernozem soils
		Noncalcic Brown or Shantung Brown soils
	Light-colored podzolized soils of the timbered regions	Podzol soils
		Brown Podzolic soils
		Gray-Brown Podzolic soils
	Lateritic soils of forested warm-temperate and tropical regions	Yellow Podzolic soils
		Red Podzolic soils (and Terra Rossa)
		Yellowish-Brown Lateritic soils
		Reddish-Brown Lateritic soils
		Laterite soils
Intrazonal soils		
	Halomorphic (saline and alkali soils)	Solonchak or saline soils
		Solonetz soils
		Soloth soils
	Hydromorphic soils of marshes, swamps, and seep areas	Wiesenböden (meadow soils)
		Alpine Meadow soils
		Bog soils
		Half Bog soils
		Planosols
		Ground-water Podzol soils
		Ground-water Laterite soils
	Calcomorphic	Brown Forest soils (Braunerde)
		Rendzina soils
Azonal soils		
		Lithosols
		Alluvial soils
		Dry sands

Source: Modified from Baldwin *et al.* (1938).

Table 8.2 | General characteristics of zonal soils of the 1938 system of classification

Soil type	Profile characteristics	Native vegetation	Soil-development processes
Zonal soils			
Tundra	Dark brown peaty layers over a grayish, mottled profile; with permafrost	Lichens, moss, flowering plants, and shrubs	Gleization and mechanical mixing
Desert	Light colored profile, low in organic matter and shallow to calcareous material	Scattered shrubby desert plants	Calcification
Red Desert	Light reddish brown A horizon and a brownish red or red B horizon with some illuvial clay; shallow to secondary carbonates	Desert plants	Calcification
Sierozem	Pale grayish soil grading into secondary carbonates at ≤30 cm	Desert plants	Calcification
Brown	Brown soil grading into a whitish Bk horizon at 30–100 cm	Shortgrass and bunchgrass prairie	Calcification
Reddish Brown	Reddish brown A horizon, grading into a red or dull red, clay-rich B horizon; at depth is a zone of secondary carbonates	Tall bunchgrass and shrubs	Calcification
Chestnut	Dark brown, friable A horizon over a brown, prismatic B horizon; secondary carbonates at 50–100 cm	Mixed tall and shortgrass prairie	Calcification
Reddish Chestnut	Dark reddish brown A horizon over a reddish brown or red B horizon; secondary carbonates at ≥70 cm	Mixed grasses and shrubs	Calcification
Chernozem	Thick, black or very dark grayish brown A horizon, grading to a lighter B; secondary carbonates at depth	Tallgrass and mixed grass prairie	Calcification
Prairie	Very dark brown or grayish brown A horizon, grading to brown (or lighter-colored) parent material	Tallgrass prairie	Calcification with weak podzolization
Reddish Prairie	Dark brown or reddish brown A horizon, grading to a reddish brown, clay-rich Bt; moderately acid	Tallgrass and mixed grass prairie	Calcification with weak podzolization
Degraded Chernozem	Nearly black A horizon, somewhat bleached, grayish E horizon, and an incipient Bt horizon; vestiges of secondary carbonates	Forest encroaching on tallgrass prairie	Calcification followed by podzolization
Noncalcic Brown (Shantung Brown)	Brown or light brown, friable A horizon over a pale reddish brown or dull red B horizon	Open deciduous forest with brush and grasses	Weak podzolization and some calcification
Podzol	Thick O horizon over a thin, dark gray A horizon; prominent, whitish gray E horizon, above a dark or coffee brown Bhs horizon and a yellowish brown BC; strongly acid	Coniferous, or mixed coniferous–deciduous forest	Podzolization
Brown Podzolic	Acidic O horizon over a thin, gray-brown or yellowish brown E horizon, over a brown Bw horizon; profile seldom >60 cm thick	Deciduous or mixed coniferous–deciduous forest	Podzolization

(cont.)

Table 8.2 | (*cont.*)

Soil type	Profile characteristics	Native vegetation	Soil-development processes
Gray-Brown Podzolic	Thin O horizon over a dark A horizon, 5–10 cm thick; grayish brown E horizon over a brown, clay-rich Bt horizon; less acidic than Podzols	Mostly deciduous forest with mixture of conifers	Podzolization
Yellow Podzolic	Thin, dark-colored O horizon over a pale, yellowish gray E horizon, 15–100 cm thick; Bt horizon is clay-rich and yellow, overlying a yellow-red-gray (mottled) parent material; acidic	Coniferous or mixed coniferous and deciduous forest	Podzolization with some laterization
Red Podzolic	Thin O horizon over a yellowish brown or grayish brown A horizon, over a deep red B horizon; acidic	Deciduous forest with some conifers	Podzolization and laterization
Yellowish-Brown Lateritic	Brown, friable, clay and clay loam horizons over yellowish brown, heavy but friable clays; acidic to neutral reaction	Evergreen and deciduous broadleaved trees	Laterization and some podzolization
Reddish-Brown Lateritic	Reddish brown or dark reddish brown, friable, clayey horizons over deep red, granular clay; C horizon reticulately mottled	Tropical rain forest to edge of savanna	Laterization with little or no podzolization
Laterite	Red-brown surface soil with a thick, red B horizon and a red or mottled C horizon; very deeply weathered	Tropical selva and savanna	Laterization and a little podzolization
Intrazonal soils			
Solonchak	Thin, gray salty surface crust with a fine granular mulch just below; more salts in lower profile	Sparse growth of halophytic grasses and shrubs	Salinization
Solonetz	Very thin, friable A horizon underlain by dark, hard columns; usually highly alkaline	Halophytic plants; much bare surface	Solonization (desalinization and alkalization)
Soloth	Thin grayish brown, friable A horizon over whitish, leached E horizon, underlain by dark brown, clayey Btn or Btz horizon	Mixed prairie or shrub	Solodization (dealkalization)
Wiesenböden and Alpine Meadow	Dark brown or black A horizon grading to a gray, mottled subsoil	Grasses and sedges	Gleization and some calcification
Bog	Brown, dark brown, or black organic materials over brown peat	Swamp forest or sedges and grasses	Gleization
Half Bog	Dark brown or black peat over gray or mottled mineral materials	Swamp forest or sedges and grasses	Gleization
Planosols	Strongly leached, acidic A and E horizons over a compact or cemented Bt "claypan"	Grass or forest	Podzolization and gleization; laterization in the tropics
Ground-water Podzols	O horizon over a very thin, acid A horizon; whitish gray E horizon, up to 100 cm thick, over a Bsm horizon that is brown or very dark brown; gleyed, gray C horizon	Forest	Podzolization and gleization

Table 8.2 *(cont.)*

Soil type	Profile characteristics	Native vegetation	Soil-development processes
Ground-water Laterites	Gray or grayish brown A horizon over a leached, yellowish gray E, over a thick, reticulately mottled, and cemented "hardpan" at ≥100 cm; pan can be >100 cm thick	Tropical forest	Podzolization and laterization
Brown Forest (Braunerde)	Very dark brown, friable A horizon, getting lighter-colored at depth; rich in calcium	Forest, usually broadleaf	Calcification with very little podzolization
Rendzina	Dark, grayish brown to black, granular A horizon underlain by gray or yellowish, usually soft, calcareous material	Usually grasses; some broadleaf forest	Calcification
Azonal soils			
Lithosols	Thin, stony/rocky A horizon; little or no illuviation; shallow to bedrock	Depends on climate	Varies
Alluvial soils	Little profile development; some organic matter accumulation; stratified, alluvial parent materials	Depends on climate	Varies
Sands, dry	Essentially no profile development; loose sands	Scanty grass, scrubby forest or bare sand	Varies

Source: Modified from Baldwin *et al.* (1938).

were weakly defined, and most definitions of class limits were qualitative (Riecken 1945). The system was based on the virgin soil profile and broke down by forcing the classifier to pigeonhole cultivated or eroded soils into categories based on what they were perceived to have been, prior to cultivation. In short, the 1938 soil classification system left much to subjective interpretation. A rigorous, more quantitative system was needed (Smith 1983). Fortunately, by the 1940s, much had been learned about the range of soil properties and processes, and thus a new system could be based on that knowledge. Major advances in soil chemistry and soil physics, better understanding of soil landscapes after years of mapping, the introduction of the catena concept (Bushnell 1942) and with it a better understanding of local-scale (within-series) soil variability, and the need for increasingly quantitative soil interpretations for agricultural purposes all led to a new, more objective system of soil classification. In 1949, James Thorp and Guy Smith declared their intent to depart from the zonal soil concepts that were the backbone of the 1938 system, "in favor of terms based on soil characteristics" (Thorp and Smith 1949, 125). This was to be a major turning point in soil classification.

This new classification, eventually called *Soil Taxonomy*, had as its primary objective the facilitation of soil survey/mapping operations. This new system had to have certain characteristics (Cline 1949a). The system was designed to improve use and management of the soil. Let no one be confused; the new system was not established to improve categorization of soils on the basis of *genesis* alone, although the categories needed to make genetic sense (Cline 1963, Bockheim and Gennadiyev 2000). The categories of the new system must have pedogenic ties or they could not be mapped, but the criteria that governed the system were based on quantifiable soil *properties*. The move away from an emphasis on soil processes was based on the assumption that soil properties were the result of processes, and properties were more quantifiable (Bockheim and Gennadiyev 2000). Thus, soil genesis would not overtly appear in the new system, but it lay everywhere behind it (Cline and Johnson 1963, Smith 1983). Importantly, the new system would not abandon already entrenched series concepts or names. And yet, it needed to link series concepts to higher taxonomic categories – hence the need for a hierarchical system of classification.

The new system, spearheaded by Guy Smith, was published as a series of working drafts or *approximations* (Thorp and Smith 1949). Each approximation was circulated to a number of practicing soil scientists and scholars and amended according to their comments and suggestions. Advice was sought at specially organized meetings, and by circulation of the approximations at national

Table 8.3 | Numbers of entries of the various categories of Soil Taxonomy

Category	Number of members[a]	Examples
Order	12	Mollisols
Suborder	70	Udolls
Great group	344	Hapludolls
Subgroup	2664	Typic Hapludolls
Family	11332	Sandy, Mixed, Frigid Typic Hapludolls
Series	23636	Binford series

[a] The number of entries in the four largest (taxonomically highest) categories is fixed by the most recent version of *Keys to Soil Taxonomy*. The numbers of entries in the two lowest categories, however, change regularly, as field soil mappers discover, describe, and formalize new families and series. Data in the table were determined in April 2012.

and international conferences. Hundreds of people were involved. Finally, in 1960, the new 7th Approximation was deemed sound enough to be published widely (Soil Survey Staff 1960). Fifteen years later, Soil Taxonomy was first published in a thick green hardcover book as USDA Agriculture Handbook 436; it was essentially the 8th Approximation, but now it would become known simply as Soil Taxonomy or, to many, "the green book" (Soil Survey Staff 1975). The second edition (a purple book) was published in 1999. The staff of the Natural Resource Conservation Service's National Soil Survey Center continue to teach, interpret, and maintain the system. As new knowledge comes forth, especially with respect to soils outside the United States (Soil Survey Staff 1999), the system is updated and improved. An abridged version known as the Keys to Soil Taxonomy is published every 2–4 years to incorporate amendments and update the system more frequently than could be done for the hardcover form (Soil Survey Staff 2014).

How Soil Taxonomy Works

There are basically two philosophies in classifying soils: genetic and morphological (Beckmann 1984). The genetic approach is based on knowledge of pedogenic processes or on correlations of inferred pedogenic factors (see Chapters 12 and 13). The purely morphological approach deemphasizes some of the genetic interpretation in favor of an emphasis on properties of individual soils. *Soil Taxonomy* (Soil Survey Staff 1999) is based on definable and measurable soil characteristics; it is thus primarily a morphological taxonomic system but with strong genetic underpinnings. The subtitle of Soil Taxonomy is "A Basic System of Soil Classification for Making and Interpreting Soil Surveys." This subtitle explains the goals of the system most clearly. For all practical purposes, the pedon is the basic conceptual soil unit used in Soil Taxonomy; it is intended to be the sampling unit. The system draws heavily upon experience of soil survey teams in the United

States, although it seeks to be general enough to be applied globally.

Above all, Soil Taxonomy is quantitative, and it is hierarchical. Its six major categories, from most general (with the fewest taxa) to most specific (with the greatest number of taxa), are order, suborder, great group, subgroup, family, and series. With each successive category, increasing amounts of information are provided about the soil. Soil orders are broadly defined and general, but small (twelve) in terms of members. Soil series, at the other end of the spectrum, are narrowly defined in terms of specificity, but large in terms of members (Table 8.3). Thus, the system is arranged from the general to the specific, just as the Linnaean system in biology is.

Taxa within the system, e.g., orders and suborders, are first and foremost based on the presence or absence of quantitatively defined *diagnostic horizons*, which can be at the surface or in the subsurface (Table 8.4). Diagnostic horizons (genetically designated O or A horizons) at the soil surface are called *epipedons*. As we noted in Chapter 3, it is important to distinguish between genetic horizons and the diagnostic horizons used in soil classification. For example, a B horizon enriched in illuvial clay would be designated Bt in a soil description, because it shows evidence of illuvial clay. Only if it were so strongly developed that it also met certain defined criteria could it be considered a diagnostic *argillic* horizon. But if it lacked the minimum amount of illuvial clay, it might be a *cambic* horizon, another diagnostic horizon but with different criteria (Table 8.4). All argillic horizons use a *t* suffix symbol in the genetic horizon designation, but *not* all horizons designated as Bt qualify as diagnostic argillic horizons.

At the highest four levels of the system, soil names are constructed by assembling formative elements or syllables. Many formative elements are taken from Greek and Latin, while others are fragments of words commonly used today (Heller 1963). For the twelve soil orders, the formative elements are listed in Table 8.5. Each order

Table 8.4 | Diagnostic horizons and characteristics used in Soil Taxonomy

Diagnostic horizon	Defining criteria	Typical minimum thickness (cm)	Common genetic horizon equivalent	Commonly found in which orders
Epipedons (use only in mineral soils)				
Histic	Organic soil material, e.g., peat, muck, characterized by saturation and reduction	20–60	Oi, Oe, Oa	Gelisols, Spodosols, Andisols, Oxisols, Mollisols, Alfisols, Inceptisols, Entisols
Folistic	Similar to histic, but not wet; commonly overlies rock	15–25	Oi, Oe, Oa	Spodosols, Andisols, Inceptisols
Mollic	Dark (Munsell color value ≤3 when moist and ≤5 when dry; chroma ≤3 when moist), humus-rich (≥0.6% organic carbon) horizon in which bivalent cations, especially Ca, are dominant on the exchange complex; base cation saturation >50%; granular structure is common; usually forms under grasses	18–25	A, Ap	Mollisols, Vertisols Andisols Alfisols Gelisols Inceptisols
Umbric	Similar to mollic but base cation saturation is <50%; usually formed in cool, humid climates, under forest	10–25	A, Ap	Ultisols, Gelisols, Spodosols, Andisols, Alfisols Inceptisols
Anthropic	Generally similar to mollic but has high P content due to long-term human occupation	10–25	Ap	Several
Plaggen	Dark surface horizon created by long-term manuring	50	Ap	Inceptisols
Melanic	Thick, dark horizon, rich in organic carbon, usually formed by continual, thin additions of volcanic ash	40	A, Ap	Andisols
Ochric	All other epipedons; typically thin, often formed under forest	None	A, Ap	Several
Diagnostic subsurface horizons and layers				
Agric	Illuvial horizon, below the Ap, in which long-term cultivation has led to translocation of silt, humus, and clay	10	Varies	
Albic	Bright-colored, leached E horizon; reflects the color of uncoated sand and silt grains; often coarse-textured	1	E	Spodosols, Alfisols, Andisols, Ultisols, Mollisols
Argillic	Horizon of illuvial phyllosilicate clay; contains ≥1.2 times as much clay as overlying horizons; has coatings of oriented clay (argillans) on ped faces; commonly, the fine/total clay ratio is greater than in overlying horizons; can exist as clay bands (lamellae) or as clay bridges between sand grains	7.5–15	Bt	Alfisols, Ultisols, Mollisols, Aridisols, Spodosols

(cont.)

Table 8.4 | *(cont.)*

Diagnostic horizon	Defining criteria	Typical minimum thickness (cm)	Common genetic horizon equivalent	Commonly found in which orders
Calcic	Horizon of secondary carbonate accumulation, commonly in dry climates but can occur as precipitation from shallow groundwater; not cemented	15	Bk	Aridisols, Vertisols, Mollisols, Inceptisols, Andisols, Alfisols
Cambic	Minimally developed B horizon; formed by various physical alterations and/or chemical transformations; cannot be extremely sandy; often called "color B"	15	Bw, Bg	Inceptisols, Gelisols, Vertisols, Mollisols
Glacic	Massive ice or ground ice (wedges or lenses)	30	Af, Bf, Cf	Gelisols
Glossic	Degraded (removal of clay and Fe) argillic, kandic, or natric horizon; albic (E) material interfingers into the horizon from above	5	E/Bt, E/Btn, E/Btx	Alfisols, Ultisols
Gypsic	Horizon of secondary gypsum accumulation, in dry climates on gypsum-rich parent materials; not cemented	15	By	Aridisols, Vertisols, Mollisols, Gelisols, Inceptisols
Kandic	Weathered, clay-rich subsurface horizon; most of the (not necessarily illuvial) clay has low CEC and is 1:1 phyllosilicates; may not exist as lamellae	15–30	Bt	Ultisols, Alfisols, Oxisols
Natric	Special kind of argillic horizon that has high levels of sodium; commonly dense and slowly permeable; columnar or prismatic structure is commonplace	7.5–15	Btn, Bn	Mollisols, Aridisols, Alfisols, Vertisols
Oxic	Usually clayey (but sometimes sandy), subsurface horizon with low CEC values and few weatherable minerals; dominated by kaolinite, oxide clays	30	Bo	Oxisols
Petrocalcic	Illuvial horizon cemented by secondary carbonates	10	Bkm, Bkkm	Aridisols, Mollisols
Petrogypsic	Horizon cemented by secondary gypsum; dry fragments do not slake in water	10	Bym, Byym	Aridisols
Placic	Thin, black to dark reddish pan, cemented by Fe oxides, or Fe and Mn oxides with organic matter; commonly wavy or convoluted	1		Spodosols
Salic	Horizon in which salts more soluble than gypsum, e.g., halite, have accumulated; high electrical conductivity	15	Bz, Cz	Aridisols, Vertisols, Mollisols
Sombric	Subsurface horizon of illuvial humus not associated with Al or Na; low CEC values and base cation saturation	None	Bh	Inceptisols, Oxisols, Ultisols

Table 8.4 | (cont.)

Diagnostic horizon	Defining criteria	Typical minimum thickness (cm)	Common genetic horizon equivalent	Commonly found in which orders
Spodic	Illuvial horizon dominated by spodic materials (amorphous Al and organic materials, sometimes with Fe); typically underlies albic horizons; dark and/or reddish colors; low (≤5.9) pH	2.5	Bs, Bh, Bhs, Bsm, Bhsm	Spodosols
Sulfuric	Horizon with high levels of sulfur, and hence, low pH; composed of either mineral or organic soil materials	15	Cj, Cse	Histosols, Entisols

Subsurface pans

Duripan	Silica-cemented subsurface horizon; commonly will not slake in weak acid but often does in a strong base; common to dry regions	None	Bqm	Mollisols, Inceptisols, Alfisols, Andisols, Aridisols
Fragipan	Dense, brittle horizon that restricts entry of water and roots, yet air-dry fragments will slake in water; commonly has coarse prismatic structure with eluviated material between prisms	15	Bx, Ex, Btx	Alfisols, Inceptisols, Ultisols, Spodosols
Ortstein	Cemented horizon composed of spodic materials; may be present as nodules or as a continuously cemented horizon	2.5	Bsm, Bhsm	Spodosols

Characteristic materials and conditions (selective list)

Aquic conditions	Soils that currently undergo continuous or periodic saturation and reduction	None	Eg, Bg, Cg	All orders
Gelic materials	Soil materials that show evidence of cryoturbation or ice segregation	None	Bf, Bff, Cf, Cff	Gelisols
Lamellae	Thin (<7.5 cm) layers of illuvial, silicate clay, commonly found in sandy soils	None	E and Bt, Bt and E	Alfisols, Spodosols, Inceptisols, Entisols, Ultisols
Permafrost	Thermal condition in which a material remains below 0°C for two or more consecutive years	None	Bf, Cf	Gelisols
Plinthite	Iron-rich, humus-poor mixture of clay, quartz, and other minerals, forming platy, polygonal or reticulate patterns; changes irreversibly to ironstone hardpan after wetting/drying	None		Oxisols, Ultisols, Inceptisols
Sulfidic materials	Waterlogged materials, mineral or organic, that contain >7.5 g kg^{-1} of sulfide-sulfur	None	Cj, Cse	Histosols, Entisols

Source: Simplified: from Soil Survey Staff (1999). The reader is referred to that source for precise definitions.

Table 8.5 | Names, formative elements, and generalized descriptions of soils in the twelve taxonomic orders

Order	Formative element	Derivation[a] of formative element	Pronunciation	General characteristics and descriptors
Gelisols	-el	L. *gelare*, to freeze	Gel	Cold soils with permafrost within 2 m of the surface
Histosols	-ist	Gr. *histos*, tissue	Histology	Organic soils without permafrost, dominated by decomposing organic matter; most are saturated at times
Spodosols	-od	Gr. *spodos*, wood ash	Odd	Soils in which translocation and subsurface accumulation of compounds of humus and Al, and sometimes also Fe, have occurred
Andisols	-and	Modified from *ando*	And	Soils that formed in parent materials with a large component of volcanic ash; dominated by short-range-order minerals
Oxisols	-ox	F. *oxide*, oxide	Oxides	Highly weathered, relatively infertile soils dominated by oxide, low-activity clays
Vertisols	-ert	L. *verto*, turn, mix	Invert	Dark soils of semiarid grasslands and savannas that develop deep cracks in the dry season; as the shrink-swell clays rehydrate and expand in the wet season, the cracks close
Aridisols	-id	L. *aridus*, dry	Arid	Soils of dry climates with some development in the B horizon, often evidenced by slight weathering or by illuvial compounds of calcium, gypsum, or other materials
Ultisols	-ult	L. *ultimus*, last	Ultimate	Acidic, leached soils, usually in warm, humid climates, that have a Bt horizon enriched in clay, mainly of the 1:1 phyllosilicate type
Mollisols	-oll	L. *mollis*, soft	Mollify	Base cation–rich soils that have a thick, dark A horizon, often formed under grasslands or savanna, and a B horizon that exhibits moderate or stronger development
Alfisols	-alf	Aluminum (Al) and iron (Fe)	Ralph	Soils, usually less acidic than Ultisols and in cooler climates, that have a Bt horizon enriched in silicate clays, or a fragipan with illuvial clay
Inceptisols	-ept	L. *inceptum*, beginning	Inception	Soils with dark, carbon-enriched epipedons, weak to strong B horizon development, or high Na content plus endosaturation
Entisols	-ent	Meaningless syllable from "recent"	Recent	All other soils, usually very weakly developed, on young surfaces or eroded/disturbed sites or forming in difficult-to-weather materials

[a] F., French; Gr., Greek; L., Latin.

Source: From Soil Survey Staff (1999). The reader is referred to that source for more precise definitions and limits.

name contains a formative element and a connecting vowel (*o* or *i*), ending with *sol* (Latin *solum*, soil). In the Gelisol order the formative element is *el*; in Aridisols it is *id*. These formative elements continue to be used as endings for the suborders, great groups, and subgroups. Thus, the names of all taxa (higher than series) that are in the Entisols order end in *ent*. Each order is strictly defined; most require the presence of at least one diagnostic horizon (Tables 8.5, 8.6). Many of the major order concepts were taken directly or in large part from the 1938 system of classification (Baldwin *et al.* 1938; Table 8.7), and many also correspond, roughly, to a traditional sequence of

Table 8.6 | Diagnostic horizons and properties, arranged by taxonomic order

Soil order	Commonly observed epipedons[a]	Commonly observed subsurface diagnostic horizons and properties[a]	Representative genetic horizon sequence(s)
Alfisols	Ochric	Albic horizon	A/E/Bt/C
		Argillic horizon	A/E/Bt1/Bt2/C
		Fragipan	Ap, Bt, C
		Duripan	A/E/Ex/Btx/C
		Kandic horizon	A/Bt/Bqm/C
			A/E/Bt/C
Andisols	Ochric, Mollic, Melanic	Albic horizon	A/Bw/C or A/Bs/C
		Andic soil properties	
		Placic horizon	A/Bsm/C
Aridisols	ochric	Natric horizon	A/E/Btn/C
		Calcic or petrocalcic horizon	A/Bkk/Bkkm/Ck/C
		Gypsic or petrogypsic horizon	A/Byy/Byym/C
		Argillic horizon	A/Bt/C
		Duripan	A/Bkqm/Bqm/C
		Salic horizon	Az/Bz/C
		Cambic horizon	A/Bw/Bk/C
		Identifiable secondary carbonates	
Histosols	None	None	Oa/Oe/Oi1/Oi2
		Placic horizon	Oa/2Bsm/C
Mollisols	Mollic	Albic horizon	A/E/Bt/C
		Calcic horizon	A/Bk/C
		Natric horizon	A/Btn/C
		Argillic horizon	A/Bt/C
		Duripan	A/Bqm/C
		Cambic horizon	A/Bw/C
Oxisols	Ochric	Oxic horizon	A/Bo1/Bo2/Cr
		Kandic horizon	A/Bo1/Bo2/C
		Sombric horizon	A/Bh/Bo/C
Spodosols	Folistic, Histic, Ochric	Albic horizon	Oa/E/Bhs/Bs/C
		Spodic horizon	Oa/E/Bs/ Bhs/Bs/C
		Albic horizon	Oa/E/Bs/C
		Placic horizon	Oi/E/Bsm/Bs/C
		Fragipan	Oi/E/Bs/E'/Ex/Btx/C
Ultisols	Ochric	Albic horizon	A/E/Bt/C
		Argillic horizon	A/E/Bt/C
		Kandic horizon	A/E/Bt/C
		Plinthite	A/E/Bt/Cv/R
		Fragipan	A/E/Bt/Btx/C
Vertisols	Mollic, ochric	Slickensides, cracks	A/Bss/Bssg
		Salic horizon	A/Bn/Bnss/C
		Gypsic horizon	A/By/Bssy/C
		Calcic horizon	A/Bk/Bkss/C
Gelisols	Ochric, histic	Gelic materials	O/A/Bgjj/Cf
Inceptisols	Histic, folistic, Ochric, umbric	Cambic horizon	A/Bw/C or A/Bg/Cg/C
		Fragipan	A/Bw/E/Bx/C
		Placic horizon	A/E/Bsm/C
Entisols	Anthropic, ochric	None	A/C or A/Bs/C or A/Bw/C
All orders	–	Redoximorphic features	A/Bg/Cg

[a] Shown here are some examples of diagnostic horizons or other features that the soils of each order usually do/can have; not all are required for a soil to classify within that order.

Source: Partially from Bockheim and Gennadiyev (2000).

Table 8.7 | Taxonomic equivalencies between the 1938 soil classification system and Soil Taxonomy

1938 classification great soil groups	*Soil Taxonomy* great groups or other taxa *mostly* included	Great groups or other taxa *partly* included
Alluvial soils	Fluvaquentic and fluventic Inceptisols and Mollisols; fluvents	Cumulic Mollisols; Cryorthents, Psammaquents, Endoaquents, and Endoaquepts
Alpine Meadow soils	Cryaquods and Cryands	Cryaquolls and Cryaquepts; some Gelisols
Ando soils	Andisols	
Bog soils	Histosols	
Brown soils	Aridic Argiustolls, Argixerolls, Haploxerolls, and Haplustolls; Ustollic and Xerollic Argids and Cambids	Frigid families of Xerollic Argids, Cambids, and Xerolls; frigid families of Aridic Haplustolls and Hapludolls
Brown Forest soils	Eutrudepts and Haplustepts	Haploxerolls and Hapludolls
Brown Podzolic soils	Cryudands; mesic families of Entic Orthods	Dystrudrepts, Eutrudepts, and Fragiudepts
Calcisols	Calcids, Calciargids, Petrocalcids, Calcigypsids, and Calcicryids	Haplocambids, Haplodurids, Haplustepts, and Haploxerepts
Calcium Carbonate Solonchaks	Calciaquolls	Aquic Calciustolls
Chernozems	Cryolls; mesic families Argiustolls and Haplustolls	Mesic and frigid families of Haploxerolls
Chestnut soils	Frigid families of Argixerolls, Durixerolls, Haploxerolls, and Palexerolls; mesic families of Aridic Argiustolls, Argixerolls, Haploxerolls, and Haplustolls	Mesic and frigid families of Calcic and Calcidic Argixerolls and Haploxerolls
Degraded Chernozems	Mollic Hapludalfs and Endoaqualfs	Alfic Argiudolls and Argiustolls
Desert soils	Mesic families of Argidurids, Haplargids, and Paleargids	Mesic families of Haplocambids and Haplodurids
Gray-Brown Podzolic soils	Hapludalfs and Glossudalfs	Aeric Endoaqualfs; Fragiudalfs and mesic families of Hapludults
Gray Wooded soils	Eutrudepts and Dystrudepts	Mesic and frigid families of Udalfs and Ustalfs; Glossudalfs
Ground-water Laterite soils	Plinthaquults, Plinthudults, Plinthustalfs, and Plinthustults	
Ground-water Podzols	Aquods	Haplohumods; Aquands; Spodic Psammaquents
Grumusols	Vertisols	Vertic Endoaquepts, Epiaquepts, and Endoaquolls; Ustertic, Vertic, and Xerertic Haplargids, and Haplocambids
Half-Bog soils	Histic Cryaquepts and Humaquepts	Histic Cryaquolls, Endoaquolls, and Epiaquolls; Histic Glossaqualfs
Humic Ferruginous Latosols	Humults	Dystrudands; Oxisols
Humic Gley soils	Argiaquolls, Cryaquolls, Haplaquolls, and Humaquepts	Aquults and Aquands; Calciaquolls, Fluvaquents; Mollic Endoaquepts and Endoaqualfs

Table 8.7 | (*cont.*)

1938 classification great soil groups	*Soil Taxonomy* great groups or other taxa *mostly* included	Great groups or other taxa *partly* included
Humic Latosols	Humults	Dystrudands and Hydrudands; Humic Ustox, Perox, and Udox
Hydrol Humic Latosols	Hydrudands	
Laterite soils	Acrotorrox, Acroperox, and Acrudox	
Latosolic Brown Forest soils	Dystrudands	Udivitrands
Latosols	Humults	Oxisols; Hydrudands, Rhodustults, Kandiustults, and Dystrudepts
Lithosols	Lithic extragrades	"Shallow" familes of any soil series
Low Humic Gley soils	Aquults, Endoaquents, Endoaquepts, and Endoaqualfs	Aquepts, Aqualfs, and Aquods; Cryaquents, Fluvaquents, and Psammaquents
Low Humic Latosols	Oxic Dystrustepts and Haplustepts	Haplustox and Kandiustox
Noncalcic Brown soils	Durixeralfs, Haploxeralfs, and Palexeralfs	Haploxerepts
Planosols	Albaqualfs, Argialbolls, and Fragiaqualfs	Glossaqualfs, Fragiaquults, and Albaquults
Podzols	Spodosols	Cryands and Aquands; Psammoturbels and Psammorthels
Prairie soils and Brunizems	Argiudolls, Haplaquolls, Argiaquolls, Hapludolls, Argixerolls, and Haploxerolls	Haplocryolls, Durixerolls, Palexerolls, and Udic Argiustolls
Red Desert soils	Haplocambids, Argidurids, Haplargids, and Paleargids	Haplocalcids
Reddish-Brown Lateritic soils	Humults and Rhodudults; Rhodic Paleudalfs and Paleudults	
Reddish-Brown soils	Haplustalfs and Paleustalfs; Typic and Ustollic Haplocalcids, Haplocambids, and Haplargids	Haplotorrands; Aridic Argiustolls, Calciustolls, and Haplustolls
Reddish Chestnut soils	Haplustalfs and Paleustalfs; Aridic and Typic Argiustolls and Paleustolls	Haplustolls
Reddish Prairie soils	Thermic families of udic subgroups of Argiustolls and Paleustolls	Torrands; thermic families of Paleudolls
Red-Yellow Podzolic soils	Thermic families of Fragiudults, Hapludults, Kandiudults, Kanhapludults, and Paleudults	Mesic families of Fragiudults, Hapludults, Kandiudults, Kanhapludults and Paleudults; thermic families of Paleudalfs; some Haplustalfs and Paleustalfs
Regosols	Psamments, Orthents	Entic subgroups of Haploxerolls, Hapludolls, and Haplustolls; Psammaquents, Xerochrepts, and Cryands
Rendzina soils	Rendolls	Xerolls and Ustolls
Sierozems	Haplocambids, Argidurids, Haplargids, and Paleargids	
Solonchak soils	Salorthids	Halaquepts
Solonetz soils	Natric great groups of Alfisols, Aridisols, and Mollisols	

(*cont.*)

Table 8.7	(cont.)		
1938 classification great soil groups	*Soil Taxonomy* great groups or other taxa *mostly* included	Great groups or other taxa *partly* included
Soloth soils		Natraqualfs; Natric subgroups of Cryolls; Duraquolls, Durixeralfs, and Haploxeralfs
Sols Bruns Acides	Dystrochrepts, Fragiochrepts, and Haplumbrepts	Udorthents and Xerumbrepts
Subarctic Brown Forest soils	Cryepts	Cryolls
Tundra soils	Gelisols	Vitricryands, Cryochrepts, and Cryumbrepts

Source: After a variety of sources, including Baldwin *et al.* (1938), Cline (1955), Harper (1957), Kellogg (1941, 1950), Kellogg and Nygard (1951), Marbut (1935), McClelland *et al.* (1959), Oakes and Thorp (1950), Simonson *et al.* (1952), Tavernier and Smith (1957) and Thorp and Smith (1949).

development: Entisols – Inceptisols – Spodosols/Alfisols – Ultisols – Oxisols (Fig. 8.3). For some orders, the influence of lithology, climate, or local site characteristics is dominant (Figs. 8.3, 8.4).

Soils are further categorized into *suborders*. Suborder names are two syllables long; the first syllable connotes something about an important property or properties of the soil, whereas the second is the formative element of the order (Table 8.8). There are currently 70 suborders (Table 8.3). For example, the Fibrists suborder is constructed by adding the formative element *Fibr* (meaning fibrous) (Table 8.8) to the *ist* for the Histosols order. Fibrists are Histosols (organic soils) in which the organic matter is minimally decomposed and high in fibers. Likewise, a Psamment is a sandy Entisol (*Psamm* means sandy, and *ent* is for Entisols). In this way, formative elements are strung together to make words that have meaning and connote an image. Because each of the taxa formed by stringing together formative elements is quantitatively defined, the words also have a strict meaning that is quantifiable; subjective interpretation is minimized.

The name of a *great group* consists of its suborder name and an additional prefix descriptor that consists of one or two formative elements that provide additional information about the soil (Table 8.9). Great group names, therefore, have three or four syllables and end with the formative element of an order. For example, a *Fragiaquod* is a soil that is in the Spodosol order (-*od*) and has aquic conditions (-*aqu-*) and a fragipan (*Frag-*). A fragipan is a diagnostic subsurface horizon that is dense and often brittle (see Chapter 13). Likewise, a *Natrargid* is an Aridisol (-*id*) that contains argillic (-*arg-*) and natric (*Natr-*) diagnostic horizons (Table 8.4).

The Soil Survey Staff stopped the process of adding onto soil taxonomic names with great groups. *Subgroup* names, therefore, consist of the name of a great group modified by one or more adjectives that precede the great group name as a separate word or words (Table 8.10). Subgroup adjectives normally take on one of three functions: (1) denoting the soils as the *central concept* for that great group, (2) indicating that the soil is an *intergrade* between its great group and another, or (3) noting that the soil has properties that are not representative of another kind of soil in a higher category of any order, suborder, or great group (*extragrade* soils). The adjective *typic* is used for "central concept" soils within a great group, e.g., Typic Haplocalcid. In other cases, the typic subgroup is simply used for all the soils of a great group that do not fit into the possible intergrade or extragrade subgroups, i.e., the typics are the leftovers. Typic subgroups have all the diagnostic properties of the order, suborder, and great group to which they belong (Soil Survey Staff 1999). *Intergrade* subgroups belong to one great group but have some properties of another order, suborder, or great group, i.e., they are transitional in some way. For example, *Vertic Torrifluvents* have some properties of Vertisols, but they display the complete set of diagnostic properties of Torrifluvents – Entisols on floodplains (-*fluv-*) in an aridic (*torric*) soil moisture regime. They are Torrifluvents that are transitional to Vertisols. *Extragrade* subgroups do not necessarily show properties of another order, but rather, they display important properties that may affect their use and management. They are not necessarily transitional to another order or suborder. Extragrade subgroups are named by modifying the great group name with an adjective that connotes something about the nature of the unique soil properties (Soil Survey Staff 1999). For example, a *Typic Haplorthod* is the central concept of Haplorthods, an *Andic Haplorthod* is an intergrade Haplorthod to an Andisol, and a *Lithic Haplorthod* is an extragrade whose noteworthy characteristic is the presence of hard, coherent bedrock at a shallow depth. In certain circumstances, multiple adjectives may be used in a subgroup name, e.g., Abruptic Haplic or Cumulic Vertic (Table 8.11).

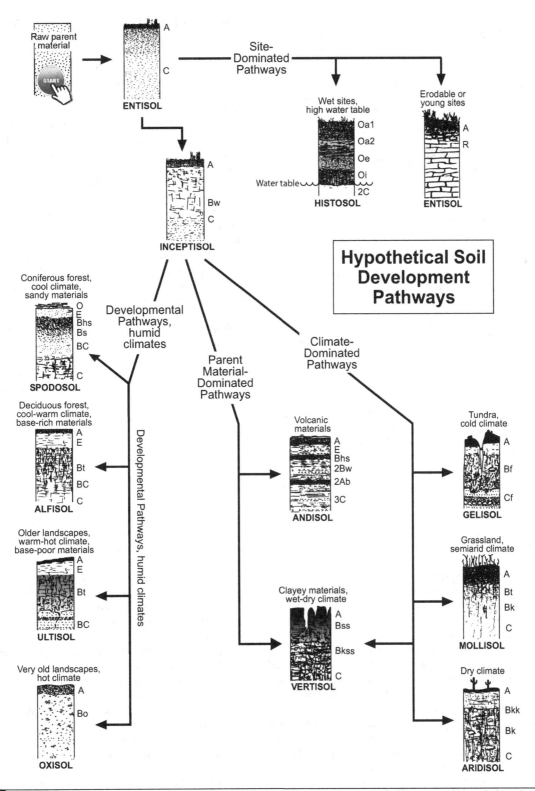

Fig. 8.3 A general scheme illustrating the major genetic linkages among the 12 soil orders in Soil Taxonomy. Although this scheme illustrates a few of the possible genetic pathways, soils could evolve along different pathways, regress, or remain within one order for millennia.

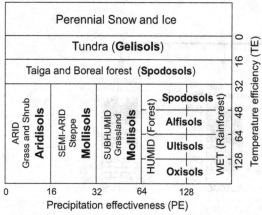

Thornthwaite (1931), Ruhe (1975a)

Fig. 8.4 Matrix showing how several major soil orders relate to each other, along two climatic axes. PE values correlate to moisture availability (>128 = superhumid, 64–128 = humid, 32–64 = subhumid, 16–32 = semiarid, and <16 = arid). TE values correlate to temperature (>128 = very hot, 64–128 = hot, 32–64 = warm, 16–32 = cool, and <16 = cold).

Preceding the subgroup name is that of the family, within which additional detail is provided. Family differentia include terms related to some of the following classes: (1) particle size, (2) mineralogy, (3) cation-exchange activity, (4) calcareous and reaction character, (5) soil temperature, (6) soil depth, (7) rupture resistance, (8) classes of coatings on sand grains, and (9) classes of permanent cracks. The names of most families have three to five of these descriptive terms that modify the subgroup name (Soil Survey Staff 1999). Table 8.11 provides a listing of a variety of family names, as examples of the diversity of soils; it also shows how the system works.

Soil Moisture Regimes

At the suborder level, one of the most common descriptors refers to the soil's long-term soil moisture or wetness status. The *soil moisture regime* is primarily a function of climate (water supply) and topography/relief as it affects the local water table (Table 8.12). Because of the importance of water in soil genesis, knowing the soil moisture regime of a soil is vital to understanding its genesis, use, and management. But the status of soil water is variable. Soils dry out and wet up on various timescales (daily, seasonally, and over long periods of climate change) and with depth (the upper horizons may be much drier than lower ones; a perched water table may exist; etc.), complicating the assignment of a soil moisture regime. Finally, soil moisture can vary tremendously across the landscape, even among pedons only a few meters apart.

In Soil Taxonomy, the moisture status of a soil is evaluated by the presence or absence of water held at a tension of less than 1,500 kPa (15 bars) in the soil. Water held at tensions >1,500 kPa is retained so tightly that most plants cannot remove it. When a soil horizon is *saturated*, the water within is retained at tensions near zero, or even at positive pressures. Consequently, a horizon is considered *dry* when the water within is held at tensions >1,500 kPa, and it is considered *moist* if the water within is held at tensions <1,500 kPa but less than zero. Soils with salty or saline water are considered salty rather than dry.

Soil moisture regimes are loosely related to macroclimate (Paetzold 1990). For example, permeable soils in humid climates have water that is available to plants most of the time. Conversely, one might think that soils in deserts are always dry. Not so. They may be dry, moist, or saturated, depending on their position on the landscape and the particular day of the year. Desert soils may be wet after rainfall events, and they may receive water from sources other than precipitation, such as groundwater. Thus, their wetness varies in time and space; the soil moisture regime concept is intended to reflect the long-term status of the soil's wetness condition. In the Northern Hemisphere, precipitation is more effective in soils on north aspects than in soils on south aspects (Franzmeier *et al.* 1969, Hunckler and Schaetzl 1997, King *et al.* 1999), possibly leading to different soil moisture regimes on opposing slopes, and again illustrating the variation in long-term soil wetness across space. Soils on steep slopes or slowly permeable materials may lose part or most of the precipitation that impinges upon them as runoff, making them drier than they would otherwise be. Thus, across vast areas, soils may have similar soil moisture regimes due to macroclimate, but on certain landscape positions soils may exist that occupy the drier or wetter regimes, on the basis of local influences.

Determining the moisture regime of a soil requires knowledge of inputs of water to it, throughputs of water vapor and liquid within it, and losses of water from it. We usually only infer soil moisture status over longer timescales, on the basis of climatic data for which we have good areal coverage and long temporal runs. Soil Taxonomy uses probabilities of long-term, mean annual precipitation to determine the climate of the "normal year" (Soil Survey Staff 1999) and bases its assessment of soil moisture regimes on that.

The moisture characteristics of a soil vary with depth, and throughout the year. A soil may be continuously moist in some or all horizons either throughout the year or for some part of the year. It may be either moist in winter and dry in summer or the reverse (Soil Survey Staff 1999). Thus, we beg the question, what *part* of the soil profile must be considered when classifying a soil's long-term moisture regime? Soil Taxonomy uses an estimate of the water status (moist or dry) in the *soil moisture control section*. The upper boundary of this section is the depth to which a dry soil will be wetted by 2.5 cm of water within 24 hours. Its lower boundary is the depth to which the same soil will be moistened by 7.5 cm of water within 48 hours. The intent

Table 8.8 | Derivations, mnemonics, and generalized meanings of the formative elements for suborders

Formative element	Derivation[a]	Mnemonic	Connotation/meaning[b]
Alb	L. *albus*, white	Albino	Has an albic (bright E) horizon
Anthr	Gr. *anthropos*, human	Anthropology	Has an anthropic epipedon, usually implying long-term human occupancy
Aqu	L. *aqua*, water	Aquarium	Has aquic conditions; wet or saturated for parts of the year as indicated by saturation and redoximorphic features
Ar	L. *arare*, to plow	Arable	Has been mixed or plowed
Arg	L. *argilla*, white clay	Argillite	Has an argillic (Bt) horizon, enriched in illuvial clay
Calc	L. *calcis*, lime	Calcite	Has a calcic (Bk) horizon, enriched in secondary carbonates
Camb	L. *cambiare*, to exchange	–	Has a cambic (Bw) horizon, slightly weathered and with pedogenic structure
Cry	Gr. *kryos*, icy cold	Crystal	Has a cryic soil temperature regime (<8°C)
Dur	L. *durus*, hard	Durable	Has a duripan (Bqm), dense and cemented with silica
Fibr	L. *fibra*, fiber	Fibrous	Dominated by organic materials that have been minimally altered by decomposition
Fluv	L. *fluvius*, river	Fluvial	Is on a floodplain or developed in alluvium
Fol	L. *folia*, leaf	Foliage	Has organic materials, not necessarily wet, with abundant foliage (leaves); often lying on bedrock
Gel	L. *gelare*, to freeze	Jell-o	Has a gelic soil temperature regime
Gyps	L. *gypsum*, gypsum	Gypsum	Has a gypsic (By) horizon, enriched in secondary gypsum
Hem	Gr. *hemi*, half	Hemisphere	Has organic materials dominated by those in a partially decomposed (hemic) state
Hist	Gr. *histos*, tissue	Histology	Dominant presence of organic soil materials
Hum	L. *humus*, earth	Humus	Has abundant organic matter, usually in the B horizon (Bh)
Orth	Gr. *orthos*, true	Orthographic	Is a common type of soil for this order
Per	L. *per*, throughout time		Has little seasonality with respect to temperature or moisture; often implies tropical climate
Psamm	Gr. *psammos*, sand	Psammite	Has sandy texture
Rend	Modified from Rendzina		Has abundant carbonates in soil and parent material
Sal	L. base of *sal*, salt	Saline	Has some type of a salic (salty) horizon (Bn, Bz), with excess soluble salts
Sapr	Gr. *saprose*, rotten	Saprophyte	Has organic materials dominated by those in a well-decomposed (sapric) state
Torr	L. *torridus*, hot and dry	Torrid	Has a torric (aridic) soil moisture regime; dry
Turb	L. *turbidis*, disturbed	Turbulent	Has been mixed by cryoturbation, usually soil ice
Ud	L. *udus*, humid	Udometer	Has a udic soil moisture regime (humid, moist)
Ust	L. *ustus*, burnt	Combustion	Has an ustic soil moisture regime (semiarid to subhumid)
Vitr	L. *vitrum*, glass	Vitreous	Has glass or volcanic ash within the profile
Wass	Ge. *wasser*, water	–	Is usually submerged by water
Xer	Gr. *xeros*, dry	Xerophyte	Has a xeric (Mediterranean) soil moisture regime; dry summers and moist winters

[a] Gr., Greek; L., Latin; Ge., German.
[b] Simplified from Soil Survey Staff (1999, 2010). The reader is referred to those sources for precise definitions.
Source: Soil Survey Staff (1975).

Table 8.9 | Derivations, mnemonics, and generalized meanings of the formative elements for great groups

Formative element	Derivation[a]	Mnemonic	Connotation
Acr	Gr. akros, at the end	Acrolith	Extremely weathered, low CEC values
Al	Modified from aluminum	Aluminum	Illuvial horizons dominated by aluminum
Alb	L. albus, white	Albino	Has an albic (bright E) horizon
Anhy	Gr. anydros, waterless	Anhydrous	Very dry
Anthr	Gr. anthropos, human	Anthropology	Has an anthropic epipedon, usually implying long-term human occupancy
Aqu	L. aqua, water	Aquarium	Aquic conditions; wet or saturated during parts of the year
Argi	L. argilla, white clay	Argillite	Has an argillic (Bt) horizon, enriched in illuvial clay
Calci, calc	L. calcis, lime	Calcium	Has a calcic (Bk) horizon, enriched in secondary $CaCO_3$
Cry	Gr. kryos, icy cold	Cryogenics	Exists in a cold, cryic soil temperature regime (<8° C)
Dur	L. durus, hard	Durable	Hard, cemented. Used with duripans (Bqm) and ortstein (Bsm) horizons
Dystr; Dys	Gr. dys, ill; dystrophic, infertile	Dystrophic	Low base cation saturation, acidic
Endo	Gr. endon, endo, within	Endoderm	Regional groundwater table present within the soil as a continuous zone of saturation at least 2 m thick, "endosaturation," as in Endoaquods
Epi	Gr. epi, on, above	Epiderm	Perched water table present within the soil, overlying unsaturated layers, "episaturation," as in Epiaquods
Eutr	Gr. eu, good; eutrophic, fertile	Eutrophic	High base cation saturation, high pH, alkaline, fertile
Ferr	L. ferrum, iron	Ferruginous	Abundant iron, usually implies "within B horizon"
Fibr	L. fibra, fiber	Fibrous	Organic soil materials dominated by raw or virtually raw (fibric) materials
Fluv	L. fluvius, river	Fluvial	Soils on floodplains or developed in stratified alluvium
Fol	L. folia, leaf	Foliage	Organic soil, not necessarily wet, with abundant leaves; often lying on bedrock
Fragi	L. fragilis, brittle	Fragile	Has a fragipan (Bx or Btx horizon), brittle and dense
Fragloss	Compound of frag- and gloss		Has fragipan (Bx or Btx horizon) and glossic (tonguing E or Ex) horizon
Frassi	Ger. frasi, fresh	Frass	Low electrical conductivity in subaqueous soils
Fulv	L. fulvus, dull brownish yellow	Fulvous	Dark brown or yellow color; presence of dissolved organic carbon; acidic
Gel	L. gelare, to freeze	Jell-o	Has a gelic soil temperature regime
Glac	L. glacialis, icy	Glacial	Presence of ice lenses or wedges
Gyps	L. gypsum, gypsum	Gypsum	Has a gypsic (By) horizon, enriched in secondary gypsum
Gloss	Gr. glossa, tongue	Glossary	Has a glossic horizon (tonguing of E into B horizon)
Hal	Gr. hals, salt	Halite	Saline or high levels of exchangeable Na
Hapl	Gr. haplous, simple	Haploid	Minimum horizonation, not a "complicated" profile
Hem	Gr. hemi, half	Hemisphere	Organic materials dominated by partially decomposed (hemic) materials

Table 8.9 | (cont.)

Formative element	Derivation[a]	Mnemonic	Connotation
Hist	Gr. histos, tissue	Histology	Presence of organic soil materials
Hum	L. humus, earth	Humus	Abundant organic matter, usually in the B horizon (Bh)
Hydr	Gr. hydor, water	Hydrolysis	Presence of water, wet
Kand, kan	Modified from kandite	Kandite	Dominated by 1:1 phyllosilicate clays
Luv	Gr. louo, to wash	Alluvium	Illuvial materials in B horizon
Melan	Gr. melas, black	Melanic	Black or dark in color; abundant organic carbon
Moll	L. mollis, soft	Mollify	Has mollic (dark, thick, base-cation–rich A horizon) epipedon
Natr	from natrium, sodium	Natron	Has natric (Btn) horizon, rich in sodium
Ombr	Gr. ombros, rain	Ombrology	Has surface wetness
Pale	Gr. palaeos, old	Paleosol	Extreme profile development, usually of B horizon; old soils
Petro	Gr. petra, rock	Petrology	Has a rocklike or a cemented horizon near the surface
Plac	Gr. plax, flat stone	Placoid	Has a thin ironpan (placic horizon)
Plagg	Ger. plaggen, sod		Has a plaggen epipedon, produced by long-time manuring
Plinth	Gr. plinthos, brick	Plinth	Has plinthite
Psamm	Gr. psammos, sand	Psammite	Sandy textures dominate
Quartzi	Ger. quarz, quartz	Quartz	Extremely high quartz content, and sandy
Rhod	Gr. base of rhodon, rose	Rhododendron	Dark red colors
Sal	L. base of sal, salt	Saline	Has salic (Bz) horizon, dominated by salts more soluble than gypsum
Sapr	Gr. saprose, rotten	Saprophyte	Organic soil materials dominated by well decomposed and rotten (sapric) materials
Somb	F. sombre, dark	Somber	Has a sombric (acidic, Bh) horizon
Sphagn	Gr. sphagnos, bog	Sphagnum	Dominated by fibers of Sphagnum moss
Sulf	L. sulfur, sulfur	Sulfur	Presence of sulfides or their oxidation products; may have a sulfuric horizon
Torr	L. torridus, hot and dry	Torrid	Torric (aridic) soil moisture regime; dry
Ud	L. udus, humid	Udometer	Udic soil moisture regime (humid, moist)
Umbr	L. umbra, shade	Umbrella	Has umbric (dark, thick, low base saturation) epipedon
Ust	L. ustus, burnt	Combustion	Ustic soil moisture regime (semi-arid to subhumid)
Verm	L. vermes, worm	Vermiform	Abundant worm activity
Vitr	L. vitrum, glass	Vitreous	Presence of glass or volcanic ash
Xer	Gr. xeros, dry	Xerophyte	Xeric (Mediterranean) soil moisture regime; dry summers and moist winters

[a] F., French; Ger., German; Gr., Greek; L., Latin.
Source: Simplified from Soil Survey Staff (1975, 1999). The reader is referred to that source for precise definitons.

Table 8.10 | Derivations, mnemonics, and generalized meanings of the formative elements for subgroups

Adjective	Derivation[a]	Connotation
Central concept		
Typic	Typical, normal	Central concept; not normally grading toward another type of profile or taxonomic group
Intergrades		
Acraquoxic		Combination of *Acr-*, *Aquic, and Oxic*; intergrade to an Acraquox
Acrudoxic		Combination of *Acr-*, *Udic, and Oxic*; intergrade to an Acruduox
Acrustoxic		Combination of *Acr-*, *Ustic, and Oxic*; intergrade to an Acrustox
Albaquultic		Combination of *Albic*, *Aquic, and Ultic*; intergrade to an Albaquult
Alfic	*Al* for aluminum, and *Fe* for iron	Intergrade to an Alfisol, often implying the presence of a Bt horizon
Andic		Intergrade to an Andisol, often implying that the soil has had additions of volcanic glass, ash, or pyroclastic materials
Aqualfic		Combination of *Aquic* and *Alfic*; intergrade to an Aqualf
Aquandic		Combination of *Aquic* and *Andic*; intergrade to an Aquand
Aquentic		Combination of *Aquic* and *Entic*; intergrade to an Aquent
Aquertic		Combination of *Aquic* and *Vertic*; intergrade to an Aquert
Aquic		Intergrade to a taxon with aquic conditions at a shallow depth; often implies that the soil is less than well drained (if no -aqu- in suborder position)
Aquicambidic		Combination of *Aquic, Cambic, and Aridic*; intergrade to an Aquicambid
Aquodic		Combination of *Aquic* and *Spodic*; intergrade to an Aquod
Aquollic		Combination of *Aquic* and *Mollic*; intergrade to an Aquoll
Aquultic		Combination of *Aquic* and *Ultic*; intergrade to an Aquult
Argidic		Combination of *Argic* and *Aridic*; intergrade to an Argid
Argiduridic		Combination of *Argic, Duric, and Aridic*; intergrade to an Argidurid
Aridic		Intergrade to an Aridisol, often implying that the soil moisture regime is borderline Aridic/Torric
Calciargidic		Combination of *Calcic, Argic, and Aridic*; intergrade to a Calciargid
Calcidic		Combination of *Calcic* and *Aridic*; intergrade to a Calcid
Cambidic		Combination of *Cambic* and *Aridic*; intergrade to a Cambid
Duridic		Combination of *Duric* and *Aridic*; intergrade to a Durid
Entic		Intergrade to an Entisol, often implying that some part of the soil is more weakly developed than is typical for the great group
Fluvaquentic		Combination of *Fluv-*, *Aquic, and Entic*; intergrade to a Fluvaquent
Fluventic		Combination of *Fluv-* and *Entic*; intergrade to a Fluvent
Folistic		Connotes a folistic epipedon in mineral soils that intergrade to Histosols (Folists)
Haplargidic		Combination of *Haplic, Argic, and Aridic*; intergrade to a Haplargid
Haplocalcidic		Combination of *Haplic, Calcidic, and Aridic*; intergrade to a Haplocalcid
Haploduridic		Combination of *Haplic, Duridic, and Aridic*; intergrade to a Haplodurid
Haploxeralfic		Combination of *Haplic, Xeric, and Alfic*; intergrade to a Haploxeralf
Haploxerollic		Combination of *Haplic, Xeric, and Mollic*; intergrade to a Haploxeroll

Table 8.10 | (cont.)

Adjective	Derivation[a]	Connotation
Histic		Connotes a histic epipedon in mineral soils that intergrade to Histosols; can also imply that the soil has an O horizon thinner than is required for a histic epipedon
Humaqueptic		Combination of *Humic, Aquic, and Inceptic*; intergrade to a Humaquept
Hydraquentic		Combination of *Hydric, Aquic, and Entic*; intergrade to a Hydraquent
Inceptic		Weakly developed; intergrade to an Inceptisol
Kandiudalfic		Combination of *Kandic, Udic, and Alfic*; intergrade to a Kandiudalf
Kandiustalfic		Combination of *Kandic, Ustic, and Alfic*; intergrade to a Kandiustalf
Mollic		Intergrade to a Mollisol, often implying that A horizon is not quite dark and/or thick enough to be a mollic epipedon
Natrargidic		Combination of *Natric, Argic, and Aridic*; intergrade to a Natrargid
Natrixeralfic		Combination of *Natric, Xeric, and Alfic*; intergrade to a Natrixeralf
Oxic		Intergrade to an Oxisol, often implying a low CEC or a dominantly oxide clay mineralogy
Paleargidic		Combination of *Pale-, Argic, and Aridic*; intergrade to a Paleargid
Palexerollic		Combination of *Pale-, Xeric, and Mollic*; intergrade to a Palexeroll
Petrocalcidic		Combination of *Petro-, Calcic, and Aridic*; intergrade to a Petrocalcid
Plagganthreptic		Combination of *Plaggen, Anthropic, and Inceptic*; intergrade to a Plagganthrept
Psammentic		Combination of *Psamm- and Entic*; intergrade to a Psamment
Salidic		Combination of *Salic and Aridic*; intergrade to a Salid
Spodic		Intergrade to a Spodosol, often implying that the soil has a spodiclike B horizon but fails to meet one or more criteria for a spodic horizon
Sulfaqueptic		Combination of *Sulfic, Aquic, and Inceptic*; intergrade to a Sulfaquept
Torrertic		Combination of *Torric and Vertic*; intergrade to a Torrert
Torrifluventic		Combination of *Torric, Fluv- and Entic*; intergrade to a Torrifluvent
Torriorthentic		Combination of *Torric, Orthic, and Entic*; intergrade to a Torriorthent
Torripsammentic		Combination of *Torric, Psamm-, and Entic*; intergrade to a Torripsamment
Torroxic		Combination of *Torric and Oxic*; intergrade to a Torrox
Turbic	L. *turbidis*, disturbed	Connotes cryoturbation in soils that intergrade to Gelisols (Turbels)
Udandic		Combination of *Udic and Andic*; intergrade to a Udand
Udertic		Combination of *Udic and Vertic*; intergrade to a Udert
Udic		Intergrade to a Udic soil moisture regime
Udifluventic		Combination of *Udic, Fluv-, and Entic*; intergrade to a Udifluvent
Udollic		Combination of *Udic and Mollic*; intergrade to a Udoll
Udorthentic		Combination of *Udic, Orth-, and Entic*; intergrade to a Udorthent
Udoxic		Combination of *Udic and Oxic*; intergrade to a Udox
Ultic		Intergrade to an Ultisol, often implying that the soil is neither weathered enough nor acidic enough to be an Ultisol
Ustalfic		Combination of *Ustic and Alfic*; intergrade to a Ustalf
Ustandic		Combination of *Ustic and Andic*; intergrade to an Ustand

(cont.)

Table 8.10 | *(cont.)*

Adjective	Derivation[a]	Connotation
Ustertic		Combination of *Ustic* and *Vertic*; intergrade to an Ustert
Ustic		Intergrade to an Ustic soil moisture regime
Ustifluventic		Combination of *Ustic*, *Fluv-*, and *Entic*; intergrade to an Ustifluvent
Ustollic		Combination of *Ustic* and *Mollic*; intergrade to an Ustoll
Ustoxic		Combination of *Ustic* and *Oxic*; intergrade to an Ustox
Vertic		Intergrade to a Vertisol, often implying that the soil cracks in a dry season, but not deep enough or for long enough periods to classify as a Vertisol
Vitritorrandic		Combination of *Vitric*, *Torric*, and *Andic*; intergrade to a Vitritorrand
Vitrixerandic		Combination of *Vitric*, *Xeric*, and *Andic*; intergrade to a Vitrixerand
Xeralfic		Combination of *Xeric* and *Alfic*; intergrade to a Xeralf
Xereptic		Combination of *Xeric* and *Inceptic*; intergrade to a Xerept
Xerertic		Combination of *Xeric* and *Vertic*; intergrade to a Xerert
Xeric		Intergrade to a Xeric (Mediterranean) soil moisture regime
Xerofluventic		Combination of *Xeric*, *Fluv-*, and *Entic*; intergrade to a Xerofluvent
Xerollic		Combination of *Xeric* and *Mollic*; intergrade to a Xeroll
Extragrades		
Abruptic	L. *abruptum*, torn off	Abrupt textural change, usually from A (or E) to B horizon
Aeric	Gr. *aerios*, air	Aeration; slightly drier, better drained, than is normal for the great group
Albic		Has an albic horizon
Alic	Aluminum	High amounts of Al^{3+}
Anhydritic	Gr. *Anudros*, anhydrous	Has an anhydritic horizon
Anionic	Gr. *anion*, a thing going up	Negatively charged colloids
Anthraltic		Presence of human-altered material
Anthraquic	Gr. *anthropos*, human and L. *alterare*, to change	Controlled flooding of these soils
Anthrodensic	Gr. *anthropos*, human and L. *aqua*, water	Presence of mechanically compacted materials
	Gr. *anthropos*, human and L. *densus*, compact	
Anthropic	Gr. *anthropos*, human	Has an anthropic epipedon
Anthroportic	Gr. *anthropos*, human and L. *portare*, to carry	Presence of human-transported materials
Arenic	L. *arena*, sand	Has a thick, sandy epipedon
Argic		Has an argillic horizon
Calcic		Has a calcic horizon
Chromic	Gr. *chroma*, color	High chroma (bright-colored, not gray); usually imples "low in organic matter"
Cumulic	L. *cumulus*, heap	Overthickened, dark epipedon
Duric		Has a duripan or a mostly cemented, duripan-like layer
Durinodic	L. *durabilis*, lasting/enduring, and L. *nodus*, knot	Has durinodes or is brittle

Table 8.10 | *(cont.)*

Adjective	Derivation[a]	Connotation
Dystric		Low base cation saturation, acidic
Epiaquic		Combination of *Epi-* and *Aquic*; surface wetness, usually implying a perched water table within the profile
Eutric		High base status or pH
Fibric		Contains organic materials dominated by raw or virtually raw (fibric) materials
Fragic		Has a fragipan or a fragiclike horizon
Glacic		Presence of ice lenses or wedges
Glossic		Has a glossic horizon, tonguing horizon boundaries
Grossic	L. *grossus*, thick	Thick layers of potentially fluid (low *n* value) mineral soil material
Grossarenic	L. *grossus*, thick, and L. *arena*, sand	Has a thick, sandy epipedon
Gypsic		Has a gypsic horizon
Halic		Has high amounts of soluble salts
Haplic	Gr. *haplos*, simple, and	Minimum horizonation, not a "complicated" profile
Haploplaggic	Ger. *plaggen*, sod	Thin, plaggenlike surface horizon
Hemic		Contains organic materials dominated by partially decomposed (hemic) materials
Humic		Has abundant organic matter; may also connote melanic, mollic or umbric epipedons
Hydric		Presence of water, often implying a layer of water in inorganic clays and humus, or simply a layer of water within the profile
Kandic		Dominated by 1:1 phyllosilicate clays; probably has a Bt horizon that nearly qualifies as a kandic horizon
Kanhaplic		Combination of *Kandic* and *Haplic*
Lamellic	L. *lamina*, thin plate	Argillic horizon that consists of lamellae (clay bands)
Leptic	Gr. *leptos*, thin	Has thin profile
Limnic	Modified from Gr. *limne*, lake	Presence of limnic materials (lake sediments) at depth
Lithic	Gr. *lithos*, stone	Shallow to hard bedrock (lithic contract)
Natric		Has a natric horizon
Nitric	Gr. *nitron*, soda	High in nitrates
Ombroaquic	Gr. *ombros*, rain, and L. *aqua*, water	Water on the soil surface; sometimes saturated in the upper solum
Oxyaquic	Modified from *oxygen*, and L. *aqua*, water	Oxygenated water occupies saturated soil horizons for at least some part of the year
Pachic	Gr. *pachys*, thick	Has a thick epipedon
Paralithic	Gr. *para*, beside, and *lithic* (stone)	Shallow to soft and/or weathered bedrock (paralithic contact)
Petrocalcic		Has a petrocalcic horizon
Petroferric	Gr. *petra*, rock, and L. *ferrum*, iron	Shallow to ironstone (petroferric contact)
Petrogypsic		Has a petrogypsic horizon

(cont.)

Table 8.10 | (*cont.*)

Adjective	Derivation[a]	Connotation
Petronodic	Gr. *petra*, rock, and L. *nodulus*, little knot	Has nodules or concretions
Placic		Has a placic horizon
Plinthic		Has plinthite
Rhodic		Dark red color
Ruptic	L. *ruptum*, broken	Intermittent or broken horizons, often over bedrock
Salic		Has a salic horizon
Sapric		Contains organic materials dominated by highly (sapric) materials
Sodic	Modified from sodium	Abundant sodium in the profile
Sombric		Has a sombric horizon
Sphagnic		Dominated by fibers of *Sphagnum* moss
Sulfic or Sulfuric		Presence of sulfides or their oxidation products; may have a sulfuric horizon
Terric	L. *terra*, earth	Organic soils with a mineral layer within 1.5 m of the surface
Thapto, thaptic	Gr. *thapto*, buried	Profile contains a buried soil or horizon
Thapto-Histic	Gr. *thapto*, buried, and *histos*, tissue	Combination of *Thapto* and *Histic*; profile contains a buried organic soil or histic epipedon
Umbric		Has an umbric epipedon
Vermic		Worms very abundant, or mixed by animals
Vitrandic		Has volcanic ash, glass, cinders, pumice, and pumicelike fragments
Vitric		Has some volcanic ash and/or glass influence
Xanthic	Gr. *xanthos*, yellow	Yellow colors
Combinations of extragrades and intergrades		
Albaquic		Combination of *Albic* and *Aquic*
Anthraquic		Combination of *Anthropic* and *Aquic*
Argiaquic		Combination of *Argic* and *Aquic*
Fragiaquic		Combination of *Fragic* and *Aquic*
Glossaquic		Combination of *Glossic* and *Aquic*
Plinthaquic		Combination of *Plinthic* and *Aquic*

[a] Derivation is included here only if not previously provided in an earlier table.
Source: Soil Survey Staff (1999). The reader is referred to those sources for precise definitions.

of these seemingly arbitrary limits is that the soil moisture control section roughly corresponds to the rooting depths for many crops. To put some numbers on these concepts, one can assume that the soil moisture control section of a finer-textured soil is between 10 and 30 cm depth. In sandy soils, the section might be 30 to 90 cm deep (Soil Survey Staff 1999).

Most soil moisture regimes correspond approximately to major climate regions, such as arid and semiarid (aridic), Mediterranean (xeric), subhumid (ustic), and humid (udic) climates (Table 8.12, Fig. 8.5). In contrast, the wettest of the soil moisture regimes, named *aquic*, is not tied to macroclimate and generally occurs in soils with a periodically reducing (anoxic) regime, as a result of periodic saturation (Soil Survey Staff 1999). It is usually associated with soils in low, wet spots on the landscape, where the water table is high. Only in vast swamps or wetlands would an aquic soil moisture regime extend across large tracts of landscape. Some of these areas are shown in Figure 8.5, although many small upland areas within what is mapped as aquic contain soils in drier soil moisture regimes. Thus, many landscapes have two types of soil moisture regimes: Upland soils are

Table 8.11	Examples of soil families within the system of Soil Taxonomy, listing some characteristics discernible from the name

Family names	Generalized description of the main characteristics of that family
Central concept (Typic) subgroups	
Fine, mixed, semiactive, acid, isomesic Typic Placaquepts	Inceptisol with placic horizon and an ochric epipedon (A horizon), located in a mesic soil temperature regime with little seasonal variation(iso-); the high water table results in aquic (aqu-) conditions and an aquic soil moisture regime at some times; it has acid pH values, has mixed mineralogy, and is fine textured
Clayey, parasequic, isohyperthermic Typic Rhodudults	Very red (Rhod-), clay-rich, and weathered soil (Ultisol) in a udic soil moisture regime; Ultiols have a B horizon enriched in clay, in this case probably 1:1 clays and some oxide clays, e.g., goethite and hematite (parasequic); its mean annual soil temperature is >22°C (hyperthermic), with little seasonality (iso-)
Fine-silty, mixed, active, mesic Typic Fragiaqualfs	Alfisol with an argillic horizon and a fragipan, probably a Btx horizon; the high water table gives the soil aquic conditions close to the surface; soil temperature regime is mesic, mineralogy is mixed, and textures are silt loam and silty clay loam (fine-silty), the latter characteristic indicating that the likely parent material is loess
Coarse-loamy, mixed, superactive, subgelic Typic Aquiturbels	Cold soil (Gelisol) with cryoturbated horizons; water perches on the permafrost below, leading to aquic conditions; has mixed mineralogy and moderately coarse textures (coarse-loamy); high organic matter content gives it a high (superactive) CEC/clay ratio
Intergrade subgroups	
Loamy-skeletal, parasequic, mesic Ultic Palexeralfs	Alfisol with a thin (ochric) A horizon and a well-developed (Pale-) argillic (Bt) horizon; soil is an intergrade to an Ultisol, probably because of its clay mineralogy and low base status due to strong leaching; its mineralogy is dominated by various iron oxide and aluminum hydroxide minerals (parasequic); it is in a xeric (Mediterranean) soil moisture regime and has a mesic soil temperature regime; soil has loamy textures and many rock fragments (skeletal)
Fine, smectitic, frigid Argiaquics Argialboll	Mollisol with a dark, thick A horizon (mollic epipedon), but also with E (albic) and Bt (argillic) diagnostic horizons; soil is an intergrade to Argiaquolls and is drier than typical, poorly-drained Albolls, but also lacks an abrupt textural change and has higher hydraulic conductivity in the argillic horizon; soil does, however, have aquic conditions (aqu-) like other Albolls; soil has a frigid soil temperature regime, has smectitic clay mineralogy, and is fine textured
Coarse-loamy, mixed, superactive, thermic Ustic Haplocambids	Aridisol (in an aridic soil moisture regime) with a cambic (Bw, or weak Bk) horizon and an ochric A horizon; soil is an intergrade to an Ustorthent, which is a weakly developed (Entisol) soil in an ustic soil moisture regime; thus, soil moisture regime of this soil is aridic but borders on ustic; the soil, in a thermic soil temperature regime, has mixed mineralogy and moderately coarse textures (coarse-loamy)
Very-fine, smectitic, calcareous, mesic Vertic Ustifluvents	Weakly developed soil (Entisol) that has a tendency to develop cracks in the dry season (hence, the Vertic intergrade), but cracks do not stay open long or often enough to qualify as a Vertisol; soil has formed in clayey alluvium that has a high content of montmorillonite and similar smectite-family clay minerals; ustic soil moisture regime tends to promote cracking, allowing the soil to dry during the warmest times of the year; soil has a mesic soil temperature regime; minimal leaching in this dry climate has led to carbonates within the profile
Euic Fluvaquentic Cryohemists	Organic soil (Histosol) in which most of the organic material is not highly acidic (euic) and only partially decomposed (hemic); soil is in a cryic soil temperature regime; it intergrades with Fluvaquents, which are wet mineral soils that form on floodplains; thus, it has thin strata of mineral soil material that were deposited as layers of alluvium between layers of organic soil materials

(cont.)

Table 8.11	(cont.)	
Family names	**Generalized description of the main characteristics of that family**

Extragrade subgroups

Loamy-skeletal, magnesic, mesic Lithic Haploxerolls	Mollisol with a dark, thick A horizon with a xeric (Mediterranean) soil moisture regime and a mesic soil temperature regime; the profile has simple horizonation (hapl-) and is shallow to serpentine bedrock (lithic contact), hence the magnesic mineralogy, connoting the presence of magnesium-silicate minerals; loamy textures of the soil include many rock fragments (skeletal)
Medial-skeletal over fragmental or cindery, amorphic over mixed, isomesic Humic Ustivitrands	Andisol formed in slightly weathered volcanic materials (tephra), which have a low to medium content of volcanic glass (vitr-); it is in an ustic soil moisture regime and has a dark, thick A horizon (humic) that is a melanic, mollic, or umbric epipedon; soil has formed in at least two distinct parent materials: the upper one with weathered ash that has a moderarely low water-holding capacity (medial) mixed with many rock fragments (skeletal), and a dominance of short-range-order clay minerals such as allophone and imogolite; the lower one mostly basaltic cinders (cindery) of mixed mineralogy
Sandy, siliceous, thermic Grossarenic Haplohumods	Spodosol with a carbon-rich Bh horizon (hum-), an ochric A horizon, and an albic E horizon; overthickened (gross-), sandy (arenic-) textured epipedon is dominated by quartz (siliceous); is in a thermic (subtropical) soil temperature regime
Fine-loamy, siliceous, semiactive, thermic Plinthaquic Paleudults	Weathered, upland Ultisol in a humid (udic) climate; soil is strongly developed (pale-) with a thick profile and probably a strong Bt horizon and plinthite (secondary iron compounds); soil is slightly wetter than typical Paleudults; it is gradational to an aquic soil moisture regime but still is within udic; although the soil material is loamy in texture, the sands are mainly silica minerals (quartz), which are resistant to weathering in a humid climate
Fine, mixed, superactive, mesic Chromic Gypsitorrerts	Vertisol that develops wide, deep cracks during a long dry season; moisture regime of this soil (torr-) alludes to a very short wet season, during which the cracks usually close; soil has a brightly colored ochric epipedon (chromic) and a By horizon with accumulated gypsum that occurs below the zone of maximum shrinking and swelling; it has a mesic soil temperature regime; although fine textured, with a high linear extensibility, it has mixed mineralogy; high organic matter content gives it a high (superactive) CEC/clay ratio
Fine, ferritic, isothermic Xanthic Acroperox	Yellow-colored (xanthic-) Oxisol (highly weathered soil dominated by oxides and 1:1 clays) in a continuously moist climate (perudic soil moisture regime);s soil is very weathered and leached, even for Oxisols, leading to a low effective CEC value (acr-); its isothermic soil temperature regime implies a continuously warm, tropical climate; it is fine textured (clay and silty clay textures) and has high iron contents (ferritic)

in the regime that corresponds to the macroclimate, e.g., ustic, udic, xeric, while aquic soils occur in lowlands (or on flat uplands) where high (or perched) water tables impact the profile. The aquic soil moisture regime is temporal and may exist for only a few days within a year or 12-month period. In these circumstances, the soil has another moisture regime for the balance of the time. Some soils with an aquic moisture regime also have a xeric, ustic, or aridic (torric) regime (Soil Survey Staff 1999). The very wettest soil moisture regime is named *peraquic*, connoting permanent wetness where groundwater is always at or very close to the soil surface. Examples of soils with a peraquic soil moisture regime are subaqueous soils, soils in tidal marshes, and soils in closed, landlocked depressions fed by perennial streams.

Soils that have an *aridic (torric)* moisture regime normally occur in arid climates, although a few occur in semi-arid climates where soil properties inhibit infiltration or facilitate runoff. Soils may be moistened after storms, but subsequently, much of this water moves upward along matric tension gradients, as unsaturated flow and evaporates. The result is a soil with a B horizon rich in

Table 8.12 | Classes and characteristics of soil moisture regimes

Soil moisture regime	Simplified definition and limits
Peraquic	Reducing regime for soils that are free of dissolved oxygen and permanently saturated, because groundwater is almost always at, or very close to, the soil surface
Aquic	Temporary reducing regime in periodically saturated soils, due to either a perched water table (episaturation) or a regional groundwater table (endosaturation); when saturated, soil is virtually free of dissolved oxygen
Aridic and torric	Typical of upland soils in dry climates; in warm season of normal years soil-moisture control section is dry more than half of the time or is moist for <90 consecutive days
Ustic	Typical of semiarid grasslands and savannas; in normal years, soil-moisture control section is dry for ≥90 cumulative days, but moist for >180 cumulative days (or >90 consecutive days)
Xeric	Typical of Mediterranean climates; in normal years, soil-moisture control section is dry for ≥45 consecutive days in the summer and fall, and moist for ≥45 consecutive days in winter
Udic	Typical of humid climates where precipitation exceeds evapotranspiration; in normal years, soil-moisture control section is not dry for 90 cumulative days; extended dryness (>45 consecutive days) in summer is not allowed
Perudic	Typical of extremely wet climates where leaching occurs, i.e., precipitation exceeds evapotranspiration, in every month

Source: Simplified from Soil Survey Staff (2010). The reader is referred to that source for precise definitions.

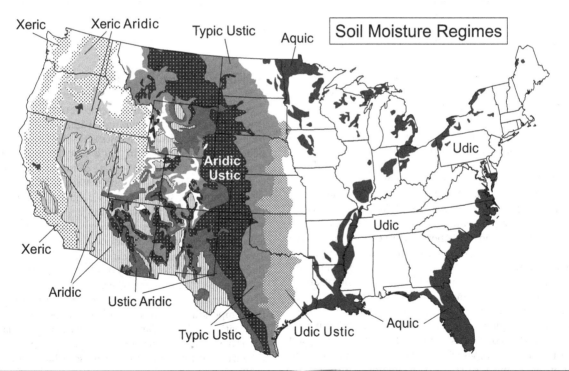

Fig. 8.5 Soil moisture regimes of the continental United States. Generalized and redrawn from various maps created by the NRCS National Soil Survey Center, Lincoln, Nebraska. Areas of complex soil moisture pattern, in the mountains of the western United States, are shown as white. Although areas designated with the aquic soil moisture regime are dominated by aquic soils, on uplands in these areas, soils with udic soil moisture regimes are also very common.

Table 8.13 | Soil temperature regimes and classes, and simplified definitions

Soil temperature regimes (STRs) and classes	Simplified definitions[a] and limits
Gelic (STR only)	MAST[a] ≤ 0°C (in Gelic suborders and Gelic great groups) or ≤ 1°C (in Gelisols)[b]
Hypergelic	MAST ≤ – 10°C; Gelisols, Gelic suborders, and Gelic great groups only
Pergelic	MAST between –4°C and –10°C; Gelisols, Gelic suborders, and Gelic great groups only
Subgelic	MAST between 1°C and –4°C; Gelisols, Gelic suborders, and Gelic great groups only
Cryic (STR only)	MAST between 0°C and 8°C, but the soil does not have permafrost[b]
Frigid	MAST between 0°C and 8°C, but warmer in summer than Cryic soils; summer minus winter soil temperature >6 C: i.e., seasonality
Mesic	MAST between 8°C and 15°C; summer minus winter soil temperature >6°C: i.e., seasonality
Thermic	MAST between 15°C and 22°C; summer minus winter soil temperature >6°C: i.e., seasonality
Hyperthermic	MAST >22°C; summer minus winter soil temperature >6°C: i.e., seasonality
Isofrigid	MAST <8°C; summer minus winter soil temperature <6°C: i.e., little seasonality
Isomesic	MAST between 8°C and 15°C; summer minus winter soil temperature <6°C: i.e., little seasonality
Isothermic	MAST between 15°C and 22°C; summer minus winter soil temperature <6°C: i.e., little seasonality
Isohyperthermic	MAST >22°C; summer minus winter soil temperature <6°C: i.e., little seasonality

[a] MAST, mean annual soil temperature.
[b] By definition, soils with permafrost within 100 cm are within the Gelisol order (Table 8.5).
Source: Simplified from Soil Survey Staff (1999). The reader is referred to that source for precise definitons.

chemical precipitates, such as $CaCO_3$ or soluble salts (see Chapter 13). Soil moisture regimes of semiarid climates, in which the concept is one of soils that are usually dry but at some times are moistened for days or weeks at a time, are split into two groups. Those with a wet season that corresponds to the warm season are considered *ustic*; those where wintertime rainfall is the rule are *xeric* (Engel *et al.* 1997). In ustic soils, the soils are wettest when conditions are suitable for plant growth, as in grasslands and savannas (Soil Survey Staff 1999). Some leaching occurs here – more than in aridic regions – although subsoil accumulations of secondary carbonates are commonplace (see Chapter 13). In xeric soil moisture regimes, precipitation is concentrated in the cool season when many plants are dormant, and thus it is more effective at translocation. To qualify as xeric, a soil must be wet for a period in winter and dry in summer (Table 8.12). Most humid climates have adequate precipitation to wet and thoroughly leach soils annually. This concept is embodied within the *udic* soil moisture regime, in which water moves entirely through the profile in most years. Thus, accumulations of secondary carbonates and salts are generally, though not always, e.g., Schaetzl *et al.* (1996), absent. In climates where the soils are almost constantly moist, and deep leaching is more or less continuous, the moisture regime *perudic* is used (Table 8.12).

Soil Temperature Regimes

At the family (and sometimes higher) level, soils are split out according to their long-term temperature characteristics (Paetzold 1990). Soil temperature is important; it regulates biological activity in and on the soil, as well as geologic and pedologic processes within and below the soil. Moreover, it can dramatically affect the state of soil water (Schmidlin *et al.* 1983, Jensen 1984, Alexander 1991, Isard and Schaetzl 1995, Mount 1998). Frozen soils behave quite differently than do soils where the water is in liquid form (Cary *et al.* 1978). Soil temperatures are variable among horizons, and among various parts of the landscape (Carter and Ciolkosz 1980, Nullet *et al.* 1990). Near-surface temperatures often fluctuate daily and seasonally (Beckel 1957). Daily soil temperature fluctuations may be very large, especially in soils of dry climates; at the other extreme, under melting snow, the temperature at the soil surface may be nearly isothermal with depth. Temperatures of wet soils fluctuate less than they do in dry soils, because the specific heat of water is nearly four times that of dry soil (see Chapter 6).

Soil Taxonomy recognizes ten *soil temperature regimes* used in the higher categories and eleven *soil temperature classes* used in the family categories (Table 8.13). The *gelic* soil temperature regime is represented at the family level with the hypergelic, pergelic, and subgelic temperature classes. The *cryic* soil temperature regime is used to name

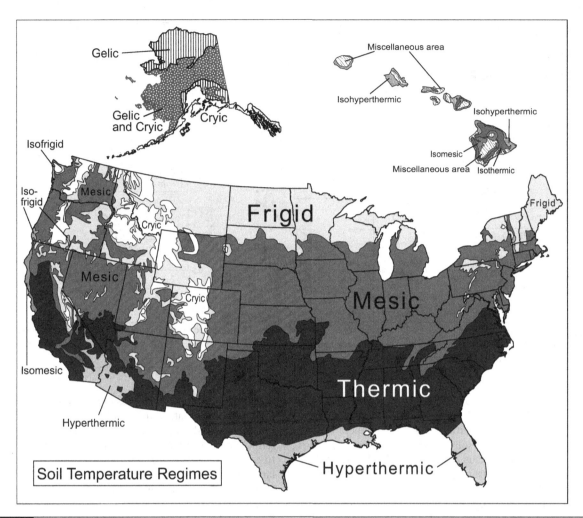

Fig. 8.6 Soil temperature regimes of the continental United States. Generalized and redrawn from various maps created by the NRCS National Soil Survey Center, Lincoln, Nebraska.

taxa at the suborder and great group categories, but it is not used in the family category because it is encompassed by the temperature ranges of the frigid and isofrigid temperature classes, and its use there would be redundant. Many of these occur in the United States (Fig. 8.6). These regimes are usually only estimated from long-term atmospheric data from nearby sites (Bocock *et al.* 1977, Reimer and Shaykewich 1980). This approach is taken because real soil temperature data that can be gathered today are reflective only of contemporary conditions. Hardly ever do long-term temperature data exist from soils, requiring that we determine long-term soil temperatures on the basis of statistical relationships between the soil and the atmosphere (Isard and Schaetzl 1995). Soil temperature regimes are difficult to map because, like atmospheric climate, soil climate changes on short and long timescales (Isard and Schaetzl 1995, Mokma and Sprecher 1995). What was mesic in one year may have been frigid in five of the previous ten years.

What, exactly, are the relationships between air and soil temperatures? Long-term mean soil temperatures change little below a certain depth, perhaps about 9 m (Soil Survey Staff 1999). According to the Soil Survey Staff (1999), a single temperature reading at 6 m is assumed to be within 1°C of the mean annual air temperature (MAAT), regardless of when it is taken. And at 10 m, a reading is usually within 0.1°C of the MAAT, although with increasing depth, soil temperatures will gradually increase as geothermal sources of energy are approached. Mean annual soil temperature (MAST) is usually simply estimated to be MAAT plus 1°C or 2°C. This relationship breaks down in areas that have thick snowpacks, as snow is an effective insulator from cold air temperatures (Hart and Lull 1963, Ping 1987, Sharratt *et al.* 1992, Isard and Schaetzl 1995). MAST is also affected by the amount and distribution of precipitation, the protection provided by shade and by O horizons in forests, slope aspect and gradient, and irrigation (MacKinney 1929, Smith *et al.* 1964, Soil Survey Staff 1999). Factors such as

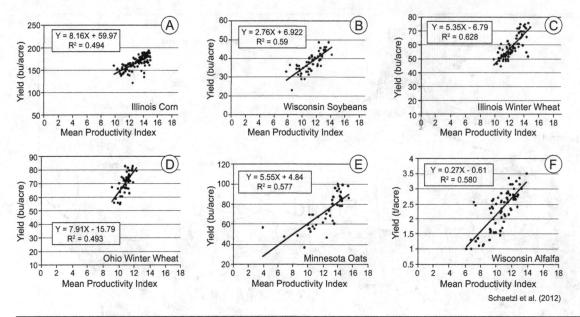

Schaetzl et al. (2012)

Fig. 8.7 Scatterplots showing the relationships between countywide yields for selected crops and midwestern U.S. states, and mean PI values for all parcels in each county that were planted to that crop in 2009. The PI works best in areas where crops are not irrigated, because irrigation alters the natural (inherent) productivity of soils, which is what the PI attempts to mimic.

soil color, texture, and content of organic matter, however, have negligible effects on MAST (Smith *et al.* 1964). In the tropics, where seasonal differences in air temperature are negligible, variation in soil temperature corresponds with elevation and wet versus dry seasons. If MAST cannot be accurately estimated from MAAT, it can be estimated by the average of soil temperature values measured once every three months, over a one-year period, at 50 cm depth. Daily temperature fluctuations are dampened out below that depth. At 50 cm or deeper, soil temperatures change more in response to short-term weather patterns and seasonal trends than they do to diurnal patterns (Soil Survey Staff 1999). If soils are shallower than 50 cm to a root-limiting layer, such as a paralithic contact, the depth for measuring soil temperature parallels that layer.

An Overview and Assessment of Soil Taxonomy

Sometimes, when classifying a soil, one may become bogged down in the details. At first, classifying a soil, i.e., keying out the description and measured characteristics of a pedon to arrive at a taxonomic class, may be cumbersome and difficult. However, using the class names, i.e., going from the name to a pedon's concept in one's mind, given just a little practice, can be simple and straightforward (Table 8.11).

Like most other soil classification systems worldwide, Soil Taxonomy is inherently logical. Its logic and rigorously defined taxonomic groupings also make it potentially useful for any number of possible applications involving taxonomic information and how it is distributed across soil landscapes. In short, the taxonomic information in the system, when *symbolized, named,* and *mapped,* should be able to convey a great deal of information about soils and soil landscapes. Indeed, there is great value in a name – a taxonomic name, in this case. A name can convey a great deal of information about the soil (Schimel and Chadwick 2013). Knowing this, Schaetzl *et al.* (2009) developed a natural soil drainage index (DI), which estimates long-term soil wetness, using only subgroup classification information, and, if desired, map unit slope (see also Schaetzl 1986a). The index is designed to represent, along an ordinal range from 0 to 99, the amount of water that a soil contains and makes available to plants under normal climatic conditions. Soils with DIs of 99 are, essentially, open water, whereas those with DIs of 1 are so thin and dry as to almost be bare bedrock. Later, Schaetzl *et al.* (2012) developed a similar index for soil productivity – the Productivity Index. Ranging from 0 to 19, the PI correlates well to land use categories and to yields of various crops (Fig. 8.7). Taken together, these indices illustrate the validity of Soil Taxonomy's classes, and they highlight just a few applications of these data to soil landscapes and ecological modeling (Kowal *et al.* 2013). More information on the DI and PI are provided at http://foresthealth.fs.usda.gov/soils.

Soil Taxonomy restricts the number of descriptive adjectives that can be used for extragrades (Table 8.10). Generally, only one or two adjectives are allowed, even when more may be applicable. If that limit were to be loosened, more

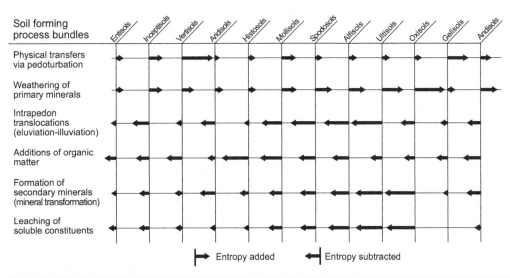

Fig. 8.8 Degree of expression of some of the major process bundles (see Chapter 13) in the 12 soil orders of Soil Taxonomy. The direction and length of the arrows suggest the extent to which these processes increase (arrow left) or decrease (arrow right) the organization of soil horizons. Inspired by Smeck et al. (1983).

extragrades could be added, and thus, more details of the soils being classified could appear in the subgroup name.

Some have argued that Soil Taxonomy relies too much on laboratory data that can take months to acquire, leaving a pedon's classification in limbo. However, this problem is inherent in any highly quantitative, property-based classification system. Others, e.g., Birkeland (1999), noted that Soil Taxonomy may focus too much on surface horizons and not enough on subsoil development. The latter often provides more information about the age of the surface the soil may have formed on, rendering it more useful from a geomorphic standpoint. Soil quality researchers, on the other hand, may believe that not enough emphasis is placed on surface horizons, which are used for food production and are also the first to be eroded or polluted. Campbell and Edmonds (1984) contended that Soil Taxonomy should have a stronger geographic focus. That is, many soil taxa, as they exist on the Earth's surface, have boundaries that do not coincide with natural features, making them difficult to map. They argued that, because Soil Taxonomy is designed to assist in making soil surveys, it should, by its very nature, consider geographic features in classification units. Also, the abstract, taxonomic entities defined by Soil Taxonomy do not always coincide with genetic soil units. Along these same lines, many have stated that Soil Taxonomy is not genetic enough – that it relies on observable characteristics more than interpretation of processes. Processes that have left little measurable imprint on the soil are not often considered in the classification. This is true. But as Guy Smith (1983: 43) noted:

Processes can rarely be observed or measured. They vary with the season and with the year. They change if climate changes or as

the result of new cycles of landscape dissection. They leave marks in the soil, but the marks may persist long after the processes that produced them have ceased to act.

Thus, Soil Taxonomy assumes that if a given set of processes has been dominant for a long period, it will have left its mark morphologically or chemically in such a way that the results will be measurable. The problem with placing too much emphasis on genetic interpretations of soils lies in the fact that interpretations can differ among observers, and they change as knowledge advances. Recall also that Soil Taxonomy was designed to further our *use and management* of soil, the knowledge of which generally outpaces advances in knowledge of soil genesis. Soil Taxonomy tries to ensure that different people will arrive at the same classification of a soil, even if they disagree on its genesis (Smith 1983). This trait ensures that the system will be somewhat independent of the day-to-day vagaries of pedogenic theory or the views of any particular soil scientist. (Note that the system *is* amended frequently, in light of advances in pedogenic understanding and theory.) Genesis does not appear so much in the *definitions* of the taxa in Soil Taxonomy, but it lies everywhere *behind* them (Smith 1983, Ciolkosz et al. 1989) (Fig. 8.8). The founders of Soil Taxonomy intended the system to keep soils of similar genesis in the same taxon, although this is not always possible because many properties can be developed by different processes, i.e., they exhibit equifinality. The deeper one proceeds into the system, however, the more genetic similarity the soils will have. Soils of a given suborder are somewhat alike genetically, but at the series level the similarities are as close as knowledge permits.

Other criticisms of Soil Taxonomy center on class limits and the perceived inflexibility of the system (Hallberg 1984). Beckmann (1984) felt the class limits of Soil Taxonomy were often arbitrary (see also Webster 1968). To that end, most diagnostic horizons are a quantitative reflection of our understanding of pedogenic processes, taken beyond a certain minimum point. For example, an argillic horizon is designed to reflect the process of clay illuviation, assuming that it has been operative for some time. Although most, if not all, soils have technically experienced some clay illuviation, making note of it within the taxonomic class of a soil is simply not worthwhile unless the process has proceeded to the point where it has dramatically impacted the profile. How dramatically is enough? That answer is found in the quantitative limits outlined by the Soil Survey Staff (1999). For a more thorough discussion of how genetic thought factored into the making of Soil Taxonomy, we suggest that you consult the Guy Smith interviews (Smith 1986).

The Canadian System of Soil Classification

Unlike Soil Taxonomy, which is a global system, the Canadian system of soil classification is intended primarily for use in Canada (Table 8.14). Provision is not made for soils in thermic and warmer soil temperature regimes or for soils in torric and xeric soil moisture regimes, because these regimes do not exist in Canada (Expert Committee on Soil Science 1987, Soil Classification Working Group 1998). A strength of the system lies in its classification of cold soils and salt-affected soils, of which Canada has many millions of hectares.

The two classification systems have many similarities, and their structure and logic are essentially the same. Some of the soil orders are essentially equivalent in the two systems: Spodosols and Podzols, Alfisols and Luvisols, Entisols and Regosols, Inceptisols and Brunisols. Perhaps because there are fewer types of soils to classify in Canada than exist worldwide, the Canadian system has fewer soil orders (ten) and one less hierarchical level than does Soil Taxonomy; below the order level, the system goes immediately to the great group level. Other taxonomic levels are similar to those of Soil Taxonomy and use the same names, e.g., subgroup, family, and series.

One main difference between the two systems lies in how wet soils and salt-affected soils are classified. The Canadian system breaks these soils out at the highest (order) level of classification: Gleysols and Solonetzic soils, respectively. In Soil Taxonomy, on the other hand, soil moisture condition is not a defining criterion until the suborder level, at which various -aqu- suborders appear, e.g., Aqualfs or Aquepts. Likewise, the presence of salts in a soil horizon does not enter into Soil Taxonomy until the great group level, e.g., Haplosalids or Natrustolls.

The World Reference Base for Soil Resources

In 1974, the United Nations Food and Agriculture Organization (FAO), in conjunction with the United Nations Educational, Scientific and Cultural Organization (UNESCO), published the first detailed soil map legend for the world at a 1:5,000,000 scale. This legend was part of the International Reference Base for Soil Classification (IRB) project, which intended to establish a framework through which existing, nation-based soil classification systems could be correlated. It was the first attempt to create a soil map for the entire world by using a uniform legend. Reaching international agreement on which major soil groupings should be recognized at a global scale, as well as on the criteria and methodology to be applied for defining and identifying them, was the difficult but laudable goal of the project.

Since the publication of the 1974 world soil map legend, it has progressed and expanded to encompass the major soils of the world in more detail. For example, in 1988 an updated map legend was published. Finally, in 2006 the International Union of Soil Sciences (IUSS) Working Group published the latest update to the system, renaming it the World Reference Base (WRB) for Soil Resources (IUSS 2006). Today, the WRB stands as the international standard for soil classification, particularly at small scales, i.e., over large areas (Gray *et al.* 2011). WRB is designed to facilitate communication among soil scientists, as they strive to identify, characterize, and name the world's major soil types. It is not designed to replace national soil classification systems, but to be a resource by which better correlation between national systems can be achieved.

Like Soil Taxonomy, the WRB is a hierarchical system, with 32 reference soil groups (RSGs) in its top tier (Table 8.15). Examples include some familiar names and categories, e.g., Histosols, Andosols, and Regosols, as well as some newer ones – Stagnosols, Albeluvisols, and Technosols. These groups are mainly used for delineating soils on small-scale maps. The second tier of soil categories, used on larger scale maps, includes adjectives called qualifiers; these are used as suffixes and prefixes to modify the RSGs and are designed to convey additional information about the soil type. Direct correlation of WRB taxonomic units with those in Soil Taxonomy is somewhat problematic, however, because the WRB does not use climate information about soil temperature or moisture as criteria in the classification system. The WRB has a flatter architecture in its hierarchy, with only two main categories versus the six categories of Soil Taxonomy. Because the WRB system lacks family and series categories, more information must be encompassed in tier 2 with the use of qualifiers to the RSG. The WRB also recognizes human-influenced soils in tier 1, with two RSGs, i.e., Technosols and Anthrosols. The reader is referred to Table 8.15 and a number of online

Table 8.14 | Approximate equivalents between the Canadian system of classification and Soil Taxonomy

Canadian system	Soil Taxonomy
Order and great group	Great groups or other taxa partly or mostly included
Chernozemic soils	Ustolls and Cryolls
Brown	Aquic and Vertic Ustolls; Calciustolls; Cryolls other than Natric
Dark Brown	Ustolls other than Aridic and Udic; Cryolls other than Natric
Black	Frigid and Cryic Argiudolls and Hapludolls; Aquic and Vertic Argiustolls and Haplustolls; Calciustolls other than Typic and Aridic; Cryolls other than Natric
Dark Gray	Cryic and Frigid Ustolls and Udolls
Solonetz soils	Alfisols and Mollisols with a natric horizon
Solonetz	Natrustalfs, Natrustolls, Natraquolls, Natricryolls; Frigid Natrargids
Solodized Solonetz	Natralbolls
Vertisolic Solonetz	Vertic intergrades of Natr- great groups
Solod	Glossic Natraqualfs, Natrudalfs, Natrustalfs, Natrudolls, Natrustolls, and Natralbolls; Frigid Natraquolls
Luvisolic soils	Udalfs
Gray Brown Luvisol	Hapludalfs and Glossudalfs
Gray Luvisol	Cryalfs and Udalfs that do not have a kandic or natric horizon
Podzolic soils	Spodosols not within an aquic soil moisture regime
Humic Podzol	Humods and Humicryods
Ferro-Humic Podzol	Cryods and Placorthods
Humo-Ferric Podzol	Cryods, Placorthods, and Frigid Orthods
Brunisolic soils	Inceptisols not within an aquic soil moisture regime
Melanic Brunisol	Eutrocryepts, Eutrudepts, and Hapludolls
Eutric Brunisol	Cryepts and Eutrudepts
Sombric Brunisol	Dystrudepts and Eutrudepts
Dystric Brunisol	Dystrudepts and Dystrocryepts
Regosolic soils	Entisols
Regosol	Entisols not within an aquic soil moisture regime
Gleysolic soils	Soils of aquic suborders that are Frigid or Cryic
Humic Gleysol	Cryaquolls, Calciaquolls, and Humaquepts; Frigid Endoaquolls
Gleysol	Aquents, Aquepts, and Fluvents
Luvic Gleysol	Albolls, Aquolls, and Aqualfs
Organic soils	Histosols
Fibrisol	Frigid and Cryic Fibrists
Mesisol	Frigid and Cryic Hemists
Humisol	Frigid and Cryic Saprists
Folisol	Frigid and Cryic Folists

Source: Simplified and combined from Leahy (1963), Soil Survey Staff (1975, 1999), Expert Committee on Soil Science (1987), and Soil Classification Working Group (1998).

Table 8.15 | An overview of the reference soil groups of the World Reference Base system (WRB)

Reference soil group(s)[a]	Characteristics
Histosols	Organic soils
Anthrosols and Technosols	Soils formed or strongly modified by human activity such as agriculture, including areas of mine soils and other forms of human waste
Cryosols and Leptosols	Soils with severe rooting limitations due to permafrost or bedrock
Vertisols, Fluvisols, Solonetz, Solonchaks, and Gleysols	Soils that have been strongly influenced by water, either through shrink-swell cycles caused by wetting and drying, by fluvial activity, by translocation of salts, or by saturation within the profile
Andosols, Podzols, Plinthosols, Nitisols, and Ferralsols	Soils in which genesis involves Fe and Al chemistry and translocation
Planosols and Stagnosols	Soils that perch water within the profile
Chernozems, Kastanozems, and Phaeozems	Soils in semiarid regions that have thick, dark A horizons and high base cation saturation values
Gypsisols, Durisols, and Calcisols	Soils of dry regions that have subsoil accumulations of secondary gypsum, silica, and calcium
Albeluvisols, Alisols, Acrisols, Luvisols, and Lixisols	Soils with subsoil accumulations of clay
Umbrisols, Arenosols, Cambisols, and Regosols	Soils with minimal development of horizons

Notes: RFGs are roughly equivalent to soil orders in Soil Taxonomy.
After Michéli and Spaargaren (2011).
[a] The soil groups are best understood if keyed out in the logical order presented here.

resources for additional information. And for an excellent overview of the many other, national-level soil classification systems in use worldwide, we urge you to consult Michéli and Spaargaren (2011).

Soil Mapping and Soil Maps

The most direct application of soil classification is in soil mapping. In the United States, the Natural Resources Conservation Service (NRCS) coordinates the National Cooperative Soil Survey Program in conjunction with other federal agencies, cooperating state and local governments, land-grant and other universities, and private organizations. The precursor to the NRCS was the Soil Conservation Service (SCS), which was established by Congress in 1935 in response the detrimental effects of soil erosion associated with the Dust Bowl of the 1930s. The SCS was founded largely through the efforts of Hugh H. Bennett, a soil conservation pioneer who had intimate knowledge of the ecological disaster of the Dust Bowl. The SCS changed its name to the NRCS in 1994, and its mission has expanded to include conservation efforts of all kinds, ensuring that lands are conserved, restored, and made more resilient to environmental challenges. Today, the NRCS is the lead agency in soil mapping efforts in the United States; its headquarters is in Washington, D.C., and its National Soil Survey Center (NSSC) is in Lincoln, Nebraska.

Utility of Soil Maps

Soil maps were originally created as a way to assess land value for taxation purposes. Today, soil maps are still used for that purpose, but they have a multitude of other functions, such as land-use zoning and planning, determining septic tank filter field suitability, environmental quality protection and management, siting highway routes, in agriculture and horticulture, and in many other applications in geology, geomorphology, and archaeology, to name a few (Saucier 1966, Lammers and Johnson 1991, Hartemink and McBratney 2008). Indeed, soil maps are excellent surficial geology maps, for quite frequently the initial parent material of the soil can be determined from the soil map (Schaetzl *et al.* 2000, Miller *et al.* 2008, Whitmeyer *et al.* 2010, Yang *et al.* 2011) (Fig. 8.9).

As discussed previously, soil maps can be also used as excellent proxies for natural landscape wetness. Large-scale maps, for example, are often used to determine the location and extent of wetlands and floodplains. Schaetzl (1986a) developed a natural soil drainage index (DI), which ranges from 0 to 99, with wetter soils having larger numbers. Once converted to this index, soil map units can then

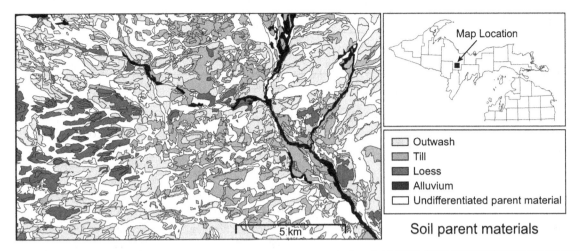

Fig. 8.9 Map of soil parent materials for a part of Marquette County, northern Michigan, created in a GIS by assigning soil series to parent material categories, as defined by their official series descriptions. In this way, standard NRCS soils data can be used to map surficial sediment. Comparable maps can easily be made of other soil attributes, e.g., texture, gravel content, slope, and wetness. See Schaetzl *et al.* (2013) for additional examples.

be examined and remapped; the resultant maps are excellent spatial representations of long-term landscape wetness (See Figure 8.24). More recent work by Schaetzl *et al.* (2009) expanded upon this initial application and provided a Web site that enables the user to determine the DI of any soil, or to download a file that, when linked to a standard NRCS soil map file in a GIS, provides DI values for all the soils on that map. Groffman and Tiedje (1989) applied such data to identify areas likely to undergo denitrification, and Nikièma *et al.* (2012) used the DI (in a GIS) to determine extent of soils of certain wetness conditions across their study area. These studies show how the DI can provide information on soil wetness across landscapes, which can then be used to map wetness-mediated processes such as weathering and leaching, not to mention biomass accumulation. One could also produce maps of other soil attributes, such as N and organic matter content, erodibility and runoff potential, soil attenuation potential, and so forth (Cates and Madison 1993).

Using a few assumptions about the relationships between taxonomic classification and natural soil productivity, Schaetzl *et al.* (2012) also developed a soil productivity index, or PI. Certain soil properties that are indicative of high productivity, e.g., the thick, dark A horizons of Mollisols or the relatively high pH values of Eutrudepts, make their way into the classification of soils. Excessive amounts of soluble salts or the dominance of low CEC clay minerals – characteristics that also can be determined from the soil's classification – can similarly be used to determine, if only in a relative sense, a soil's productivity. The PI correlates well with various land uses, illustrating that farmers are utilizing the best soils for the most intensive

Table 8.16 | Mean Productivity Index (PI) values for sites of different crops and land uses in Lower Michigan, in 2009

Crop or land use	Productivity Index (mean ± standard deviation)
All field crops	10.94 ± 2.4
Alfalfa	10.35 ± 2.1
Corn	10.85 ± 2.4
Dry edible beans	12.25 ± 2.1
Potatoes	10.33 ± 2.5
Soybeans	11.27 ± 2.1
Sugar beets	14.00 ± 0.0
Winter wheat	10.43 ± 3.4
All forest	7.77 ± 3.2
Deciduous forest	8.17 ± 3.2
Evergreen forest	5.87 ± 2.9
Mixed forest	6.18 ± 2.8

Note: Based on the sample of 1,000 random points, taken with a GIS. The PI ranges from 0 (least productive) to 19 (most productive). See Schaetzl *et al.* (2012) for more information

land uses (Table 8.16). Similarly, crop yields in 2009 correlated well with the PI (Fig. 8.7).

In the past, soil maps were produced as paper products only. Today, most are available as digital files, which allow for a multitude of additional uses in a geographic

Table 8.17 | The five orders (scales) of soil maps

Map order	Minimum map unit size (ha)	Typical map scale	General uses	Example
First order	≤1	<1:4800	Intensive uses, building lots, experimental farm plots	Soil map of the Davis Natural Area
Second order	0.6–4.0	1:12000–1:31,680	Intensive, general agriculture and zoning, urban planning	Soil Survey of Grant County, Wisconsin Soil Survey Geographic database (SSURGO)
Third order	1.6–16	1:20,000–1:63,360	Extensive, range land determinations, broad-scale zoning	State Soil Geographic Data Base (STATSGO) map
Fourth order	16–252	1:63,360–1:250,000	Extensive, for broad land-use potential	Major Land Resource Area (MLRA) map
Fifth order	252–4,000	1:250,000–1:5,000,000	Very extensive, regional and global soil maps for general inventory	Digital soil map of the world

information system (GIS) (Schaetzl 2002, Schaetzl et al. 2002). Downloads of county-scale soil maps for the United States can be accomplished through the NRCS's Web Soil Survey site: http://websoilsurvey.sc.egov.usda.gov/App/HomePage.htm. Because the soils are identified with map units, or polygons, the GIS data are created and served out as vector data. Nonetheless, converting these vector data to raster grids is commonplace and serves to reduce the large file sizes associated with vector data.

Although the possible applications of soil information within a GIS are myriad, we mention a few here. Kudryashova et al. (2011) used soil data in a GIS to estimate soil carbon stocks across a region. Wu et al. (1997) used NRCS soil map data in a GIS to estimate soil erodibility across a landscape in Kansas. Both Jacobs et al. (2011) and Scull and Schaetzl (2011) mapped loess thickness across landscapes using soil maps, by coding each map unit to its typical loess thickness, as described in their official soil series description. Miller et al. (2008) coded soil map units to parent material, enabling them to map in great detail the parent materials and landforms of the Des Moines glacial lobe in Iowa and Minnesota. A similar soil geomorphology application was reported by Millar (2004), in which ice margins were mapped using soils and terrain data, in a GIS.

Applications using GIS and soil maps will continue to expand as the need to understand soils, sediments, and landforms better at a landscape scale grows. Indeed, because of the increased demand for expertise and knowledge in GIS-soils applications, in 2002 the NRCS established a Geospatial Research Unit (GRU) within the National Soil Survey Center (NSSC). The GRU focuses on innovative applications of GIS and related technologies, emphasizing natural resource inventory and mapping activities.

Components of Soil Maps

What is a soil map and what are it components? The most fundamental element of a soil map is the *map unit* (Soil Survey Division Staff 1993). When drawn on a map, a map unit is called a *delineation*, and the map unit name generally corresponds to the dominant soil therein, e.g., Drummer silty clay loam, a Typic Endoaquoll.

The amount of detail that can be shown on the soil map – and its accuracy – is primarily a function of scale; large-scale maps, such as county soil maps, can show a great deal of detail, while small-scale maps of states or regions must generalize significantly (Lyford 1974). Even on the largest-scale map, some degree of generalization is required; it is not possible to map every pedon. Generally, five types or orders of soil maps, based on scale and detail, are recognized (Table 8.17). Second-order maps, like those of county soil surveys, keep errors and generalization to a minimum. As the area mapped becomes larger, however, the amount of detail that can be shown decreases, such that similar map units must be combined. When this happens, the mapmaker can rename the units using both of the original names, e.g., Miami-Conover soils, or use only the name of the dominant soil and report that minor components of the other exist in these map units.

Almost all soil map units have *minor components* (previously called *inclusions*) of similar and dissimilar soils. For example, a Drummer map unit would be dominated by soils that fall within the limits of the Drummer series, along with minor components of some other soils. What can be done to minimize the area or number of minor components on smaller-scale maps? More time could be spent mapping, and this mapping could be done at larger scales. But time is money, and thus, this solution is often unrealistic.

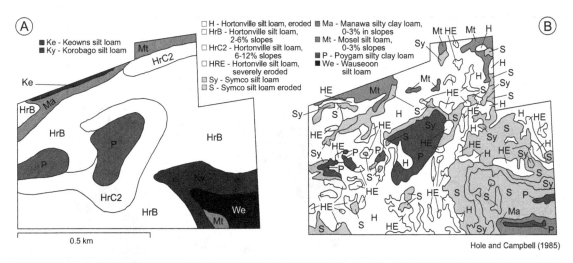

Fig. 8.10 Two soil maps of the same 50-ha area in Winnebago County, Wisconsin. (A) Published county soil map (Mitchell 1980). (B) A larger-scale soil map used for research, based upon intensive field mapping and air photo interpretation.

Another solution to landscape complexity issues – and how to map them – is to define the soil map units differently. On a large-scale map of a pasture, for example, units of Drummer silty clay loam may be readily mappable with a minimum percentage of minor components and a high level of taxonomic purity. On a smaller-scale (countywide) soil map, however, the soil scientists might not be able to delineate relatively pure Typic Endoaquoll (Drummer) units, because of cartographic and/or time constraints. However, they may be able to delineate and name map units that are less taxonomically restrictive, e.g., Endoaquolls, and that may include similar soils that classify in Typic, Aeric, and Cumulic subgroups. That is, broadening the definition of the taxon that is being mapped enables the mapper to delineate areas that are still reasonably pure, in terms of the map legend, and may be able to be differentiated at larger, more detailed map scales.

The amount and character of minor components in a map unit vary from place to place. In some complex landscapes, all of the minor components together may compose half or more of the map unit. And on small-scale maps, these minor components cannot easily be shown, even if one knows of their existence and location (Figs. 8.1, 8.10). Spot symbols are often used in this case.

As explained, a map unit with no minor components would be taxonomically pure. Such map units are extremely rare.

Types of Map Units

The NRCS recognizes three main types of map units: *consociations*, *complexes*, and *associations*. Consociations are dominated by one soil series. A *soil series* is the lowest and most homogeneous taxonomic category, representing a tightly defined group of soils that have a similar arrangement of horizons and other differentiating characteristics,

e.g., texture, color, stoniness, or depth to carbonates (Soil Survey Division Staff 1993). Soil series are defined by restricting the range of allowable values for each of the differentiating characteristics. Series are usually named for a geographic location near where they were first identified and mapped, e.g., Antigo, Portsmouth, or Islandlake, but some series names are invented. On a soil map, most consociations are named for the series (Antigo) and its typical surface-horizon texture (silt loam). The map unit name may also refer to the *phase* of a series when distinctions are desired within the normal range of characteristics. Examples might be Antigo silt loam, stony phase; Antigo silt loam, deep substratum; or Antigo silt loam, severely eroded.

Normally, at least half of the area of a consociation map unit represents the series for which the map unit is named, e.g., Milaca (Campbell and Edmonds 1984). Minor components of dissimilar soils are permitted, but only within predetermined limits (Soil Survey Division Staff 1993). The consociation map units of most county-scale NRCS surveys are designed to represent a single series with as few minor components as possible (Fig. 8.11). Map units called *complexes* and *associations* consist of two or more dissimilar soil taxa. Complexes are map units in which the components occur in a pattern that is, although often a repeating and often understandable one, so complex that it cannot be resolved at the map scale being used. Alternatively, in a soil association, the pattern is known and could be resolved (and hence, portrayed) on a larger-scale map, e.g., 1:24,000 or 1:15,840 (Fig. 8.11). Hence complexes and associations are named with at least two soil names, e.g., Rubicon-Grayling association, or Catlin-Flanagan complex. Generally, complexes are used for map units at scales of 1:24,000 or larger. They represent patterns of soils that are so intricately mixed on the landscape that mapping them

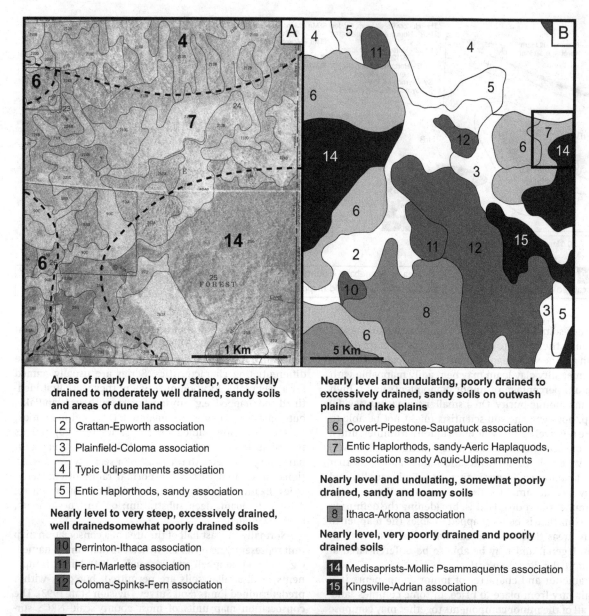

Areas of nearly level to very steep, excessively drained to moderately well drained, sandy soils and areas of dune land

2   Grattan-Epworth association

3   Plainfield-Coloma association

4   Typic Udipsamments association

5   Entic Haplorthods, sandy association

Nearly level to very steep, excessively drained, well drainedsomewhat poorly drained soils

10   Perrinton-Ithaca association

11   Fern-Marlette association

12   Coloma-Spinks-Fern association

Nearly level and undulating, poorly drained to excessively drained, sandy soils on outwash plains and lake plains

6   Covert-Pipestone-Saugatuck association

7   Entic Haplorthods, sandy-Aeric Haplaquods, association sandy Aquic-Udipsamments

Nearly level and undulating, somewhat poorly drained, sandy and loamy soils

8   Ithaca-Arkona association

Nearly level, very poorly drained and poorly drained soils

14   Medisaprists-Mollic Psammaquents association

15   Kingsville-Adrian association

Fig. 8.11   Maps taken from the NRCS county-level soil survey of Mason County, Michigan (Johnson 1995), showing the level of generalization common to small-scale soil association maps. (A) Part of map page 16, originally mapped at 1:15,840 scale, showing mainly soil consociations. (B) Part of the much smaller scale (1:190,080) soil association map for the county. The inset map (rectangle) in part B illustrates the extent of the map shown in part A.

as consociations is impractical, if not impossible. Usually, estimated percentages of each series within the complex are provided in the map metadata, and sometimes the pattern of complexity is explained in a soil survey report. For example, Catlin soils may occur as isolated patches (on high spots) within the Flanagan background or in a linear pattern. Fridland (1965) noted that regularly occurring soil complexes (or, as he called them, soil combinations) can be divided into three types: (1) microcombinations, often

due to microrelief such as treethrow pits or mounds; (2) mesocombinations, linked to patterns of mesorelief like swell-and-swale topography; and (3) macrocombinations, which relate to macrorelief, and thus are similar to catenas. Soil combinations of types 1 and 2 are usually so complex that they would be mapped as map unit complexes. Type 2 and 3 combinations are sufficiently large that they may be delineated as consociations. Soil associations are commonly mapped by the NRCS at scales of ~1:200,000 and are often simply generalized versions of more detailed maps drawn to a larger scale.

In creating a soil map, the mapper tries to delineate map units that are as taxonomically pure as possible, within the scale constraints of the job he or she is undertaking (Amos and Whiteside 1975). This is often a Herculean task, given the great amount of complexity that exists on the landscape. For example, in a large-scale soil map of an agronomy research plot, where high accuracy is required, map units might be as much as 95% pure with respect to a single soil series. That is, minor components of dissimilar series would constitute less than 5% of the map units. Small-scale maps, like those used for general planning purposes, allow higher percentages of soils that are not of the same series (Fig. 8.10). In such a map, the legend not only should mention what percentage of the map unit is a different series, but also how much of that map unit is composed of similar versus dissimilar soils. *Similar* series may differ only in subtle ways, such as slight changes in texture, depth to carbonates, or a different but similar taxonomic classification. Their use and management are not expected to differ from those of the dominant soil. *Dissimilar* soil series may be in another order, as, for example, when pockets of Histosols occur on a sandy till plain, or when Natraquolls dot an otherwise dry, shortgrass landscape. Amos and Whiteside (1975) took this exercise even further, determining not only the number of minor components in a Michigan landscape, but also the degree of contrast they had with the soil the map unit was named for. Their work pointed out that it is not just the amount of minor components that is important in assessing soil map quality, but how *different* these are from the soil that the map shows on the landscape.

Making a Soil Map: Fieldwork

Soil mapping is an exercise in identifying patterns of soil variation and representing those patterns in a two-dimensional model – the map. The initial phases of mapping include an assessment of what kinds of soils are on the landscape, and what kinds of repetitive and predictable patterns they occur in. Often, this is accomplished by walking transects across the various segments of the landscape and making observations at predetermined intervals along the way. Roadcuts and other exposures are also examined carefully. Any other previously mapped or remotely sensed data for the area, including bedrock maps, surficial geology maps, and topographic maps, are also inventoried and studied to familiarize the mapper with the soil landscape. The mapper then examines detailed aerial photographs of the area and looks for places where vegetation, slope, or some landscape phenomena change abruptly; these areas may signal a significant change in soil properties. The mapper then draws preliminary lines on the air photo to mark those locations. Because these lines *could* correspond to soil boundaries, they must be field-checked. With all this preparation, the mapper rarely goes into the field "blind." To do so would be a learning experience, but little more.

The best place to start mapping in the field, of course, is at the beginning. The first pedon encountered, like so many others that will follow, must be exposed and classified, usually to the series level. Exposing the soil/pedon is often done with an auger, shovel, or push probe (Figs. 8.1, 8.12). If time is available or more detail is needed, a pit may be opened by hand or with a backhoe. In rocky areas where bucket augers are not optimal or where soils are shallow to bedrock, the mapper may simply use a shovel or spade to open a small, shallow pit and determine the soil series based on the upper few decimeters of the profile. In any event, a classification decision must eventually be made about the pedon – usually to the series level.

The mapper uses morphological clues to determine which soil series best fits the soil at the site. For every soil series, ranges have been defined for each major property, such as types and arrangement of horizons, thickness, texture, and color or horizons, and depth to redoximorphic features (mottles), among others (Lyford 1938; Fig. 8.13). Usually, the mapper knows the major series of the region so well that only a few key clues are enough to ascertain which soil series is appropriate. For example, the mapper may know that, most of the time, soils of only one of three series occur on uplands in the mapping area. One soil may typically be found in association with a very gravelly and rocky surface, while the others lack coarse fragments. Each may be associated with a particular vegetation type. Land use can also help. One soil may be more erodible than the others; the list of possible clues is long, but with experience, the mapper is able to make reliable distinctions.

Of the possible mapping clues, the variation of major horizons across the landscape is certainly the most important. The mapper must determine, for example, the types of epipedons, e.g., mollic, ochric, and the types of subsurface horizons, e.g., Bt, Bh, Bg, that are present. At a given site, the mapper must also decide whether the genetic horizons described meet the criteria for a particular type of diagnostic horizon. For example, Milaca loam, a coarse-loamy, mixed, superactive, frigid Typic Hapludalf, must have an ochric epipedon, an argillic horizon, and no other unique or extragrade features such as shallow bedrock or a fragipan (although it can have an albic horizon). It must also have coarse-loamy textures in the textural control section. If the pedon examined is similar but has, for example, a fragipan, the pedon would classify as a Fragiudalf. Alternatively, if the pedon were similar but had carbonates at a shallow depth, it would still classify as a Typic Hapludalf, but within a series called Emmet. Each soil series has a suite of related but different series, all of which differ by one or more aspects – texture, thickness, presence or absence of a diagnostic horizon, drainage class, etc. The mapper must be aware of both the geographically associated series as well as the competing series, to settle on the correct series for a single pedon. In the process of sampling soils across the landscape, the mapper will gain insight

Fig. 8.12 An NRCS soil scientist using a bucket auger to determine the soil series at that site. He augers (A) and then examines the auger shavings (B) for color, texture, consistence, and so on. Photos by RJS.

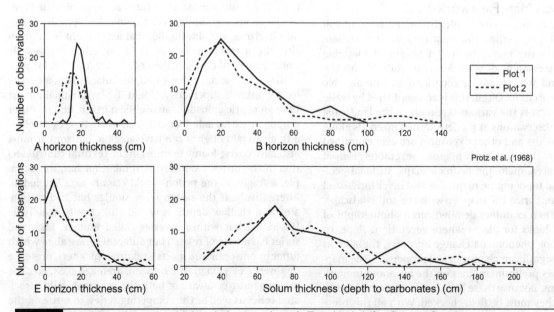

Protz et al. (1968)

Fig. 8.13 Variability in horizon and solum thickness in plots of a Typic Hapludalf in Ontario, Canada.

Table 8.18 | Examples of soil series that form a catena or drainage sequence

Natural drainage class	Soil series[a]	Family taxonomic designation
Well drained[b]	Milaca	Coarse-loamy, mixed, superactive, frigid Typic Hapludalfs
Moderately well drained	Mora	Coarse-loamy, mixed, superactive, frigid Aquic Hapludalfs
Somewhat poorly drained	Ronneby	Coarse-loamy, mixed, frigid Udollic Epiaqualfs
Poorly drained and very poorly drained	Parent	Coarse-loamy, mixed, frigid Typic Epiaquolls
Very poorly drained	Twig	Coarse-loamy, mixed, acid, frigid Histic Humaquepts

[a] All the soils in this Table (1) have developed in the same parent material (dense loamy glacial till on drumlins or moraines), were exposed by deglaciation in the Late Pleistocene, and (2) formed under forest vegetation.
[b] "Drier" drainage classes are defined, but only in sandy parent materials.

about how the diagnostic horizons thicken and thin as topography and parent material vary.

The natural soil drainage class of a soil is determined by first locating oneself on the landscape; poorly drained soils are not likely on steep slopes or on the tops of sandy hills. This knowledge gives the mapper clues about the soil's natural drainage class well before it is even examined. The information that the soil provides about its natural drainage class includes presence or absence of a water table and, most importantly, presence, character, and depth to redoximorphic features, or gleying. Although we discuss the major types of natural soil drainage classes in Chapter 14, we note here that the shallower the soil's water table, the more likely it is to undergo oxidation-reduction processes, which leave behind a telltale morphology of gray and reddish markings, called redoximorphic features, or mottles. Generally, the more poorly drained the soil, the higher in the profile these features will occur. For most soil series in humid regions, a complete suite of drainage class counterparts exist, all of which have formed in similar parent materials and exist as a drainage sequence (or catena) (Table 8.18). An experienced mapper can draw a sample from the appropriate soil depth, interpret the morphology, and quickly determine the natural drainage class, thereby eliminating many series from the list of possibilities.

Once the soil at a site has been identified to the series level, the mapper reexamines the air photo and topographic map, surrounding landscape, vegetation, slope, and other indicators. The mapper uses all this evidence to project a pedogenetic model onto the landscape and predict the most likely patterns of soil variation. The model should predict locations where soil properties might change significantly, and those predictions become provisional map unit boundaries. Boundaries are drawn where the rate of change of one or more soil properties is expected to be the greatest (King et al. 1983). Delineations on the map are tested and further refined by systematically traversing the landscape and sampling soil properties on either side of projected map unit boundaries. In addition, systematic transects with regular sampling points within mapping units are especially important for quantifying the contributions

that similar and dissimilar minor components make to the mapping unit. Determining the location of the map unit boundaries, as well as the correct taxonomic identification of soil components in the unit itself, is the most important part of soil mapping.

To repeat an important point, soil mappers locate the place on the landscape where the *rate of change* from one soil to another to another is most rapid, and they draw the map unit boundary there. As a result, a map unit boundary does not provide an exact location where soil series A yields to soil series B (Campbell 1977). Rather, the boundary is an estimate of the place where, on the landscape, the change from soil A to soil B is most likely. Soil A may still occur as an impurity in map unit B, and vice versa (Fig. 8.14).

There are many tricks to discerning possible soil boundaries in the field (Scull et al. 2003). Recall that the mapper gains a sense of where soil boundaries might be placed by interpretation of aerial photographs of vegetation and terrain, prior to entering the field. But these possible boundary locations must be field-checked. The mapper is armed with knowledge of the five soil-forming factors (Jenny 1941b) and knows that as one factor changes on the landscape, soil characteristics are also likely to change (Muckenhirn et al. 1949, Gile 1975a, b). Thus, the mapper looks for changes in vegetation, topography, or parent material as clues to possible soil variation, and thus a soil map unit boundary. Usually, at local scales, one of the most reliable of these genetic factor clues is topography. Subtle differences in slope curvature (concave vs. convex), steepness, or location can affect soil processes and therefore soil properties (King et al. 1983; Fig. 8.15). Downslope, the soil mapper may be able to find where soil properties change as a result of an increasingly shallower water table. This type of pattern is so predictable that in some landscapes one could almost map soils, or at least some attributes of soils, on the basis of terrain alone (Klingbiel et al. 1987, Moore et al. 1988, 1993, Bell et al. 1992). In the office, the mapper will judge where soils change and draw preliminary lines on the map (air photo) in the appropriate locations. In the example shown in Figure 8.16A, most of the map unit boundaries correspond with changes in topography

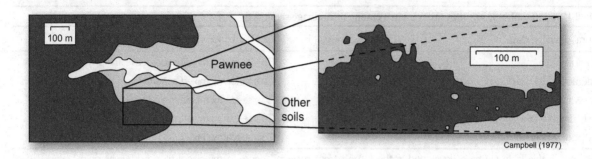

Campbell (1977)

Fig. 8.14 A map unit boundary in Kansas. (A) As drawn on a county soil survey map (1:20,000 scale), and (B) closer to reality, based on detailed grid sampling across the boundary.

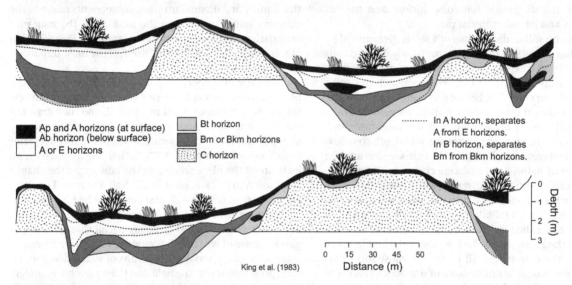

- ■ Ap and A horizons (at surface)
  Ab horizon (below surface)
- □ A or E horizons
- ▨ Bt horizon
- ▨ Bm or Bkm horizons
- ⣿ C horizon

- ⋯⋯ In A horizon, separates A from E horizons.
  In B horizon, separates Bm from Bkm horizons.

King et al. (1983)

Distance (m)

Depth (m)

Fig. 8.15 Two transects showing soil profile variability across a gently rolling Saskatchewan landscape where the soils are developed in till and till-derived sediments.

and, hence, drainage class. Spodic horizons have formed in upland soils that are well drained and somewhat poorly drained (Rouseau and Wainola soils), whereas both excessively drained and poorly drained soils lack this diagnostic horizon. In very poorly drained landscape positions, muck has accumulated and a Histosol is mapped.

In dry climates, most of the landscape is well drained. Soil map unit boundaries may then be associated with changes in slope gradient or aspect or differences in surface age, because the different slopes will reflect younger surfaces or will have affected the genesis of the soils (Gile 1975a, b). For example, on steeper slopes on a North Dakota prairie, most water runs off instead of infiltrating and argillic horizons do not form, while on more gently sloping parts of the landscape argillic horizons *have* formed. Likewise, at the base of slopes, A horizons may be overthickened.

On many landscapes, changes in topography also accompany changes in parent material or perhaps signal a change to a surface of different age, e.g., a lower, eroded surface or a higher, constructional surface such as a sand dune (Gile 1975a, Nettleton and Chadwick 1991). This is particularly true in "rock country," where certain types of bedrock may stand up as ridges or maintain steeper slopes than other kinds of rock (Fig. 8.16C). But this system also works in glaciated terrain; outwash plains tend to be flat while moraines and dunes are more rolling. Floodplains tend to be flat while uplands tend to be undulating, and the contact between the two, i.e., the valley wall, is often sharp. Moving from a plain onto a sloping surface, therefore, often indicates a change in parent material and drainage class, and hence, soils. On older landscapes, deposition of a new parent material may lead to a soil boundary at the contact of the two sediment

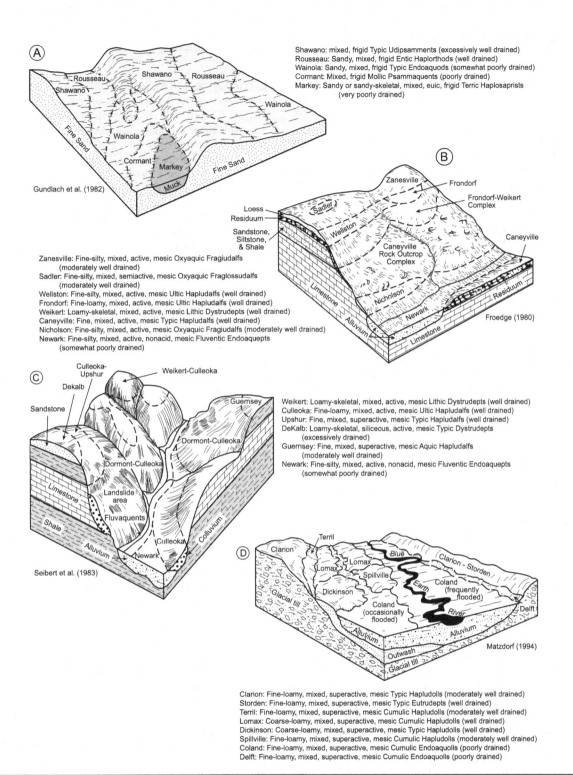

Fig. 8.16 Typical patterns of soils on various landscapes, as shown on block diagrams. (A) A glaciated landscape (udic, frigid) in Wisconsin. Because this landscape is dominated by a single (sandy) parent material, most of the soil map unit boundaries here are associated with changes in natural soil drainage class, which are, in turn, directly affected by slope position. (B) An unglaciated landscape (udic, mesic) in Kentucky. Most of the soil map unit boundaries here are associated with changes in parent material, although drainage class does change slightly along the slopes, and soils are considerably wetter in the valley bottoms. (C) A steep, mountainous, unglaciated landscape (udic, mesic) in southwestern Pennsylvania. Most of the soil map unit boundaries here are associated with changes in parent material or, less commonly, natural soil drainage class. (D) A glaciated landscape (udic, frigid) in Minnesota. Most of the soil map unit boundaries here are associated with changes in parent material and natural soil drainage class, although vegetation covaries significantly with these factors as well.

types (Gile 1975a). Soils also vary as a function of slope concavity and convexity, at sediment-accumulating sites versus sites where sediment is lost (Walker *et al.* 1968; see Chapter 14). Similarly, and often very importantly, the mapper looks for changes in vegetation, which may signal a different parent material or drainage class (Host and Pregitzer 1992). Landscapes of low relief and uniform parent materials have very subtle soil boundaries (Gile 1975b). On such landscapes, map unit boundaries may be placed to correspond with subtle changes in relief that indicate a change in drainage class. On some flat lacustrine plains, all soils are in the same drainage class but parent materials may change laterally. These types of changes may only be manifested as subtle differences in vegetation. Other types of map unit boundaries can occur as slopes change from steep to gentle, or as slope aspect changes from cool, northeast-facing slopes to warm, dry slopes that face to the southwest.

Fundamentally, soil mapping boils down to deciding which map units correspond best to the pattern of soil properties on a given landscape and then determining where boundaries between those units should be placed on the map. Soil mappers use information about the factors of soil formation (climate, organisms, relief, parent material, and time) (see Chapters 12 and 13) to hypothesize how soils may change across the landscape. Soil samples provided by their shovels or augers support or reject their hypotheses. In the end, the degree to which actual variation of soil properties can be represented on the map depends on the time the mapper has to do the job, the scale at which the map is being drawn, the complexity of the landscape, and, of course, the skill and experience of the mapper.

Making a Soil Map: After the Field

Making a soil map obviously does not stop in the field. In the office, the mapper will review the map unit boundaries drawn in the field and transpose them onto an aerial photograph that has been corrected to remove distortion due to camera angle and topography. Each map unit will be checked for accuracy and proper labeling. After that, the lines will be digitized, along with attributes for each map unit (series, slope, etc.).

Most county-scale soil maps have map sheets that span 6–8 square miles (15.5–20.7 km²). In general, a *team* of soil mappers works on a county soil map; the soils on a given map sheet are not mapped solely by one person. Rather, the sheet is divided into sections and different people map different sections. Then, the edges of the sections (or sheets) are matched or joined, as a mean of quality control and error checking. Soil map units rarely edge-match perfectly, and the mappers are left to work out their "joins." Sometimes two *different* soil series will meet at a map sheet edge, requiring a reconciliation. Often, additional fieldwork will be required to rectify such a situation, in the end leading to a better map.

Error, Uncertainty, and Variability in Soil Maps

Landscapes with high amounts of natural soil variability provide a significant challenge to mapping. To produce a good soil map, the mapper seeks to delineate map units that can be consistently and explicitly described (Edmonds *et al.* 1985). The difference between the range of soil properties predicted by a delination on a soil map and the actual soil properties that are found on the corresponding landscape may be called soil map *error*. Map error may result from human error, for example, applying an inappropriate pedogenetic model to represent variation in soil properties. But another kind of error is more common. It occurs because the scale of the map does not allow for all the complexities to be mapped (Ball and Williams 1968, Walker *et al.* 1968, Campbell 1979). Error may vary as a function of sedimentological and geomorphic properties. For example, variability in horizon thickness might be much greater on some landscape positions than on others (Lepsch *et al.* 1977a, Ovalles and Collins 1986), or in some types of parent materials (Drees and Wilding 1973). Parent materials such as loess or eolian sand have low within-unit variability, while glacial deposits and some types of lacustrine sediments are highly variable from place to place (Campbell 1979).

Map units are not taxonomically pure, and, thus, knowing the amount of variability or potential error *within* a map unit is important when making many land-use and soil management decisions (Mader 1963, Cipra *et al.* 1972, Campbell 1978, Campbell and Edmonds 1984) (Fig. 8.17). *Within*-map unit variability is usually great enough that contrasting, minor components are explicitly recognized in map unit descriptions (Lark and Beckett 1995). Many of these minor components would be extragrades, e.g., a minor component of Lithic Hapludults within a Typic Hapludults map unit. Similar soils are not considered minor components; only dissimilar soils are listed as minor components in map unit descriptions. Obviously, the mapper tries to define map units that will contain as few highly contrasting minor components as possible (Wilding *et al.* 1965, Beckett and Webster 1971).

In addition to the variability within existing soil map units, soil map error and uncertainty can be associated with the *location* of the map unit boundaries, as on complex landscapes of undulating relief where parent materials change frequently over short distances. Like all soil landscape characteristics, it may increase or decrease with time (Barrett and Schaetzl 1993, Lagacherie *et al.* 1996, Phillips 2001a). Map unit boundaries misrepresented spatially are an example of one kind of map *uncertainty*. That is, the map unit location only approximates the actual location of the boundary between the two unlike soils. Someone standing on the landscape near a map unit boundary would express the uncertainty as "I am *probably* in map unit A." The spot was mapped as soil A but there exists some degree of uncertainty in the map. Uncertainty may or may not be manifested on the surface, making it difficult to map

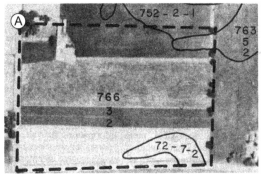

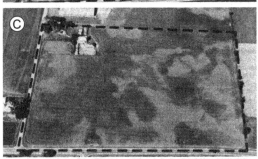
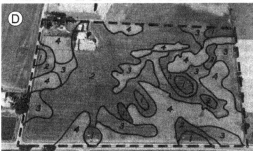

Milfred and Kiefer (1976)

Fig. 8.17 Aerial photographs of a field in Dane County, Wisconsin. (A) The field with soil boundaries overlaid on it, as depicted in the Dane County soil survey. (B) The same field in July, 3 days after receiving 2.5 cm of rain; these soil patterns primarily reflect soil water content. (C) The same field in September, showing patterns in the corn crop, which are indicative of soil characteristics that affect productivity. (D) The same field, mapped in more detail on the basis of data obtained from repetitive aerial photographs. This level of detail is not possible with most county soil surveys but does illustrate the complexity of soils that is not always captured on standard NRCS soil maps.

the boundary with precision (Fig. 8.18). Additionally, soil boundaries have a degree of *fuzziness*, i.e., their distinctness varies. Fuzziness reflects variability in the rate at which soil properties vary close to a map unit boundary. One map unit may grade into the other, or the change can be more abrupt. Map unit boundaries along catenas are often fuzzy (Fig. 8.18A and B). Explaining your location vis-à-vis a fuzzy soil boundary could go like this: "The soil here, at the edge of map unit A, is more like series A than series B." Interpreting soil maps, especially digital maps in a GIS, is done more accurately when the degrees of uncertainty and fuzziness are taken into account. Knowing the *distinctness*, gradual or sharp, of a map unit boundary can be useful to the correct interpretation of the soil landscape.

Digital Soil Maps and Mapping

By using the labor- and capital-intensive methods described, soils in the United States have largely been mapped at larger scales than most places in the world. With most of the United States mapped, using the traditional methods described earlier, and given the dwindling resources to continue this effort at even larger scales, the obvious question that arises is "What's next?"

Because soil maps are so important for resource inventory and modeling, simply living with the existing maps is not a realistic option. These maps can and always must be made better, to serve societal and scientific needs. The NRCS has several options to upgrade its storehouse of soil map data, as listed here. (1) Data enrichment: retain existing map unit boundaries but fill in the gaps in the attribute tables that lie behind them, i.e., ensure that ancillary data associated with the map units are as complete and accurate as possible. This effort is currently under way, providing important data on ecological characteristics, land use limitations, and physical properties, for soils around the nation. (2) Remapping: acquire additional field and lab data and map some areas again – where the original mapping may be poor or dated, or where soils are under intensive development pressures. Often, remapping can be justified because new understandings of soil-landscape or soil-parent material relationships within the study area have emerged. (3) Disaggregation: using field and remotely sensed data, split apart some of the map unit *complexes* that are composed of two or more soils, creating smaller map units that are more internally homogenous, i.e., *consociations*. (4) Adjustment: use primarily remotely sensed data to adjust (move) some of the map unit lines/boundaries to reflect the soil landscape better. This effort is low priority, although in some areas where the initial mapping effort was inadequate, it is necessary. Given the budgetary constraints for field data collection, accomplishing options (3) and (4) is likely to use modeling approaches. Models use correlation with remotely sensed data and interpolation to drive digital soil map predictions. To many, this approach represents a soil mapping future with vast potential. For this reason, we expand on it in the Landscapes box.

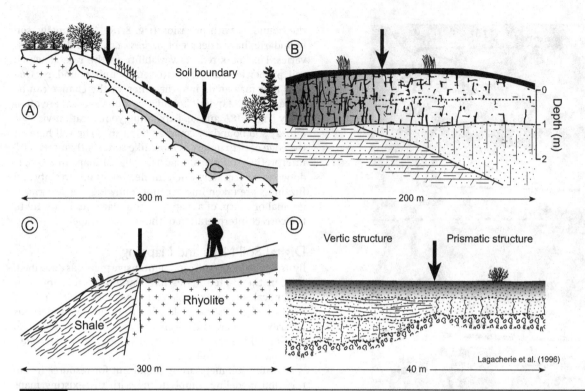

Fig. 8.18 Fuzziness and uncertainty in soil boundaries (indicated at the arrows). (A) High fuzziness, low uncertainty. The gradual variation in soils along this catena is expressed clearly by changes in vegetation. (B) High fuzziness, high uncertainty. The gradual variation in stoniness of a subsurface horizon is not expressed on the surface. (C) Low fuzziness, low uncertainty. The abrupt variation in the soils, due to a subsurface geological boundary, is clearly manifested at the surface as a slope break and changes in stoniness. (D) Low fuzziness, high uncertainty. The undetectable but abrupt variation in subsoil structure is not manifested on the surface.

Landscapes: Digital Soil Mapping (DSM)

The past few decades have witnessed a proliferation of digital data about the Earth's surface, most of it remotely sensed by satellite or other types of airborne sensors. Increasingly, better and more accurate data are becoming available for land cover, soil water, and organic matter contents, as well as various other properties correlated with spectral signatures, obtained from both passive and active sensors. Most importantly, detailed and accurate terrain data are now widely available, often as 10-m-resolution grid files, or even better, as LiDAR (Light Detection and Ranging) data. LiDAR data can have horizontal resolutions of 1 to 3 m, and submeter vertical accuracies. With the help of enhanced computing capabilities, scientists have been combining existing soil map legacy data, which were created using traditional methods, with new types of environmental data as inputs to computer models that generate detailed digital soil maps that are, presumably, an improvement on what was previously available (Shi et al. 2009, 2012, Yang et al. 2011, Thompson et al. 2012). DSM has, therefore, evolved out of more traditional soil survey techniques to take advantage of these advances in computing and geographic data handling, as well as the increased availability of environmental covariates from DEMs and remotely sensed imagery (Thompson et al. 2012). The importance and relevance of such an approach are underscored by the fact that

a team of scientists is now attempting to create a digital soil map of the world for major soil properties at a 90-m grid resolution (Sanchez et al. 2009, Global Soil Map 2012).

Although digital mapping models have different components, approaches, and assumptions, they all use various environmental parameters as inputs, with the goal of predicting the soil classes and properties for a given location, or cell (Scull et al. 2003). The model is guided by a set of rules. These rules are, essentially, quantitative relationships among soil classes, soil properties, and environmental covariate input data, e.g., terrain characteristics or spectral signatures of the soil surface. They are derived directly by the soil mapper or automatically based on various algorithms such as random forest (RF) (Breiman 2001), artificial neural network (ANN) (Rumelhart and McClelland 1986, Hastie et al. 2009), and clustering (Rubin 1967). Unlike a traditional soil mapper, however, who only delineates polygon soil map units, most digital mapping models (1) do not focus on map soil taxonomic units, but instead can map soil classes and soil properties such as wetness, organic matter content, or depth to bedrock and (2) can generate gridded maps and/polygon maps from aggregated gridded maps.

Additionally, many digital soil maps do not create cells that are fully defined as Soil A or Soil B, but instead, define fuzzy membership classes for each cell. In this situation, for each cell there exists a number (its membership class) that denotes how much that cell is like Soil A and B (Qi et al. 2006, Zhu et al. 2001, 2010). At any time, raster cells in the grid can have their memberships hardened and aggregated, so that the digital grid soil map can be aggregated into polygon maps.

Some of the most commonly used digital soil mapping approaches are SoLIM (Zhu et al. 1997, Shi et al. 2009), SCORPAN (McBratney et al. 2003), and ConMap (Behrens et al. 2010). More and better models will certainly appear in the near-future. Much work needs to be done to improve and apply these models, and to make the digital products that result from them more accurate and useful. Nonetheless, digital soil maps do have one advantage over traditional maps – they are repeatable, are consistent, and can be refined as more detailed and accurate data become available.

Digital soil map products, like the soil maps in many other nations, are grid-based products. Traditional soil maps, created by a field soil mapper, show map units as polygons. Each map unit polygon has an inherent suite of attributes. However, the *nature* of these attributes, much of which centers on soil variation *within* the polygons, is not represented on such a map. To obtain additional data, e.g., carbon content, CEC, water table depth, degree of past erosion, one could either disaggregate (cut up) the polygons of soil associations into their respective soil components or grid the polygons into raster cells. The gridded polygon approach is common in countries outside the United States, where the manpower does not exist to map the landscape using traditional methods. This polygon disaggregation approach, however, can provide better spatially distributed soil properties, as compared to traditional soil association polygons. This is especially important for associations that contain many soil components. However, both the traditional polygon maps and disaggregated polygon maps still provide ranges for soil properties with no indication about the spatial distribution of the range within the soil polygon or soil components. In contrast, digital soil maps, especially those based on fuzzy logic, have the advantage of being able to provide (1) a more continuous representation of soil properties across the landscape, (2) soil data of finer resolution, and (3) information about uncertainty. Raster data are more amenable to various environmental modeling applications and site-specific natural resource management than are polygon data. Mapping soils and displaying their characteristics as raster grids are important to the future of soil mapping and digital soil databases.

Pedometrics

In the last 30 years, soil scientists have been developing and applying a host of new mathematical and statistical approaches for predicting soil properties and processes (Webster 1994, McBratney et al. 2000). The field of *pedometrics* deals with topics such as integrating spatial data obtained at multiple scales, characterizing and comparing complex spatial patterns in soil properties, computer simulation of pedogenic processes, designing robust sampling schemes for soil characterization, and using remotely sensed data to predict soil properties, among many other topics. Pedometric analyses rely on both numerical and statistical theory. Some pedometrics methods are especially useful for generating digital maps of soil properties.

Recall that the basic purpose of a soil map is to predict soil properties on the landscape. When soil maps are made as described earlier in this chapter, they are based on the assumption that well-known factors of soil formation, such as parent material, climate, or landscape position, have converged to produce a predictable pattern of soil properties. A pedogenic model (see Chapter 12) guides the mapper in choosing where soils are sampled and where map unit boundaries are to be placed. Another approach to mapping soil properties is based primarily on the assumption that the similarity of soil properties at any two points on the landscape depends simply on how close the points are to one another. This tenet is Waldo Tobler's first law of geography – "Everything is related to everything else, but near things are more related than distant things." It follows that, the closer two soils are, the more likely their degree of similarity. This simple idea is also a foundational principle of *geostatistics*, an approach that has many applications in soil science.

Geostatistical analyses offer the opportunity to assess the spatial structure of soil properties quantitatively. The soil scientist can look for correlations in the spatial structures of multiple properties and test dependencies among them. For example, is the quantitative distribution of soil organic C across a three-dimensional nose slope correlated with slope gradient, slope aspect, or both, or neither? Structural analysis of spatial variability requires sampling many (usually >100) points on a landscape and plotting an index of variability as a function of distance between points. The plot is called a semivariogram, and, when it is coupled with a variety of sophisticated interpolation techniques, information on spatial structure obtained from the semivariogram can be used to predict values of soil properties at unsampled points. In addition to predicting the value of a soil property, e.g., clay concentration in the A horizon, one can make quantitative predictions of the uncertainty of that value, e.g., $25 \pm 5\%$. Predictions of soil properties at unsampled points can also be improved by incorporating the correlation of properties that are easy to measure, e.g., elevation, with those that are expensive to measure, e.g., organic carbon concentration, into the interpolation algorithms (Fig. 8.19).

Reliable geostatistical analysis of spatial structures and meaningful interpolations usually require many samples, and a well-conceived sampling plan. Therefore, to justify the expense of intensive sampling, geostatistical techniques have their greatest application in contexts where accurate and precise predictions are necessary. A more complete discussion of geostatistical analyses is beyond the

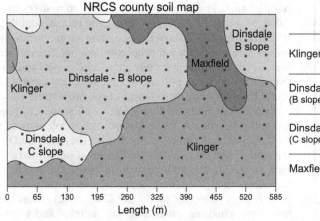

NRCS county soil map

	Subgroup classification	Drainage class	Landscape element	Slope gradient
Klinger	Aquic Hapludoll	Somewhat poorly drained	Interfluves and concave slopes	0 - 2%
Dinsdale (B slope)	Typic Argiudoll	Well drained	Level to convex interfluves	2 - 5%
Dinsdale (C slope)	Typic Argiudoll	Well drained	Sideslopes	5 - 9%
Maxfield	Typic Endoaquoll	Poorly drained	Concave headslopes	0 - 2%

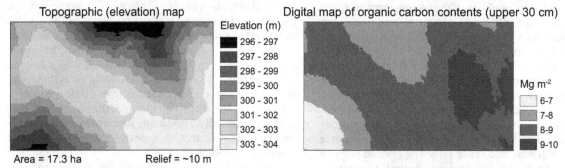

Topographic (elevation) map

Digital map of organic carbon contents (upper 30 cm)

Elevation (m)
- 296 - 297
- 297 - 298
- 298 - 299
- 299 - 300
- 300 - 301
- 301 - 302
- 302 - 303
- 303 - 304

Area = 17.3 ha   Relief = ~10 m

Mg m^{-2}
- 6-7
- 7-8
- 8-9
- 9-10

Fig. 8.19 Application of geostatistics to predict the organic carbon distribution in an agriculturally managed field in Black Hawk County, Iowa. For the analysis, 143 samples were recovered to a depth of 30 cm and analyzed for organic carbon content. The map in the lower right was developed by ordinary kriging, based on a spherical model of the semivariogram. The greatest concentration of organic carbon occurs at the head of a drainageway and roughly corresponds with portions of the poorly drained map unit (Maxfield) and the somewhat poorly drained map unit (Klinger).

scope of this book; readers are referred to several excellent sources for an introduction to these powerful techniques (Webster 2000, Yates and Warrick 2002, Webster and Oliver 2007, Webster and Lark 2013).

Soil Landscape Analysis

As discussed, geographers and soil scientists are interested in understanding soil patterns, or the distribution of soils on landscapes (Fridland 1974; Fig. 8.20). Many decades ago, Dokuchaev noted that soils exist on the landscape in distinct geographic combinations, later called soil combinations or *areals* (Neustruyev 1915, Prasolov 1965). Closely related are the concepts of soil catenas and soil associations, both of which are highly geographic in nature and deal with the spatial relationships among soils (Fridland 1965). The science of *soil landscape analysis* explores such issues (Hole 1978, Hole and Campbell 1985). Francis Hole (1953) was perhaps the first to draw attention to the fact that soils should be studied as three-dimensional bodies,

at a time when the focus was primarily on the two-dimensional profile.

Soil landscape analysis places soils within their landscape context and in so doing provides a basis for characterizing and explaining soil patterns. Soil landscape analysis uses as its basic unit the soil body – an abstract concept that is more analogous to a polypedon than a map unit. Fridland (1965) refers to a similar concept as that of the *elementary soil areal*. Think about the boundaries of a soil body, the basic unit of study in soilscape analysis. The upper boundary is clear – where the soil surface meets the atmosphere. The lower limit of the soil is sometimes obvious but also can be diffuse, depending on how we define B, C, and D horizons (see Chapter 3). The lateral boundaries of this same soil body may be defined by its taxonomic description, i.e., also by human constructs. In other words, except where the soil body abuts nonsoil material such as a lake or bare bedrock, the lateral and lower edges are ultimately conceptual (Hole 1953). Other soil bodies abut it, and all fit together like a jigsaw puzzle. Explaining the nature of this soil puzzle is a part of pedology called soil landscape analysis.

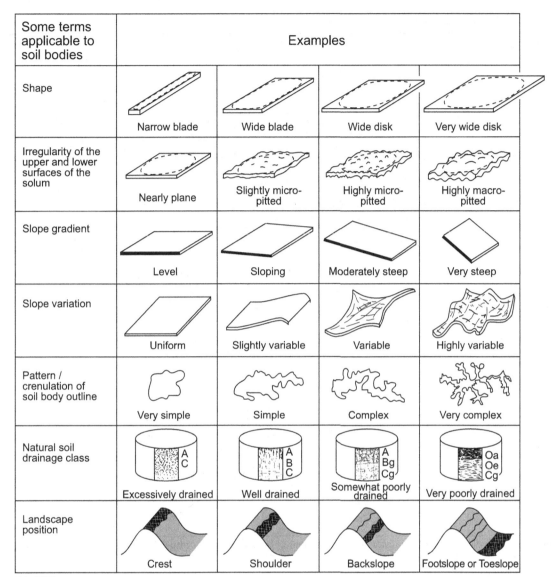

Some terms applicable to soil bodies	Examples			
Shape	Narrow blade	Wide blade	Wide disk	Very wide disk
Irregularity of the upper and lower surfaces of the solum	Nearly plane	Slightly micro-pitted	Highly micro-pitted	Highly macro-pitted
Slope gradient	Level	Sloping	Moderately steep	Very steep
Slope variation	Uniform	Slightly variable	Variable	Highly variable
Pattern / crenulation of soil body outline	Very simple	Simple	Complex	Very complex
Natural soil drainage class	Excessively drained	Well drained	Somewhat poorly drained	Very poorly drained
Landscape position	Crest	Shoulder	Backslope	Footslope or Toeslope

Hole (1953)

Fig. 8.20 Examples of terms used to describe the various aspects of soil map units, when examined from a soil landscape analysis perspective.

Although the concept of a soil body is related to that of the polypedon, our best information about soil bodies is gained from soil maps. Hence, for purposes of discussion we will use the terms synonymously. We recognize, however, that a soil map is simply our best representation of the net or fabric of soil bodies that exists on the landscape. Soil landscape (soilscape) analysis places emphasis on describing and explaining the *patterns* that soil bodies make on the landscape. In this way, soil landscape analysis is a core part of soil geography. The basic premise of soilscape analysis is that the *geographic distributions of soil*

characteristics are the criteria that can be used to distinguish one *landscape* from another, as a profile description can be used to distinguish one *pedon* from another (Pavlik and Hole 1977). One attempts to determine the nature and patterns governing the distribution of soils as reflected in the makeup of the soil cover (Fridland 1965; Fig. 8.21).

Describing Soil Bodies and Soil Landscapes

In soil landscape analysis, each soil body is described by a number of semiquantitative and quantitative terms – shape, surface, and subsurface irregularity; slope gradient

Table 8.19 | A selection of soil landscape descriptors

Type of soilscape measure[a]	Description of method and what it portrays
Soil body shape	Shapes (random, elongate, circular, anastomosing, complex, etc.) of soil bodies on the landscape (see Fig. 8.21)
Alignment of soil boundaries	Compass orientations of the boundaries of soil map units (see Fig. 8.21)
Number and arrangement of nodes	Distribution of the points on the landscape where map unit boundaries join; also of interest is the sheer number of nodes, as a measure of soilscape complexity
Density (or size) of soil bodies	Density is an indirect measure of the size of soil map units (mean, median, range, standard deviation, etc.) (see Fig. 8.23)
Composition	Proportionate amounts of the various types of soil taxa (series, subgroups, great groups, etc.) on a landscape
Taxonomic contrast	Relative numbers of taxa on a landscape
Soil heterogeneity	Number of soil bodies per unit area (i.e., the density) multiplied by the taxonomic contrast; provides an indication of taxonomic complexity
Number of landscape positions	Assessment of which of the five major landscape positions are represented on the soil map
Soil moisture regime (Drainage Index)	Assignment of a numeric value to each mapping unit, based on its natural soil drainage class or productivity class (See Fig. 8.24)
Distinctness of soil boundaries	Estimation of the pedogenic or taxonomic contrast/sharpness of map unit boundaries

[a] Most measures presented are readily quantifiable.
Source: Hole and Campbell (1985).

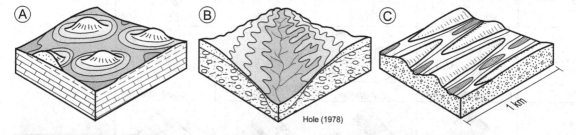

Hole (1978)

Fig. 8.21 Examples of three different types of landscapes and the shapes of the soil bodies (or map units) that might be observed on them. **(A)** Isolated coppice dunes on a plain. **(B)** A fluvially dissected plateau. **(C)** A drumlin field.

and variation; outline crenulation/pattern; wetness; and position on the landscape (Pavlik and Hole 1977; Table 8.19, Figs. 8.20, 8.21). For example, one could quantify the distinctness of the boundary (sharp, gradual, etc.) between two soil bodies in several ways. The shape, slope, pattern, surface irregularity, and natural drainage class of a soil body can all be determined, and after a number of soil bodies on a landscape have been studied, generalizations made about the spatial character of the soilscape (Hole 1953, Schaetzl 1986a). In sum, these properties are a landscape's *pedogeomorphic fabric* (Habermann and Hole 1980).

Shapes of soil bodies on a landscape can be a very useful indicator of the physiography of that landscape and how it may have evolved (Table 8.19, Fig. 8.21). One can describe the general shapes of soil bodies in qualitative terms such as disks, spots, stripes, and dendritic patterns, or they can be quantified in terms of length/width ratios, area/perimeter ratios, degree of boundary crenulation, and so on (Fridland 1965, Hole 1978). The *orientation* of *soil boundaries*, regulated by the arrangement of the soil bodies on the landscape, can be examined, as a way of determining the degree of linearity of soil bodies on the landscape. Fluted landscapes with drumlins would exhibit high linearity, while an end or ground moraine might not (Fig. 8.22). Density of soil bodies per unit area can also be examined (Fig. 8.23), although this will vary according to map scale and the tendency of the mappers to be either "splitters" or "lumpers." Likewise, the size of map units, which determines map unit density, is affected by access (forested areas are difficult to access and hence may contain larger map units) and whether the mappers choose to combine some difficult-to-map consociations into map unit complexes. Thus, map unit size is

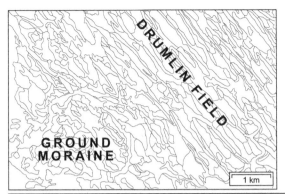

Fig. 8.22 Soil map unit boundaries in a part of Presque Isle County, Michigan, soil survey (Knapp 1993). Note the high degree of linearity in the drumlin field versus the more random shapes of map units on the ground moraine.

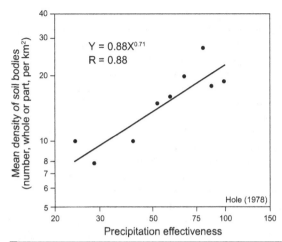

$$Y = 0.88X^{0.71}$$
$$R = 0.88$$

Hole (1978)

Fig. 8.23 Relation between the density of soil bodies and precipitation effectiveness, a measure of leaching, from the dry climate of Colorado to the humid climate of Indiana.

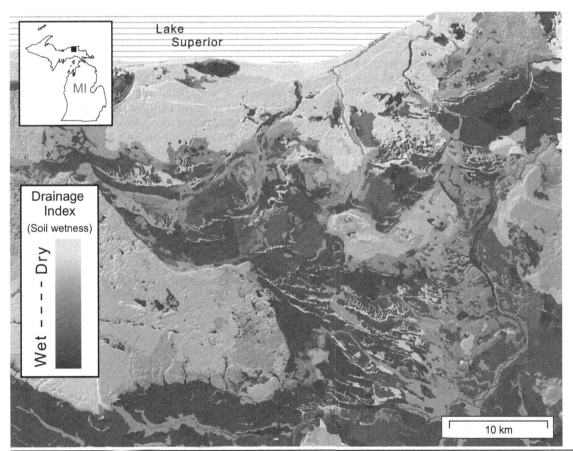

Fig. 8.24 Grayscale diagram showing the variation in natural soil wetness, as indicated by the Natural Soil Drainage Index (DI) of Schaetzl *et al.* (2009), across a landscape in northern Michigan, near the Lake Superior shoreline. Black areas (DI = 99) are open water. Swamps with Histosols (DI ~ 80) show as dark gray. Dry uplands dominated by excessively drained sandy soils (DI ~ 20) show as light gray, while well-drained uplands, on which soils have formed in loamier tills, are intermediate in tone. Note the high amount of contrast between the excessively drained sand dunes that occur within otherwise poorly drained swamps.

a difficult and potentially misleading soilscape parameter and must be used with caution. One possible way to avoid this problem is to examine patterns that are more geomorphic than pedologic in origin, such as slope or drainage class. In this way, explainable patterns are more likely to emerge, regardless of the soil classification scheme used, accessibility of the landscape, or biases of the mapper (Habermann and Hole 1980).

Information on *pedologic diversity*, a measure of the heterogeneity of a soilscape, may assist biologists interested in *biodiversity* and provide important geological/geomorphological information as well (Hole 1978). It may be quantified by examining the suite of different soil taxa on a landscape, or by examining the taxonomic contrast across a number of soil boundaries (Schaetzl 1986a). Contrast may be defined on the basis of natural soil drainage class, slope, and productivity, or taxonomically. Taxonomic contrast can take all forms, from high-level contrast (a Mollisol abutting an Entisol or an Andisol) to contrast at the subgroup level (Typic vs. Lithic Hapludolls).

Although difficult to generalize, many soilscapes, especially recent (postglacial) ones and those with high local relief, have great pedologic diversity, while old, flat landscapes may be dominated by fewer numbers of taxa. Schaetzl (1986a) has shown that pedologic diversity is not necessarily related to readily observable landscape qualities such as local relief. At the very lowest level of taxonomic contrast are soil boundaries where the same series is mapped on each side of the boundary, but the slope of each map unit is different. This map unit boundary exhibits no taxonomic contrast whatsoever, only contrast in slope gradient; many would consider these two map units within the same soil body.

Some soilscapes may be primarily composed of well and somewhat poorly drained soils, while others are mainly wet, and still others have nearly equal proportions of soils from four or more drainage classes – this is landscape contrast according to soil wetness. With respect to natural soil drainage class and soilscape diversity, Hole (1953) developed, and Schaetzl (1986a; Schaetzl *et al.* 2009) expanded upon, a type of soil drainage index designed to capture the essence of the soilscape's diversity with respect to wetness. Variability in drainage index (0–99) across various types of landscapes can then be plotted and visualized (Fig. 8.24).

As a logical next step, data on soil bodies from entire landscapes can be combined and interregional comparisons made (Habermann and Hole 1980). For example, on the plains of Kazakhstan are very large soil bodies, while areas of similar Mollisols in North Dakota may be more intricate, with patches of small soil bodies interspersed in glacial kettles or along drainageways. Hole (1978) reported on one such comparison, in which the density of soil bodies increased in direct relation to precipitation effectiveness, i.e., leaching potential (Fig. 8.23). Pavlik and Hole (1977) illustrated how soil information can be used to discriminate one type of landscape from another, and to determine the degree of difference between adjoining soilscapes. Schaetzl *et al.* (2013) mapped the physiography of Michigan in some detail and found that soils and soil characteristics provided many of the most important criteria for distinguishing one physiographic region from another. In sum, soilscape analysis provides a refreshing and open-ended way of examining soil landscapes and, in so doing, gleaning landscape-based information that might otherwise have remained hidden.

Chapter 9

Weathering

An important step in the formation of *soil* from *rock* involves *weathering* of the rock into smaller and/or chemically altered parts (Yatsu 1988). Weathering is the physical and chemical alteration of rocks and minerals at or near the Earth's surface, produced by biological, chemical, and physical agents (in actuality, by their *combination*), as they adjust toward an equilibrium state in the surface environment (Pope *et al.* 2002). Few soils form directly from bedrock. More often soils develop after an intermediate step in which weathering processes break rock down in situ, or geomorphic processes comminute and erode the rock. This intermediate step forms various types of regolith, or rock overburden, which are then acted upon by pedogenic processes to form soil (see Chapter 13). In the end, rocks become discolored, are structurally altered, acquire precipitates of weathering by-products, and experience collapse as a result of weathering. In this section we examine the main components and processes of weathering – a discussion that logically precedes our discussion of soil parent materials in Chapter 10. Excellent reviews of weathering are found in Ollier (1984), Yatsu (1988), Pope *et al.* (1995), Bland and Rolls (1998), and Hall *et al.* (2012).

In various degrees, rocks are physically and chemically unstable at the Earth's surface, and hence they *weather*, because the surficial environment is far different from the one in which they formed. For most rocks, the surface (soil) environment is colder, with less pressure and increased amounts of oxygen, water, and biota, than their formative environment, be it in a volcano's cooling magma, below the seafloor, or deep within the crust. For this reason, rocks, minerals, and soils are typically the most weathered at the surface and progressively less weathered with depth (April *et al.* 1986).

As noted in Chapter 4, primary minerals are those that crystallize as magma cools from high temperatures. Over long periods, primary minerals are unstable in soils and weather to secondary minerals, commonly clay minerals. In short, the essence of weathering is the breakup of rock and the formation of secondary minerals from the inherited (primary) minerals, as rocks are changed into forms that are more stable at the Earth's surface.

Several other processes are almost always associated with weathering, e.g., erosion (the wearing away of rocks or sediments/soils) and transport (the movement of those same materials), which collectively are termed *denudation*. Weathering, erosion, mass wasting, and transportation are all included within denudational processes. During transport, sediments may also be *corraded* (eroded by friction or ablation).

Rocks (Table 9.1) and minerals (Fig. 9.1) vary in their resistance to weathering. Minerals that weather slowly or difficultly are termed "resistant" minerals, whereas those that weather more quickly are variously referred to as "weatherable" or "weak" minerals. The susceptibility of various minerals to weathering is largely captured by Bowen's series, which illustrates that minerals that crystallize first from a molten state are most susceptible to weathering, and those that crystallize later are more resistant (White and Brantley 1995; see Chapter 4). The structure of minerals also dramatically affects not only the susceptibility of a mineral to weathering, but the manner in which it weathers (Fig. 9.2).

Rocks with many weathering-susceptible minerals are vulnerable to weathering. However, rocks with many individual minerals that are only loosely held together will also be highly weatherable, regardless of their mineralogy. Weathering susceptibility of rocks is also a function of the size of the individual minerals as well as the degree to which they are cemented together. In general, coarse-textured rocks with larger mineral crystals, such as granite, weather more rapidly than do finer-textured intrusive rocks. It is also easier for physical weathering processes to separate the mineral particles in the intrusive, igneous and acidic, metamorphic rocks, e.g., gneiss, than in finer-textured, extrusive rocks such as basalt. For many clastic sedimentary rocks, e.g., sandstone, the degree of cementation of the grains is a primary factor determining their resistance to weathering.

Table 9.1 | General classification of the physical strength of some rock types

Description	Examples
Very weak rocks	Chalk, rock salt, lignite
Weak rocks	Coal, siltstone, schist
Moderately strong rocks	Slate, shale, sandstone, ignimbrite
Strong rocks	Marble[a], limestone[a], dolomite, andesite, granite, gneiss
Very strong rocks	Quartzite, dolerite, gabbro

[a] Marble and limestone are *physically* resistant rocks but are prone to rapid dissolution by carbonic acid present in the rain and groundwater of humid climates. As a result, in many humid landscapes, limestone and marble form valleys.
Source: Selby (1980).

Physical Weathering

Traditionally, weathering has been subdivided into physical and chemical components, with the former being a necessary precursor to the latter. In physical weathering, also known as *mechanical weathering* or *disintegration*, physical stresses combine to break rock apart. Chemical weathering is driven by processes that decompose geologic materials into substances that have different chemical compositions, usually resulting in a loss of mass, as some of the weathering by-products are flushed from the system. We recognize that this bipartite categorization of weathering may be simplistic (Hall *et al.* 2012), but it nonetheless does provide an easily conceptualized framework from which to start this discussion. We also stress that neither form of weathering is purely abiotic; organisms and biological processes play a role in almost all forms of weathering.

In physical weathering, little or no chemical alteration is implied; the rock is simply broken into smaller pieces. These breaks or fractures can occur at seemingly random locations across the rock, along joints or bedding planes, as in a limestone, or between mineral grains, as in a sandstone or granite (Fig. 9.3). The size of fragments formed by physical weathering depends on the degree of fracturing and the spacing of fractures, as well as grain-to-grain contacts within the rock. A rock with large phenocrysts, e.g., granite, will weather into larger mineral fragments than will a rock composed of small mineral grains, e.g., schist. A common form of weathered, coarse-grained rock is *grus* – a sandy and gravelly, slightly weathered version of the parent rock, which, however, crumbles easily (Migoń 1997, Migoń and Thomas 2002).

To a soil geomorphologist, the primary importance of physical weathering centers on the formation of additional surface area and pore space. As rocks become fractured, their surface area increases geometrically (Fig. 9.4). Chemical weathering (discussed later) reactions operate on *surfaces*, and thus as a rock becomes increasingly fractured, the surface area upon which chemical reactions can occur increases. Likewise, fractured rocks with more pores allow more oxygen and water to have direct contact with the rock, again facilitating chemical weathering. Water and oxygen are important to many chemical weathering reactions. Highly fractured rock tends to have increased porosity and permeability, which when combined with abundant water facilitate the transport of by-products of chemical weathering reactions. If those by-products are not flushed from the system, the weathering reactions slow and eventually stop. Pore spaces also facilitate the growth of roots, which in turn can cause further physical breakup. Roots promote chemical breakup by their exudates and various fungal hyphae associated with them (Bornyasza *et al.* 2005). Roots also release CO_2, which combines with water to form carbonic acid, further enhancing weathering.

Most forms of physical weathering are incapable of breaking up rock into particles smaller than sand or coarse silt size (Stromsoe and Paasche 2011). Thus, areas dominated by physical weathering tend to have coarse-textured regolith, residuum, and soils. In order to break down the sand and silt grains produced by physical weathering, various chemical agents are usually required (see later discussion).

Forms of Physical Weathering

Physical weathering occurs as stress is *added to* a rock (usually from along an existing plane or point of weakness) or as pressure is *released from* the rock, allowing it to expand and crack. The latter process is called *unloading* or *dilation*. Unloading occurs when weight is gradually or suddenly removed from dense rock masses, causing the rock to expand and sometimes resulting in sheets (sheeting) or flakes (flaking) that appear to peel from the rock. Dilation (expansion) can be up to 1% in some granites. Rocks undergoing unloading typically expand along planes that are at right angles to the direction of pressure release. When this occurs on a large scale and the sheets are thick and/or curved, the term *exfoliation* is used (Bradley 1963). Exfoliation is often restricted to massive, otherwise unjointed rocks; slaking or sheeting is used for smaller-scale versions of the same. Rock corners and edges are the most susceptible to sheeting and flaking, as corners have more surface area to attack, and therefore potential for rounding. Thus, one major impact

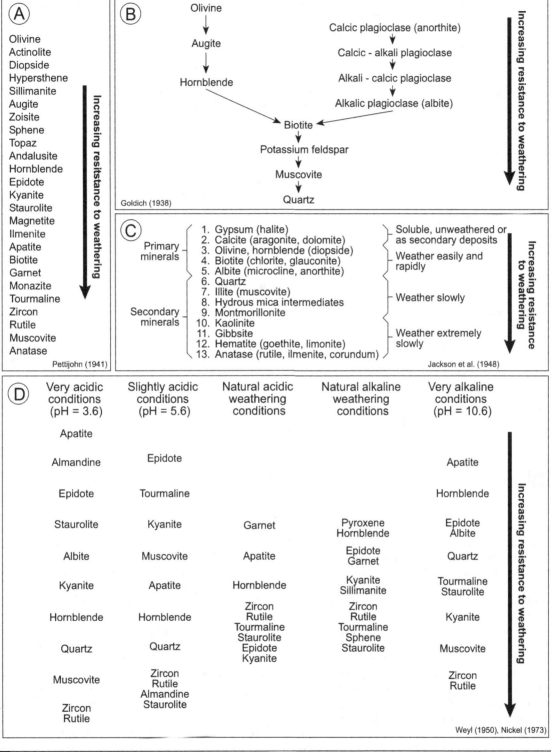

Fig. 9.1 Various rankings of common rock-forming minerals with respect to their general resistance to weathering. Because they have been compiled from various sources, exact correspondence among the sequences is not implied. (A) On the basis of their frequency in sedimentary rocks of increasing age, which he interpreted as their degree of stability. (B) Approximating the inverse sequence with which minerals crystallize from a melt, i.e., Bowen's series. (C) Weathering sequence of clays and clay-sized minerals. (D) Mineral resistances at various acidity/alkalinity conditions.

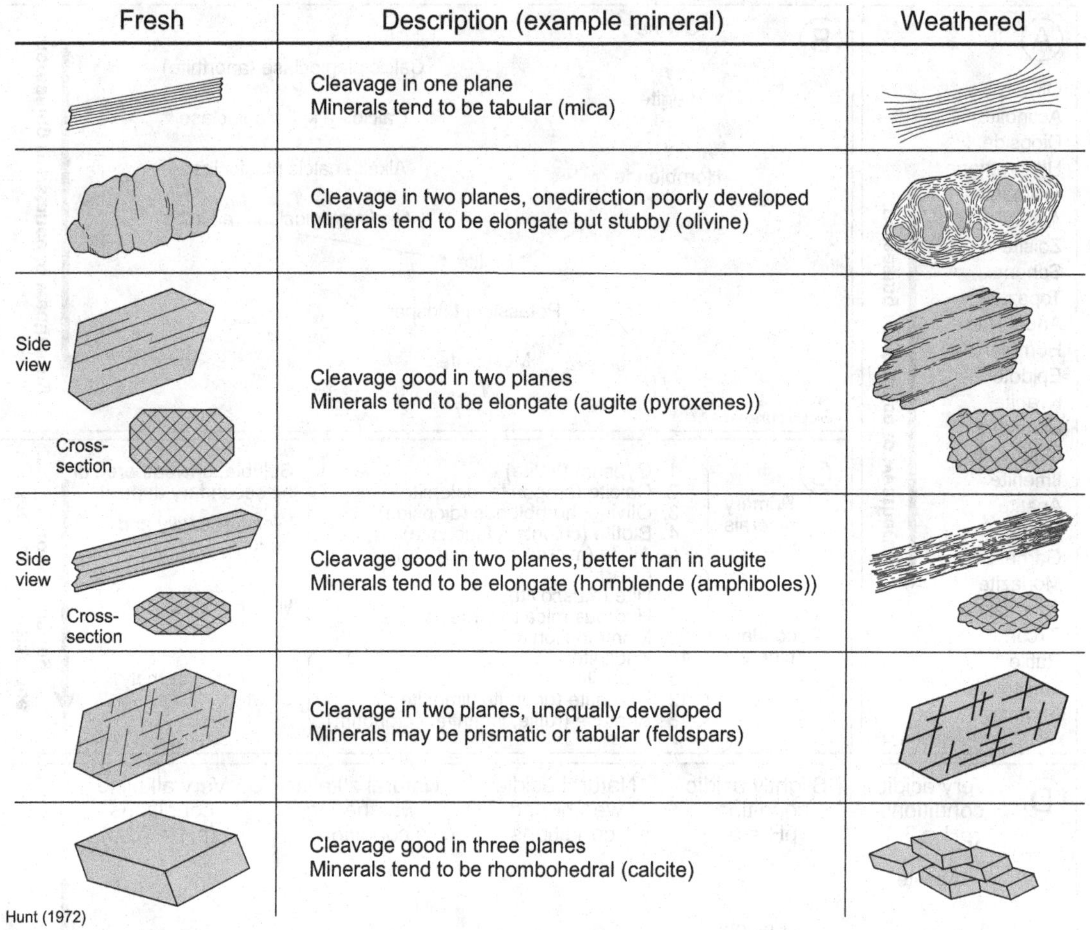

Fresh	Description (example mineral)	Weathered
	Cleavage in one plane Minerals tend to be tabular (mica)	
	Cleavage in two planes, onedirection poorly developed Minerals tend to be elongate but stubby (olivine)	
Side view Cross-section	Cleavage good in two planes Minerals tend to be elongate (augite (pyroxenes))	
Side view Cross-section	Cleavage good in two planes, better than in augite Minerals tend to be elongate (hornblende (amphiboles))	
	Cleavage in two planes, unequally developed Minerals may be prismatic or tabular (feldspars)	
	Cleavage good in three planes Minerals tend to be rhombohedral (calcite)	

Hunt (1972)

Fig. 9.2 Examples of how mineral structure and cleavage affect the way they break up by physical processes.

of these processes is to round the edges of rocks. Continued sheeting and flaking wear away the sharp edges of the rock, producing a round or spheroidal shape, hence the name "spheroidal weathering" (Fig. 9.5).

Most forms of physical weathering occur when internal stresses overcome the strength of the rock, causing it to disintegrate. These stresses can result from a number of processes: (1) salt crystal growth, occurring as salty water within cracks in the rock evaporates, leading to the growth and expansion of salt crystals in the void (Amit et al. 1993, Smith et al. 2005, Prikryl et al. 2007, Buj et al. 2011); (2) ice crystal growth as water within cracks freezes (Fahey and Lefebure 1988, Matsuoka 1995, 2008, Matsuoka and Murton 2008, Ruedrich et al. 2011); (3) root growth within cracks (Johnson-Maynard et al. 1994); and (4) variable expansion stresses set up during wet-dry and thermal cycles (Matsuoka and Murton 2008, Dorn 2011). All of these stresses are more effective when they occur in numerous and repetitive cycles (Smith et al. 2005, Viles and Goudie 2007, Eppes et al. 2010, Viles et al. 2010).

Although most common there, salt weathering is certainly not restricted to deserts (Goudie and Viles 1997). Weathering by salts and gypsum is most pronounced in arid regions and near sea coasts, where salty water is abundant (Reheis 1987b, Doornkamp and Ibrahim 1990, Smith et al. 2000). In some urban areas an overabundance of salts due to ice removal, fly ash dust, general traffic dust, and other materials occurs. Water containing these dissolved substances penetrates crevices in rocks. Upon drying, evapoconcentration processes facilitate the growth of salt crystals, which can force apart the rock (Holmer 1998). Water vapor is later drawn to the dry salts, hydrating and expanding them, further enhancing the salt weathering process (Amit et al. 1993). In some salty soils, the lack of rocks in the upper solum has been attributed in part to physical shattering of clasts by salts and other forms of mechanical weathering (Dan et al. 1982, Gerson and Amit 1987). Amit et al. (1993) stressed, however, that salt concentrations alone do not cause shattering of clasts, for within petrosalic (Bzm) horizons, shattering does not

Fig. 9.3 Images of physically weathered rocks. Photos by RJS. (A) Seemingly random cracking of a rock, probably driven by frost crystal growth within microfissures. (B) Exfoliation of granite in Yosemite National Park, California, driven mainly by unloading weight release. (C) Sandstone breaking apart along bedding planes, Arches National Park, Utah. (D) Granular disintegration of a granite, due mainly to wetting-drying and salt crystal growth pressures, Anza-Borrego Desert State Park, California. Knife for scale.

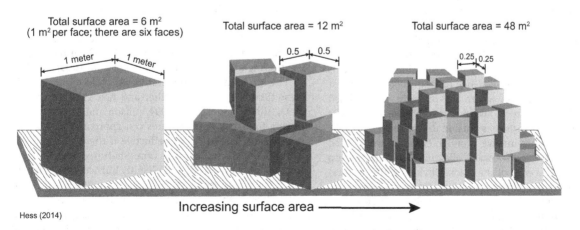

Hess (2014)

Fig. 9.4 Change in surface area as a cube-shaped rock is repeatedly fractured.

Fig. 9.5 Spheroidal weathering of soft siltstone, caused by sheeting and flaking, Goblin Valley State Park, Utah. Photo by RJS.

occur. Rather, changes in moisture content and temperature above, in the upper solum, coupled with the presence of salts and rocks with microcrystalline fractures, lead to rapid and complete rock shattering. The process is as follows: (1) Salt crystal shattering relies upon the formation of a salt crust on the rocks; (2) these crusts plug the openings of the cracks, confining saline solutions to the interiors; and (3) upon further drying and/or heating, the pressures inside the rocks result in shattering.

Sometimes, weathering is a combination of chemical and physical processes. For example, physical weathering, initiated by additions of rain rich in sulfuric acid (H_2SO_4), can weather rocks physically, especially carbonate rocks, as follows:

$$H_2SO_4 + CaCO_3 + H_2O \rightarrow CaSO_4 \cdot 2H_2O + CO_2$$

Gypsum ($CaSO_4 \cdot 2H_2O$) crystals form just below the rock surface, occupying more volume than the original calcite crystals. The newly formed gypsum crystals can flake the limestone away – physically.

The role of freeze-thaw activity and ice crystals in weathering is undeniable and yet, exactly how it functions – and over what temperature range – is debatable (Hall 2007). In theory, as water in cracks in rocks freezes, it expands by 9%, potentially forcing the rock apart and shattering it. This process, also called *frost shattering* or *gelifraction*, works best in deep, narrow cracks where the pressure is contained wholly within the rock and not readily released as the developing ice mass pushes outside it. Optimal conditions for frost shattering involve rapid freezing of water-filled cracks from the top downward, allowing the ice to act as a seal on the outer edge of the crack. Obviously, this process can only operate in climates that at least occasionally have freezing temperatures, and the more cyclicity there is, e.g., in the high alpine or high midlatitudes, the better (Thorn 1979, Dredge 1992, Matsuoka 2008). Recent work on this

process has, however, suggested that the main agent of disintegration may not be expansion by ice but the migration of thin films of water along microfractures. These films may be adsorbed so tightly that they cannot freeze, but they can exert pressure on the rock and cause it to fracture (White 1976, Walder and Hallet 1985, Bland and Rolls 1998).

Thermal expansion of rocks is driven by repeated cycles of insolation (solar heating) and by fire (Dragovich 1993). The former type is also called *insolation weathering* and is most efficient in sunny climates where rocks are exposed at the surface, e.g., deserts of all kinds (McFadden *et al.* 2005, Moores *et al.* 2008). In theory, expansion occurs – and microfractures form – as the outer layers of a rock are heated by the sun (or a fire), only to contract as they cool (Peel 1974, Luque *et al.* 2011). This expansion and contraction – usually on a diurnal basis – lead to thermal gradients and compressive-tensile stresses, i.e., thermal fatigue (Halsey *et al.* 1998, Hall 1999, Sumner *et al.* 2007, Eppes *et al.* 2010). Small-scale differences in the rate of thermal expansion between minerals or weathering rinds may also contribute to the physical breakdown of rocks. In some sedimentary rocks that contain clay minerals, e.g., shale and some limestones, thermal effects enhance the wetting and drying expansion pressures within the rock, adding another dimension of stress (Weiss *et al.* 2004). Overall, these fatigue effects reduce the cohesive strength of intergranular bonds in the rock and initiate microfracture development (Warke *et al.* 1996, Hall and Andre 2001). Cracks formed by thermal expansion and contraction can be further expanded as water and dust enter and undergo successive dissolution/precipitation cycles (Dorn 2011; Fig. 9.6).

The degree to which minerals in the rock expand with temperature is a function of their coefficient of thermal expansion, and in sunlight, their albedo. Minerals vary in these properties, setting up a condition in which the parts of the rock expand and contract at different rates and by different amounts, all contributing to stresses and breakdown. For breakage to finally happen, however, the elastic limit of the rock must be exceeded, and this also varies among rock types (Bland and Rolls 1998). This type of weathering, once thought to be limited to hot, dry climates, is actually also a very important form of physical weathering in cold, dry environments (Hall 1999).

Fire has long been known to be an agent of rock weathering (Blackwelder 1926, Ollier and Ash 1983, Dragovich 1993, Allison and Goudie 1994, Allison and Bristow 1999). During fire, rock temperatures can approach 500°C. "Fire weathering" is particularly effective if the fire is immediately followed by a cooling rain. Shattered rocks on the surface of hot deserts and in recently burned areas attest to the efficacy of thermal expansion as a weathering process, in addition to shattering by salt crystals (Dorn 2003).

Biotic forms of physical weathering have long centered on the role of plants. Roots penetrate fissures in rock and

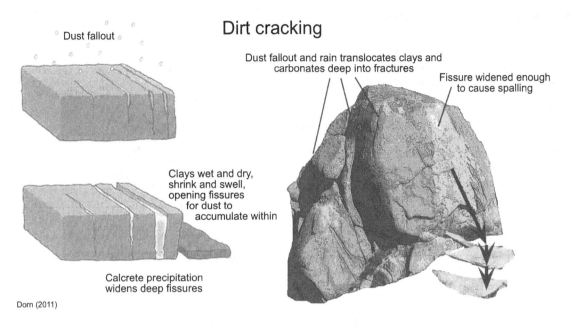

Dust fallout

Dirt cracking

Dust fallout and rain translocates clays and carbonates deep into fractures

Fissure widened enough to cause spalling

Clays wet and dry, shrink and swell, opening fissures for dust to accumulate within

Calcrete precipitation widens deep fissures

Dorn (2011)

Fig. 9.6 The process of dirt cracking – a type of physical weathering. (A) Diagrammatic sketch of the process. (B) Electron-microscope image of assumed eolian in-fill within a rock crack. Carbonates, introduced as eolian particles, have dissolved and reprecipitated onto the walls of the crack as laminar calcrete. This layer thickens with time. Silt- and clay-sized dust particles have since accumulated in the crack, as the process begins again. These particles may eventually reorient and align themselves with the walls of the cracks and, if they contain expansive clays, lead to additional stresses during wetting-drying cycles.

can exert enough pressure to break them apart or move them within regolith (Fig. 9.7). However, fauna also play a large role in disintegration processes, usually functioning at the smaller end of the scale, where their impacts are less noticed (Avakyan *et al.* 1981). Many burrowing animals ingest sand and gravel, and as this material passes through their gut it is physically comminuted (Suzuki *et al.* 2003).

Fig. 9.7 The power of root expansion. This rock, in Michigan's Upper Peninsula, is known locally as Split Rock. Photo by RJS.

Large mammals such as badgers and wombats, however, are more than able to break apart rock or cemented soil horizons as they tunnel.

Last, we note that wetting and drying cycles are also effective agents of physical weathering, especially in rocks that are weak initially or are able to absorb large amounts of water (Hall and Hall 1996, Gokceoglu *et al.* 2000, Paradise 2002). Wetting can dissolve substances that are within cracks, which then precipitate upon drying. For example, if $CaCO_3$ is present in dust, it can dissolve and reprecipitate, causing further expansion, as does the wetting of expansive clay minerals. Repeated cycles of dissolution-precipitation can create expansion pressures and physically break rocks apart.

Ultimately, most physical weathering processes, and many geomorphic processes as well, e.g., glacial grinding, fluvial corrosion, and eolian abrasion, combine to produce regolith that comprises silt-sized or larger particles. Although some coarse clay (>1.0 μm) in soils is composed of primary quartz particles that may have been formed via physical weathering, these are the exception. Most clay-sized (< 2.0 μm) particles are secondary minerals that have been formed via chemical weathering processes at and near the Earth's surface (Fig. 2.6). The large surface area created by physically weathering rocks into smaller particles then facilitates chemical weathering, further reducing the particles in size and, eventually, leading to their dissolution.

Chemical and Biotic Weathering

Chemical weathering, also known as *decomposition*, involves the chemical and biochemical breakdown of rocks and minerals into different products and compounds. Common end products of chemical weathering are clay minerals, soluble acidic compounds, and various ionic species. The intensity of chemical weathering is largely governed by climatic conditions that govern temperature and the movement (or lack thereof) of water in the soil and rock, water quality, and the mineral and chemical composition of the rocks themselves. As noted, rock permeability and porosity, as well as surface area, are also important, as they facilitate or retard flushing of reaction by-products. Indeed, in areas where chemical weathering is abnormally active or inactive, in terms of what would be climatically "expected," one can usually look to parts of the chemical weathering environment not governed by climate, e.g., permeability, rock chemistry, strength of weathering agents, for an explanation (Pope *et al.* 1995).

Chemical weathering is the primary means by which rock material, and the gravel, sand, and silt particles that form from it by *physical* weathering, are ultimately decomposed into (1) ionic forms and (2) chemical and mineral structures that are highly stable at the Earth's surface and in soils. Chemical weathering of primary minerals results in the release of base cations (Ca, Mg, K, and Na) to the soil solution. Equally importantly, chemical weathering is the engine that drives the formation of secondary clay minerals, which are more stable in the regolith than are primary minerals, which form from molten magma or lava.

Chemical weathering can be subdivided into congruent and incongruent dissolution. Congruent dissolution occurs when a mineral dissolves completely in solution and nothing is immediately precipitated from it. Many highly soluble minerals, e.g., halite, weather congruently – i.e., they dissolve – in water. Incongruent dissolution means that all or some of the ions released by weathering precipitate at or near the site of weathering to form new minerals such as phyllosilicates or oxyhydroxides.

Enhancing this twofold generalization, Bland and Rolls (1998) divided chemical weathering into three categories, i.e., focused on its three main outcomes:

1. Solution of ions and molecules
2. Production of new minerals, e.g., clay minerals, oxides, and hydroxides
3. Release and residual accumulation of insoluble or otherwise unweatherable materials

Any or all of the preceding situations results in chemical changes in the preexisting mineral and the formation of new minerals or parts (ions) that could be used to "construct" them. Hence, although chemical weathering by definition involves rock decomposition, an equally important part of the process is the synthesis of new compounds

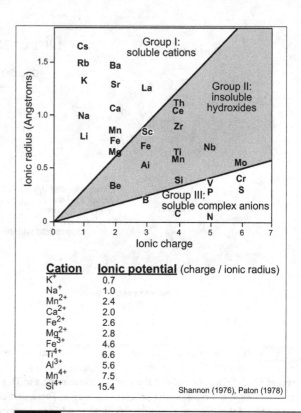

Cation	Ionic potential (charge / ionic radius)
K^+	0.7
Na^+	1.0
Mn^{2+}	2.4
Ca^{2+}	2.0
Fe^{2+}	2.6
Mg^{2+}	2.8
Fe^{3+}	4.6
Ti^{4+}	6.6
Al^{3+}	5.6
Mn^{4+}	7.5
Si^{4+}	15.4

Shannon (1976), Paton (1978)

Fig. 9.8 The ionic potentials of various elements common to soils and regolith. The ion of each element can be placed into one of three categories that roughly coincide with their relative potential to be leached from soils.

that occurs concomitantly or long after the main reaction has stopped. Some of the more important and commonly mentioned mechanisms of chemical weathering are discussed later.

Chemical weathering reactions can be described on the basis of their overall stoichiometry or at the molecular scale. The molecular-scale steps depend on the mineral and its purity as well as the composition and movement of water at the particle surface. At the interface between mineral crystal and water, the kinds and rates of chemical reactions are regulated by pH, redox potential, ionic strength, presence of complexing agents, and potential for precipitation of additional minerals.

Mineral Solubility in Water

Minerals in evaporite rocks form by crystallization from a saturated or supersaturated solution. Some evaporite minerals, such as halite (NaCl) and sylvite (potash, KCl), contain alkali or alkaline earth metals and are readily soluble in water. The ability of water molecules to solvate alkali and alkaline earth cations reflects the relatively low ionic potential or charge density (ratio of charge to ionic radius) of these ions (Fig. 9.8). For example, the low ionic potential of Na^+ in halite allows the ionic bonds linking Na^+ and Cl^- to be disrupted by water molecules that compete with and replace the Cl^- ions around each Na^+. Therefore, halite is very soluble.

In general, the lower the ionic potential of an ion, the more likely it is to be soluble. Still, the solubility of evaporite minerals is not uniformly high. For example, calcite ($CaCO_3$), dolomite ($CaMg(CO_3)_2$), and hydroxyapatite ($Ca_5(PO_4)_3(OH)$) are sparingly soluble in pure water, whereas gypsum ($CaSO_4 \cdot 2H_2O$) is intermediate in solubility. The solubility of these minerals is increased in acidic solutions, as described later.

Destabilization by Charge Imbalance

Chemical weathering of minerals occurs first at the exterior surfaces of mineral particles, often where broken bonds or lattice distortions occur at points of crystal defects. This explains why physical weathering is an important precursor to chemical weathering; physical weathering creates more surface area (Fig. 9.4). Molecular-scale weathering processes create local charge imbalances that weaken and break bonds, followed by transfer of ions to the solution that surrounds the particle. The exact sequence of steps varies; it usually includes (1) ion exchange, (2) protonation, and (3) hydration reactions. These reactions depend significantly on solution pH because hydronium (H_3O^+) ions or hydroxyl ions are key reactants. The term *hydrolysis* (Greek, *hydro*, water, *lysis*, split) is often used to describe the breakdown of minerals by chemical reactions that involve water, but it can also refer to the breaking of specific bonds in a mineral, as described later.

As noted previously and in Chapter 5, carbonic acid derived from dissolution of CO_2(g) in water is a key source of hydronium ions. A portion of the H_2CO_3 thus formed dissociates in the water to produce H_3O^+ and HCO_3^-. Under acidic conditions, hydronium ions are capable of *exchanging* with monovalent and divalent cations near a crystal's surface. Consider the feldspar albite, for example. In this mineral, Na^+ balances the excess negative charge that results where Al^{3+} has substituted for Si^{4+} in the silica framework. Exchange of H_3O^+ for Na^+ releases Na^+ to the surrounding solution. It also distorts the distribution of charges associated with nearby $\equiv Al-O-Si \equiv$ bonds, where $\equiv$ represents bonds to adjacent O^{2-} bridges. One resolution to this distortion is for H_3O^+ to donate a proton to an adjacent O^{2-} ion, a process called *protonation*. Protonation of the bridging O in an $\equiv Al-O-Si \equiv$ group in the framework weakens the Al–O bond to the point that it breaks, forming $\equiv Si-OH$. A water molecule is then added to the Al^{3+} ion (this process is called *hydration*) to form $\equiv Al-OH_2^+$ at that point on the surface.

This suite of reactions, taking place at many points on a mineral's surface, leads to many broken bonds and a weakened crystal structure. It is nonetheless often observed that some facsimile of the structural framework remains intact. Near the particle surface, a cation-depleted crust (10s to 100s of nms thick) forms (Brantley 2003). Ultimately, this

weathering rind disintegrates, too, releasing Al and Si to the surrounding solution. To summarize the exchange, protonation, and hydration steps, the overall stoichiometry of albite weathering may be written as a hydrolysis reaction:

$$NaAlSi_3O_8(s) + 4H_2O + 4H^+(aq) \rightarrow Na^+(aq) + Al^{3+}(aq) + 3Si(OH)_4(aq)$$

Protonation is also an important first step in the weathering of calcite ($CaCO_3$), the dominant mineral in limestone. The primary source of protons depends on the pH of the system because that determines whether the most likely species near the calcite surface will be H_2CO_3, H_3O^+, or H_2O. When one of these species protonates CO_3^{2-} groups at the crystal surface to form HCO_3^-, the local structure is electrically imbalanced and destabilized, allowing Ca^{2+} to be solubilized by H_2O molecules (Plummer *et al.* 1978).

Dissolution of calcite by carbonic acid creates a plethora of weathered forms in limestone. The natural fracture patterns in carbonate rocks are often enlarged, leading to increased porosity and permeability. Thus, surface waters can percolate through these rocks rapidly, and there exist few surface streams. Underground rivers, collapsed caves, swallowholes, and sinkholes are common features of *karst* landscapes that are underlain by this type of weathered limestone (Fig. 9.9). This type of weathering is sometimes referred to as *carbonation* – a general term for weathering processes associated with CO_2 in aqueous solution.

In addition to hydration reactions, oxidation and reduction reactions play important roles in mineral weathering. Many primary silicates, e.g., olivines, hornblendes, and pyroxenes, contain the redox-sensitive element Fe, and oxidation of Fe leads to charge imbalance and crystal destabilization. Biotite is an Fe-bearing, primary layer silicate (a mica) that crystallizes from the magma when oxygen is not present. However, when biotite is in an oxidizing soil environment, the oxidation of Fe ($Fe^{2+} \rightarrow Fe^{3+} + e^-$) in the octahedral sheet of the crystal results in decreased layer charge and weakens the electrostatic attraction for K^+ in the interlayer position. For this reason, biotite weathers more readily than muscovite, another mica that does not contain Fe (Thompson and Ukrainczyk 2002).

Another example of a mineral affected by oxidation-reduction processes is the accessory mineral pyrite, composed of reduced forms of Fe and S. Although it is a complex, multi-step process, the oxidation of S in pyrite can be described by the following summary reaction:

$$4FeS_2(s) + 14O_2(g) + 4H_2O \rightarrow 4Fe^{2+}(aq) + 8SO_4^{2-}(aq) + 8H^+(aq)$$

This reaction can proceed abiotically, although it is much faster when S-oxidizing bacteria are present. The Fe (II) released from the mineral is oxidized to Fe (III), as shown in the following reaction:

$$4Fe^{2+}(aq) + O_2(g) + 4H^+(aq) \rightarrow 4Fe^{3+}(aq) + 2H_2O$$

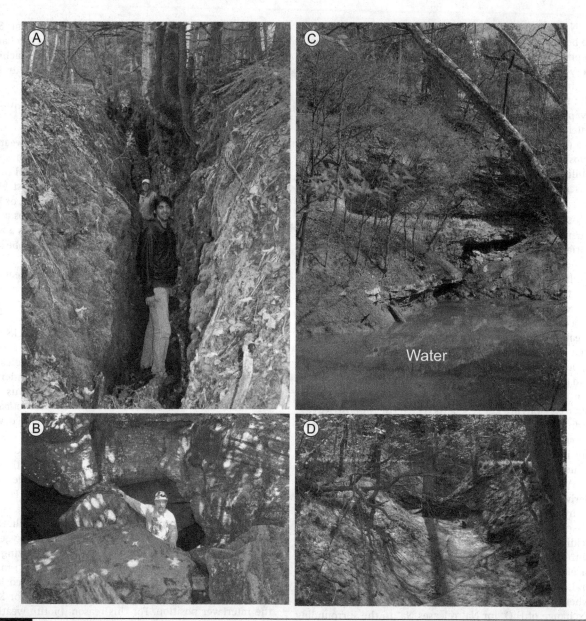

Fig. 9.9 Geomorphic features associated with dissolution of limestone rocks, i.e., karst. Photos by RJS. (A) Carbonation is focused along cracks in this limestone bedrock, where water and its associated carbonic acids flow, and through which reaction by-products can be flushed. Eventually, these cracks are widened, as is shown here, in northern Michigan. (B) In karst landscapes, long-term underground dissolution of limestone by carbonation results in caverns, which are manifested as sinkholes and caves, like this one in southern Pennsylvania. (C) After rivers flow into swallowholes, like the one shown here, they often flow underground for some distance. This swallowhole occurs in the valley of the Lost River in southern Indiana. (D) The valley of the Lost River, downstream from a swallowhole; all that is visible is a dry channel that only carries water during flood events, i.e., when the swallowhole upstream cannot handle all the surface water.

leading to further oxidation of S in the solid phase, carried out now by Fe^{3+} in the aqueous phase:

$$FeS_2(s) + 14Fe^{3+}(aq) + 8H_2O \rightarrow 15Fe^{2+}(aq) + 2SO_4^{2-}(aq) + 16H^+(aq)$$

Reduction reactions can also advance mineral dissolution. Reducing conditions brought about by microbial activity in anaerobic, waterlogged soils promote the dissolution of ferric oxyhydroxides such as goethite and lepidocrocite. When the redox potential is sufficiently low, Fe^{3+} in those

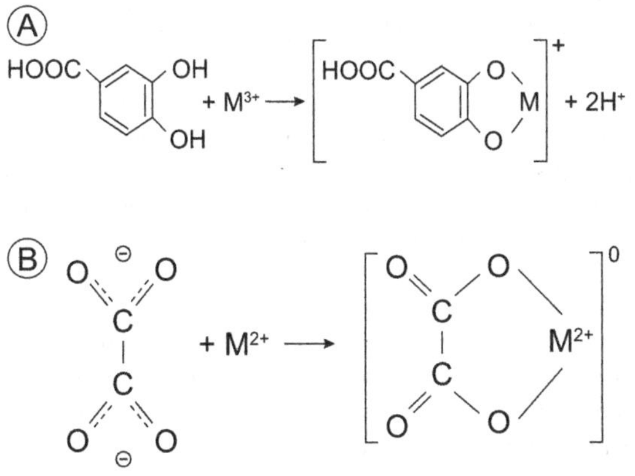

Fig. 9.10 Examples of chelation of metal cations by organic acids at pH ~4.5. Water molecules associated with the metal ions are omitted for clarity. (A) Complexation of metal, e.g., Fe^{3+} or Al^{3+}, with phenolic groups of protocatechuic acid. After Vance et al. (1986). (B) Chelation complex of a divalent metal, e.g., Ca^{2+}, with oxalate. After Bland and Rolls (1998).

minerals is converted to Fe^{2+}, destabilizing the crystal structures, and leading to soluble Fe^{2+} (see Chapter 13):

$$FeOOH(s) + 3H^+(aq) + e^- \rightarrow Fe^{2+}(aq) + 2H_2O$$

Preventing the Backward Reaction

If the products of a chemical weathering reaction were to remain near the reaction site, they might react with one another in the reverse reaction. Therefore, chemical weathering is promoted by processes that prevent reversible chemical reactions, such as exchange reactions, from going backward. For example, when weathered cations are transported away from the solid surface by water, their concentrations in the solution at the solid-liquid boundary decline, allowing the forward reaction to proceed further. Hence, weathering continues. Alternatively, the ionic products of weathering reactions may precipitate into secondary minerals. For example, Fe^{3+} released from Fe-bearing silicates commonly precipitates as an oxyhydroxide solid that ultimately crystallizes to goethite, α-FeOOH. Similarly, Al and Si ions released during the weathering of feldspar may precipitate to form kaolinite, $Al_2Si_2O_5(OH)_4$, as shown here in a summary reaction for albite weathering:

$$NaAlSi_3O_8(s) + \tfrac{9}{2}H_2O + H^+ \rightarrow Na^+(aq) + 2Si(OH)_4(aq) + \tfrac{1}{2}Al_2Si_2O_5(OH)_4(s)$$

Although precipitation of weathering by-products does remove them from participating in back reactions, precipitation of secondary minerals close to a weathering surface may also slow the forward reaction by limiting the diffusion of solutes to pores where mobile water could remove them from the reaction zone.

Finally, we note that complexation reactions, in which metal ions are bound in aqueous complexes with organic ligands, are particularly effective in promoting mineral weathering (Fig. 9.10). The organic ligands are normally low-molecular-mass conjugate bases of organic acids that have been released to the solution by microbial decomposition of organic matter, or directly from plant roots. Chelation complexes, in which more than one ligand from the same anion coordinates with the central metal ion, are especially stable. Not only are complexed metals such as Ca, Fe, or Al effectively removed from participation in back reactions by this process, the complexed species also tend to remain in solution and thus can be more readily moved away from the dissolving mineral surface by mass transport in the surrounding solution.

In summary, silicate weathering reactions are typical in that (1) a primary silicate mineral is weathered in the presence of water, carbonic acid, and/or commonly occurring complexing agents; (2) a secondary mineral forms as a result; (3) some of the reaction by-products are soluble and, hence, can be removed from the system; and (4) in the end, the pH of the soil solution is lowered as pH-buffering components are dissolved and leached. To the extent that the reaction by-products, e.g., base cations and other metals, are soluble, a portion of the original mass is lost. In some environments, e.g., the humid tropics, the silica remains soluble and leaves the system by leaching (*desilication*, see Chapter 13), while in others it may precipitate as quartz or other microcrystalline silicate minerals or amorphous compounds.

Biological Agents

Almost all weathering reactions are enhanced or mediated by biota. Plants and their associated microbiota directly impact weathering by generating complexing agents and other chemical compounds capable of weathering rock. Biota modify the pH of the rock-soil interface, through production of CO_2 and organic acids. Richter and Markewitz (1995) provided abundant data on the importance of carbonic acid as a weathering agent, particularly in the C horizon, where organic and other acids are less active. Respiration from microbes and plant roots provides the CO_2 to the deep subsoil, fueling the formation of carbonic acid and driving weathering processes. They also stress that, because most of the carbonic acid in soils is derived from plants and other biota, weathering by carbonic acid should be considered a biochemical process, initiated and strongly mediated by biota.

Production of organic acids may be the most important contributor to biochemical weathering in soils. These acids are released directly from the organism and also are derived as by-products of organic matter decomposition. Bland and Rolls (1998) described two ways in which organic acids contribute to chemical weathering:

(1) Monovalent cations in the mineral are replaced by hydronium ions and then removed in solution, and (2) di- and trivalent cations are made soluble, either after redox reactions involving dissolved organic compounds have occurred or by chelation by organic acids (Schatz 1963). In either event, minerals are decomposed and their soluble by-products are flushed from the system. Obviously, biota also play a critical role in the fate of weathering by-products by biocycling them.

Recently, research on rock weathering has focused on the role of fungi, which often form symbiotic relationships with plant roots (Etienne and DuPont 2002, Hoffland *et al.* 2002). Particularly, *ectomycorrhizal fungi* mobilize essential plant nutrients directly from minerals in rocks, through excretion of organic acids (Landeweert *et al.* 2001). In return, the fungi receive carbohydrates from the plant. Chemical weathering of clay minerals in termite mounds has also been associated with the presence of fungi that are deliberately cultivated by some species of termites.

Lichens – composite organisms whose body (*thallus*) consists of a fungus and an alga growing symbiotically – are early colonizers on bare rock surfaces and are therefore instrumental in the incipient stages of both physical and chemical rock weathering (Williams and Rudolph 1974, Ascaso *et al.* 1982, Chen *et al.* 2000, Aghamiri and Schwartzman 2002, de la Rosa *et al.* 2012). Their nutrient-absorbing and transmitting bodies, *mycelia*, are composed of a complex web of fine, threadlike *hyphae*. The hyphae can penetrate the rock surface along microcracks and generate enough stress to fracture off parts of it. Likewise, as water is absorbed into the hyphae they expand and create additional pressure.

On bare rock surfaces, lichens enhance the water-holding capacity of the surface, which may increase chemical dissolution and, climate permitting, frost wedging (Bjelland and Thorseth 2002). Lichens also emit organic acids, especially oxalic acid, which are very effective chemical weathering agents (Ascaso and Galvan 1976). The swelling action of the organic and inorganic salts originating from lichens may also add to physical breakup of rocks (Erginal and Ozturk 2009). The exact nature of the chemical weathering process associated with lichens remains unclear, but it is suspected that the acids from lichens (1) donate protons (H^+) that cause hydrolysis of the minerals and (2) act as agents of chelation. Last, the CO_2 emitted by the lichen plays a role in weathering via its generation of carbonic acid.

Although lichens invade bare rock surfaces and have an important role in the initiation of rock weathering, they also act to protect the rock and its regolith (Mottershead and Lucas 2000). The relative importance of lichens vis-à-vis rock protection versus rock weathering is still being debated (Arino *et al.* 1995, Di-Bonaventura *et al.* 1999, Carter and Viles 2005).

Products of Weathering

Weathered rock often has an intermediate stage before it becomes soil – *saprolite* (see Chapter 3). Saprolite is chemically weathered rock that retains most of its original physical structure. Many of the more weatherable minerals have been lost from saprolite, but the skeleton of the rock remains. The fate of the various ions, weathered chemically and released into the soil solution, varies. They may be (1) taken up by plants and cycled through biota, (2) precipitated in secondary minerals, (3) retained in the soil by exchange or adsorption reactions, or (4) leached, ending up in surface waters or groundwater.

Focused on the potential of leaching of weathering by-products, a research stream that is concerned with measuring rates and types of weathering, usually on watershed scales, has emerged. The technique applies data on the concentrations and types of cations in ground and surface waters exiting the basin to determine the types and rates of weathering occurring within (Velbel 1985, April *et al.* 1986, Pecher 1994, Richards and Kump 1997, Furman *et al.* 1998). The chemical signature of stream water is influenced by inputs from the atmosphere and from rock weathering inputs and is mediated by biological uptake, soil storage, and fluctuations in the stores of organic matter in the watershed (Gassama and Violette 2012). The importance of weathering and biological activity can be assessed – even seasonally – by subtracting atmospheric inputs.

The assumption in these studies is that the soluble cations released by weathering reactions eventually leave the system via flux from springs, streams, and groundwater, and if the geology is known, a mass balance approach can provide information about which minerals are weathering and at what rates (Duan *et al.* 2002, West *et al.* 2002, Soumya *et al.* 2011). Rates of landscape denudation due to chemical weathering can even be calculated using such data (Pope *et al.* 1995, Yuretich *et al.* 1996, White *et al.* 1998).

Controls on Physical and Chemical Weathering

Many factors control and impact weathering rates and govern the fates of weathering by-products. Not the least of these is macroclimate, because of its impact on moisture availability and temperature (Büdel 1982). Temperature impacts physical weathering in that it can affect freeze-thaw processes and, in hot deserts, thermal and salt weathering. It may be equally or more important for chemical weathering because reaction rates increase at higher temperatures (White *et al.* 1999). This concept is captured in the general rule of thumb that, for every 10°C increase in temperature, the reaction rate is doubled. Obviously, this

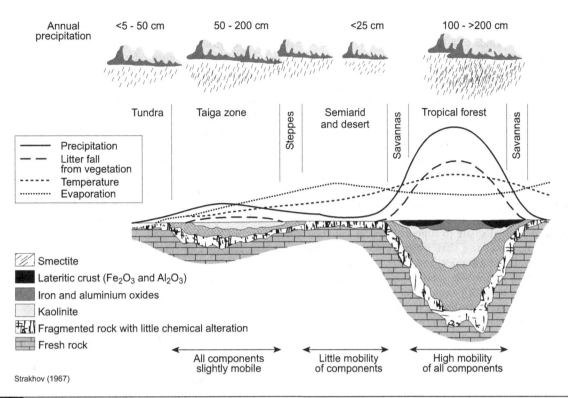

Strakhov (1967)

Fig. 9.11 Generalized relationships among depth of weathering, i.e., regolith thickness, and soil mineralogy along a latitudinal transect from the equator to the pole. Although generally correct, this simplified view of weathering across the globe creates misconceptions that ignore denudation rate adjustments and climate change.

"rule" only holds within certain temperature limits, usually taken to be 5°C–35°C. Nonetheless, the implications of this rule are huge; in the humid tropics many chemical reactions operate more rapidly than in midlatitudes, and where water supply is adequate, deep weathering profiles of highly weathered regolith can result (Fig. 9.11). We stress that temperature affects reaction *rates*, but the *types* of reactions are determined by the minerals present as well as solution variables such as pH, redox potential, and ionic strength.

Physical weathering and its products have traditionally been viewed as being dominant in rocks undergoing weathering in cold and/or dry climates. Peltier's (1950) classic work, although now dated, retains its value in pointing out that physical weathering is best expressed in cold, dry climates, especially those with frequent freeze-thaw cycles (Thorn 1979, Matsuoka 2001; Fig. 9.12). Physical weathering is also important in many deserts, and along sea coasts, where salts and water commingle.

A major control on chemical weathering involves the degree to which water can participate in the reaction, including the ability to flush the system of weathering

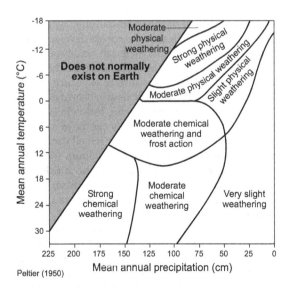

Peltier (1950)

Fig. 9.12 Relative importance of the various types of physical and chemical weathering in different environments.

by-products. In stagnant but wet settings, chemical weathering quickly grinds to a halt, as weathering by-products accumulate. Because water availability is a function of not only local factors but also macroclimate, it has long been assumed that chemical weathering is dominant in hot, wet (humid) climates where the reactions involved are assisted by warm temperatures.

Although largely correct, these characterizations of large-scale climate-weathering relationships are coming under increasing scrutiny (Pope *et al.* 1995). Hall *et al.* (2002) pointed out that, contrary to long-standing belief, weathering, including chemical weathering, is not strictly temperature-limited but is, instead, often limited more by moisture availability. Even in polar regions, temperatures are more than adequate to support physical *and* chemical weathering, providing that water is present. Wherever water is available, chemical and biotic weathering can be an important part of the weathering regime (Thorn *et al.* 2001). Pope *et al.* (1995) argued that the macroclimatically driven approach to weathering is flawed because it is based on large-scale qualitative observations, and that molecular-scale mechanisms of weathering do not support the broad-brush, macroclimate-zonal categorization of Peltier (1950). In short, the climatic-zonation models of weathering fall apart because the process link between tropospheric climatic conditions and boundary layer weathering processes is often disjunct.

Weathering versus Soil Production

Many soils form in thick sequences of unconsolidated parent materials that had been previously transported to the site by glaciers, rivers, or other geomorphic agents. Soils that form in this kind of *transported regolith* develop quickly, down into the unconsolidated materials, and profile thickness at any one time is mainly a function of the (1) rate of pedogenesis and (2) soil, or surface exposure, age (see Chapter 15). In other areas, the soils are formed directly in *residuum* derived from the underlying bedrock. Here, profile thickness is mainly governed by the rates of soil (or regolith) *production* versus *erosion*. That is, soil cannot form directly *in* bedrock; it must form in residuum. Therefore, in many shallow-to-bedrock soils, regolith thickness is equivalent to the rate of regolith production from bedrock, minus the rate at which this material is lost via slope processes and leaching.

Knowing this, an area of research that focuses on "soil production" rates has recently emerged (Humphreys and Wilkinson 2007, Almond *et al.* 2007). A goal of this research is to establish the relationship between soil production and soil thickness: How does soil production from bedrock vary as a function of the thickness of regolith? Does this relationship vary with slope position? Does the production of soil from weathering diminish over time? How does denudation of the surface affect the weathering rates of the materials below (Schaller *et al.* 2009)?

Geomorphologists who study and model the evolution of hillslopes are also concerned about soil production rates, because they relate directly to landscape downwasting, slope stability, and sediment transport (Almond *et al.* 2007). Additionally, knowledge of erosion versus production rates will help constrain the limits of sustainable soil loss by land use practices such as deforestation and tillage (Heimsath *et al.* 2001, McNeill and Winiwarter 2004, Dosseto *et al.* 2011). When erosion outpaces soil production, soils thin and bare bedrock can become exposed, or even dominate a landscape. In short, it is useful and practical to improve understanding of not only soil production, but soil erosion as well.

This field of research is based on the concept of a *soil production function* (SPF) that approximates the production of soil (residuum) from bedrock. On the basis of the pioneering work of G. K. Gilbert (1877), the rate of soil – or regolith – production on bedrock surfaces has been assumed to depend largely on the depth of overlying soil or regolith (Humphreys and Wilkinson 2007). Regolith production on bare bedrock is thought to proceed mainly by physical weathering processes, with freeze-thaw processes and crystal growth leading the way (Fig. 9.13A). Chemical weathering is slower on bare bedrock than where the rock is overlain by some thickness of weathered material (soil/regolith), because the regolith retains additional water and fosters the growth of plants. The increased frequency and duration of wetness at the soil-bedrock surface (often proportional to regolith thickness) enhance chemical weathering at the interface (Merrill 1906). Plants and animals, repelled by bare bedrock sites but attracted by regolith, foster rock breakdown both chemically (mainly by additions of $CO_2(g)$ and organic acids) and physically (mainly by rooting of plants and burrowing of animals). Also, water that percolates to the regolith-bedrock interface through only a thin soil cover is relatively fresh and chemically aggressive, i.e., not saturated with soluble weathering by-products. As a result, the underlying bedrock weathers more rapidly and soil production is maximal under this thin regolith (Gilbert 1877; Fig. 9.13A, B). However, as regolith thicknesses increase beyond some thickness optimal for weathering, the rock below is exposed to fewer wet-dry and freeze-thaw cycles, and the influence of biota and macroclimate is diminished. Water that affects the regolith-bedrock interface has also, by this time, percolated through such thick sequences of weathered material that it has become enriched in weathering by-products, and as a result it does not promote weathering at the bedrock-soil interface. Instead, it shuts down (or slows) chemical weathering reactions occurring there. As a result, as regolith thickens, production of additional regolith from the bedrock also slows (Burke *et al.* 2007). Theoretically, then, soil production is maximized at thin to intermediate

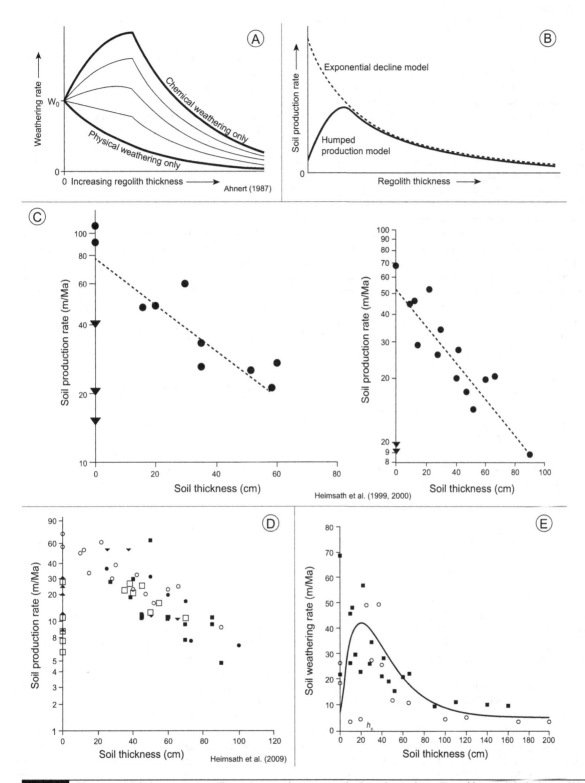

Fig. 9.13 Various modeled permutations illustrating weathering rates (soil production), as affected by regolith thickness. (A) Weathering rates as a function of regolith thickness. (B) Schematic representation of the humped soil production model of Gilbert (1877), as first depicted by Carson and Kirkby (1972), and the inverse exponential soil production function. (C) Various data compilations that support the exponential decay model of soil production. (D) Soil production data for a number of sites in Australia, each indicated by its own symbol, showing some support for a humped production model. (E) Data from Heimsath et al. (2000, 2001), as graphed by Minasny and McBratney (2006), again showing support for a humped production model.

thicknesses of regolith, and the soil production function takes on a polynomial form (Fig. 9.13B). Although Gilbert (1877) did not explicitly name it as such, the relationship that he espoused – that soil production is maximal under thin regolith and decreases as it thins and thickens – has come to be known as the *humped production model* or function (Carson and Kirkby 1972, Cox 1980, Gabet and Mudd 2010; Fig. 9.13B). In this model, soil production is maximal where regolith is present, but thin (Humphreys and Wilkinson 2007). But what is the optimal thickness, and does this humped function really mimic the rock-to-soil production system? Answering those questions has been the focus of much research, which we discuss later.

Data compiled to test this model empirically began appearing in the 1990s. First, soil production rates on slopes were estimated by using measurements of sediment accumulation and transport, taken from downslope sites. Recall that if a slope is in steady state, losses of soil from the slope should equal the production of soil from bedrock. To that end, Reneau *et al.* (1989) and Reneau and Dietrich (1990) estimated rates of production of eroded sediment in small watersheds, using ^{14}C ages of charcoal deposited in colluvial fills. Although innovative, this approach only indirectly estimated soil production rates for upslope sites on bedrock. Soon thereafter, Monaghan *et al.* (1992) and McKean *et al.* (1993) used a new method to estimate soil production *at the site* of production – on the slope itself. They measured fallout ^{10}Be concentrations in vertical profiles of soil pits (see Chapter 14) excavated into the hillslope and used those data to estimate rates of bedrock-to-soil conversion. Even better estimates of soil production surfaced after in situ cosmogenic nuclide data became available. Using measurements of in situ–produced ^{10}Be and ^{26}Al concentrations in saprolite and bedrock, sampled below soils of different depths, Heimsath *et al.* (1997) determined that an *exponential* decay equation, not a humped model, best explained soil production (Fig. 9.13C, D). This work supported earlier modeling research by Dietrich *et al.* (1995), which had found that the humped production model led to landscapes with a large proportion of bare bedrock on convex crests. When they applied an exponential soil production function, fewer instances of bare bedrock occurred, providing a better match to the actual landscape. In an exponential decay function, soil production is maximal on bare bedrock and decreases rapidly as regolith thickens (Fig. 9.13C, D).

Further work by Heimsath *et al.* (1999) yielded an integrated soil production function – of sorts – that appeared to optimize soil production at ~30 cm depth. Continued research on soil production, using in situ cosmogenic isotopes but informed by knowledge of pedoturbation, suggested that each model – humped and exponential – has merit and can be applied in different settings (Furbish and Fagherazzi 2001, Heimsath *et al.* 2000, 2001, 2009, Wilkinson *et al.* 2005, Minasny and McBratney 2006). In those studies that do support a humped production

model, soil production appears to be maximal at regolith thicknesses of 30–40 cm.

The humped production model illustrates a potential negative feedback mechanism that helps explain the relationships among weathering, soil production, and denudation (Ahnert 1987). When regolith is thin, weathering operates freely, within climatic and geologic controls, and soils gradually thicken. But as regolith thickens, its impact is to force weathering reactions at the top of the bedrock to slow, and eventually to cease. The negative feedback here is clear – an increase in one system component (regolith thickness) causes a decrease in another (soil production). Alternatively, when regolith becomes too thin – thinner than some optimal value – production may not be able to keep pace with natural erosion processes, and the system could go into a negative feedback loop in which the soil is completely eroded, resulting in bare bedrock. Because of this interaction, weathering and denudation rates tend to adjust mutually in a type of dynamic equilibrium, resulting in a reasonably constant thickness of regolith across slopes.

Assessing Weathering Intensity

It is useful to know the degree to which a soil or regolith is weathered. Often, degree of weathering parallels degree of soil development. In theory, the older the sediment or landform (surface) is, the more weathered it should be. Be aware, however, that soils and sediments are not immobile. Materials may weather in one place and then be transported to another. For this reason, a geomorphic surface (and its associated soils) may be much younger than the material it is composed of. That material may have characteristics of long or intense weathering periods, e.g., lack of primary minerals and accumulation of secondary minerals, that are unrelated to the present pedogenic cycle (see Chapter 14).

Thus, weathering data tend to focus (minimally) on *relative* degrees of weathering at different sites, but if possible, more rigorous numerical treatments of weathering intensity can be achieved. *Absolute rates* of weathering may be estimated by using mass balance equations of fluxes of weathering by-products that exit a watershed in surface water and groundwater.

The methods used to determine the relative degree of rock weathering range from qualitative ones involving color or hardness to highly quantitative measures involving mineralogy or chemistry. Many of these rock weathering methods are discussed in Chapter 15, where their application is fit into the context of using weathering to assess the age of the sediment or the landform within which they reside. Examples, particularly those associated with Robert Ruhe's work in Iowa, are provided in Chapter 14.

Chapter 10

Soil Parent Materials

The influence of parent materials on soil properties has long been recognized. Early pedologists and soil geographers based their taxonomic concepts of soils largely on their presumed parent material. Later, parent material became viewed as simply one of five factors that influence soil development – an influence that diminishes in importance with time. Parent material provides the framework for the developing soil profile and, to a large extent, determines the limited array of soils that can possibly form at a place (Fig. 10.1). In fact, the more holistically a soil is examined, including the many deep horizons below the upper meter or two, the more important parent material seems.

One goal of soil geomorphology is to identify the types and origins of a soil's parent material, because such knowledge helps to explain the evolution of the soil and the landform on which it has developed. Soil genesis studies must almost always ascertain the origin and characteristics of the parent material. This is particularly true for soils on Pleistocene or younger surfaces (Roy et al. 1967, Schaetzl et al. 2000). Many times, the role of the soil geomorphologist is to produce a surficial geology map from a soil map (Fig. 8.9). From this surficial sediment map a better understanding of the past sedimentological and surficial processes that were operative across this landscape is possible. Such an exercise may be relatively easy for young soils on young landscapes, but it is an extreme challenge on older landscapes. Indeed, making a soil survey is nearly impossible without detailed information about surficial geology, illustrating the interrelatedness of soils and parent materials in almost all landscapes.

At least two potential problems may arise when trying to identify the parent material of a soil. First, it may have been highly weathered and altered prior to the current period of pedogenesis. In many such cases, the parent material is actually a previous soil and the soil is, by definition, polygenetic, e.g., Waltman et al. (1990) (see Chapter 13). Second, the soil may be formed in more than one parent material, as either discrete or intimately mixed layers, e.g., Frolking et al. (1983) or Rabenhorst and Wilding

(1986a), some of which may be so thin that pedogenesis and pedoturbation have blurred them.

This chapter focuses on the discernment and characteristics of the major types of soil parent materials, and what they might be able to tell us about the geomorphic and sedimentological history of a site.

Effects of Parent Material on Soils

Parent material is one of the five state factors of soil development (Jenny 1941b; see Chapter 12), and for young soils, it is often the most important of the five. Soils start out as parent material and gradually develop, as various pedogenic, geomorphic, and biological processes act on these initial materials, changing them into a soil that is unique to that location in time. One way to look at pedogenesis is that it constitutes the sum of processes that form soils from parent materials. Over time, soils diverge more and more from their inherited characteristics, i.e., the characteristics of the parent material. In extremely old soils, parent material influence may be minimal (Chesworth 1973).

Although the standard definition of a C horizon is unweathered parent material, most C horizons have been somewhat altered by pedogenic processes, at least in their upper part. The C horizon may be very similar to the original parent material, but unless it exists in a very young soil, it has been altered from its initial state. Completely unaltered parent material, which occasionally occurs below the C horizon, does exist. It is informally called the D horizon (see Chapter 3). Not always obvious or easily recognized, D horizon material may be buried deeply beneath water, sediment, or rock; it may be frozen; or it may have been recently deposited, such as volcanic ash or fresh beach sand. Buol et al. (1997: 136) considered parent material to be "that portion of a C or R horizon which is easily obtainable but reasonably distant" below the solum. This approach will work for soils developed in thick, uniform parent materials, but will not suffice for soils that have formed in multiple layers of sediment;

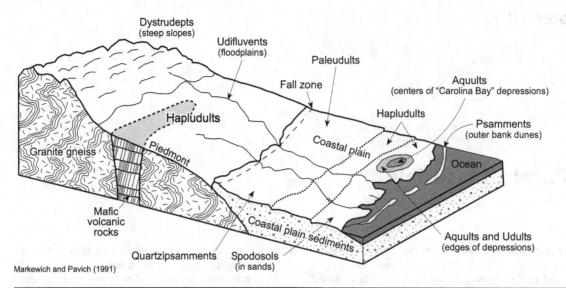

Markewich and Pavich (1991)

Fig. 10.1 Block diagram of the major soils and parent materials in the Piedmont and Coastal Plain regions of the southeastern United States, illustrating the strong association between parent material and soil development at the highest levels of classification.

the upper materials may be so altered by pedogenesis that it is impossible to characterize the original parent material. For this reason and more, ambiguities as to what exactly constitutes a given soil's parent material are commonplace.

For many soils, especially weakly developed Entisols, Gelisols, and Inceptisols, discernment and evaluation of their parent material may be relatively easy, e.g., Ehrlich *et al.* (1955), Joseph (1968), Schaetzl (1991a), and Mason and Nater (1994). The solum may be thin enough and the soil young enough that a pit a few meters deep exposes minimally altered parent material. In such soils, some characteristics of the original parent material, e.g., color, texture, and mineralogy, can be surmised by examining the profile alone (Oganesyan and Susekova 1995). For thicker profiles and older soils, the solum may be so different from the original parent material that only generalizations can be made about the original C horizon.

With time, soils increasingly differentiate from their parent materials (Muckenhirn *et al.* 1949, Yaalon 1971, Chesworth 1973). The effects of climate and relief often become dominant over that of parent material (Short 1961), such that adjacent or neighboring soils formed from different materials will seem to converge, morphologically. Color is quickly changed under different oxidizing or reducing conditions, by translocation of clay-humus complexes or salts, and by melanization and rubification (see Chapter 13). Pedoturbation can quickly destroy stratification or create layering where none previously existed (Johnson 1989, 1990; see Chapter 11). Leaching, weathering, and/or additions of material on the top of the soil surface may change soil pH, such that the original pH may not be discernible.

Conversely, many properties of parent material are persistent and continue to exert an impact on the soil for long periods (Ehrlich *et al.* 1955). Clayey parent materials will develop into clayey soils, and sands generally produce sandy soils; in other words, texture has a way of persisting (Yaalon 1971). Some sandy parent materials may weather into a loamy soil, however, if many of the sand-sized minerals are weatherable, as opposed to quartz. A soil horizon may have more clay than it inherited from the parent material either because larger particles have weathered to clay, or because clay has been translocated into that horizon. Conversely, clay can be lost from the horizon or the profile, leaving a sandier profile behind. Similar statements could be made for a suite of soluble substances common to soils.

Parent material, as a consequence of texture and surface area, also affects rates of pedogenesis. In general, pedogenesis proceeds faster in sandier materials, because of its lower surface area. Color changes, which mainly reflect particle *coatings*, occur on the *surfaces* of particles, implying that coarser-textured sediments can change color faster, all else held constant (see Chapter 13). Sandier sediments promote deeper wetting, and hence, deeper weathering and translocation of soluble materials and other particles, all else held constant. As a result, sola may be thicker – and minimally altered parent materials deeper – on coarser-textured parent materials (Harris *et al.* 2005).

The Mutability of Time$_{zero}$

In the simplest case, unweathered parent materials are deposited by sedimentation, or exposed by catastrophic erosion. Then, pedogenic processes begin to operate. This moment in time is called time$_{zero}$. Rarely, however, is pedogenesis that simple (Johnson 1985). Two kinds of complicating scenarios may occur.

First, erosion may slowly or rapidly remove material, even as the soil forms in it, lowering the surface through time. In this situation, the lower boundary of the solum may continually grow downward and as a result, encounter fresh parent material (Pavich 1989). If the soil contained coarse fragments, the soil surface may become enriched in them as a surface *lag concentrate* or stone line, because erosion may preferentially remove the finer sediment, leaving the coarser materials behind (see Chapter 14). In this scenario, the eroding (unstable) surface lessens the utility of the time$_{zero}$ concept, because the soil is continually being eroded.

Second, soils can receive slow, steady or intermittent, surficial additions of fresh mineral material, thereby aggrading the soil surface (Johnson 1985, McDonald and Busacca 1990). Small, slow additions of sediment allow soil processes to adjust; gradually the soil may "grow upward" and thicken as the new sediment plus the weathering profile become the parent material for the "new" soil (Riecken and Poetsch 1960, Wang and Follmer 1998). Johnson (1985) called this process "developmental upbuilding." It is also known as cumulization (see Chapters 13 and 14). Cumulization can result from eolian, fluvial, colluvial, or anthropogenic additions to the soil surface. It often occurs on alluvial surfaces, e.g., floodplains, although it can also occur when loess or eolian sand is slowly added to upland soils. Small inputs of sediment – slow relative to the rate of horizon formation – epitomize this process.

All soils must have at least one time$_{zero}$, or a starting point. At time$_{zero}$, pedogenic processes begin their work on unconsolidated surficial material or weathered rock. For many soils, that starting point represents only one of many such time$_{zeros}$, as the soil may form, be eroded, and begin forming again in the remaining parent material. In other areas, e.g., those with hard bedrock near the surface, alteration of the rock to saprolite may need to occur before soil development, and thus time$_{zero}$, can begin.

Polygenetic soils are those that have evolved along more than one pedogenic pathway (see Chapter 12). Thus, many polygenetic soils have more than one time$_{zero}$. For example, when an Aridisol encounters a suddenly wetter climate, the new, humid-climate pedogenic pathway uses the old soil as parent material (Bryan and Albritton 1943, Chadwick *et al.* 1995). Similarly, a soil may encounter a change in vegetation, forcing modifications to its pedogenic processes. In both cases, the preexisting soil is viewed as the parent material for the one that will henceforth develop, i.e., parent material may not necessarily be pedogenically *unaltered* sediment.

In sum, soil parent material should be evaluated in light of surface stability and pedogenic pathways. Soils can develop steadily in a single parent material on a stable geomorphic surface, and thus have a clear and well-defined

time$_{zero}$, but such situations may occur only for short periods. Such "stability" is never retained over geologic timescales (see Chapter 14). Thus, the concept of parent material is one that also has short-term meaning. What is "parent material" to a soil today will be part of the solum, eroded or buried at some time in the geologic future. This line of reasoning also reinforces the importance of being able to determine what the parent material of a soil is/was, for in so doing we can learn a lot about surface stability, soil development, and landscape evolution. For this reason, obtaining accurate information about soil parent material is often one of the most important and challenging tasks of the soil geomorphologist (e.g., Chadwick and Davis 1990, Dahms 1993, Schaetzl *et al.* 2000, Wysocki *et al.* 2005).

Types of Parent Materials and Their Characteristics

Early approaches to soil survey, mapping, and classification relied heavily on interpretations of a soil's parent material, largely because soils were thought of as little more than disintegrated rock mixed with decaying organic matter (Simonson 1952, 1959, Kellogg 1974, Tandarich 2001, Brevik 2002). For example, terms such as "granite soils" and "loessial soils" were common, because pedogenesis was poorly understood. As our understanding of soils progressed, it became clear that soils formed in similar parent materials can vary considerably, spatially and temporally (Phillips 2001a). Today, we acknowledge that understanding the parent material of a soil goes a long way toward knowing its initial state, or time$_{zero}$ condition, and this information provides certain limits to what the soil can eventually develop into, even though we seldom use a parent material type as the only soil name modifier. As with any broad topic, understanding all of its complexities may be best done by categorization and classification, which are what we attempt in the following.

Residual Parent Materials

All loose, unconsolidated materials that overlie bedrock are referred to as *regolith* (see Chapter 3). First coined by Merrill in 1897, the term "regolith" is derived from the Greek words *rhēgos* (blanket) and *lithos* (stone). In Canada and some other parts of the world, minimally developed soils that still largely resemble their "regolithic" parent materials are classified as Regosols (VandenBygaart 2011).

Either regolith may be *residual* (formed in place) or it may have been *transported* to its location by water, gravity, wind, and so on. Residual regolith is called *residuum* (Short 1961). Often, it is impossible to know with certainty whether regolith has been transported at some point in the past, or whether it is "true" residuum. For example, on limestone terrain in Mediterranean regions, a red sediment often interpreted to be residuum overlies carbonaceous

Fig. 10.2 Soils formed in residual parent materials. Photos by RJS. (A) A shallow, residual soil (Lithic Udorthent) on fractured sandstone bedrock in southwestern Wisconsin. Small additions of loess (wind-blown silt) may also be present in the soil. (B) Saprolite formed from gneiss bedrock, Colorado Front Range.

bedrock. However, this sediment is often overthickened in depressional areas and on footslopes (Wieder and Yaalon 1972, Yassoglou et al. 1997), and thinner on steeper slopes (Benac and Durn 1997). Thus, what may initially have been residuum has undoubtedly been transported to some extent. Similar relationships exist in the residual bauxite areas of Arkansas. Saprolite rich in aluminum but relatively Fe-poor is termed *bauxite* (Sherman et al. 1967, Mutakyahwa et al. 2003, Liu et al. 2010). Here, long-term weathering of nepheline syenite bedrock on uplands has led to thick, residual deposits of bauxite ore, rich in gibbsite and kaolinite (Gordon et al. 1958). Thick bauxite deposits on lower

slopes exist because of transportational (colluvial and alluvial) processes.

The thickness of the residual cover on a slope is a function of the interplay between the weathering processes that produce it and the transportation processes that erode and carry it away. On slopes with thick residual accumulations, long-term weathering rates exceed transportation rates; the slope is considered *transport-limited*. On other slopes, residuum may be thin; these may be *weathering-limited* slopes where the rate of weathering, i.e., residuum formation, cannot keep pace with the transport processes (Paradise 1995; Fig. 10.2). The terms do not imply that either process is, in and of itself, proceeding slowly or rapidly. They simply refer to the *balance* between the two opposing processes.

Saprolite

Saprolite is in situ weathered rock that retains some of the original rock structure in the form of veins or dikes (Aleva 1983, Whittecar 1985, Stolt and Baker 1994, Ollier and Pain 1996; Fig. 10.2B). It forms a transition between the solum and hard bedrock. Corestones of solid bedrock may occur within its lower parts (Berry and Ruxton 1959, Fletcher and Brantley 2010; Fig. 3.12). Saprolite contains both primary minerals and their weathering products (Hurst 1977) and is usually soft enough to be penetrated with a sharp shovel blade or knife (Hirmas and Graham 2011). Rock strength is weaker and porosity is greater in the saprolite, compared to the original bedrock, thereby promoting processes that act to break up the rock further, such as shrinking and swelling (Frazier and Graham 2000). The presence of thick saprolite usually implies that weathering has been dominated by chemical, rather than physical, processes. Weathering of hard bedrock into saprolite is often called *saprolitization* or *arenization* (Eswaran and Bin 1978b).

More often than not, the presence of thick saprolite implies long-term, deep weathering, dominated by chemical rather than physical processes, on old geomorphic surfaces and under warm, humid climatic conditions. On these old geomorphic surfaces, such as in parts of the Appalachian Mountains, ample time has passed to produce saprolite that can be >30 m in thickness (Bouchard and Pavich 1989, Cremeens 2000). Saprolite can form quickly and can persist for several thousands of years, with some examples dating to the Tertiary.

Saprolite is best expressed, i.e., thickest, on old, gently sloping, transport-limited surfaces, such as stable bedrock uplands (Aleva 1983, Whittecar 1985, Pavich 1989, Stolt et al. 1992, Ohnuki et al. 1997). Here, erosion of clastic materials by running water is minimal, and the main losses from the system have occurred in solution, by deep percolation. Soluble weathering by-products have mainly been flushed from the system via fractures, which may today be few and/or filled with illuvial clay.

Depending on the degree of weathering and composition of the parent rock, saprolite either can weather

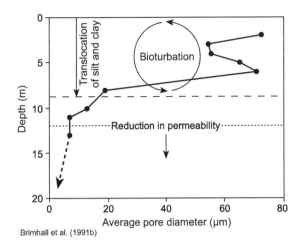

Brimhall et al. (1991b)

Fig. 10.3 Average pore diameter in a core from an Oxisol and its subjacent saprolite in Mali.

isovolumetrically or can be "collapsed" compared to the original rock volume (McFarlane and Bowden 1992). Collapse occurs as minerals are weathered and their weathering products are removed in solution. This process mainly occurs at the top of the saprolite, where it is being converted to soil; Pavich (1989) reported a mass loss of about one-third.

Alternatively, the formation of saprolite from crystalline rocks such as gneiss and schist is commonly isovolumetric, i.e., mass is lost but with little or no volume change (Stolt and Baker 2000). In these cases, the saprolite retains the original rock structure. Velbel (1990) listed the three mechanisms by which isovolumetric weathering can occur:

1. Minerals weather in different sequences and "take turns" maintaining the rock structure. For example, hornblende weathers while plagioclase alters to clay; later the feldspar weathering products (clay) maintain the structure while the hornblende weathers.
2. A rigid framework that resists collapse is produced during weathering, maintained largely by resistant minerals such as quartz and muscovite.
3. Etch pits form in minerals as portions of these minerals weather at different rates. Material is removed in solution from etch pits, but because the general outline and bulk volume of the mineral remain unaltered, the rock does not collapse.

Another process at work to inhibit or counteract collapse involves soil infauna. They, along with roots, facilitate the "opening up" of the saprolite, forming the pores that lead to the decreased bulk density of soil vis-à-vis saprolite, but at the same time, the rock structure can be retained.

Rock and saprolite weathering can only continue if the rock is porous enough to allow water to remove weathering by-products, e.g., CaO, Na$_2$O, and SiO$_2$ (Pavich 1989).

Permeability also facilitates the translocation of clay and dissolved substances; hence, saprolites may have argillans, even at depth. Many saprolites are more clay-rich near the surface than at depth, however, because clay particles are arrested before they can move to great depths. A follow-up to this involves the examination of stream chemistry in landscapes where bedrock weathering is the main producer of ions, for such knowledge allows calculation of rates of saprolite production (Velbel 1990). Studies have reported that as much as 1 million years is required to form 4 m of saprolite in some areas (Pavich et al. 1986, Pavich 1989, Cleaves et al. 1970).

Many studies of saprolite formation approach the problem because of a geomorphic interest. By knowing the rates of saprolite formation and its thickness on the landscape, a degree of understanding of landscape age and evolution can be achieved.

Hunt (1972) described the typical saprolite profile in the southern Appalachians, on crystalline rocks such as schist, gneiss, and granite (see also Calvert et al. 1980a, b, and Pavich 1989). Overlying, hard bedrock is discolored, partially weathered bedrock with a bulk density 10–20% lower than that of the bedrock. The weathered rock does not ring when hit with a rock hammer, but instead produces a dull thud. Cracks within are coated with yellow and brown hydrated iron oxides. *Structured saprolite*, which makes up most of the regolith volume, overlies the weathered bedrock (Pavich et al. 1989). Argillans and iron stains are common, and although its density is only half that of the bedrock, the original rock structure and fabric are preserved. Weathering and translocation have resulted in losses of many basic cations and any soluble salts, especially Na$_2$O and CaO (Pavich 1989). In places where erosion has been slow and weathering rapid, structured saprolite is overlain by *massive saprolite*, implying that rock structure has been lost, mostly through bioturbation, wet-dry, creep, and other near-surface mixing processes. Thus, quartz veins found in the bedrock and structured saprolite end abruptly at the base of the massive saprolite, and in its place may be scattered quartz fragments.

To many pedologists, *massive saprolite* is simply the lower solum, in which pedogenic and pedoturbative processes have destroyed the original saprolite structure. The transition from saprolite to the soil profile, which can be abrupt or gradual, is largely controlled by the following processes: (1) mixing and formation of pedogenic structure by fauna and plant roots, (2) accumulation of organic matter, and (3) additions of eolian materials and translocated substances, e.g., clay or carbonates (De Villiers 1965, Brimhall et al. 1991). The change from rock structure with very small pores to the blocky and granular pedogenic structure with much larger pores is a fundamental part of the saprolite-soil transition (Fig. 10.3). This transition usually occurs at the maximum depth of rooting, about 1 m according to Fölster et al. (1971), but often deeper.

Formation of soil from saprolite is often *not* isovolumetric. The volume increase (*dilation*) that occurs in the bedrock-to-soil morphogenesis can be 200–300% (Brimhall *et al.* 1991). Mass may have been lost in the saprolite-to-soil conversion, but the addition of pore space leads to expansion of the matrix.

Flach *et al.* (1968) coined the term *pedoplasmation* for the collection of physical and chemical changes that alter saprolite to soil. It is particularly important in many Oxisols and Ultisols on old, bedrock landscapes. Literally, this term means the formation of soil plasma, or what Fölster *et al.* (1971) referred to as a *homogenization horizon*. The processes responsible for pedoplasmation include bioturbation and other mechanical disturbances such as shrink-swell and root activities. Brimhall *et al.* (1988) were able to distinguish this homogenized zone from the saprolite below and noted that zircon minerals in the saprolite resembled those in the bedrock, while zircons in the soil above were smaller and more rounded. Because the upper zircons had presumably been deposited by wind, they must have been mixed to depth by homogenization processes such as pedoturbation and root growth. Zircons in the saprolite, below the homogenization horizon, had not yet been exposed to these processes.

Saprolite Formation in the Humid Tropics

The main differences in saprolitization between the humid tropics and elsewhere revolve around the much greater rates at which it occurs, the types of minerals affected by weathering, and the different mobilities of the ions that result. Given the intense heat, the availability of water (at least during parts of the year), and the great age of many tropical landscapes, residual parent materials there are highly weathered and frequently underlain by thick saprolite (Migon 2009). During saprolitization in this environment, chemical weathering is especially intense, leaving many soils with a high proportion of resistant materials and ions, such as Ti, Zr, and various metal oxides and oxyhydroxides, mostly as fine clay (Soileau and McCracken 1967, Bigham *et al.* 1978, Burke *et al.* 2007). In tropical soils, bases and silica are translocated from the profile (see Chapter 13), leaving behind large amounts of free Fe and Al; what happens to these elements markedly affects soil evolution. Some may stay in the profile to form nodules, concretions, and crusts, while others may be transported laterally into lower, wet areas at the bases of slopes, where they can form massive ferruginous crusts (Fölster *et al.* 1971, Beauvais and Colin 1993; see Chapter 13).

Kaolinite, gibbsite, and other sesquioxides are generally regarded as products of chemical weathering in the humid tropics, with gibbsite often seen as the ultimate end product (Aleva 1983). Gibbsite is most common in dry, hot soils and may be almost absent in soils with high water tables, where it cannot crystallize (Cady and Daniels 1968). In aluminum-rich bedrock, boehmite and gibbsite may be dominant secondary minerals.

Bauxite (Al-rich saprolite) tends to form more often in humid climates than do many Fe-rich saprolites. In order to concentrate Al at the expense of Fe, two factors must also occur: (1) The parent rock must be iron-poor, such as in basic volcanics and phyllites, and (2) the site must have high biotic productivity so that the Fe can be chelated by organic acids and removed from the system (Thomas 1974). Desilication must also be a strong process in order to concentrate aluminum to the extent that the end product is considered bauxite.

In the humid tropics, saprolites may exceed 100 m in thickness (Aleva 1983). Saprolite thickness depends on many factors, primarily age and stability of the surface, type of rock, location on the landscape, and climate. For example, no real saprolitic zone may overlie many basic and ultrabasic rocks. Here, the base of the soil rests directly on rock, or on a thin weathering crust (Soil Survey Staff 1999).

Using a mass balance approach to soil formation and saprolitization on an Oxisol in Mali, Brimhall *et al.* (1991) were able to show that much of the Fe and Al in these soils was actually not residual; these elements were added to the soil system from upslope or by eolian dust. The eolian dust can only penetrate to a certain depth, depending on soil pore size and depth of pedoturbation. Thus, it appears that purely residual accumulations of sesquioxides may be restricted to the basal parts of some saprolites. Many of the chemically mature sediments in the tropical soils may have been imported as dust from nearby highly weathered landscapes. Much of that dust ends up in oceanic areas immediately downwind (Fig. 10.4).

Saprolite often displays discrete *zones* that vary in degree of weathering and in amount of pedogenic imprinting. Acworth (1987) described four main saprolite zones below the solum; Jones (1985) described five (Fig. 10.5). The bedrock-saprolite interface, often tens of meters below the surface, is usually diffuse and shaped mostly by fracture patterns in the bedrock (Fig. 10.7). Here, soluble constituents and weathering by-products are removed in groundwater, but rock structure is preserved. Corestones of bedrock are common well up into the saprolite (Berry and Ruxton 1959, Brantley *et al.* 2011). The permeability of the saprolite zone with corestones is high; solutes are able to move between corestones and within fractures. In a description of the typical rock-saprolite-soil sequence in the Gold Coast (Africa), Brückner (1955) noted that in many sections it was possible to distinguish a lower layer of saprolite from an upper layer (Fig. 10.6). Roots proliferate only in the upper, redder layer. Eswaran and Bin (1978a) described six soil-saprolite zones in humid Malaysia – the uppermost of which was considered soil. Below this zone was a stone line or stone zone. Below this zone were three zones of decreasing amounts of weathering and increasing rock structure.

Weathering intensity, which increases toward the surface, follows a pH gradient; soils are often most acidic at the surface and less so at depth, because of accumulations

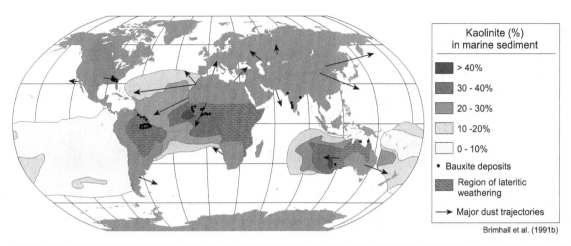

Kaolinite (%)
in marine sediment

■ > 40%

■ 30 - 40%

■ 20 - 30%

□ 10 -20%

□ 0 - 10%

• Bauxite deposits

▨ Region of lateritic weathering

→ Major dust trajectories

Brimhall et al. (1991b)

Fig. 10.4 Map showing the locations of Precambrian shields (cratons), lateritic and oxic soils, and kaolinite-enriched marine sediments. Global dust trajectories are from Pye (1988).

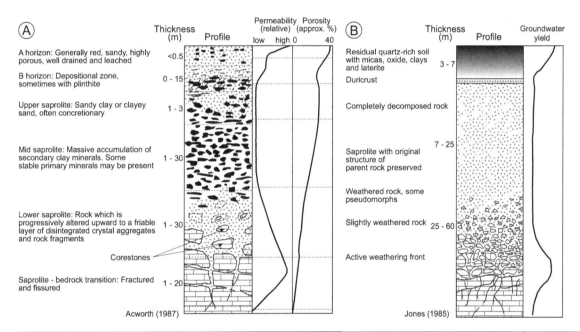

Fig. 10.5 Typical soil-saprolite (weathering) profiles developed on crystalline bedrock in the humid tropics.

of weathering by-products (van Wambeke 1962). The increase in weathering may be either gradational or abrupt, largely depending on water flow pathways and parent material mineralogy. Fingers of more highly weathered materials often protrude down into less weathered rock, reflective of infiltration conduits. Likewise, corestones, protuberances, and diapirlike structures of less weathered rock may commonly be found higher in the weathering profile than one would expect; they are preserved because infiltrating water is steered around, rather than into, them (Fig. 10.8). The mineralogy changes that are associated with increased

near-surface chemical weathering can be observed in the transition from rock to saprolite to soil (Fig. 10.7). Conditions also become less oxidizing with depth, leading to mottled or reduced zones. If large amounts of soluble iron are present, ironstone fragments or *plinthite* will form (Fölster *et al.* 1971, Eswaran and Bin 1978a, Aide *et al.* 2004; see Chapter 13).

Although it is difficult to make general statements about the changes in mineralogy from the base of tropical saprolites to the soil profile, it is clear that weatherable minerals decrease in near-surface layers. One of the first secondary

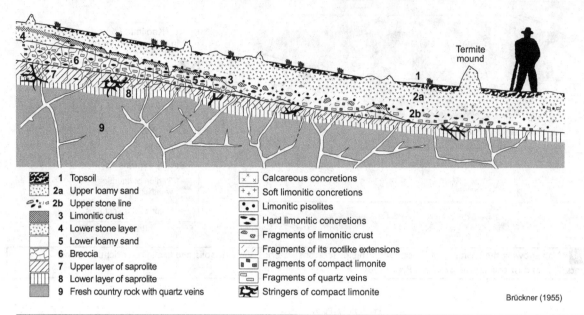

1 Topsoil
2a Upper loamy sand
2b Upper stone line
3 Limonitic crust
4 Lower stone layer
5 Lower loamy sand
6 Breccia
7 Upper layer of saprolite
8 Lower layer of saprolite
9 Fresh country rock with quartz veins

Calcareous concretions
Soft limonitic concretions
Limonitic pisolites
Hard limonitic concretions
Fragments of limonitic crust
Fragments of its rootlike extensions
Fragments of compact limonite
Fragments of quartz veins
Stringers of compact limonite

Brückner (1955)

Fig. 10.6 Idealized geologic section through the "mantle rock" of the Gold Coast, Africa.

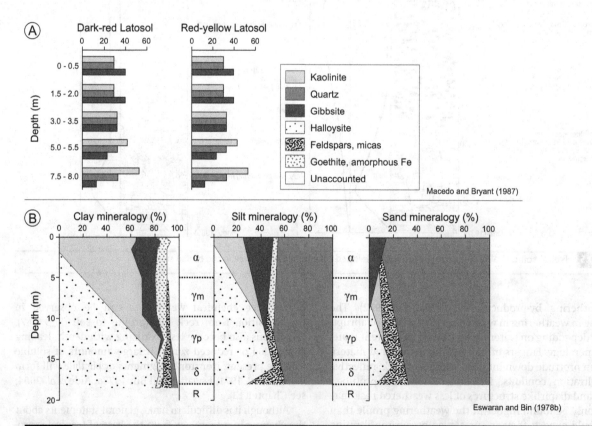

Macedo and Bryant (1987)

Eswaran and Bin (1978b)

Fig. 10.7 Variation in mineralogy, with depth, of the fine earth fraction of some weathering profiles in the humid tropics. (A) Mineralogy of two Acrustox in Brazil. (B) Mineralogy of a deep weathering profile in Malaysia.

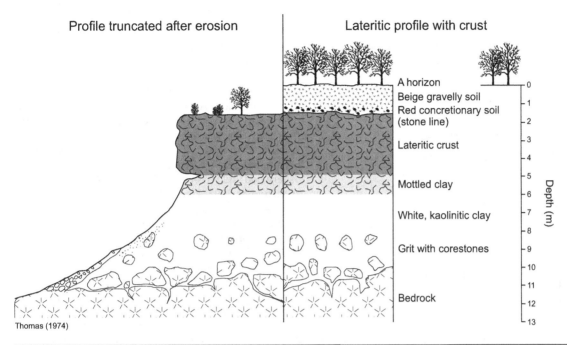

Profile truncated after erosion   Lateritic profile with crust

A horizon
Beige gravelly soil
Red concretionary soil
(stone line)

Lateritic crust

Mottled clay

White, kaolinitic clay

Grit with corestones

Bedrock

Depth (m)

0
1
2
3
4
5
6
7
8
9
10
11
12
13

Thomas (1974)

Fig. 10.8 Generalized bedrock-saprolite-soil sequence under forest and savanna in Sudan.

minerals observed in deep saprolite in the humid tropics is kaolinite, which weathers from feldspars (Fig. 10.7). The abundance of kaolinite is often (although not always) expressed as whitish layers in the upper saprolite (Fig. 10.8). Moss (1965) and Thomas (1974) referred to this zone as the pallid zone. Nearer to the surface, the colors change to reds and browns, as kaolinite is increasingly weathered to gibbsite (Aleva 1983, Macedo and Bryant 1987; Fig. 10.7). The decrease in kaolinite in near-surface layers may be due to weathering, or to eluviation in particulate form (Aleva 1987). However, in highly weathered soils, the loss of kaolinite is usually due to the former. In some highly leaching environments, the amount of aluminum-rich weathering products, i.e., boehmite, is greater near the surface because Fe has been preferentially leached (Aleva 1965).

Another feature typical of humid tropical weathering profiles is the presence of secondary minerals in the sand and silt fractions – as small ironstone fragments or parts of partially altered primary minerals surrounded by an alteration product. According to Curi and Franzmeier (1984), high hematite and gibbsite contents reflect a leached, dry soil environment with high Fe^{3+} concentrations in solution, while moist sites typically have more goethite and kaolinite, reflective of lower levels of desilication. The predisposition for goethite to form in wetter horizons is supported by the work of Fölster et al. (1971), who noted that while goethite and hematite can form in the same horizon, in wetter horizons goethite is dominant.

The red, brown, and white masses of kaolinite and gibbsite, so common in tropical saprolite, become mixed and merge into a more uniform, almost isotropic, reddish brown material in the solum (Flach et al. 1968). In the upper B horizon, this uniform brown material will possess strong blocky or granular structure. Its clay contents will have increased almost threefold, and it will have lower bulk densities (Fig. 10.9). Silt and even sand-sized masses of kaolinite, common in the saprolite, are weathered in near-surface horizons such that in the upper solum all the kaolinite is in the clay fraction. Thus, clay in the saprolite becomes increasingly disaggregated and "finer" as weathering proceeds in near-surface layers.

Effects of Rock Type on Residuum Character

The character (mineralogy, texture, porosity, base status, etc.) of the rock largely determines the nature of its residuum (Stephen 1952, Plaster and Sherwood 1971). In many instances, this relationship is obvious. For example, sandstone will produce sandy, porous residuum; shale leads to clayey residuum. Quartzite is so difficult to weather that it produces little residuum, so soils on quartzite tend to be thin. Limestone weathers so completely that its residuum mainly comprises insoluble components from the parent rock (Moresi and Mongelli 1988). Basic rocks, such as basalt, diorite, and gabbro, weather to a dark, clay-rich residuum that is high (initially) in pH. Soils formed in this type of residuum tend to be more fertile and balanced with respect to cations than are the soils formed from acidic rocks. Smectite may form from Mg- and Na-rich, acidic rocks, if the climate is not so humid that these ions are leached from the profile. Ultramafic rocks such as peridotite, which are very high in Mg and Fe and low in Al, K, Ca, and Na, produce silty and clayey residuum (Rabenhorst et al. 1982).

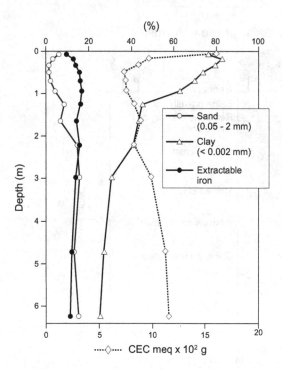

Table 10.1 and Figures 10.10 and 10.11 provide baseline information on some common rocks and minerals.

The discussion that follows focuses on residuum from different rock types. The emphasis is on those that do *not* develop thick sequences of saprolite, e.g., many sedimentary rocks, gabbros, and basalts. Rocks that do form saprolitic residuum, such as granite, gneiss, and schist, are discussed only briefly, because saprolitization has been discussed previously.

Residuum from Siliceous Crystalline Rocks

Granite and granitic gneiss are common, coarse-grained, acid crystalline rocks; other examples include tonalite, quartz monzanite, and granodiorite. These rocks form deep within Earth's crust, where slow cooling allows for the formation and development of large mineral crystals. The magmas that these rocks are derived from are termed *felsic*, implying that they are rich in silicon (si), but with some iron (fe) as well (Fig. 10.11). Quartz and feldspar are common minerals in these rocks, which is why they are typically light-colored. Biotite, distinctive because of its shiny appearance and platelike cleavage, may be a weak link in these rocks, from a weathering standpoint (Isherwood and Street 1976, Dixon and Young 1981). It tends to be the least resistant to chemical attack, weathering to vermiculite in humid, leaching environments (Borchardt *et al.* 1968).

Because of their high quartz content and large crystals, acid igneous rocks tend to weather primarily by physical means, to sandy and gravelly residuum, often with low base status and poor nutrient reserves (Muckenhirn *et al.* 1949, Stephen 1952, Plaster and Sherwood 1971, Eswaran and Bin 1978a, Wang *et al.* 1981, Evans and Bothner 1993, Oganesyan and Susekova 1995). Dixon and Young (1981) referred to the weathering of granite to coarse-textured soils as *arenization*, or the formation of sandy mantles. Weathered granite saprolite is coarse-textured and has been called *grus*; its formation from bedrock is *grusification* (Isherwood and Street 1976, Dixon and Young 1981, Migon and Thomas 2002; Fig. 10.12). Sand grains in grus might contain a significant amount of weatherable minerals, such as mica, chlorite, and feldspar, while in the resulting soil profile these minerals have often been weathered away. For example, kaolinite and Fe oxides are derived from weathering of micas and feldspars (Short 1961, Rebertus and Buol 1985). As a result, acidic, light-colored or reddish soils that have lost silica often form in this type of residuum (Ehrlich *et al.* 1955, Eswaran and Bin 1978a).

Gneiss and schist, metamorphic equivalents of granite, often have high amounts of mica (Fig. 10.10). Their saprolite tends to be silty, but still coarser-textured and more permeable than residuum from other metamorphic rocks. Where muscovite was the primary mica in the rock, illite and vermiculite may occur in the clay fraction of the saprolite. Residuum from chlorite schist may be more clayey, with high amounts of Mg and perhaps smectite.

Residuum from Base-Rich Crystalline Rocks

Subsilicic rocks such as basalt, diabase, dolerite, gabbro, and metagabbro are dark-colored and rich in base cations, especially Fe and Mg. Also included in this category are andesites, diorites, and hornblende gneisses (Fig. 10.10). These rocks are dominated by ferromagnesian minerals such as amphiboles, pyroxenes, biotite, olivine, and chlorite (Rice *et al.* 1985; Fig. 10.11). Magmas and lavas that gave rise to these rocks are termed *mafic*, because they are rich in magnesium- (ma) and iron- (fi, fe) bearing minerals. Hornblende, an amphibole, may constitute a significant component of the sand fraction of C horizons in this type of saprolite, but in the solum it may have been weathered away. Plagioclase feldspars in these rocks are rich in Ca, and like many of the other minerals that compose these rocks, weather rapidly. Generally speaking, minerals that crystallize early from molten materials under high temperatures and pressures tend to weather more rapidly than those that crystallize later, as they are more out of equilibrium with the cool, and often wet, surficial environment.

Soils formed from base-rich crystalline rocks tend to have dark colors, clayey textures, and high pH values (Fig. 10.13). Dark primary minerals such as hornblende and olivine contribute to dark colors in residuum and soils. Nevertheless, most of the darker colors result from organic matter, which tends to be found in large amounts in these base-rich, fine-textured soils, where it strongly binds to the

Table 10.1 | Classification and composition of sedimentary rocks

	Texture	Composition	Rock name
Clastic	Coarse-grained	Rounded fragments; quartz, quartzite, and chert dominant	Conglomerate
		Angular fragments; quartz, quartzite, and chert dominant	Breccia
	Medium-grained	Quartz and others	Sandstone
		Quartz and feldspars	Arkose sandstone
	Fine-grained	Quartz and clay minerals	Siltstone
	Very fine-grained	Quartz and clay minerals	Shale, siltstone, or mudstone
Chemical or organic	Medium- to coarse-grained	Calcite (CaCO$_3$)	Limestone
	Microcrystalline, conchoidal fracture		Micrite
	Aggregates of oolites		Oolitic limestone
	Fossils and fossil fragments, loosely cemented		Coquina
	Abundant fossils in calcareous matrix		Fossiliferous limestone
	Shells of microscopic organisms, soft		Chalk
	Banded calcite (cave or "hot spring" deposits)		Travertine
	Textural varieties similar to limestone	Dolomite (CaMg(CO$_3$)$_2$)	Dolomite
	Cryptocrystalline, dense, conchoidal fracture	Chalcedony (SiO$_2$)	Chert
	Fine to coarse crystalline	Gypsum (CaSO$_4$ · 2H$_2$O)	Gypsum
	Fine to coarse crystalline	Halite (NaCl)	Rock salt

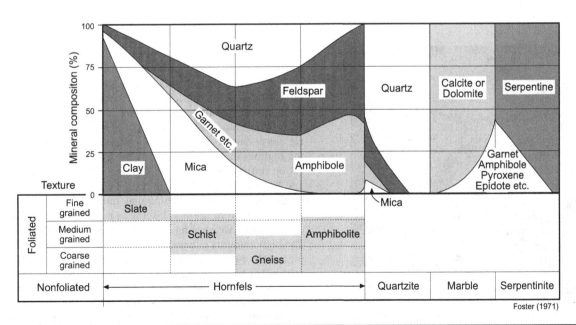

Foster (1971)

Fig. 10.10 A textural/mineralogical classification of the common metamorphic rocks.

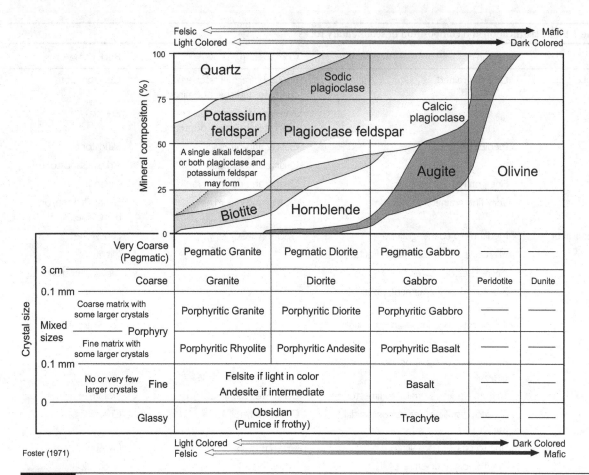

Felsic ⟵ — Mafic
Light Colored ⟵ — Dark Colored

Light Colored ⟵ — Dark Colored
Felsic ⟵ — Mafic

Foster (1971)

Fig. 10.11 Classification of common igneous rocks, based on mineral composition and crystal size. Minor accessory minerals are omitted for brevity. The name "granite" is used in the widest sense; several names are, however, in common usage for this type of rock. The modifier "porphyritic" is used only when crystals are large and prominent.

Fig. 10.12 A blue-bellied lizard (*Sceloporus occidentalis*) scurries along the surface of a weathered granite upland in Yosemite National Park, California. Coarse-textured granite like this produces sandy, acidic residuum, or grus. Photo by RJS.

Ca ions released from the rock. Many of the red colors are derived from oxidized iron compounds.

These soils generally lack sand in the saprolite and soil profile because of the low quartz content in the rocks, although exceptions occur (Jackson *et al.* 1971, Graham and Franco-Vizcaino 1992). As long as Ca, Mg, and other base cations persist in the residuum, because of either low leaching regimes or eolian influxes of Ca-rich dust, a high base cation status can be maintained in the soil above (Graham and Franco-Vizcaino 1992). In humid climates where base cycling is not strong, leaching may quickly remove these base cations from the soil, leading to more acidic soil conditions, rapid weathering, and even more clay production. The high amounts of base cations in these soils, when coupled with organic matter in a wet-dry climate (or an intermittently perched water table), often lead to the formation of smectite clay minerals and Fe-Mn concretions (Plaster and Sherwood 1971, Rice *et al.* 1985). Otherwise, kaolinite, vermiculite, and halloysite are common secondary clay minerals; smectites are more common in drier climates.

The conditions described previously apply to mafic rocks. Ultramafic rocks are characterized by even lower (<40%) silica contents and high amounts of Mg, Fe, Cr, and Ni (Mansberg and Wentworth 1984, Lee *et al.* 2001). They are poor in Ca, Al, K, and Na. The most common ultramafic rocks are peridotite and serpentinite, both of which are rich in the 1:1 clay mineral serpentine; others are dunite and soapstone. According to folklore, serpentine gets its

name from the resemblance of the soils formed on it to a mottled, greenish-brown snake.

In soils formed in ultramafic residuum, quartz, feldspar, and mica minerals are rare, while normally uncommon minerals such as talc, magnetite, and olivine are found in abundance. Residual soils and their ultramafic residuum tend to be thin, droughty, nutrient-poor, erodible, stony, and fine-textured (Parisio 1981, Bulmer and Lavkulich 1994, Alexander and DuShey 2011, Lesovaya *et al.* 2012; Fig. 10.14). The shallowness of these soils has been attributed to (1) the lack of resistant minerals such as quartz and feldspar, which facilitates nearly complete weathering; (2) limited clay mineral formation due to a lack of Al; (3) accelerated erosion due to sparse vegetative cover; and (4) removal of mass by Mg leaching (Parisio 1981, Alexander and DuShey 2011).

Because serpentine clays weather to release high amounts of Mg and metal ions, ultramafic residuum and soils have very low Ca:Mg ratios (Walker *et al.* 1955, Rabenhorst and Foss 1981). Roberts (1980) reported Ca:Mg ratios for some serpentine soils in Newfoundland that typically were <0.012. Nickel, cobalt, and chromium levels are commonly high as well (Rabenhorst *et al.* 1982) and can even reach toxic levels. Soils formed on serpentine are therefore notoriously infertile because of the very low Ca:Mg ratios, metal (especially Ni) toxicities or macronutrient deficiencies, and often, negligible concentrations of K (Gordon and Lipman 1926, Spence 1957, Mansberg and Wentworth 1984, Kram *et al.* 2009, Rajakaruna *et al.* 2009, Weerasinghe and Iqbal 2011). In most horizons, ≥80% of the exchange sites are occupied by Mg^{2+}. The high magnesium contents block the plant's ability to take in nutrients, especially Ca. With time and continued leaching, Mg and Ni can, however, become depleted from serpentine soils (Fig. 10.14). Si and Al, which were present

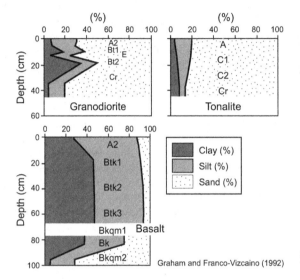

Fig. 10.13 Particle size distribution, with depth, in three residual soils in Baja California. Basalt is a fine-grained, base-rich rock; tonalite is a coarse-grained, acidic rock; and granodiorite is intermediate between the two in many respects.

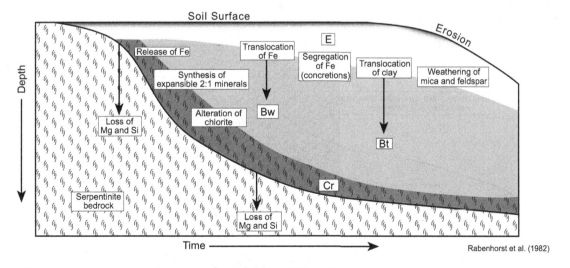

Fig. 10.14 Generalized model of soil genesis in serpentine and serpentinite residuum. Modified with data from Bulmer and Lavkulich (1994).

in very low amounts in the parent material, then become relatively enriched (Wildman *et al.* 1968, Rabenhorst *et al.* 1982, Bulmer and Lavkulich 1994). Nonetheless, even when leached, the soils retain high amounts of exchangeable Mg, especially in the lower profile; pH values remain high.

Residuum from Sedimentary Rocks

Residuum formed from clastic sedimentary rocks is texturally much like the clastic particles within the rock, e.g., sandstones weather to sandy sediment. Clay mineral assemblages, however, can vary from one rock unit to another.

Shale is perhaps the most easily weathered of the common sedimentary rocks. For the most part, weathering of shales (and sandstones) tends to be dominated by physical (disintegration) processes, rather than chemical weathering. Shales tend to be high in layer silicates and feldspars; soils developed in shales tend to be clayey, thin, dark, and slowly permeable (Oganesyan and Susekova 1995). Some shales contain pyrite and other sulfurous minerals, leading to very acidic soils that are dominated by kaolinite. Other shales are high in carbonates and bases, with illite and smectite clays. Hence, a soil that develops on acid shale may be quite different from one on calcareous shale (Table 10.1). Weathering, lessivage, and other pedogenic processes may begin earlier in acid residuum, prompting thicker sola and more clay formation and translocation.

Sandstones, commonly >60% sand and usually dominated by quartz, weather to sandy residuum. The rate of weathering largely depends on the nature of the cementing agent. Most commonly, sandstone residuum is acidic and thick, although erosion may, in places, prevent thick residuum from accumulating. Arkose has high feldspar contents and weathers to a more clayey residuum. Greywacke consists of angular fragments of quartz, feldspar, and other minerals set in a fine-textured base. Soils formed in greywacke residuum are often loamy. Quartz and feldspars are common minerals in residual soils developed in pure sandstone.

Landscapes: The Mesa-and-Butte Desert Landscape of the Colorado Plateau

On bedrock landscapes in humid regions, saprolite, regolith, and soils often completely cover the landscape, occurring on all but the steepest slopes. In dissected, arid landscapes, however, bedrock outcrops are often much more common, with the Grand Canyon area of the Colorado plateau being the classic example. Sandstone and limestone tend to be resistant rock units in dry climates, and thus they tend to stand up in nearly vertical walls. Rock outcrops occur in desert landscapes because, for many of the most resistant rocks, weathering is so slow that the production of saprolite and residuum cannot keep pace with the forces of erosion. Hence, soils are thin or nonexistent on such rocks, and slopes steepen as erosion continually tears away at them. Additionally, erosion rates are high on these sites because of the sparse vegetation cover. And the vegetation is sparse there because the soils are thin. Clearly, such sites exhibit a strong positive feedback setup, with the end result being expansive areas of bedrock outcrop.

Alternatively, on areas of softer rock, e.g., shale, weathering is rapid and saprolite can form, even in this dry climate. Shale is a weaker rock that retains more moisture after precipitation events and weathers quickly by wetting and drying processes. Thus, it crops out as gently sloping surfaces, and shallow Entisols occur on such residuum.

The southern Arizona landscape shown in the photo is typical of this scenario. The steep slope on sandstone has no associated soil. Weathering proceeds slowly here, and when the rock does break apart, rockfalls occur, exposing more fresh bedrock. On the gentler slope in the foreground, weathered shale has developed into Ustorthents. Vegetation can get a foothold here, further stabilizing the slope, facilitating soil formation, and accelerating weathering. Soils form further, and regolith thickens even more.

This example illustrates how weathering rates, slope processes, residuum formation, and soil development all interact with and feed back on each other, within the landscape context. It sets the stage for the sections on soil geomorphology that follow.

Figure: Plateau and escarpment landscape of the Hualapai-Havasupai region, northwestern Arizona.

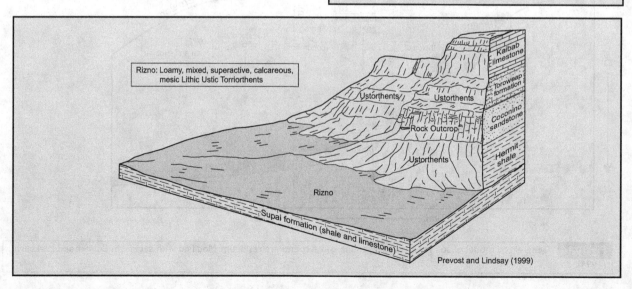

Rizno: Loamy, mixed, superactive, calcareous, mesic Lithic Ustic Torriorthents

Kaibab limestone

Toroweap formation

Coconino sandstone

Hermit shale

Ustorthents

Rock Outcrop

Rizno

Supai formation (shale and limestone)

Prevost and Lindsay (1999)

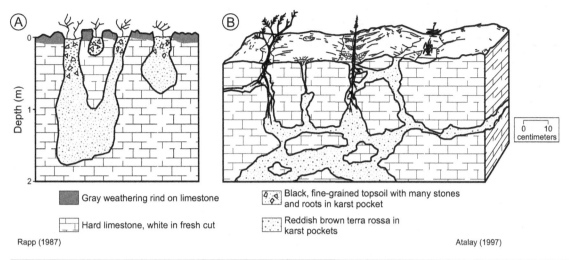

Rapp (1987)

Gray weathering rind on limestone

Hard limestone, white in fresh cut

Black, fine-grained topsoil with many stones and roots in karst pocket

Reddish brown terra rossa in karst pockets

Atalay (1997)

Fig. 10.15 Typical patterns of soil-residuum-rock in fissured outcrops of limestone.

Within the carbonaceous family of sedimentary rocks (Table 10.1), chemical weathering (carbonation and hydrolysis) is dominant over physical weathering. Limestone dominated by calcite is more weatherable and soluble than limestone dominated by dolomite, but both produce little residuum, as insoluble residue contents are commonly <1% (MacLeod 1980, Danin et al. 1983). Either most, if not all, of the calcite or dolomite in the limestone is leached as soluble bicarbonate (HCO_3), Ca^{2+}, and Mg^{2+}, or the Ca and Mg become biocycled. Usually, these rocks contain some chert fragments and smaller amounts of insoluble residue, which ultimately become major components of the residuum (Barshad et al. 1956, Rabenhorst and Wilding 1984, Moresi and Mongelli 1988). Residue from calcitic and dolomitic limestone is often clay- and silt-rich, because clay and silt particles were components of the seafloor when the carbonates were precipitated or deposited (Plaster and Sherwood 1971, Rapp 1984, Crownover et al. 1994). Limestone residuum is typically dominated by illite clays, although aragonite, siderite, and quartz clays may also occur (Plaster and Sherwood 1971, Oganesyan and Susekova 1995). Some of the quartz may have been deposited as a silica gel and later dehydrated to form chert, jasper, and flint (Buol et al. 1997).

Limestone weathers along preferential joints, seams, or pockets, where water and carbonic acid can easily penetrate. Once developed, these pathways become preferred sites for additional weathering. Thus, the classic red residuum of limestone typically occurs in pockets, deep cracks, and fissures, often with bare rock between (Atalay 1997). In Europe, the rocks that stand between wide, deep cracks in limestone are called lapiès (Danin et al. 1983). Frolking et al. (1983) described a brown clay zone, several millimeters thick, that lay between the reddish clayey residuum and unweathered limestone bedrock in Wisconsin. In other cases, the contact is clean and abrupt (Atalay 1997). It is often assumed that clayey, red, weathered material washes off the rock into the fissures, keeping the rock relatively free of residue, while facilitating the accumulation of residuum in pockets (Fig. 10.15). On marls and other carbonaceous rocks that have few fissures and cracks, more water runs off, leading to erosion and thinner residuum (Verheye 1974).

The production of red colors, common in limestone residuum, is termed *rubefaction* or *rubification*. It develops as a result of various iron-bearing minerals and compounds (Benac and Durn 1997; see Chapter 13). Rubification is enhanced by aeration, warm temperatures, a wet season sufficient to permit growth of microorganisms (specifically iron bacteria), and rapid turnover of organic matter (Glazovskaya and Parfenova 1974, Boero and Schwertmann 1989). Fractured limestone uplands in Mediterranean climates are well suited to provide these factors. Here, iron released by weathering is precipitated as iron oxides and hydroxides, coating other clay minerals and imparting red colors to the soil and residuum (Glazovskaya and Parfenova 1974, Bech et al. 1997). Once formed, these compounds tend to persist.

Many studies of the red, clay-rich residuum that overlies limestone focus on the extent to which it is pure residuum versus material introduced from outside, most notably via eolian influx. To resolve this question, researchers have examined silt and clay mineralogy, as well as the texture of the insoluble fraction in the parent rock, and compared them to those of the residuum (Glazovskaya and Parfenova 1974, MacLeod 1980, Frolking et al. 1983, Muhs et al. 1987; Table 10.2). Questions asked include the following: Do certain minerals increase or decrease in the residuum vis-à-vis the rock? Are some minerals in the regolith not present in the rock? If not, could they be weathering products of minerals in the rock, or must they have originated outside? Other clues also exist. For example, chert layers in

Table 10.2 Characteristics of dolomite residuum, overlying red clay and loess in Wisconsin uplands

Characteristic (units)	Dolomite residuum[a]	Red clay	Loess
Sand content (mean%)	10.1	5.2	1.4
Silt content (mean%)	43.0	14.7	70.0
Clay content (mean%)	47.0	82.0	28.0
Fine clay content (mean%)	34.0	70.0[b]	16.0
Smectite content[c]	1.1	4.5	4.7
Vermiculite content[c]	1.1	1.9	2.7
Mica content[c]	3.6	1.5	1.5
Kaolinite content[c]	1.0	2.4	1.7
Quartz content[c]	4.2	2.0	1.7
I_{100}:I_{101} ratio[d]	0.70	1.02	0.23

[a] Dolomite residuum represents material taken directly from the rock
[b] Much of the fine clay could have been illuvial
[c] Mineralogy is an average of the fine and coarse clay fractions, based on XRD peak areas. 5 = >50%, 4 = 31–50%, 3 = 16–30%, 2 = 5–15, and 1 = <5%
[d] Ratio of the 100 and 101 XRD peaks for the 1–10 μm quartz fraction. This ratio measures the relative concentration of euhedral quartz grains in a sample. Assumedly, chert weathers to euhedral grains, whereas eolian loess has primarily angular quartz grains with pitted surfaces. Higher I_{100}:I_{101} ratios indicate more chert weathered from the dolomite, and less loessial influence
Source: Frolking *et al.* (1983).

limestone bedrock may also exist in the regolith, providing strong evidence for a purely residual origin (Frolking *et al.* 1983). Quartz grains from eolian sources may have different shapes and surface textures than those taken directly from the bedrock; these data can also assist in identifying the source of the sediment (Frolking *et al.* 1983, Rabenhorst and Wilding 1986a, Levine *et al.* 1989). Figure 10.16 illustrates this point with scanning electron microscope (SEM) photomicrographs of silt grains removed directly from limestone and from the A horizon of the same profile, which has formed in limestone residuum *and* loess. The former grains have an obvious euhedral shape, typical of authigenic quartz crystals formed in voids in limestone (Rabenhorst and Wilding 1986a). The latter, however, are pitted and rounded, presumably by eolian transport (see also Frolking *et al.* 1983).

Limestone residuum, like that of calcareous shales and marls, is saturated with base cations, especially Ca, when it weathers out of the parent rock. Although this residuum may eventually become leached, depending on climatic and biotic conditions at the site, many pedogenic processes, e.g., lessivage, will be inhibited until some of the base cations are removed and the pH falls (see Chapter 13). Melanization will be more restricted to the upper profile, because of the strong Ca-humus bonds that

develop, rendering organics relatively immobile. For these reasons, soils formed from limestone residuum often are less developed and have thinner sola than comparable soils developed on other parent materials (Ehrlich *et al.* 1955). Although limestone is a highly weatherable rock in humid climates, fragments may persist for long periods in the residuum as infiltrating water takes preferred pathways around them. Infiltrating, acidic water is buffered by the continual release of bicarbonate at the limestone-residuum contact and from limestone fragments within the soil, as the rock weathers (Ehrlich *et al.* 1955). If clay translocation does occur, much of the illuviated clay may be located in a thin layer near or immediately on top of the limestone.

Landscapes: Terra Rossa Soils of the Mediterranean Region

Terra rossa is an Italian term for soils that are red, shallow, clayey, and undifferentiated and that usually overlie calcareous bedrock (Stace 1956, Isphordi 1973). Commonly, they have sharp contacts with the underlying bedrock and often fill solution cavities within the rock (van Andel 1998, Muhs *et al.* 2012). Although terra rossa soils are

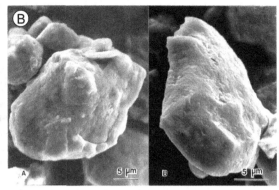

Rabenhorst and Wilding (1986a)

Fig. 10.16 Photomicrographs of quartz grains from a Texas soil and bedrock. Photomicrographs by M. C. Rabenhorst. (A) Euhedral prismatic quartz grains from the medium silt (5–20 μm) fraction of the noncarbonate limestone residue. (B) Silt-sized quartz grains from the upper solum, showing rounded edges and pitted surfaces, suggesting an eolian origin.

often called red Mediterranean soils, in areas where they dominate, most limestone soils are, in fact, not red (Verheye 1974, Benac and Durn 1997). The term has also been used to describe the red residual material that commonly overlies carbonatic rocks (Darwish and Zurayk 1997). Most terra rossa soils classify as Xeralfs, Ochrepts, or Xerolls; many are Rhodoxeralfs (Soil Survey Staff 1999). The red or brown colors are due to iron oxides (Jackson and Frolking 1982). Many have a significant clay increase at the bedrock interface, where base cations released from the rock cause clays to coagulate (Durn et al. 1999).

It has often been presupposed that these soils formed in residuum from calcareous bedrock, i.e., the clay and other silicate minerals are acid-insoluble residues (Barshad et al. 1956, Moresi and Mongelli 1988, Feng et al. 2009). This theory implies that the soils are very old and exist on exceedingly stable surfaces. The presence of weakly developed soils on limestone, whose mineralogy matches that of the limestone residuum below, tends to support this model (Glazovskaya and Parfenova 1974). Verheye (1974) noted that terra rossa soils are primarily found on stable uplands, where erosion is minimal; on other landscape positions less-developed brown and yellow soils occur. Some, however, observed that the red residual sediments are also accumulating in karst depressions as thick deposits in accretionary slope positions (Benac and Durn 1997).

Recently, the residuum theory of terra rossa development has fallen into disfavor. The arguments against it are the following:

1. The sharp contact between terra rossa and bedrock, usually but not always observed (Glazovskaya and Parfenova 1974), is inconsistent with a residual origin (Olson et al. 1980). That is, transitional regolith between the soil and bedrock is absent.

2. Typically terra rossa soils contain more mass than can be accounted for by weathering alone (Olson et al. 1980, Muhs et al. 1987, Herwitz and Muhs 1995, Durn et al. 1999). For example, at a site in Greece it was determined that ~130 m of limestone would have to have weathered (and all the residuum stay in situ) to produce only 40 cm of red clay (Macleod 1980).

3. Cyanobacterial weathering patterns on the limestone indicate that weathering essentially ceases after residuum reaches a few decimeters thick (Danin et al. 1983).

4. Clay contents and silt:clay ratios in the soils are often different from those of the acid-insoluble limestone residue (Macleod 1980, Durn et al. 1999).

5. Iron contents in the limestone residue are often much lower than would be expected, given the red color and high iron contents of the terra rossa soils.

6. Clay minerals and other elemental oxides in the rock and residue do not match those in the red sediment (Olson et al. 1980, Moresi and Mongelli 1988).

7. The angularity of the silt and sand grains in the residuum matches classical eolian particles more closely than those encased in the limestone (Rapp 1984).

An additional line of evidence arises from geomorphology. Over-thickened silty sediments in depressional areas and on footslopes attest to the instability of terra rossa landscapes (Wieder and Yaalon 1972, Yassoglou *et al.* 1997); residuum does not accumulate to great thicknesses on such weathering-limited slopes. Yassoglou *et al.* (1997) noted that most terra rossa soils in the Mediterranean region have undergone such severe erosion that the complete profile is rarely found (see also Boero and Schwertmann 1989). Even in the lower-relief landscapes of the midwestern United States, terra rossa is primarily debris that has been transported by slope processes (Olson *et al.* 1980).

Today, most agree that the clay and silt in terra rossa soils mainly have eolian origins (Rapp 1984). In the Mediterranean region, eolian clay and silt from the Sahara Desert and Ukraine fall out as "mud" or "blood" rain and snow (Lundqvist and Bengtsson 1970, Macleod 1980, Yaalon 1987, 1997a). Dust from the Sahara affects locations as far north as Great Britain (Pitty 1968, Prodi and Fea 1979) and as far west as Bermuda, the Bahamas, and the Caribbean (Prospero and Lamb 2003, Muhs *et al.* 2012). Limestone weathering contributes *some* residuum to the terra rossa soils, but the majority of the parent material is eolian silt and clay.

Figure: Terra rossa soils. (A) Soil differences on marl versus fractured, hard limestone, in the Mediterranean. (B) Particle size data, plotted on a standard textural triangle, for some terra rossa soils, Indiana, United States, and Croatia. (C) Relationship between clay content and dithionite-extractable Fe content of some terra rossa soils in Greece.

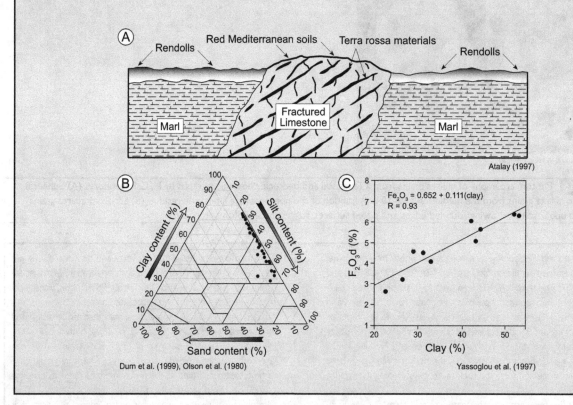

Atalay (1997)

$Fe_2O_3 = 0.652 + 0.111(clay)$
$R = 0.93$

Durn et al. (1999), Olson et al. (1980)

Yassoglou et al. (1997)

Organic Parent Materials

In some places, organic materials released from vegetation accumulate on the surface faster than they can decompose. This scenario usually occurs where decomposition rates are slow, and where the organic materials are only minimally mixed with mineral materials below. When organic materials accumulate to thicknesses exceeding 20 cm, the soil may be classified as a Histosol (Soil Survey Staff 1999). Organic soil materials occur in various stages of decomposition, from raw vegetable matter, through peat, mucky peat, peaty muck, and muck (Malterer *et al.* 1992, Soil Survey Division Staff 1993; Table 3.2). The degree of decomposition is, ultimately, dependent upon a number of factors, including (1) supply of oxygen to the organic materials; (2) water chemistry, especially pH; (3) temperature; (4) biodegradability of the plant matter; (5) amount of faunal activity, such as worms and beetles; and (6) time available for decomposition.

Although thick accumulations of organic matter develop fastest and occur most commonly in wet, saturated sites, they can also form on uplands (Ciolkosz 1965, Daniels *et al.* 1977; see Chapter 13). Most Histosols occur in wet, cool sites such as bogs, fens, swamps, marshes, and tidal backwaters (Gates 1942, Heinselman 1970, Bockheim *et al.* 1999) because the rate of decomposition is slowed by wet, waterlogged, and sometimes anoxic (anaerobic) conditions

(Rabenhorst and Swanson 2000). Although many wet sites are highly productive from a biological perspective, they also tend to have low rates of decomposition and, thus, are well suited for organic matter accumulation.

Depending on the site, organic materials in Histosols may be a mix of many different kinds of plant material, or they may be dominated by woody debris, *Sphagnum* moss, heather, grasses, and sedges. Over time, organic materials may continue to accumulate and thicken, or they may slowly decompose as site conditions change. The latter situation is likely to occur as the climate warms or dries, or as water tables fall. Many Histosols burn when they are dry, retaining layers of charcoal indicative of dry paleoclimatic periods. Thick organic soil materials often preserve within them very detailed records of past plant occupation, fire history, and paleoclimate, in the form of plant macrofossils and pollen, which accumulate over time and are only minimally mixed or turbated within the accumulated organic materials (see Chapter 16).

On bare bedrock surfaces, even on uplands, organic materials can also accumulate and thicken, providing that the climate is cool (Fig. 3.7). How and why would this occur? Recall that one of the factors that facilitate organic matter accumulation is minimal mixing with mineral materials below. On most dry, upland soils, organic matter falls on the surface, decomposes, and then is mixed with the mineral material below; this process forms A horizons (see Chapter 13). However, if the organic material falls onto bare bedrock, the mixing is thwarted, and organic matter accumulates on the rock. Of course, it will accumulate more rapidly if the climate is cool, or if the rock is rich in base cations (the Ca^{2+} cations will bind organic matter and slow its decay), but it will, nonetheless, accumulate. If it accumulates to >15 cm, the soil is considered a Folist in Soil Taxonomy (organic soils that are rarely saturated with water under natural conditions). Folists occur mainly in cool, high mountainous or upland areas, where they form directly on bedrock or fragments of bedrock (Witty and Arnold 1970; see Chapter 13).

A historian once described the efforts of firefighters on Isle Royale in Lake Superior, in 1935, where Folists are common on basalt bedrock: "The soil is leaf mold and humus, lying in a shallow layer over clay and rock. The soil itself burns." The lack of mineral soil material below a Folist's O horizon precludes significant mixing with mineral material below and can actually keep the litter mat drier than it would otherwise be, thereby slowing decomposition. On surfaces that are not bare rock, Folists can occur as organic drapes on large boulders that stand above the mineral soil (Fig. 3.7). Schaetzl (1991a) reported on Folists on Bois Blanc Island, a cool site in Lake Huron. Here, acid organic materials have accumulated in and on top of coarse beach gravel dominated by limestone fragments. The high pH assists in the formation of Ca-humus precipitates (specifically, Ca precipitates the polyphenolic precursors of humic and fulvic substances), slowing decomposition of the organic materials (Bochter and Zech 1985). Although organic materials are, technically, the parent material of Folists, most plants are affected by and are able to utilize the mineral material below, especially if it is fractured bedrock or skeletal materials (Alexander 1990).

Transported Parent Materials

Worldwide, there are far more transported parent materials than residual parent materials (Hunt 1972). During transport, parent materials become considerably altered in character. Many become more intimately mixed than the material from which they were derived, whereas others are better sorted by particle size and/or mineralogy. Abrasion during transport can reduce the size of particles within the sediment, and change their surface texture.

Most transported parent materials are categorized on the basis of their mode of transport, e.g., water or wind. The origins of transported parent materials can often be discerned by examining their sedimentology, mineralogy, and geomorphology. For example, eolian sand has a distinctive texture and sorting and is typically found in dunes. Alluvium tends to be stratified. The categorization of transported parent materials below illustrates one possible genetic grouping.

Transported by Water

Materials transported and deposited by running water in streams are called *alluvium*. All alluvium is first entrained (picked up) by the stream, transported, and then deposited. Understanding these processes can be of benefit in the interpretation of alluvial parent materials (Friedman 1961). Other sediment that has some connection to water includes marine sediment (oceans) and lacustrine deposits (lakes). Most alluvium is deposited within stream channels, on floodplains, and in deltas and alluvial fans. Thin, scattered alluvial deposits may also occur at the base of slopes, the result of sheetflow (nonchannelized water).

All transported sediment has a certain amount of common history – it must first be entrained by the agent of transport, then it is transported, and finally, it is deposited when gravity overcomes the ability of the transporting agent to keep the sediment moving. Transport of sediment is primarily a function of the sediment size and the energy (velocity) of the transporting medium. Transport by water follows the same rules, although other factors such as the water's viscosity and turbulence are also important. Coarse sand, gravel, and larger stones are difficult to entrain because of their mass. Clay particles are difficult to entrain because they must be chemically dispersed to remain suspended (Fig. 10.17). Thus, it is easier for a stream to entrain silt than either coarse sand or fine clay. Transportation, on the other hand, is more directly proportional to the size of the sediment than is entrainment. Depending on the water's ionic strength, most clays will settle out of suspension only in calm water, whereas even the finest sand

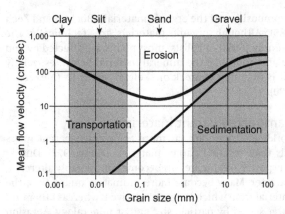

Fig. 10.17 The Hjulstrom diagram, shown here, depicts water/current speeds necessary for erosion, transportation, and/or deposition of various grain sizes.

Fig. 10.18 A small floodplain at the base of a sandy slope in Wexford County, Michigan. Udifluvents occupy the floodplain; Haplorthods dominate on the slope. Photo by RJS.

requires comparatively fast velocities of 1 cm s⁻¹ to avoid deposition.

To some extent, most alluvial deposits are stratified (layered) during deposition. Lack of stratification implies either deposition in turbid, swirling waters and subsequent reentrainment, or, commonly, some form of postdepositional mixing. Coarse-textured sandy deposits, such as midchannel bars, are often stratified and cross-bedded. Silty to clayey overbank deposits are finely stratified in more horizontally lying beds. Levee sediments are also stratified, but somewhat coarser than are overbank sediments on the floodplain proper. In oxbow lakes and backswamp areas on floodplains, clays and silts can settle out of calmer waters. Point bar deposits, usually coarser-textured, form in the bends of stream meanders. Fining upward sequences, typical of point bar deposits, occur because larger particles are deposited first, and as the energy of the river decreases during a depositional episode, continually smaller particles are deposited, until at the top of the sedimentary sequence, only the finest materials remain (Baker *et al.* 1991). Coarsening upward sequences are rare, for they would imply that the river was gaining in energy and velocity while building its channel. The young age of the soils on most upwardly aggrading fluvial landforms, such as floodplains, explains their common classification as Entisols (Fig. 10.18). Many soils on floodplains classify as Fluvents, implying that they are relatively weakly developed and that their parent materials have a fluvial origin. Because of their aggradational origins, Fluvents have variable depth functions for organic carbon and other soil parameters.

Like almost all transported parent materials, alluvial deposits are progressively younger toward the top of the deposit. They also can exhibit many sudden changes or breaks in sediment texture, packing or bedding, along strata. These sedimentologic discontinuities can dramatically affect pedogenesis. Water (wetting fronts) carrying dissolved and suspended materials tends to stop or hang at such discontinuities, especially where fine material overlies coarser sediment (see Chapter 13). In such situations, the infiltrating water may not enter the coarser material below until the fine material above is saturated, or nearly so. Thus, preferred sites for deposition of materials translocated in percolating water are along the contact (discontinuity) and within the sediment above it.

Alluvial facies can be continuous for many kilometers, or they may be highly discontinuous. Overbank flood deposits can extend for great distances, whereas a midchannel bar deposit might only be a few meters wide or long.

Many streams in North America and Europe carried glacial meltwater during the waning phases of each of several glaciations during the Pleistocene Epoch. Many of these streams filled their valleys with sediment, only later to deepen and incise them when the large influx of glacial debris no longer was a factor. Postglacial incision of valley-fill sediments resulted in the abandonment of former floodplains, leaving terraces behind (Fig. 10.19). Soil development on terraces can provide clues to the timing of terrace abandonment and, hence, the age of the soils (Howard *et al.* 1993, Alonso *et al.* 1994, Shaw *et al.* 2003, Tsai *et al.* 2006). Large valleys, once choked with coarse sediment, these streams are now *underfit*, carrying much less water and sediment than their large valleys would suggest.

Lacustrine (lake) deposits reflect depositional environments that range from deep, calm conditions through those of the active shore zone. As a result, they can be highly variable in texture, although most lacustrine deposits have excellent stratification. In the deeper parts of lake basins, stratified clay- and silt-rich sediments are typical (Fig. 10.20). This is especially common in proglacial lakes, which had a constant supply of silt- and clay-sized material from meltwater. In glacial lakes, and some nonglacial lakes, freezing in winter coupled with a lack of sediment influx allows the finest clays to settle out of suspension.

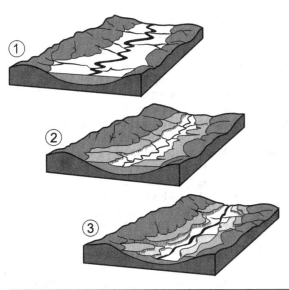

Fig. 10.21 The flat and nearly featureless floor of Glacial Lake Saginaw, in southern Michigan. Silty and clayey lacustrine sediment composes the parent material for the Endoaquolls that dominate this landscape. Photo by RJS.

Fig. 10.19 Examples of a typical valley cross section, illustrating how a valley fill deposit (alluvium) can later become incised to produce alluvial terraces of varying ages.

Fig. 10.20 Clayey lacustrine sediments from a Glacial Lake Oshkosh, in eastern Wisconsin. The stratified nature of these sediments is strikingly obvious. Photo by RJS.

In summer, the deeper waters of the lakes are more turbid and have slow currents, allowing only silts and coarser materials to be deposited, while clay remains suspended or leaves the lake basin via an outlet. Thus, an annual couplet of fine, dark clay (the dark colors originate from finely comminuted organic matter) and lighter silts (with some quartz) are deposited annually. These annual couplets are called *varves* (Bradbury *et al.* 1993, Leonard 1997, Andren *et al.* 1999, Schaetzl *et al.* 2000).

In the shallower parts of lakes and near the coastline, wave action and currents produce higher energy environments, and as a result, coarser-textured sediment is more commonly deposited there. Waves and currents can be major erosive forces, beveling off any high spots on the bottom of the lake and eroding headlands along the coastline. Eventually, the bottoms of many lakes become quite flat, as a result of waves constantly shoaling, and thus, eroding these areas (Fig. 10.21). Sediment also infills the deeper sections of the lake, resulting in flat, nearly featureless lake floors (Clayton and Attig 1989).

Beach sands and offshore sand bars are an important component of many lacustrine environments – if for no other reason than that they provide pedologic diversity to the otherwise monotonous lake floor landscape. On the floor of Glacial Lake Saginaw, a now-drained paleolake in southern Michigan, these types of sandy deposits represent some of the only coarse-textured sediment on the lake plain. Many of these sands have since been deflated and deposited as dunes (Arbogast *et al.* 1997). Often, these sandy soils are the only locations where well-drained soils occur on lake plains.

Lakes and some floodplains, rich in sediment derived from carbonate rocks, may develop thick deposits of marl, a light-colored, carbonate-rich form of organic-limnic sediment (Haile-Mariam and Mokma 1990, Chambers 1999). Shells and shell fragments are common indicators of marl, which is primarily composed of finely divided calcite (Johnston *et al.* 1984).

Ephemeral desert lakes (playas) typically are also very flat and usually contain salty deposits and saline soils (Peterson 1980). Eolian dust that blows out of these playas during dry periods is important to soil development on the nearby uplands (Chadwick and Davis 1990).

One group of widespread parent materials associated with coasts is coastal plain deposits. These sediments form in association with current and former coastlines; they include former barrier islands, deltas, beaches, and bars. Most coastal plain sediments are sandy. These low relief

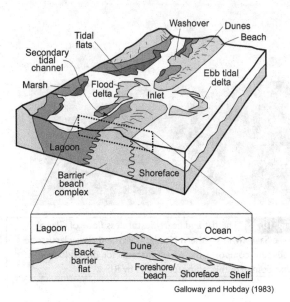

Galloway and Hobday (1983)

Fig. 10.22 Depositional environments associated with oceanic coasts, with barrier islands.

landscapes result from a fall in sea level that exposes the sediments subaerially.

In the coastal system, many additional sedimentary environments can be isolated, each of which will have variously saline, brackish, or freshwater conditions; tidal activity adds further complications. Estuaries are drowned river valleys usually caused by Holocene rises in sea level. In order to prevent estuaries from filling with alluvium, rivers must deposit sediment at rates that do not exceed the transportational abilities of waves and currents from the ocean. Otherwise, the estuary will fill and a delta will slowly form. At the heads of estuaries, fluvial sands grade into finer-textured sediments in the estuary proper. Estuarine facies include muds, silts, and sands; farther seaward, other forms of tidal deposits are common.

In certain areas along coastlines, sediments can become rich in sulfur and the sulfur-bearing mineral pyrite. Through a process called *sulfidization*, these sediments develop into acid sulfate soils, with pH values commonly <3.0 (Fanning and Fanning 1989; See Chapter 13). The process is initiated as sulfur-bearing sea water has contact with tidal marsh sediments, usually in backwater areas and not on the open coast proper (Darmody and Foss 1979; see Chapter 13).

Deltas are perhaps some of the more complicated fluvial depositional systems. Their character can vary from sandy through clayey, not only across the delta but vertically within it. Ages of the sediment also vary greatly, as the major distributary channels shift across the delta over time. In general, delta sediments are deposited in wedges that are thinner and finer-textured farther from the delta head (or river mouth), to the prodelta. Nonetheless, splays of coarser material can occur at any place, leading to sandy deposits overlying silts, or any imaginable combination.

Oceanic coasts are complex systems of deposition and erosion. On actively eroding coasts, sea cliffs and bedrock stacks provide unmistakable evidence of the power of the sea. Other coasts show evidence of ongoing deposition and reworking of sands by waves and currents. One variable common to all sea coasts and nearby environments is the recent geologic changes in sea level. Eustatic (global) sea level rise following the melting of the last ice sheets was on the order of 120 m (Gehrels 2010). These changes forced coasts to retreat, and with them offshore barrier islands and their sandy sediments, often with dunes at the landward side (Fig. 10.22). Barrier islands separate the ocean from the brackish lagoon behind them. Lagoons are important sedimentation systems for not only fine sediment but also organic materials. They often contain organic-rich muds over (deep) sands, which may be interbedded with sands washed over the barrier during storms. On the landward side of the lagoon, brackish marshes and tidal flats accumulate various types of sediments from streams that feed into the coastal system, and acid sulfate soils may form (see Chapter 13). Location within the marsh generally controls grain size. On tidal flats, sediments tend to be coarser near low-tide lines and fine near high-tide lines (Daniels and Hammer 1992).

Transported by Wind

A variety of soil parent materials are deposited from the atmosphere. Entrained by the wind, the distances that sand, silt, and clay are transported depends on wind velocity, atmospheric turbulence, and grain size. Pyroclastic materials (fragments ejected during a volcanic explosion) are blown up into the atmosphere and then either settle out of the air directly or are moved still farther by wind (Kieffer 1981).

Wind is best capable of entraining and moving small particles, typically sand size or less. Strong winds coupled with dry sediment of the appropriate size – if not stabilized by vegetation – facilitate eolian entrainment and transport (Fig. 10.23). Once entrained, larger particles will *saltate* (bounce) along the surface, while smaller particles may be lifted up and move in suspension. As saltating grains impact the surface, they dislodge others, which may then begin to move by creep, by saltation, or in suspension (Fig. 10.24). Erosion by wind, i.e., *deflation*, may remove most of the fine particles, sometimes resulting in a lag concentrate, or pavement, at the soil surface (Parsons *et al.* 1992). Stones eroded, grooved, and shaped by this type of sand blasting are called *ventifacts* (Sharp 1949, Whitney and Splettstoesser 1982). They are commonly observed on the surfaces of dry, windy desert areas (Fig. 10.25).

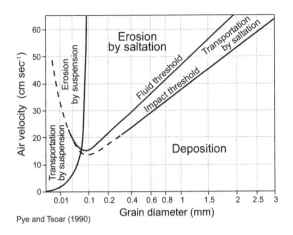

Pye and Tsoar (1990)

Fig. 10.23 Threshold velocities required for wind to erode/entrain, transport, and deposit dry particles of various sizes.

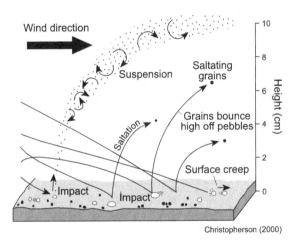

Christopherson (2000)

Fig. 10.24 Idealized example of particle movement by the wind, i.e., by creep, saltation, and suspension.

Fig. 10.25 Two small ventifacts on the surface of the Anza-Borrego Desert, near Palm Springs, California, attest to the strong winds in this area. Photo by RJS.

When wind velocities drop below a critical value, deposition occurs. The coarsest particles will settle out of suspension (or saltation) first, and as wind speeds drop increasingly more of the smaller particles will also be deposited. For this reason, eolian (wind) deposits are usually fairly well sorted.

EOLIAN SAND PARENT MATERIALS

Distinct landforms made of eolian sand are called dunes (Muhs and Wolfe 1999); these landforms are called sand sheets or cover sands if the sand resembles a "blanket" on the landscape (Cailleux 1942, Kocurek and Nielson 1986, Thorson and Schile 1995, Frechen *et al.* 2001, Holliday 2001, Haase *et al.* 2007, Semmel and Terhorst 2010). Dune fields form when winds above a threshold velocity interact with sandy surfaces that lack adequate stabilizing vegetation. The instability is usually caused either by fresh deposition of sand (and hence no vegetative cover) or by fire, drought, or other factors destroying stabilizing vegetation. Many coastal dunes are associated with the first scenario.

Similar to soil production on slopes, erosion over entire landscapes can be considered either sediment-limited or transport-limited with respect to eolian processes (Kocurek and Lancaster 1999). Sediment-limited situations occur where either (1) the available sediment is of the wrong size for eolian transport or (2) the grains are not susceptible to entrainment and transport, as may happen when the sand is wet, cemented, covered with lag gravel, or vegetated. Transport limitations primarily involve wind speed, which may also be affected by a cover of vegetation. Many of the world's dune fields are not active today because of limitations with respect to sediment supply or, more commonly, transport vectors. Thus, they represent a time in the past when conditions were different, and sand was mobile (Muhs and Zárate 2001). Small shifts in climate (moisture

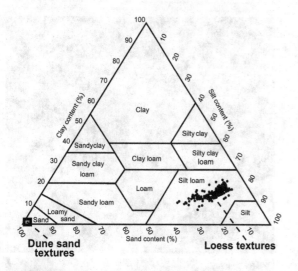

Fig. 10.26 Textural data from two types of eolian sediment (dune sand and loess), as plotted on a textural triangle. The loess data are from various sites in Wisconsin where loess thickness is > 1 m. Sand dune sample data were derived from small dune fields scattered throughout Michigan and Wisconsin. Internal data generated at the Michigan State University Geomorphology lab, using laser particle size diffraction techniques.

balance, wind strength, or land use) might be enough to initiate or reinitiate sand transport (Keen and Shane 1990, Muhs and Holliday 2001). Soils that are forming on stabilized dunes can provide information about when the dunes stabilized and pedogenesis began (Arbogast *et al.* 1997, Arbogast 2000).

Eolian sands are often yellow, buff, or brown in color, as a result of thin coatings of oxidized iron and clay on the quartz grains. Sand grains may appear "frosted" and chipped as a result of impact in transit, and many are well rounded. Well-sorted, medium and fine, quartz-rich sands dominate many dunes (Ahlbrandt and Fryberger 1980, Dutta *et al.* 1993, Arbogast 1996, Muhs *et al.* 1999; Figs. 10.26, 10.27). Often, little or no measurable clay fraction is found in dune sand (Arbogast *et al.* 1997). For obvious reasons, therefore, most soils formed in dune sand are quartz-rich, although many have illuvial clay lamellae at depth (Schaetzl 1992b, 2001, Rawling *et al.* 2008, Bockheim and Hartemink 2013a; see Chapter 13). The kinds of soils that form in dune sand, and the rate at which various ions and minerals are released to the soil by weathering, are largely a function of climate and initial dune sand lithology, age of the deposits notwithstanding.

The term *mineralogical maturity* is often used to refer to weathered eolian sands. Mineralogical maturity implies that the grains are more quartz-rich, and often more rounded,

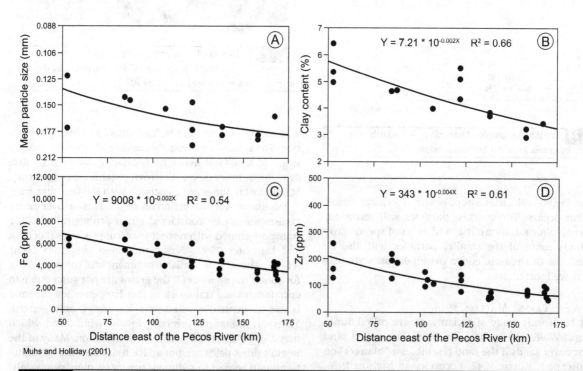

Muhs and Holliday (2001)

Fig. 10.27 Variation in dune sand characteristics away from the Pecos River, Texas, the sand source.

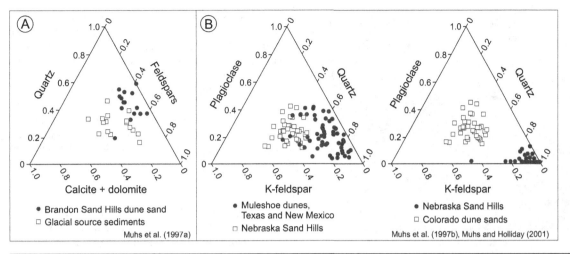

Fig. 10.28 Ternary plots of dune sand mineralogy. (A) Mineral composition of dune sands versus the presumed source material. (B) Comparisons of the mineral composition of dune sands from different dune fields.

frosted, and abraded. This characteristic may be due to the great age of the eolian sand, facilitating the loss of many of the more weatherable minerals (Dutta *et al.* 1993; See Figure 10.28 for examples). Alternatively, repeated episodes of transport, during which time abrasion and ballistic impacts deplete the carbonate minerals and feldspars from the dune sand, can also lead to mineralogical maturity (Muhs *et al.* 1997a, b, Muhs 2004). Experimental evidence suggests that sand grains are continually comminuted as they saltate. It is important to note, however, that mineralogical maturity may represent either or a combination of these two factors, but it also may simply be inherited from the source rock, perhaps sandstone that is dominated by quartz grains (Muhs 2004). For that reason, information on mineralogy and maturity of eolian sands is also useful in identifying the ultimate source material and age of the dunes (Muhs *et al.* 1997a, b, Arbogast and Muhs 2000, Garzanti *et al.* 2012; Fig. 10.28).

For mature eolian sands, determining whether they are low in weatherable minerals because they have experienced long-term weathering or because the original parent material was quartz-rich is often an important question, because it speaks to the dynamics of the eolian system, and the age of the sediment (Dutta *et al.* 1993, Muhs *et al.* 1997b). Sand dunes are dynamic landforms, being stable for periods of time, but then, as conditions change, being subject to erosion and remobilization, or burial by sand being transported from upwind sources. As a result, many sand dunes show evidence of buried soils, formed at times in the past when they were at the surface of the dune (Anderton and Loope 1995, Arbogast *et al.* 2004, Halfen *et al.* 2010). Organic matter or other forms of buried organic materials,

e.g., wood, charcoal, can be dated using radiocarbon dating techniques to determine when the burial event may have happened (Anderton and Loope 1995, Dalsgaard and Odgaard 2001, Arbogast *et al.* 2004, Goble *et al.* 2004, Hansen *et al.* 2004). If one assumes that there has been no translocation of organic matter from younger, overlying soil horizons, then such dates provide maximum-limiting estimates for the burial event, i.e., the burial event must have occurred after the date established by ^{14}C (Arbogast and Loope 1999; see Chapter 15).

Evidence for small amounts of eolian sand or silt in soils or sediments can be highly informative as to the depositional history of the site, but such evidence is often elusive and difficult to evaluate without concomitant sedimentological analysis (Schaetzl and Luehmann 2013). Crawford *et al.* (1983) were able to differentiate eolian sand grains in soils from those that had weathered out of bedrock in situ; eolian sand grains were more rounded (Fig. 10.29). Indeed, detailed analysis of sand grain shape and texture can prove quite useful in determining whether the particles have undergone eolian transport at some time in the past (Krumbein 1941a, b, Cailleux 1942, Elzenga *et al.* 1987). Mycielska-Dowgiallo and Woronko (2004) named the process whereby grains acquire certain shape and surface characteristics due to eolian transport *eolitization*. Leaning on methods originally used by Krumbein (1941a), Cailleux (1942), Gozdzik (1980), and Mycielska-Dowgiallo and Woronko (1998), they examined various textural features of medium sand grains from cores of Quaternary deposits in Poland to determine the extent to which eolian versus other types of processes had been involved in their emplacement, i.e., how much eolian influence the grains displayed. Particular attention

Sandy soils, such as those that form on dunes, tend to develop rapidly and their pH can adjust quickly to various state factor inputs, largely because of the low surface area of sand (see Chapter 13). Podzolization is a common process on dunes in cool, humid climates; in grasslands melanization is dominant in dune soils. Because dune sand is so quartz-rich and clay-poor, lessivage is seldom a dominant process, although if the dune sand has some clay, textural clay bands known as *lamellae* can, and often do, form (Oliver 1978, Gile 1979, Berg 1984, Schaetzl 1992b, 2001, Rawling 2000).

PARNA PARENT MATERIALS

Clay-sized particles, if flocculated into sand-sized aggregates, can saltate and form "clay dunes" (Butler 1956, Beattie 1970, Bowler 1973, Munday *et al.* 2000). *Parna*, an eolian sediment composed of dominantly clay size aggregates, is common in parts of Australia, where it can attain thicknesses > 8 m. Many of these deposits have deflated from dry, clay-rich lake basins, whose salts assist in the flocculation of the clays. Dare-Edwards (1984) suggested that the term "parna" be dropped, suggesting instead that these deposits be referred to as loessic clay or sand loess.

LOESS PARENT MATERIALS

One of the most widespread types of eolian sediment is wind-blown silt, or loess. Loess is dominated by silt-size, usually quartz, particles. In raw form it is usually tan or light brown in color, with 10YR hues dominating. Thick loess sheets are widespread in the midcontinent United States, China, Argentina, and eastern Europe (Smalley 1975, Pye 1984, 1987, Eden *et al.* 1994, Haase *et al.* 2007; Fig. 10.31A). In the central United States, loess sheets are thickest near major meltwater valleys, which were loess sources as continental glaciers receded during multiple Pleistocene glacial periods (Figs. 10.31, 10.32). Here, loess deposits can exceed several meters in thickness (Roberts *et al.* 2003). When freshly exposed in roadcuts or excavations, loess stands vertically, with pronounced vertical cleavage (Ruhe 1975b; Fig. 10.33). Loess that was deposited just downwind from the source is coarse-textured and may be slightly stratified, but most loess is essentially massive and unstratified; often it is calcareous when deposited. Loess deposits are readily erodible by water and wind; gullying in loess soils is not uncommon. All manner of gradational loess-sand deposits exist where loess sheets interfinger with sand, and vice versa (Mason 2001, Muhs and Zárate 2001, Iriondo and Krohling 2007, Zarate and Tripaldi 2011, Vandenberghe 2013).

Loess has two main origins: nonglacigenic (desert) loess and glacial loess. Nonglacigenic loess is deposited downwind of dry, minimally vegetated areas, where loose sands may build into dunes. The smaller silt grains are entrained, winnowed, and deposited farther downwind,

Fig. 10.29 Photomicrographs of coarse sand from a soil near Point Reyes, California. (A) Rounded grains transported to the site by wind. (B) Angular grains weathered from bedrock, present in the residuum.

Crawford et al. (1983)

was paid to the degree of rounding and surface frosting displayed on quartz grains, using light and electron microscopy (Fig. 10.30). Interestingly, degree of rounding was based on percentage of grains that were spherical enough to roll down an inclined glass plate. Rounded grains typify eolian sediment, and the percentages of such grains in a sediment increase with the duration of the process (Kowalkowski 1995; Fig. 10.30). Last, Mycielska-Dowgiallo and Woronko (2004) used mineral ratios as indicators of duration of eolitization, because resistant mineral contents, e.g., quartz, garnet, and tourmaline, relatively increase during eolian transport, because of abrasion. Less resistant minerals, e.g., feldspars, biotite, and muscovite, tend to show relative decreases for the same reason (Kobojek 1997). In short, this work illustrates that much information can be gleaned about sedimentological history from surface texture and grain shape alone.

Another indicator of eolian materials (sands) in soils is the particle size distribution within the sand fraction; eolian sand is often quite well sorted, whereas sand from other sources may be less sorted. Similar statements could be made for small additions of loess to soils.

  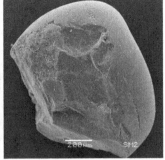

Angular, coated grain.
Weathered directly from bedrock or in soil profile.
Not transported.

Fresh, angular, and cracked grains.
Slightly modified in glacial, fluvioglacial, periglacial, and fluvial environments.

 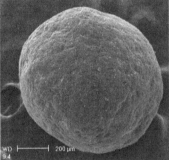 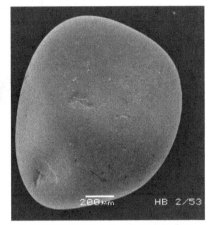

Short distance transport   **Long distance transport**
Rounded grains. Transported by wind and indicative of "eolitization"

Rounded grain with a smooth, shiny surface.
Greatly modified in fluvial/coastal environments.

Fig. 10.30 Scanning electron microscope (SEM) images of sand grains, illustrating how their surface textures can be used to infer different types of depositional environments. Photos by B. Woronko.

as loess (Smalley and Vita–Finzi 1968, Yaalon 1969, 1987, 1997b, Smalley and Krinsley 1978, McTainsh 1987, Tsoar and Pye 1987, Iriondo and Krohling 2007). Some of the thickest and most rapidly accumulating loess in the world has just such an origin; it lies just downwind of the Nebraska Sand Hills (Roberts *et al.* 2003). Nonglacigenic loess in parts of Colorado and Nebraska is presumed to have been eroded from shales and siltstones, exposed during cold, dry periods in the Pleistocene (Aleinikoff *et al.* 1999, Muhs *et al.* 1999, Rawling *et al.* 2003). The deposition rates and transport directions of nonglacigenic loess may be influenced by glacial-interglacial climate cycles, suggesting that the term "nonglacigenic" is a misnomer (Mason 2001).

The largest accumulation of loess in the world is nonglacigenic, desert loess. Chinese Loess Plateau sections are commonly >150 m thick (Lu and An 1998), and in some places >330 m thick (Derbyshire 1984). The loess is so thick that many people build their homes directly into

the deposits (Fig. 10.34). The Chinese Loess Plateau exists downwind from the deserts of Mongolia and China (Kukla and An 1989, Ding *et al.* 1992, 1999). Over the past 2.5 million years, this loess was transported from deserts that lie to the northwest (Ding *et al.* 1993, Eden *et al.* 1994) by the winds of the winter Asian monsoon (Lu and Sun 2000; Fig. 10.35). As noted, this nonglacigenic loess also has an indirect glacial origin. During glacial periods, increased aridity and expansion of the deserts led to dust influxes onto the plateau; loess input was much less during cool, moist glacial periods (Ding *et al.* 1995, Porter 2001, Pullen *et al.* 2011).

In the midlatitudes, near the margins of the various Quaternary ice sheets, loess is particularly commonplace (Markewich *et al.* 1998, Haase *et al.* 2007). Much loess originated as glaciers ground rock into silt-sized sediment, which then was directed down spillways and then spread onto floodplains as meltwater deposits in river valleys and similar landforms (Smalley 1972, Leigh 1994, Sun *et al.*

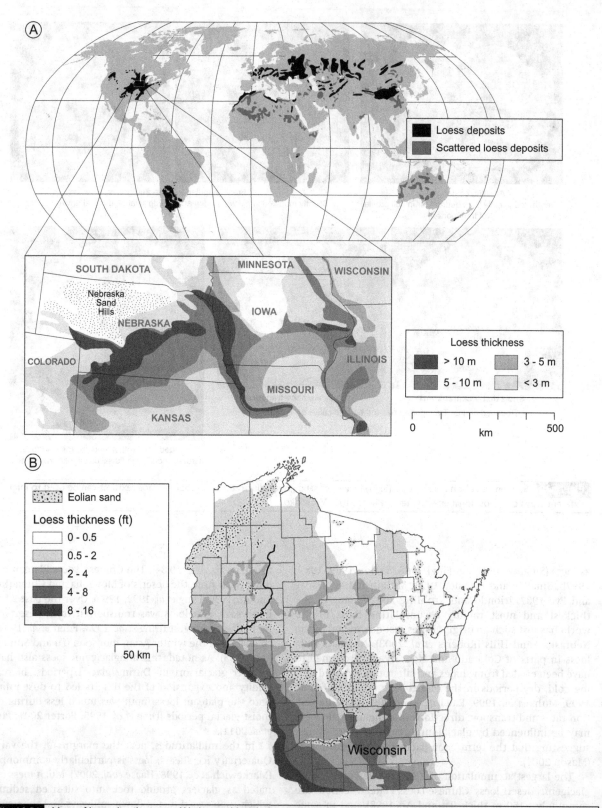

Fig. 10.31 Maps of loess distribution in (A) the world, with detail on loess thicknesses in the central Great Plains, and (B) Wisconsin. Compiled from numerous sources, but most notably, Hole (1950).

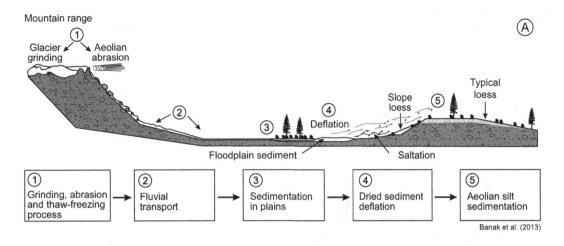

Banak et al. (2013)

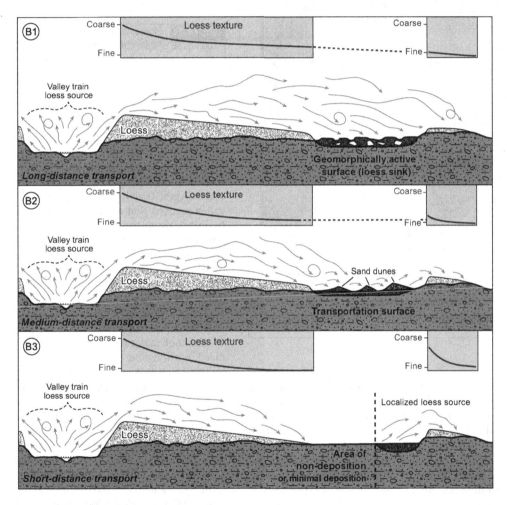

Fig. 10.32 Mechanisms by which nonglacigenic loess deposits are generated, typical of loess deposits in glaciated, midlatitude locations. (A) The loess production-conveyor belt, from glacial source to thick loess deposits, as assisted by fluvial transport of silt-laden meltwater. (B) Three models that detail the possible options for loess transport across glaciated terrains. (1) A distant, valley train, loess source coupled with long-distance transport. Stable sites far from the source continue to receive loess, but that loess is undetectable in areas that are geomorphically unstable. (2) Same as "A," but involving short-distance transport. Here, however, a surface of transport enables loess to be transported beyond its normal limits. (3) As in loess transport from the valley train is also spatially restricted. Here, however, local outwash plains or lake plains, all unrelated to the larger, valley train, source, act as additional loess sources. Note that the textural and thickness characteristics of the loess provide key information as to which model is relevant.

Fig. 10.33 An exposure of thick loess near the Danube River, in Serbia. As shown here, loess typically will stand in vertical faces that can be many meters tall. Note the prominent buried soils (paleosols). At least two additional, better-developed, paleosols exist above the one that is being pointed out. Photo by RJS.

Fig. 10.34 Homes built directly into thick deposits of loess, 100 km SE of Yulin City, Shaanxi Province, Chinese Loess Plateau. Raw loess is heated to create the hard bricks also used in making the exterior walls of these homes. Photo by X. Miao.

2007; Fig. 10.32A). Silty glacial debris was carried from the ice sheets, whose grinding was very effective at comminuting larger rocks and sand into silt-sized particles (Pye 1984, 1995, Assallay *et al.* 1998). Loess is often called glacial flour, because, like wheat, it was formed by physical grinding (in this case, by ice) into smaller sizes, and because it has the same consistency when dry. The meltwater valleys were filled with milky-colored, silt-rich meltwater, particularly in summers.

Loess in the U.S. Midwest has been studied intensively for more than 75 years. The valleys of meltwater rivers, such as the Mississippi, Missouri, Wabash, and Illinois Rivers, would have been dry in winter, and because of the great volumes of meltwater they carried in the warm seasons, completely unvegetated. Strong winds associated with the ice sheet were able to entrain the silt from the dry valleys and transport it to the adjoining uplands, where it often accumulated to great thicknesses, especially near the river valleys (Smith 1942, Wascher *et al.* 1947, Rutledge *et al.*, 1975a, Putman *et al.* 1989, Bettis *et al.* 2003). Data from Fehrenbacher *et al.* (1965b) and Putman *et al.* (1989) even suggest that loess is both thicker and coarser near valley sections that are wider. Coarser sands, too large to be moved from the valleys, occasionally accumulated as dunes on the valley floors, only to be destroyed by the next year's floods. Areas that lack loess today either were under the ice sheet during the period of loess deposition, are too distant from source areas to have received large eolian influxes, or were geomorphically unstable sites, such that loess falling there could not be retained (Schaetzl and Attig 2013, Luehmann et al. 2013).

Although major meltwater valleys have long been recognized as prodigious loess sources, recent work is bringing to light other loess sources from the postglacial landscape (Fig. 10.32B). Indeed, any area that lacked vegetation (which is normally the case on recently deglaciated landscapes) and had ample amounts of silt could have served as a loess source. Wind velocities sufficient to entrain and transport the loess were usually present, as winds near the ice sheet were strong and variable (Hobbs 1943, Thorson and Schile 1995, Krist and Schaetzl 2001, Schaetzl and Attig 2013). For example, Stanley and Schaetzl (2011) reported on an end moraine that served as a loess source, as it contained many ice-walled lake plains, rich in silty lake sediment. Schaetzl and Attig (2013) provided evidence that proglacial outwash plains were also important loess sources, with the loess being transported mainly by katabatic (directly off the glacier) winds. These examples illustrate that wherever winds blow across dry sediment on unstable landscapes, eolian sediment can accumulate in downwind areas.

Fig. 10.35 Map of the Chinese Loess Plateau and nearby desert and mountainous areas. After several sources.

Although loess can be preferentially deposited in some areas to form hills (loess dunes), it usually covers the landscape more or less evenly, like a blanket – at least initially. Loess is usually thickest on flat summit positions (Ruhe 1954). On rolling landscapes, however, some of the loess eventually is transported downslope by various surficial processes, producing variable thicknesses of loess across the landscape (Jacobs et al. 2012). Some evidence has also shown that loess is slightly thicker on lee sides of hills, rather than being of uniform thickness in all landscape positions (Simonson and Hutton 1942).

At landscape scales, loess sheets are thickest and coarsest-textured near the source, whether it be a desert margin, a meltwater valley, or an outwash plain (Smith 1942, Frazee et al. 1970, Kleiss 1973, Fehrenbacher et al. 1986, Roberts et al. 2003, Scull and Schaetzl 2011). The progressive thinning of loess – and its textural fining – with distance from the source areas is so predictable that it has often been described mathematically (Simonson and Hutton 1942; Figs. 10.36, 10.37) and is a reliable ways to trace loess to its source area (Ding et al. 1999, Mason 2001, Zarate 2003, Schaetzl and Hook 2008, Stanley and Schaetzl 2011 Zarate and Tripaldi 2011). In the northern midlatitudes, loess typically thins to the east and southeast of source areas, although in the case of meltwater valleys, a secondary but more subtle thinning trend to the west and northwest is common (Frazee et al. 1970, Ruhe et al. 1971, Ruhe 1973, Muhs and Bettis 2000) (Figs. 10.36, 10.37). This secondary

thinning pattern is usually ascribed to infrequent easterly winds, while the characteristic west-to-east thinning pattern of the loess suggests that the dominant winds were from the west (Fehrenbacher et al. 1965b, Muhs and Bettis 2000). Smaller particles increase in relative abundance away from source areas, while larger particles decrease predictably, along the same transect (Fig. 10.37). Often, the "break-even" point occurs at roughly the 40–50 μm fraction – very coarse silt. Some evidence also suggests that certain minerals in the loess decrease as a function of distance from the source, all of which point to the exceptional sorting abilities of wind, especially when the transportation is via suspension (Fig. 10.37).

Loess deposition rates vary greatly in space and time (McDonald and Busacca 1990, Roberts et al. 2003). For glacial loess, accumulation rates were maximal when the ice sheet was rapidly melting, when winds were strong, and when vegetation cover was minimal (Muhs and Bettis 2000). It is also commonly assumed that loess is deposited earlier and more rapidly on uplands nearer to the source area than on those farther away. For example, in Illinois and Iowa, it has been postulated that the earliest additions of loess associated with Pleistocene ice sheets occurred near the rivers (Ruhe et al. 1971). Later, loess deposition enveloped more and more of the landscape, so that at the time of maximal deposition, sites far away were also receiving silt inputs (Ruhe 1969). As loess deposition slowed near the end of the glacial period, sites farthest from the source ceased

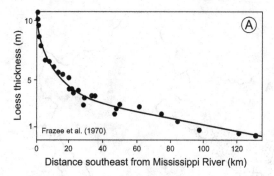

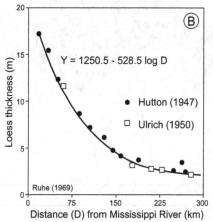

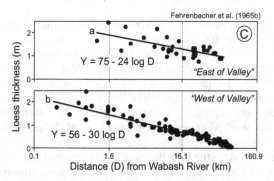

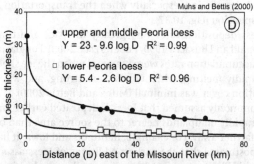

Fig. 10.36 Mathematical relationships between loess thickness and distance from the loess source. (A) In southern Illinois, east toward southern Indiana. (B) In western Iowa, east toward south-central Iowa. (C) In southern Illinois and Indiana, to the east and the west. (D) In western Iowa.

receiving loessial additions and soils began to form. The last sites to receive loess were those nearest the river. Thus, the age of the *base* of the loess is progressively younger as distance from the river increases, and the window during which the loess was deposited is progressively smaller, farther from the river (Ruhe 1969). Slower loess deposition rates farther from the source areas presumably allowed for syndepositional weathering of the loess, facilitating increased leaching and pedogenesis (Ruhe 1973, Muhs and Bettis 2000).

Near source areas like the ones described, soils developed on younger surfaces and in coarser loess, creating a complex space-time condition for soil formation (Ruhe 1969). Postdepositional processes on steep slopes near major river valleys were particularly effective at eroding the coarser loess, leaving behind gullied and eroded landscapes, with steep slopes and minimally developed soils. Farther from the river valleys, where erosion was slower and landscapes were flatter and more stable, the loess is thinner, finer-textured, and more weathered. Also, if the loess covered a slowly permeable surface containing an acidic paleosol, this surface would have facilitated weathering and soil development (Kemp *et al.* 1998). Increased weathering in soils of the thin loess areas occurs for two main reasons: (1) a "wetness effect" in which the buried soil perches water and facilitates weathering in the loess, via ferrolysis, and (2) an acidity effect, because the buried soils are commonly leached and acidic, promoting loss of weathering products from the (often) calcareous loess above. In sum, soils tend to be better developed in thin loess areas than they are nearer the loess source, other factors being equal (Ruhe 1973, Harlan and Franzmeier 1977).

Spatial trends in loess deposition rate, texture, and thickness all impact subsequent soil development (Ruhe 1954). If loess deposition is slow, the new silty sediment can be readily incorporated into the soil profile. Under this scenario, the soil profile, especially the A horizon, slowly thickens as the loess is mixed into it (McDonald and Busacca 1990). This process is called *cumulization* (see Chapter 14). In thin loess areas, slow aggradation of loess upon preexisting soils will eventually even lead to thick, welded soils, in which a surface soil slowly grows upward via cumulization, thickening as increments of loess are added (Ruhe and Olson 1980; see Chapter 16). In cases where loess deposition is rapid, sedimentation outstrips the soil's ability to incorporate the loess into the profile, and the soil becomes buried. Buried paleosols are common in thick loess columns (Ruhe *et al.* 1971, Gerasimov 1973, Valentine and Dalrymple 1976, Wintle *et al.* 1984, Busacca 1989, Olson 1989, McDonald and Busacca 1992, Feng *et al.* 1994a, Maher and Thompson 1995, Bronger 2003, Jacobs and Mason 2007, Mason *et al.* 2008, Pierce *et al.* 2011). The length of the pause in sedimentation and the rate of soil formation during that time interval will determine whether the buried soil is strongly developed or simply a

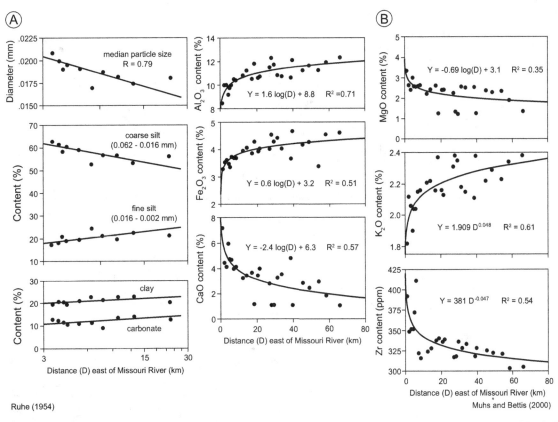

Fig. 10.37 Mathematical relationships between several loess characteristics and distance from the loess source, eastward from the Missouri River in southern Iowa.

dark band enriched in organic matter (Daniels *et al.* 1960, Jacobs and Mason 2004).

Silt loam and silty clay loam textures dominate loess soils. Loess-derived soils are also quite porous. Their porosity renders them susceptible to erosion, but also facilitates their use as habitat for some species of nesting birds (Smalley *et al.* 2012, 2013)! Loess soils are generally fertile and high in pH, unless underlain by a shallow, acid paleosol. Many loess-derived soils are relatively young, having formed in the Late Pleistocene or early Holocene. Thus, weathering of primary minerals is typically not advanced, and their mineralogy is related closely to the mineral suite inherited from the parent material (Jones *et al.* 1967, Jeong 2008). The clay mineralogy of loessial soils varies, primarily on the basis of what clay minerals were originally in the parent material and on the degree of postdepositional weathering.

Because loess is such a well-sorted deposit across large tracts of landscape, many of its physical and mineralogical characteristics have long been used to isolate its source area, and hence, to improve understanding of wind direction and speed, i.e., paleoclimate. Grain size (texture), silt

and clay mineralogy, magnetic susceptibility, contents of various heavy minerals, isotopes and rare earth elements, and carbonate content, among several other types of geochemical data, have all been used to differentiate loess deposits from each other and to pin particular loess deposits to their source area (Rutledge *et al.* 1975a, b, Beer *et al.* 1993, Frakes and Jianzhong 1994, Thompson and Maher 1995, Aleinikoff *et al.* 1999, 2008, Muhs *et al.* 1999, Muhs and Bettis 2000, Schaetzl and Loope 2008, Wu *et al.* 2011, Li *et al.* 2012; Fig. 10.38). Of these, texture is most commonly used, and most easily determined (Smith 1942, Frazee *et al.* 1970, Fehrenbacher *et al.* 1965b, Rutledge *et al.* 1975a, Khadjeh *et al.* 2004, Schaetzl and Hook 2008, Stanley and Schaetzl 2011, Schaetzl and Attig 2013). Nonetheless, the use of multiple kinds of data provides the best and most robust results as to loess provenance (Muhs *et al.* 2008a, Ujvari *et al.* 2012).

It is often useful to determine whether a soil – one not formed entirely in loess – has nonetheless had some loess admixed into its surface horizons (Dixon 1991, Dahms 1993, Kowalkowski 1995, Mason and Jacobs 1998, Muhs *et al.* 1999). In some soils, identification of a thin loess cap is

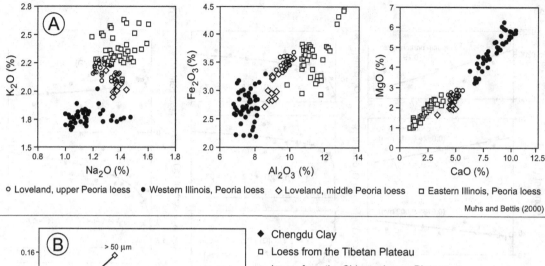

○ Loveland, upper Peoria loess   ● Western Illinois, Peoria loess   ◇ Loveland, middle Peoria loess   □ Eastern Illinois, Peoria loess

Muhs and Bettis (2000)

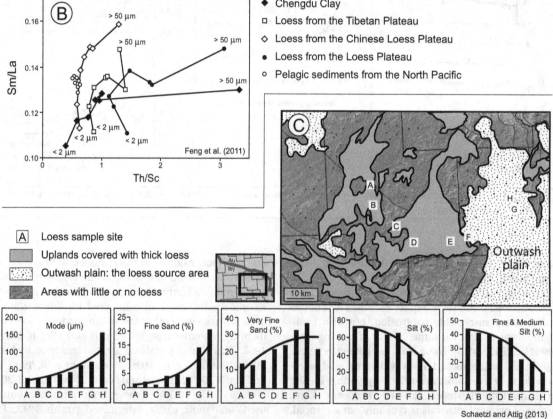

Schaetzl and Attig (2013)

Fig. 10.38 Examples of various mineralogical and textural data that have been used to characterize loess deposits and determine their provenance (source area). (A) Elemental oxides in loess deposits from eastern Nebraska and western Illinois. The clustering of the loess data indicates different source regions and environments of deposition. (B) Comparative ratios of rare earth elements for loess from China, and residual components of North Pacific sediments, sorted by size fraction. Note that the ratios for the four loess samples are distinctly different from each other, and that the eolian-dominated North Pacific sediments are similar to the fine fractions of loess from the Chinese Loess Plateau. (C) Textural data from loess deposits in NE Wisconsin, United States. The loess source area, on the eastern edge of the map, was a Late Wisconsin outwash plain. Loess was transported east to west in this region. Texture data along a transect of points illustrate that the loess becomes progressively siltier and less sandy, away from the outwash plains.

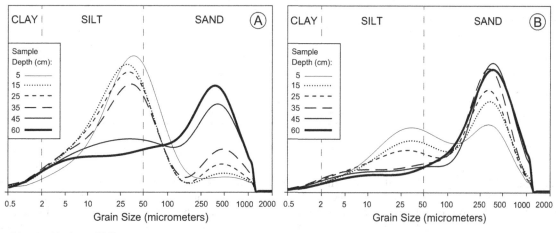

Schaetzl and Luehmann (2012)

Fig. 10.39 Detailed particle size curves for two soils, each of which has a thin, silty loess cap over sandy glacial sediment. Each curve represents the particle size characteristics for a given depth. The soil in A has ~40 cm of loess over the till, whereas in soil B the loess is so thin (<15 cm) that it has been mixed into the sandy till below. Note that even in the uppermost sample from soil A, some sand from below has been mixed into the loess. The curves illustrate the distinct textural character of each sediment, the clear presence of a mixed zone between the two, and the utility of continuous particle size data in studies of sediment and soil origins.

manifested as an increase in silt content nearer the surface, and often with a mixed zone near the sedimentological break (Stanley and Schaetzl 2011, Schaetzl and Luehmann 2013). For such soils, textural data can help determine whether a mixing zone exists and, if it does, determine its thickness (Fig. 10.39). Even in soils formed in thick loess, a mixing zone between the loess and the sediment below can often be determined (McSweeney *et al.* 1988).

Dust is an increasingly popular term used to refer to silt-size or smaller sediment that has an eolian origin (Dan 1990, An 1991, Simonson 1995, Offer and Goossens 2001, Jacobs and Mason 2005, Jeong 2008, Hladil *et al.* 2010, Pullen *et al.* 2011). The importance of eolian dust to soil genesis in arid landscapes is now unquestioned (Jenkins and Bower 1974, Chester *et al.* 1977, Rabenhorst *et al.* 1984a, Amit and Gerson 1986, Yaalon 1987, Levine *et al.* 1989, Reheis *et al.* 1995, Blank *et al.* 1996, Naiman *et al.* 2000; see Chapter 13). More recently, global effects associated with long-distance transport of silt and smaller sized particles, many of which have traversed entire oceans, are becoming increasingly recognized. For example, unambiguous evidence has documented dust transport from California to the nearshore Channel Islands (Muhs *et al.* 2008b), and from the Sahara Desert to islands in the Caribbean Sea (Muhs *et al.* 1990, 2007, Muhs and Budahn 2009). Dust is frequently transported from the Sahara to the Canary Islands, 100 km off the African coast (Menendez *et al.* 2009b; Fig. 10.40). Here, Menendez *et al.* (2009b) have documented deposition rates on the order of 20 g m^{-2} y^{-1} (Menendez *et al.* 2009b). Cores from the ocean basins and ice sheets, direct observations of dust storms and their trajectories, as well as soils data

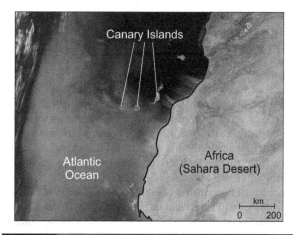

Fig. 10.40 Satellite image of a major dust storm over northwest Africa and the Sahara Desert on February 11, 2001. Note the eddies and waves on the lee side of the mountainous Canary Islands (center). Image provided by the Sea WiFS Project, NASA/Goddard Space Flight Center, and ORBIMAGE.

confirm that dust transport is truly global in scale and that it has been ongoing for millennia (Pitty 1968, Biscaye *et al.* 1974, Prodi and Fea 1979, Glaccum and Prospero 1980, Pokras and Mix 1985, Dansgard *et al.* 1989, Rea 1990). As a result, the importance of fine-grained eolian sediment has now transcended its effects on soils and is becoming a critical link in our understanding of climate change, biogeochemical cycling, and nutrient supplies to ecosystems worldwide.

Table 10.3 | Grain-size-based nomenclature for some common pyroclastic deposits

Grain size	Unconsolidated tephra	Consolidated pyroclastic rock
<1/16 mm	Fine ash	Fine tuff
1/16–2 mm	Coarse ash	Coarse tuff
2–64 mm	Lapilli	Lapillistone
>64 mm	Bomb (fluidal shape) and block (angular)	Agglomerate or pyroclastic tephra

Source: Orton (1996).

VOLCANIC PARENT MATERIALS: ANDISOLS

Volcanic materials are typically classified by their size and elemental content (Table 10.3). Low-viscosity *magmatic lavas* tend to effuse slowly out of volcanic vents and harden to basalt. On the other hand, explosive eruptions tend to be associated with *silicic lavas* that are more viscous and generally originate as slightly cooler magmas. On the far end of the silicic spectrum are various airborne, ejected, pyroclastic materials, classified primarily on size: dust (<63 μm), ash (63 μm–2 mm), lapilli (2–64 mm), and blocks and bombs (>64 mm). Scoria and pumice are ill-defined terms for glassy, vesicular fragments in the lapilli and bomb size range (Orton 1996). Ashfall deposits tend to cover large tracts of landscape and, thus, are useful stratigraphic marker beds, providing that their age and correlation can be determined.

The distribution and spread of pyroclastic materials, or *tephra*, are largely determined by their size and the direction of the prevailing winds. Large particles, e.g., lapilli and bombs, follow trajectories that are more or less ballistic, little affected by wind. Few travel farther than a few kilometers from the vent. However, some pyroclastic materials are significantly impacted by local and regional winds. Particles smaller than 250 μm that penetrate the tropopause and enter the upper levels of the atmosphere may circle the Earth many times before falling (Orton 1996). Because winds vary in speed and direction at different levels of the atmosphere, dispersal of the ash cloud can be complex. Nonetheless, ash deposits usually thin with distance from the source, and at any one site ash deposits usually exhibit a fining-upward sedimentology. Obviously, many sites that have soils developed in ash also record a history of episodic ash additions, such that layers of "fresh" ash often overlie paleosols developed in an older ash, and so on (Hay 1960, Inoue *et al.* 1997). This type of parent material stratigraphy is typical of Andisols – soils formed in tephra and related deposits (McDaniel *et al.* 2011). Multiple ash deposits of varying thickness may exist at a single site, each deposit constituting a distinct layer. Many of the volcanic materials associated with ash-derived soils have undergone subsequent transport as mudflows, filling canyons and low-lying areas, or subsequent remobilization as eolian sediments (Arnalds *et al.* 1995). The dating of ash deposits and their placement within their proper chronostratigraphic context is called *tephrochronology* (see Chapter 15).

Glass and other amorphous, glasslike materials dominate tephra. Their lack of crystallinity is caused by the rapid cooling of the molten ejecta; crystals do not have enough time to form before cooling hardens the molten material. Light-colored ash is acidic, with quartz, hornblende, and rhyolite the main crystalline minerals. Darker ashes tend to be basic (mafic) and enriched in olivine and other ferromagnesian minerals. Amorphous materials dominate the mineralogy of younger soils developed in ash deposits. With time, however, the amorphous glassy minerals weather, and minerals such as allophane, imogolite, and halloysite take their place.

Volcanic parent materials and their weathering products impart distinct and unique properties to soils. For this reason, acid soils formed in volcanic materials, with their thick, dark A horizons; greasy feel; and abundant organic acids, have been given special names such as Ando soils (Japanese *an*, dark, and *do*, soil), Humic allophane soils, and Andosols (Thorp and Smith 1949, Simonson and Rieger 1967, Wada and Aomine 1973, Simonson 1979, Tan 1984). In 1990, the Soil Survey Staff established the Andisol order for these kinds of soils (Parfitt and Clayden 1991, Shoji *et al.* 1993, Kimble *et al.* 2000, Ping 2000, Takahashi and Shoji. 2002, Dahlgren *et al.* 2004, McDaniel *et al.* 2011). The dark A horizon that is characteristic of these soils was also deemed unique enough that it was also eventually separated out as a melanic epipedon (Table 10.3).

The central concept of Andisols is that of soils, commonly with depositional stratification, developing in loose volcanic ejecta, e.g., volcanic ash, pumice, scoria, and cinders. After only a minimal amount of weathering of these types of parent materials, the colloidal fraction tends to become dominated by amorphous materials and short-range-order Al- and Si-rich minerals, such as allophane, imogolite, and ferrihydrite (Yoshinaga and Aomine 1962a, 1962b, McDaniel *et al.* 2011). Secondary Al-, Fe-, and Si-humus complexes and chloritized 2:1 clay minerals are also common, depending on the nature of the volcanic material (Shoji and Ono 1978, Parfitt and Saigusa 1985). Typical characteristics of Andisols include (1) low bulk densities, (2) very high macroporosity with rapid drainage, (3) deep rooting, (4) low strength with regard to mechanical disturbance, (5) high potential for wind and/or fluvial erosion, (6) high available water contents, and (7) moderate to high retention capacities for phosphate and sulfate (Shoji

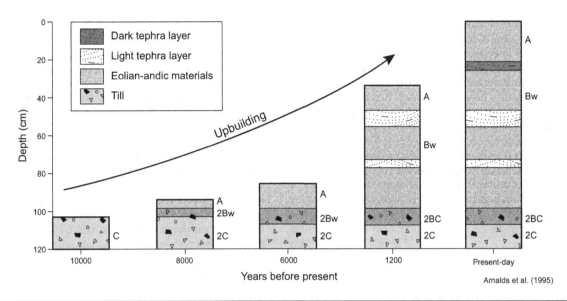

Fig. 10.41 Chronostratigraphic development of a typical Andisol in Iceland that has formed in tephra and volcanic materials reworked by wind.

and Ono 1978, Wada 1985, Parfitt and Clayden 1991, Ping et al. 2000). Other common but not requisite characteristics include low pH and base saturation values, and high amounts of exchangeable aluminum, much of which is complexed with organic matter (Aran et al. 2001). One of the outstanding features of Andisols is their generally high natural productivity, despite the fact that many are on steep, unstable slopes that preclude intensive agriculture.

In Andisols, rapid weathering processes tend to dominate. The Andisol order is defined by characteristics inherited from weathering products of volcanic materials, not on the assumed presence of volcanic ejecta – a critical distinction. Weathering of parent minerals that are not volcanic can, at times, lead to the formation of short-range-order minerals; some of these soils also are included in the Andisol order (Soil Survey Staff 1999). Assisted by the noncrystalline nature of the primary minerals and the high porosities and water-holding capacities of the parent materials and soils, weathering is often exceedingly rapid (Hay 1960).

Andisols are usually found in humid, volcanically active areas – typically in or near volcanic mountain ranges. Areas with abundant Andisols include Iceland, Alaska, Japan, Indonesia, New Zealand, the Andes Mountains of South America, and Russia's Kamchatka peninsula (Seki 1932, Ulrich 1947, Tan 1965, Simonson and Rieger 1967, Wada 1986, Arnalds et al. 1995, Arnalds and Kimble 2001, Arnaulds 2004). In these regions, the likelihood of encountering Andisols tends to increase with altitude.

Many Andisols are pedogenically young, because they have formed either in recently deposited ash or in parent materials with thin increments of geologically young ash embedded in their upper profile (Tan and Van Schuylenborgh 1961). As a result, many Andisols display geologic and pedogenic stratification (Fig. 10.41). Indeed, understanding their genesis requires a pedostratigraphic approach, combined with an appreciation of the possibility that some of the horizons in the profile might actually be buried horizons from an older soil. Additionally, Andisols tend to occur on unstable slopes in mountainous terrain, where erosion continually removes the old, more weathered materials in the upper profile, exposing new, relatively unweathered parent materials. In sum, they frequently become pedogenically rejuvenated (Bakker et al. 1996). Thus, most Andisols are highly polygenetic soils.

Andisols are unlike almost all other soils, in that their surface horizons are often less weathered than the B horizons below because of the additions of new parent materials to the surface (McDaniel et al. 1993, Bakker et al. 1996). Breaks between ash layers may be identified as lithologic discontinuities. They are so common, however, that Arnalds et al. (1995) did not recommend distinguishing individual ashfall layers as separate parent materials. Thin ashfalls may lead to cumulization of the A horizon, whereas thick ashfalls bury the preexisting surface horizons (Shoji and Ono 1978, Smith et al. 1999). Volcanic glass – in its unweathered state – is a common component of the silt and sand fractions of Andisols.

Most Andisols have formed in *acidic* volcanic deposits, and because these deposits contain Al- and Si-rich materials, they weather to form allophane and other short-range-order minerals. Weathering of primary aluminosilicates, where they occur, has proceeded only to the point of formation of short-range-order minerals (Dahlgren and Ugolini

1991). In humid climates, silicic tephra readily weathers to soils rich in allophane, ferrihydrite, imogolite, halloysite, and hydroxy-interlayered vermiculite (Vacca *et al.* 2003, Dahlgren *et al.* 2004). In many Andisols, the soil solution is rich in silica, weathered from the tephra components; some of this silica eventually precipitates as opaline silica. Basic (nonacidic) materials tend to weather to soils richer in halloysite. Volcanic materials in dry climates weather more slowly and soils may take much longer to develop into Andisols (Dubroeucq *et al.* 1998). When wet, Andisols feel greasy and smeary, largely because of the dominance of these types of amorphous weathering products. The active Al and Fe compounds, so typical of Andisols, also have the ability to sorb and strongly bind anions such as fluoride and phosphate, rendering much of the P unavailable for plant growth. Indeed, P retention is used as one of the defining criteria for Andisols. In the end, the formation of immobile Al-organic matter complexes leads to high organic matter contents in near-surface horizons and dark soil colors.

Hydrous oxides of Fe, Al, and Si, along with Fe- and Al-organic matter complexes, readily form in Andisols, as these cations are released from the noncrystalline glass or from primary minerals (Shoji *et al.* 1985, Hunter *et al.* 1987). Once weathered, the cations (primarily Si, Al, and to a lesser extent Fe) released are usually quickly complexed by organic acids, producing stable organometallic complexes when the soil pH is low (Parfitt and Saigusa 1985, Shoji *et al.* 1993). With time, additions of more free silica, dehydration, and structural rearrangement of the allophane and imogolite can lead to the formation of halloysite (Dubroeucq *et al.* 1998).

Unlike podzolization (see Chapter 13), in which metal-cation complexes are mobile and become translocated to form E-Bhs sequa, little such translocation occurs in Andisols. Why? One hypothesis presupposes that the organic matter immobilization is caused by such high aluminum contents in the soil that the humic compounds become saturated in the A horizon and cannot be readily translocated (Ugolini *et al.* 1988, Aran *et al.* 2001). Weathering of volcanic materials in Andisols can indeed provide large amounts of Al cations to the soil solution. Indeed, some Andisols even have aluminum toxicity problems (Wada and Aomine 1973). Another likely cause has to do with the types of organic compounds involved: Larger, humic-type organic acids are not as likely to become mobile when chelated with Al cations. Aliphatic OM is also much more dominant in Andisols (as are microbial products); that may be why they retain a lot of OM in the A horizon ().

Andisols often mingle with Spodosols on ash-dominated landscapes; the two kinds of soils have much in common genetically. Both orders are dominated by soil horizons whose active components consist mainly of amorphous materials (Flach *et al.* 1980, Lindeburg *et al.* 2013). Whereas Andisols are associated with the *immobilization* of Al and Si by organic matter complexation, in podzolization the organic acids assist in the *mobilization* of metal cations. Although humic acid colloids tend to dominate allophane-rich Andisols (Wada and Aomine 1973, Flach *et al.* 1980), podzolization is triggered by smaller organic acids with greater complexation capacities (Ugolini *et al.* 1988, Aran *et al.* 2001). Thus, as one progresses from warmer, drier environments to cooler, humid, forested conditions, Andisols grade into soils with spodiclike profiles (Simonson and Rieger 1967, Shoji *et al.* 1988). Along such an Andisol-Spodosol transect the amounts of organic materials, Fe, Al, and Si that have been translocated to the B horizon increases. E horizons become more acidic, and smectite, rather than allophane, comes to dominate their mineralogy (McDaniel *et al.* 1993). In drier pedogenic environments, organic matter production is less. Under strong leaching conditions, desilication will facilitate the formation of gibbsite.

Transported by Ice

Geologists sometimes use the term *drift* to refer to material carried by glacial ice. The glacial sedimentary environment is unique and complex because sediments are often transported both by solid ice and, later, by meltwater (Ashley *et al.* 1985). Indeed, according to Ashley *et al.* (1985: 2), "If any generalization can be made concerning the distribution of glacial sedimentary facies, it is that few generalizations can be made." Glacial sediments vary from unsorted, boulder-rich till to stratified lacustrine silts and clays. Transitions from one deposit to another may occur abruptly, or they may imperceptibly grade together. Multiple deposits, stacked one upon another and all dating to the same ice advance, are common. Many of the world's glacial deposits are covered by eolian silt (loess), which was usually derived from glacial deposits. It is little wonder that perhaps the most complex soil landscapes are those that have been recently deglaciated.

Material deposited directly by the ice, glacial *till*, is often unstratified and unsorted (Dreimanis 1989). Till tends to run the gamut from very clay-rich material to coarse, sandy sediment, depending upon the rocks and landscapes the ice most recently overrode. Large clasts may be concentrated in some parts of a till deposit while other tills may be nearly gravel and stone-free. From a sedimentological perspective, perhaps the most distinguishing characteristic of till its unpredictability (Hartshorn 1958, Attig and Clayton 1993, Khakural *et al.* 1993; Fig. 10.42). Till can be sandy, clayey, silty, cobbly, dense or porous, acidic or calcareous (Willman *et al.* 1966, Levine and Ciolkosz 1983, Mohanty *et al.* 1994, McBurnett and Franzmeier 1997). In many areas, however, a critical distinction can be made between basal till (deposited under the ice sheet) and ablation or superglacial till. The former is often dense, while the latter is more porous and often shows evidence of the influence of running water, as stringers of sorted sand lenses. In some soils with dense fragipans in the subsurface, the close-packed nature of the fragipan has been ascribed to inheritance from compact till (Lyford and Troedsson 1973).

Fig. 10.42 Examples of glacial till from the Midwest, United States. (A) Fine-textured till with a few scattered coarse fragments. (B) Sandy till with abundant stones and boulders. Photos by RJS.

Soils that have formed in till deposited during the last (Wisconsinan, Pinedale, Würm) glacial advance are generally not strongly developed, whereas the soils on older deposits (Illinoian, Bull Lake, Riss) can sometimes be extremely well developed (Allan and Hole 1968, Dalsgaard *et al.* 1981, Franzmeier *et al.* 1985, Berry 1987, McCahon and Munn 1991, Applegarth and Dahms 2001). Soils in many of the older deposits, however, have since been buried or eroded.

Another common aspect of till soils and landscapes is their layered stratigraphy, whereby one till sheet is deposited upon another. Buried soils (paleosols) are, therefore, routinely found in the upper parts of the older till deposits (Bushue *et al.* 1974, Griffey and Ellis 1979, Hughes *et al.* 1993). In many landscapes, the till itself does not contain a soil but is buried by loess that contains the modern soil at its surface. In these cases, loess deposition occurred immediately after the glacier receded, before a soil could form in the till.

Transported by Meltwater

Glacial meltwater usually entrains, transports, and sorts some of the sediment deposited by melting glaciers, carrying away the silt and clay (some of which will eventually become loess) and leaving behind the larger sand and gravel particles as glaciofluvial sediment, or outwash. *Outwash* is sandy and gravelly sediment that is affected by pedogenesis in many of the same ways as are sandy alluvium or dune sand (Fig. 10.43). It is unlike dune sand in that outwash (1) contains coarse fragments and (2) is usually stratified. The many sedimentologic contacts between the strata dramatically affect water flow in the soils and facilitate deposition of dissolved and suspended materials that may be moving in soil water, i.e., Fe, humus, and clay (see Chapter 13). The tendency for illuvial materials to accumulate at points of textural change has been demonstrated numerous times (Bartelli and Odell 1960a, Bouabid *et al.* 1992, Schaetzl 1998). One of the most common

occurrences of this phenomenon happens when illuvial clay tends to accumulate at the contact between a sandy zone and an underlying gravelly zone.

Outwash soils, like all sandy soils, tend to form rapidly because they have low surface area and because water (wetting fronts) can penetrate deeply. Particle coatings can be stripped quickly, forming E horizons much more rapidly than in loamy or fine-textured parent materials. Most outwash has only small amounts of clay, which can be translocated comparatively quickly, sometimes forming Bt horizons, but more commonly, illuvial clay lamellae (Wurman *et al.* 1959, Dijkerman *et al.* 1967, Host and Pregitzer 1992, Schaetzl 1992b, Khakural *et al.* 1993, Cooper and Crellin 1996). Because so little clay needs to be translocated, and because water flows so readily through these sandy soils, lamellae can form in matters of decades (Schaetzl 2001).

Transported by Gravity

Material moved downslope, primarily under the influence of gravity but usually assisted by water, is called *colluvium* (Fig. 10.44), and the process is called *colluviation* (Goswami *et al.* 1996). Slumps, slides, flows, and heaves all represent such movement of regolith (Moser and Hohensinn 1983, Vaughn 1997, Leopold and Volkel 2007). Colluvium that has been reworked by running water or slope wash is called *coalluvium* (Fig. 10.44B). Colluvium can range in texture from fine to extremely cobbly and bouldery, depending on the source material composition, and this variability can occur in both the lateral and vertical dimensions (Fig. 10.45). Although colluvium is usually quite heterogeneous, it often does have crude textural zones or strata, which can impact pedogenesis at a later point (Ciolkosz *et al.* 1979, Kleber 1997, Porter *et al.* 2008), and the contacts between such zones can be quite abrupt.

When regolith on a slope moves, the slope is said to have failed. Slope failure occurs because the shear strength

Fig. 10.43 Examples of glacial outwash. (A) Proximal (near to the ice margin) outwash sediment, showing alternating fine- and coarse-textured strata, possibly reflecting interannual deposition. (B) Coarse-textured, gravelly outwash, with fine sand strata. (C) Fine-textured, sandy outwash with few coarse fragments, displaying cross-bedding and ripples. (D) Thick sequences of gravelly outwash being mined for road aggregate. Photos by RJS.

of the material is compromised, or because shear stresses become too great. For example, shear stress can increase when a slope is overloaded with heavy, wet snow, or by tectonic stresses and faulting (Forman *et al.* 1991, Amit *et al.* 1995). Loss of vegetation by fire or drought, undercutting of the lower slope section by fluvial erosion, jointing in the parent rocks, and saturated soil conditions after long rain events all act to weaken the cohesiveness (shear strength) of the sediment/soil. As a result, these events can initiate slope failure (Eschner and Patric 1982, Riestenberg and Sovonick-Dunford 1983, Graham *et al.* 1990a). Other situations in which slope failure is common include thin soil areas over impervious bedrock, areas of convergent subsurface flow, areas with shallowly rooted vegetation or lack of vegetation, areas that contain impermeable layers near the surface, and of course, steep slopes (Neary *et al.* 1986; Fig. 10.46). All of these situations tend to create low shear strength, or add to the shear stress, for the slope materials.

Colluvial deposits may be very recent, or they may date to Pleistocene events and climates (Ridge *et al.* 1992, Bettis 2003, Geertsema *et al.* 2006). In the southern Appalachian Mountains, many debris flow deposits date back to hurricanes in the geologic past (Liebens and Schaetzl 1997) or to infrequent, intense storms in recent history (Williams

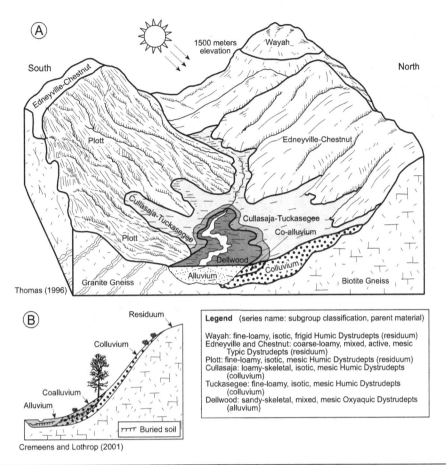

Fig. 10.44 Parent materials and landscapes in the Appalachian region, United States. (A) Block diagram of the soil-landscape-parent material relationships in Macon County, North Carolina. Residuum occurs on upper slopes while colluvium, coalluvium, and alluvium dominate the footslopes. (B) Schematic valley cross section in a first-order drainage basin, illustrating the typical distribution of regolith types along the slope.

and Guy 1973, Neary *et al.* 1986). Elsewhere, many colluvial deposits are associated with permafrost, solifluction, and much colder climates that occurred during glacial periods (Denny 1951, Mills 1981, Waltman *et al.* 1990, Mason and Knox 1997). In those past, colder climates, widespread permafrost facilitated slope failure by restricting infiltration into the subsoil, resulting in saturated conditions in the upper few meters (Ciolkosz *et al.* 1979).

Perhaps the two factors that contribute most to slope instability, and are most related to colluvial thickness, are slope gradient and vegetative cover (Salleh 1994, Goswami *et al.* 1996). Steep slopes with minimal vegetation are the most prone to failure, other factors being equal. The effect of slope gradient is obvious. Vegetation cover has clear impacts as well. The presence of less vegetation leads to more slope failures, and conversely, vegetation cover helps to retain existing regolith, so that it does not move downslope.

Colluvial soils, usually most common on the lower parts of slopes (Fig. 10.44), have primarily been studied in areas of high relief (Hoover and Ciolkosz 1988, Liebens 1999, Ogg and Baker 1999). Colluvium and colluvial soils tend to be thickest in the low-gradient deposition zones at the bases of slopes, in depressions, and at the ends of small valleys; they are thinnest on steeply inclined, upslope areas (Goswami *et al.* 1996, Ohnuki *et al.* 1997; Fig. 10.47). Thus, landscapes with frequent slope failure often are a complex of thick, colluvial soils and fans in coves and valleys, while upland areas contain residual soils, saprolite, and scars where soils are thin or nonexistent (Whittecar 1985, Mew and Ross 1994; Fig. 10.48). Most areas of intense and rapid slope failure have *initiation zones*, which are often steep shallow to bedrock areas, *runout zones*, valleys where colluvium moves rapidly downslope, as well as fan-shaped *deposition zones*, where the transported material comes to rest. Scar areas (initiation zones) are often prone to subsequent

Cremeens and Lothrop (2001)

Fig. 10.45 Colluvial soils and paleosols in the unglaciated Allegheny Plateau of West Virginia. (A) Colluvium and a colluvial soil (Typic Hapludult) above sandstone bedrock. Photo by D. Cremeens. (B) Excavation into coalluvium on a toeslope, showing two buried paleosols on older, previously stabilized surfaces. The lower paleosol contains charcoal that dates between 2470 and 3450 BP, and is buried by coalluvium. The upper paleosol is buried by more recent alluvium and slopewash deposits and dates to about 600 BP.

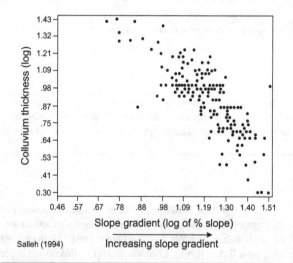

Salleh (1994)

Fig. 10.46 Scatterplot showing the relationship between thickness of colluvium and slope gradient.

slope failure, because the thin soils over hard bedrock are only weakly retained on the slope (Fig. 10.48). Thus, a cycle often ensues in which areas of slope failure tend to repeat this process (Liebens and Schaetzl 1997). Many colluvial deposits bury or truncate preexisting soils (buried paleosols) or bury still older colluvium (Waltman *et al.* 1990, Crownover *et al.* 1994, Kleber 1997, Yassoglou *et al.* 1997). These contacts represent lithologic breaks, which also can occur within the same colluvial mass (Whittecar 1985). Finally, we emphasize that, unlike many other soil

parent materials, colluvium is partly or entirely composed of preexisting, weathered, and pedogenically altered material (buried soils), reflecting the episodic nature of its formation, transport, and deposition.

Lithologic Discontinuities in Soils

Soils form in parent materials that may be uniform, layered, or simply random mixes of various types of sediment. Roughly horizontal layers of sediment can form in several ways: (1) by the geologic depositional processes prior to soil formation; (2) by depositional upbuilding (additions of new sediment) during soil formation; (3) by weathering and vertical or lateral translocations, i.e., pedogenesis; and (4) by bioturbation (Phillips and Lorz 2008). In this chapter on soil parent materials, we are most concerned with the detection and implications of layering prior to soil formation, i.e., geologic layering and breaks in the sedimentary column. These breaks, termed *lithologic discontinuities*, represent zones of abrupt or gradual change in the lithology of soil parent material. Typically, they have been restricted to situations in which the contact reflects a clear lithologic change, i.e., they have not formed via pedogenesis. Lorz and Phillips (2006) took a much broader and more inclusive approach, interpreting LDs as abrupt changes in texture, structure, fabric, geochemistry, and/or mineralogy. In this discussion, however, we restrict LDs to those abrupt changes in the soil profile that can be attributed to initial parent material layering. So defined, lithologic

Fig. 10.47 Colluvial landscapes in the Appalachian Plateau of Pennsylvania. Photos by D. Cremeens. (A) Colluvium is several meters thick on these gentle slopes (Greene County), mostly underlain by Oxyaquic Hapludalfs. (B) Colluvial material creeping into the base of this small stream valley, even as the stream attempts to remove it (Nicholas County). Soils on the slopes classify as Typic Haplumbrepts.

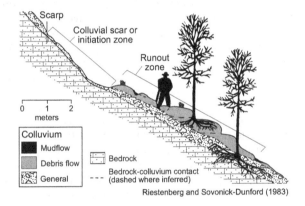

Riestenberg and Sovonick-Dunford (1983)

Fig. 10.48 Diagrammatic cross section through a wooded slope in Ohio, showing various colluvial deposits and their relationship to slope position, bedrock, and standing trees.

discontinuities (LDs) can provide critical information on parent material origins or breaks in the past sedimentation system, and thus are important in most soil geomorphic studies, e.g., Kowalkowski (1995) and Liebens (1999).

Knowledge of lithologic discontinuities can be vitally important to interpreting the evolution of a soil. For a soil that has a clear LD, one can almost always assume one of two scenarios: (1) that the sedimentation *system* that deposited the parent material underwent a dramatic and often sudden change at some point, e.g., from deposition of glacial outwash by meltwater to deposition of windblown sediment (loess), or (2) that an *erosion episode* occurred, marking the LD between two sediments that may have had similar genetic origins. Only rarely do LDs indicate some other type of scenario. Thus, discerning the presence and characterizing an LD are key components that soil geomorphologists use to interpret the geomorphic and sedimentologic

history of a landscape. The importance of lithologic discontinuities cannot be overestimated. Knowledge of LDs is also important in geoarchaeology, as their character and presence influence the interpretation of cultural materials contained within the layers (Johnson 1989, 1990). From a soil genesis perspective, breaks in the sediment column have dramatic effects on subsequent pathways of soil development, e.g., Kuzila (1995).

Detecting LDs and determining their nature (abruptness, topography, etc.) are essential starting points for any pedogenic study (Parsons and Balster 1966, Chapman and Horn 1968, Raad and Protz 1971, Evans 1978, Meixner and Singer 1981, Norton and Hall 1985, Arnold 2005). Additionally, any attempt at quantification of soil development requires that the uniformity of the soil parent material be established, i.e., that it lacks lithologic discontinuities (Haseman and Marshall 1945, Evans and Adams 1975a, Santos *et al.* 1986, Chadwick *et al.* 1990, Anda *et al.* 2009). Often, the term "uniformity" is associated with, or implies, the lack of any kind of lithologic discontinuities.

By definition, lithologic discontinuities are inherited from their parent material. Nonetheless, they may be confused with textural or mineralogical breaks that have developed pedogenically (Paton *et al.* 1995, Arnold 2005). Given enough time, soils can develop abrupt textural breaks that are identified as horizon boundaries. Because they were not inherited from the parent material they would not be considered true LDs. Many are referred to as "abrupt textural changes" by the Soil Survey Staff (2010). Although the focus in this section is on *inherited geologic discontinuities*, the principles put forth are applicable in either case. In the sections that follow, the methods by which lithologic discontinuities can be detected, and their importance to pedogenesis, are based largely on Schaetzl (1998).

Lithologic discontinuities are very common. Schaetzl (1998) sampled a listing of U.S. soil series and determined that about a third (33.5%) of them have lithologic discontinuities, and nearly 7% have two or more. Although soils with discontinuities are common, many LDs are subtle and their detection may be difficult. Even if detected, it is often unclear whether the discontinuity is inherited or if it developed by pedogenesis. Additionally, many sedimentary deposits, assumed to be homogeneous throughout, are crudely stratified, e.g., coalluvium or tephra, forcing the investigator to decide to what extent contacts between individual strata should be considered discontinuities (Asamoa and Protz 1972, Cremeens and Mokma 1986, Schaetzl et al. 1996, Liebens 1999). In the end, deciding whether or not a sediment or soil has a lithologic discontinuity often has an element of subjectivity.

Types of Discontinuities

As mentioned, distinct breaks in the lithology, texture, or mineralogy of soils can occur in two ways: (1) geologically/sedimentologically or (2) pedologically (Raad and Protz 1971, Follmer 1982, Paton et al. 1995). Most discontinuities are of the former type. In principle, a geologic/sedimentologic discontinuity, i.e., a lithologic discontinuity, is the physical manifestation of either (1) a break/change in sedimentation or (2) an erosion surface. In the first instance, sediments accumulating at a site abruptly changed in character during the depositional process, creating an LD (Fig. 10.49). Soils that formed in these materials would, then, have a clear LD. In the second instance, after an initial period of sedimentation, a period of soil formation (and hence, nondeposition) and/or erosion (Imbellone and Gimenez 1998) occurred. Subsequently, a second period of sedimentation ensued on top of the soil or its eroded remnant. The break in sedimentation may indicate changes in the sedimentation regime, such as from deposition of sands by running water to deposition of silts by wind (Kuzila 1995, Soil Science Society of America 1997). Follmer (1982) suggested that LDs are typically expressed by a departure in depth trends between zones of otherwise relative uniformity. Arnold (1968) used the term "abrupt changes" to describe loci within depth functions where LDs might be indicated (see also Raad and Protz 1971).

Pedogenically formed discontinuities, such as biomantles (see Chapter 11), usually involve pedogenic processes that, in essence, sort near-surface sediments. Although common, they are often conceptually ignored (Johnson 1990, Humphreys 1994, Nooren et al. 1995). Because they are formed by pedogenic, and not geologic, processes, these textural or mineralogical "breaks" are not lithologic discontinuities. Nonetheless, pedogenically formed discontinuities affect water flow and many other types of pedogenic processes, and, as with LDs, documenting their existence can help in understanding the soil system (Bockheim 2003). Most pedogenic discontinuities involve two or more distinct layers within the solum, and the discontinuity is usually placed near the depth at which the influence of a surficially driven sorting process, e.g., splash

erosion, bioturbation, or tree uprooting, diminishes to zero (Johnson 1990). Pedogenically formed discontinuities usually exhibit more gradual changes than do geologically formed ones (Smeck et al. 1968). In tropical regions, where such discontinuities and their associated stone lines exist in soils on old landscapes, their detection and accurate interpretation have greatly enhanced our understanding of soil geomorphic systems and their evolution (Bishop et al. 1980, Johnson and Balek 1991, Johnson 1993b, Paton et al. 1995, Brown et al. 2004b, Fey 2010) (see Chapters 11 and 14). Pedogenically formed discontinuities are more common on older landscapes, where the surficial and pedogenic processes have acted in concert for many years to form, for example, relatively stone-free biomantles, stone lines, and texture-contrast soils (Johnson 1993b, 1994, Schwartz 1996, Ande and Senjobi 2010).

Because lithologic discontinuities reflect changes in past sedimentation/erosion systems, their existence is highly applicable to studies of near-surface sedimentary history (Fig. 10.49). Applications include detecting and distinguishing among deposits of loess, colluvium, alluvium, till, and lacustrine sediments (Raukas et al. 1978, Bigham et al. 1991, Karathanasis and Golrick 1991, Schaetzl 1998). Identification of a paleosol in the upper part of a sedimentary layer, e.g., just below an LD, implies that a soil forming interval has occurred before the sedimentary layer was deposited (Ruhe 1956b, 1974a, Foss and Rust 1968, Schaetzl 1986c, Ransom et al. 1987b, Olson 1989, Tremocoldi et al. 1994, Anderton and Loope 1995). Identification of a stable geomorphic surface buried within the near-surface stratigraphic column also has implications for climatic change, landscape stability, and archaeology, among many others (Holliday 1988, Cremeens and Hart 1995, Curry and Pavich 1996).

Detection of Discontinuities

Examination and sampling of the soil from the surface downward, at numerous, closely spaced intervals, is the first step in detecting possible lithologic breaks. If the data are then plotted as depth functions, distinct breaks in some, but not necessarily all, of the depth trends will suggest the presence of lithologic discontinuities (Tejan-Kella et al. 1991, Lorz and Phillips 2006; Fig. 10.50). "Breaks" in depth functions for pedogenic data may not always coincide with the lithologic discontinuity, because of (1) the subtlety or diffuse nature of the discontinuity or (2) large amounts of depth variability within the presumed homogeneous lithologic units above and below, such as might be common in glacial till (Oertel and Giles 1966, Busacca and Singer 1989). Alternatively, the depth function being examined may not be appropriate for the detection of LDs, e.g., the depth distribution of Fe in a Spodosol. In this case, an LD may be indicated where, in reality, none exists, because of the choice of inappropriate evaluative tools. For example, Oertel and Giles (1966) observed that the disagreement among three studies on the locations and/or existence of

Fig. 10.49 Soil profiles with obvious and abrupt lithologic discontinuities, shown by dashed lines. (A) Loess overlying outwash sand in a Glossudalf from northern Wisconsin. (B) Sandy outwash over gravelly outwash in a Haplorthod from northern Michigan. Note the wavy nature of the discontinuity. (C) Glacial till over gravel-free outwash sand, in a Dystrudept from northern Michigan. (D) Dune sand over lacustrine silt and clay, in a Haplorthod from northern Michigan. Photos by RJS.

LDs within the same soil was due to the use of different criteria for their detection, some of which were probably inappropriate. Beshay and Sallam (1995) could not reconcile depth function data, designed to detect LDs, because some of their parameters were developmental, i.e., acquired via pedogenesis, and some were geologic. Thus, it is important to use only certain, carefully chosen parameters for the identification of LDs.

Which parameters, then, should be used to detect lithologic discontinuities, from depth function data? Table 10.4 lists some of the many parameters that have been successfully used to detect lithologic discontinuities in soils (Schaetzl 1998). Taken collectively, these studies show that that only data from *immobile fractions* – the soil skeleton – should be used to detect LDs. That is, use of any soil constituent that is pedologically mobile, i.e., part of the soil plasma (particles <2 μm or <4 μm, cf. Meixner and Singer 1981, Asadu and Akamigbo 1987), can produce erroneous results. Generally, concentrations of sand and silt fractions, part of the soil skeleton, can be determined and then plotted as ratios, e.g., sand:silt or medium silt:fine silt, to determine where LDs might be located. Considered individually, the following types of soil characteristics are rarely useful for detection of LDs: pH, clay content or mineralogy, organic matter, CEC, base saturation, contents of various iron oxides, concentrations of soluble species, certain forms of magnetic susceptibility, and content data for mobile elements, as well as most soil morphological properties, e.g., color, structure, consistence, bulk density, or electrical conductivity (Table 10.5). Depth functions for one or more of these parameters can and often do change

abruptly at an LD, e.g., Karathanasis and Macneal (1994) and Kuzila (1995), but they should not be used to *detect* one. Pedogenic properties, as well as any properties related to the soil plasma, change at LDs because the discontinuity affects the flow of infiltrating water, which then impacts the deposition of plasma components and the rate of weathering above and below the LD. Detection of LDs in soil is more reliable when using immobile fractions.

Because they are designed to capture the effects of pedogenesis, soil development indices, such as Bilzi and Ciolkosz's (1977) relative horizon distinctness (RHD) index, also should not be used to detect LDs. As stated previously, these indices often rely heavily on secondary (pedogenic) data. At best, pedogenic data (or indexed compilations of acquired pedogenic characteristics) may offer *permissive* or supportive evidence for an LD, but not *conclusive* evidence.

In most environments that lack permafrost, ongoing pedoturbation, or tillage, particles coarser than about 30 μm in diameter can be considered immobile (Karathanasis and Macneal 1994). Caution must be exercised, however, in situations in which large particles could have formed pedogenically, e.g., concretions, or, more commonly, where they could have been moved by pedoturbation (Wood and Johnson 1978, Johnson et al. 1987, Johnson 1989, Cox 1998, Humphreys and Adamson 2000; see Chapter 11). Biomantles (Johnson 1990, Humphreys 1994) could be mistaken for a second parent material, because they commonly overlie a stone line that itself may be mistaken for an LD (Johnson 1989). Because the shape and sphericity of sand grains can be diagnostic of their depositional environments (Patro and Sahu 1974), sand grain

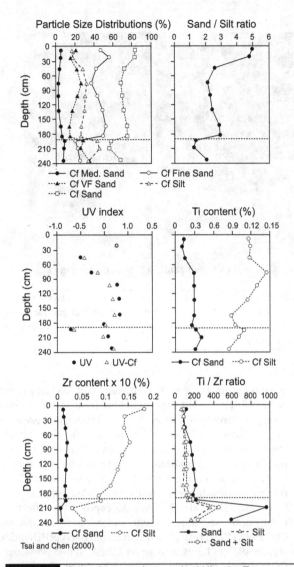

Particle Size Distributions (%)   Sand / Silt ratio

UV index   Ti content (%)

Zr content x 10 (%)   Ti / Zr ratio

— Cf Med. Sand   —○— Cf Fine Sand
···▲··· Cf VF Sand   ···△··· Cf Silt
···□··· Cf Sand

● UV   △ UV-Cf

—●— Cf Sand   ···○··· Cf Silt

—●— Cf Sand   ···○··· Cf Silt

—●— Sand   ···△··· Silt
···○··· Sand + Silt

Tsai and Chen (2000)

Fig. 10.50 Depth functions for an upland Ultisol in Taiwan that has a lithologic discontinuity (dashed line) at 190 cm.

morphometry could be useful for detecting LDs (Schaetzl 1998). The *standard deviation* of sand grain roundness and sphericity data also may be more efficacious in differentiating parent materials than mean values alone (Patro and Sahu 1974, Schaetzl 1998).

To be even more confident in the identification of an LD, one may use data from an immobile *and* slowly weatherable soil fraction (Langohr *et al.* 1976, Tejan-Kella *et al.* 1991; Schaetzl 1998, Aide and Smith-Aide 2003, Stiles *et al.* 2003). Data from the immobile fraction (skeleton) should be excellent indicators of LDs because they reflect sedimentology better than do mobile, or plasma, elements (Washer and Collins 1988). However, some of these immobile constituents may be quite weatherable, and thus may

have been lost from the upper solum, even if the soil has formed in a single parent material (Nikiforoff and Drosdoff 1943). Thus, it is even better to use data from an immobile *and* slowly weatherable fraction (Table 10.4). In a uniform profile these data (and their ratios) should plot along a line that has no major "breaks" with depth (Santos *et al.* 1986; Fig. 10.51). Examples of slowly weatherable minerals include tourmaline, beryl, rutile, anatase, and zircon. Fortunately, many of these minerals are associated with relatively rare elements, allowing for quick determinations of the amounts of these minerals by analyzing for their elements, e.g., Zr for zircon, and Ti for rutile or anatase (see Chapter 14). Although this is a time-tested method for ascertaining the location of LDs, it has one drawback: Many soils, especially highly weathered ones, contain very low amounts of such minerals (Tsai and Chen 2000).

Finally and most importantly, calculation of depth functions for immobile and/or inert components is best done on a *clay-free basis*, which removes the effects of the mobile element, clay, from the result (Rutledge *et al.* 1975a, Asady and Whiteside 1982, Bigham *et al.* 1991, Ogg *et al.* 2000). Clay-free silt and sand contents are routinely used to detect LDs (Table 10.4, Fig. 10.51), and their use is recommended. Clay-free data are simply those in which the clay has been mathematically removed, resulting in proportional increases in the amounts of the remaining fractions, that is:

Clay-free sand = 100 * [(% sand) / (% silt + % sand)].

Even when skeletal (sand and silt) data are used, as shown in Figure 10.52, if they have not been calculated on a clay-free basis, the resulting conclusions can be misleading. Figure 10.53A illustrates how, using raw skeletal data, one might mistakenly assume that an LD exists between the E and Bw1 horizons, when, in fact, none does, according to the clay-free data. In Part B of Figure 10.53, the raw data suggest that this soil lacks an LD, when, in fact, one exists at 42 cm, on the basis of the >6% changes in clay-free sand and silt, across this boundary. Where necessary, one might take this precaution one step further and use clay-free *and* carbonate-free data (Raad and Protz 1971).

We add two last points of clarification and guidance. In depth functions, LDs are assumed to exist at areas where the break in slope is sharp (Figs. 10.50, 10.51, 10.52). First of all, we caution that these kinds of data should not be used, singularly, to suggest a lithologic discontinuity; field data and observations should also support the data. Depth function data should be used to confirm or refute an LD that was previously thought to exist on the basis of field observations. Second, how large should the break be, in order to confirm an LD? Although this question has no clear answer, content changes of 5–10% or more, across a suspected LD, should normally exist. Additionally, such changes should be present in *more than one kind* of data, e.g., clay-free silt- and sand-sized Zr content.

Table 10.4 Parameters that have been successfully used to detect (or refute) the presence of lithologic discontinuities in soils

Indicator of lithologic discontinuity	References
Mineral content (mineralogy) Presence or absence of a detrital fossil	Barnhisel et al. 1971, Raad and Protz 1971, Karathanasis and Macneal 1994, Kuzila 1995 Karathanasis and Macneal 1994
Resistant mineral content of a nonclay fraction	Chapman and Horn 1968, Wascher and Collins 1988, Tsai and Chen 2000, Stiles et al. 2003
Elemental composition of a nonclay fraction	Alexander et al. 1962, Oertel and Giles 1966, Arnold 1968, Barnhisel et al. 1971, Foss et al. 1978, Norton and Hall 1985, Tejan-Kella et al. 1991, Karathanasis and Macneal 1994
Rare earth or elemental content, on a whole soil basis	Fernandes and Bacchi 1998, Ahr et al. 2013
Heavy mineral content or composition	Chapman and Horn 1968, Khangarot et al. 1971, Cabrera-Martinez et al. 1989, Kleber 2000
Coarse fragment content	Follmer 1982, Arnold 1968, Raad and Protz 1971, Asamoa and Protz 1972, Meixner and Singer 1981, Schaetzl 1996, 1998, Liebens 1999, Kleber 2000, Ande and Senjobi 2010
Content of sand, or a sand fraction, or mean sand size	Oertel and Giles 1966, Arnold 1968, Borchardt et al. 1968, Gamble et al. 1969, Caldwell and Pourzad 1974, Langohr et al. 1976, Meixner and Singer 1981, Follmer 1982, Santos et al. 1986, Schaetzl 1992b, 2008
Content of silt, or a silt fraction	Oertel and Giles 1966, Caldwell and Pourzad 1974, Price et al. 1975, Langohr et al. 1976, Meixner and Singer 1981, Follmer 1982, Santos et al. 1986, Kleber 2000, Schaetzl 2008
Content of clay-free sand, or a clay-free sand fraction	Wascher and Collins 1988, Busacca 1987, Busacca and Singer 1989, Karathanasis and Macneal 1994, Schaetzl 1996, 1998, Tsai and Chen 2000, Ahr et al. 2013
Content of clay-free silt, or a clay-free silt fraction	Chapman and Horn 1968, Barnhisel et al. 1971, Asamoa and Protz 1972, Price et al. 1975, Rutledge et al. 1975b, Wascher and Collins 1988, Busacca 1987, Busacca and Singer 1989, Karathanasis and Macneal 1994, Schaetzl 1996, Wilson et al. 2010
Content of a clay-free and carbonate-free sand fraction	Raad and Protz 1971, Schaetzl 1998
Content of a clay-free and carbonate-free silt fraction	Raad and Protz 1971
Clay mineralogy	Follmer 1982, Dixon 1991, Kuzila 1995, Kleber 2000
Oxygen isotope composition ($\delta^{18}O$) of quartz	Mizota et al. 1992
Magnetic susceptibility	Singer and Fine 1989, Fine et al. 1992, Wilson et al. 2010
Shape of sand particles	Schaetzl 1998
Ratio of one sand fraction to another	Fiskell and Carlisle 1963, Oertel and Giles 1966, Wascher and Collins 1988, Cabrera-Martinez et al. 1989, Hartgrove et al. 1993, Beshay and Sallam 1995, Liebens 1999
Ratio of one silt fraction to another (in some cases, clay-free)	Follmer 1982, Sobecki and Karathanasis 1992, Kuzila 1995, Wilson et al. 2010
Ratio of sand/silt or silt/sand (in some cases, clay-free)	Chapman and Horn 1968, Raad and Protz 1971, Smith and Wilding 1972, Asady and Whiteside 1982, Busacca 1987, Busacca and Singer 1989, Sobecki and Karathanasis 1992, Tsai and Chen 2000
Ratio of contents of two minerals in a nonclay fraction	Chapman and Horn 1968, Price et al. 1975, Follmer 1982, Busacca and Singer 1989, Bigham et al. 1991, Beshay and Sallam 1995
Ratio of an element to a resistant mineral in the nonclay fraction	Santos et al. 1986

(cont.)

Table 10.4 | *(cont.)*

Indicator of lithologic discontinuity	References
Ratio of two or more elements in a nonclay fraction	Smith and Wilding 1972, Foss *et al.* 1978, Amba *et al.* 1990, Bigham *et al.* 1991, Sobecki and Karathanasis 1992, Karathanasis and Macneal 1994, Tsai and Chen 2000, Tejan-Kella *et al.* 1991, Stiles *et al.* 2003
Formulas involving particle size fractions: uniformity value	Cremeens and Mokma 1986, Schaetzl 1998, Tsai and Chen 2000
Comparative particle size distribution index	Langohr *et al.* 1976

Note: Parameters listed are of the kind that can be displayed as depth functions; qualitative and/or morphological indicators of discontinuities are not addressed in this table. See Raukas *et al.* (1978) for a list of parameters used to study tills, some of which could have application within pedology.

Source: Modified from Schaetzl (1998).

Table 10.5 | Utility of various parameters used to detect lithologic discontinuities in soils

Very useful in most instances[a]	Useful in some instances	Rarely useful; should be used only in conjunction with instances parameters	Not generally recommended
Content of a clay-free, immobile particle size fraction	Content of a skeletal particle size fraction, whole-soil basis	Clay mineralogy	Content of organic matter
Elemental composition of an immobile and difficultly weatherable, skeletal particle size fraction	Content of a mineral or element, whole-soil basis	Magnetic susceptibility	Content and mineralogy of clay fraction
Content of a resistant mineral in a skeletal particle size fraction	Shape (roundness, sphericity, etc.) of a skeletal particle size fraction	Paleobotanical or paleontological data	Pedogenic physical features, e.g., structure, consistence, color, bulk density, and soil "development indices"
Indices specifically developed and designed to detect lithologic discontinuities	Presence of a stone line or zone		Soil chemistry, e.g., pH, electrical conductivity, CEC, base saturation
			Certain kinds of magnetic susceptibility
			Contents of most elements, except those associated with difficultly weatherable minerals (Zr and Ti)

[a] Application of ratios of and between useful parameters is generally encouraged.

Source: Schaetzl (1998).

Ratios of some of the preceding parameters are also widely used in the detection of LDs, and their use is recommended (Santos *et al.* 1986; Table 10.4). The ratio of two sedimentologic parameters has an advantage over data of just one, because when more data are incorporated, the between-material differences become more apparent (Foss *et al.* 1978). Ratios also have, however, a distinct mathematical disadvantage over other parameters. Extremely small values, when positioned as the denominator, can inflate the ratio and make interpretation difficult. Many ratios of resistant mineral contents encounter this problem, since the values are small to begin with and can even

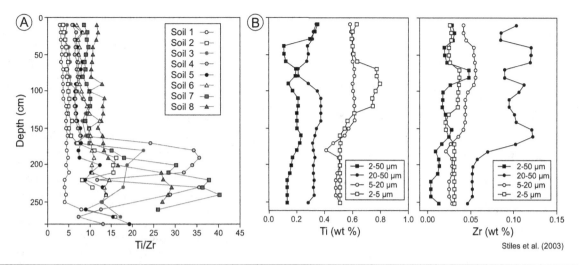

Fig. 10.51 Depth plots for Ti and Zr in some Vertisols in Texas. These depth patterns may be more indicative of weathering or pedogenic trends than lithologic changes per se. (A) Ti/Zr ratios on the whole soil fraction for eight different soils. (B) Ti and Zr contents for various size fractions of two different soils.

Stiles et al. (2003)

be zero (Beshay and Sallam 1995). Molar ratios of elements and resistant/weatherable mineral ratios have been used to determine the age or weathering status of soils (e.g., Beavers *et al.* 1963, Busacca and Singer 1989, Bockheim *et al.* 1996) and often do change abruptly at discontinuities (Foss *et al.* 1978). But because they also reflect pedogenesis, they should not be relied upon solely for detecting LDs.

In sum, when *many lines of evidence* point to a discontinuity at a given depth, one can have confidence that a discontinuity exists (Evans 1978, Raukas *et al.* 1978, Beshay and Sallam 1995). Taking off on this generalization, several researchers have developed mathematical indices that incorporate a multiplicity of parameters, so as to discern the presence or absence of a lithologic discontinuity. These indices usually compare data between the horizon in question and the one immediately above. The Comparative Particle Size Distribution (CPSD) Index of Langohr *et al.* (1976) uses data from the particle size fractions between 2 μm and 2 mm, i.e., it uses clay-free data. The CPSD determines the similarity of particle size fractions of two samples, one above and one below the perceived discontinuity, such that the higher the CPSD the more similar are the samples. Rindfleisch and Schaetzl (2001) developed a modified version of the CPSD, including data on the 2–4 mm, 4–8 mm, and total coarse fragment contents, expressed in percentages of the entire sample (not just the fine earth fraction). In its original form, perfect similarity will result in CPSD values of 100; in this modified index, values can exceed 100.

Cremeens and Mokma (1986) successfully used another index, the uniformity value (UV), to detect LDs (Tsai and Chen 2000). As with other indices, the UV compares particle size data from horizons above and below a perceived LD. It is calculated as

$$UV = \frac{[(\%silt + \%very\ fine\ sand)/(\%sand - \%very\ fine\ sand)]\ in\ upper\ horizon}{[(\%silt + \%very\ fine\ sand)/(\%sand - \%very\ fine\ sand)]\ in\ lower\ horizon} - 1.0.$$

As with the CPSD, the closer the UV is to zero, the more likely that the two horizons formed from similar parent materials; Cremeens and Mokma (1986) assumed that UV values >0.60 indicated nonuniformity, i.e., a discontinuity.

Pedogenic/morphological indicators, if chosen carefully, may help locate LDs, or they may substantiate LDs suggested by other methods (Table 10.5). For example, horizon boundaries often coincide with LDs (Bartelli and Odell 1960a, b, Bockheim 2003). Schaetzl (1996, 1998) noted that the depth of leaching in some Michigan Udalfs is usually coincident with an LD (Fig. 10.52B). Unusually thick or thin horizons may be indicative of a discontinuity, as are some types of abrupt horizon boundaries. Depth functions of physical parameters such as bulk density or porosity have also been used to detect LDs (Arnold 1968), although they should be used cautiously. Stone lines within a profile may indicate the presence of an erosional episode, and thus, can be at or near a discontinuity (Ruhe 1958, Parsons and Balster 1966, Ande and Senjobi 2010). Care must be taken, however, since many stone lines have pedogenic, not geologic, origins (Johnson 1989, 1990, Johnson and Balek 1991, Johnson *et al.* 2005).

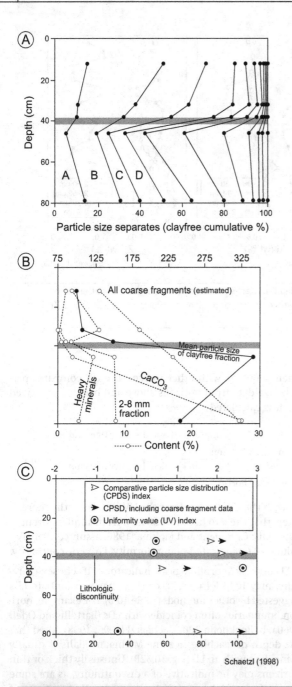

Schaetzl (1998)

Detection of LDs becomes increasingly more difficult as soils develop (Fine *et al.* 1992). Translocation of mobile soil components, mixing processes, and transformations within the profile may blur and confound any preexisting sedimentologic evidence of the discontinuity. Thus, in situations in which an LD is suspected but the soil is highly developed, it becomes increasingly imperative to focus the interpretation of LDs on immobile elements.

Although most LDs can be detected by the examination of depth functions of carefully selected parameters, this method has the disadvantage of being somewhat subjective. More quantitative methods, involving discriminant analysis and principal components analysis, have been reported (Norton and Hall 1985, Litaor 1987, Ogg *et al.* 2000). However, these often require some sort of a priori reasoning regarding the possible location and/or existence of discontinuities. Like depth functions, discriminant functions are best calculated using data on an immobile and slowly weatherable soil fraction.

Pedogenic Importance of Discontinuities
Pedogenic processes that involve translocation are variously affected and altered at (and by) the discontinuity (see Chapter 13). Discontinuities where fine material overlies coarse layers dramatically affect eluviation-illuviation processes because of higher matric tension in the upper, fine-textured material. Unsaturated flow cannot enter the lower, coarser-textured sand until the overlying material is nearly saturated (Bartelli and Odell 1960a, b, Clothier *et al.* 1978). Thus infiltrating water may hang at the fine-over-coarse discontinuity (Asamoa and Protz 1972, Busacca and Singer 1989, Khakural *et al.* 1993, Bockheim 2003), leading to less frequent wetting of the materials below the discontinuity and deposition of illuvial materials immediately above (and at) it. When water does break through the contact, it often moves through the lower material rapidly and as finger flow (Liu *et al.* 1991, Boll *et al.* 1996, Dekker and Ritsema 1996). In these situations, LDs essentially cause the zone

Fig. 10.52 Depth functions for an Alfisol in northern Michigan. The lithologic discontinuity, as judged in the field, is shown by the broad, stippled band. (A) Clay-free, cumulative percentages of particle size separates of the fine earth (<2 mm diameter) fraction. The particle size breaks begin with fine silt (0.002 to 0.016 mm, labeled "A") and becomes coarser to the right. The rightmost line represents the 1.4–2.0 mm fraction. (B) Depth functions of various pedological and sedimentological data. All are in units of percentages, except mean particle size, which is in micrometers. (C) Data from three indices designed to show lithologic discontinuities, based on particle size data: (1) the comparative particle size distribution (CPSD) Index, (2) the CPSD including coarse fragment data, and (3) the uniformity value (UV) index. Plotted values show the index value, compared to that of the horizon *immediately above*. For the UV index, greater deviations, either way from zero, indicate a stronger likelihood that the two horizons are separated by a discontinuity. Higher values for the CPSD index indicate a lower likelihood of a discontinuity between the two horizons; the farther to the right the value plots, the lower the likelihood of a lithologic discontinuity between that horizon and the horizon above.

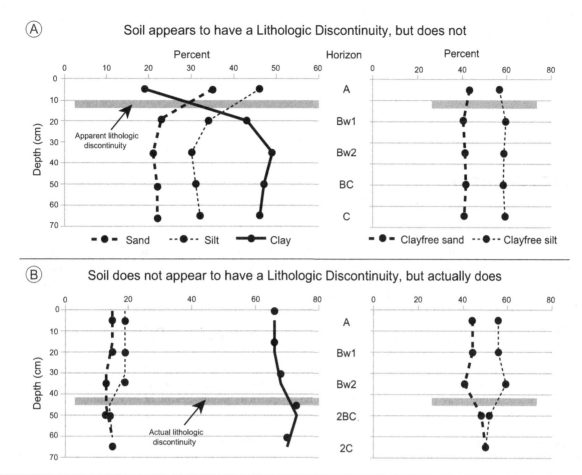

Fig. 10.53 Depth functions of textural components for two fictitious soils, illustrating the utility of clay-free particle size data in determining the locations of lithologic discontinuities. Soil A *lacks* a lithologic discontinuity but may appear to have one, on the basis of raw sand and silt data only. Clay-free particle size data show that this soil lacks a lithologic discontinuity. Conversely, Soil B *has* a lithologic discontinuity at 42 cm, which is only apparent from the clay-free data.

of illuviation to be vertically compressed; soil morphology and horizonation are dramatically affected.

Where coarser materials overlie fine-textured sediment, infiltrating water can also be perched at the contact (Rozanov 1957), promoting gleying and deposition of illuvial materials at and above the contact. Eventually, this water can be "pulled" into the lower material as a result of increased matric tension. In the latter case, the depths at which soluble and suspended materials are deposited in the soil are affected by the discontinuity.

Chapter 11

Pedoturbation

Pedoturbation, a word popularized by Francis Hole (1961), is synonymous with soil mixing. The mechanisms and vectors by which physical mixing is accomplished are many, and they operate from microscopic scales at which crystals grow and deteriorate, to larger mixing processes associated with uprooted trees, to massive termite mounds and solifluction lobes. The importance of pedoturbation has traditionally been underemphasized by many soil and earth scientists. Nonetheless, it is ubiquitous, and its importance to soil and landscape genesis is becoming increasingly documented. Pedoturbation may promote horizonation and physical organization in a soil, but it can also be a regressive (mixing) process that promotes disorder. In short, although it is a form of *mixing*, pedoturbation is not, as we shall see, always synonymous with *homogenization*.

Pedoturbation affects soil genesis and its developmental pathways almost continually, but it often is little noticed. Knowledge of pedoturbation is vital for the study of preexisting stratification, such as in archaeology (Wood and Johnson 1978, Rolfsen 1980, Stein 1983, Bocek 1986, McBrearty 1990, Balek 2002) and sedimentology, as well as for those who study pedogenic layering, e.g., soil scientists or geomorphologists (Johnson *et al.* 1987). For example, geological lithologic discontinuities can be blurred or completely masked by pedoturbation processes. Alternatively, layering formed by pedoturbation processes can be mistakenly attributed to geological processes that operated prior to pedogenesis. Pedoturbation plays a critical role in maintaining macroporosity in most soils, which in turn aids in infiltration and retards runoff and erosion. Physical mixing of organic matter in surface litter layers into the underlying soil is largely accomplished via pedoturbation. In short, most pedogenic pathways are in some way affected by pedoturbation.

Many traditional soil genesis studies have focused on ascertaining how and why the soil profile (an ordered state) has developed. Comparatively few studies, however, have stressed or examined the converse – the formation of disorder (haploidization, simplification) from an otherwise pedologically ordered soil, i.e., one with horizons (see Chapter 12). Still fewer have examined the preservation of disorder by pedoturbation, or the formation of soil order by pedoturbation. This oversight is the impetus for this chapter.

Expressions of Pedoturbation

In its various forms (Table 11.1), pedoturbation is studied either by observing the process, such as ants moving soil particles (Pérez 1987b), or by examining and interpreting the end products of pedoturbation (Baxter and Hole 1967, Schaetzl 1986b, Cox *et al.* 1987a, Johnson 2002, Robertson and Johnson 2004). In the latter case, morphological or chemical signatures of the pedoturbation are observed and measured, e.g. Southard and Graham (1992). Because many forms of pedoturbation move *coarse fragments*, this type of evidence is commonly used to infer pedoturbation (Table 11.2).

Signatures of previous pedoturbation are expressed as within-profile morphological imprints, and as surface topographic features. Within-profile imprints include slickensides, faunal fecal pellets, reorganization and reorientation of the soil microfabric, stone lines, broken and disrupted horizons, and infilled burrows, among others (Wang *et al.* 1995). Surface expressions of pedoturbation occur as microrelief, e.g., gilgai, treethrow mounds and pits, anthills, termite mounds, patterned ground, earthwork kommetjies, and caved-in animal burrows (Haantjens 1965, Fey 2010). The scale at which these features are manifested ranges from microscopic to large features observable from airplanes.

Proisotropic versus Proanisotropic Pedoturbation

Pedology is replete with innumerable studies that examine how *pedologic order* (anisotropy, soil horizonation, layering) evolves from sediments with *geologic disorder* (isotropic

Table 11.1 | Major types of pedoturbation vectors

Form of pedoturbation	Soil mixing vectors (and representative references)
Aeroturbation	Gas, air, wind (McFadden et al. 1987)
Anthroturbation	Humans (Eidt 1977, Short et al. 1986)
Aquaturbation	Water (Abrahams et al. 1984)
Argilliturbation Bombturbation	Shrinking and swelling of clays, such as smectite (Graham and Southard 1983, Wilding and Tessier 1988, Coulombe et al. 1996, Nordt et al. 2004) Explosive munitions and ordnance (Hupy and Schaetzl 2006, 2008, Hupy and Koehler 2012)
Cryoturbation	Freeze–thaw activity, ice crystals (Benedict 1976, Manikowska 1982, Van Vliet-Lanoë 1985, Tarnocai 1994)
Crystalturbation	Crystals, such as ice and various salts
Faunalturbation[a]	Animals, including insects (Webster 1965a, Langmaid 1964, Cox et al. 1987a, Johnson 1989)
Floralturbation[a]	Plants (Schaetzl 1986b, Small et al. 1990, Samonil et al. 2010a)
Graviturbation	Mass movements, such as creep (Moeyersons 1978)
Impacturbation	Extraterrestrial impacts such as comets and meterorites, and human-generated impacts, e.g., artillery shells and bombs
Seismiturbation	Earthquakes

[a] Collectively, floralturbation and faunalturbation are referred to as *bioturbation*.
Sources: Hole (1961), Johnson et al. (1987), and Hupy and Schaetzl (2006).

parent materials such as loess or dune sand) or *geologic order*, such as stratified alluvium or saprolite. This mind-set has proliferated, deliberately or inadvertently, the notion that pedoturbation is a regressive soil process – one that blurs soil horizons or prevents them from forming (see Chapter 12). Fewer studies observe that pedoturbation can actually create, or even preserve, the anisotropy that often directly results from pedogenesis.

With this in mind, we follow the classifications of others (Hole 1961, Johnson *et al.* 1987), who placed pedoturbation into one of two categories: *proisotropic* or *proanisotropic*. The former term implies a condition tending toward (pro) isotropy (uniformity, disorder, randomness, etc.), while proanisotropic pedoturbation means a tendency toward layering, order, nonrandomness, and sorting (Fig. 11.1). More rigorously defined (Johnson *et al.* 1987; see also Hole 1961: 375), *proisotropic pedoturbations* include processes that *disrupt, blend,* or *destroy* horizons, subhorizons, or genetic layers and/or *impede their formation*. These processes cause morphologically simplified profiles to evolve from more ordered ones. Proanisotropic pedoturbations include processes that *form* or *aid in forming/maintaining* horizons, subhorizons, or genetic layers. They cause an overall *increase in profile order*. In most cases, soil mixing has components of each of these two sets of interacting processes.

Pedoturbation processes can be classified as proanisotropic (progressing toward the development of soil horizons) or proisotropic (progressing toward the destruction of horizons). That is, pedoturbation can promote a simpler or more complex soil profile, or do neither. Proisotropic pedoturbation is akin to homogenization, although the homogenization may only occur in part of the profile and only for a brief period. The literature contains many examples of proisotropic pedoturbation, such as the destruction of soil horizons by burrowing animals or tree uprooting (Schaetzl *et al.* 1990). Borst (1968) described how ground squirrel burrowing activities had destroyed incipient horizons and prevented the development of argillic horizons. By continually mixing soils in the same area, the squirrels were capable of affecting the soils to a depth of 75 cm every 360 years. Worm activity can have similar results (Buntley and Papendick 1960). For example, Baxter and Hole (1967) studied ant mounds on a prairie remnant in Wisconsin, where the mounds occupied 1.7% of the surface. They assumed a 12-year occupancy for each mound, implying that each point on the landscape might be occupied by a mound every 600 years. At this intensity of faunalturbation, Baxter and Hole (1967) argued that strongly horizonated (anisotropic) Alfisols had been changed into more isotropic Mollisol profiles in the 3,500 years since prairie replaced forest. One of the more interesting ways in which the ants had accomplished this was their selective mining of clayey B horizon material, leading to the formation of a humus- *and* clay-rich A horizon in the resulting Mollisols. This example, illustrates that, in many soils with diffuse horizon boundaries, this morphology can be ascribed to long-term faunal activity, which is here defined under proisotropic pedoturbation. Proanisotropic processes act

Table 11.2 Pedoturbation processes that can move coarse fragments

Process	Representative sources
Faunalturbation: burrowing, tunneling, or mounding	
Earthworms	Darwin 1881, Webster 1965b, Johnson et al. 1987, Ponomarenko 1988
Pocket gophers	Murray 1967, Cox and Allen 1987a, Johnson et al. 1987, Johnson 1989, Cox and Scheffer 1991
Ground squirrels	Borst 1968, Johnson et al. 1987
Mole rats	Cox et al. 1987b, 1989, Cox and Gakahu 1987
Prairie dogs	Carlson and White 1987, 1988
Aardvarks	Melton 1976
Crayfish	Stone 1993, Robertson and Johnson 2004
Termites	Lee and Wood 1971a, b, Gillman et al. 1972, Holt et al. 1980, Bagine 1984, Nutting et al. 1987, Lobry de Bruyn and Conacher 1990, Grube 2001
Ants	Baxter and Hole 1967, Salem and Hole 1968, Humphreys 1981, Mandel and Sorenson 1982, Levan and Stone 1983, Cox et al. 1992
Floralturbation: tree uprooting	Lutz 1960, Johnson 1990, Schaetzl 1990, Schaetzl et al. 1990, Small 1997, Small et al. 1990, Phillips and Marion 2006, Samonil et al. 2010a, b
Cryoturbation (also congelliturbation): freezing and thawing	Bryan 1946, Inglis 1965, Pissart 1969, Manikowska 1982, Mackay 1984, Van Vliet-Lanöe 1985, Pérez 1987a
Argilliturbation: swelling and shrinking of clays	Springer 1958, Johnson and Hester 1972, Cooke et al. 1993
Graviturbation: mass wasting	
Soil settling	Moeyersons 1978
Creep	Schumm 1967, Williams 1974, Moeyersons 1978, 1989
Debris flow	Van Steijn et al. 1988
Aeroturbation	
Wind erosion	Wilshire et al. 1981
Eolian accumulation	McFadden et al. 1987
Aquaturbation	
Splash creep	Moeyersons 1975
Runoff creep	De Ploey et al. 1976
Surface wash	Kirkby and Kirkby 1974
Hydraulic erosion	Abrahams et al. 1984, Poesen 1987
Crystalturbation: growth and wasting of crystals	Cooke et al. 1993
Seismiturbation: earthquakes and vibrations	Clark 1972, Hole 1988
Anthroturbation	
Trampling	Barton 1987
Tillage	Kouwenhoven and Terpstra 1979
Off-road vehicle traffic	Elvidge and Iverson 1983, Pérez 1991
Bombturbation	Hupy 2006, Hupy and Schaetzl 2006

Source: Modified from Poesen and Lavee (1994).

in the opposite fashion; they create order (horizonation) in soils and/or strengthen existing order.

Seldom are pedoturbation processes entirely proisotropic or entirely proanisotropic. Instead, they usually have elements of both, with one form of pedoturbation often more strongly expressed than the other (Fig. 11.1). For example, earthworms may mix O horizon material into the A horizon, and in so doing isotropically blur the

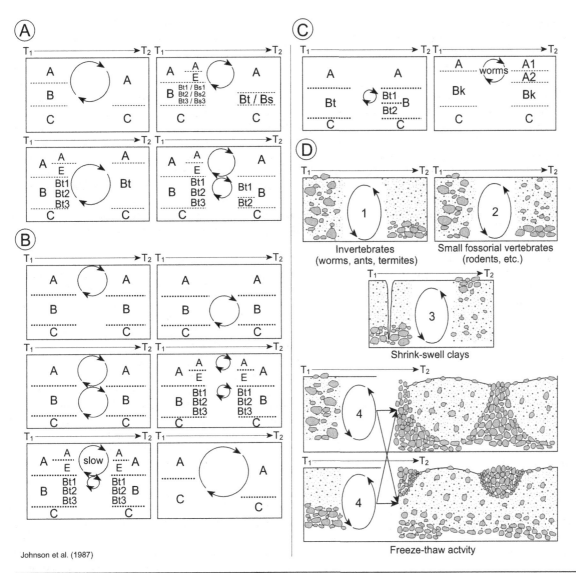

Johnson et al. (1987)

Fig. 11.1 Hypothetical examples of proisotropic and proanisotropic pedoturbation, during the theoretical period T_1 to T_2. (A) Proisotropic pedoturbation, in which a horizon or horizons are simplified or destroyed by pedoturbation. (B) Proanisotropic pedoturbation, in which a horizon or horizons are maintained despite the mixing. (C) Proanisotropic pedoturbation, in which a new horizon is produced by pedoturbation, showing examples of geologically deposited sediment becoming pedogenically organized. In the right diagram, a Vermdudoll with two distinct A horizons is formed by intense faunalturbation. (D) Proanisotropic pedoturbation, in which geologically organized and/or unorganized sediment is reorganized by various types of pedoturbators.

two horizons (Nielsen and Hole 1964). But in the process, the worms thicken the A horizon at the expense of the O, thereby promoting horizonation. Alternatively (and probably concurrently), they may mix small amounts of mineral matter from the A horizon with raw litter and deposit organic-rich casts on top of the O horizon; the O horizon thus thickens – a form of anisotropic mixing (Bernier 1998). Nooren *et al.* (1995) discuss a different way that earthworms can create anisotropy in soils. In the Ivory Coast of Africa, earthworms are credited as the main force

behind the formation of soils with relatively sandy surface horizons over clayey subsurface horizons. Deep burrowing moves clay-rich casts to the surface, where they are then disintegrated by heavy rains, leaving mainly sandy sediment behind. The displaced, finer particles end up in nearby rivers. Thus, the surface horizons become increasingly sandy. This example points out why many of the morphological signatures of pedoturbation in soils are mainly evident *texturally*, because they result from a biomechanical, not a biochemical, process.

Archaeology provides examples of how both forms of pedoturbation can operate synchronously. Human artifacts, such as tools, dropped on the soil surface become buried as worms deposit their casts on top of the artifacts (Johnson 1990, 2002, Cremeens 2003, Feller *et al.* 2003). After a few thousand years, the artifacts are lowered into the soil, but their layering and even their spatial organization are often preserved (Darwin 1881, Atkinson 1957, Frolking and Lepper 2001, Van Nest 2002, Alexandrovskiy 2007). To most observers, *before* and *after* snapshots of the upper horizons in such a soil would be indistinguishable, even though the soils would have been undergoing continuous pedoturbation. The main difference lies not in the fine earth fraction, but in the coarse materials – the artifacts. Similarly, the burrowing of cicada nymphs in grassland soils can be viewed as mainly proanisotropic. Hugie and Passey (1963, 79) observed mixing of soil horizons by cicadas only at abrupt and clear soil boundaries, suggesting that the cicada nymphs continually backfill their burrows with local soil materials, keeping open only a short lead tunnel. Local horizonation is largely preserved.

These examples illustrate that pedoturbation is rarely wholly isotropic. It may be mostly anisotropic, or it may be partly anisotropic and partly isotropic, depending on such factors as time of year, which part of the profile the activity occurs in, which size fraction is being examined, what the mixing vector is, and other factors.

Pedoturbation is pervasive. It is a suite of pedogenic processes that are as important as other, more traditional pedological process suites, e.g., eluviation, illuviation, weathering, and leaching. Throughout much of the twentieth century, however, the agronomic focus of soil scientists, as well as many geomorphology paradigms, did not fully appreciate the importance of pedoturbation. Only recently has a reevaluation of its importance commenced. Pedoturbation is no small matter!

Biomantles and Stone Lines

The spatial expression of pedoturbation and whether it is proisotropic or proanisotropic often depend on the form of the process (mixing by tree uprooting vs. ant mounding vs. freeze-thaw activity), its intensity and duration, and the *particle size fraction* that is being examined. That is, pedoturbation can be regressive or progressive, isotropic or anisotropic, depending on the point of view – is the focus the fine earth fraction, or the gravel and stones? Pedoturbation, in its many forms (see later discussion), is capable of mixing the fine earth fraction, while concomitantly sorting and creating order in the larger size fractions, such as gravel or stones, by slowly allowing them to settle to the bottommost level of mixing. For example, many insects, arthropods, and small mammals burrow in and through the soil profile, moving soil material to the surface and/or reorganizing it (Wiken *et al.* 1976, Johnson 1990, Munn 1993, Robertson and Johnson 2004).

As discussed later, mixing of soils by plants (flora) and animals (fauna) is called *bioturbation*. Soil fauna that reside within the soil are called *infauna* (Fig. 7.2). Some fragments in soils are too large for infauna to carry to the surface or even move upward in the soil. Rather, in the process of burrowing around such fragments, they eventually settle downward, as tunnels collapse and smaller materials nearby are moved to the surface. In time, these fragments become concentrated at the approximate maximum depth of burrowing and are often collectively called a *stone line*, stone zone, or stone layer (Williams 1968, Johnson 1989, 2002). Overlying the stone line is a *biomantle*, which is a layer of relatively stone-free material, sorted and moved to the surface by biota, via bioturbation (Soil Survey Staff 1975: 21, Johnson 1990, Humphreys 1994, Balek 2002, Johnson *et al.* 2005, Horwath and Johnson 2006, Johnson and Horwath Burnham 2012).

Because they have such a distinctive mode of formation, many biomantles also have a characteristic *biofabric*. For example, a worm-worked biomantle will have a strong crumb or granular fabric associated with worm casts and voids (Buntley and Papendick 1960, Peacock and Fant 2002). Horizons within the biomantle can be thoroughly mixed, or they can retain some stratification. Van Nest (2002) discussed how human artifact patterns, rock foundations of buildings, and brick patio arrangements can be preserved beneath a biomantle, as long as the burial processes are slow and spatially uniform. The large clasts are lowered so slowly and evenly that their original spatial arrangement is preserved. She used this knowledge to explain why artifacts >3,500 years old are often found several centimeters beneath the soil surface and may thus be overlooked by routine archaeological surveys. In such cases, the geological principle of superposition (see Chapter 15) obviously does not apply, as the artifacts are older than the biomantle, but lie below it. This fact *must* be taken into consideration whenever interpretations based on stratigraphy and superposition are made in soils (Piperno and Becker 1996, Balek 2002, Cremeens 2003).

Charles Darwin, in his classic (1881) book *On the Formation of Vegetable Mould through the Action of Worms*, showed how bioturbation texturally differentiates soils and sediments (Johnson 2002, Meysman *et al.* 2006). In particular, Darwin described how a loamy, worm-worked topsoil was formed above a subsurface layer of flints and cinders that had, years before, been spread on the surface. As a result of worm bioturbation, the coarse fragments had migrated downward to form a subsurface stone line. Darwin thought of his book as "curious" and "of small importance." He referred to bioturbation as a "subject that I have perhaps treated in foolish detail" (Meysman *et al.* 2006). Nonetheless, Darwin was the first to document that reworking activities by soil fauna could have dramatic consequences at far larger scales, e.g., in landscape formation, and could be vital to

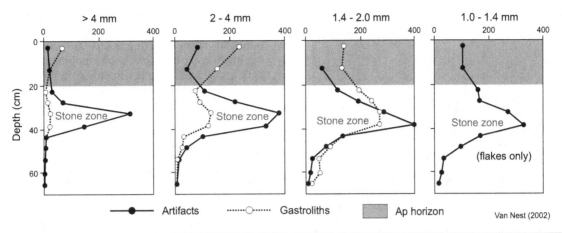

Fig. 11.2 Distributions of artifacts and gastroliths with depth in an upland soil in Warren County, western Illinois. Note the presence of a developing stone zone at depth.

an understanding of pedogenesis at the smallest of scales. And his work on earthworms showed the potential importance of carefully observing natural processes that might at first seem trivial (Tsikalas and Whitesides 2013, Johnson and Schaetzl 2014).

Stone lines are omnipresent in areas in situations where burrowing infauna have occupied the soil for a long time, provided the soil has coarse fractions. In older pedological literature, they have often been erroneously interpreted to represent erosion-derived lag concentrates at a former soil surface. If the stone line contains human artifacts, a false impression of a paleo land surface can also be construed. By understanding these processes and their importance, Atkinson (1957: 222) concluded that many archaeological materials have been displaced downward, and that in some cases the amount of displacement may have been sufficient to alter their apparent stratigraphic relationships (Erlandson 1984). Moreover, the upper limit on the size of materials or structures that can be buried by bioturbation is quite large. For example, in his book on earthworms, Darwin also showed that some of the fallen stones of Stonehenge had sunk several centimeters into the soil since they had originally toppled.

The depth of the stone line is primarily a function of the depth to which the burrower operates; coarse fragments tend to settle to the maximum depth of burrowing. The size of fragments in the stone line also depends on the burrower. Fragments that are moved by the burrower stay in the biomantle, whereas larger fragments settle into the stone line. Where stone-free sediments, like loess, overlie sandy and gravelly sediment, and the gravels are so small that they can all be moved by infauna, the mixing may be mainly anisotropic; no stone line may form. For example, Munn (1993) discussed how prairie dog burrowing in soils developed in alluvium over gravel thoroughly mixes the two zones. Ants may form a stone zone at 2 m, composed primarily of particles > 3 mm in diameter. But for gophers,

the stone zone may be deeper and limited to much larger stones. Johnson (1989) noted that a certain species of pocket gopher in California does not transport stones that are larger than ~6–7 cm long. In summary, if the largest clasts in the soil are smaller than the maximum size of particles that the infauna are capable of moving, a stone line will not form.

Often, contained within the stone line or zone are *gastroliths* – gravel-sized fragments that are ingested by birds to facilitate digestion (Cox 1998). After they pass through the bird's gut or are coughed out, their fate is similar to that of any other stone, human artifact, or coarse clast (Fig. 11.2).

In cases where bioturbation is intense and widespread, most of the original soil horizonation is destroyed, but the coarse fragments either retain their stratification or develop stratification from a less-ordered state. In other instances, where the burrowing activities are scattered in time and space, soils take on a form of patchiness and overall landscape complexity increases; vestiges of previously formed soil horizons can therefore be maintained. Indeed, the patchiness or pattern of entire landscapes is often due simply to long-term pedoturbation, usually by mounders such as prairie dogs and the South American vizcacha (Löffler and Margules 1980, Branch 1993).

Although many forms of pedoturbation create stone-free biomantles, other forms of pedoturbation can move coarse fragments, including archaeological artifacts, upward, even as far as the surface (Holmes 1893). Tree uprooting is a major cause of large rocks and stones being thrust upward (Lutz 1960, Daniels *et al.* 1987, Schaetzl *et al.* 1990, Small *et al.* 1990, Phillips *et al.* 2008). Then, when fine materials are washed off the root + soil mass, coarse fragments can become concentrated (Denny and Goodlett 1956, 1968, Beatty and Stone 1986, Osterkamp *et al.* 2006) (Fig. 11.3). Freeze-thaw processes, i.e., cryoturbation, move large clasts to the surface and sort them (see later discussion). Argilliturbation – mixing by shrink-swell clays – can

Fig. 11.3 An eroding root mass from a tree that was uprooted about 2–3 years previously. Erosion by water and wind has led to the formation of rock pedestals on the surface, a form of lag concentrate. Thus, a form of pedoturbation has also led to sedimentologic sorting. Note knife for scale. Photo by RJS.

function similarly (Johnson and Hester 1972, Muhs 1982), as can bioturbation by some larger mammals and reptiles. Thus, pedoturbation can be viewed as both proisotropic (destroying surface horizons and mixing the fine earth and small gravel fraction) *and* proanisotropic (forming a coarse-fragment-enriched stone zone at depth – or at the surface – and a stone-free biomantle). It simply depends on the particle size fraction that is being considered (Johnson *et al.* 1987).

Pedoturbation Vectors

Previously, we classified pedoturbation along the proisotropic-proanisotropic continuum, depending on the degree to which the mixing processes increased or decreased the physical organization of particles in the profile. Early classifications of pedoturbation, however, focused on the *vectors* that drove the process (Hole 1961, Wood and Johnson 1978, Johnson *et al.* 1987; Table 11.1). Most of the remainder of this chapter is a discussion of the various pedoturbation vectors – mixing classified on the basis of the driving force behind it, e.g., ice crystals, animals, flora (plants), shrink-swell clays.

Aeroturbation

Mixing of soils by gases occurs on a macroscale as soil particles are transported by wind and on a microscale as gases, emitted as by-products of organic or inorganic reactions, move soil particles micromillimeters at a time. Erosion and transportation of sediments by *wind* are not, strictly speaking, pedoturbation, although they clearly can and will affect pedogenesis. Only when the particles are translocated within (more likely, from place to place on top of)

the same pedon can wind transport be considered aeroturbation sensu stricto. Because aeroturbation includes minuscule forms of mixing as gases are given off from various microscopic life-forms, roots, and small faunal inhabitants, it may have more in common with Brownian motion than with the other pedoturbation vectors. Little work has been done on aeroturbation, and little is known about its pedogenic impact.

Aquaturbation

In most instances, water flow in soils is viewed as an organizing, rather than a mixing vector (see Chapter 13), e.g., the translocation of colloids and ions to form illuvial B horizons. Aquaturbation, *mixing* by water, is nonetheless ubiquitous. All water that moves into a soil is capable of dissolving soluble substances and entraining clastic particles, and this process usually occurs on microscopic or otherwise very small scales. As water tables rise and fall, and as wetting fronts penetrate the soil, some soil particles are indisputably moved, if only a few micrometers. Water can also inhibit organization. A spring seep, for example, where water is slowly flowing upward, can inhibit the formation of some types of subsurface horizons. However, our knowledge of aeroturbation is still in its infancy.

Bioturbation: Faunalturbation and Floralturbation

Recently, Johnson *et al.* (2005), inspired by contemporary research of others, reported on a unique way to classify *bioturbation* processes (Table 11.3). Bioturbation is a composite term, commonly used to refer to soil mixing by biota – animals, plants, fungi, protists, and even microbes. Although all play a role, we focus on animal and plant bioturbation. But first, it seems appropriate to make a few comments about bioturbation in general.

Wilkinson *et al.* (2009: 257) noted that the primary effects of bioturbation include soil production from saprolite, the formation of surface mounds, soil burial, and downslope transport of sediment. Johnson *et al.* (2005) focused on the *styles* and *expressions* of bioturbation, asking questions like (1) What are the results and what are the physical manifestations of the biotic mixing processes? (2) What are the biota doing in (and to) the soil? and (3) How and where do they move sediment? They started by asserting that all soil organisms are motile; they all wriggle, grow, and/or move in some way within or on the soil. In the process, soil biota have the potential to mix and stratify soil particles, thereby forming vertical and horizontally layered anisotropy. Some – the *moundmakers* – make mounds by mainly upward, focused transfers of soil and sediment, a process noted by many (Humphreys 1981, Cox 1984, Cox and Allen 1987a, b, Carlson and White 1988, Eldridge 2004, Richards 2009, Johnson and Horwath Burnham. 2012). Collapse of burrow tunnels is an obvious and eventual countereffect, but the mounds nevertheless are a major

Table 11.3 | Major styles and expressions of bioturbation

Bioturbation style	Operative term	Morphological expression in the soil	Example biovectors
Upward biotransfers of sediment	Conveyor belt species (and moundmakers)	Loosened, texturally anisotropic biomantle; surface mounds, heaps, small tumuli, and biofabric	Ants, termites, worms, crustaceans, marine and terrestrial invertebrates, wombats, ground squirrels, badgers, prairie dogs, tuco-tucos, mole rats, moles, and armadillos
Biomixing	Mixmaster species (and moundmakers)	Loosened, texturally anisotropic biomantle; surface mounds and vuggy biofabric	Moles, pocket gophers, tuco-tucos, molerats, armadillos, marsupial moles, marine and terrestrial invertebrates, and humans
Cratering	Cratermakers	Surface craters and depressions, shallow licks, scratched and scraped areas, burrow collars, surface rubble, vuggy spoil heaps and piles, excavations, and furrows	Badgers, aardvarks, wombats, viscachas, armadillos, pigs, birds, skunks, trees, aquatic invertebrates, fish, and humans
Soil volume increases, mainly within a biomantle	—	Loosened biomantle, soil microstructural features, biopellets, biopores, biochannels, and biovugs (biofabric)	Growth structures of plants, fungi, algae, and free-living protoctists, and bacteria

Source: Modified from Johnson et al. (2005).

factor in constantly replenishing the biomantle. Initially, biomantles formed by moundmakers will be spatially discrete and patchy, but over time more of the surface will become affected, and the biomantle will expand laterally. Its thickness will remain steady, increase, or become thinner as erosion and other process vectors dictate.

Other biota move soil material upward, downward, and in various other directions; these are termed *mixmaster* species, and like the moundmakers, they are primarily burrowers. Most earthworms are classical mixmasters – moving throughout the upper soil, or biomantle, and mixing it as they go (Dexter 1978, Sveistrup *et al.* 1998, Edwards 1999). The actions of mixmasters will often be felt more widely across landscapes than would the more focused activities of moundmakers. And they may be more subtle. In both situations, however, stone lines may form below the biomantle at the maximum depth of burrowing.

Cratermakers dig, scratch, and furrow the soil surface, often as a means of finding food or shelter (Eldridge and Rath 2002, Eldridge *et al.* 2002, 2012, Eldridge and Mensinga 2007, Richards *et al.* 2011). In so doing, they move soil material laterally, and some sorting can occur. Like mound making, cratering is a point-centered style of mixing that has clear lateral edges. Therefore, biomantles augmented in this way will take considerable time to become spatially continuous.

Last, Johnson *et al.* (2005) turned to the smaller-scale and less obvious effects of bioturbation – volume increases and collapse. As roots and infauna grow, die, and collapse, soil

materials are moved and displaced (Hole 1988). Although this type of mixing is less obvious and often occurs at small scales, its impact is substantial because of the ubiquity of the process. In all, the work of Johnson *et al.* (2005) and others cited in this chapter encourages an examination not only of the bioturbator type, but also the direct effects of that mixing, the extent to which they impact the soil, and the role that bioturbation plays in soil formation generally.

Faunalturbation

Within bioturbation research, faunalturbation is more often studied, probably because the mixing processes and outcomes of soil fauna are more easily observed and measured than are the effects of plants. Faunal processes usually operate faster than do those related to plants (Thorp 1949, Heath 1965, Hole 1981, Stein 1983, Lobry de Bruyn and Conacher 1990). Mammals such as wombats, badgers, gophers, and moles; invertebrates such as earthworms; and many species of insects such as ants and termites are all burrowers. They burrow as a means of finding food and providing for shelter, hibernation, estivation, or reproduction (Thorp 1949, Hole 1981, Carpenter 1953, Whitford and Kay 1999, Van Nest 2002, Fey 2010). Many fauna pass soil particles through their guts as they burrow, while others simply push soil aside or move it to a preferred location. Sediment moved by soil fauna often has distinct textural characteristics, and which are often different from those of the native soil (Table 11.4). Mound material may

Table 11.4 | Characteristics of material transported by ants and termites versus the native soil

Location	Rate of accumulation (mm yr^{-1})	Soil mass moved to suface	Soil turnover rate (t ha^{-1} yr^{-1}) (unless otherwise noted)	Mound longevity (yr)	References
Termites					
Northern Australia	0.0125	–	–	10	Williams 1968
Northern Australia	0.02–0.10	65	470 g m^{-2} yr^{-1}	–	Lee and Wood 1971a
Northern Australia	0.05–0.025	20 in 25–50 years	–	–	Holt et al. 1980
Northern Australia	–	11–26	–	–	Spain and McIvor 1988
West Africa	0.02	–	1.25	80	Nye 1955c
Africa	–	17.6	–	–	Maldague 1964
Africa	–	20–25 m^3 ha^{-1}	–	–	Lepage 1972, 1973
Africa	–	–	0.35	–	Nel and Malan 1974
Africa	0.04–0.115	–	–	–	Pomeroy 1976
Africa	–	–	0.3	–	Wood and Sands 1978
Africa	–	–	4.0	–	Aloni et al. 1983
Africa	–	3.7	–	–	Akamigbo 1984
Africa	0.06	–	1.06	–	Bagine 1984
Africa	1.5–2	–	–	20–25	Lepage 1984
India	–	–	15.9 g m^{-2} day^{-1}	–	Gupta et al. 1981
South America	–	–	0.78	–	Salick et al. 1983
Africa	0.19	3.0 t ha^{-1} yr^{-1}	3	8	Aloni and Soyer 1987
Africa	–	–	–	15–20	Janeau and Valentin 1987
North America	–	–	0.07	–	Nutting et al. 1987
Ants					
South Australia	–	–	400 cm^3 ha^{-1} yr^{-1}	100	Greenslade 1974
Eastern Australia	–	–	8.41	months	Humphreys 1981
Southeast Australia	0.03	–	0.35–0.42	–	Briese 1982
Eastern Australia	–	–	<0.05	–	Cowan et al. 1985
Michigan	–	85.5 g m^{-2}	–	–	Talbot 1953
Wisconsin	–	2500 kg in 1040 m^2	11.36	–	Salem and Hole 1968
USA	–	0.28–0.8 g m^{-2}	–	–	Rogers 1972
USA	–	–	11.36	–	Wiken et al. 1976
Wisconsin	–	948 kg ha^{-1}	–	–	Levan and Stone 1983
England	–	–	8.24	–	Waloff and Blackith 1962
England	–	40–400 mg day^{-1} nest^{-1}	–	–	Sudd 1969
Argentina	0.085	300	11	–	Bucher and Zuccardi 1967

Source: Lobry de Bruyn and Conacher (1990).

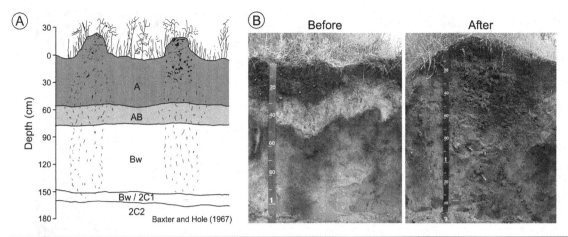

Fig. 11.4 Ant mounds formed by a species of the genus *Formica*. (A) Diagram of mounds formed in a Mollisol in Wisconsin. (B) Before and After views of a Spodosol in northern Michigan. Tape increments in centimeters. Note the slight discrepancy in scale. Before: a well-horizonated Haplorthod. After: the same soil, 3 meters away, showing abundant evidence of mixing and haploidization by ants. Photos by RJS.

become more sorted by the fauna during burrowing, and then again as the mound collapses and water washes over it (Wiken *et al.* 1976, Breuning-Madsen *et al.* 2004). The degree of particle size selectivity with which soil fauna move particles, and the locations of their burrows, often determine whether the faunalturbation is predominantly proanisotropic or proisotropic (Humphreys and Mitchell 1983, Johnson 1989, 1990).

In addition to the most obvious mounding, mixing, and cratering activities of soil fauna (Fig. 11.4), they form voids, form and destroy peds, regulate soil erosion, impact plant and animal litter, facilitate the movement of water and air within the soil, and regulate many biogeochemical cycles (Hole 1981, Lobry de Bruyn and Conacher 1990). Burrowing, whether by ants (Lokaby and Adams 1985) or by other fauna, creates large voids in the soil (Fig. 11.4B). These voids provide conduits for rapid passage of gases and fluids and contribute to lower bulk density in the soil (Salem and Hole 1968, Lobry de Bruyn and Conacher 1990). Many plants rely on recently pedoturbated soils as seedbeds; without this type of disturbance their germination success would be greatly diminished (Dean and Milton 1991, Schiffman 1994).

Animal burrows, termed *krotovinas* or *crotovinas* (Borst 1968, Pietsch 2013), provide unmistakable evidence of faunalturbation. This sediment often contrasts in color and density to its surrounding matrix, making for easy recognition and providing a valuable archive of paleobiological activity. When they become filled with A horizon material, as is often the case, krotovinas are darker than the surrounding matrix (Fig. 11.5A). In some Aridisols, krotovinas are recognizable because the matrix is rich in light-colored, secondary carbonates (Fig. 11.5B).

Landscapes: Rapid Soil Changes Due to Invasive Earthworms

Everyone knows something about earthworms. Pedologists and fishermen love them, but for different reasons. Some children fear them. But we all will have to deal with them increasingly as time goes on.

You see, earthworms are becoming more common, in places and in soils that have not had them before. For example, in glaciated parts of the midwestern United States, earthworms were absent until recently. Human activities, however, have introduced earthworms to many of these forested areas. Scientists have been able to identify the edges of these colonies, i.e., the invasion fronts, and the direction of their movement (Seidl and Klepeis 2011). Knowing this, the effects of the worms on soil properties can be determined by examining soils ahead of, at, and behind the front (Suarez *et al.* 2006a, b, Eisenhauer *et al.* 2011).

Humans are the principal agents of their dispersal, spreading the worms accidentally via horticulture and land disturbance, and through composting and the improper disposal of fish bait (Seidl and Klepeis 2011). Forests that have, for millennia, lacked earthworms now have them, and the wave of these invaders can be readily tracked in the forest (Lyttle *et al.* 2011). As a result, their effects on native soils can be easily ascertained by examining soils that lack worms and comparing them to soils that have had worm populations for 5, 15, 25, or 50 years or more. In essence, by introducing worms to these landscapes, humans have created a natural laboratory for soil scientists. Indeed, determining the effects of these invasions on soils has recently become a topic of active research (Alban and Berry 1994, Bohlen *et al.* 2004, Frelich *et al.* 2006, Karberg and Lilleskov 2008). In Minnesota, where the wave of invading worms has been tracked for years, pedologists have concluded that O horizons in native forests with invasive earthworms are thinning or even disappearing at some times of the year (Hale *et al.* 2005, Resner *et al.* 2011; see figure).

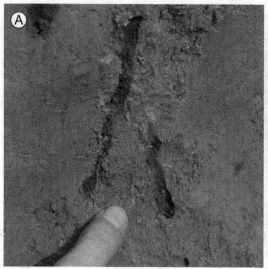 

Fig. 11.5 Krotovinas. Photos by RJS. (A) Earthworm krotovinas in an Alfisol in Wisconsin. Note the dark color of the walls of the krotovina, caused as the worms drag humus into the burrow. (B) Cicada krotovinas (dark-colored infillings, from clay-rich Bt horizon above) in an otherwise light-colored, carbonate-impregnated Bkm horizon of an Aridisol, New Mexico.

Although the effects of the introduced earthworms on soils vary from place to place, most studies have found that they change soils dramatically and quickly. Bioturbation by introduced worms affects O and A horizon thickness, structure, porosity, and permeability, as well as the biogeochemical cycles of nutrients and carbon. Habitats for litter-dwelling organisms are affected, and plant species richness often declines, as worms invade a forest (Holdsworth et al. 2007, Hale 2008).

What influences these new invaders, and their rate and direction of spread? Most research suggests that soil moisture and texture, as well as litter quality and thickness, are the major determinants (Hale et al. 2005, Ammer et al. 2006, Tiunov et al. 2006, Bernard et al. 2009). Therefore, some areas are more vulnerable to new earthworms than are others, and their effects will not be distributed equally, even though the average dispersal rate is about 10 m/year.

Like it or not, we live in a world of invasive species, with new insects and fish usually capturing the headlines. Invasions of mammals can be equally dramatic. Burrowing activities of pocket gophers in the Siskiyou Mountains of California have led to thicker A horizons than in adjacent forested areas where the gophers are absent (Laurent et al. 1994). Forested soils have O horizons, whereas the areas populated by gophers generally do not.

As the saying goes, the times they are a-changing! So are the soils.

Burrowers can affect soil morphology and chemistry over very short periods. Nielsen and Hole (1964) studied the middens formed as earthworms drag raw litter into their vertically oriented burrows (Fig. 7.8). Worm burrows were filled with litter to an average depth of 11 cm, which coincidentally was also the average A horizon thickness. Annually, an average of 46 leaves were gathered into each midden, meaning that 2.8 million leaves per hectare could have been dragged into the subsurface by earthworms in this Wisconsin forest.

Mounders and Nonmounders

In Table 11.3 we recognized that most macroscale infauna are mounders, mixmasters, or cratermakers. Ants, termites, certain large earthworms and insects, other arthropods, rodents, gophers, and foxes are usually mounders (Wood and Sands 1978, Reichman and Smith 1991, Whitford and Kay 1999, Richards 2009; Tables 11.4, 11.5, Fig. 11.6). Their pedogenic role has a proisotropic component, by moving mixed, unsorted soil materials to the surface and depositing them in mixed piles, and a proanisotropic component, by forming stone lines at depth, and via the later collapse of tunnels. Alternatively, mixmasters, such as the *Citellus* ground squirrel of California (Borst 1968), tend simply to redistribute material in the subsoil by continuously burrowing, followed by tunnel collapse. Both types of burrowing activity are capable of lowering objects within the soil, although mounders perform this function faster and better.

Mounds are obviously built by soil fauna for protection and housing, but they are sometimes also built as an adaptation to high water tables or shallow-to-bedrock conditions. Some moundbuilders never really occupy their mounds; they are simply waste piles of soil created during burrowing. Eventually the mounds collapse and form a biomantle. Many mounds are occupied for only a period of time, up to a few decades, and then abandoned (Watson 1962). Abandoned mounds provide homes for other fauna, which invade and establish residence there (Anduaga and Halffter 1991), while others (after collapse) become permeable depressions that act as loci for infiltration.

Table 11.5 | Characteristics of termite mounds built by the three Australian termite species at "Redlands," North Queensland, Australia

	Density (mounds ha⁻¹)	Bulk density (Mg m⁻³)	Maximum (and mean) height (m)	Mean basal diameter (m)	Mass of soil in mounds (Mg ha⁻¹)	Total basal area (m² ha⁻¹)	Mean depth of termite activity (m)
Amitermes vitiosus	173.2	1.45	1.35 (0.54)	0.50	15.3	59	0.4
Drepanotermes perniger	25.6	1.36	0.90 (0.53)	0.73	3.0	15	0.4
Tumulitermes pastinator	28.1	1.02	0.93 (0.57)	0.81	2.4	16	0.2

Source: Holt *et al.* (1980).

Fig. 11.6 Mounds built by soil fauna. Photos by RJS. (A) A large ant (*Formica* spp.) mound built in the dry, sandy Entisols near Grayling, Michigan. Note the new mound materials, piled upon an older, lower mound. The measuring stick is a standard 12-inch (30-cm) ruler. (B) A crayfish mound (or chimney) built in silty soils in North Carolina. The mounds are composed of material excavated from a vertical tunnel, built to provide access for the crayfish to the water table.

Mounds of many soil fauna represent deliberate or inadvertent concentrations of soil constituents, e.g., a certain clastic size fraction, organic matter, or a chemical fraction (Lee and Butler 1977, Levan and Stone 1983, Coventry *et al.* 1988, Moorhead *et al.* 1988, Abe *et al.* 2011). For example, many ant and termite mounds have higher pH values, N and organic matter contents, and base cation concentrations than do surrounding soil materials (Wiken *et al.* 1976, Salik *et al.* 1983, Nutting *et al.* 1987, Lobry de Bruyn and Conacher 1990, Ackerman *et al.* 2007, Vele *et al.* 2010, Arveti *et al.* 2012). The organic matter enrichment in these mounds is derived from feces, salivary secretions, corpses, and dead predators, as well as organic material deliberately carried in as food sources and construction materials (Lee and Wood 1971a, Lee and Butler 1977). Often, the enhanced nutrient stores in most mounds are inaccessible to plants while the mound is active, as most insects keep the mounds root- and plant-free. However, these nutrients are released to the soil after the mounds are abandoned, enriching the soil and making mound occupancy periods important from a nutrient-cycling perspective. Indeed, in many ecosystems where soils are inherently infertile, former mounds represent nutrient pools and loci of porous, fertile soil materials that can be exploited by plants as preferred germination sites, leading to ecosystem patchiness and heterogeneity that are a direct result of bioturbation (Sileshi *et al.* 2010). Termite mounds are even exploited by indigenous peoples as nutrient sources for crops (Watson 1977, Tilahun *et al.* 2012), and as human food supplements; the latter practice is called geophagy (Mills *et al.* 2009; see Chapter 4). Indeed, because of their worldwide extent and local importance, there may be no more important mounders than termites and ants.

Earthworms

Earthworms are a very important group of infauna, especially in finer-textured soils and in humid climates (Darwin 1881, Atkinson 1957; Fig. 11.7). Most are nonmounding, mixmaster-type bioturbators. Earthworms occupy different ecological niches in soils; some burrow deeply and mainly

Fig. 11.7 An earthworm in its vertical burrow, and a nearby krotovina. This worm, probably *Lumbricus terrestris*, is the common nightcrawler and is an introduced species in North America. Do not feel sad; he gave his life in the interest of science. Many of his relatives went the same route for something much less important, i.e., as fish bait!

vertically, while others live in O horizon materials only and seldom enter the mineral soil. Figure 7.9 illustrates the various habits and names of the different types of earthworms. Casting on the surface by vertically burrowing earthworms is a particularly effective form of anisotropic pedoturbation. It dilutes any illuvial clay concentrations that may have existed, resulting in more isotropic textures than would otherwise be expected. Indeed, one of the explanations given by Buol *et al.* (1997) for the occurrence of nearly equal amounts of clay in A and B horizons in some soils is that B horizon clay is continually being translocated upward by ants, worms, and other biota. This hypothesis is substantiated by Baxter and Hole (1967), who found that clay contents of ant mounds in a prairie soil were similar to that of the B horizon.

Most earthworms are very important in the production of soil structure (via casts) and macroporosity (via biopores), and in the consumption and mixing of organic materials into the mineral soil (Kladivko and Timmenga 1990, Zhang and Schrader 1993, Blanchart *et al.* 1999). The long-term implications of this important activity center on soil permeability and carbon storage in soils. Large tropical earthworms can even create microrelief, as their casts can be on the order of decimeters in size (Haantjens 1965).

The mixing activities of earthworms are usually beneficial to crops (Mackay *et al.* 1982, Stockdill 1982) and very efficient at sorting finer from larger materials, i.e., those too large to pass through the worm's gut. Vermudolls and Vermustolls have an A horizon dominated by "wormholes, worm casts, or filled animal burrows" (Soil Survey Staff 1992). Many other types of Mollisols that lack discrete

horizon boundaries often owe such morphologies to long-term mixing of their upper layers by infauna, especially worms and ants (see also Langmaid 1964). Worms are generally rare or absent in sandy soils or soils of the drier and colder climates. They prefer to inhabit soils that have high pH values, and readily digestible litter (Vahder and Irmler 2012).

Arthropods

Arthropod fauna, such as ants and termites, occur worldwide. They bioturbate differently, however, because of their different feeding and burrowing habits. Termites feed on wood and plant material containing cellulose, such as seeds and grasses, whereas ants ingest fluids from other insects, and mine flora for seeds and other edible parts. Most ants excavate subterranean nests and build mounds, but overall may do less to alter the soil than do termites. Most ant activity involves moving soil out of the way as they dig new tunnels, and many ant mounds are small. Some ants and termites do not build mounds at all. On the other hand, many species of termites deliberately move soil *into* large mounds, or termitaria. These structures contain runways, shelters, and sheeting, cemented together with feces, saliva, and undigested wood fibers (Gillman *et al.* 1972, Nutting *et al.* 1987). Conversely, ants are more likely to enrich their mounds in *raw* organic matter, like twigs, seeds, and grass shards.

The amount of soil mixed and otherwise influenced by mound-building arthropods is a function of the size of the mounds, as well as their density and longevity (Table 11.6). Densities > 500 termite mounds per hectare have been reported in Australia (Spain *et al.* 1983), and Wiken *et al.* (1976) reported ant mound densities exceeding 1,100 ha⁻¹. In time, as burrows, mounds, and hibernacula are created and destroyed, a larger and larger proportion of the surface and subsurface comes to exhibit evidence of bioturbation. On landscapes where the mounds are inhabited and maintained for long periods, as with some species of termites, the impact of pedoturbation is so intense in these localized spots (Fig. 7.7) that it may take centuries to obliterate. Likewise, some ants move to a new mound annually, allowing them to affect more of the landscape, but the net impact of each mound is less (Greenslade 1974).

Literature concerning the effects of arthropod burrowing on soils could fill volumes; only a few examples are provided here. Changes in soils brought about by arthropod activity include nutrient content, pH, bulk density, and organic matter content, as well as the many impacts that the mounds have on aboveground vegetation and ecosystems (Watson 1962, King 1977, Mandel and Sorenson 1982, Levan and Stone 1983, Litaor *et al.* 1996, Lafleur *et al.* 2002, Asawalam and Johnson 2007, Barton *et al.* 2009). They are true ecosystem engineers (Jouquet *et al.* 2006). Baxter and Hole (1967) found that ant mounds had much higher amounts of available P and K than did the soil below, or beside, the mounds. Termite mounds, and the soils immediately below, are

Table 11.6 | The effects of termites on physical soil properties

Termite species	Sample location	Coarse sand content (%)	Fine sand content (%)	Silt content (%)	Clay content (%)	References
Macrotermes bellicosus	Mound	30	38	15	17	Watson 1969
	Nearby soil	67	23	5	5	
Amitermes laurensis	Mound	19	49	5	24	Lee and Wood 1971a
	Nearby soil	26	65	4	4	
Nasutitermes exitiosus	Mound	23	37	5	33	Lee and Wood 1971a
	Nearby soil	36	50	6	7	
Macrotermes bellicosus	Mound	9	26	10	54	Boyer 1973
	Nearby A horizon	26	40	13	18	
	Nearby deep subsoil	21	23	6	47	
Macrotermes bellicosus	Mound	5	23	14	55	Boyer 1975
	Nearby soil	31	26	6	36	
Macrotermes falciger	Mound	22	37	12	29	Watson 1977
	Nearby A horizon	52	38	5	5	
Amitermes vitiosus	Mound	31	34	8	28	Holt et al. 1980
	Nearby soil	39	35	8	18	
Trinervitermes trinervoides	Mound	2	46	14	37	Laker et al. 1982
	Nearby soil	6	68	6	20	
Cubitermes oculatus	Mound	61		20	19	Wood et al. 1983
	Nearby A horizon	77		12	11	
Nasutitermes sp.	Mound	25	26	13	37	Akamigbo 1984
	Nearby soil	31	37	7	24	
Macrotermes michaelseni	Mound	48		14	38	Arshad et al. 1988
	Crust Nearby A horizon	67		15	18	
Iridomyrmex purpureus	Mound	29	56.0	9	5	Greenslade 1974
	Nearby soil	30	52.0	10	6	

Source: Lobry de Bruyn and Conacher (1990). Values rounded to the nearest whole number.

also commonly enriched in many nutrients, even precious metals (Brossard et al. 2007, Stewart et al. 2012; Fig. 11.8). Termite mounds studied by West (1970) had anomalously high zinc, gold, and silver contents. They were underlain at 10–25 m by rocks rich in those metals, pointing not only to the enriching tendencies of termite activity but also to their extreme depth of burrowing. Some mounds are so nutrient-rich, and the nearby soils so infertile, that some have even suggested they could be crushed and used as a soil amendment (Watson 1977, Tilahun et al. 2012). This type of surface enrichment is particularly important to the long-term health of ecosystems that would otherwise have low productivity, e.g., the dry tropics (Coventry et al. 1988). Because of their foraging activities, termites play important roles in the ecosystem, as decomposers. Indeed, Jouquet et al. (2011) argued that ecosystems services provided by termites, particularly their influence on the distribution of natural resources, e.g., water and nutrients, in the landscape, and consequently their influence on the diversity of soil microbes, are not sufficiently appreciated.

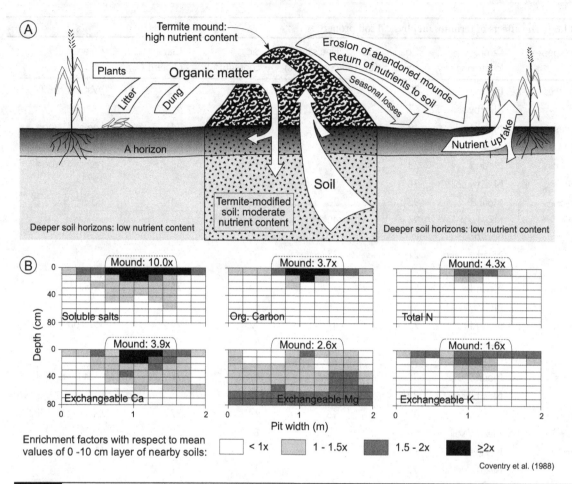

Fig. 11.8 Effects of termite mounds on soil fertility below and near the mound. (A) Generalized flow patterns of nutrients through termite mounds. Similar flow paths could be fashioned for ant mounds. (B) Distribution patterns and enrichment factors for various soil chemical components within and beneath termite (*Amitermes vitiosus*) mounds versus nearby soils.

The effects of faunalturbation on the formation and maintenance of soil porosity cannot be overestimated, as the burrows (galleries) provide rapid conduits for infiltrating water (Green and Askew 1965, Stone 1993). Soils bioturbated in this manner can "wick up" water from deep below (Watson 1962), and the burrows provide important entryways for gases. Infauna also dramatically influence the development of soil structure, from the granular structure formed by worm casts to the subangular blocky structure of ant hills to the cylindrical peds formed as cicada krotovinas are infilled (Buntley and Papendick 1960, Hugie and Passey 1963, Wiken *et al.* 1976).

Landscapes: Worm Bioturbation of Artifacts on the Hillslopes of Western Illinois

For most of his life Charles Darwin was fascinated by earthworms, and his last book (1881) was devoted to synthesizing his lifelong study of them. Darwin had become well acquainted with the fact that continued bioturbation by earthworms resulted in the burial of objects larger than the worms were capable of moving (2.1 mm as measured by Darwin). Included among these objects were ancient artifacts and architectural features, such as coins, paving stones, and the remains of stone foundations. The process Darwin described involved earthworms' moving fine sediment to the surface, facilitating the slow, downward settling of objects that were originally deposited. Given sufficient time, a fine-grained, humus-rich "vegetable mould" (A horizon) would form, over a subsurface layer of stones a few decimeters in thickness. For Darwin, the known age of the archaeological materials provided important insights into the rates at which this process might operate over geological timescales. At first, the processes were relatively rapid in shallow portions of the soil where the worms were most active. But they were much slower for objects being buried to greater depths, eventually reaching a maximal depth to which the worms would burrow.

For years, clastic objects, e.g., pots, fragments, and projectile points, found at archaeological sites and buried in this manner have been overlooked or undervalued. One archaeological landscape where this process has been important is the loess-mantled uplands of the midwestern United States, where for generations archaeologists have relied almost exclusively on surface reconnaissance surveys to locate sites. Recent research shows that burial of artifacts to subplow zone depths in biomantles has occurred even on relatively young (Late Pleistocene) surfaces. Without a full appreciation of bioturbation processes and a theoretical framework within which to study them (Johnson 1993a, 2002), archaeologists struggled to explain why so many Holocene artifacts were buried on these upland surfaces. In a case study from western Illinois, Van Nest (2002) showed how Archaic Period (>2,500 years), but not younger, artifacts had become buried in stone zones 30–40 cm beneath the surface. In these prairie-influenced soils, burial was accomplished mainly by earthworm faunalturbation.

Van Nest (2002) observed that archaeological sites with artifacts at the surface often occur in ribbonlike patterns on shoulder and upper backslopes along headwater valleys. Where present, artifacts found upslope or downslope from these sites are usually buried. Using biomantle-stone-zone theory, Van Nest easily explained this type of distribution. On hilltops, soil loss by creep was effectively nil, allowing artifacts to become buried beneath a worm-worked biomantle, ultimately settling to the depth of bioturbation. On steep backslopes, artifacts were commonly found near the surface and within the plow zone. Here, the rate of soil creep exceeds the rate of burial by soil fauna; as the artifacts are buried, erosion exposes them and they remain near the surface. Artifacts at these sites are readily found by survey archaeology teams. At the bases of the slopes, artifacts may be buried by sediment washed from upslope, by alluvium, or by biomantle materials. Thus, they are more difficult to locate.

These findings counter the once-prevalent assumption that upper slopes, where cultural remains occur at the surface, are eroded and degraded. Upland archaeology can begin anew, with this new conceptual framework.

Termites

In this section we focus on the role of *termites* in the soil systems of the humid tropics, recognizing that they are also important in other warm climates.

Landscapes: The Mima Mound Mystery – Solved!

Mima mounds, named for the Mima Prairie in Washington State, are dome-shaped earth mounds, often >2 m high and 10–50 m in diameter, that are widespread on many grassland landscapes (Cox 1984, Cox and Roig 1986, Lovegrove and Siegfried 1986; see figure). Known by many names, e.g., pimple mounds or prairie mounds, they sometimes cover vast acreages at densities exceeding 100 ha^{-1}. The mounds occur only on soils where bedrock or a subsurface pan, such as a duripan, is near the surface or in soils that have high water tables (Cox and Scheffer 1991). The origin of these features has long been controversial (Scheffer 1947, Horwath Burnam and Johnson 2012). One hypothesis had them as erosional features – relict landforms left behind after wind and water eroded areas between (Waters and

Flagler 1929, Melton 1935, Ritchie 1953). Other evidence suggested that the mounds formed as vegetation (shrubs or clumps of grass) trapped eolian or fluvial sediment, making them purely constructional features that date back to a past time when wind erosion and sediment transport were more active (Barnes 1879). A third hypothesis gave them a paleoperiglacial origin, like Arctic stone circles, in large part because areas between the mounds were often bare of vegetation and strewn with large rocks (Masson 1949, Newcomb 1952, Malde 1964). Berg (1990) proposed a seismic hypothesis.

The fossorial rodent hypothesis of Mima mound origin, widely accepted today, was once ridiculed. To quote Ritchie (1953: 41), "This novel idea left the geologist in a position of arguing on what a gopher might or might not do (to form such mounds)." Its main proponent is George Cox, a biologist at San Diego State University (Cox and Gakahu 1986, Cox and Allen 1987a, Cox 1990, Cox and Hunt 1990, Cox and Scheffer 1991). This hypothesis postulates that Mima mounds are formed as pocket gophers or similar burrowing rodents, e.g., moles or tuco-tucos, translocate soil material centripetally from their subsurface homes (Price 1949, Horwath and Johnson 2006). As the gophers tunnel outward they push soil material behind them, building up a mound. The mounds serve as nest chambers in the thin or wet soils, providing the increased soil thickness necessary to protect them from predation, winter cold, or high water tables (Cox and Scheffer 1991). Because these animals are so territorial, their mounds come to be located almost in perfect, regular arrangements on the landscape. Where soils are thick, gophers are not restricted in siting their nests and thus move from place to place, and mounds are not formed. The fossorial rodent hypothesis is strongly supported by association (wherever mounds are found, rodents are always present in or near the mound fields) and by various types of data from mound soils. Between the mounds is commonly a zone where soil is thin and rocks are numerous, largely because the rodents have removed the soil from these areas (Cox and Allen 1987b). The mounds contain small stones only (Cox and Gakahu 1986), and stone lines containing stones too large for gophers to move (> ~6 cm diameter) commonly underlie and ring the mounds (Ritchie 1953). Thus, the mounds are essentially overthickened biomantles formed by point-centered burrowing (Horwath and Johnson 2006). Small stones are not concentrated on mound tops, as would occur if the mounds were erosional. The limited amount of soil available to form mounds explains why larger mounds are generally located farther from their nearest neighbors than are smaller mounds (Cox and Gakahu 1986). Finally, many mounds continue to be occupied by gophers (Cox and Gakahu 1986), and those that lack gophers contain krotovinas and collapsed nest chambers.

Mima mounds provide a unique look into the world of biogeomorphology – the study of landscapes that result from the interaction of abiotic and biotic processes (Viles 1988). For a more thorough examination of these fascinating features, we urge the reader to consult Horwath Burnham and Johnson (2012).

Figure: Mima mounds. (A) Map of Mima mound terrain in North America. (B1) The Mima Prairie landscape, Washington State, United States. (B2) Interior of a Mima mound, showing the dark, organically enriched biomantle above a stone line. Photos by D. Johnson.

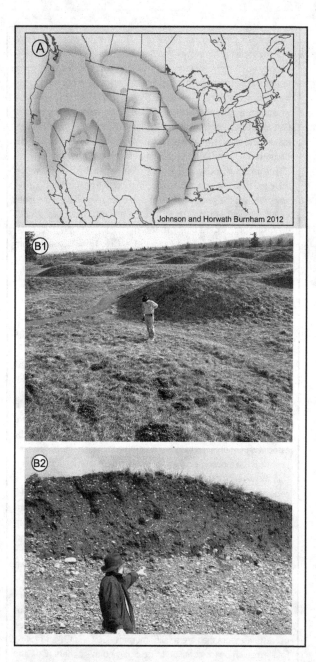

Johnson and Horwath Burnham 2012

exceed 10 m in height (Pullan 1979; Fig. 11.9). By moving up only sand-size and smaller-size fractions, sometimes from depths of 20 m or more (Eggleton and Taylor 2008), termites create a stone-free mound that will eventually become a texturally distinct, stone-free biomantle of less than 1 m to more than 10 m in thickness (Pomeroy 1976, Holt *et al.* 1980, Breuning-Madsen *et al.* 2004). Many tropical landscapes with high densities of termites have stone lines at depth, suggesting that the stone lines, and most of the stone-free biomantles above them, are due to these arthropods (Nye 1955c, Johnson 1990, Paton *et al.* 1995; Figs. 11.10, 11.11).

Termite nests and mounds consist mainly of soil and wood particles cemented together with feces and saliva (Wood *et al.* 1983, Martius 1994). A typical mound is assembled from sand-sized pellets of clay and has a hard, compact, clay-rich casing that is resistant to erosion (Nye 1955c; Fig. 11.10A). Evaporation of Ca-rich soil water from the mound can lead to carbonate precipitation in the casing, further hardening it (Fig. 11.12). Inside the casing is the nest framework, which is essentially a mass of tunnels, nurseries, and the royal chamber – all composed of fine materials (Matsumoto 1976). Nye (1955c) suggested that the nest is composed mainly of material that has passed through the bodies of the termites, necessitating that there be no particles in it larger than fine sand.

Termites consume large amounts of plant material, residues of which end up in mounds as humified organic matter (Pomeroy 1976, Wood and Sands 1978, Brener and Silva 1995). However, even when termite populations are high, they still consume less than 3% of the dead plant biomass (Matsumoto 1976). Much of the biomass they ingest and redeposit in the mound is so thoroughly consumed and reconsumed by cannibalism and consumption of feces, however, that often little is left behind to enrich the soil, as soil organic matter (Fig. 11.13). In fact, one of the reasons that many tropical soils have so little organic matter is a result of efficient consumption by termites. In the nest, many species of termite cultivate fungus gardens or *combs*. Biomass is fed to the fungi by the termites; this is another way that organic materials associated with termite mounds become so well digested. Last, many termites eat their own excreta and dead, keeping the nutrient cycle tight and within-mound (Hesse 1955).

With a few exceptions (Holt *et al.* 1998), soils near termite mounds are notoriously humus-poor and on a longer, slower nutrient cycle than the mounds themselves, which become enriched in nutrients and organic matter (Lee and Wood 1971b, Pomeroy 1976, Salik *et al.* 1983, Liu *et al.* 2002, Ruckamp *et al.* 2012; Table 11.7, Figs. 11.13, 11.14). Because mounds can remain viable and occupied for more than a century (Watson 1972), and most are occupied for 10–20 years (Holt *et al.* 1980), the potential for nutrients to be tied up in them is considerable (Pomeroy 1976). Even when abandoned, it

Most termites are social insects that live in discrete mounds or nests and access remote food resources via covered runways or galleries (Martius 1994). Some species of termites build tall mounds while others create numerous covered runways – covered tunnels on the soil surface – that can extend more than 50 m beyond the central nest (Bagine 1984). These "forage soil sheetings" are a very effective way of transporting stone-poor, fine-textured soil materials to the surface and, unlike mounds, are widely dispersed and fairly ephemeral. Termite mounds can occur in great densities and some

Fig. 11.9 Examples of termite mounds. (A) A mound near Niamey, Niger. Photo by J. Olson. (B) A mound in Ghana that has been mined for geophagical clays. Photo by J. Hunter. (C) A termite mound in Malawi. Photo by L. Zulu.

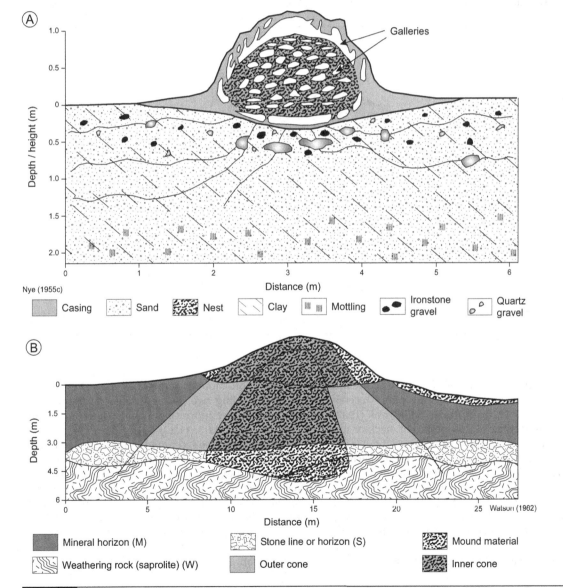

Fig. 11.10 Termite mound structures. (A) Diagram of a typical mound of *Macrotermes nigeriensis*. (B) Diagram of a termite mound, showing its components and relationship to M, S, and W horizons.

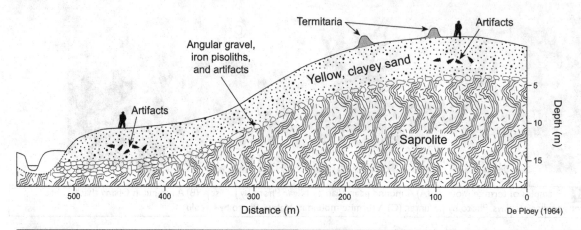

Termitaria   Artifacts

Angular gravel,
iron pisoliths,
and artifacts

Yellow, clayey sand

Artifacts

Saprolite

Distance (m)

De Ploey (1964)

Fig. 11.11 Cross section through a hillslope in Zaire, showing a stone line at the base of a biomantle – both a result of long-term termite bioturbation.

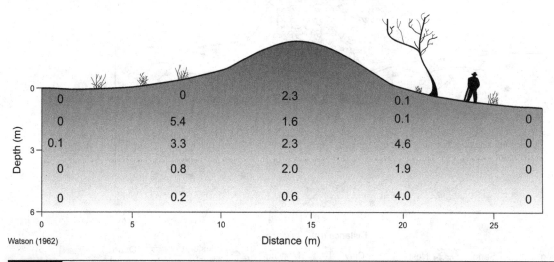

Watson (1962)

Distance (m)

Fig. 11.12 The distribution of carbonates below a termite mound. Units shown are percentages of CaCO$_3$ in the fine earth fraction.

may be a few years to a few decades before the mound disintegrates and the nutrients become plant accessible (Watson and Gay 1970). In Brazil, Salik *et al.* (1983) reported that an average of 165 nests per hectare are abandoned annually in a tropical rain forest where more than 1500 nests occur per hectare, pointing to a perhaps more rapid turnover of soil and nutrients than is typical. In a part of Queensland, Australia, termite mounds contained only about 1% of the A horizon soil mass but 5%–7% of the nutrients (Coventry *et al.* 1988). Coventry *et al.* (1988) calculated "enrichment factors" for various nutrients, where the factors are the relative enrichment over and above nutrients in the upper 10 cm of

unaffected soil. Mounds sometimes showed enrichment factors exceeding two, i.e., nutrient concentrations in the mounds were more than double those in the unaffected soil. (Figs. 11.8, 11.13).

Archaeologists are interested in the effects of termites on soils not only because they can lower artifacts to the depth of burrowing but because they change the pH and alter the aeration status of the soil, affecting bone preservation (Watson 1967, McBrearty 1990). In general, the effects of increased aeration more than offset the higher pH values created within mounds, and thus termites do more to destroy bones than to preserve them (Lee and Wood 1971b).

Table 11.7 | Nutrient contents and other parameters associated with termitaria and nearby lateritic soils in Brazil

Parameter	Termitaria	"Laterite" soil
pH	3.9	3.9
Carbon (%)	22	2.9
Total N (%)	1.5	0.3
Total P (%)	0.08	0.02
Exchangeable K (%)	0.03	0.003
Exchangeable Ca (%)	0.02	0.002
Exchangeable Mg (%)	0.1	0.001
Exchangeable Na (%)	0.004	0.002

Source: Salik *et al.* (1983).

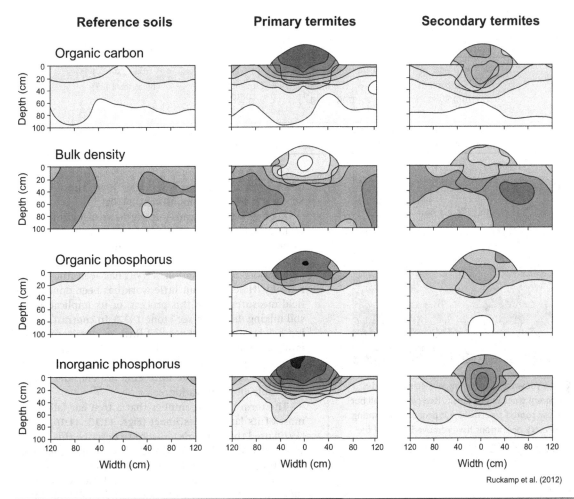

Ruckamp et al. (2012)

Fig. 11.13 Modeled pedogenic characteristics of soils in termite (*Cornitermes silvestrii* and *Nasutitermes kemneri*) mounds in Brazil versus nearby (reference) soils. Darker colors indicate higher values.

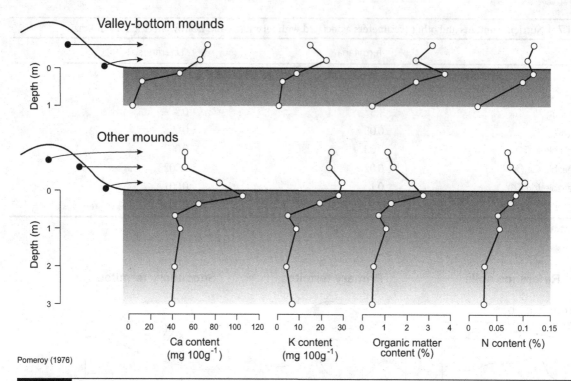

Fig. 11.14 Depth functions of soil constituents within and below termite mounds in Uganda. Data for soils below the mound are actually from soils adjacent to the mound, so as to reflect the native soil prior to mound building.

Fig. 11.15 Blowdown and uprooting of a stand of red pine (*Pinus resinosa*) trees in Wisconsin. Red pine is a tall but relatively shallow rooted tree, making it prone to uprooting. Subsequent forestry operations have removed most of the downed wood from this area. Photo by RJS.

Floralturbation

Floralturbation or *phytoturbation* (Vasenev andTargul'yan 1995) is the mixing of soil by plants, usually via (1) root expansion, (2) decay and infilling of root channels, (3) root movement/shaking caused by wind moving the above-ground plant (Hole 1988), and (4) uprooting, especially of larger plants such as trees (Lutz 1940, Hintikka 1972, Schaetzl *et al.* 1990, Samonil *et al.* 2010a).

Root growth may cause upward expansion of the soil surface. Later, as the roots die and decay, the infilling of the former root channels and stumps allows materials from upper horizons to slump and mix (Roberts 1961, Phillips and Marion 2006). Agitation of roots has been studied theoretically (Hole 1988) but little work has been done on the field measurement of this process, or its implications for soil mixing (however, see Stone 1977). In contrast, uprooting has been the most studied form of floralturbation. It is most common in forested areas (Schaetzl *et al.* 1990). Because it is often manifested as a form of microtopography, the effects of tree uprooting are considerably less noticeable after cultivation.

The term *uprooting* implies that a tree has fallen with most of its larger roots intact (Figs. 11.15, 11.16). Brown (1979) used the term *arboturbation* to describe this process. The terms "treefall," "tree-tip," and "treethrow" have also been used, but these terms may also refer to trees whose trunks are broken near the base, with little or no soil disruption. Uprooting may disrupt considerable amounts of soil to the depth of rooting, and in many areas it is a major cause of downslope movement of regolith (Norman *et al.* 1995, Walther *et al.* 2009). Schenck (1924) referred to the process as "natural plowing" of the soil. By moving otherwise deeply buried rocks and nutrient- or clay-rich subsoil

materials to the surface, uprooting provides a mechanism for soil rejuvenation and production (Roering *et al.* 2010). Indeed, this process is one of the few forms of pedoturbation that readily transport coarse fragments up to the surface (Fig. 11.17). As a result, gravelly biomantles or lag deposits may form on mounds that are created as soil slumps from the root plate (Small *et al.* 1990, Phillips *et al.* 2005, Phillips and Marion 2006; Figs. 11.3, 11.17).

Although the effects of uprooting may persist for millennia (Samonil *et al.* 2013), they are often quite localized. Soil and rocks falling and washing off deteriorating root plates tend to create irregular patches of mixed and discontinuous horizons, often within a mound (Pawluk and Dudas 1982, Vasenev and Targul'yan 1995). Certain pedogenic characteristics, such as the interrupted and cyclic horizon character of many Spodosols and Alfisols, sometimes expressed as E/B and B/E horizons, can be used to infer past floralturbation events. In some cases, as on steep slopes where the root plate is overhanging the vertical, soil horizons may literally fold over each other (Veneman *et al.* 1984, Schaetzl 1986b, Pawlik *et al.* 2013; Fig. 11.18). If intense and frequent, uprooting will regress well-horizonated forest soils such as Spodosols or Alfisols to lesser-developed orders such as Inceptisols or Entisols, or keep weakly developed soils in that state by persistent, proisotropic pedoturbation.

The surface expression of tree uprooting is a landscape with scattered pits (at sites where the roots once were) and adjacent mounds, formed as soil slumps off the roots (Figs. 11.3, 11.17). Many forested (or previously forested) landscapes have pits and mounds at varying densities and of various sizes. These features are often known as *cradle knolls*, because babies could be placed in the pits – a kind of cradle – while loggers worked nearby: hence the commonly used term "cradle-knoll topography" (Lyons and Lyons 1979, Kabrick *et al.* 1997a). Once thought to be rather ephemeral features, lasting only a few centuries at best (Stephens 1956, Denny and Goodlett 1968, Stone 1975 Beatty and Stone 1986), pits and mounds are now known to have longevities exceeding several thousand years, especially in sandy soils where runoff and erosion are minimal

Fig. 11.16 A sequence of uprooted trees whose root plates are in various stages of disintegration. Photos by RJS. (A) An uprooted red pine (*Pinus resinosa*) tree. The event is less than 4 weeks old. Potter County, Pennsylvania. (B) Another very recently uprooted tree. This one fell because it was rooted in a Histosol, which provided very little support against strong winds. Note the white marl deposit that was also torn up by the tree. (C) An uprooted red pine tree, about 10 years old. Note how much soil has fallen from the root plate in this short period. (D) A root plate in an advanced state of disintegration, near L'Anse, Michigan.

Fig. 11.17 Large rocks caught in the root plate of an uprooted tree in southern Michigan. Uprooting is one of the few forms of pedoturbation that can effectively move large clasts upward in the soil profile, depositing them on the soil surface. Photo by RJS.

(Schaetzl and Follmer 1990, Samonil *et al.* 2013). Greater mound/pit longevities imply that rates of tree uprooting (pedoturbation) may be less than previously thought, and that the pedogenic effects of the microtopography will be focused on small areas for longer periods, increasing the spatial variability of soils much more than if these features were short-lived (see Chapter 14). In summary, in forested landscapes, tree uprooting is responsible for mixing soils and creating patterns of microtopography, and these effects may persist for millennia, making the soils of such areas very difficult to map at scales smaller than ~1:3,000 (Alban 1974, Meyers and McSweeney 1995, Kabrick *et al.* 1997b).

Cryoturbation and Gelisols

Many areas of the Earth's surface, mostly in the Arctic, Subarctic, and boreal ecozones and in high mountain areas of the globe, are associated with permafrost and undergo annual or more frequent freeze-thaw cycles (Tedrow 1962, 1977, Bockheim and Tarnocai 2011). Cryoturbation, a subset

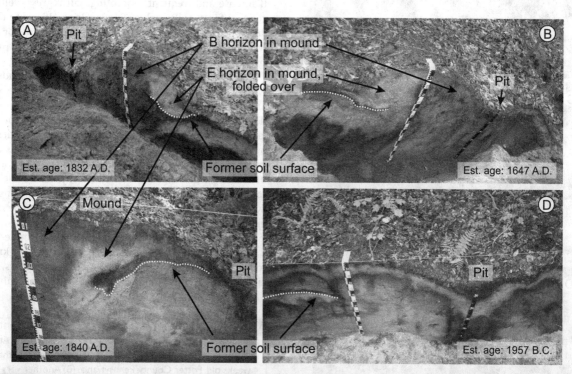

Fig. 11.18 Photos of the interiors of treethrow mounds of various ages from northern Michigan. The native soils here are sandy Spodosols, with bright white E horizons and dark, brown-red B horizons below. Note how, in the young mounds, this horizon sequence is often inverted or folded over, and yet, otherwise largely undisturbed. Buried soils, containing charcoal and wood, both datable by ^{14}C dating, lie beneath the treethrow mounds and are visible in all four photos. The maximum ages of the uprooting events (as shown) were determined using radiocarbon dates on this buried wood and charcoal. Note the amount of additional soil development in the oldest mound, vis-à-vis the other, younger mounds. Increments on the tapes are in centimeters. Photos by P. Samonil.

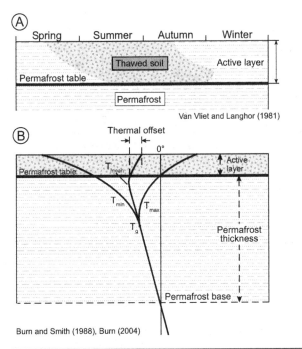

Van Vliet and Langhor (1981)

Burn and Smith (1988), Burn (2004)

Fig. 11.19 Schematic representations of temperatures in soils with permafrost. (A) The annual, spatiotemporal cycle of freezing and thawing in soils with permafrost. The typical depth to the permafrost table varies from 20 to 250 cm. (B) The famed "trumpet diagram" that illustrates temperature variation with depth, throughout the year, in soils with permafrost.

of the larger term for frost action in sediments of all kinds, is an important pedogenic and geomorphic process in such landscapes (Bryan 1946, Washburn 1980a, Bockheim *et al.* 2006). Older terms for cryoturbation include "geliturbation" and "congelliturbation." Bryan (1946) suggested the term *cryopedology* for science that deals with cryogenic processes in soils and near-surface sediments. Processes of vertical translocation, commonplace in mid- and low-latitude soils, are slowed and inhibited by long cold seasons and frozen substrates. Add to this the mixing effects of cryoturbation, and it becomes obvious that freezing is an important pedogenic process, and soils affected by these processes are unique.

In this section, we discuss not only cryoturbation, but the characteristics and pedogenesis of Gelisols, a soil order that is defined on the basis of cold soil temperatures and the processes associated with these types of conditions.

Soil Freezing and Cryoturbation

Water expands about 9% in volume upon freezing; this expansion can create significant amounts of heaving, cleavages, partings, and segregations in soils, especially if repeated frequently (Konrad 1989, Sveistrup *et al.* 2005,

Azmatch *et al.* 2012). In some soils, the expansion caused by freezing is vital – it creates voids that enhance infiltration and increase hydraulic conductivity (Chamberlain and Gow 1979, Schaetzl and Tomczak 2002). The permeability of frozen soils depends on the type of frost present. Four main types of soil frost are generally recognized: (1) concrete, (2) honeycomb, (3) stalactite, and (4) granular (Post and Dreibelbis 1942, Pierce *et al.* 1958). Of these, concrete frost, which forms in wet sands and soils with weak structure grade, is the least permeable. On smaller scales, needle-ice formation may cause significant heaving and mixing in the uppermost O and A horizons and can move large fragments (Pérez 1986, 1987a, 1987b).

Freezing fronts develop in soils just as wetting fronts do – from the top down, as the soil loses heat to the cold atmosphere. If permafrost (permanently frozen ground) exists at depth, ice and subfreezing conditions may also grow upward from it, while the freezing front from the surface grows downward (Fig. 11.19).

In dry or coarse-textured soils, little disruption or mixing is due to freezing. However, freezing in wet or moist soils can cause *ice segregations* to form, break up the soil, and cause mixing – both when they form (by expansion) and when they melt (by collapse). The ice segregation process is so important that the terms *ice lens* and *ice segregation* not only are descriptive, but also imply a genesis, i.e., ice lens formation can usually (but not always) be attributed to segregation processes. If enough water moves to the freezing front, large, relatively pure ice segregations will develop and grow, starting out as crystals and growing into lenses, vein ice, and ice wedges (Radke and Berry 1998). These features are so common in cold climate soils that the Agriculture Canada Expert Committee on Soil Survey (1987) has designated a major type of soil horizon (Wz horizons) for thick ice segregation features. W may be used for not only ice lenses/segregation features, but also ice wedges, intrusive ice, buried glacier ice, or any other type of ground ice not resulting from ice segregation. Both the U.S. Soil Survey Staff (1999) and the Agriculture Canada Expert Committee on Soil Survey (1987) define some types of ground ice features as *glacic layers*.

Ice segregations form mainly in moist and fine-textured soils, as water in vapor and liquid forms moves from unfrozen soil nearby, toward the freezing front, and freezes there as ice (Mackay 1983, Smith 1985). This type of water migration to the freezing front is negligible in sandy and gravelly soils. The movement of water toward the freezing front is regulated by two forces: (1) suction gradient (controlled strongly by the unfrozen water content, which is higher in clays and lower in sandy soils), and (2) hydraulic conductivity (controlled by the tortuosity and connectivity of unfrozen water films). Bronfenbrener and Bronfenbrener (2010) refer to the former mechanism as the *cryostatic suction effect*. Clays have the strongest potential gradients, but their conductivity is quite limited. Silty

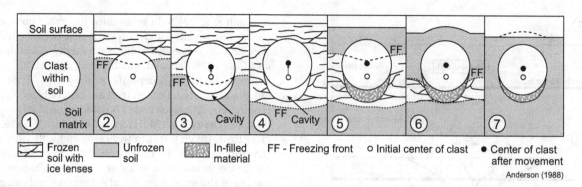

Anderson (1988)

Fig. 11.20 The frost-pull process, which results in upfreezing of clasts within soils. In panels 1–4, ice lenses in the soil cause frost heave and expansion of the soil. By panel 3, the clast is frozen into the overlying soil and pulled upward as the soil is heaved, forming a cavity below the clast. Settling of the soil around the clast into this cavity partially or completely infills it, in panels 5–7. After the soil thaws, the clast settles slightly but cannot return to its original position because of the infilled material below, resulting in a net upward displacement.

soils are generally the greatest transmitters of water to the freezing front, and hence are often the most prone to cryoturbation.

It is important to get the terminology correct. Segregated ice refers to ice resulting from the ice segregation process and, as such, describes not only the form of the ice but also the genesis (in the case of segregated ice, by cryosuction). A preferred term to describe a discrete ice body in the soil might be *ground ice*, or *massive ice*, in the case of large bodies of ice, e.g., ice wedges, buried glacier ice, pingo ice, or very large ice lenses.

Soil and other debris are pushed aside as ice features expand in soils, heaving the soil surface upward and laterally and adding stress and pressure to subsoil layers (Van Vliet-Lanoë 1988, Konrad 1989, Chamberlain *et al.* 1990). Over time, these ice segregation features may grow into areally extensive ice lenses, usually but not entirely parallel to the soil surface, but typically oriented perpendicularly to the direction of heat flow. They are more prevalent in wet soils than in dry soils because of the greater water supply and, at least for segregation features, more common in fine-grained soils than sandy soils.

A second type of freezing-related displacement occurs in clay-rich soils, below the freezing front. Here, as a result of freezing and migration of water to the freezing front above, progressive desiccation develops in the unfrozen soil below. Large negative pore water pressures develop and can form vertical *shrinkage cracks* (Chamberlain *et al.* 1990) that create additional pore space within the soil. Soils may even collapse in on themselves as a result of this shrinkage.

Cryoturbation can intensively mix soils (see later discussion). But it can also sort them. Intense frost action in soils can effectively sort coarse fragments, which are often pushed or pulled to the surface and concentrated in various patterns (Corte 1963, Inglis 1965). This process is termed *upfreezing* (Vilborg 1955, Anderson 1988, Gozdzik and French 2004) or *frost jacking*. Archaeologists are aware that upfreezing can displace artifacts and destroy stratigraphy (Wood and Johnson 1978, Mackay and Burrous 1979). The process begins as the soil develops a freezing front that advances from the surface downward (Fig. 11.20). Because the thermal conductivity of rocks in the soil is greater than that of the soil matrix, the freezing front advances more rapidly through the rocks. The frozen soil expands (heaves) as it develops ice lenses. At some point, each rock is frozen to the frozen soil above it, and as that soil heaves upward, the rock is *pulled* upward, because the forces holding it to the unfrozen soil below are minimal (Inglis 1965). (Early, competing hypotheses that involve ice lenses forming below rocks and pushing them upward have since been rejected [Taber 1943].) As one might envision, upfreezing is most efficient when the rate of frost penetration is slow.

During thaw, the upfrozen rock remains anchored in the frozen soil above, while thawed soil around it settles. Some of the unconsolidated soil slumps and washes into the cavities below it. The partially infilled cavities prevent the rock from returning to its original location, resulting in a small increment of upward movement. Factors that maximize ice lens growth, such as high moisture contents or slow advance of the freezing front, favor upfreezing. Therefore, upfreezing is potentially greatest in fine-textured soils because they hold more water than do coarse-textured soils, and because water has greater mobility in fine pores, when it is near freezing (Anderson 1988).

When the ground surface is not even or level, freezing fronts penetrate the soil at angles, pulling clasts upward at acute angles to the surface (Vilborg 1955). Movement of clasts in a nonvertical dimension by upfreezing can actually result in larger net displacements, because the clast cannot return as easily to its original position.

Upfreezing should not be confused with frost heave, which is simply the expansion of wet soil during freezing. Heave is driven largely by ice lens growth (see later discussion), providing that temperature and moisture conditions allow. When these lenses melt, the large pores/cavities that remain allow for the translocation, in suspension, of silt and clay. Silt is not normally mobile in soils because pore sizes are not large enough. But in circumstances like these, silt particles can wash or flow downward. *Pervection*, the name given to the translocation of silt grains in this manner (Table 13.1), is only common in soils that undergo frequent freezing and thawing, and it is typically manifested as silt caps on stones and as silt-enriched horizons (Fedorova and Yarilova 1972, Frenot *et al.* 1995). Silt caps form as silt-rich, percolating meltwater replaces ice sheaths around stones by infilling the voids that appear above them as they settle into cavities left after melting (Payton 1993a).

Permafrost

Permafrost is rock or soil material that remains at or below freezing for at least two consecutive years; it is defined by the temperature, not composition of the material. The top of the permafrost zone is called the permafrost table (Fig. 11.19). Permafrost can extend to many hundreds of meters, although most areas with permafrost have a surface zone, i.e., an *active layer* that thaws seasonally (Miller *et al.* 1998). According to Bockheim and Tarnocai (2000), permafrost may underlie more than one-fourth of the Earth's land surface. Near the margins of continuous permafrost its distribution becomes discontinuous or sporadic, usually being found only within organic soils with high insulating abilities (Zoltai and Tarnocai 1975).

Some permafrost is pure ice; other types may include layered ice and clastic or humic materials. Dry permafrost contains insufficient moisture to be cemented, and it is found at only the very driest sites, e.g., Antarctica (Ferrians *et al.* 1969, Mckay 2009). Cryoturbation processes are minimal in soils with dry permafrost but become dramatic in wetter and finer-textured soils, where frost boils and intrusions of varying kinds cause mixing of ice and soil materials (Bockheim 1980a, 2007; see later discussion).

The active layer – the compartment within which pedogenesis and cryoturbation occur – is strongly influenced by the permafrost below (Miller *et al.* 1998). Water perches on the permafrost and the soils above can become extremely wet, placing the soil on a much different pedogenic pathway, generally with more pronounced mixing. Soils within the active layer are often within these perched water tables and, thus, exhibit wet conditions during the warm season. Only on sloping surfaces and in the driest polar deserts is the active layer not saturated for much of the warm season.

In polar regions, the thickness of the active layer generally decreases poleward, and on colder (poleward-facing) aspects. On a more local scale, the thickness of the active layer is a function of the nature of the substrate, insulating

Fig. 11.21 An excellent example of patterned ground on the Mackenzie River delta, Northwest Territories, Canada. Low-centered polygons dominate the foreground, while high-centered polygons are in the distance. All of the cracks between the polygons are associated with ice wedges, and in the summer, they fill with water, as shown here. Photo by C. Tarnocai.

cover (e.g., a histic epipedon), vegetative cover, relief, aspect, or other factors. Well-drained sites tend to thaw to greater depths than do wetter sites. Sites that have patches late in the season may thaw only a few centimeters into the mineral soil before the next cold season sets in. Likewise, many organic, permafrost-affected soils never thaw into the mineral soil below because of the insulating effects of the peat.

Gelisols occur on geomorphic surfaces that are, despite their low slope, inherently unstable. On sloping surfaces, the active layer will commonly "flow" slowly downslope (Benedict 1976). This process is called *gelifluction* or *solifluction*. Usually involving saturated masses (lobes) of soil whose cohesive forces have been weakened by repeated freeze-thaw activity, solifluction can occur on slopes of even very gentle inclination, at rates up to 1 m yr^{-1}. Lobes, stripes, terraces, and steps are common, as the sediment in the active layer continues, albeit episodically, to move downslope (Benedict 1976, French 1993). Along with cryoturbation, solifluction is an important mechanism by which organic matter becomes incorporated into soils. Organic matter on the soil surface is buried by slowly oozing, geliflucting materials that override soil surfaces on lower slopes (Smith *et al.* 2009).

Patterned Ground

A very common morphological outcome of long-term cryoturbation is *patterned ground* – often taking the form of sorted and unsorted circles, nets, steps, stone stripes, and polygons (Washburn 1956, 1980b, Black 1969, Fitzpatrick 1975, Tedrow 1977, Thorn 1976, Van Vliet-Lanoë 1985, 1998, Ugolini *et al.* 2006; Figs. 11.21, 11.22). The forms of patterned ground are largely due to a pedoturbative process. In fact, the reason that patterned ground is so common

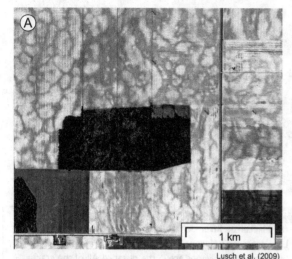

1 km

Lusch et al. (2009)

Dry suture area in forest

Water-filled suture area

Sutures intersecting a roadside ditch

and persistent on polar landscapes is that cryoturbation is omnipresent, while other forms of pedoturbation *are not*. If faunalturbators were common in this landscape, patterned ground would be less dramatic and in some areas, absent (Smith 1956).

Patterned ground results from repeated freezing and thawing in the active layer. The details of the type and formation of patterned ground are varied and multifaceted (Washburn 1956, 1980b, Goldthwait 1976, Hallet and Prestrud 1986, Van Vliet-Lanoë 1985, 1988, 1998). But, in order to understand the formation of patterned ground, one must first understand *ice wedge* formation. Frost or ice wedges are V-shaped bodies of ground ice, usually <4 m deep and ~2 m wide, that typically form in areas of continuous permafrost, where mean annual air temperatures are < 6°C (Péwé 1966, Romanovskij 1973; Fig. 11.23). Ice wedges begin forming in winter when the soil freezes at the surface, forming fractures and cracks by thermal contraction. The cracks form in a somewhat random, although generally polygonal, pattern (Figs. 11.21, 11.22, 11.24). Indeed, patterned ground is often formed – or at least starts forming – by intersecting frost wedges (Fig. 11.24B). Repeated cracking in the same location causes the frost wedge to grow in size as soil water, in vapor and liquid form, migrates to the site and freezes, and as meltwater seeps into an open crack. Later, as this water freezes, the wedge expands and grows. Upturned, displaced, and distorted soil and sediment layers are telltale indicators of this process, even after the ice is gone (Figs. 11.25, 11.26).

Some wedges, called sand wedges or ground wedges, do not fill with ice; the open cracks fill with blowing sand and snow. Sand wedges form in cold, dry landscapes with strong winds, minimal snow cover, and an abundant supply of sand. They eventually take on the same form as ice wedges (Fig. 11.25), but because they are infilled with sandy material and not ice per se, they exhibit minimal slumping when the climate warms and the permafrost thaws completely (Mackay and Matthews 1983, Wayne 1991).

Fig. 11.22 Evidence for Late Pleistocene permafrost and patterned ground on the Glacial Lake Saginaw Plain, southern Michigan. (A) Aerial photograph of the landscape today, showing the darker streaks associated with the wet, organic matter–rich soils of the polygon edges. (B) On-ground photos showing these same "suture" areas. They exist as linear depressions at former polygon edges, which presumably were formerly occupied by ice wedges. These interpolygon suture areas sometimes retain surface water.

Fig. 11.23 Photographs of active, well-formed ice wedges. (A) A large ice wedge, showing the typical vertical foliations, indicative of the lateral growth of the ice through time. Note shovel for scale. Photo by J. Munroe. (B) and (C) Ice wedges near the Beaufort Sea coast of Alaska. Photos by M. Kanevskiy. (D) Ice wedges in silty Late Pleistocene deposits exposed along a 35 m high bluff of the Itkillik River, Alaska. These wedges are up to 7 m wide and more than 35 m tall. Photo by M. Kanevskiy.

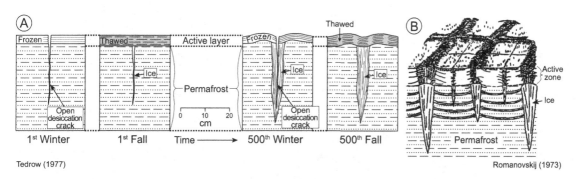

Fig. 11.24 Frost wedge morphology and genesis. (A) Idealized diagram showing the formation of a frost wedge over time. (B) Frost wedges and patterned ground, in three dimensions.

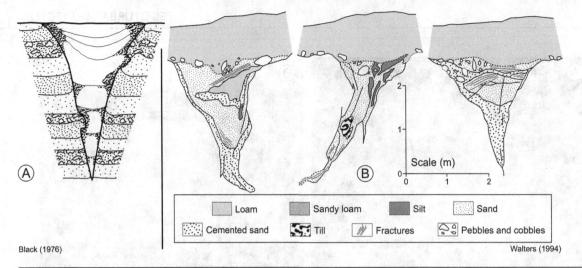

Loam   Sandy loam   Silt   Sand

Cemented sand   Till   Fractures   Pebbles and cobbles

Scale (m)

Fig. 11.25 Drawings of ice wedge casts. (A) Idealized ice wedge cast, showing upturned strata at its edges, and slumpage within the fossil cast. (B) Cross-sectional drawings of ice wedge casts from the Iowan erosion surface, northeast Iowa.

Fig. 11.26 Typical cross sections through ice wedge casts. (A) Photograph and diagram of an ice wedge cast in glacial deposits in Chippewa County, Wisconsin. Note the upturned sediment beds near the margins of the cast, as are typical. (B) Photographs of ice wedge casts in glacial outwash and sandy, stratified glacial sediment associated with the Beaupré Moraine, near Quebec City, Quebec, Canada. Photos by J. Munroe.

Landscapes: The Boreal Forest of Interior Alaska

Near the boreal forest–tundra ecotone in central Alaska, permafrost is an important component of the soil landscape (Swanson 1996). Here, as in many other sites where the landscape is not a uniform plain, meso- and microtopography have a highly important influence on pedogenesis, largely because they affect snow patch thickness and soil wetness, which in turn affect soil temperature and rates of cryoturbation (Mueller *et al.* 1999).

On a typical slope, gravelly bedrock knobs have the driest soils, usually Haplorthels (simple Gelisols), with Oe-Bw1-Bw2–2Bc-2Cr horizonation, beneath spruce-birch forest. Here, the permafrost table is deep and cryoturbation is minimally expressed. The wettest, coldest sites are within microtopographic lows; Hemistels (Gelisols with thick, peaty surface horizons) occur here. They typically have permafrost in their lower sola.

Feedback mechanisms create linkages between cold, wet sites with thick O horizons and the shallow active layer that underlies them. Summer thaws penetrate less deeply in such soils because – first – of the high heat capacity of water. Once frozen in winter, these soils give off more heat than do dry soils, because ice has high thermal conductivity values. This leads to colder soil temperatures and thinner active layers in the ensuing summer, both of which enhance organic matter accumulation, setting up a positive feedback loop. The thinner active layer then causes water to perch, perpetuating the hydric conditions.

Also pertinent to this discussion is the importance of latent heat. The latent heat required for thawing of wet organic matter is orders of magnitude larger than the thermal capacity of wet organic matter. Thus, it is not the heat required for warming the wet O horizon that is the greatest limit to thaw penetration, but rather the heat required for thawing the ice in it. The low thermal conductivity of the wet peat slows thawing in the summer, setting up the same feedback loop.

Alternatively, dry sites on bedrock knobs warm quickly in summer, facilitating the mineralization of any organic matter that accumulates there. Dry conditions inhibit peat formation, allowing heat to penetrate deeply into the soils, driving the permafrost table deeper. These soils have brown, oxidized Bw horizons and less evidence of cryoturbation. Thus, although many site factors influence depth to permafrost, subtle microtopographic variation can be enough to cause major differences in soil properties across very short distances.

Figure: Soil-landscape diagram for the area near Hughes, central Alaska. Soil horizonation is vertically exaggerated, relative to topography.

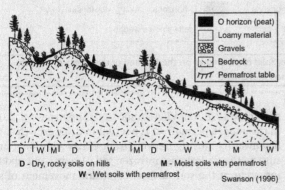

O horizon (peat)
Loamy material
Gravels
Bedrock
Permafrost table

D - Dry, rocky soils on hills   M - Moist soils with permafrost
W - Wet soils with permafrost

Swanson (1996)

When temperatures are no longer cold enough, ice wedges become fossilized and inactive. They are then variously called frost wedges, fossil ice wedges, ice wedge casts, or relict ice wedges (Billings and Peterson 1980; Fig. 11.26). Frost wedges undergo morphological changes when they thaw, making their unequivocal identification somewhat subjective (Harry and Gozdzik 1988). Like patterned ground, evidence from frost wedges is, nonetheless, variously retained after the cold conditions have ended and can be used to infer a period of permafrost (Sharp 1942, Smith 1942, 1962, Fosberg 1965, Péwé 1966, Kozarski 1974, Walters 1994, Mader and Ciolkosz 1997). Ice wedge casts fill with sediment from nearby and overlying materials and usually retain an irregular wedge shape (Péwé 1966, W. Johnson 1990). The sediment in the cast typically shows synclinally curved stratification and normal faulting, which form as the surrounding sediment slumps down when the ice in the wedge melts. (Thaw is usually from the top down.) Sediment near the cast then becomes even more distorted. The end structure is usually wider than the original ice wedge (Black 1976). The sediment *within* ice wedge casts may not be distorted, however, since there would have been little ice to melt and create additional volume.

Both upfreezing and outfreezing from polygon centers help drive the formation of patterned ground. Often, evidence for permafrost – paleo- or contemporary – is manifested as patterned ground. It often occurs as networks of *linear or curvilinear depressions* or *shallow ridges* separated by troughs (Goldthwait 1976, Harry and Gozdzik 1988, Lusch *et al.* 2009). Troughs/cracks between polygons are wet sites that accumulate organic materials, leading to a polygonal pattern of soils with overthickened O and A horizons (Ugolini 1966, Lev and King 1999), while the intrapolygon soils are higher and better drained (Figs. 11.21, 11.22).

Models of Cryoturbation

At least five general models describe how soils become cryoturbated (Vandenberghe 1992, Phillips 2011; Fig. 11.27). In the *cryohydrostatic model* (Shilts 1978, Coutard and Mücher 1985, Van Vliet- Lanoë 1991), increasing amounts of water are trapped between an underlying, impermeable layer (usually permafrost) and the freezing front, creating a hydraulically closed system (Fig. 11.27A). As the freezing front penetrates deeper, pore water pressures in the unfrozen pocket progressively increase. Eventually, *liquefaction* (behaving like a liquid) occurs in the unfrozen pocket. As a result, sediment is squeezed and eventually expelled, under pressure, into weak areas in the surrounding frozen soil. Frost or desiccation cracks, often existing in polygonal patterns, may act as zones of obvious weakness in the overlying frozen soil. The ensuing cryohydrostatic movement occurs in a fluid state. In soils, characteristic forms expected from cryohydrostatic pressure are dikelike, irregularly shaped extrusion features oriented in all directions,

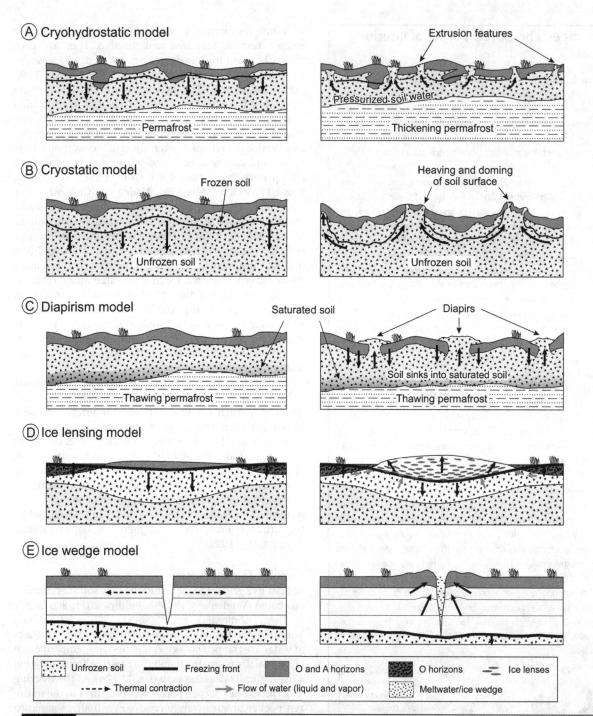

(A) Cryohydrostatic model

Extrusion features

Pressurized soil water

Permafrost

Thickening permafrost

(B) Cryostatic model

Frozen soil

Heaving and doming of soil surface

Unfrozen soil

Unfrozen soil

(C) Diapirism model

Saturated soil

Diapirs

Thawing permafrost

Soil sinks into saturated soil

Thawing permafrost

(D) Ice lensing model

(E) Ice wedge model

| Unfrozen soil | —— Freezing front | O and A horizons | O horizons | ⫶ Ice lenses |
| ----> Thermal contraction | ——> Flow of water (liquid and vapor) | Meltwater/ice wedge |

Fig. 11.27 Diagrams illustrating five general models of cryoturbation. Solid black arrows on the left side of the diagrams indicate freezing front advancement, whereas those on the right side indicate soil movement.

but mainly upward. Only the lower, unfrozen sediments are deformed, while the upper, overlying and frozen soil is not deformed significantly.

In the *cryostatic model* (Fig. 11.27B), unfrozen soil is trapped below a frozen layer at the soil surface (Tarnocai

and Zoltai 1978, Van Vliet-Lanoë 1991, Superson *et al.* 2010). In this model, permafrost is not required. Ice growth at the downwardly moving freezing front exerts additional pressure on the underlying unfrozen soil. Heaving and updoming occur at the soil surface. Cryostatic movement of soil material occurs in a plastic state, unlike in the cryohydro-static model. Vandenberghe (1992) described the upheaval as occurring in celllike patterns, leading to a hummocky surface.

In the *diapirism* (or *load casting*) *model* (Vandenberghe and Van den Broek 1982, Murton and French 1993, Swanson *et al.* 1999, Harris *et al.* 2000, Bertran *et al.* 2009), soil above the permafrost table flows upward by viscous flow – like diapirs – as denser overburden sinks into it (Fig. 11.27C). For this model to function, low-density soil must exist near the permafrost table, overlain by denser soil. Swanson *et al.* (1999) suggested that this type of density profile may develop as water is released into the active layer from thaw-ing of ice-rich upper permafrost. As it thaws, large amounts of water move upward from the permafrost, into the lower part of the active layer, which rapidly becomes saturated. If the water volume exceeds the pore volume, the density of the lower active layer is effectively lowered relative to that of the overburden, allowing the denser overburden to sink into the less dense soil below, as droplike structures, while saturated soil from the lower active layer moves up and into the overlying soil, as diapirs (Harris *et al.* 2000, Swanson *et al.* 1999, Superson *et al.* 2010). Upward diapir movement will cease if the diapir encounters another layer of low den-sity, such as an organic layer, at or near the soil surface. The result of the movement is a celllike pattern with polygonal form (Vandenberghe 1992). Mushroom- or pillar-shaped forms of rising lower sediment alternate with pockets of downsinking upper sediment (Dzulinski 1966). Some soils have alternating downsunk pockets and updomed diapirs. This form of cryoturbation is the most common in loose Pleistocene deposits (Vandenberghe 1992). In cases when diapirs breach the soil surface, they destabilize the vegeta-tive cover and alter the thermal regime of the soil. This can potentially initiate a positive feedback loop, where reduced insulation from vegetation and organic matter causes deeper thawing and increased release of water from the ice-rich upper permafrost (Swanson *et al.* 1999).

In the *ice-lensing model*, the freezing front penetrates into the soil, but more quickly and deeply where the O horizon is thin or absent (Fig. 11.27D). Water migrates from organic-rich areas to organic-poor areas, as a result of lateral suction gradients caused by the convoluted freezing front. As a result, ice lensing is more pronounced in the organic-poor soil, heaving the soil surface more than the adjacent organic-rich soil and creating a mound. Additional movement and slumping occur when ice lenses thaw.

Last, in the standard *ice wedge model* (Fig. 11.27E), the frozen soil drives thermal contraction processes. Cracks open from the surface downward, to release the tension. Meltwater fills the cracks and freezes. Although relax-ation occurs as the temperature occasionally increases, the expansion is constrained by the ice in the wedge, and hence, the soil must deform to release the excess pressure. The effect is cumulative over many cycles of contraction,

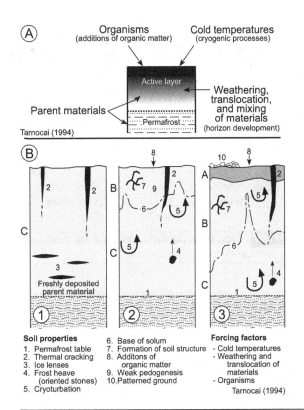

Soil properties
1. Permafrost table
2. Thermal cracking
3. Ice lenses
4. Frost heave (oriented stones)
5. Cryoturbation

6. Base of solum
7. Formation of soil structure
8. Additons of organic matter
9. Weak pedogenesis
10. Patterned ground

Forcing factors
- Cold temperatures
- Weathering and translocation of materials
- Organisms

Tarnocai (1994)

Fig. 11.28 General factors and processes involved in the formation of Gelisols. (A) Factors and processes important to the genesis of Gelisols. (B) Stages (1, 2, and 3) in the development of a Gelisol (Turbel).

cracking, and wedge growth, leading to ridges and troughs over ice wedges and intense cryoturbation adjacent to them (Figs. 11.24, 11.28).

Pedogenesis Dominated by Cryoturbation

Permafrost, cryoturbation, and the pedogenic influence of ice lenses and frozen sediment are concepts central to the Gelisol order of the U.S. Soil Taxonomy. Gelisols are (broadly) defined as soils that have permafrost within a meter of the surface (Soil Survey Staff 2010). The Soil Survey Staff (1999) defines materials that show evidence of cryo-turbation and/or ice segregation in the active layer and/or the upper part of the permafrost as *gelic materials*. Canadian soil scientists have long recognized the importance of cryo-turbation as a pedogenic process. In their system, soils sim-ilar to Gelisols are classified as Cryosols (Soil Classification Working Group 1998, Tarnocai and Bockheim 2011). Soils can be classified within either of these orders even if the permafrost is slightly deeper, as long as they also have evi-dence of cryoturbation.

Pedogenesis in Gelisols is set within a context of very cold temperatures, commonly with mean annual soil tem-peratures below 0°C. Pedogenesis is normally restricted

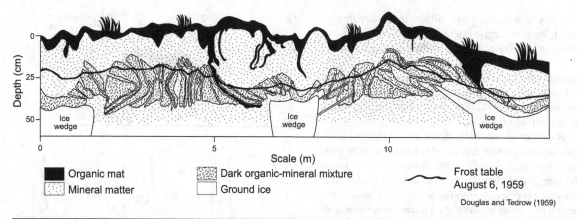

Fig. 11.29 Cross section of Gelisols on the Arctic coastal plain of Alaska, showing microtopography and distorted horizon boundaries, both due to cryoturbation.

to the active layer of the permafrost, and it is slowed by the cold soil temperatures (Fig. 11.28). Physical processes are driven primarily by the volume changes – expansion – associated with ice formation, and by thermal contraction (Bockheim *et al.* 1997, Bockheim and Tarnocai 2000). In addition to cryoturbation, Ping (1997) and Höfle *et al.* (1998) listed oxidation-reduction and organic matter accumulation as important subsidiary processes associated with Gelisols.

Gelisols are found mainly in polar, subarctic, and boreal regions, as well as localized high alpine areas. Low soil temperatures cause pedogenic processes, especially weathering, humification, and mineralization, to proceed very slowly, making the influence of parent material and natural soil drainage stronger than in most landscapes (Oganesyan and Susekova 1995). Because Gelisols are frozen for much of the year, they have the potential to preserve pedogenic features for long periods (Gubin 1994). Wet conditions prevail in many Gelisols because of perched water above the permafrost (Tarnocai 1994). Organic matter production is higher in polar landscapes than one might assume, because the soils become warmer in summer than does the atmosphere (Blume *et al.* 1997).

Stagnant, perched water and thin active layers also allow for the accumulation and preservation of thick peat layers at the surface, many of which are subsequently cryoturbated and buried (Zoltai and Tarnocai 1975, Tedrow 1977, Swanson 1996, Becher *et al.* 2013). Decomposition of organic matter is impeded not only by standing water but also by cold temperatures and decomposition-resistant plant materials (Höfle *et al.* 1998). Hence, the organic carbon content of many Gelisols is often high and extremely variable with depth. In some soils, it is maximal at the surface, and in others it is maximal at depth, or at the permafrost table. The type of cryoturbation model that dominates the soil largely determines how intimately mixed the organic matter becomes, and the resultant depth distribution patterns (Fig. 11.27).

Horizon boundaries in most Gelisols (and especially in Turbels) are contorted, broken, and often vertically aligned (Figs. 11.29, 11.30). Frequently organic material is incorporated into the subsurface, where it can remain for many years (Ugolini 1966, Brown 1969, Becher *et al.* 2013). Much of the incorporated organic material eventually resides at the top of the permafrost or in its upper few decimeters (Michaelson *et al.* 1996). In short, mixing and burial are slow but omnipresent in Gelisols, and in soil landscapes undergoing cryoturbation.

Worldwide, Gelisols contain massive amounts of organic carbon. It is estimated at between 1,400 and 1,700 Pg, almost twice the present atmospheric carbon pool (Lal 2004, Tarnocai *et al.* 2009, MacDougall *et al.* 2012) and almost half of the estimated global belowground organic carbon pool. Loss of carbon from these soils will have potentially important long-term, global climate implications (Stokstad 2004). As a result of increasing greenhouse gas emissions and climatic warming, Gelisols are expected to release large amounts of CO_2 and methane, setting off a complex positive feedback loop that may further accentuate global warming (Schaefer 2011, von Deimling *et al.* 2012). Specifically, processes in the high latitudes that will likely change, either setting off or slowing this feedback loop, include (1) increases in C sequestration due to CO_2 fertilization of ecosystems, (2) permafrost thawing and subsequent C mineralization, and (3) warming-induced increased CH_4 fluxes, which may be partially offset by a reduction in wetland extent (Koven *et al.* 2011). Also likely to occur are increased mineralization rates, the drying of previously anaerobic soils (especially peat), increases in C lost to fires, and increases in C:N ratio as plant communities change. It is a complex situation that is currently under intense study and merits watching (Zhuang *et al.* 2006).

Cryoturbation is manifested in soils as (1) irregular or broken horizons; (2) involutions; (3) highly irregular depth distributions of organic matter above, and within, the permafrost; (4) oriented rock fragments; and/or (5) silt-enriched

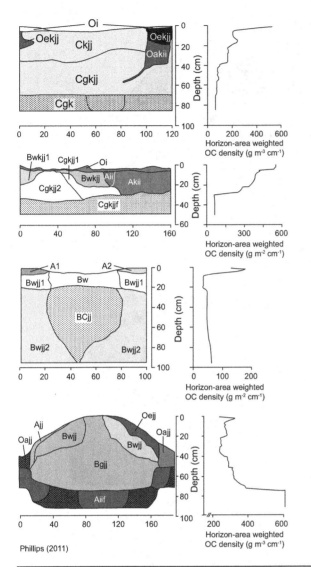

Phillips (2011)

Fig. 11.30 Horizonation patterns and corresponding depth distributions of organic carbon in four representative Gelisols in the Yukon and in Nunavut, Canada. Colors and shading of horizons are not intended to imply process or organic matter content, but simply to help discriminate them visually.

layers. Ice segregation is manifested as (1) ice lenses, (2) vein ice, (3) segregated ice crystals, and/or (4) ice wedges (Frenot et al. 1995, Michaelson et al. 1996, Bockheim et al. 1997, Bockheim 2007; Figs. 11.31, 11.32).

Cryoturbation also leads to repeating, complex, and seemingly random patterns of microtopography, and when combined with the below-ground mixing results in highly variable soils at small and medium scales (Everett 1980, Van Vliet-Lanoë 1998, Lev and King 1999, Mueller et al. 1999, Bockheim 2007; Fig. 11.27B). Earth hummocks and frost boils create specialized niches and microsites

for plant growth, peat formation, and oxidation-reduction (Ping 1997; Fig. 11.31). Soil horizons, already mixed and contorted by cryoturbation, are further pushed toward nonhorizontality by microtopography, which creates preferred areas of sedimentation, saturated areas where peat can form, and highs that are prone to wind erosion. Thus, Gelisols can be highly spatially complex on scales of a few meters, but across large tracts of landscape the overall soil pattern gives the appearance of monotony.

Cryogenic processes are the dominant feature and defining characteristic of Gelisols, but the expression of these processes varies among the suborders. Gelisols have three suborders (Soil Survey Staff 1999, Bockheim and Tarnocai 2011). Turbels are the most common suborder, and they have the strongest evidence of cryoturbation. They are often found in association with patterned ground (Pettapiece 1975, Douglas and Tedrow 1979, Bockheim and Hinkle 2007). Cryoturbation and ice segregation are key processes in the genesis of Turbel morphology. Histels are wet Gelisols dominated by organic materials (Tarnocai 1972, Zoltai and Tarnocai 1975). They meet the requirements of a Histosol, except that permafrost is also present. Orthels are mineral soils that have minimal evidence of cryoturbation and are the most common suborder in areas of dry permafrost, such as polar deserts, or in other drier, cold landscapes. They are relatively uncommon.

Particularly in Turbels, cryogenic processes are driven by the mobility of unfrozen water as it moves along thermal and pressure gradients, into and through the frozen system, feeding growing ice bodies and segregations (Tarnocai and Bockheim 2011). The continuous growth of ice segregations and the concomitant volume increases associated with them lead to differential heave and cryostatic pressures, producing the contorted and imperfectly sorted sediment units that these complex soils comprise. Not all Gelisols are as mixed and contorted as are Turbels. In Histels and Orthels, however, migration of unfrozen water to growing ice bodies is a less important process, because they are either not frost-susceptible (and thus not subject to the ice segregation process) or too dry. Indeed, cryostatic pressures, per se, are not always present in cryoturbation-affected soils (Mackay and Mackay 1976, Mackay 1980). Cryoturbation can also occur by differential frost heave or by the deformation of sediments beside growing ice wedges, or by a number of other processes.

Bockheim et al. (2006) stressed that cryogenic processes should be viewed as distinct processes that dominate soils in cold climates, producing soil properties and microrelief that are typically not found in other orders. Cryopedogenic processes are not restricted to Gelisols; they occur in all soils that freeze. Likewise, processes associated with warmer soils, such as podzolization, calcification, salinization, weathering, and humification (see Chapter 13), also occur in Gelisols, although usually at reduced rates and intensities (Bockheim 1982, Claridge and Campbell 1982,

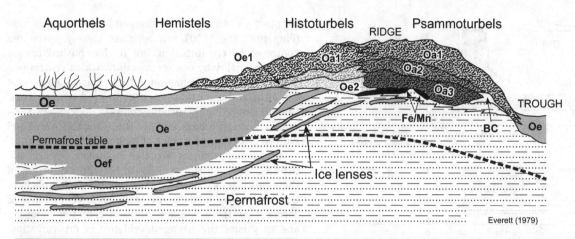

Fig. 11.31 Schematic cross section of the soils in and near a low-centered ice wedge polygon near Barrow, Alaska.

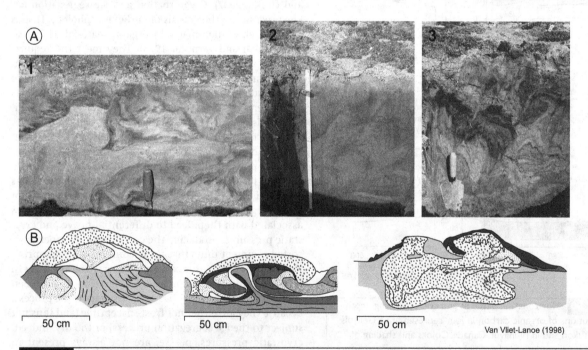

Fig. 11.32 Gelisols (Turbels) with involuted and contorted horizonation. (A) Photos. 1. Soil associated with a nonsorted circle near Bathurst Island, Nunavut, Canada. The lighter material is parent material pushed up by cryoturbation. The dark colored B horizons are strongly contorted by cryoturbation. 2 and 3. Similar soils, located on Ellesmere Island, Canada, again with convoluted bands of organic matter, due to cryoturbation. Photos by C. Tarnocai. (B) Diagrams. Various types of frost-heave hummocks, typical of strongly cryoturbated, moderately well drained, and somewhat poorly drained Gelisols. Colors and patterns imply different types of horizons or materials.

Forman and Miller 1984, Ugolini 1986a, Ugolini *et al.* 1987, Sokolov and Konyushkov 1998). Lessivage involving clay is rare in polar landscapes (Ugolini 1986a, Bockheim and Tarnocai 2000), although pervection of silt is a signature process in many Turbels.

Gelisol pedogenesis – again, particularly for Turbels – is perhaps best described by Tarnocai (1994) and Bockheim

et al. (2006; Fig. 11.28). Permafrost develops quickly after time$_{zero}$. Granular and platy structure forms quickly in surface horizons, aided by heave and shrinkage processes and promoted by ice lens and vein ice formation. Establishment of vegetation leads, eventually, to the formation of O and A horizons, which may be cryoturbated at any time. Soil horizon boundaries become contorted (Fig. 11.28). The

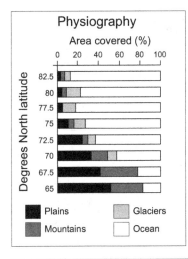
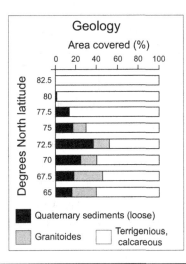
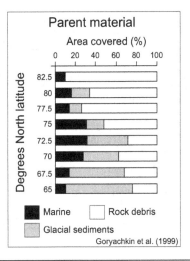

Fig. 11.33 The distribution of terrain and parent material features in the Arctic, as a function of latitude.

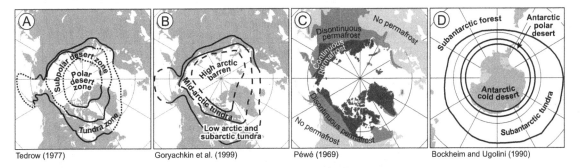

Fig. 11.34 Zonation in the Arctic and Antarctic, as it pertains to soils and permafrost. See also Tedrow (1968). (A) Soil zonation in the Arctic. (B) Soil zonation in the Arctic. (C) Permafrost zones in the Arctic. (D) Soil zonation in the Antarctic.

growth of the freezing front, downward, into the soil each autumn leads to extreme desiccation and high cryostatic pressures in the soil between it and the permafrost below (Bockheim and Tarnocai 2000). The pressures placed on the soil between the two freezing fronts squeeze it, creating subsoil horizons that are often dense, massive, and subject to desiccation. Patterned ground develops at the surface, and the B horizon, although distorted, may eventually show evidence of podzolization, gleization, or brunification. And all of these processes tend to be better expressed in wetter and siltier soils (Bockheim 2007).

Soils of Polar Landscapes
Polar landscapes are highly diverse pedologically, and especially hydrologically, containing some of the driest and wettest soils on the planet. Surprisingly, many soils here are not dominated by cryoturbation processes. Parts of the flat, permafrost-laden landscapes are wet and waterlogged for much of the short warm season. Conversely, large expanses of Antarctica are extremely dry; soils accumulate salts, and oxidation and salt weathering are dominant processes (MacNamara 1969, Bockheim 1982, Beyer *et al.* 2000).

Several attempts have been made to classify, map, and otherwise subdivide the polar landscapes into pedogenically homogeneous subregions (Tedrow 1968, Tuhkanen 1986, Ugolini 1986a, Bockheim and Ugolini 1990, Campbell and Claridge 1990, Blume *et al.* 1997, Sokolov and Konyushkov 1998). Soil distribution in the highest latitudes is primarily dependent upon terrain and parent material, which often vary by latitude (Fig. 11.33). Goryachkin *et al.* (1999) subdivided the Arctic landscape into three soil zones: High Arctic barren, Mid-Arctic tundra, and Low Arctic and Subarctic tundra (Fig. 11.34). Pedogenic processes change markedly across these regions (Fig. 11.35).

Little information exists, however, on the importance of geographic variation in the strength or intensity of cryoturbation across these regions. In his review of the major

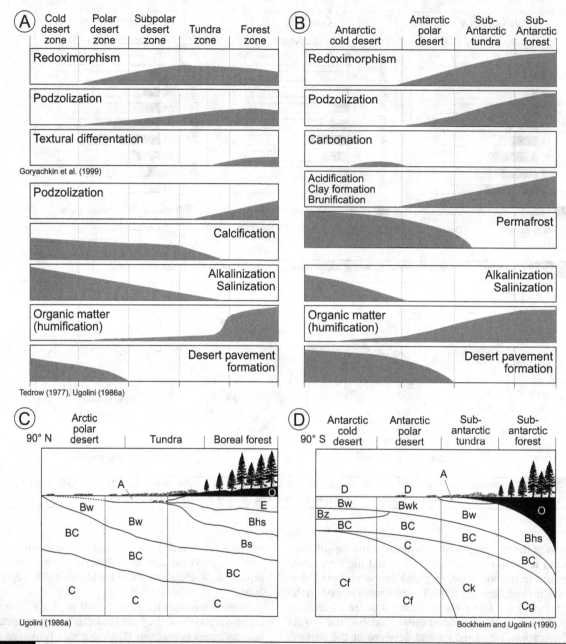

pedogenic processes in polar regions, Ugolini (1986a) scarcely mentions cryoturbation. We suspect that cryoturbation is at a maximum in the mid-Arctic. Equatorward of the High Arctic barren, one gradually encounters less permafrost and more trees, and podzolization becomes increasingly dominant. Wet, gleyed soils, many with histic (O) epipedons, become common. Poleward, increasingly larger

parts of the landscape are dry, windswept islands where inputs of salt aerosols are high and rocky parent materials experience little leaching and pedogenesis (Fig. 11.33). Here, high winds and scarce precipitation create conditions where cryoturbation is subsidiary to salinization and calcification. In short, soil development, but not necessarily cryoturbation, decreases poleward (Goryachkin et al.

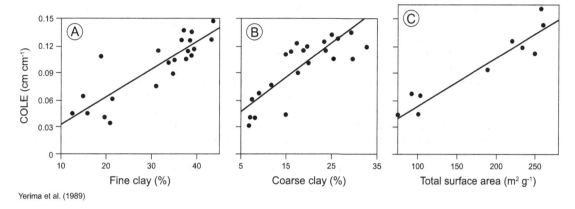

Yerima et al. (1989)

Fig. 11.36 Relationships among coefficient of linear extensibility (COLE), clay content, and surface area, for some Vertisols and Alfisols in Cameroon.

1999). Indeed, many of the very coldest, driest (ultraxerous) soils in Antarctica and in the polar deserts of the Northern Hemisphere have virtually no pedogenic signature, except for salt concentrations at or near the surface and a cover of desert pavement (Claridge and Campbell 1968, Blume et al. 1997). Their lack of an appreciable amount of organic matter has earned these soils the name *ahumic soils* (Ugolini and Bull 1965, Claridge and Campbell 1982, Bockheim and Ugolini 1990), and cryoturbation here is also minimal.

Argilliturbation and Vertisols

Some 2:1 silicate clays (see Chapter 4), mainly smectite, have high coefficient of linear extensibility (COLE) values; they expand upon wetting and shrink when dry. As a result, soils dominated by these types of clays undergo significant volume changes upon wetting (Jayawardane and Greacen 1987). This type of expansion-contraction causes soils to mix and is the central tenet of *argilliturbation* – soil mixing by shrink-swell, usually smectite-rich, clays. In this section, we examine the nuances of argilliturbation and discuss the soil order that is conceptually interlinked to argilliturbation – Vertisols.

Strong argilliturbation occurs where (1) soils are rich in smectitic clay and (2) the climate has a distinct wet and dry season, so that the soil undergoes wet-dry cycles that are frequent (and intense) enough for the process to operate fully. Not coincidentally, the formation and preservation of smectite clay are promoted in wet-dry climates where pH values and base saturation remain high, and where silica is retained in the profile (Folkoff and Meentemeyer 1987, Southard et al. 2011). It is a perfect storm – wet-dry climates promote the formation of smectites, and, in turn, the smectites and the nature of the climate facilitate seasonal cracking.

Smectites usually occur in the fine clay fraction of soils, and COLE is more strongly correlated with fine clay than with coarse clay content (Yerima et al. 1989). Soils with abundant fine clay have a greater volume of small pores that separate smectite microaggregates than do soils with more coarse clay and silt (Coulombe et al. 1996). As water moves into these pores, the local volume swells (Wilding and Tessier 1988, Coulombe et al. 1996).When the pores lose water upon drying, the soil volume shrinks. The combination of smectite mineralogy *and* high clay content leads to high COLE values.

Pedogenesis Dominated by Argilliturbation
Vertisols – Vertisolic soils in Canada – are defined in terms of the dominance of argilliturbation (Brierley et al. 2011, Southard et al. 2011). The name is derived from the Latin *vertere*, "to turn over or invert." They are rich in smectite clays, especially fine (<0.2 µm diameter) clays (Nelson et al. 1960). Plastic and sticky when wet, Vertisols may become very hard when dry (Jayawardane and Greacen 1987). Clay contents can approach 90% or more in some Vertisols (Hallsworth and Beckmann 1969). Although exceptions do occur, e.g., Isbell (1991), the high COLE values of Vertisols typically increase with increasing amounts of fine clay (Anderson et al. 1973, Kishne et al. 2009), which are correlated with increasing amounts of smectite clays, particularly montmorillonite and beidellite (Nelson et al. 1960, Roy and Barde 1962, Ahmad and Jones 1969, Fadl 1971, Graham and Southard 1983, Newman 1983, Hussain et al. 1984, Yerima et al. 1985; Fig. 11.36). For lack of fine clay, many soils that are silty clay, silty clay loam, or coarser in texture do not meet the cracking requirements of Vertisols (see Landscapes: The Blackland and Prairies of East Texas).

Southard et al. (2011) estimated that about 2% of the ice-free land area on Earth is covered with Vertisols. These soils dominate parts of the Deccan Plateau of India, where they have developed on gently sloping plains underlain by weathered basalt (Simonson 1954b, Roy and Barde 1962, Sehgal and Bhattacharjee 1990). In Sudan and other parts of semiarid Africa, Vertisols and similar soils are widespread

(Stephen *et al.* 1956, El Abedine *et al.* 1971, Acquaye *et al.* 1992, Favre *et al.* 1997). In Morocco and Algeria they are known as Tirs (Del Villar 1944). They are also dominant on the prairies in eastern Texas (Nelson *et al.* 1960, Kunze *et al.* 1963, Newman 1983, Nordt *et al.* 2004). Australia has the largest acreage of Vertisols of any nation (Costin 1955a, Hallsworth *et al.* 1955, Isbell 1991).

Vertisols are usually found on flat to gently rolling surfaces that have formed in clayey parent materials (Roy and Barde 1962, Ahmad 1983, Isbell 1991). On steep slopes erosion may outpace profile formation, leading to shallow sola and weakly developed Entisols, while Vertisols occupy adjoining lowlands where the eroded clays accumulate (Ahmad 1983; see Landscapes: The Blackland and Prairies of East Texas). Additionally, sloping sites dry faster as a result of runoff, reducing the full cycle of cracking. Because of the inherently slow permeabilities of Vertisols, low slopes tend to promote wetting (De Vos and Virgo 1969). Many of the best-expressed and deepest Vertisols are in shallow depressions, which act as settling basins for clays eroded from nearby and where water can pond in the wet season. Extended periods of ponding further assist in smectite neoformation and virtually assure complete wetting of the profile, accentuating the wet-dry seasonality of the site.

Landscapes: The Blackland and Grand Prairies of East Texas

The Blackland and Grand Prairies of East Texas occupy nearly level to gently rolling plains. The Vertisols here have developed on calcareous marine shales, marls, and clays (Diamond and Smeins 1985). Large rivers crossing the area have broad but shallow valleys. The Blackland Prairie once supported a tallgrass prairie, and deciduous bottomland forests were common along rivers and creeks. The high soil fertility is ideal for row crop agriculture, although only a few hay meadows and ranches remain. This area receives about 750–950 mm of precipitation during the year and is known for its extreme heat and dryness during midsummer, when wide, deep cracks form and open in the clayey soils. Most of the soils here are classified in the Ustic soil moisture regime, or as Udic intergrades. This is the land where the famous Houston Black soil series (Udic Haplustert) – one of the first Vertisols to have been studied anywhere (Templin *et al.* 1956) – is a mainstay.

The overwhelming importance of clay-rich, smectitic parent materials in a semiarid climate like this cannot be missed. In McClennan County, where the parent materials have been weathered from shale and marl and are therefore *initially* clayey and smectitic, Vertisols with clay textures throughout the profile form quickly and dominate the landscape. Only in the moist (Udic) microsites of the river valleys are Vertisols not mapped. Here, cracks in the Tinn series open less frequently; it is a Vertic intergrade.

In Williamson County, however, Houston Black soils are found only the lowlands, where enough of the fine clays have accumulated by runoff. Uplands here are shallow to soft chalk bedrock, and soil textures are slightly coarser – silty clay. Therefore, soils on the remainder of the landscape do not have enough clay to crack deeply or widely enough to be classified as Vertisols; they are shallow Inceptisols and Entic Mollisols. However, weathering of the chalk, coupled with the semiarid climate, keeps soil pHs high. Weathered materials from the chalk bedrock are rich in expansible clays, and they provide the cations necessary for the continued neoformation of smectites. In short, this landscape is not likely to lack Vertisols on uplands forever!

Figure: Block diagrams of soil landscapes on the Blackland Prairie, McLennan and Williamson Counties, Texas.

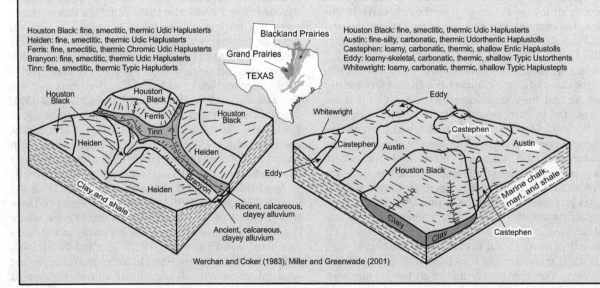

Houston Black: fine, smectitic, thermic Udic Haplusterts
Heiden: fine, smectitic, thermic Udic Haplusterts
Ferris: fine, smectitic, thermic Chromic Udic Haplusterts
Branyon: fine, smectitic, thermic Udic Haplusterts
Tinn: fine, smectitic, thermic Typic Hapluderts

Blackland Prairies
Grand Prairies
TEXAS

Houston Black: fine, smectitic, thermic Udic Haplusterts
Austin: fine-silty, carbonatic, thermic Udorthentic Haplustolls
Castephen: loamy, carbonatic, thermic, shallow Entic Haplustolls
Eddy: loamy-skeletal, carbonatic, thermic, shallow Typic Ustorthents
Whitewright: loamy, carbonatic, thermic, shallow Typic Haplustepts

Werchan and Coker (1983), Miller and Greenwade (2001)

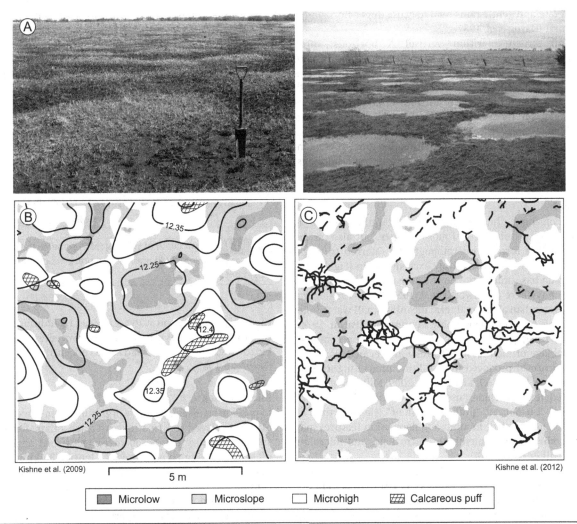

Microlow   Microslope   Microhigh   Calcareous puff

Fig. 11.37 Gilgai microrelief. (A) Photos of circular gilgai from the Texas prairies, by W. Miller (left) and C. Morgan (right). (B) Detailed topographic map of gilgai, with a 5 cm contour interval, showing microhighs and microlows. (C) Locations of cracks, as mapped in three different years, on the same plot. Note that most of the cracking occurs in association with microhighs.

As in many soils, when a Vertisol dries, cracks form at the surface and expand downward. Cracks start to develop first on the microhighs, which dry out first. With continued drying, they extend across all of the surface microrelief features (Kishne *et al.* 2009, 2012, Miller *et al.* 2010). During several dry-wet-dry periods within a single season, the cracks will open-close-open in the same locations (Miller *et al.* 2010). On interannual timescales, cracks may reopen in the same places or their locations will shift slightly (El Abedine *et al.* 1971). Nonetheless, over time, the cracks still cluster in the same general microtopographic locations.

Deepening cracks facilitate the continued drying of the lower solum. The largest cracks tend to be in surface lows (or microlows), sometimes with one primary crack in the center (Newman 1983). Soil material sloughs off the sides of open cracks and fills them; this is expected because the clay-rich subsoil often shrinks more than the surface soil, resulting in cracks that sometimes widen with depth and overhang near the surface (Hallsworth *et al.* 1955). We cannot overemphasize this point, because cracks that are widest at the top will fill with less soil material than those that widen with depth. Soil can also be blown in by wind, knocked in by biota, or washed in by rain (Buol *et al.* 2011). Cracks that remain open after swelling are soon filled as surface soil washes in (Hallsworth *et al.* 1955). Direct evidence of this process is present in most Vertisols as darker, humus-rich masses in lower or midprofile positions (El Abedine *et al.* 1971, Hussain *et al.* 1984). Rewetting, which causes the cracks to close, can occur from the bottom up, from the top down, or from side to side (Graham and Southard 1983, Coulombe *et al.* 1996). Heavy rains or flooding can cause the soil to wet from the bottom up, while moderate and gentle rains tend to cause upper horizons to wet first (Dasog *et al.* 1987).

Soil structure within Vertisol A horizons, often described as a "surface mulch," is typically strongly granular (Dasog *et al.* 1987). The hard, nutlike structure resembles loose gravel on the surface (Dudal and Eswaran 1988). These clay- and humus-rich horizons undergo even more frequent wet-dry cycles than the Bss horizons (with slickensides, see later discussion) below, continually fracturing peds into smaller granules. Additionally, grass roots contribute to the integrity of granular peds in the A horizons, and worm casts are also abundant.

Gilgai

The continual churning of soils by argilliturbation is manifested within Vertisol morphologies, but also on the surface as a form of microtopography referred to as *gilgai*. Relief on gilgai commonly exceeds 15 cm (Fig. 11.37). Australia has some of the most extensive and varied gilgai forms anywhere (Hallsworth *et al.* 1955, Isbell 1991); *gilgai* is an Australian aboriginal term meaning small water hole, referring to microtopographic depressions (Paton 1974). The term "gilgai" has since come refer to the complete assemblage of mounds *and* depressions (termed *microhighs* and *microlows*) produced by argilliturbation (Knight 1980, Soil Survey Staff 1999, Kishne *et al.* 2009, Miller *et al.* 2010). Related terms include "melonhole," "crabhole," "Bay of Biscay," "hushabye," "puff," and "shelf" (Hallsworth *et al.* 1955). Gilgai surface mounds and depression features may be subdivided into microhigh, microslope, and microlow sites (Miller *et al.* 2010).

The spatial variability of gilgai is generally predictable across the landscape (Milne *et al.* 2010). Gilgai are not just random mounds and depressions, although they are rarely the same from one landscape to the next. Newman (1983) argued that each Vertisol series has its own signature gilgai, much like a fingerprint, and various types have been recognized on the basis of the dominant shape of microhighs and microlows, e.g., circular, elliptical, and linear gilgai (Hallsworth and Beckmann 1969, Paton 1974, Bhattacharyya *et al.* 1999). Regularity of mounds and depressions can be such that amplitudes can be readily determined (Jensen 1911, Hallsworth *et al.* 1955, Stephen *et al.* 1956). On steep slopes gilgai may align down the slope (White and Bonestell 1960).

Gilgai are thought to form by a number of stress/heave configurations (Knight 1980). Most common is the assumption that they form in areas between deep soil cracks, i.e., the swelling pressures induced by additional soil material within cracks are transferred to the microhighs between. Soils in the depressions crack most often and are most likely associated with large cracks or former crack locations (Graham and Southard 1983). Alternatively, microhighs often have the greatest crack densities because the cracks form in new locations (Kishne *et al.* 2009). Establishment of a gilgai pattern, especially depressions

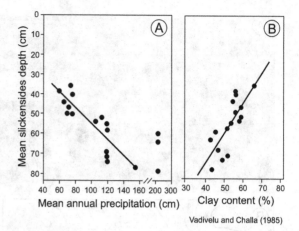

Vadivelu and Challa (1985)

Fig. 11.38 Mean depth of slickensides in some Vertisols of India, as a function of (A) mean annual precipitation and (B) clay content.

that occasionally pond water, can lead to additional pedoturbation when crayfish invade the wet sites and begin burrowing. Ant mounds tend to be concentrated on knolls (Newman 1983).

Slickensides

Expansion stresses caused by argilliturbation force masses of moist, clayey soil to slide past each other, forming a subsoil structure composed of wedge-shaped aggregates, often resembling *parallelepipeds* or lentils (Blokhuis 1982). These structural units, found only in Vertisols, are most common in the B (Bss) horizons. De Vos and Virgo (1969: 199) described these features: "Individual peds have triangular or trapezoidal faces tapering to points at each end in the form of a 'double wedge,' but forming an obtuse-angled dome at the upper and lower sides." They argued that the term *bicuneate* should be used to describe these double wedge-shaped aggregates in Vertisol B horizons (Latin *bi*, two, *cuneus*, wedge). Found closer to the surface in drier climates, they are better expressed where the wet-dry seasonality is extreme (Ahmad 1983, Vadivelu and Challa 1985; Fig. 11.38). As the soil repeatedly expands and contracts, the structural units are continually recast and remolded. In time, they acquire blocky or wedge shapes, with smooth edges and facets. Most are wider than they are tall (Ahmad 1983).

Slickensides, which are highly characteristic of Vertisols, are the shear planes formed along the edges of the wedge-shaped aggregates. They represent the plane along which the expanding soil matrix has moved, much like a thrust fault. Often shiny when moist, slickensides appear polished or grooved when freshly exposed. They can be striated if the aggregates that slid past each other had gravel or sand grain protrusions (Lynn and Williams 1992; Fig. 11.39). The shear surface is seldom planar, but rather is slightly curved

Fig. 11.39 Slickensides in Vertisols. (A) Large slickenside exposed in a Texas Vertisol. Photo by W. Lynn. (B) A slightly fluted slickenside. Photo by RJS.

like a flattened spoon (White and Bonestell 1960, Lynn and Williams 1992). Larger slickensides may even be fluted (Fig. 11.39). Once formed, a slickenside is the surface of least resistance to shear for the next wetting-expansion event; thus they persist. In situ slickensides often align along a plane that is 20° to 60° from the horizontal, indicating that most of the shear is occurring out and up, rather than vertically or entirely horizontally (Knight 1980). Lynn and Williams (1992) noted that many slickensides represent planes of movement in two directions, upward upon wetting and downward upon drying, and that many vertical surface cracks lead down to a slickenside in the subsurface.

Models of Argilliturbation

Early researchers described Vertisols as self-plowing, self-swallowing, or self-churning soils. As described earlier, the self-plowing concept has traditionally involved (1) shrinking of the soil matrix and formation of desiccation cracks; (2) partial infilling of the cracks with material from outside the profile, or from higher up within the same profile; (3) wetting and expansion of the soil matrix; resulting in (4) soil material moving (churning) horizontally and vertically to accommodate the additional volume (Simonson 1954b, Bronswijk 1991). Although this model still has considerable

merit, contemporary research has shown it to be rather simplistic and not fully explanatory.

In the current pedogenic model of Vertisol genesis, argilliturbation occurs because of (1) swelling pressures and mixing within the solum and (2) self-swallowing processes that lead to volume disruption, because surface materials accumulate in the lower profile. Although self-swallowing is ongoing in Vertisols (El Abedine *et al.* 1971, Graham and Southard 1983), recent research has deemphasized its role, in favor of simple expansion-shrinkage processes (Beckmann *et al.* 1984). Evidence is provided by Vertisols with pronounced gilgai that have protrusions of subsoil material into near-surface locations (Paton 1974). Newman (1983) called these subsurface structures chimneys, while other have used the terms *diapir* and *mukkara*. These features may extend completely to the surface, where they form gilgai mounds or *puffs*, although most do not (Figs. 11.40, 11.41). Sometimes, they are expressed at the surface only as patches of light-colored soil. Vertisol sola, i.e., defined by depth to the C horizon or to carbonates, are much thinner above these upthrust chimney areas (Kunze *et al.* 1963, Miller *et al.* 2007). Shallow depressions in the surface, occurring between chimneys, are often referred to as *microlows*; they range from 2 to 5 m across and are 1–2 m deep (Williams *et al.* 1996). Do not be confused: The terms "microhigh" and "microlow" describe surface topography, whereas the terms "bowl" and "chimney" refer to subsurface features (Figs. 11.41, 11.42).

The widespread presence of chimneys and bowls in Vertisol landscapes has forced a rethinking of the traditional self-swallowing model of Vertisol genesis, for if cracks developed in *random* locations, and material was deposited in them evenly each year, chimneys and bowls would not exist. The traditional self-swallowing model also does not explain certain features of Vertisols, such as

systematic depth functions of organic carbon, salts, and carbonates, and the presence of Bt horizons, which should be destroyed by repetitive and randomly located cracking and argilliturbation (White and Bonestell 1960). Instead, what is observed are repetitive patterns of highs and lows, with predictable characteristics across this sequence (Wilding *et al.* 1990). Material in chimneys often has a different texture *and* color than the surrounding material; bowl areas are traditional black clay, with thick A horizons, whereas chimneys consist of lighter-colored, brown clay, like subsoil material under much thinner A horizons (Hallsworth *et al.* 1955, Hallsworth and Beckmann 1969, Paton 1974, Coulombe *et al.* 1996, Kishne *et al.* 2009). Years ago, Paton (1974) hypothesized that, in chimneys, subsoil material had been thrust upward as a result of differential loading pressures on wet clay, which can flow upward through overlying soft sediments, as in salt domes. Hirschfield and Hirschfield (1937) had shown that, upon wetting, the subsoil under gilgai chimneys expanded considerably more than did the surface layer. These data support Paton's (1974) model, expanded upon and essentially codified by Wilding *et al.* (1990), which involves subsoil materials pushing upward in chimneys, through a darker, thin surface mantle that periodically cracks and swells. This hypothesis explains why material in the chimneys is lighter in color, and why A horizons there are thinner (Figs. 11.40, 11.41, 11.42). Miller *et al.* (2010) and Kishne *et al.* (2012) found that cracking

Fig. 11.40 Photo of a Vertisol profile, showing a chimney and bowl. Photo by W. Miller.

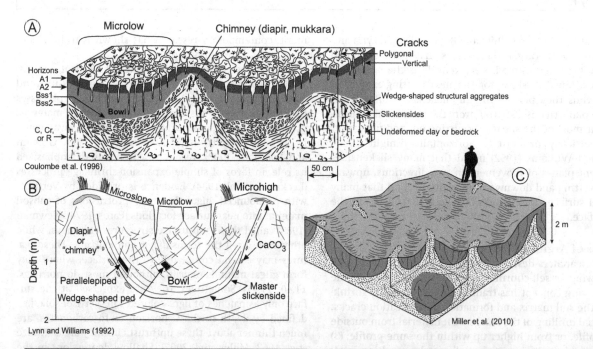

Fig. 11.41 Diagrams of the surface and subsurface morphology of Vertisols. (A) Transect through several Vertisol pedons, showing morphology typical of the order. (B) Diagrams through a single Vertisol bowl and chimney pair, showing the relationships among slickensides, soil structure, and CaCO$_3$ content. (C) Schematic block diagram showing the microvariability of a Vertisol, and its relationship to gilgai, on the Texas Gulf Coast Prairie.

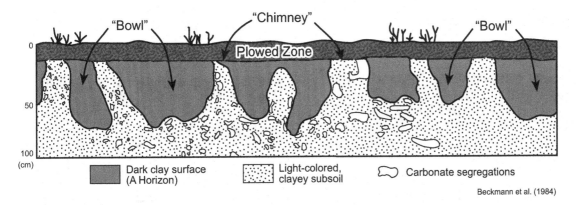

Fig. 11.42 Cross section (true to scale) along an exposure of Vertisols in Queensland, Australia, showing chimney and bowl structures. Note that the surface microtopography has been altered by cultivation.

began, and crack density was greater on, microhighs or the upper part of microslopes (Fig. 11.37). Under prolonged periods of drying, cracks formed in microlows as well, and these cracks were typically deeper than on microhighs (Fig. 11.41). Also in support of this model is the observation by Lynn and Williams (1992) that the outer surfaces of bowls often have large, master slickensides that essentially form the boundary between chimneys and bowls (Knight 1980, Wilding *et al.* 1990; Fig. 11.39A). These can be traced from nearly vertical orientations near the surface (beside the chimney) to nearly horizontal orientations at the base of the bowl, about 1–2 m below the surface.

Wilding and Tessier (1988) expanded upon Paton's model. They noted that the area between the chimneys forms a type of syncline, or synclinorum, with the chimneys occupying an anticline-like area between (Beckmann *et al.* 1984, Williams *et al.* 1996) (Fig. 11.41). In the Wilding and Tessier (1988) model, dry, cracked soil rewets from the base of the cracks upward. The lighter-colored subsoil, which may have higher COLE values than surface horizons, expands in all directions (Beckmann *et al.* 1984). The only way for the increased volume to be accommodated is by buckling of the surface, similar to the way folds develop in a carpet (Fig. 11.42). This buckling *starts* in the microlows, thrusting the soil outward and upward. Once the folds are initiated, further expansion is accommodated by vertical motion of the subsoil. Slickensides form at this point and induce the formation of *thrust cones* or bowls, between which are chimneys (Fig. 11.43). Within the thrust cones, shrink-swell processes, slickensides, self-churning, and argilliturbation are maximal, while between the bowls (in the chimneys) the main processes are the slow upward movement of subsoil material. Upward and outward thrusting in microhighs allows them physically to "shed" materials into the bowllike depressions of the microlow (Stiles *et al.* 2003). This situation is enhanced when COLE values in the subsoil exceed those of the upper sola, a situation common to landscapes with gilgai (Hallsworth and Beckmann

1969, Beckmann *et al.* 1984). The plastic limits of the lighter subsoil material versus the darker bowl material are such that, at certain moisture contents, the subsoil might still be deformable, while the dark soil above would be drier and more rigid (Beckmann *et al.* 1984). This juxtaposition assists in chimney formation, as the subsoil buckles upward. As expected, because of the microtopography and the depth of cracking (deeper in bowls), leaching is maximal within bowls and minimal on chimneys (Table 11.8; Fig. 11.44). Often, within chimneys, secondary carbonates and gypsum occur very near the surface (Hallsworth *et al.* 1955, Lynn and Williams 1992, Miller *et al.* 2007).

The preceding processes are typical of flat, clay-rich landscapes. At the bases of long slopes, however, downslope creep pressures force the entire sequence to be compressed, forcing chimneys to fold over each other in the downslope direction, like waves breaking on a beach. Some of the subsoil (lighter brown) clays of the chimney are washed into the bowls, essentially completing the circuit. Because argilliturbation processes lead to pedons (bowls) with overthickened A horizon material between chimney soils that are much shallower, the entire mixing suite probably could be considered proisotropic *and* proanisotropic (Johnson *et al.* 1987).

All in all, strongly expressed argilliturbation and shrink-swell processes in Vertisols lead to intimately mixed horizons and short-range – but usually predictable – spatial variability. In soils where bowl-and-chimney morphology is expressed in the subsurface, surface expression, in the form of gilgai, is also found. Some Vertisols lack strong evidence of gilgai, and in these, churning may not be as ordered, and cracking patterns may be more random. For these soils, a more traditional, self-swallowing model may apply.

The development of bowls/thrust cones potentially leads to a series of interesting feedback mechanisms. First, we must assume that the finest clays have the most shrink-swell potential, regardless of their mineralogy. Agents that flocculate those clays into larger masses with

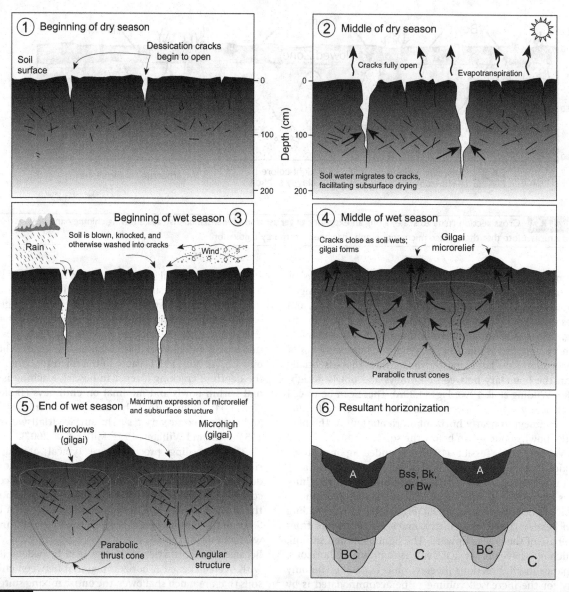

Fig. 11.43 Contemporary views, compiled from several sources, of the formation of Vertisols by argilliturbation and shrink-swell processes, into bowl-and-chimney subsurface morphology, with resultant gilgai.

lower shrink-swell capacity include organic matter, gypsum, and calcite. In the lower, wetter bowls, organic matter tends to accumulate, leading to flocculation of clays and potentially less shrink-swell activity. Working against this trend, however, is the tendency for microlows to be more intensively leached, allowing clays near the surface to be less flocculated and thereby enhancing shrink-swell processes. Although these two process bundles might offset each other, Williams *et al.* (1996) suggested that the leaching of carbonates and gypsum has more influence on the system than does the accumulation of organic matter. In carbonate-rich chimneys, argilliturbation processes will be discouraged by the tendency for smectites there to be

flocculated. The end result is a system that tends to perpetuate argilliturbation in the bowls, opening up the soil with cracks and facilitating leaching. At the same time, carbonates and gypsum (which are not translocated from the chimneys) work chemically to mitigate deep cracking there, which in turn minimizes wetting and leaching processes.

Although the general processes at work within Vertisols have been known for decades, details of the process are still being sorted out. However, current research appears to indicate that Vertisols are not as isotropic as once thought and that *thrusting* due to swelling may be more important than *mixing*.

Table 11.8 | General morphological trends between microbasin (bowl) and microhigh (chimney) positions in Vertisols

Property	Microlow	Microhigh
A horizon thickness	Thicker	Thinner
Solum thickness	Thicker	Thinner
Color	Darker	Lighter
Organic carbon content	Greater	Lower
pH	Lower	Higher
Structure grade	Finer	Coarser
Consistence	Very firm	Extremely firm
Cracking characteristics	Many, largest	Earliest. Narrower and shallower
Shrinkage	Higher	Lower
Slickenside expression	Anticlinal	Synclinal
Moisture content (mean)	Higher	Lower
Depth of leaching	Greater	Lesser
Carbonates	Less, deeper	More, higher in profile

Source: Wilding *et al.* (1991).

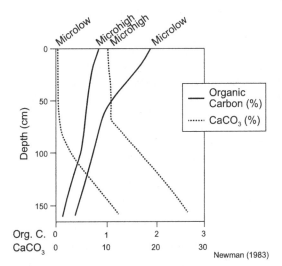

Fig. 11.44 Depth distributions of organic carbon and CaCO$_3$ for a bowl (microlow) and chimney (microhigh) in a Texas Vertisol.

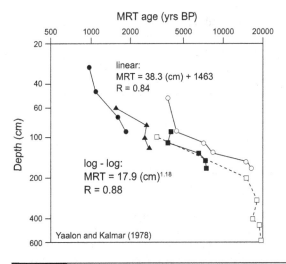

Fig. 11.45 Mean residence time (MRT) (see Chapter 15) of organic matter in five Vertisols and Vertisol-like profiles from Israel.

Rates of Argilliturbation

Rates of argilliturbation in Vertisols largely depend on the dynamics of surficial materials being incorporated into cracks, and exactly *where* within the profile one examines it, i.e., bowl or chimney. Nonetheless, Simonson (1954b: 201) postulated that mixing of a Vertisol could occur "in a matter of centuries." El Abedine *et al.* (1971) measured the volume of infilled soil material in closed cracks in a Vertisol in Sudan and determined that the entire profile could be overturned in 8,700 years. Yaalon and Kalmar (1978)

recalculated mixing rates, using the data of El Abedine *et al.*, and determined turnover times of 700–1,250 years for the *upper* solum.

Radiocarbon ages of soil organic matter tend to confirm that rates of mixing are less than the rates of organic matter additions, explaining why distinct A horizons form so readily in these soils. In Vertisols with continuous mixing to great depths, ^{14}C ages should be fairly similar with depth, but they seldom are (Scharpenseel 1972c). Rather, ^{14}C ages increase with depth (Fig. 11.45), implying that organic matter additions proceed more rapidly than does mixing. Mean

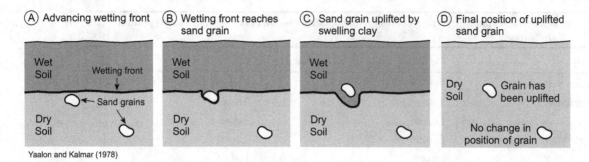

(A) Advancing wetting front
(B) Wetting front reaches sand grain
(C) Sand grain uplifted by swelling clay
(D) Final position of uplifted sand grain

Yaalon and Kalmar (1978)

Fig. 11.46 Diagrammatic representation of how argilliturbation can move small objects upward in a soil profile.

residence times of ^{14}C reported by Yaalon and Kalmar (1978) also suggested that most mixing in Vertisols is confined to the upper solum. Southard and Graham (1992) used an isotopic tracer, ^{137}Cs, produced from open-air nuclear testing and, which is strongly adsorbed by some clay minerals, to estimate the rates at which surficial materials had become incorporated into the subsoil of a Vertisol. Within 25 years since the peak ^{137}Cs fallout, some mixing had occurred to the maximum depth of cracking, but most mixing was limited to the upper solum. The ^{137}Cs data by Southard and Graham (1992) indicated that sloughing of soil into the bottoms of *deep* cracks occurs only in the very driest years, and that argilliturbation decreases markedly with depth. This work may help explain why the Bss horizon and its slickensides are maximally developed at the *mean*, rather than the *maximal*, depth of cracking. Over time, more soil material is likely to fall to the mean depth of cracking than to any other depth, and, therefore, expansion pressures will, on average, be greatest there.

Argilliturbation and Coarse Fragments
Argilliturbation is one of the few forms of pedoturbation that can transport gravel, gastroliths, carbonate nodules, and other large clasts upward in soils, making it a proanisotropic pedoturbation vector for the larger size fractions (Blokhuis 1982, Miller *et al.* 2007). Some Vertisols even have sand maxima at the surface (Yaalon and Kalmar 1978). For fine particles like sand the process functions as shown in Figure 11.46. As a wetting front advances into a soil with abundant smectites, water rapidly flows around sand or small gravel particles, ahead of the front (Fig. 11.46B). Clays below the grain expand, pushing it upward. Upon drying, subsequent movement of the grain downward, into the clay, is not as likely as is upward thrusting during wetting. Instead, drying of the wet soil occurs rather uniformly, causing the grain to remain at its new position, or to fall back only a fraction of the distance it was lifted.

For clasts that are larger than most cracks, another process is at work (White and Bonestell 1960). As larger clasts are moved upward by thrusting, they are unable to fall back into the now-too-small cracks and, thus, become

concentrated nearer the surface (Springer 1958, Mabbutt 1965). In the end, this coarse material comes to rest on the surface, or at least moves upward slowly, in the soil (see also Jessup 1960). Because of this process, the bedrock-soil contact in Vertisols can be abrupt; as soon as a rock is dislodged from the bedrock it is swept up and into the solum (Johnson *et al.* 1962: 393). In extreme situations, argilliturbation can result in a surface stone pavement (Muhs 1982). Johnson and Hester (1972) pointed out how all clasts can be moved to the soil surface by upward heave, but only the largest clasts remain there, as others fall into cracks and are cycled through the profile (Fig. 11.47). Vertisols and vertic Mollisols in Iraq have limestone fragments in their upper sola, but are stone-free below, suggesting that argilliturbation has moved them upward (Hussain *et al.* 1984). Similarly, in Australia, Mabbutt (1965) described Aridisols with pronounced stone pavements, stone-free B horizons, and duricrusts at depth (Fig. 11.48). He hypothesized that argilliturbation had moved these stones upward from the duricrust below. Since that time, the climate has dried, argilliturbation has ceased, and an ordered A-E-B-C profile has developed, while retaining the inherited rock depth trend. A second possibility Mabbutt (1965) considered is that, after a certain number of rocks are moved to the surface, an intrinsic threshold is crossed. The rocks may act as a mulch, effectively minimizing surface cracking by limiting evaporation.

Graviturbation
Downslope transport of soil and sediments under the influence of gravity is termed *mass wasting, mass movement*, or soil erosion. Usually, interstitial water aids in this movement by reducing the shear strength of the materials. In-transit mixing of regolith is typically a function of the category of mass movement, of which there are four: slides, flows, falls, and heaves. In slides, cohesive blocks of material move downslope along a distinct plane of failure. Mixing within the sliding material may be minimal. Potential for mixing within falls and flows, however, is considerable, either while the soil material is in transit or as a result of settling and compaction/rearrangement immediately

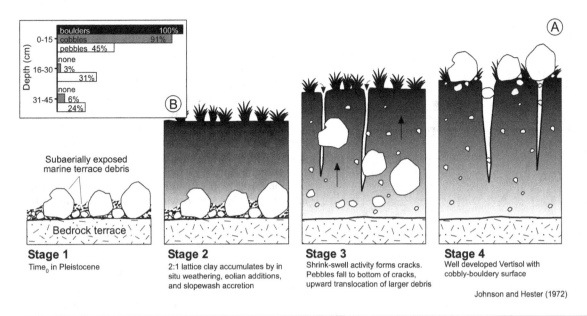

Fig. 11.47 Illustrations of the sorting effects of argilliturbation on coarse fragments. (A) Hypothetical model showing how clasts on the surface of a marine terrace on San Clemente Island can be argilliturbated to the surface. (B) Content of coarse fragments of varying sizes in a Vertisol on San Clemente Island, California.

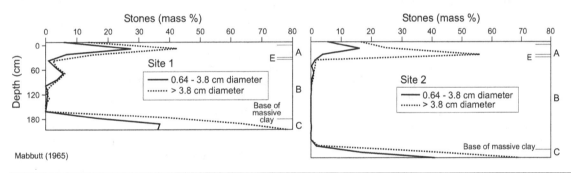

Fig. 11.48 Depth distribution of stones in an Australian Aridisol.

upon cessation of movement. Nonetheless, because most of the aforementioned mass movements are generally catastrophic or of moderate size, they are usually not considered within the rubric of pedoturbation. Indeed, most of the soil may be lost and destroyed during falls, slides, or flows, rather than mixed in place.

In many cold climates, particles near the soil surface are heaved upward as a result of ice crystal growth (crystalturbation) as wet soils freeze. Heave contributes in large part to the slow but persistent process of soil creep, which is generally confined to the upper 20–40 cm of soil and diminishes rapidly in intensity with depth (Young 1960, Finlayson 1981, Heimsath et al. 2002). Frost heaving is most efficient in silty materials and in climates where freeze-thaw cycles are frequent.

Graviturbation processes are more active as slope angle increases and as the effects of stabilizing vegetation decrease (Schumm 1967). Slope aspect, via microclimate, may also affect slope movement processes. Since graviturbation has been little studied from a pedological perspective (but see Pérez 1984, 1987a, 1987c), it is difficult to be specific about the type and amount of mixing caused by this vector of pedoturbation.

Anthroturbation

Humans are perhaps the most obvious contemporary soil mixers and modifiers, and yet our activities are often ignored, considered as unnatural. At this point, we draw the necessary distinction between soil *mixing* due to human action, mainly due to agriculture and cultivation, and soil *disturbance*, more often related to construction and mining activities. Our focus here is on mixing by humans.

Richter and Yaalon (2012) argued that one of the major, ongoing changes in the genetic model of soil

Ap horizon

Fig. 11.49 The clear, sharp lower boundary of the Ap horizon is evident (dashed line) in this Alfisol from northern Michigan. Most of the preexisting E horizon has been plowed up, into the Ap. Increments on tape are 10 cm. Photo by RJS.

involves the notion that soil is being transformed globally from a natural to a human-natural body. Although many "natural" soils remain on Earth, increasingly more soils are being modified and mixed by human action. Many kinds of anthropogenic soil changes exist, e.g., erosion, compaction, salinization, terracing, fertilization, and the effects of drainage and irrigation, to name a few (Bidwell and Hole 1965). Indeed, research focusing on anthropogenic soils is increasing worldwide (Glaser *et al.* 2001, Grieve 2001, Blume and Leinweber 2004, Richter 2007, Yaron *et al.* 2008, Certini and Scalenghe 2011, Chendev *et al.* 2012).

Anthroturbation includes all forms of soil mixing by humans, mainly involving agricultural and mining activities but also occurring in cities and in areas of obvious soil engineering (Ackermann *et al.* 2004). Humans plow the soil and in so doing quickly destroy all vestiges of previous horizonation in the plow layer, or Ap horizon. The telltale indicator of an Ap horizon is its planar base, usually about 25–30 cm deep, produced because moldboard plows have plowsoles that pull soil up from a specified depth (Fig. 11.49). Plowing and subsequent cultivation practices change the character of the plowed horizon by crushing structural units (peds) and by mixing together (often) several horizons into one. They also affect pedogenic processes deeper in the profile, e.g., plowing and tillage can

lead to the formation of an *agric horizon* immediately below the mixed zone. Agric horizons are subsurface horizons enriched in illuvial silt, clay, and humus (Soil Survey Staff 1999). Plowing renders the silts and clays in the Ap horizon mobile by breaking peds apart, allowing individual particles to be translocated. Facilitated by these large pores and by the absence of roots, turbulent flows of muddy water move into and through the base of the Ap horizon (Soil Survey Staff 1999). This water enters wormholes or fine cracks below the Ap horizon, depositing suspended materials there. Eventually, characteristic features of the agric horizon form: Worm channels, root channels, and surfaces of peds become coated with a dark-colored mixture of organic matter, silt, and clay, which can become so thick that it fills the holes.

Effland and Pouyat (1997) defined a subset of anthroturbation, *urbanthroturbation*, which *excludes* agriculture-related mixing. It includes activities one might find in cities, at mines, and wherever soils are scalped, filled, or otherwise modified for human use (Fanning and Fanning 1989). Names given to soils that have been markedly affected by urbanthroturbation include "Anthrosols," "urban earths," and "urban soils" (Sjöberg 1976, Woods 1977, Stroganova and Agarkova 1993, Howard and Olszewska 2010; see Chapter 12).

Bombturbation

In 2006, Hupy and Schaetzl introduced a 12th type of pedoturbation – *bombturbation* – coined for mixing of soils by explosive munitions, usually during warfare or military training activities (Table 11.1). Unlike the rare instances of extraterrestrial (meteoroid) impacts (*impacturbation*), mixing of soils by bombs and munitions is common worldwide (Certini *et al.* 2013). Deep and often overlapping cratering is a common microtopographic expression of bombturbation (Fig. 11.50). Although it could be argued that bombturbation is a type of anthroturbation or impacturbation (Hupy and Koehler 2012; Table 11.1), its effects are so dramatic that Hupy and Schaetzl (2006) felt it logical to single it out.

Bombturbation usually results in craters, marking the epicenter of an explosive ordinance, with rims of disturbed and mixed soil/debris surrounding them (Fig. 11.50). Because such explosions are nonselective, i.e., all of the soil is removed, mixed, and redistributed to the depth of the explosion – bombturbation is a proisotropic form of pedoturbation. Existing soil horizons are entirely destroyed or intimately mixed, and nearby soils become buried by the blasted-out debris. On some battlefields, such as near Verdun, France (Hupy 2006), and in parts of Vietnam (Hupy 2011), it is so prominent that, in places, little or none of the original soil surface remains undisturbed (Fig. 11.50B).

Craters left behind by bombs can dramatically influence the pathways of soil development, by becoming focal points for infiltration, litter accumulation, and pedogenesis

Fig. 11.50 Illustrations of bombturbation from the Verdun, France (World War I) battlefield. (A) Diagrammatic sketch of the soils (shallow to limestone bedrock) before and after an explosive munitions blast. (B) Photos of the land surface, showing the impact craters. Photos by J. Hupy and A. Arbogast.

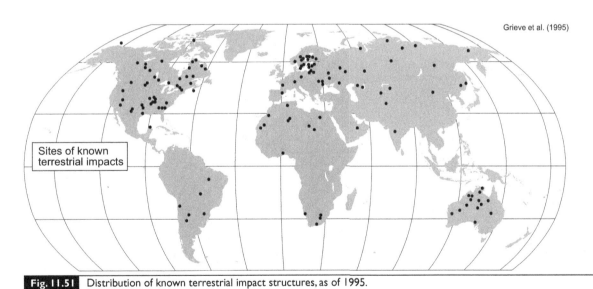

Fig. 11.51 Distribution of known terrestrial impact structures, as of 1995.

(Hupy and Schaetzl 2008). Subsoil materials, moved to the surface by the explosions, can dramatically affect the nutrient cycles and soil chemistry of the soils and surfaces where they land. In short, the soil landscape, at least when examined at small scales, can often bear the marks of warfare for extremely long periods.

Less-Studied Forms of Pedoturbation

Crystal growth, other than ice, has been shown to be an effective agent of weathering (Winkler and Wilhelm 1970, Ugolini 1986b, Theoulakis and Moropoulou 1999, Torok and Rozgonyi 2004). However, it has been less studied as an agent of pedoturbation, wherein it is referred to as *crystalturbation*.

It commonly occurs as the water in an ion-rich soil solution is lost to evaporation or root uptake. The resultant evaporite crystals grow and expand into the surrounding matrix, disrupting some soil material. The more common precipitates, some crystalline, that form authigenically in soils include sodium chloride (NaCl), gypsum ($CaSO_4 \bullet H_2O$), calcite ($CaCO_3$), pedogenic dolomite $CaMg(CO_3)_2$, opaline silica ($SiO_2 \bullet 2H_2O$), and jarosite ($KFe_3(OH)_6(SO_4)_2$) (Kittrick *et al.* 1982, Dixon and Weed 1989, Nettleton 1991). Most of these precipitates are too small as to be significant as pedoturbation vectors. However, when the volume of minerals is great relative to the soil matrix, or when their formation/dissolution frequencies are high relative to horizonation processes, they can be disruptive. Salt playas in desert regions seldom have any appreciable horizon development, primarily because of their unique hydrology but also because of salt crystal growth during repeated periods of hydration and drying.

Seismic activity is capable of moving and crushing solid rock; the soil that exists above that rock also is exposed to mixing processes as the seismic waves traverse through the crust. *Seismiturbation* is usually associated with small-scale forms of graviturbation and mass wasting, as the disrupted soil moves, slumps, slides, and flows into more stable positions.

Impacturbation of soils, a term coined by Johnson *et al.* (1987), has been little studied. At least 150 terrestrial astroblemes are known, and new ones are discovered annually, e.g., Dort and Dreschhoff (2002). These 150 astroblemes represent a small, biased sample of a much larger population (Grieve *et al.* 1995; Fig. 11.51). Meteors continually impact the Earth, and many occur on land that retains a cover of soil. Their effect on soils is small in comparison to that of other forms of pedoturbation, but nonetheless is large at the impact site.

Chapter 12

Models and Concepts of Soil Formation

Soils are complex. They exist at the interface of the lithosphere, atmosphere, and biosphere and function as integral components of the hydrosphere. They inherit, react to, and affect (seemingly simultaneously) all of these realms. Erosion, burial, climate change, biomechanical movement and mixing processes, water table effects, inputs of eolian dust, microclimatic effects of aspect and topography, and innumerable other nuances of the soil-forming environment all interact to form the most complex of natural systems – soil. Soil plays a key role in global energy, water, and geochemical cycles (Bockheim and Gennadiyev 2000; Fig. 12.1). To top it off, soils have no single end point toward which they are developing. Every one of Earth's soil bodies is on its own individual journey to a destination that may be impossible to envision. How fun!

Question: How can we possibly make sense of this commotion and sensory overload? The answer: By using *conceptual models* that help us understand the soil system and distinguish a signal from all the noise (Minasny *et al.* 2008, Bockheim and Gennadiyev 2010). Conceptual models are essential tools of science; they are simplified descriptions of natural systems (Drury and Nisbet 1971). Rather than being precise mathematical descriptions that can be solved with ample data, they are used to help put soil information into perspective and provide insight into the system interrelationships, process linkages, and nuances of pedogenesis and soil geomorphology (Jenny 1941b, Cline 1961, Dijkerman 1974, Conacher and Dalrymple 1977, Burns and Tonkin 1982, Phillips 1989, 1993b, Hoosbeek and Bryant 1992, Johnson *et al.* 1990). Models provide a way to organize, simplify, and enumerate the factors that affect soil systems and processes (Smeck *et al.* 1983, Bockheim *et al.* 2005). They help to view things in ordered ways, organize our thoughts, and provide a conceptual framework within which to consider facts (Johnson and Watson-Stegner 1987). Some models are complex, but most try to take the complex and make it simple. They help us see the big picture. By their very nature, models are simplifications of reality. In fact, sometimes the simpler the model, the better.

Unfortunately, models can also serve as mental blinders, preventing us from seeing aspects of reality that do not necessarily fit with the model. They can constrain our viewpoint on the world and force us to see it through a certain type of conceptual lens.

Conceptual pedogenic models have been critical in driving advances in the fields of soil science and soil geomorphology (Runge 1973, Huggett 1975, 1976a, Simonson 1978, Johnson 1993b, 2000). History is full of examples in which the application of certain pedogenic models influenced perceptions of soil genesis. Nonetheless, there are also examples in which models were ignored, despite being published, or were otherwise unknown (Johnson 1999, 2002). Had these models been employed, the ways in which we understand soil might have been very different.

In this chapter we present, apply, and evaluate several of the major conceptual models and explain how each provides a unique perspective on soil genesis and geomorphology. Our focus is on models that apply across all of soil science, rather than on models that may help understand the formation of only a certain suite, or type, of soils, e.g., Weisenborn and Schaetzl (2005b), Rabenhorst and Wilding (1986b). Those models are better discussed and evaluated in Chapter 13, which focuses on pedogenesis proper. By devoting a chapter to pedogenic models, we underscore the fact that they form the basis and foundation for soil research and explanation.

The conceptual pedogenic models presented in this chapter appear in their historical sequence, to allow us to observe how one model often was inspired by another or developed because of a shortcoming of a predecessor. Nonetheless, many models have done more than just build on a previous one. They have taken pedogenic theory in entirely different directions. Thus, just as soil development pathways are not always progressive, the models used to describe them, if examined temporally, also ebb and flow in their approach to pedogenesis.

It is our hope that the models discussed in this chapter will, as suggested by Smeck *et al.* (1983), lead to the formulation and development of even more advanced and

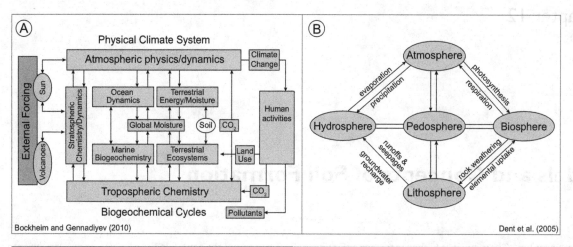

Fig. 12.1 Examples of linkage models showing interrelations among the various Earth systems, and how soils fit into these systems. (A) The Bretherton Diagram model for earth-system science. (B) Interactions among Earth's major spheres.

testable models. Because no model is perfect, we must always strive to develop better models – ones that help us better organize and understand our observations about the complex reality that is the soil landscape.

Dokuchaev and Jenny: Functional-Factorial Models

When it comes to longevity and perhaps applicability, the functional-factorial or state-factor model is the granddaddy of them all. It is often ascribed to Hans Jenny but was actually developed by Russian scholars. Detailed reviews of these models are found elsewhere (Jenny 1961, Dijkerman 1974, Smeck *et al.* 1983, Phillips 1989, Hoosbeek and Bryant 1992, Tandarich and Sprecher 1994, Wilding 1994, Minasny *et al.* 2008). In a functional-factorial categorization, the model envisions soil to be formed by the interplay among several *state factors*, and the model is written much as a mathematical *function* is:

$$S = f(\text{factor}_1, \text{factor}_2, \text{factor}_3, \ldots)$$

Hence the name.

Each factor is assumed to influence the soil through its effects on a variety of pedogenic processes, i.e., factors are not processes and do not *directly* cause soils to form. Functional-factorial models do not include processes directly within their framework. They leave this link to the person applying the model.

During the 1870–1880s, the Russian geologist/agricultural chemist V. V. Dokuchaev, largely under an agronomic and tax mandate, attempted to unravel and make sense of the distributions of, and genetic pathways within, soils on the loess-mantled Russian plains (Zavalishin 1958, Johnson and Schaetzl 2014). Dokuchaev viewed soils not just as an

amalgamation of weathered rock debris, as did others of his time, but as natural bodies worthy of study in their own right. His effort culminated in his treatise on Russian Chernozem soils (Dokuchaev 1883, Dobrovolskii 1996, Johnson 2000). Eventually, Dokuchaev and his colleagues, N. M. Sibirtsev and K. D. Glinka, formulated the first factorial model of soil formation (Dokuchaev 1886, Afanasiev 1927). The people who influenced him have been meticulously reviewed by Tandarich and Sprecher (1994) and are not discussed here. His model is formulated thus:

$$P = f(k, \Phi, g, v)$$

where P is soil (*pochva*) or soil properties, k is climate (*klimatz*), Φ is organisms (*organism*), g is subsoil (*gornaya poroda*), and v is age of the soil (*vosrast*) (Dokuchaev 1899). According to Florinsky (2012), the topography factor was n ot included, owing to a stenographer's mistake. In the original Russian publication, the equation shown here was actually preceded by two sentences discussing the important role of topography in soil formation. Later, Dokuchaev did formally add topography, or relief, as the fifth factor (Nikiforoff 1949).

Soil is clearly the *dependent variable* in this and most other models that we will discuss. Thus, in Dokuchaev's view, soil properties and the processes that interact to form soils are a function of interacting factors. Other scholars of the same era also developed similar state-factor equations of soil formation, all of which were expansions on, or modifications of, those originally proposed by Dokuchaev. Although these concepts were slow to filter out to other parts of the world, they eventually did. For example, the idea/equation/model was included in a German soils text by K. Glinka (1914), in a Russian soils text by Zakharov (1927), and then introduced to the Western world in C. F. Marbut's (1927b) translation of Dokuchaev's work and in a journal article by Shaw (1930). Wilde (1946b) later expanded on

Dokuchaev's model and defined soil (S) as the integration of three, somewhat reorganized soil-forming factors, acting through time:

$$S = \int (g, e, b)dt$$

where g is geologic parent material, e is environmental influences, b is biological activity, and t is time. The dt implies that the factors are integrated over time.

Factors of Soil Formation

Dokuchaev's model set the stage for the development of the most historically influential of soil genesis models – the state-factor model of Hans Jenny. A Swiss-born soil chemist, Jenny spent much of his career at the University of Missouri, before moving to the University of California at Berkeley. He was influenced, though not directly, by the Dokuchaev functional-factorial school of thought, developing his state-factor model in the 1930s and 1940s (Tandarich *et al.* 1988). Even though Jenny's ideas about soils as functions of various factors eventually were *applied* in soil survey and mapping, it was Jenny who encouraged this approach in the *study* of how soils formed and varied across the landscape. Jenny's publications in journals spoke to his early interest in factorial analysis of soil development (Jenny and Leonard 1934, Jenny 1935, 1941a, 1958). For historical background on Jenny and the development of this model, see Tandarich *et al.* (1988), Tandarich and Sprecher (1994), and Arnold (1994).

Jenny focused on attempts to clarify and quantify Dokuchaev's *five factors model* further (Johnson 2000). The model has, obviously, had a large and long-lasting theoretical impact on the pedologic community. As Johnson and Hole (1994) observed, the intellectual ambience that prevailed in 1941 was probably optimal for the formalization of the model that many had been using informally for decades, but mainly in Russia. By 1941, the factorial approach had become the principal paradigm of pedology. Its rise to prominence was based on its utility in soil mapping, classification, and human use (Johnson *et al.* 1990), appearing in a number of major American and international publications (Dokuchaev 1893, 1899, Glinka 1914, Neustruev 1927, Marbut 1927b, 1935, Joffe 1936, Byers *et al.* 1938, Thorp 1947, 1948). To recapitulate, the largely Russian pedological approach had, by the late 1930s, been accepted, endorsed, and widely promulgated in the Western world. No competing theory existed or was widely known (Johnson and Hole 1994). Thus, the stage was set for Jenny's 1941 book on the state-factor model – *Factors of Soil Formation* (Fig. 12.2) – to be a success, and the clear and simple language and elegant illustrations within did not disappoint. Still, the newness of the five factors concept was apparent. In fact, when Jenny first submitted his now-classic book, it was rejected for publication. It was five years before the book was eventually published, in 1941 (Yaalon 2000).

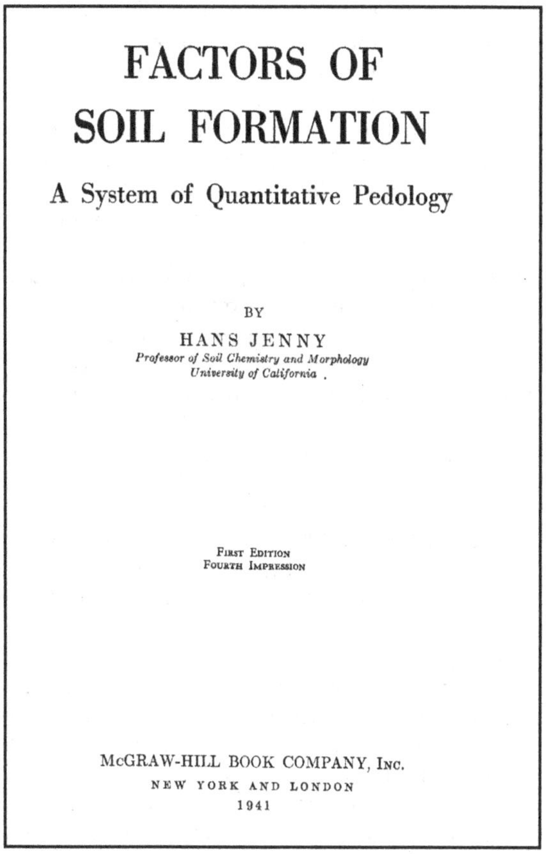

FACTORS OF SOIL FORMATION

A System of Quantitative Pedology

BY

HANS JENNY
Professor of Soil Chemistry and Morphology
University of California .

FIRST EDITION
FOURTH IMPRESSION

McGRAW-HILL BOOK COMPANY, Inc.
NEW YORK AND LONDON
1941

Fig. 12.2 The faceplate of Han Jenny's 1941 book, Factors of Soil Formation.

Out of the great complexity that was the soil environment, others had tried before (and after, see Stephens [1947]) to assemble various combinations and definitions of factors, but Jenny's model represented a final settling on *five* factors. The five state factors seemed to resonate, and like many successful models, it was sold well, in this case, in Jenny's 1941 book. Each of the factors is meant to define the state and history of the soil system (Wilding 1994) and are thus referred to as state factors. They are not *forces* or *causes* but rather *factors* – independent variables that define the soil system. The model, often referred to as the *clorpt model*, is

$$S = f (cl, o, r, p, t, ...)$$

where S is the soil or a soil property, cl is the climate factor, o is the organisms factor, r is the topography factor (r stands for relief, but the original intent was that of *topography*), p is the parent material factor, t is the time factor, and the string of dots represents other, unspecified factors that may be important locally but not universally, such as inputs of eolian dust, sulfate deposition in acid rain, or the effects of humans (Phillips 1999, Richter and Yaalon

2012) or fire (Ulery and Graham 1993). The factors define soil in terms of the *controls* on pedogenic processes and ultimately the soil patterns that evolve on the landscape (Wilding 1994). Jenny's *clorpt* model was his way of conveying the idea that if we could specify the state of the entire soil system, we would be able to predict soil properties at unsampled locations. In order to define the state of the system precisely, however, we would need to address and define at least five aspects of it. The five factors define the state of the system; hence they are referred to as *state factors*. They were not meant to explain *how* these particular conditions influenced soil properties, only that a given set of environmental conditions would *result in* a particular soil property. That is, the initial state and the end points were the foci of the model, rather than the processes that led from one point to the other. This was not a process model. According to Florinsky (2012, 450), the Jenny model, simply yet eloquently expressed in *Factors of Soil Formation*, was first used in this form by a contemporary of Dokuchaev named Sergey Zakharov. For that reason, he argued that the equation Jenny used "should be called the Zakharov equation."

Climate and organisms are considered the more active factors, whereas topography (relief), parent material, and time are passive, i.e., they are being acted upon by active factors and pedogenic processes. Johnson *et al.* (1990) discussed in some detail a very complete listing of passive and active *vectors* in pedogenesis. Vectors are different from factors, but the distinction between active and passive is nowhere drawn more clearly than in the aforementioned paper.

In this light, the parent material factor should be viewed as the initial state of the system, including its physical, chemical, and mineralogical characteristics, as well as all other inorganic and organic components. The model defines the soil in terms of the *controls* on pedogenesis and soil distribution factors, which Jenny called an environmental formula. By its presentation as an equation, the model implied that the soil system could be quantitatively investigated, at a time when the scientific community was hungry for quantification of physical systems.

The five factors model remains today the primary pedogenic model throughout the world. It is the standard against which all other pedogenic models are judged – the main model used to explain soil distributions at most scales. The model has allowed us to view the soil as part of a larger environmental system. This viewpoint facilitates descriptions and connections of soils with the rest of the physical world, providing a conceptual framework on which to hang all this information, which might otherwise seem impossibly complex and unrelated.

The model has perhaps its most utility at intermediate scales, e.g., in soil mapping (Ciolkosz *et al.* 1989). Field mappers use the model, mentally, to explain and predict soil variation *on the landscape*, as a function of the five factors (Jenny 1946, Johnson and Hole 1994, Scull *et al.* 2003).

Indeed, soil mapping may be viewed as a field solution to the state-factor equation. As Birkeland (1999: 142) put it:

In the field, coring and mapping the soils, one has to wonder why the soils differ. The differences in soil may be due to differences in parent material, topographic position, slope steepness, redistribution of moisture, vegetation, age of the associated landscape, etc. Because all of these can be seen or visualized in the field, in time a correlation of factors with mapping boundaries develops. Thereafter, the mapping proceeds at a more rapid pace and one can predict the location of contacts better.

The state-factor model is also a valuable teaching tool, still used in most introductory courses, since it is elegant and easy to understand. The value, utility, and comprehensibility of the factorial approach to the understanding of natural systems are underscored by the fact that it is used in related disciplines such as geoarchaeology (Holliday 1994), Quaternary geology (Birkeland 1974), paleopedology (Retallack 1990), and Earth surface systems (Huggett 1991). The state-factor approach has the distinct advantage that the factors/controls are generally observable and measureable. They vary spatially, as do soils, making this model a favorite of soil mappers and soil geographers.

The model attempts to explain the soil system characteristics in terms of external variables. But by doing so, the model actually reveals little about soil system dynamics or pedogenic processes. In fairness, however, that was never Jenny's aim (Yaalon 1975, Huggett 1976a, Wilding 1994). The link – the conceptual leap – between factors and processes must be made by the person using the model. And herein lies a shortcoming of Jenny's and every other factorial-type model. For example, one might use the functional-factorial model to surmise correctly that soils at the base of a hillslope will be different from those at the summit, because the relief factor (at least) is different. However, only through a *knowledge* of oxidation-reduction *processes* will one be able to hypothesize correctly *what* those differences should be, *how extensive* they might be, and *why* they occur. The state-factor approach suggests to us that soil A might be different from soil B (because at least one state factor is different), but not why this is so. In this sense, the model is focused on correlation, not causation.

Because the state-factor equation is a semiquantitative formula, it can be used to facilitate quantitative relationships among soils and the state factors. Many have attempted to "solve" the state-factor equation itself and have made incremental advances in the process. A solution would imply the ability to predict or describe the state of the soil system or a property thereof, in terms of the state factors (Phillips 1989). Mathematically *solving* the entire equation involves determining what the numerical coefficients of each factor in the equation are, and using them to define the nature of the environmental system and its interrelationships more quantitatively. The equation cannot be fully solved, however, partly because the soil system is so complex that we cannot ever define it well enough in terms of state factors, but primarily because the factors *are not independent*, e.g., organisms and soil covary (Stephens 1947, Yaalon 1975, Phillips 1993c). Indeed, time may be the only truly independent state factor (Chesworth 1973), and Jenny (1941b: 16) clearly acknowledged the lack of factor independence in his initial model formulation. Many view this as a conceptual shortcoming of *the model*, when it may, in fact, be a shortcoming of *our* abilities to gather enough of the right kinds of data! Because one might never be able to collect all the data necessary actually to solve the equation, what practical value would solving it have? As Phillips (1993c) put it, the inability to solve the state-factor model does not diminish its utility as a conceptual framework.

Johnson and Hole (1994) pointed out that the state-factor model falls short on two additional theoretical elements, both of which had been mentioned in pre-1941 publications. One element is the notion that soil morphological properties and conditions can *evolve* such that they, by their very presence, can cause profiles to change – more or less independently of any external environmental factors. These *pedogenic accessions*, important aspects of soil development, are not captured in the model (see later discussion). An example might be a soil horizon that continues to accumulate illuvial clay and develops to the point where it becomes an aquitard. This pedogenic accession (the slowly permeable Bt horizon) dramatically changes soil development by perching water and facilitating oxidation-reduction cycles in the overlying horizon. Ferrolysis acts on the clays and weathers them, leading to a soil that is clay-poor, acidic, and sandy (see Chapter 13). All these changes occur because of intrinsic processes within the soil, and they are not captured by the extrinsic environmental factors in the state-factor model. Soil properties and processes can and do emerge without any changes in the external soil-forming factors other than the passage of time.

Another missing element in the *clorpt* model is the absence of *biomechanical processes*, as originally defined in the *o* factor. Jenny's *o* factor concept was primarily centered on plants and their *biochemical* impacts on soil; this is understandable as Jenny was a soil chemist. The effects of animals on soil formation, now known to be dramatic in many different contexts (Hole 1981, Hall and Lamont 2003, Johnson *et al.* 2005, Richards 2009, Bartlett and Ritz 2011), are essentially dismissed in the state-factor model. To sum up, these are Johnson's (2000) concerns:

The five factors model is fine for students, and is useful as an explanatory-mapping model in soil survey work. Certainly scholars should be aware of the "whys" and "wherefores" of its impacts on the field. But as a major theoretical tool in research and graduate training, it is visionally restrictive.

We do not imply that the state-factor model *cannot* incorporate biomechanical processes, but rather that it *did not*, and those using it generally also *chose not* to include them

in application of the model. Thus, the lack of recognition of biomechanical processes in the myriad studies that have used the five factors model is not so much a major conceptual flaw, but rather an omission by the scientists applying the model.

Although the state-factor equation cannot be solved in a strict mathematical manner, and this is viewed by some as a serious shortcoming (Stephens 1947), many studies have attempted to isolate each factor or variable to assess its effect on the soil system evaluated in a quantitative manner (Richardson and Edmonds 1987). This approach leads to five equations, each of which describes a pedogenic functional relationship (Stephens 1947). Indeed, the functional-factorial model has been applied mainly through such pedogenic functions. Jenny (1941b: 17) anticipated this. He noted that, to ascertain the role or impact of a soil-forming factor X on pedogenesis, several soils must be examined in which factor X is allowed to vary, while all the remaining factors must be held constant. Thus, he obtained the following set of equations to describe the five possible scenarios and called each a *function*:

$$S = f_{cl} \text{ (climate)}_{o,r,p,t} \dots \qquad \text{climofunction}$$

$$S = f_o \text{ (organisms)}_{cl,r,p,t} \dots \qquad \text{biofunction or floral-function (plants only)}$$

$$S = f_r \text{ (topography/relief)}_{cl,o,p,t} \dots \qquad \text{topofunction}$$

$$S = f_p \text{ (parent material)}_{cl,o,r,t} \dots \qquad \text{lithofunction}$$

$$S = f_t \text{ (time)}_{cl,o,r,p} \dots \qquad \text{chronofunction}$$

Combined functions are also possible, though uncommon, e.g., a topolithofunction: $S = f_{r,p}$ (relief, parent material)$_{cl,o,t}$. ... Each state-factor function thus provides a formula within which to evaluate and explore the effects of that factor on soil development, or, theoretically, to predict S on the basis of one or more of the independent variables (Muckenhirn *et al.* 1949, Phillips 1993c). To do so, the investigator routinely examines a group of soils in which four (or less commonly, three) of the state factors are held constant and allows one (or, at most, two) to vary. The series of soils so examined is called a *sequence*, whereas the equation(s) derived from the soils is the *function*. For example, a series of soils that have all formed in the same parent material, are of the same age, and formed (presumably) under the same climate and vegetation would compose a *toposequence*. Only the factor of topography (relief) has been allowed to vary within this group or set of soils. The *equation* that describes clay content in the B horizon as a function of, for example, feldspar content of the parent material would be the *lithofunction*. The functional-factorial model of Jenny (1941b) has been used many times to examine climosequences, biosequences, toposequences, lithosequences, and chronosequences. In fact, therein lies the primary research utility of the model. Examples of each are briefly presented in the following.[1] Obviously, holding the other four factors constant, as much as is possible, is paramount to the formulation of any and all functions. It is also a major challenge, because the factors almost always covary, e.g., climate and organisms (Kohnke et al. 1968).

The Time Factor

Perhaps the most common of the sequences derived from the state-factor equation is the *chronosequence*. In a chronosequence, a series of soils of varying age are examined, providing valuable information about soil development pathways and rates, particularly regarding the various developmental stages through which soils may pass. Although chronosequences are discussed in more detail in Chapter 15, we emphasize some points here.

Critical to the development of a chronosequence is the notion of time$_{zero}$, or the time when soil formation started. More exactly, this notion equates to the time when parent material was exposed and a geomorphic surface stabilized, i.e., after any periods of erosion or burial. For many soils, this can be ascertained with reasonable certainty, e.g., those developing on a mid-Holocene ashfall deposit. In other situations, it is less certain, as with soils on a glacial moraine: You know the general age of the feature but are less certain about exactly when the surfaces on that moraine stabilized so that soils could start forming. For still others, such as an old plateau in the humid tropics, it might be acceptable simply to be within an order of magnitude of the age. Confounding all this is the notion that most soils are polygenetic (see later sections of this chapter for definitions) (Bryan and Albritton 1943, Johnson et al. 1990, Fuller and Anderson 1993, Mossin et al. 2001). In others, development is arrested at a certain moment in time by an erosion or burial event, resetting the clock and forcing a new time$_{zero}$ to begin (Mausbach et al. 1982). Finally, the accuracy of the chronosequence is always dependent upon the accuracy and availability of information about the ages of the soils or surfaces on which they have formed (Yaalon 1975). In many instances, relative ages of soils may be known, but without accurate numerical dating a chronofunction cannot be constructed, and hence, the rates of soil development cannot be ascertained.

Phillips (1993b, 2001a) raised another problem: Soil development can be so sensitive to initial conditions and can be so affected by small perturbations in the system that variability in development increases dramatically through time. Thus, soils of similar age can be radically different, even on the same site, further reducing the predictability of the *t* factor. Yet another problem with chronosequences is that a certain amount of constancy

of the other four factors must be assumed. Although this assumption may be acceptable for the *p* and *r* factors, it is highly unlikely that *cl* and *o* have been immutable, except in perhaps the very shortest chronosequences (Stevens and Walker 1970). Despite all these potential complications, chronofunctions have been valuable in comparing (1) rates of soil development, (2) how these rates may vary through time, and (3) the amount of time necessary for certain pedologic features or characteristics to develop (Birkeland 1978, Machette 1985, Mellor 1987, Birkeland et al. 1991, Helms et al. 2003).

In a chronofunction, time is the independent variable and some soil property or soil development index is used as the dependent variable (Fig. 12.3). The chronofunction is therefore usually presented as a best-fit regression line, fit to a scatter of points where each point represents a soil or soil property at a particular time. This tendency has led to a high degree of quantification of pedogenic properties as a function of time; chronofunctions tend to be presented statistically. Depending on the soil property, the data may be best fit to linear, logarithmic, exponential, or other types of functions (Schaetzl et al. 1994; see Chapter 15).

Some chronofunction data can be used to determine the *amount of time that was necessary* for certain pedogenic features or horizons to form (Fig. 12.4). This is a major goal of much chronofunction research. Another goal or group of goals centers on determining how soil development *rates change through time*, rather than just the amount of morphologic change from a certain time in the past, as in the first goal mentioned. For example, one might be able to determine whether soils reach steady state after a while, or whether, beyond some threshold, the rate of change increases or decreases markedly (Egli et al. 2001; Fig. 12.3).

As will be noted later in this chapter, continuous chronofunctions assume smooth pedogenic change through time. They often do not or cannot take into account perturbations in development that may have occurred in the past. For example, soils can regress briefly or endure periods of little or no development change, and morphologic evidence for this may be lacking.

The Parent Material Factor

Lithosequences represent soils that have developed in parent materials that sequentially change along some gradient. Parent material differences could be related to a number of factors, such as texture, mineralogy, coarse fragment content, and even parent material layering/stratification (Schaetzl 1991a, Graham and O'Geen 2010, Werts and Milligan 2012). A traditional type of lithosequence occurs in residual soils, because they form a natural laboratory of soil variation along a sequence of rocks (parent materials) that vary in mineralogy or texture (Cady 1950, Barnhisel and Rich 1967, Parsons and Herriman 1975). Changes in parent material can influence many soil properties. For example, coarser-textured soils generally develop faster

---

[1] The discussion of the various state factors, below, may seem uneven because others are discussed in more detail in other chapters. For example, relief and water table effects are discussed in Chapter 13 and paleoclimate is a focus in Chapter 15.

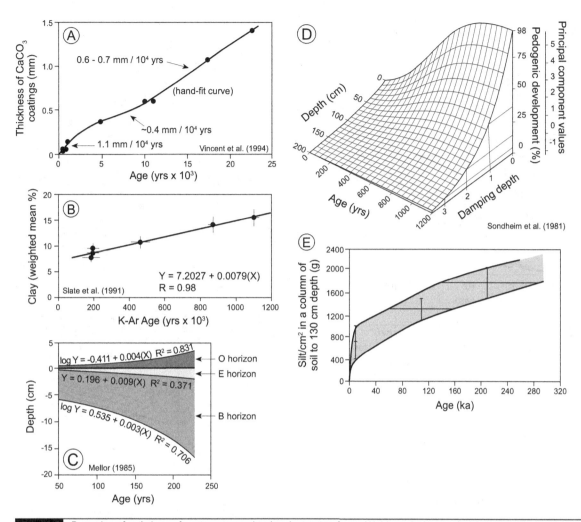

Fig. 12.3 Examples of soil chronofunctions, in graphical and equation form. (A) Mean thickness of pedogenic carbonate coatings on the bottoms of pebbles in some soils in Idaho. The chronosequence was developed for soils on a series of fluvial terraces. Note that the hand-drawn chronofunction line suggests that the rate of accumulation changes subtly through time. (B) Profile-weighted mean clay contents in some soils in Sonora, Mexico. The chronofunction was developed for soils on a series of dated lava flows. (C) Graphical display of horizon thicknesses versus time, for some soils in Norway. The chronofunction was developed for soils on a series of glacial moraines. (D) Three-dimensional plot of pedogenic development, as indicated by principal component scores versus age *and* depth. (E) Chronofunction showing the accumulation of silt over time in a series of soils on raised beach ridges in Spitzbergen. The envelope shows the total weight of silt in a 1-cm² column, to a depth of 130 cm. The width of the envelope was determined by error bars; horizontal error bars represent the age ranges for the soils, whereas the vertical bars are the range in silt weights.

and to greater depths because they have less soil volume and less surface area that needs to be pedologically altered (Schaetzl 1991b; Fig. 12.5).

In order for lithofunctions to be constructed and then readily interpreted, we must be able to assign numerical values to parent material characteristics, e.g., % sand, % gravel, $CaCO_3$ content, weathering potential, or porosity (Jenny 1941b, Pedro 1966). If this cannot be done, the default is to generate nonparametric lithofunctions, or to use some sort of parent material index for the dependent

variable – one that integrates a variety of parent material characteristics into one value, such as a principal component score (Yaalon 1975, Campos *et al.* 2012; Fig. 12.6). For this reason, lithosequences are commonly portrayed graphically, rather than as a scatter of points fit to a regression line (Fig. 12.7).

The Topography (Relief) Factor

Relief is not the same as topography. Technically, relief is the *relative difference in elevation* between an upland and a

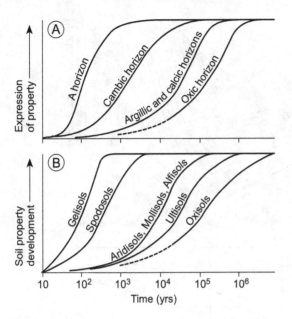

Theoretical curves, compiled from a variety of chronosequence studies, illustrating the time necessary for certain horizon types or soil orders to develop, assuming that their pedologic "clocks" are not reset. In part, after Birkeland (1999).

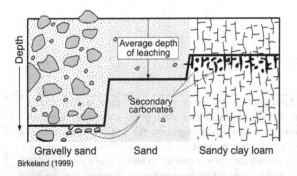

Birkeland (1999)

Hypothetical depth of leaching and soil development (as partially exemplified by carbonate translocation) in soils that have varying porosity and surface area. Texture and coarse fragment content directly affect porosity and surface area.

lowland within a specified region. Topography – which is what is implied in this state factor – is the *relative positions and elevations* of the land surface in an area, used to describe the configuration of its surface. Jenny used the letter *r* for relief, but he probably *meant* topography. Because of long-standing convention, in this book we will do the same. We will use *r*, but we will mean topography.

Quantitative functions examining the effects of relief on soils are termed *toposequences* or *catenas*, emphasizing the influence of topography. The relief factor carries with

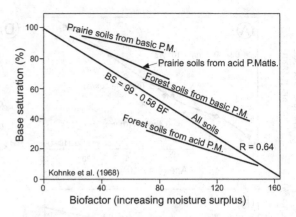

Quantitative relationships between a biofactor (a measure of precipitation excess, over and above environmental water demand) and base saturation, across an array of different parent materials and vegetation types.

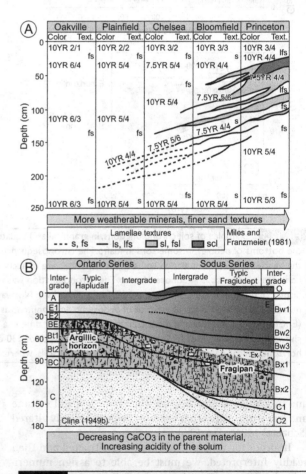

Graphical illustrations of the changes that occur in soils as a function of lithology. (A) Soil morphology along a sandy lithochronosequence in Indiana. The soils vary in age but also in texture and mineralogy. (B) Soil morphology along a lithosequence in New York State. The parent materials of the soils vary mainly in $CaCO_3$ content.

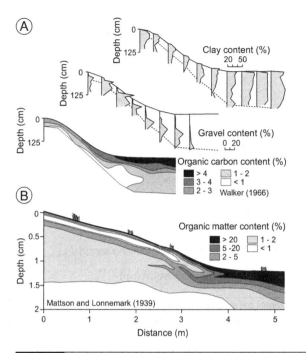

Toposequences – changes in soil properties along hillslopes. (A) Depth functions of clay content, gravel content, and organic carbon content for a series of soils on a slope in Iowa. (B) Organic matter in soils along a hillslope in Sweden. Large data sets provide the ability to draw isolines and therefore present a somewhat continuous surface.

it a number of subfactors, mainly slope and water table effects. Toposequences are often correlated with hydrosequences in which depth to water table varies, along with concomitant subfactors such as soil moisture, degree of oxidation within the groundwater, and vegetation differences (Boersma *et al.* 1972, Knuteson *et al.* 1989, Daniels and Buol 1992). Also included within the relief factor are the concepts of elevation, important in mountainous terrain; slope aspect (compass direction that the slope points toward – see Chapter 14); and slope gradient or steepness (Acton 1965). Finally, the three-dimensional slope curvature of the surface must also be considered, for certain slopes are water-concentrating slopes while others are divergent, e.g., nose slopes (see Chapter 14).

Variability in soils across short distances is often best explained by the relief factor. Within a few meters horizontally soils can be radically different because of varying water tables, location on steeply sloping erodible sites, or location on sites at the base of the slope where sediment from upslope accumulates. Microrelief, in particular, causes great amounts of short-distance variability in soils (see Chapter 14). The diagrammatic portrayal of soil properties in toposequences can take many forms; the most useful is perhaps the kind in which variations in soil morphology or

chemistry are portrayed along them (Fig. 12.8). Additional details about the effects of relief and topography on soils are discussed in Chapter 14.

The Biotic Factor

Both Jenny (1958) and Crocker (1952) defined the biotic factor as the potential floristic list, or the potential natural vegetation of the site. What *actually grows there*, according to Jenny, is dependent on the other state factors. As mentioned, this factor had, in Jenny's (1941b) original interpretation, primarily a biochemical meaning and was more or less restricted to flora. Jenny's (1941b) research emphasis with regard to the *o* factor was on the nitrogen and organic matter contents of soils, and on the parallelism between plant succession and soil development. Jenny even referred to the *o* factor as the "plant factor" (Jenny 1958). Biomechanical effects resulting from the burrowing activities of soil fauna were largely ignored. Jenny's research agenda on biochemical functions, as well as the paucity of information on bioturbation at this point in time, probably led to his interpretation and bias. The biomechanical component, which can occur as animals and plants move and rearrange soil, has only recently become a focus of soil research (Johnson and Hole 1994). It is not that Jenny did not consider animals as being part of the *o* factor, but, as he stated, "because of lack of sufficient observational data …, the discussion of animal life is omitted" from his book (Jenny 1958: 203). Although many studies on the effects of animals on soil development had been published in the early years of pedology and soil science, they were by and large ignored by both Dokuchaev and Jenny. Only later would the literature catch up (Johnson 2002, Johnson *et al.* 2005), as scientific papers on the effects of animals became increasingly prominent, e.g., Thorp (1949), Hole (1981), Johnson (1989, 1990, 1993a), and Cox and Scheffer (1991).

The biochemical/biotic factor is one of the more difficult to isolate and tease out within the functional-factorial framework, across small or medium-sized areas, because plants (and animals) rarely function independently of the other factors. In fact, it is more likely that vegetation is a function of soil, rather than that vegetation is the independent variable that is stated in Jenny's equation (Noy-Meir 1974). As a result of these interdependencies, Yaalon (1975) felt that true biotic functions were rare and suggested that biotic attributes should be treated more as dependent, rather than independent, variables.

Plant communities continually change through time, naturally by succession but also by disturbance events, adding another level of complexity to the biofunction. These interdependencies again make it difficult to generate a true biosequence, which assumes generally uniformity in the plant factor through time (Crocker 1952). Plus, even when climate, relief, and time are controlled for (usually within a small area), it is difficult to find vastly different plant communities while holding parent material constant (Curtis

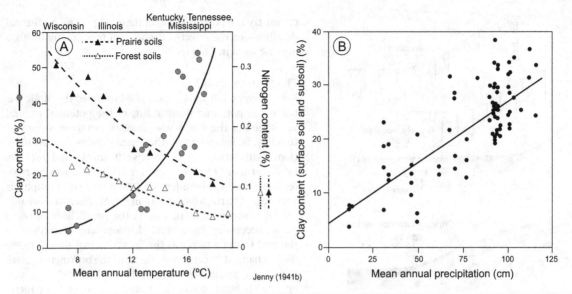

Fig. 12.9 Quantitative relationships between climate and soils. (A) Mean annual temperature versus clay content (probably of the most clay-rich horizon), and versus N content of the A horizon. (B) Mean annual precipitation versus mean clay content (to 100 cm depth).

1959). Thus, most biosequences cover large areal expanses, often from forest into grassland or from forest into tundra (Liptzin and Seastedt 2010). They are rarely pure biosequences, because climate covaries with biota (Shrader 1950, White 1955, White and Riecken 1955, Bailey *et al.* 1964, St. Arnaud and Whiteside 1964, Severson and Arneman 1973).

Another way to approach biosequences commonly analyzes how a *change in vegetation* affects existing soils (Fuller and Anderson 1993). The researcher assumes that, at some point in the past, there was a vegetation change in some areas but not in others. Analysis of the effects of these varying biotic changes can present a useful type of biosequence (Geis *et al.* 1970, Barrett and Schaetzl 1998, Mossin *et al.* 2001).

The Climate Factor

After early soil scientists began to realize that soil was more than just weathered rock, climate replaced parent material as the most important soil-forming factor. Russia has a suite of east-west trending climate zones that are uninterrupted by mountains or oceans, such that early pedologists traveling on a north-south transect could not help being swayed by the effect of climate on soils – especially across large areas (Kohnke *et al.* 1968). Hilgard's (1906) early classification of U.S. soils initially used the division between soils of arid and humid climates. For these reasons, the *cl* factor was originally defined as the regional climate, although clearly the climate that the soil reacts to is influenced by the biotic cover that lies between the soil and the atmosphere, by slope gradient and aspect, snow cover, and many other microclimatic factors. Thus, *soil climate* is often

quite a different thing from regional atmospheric climate (Schaetzl and Isard 1991, Johnson-Maynard *et al.* 2004).

Although climate is clearly one of the most important state factors (Yaalon 1983), climosequences generally, by their very nature, are difficult to isolate. First of all, climate is seldom independent of biota; therefore, climobiosequences are more common and easier to study than are pure climosequences (Brye *et al.* 2004). Second, by its very nature, climate is variable and changing. Thus it is difficult to know how the various aspects of climate along the modern climosequence compare in space and magnitude to climates of the geological past. What is a dry tropical to humid tropical sequence today may have been very different in the past. Deserts may have been wetter; the seasonal timing of precipitation, so important to pedogenesis, may have changed even if the annual precipitation totals did not. Although climosequence data must be viewed cautiously, where a good climosequence has been teased out or modeled, the results can be very useful in the interpretation of pedogenic properties (Arkley 1963).

In most climosequences, the primary climatic variables (independent variables in a regression equation) are mean annual precipitation, mean annual temperature, or some index that measures water need or evapotranspirative demand, of which many exist (Lang 1915, Thornthwaite 1931, Yaalon 1975, 1983, Joshi *et al.* 2003, Tsai *et al.* 2010). Only one of many possible climatic inputs is regressed against soil properties such as nitrogen content of the A horizon, depth of leaching, clay content, and mineralogy, and translocation depths of any of a number of soil components (Kohnke *et al.* 1968; Figs. 12.6, 12.9). One can immediately see the difficulty of capturing the essence of "climate" in one or even a few variables, which is why well-crafted climofunctions are a challenge.

Note the complexity of the arena – and potential – of climate-soil relationships. In this era of rapidly changing climate, it has emerged as a clear research focus. Effects of a changing climate on nutrient cycling, food production, soil quality, and carbon dynamics are areas of great research interest, and understandably so. Some of this research is informed by traditional climosequence studies, but increasingly more of it is based on modeling. Regardless, models of soil change under a changing climate must start with accurate base data – the kind gleaned from soil characterization and functional studies. We must know what soils are like, why they are distributed where they are, and how they relate to the soil-forming factors, before we can accurately model them. An alternative track to soil-climate studies has traditionally been to examine soils formed under past climates as analogs – studies normally cataloged under the rubric of paleopedology, or the study of paleosols (see Chapter 16). Perhaps no area of soils research offers so many challenges, or so many potential rewards, as the interrelationships between soils and climate.

Zonation of Soils in Alpine Regions: Where Climo-, Topo-, and Biosequences Meet

Like biosequences, climosequences tend to be studied over larger areas, e.g., from a desert to a humid climate, from the core of a desert to its fringes, or across the tundra-forest ecotone (Buntley and Westin 1965, Yair and Berkowitz 1989). However, a common form of climosequence is one that examines soils (recognizing, as always, that vegetation often parallels climate) up a mountainside, or within a mountain range (Retzer 1956, Whittaker *et al.* 1968, Hanawalt and Whittaker 1976, Nettleton *et al.* 1986, Dahlgren *et al.* 1997, Darwish and Zurayk 1997, Trifonova 1999, Tsai *et al.* 2010). Generally, air temperatures become predictably cooler and evapotranspirative demand diminishes as elevation increases, and moisture/precipitation changes as well (usually increases). If well-drained soils can be chosen at all sites, these predictable trends provide a natural laboratory for the establishment of an areally tight climosequence that often also has good control on lithology and relief (Cortes and Franzmeier 1972). The environmental lapse rate – the change in air temperature with elevation – varies as a function of water vapor content, but is generally about 6.4°C per 1,000 m. Thus, one can assume that a location at 3,000 m will be, on average, about 19°C cooler than one at the base of the mountain at sea level. As air temperatures become cooler with elevation, precipitation usually increases as well, although in very high mountains a drop-off in precipitation occurs at the highest elevations, because the cold air at this height cannot hold as much moisture. However, the continued lowering of evapotranspirative demand with elevation may act to offset the drier conditions in the high alpine areas, rendering them climatically humid.

The sequences of soils that occur along an alpine climosequence (or climobiosequence) vary greatly from region to region. It is difficult to generalize. The soils and vegetation at the base of the sequence are dependent upon the general macroclimate of the region; the soils could be Aridisols, Mollisols, and so on, with variations in elevation, climate, and vegetation. The units of the sequence are usually expressed as discrete vegetation *zones* that wrap around the mountain but are lowest on the warmer, drier southwest slopes (Fig. 12.10). In all but the lowest and highest mountain ranges, a forested zone occurs somewhere along the sequence (Amundson *et al.* 1989a). In mountain ranges with multiple forested zones, the uppermost one is usually a coniferous zone, which gives way to the treeless *alpine zone* at height (Fig. 12.10). The ecotone between the uppermost forest and the alpine zone is called the *alpine treeline*; it is usually not a distinct line but a series of disjunct patches and outliers/inliers of stunted trees. As the climate becomes more severe with increasing elevation, trees become limited to sites that are sheltered, appearing only in patches or groups.

Soils change markedly along with vegetation and climate in alpine climosequences. Generally, effective moisture and organic matter contents increase with elevation, to a point (Chadwick *et al.* 1995, Djukic *et al.* 2010; Fig. 12.11). This trend occurs because biomass production increases (to a point) and then decreases along the elevation transect. But concomitant with that, litter mineralization decreases steadily as temperatures decrease with elevation. The ebb and flow of the intensity of the various pedogenic processes along alpine transects generally correspond to local climate, which not only changes with elevation but also is affected by aspect (see Chapter 14; Fig. 12.12). For example, calcification in soils at the base of mountains (in desert or dry grassland areas) diminishes with elevation as leaching increases. Leaching, however, then becomes less pronounced in the high alpine zones as precipitation decreases.

For the remainder of this section, our focus will be on the soils of the high alpine, i.e., the areas above treeline, as soil patterns here are easier to generalize than they are along the entire mountain slope. Costin (1955b) defined *alpine soils* as occurring above treeline where the ground is snow-covered for more than half of the year (Retzer 1956, 1965). Although pedogenesis is typically slower here than in warmer climates, many of the same processes are operative, providing that a similar suite of soil-forming factors exists (Stützer 1999).

By many definitions, alpine soils are weakly developed, mainly because there is a short growing season, but also because they form on slopes that are often erodible and unstable (Grieve 2000, Birkeland *et al.* 2003; Fig. 12.13). Sola are increasingly thinner and soils are less developed with

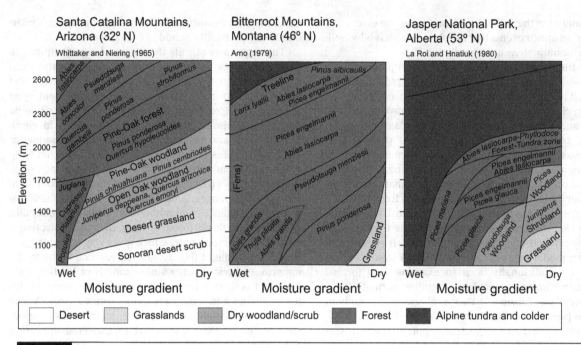

Fig. 12.10 Vegetation-elevation mosaics for three mountain ranges in western North America. *Wet* and *Dry* sides imply the NE and SW slopes of the ranges, respectively.

elevation in the alpine zone, until they are little more than regolith with interspersed patches of bedrock, covered for much of the year with snow and ice. This highest/coldest area of the alpine (except for the areas of perennial snow and ice) is often called the shatter zone or blockfield zone (Rudberg 1972). Many alpine soils are shallow to bedrock, and most are coarse-textured, rocky, and gravelly, reflecting the dominance of physical weathering and freeze-thaw activity. Cryoturbation is dominant here, not just because this is a cold place but also because it has strong variations in diurnal temperature (Harris 1982). The thin air and high elevations promote rapid nocturnal cooling, and at the same time the intense insolation at high altitudes allows for rapid heating during the day. Thus, freeze-thaw cycles are commonplace, and physical weathering by ice crystal growth is a dominant weathering process. Because physical weathering processes dominate over chemical weathering, many alpine soils are coarse-textured (Gao and Chen 1983). Here, water, which is so necessary for chemical weathering, is minimal and frozen for much of the year. In summer, however, small amounts of hydration and chelation are operative (Ellis 1980a).

Alpine soils are unique enough that they are often studied and discussed individually. Nonetheless, they vary dramatically across short distances, from Histosols to Gelisols to Entisols and Inceptisols. Topography is the most important soil-forming factor affecting soil patterns. Many of the steeper slopes are unstable; solifluction and processes associated with patterned ground are common (Benedict 1970,

Grieve 2000). Sags and cols are areas of accumulation of snow and sediment.

The alpine zone is also a very windy place. Wind direction and speed are, of course, affected by topography (Thorn and Darmody 1980, Dahms 1993). Many fine materials, like silt and clay, are transported across this landscape by wind and removed in the same manner. Thus, thin eolian caps are common in alpine soils (Litaor 1987, Dixon 1991, Muhs and Benedict 2006).

Largely because of the persistent winds – in all seasons – and rugged topography, soil patterns become patchy within the alpine zone and each of its subzones. Long-lasting snow patches form in the lee of wind obstructions, e.g., tree islands or large rocks. These features, both the snow patches and the tree islands, have a dramatic effect on soil development (Stepanov 1962, Ellis 1979, Grieve 2000; Fig. 12.14). In protected areas, snow and litter can accumulate; soils there are some of the deepest and most leached of anywhere in the alpine zone (Stepanov 1962). Snowmelt water facilitates weathering and soil development, and eolian material within the snow is added to the soil (Thorn and Darmody 1980), enhancing any fine-textured mantle that may have been present. Then, after the fine-textured mantle has been established, the long-lasting snow patches help to reduce its subsequent erosion by wind and water. Other soils, in exposed locations, are desiccated and eroded by the strong winds.

One way to make geographic sense of the alpine is to use geomorphic models to subdivide the area into areas

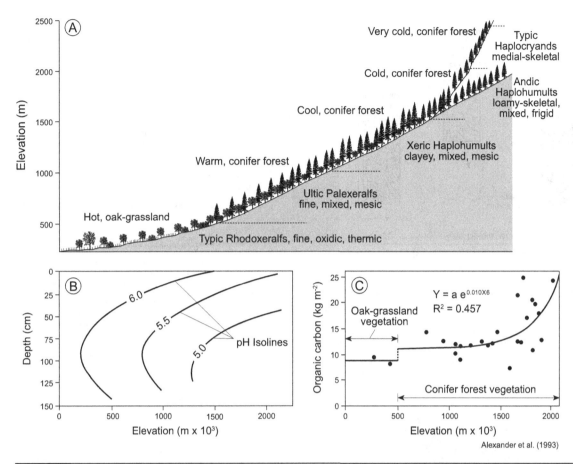

Fig. 12.11 Trends in vegetation and soil properties with elevation, in the Cascade Mountains, California. (A) Dominant soil types and plant communities, idealized. (B) pH trends with depth and altitude. (C) Organic carbon contents in the upper meter of soil. Note the change due to vegetation.

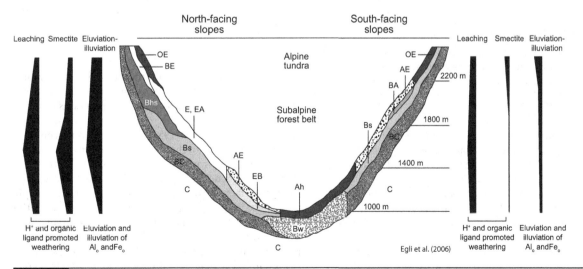

Fig. 12.12 Idealized altitudinal sequence of soil characteristics, summarized by elevation and aspect (N versus S), for the Alps of northern Italy. Broader areas on the graphs indicate higher concentrations or more intensive processes.

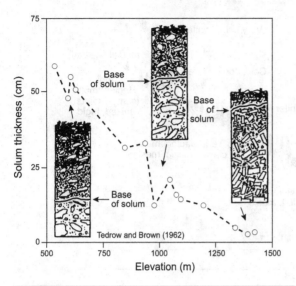

Fig. 12.13 Changes in soil thickness and morphology with elevation in the Brooks Range of Alaska. Most of the soils are Cryepts. The dotted line represents maximum solum thickness observed at that altitude. The arrows show the base of the solum.

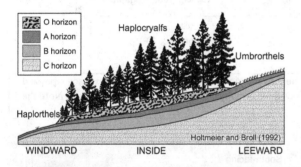

Fig. 12.14 Soil horizonation and classification as affected by an alpine tree island. Snow accumulates on the leeward side of the island, whereas at windward sites soils become eroded and are thinner. Soil development is strongest within the wind-protected tree islands, where thick O horizons can form.

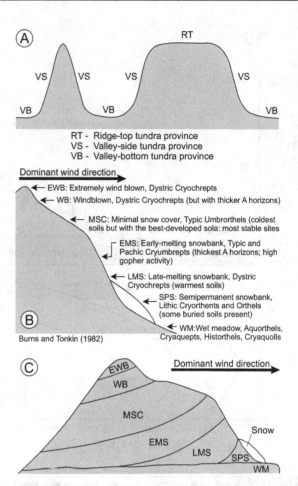

Fig. 12.15 Alpine geomorphic/soil provinces. (A) The major alpine geomorphic provinces. (B) Microenvironmental sites on the lee slopes, based on the Synthetic Alpine Slope (SAS) model. Dominant soil types and their characteristics (in the southern Rocky Mountains) are listed. (C) The SAS model for both windward and lee slopes in alpine environments.

of relative uniformity. Burns and Tonkin (1982) developed a classification scheme for alpine regions, based on work done in the southern Rockies; it has found widespread use (Fig. 12.15). On a gross scale, the alpine can be divided into ridge tops, valley sides, and valley bottoms. Periglacial activity dominated soil genesis on the ridge tops, which often generally escaped Late Pleistocene glaciation. The slopes here are fairly stable and, although heavily cryo-turbated, have thicker profiles than at lower elevations. Snowpack duration and distribution are key determinants of soil character and vegetation is this zone (Isard 1986, Litaor *et al.* 2008). The steep slopes of the valley side tundra

areas (Fig. 12.15), coupled with recent glacial activity, interact to create slope instability, leading to thin, rocky soils and rock outcrops. Glacial and fluvial processes dominate the valley bottoms.

Burns and Tonkin (1982) developed an even more detailed assessment of the alpine zone, taking into account windswept ridge top areas, called the synthetic alpine slope (SAS) model (Fig. 12.15). The model incorporates the major factors that control the distribution of alpine soils: aspect, topography, seasonal snow accumulation, distribution of plants, and alpine eolian sediments. All of these components are rarely seen on one slope but form a mosaic within the alpine zone. The SAS model is based on the assumption that spatial variations in soil characteristics follow topographically controlled variations in snow cover, the distribution of which dramatically affects soil development. Because wind and topography influence snow cover and melting more than other local-scale factors, the SAS is largely based on how topography impacts wind and drifting snow. The SAS model illustrates that the highest alpine sites are often the most windblown and snow-free and, hence, the driest and most prone to freeze-thaw cycles. Areas with long-lasting snowpacks tend to have minimal vegetation and rocky soils. From the perspective of this soil-geomorphic model, alpine soil patterns and characteristics, especially A horizon thickness and degree of eolian contribution, become more predictable and understandable.

Humans as a Soil-Forming Factor: Anthrosols

Early pedological ideas, developed in the eighteenth and nineteenth centuries, centered on soil as a natural body worthy of its own scientific investigation (Table 1.1). For more than a century, scientists examined the soil system as one that had developed from a complex of natural processes (Richter 2007). However, by the mid twentieth century, the effects of humans on soils became increasingly evident, and today, it has become a focus of considerable discussion and research (Tugel *et al.* 2005, Certini and Scalenghe 2011). More than half of Earth's soils are cultivated, grazed, or periodically logged for wood. Ever-increasing are tracts developed for residential, industrial, transportation, and recreation or used for waste disposal.

The human impact on soils can no longer be ignored or considered aberrant, as pedologists are becoming increasingly focused on the science and management of human-affected soils (Richter 2007). Indeed, many now consider the last 200–2,000 years a period of the Holocene that is so unique as to merit its own name – the *Anthropocene* (Crutzen and Stoermer 2000). This designation singles out the current epoch of Earth's history, during which humankind's influence has met or exceeded those of the other environmental forces that have shaped the planet and its soils (Certini and Scalenghe 2011).

How does the ever-increasing impact of human agency fit into the state-factor model? Although Jenny (1941b) recognized that humans have an effect on soils, as part of the *o* factor (Bidwell and Hole 1965), we argue that human agency influences soil development through any or all of the state factors. Jenny did, however, note that "a considerable number of human influences on soil appear to stand in no direct relationship to soil-forming factors" (Jenny 1941b: 203). Humans certainly influence the *t* factor by resetting the clock in the same way that geologic or biologic events might, and by forming new materials and surfaces for pedogenesis. Effland and Pouyat (1997) noted that human effects can be incorporated into the state-factor model in either of two ways: (1) They can be considered on an equal basis with other organisms and hence included within the *o* factor, or (2) they can be given "factor status" as

$$S = f(a, cl, o, r, p, t, ...)$$

where *a* is the anthropogenic factor, or as Yaalon and Yaron (1966) called it, the *metapedogenic factor*. Adding the sixth factor, *a*, increases the interdependence of the factors, because humans can and do alter four of the remaining five factors. Another problem associated with the *a* factor is related to the length of time it has been operative. Because of the relatively brief time span over which humans have altered soils, the *a* factor can rarely be given equal status with the other five factors. That is, humans may be better termed soil *modifiers* than soil *formers* (McPherson and Timmer 2002; Table 12.1). Last, when the human influence is strong, the other state factors are usually forced to change, e.g., to cultivate a soil one must remove the natural vegetation or drain it (Montagne *et al.* 2008, 2009).

The effects of humans are often treated only minimally in many pedology books. Although we examine some aspects of the *a* factor, we focus only minimally on, for example, the pedogenic effects of agricultural practices. Certainly, however, long-standing agriculture (and its associated soil drainage and irrigation) dramatically affects soil properties and processes (Anderson and Browning 1949, Skidmore *et al.* 1975, Martel and Mackenzie 1980, David *et al.* 2009, Montagne *et al.* 2009, Chendev *et al.* 2012). In Europe, this is well known, as long-term manuring and cultivation lead to the formation of plaggen and mollic epipedons (Pape 1970, Conry 1971, Blume and Leinweber 2004). In the humid tropics, indigenous peoples have long been deliberately altering the naturally infertile soils to make them more productive, mainly through additions of biochar (Fraser *et al.* 2011). The resultant soils, called Amazonian Dark Earths, or *terra preta*, are currently an object of considerable study by scientists interested in sustainable agriculture in the tropics (Glaser *et al.* 2001, Woods *et al.* 2008, Glaser and Birk 2012). Indeed, anthropologists and archaeologists have long referred to human-impacted soils (informally) as anthropogenic soils or *Anthrosols* (Sjöberg 1976, Woods 1977, Delgado *et al.* 2007). The term has caught on with soil scientists as well (Dazzi *et al.* 2009), although it has not yet received formal recognition as a taxonomic class.

Records of human impact on soils, dating back more than 7,000 years (Zi-Tong 1983), could fill volumes, but here we only summarize some direct and indirect anthropogenic effects (Bidwell and Hole 1965; Table 12.1). Directly, humans till and terrace the soil, facilitating erosion and/or leading to overthickened profiles (Chang 1950, Sandor *et al.* 1986a, Smith and Price 1994). Warfare and its associated activities dramatically affect soils (Certini *et al.* 2013). People construct burial and ceremonial mounds (Bettis 1988, Cremeens 1995; Fig. 12.16), and in urban areas, fill materials are often added to the soil surface (Delgado *et al.* 2007, Howard and Olszewska 2010). Amendments, such as manure, lime, and fertilizers, are added (Gaikawad and Hole 1965). Although long-term

Table 12.1 | Examples of anthropic soil modifications

Manipulation	Principal processes observed in the soil
Topographical features	
Terracing or land leveling	Reduction of erosion; humus content increase; rejuvenation of pedogenic processes; altered catenary slope differentiation
Dam construction	Cessation of sedimentation and leaching; water table rise; salt accumulation
Draining of wetlands and mining	Surface subsidence
Hydrological factors	
Drainage, lowering of water table	Improved oxidation; structure formation; permeability change
Planting of windbreaks	Changed moisture regime; altered base cation saturation; carbonate leaching
Dredging of bays, harbors, and wetlands	Additions of dredged materials onto otherwise undisturbed soils
Flooding of paddy fields	Hydromorphic water regime; reduced oxidation; gleying
Changing the hydrologic cycle by cloud seeding and irrigation	Potentially increased leaching or translocation; additional soluble salt accumulations
Chemical factors	
Irrigation with sodic water	Adsorption of sodium; structural degradation; decrease in permeability
Clay marling or warping	Textural change in upper horizons; moisture regime change; altered base cation saturation
Cultivation and cropping factors	
Deforestation and cultivation in temperate areas	Mixing of upper horizons; change in pH and nutrient status; retardation of natural pedogenesis
Deforestation and shifting cultivation in tropical areas	Erosion; drainage and drying of soil materials
Overgrazing	Erosion of surface horizons; reduction in infiltration

Sources: Yaalon and Yaron (1966), Bidwell and Hole (1965).

manuring can lead to overthickened, organic-rich A horizons, cultivation without large additions of organic materials usually results in decreased organic matter contents in soils (Fig. 12.17). Continued human occupation, especially in prehistoric time, often creates elevated phosphorus levels in the soil, which if high enough qualify the epipedon as anthropic (Eidt 1977, Sandor *et al.* 1986a, Leonardi and Miglavacca 1999, Soil Survey Staff 1999, Lima *et al.* 2002, Migliavacca *et al.* 2012; Table 8.4). The sources of the P are commonly associated with bone and kitchen middens – residue related to mealtime activities (Kaufman and James 1991). For this reason, elevated P levels in soils are commonly used by archaeologists as being diagnostic for (paleo)human habitation (Eidt 1977, Sandor *et al.* 1986c, Schlezinger and Howes 2000).

Human impacts on soils vary considerably. They may be beneficial or they may cause degradation. They may be so dramatic that in some cases soil classification and soil maps must be changed to reflect the new soil characteristics (Veenstra and Burras 2012). In other cases, cultivation

and human action cause little change, even over long timescales (Table 12.2). Thus, some aspects of soil are more fleeting under human influence; some are more persistent. Research on this topic continues.

Urban soils, urban earths, and other forms of disturbed land are emerging expressions of Anthrosols (Indorante and Jansen 1984, Stroganova and Agarkova 1993, Marcotullio 2011). In cities, the soilscape is often a mix of natural soils and those affected by human actions (Alexandrovskaya and Alexandrovskaya 2000). Effland and Pouyat (1997) described an *anthroposequence* as a series of soils, often from the city center to the countryside, in which the human effects change along a predictable gradient (McDonnell and Pickett 1990). They went on to coin the term "anthropedogenesis" for the role of humans in changing the pathway of soil formation.

Urban Anthrosols are directly impacted by human activity in many ways. Nonpoint source pollution is a particularly important human impact in large urban areas, leading to high concentrations of metals and other contaminants

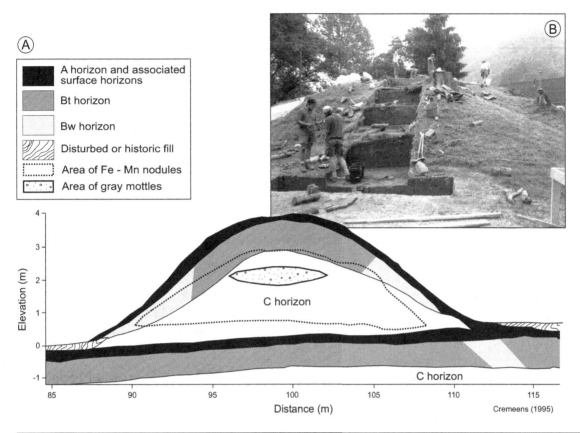

Fig. 12.16 Soil development in the sediments of Cotiga Mound, a mortuary mound from the Early Woodland Period, West Virginia. (A) Soil development in Cotiga Mound, as seen in cross section. (B) Mound excavation. Photo by D. Cremeens.

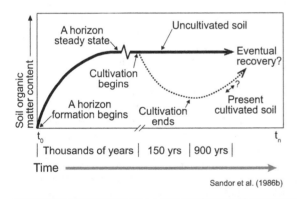

Fig. 12.17 Schematic diagram of changes in organic matter content in soils under natural and cultivated conditions, for a site in New Mexico.

(Laidlaw and Filippelli 2008, Wei and Yang 2010). They are modified by compaction, additions and removals of chemicals, and overthickening where spoil or fill material is added to a site to raise its elevation or to provide foundation materials (Parker *et al.* 1978, Kahle 2000, Scalenghe

and Marsan 2009). Soils intercept pesticides, contaminants, and other toxic substances and often contain fragments of human activity such as glass, bricks, wood, nails, and ash (Effland and Pouyat 1997). Large parts of many cities actually are built on spoil, rather than natural soils and sediments (Short *et al.* 1986a, b, Buondonno *et al.* 1998). Humans also mine, excavate, and backfill areas outside cities, sometimes creating immense areas of disturbed land, and, in so doing, reset the pedogenic clock (Shafer 1979, Ciolkosz *et al.* 1985, Strain and Evans 1994).

Indirectly, human alteration of soils is ongoing and can reach well into the hinterland, beyond direct human habitation (Stottlemyer 1987, Burghardt 1994). Exhaust from motor vehicles and emissions from factories add metals to soil (Parker *et al.* 1978, Coester *et al.* 1997, Hiller 2000, Perkins *et al.* 2000). Road dusts settle on nearby soils. Acids from coal-burning factories introduce inordinate amounts of sulfur into the soil system, far from the cities that are the sources. Changes in plant communities, for example, by reforestation practices, can also change pedogenic pathways (Mossin *et al.* 2001, Bonifacio *et al.* 2006). The list could go on, but suffice it to say that disturbed, urban, and anthropogenic soils are increasing in area and are rapidly

Table 12.2	Soil properties grouped according to their rate of change in response to human activity, e.g., agriculture, forestry, air pollution, or human-caused alterations of hydrology or climate		
Dynamic: change considerably over decades	Slowly dynamic: change considerably over centuries	Persistent: change considerably over millennia	
Organic carbon content	Fe and Al oxide contents	Non-pH-dependent charge	
Acidity	Humic substances	Texture	
Salinity	Illuvial clay accumulation	Coarse fragment content	
pH-dependent surface charge		Cemented pans, e.g., plinthite, duripan, petrocalcic horizon	
Bulk density			
Porosity			
Structure and aggregation			
Bioavailability of nutrients and contaminants			
Redoximorphic features			
Porosity			
Hydraulic conductivity			
Rooting depth and volume			

Source: Richter (2007).

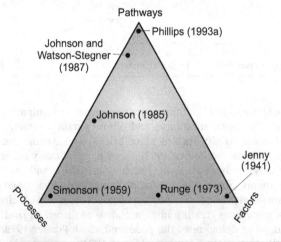

Pedogenic Model Emphases

Fig. 12.18 Schematic representation of the emphases of a few of the major pedogenic models discussed herein. Interpretation is that of the authors and does not imply judgment or value.

becoming an important focus of pedological research (DeKimpe and Morel 2000). Humanity's transformation of soils worldwide challenges scientists to develop a pedological viewpoint that has a broad purview and a decades-long timescale, one that supports the science and management of both the natural and the human-modified soil environment (Richter 2007).

Simonson's Process-Systems Model

Roy Simonson developed and presented a process-systems model of soil formation in his classic 1959 paper and refined it 19 years later (Simonson 1978). Unlike the state-factor model of Jenny (1941b), which was long on factors that affected soil development but said nothing directly about processes that actually formed the soil, Simonson's model was entirely process-based (Fig. 12.18).

The background for Simonson's model is worth exploring. In his 1959 paper, Simonson drew attention to the paradigm shifts that had been occurring in soil science. The functional-factorial model, born in Russia 75 years earlier, was yielding to a point of view that held that soils evolve continuously. Each soil stage may appear, disappear, recur, and fade away, leading to changing soil types and patterns. This ebb-and-flow viewpoint placed an emphasis on process – the processes that were causing this to happen – setting the conceptual stage for Simonson's process-systems model. Simonson noted that all soils have similarities and differences with one another. This observation was important. Recognizing the commonality among soils, we are forced us to explain their differences as the result of different strengths of the *same types of processes*. Simonson argued that soils differed because the processes they shared varied in *degree*, not in *kind* (Fig. 12.18).

In Simonson's process-systems model, soil genesis is viewed as consisting of two steps: (1) the accumulation of parent material and (2) the differentiation of that parent material into horizons (Arnold 2005). Yaalon (1975) noted that the general operation of any process-response model,

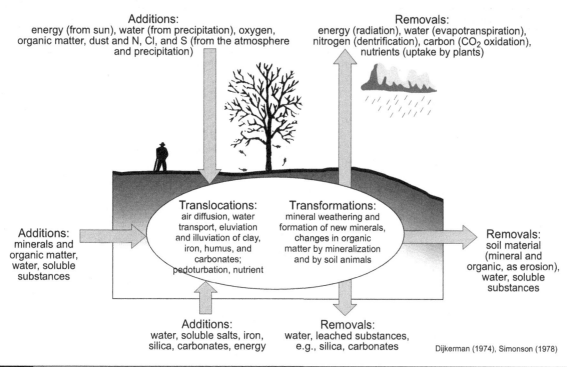

Additions:
energy (from sun), water (from precipitation), oxygen,
organic matter, dust and N, Cl, and S (from the atmosphere
and precipitation)

Removals:
energy (radiation), water (evapotranspiration),
nitrogen (dentrification), carbon (CO_2 oxidation),
nutrients (uptake by plants)

Translocations:
air diffusion, water
transport, eluviation
and illuviation of clay,
iron, humus, and
carbonates;
pedoturbation, nutrient

Transformations:
mineral weathering and
formation of new minerals,
changes in organic
matter by mineralization
and by soil animals

Additions:
minerals and
organic matter,
water, soluble
substances

Removals:
soil material
(mineral and
organic, as erosion),
water, soluble
substances

Additions:
water, soluble salts, iron,
silica, carbonates, energy

Removals:
water, leached substances,
e.g., silica, carbonates

Dijkerman (1974), Simonson (1978)

Fig. 12.19 Following on the Simonson (1959) process-systems model, this diagram illustrates some common additions, removals, translocations, and transformations in soils.

like Simonson's, can be characterized in terms of three determinants: (1) the initial state of the system (step 1, (2) the processes it is subjected to (step 2), and (3) the stage or duration of the processes (time). The focus of Simonson's model is on the second step. The four major kinds, or bundles, of processes in the model were designed, by necessity, to be very general in order to cover the full range of pedogenic processes (Smeck *et al.* 1983). Although not originally conceived as an equation, the model can be written as

$$S = f(a, r, t_1, t_2)$$

where S is the soil, a is additions, r is removals or losses, t_1 is transfers/translocations, and t_2 is transformations. Simonson (1978) envisioned that losses and additions are to *the soil as a whole*, whereas translocations are losses and additions, i.e., movements, *between horizons within a single pedon* (Fig. 12.19). These four sets of processes operate simultaneously and in all soils. Their *balance* and *character* govern the actual expression of the soil (Simonson 1978). Simonson's model might be considered part of a broader class of mass balance/budget models, to which sediment, water, nutrient, and geochemical balance models also belong. In this model, soil horizons and the profile develop their distinctive character because of a unique blend/suite of additions, removals, translocations, and transformations Table 12.3). A horizons, for example, are dominated by additions of organic material, whereas salts, clay, and other materials may be removed from them.

The process-systems model of Simonson was not developed for, nor has found strong usage in, the interpretation of soil variability, or in soil mapping applications; for this purpose the functional-factorial model of Jenny (1941b) is more appropriate. Users of Simonson's model need some knowledge of processes in order to apply it effectively to explain soil spatial variability. Phillips (1989: 167) summed it up best:

The process approach is more useful and appropriate for analyzing and describing the properties and development of a particular soil profile. However, the state-factor approach is often more useful for understanding the geographical variation of soils and ecosystem-level relationships with other environmental components.

The model is popular in the classroom, where it is used to teach students about soil process bundles and the flows of materials and energy through soils. It is used as a framework for many studies that center on pedogenic processes, e.g., Cutler (1981), Vreeken (1984b), Hoosbeek and Bryant (1992), Gessler *et al.* (2000), and theory, e.g., Conacher and Dalrymple (1977), Johnson and Watson-Stegner (1987). Unlike some other soil genesis models, it can be employed to understand both the development of individual horizons as well as that of whole soil profiles. It encourages the user to identify clearly the boundaries across which additions, losses, and translocations occur and within which

Table 12.3 | Major soil horizons as they pertain to components of Simonson's (1959) process-systems model

Horizon	Additions	Removals	Transformations
O	Organic matter (OM) and materials Atmospheric fallout (ions, dust) Mineral matter moved up from below (by burrowing organisms)	Decomposed OM and humus	Raw OM to humus (humification) Raw OM to its constituent ions (mineralization)
A	OM and humus Mineral matter moved up from below (by burrowing organisms)	Base cations (Ca, Mg, K, Na) (unless biocyled), carbonate Clay Fe and Al Salts	Primary mineral weathering Humus to a more decomposed state
E	None permanently	Base cations and carbonate Clay OM and humus Fe and Al	Primary mineral weathering Clays degraded
B	t = clay s = Fe and Al h = humus y = gypsum z = soluble salts n = sodium k = $CaCO_3$ w = few or none g = None	g = iron Carbonate	w = primary mineral weathering g = Fe^{3+} to Fe^{2+}
C	None, technically, although some clay and dissolved substances added, in small amounts	None	None, technically, although some weathering expected

Note: Compiled as representative examples. Translocations are omitted because they are captured by the essence of removals + additions when the soil is evaluated on a horizon-by-horizon basis. If examined on a whole-soil basis, translocations become horizon-to-horizon movements of materials and energy, whereas additions and removals are considered to be from (or to) the soil as a whole.

transformations occur. Because of the way it compartmentalizes each of the four sets of processes, the model is also adaptable to computer simulations of soil genesis (Kline 1973, Levine and Ciolkosz 1986, Gaston *et al.* 1992, Hoosbeek and Bryant 1992, 1994). The model has proven most useful when evaluating the gains and losses of materials from profiles and horizons to improve understanding of their genesis. In this regard, the process-systems model has formed the conceptual backbone of an emerging branch of pedology that is concerned with mass balance calculations. In this field, actual gains and losses of various constituents from soil horizons (and the profile) are determined by comparing their amounts with those of an immobile, slowly weatherable mineral, usually zircon, tourmaline, or quartz (Brewer 1976, Brimhall *et al.* 1988, 1991, Sohet *et al.* 1988, Olsson and Melkerud 1989, Chadwick *et al.* 1990, Merritts *et al.* 1992, Jersak *et al.* 1995). This topic is discussed more thoroughly in Chapter 15.

Runge's Energy Model

Ed Runge, a soil scientist then at the University of Illinois, developed a factorial model of soil development that is somewhat of a hybrid between the state-factor model of Jenny and Simonson's process-systems model (Runge 1973; Fig. 12.18). In formulating the model, Runge found a way to merge a considerable amount of process into Jenny's factorial framework, although Yaalon (1975: 199) called it nothing more than a "new verbal dress for the same model."

Runge's energy model, initially suggested by Smeck and Runge (1971), characterizes water as the primary energy source influencing entropy in soils (Brye 2004). The gravitational potential energy of water, as it percolates through the soil, decreases the entropy, i.e., increases the organization of the whole soil system, organizing the soil via processes of translocation.

Runge emphasized two *priority factors* from Jenny's model, climate and relief, which he regarded as most important. He combined them into a single *intensity factor* that he defined as the amount of water available for leaching (*w*), which was governed by climate, through the balance of precipitation and evapotranspiration, and topography, because certain sites are run-on sites and others are runoff sites. Thus, when examined together, the two factors produce a process vector that is roughly comparable to the potential for water to enter and percolate through the profile. This makes sense, for as we will see in Chapter 13, percolating water drives many, if not most, pedogenic processes. Runge therefore saw the *w* factor as an *organizing vector* that utilized gravitational energy to organize the soil profile, decrease parent material entropy, and create horizons, i.e., make it more anisotropic. Water that runs off the surface is not available for leaching and represents gravitational energy that is lost to the system.

He then combined the parent material and organisms factors (once again assuming that the organisms factor was primarily concerned with plants and hence biochemical in nature) into another intensity factor, called organic matter production (*o*), or lack of mineralization. The rationale for the *o* factor is less obvious. Basically, Runge knew that plants were the source of organic matter (humus) in the soil and assumed that their ability to grow was governed largely by parent material. For example, if the parent material is infertile, little organic matter can be produced. Thus, although a number of environmental factors govern the production of biomass and soil organic matter, Runge argued that many of them were captured by the parent material factor. Unlike water available for leaching (*w*), the *o* factor was seen as a renewing or rejuvenating vector. How or why? Humus coats mineral soil particles (melanization) and thus inhibits weathering. The prairie grasses near Runge's home (Illinois) are also excellent base cation cyclers, and thus the better they grow the higher the pH of the soil remains, further inhibiting weathering and lessivage (clay translocation). As a result, the more humus-rich the profile, the less weathered and the more isotropic it often is. Runge, consequently, viewed the *o* factor as offsetting the *w* factor. He also included a time factor (*t*). Because the model relies heavily on (1) the gravitational energy that drives infiltrating water and in turn causes horizonation and (2) (indirectly) on radiant solar energy for organic matter production, it has come to be known as the *energy model* (Smeck et al. 1983):

$$S = f(o, w, t)$$

where *S* is the soil, *o* is organic matter production, *w* is water available for leaching, and *t* is time.

Each of Runge's two intensity factors is conditioned by a number of *capacity factors*. *W* is conditioned by such factors as duration, intensity, and temporal distribution of precipitation (P); runoff versus. run-on; soil permeability; evapotranspirative demand; and others. The *o* factor

is conditioned by nutrient (especially P) and water availability, porosity, available seed sources, fire, and sother factors.

As shown in Figure 12.18, the energy model is a hybrid of the factorial model of Jenny (1941b) and the pure process-based model of Simonson (1959). It combines many of the positive attributes of factor-based models and process-based models. Nonetheless, it is just as difficult to quantify (Huggett 1975). It also does not actually quantify energy, even though it is an energy model. Nonetheless, as a conceptual model, it is simple, easily comprehended, and process-oriented.

The energy model is applicable in a number of settings where lowland or depressional sites with excess water due to run-on tend to have more strongly developed and horizonated soils, i.e., it works best in humid (leaching) soil environments. Runge noted that it is probably limited in application to soils on unconsolidated surficial deposits such as loess or till, in which leaching can operate freely. In other words, the model does not work well on residual soils. On the prairies of Saskatchewan, Miller et al. (1985) reported on depression-focused recharge by snowmelt that led to greater leaching in those areas. Sola were thicker in these sites, where water available for leaching was greatest (see also Pennock et al. 1987). In areas where a high water table inhibits leaching (ground water discharge depressions), soil development is inhibited. This last example shows the efficacy of the energy model – where water is not likely to leach vertically (because of a high water table), soil development is not accelerated, even if the site is wet. These and similar studies, e.g., Sinai et al. (1981), Donald et al. (1993), Fuller and Anderson (1993), Manning et al. (2001), and Schaetzl and Schwenner (2006), continue to show that infiltrating water is a source of organizational pedogenic energy and in so doing validate Runge's energy model. Applications of this general approach to landscape-scale patterns of denitrification are also common, e.g., Elliott and DeJong (1992), Brye (2004).

Perhaps the strongest criticism of the energy model centers on the *o* factor. In Illinois, where the model was developed, soils are mostly Mollisols that are rich in humic materials. In nonacidic soils, large amounts of humic materials retard mineral weathering, slow translocation processes, and enhance bioturbation. However, in forested soils, especially in forests with a coniferous component, organic matter is more acidic and more likely to promote mineral weathering by forming soluble complexes with dissolved metals. In this way, horizonation is promoted. Even in such landscapes, however, the model has been found useful (Hupy and Schaetzl 2008). Schaetzl (1990) reported accelerated soil development in pits formed by tree uprooting; the pits are loci of infiltration. They also have higher organic matter accumulations, which in the forests of Michigan also act as an organizing vector. It is likely that the increased soil development in pits is due to both increased infiltration and higher concentrations of mobile organic acids. In conclusion, whereas the model is most applicable to grassland environments developed on unconsolidated parent materials, as originally stated by Runge (1973), it is still useful for forested sites and sites where soil development is driven mainly by percolating water (Schaetzl and Schwenner 2006).

Johnson's Soil Thickness Model

Pedogenic models generally focus on the formation of the profile, the development of horizons, the loss or degradation of those horizons, and similar topics. Most implicitly assume, for ease of comprehension, that the soil surface is static and that parent material has already been in place. Almond and Tonkin (1999) called this concept *top-down pedogenesis*, in which the depth of pedogenic alteration increases with time, but one where the main impacts are from the external, subaerial environment on top of the soil.

This situation, however, does not hold for aggrading surfaces where soils become buried (rapid aggradation) or upbuild slowly (slow aggradation) by additions of loess, alluvium, tephra, dust, and so forth (Nikiforoff 1949, Phillips *et al.* 1999, Phillips and Lorz 2008). Almond and Tonkin's (1999) study highlighted upbuilding and soil thickness concepts that had been addressed by Donald Johnson more than a decade earlier: *Soil thickness* is an important soil geomorphic attribute in its own right. Additions to, and removals from, the soil surface are an integral part of pedogenesis, not to mention soil classification. In a landmark paper, Johnson (1985) isolated soil thickness and set about examining the processes that affect it. It is an interesting twist on previous pedogenic models that focused on the totality of soil development as the dependent variable. Here, thickness is considered as a dependent variable. In essence, the soil thickness model is an outgrowth of Simonson's process-systems model but it focuses on additions and removals from the soil surface.

In the model, the thickness (T) of a mineral soil is viewed as a dynamic characteristic involving processes of profile deepening (D), upbuilding (U), and removals (R):

$$T = D + U + R$$

In this model, removals are similar to their formulation in Simonson's model. They refer mainly to losses of material from the surface through erosion and mass wasting, although subsurface removals by through flow, leaching, and biomechanical processes are also included. Oxidation and mineralization of organic matter are also considered removals. Upbuilding occurs as allochthonous materials are added to the soil surface by eolian or slope processes or enter the soil laterally via through flow. Organic matter additions from biota are also included in this category. Profile deepening occurs as the base of the solum grows

downward into fresh parent material, via leaching and weathering. Soils become thinner when $D + U < R$, and they become thicker when $D + U > R$, $D > U - R$, or $U > D - R$. Phillips *et al.* (1999) defined *truncation* as an instance when $D + U < R$, i.e., the soil becomes thinner and the surface is lowered – but not fast enough to remove all horizons completely.

Upbuilding can take two forms (Fig. 12.20). In *developmental upbuilding*, surface additions are slow enough that pedogenesis can keep pace (McDonald and Busacca 1990, Almond and Tonkin 1999, Eger *et al.* 2012). The new sediment becomes soil as fast as it is added. This concept is equivalent to the term "cumulization" (see Chapter 13). Examples of developmental upbuilding include slow accumulations of loess in glacial environments, dust in deserts, and overbank alluvium. In *retardant upbuilding*, materials are added to the surface faster than pedogenic processes can effectively incorporate them, and the soil surface, momentarily at least, becomes buried (Schaetzl and Sorenson 1987, Eger *et al.* 2012). In this case, the soil per se is no thicker, only buried by sediment that is not considered a part of the soil profile.

Conceptually, this model (1) draws attention to a concept (soil thickness) that had heretofore not been examined per se and (2) emphasizes the dynamic nature of the soil surface as one that is constantly experiencing erosion and/or deposition (Haynes 1982, Sunico *et al.* 1996). The model introduced concepts that eventually led to Johnson and Watson-Stegner's (1987) soil evolution model, just two years later.

Johnson and Watson-Stegner's Soil Evolution Model

Just as soil thickness, as a soil property, can be shown to be either static, regressive (getting thinner), or progressive (getting thicker), soil development be viewed in the same manner. For most soils, thickness is a dynamic condition that ebbs and flows through time (Fig. 12.20); so it is with soil development. The thrust of Johnson and Watson-Stegner's (1987) soil evolution model is that soils *evolve*, ebb, and flow, rather than unidirectionally develop and progress from not-soil to some theoretical, steady-state end point. Indeed, Nikiforoff (1959) quoted Dokuchaev's statement that "soil, like any other plant or animal organism, eternally lives and changes, now progressively and then regressively."

The Normal Soil Concept
The soil evolution model was developed, in part, as a reaction to perceived flaws in the normal soil concept of unidirectional soil development. In the early days of soil science, soils were viewed as developing steadily toward

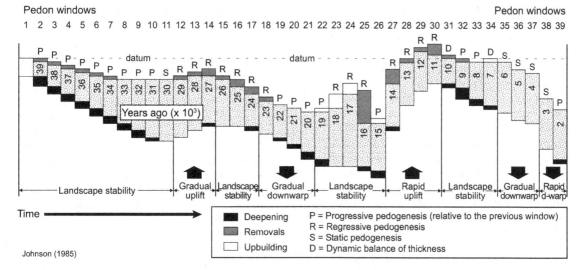

Fig. 12.20 Conceptual diagram showing the relationships of various soil thickness processes in a hypothetical soil, on a highly dynamic and tectonically unstable landscape. Progressive, regressive, and static pedogenesis are illustrated. Permutations of soil thickness and elevation changes are made, relative to a constant datum, by means of 39 pedon time windows, each arbitrarily specified at 1,000 years. Time$_{zero}$ was 39,000 years ago, implying that pedon 39 is the same soil after 39,000 years of pedogenesis. Pedon windows 1–7 show the changing soil thicknesses from time$_{zero}$ to 33,000 years ago. Although surface removals occur, they lag behind the pace of soil deepening, and surface additions are negligible. The soil becomes thicker via deepening and, in the absence of appreciable haploidization, experiences progressive pedogenesis. During the period 32,000 to 29,000 years ago (pedon windows 8–10), removals and additions are negligible but deepening continues, as the soil continues to reflect progressive pedogenesis. From pedon windows 10–11 the soil shows no change in thickness, because no appreciable additions or removals occur, and deepening processes are negligible; this state is called *static pedogenesis*. Pedon windows 12–18 represent a period of regressive pedogenesis in which removals exceed deepening, as a result of tectonic uplift and, hence, increased surface erosion. Uplift ceases at the end of pedon window 14, leading to landscape stability in pedon windows 15–17. However, regressive pedogenesis continues in pedon windows 15–17 because surface removals exceed subsurface deepening. Tectonic downwarping begins at pedon window 18, but the soil continues regressing because of removals and negligible deepening. Pedon windows 19–30 are self-explanatory, as the thickness of the soil continues to change. A severe erosion event, perhaps induced by a large storm, occurs at pedon window 26, thinning the soil considerably. The rapid uplift that follows during pedon windows 27–30 adds more fuel to the fire of soil thinning and regression. Pedon window 31 is characterized by structural stability of the landscape. Removals are balanced by profile deepening processes, so that while the soil actively lowers relative to the datum, it neither thins nor thickens. This thickness condition is a case of *dynamic balance*. Pedon windows 32 and 33 reflect a return to progressive pedogenesis with modest or negligible removals and active profile deepening. In pedon window 34, rates of upbuilding match those of deepening, which thickens the soil and causes it to rise relative to the datum; this is another example of dynamic balance of thickness. Static pedogenesis typifies pedon windows 35–38, despite the slow tectonic downwarping that occurs during this 4,000-year period. At pedon window 38, rapid downfaulting occurs. The thickness of the soil remains constant, and by the end of pedon window 38, the soil surface is again below datum. Pedon window 39 marks a return to progressive pedogenesis, with some deepening in the absence of appreciable removals or additions. Rapid episodic downfaulting of the landscape continues during this last (39th) millennium of pedogenesis.

an end point – the normal or *zonal soil* (Bockheim *et al.* 2005, see Chapter 8). Beginning with Dokuchaev and continuing with those used in the United States until ca. 1960 (Marbut 1927a; Baldwin *et al.* 1938), soil classification schemes were based on this zonality concept. Zonal soils displayed well-developed characteristics that reflected the broad, regional influence of climate and vegetation, on stable, well-drained sites. Most soils were viewed as eventually developing toward that end point; soil development was a one-way street. Johnson's and Watson-Stegner's soil

evolution model essentially demands that we view soil development as a two-way street; soil change that has no clear-cut or preconceived end point. Let us explore this concept further.

Normal soils, synonymous with zonal or mature soils, were thought to be the end point of soil development. Each kind of normal or zonal soil was linked to its particular climate-vegetation assemblage (Marbut 1923, Baldwin *et al.* 1938, Cline 1949b, Frei and Cline 1949, McCaleb 1954, Johnson *et al.* 1990). For example, Chernozems were the normal soils associated with grasslands of semiarid climates, and Podzols were the normal soils for cool, humid, coniferous forest biomes. Because both Dokuchaev and Jenny emphasized climate as a dominant, active soil-forming factor, it came to be assumed that in each climate there existed a profile morphology that soils would develop toward, unless thwarted by excess wetness, rockiness, sandy or clayey textures, or other factors. Indeed, the first major treatise on soil development, Dokuchaev's (1883) *Russkii Chernozem*, was essentially a mandate for the zonal soil concept and the dominant influence of climate on soil formation. Thus, from its very beginnings, the science of pedology has been influenced by the notion that each climate or bioclimate has a dominant, almost climaxlike, soil type. The one climate–one soil idea was reinforced by Bryan and Albritton's (1943) paper, influential though conceptually flawed, on monogenetic soils. In a way, our book follows some of those same lines of reasoning in Part III, although we approach it with a wider conceptual and theoretical bag of tricks.

Belief in the normal soil concept during the early and mid twentieth century was almost a doctrine (Johnson and Watson-Stegner 1987), espoused and supported by all the major soil scientists of the day: Marbut, Jenny, Thorp, Baldwin, and Kellogg. The science of geomorphology was, at that time, conceptually no different; the zonal implications of climatic geomorphology (1950–1990) initiated by Peltier (1950) were showcased by Büdel in his 1982 book *Climatic Geomorphology* (Johnson 2000). Ecologists were also focusing on the progression of vegetation communities through a series of stages, to a stable and unchanging climax community (Cooper 1913, Clements 1936, Graham 1941, McComb and Lomis 1944, Drury and Nisbet 1971). Thus, all the major soil, landscape, and vegetation successional-climax models and theories had come to emphasize progressive and unidirectional evolutionary change (Johnson 2000).

Although the Russian school did not entirely accept the normal soil concept, they acknowledged that there was little opportunity for soils to regress. Instead, they subscribed to the idea of a series of waves of soil development, governed largely by primary mineral weathering (Nikiforoff 1949). Thus, the Russians emphasized (and correctly so) that some aspects of soil formation simply cannot revert. Weathering is one of them. Rode, a prominent Russian pedologist, supported the idea that some soil processes are cyclic – but some, like weathering, are not.

Emphasis within pedology at the time was, therefore, placed on determining the developmental processes that led to formation of the normal soil for its given climate. Although this pervasive view may have been imbalanced with respect to soil genesis, it was probably necessary at the time; pedologists needed to know how horizons became differentiated and how soils formed before they could focus on how they regressed.

Background and Development of the Soil Evolution Model

Although the background for the soil evolution model is complicated, it was certainly influenced by the Russians, especially Constantin Nikiforoff (1949), whose lesser-known model or concept of soil evolution was based on three assumptions: (1) that soil development is continuously affected by certain progressive processes, (2) that these processes do not operate steadily, and (3) that many pedogenic processes are terminal and irreversible, i.e., they operate to completion, and then are no more. Nikiforoff provides weathering of minerals as an example of assumption 3. A soil may experience successive waves of these beginnings, peaks, and endings, each of which varies for the different processes. For example, decalcification may be followed by lessivage, which may then be followed by podzolization and fragipan formation. Although the Russians, Nikiforoff in particular, conceived of some overlap among these waves, the process waves nonetheless maintained their identity throughout the soil development timescale. Thus, the Russian concept of soil evolution was like that of human evolution – generally progressive, occurring in waves of distinct species (soils). Although it uses the same name (evolution), the Nikiforoff model is not quite the same as the concept embodied in the soil evolution model of Johnson and Watson-Stegner (1987).

Johnson and Watson-Stegner, two geographers then at the University of Illinois, argued that the term *soil genesis* had, by the 1980s, become synonymous with soil formation, and that soil formation implies an increase in soil organization, or a loss of entropy and isotropism. Thus, many began to perceive soils, by virtue of the language that was being used, as systems that progressed along a unidirectional *pathway* of continued development, i.e., a holdover of the zonality concept and one that implies a certain degree of pedogenic constancy (Bryan and Teakle 1949). This was the prevailing view, despite the many position papers and empirical research products that pointed to the importance of soil regression, mixing, and even erosion and burial as normal, ongoing processes in soils (Hole 1961, Runge 1973, Vreeken 1975, Wood and Johnson 1978, Humphreys and Mitchell 1983, Johnson *et al.* 1987).

Why did scholars continue to view soil genesis as strictly the organizing half of the equation, despite a significant body of evidence for contrary (regressive) processes (Nikiforoff 1949)? Much of the answer is related to the application of the state-factor model to enhance understanding of soil *development*. Jenny's state-factor model (paraphrased and quoted from Johnson and Watson-Stegner 1987: 349)

encouraged viewing soil genesis through a formational/developmental filter. Its main message is that soils form and progressively develop under the influence of environmental factors. The time factor became implicitly linked to the formational/developmental model, in that soils were seen as progressing from simple to complex and ultimately toward a stable "mature" state.

Table 12.4 | Vector/process components of soil genesis and its two major soil evolution pathways

Progressive pedogenesis	Regressive pedogenesis
Horizonation	**Haploidization**
Proanisotropic conditions and processes that promote organized profiles; profile differentiating aspects of additions, removals, transfers, transformations, intrinsic feedbacks, and proanisotropic pedoturbation	Proisotropic conditions and processes that promote simplified profiles; profile rejuvenating aspects of additions, removals, transfers, transformations, melanization, nutrient biocycling, enrichment, intrinsic feedbacks, and proisotropic pedoturbation
Developmental upbuilding	**Retardant upbuilding**
Pedogenic assimilation of surface-accreted materials	Pedogenic impedance produced by surface-accreted materials
Soil deepening	**Soil thinning**
Downward migration of the lower soil boundary into fresh, relatively unweathered material	Surface erosion and mass wasting

Source: After Johnson and Watson-Stegner (1987).

Studies had continually pointed out that the morphology of many soil profiles reflected not just organization and progressive formation, but also disorder, mixing, regression, and haploidization (simplification). Indeed, entire soil orders (Vertisols) are centered on the concept of mixing and regression (Brierley *et al.* 2011, Southard *et al.* 2011). And yet, processes that might lead to profile simplification include not just pedoturbation, but also melanization, biocycling of nutrients, erosion, and high water tables, to mention a few. The pedogenic implications of these sorts of processes did not fit easily into a unidirectional model, and so the processes were effectively ignored in discussions of "whole-soil" genesis. The pedogenic community has commonly focused on soil development as being progressive. A model that could see both sides of the coin was clearly needed. Such a model needed to incorporate both progressive and regressive soil development processes, as well as thickness concepts. In this light, Johnson and Watson-Stegner (1987) presented their soil evolution model, as

$$S = f(P, R)$$

where S is the soil or a soil property, P is progressive pedogenesis, and R is regressive pedogenesis. The soil evolution model stresses that soils proceed along two *interacting genetic pathways* that reflect variable exogenic (from the outside) and endogenic (from the inside) processes, factors, and conditions (Fig. 12.18). Every soil has a progressive pathway along which the soil moves forward or develops horizons, and a regressive pathway that typifies a reversion to an earlier or simpler form. Whether a soil develops or regresses depends on which pathway is stronger at the moment or in the recent past. Even though a soil may display morphological and/or physicochemical order and stability, and thereby reflect the predominant strength of the horizonation (progressive) vectors, the subsidiary haploidization vectors continue to operate. Static pedogenesis, which is only infrequently realized, occurs when the two pathways are essentially equal in strength.

Each pathway has *three components*, and each of these has a set of opposing vectors: (1) horizonation/haploidization, (2) retardant upbuilding or developmental upbuilding, and (3) profile deepening or profile thinning (Table 12.4). The *progressive pathway* is composed of horizonation processes, developmental upbuilding, and soil deepening/thickening (Tables 12.4, 12.5, Fig. 12.20). Progressive pedogenesis, or soil progression, is synonymous with soil formation, development, and organization. It includes those processes and factors that lead to organized and differentiated, i.e., more anisotropic, profiles. When progressive pedogenesis dominates, a soil usually develops more, thicker, and better-expressed genetic horizons. The *regressive pathway* includes haploidization (simplification) processes, retardant upbuilding, or soil thinning (Tables 12.4, 12.5). Regressive pedogenesis, or soil regression, reverses, stops, or slows soil progression and development. It includes those processes and factors that lead to simpler and less differentiated, i.e., more isotropic, profiles. When regressive pedogenesis dominates, soil horizons become thinner, blurred and/or mixed, and even eroded. Figure 12.21 provides examples of how the two soil evolution pathways interact, including a breakdown of the three components included in each pathway.

An important aspect of the model is inherited from Johnson *et al.* (1987) – that mixing (pedoturbation) processes are not exclusively haploidizing and that some (proanisotropic pedoturbations) may actually promote solum order. Much of the older literature focused only on the "blurring" or proisotropic aspects of pedoturbation (Blum and Ganssen 1972). As discussed in Chapter 11, pedoturbation processes are typically thought of, even today, as being mainly proisotropic, i.e., tending to promote profile simplification. This is often the case when the mixing is between horizons, such as when ants burrow through the entire profile. But within-horizon mixing need not destroy horizons, and it may even promote their identity (Johnson *et al.* 1987).

Table 12.5	Components of the progressive and regressive pathways of pedogenesis
Major components	Examples
Horizonation vectors (progressive pathway)	
Additions (to one or more horizons)	Energy and heat
	Water, as on-site rain and snow or ice melt, fog condensation and leaf drip, and surface water run-on
	Gases, via diffusion and as mass flow via wind and pressure gradients
	Solids, e.g., eolian dust and ash, loess, ions in rain and snow, sediment added via fluvial and mass wasting processes, organic matter production in place, feces and urine, pollen and seeds, microbes
Removals (from one or more horizons)	Leaching of soil constituents in solution, out of the profile
	Surface erosion of solid and dissolved substances
Translocations (from one horizon or portion of the profile to another)	Humus, clay, silt, Al, Fe, soluble salts, carbonate, silica, gypsum, etc.
Transformations (of materials in place and of the soil chemical environment)	Weathering of primary and secondary minerals
	Organic matter synthesis, resynthesis, and humification
	Oxidation-reduction and cation exchange reactions
	Alkaline to acid pH changes
Developmental feedback processes induced by intrinsic thresholds and pedogenic accessions	Time-delayed formation of Bhs horizons consequent upon sufficient accumulation of Fe^{3+} and Al^{3+} in Bs horizon to effect immobilization of humus
	Lessivage, following decarbonation
	Dispersion and translocation of clay, induced by accumulation of Na
Proanisotropic pedoturbations (mixing that acts to form and/or maintain horizons)	Faunalturbation, forms some A horizons, stone lines, stone zones, and biomantles
	Argilliturbation, forms surface stone pavements
	Aeroturbation, forms vesicular horizons and desert pavements
	Cryoturbation, forms surface stone pavements, polygonal ground, stone circles, garlands, etc.
Deepening	Downward migration of lower soil boundary into fresh parent material
Developmental upbuilding	Pedogenic assimilation of relatively small amounts of clastic materials into the upper profile
Haploidization vectors (regressive pathway)	
Additions	As above
Removals	As above
Translocations	As above
Transformations	As above
Nutrient cycling (soil enrichment and rejuvenation)	Biocycling, eolian enrichment, melanization, etc.
Simplifying feedback processes, induced by intrinsic thresholds and pedogenic accessions	Retardation or cessation of lessivage and vertical water percolation due to pore reduction and clogging, promoting increased lateral throughflow, runoff, and erosion; in some cases, a smectite-rich Bt horizon becomes vertic, engulfs its A horizon, and evolves to a Vertisol with concomitant profile simplification
Proisotropic pedoturbations (horizon disruption and/or destruction processes)	Potentially, all forms of pedoturbation; see Table 11.1

Table 12.5 (cont.)

Major components	Examples
Removals	Deflation, sheetwash, stripping, piping, and mass movements
Retardant upbuilding	Impeding or retarding effects of relatively large deposits of eolian- and slope-derived materials, added to the soil surface; occurs on some uplands by eolian processes, on many floodplains by fluvial processes, and on footslopes, toeslopes, and depressional area by slopewash and mass movements

Note: Shown here is a condensed listing. See Johnson and Watson-Stegner for a more inclusive list.

Source: Johnson and Watson-Stegner (1987).

The soil evolution model places an emphasis on *vectors* or *pathways* (plural!) of soil development, and these pathways can be regressive *and* progressive. The continual give and take between the two pedogenic pathways is the essence of soil evolution (Fig. 12.22). As soils evolve, their morphologic and physicochemical profiles progress and regress through time; they do not simply advance or develop in one direction.

Changes in Pedogenic Pathways: How and Why?

Knowing that pedogenic pathways can be progressive or regressive, we next explore how and why these pathways ebb and flow. Pedogenic pathways are affected by many intrinsic and extrinsic factors. Extrinsically, the possible pathways a soil can follow are set initially by parent material, topography, and climate, e.g., Nettleton *et al.* (1975). Parent material and topography *precondition* and *constrain* the pedogenic system to only a certain number of initial pedogenic pathways. As topography changes and as parent material is eroded from or added to the soil, however, other options may open.

Pedogenic pathways can be altered and changed by externally or internally driven, active or developmental *forces.* These forces vary along a continuum from extremely subtle to major, e.g., as shown in Figure 12.23. Subtle *external* forces, such as a drier climate for a few decades, might slow clay translocation, but not change the overall *direction* of the pedogenic pathway; it still is ongoing but it has simply changed its rate and probably decreased in strength. Plant succession from aspen forest to hemlock-pine may accelerate podzolization in a soil that was already on that same pathway. McPherson and Timmer (2002) studied the effects of deforestation, cultivation, and reforestation on soils and found clear evidence for morphological changes associated with these shifts in extrinsic state factors (Fig. 12.24). Additions of thin increments of loess or dust can slow or even change the pedogenic pathway of a soil (Sprafke *et al.* 2013). Similarly, subtle *internal* changes in a

pedogenic pathway or vector may occur as a soil exhausts its primary minerals via weathering, slowing the production of clay. Pedogenic pathway changes can occur as the B horizon of a Spodosol becomes continually enriched in iron and humus. Root proliferation within it is, therefore, enhanced, and thus, the rate at which that horizon gains humus continues to increase via increased root decay. Initially, roots had sought out the illuvial B horizon as a source of nutrients and soil water. Later, the horizon becomes even more attractive in this regard. The change was a matter of degree, not kind.

Chief among the criticisms of the soil evolution model is the ambiguity of the terms, especially "pathway." Intuitively easy to understand for some, it is nonetheless a difficult concept to define rigorously. The model, although theoretically sound, is also difficult actually to *apply.* Most would agree that the state-factor model is better for soil science (mapping, classification, etc.), whereas the soil evolution model is perhaps equally or more useful for pedogenic and soil geomorphological research and higher-level theoretical training. Thus, although the soil evolution model is conceptually rigorous, its use may be generally restricted to pedologists and those with a thorough understanding of the pedogenic system, for it requires that the user be able to elucidate a number of possible scenarios – past and future – to explain soil evolution on site. Without a solid pedogenic background, it is difficult to do that. The soil evolution model also focuses on soil as a whole, rather than its properties. This is probably on purpose, since it is difficult to imagine that soil properties regress or progress; the soil as a whole regresses or progresses. Thus, the model takes a larger view of soil genesis, perhaps at the expense of the pieces. Finally, because the soil evolution model is phrased in generalities, it is much more difficult to test than are some others. Designing a study to test the pathway shifts that a soil may or may not have taken, its degree of regression and/or progression, is a Herculean task. Perhaps it is one model we are destined to speak of in generalities only.

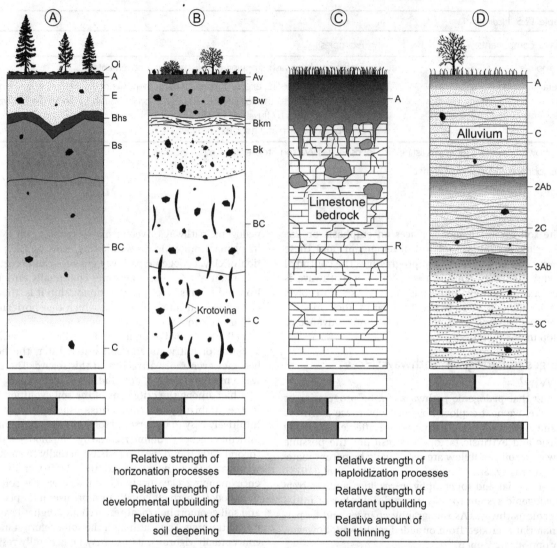

Fig. 12.21 Examples of hypothetical soil profiles operating under various permutations of horizonation-haploidization vectors. (A) A Typic Haplorthod developed in sandy glacial outwash in a Canadian coniferous forest. Horizonation vectors dominate, as almost no soil fauna are present to mix the soil. Runoff is minimal because of the coarse textures, and thus soil thinning processes are weak. The profile continues to deepen via leaching and translocation processes. The only major regressive pathway is provided by infrequent tree uprooting (Schaetzl 1986b, Samonil et al. 2010a). (B) A Typic Petrocalcid developed on an alluvial fan. Horizonation vectors dominate, as translocation of $CaCO_3$ carried in by eolian dustfall has helped form a thick petrocalcic horizon. A pedogenic threshold is crossed as the Bkkm horizon forms. Less water reaches the lower solum because of the almost impermeable nature of the Bkkm horizon, and hence profile deepening is slowed or even stopped. Additions of eolian dust have led to some profile thickening and developmental upbuilding, as the dust is quickly incorporated into an Av horizon (McFadden et al. 1987), although slight deflation losses are also occurring. Generally vertically oriented cicada burrows in the subsoil attest to some mixing at depth (Hugie and Passey 1963), although this process subsides after the petrocalcic horizon forms. (C) A Typic Rendoll developed on limestone bedrock, on a steep slope in Greece. Dissolution of the pure limestone bedrock is slow in this climate, and even when it does weather it provides little mineral material to the soil; the main process affecting soil thickness, therefore, centers on episodic inputs of dust from the Sahara (Prodi and Fea 1979). However, runoff on the steep slope essentially offsets these eolian additions. Melanization is dominant, and given the high pH values of the soil, clay translocation is minimal. The limited profile space and high organic matter contents facilitate worm and small insect bioturbation, leading to a homogenized (but thin) profile. The soil approaches static pedogenesis. (D) A Typic Ustifluvent on the floodplain of a medium-sized river in India, which drains a rugged, wetter, upland area. The stream frequently floods. During these floods, alluvium buries the soil. In the intervals between floods, however, horizonation processes operate and form a thin A horizon, which becomes buried by the next flood event. Hence, buried paleosols occur at depth. The alluvium is base-rich, preventing clay translocation, and its coarse textures limit faunalturbation by earthworms.

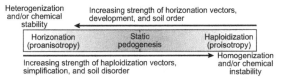

Johnson and Watson-Stegner (1987)

Fig. 12.22 Theoretical relationships between the horizonation and haploidization vectors – part of the progressive and regressive soil evolution pathways (see Table 12.4). The diagram illustrates how each vector, while opposing the other, remains operative at any given time. Their relative strength determines whether the soil progresses or regresses.

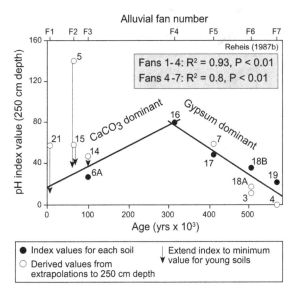

Fig. 12.23 A chronosequence of soils on alluvial fans in Wyoming that shows evidence for a change in pedogenic pathway at ca. 300,000 years ago. Early on, pedogenic carbonate accumulated from eolian sources. Later, despite continued additions of carbonate, accumulation of secondary gypsum was the main pedogenic process. This change in pathway is illustrated by a pH index (y-axis).

Thresholds and Feedbacks

In order to examine fully what factors, internal and external, affect pedogenic pathways, and how this fits into the soil evolution model, we must briefly digress into geomorphology, where the concept of *thresholds* is entrenched and has been increasingly utilized, e.g., Langbein and Schumm (1958), Graf (1978), Coates and Vitek (1980), Chappel (1983), Carson (1984), Wescott (1993), Church (2002), and Phillips (2006).

Let us consider an example of a threshold. Sediment in a streambed will remain stationary until a critical (threshold) flow velocity is reached, after which it will freely move. The threshold is a value, unique to the system, beyond which the system adjusts or changes, not just in rate but in kind. Similarly, sediment yield per unit area in dry climates has been found to be directly proportional to precipitation and runoff, but only up to a point, a threshold point, beyond which sediment yield drops significantly, even with increased precipitation (Langbein and Schumm 1958). Why? Because vegetation reaches a threshold coverage level at some value of precipitation, holding more soil in place, thereby reducing surface erosion and sediment yield.

Stanley Schumm (1980), credited by many as introducing the geomorphology community to the importance of thresholds, stressed their importance by emphasizing that knowing *average rates* of a process over time is misleading, because they do not display the complexities and variabilities of their development. The same could be said for the soil system. Average rates may tell us little about the system, if most of the change occurs when and immediately after a threshold is crossed.

Schumm's classic 1979 paper first called attention to the concept of thresholds in geomorphology. In it, he defined two main types of thresholds: *extrinsic* and *intrinsic*. In an extrinsic threshold, change in an external variable or factor triggers abrupt changes or failure within the affected system (in this case, either geomorphic or pedologic). The system responds to the external change at an extrinsic threshold level; below that level no (or minimal) change occurs (Schumm 1979). Many extrinsic thresholds are caused by climatic or geomorphic change, as mentioned in the preceding paragraph. Intrinsic thresholds, on the other hand, occur when a system changes (internally), without a change in an external variable. Schumm (1979) provided an example from geomorphology: Long-term weathering of rock reduces the strength of slope materials until eventually the slope fails. The critical shear strength value of the rock, in this case, sets the threshold level, below which the system is forced to readjust or change its pathway. Another commonly referenced example of an intrinsic threshold is in fluvial systems (Schumm and Parker 1973). Sediment in small stream catchments, carried in by tributaries with steeper gradients, accumulates in the tributary valleys because the trunk stream is unable to remove it all. This steepens the valley floors, but no real system change occurs until a steepness (gradient) threshold is achieved (Fig. 12.25). Then, the trunk stream, responding to the increase in stream power when its gradient has exceeded the threshold value, begins incising its valley and removing the accumulated sediment. The system has gone from an aggrading one to one dominated by downcutting (Patton and Schumm 1975), i.e., the change in pathway was initiated because a threshold was crossed.

Soil systems, like all others, have both intrinsic and extrinsic thresholds (Chadwick and Chorover 2001). Muhs

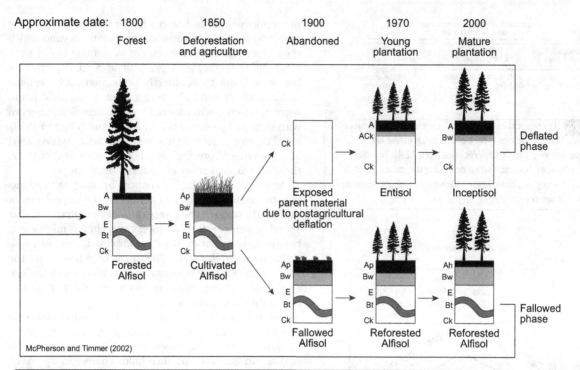

Fig. 12.24 Conceptual chronosequence of soil degradation and restoration on sandy loam soils in southern Quebec, Canada. The pedogenic pathways were altered by presettlement deforestation, followed by agriculture, land abandonment, and fallowing (upper pathway) or deflation (lower pathway). Soil restoration was initiated by reforestation with red pine.

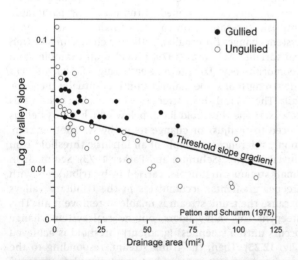

Fig. 12.25 Relation between valley slope and drainage area for the Piceance Creek Basin, Colorado. The line defines the intrinsic threshold gradient, beyond which the fluvial system begins to incise and becomes gullied.

(1984: 100), who is often credited with introducing the concept to soil systems, defined a pedologic threshold as

the limit of soil morphologic stability that is exceeded either by *intrinsic change* of the soil morphology, chemistry, or mineralogy, or by a subtle but progressive change in one of the *external* soil-forming factors (of) parent material, climate, relief, and organisms. [emphasis added]

By way of example, Paradise (1995) showed that sandstone weathering rates vary considerably as a function of iron content (an internal threshold) and solar angle (an external threshold). Thresholds in this type of system are manifested when rates of weathering (the dependent variable) change abruptly across a range of values associated with the independent variable, in this case, iron contents and solar insolation (Fig. 12.26). Threshold-induced changes in the rate of weathering show clearly as distinct breaks in slope (Coates and Vitek 1980).

Extrinsic thresholds in soils commonly involve changes in texture, almost like a chronolithosequence. As parent material texture changes, pedogenic processes also often change, sometimes abruptly (Pedro *et al.* 1978, Duchaufour 1982). Likewise, climate or vegetation continually shifts,

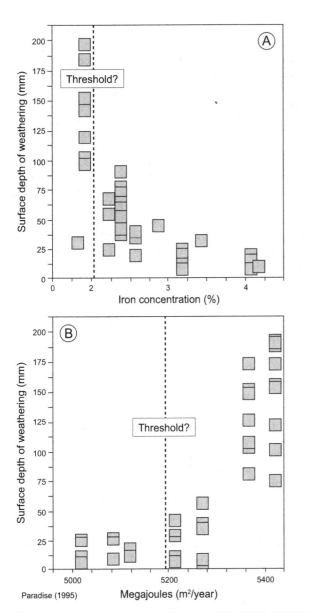

Fig. 12.26 Thresholds in the weathering of sandstone in an arid climate, as indicated by depth of recession of the rock over a 2,000-year period. (A) Relationship between depth of weathering and iron content of the sandstone, with a possible threshold at ~2% iron content. (B) Relationship between depth of weathering and accumulated annual insolation, with a possible threshold near ~5,200 megajoules/m², representative of an exposed, southern aspect.

cause a change in the pedogenic pathway (cf. Graham and Southard 1983, Reheis 1987a).

Internal drivers that can cause pedogenic pathways to change can also occur. Many intrinsic thresholds in soils center on illuvial or eluvial processes. These changes can be (but often are not) manifested completely independently of the state factors. For example, clay translocation (lessivage) probably does not begin until a threshold amount of base cations has been removed from exchange sites (Muhs 1984). Often, slow but continuous additions of clay or other illuvial substances to a B horizon change its soil water status. For example, an illuvial Bt horizon may accumulate so much clay that it becomes an aquitard, perching water and dramatically changing the redoximorphic conditions in the soil, and possibly inducing ferrolysis, which destroys clay (see Chapter 13). Hence, a pathway change can be forced by this rather straightforward type of pedogenic *accession*. Or, as soils in wet-dry climates weather and attain a threshold level of smectite clay, argilliturbation begins and offsets lessivage that may have previously been dominant (Muhs 1982). Understanding intrinsic thresholds is important, so that we do not always ascribe changes in soil morphology and pedogenic pathways to external forcing (McFadden and Weldon 1987).

Landscapes: Pedogenic Thresholds in the Buckskin Range, Nevada

One might expect thin soils on the arid, talus slopes of the Buckskin Range in Nevada, but many soils here are rocky only in near-surface horizons (Blank et al. 1996). Typical pedons (Xeralfic Paleargids) formed on talus slopes consist of large rocks, but with the interstices filled with vesicular sandy loam or loam material. Below this material lies clayey, prismatic- and blocky-structured horizons that are nearly rock-free. As one might expect in a desert, the development of these arid soils begins as raw talus acts as a trap for eolian dust. (Sites on these mountains that lack talus do not trap dust and have very thin to bedrock soils.) Rainfall washes the dust deep into the rock crevices. After a minimum amount of dust has accumulated, an intrinsic pedogenic threshold is passed and primary minerals in the dust begin to weather to smectitic clays. This does not occur in near-surface horizons, which are too dry. Continued additions of dust, coupled with its transformation to smectite, pushes the soil past yet another threshold, and argilliturbation begins. The talus is rafted upward in the incipient Vertisols (Fig. 11.47), producing a soil that has a boulder zone near the surface over a clay-rich, relatively rock-free subsoil. Muhs (1982) and Johnson and Hester (1972) described a similar sequence of development on the Channel Islands of California. The talus then protects the underlying soil from erosion.

Figure: Landscape image from the Buckskin Range, Douglas County, Nevada. Widespread occurrence of rocky talus is obvious. Photo by R. Blank.

---

forcing pathway changes. Or a group of burrowing animals moves onto a site, promoting mixing and regressive pedogenesis (Eisenhauer *et al.* 2011). Human activity can also trigger threshold changes in soil systems (Chen *et al.* 2011). These examples are all *external* drivers that can

Top of slope

Looking upslope

In a similar vein, Torrent and Nettleton (1978) introduced to the soil community the importance of pedogenic *feedback mechanisms*. Feedback is the returning of a part of the effects of a given process to its beginning or to a preceding stage, so as to reinforce that process. Thus, many feedback mechanisms or processes are self-accelerating (to a point), beyond which they decelerate and possibly terminate, as some aspect of the process becomes limiting (Vidic and Lobnik 1997). Examples include the accumulation of organic matter in A horizons, clay accumulation in Bt horizons, and the development of windows (breaks) in Bkkm horizons. Each of these processes tends to accelerate to a point, as a result of positive feedbacks set up within the soil, and then decelerate and stop changing after a time, sometimes only after a threshold is passed.

Vertisols provide excellent examples of pedogenic thresholds. They can develop directly from preexisting soils, and likewise, other soils can develop from Vertisols. Graham and Southard (1983) described a situation in which eolian materials overlie the smectite-rich argillic horizon (2Bt) of a Mollisol. The climate of the Mollisol was dry enough for the soil to develop subsoil cracks. Erosion of the upper solum of the Mollisol exhumed the clayey subsoil and allowed cracks to extend to the surface; eventually the profile developed into a Vertisol. Nettleton *et al.* (1969) had previously argued that argillic horizons could become so clay-rich and develop sufficient shrink-swell tendencies

(and cracks), over time, that they could engulf overlying materials. Although Graham and Southard's 2Bt horizon did not entirely engulf the overlying horizons into its subsoil cracks, this example nonetheless provided an example of one possible Vertisol development pathway. Graham and Southard's (1983) study provided an excellent example of polygenesis and illustrates how erosion (or any form of geomorphic instability), as an external forcing vector, can lead to a shift in pedogenic pathway. In this case, some threshold of erosion had to have been crossed before the pedogenic pathway could shift to one in which argilliturbation overwhelmed the preexisting pedogenic processes. Additionally, it should be noted that this pathway shift may be difficult to reverse, since it is unlikely that the Vertisol could develop back into a Mollisol without some sort of external forcing, such as deposition of smectite-poor sediment on top of the profile, or a change in climate.

Another Vertisol example, on San Clemente Island off the California coast, illustrates how Vertisols can develop from preexisting soils without an external forcing mechanism, but instead, by crossing an intrinsic, pedogenic threshold. The Mediterranean climate of the island has dry summers and wet winters, and the parent materials are coarse marine sands and gravels on uplifted marine terraces. Muhs (1982) described a chronosequence of soils that develop from Mollisols to Alfisols to Vertisols. Older soils contain more smectite, presumably added by eolian influx, such that eventually, when older than ~200,000 years, cracking and vertic properties become evident. Eventually, on San Clemente Island, Vertisol morphology is attained. Johnson and Hester (1972) expanded on this story by arguing that the argilliturbation processes of the Vertisols eventually force any rocks that once existed on the terrace surfaces upward, forming surface boulder and rocks fields (Fig. 11.47).

One of the many factors that cause intrinsic thresholds to be crossed centers on *pedogenic accessions* – pedogenically acquired or evolved features – that alter the genesis of the soil in some way and that can force feedback processes. Accessions include pedogenically acquired pans, fabric, concretions, pH, and various plasma features; see Johnson *et al.* (1990) for more examples. The feedback processes regulated by these acquired accessions may be more or less independent of any external conditions or vectors. For example, many desert soils accumulate secondary $CaCO_3$ in their B horizons. Eventually, these carbonates can plug the soil pores, creating an aquiclude in the B horizon. This pedogenic accession – the Bkkm horizon – forces the soil to cross an intrinsic threshold. The pedogenic pathway drastically changes from one in which water flows primarily downward, with vertically oriented processes prevailing, to one in which water infiltrates and then runs laterally. Pedogenic accessions and the feedback mechanisms they promote (Yaalon 1971, Torrent and Nettleton 1978) are captured within the soil evolution model of Johnson and Watson-Stegner (1987) but are not included, or at least directly mentioned, in any of the major pedogenic models previously discussed.

The soil evolution concept of Johnson and Watson-Stegner casts doubt on the validity of some long, strictly linear, soil chronofunctions. Linearity can occur in chronofunctions that capture only a fragment of a soil's entire development. But this linearity may not persist (Vidic and Lobnik 1997). The rate of soil development may change for times that lie outside the limits of the chronofunction, i.e., in the future. And these changes could not be predicted from the evidence presented in the chronofunction, by the soil's morphology and chemistry. We also might not be able to discern the nonlinearities and changes in development that do exist, or may occur, because we normally are operating on incomplete data; any real-world data set is of necessity incomplete. In short, the soil evolution model forces us to look at soil chronofunctions in a richer, different light, and to explore other possibilities of development – for all soils.

The soil evolution model has redefined pedogenesis, giving equal definitional rank to regressive and simplifying soil processes (Table 12.4). It recognizes and emphasizes that soils are complex, open process-response systems that are continually adjusting, by various degrees, scales, and rates to changing energy and mass fluxes, thermodynamic gradients, and environmental conditions. Disturbance and change within the soil system are, at last, viewed as natural and expected.

Steady-State Conditions

Although soils are now known to evolve and progress/regress through time, individual soil properties can eventually achieve steady-state conditions, at least for a given length of time (Egli *et al.* 2001; Fig. 12.27). Many soil properties change rapidly at first, but the rate of change decreases with time, approaching an asymptote; this asymptote condition is assumed to be a kind of steady state. At that point, the soil or some of its properties are assumed to be in a quasi-equilibrium state, exhibiting negligible change over time. Vidic and Lobnik (1997) suggested that steady state occurs when the rate of progressive soil development equals that of regressive soil development. Any soil – or any soil property – can be in a steady state, for a while. And just as surely, they all eventually will change over time. The determination of steady state or *not* steady state within a soil is dependent entirely on the timescale over which the change is being evaluated (Howard *et al.* 1993). Alexandrovskiy (2007) found that both static soil *properties* and pedogenic *process rates* change quickly and regularly in young soils, but eventually these rates slow as a steady state or period of dynamic equilibrium is approached (Fig. 12.28). He showed that this statement even held for rates of artifact burial by earthworms. Similarly, Eisenhauer *et al.* (2007, 2011) described the changes that occur in soils after earthworms – an

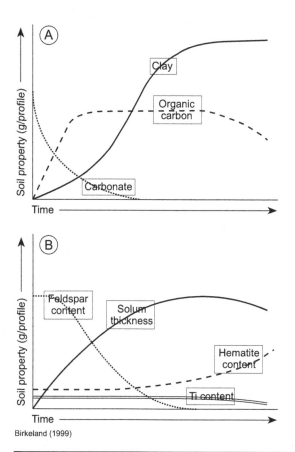

Birkeland (1999)

Fig. 12.27 Hypothetical variation in some soil properties with time, under a humid climate. Eventually, carbonates are leached from the soil. Organic carbon contents increase to a steady-state value, as does clay.

exogenous change to the soil system – entered a forest (Straube *et al.* 2009). The soil system showed the most pronounced effects of the invasion at the peak of the wave. After experiencing this wave, however, their preliminary data suggested that the ecosystem began to enter a new, steady-state condition with "altered biotic compositions and functions."

When a soil is in a steady-state condition, *inputs* of matter and energy to the system may continue, but the *net reactions* of the system, as ascertained by measurement and observation of the resultant *properties*, may be slow. If the soil or its properties *are* reacting and changing, their net change is immeasurably slow (Fig. 12.28). That is, the value (however expressed or determined) of the soil property changes little over time, and the small changes that occur do not reflect a shift in pedogenic pathway. An example of a steady-state condition in soils is the organic matter content of most A horizons. Even though the horizon continually has varying inputs of raw litter and humus, and concomitant losses via mineralization, the overall amount

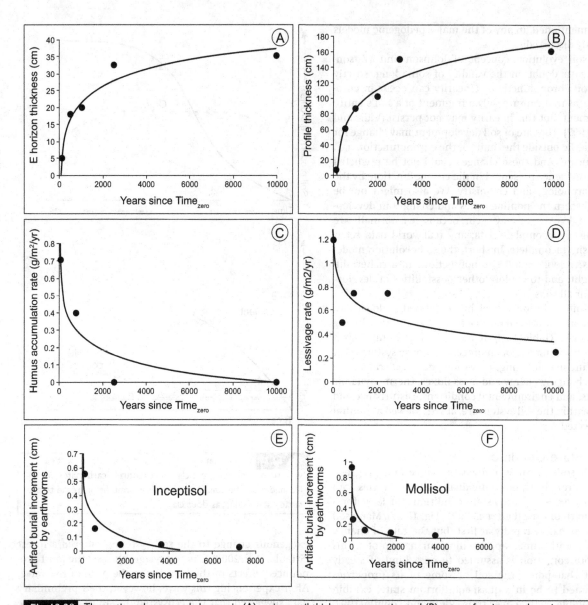

Fig. 12.28 Theoretic and measured changes in (A) various soil thickness properties and (B) rates of various pedogenic processes over time. Each set of data illustrates pedogenic changes that approach a steady state. The data are from Alexandrovskiy (2007), but the plots were prepared by the authors.

of organic matter in the horizon remains roughly the same, over medium to long timescales (Fig. 12.28).

As should be obvious, the judicious use of data-rich chronofunctions is an appropriate, and perhaps the best, way to isolate and identify potential steady-state conditions in soils, and to assess how long they may take to form (Fig. 12.27). The soil evolution and thickness models do not preclude the attainment of steady-state conditions. However, they suggest that either the condition will be ephemeral or that all aspects of the soil cannot be in a steady-state condition at the same time. That is, some

aspects of the soil will always be changing even if some are deemed to be in a quasi-equilibrium (Howard *et al.* 1993).

Polygenesis and Polygenetic Soils

The original, long-established concept of a monogenetic soil coincides well with that of the normal or zonal soil. Monogenetic soils have long been seen as developing toward some end point in equilibrium with the present climate and vegetation assemblage (Bryan and Albritton 1943). Thus, climatic stability was the basis and underlying tenet of monogenesis (Johnson *et al.* 1990). Monogenetic

Fig. 12.29 A highly polygenetic soil developed in glacial till in west-central Wisconsin, probably of Illinoian age (MIS 6), but possibly older (Syverson 2007). If so, this soil has been developing during at least two major interglacial periods and one major glacial period, and the soil remained subaerially exposed throughout all of these dramatic climate/vegetation shifts. During its development, the soil experienced minor amounts of erosion and small increments of loess deposition. The unaltered parent material of this soil, the C horizon, begins roughly at the waistline of the man holding the pick. In paleopedology, this soil would be considered a relict paleosol. Photo by RJS.

profiles were viewed as being the result of pedogenic processes that can be described by a set of variables that had constant relationships with each other for an interval of time sufficient to cause the soil attributes to form.

By definition, then, polygenetic soils have experienced at least two such episodes of monogenesis, and they usually exhibit a complex assemblage of attributes that developed in response to changes in state factors, often climate (Chadwick *et al.* 1995, Blank *et al.* 1998). Thus, if it could be demonstrated that the climate or some other external factor/variable had changed during the period of its formation, the soil could surely be considered polygenetic (Reheis 1987a, Bruce 1996; Fig. 12.29). Perhaps for this reason, Beckmann (1984) opined that a soil has a *heritage*, rather than an origin. Polygenetic soils clearly have rich heritages!

In some cases, however, the term "polygenetic soil" is also used to connote pedogenesis in a stacked sequence of parent materials; one or more of the lower materials is a buried soil (paleosol) with its own heritage, and the upper materials are thin enough that the modern soil has developed through them and into the lower sediment/soil (Karathanasis and Golrick 1991, Woida and Thompson 1993, James *et al.* 1995, Gugalinskaya and Alifanov 2000). The concept of layered parent materials and their effects on pedogenesis is discussed in Chapters 13 and 15.

Bryan and Albritton's (1943) polygenetic concept was restrictive, in that a *major* change, i.e., glacial-interglacial magnitude, was envisioned for polygenetic soils. But then, should minor climatic oscillations be handled? Johnson and Watson-Stegner (1987) (and Johnson *et al.* 1990) had an easy answer. Because pedogenic pathways are always being affected to varying degrees by changes in external and internal forces, all soils should be considered polygenetic. That is, all manner of external and internal changes impart a rich heritage *onto all soils*! Their definition lies at the other end of the continuum from that of Bryan and Albritton (1943) and suggests that even the normal, natural variability in state factors – however subtle – is adequate to render a soil polygenetic. Thus, because all soils have certainly undergone subtle process changes, they must all be polygenetic. In this vein, Johnson *et al.* (1990: 309) offered this definition of polygenesis:

> Polygenetic soil (and polygenesis) connotes multiple genetic linkages of exogenous and endogenous processes, factors and conditions, including evolved accessions, thresholds, and feedbacks, that vary with time. Therefore, all soils are polygenetic, and the older the soil, other things being equal, the more polygenetic it is.

Using this mindset, because pedogenic pathways have always been subtly shifting and evolving, the notion of the monogenetic soil is moot, and the term no longer has utility.

We take the middle ground by arguing that polygenetic soils must have demonstrably undergone some sort of major environmental change, be it extrinsically driven by changes in a state factor(s), or by intrinsic changes brought about via pedogenic accessions or the passing of a system threshold. And we assert that this change must have affected the soil in a measurable way, although the imprints may no longer be present. Many soils, especially late Holocene soils, might still be best considered monogenetic. For if we throw out monogenesis, then we are back to where we were previously and will simply have to adopt another term for soils that have undergone major vs. subtle changes in pedogenic pathways (Follmer 1998b). Each term has value, and sends a message.

Birkeland (1999) asserted, and we agree, that the pedogenic effects that confirm polygenesis, i.e., the chemical and morphologic signatures within the profile, must be *detectable*; if they are not, then the polygenetic impact may be so small that it is not worth noting (Ugolini and Schlichte 1973, Thompson and Smeck 1983, Eghbal and Southard 1993a, Fuller and Anderson 1993, Amundson *et al.* 1996, Connin *et al.* 1997, Hall 1999b, Kleber 2000, Khokhlova *et al.* 2001, Espejo *et al.* 2008, Brock and Buck 2009). And if polygenesis need not be physically or chemically expressed, the meaning of the term is devalued because it cannot be scientifically verified.

Phillips's Deterministic Chaos and Uncertainty Concepts

Soil evolution taken one step, or perhaps n steps, further epitomizes the deterministic chaos concepts espoused by Jonathan Phillips, a soil geomorphologist at the University of Kentucky. In several papers Phillips (1993a, b, c) supported and expanded upon the idea of soil evolution. Phillips's ideas focus on a mathematical and theoretical explanation of the extreme soilscape complexity that exists in the real world, especially with respect to soil pattern, using the soil evolution model and chaos and nonlinear dynamical system theory for support.

The soil evolution model, leaning as it does on pedogenic ebbs and flows, suggests that soil development is inherently nonlinear (Phillips 1993c; Fig. 12.27). The presence of thresholds also, by definition, indicates nonlinearity. Although nonlinearity does not always involve or lead to complexity, it does create the possibility of complex dynamics that cannot occur in linear systems. Pachepsky (1998) described nonlinear dynamic soil systems as those in which almost the same environmental conditions can lead to different developmental pathways and different pedogenic outcomes. Deterministic chaos theory adds that patterns, in this case, soil patterns, arise from the complex interactions of many elements behaving in an apparently random yet deterministic manner, via iterative nonlinear systems. Deterministic uncertainty is a perspective on soil spatial variability that blends two fundamental axioms: (1) the reductionist view that variability can be explained with more and better measurements of the soil system and (2) the nonlinear dynamics view that variability may be an irresolvable outcome of complex system dynamics (Phillips *et al.* 1996). It includes concepts such as dynamic instability, chaos, and divergent self-organization (Phillips 2001c). Deterministic uncertainty takes these two fundamental axioms, which might be viewed as irreconcilable, and attempts to merge them into a unifying theory.

Chaos in a system is characterized by sensitive dependence of a system on initial conditions and small perturbations. Applied to soil processes, chaos theory suggests that even minute differences at some point along a soil's evolution can lead to large and increasing differences as the soil system evolves (Huggett 1998a). The theory suggests that increasingly divergent soil development may occur over time and space, i.e., multiple soils on a common geomorphic surface may become increasingly variable with time, even if they have all been subject to the same suite of pedogenic factors and extrinsic inputs (Ibanez *et al.* 1990, Barrett and Schaetzl 1993, Huggett 1998a). Why? Combinations of minor perturbations emerge as complex patterns (Phillips *et al.* 1996). In short, the effects of minor perturbations can become large and long-lived (even if the perturbations themselves are not), when the soil system is *dynamically unstable*. Or, they may fade away if the system is stable. At least some of this persistence is due to feedbacks that are created when they first occur (Phillips 2001c). If small perturbations do not persist and grow, Phillips (1993a) argued that soil variability probably would not increase over time. Thus, the complexity of the soil fabric does not necessarily require there to have been complex external forcing vectors, such as vegetation or microtopography; the latter type of complexity is called stochastic complexity (Phillips 1993b, 2001c). Rather, underlying constraints or influences tend to *persist and grow* with time until they are disproportionately large compared to the magnitude of the original perturbation. External forces are significant to soil evolution, but Phillips's work asserts that they are not always *necessary* and that in situations when divergence occurs in the absence of variations in environmental controls and external forcings, then nothing but dynamic instability, i.e., chaos, can explain it. In later work, this framework was used to explain divergent evolution of soil morphology, i.e., increasing differentiation over time, within areas of constant soil-forming factors, on the basis of dynamical instabilities associated with the effects of individual trees, and minor initial variations in properties of underlying bedrock (Phillips and Marion 2005, Phillips *et al.* 2005).

Small local perturbations or soil variations at an indeterminate earlier time, such as any hypothesized $time_{zero}$, may be magnified and lead to significant spatial variability. Although these patterns can appear to be chaotic and random, Phillips argues for a modeling and theoretical paradigm that can address inherently nonlinear processes. Much of his work centers on unraveling and disentangling which parts of soil patterns have been caused by deterministic uncertainty versus the natural heterogeneity of the external pedogenic factors.

To explain and model this concept, Phillips (1993b) began with the soil evolution model equation

$$S = f(P, R)$$

defining S = soil development, P = progressive pathways, and R = regressive pathways. Then he differentiated the equation with respect to time:

$$dS/dt = dP/dt - dR/dt$$

expressing all the variables as a function of time, or as a rate. Written in its discrete form, this equation is

$$S_t = S_{t-1} + \Delta P - \Delta R$$

suggesting that the state of soil development at any given $time_t$ is dependent on its state at a previous, $time_{t-1}$, and on changes in the net effects of the various pedogenic pathways since that previous time. Phillips (1993a) assumed that there is a maximum rate at which progressive development occurs and that it decreases exponentially with the degree of development in the soil. Similarly, he assumed that there is a maximum rate at which regressive development occurs and that it also decreases exponentially over time. These

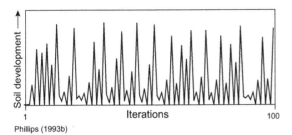

Phillips (1993b)

Fig. 12.30 Results of a soil development simulation model (Phillips 1993a) based on the evolution model of Johnson and Watson-Stegner (1987). The state of soil development (y-axis) at any given time (x-axis, iterated over 100 time steps) is assumed to be dependent on the state of development at the previous time step and on the net balance of progressive or regressive soil genesis during that step. It is further assumed that both progressive and regressive processes have maximum rates and that the rate of progressive pedogenesis decreases exponentially as the soil becomes more developed. In the simulation shown, the maximum regressive rate was 80% of the maximum progressive rate and the decline of the rate of progressive processes was sufficient to ensure that when soil development reached a steady, mature state, the progressive processes had declined to <5% of their maximum rate. The simulation illustrates that progressive processes dominated, but in an irregular pattern.

assumptions were formulated numerically and combined with the preceding equation to create a simulation model of soil development. By running this model iteratively over many time steps, a deterministic, complex pattern of soil development is produced (Fig. 12.30). The simulation model can be run iteratively for multiple pedons on a landscape, where each pedon differs only slightly from all the others with respect to its initial degree of development. Depending on the values chosen for the maximum rates of change and how fast the rates decline over time, the resulting realization may show soil properties chaotically distributed across the landscape, i.e., not all pedons will have attained similar degrees of development – even though the general soil-forming factors were the same. The concept was later applied to specific pedogenic processes and phenomena, such as parent material weathering, vegetation effects, unstable wetting fronts, and soil creep (Phillips *et al.* 1996, 2005, Furbish and Fagherazzi 2001, Phillips and Marion 2005, Toomanian *et al.* 2006, Phillips 2008).

In sum, Phillips's work shows that rich and complex patterns of soil development can ensue even if environmental inputs are reasonably uniform, as a result of *deterministic uncertainty*. Thus, the apparent chaos that one often sees on the soil landscape may be unrelated to the age of the surface, as even subtle variations in initial conditions can lead to vast differences in soils on the same surface. To quote Phillips (1993b: 175):

In the case of dramatic local spatial variability where no variation in soil-forming factors can be discerned, the chaos model offers a plausible explanation where no other explanation is available.

One way to understand the implications of Phillips's model is to take one of three perspectives on the variability of the soils of a place. The chaos theorist might throw her hands in the air and attribute the variability, at least in part, to minor, seemingly insignificant differences in soil properties at time$_{zero}$, differences that are enhanced over time as the rates of progressive and regressive processes vary. The traditional pedologist might argue that the pattern would be understandable and decipherable if the pedologic processes were better understood or scaled appropriately, and that we need more data. The deterministic uncertainty theorist would take the middle ground (Phillips *et al.* 1996) and argue that known and measurable causes can often account for the variability, but that unstable, chaotic dynamics could also be invoked to explain some degree of soil variability.

Obviously, the implications of this model cast a shadow on the use of soil chronofunctions as a surface age dating tool, especially when the number of observations is low. Chronofunctions tend to be based on surfaces of known age; anomalous pedons are ignored (Barrett and Schaetzl 1993). Chaos theory assists in interpreting these spatially complex soil landscapes and provides a powerful theoretical framework for the understanding of order and disorder in soils. The approach helps interpret and explain situations in which soil variability occurs in the absence of observable variation in soil-forming factors. Deterministic uncertainty has also proven useful in modeling of field- and hillslope-scale spatial variability and pedodiversity (Phillips 2001c, Svoray and Shoshany 2004, Toomanian *et al.* 2006, Milan *et al.* 2009). Nonlinear systems theory may be most helpful in explaining and predicting the genesis of those soils where the familiar state-factor model fails.

Final Words

It is not our intention to suggest that all of pedogenic or soil geomorphic theory is captured in the models outlined here. It is not. But because of space limitations, we have had to restrict our discussion. Other models are certainly worthy of note, including those by Stephens (1947), Hole (1961), Chesworth (1973), Huggett (1975), Johnson *et al.* (1987, 1990), Sommer and Schlichting (1997), Minasny and McBratney (2001). For those interested, Minasny *et al.* (2008) and Bockheim and Gennadiyev (2010) provide some excellent reviews of models in soil science, particularly with respect to the more quantitative models. We finally note that models of Butler (1959) and Paton *et al.* (1995) are discussed in later chapters and that soil production models were introduced in Chapter 9.

Chapter 13

Soil Genesis and Profile Differentiation

The formation of a soil from raw parent material (or a pre-existing soil) defines soil genesis, or *pedogenesis*. The study of pedogenesis is called *pedology*. Amundson (1998) noted that during its formative period, pedology focused on observation of soil properties, so as to unravel the causes and processes that formed those soils. Wilding (1994: 16) provided perhaps the most complete definition of pedology, as

the component of Earth science that quantifies the factors and processes of soil formation, including the quality, extent, distribution, spatial variability, and interpretation of soils from microscopic to megascopic scales.

Considering *scale* in pedology is important, for soils can be, and are, studied at a wide range of scales (Fig. 3.14). Today, we often think of pedology as simply the study of soil genesis (the focus of this chapter) and classification (the focus of Chapter 8).

Landscapes: Can a Soil Become Extinct?

Although soils seem to be everywhere, some soil types are rare. They occur only in isolated areas where unique combinations of state factors converge, thereby epitomizing the concept of *soil endemism* (Bockheim 2005). The loss of these soils often accompanies the loss of endemic vegetation types, as is the case with many serpentinitic soils (Spence 1957, Proctor and Woodell 1975, Mansberg and Wentworth 1984, Kram *et al.* 2009, Rajakaruna *et al.* 2009, Weerasinghe and Iqbal 2011), marl soils (Bartgis and Lang 1984), and the soils at Hans Jenny's famous pygmy forest (Jenny *et al.* 1969). Drohan and Farnham (2006) went so far as to define rare versus threatened soils. Simply put, a rare soil is of limited areal extent, e.g., Guo *et al.* (2003). A threatened soil may have a greater areal extent but is undergoing a transformation that alters its characteristics and function, making it less able to carry out that function, e.g., growing food or supporting a native ecosystem.

Some, more extensive soils are also at risk due to human activities or climate change, i.e., unintended changes in the state factors. Many Gelisols are in danger of losing their underlying permafrost, because of our globally warming climate (Stokstad 2004, Pearce 2006). The result would be a soil system that is dramatically different from the one in which the soils formed. Other soils are in danger of "going extinct" because of more direct, human-induced, land use change, e.g., mining and suburbanization (Anonymous 2004, Kaiser 2004). Soils are being plowed up, paved over, or dug up at alarming rates, prompting the concern that even many geographically extensive soil types may be forever lost (Amundson 1998). Soil maps may suggest that particular soils are widespread, but natural examples – suitable for pedological investigation – may be almost impossible to find.

By protecting some rare and endangered plants and animals, their extinction can perhaps be prevented, and their populations can grow. But the same cannot be said for soils; even if protected, rare soils will not expand in area or abundance over time. And when they are lost, they are lost forever. Unlike living organisms, soils do not reproduce; nor can they be formed again. Once they are dramatically altered by agriculture, urbanization, drainage, or other human-created land uses, they are often lost forever.

If we are concerned that soils might become endangered or extinct, are we implicitly assuming that the proper baseline for pedogenic conservation is modern, i.e., today's, soils? In most cases, contemporary soil morphology is the synthesis of many different natural processes that have occurred through time. Climatic and biotic inputs have changed dramatically in the last several thousand years. Should we mourn the Inceptisols that were likely extensive over Wisconsin and Michigan during the early Holocene, and that were replaced by Spodosols as the boreal forest replaced tundra vegetation? Can a baseline even be drawn, knowing that the state factors have changed since $time_{zero}$ and will continue to change into the future?

Do these questions cause you to think about soils differently?

Pedogenenic Processes

As discussed in Chapter 12, soil genesis involves both progressive and regressive processes (Figs. 12.21, 12.22). It includes all processes operative within the soil, whether they act to promote horizonation, preserve it, or even destroy it. Pedogenenic principles and assumptions underlie many soil classification systems, which endeavor to group soils of similar formative history and are based on observable morphological or chemical properties (Cline and Johnson 1963, Smith 1986). The links between soil genesis and classification, shown in some respects in Figures 8.3 and 8.8, can also be envisioned as follows: soil-forming factors → soil-forming processes → soil morphology and chemistry → soil classification (Bockheim and Gennadiyev 2000).

We emphasize pedogenic processes in this book for several reasons First, we must understand soil genesis if we are to classify and manage soils in a rational manner. Second, only by knowing the processes that formed soils can we hope to predict how they may change as environmental inputs, e.g., precipitation, heat, ions, and human uses, also change (Bockheim and Gennadiyev 2000). Third, the better we understand how soils form, the better we can understand how landforms evolve; the two are intimately linked – through the science of *soil geomorphology* (see Chapter 14). Last, only by knowing how soils have formed can we map them or explain their past, present, or future distributions. Thus, *soil genesis* is an integral part of *soil geography*.

Forming, maintaining, or perhaps even destroying a soil involves an extremely complex set of processes, much akin to building, maintaining, or destroying a house. Both can be compartmentalized into discrete, subcontracted blocks or bundles like (1) excavating the basement, (2) pouring the cement walls, (3) framing in the aboveground walls and roof, (4) installing plumbing and electrical components, (5) finishing and trim work, and so forth. When examined in a soil genesis context, the project might be viewed as a series of *process bundles* or macroprocesses, e.g., weathering, dealkalinization, lessivage, podzolization, and ferrolysis (Arnold 1983, Bockheim and Gennadiyev 2000; Fig. 8.8, Table 13.1). Each of these bundles or suites of macroprocesses has a certain degree of internal cohesiveness but is composed of more precise and detailed parts (microprocesses). For the home project, the names given to the more detailed jobs might include (1) hanging dry (gypsum) wallboard, (2) installing windows, (3) adding trim to window frames, and (4) painting. For the soil, examples of discrete process names might include (1) feldspar weathering, (2) complexation of metals, and (3) translocation of amorphous silica. Our point is that soil genesis, like building a house, can be viewed at a number of nested process suites, operating at a variety of scales (Zonn 1995). Be aware, however, that just as all houses are different, so are all soils different. Some houses will have stucco exteriors, while others are brick or wood, just as some soils have argillic B horizons, while others have spodic or cambic B horizons. Similar structures, but the differences lie in the details.

Pedro (1983) viewed pedological evolution as consisting of two process bundles: (1) processes associated with *weathering* of primary minerals, the subsequent *release* of various elements, and their possible *recombinations* to produce new, more stable compounds and structures; and (2) processes relating to the *arrangement* or *redistribution* of these soil constituents into soil horizons. Weathering is an important precursor to many pedogenic processes, and it is ongoing in all soils (see Chapter 9). Weathering alone can form recognizable soil horizons. For example, Bw horizons (Soil Survey Staff 1999) are slightly weathered horizons that have formed primarily in response to in situ weathering, with minimal illuvial gains or eluvial losses of constituents. Pedro (1983) called this type of pedogenesis "associative pedogenesis without plasma transfer into the solid phase," and we might call it *braunification*.

Nonetheless, most of the major pedogenic processes involve mobilization, transportation, and immobilization (precipitation) of a constituent or constituents. Pedro (1983) referred to these as dissociative pedogenic processes. Within the context of soil genesis, we would also give equal weight to pedoturbation processes, noting that they are not necessarily regressive in nature and that they occur during the genesis of every soil (see Chapter 11 and Table 13.1).

Materials in soils that can be mobilized, transported by water, and eventually deposited within the profile are referred to as *plasma*. Plasmic components, e.g., ionic species, organic compounds, and clay and organic colloids, are mobile within soils. The parts that remain behind are referred to as *skeleton* (Pedro 1983). In most cases, skeletal components include very fine sand and larger particles. Over time, however, the skeleton can be moved by pedoturbation or it can be weathered, with the resulting ionic components translocated within or out of the profile.

Eluviation-Illuviation

An important first step in the discussion of pedogenic processes is their organization into a logical scheme, as we have done in Table 13.1. There, and in this chapter, we adopt the process-systems approach of Simonson (1959; see Chapter 12), i.e., the formation of soil horizons can be ascribed to *additions to* and *removals from* the soil system, as well as *transfers* and/or *transformations* within it. Most transformations that affect soils involve primary and secondary mineral weathering (see Chapters 4 and 9). This chapter examines transformations from a pedogenic perspective but mainly focuses on interhorizon transfers. In certain instances, additions to the soil surface, e.g., dust, organic matter, are also discussed where they are viewed as being pedogenically important.

The transfer of material out of a horizon in percolating water, whether in solution or in suspension, is called *eluviation*. That is, eluviation represents a net *loss* of material from a horizon. Its alter ego, *illuviation*, refers to the *gain* of material by a horizon, usually either from an overlying horizon (vertical illuviation) or from a horizon upslope (lateral illuviation), by percolating water. Thus, eluviation and illuviation usually occur as paired couplets, or bundles. What is lost from one horizon is usually gained by another, unless it exits the profile entirely. Transfer of material

Table 13.1 Processes of soil formation: A synopsis

Term/Process	Definition, description, some pertinent references
Soil genesis; pedogenesis	Surficial processes – both progressive and regressive – that aid in the formation of soil; processes operate during and after weathering and act to arrange and reorganize the soil plasma and other constituents (Pedro 1983)
Subsets thereof (Pedro 1983): Associative pedogenesis	In situ weathering and pedogenesis; little disruption of plasma or skeleton fabric
Dissociative pedogenesis	Dissociation between skeleton and plasma, typically involving translocation of plasmic materials
Soil development; soil formation	Surficial processes that aid in the formation of, or otherwise do not destroy or weaken, soil horizons
Soil evolution	Concept that soils change, both progressively and regressively, and all combinations thereof, through time (Johnson and Watson-Stegner 1987, Huggett 1998b, McPherson and Timmer 2002)
Horizonation	Differentiation of parent material into discrete soil horizons, increasing pedogenic anisotropy and complexity (Johnson and Watson-Stegner 1987)
Haploidization	Processes involved in the destruction or blending of existing soil horizons, increasing profile isotropy and simplification (Johnson and Watson-Stegner 1987, McPherson and Timmer 2002)
Eluviation	Movement of material out of a portion of a soil profile, resulting in a net loss (Daniels et al. 1968, McKeague and St. Arnaud 1969, van Wambeke 1972, Muir and Logan 1982)
Illuviation	Movement of material into a portion of a soil profile, usually from overlying horizons, resulting in a net gain (McKeague and St. Arnaud 1969, van Wambeke 1972, Muir and Logan 1982, Fedoroff 1997, Buurman et al. 1998)
Biocycling; phytocycling	Movement of ions from the soil to the biosphere, via root uptake, and back to the soil, via litter decomposition, humification, and mineralization (Howitt and Pawluk 1985b, Quideau et al. 1996, Giesler et al. 2000, Jobbagy and Jackson 2004, Opdekamp et al. 2012, Cornelis et al. 2014)
Biogeochemical cycling	Movement of various nutrients, particles, and ions, at large scales, among the soil, biosphere, hydrosphere, and atmosphere (Likens et al. 1998, Johnson et al. 2009)
Leaching; depletion; base cation leaching	Washing (eluviating) of soluble materials from the soil profile and into the parent material (Felixhenningsen et al. 1983, van Praag et al. 2000, Egli and Fitze 2001)
Decarbonation, decalcification	Weathering, translocation, and net loss of Ca and carbonates from a soil, usually by leaching but also by plant uptake (van Breeman et al. 1983, van Breeman and Protz 1988, Rubio and Escudero 2005)
Recarbonation	Adding carbonates to a soil that had previously been leached, usually by increased Ca and Mg cycling or by additions of calcareous dust (Fuller et al. 1999)
Calcification; calcosiallitization	Accumulation of secondary carbonates in soils (Buol 1965, Gile et al. 1966, Sobecki and Wilding 1983, Machette 1985, Schaetzl et al. 1996, Douglass and Bockheim 2006, Brock and Buck 2009, Catoni et al. 2012, Gocke et al. 2012)
Sparmicritization	Dissolution of large $CaCO_3$ crystals and reprecipitation of micritic crystals of $CaCO_3$ in the resulting voids, usually facilitated by soil microorganisms (Kahle 1977, Phillips and Self 1987, Zhou and Chafetz 2009)
Surficial erosion; soil removals; superficial impoverishment	Removal of material from the surface layer(s) of a soil (Roose 1980, Johnson 1985)
Alluviation	Additions to a soil by overbank sedimentation, as occurs on floodplains (Faulkner 1998, Alexandrovskiy 2007), or preferential erosion and loss of clay from the surface horizons in soils on a hillslope, by laterally flowing water (Jackson 1965)
Upbuilding	Allochthonous surficial additions of mineral and organic materials to a soil (Johnson 1985, Lowe 2000, Eger et al. 2012)

(cont.)

Table 13.1 (cont.)

Term/Process	Definition, description, some pertinent references
Enrichment	General term for increases in the concentration of a soil constituent, by either additions or residual accumulation as other other constituents are lost
Developmental upbuilding	Additions of mass to a soil that are slow enough to allow pedogenesis to keep pace and incorporate the new material into the profile (Johnson 1985, McDonald and Busacca 1990, Almond and Tonkin 1999, Alexandrovskiy 2007)
Retardant upbuilding	Additions of mass to a soil that are faster than pedogenic processes can assimilate them into the profile, so that the soil momentarily becomes buried (Johnson 1985, Schaetzl and Sorenson 1987, Alexandrovskiy 2007, Carter et al. 2009)
Cumulization	Overthickening of the A horizon and associated upbuilding of the soil surface, caused by prolonged developmental upbuilding, as on floodplains and alluvial toeslopes (Riecken and Poetsch 1960, Hole and Nielsen 1970, McFadden et al. 1987, Alexandrovskiy 2007, Carter et al. 2009, French et al. 2009)
Noncumulative genesis; negative strain	Soil collapse as suspendable or soluble materials, e.g., carbonates, are removed; includes subsurface, lateral, and vertical removals through pervection, lessivage, and leaching (Hole and Nielsen 1970, Johnson 1985, Bryant 1989, Chadwick et al. 1990)
Loosening; positive strain; dilation	Increase in volume of voids per unit volume of soil by the activity of plants, animals, and humans; by freeze-thaw or other physical processes; or by removal of material by leaching (Lichter 1998, Mathé et al. 1999, Oh and Richter 2005)
Hardening	Decrease in volume of voids (porosity) by collapse and compaction, and by in-filling of voids with fine earth, carbonates, silica, and other materials, often leading to the formation of surface crusts and subsurface pans (Fox et al. 2004); also cementation of soil materials that leads to hard consistence
Soil deepening; profile deepening	Slow, downward migration of the lower soil boundary into relatively unweathered material below (Johnson 1985)
Acidification	Reduction of soil pH by additions of acidic materials, as in acid rain, or by hydrolysis of metal cations such as Fe^{3+} or Al^{3+}; usually accompanied by loss of base cations via weathering and leaching (Kikuchi and Gorbacheva 2005, Blaser et al. 2008a, b, Titeux and Delvaux 2010, Hedl et al. 2011, Yang et al. 2012)
Lixivation	General term for the movement of soluble salts within soils (Duchaufour 1982)
Salinization	Accumulation in soils of soluble salts such as sulfates and chlorides of Ca, Mg, Na, and K (Szabolcs 1989, Hussein and Rabenhorst 2001, Schofield et al. 2001, Rengasamy 2006, Kanzari et al. 2012)
Desalinization	Translocation of salts from the upper profile, resulting in a leached upper profile; upper profile sodication may be a by-product of this process (Kellogg 1934, Munn and Boehm 1983, van Breeman et al. 1983)
Sodication; solonization; alkalization	Accumulation of sodium in soils, resulting in an increasing ratio of Na to divalent cations in solution and on the exchange complex (van Breeman and Buurman 1998)
Dealkalization; solodization	Leaching of Na ions and salts from (out of) natric horizons (Pedro 1983)
Podzolization; spodosolization	Migration of Al and organic matter, with or without Fe, from O, A, and E horizons to accumulate in the B horizon, resulting in the relative concentration of silica (i.e., silication) in an eluviated layer (DeConinck 1980, Ugolini and Dahlgren 1987); see also acidocomplexolysis
Depodzolization	Gradual erasure of podzolic morphology from a soil, usually because of a change in climate or vegetation that weakens or stops the podzolization process (Hole 1975, Barrett and Schaetzl 1998)
Andisolization	Processes operative in soils that contain a large proportion of volcanic materials such as ash; similar to podzolization, in which fine earth fraction of soils becomes dominated by amorphous compounds
Latosolization; ferrallitization	Residual accumulation of sesquioxides in soils, as base cations and silica are leached, under long-term weathering in a hot, humid climate, and unlike laterization, little internal translocation of Fe occurs (Waegemans and Henry 1954, Sherman and Alexander 1959, Sherman et al. 1967, Lepsch and Buol 1974, Martini and Macias 1974, Beinroth 1982, Buol and Eswaran 2000)

Table 13.1 (cont.)

Term/Process	Definition, description, some pertinent references
Fersialitization	Suite of pedogenic processes involving inheritance and neoformation of smectitic clays, and immobilization of iron oxides due to alkaline soil conditions (Duchaufour 1982).
Desilication; allitization; siallitization	Leaching of silica from soils, leading to the concentration of sesquioxides, with or without formation of ironstone (laterite, hardened plinthite) and concretions (Eswaran and Bin 1978a, Latham 1980, Pedro 1983, Righi et al. 1990, Yaro et al. 2008, Schaefer et al. 2008)
Degrees and types of desilication:	(Pedro 1983)
Allitization	Total desilication and dealkalinization, forming gibbsite and ferric hydrates
Monosiallitization	Partial desilication; total dealkalinization, forming 1:1 phyllosilicates and ferric hydrates (Betard 2012)
Bisiallitization	Partial desilication; partial dealkalinization, forming 1:1 and 2:1 phyllosilicates (Betard 2012)
Laterization; plinthization	Migration of Fe compounds within soils, forming Fe concentrations at preferred sites, often producing plinthite; much of the Fe probably from outside the profile (Wood and Perkins 1976, Pfisterer et al. 1996, Aide et al. 2004, Yaro et al. 2008)
Kaolinitization	Pedogenic formation of a kaolinite-dominated mineral suite (Sandler 2013)
Silicification	Accumulation of secondary silica, either neoformed or translocated, into a soil or soil horizon (Milnes and Twidale 1983, Chadwick et al. 1987, 1989a, Evans and Bothner 1993)
Lessivage; argilluviation; illimerization	Mechanical migration of clay particles from A and E horizons to the B horizon, producing Bt horizons enriched in clay (Fridland 1958, Gorbunov 1961, Rode 1964, Smith and Wilding 1972, Howitt and Pawluk 1985a, b, Fedoroff 1997, Bockheim 2003, De Jonge et al. 2004, Quénard et al. 2011, Cornu et al. 2014)
Neoformation; argillation	Formation of clay in situ (Fridland 1958, Alekseyev 1983)
Pervection	Mechanical migration of silt (but also including some clay) particles in soils (Paton 1978, Bockheim 1979a, Harris and Ellis 1980, Forman and Miller 1984, Bockheim and Ugolini 1990)
Self-weight collapse	Process whereby wet soil or sediment collapses, increasing its overall density and reducing pore space (Bryant 1989)
Pedoturbation	Biological and physical mixing, churning, and cycling of soil materials (Hole 1961, Lee and Wood 1971b, Wood and Johnson 1978, Johnson et al. 1987, Poesen and Lavee 1994); see Table 11.1 for subtypes of pedoturbation
Proisotropic pedoturbation	Regressive pedoturbative processes that disrupt, blend, destroy, or prevent formation of horizons or subhorizons, resulting in simplification of the profile or inhibition of further horizon development (Johnson et al. 1987)
Proanisotropic pedoturbation	Progressive pedoturbative processes that form or aid in the formation and maintenance of horizons or subhorizons; results in, or promotes, increased profile order (Johnson et al. 1987)
Disintegration; physical weathering	Physical breakdown (weathering) of mineral and organic compounds into smaller pieces, with little or no change in composition (Jackson 1965)
Arenization	Weathering of and loss of fine-grained minerals in granite, leaving behind a sandy, quartz-rich residual product (Eswaran and Bin 1978b, Sequeira Braga et al. 2002)
Decomposition; chemical weathering	Chemical breakdown of mineral and organic materials, with a concomitant change in chemical composition and structure
Synthesis, neoformation	Formation of new particles of mineral and organic species from ionic components (Jackson 1965, Kloprogge et al. 1999, Walsh and Humphreys 2010)
Main pedoweathering pathways:	(Pedro 1983)
Acidocomplexolysis	Strongly acid and complexing attack on minerals by organic acids, affecting the extraction and elimination of Al, Fe, and basic cations from soils (see also podzolization)

(cont.)

Table 13.1 (*cont.*)

Term/Process	Definition, description, some pertinent references
Alkalinolysis	Strongly alkaline attack due to the presence of Na carbonates in solution, affecting alumina and silica (see also solodization)
Xerolysis	Attack on clays due to alternating periods of wetness and dryness (see also ferrolysis) (Chauvel and Pedro 1978, Ducloux *et al.* 1998, 2002)
Littering	Accumulation of raw organic materials and litter on soil surface (Hart *et al.* 1962, Lawrence and Foster 2002)
Humification; maturation of organic matter	Transformation of raw organic material into highly decomposed humus (Duchaufour 1976, Anderson 1979, Zech *et al.* 1990, Malterer *et al.* 1992, Rumpel *et al.* 1999, Prescott 2010, Schnitzer and Monreal 2011)
Phosphatization	Accumulation, translocation, and reprecipitation of P in soils due to thick accumulations of bird droppings in cold climates (Tatur and Barczuk 1985, Simas *et al.* 2007)
Paludification; paludization	Processes, dominant in wet soils, regarded by some as geogenic rather than pedogenic, including accumulation of deep (>30 cm) deposits of organic matter, as in Histosols (Tarnocai 1972, Viereck *et al.* 1993, Fenton *et al.* 2005, Lavoie *et al.* 2005, Simard *et al.* 2007, 2009, Kroetsch *et al.* 2011)
Terrestrialization	In-filling of a shallow, depressional water body by long-term accumulation of organic materials, e.g., litter and mosses (Kroetsch *et al.* 2011)
Ripening	Chemical, biological, and physical changes in organic soil material after air (oxygen) penetrates previously waterlogged material; usually accompanied by humification
Mineralization of organic matter	Release of C, N, P, S, and other elements in plant residues due to decomposition (Finzi *et al.* 1998, Jarvis *et al.* 2007, Manzoni and Porporato 2009)
Melanization	Darkening of light-colored, unconsolidated mineral materials by admixture of organic matter and humus (Wilde 1946b, Fenton 1983, Carter *et al.* 1990, Schaetzl 1991b, Douglass and Bockheim 2006)
Leucinization	Paling or lightening (in Munsell color value) of soil horizons, mainly by loss of humus, either through transformations to lighter-colored compounds or via removal (eluviation)
Brunification; braunification	Browning of soils caused by weathering and release of iron oxides from primary minerals; Fe minerals coat soil particles and form stable humus-Fe complexes, producing yellowish brown colors normally associated with goethite (Ugolini *et al.* 1990, Vodyanitskii *et al.* 2005)
Rubification; ferrugination	Reddening of soils caused by weathering and release of iron oxides from primary minerals; Fe minerals produce deep red colors normally associated with precipitation of hematite (Glazovskaya and Parfenova 1974, Torrent *et al.* 1983, McFadden and Hendricks 1985, Boero and Schwertmann 1989, Schwertmann *et al.* 1982, Beauvais and Colin 1993, Blume *et al.* 1997, Li *et al.* 2003)
Gleization; gleyzation; gleyification; hydromorphism	Chemical reduction of iron under anaerobic (waterlogged) conditions, with production of bluish, greenish-gray or whitish-gray matrix colors (Allan *et al.* 1969, Schlichting and Schwertmann 1973, Vepraskas 1999)
Sulfidization; sulfidation; pyritization	Accumulation of sulfides in soils and sediments, as sulfur-reducing bacteria change sulfate SO_4^{2-} in water and sediment to sulfide and secondary ferrous sulfide (FeS) or pyrite (FeS_2) precipitate; occurs primarily in wet soils along seacoasts, where a source of sulfur exists (Fanning and Fanning 1989, van Breeman 1982, Rabenhorst and James 1992, Glover *et al.* 2011, Johnston *et al.* 2012)
Sulfuricization	Oxidation of sulfide-bearing minerals in soils, producing jarosite and H_2SO_4 (Fanning and Fanning 1989)
Ferritization	Process whereby dissolved ferrous sulfates produced by sulfuricization are oxidized and recrystallized to form ferrihydrite and ferric sulfates (Fitzpatrick *et al.* 1996)
Chelation	Complexation of metal ions by multiple electron-donating ligands associated with single organic molecule; stability of chelated complexes promotes their mobility in the soil system (Atkinson and Wright 1957, Buurman and van Reeuwijk 1984)
Cheluviation	Combination of chelation and eluviation; translocation of complexed metals an important process in many podzols, but not all metal complexes are chelation complexes

Table 13.1 (cont.)

Term/Process	Definition, description, some pertinent references
Ferrolysis; xerolysis	Process that occurs in soils with a B horizon that perches water and also exist in wet-dry climates; pH fluctuates as redox potential fluctuates; clays destroyed in upper solum under acid conditions, forming clay-poor, acidic upper profile above clay-rich Bt horizon (Brinkman 1970, 1977b, Ransom et al. 1987a, McDaniel et al. 2001, Van Ranst and DeConinck 2002, Barbiero et al. 2010)
Chloritization	Misnomer of formation of Al-hydroxy interlayers in existing 2:1 clays under acidic weathering; clays have very low CECs and high amounts of exchangeable Al; some interlayered clays may have XRD patterns resembling chlorite, but no chlorite is formed (Dijkerman and Miedema 1988)
Gypsification	Accumulation of secondary gypsum, either neoformed or translocated, in soils (Reheis 1987b, Carter and Inskeep 1988, Buck and Van Hoesen 2002)

Source: Compiled by the authors and patterned after Buol et al. (1997).

completely out of the profile, by eluviation, is termed *leaching.* Commonly observed eluvial-illuvial couplets include those associated with clay, Fe, Al, humus, carbonates, salts, and silica. Obviously, for eluviation-illuviation to occur, enough water must be available to percolate through the profile, and substances must be made available for transport, either in solution or in suspension (Egli and Fitze 2001). Nonetheless, even in the driest of climates, soils may show some evidence of eluviation and illuviation, because the occasional storm will wet the soil thoroughly enough to initiate these processes.

Rigorously establishing the amount of material that has been translocated in soils can be time-consuming and is discussed in more detail in Chapter 15. Alternatively, a semi-qualitative assessment of illuviation can be used, e.g., the areal coverage of cutans or of a color that distinguishes ped coatings from the matrix material in the illuvial horizon. An example of the latter might be iron oxide (hematite) coatings that make a soil red, organic coatings that color it black, or a determination of the volume of cutans based on a point counting in micromorphological thin sections. In short, quantifying illuviation can be done in a variety of ways, e.g., Holliday (1998), Schaetzl and Mokma (1998).

Sometimes we wish to know the overall long-term *intensity* or *rate* of illuviation in a soil. To make this determination, we also need to know the age of the soil – the number of years since time$_{zero}$. The intensity (or strength) of the eluvial-illuvial process bundle can be assessed by determining the amount of material translocated, e.g., clay, per unit time of soil age. For example, clay translocation in a soil may have occurred at an average rate of 2 g cm^{-2} of soil area per 1,000 years. Put another way, two grams of clay were translocated downward through a column of soil one square centimeter, for each millennium since time$_{zero}$.

Effect of Surface Area

Most illuvial substances in soils are manifested as *coatings, or cutans,* on particles or ped faces. Illuvial substances are like paint that coats a wall. For larger rooms with more wall area, the paint is spread thinner. Even if the same amount of paint is used on two different rooms, it will be thicker and will coat the wall better in the room that has less surface area, less wall space. Because coarse-textured horizons have less surface area than do fine-textured horizons, illuviation in coarser horizons is more readily apparent because there is less surface area to coat/cover. In soils, it is mainly the ped and particle *surfaces* that become coated, whether these surfaces are mineral grains or ped surfaces.

Soils with abundant clay and silt have more surface area to become coated (often an indication of illuviation) or stripped of existing coatings (an indication of eluviation) than do sandy and gravelly soils (Fig. 2.8). Thus, eluvial-illuvial processes take longer to become morphologically evident in fine-textured soils. The effect of surface area is most obvious in cobbly and rocky soils; rocks and other coarse fragments occupy considerable volume, forcing the remaining processes to be compressed into less space than if the same soil had no coarse fragments (Schaetzl 1991b). Coarse fragments appear to strengthen the expression of eluvial-illuvial processes by making the products of these processes manifest more quickly. Surface area is also important to soils because fine particle surfaces retain water, cations, anions, and nutrients, forcing slower percolation and, potentially, shallower eluviation.

Examples of surface area effects abound (Broersma and Lavkulich 1980). Secondary carbonates accumulate more rapidly in gravelly parent materials, as compared to gravel-poor sediment (Gile et al. 1966). E horizons also form more quickly in coarse-textured soils than in fine-textured soils. For example, the intention of the *cambic horizon* concept (diagnostic for Inceptisols) is one of slight to very strong alteration of primary minerals and of structure development, but with some weatherable minerals present. Because these kinds of pedogenic features can form quickly in sandy materials, the Soil Survey Staff (1999) does not allow cambic (Bw or Bg) horizons to be sandy.

Schaetzl (1991b) observed that melanization and the accumulation of organic matter with depth in some gravelly soils are promoted by high contents of coarse fragments, as indicated by the strong correlation between organic carbon content and coarse fragment content ($r = 0.93$; $p < 0.001$). Holding other factors constant, as the amount of coarse fragments increases, the soil *volume* in which melanization can operate, and the surface area on which organic coatings can form, decreases. Thus if similar amounts of organic matter are incorporated into two soils, one with abundant coarse fragments and gravel, and one with little gravel or few cobbles, two hypothetical scenarios are envisioned:

A. Organic matter will accumulate to approximately similar depths in both soils (Fig. 3.8A). However, in the gravelly soil, which has less surface area on which to accumulate organic coatings, the amount of organic matter in intergravel voids and the thickness of organic coatings on skeletal grains will be greater. In effect, the same amount of organic matter is "crammed" into a smaller space. The result will be *higher concentrations* of organic matter in the upper horizons of the gravelly soil.

B. Organic matter will accumulate to greater depths in the gravelly soil (Fig. 3.8B). In this scenario, organic matter is diffused more deeply into the soil, although the soil itself contains no more organic matter than in scenario A. The amount of organic matter in intergravel voids of the upper horizons of this gravelly soil will not differ markedly from that in the intergravel voids of the gravel-free soil, but the organic matter will be *incorporated more deeply* into the gravelly soil because of its lower pore space, caused by the higher contents of coarse fragments.

In summary, the effect of eluvial or illuvial processes in soils is dependent on the *strength* and *duration* of the processes as well as on the amount of *surface area* in the receiving horizons. In the following sections, we examine the many pedogenic process suites or bundles that combine to form soils, as listed in Table 13.1.

Processes Associated with Soil Organic Matter

O horizons are formed as organic materials such as leaves, grass fragments, seeds, needles, and wood accumulate on the soil surface (Bray and Gorham 1964). Litter accumulates as it falls from living and dead vegetation, is blown onto the site, and/or washes in from upslope (Mitchell and Humphreys 1987). *Littering* is the term applied to this process (Table 13.1, Fig. 13.1A).

As litter decomposes, it forms an array of intermediate decomposition products, the most decomposed of which

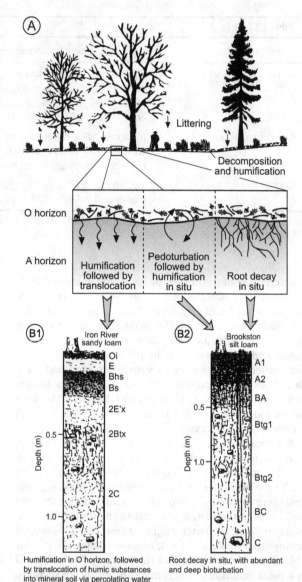

Fig. 13.1 Pathways of litter incorporation into the mineral soil. (A) The three conceptual pathways whereby litter in the O horizon eventually becomes incorporated into the A horizon as humus. (B) Examples of soils with vastly different humification pathways. (1) A sandy Spodosol (Iron River sandy loam). The acidic, nutrient-poor litter decomposes slowly, and mainly within the O horizon, entering the soil with infiltrating water, as dissolved organic substances. The paucity of soil fauna inhibits mixing between the O horizon and the mineral soil, and hence, an A horizon does not form. The B horizon becomes enriched in illuvial humus by deeply percolating water. (2) A silty Mollisol (Brookston silt loam). Abundant soil infauna in this fertile soil drag raw litter into the mineral soil and the dense root network decomposes in place; thus, a thick A horizon forms in the mineral soil. Wet conditions further facilitate litter decomposition, but clay-humus and Ca-humus complexes inhibit its complete mineralization. Note the different depth scales on the profile drawings, which are from Hole (1976).

is a substance called *humus*. The transformation of raw organic matter (OM) into humus is, therefore, called *humification* (Ganzhara 1974, Zech *et al.* 1990, Malterer *et al.* 1992; Fig. 13.1A) or *maturation* (Duchaufour 1982). This decomposition process begins as litter is fragmented and broken up by soil macro- and mesofauna such as millipedes, earthworms, springtails, and isopods. As the litter becomes more fragmented, the surface area made available to soil microorganisms for humification is greatly increased. It may be useful to envision this breakdown/decomposition process as one wherein both macro- and microorganisms are feeding on the litter simultaneously. Sometimes, the most palatable and easily decomposed materials are the first to be humified, while some of the larger and more resistant materialsmust first be variously fragmented by macrofauna.

Research into organic matter decomposition in soils is rapidly enabling our understanding of this process suite. It seems that the fate of OM in soils is not solely dependent on chemistry, but is affected by (1) the amount and solubility of the OM, (2) the decomposer community, (3) the characteristics of the microenvironment that influence the partitioning between solution and solid states (physical protection), and (4) decomposer activity.

Traditionally, soil scientists have used chemical fractionation methods based on solubility of organic compounds in a strong base and a strong acid to separate soil organic matter (SOM) into humic and fulvic acid. Humic acid – usually the larger colloids – has higher C concentrations, lower O concentrations, and higher molecular weights and it is more highly polymerized than fulvic acid (Torn *et al.* 2009). This paradigm of acid-base solubility (as a chemical property inherent to OM) has been replaced by a new paradigm for soil organic matter that emphasizes the chemical composition of SOM before acid or base extractions. In addition, interactions with mineral surfaces are thought to regulate accessibility of organic matter to microbial enzymes (Kleber and Johnson 2010, Hedges and Keil 1999, Kleber *et al.* 2007).

At pH values > 5.0, bacteria and actinomycetes are the main decomposers, producing a type of mull humus, whereas at pH values < 5.0, fungi are the main decomposers, leading to a greasy type of mor humus (see Chapter 3). With time, the mass of organic material becomes increasingly decomposed, and its C:N ratio lowered. The litter layer changes gradually, with increased amounts of fecal material, encasing fragmented and humified organic materials. Throughout this process, some of the more soluble decomposition products are eluviated from the O horizon, into the mineral soil below, as dissolved organic substances. Eventually, the litter is completely humified, and most of its decomposition products have been mixed intimately with minerals or translocated to a mineral horizon below.

A variety of factors affect the rate of litter decomposition, including soil nutrient and water-holding status (Aerts 1997, Gholz *et al.* 2000, Prescott 2010). For example,

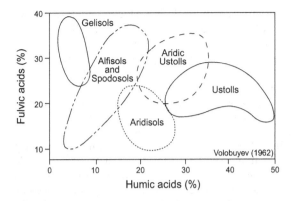

Fig. 13.2 Variation in humic and fulvic acid concentrations in soils of various types, reflective of different climate conditions.

decomposition is usually most rapid in warm, moist, nutrient-rich environments. At the seasonal to annual scale, litter characteristics become increasingly important as determinants of the rates and pathways of litter decomposition. One of the more intriguing recent findings about litter decomposition is called home-field advantage; litter decomposes more rapidly beneath the plant species it is derived from (home), than it does beneath different (away) plant species (Ayres *et al.* 2009). In certain circumstances, humification is favored in rocky materials, under grass vegetation, in wet soils, and in parent materials with high carbonate contents (Smith *et al.* 1950, Gaikawad and Hole 1965, Anderson *et al.* 1975, Schaetzl 1991b). But usually, adsorption of organic residues by clay generally slows the process. Soils with higher clay contents may retain moisture longer, thereby facilitating litter decomposition, unless the high clay contents result in waterlogging. Clay-humic acid complexes are especially common in undisturbed grassland soils where the pH is high (Fig. 13.2), a phenomenon that also helps to explain why decomposition of SOM in these soils may be slow or incomplete. In contrast, after cultivation, grassland soils quickly lose organic matter as the less decomposed humic fraction is otherwise destroyed (Martel and Paul 1974, Tiessen and Stewart 1983, Gregorich and Anderson 1985, Zhang *et al.* 1988). Organic residues mineralize more quickly after cultivation because there is increased aeration (more air = higher decomposer activities), because there are higher soil temperatures, and because disturbance decreases physical protection of SOM by disrupting aggregates. However, after long periods of cultivation, the organic matter content of these soils tends to stabilize, as the inputs of organic matter balance out the losses via decomposition, and because of long-term stabilization due to protection within microaggregates (Carrington *et al.* 2012). This interpretation is supported by two observations: (1) Easy-to-decompose sugars can be found inside microaggregates, and (2) lignin, when not physically protected by organomineral reactions, can still be decomposed (Baldock and Skjemstad 2000, Dittmar et al. 2011, Carrington et al. 2012). Eventually, increasing amounts of the SOM fraction become composed of *humic substances*, which are chemically complex, unrecognizable compounds dominated by aromatic carbon molecules (Fig. 13.2) (Stevenson 1994). These substances are thought to be enriched in decomposition-resistant compounds because of their old ^{14}C ages and prevalence in soils (Hatcher et al. 1989, Schnitzer and Monreal 2011).

In soils with high pH values and abundant exchangeable Ca, humic substances form *Ca-humate complexes*, which render the organic materials resistant to further decomposition. Calcium is an effective stabilizing agent for organic matter humic substances. In calcareous soils, Ca-humate complexes can be adsorbed to $CaCO_3$ in a carbonate film (Duchaufour 1976, Zech et al. 1990). Additionally, much of the humic fraction binds the surfaces of clay minerals via Ca bridges, resulting in *clay-humus complexes* that are highly stable and resistant to decomposition. Clay-humus and Ca-humus complexes form readily in dry, grassland soils, where pH values are high. Here, they reside mostly in the fine-clay fraction (Dudas and Pawluk 1969, Oades 1989). Once formed, these complexes can greatly extend the turnover time of soil organic matter, as the $CaCO_3$ and fine clay protect the organic matter from enzymatic attack.

Some of the humus formed in the O horizon is eventually incorporated into the mineral soil, either by translocation in percolating water or by pedoturbation (Fig. 13.1A). Once there, it will form organic matter–rich coatings (Table 2.6) or reside as humus-rich agglomerates in fecal pellets. Organic matter, in varying states of decomposition, can move by at least three pathways from the O horizon into the upper part of the mineral soil (Fig. 13.1A):

1. The litter will decompose (humify) within the O horizon and, because of its colloidal size, be translocated into the A horizon by infiltrating water.
2. The litter (in a more raw state, although it need not be entirely raw) will be translocated into the mineral soil and decompose there. Translocation may occur as soil fauna drag the litter into the soil (Fig. 7.10) or as it washes or falls into open krotovinas or cracks.
3. Organic matter will be added to the mineral soil by decomposition of roots and dead soil organisms in situ; this pathway is most important in soils where root density and turnover are high, such as grasslands. Ponomareva (1974) provided evidence that humus accumulation in Ustolls (Chernozems) can also be contributed as water-soluble root excretions, directly from plants.

In most soils, all three pathways are operational, although in acid soils forming under coniferous litter, pathway (2) is less effective because of the relative paucity of soil fauna. In forest soils, pathway (3) is less important, as root turnover is slow. Where the humification-translocation pathway (1) is dominant, e.g., sandy Spodosols, humus is incorporated into the A horizon more shallowly than

in soils where pedoturbation is active, e.g., in silty Udolls (Fig. 13.1B). Thus, in many acid Spodosols, the A horizon is thin or almost nonexistent, and the lower O horizon boundary is sharp. Where bioturbation-humification is the dominant humification pathway, i.e., pathway (2), the A horizon tends to be thicker, the rate of humification is increased, and the O-A boundary is blurred (Fig. 13.1B).

In most soils, the rates of littering, root decomposition, and humification eventually achieve a steady state, as inputs from plants balance losses due to (1) decomposition and humification, (2) mineralization, and/or (3) translocation. Despite the fact that litter production is generally rapid in warm, moist climates, soils there tend to have thinner O horizons because humification is also rapid. Disturbances such as fire will occasionally impact almost all ecosystems, temporarily upsetting this equilibrium, but in the case of O horizons, it can be reestablished in a few decades (Fox et al. 1979, Jacobson and Birks 1980, O'Connell 1987, Schaetzl 1994, Fenton et al. 2005; Fig. 3.5).

Histosols

Histosols are organic soils that have at least 40 cm of O horizon material above a mineral deposit (Everett 1971, D'Amore and Lynn 2002, Soil Survey Staff 2010). In Canada, these soils are classified within the order "Organic soils" (Kroetsch et al. 2011).

The pedogenic processes associated with the formation of organic soils seem straightforward: the continued accumulation and gradual decomposition of organic materials (litter). Histosols are almost always associated with wet and/or cool conditions, both of which slow decomposition and allow for the accumulation of thick sequences of organic materials above the mineral soil surface. Decomposition and mineralization of those organic materials are slow when soils are wet and waterlogged, because this condition excludes much oxygen from the decomposition process. However, the rate at which organic matter accumulates in wet Histosols also is affected by temperature, litter production, and litter quality (acidity, resistance to breakdown, decomposer community activity). Availability of oxygen and water matters most; the more oxygen that contacts the litter (as long as water content is not limiting), the faster it will decay. Because the main source of oxygen in soils is the atmosphere, decomposition will often proceed more rapidly in near-surface organic layers, which are usually more aerated, especially where water table levels fluctuate greatly (Collins and Kuehl 2001). Most Histosols have low bulk densities and lose most of their water at relatively high matric potentials.

Histosols form mainly in swamps, marshes, and other locations where the water table is above the mineral soil surface, i.e., in lowland sites (Transeau 1903, Gates 1942, Gorham 1957, Heinselman 1963, 1970). Kroetsch et al. (2011) neatly compartmentalized the processes that lead to the formation of Histosols in *upland* (but wet) versus lowland (open water) settings into two process bundles: *paludification* and *terrestrialization* (Fig. 13.3, Table 13.1).

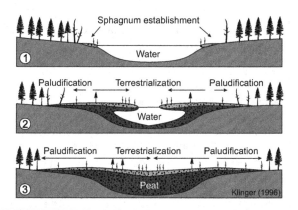

Fig. 13.3 Model of bog in-filling and succession, whereby development of the climax bog communities occurs via both terrestrialization and paludification, beginning at the margins of a small water body. Key stages include (1) establishment of peat-forming, e.g., Sphagnum, mosses along shorelines; (2) peat accumulation via both terrestrialization and paludification, causing a rise in water levels; and (3) continued terrestrialization and paludification resulting in the formation of an ombrotrophic bog in which all nutrients added to the system are from precipitation.

Paludification, common to boreal climates, is the accumulation of organic materials over the mineral soils of wet, cold, nutrient-poor uplands. Over time, moisture contents in these nutrient-poor O horizons continue to increase, often as a result of the colonization of the site by *Sphagnum* moss (van Breeman 1995). At sites undergoing paludification, litter can accumulate to great thicknesses, as soils become increasingly colder, wetter, and nutrient-poor (Gates 1942, Krause *et al.* 1959, Frazier and Lee 1971, Miller and Futyma 1987, Brugam and Johnson 1997, Rabenhorst and Swanson 2000, Crawford *et al.* 2003; Table 13.1). Eventually, cool forest soils can develop into Histosols via long-term paludification (Turunen *et al.* 1999, Moore and Turunen 2004). Paludified soils tend to be waterlogged and cold, and they have a considerable proportion of their nutrients locked up in thick, poorly decomposed O horizons (Bonan and Shugart 1989, Heinselman 1963). They are often acidic and have low microbial activities. Conditions that favor paludification include (1) cool, humid climates; (2) mineral soil textures (or pans) that inhibit percolation; (3) slightly depressional topographic sites; and (4) the presence of mosses (Fenton *et al.* 2005, Bonifacio *et al.* 2006, Simard *et al.* 2009, Kroetsch *et al.* 2011). Beaver dams can raise water tables locally and thereby induce paludification (Lavoie *et al.* 2005). Because nutrients are tied up in the decaying litter, forest productivity can decline and seedling establishment can suffer as soils undergo paludification (Greene *et al.* 1999). Fires can destroy some or all of the accumulating O horizon, however, retarding or even halting paludification (Lavoie *et al.* 2005).

Kroetsch *et al.* (2011) define *terrestrialization* as the slow in-filling of a shallow, depressional water body by long-term accumulation of organic materials (litter and mosses) (Schwintzer and Williams 1974, Podniesinski and Leopold 1998, Pokorny and Jankovska 2000, Swinehart and Parker 2000, Cai and Yu 2011, Roach *et al.* 2011). Other definitions are broader – the closing up of a water body by or with sediment of any kind. Many recently glaciated areas have wetlands and kettle lakes that have in-filled in this manner. In deeper water bodies, terrestrialization may lead only to a ring of Histosols around the shallow margins of a lake.

Both Soil Taxonomy and the Canadian System of Soil Classification recognize a unique type of Histosol that forms directly on top of rock. Folists (Soil Taxonomy) or Folisols (Canadian System) are composed of little more than decomposing litter above bedrock (Fox *et al.* 1987, 1994, Fox and Tarnocai 2011). In both systems, a minimum O horizon thickness must be met to classify as a Histosol. Unlike litter that falls onto mineral soil, few bioturbators live on rock, providing no opportunity to mix the litter into the underlying mineral material. Thus, the litter continues to accumulate, as long as rates of accumulation exceed rates of decomposition. Often Folists are thicker in rock depressional areas or crevasses, where litter can preferentially accumulate, erosional processes are minimal, and the soil can occasionally become saturated and anaerobic (Turcotte and Butler 2006, Vaughan and McDaniel 2009). The accumulation of organic materials and formation of organic soils on rocks are often considered the primary stages of soil and plant succession.

In Folists, decomposition by microbes, not macrofauna, dominates (Bochter and Zech 1985), and the overall microbial diversity is lower than for litter accumulations on top of mineral soils. As a result of all these characteristics, decomposition is slowed, and the litter continues to accumulate. Therefore, Folists are most common where litter decomposition rates are low, i.e., in cool environments (Alexander 1990, Witty and Arnold 1970, Schaetzl 1991). Because they are underlain by rock, many Folists are comparatively dry, further reducing decomposition rates (Vaughan and McDaniel 2009). They are also particularly thick when they reside above carbonate rocks, e.g., limestone. The high pH values may initially hasten decomposition, but eventually, the Ca-humate complexes that form will minimize complete mineralization of the litter.

Enloe *et al.* (2006) reported that Folist-like accumulations of litter – up to a meter in thickness – can accumulate in the crotches of trees in old-growth redwood forests, where the cool, moist conditions slow decomposition of the acidic litter. Like most Folists, these "soils" have distinct horizons that can be differentiated by degree of decomposition, even 50 meters above the forest floor! Nadkarni *et al.* (2002) referred to such litter accumulations, high in the forest, as arboreal soils, which have important ecological functions (Nadkarni and Longino 1990, Vance and Nadkarni. 1990, Veneklaas *et al.* 1990, Freiberg and Freiberg 2000). Enloe *et al.* (2006) make a strong case for the fact that arboreal litter accumulations *are* soils; they have pedogenic structure and horizons, they support plant life (epiphytes), and they develop and change over time.

Melanization: Mollisols

As humification proceeds, organic matter is increasingly decomposed, i.e., mineralized. *Mineralization* (Table 13.1) results in the release of C, N, P, and other ions that were contained in the humus, to the soil, making them available for leaching, translocation, neoformation, and biocycling.

After organic materials are transformed into humic substances, these colloidal materials may move in soil water or end up as grain coatings. These brown-black coatings, called *organs* (Table 2.6), darken the soil. *Melanization* is the development of dark, humus-rich coatings on ped faces and mineral grains, rendering the horizon a dark brown or black color (Table 13.1). Successive coats leave a more lasting and deeper color. The degree to which a soil becomes melanized is a function of the rate and duration of humus production, the types of humus produced, and its surface area. Slow weathering of primary minerals in some soils can be explained by humus coatings on the mineral particles. Ca-humate materials are particularly effective as protective coatings, especially on calcite. Think of the humus coating as a type of protective paint on mineral grains (Anderson *et al.* 1974).

Melanization can occur in any horizon where organic matter is added and retained, e.g., Bh horizons, although it usually is a dominant process only in A horizons. Melanization is, essentially, the hallmark and distinguishing process of A horizons, especially those that have formed under grassland vegetation. The imprint of melanization – the A horizon – may be all that remains after a soil is buried, providing valuable information about the presence and character of such soils (Li *et al.* 2001; see Chapter 16).

Melanization is most pronounced under grass vegetation, where in situ root decomposition provides large amounts of humus to the mineral soil, and to considerable depths (Fig. 13.1A). *Mollisols* epitomize soils that have undergone strong and long-lasting melanization and, thus, have developed a thick, organic matter–rich, A horizon. If the A horizon is thick enough and contains ample amounts of base cations, it classifies as a *mollic epipedon*. As originally conceptualized, the mollic epipedon is the morphologic expression of pedogenesis on grasslands, i.e., melanization (Smith 1986). Nonetheless, mollic epipedons occasionally develop beneath forest vegetation, although usually in frigid or colder soil temperature regimes (Nimlos and Tomer 1982; Anderson *et al.* 1975) or on the cool side of mesic temperature regimes (Gaikawad and Hole 1965). Here, mineralization rates are slow. Dark, thick epipedons also occur beneath forest vegetation where the sites are wet, as the wetness inhibits decomposition (Fig. 13.1B).

Landscapes: Soils, Carbon Sequestration, and Global Climate Change

The Earth is growing warmer (see Figure). Much has been studied and written about global climate change in recent years – enough to fill volumes. The Intergovernmental Panel on Climate Change (IPCC) recently concluded that *human activities* are largely responsible for modifying the concentrations of several key atmospheric constituents, i.e., greenhouse gases, and as a result, most of the observed atmospheric warming over the last 50 years can be blamed on increases in these gases (McCarthy 2001). Indeed, a study in 2004 of 928 journal abstracts on contemporary climate change, from papers published between 1993 and 2003, found that every one of these studies either explicitly or implicitly accepted the consensus view that climate change due to human actions is occurring (Oreskes 2004).

The soils research community is necessarily weighing in on this discussion. Soils contain about 2,500 gigatons of carbon (C) – both organic and inorganic – making the soil pool 3.3x the size of the atmospheric C pool and 4.5x the size of the biotic C pool (Lal 2004). As humans modify soils, more often than not the soils lose C to the atmosphere as CO_2, a greenhouse gas. Conversion of soils to agriculture is one of the main reasons for this loss (Reicosky *et al.* 1995, Rounsevell *et al.* 1999). Cultivated soils usually retain only ≈25–40% of their original C stores (Lal 2004). When soils become degraded, they lose even more C to the atmosphere. Still more C is being lost from soils in cold regions, as they warm and the underlying permafrost melts (Schaefer 2011, von Deimling *et al.* 2012). The potential for a positive feedback loop, in which the warming atmosphere causes even more C to be respired from these cold soils, fueling still more atmospheric warming, is a grave concern to many (Walbroeck 1993, Zhang *et al.* 2003).

For these reasons, the C cycle and *carbon sequestration* have recently become intensively studied (Guo *et al.* 2006a, b; see Figure). Carbon sequestration implies the transfer of atmospheric CO_2 into long-lived pools and the secure storage of it, so that it is not immediately returned to the atmosphere (Bruce *et al.* 1999, Moore and Turunen 2004). The total potential C sequestration in world soils is estimated to vary between 0.4 and 1.2 gigatons/yr (Sauerbeck 2001, Lal 2003). Humification of plant residues is one way that C in the biotic pool is stored in soils, although this form of storage is not necessarily permanent. C sequestration in soils means increasing soil organic and inorganic stocks, generally through judicious land use and recommended management practices (RMPs). While the potential for soils to store C is finite in capacity and time (Lal 2004), that capacity and period are very uncertain.

Lal (2004) listed several RMPs that can enhance soil organic C sequestration: mulch farming, conservation tillage, agroforestry, diversification of cropping systems, use of cover crops, and integrated nutrient management, e.g., use of manure, compost, biosolids, improved grazing, and forest management. There is also increasing evidence that C may be stored in soil for millennia in the form of land-applied biochar, a product of the incomplete combustion of organic residues by pyrolysis (Woolf *et al.* 2010). Obviously, soil organic C sequestration will also be enhanced as we restore degraded soils and ecosystems. There is much to do, and there is not a great deal of time to do it before climate change has significant negative impacts on global human societies. But most scientists agree that the soil system is a critical link in understanding, predicting, and managing global C dynamics and climate change.

Figure: (A) Land surface temperatures from 1753 to 2011, showing the unprecedented warming that the Earth is currently undergoing. (B) Modeled representations of Earth system interrelationships that affect organic C dynamics. (B1) The dynamics of C transformations and transport within soils. (B2) Flows of C among various environmental systems, emphasizing CO_2. This model does not include the flow pathways of CH_4 (methane) from the soil, which occur under anaerobic conditions. Nonetheless, most well-drained soils are sinks for CH_4.

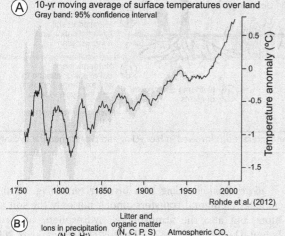

(A) 10-yr moving average of surface temperatures over land
Gray band: 95% confidence interval

Rohde et al. (2012)

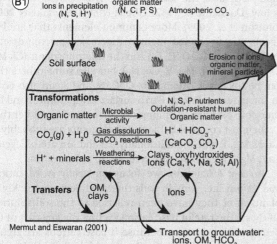

Mermut and Eswaran (2001)

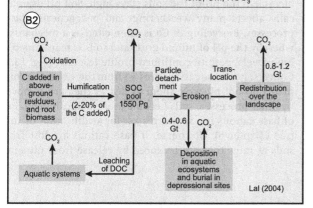

Lal (2004)

Formation of Granular Structure

A horizons are dominated by melanization, which often goes hand in hand with the formation of granular structure. Except in the sandiest soils, uncultivated A horizons usually have granular structure (Fig. 2.11A). Granular peds are small, roughly spherical or polyhedral, and bounded by curved or very irregular faces that do not exactly match those of adjoining peds (Sideri 1938; see Chapter 3). The peds have strong internal cohesiveness, yet are fairly porous. Many of these peds form as soil passes through earthworms and other fauna and exits their bodies as casts, giving them a biogenic origin (Buntley and Papendick 1960, Lee and Foster 1991). The mixing action that occurs in the guts of the animals multiplies the chemical bonds between organic and inorganic materials and mixes the entire mass with microbial gums, resulting in internally cohesive, crumb-shaped fecal pellets (Duchaufour 1976, Jongmans *et al.* 2003). Vermudolls and Vermustolls are dominated by earthworm activity and have very strong granular structure. Formation of granular structure is also aided by wet-dry and freeze-thaw activity.

Once formed, fine plant roots tend to follow interped voids, enhancing the integrity of the peds as the roots grow and expand. Root growth on the outsides of the peds slightly constricts the soil material within the ped, making it even stronger relative to forces that might destroy it. Hole and Nielsen (1970) refer to this phenomenon as interlacing roots. Fine roots that pierce granular peds add to their porosity.

Once formed, humic substances bind together, as well as coat, peds. Some of this glue is simply water-soluble root exudates (Ponomareva 1974). Also contributing to cohesiveness of the peds are microbial gums and other decomposition products of plant and animal biomass (Hole and Nielsen 1970).

Biocycling and Acidification

In all soils, even in the driest of climates, many pedogenic processes involve downward translocation of pedogenic plasma materials. Upward translocation is generally only possible via biocycling, pedoturbation, and capillary movement of soluble materials in soil water.

In *biocycling*, nutrients are taken up by plants, used for growth, and returned to the soil as litter or as root decay (Bormann *et al.* 1970, Gosz *et al.* 1976, Helmisaari 1995, Giesler *et al.* 2000). Most of these nutrient were in the soil initially, were in the parent material, or were released from primary minerals by weathering (Kolka *et al.* 1996). Biocycling is one of the main ways by which the downward transfers of soil constituents by leaching are mitigated by upward movement (Duchafour 1982). A discussion of biocycling is appropriate here, as almost all organic matter and soluble ions in the soil system have the potential to be cycled (Fig. 13.4).

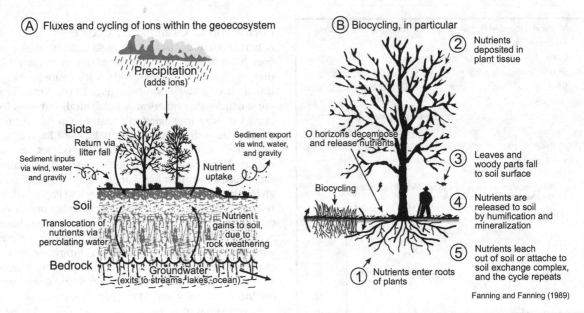

(A) Fluxes and cycling of ions within the geoecosystem

Precipitation
(adds ions)

Biota
Return via
litter fall
Sediment inputs
via wind, water
and gravity
Sediment export
via wind, water,
and gravity
Nutrient
uptake

Soil
Translocation of
nutrients via
percolating water
Nutrient
gains to soil,
due to
rock weathering

Bedrock
Groundwater
(exits to streams, lakes, ocean)

(B) Biocycling, in particular

② Nutrients
deposited in
plant tissue

O horizons decompose
and release nutrients

Biocycling

③ Leaves and
woody parts fall
to soil surface

④ Nutrients are
released to soil
by humification and
mineralization

⑤ Nutrients leach
out of soil or attache to
soil exchange complex,
and the cycle repeats

① Nutrients enter roots
of plants

Fanning and Fanning (1989)

Fig. 13.4 Generalized diagrams of some of the pathways, processes, and interactions involved in biocycling and biogeochemical cycling.

Plant nutrients can be returned to the soil in various ways, indirectly as through fall, stemflow, root exudates, and litter fall and directly as the plant itself decomposes, either aboveground or below (Eaton *et al.* 1973, Sanborn and Pawluk 1983). Animals biocycle materials as well, passing those materials back to the soil as excrement and as their bodies decompose. In this regard, Francis Hole, one of three scholars to whom this book is dedicated, often referred to things as either "soil" or "not yet soil." He even appended the honorary title "TNS" (temporarily not soil) to his name during the later years of his life! He knew that all things in nature, whether they were trees, people, or houses, were only temporarily not yet soil. His point was that parts of almost everything have come from the soil and will end up as soil. The text that follows was once displayed in front of a display case at the Soils Building on the University of Wisconsin campus, where Hole taught. It was probably written by him:

Structure built by an avian engineer using solid waste (plastic, paper, tin foil); organic debris from vegetation; and mineral soil. The soil was compacted into a platy, stratified deposit which is essentially a series of crusts of reduced hydraulic conductivity. The structure, a segment of a concretion, is formed on a tree branch close to the canopy. It is an epiphytic pedological feature, whose fate is to be translocated by free fall to the soil surface, where it will eventually be incorporated into the soil, except that part which decomposes first.

Displayed in the glass case was a robin's nest.

Biogeochemical cycling is a broader term (Fig. 13.4). It includes elemental transfers among not only the soil and biota, but also the atmosphere, hydrosphere, and rock below (Duchaufour 1982, Likens *et al.* 1998, Likens 2001, Johnson *et al.* 2009). More common elements that are biogeochemically cycled include the major macronutrients (N, P, K, Mg, S, Ca, O, C, Fe) and the micronutrients (Cl, Mn, Zn, Ni, B, Cu, Si, Mo, V, Co, Na), but also some others that can be toxic, e.g., Al and Pb. O, C, N, and S can also occur in the gaseous state at Earth-surface temperatures and are therefore exchanged among soil, plant, and air. Other than C, the most commonly biocycled elements are probably N, Ca, K, Mg, P, and S; additionally, Fe and Mn are biocycled. Ca^{2+}, Mg^{2+}, K^+, and Na^+ are referred to as *base cations*.

In this discussion, we focus on *tightly cycled* cations and anions, i.e., ions in soils that are taken up quickly by plants after they have been released to the soil solution. Of these, the rate of base cation cycling is a key part of the pedogenesis equation, because if base cations are tightly biocycled, the pH of the soils stays high. Soil pH dramatically affects many weathering- and pedogenesis-related processes. Biocycling of Ca is often cited as a mechanism whereby the pH of humid grassland soils is maintained at high levels, despite continued profile leaching (Fig. 13.5). Conversely, if base cations are lost from the system because plants do not biocycle them as fast as they are removed by other processes, soils become more acidic. Indeed, loss of base cations is a major reason for acidification in soils (see Chapter 5). Of course, if base cations are lost from soils at rates that are balanced by release from minerals

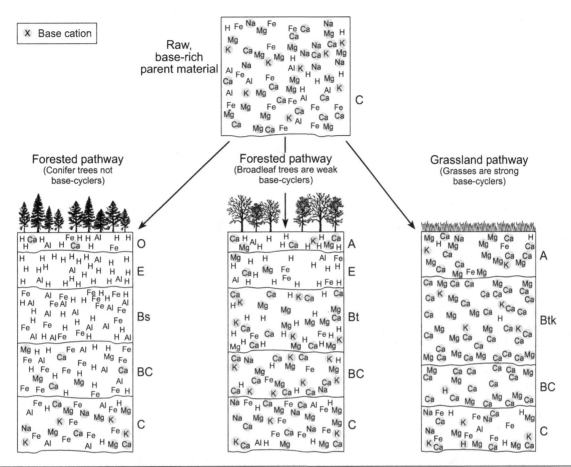

Fig. 13.5 Theoretical pedogenic pathways associated with a base cation-rich (but probably not calcareous) parent material. Soil A develops under a non-base-cycling, leaching pathway associated with coniferous trees. Soil B develops under mildly base-cycling sugar maple trees, while Soil C forms in association with strongly base-cycling tallgrass prairie. Note the distributions of cations within and between the soils, as mediated by the degree of leaching, acidification, and base cycling. Base cations are shown with a gray halo.

via weathering, then the pH may remain unchanged. And so, a delicate base cation balance may exist in soil among (1) losses due to deep leaching, driven by percolation; (2) gains from weathering of rocks, sea spray, and wet and dry deposition of ions and mineral particles in precipitation; and (3) plant biocycling (Van Leeuwen *et al.* 1996, Draaijers *et al.* 1997).

Rates of base cation biocycling are dependent upon several factors: (1) the rate at which base cations are taken up by the plants (this is species-dependent, but also is a function of availability, rates of percolation, etc.), (2) the rate at which these cations are returned to the soil (deciduous species return more foliage annually than do coniferous trees; grass litter and roots are almost completely turned over annually), and (3) the decomposition or mineralization rates of the litter (low in cold climates, rapid in warm, wet climates (see earlier discussion of mineralization). Fire

can rapidly release base cations otherwise trapped in O horizons (Fenton *et al.* 2005), although the base cations (in the ashes) are then prone to removal by wind and water. Across a variety of forest types in the southeastern United States, Sharpe *et al.* (1980) determined the amounts of four elements that are biocycled and returned to the soil as litter. Their data indicate that P and K are more rapidly biocycled (higher turnover rate) and that the litter is a large storehouse of base cations that biocycle more slowly (Table 13.2). The lower turnover rate for Ca may be due to its immobilization by fungi in the litter (Lawrey 1977, Lousier and Parkinson 1978).

In the low leaching regimes of many grasslands, cycling of Ca facilitates the formation of stable Ca-humates (Rubilin 1962). These humates bind to mineral grains, forming clay-humus complexes that not only protect the mineral particle from many forms of weathering, but also

Table 13.2	Amounts of biocycled Ca, P, Mg, and K in some forests in the southern Appalachian Mountains		
Nutrient	Amount contributed by litter fall (kg ha⁻¹ yr⁻¹)	Amount in the Oi horizon (kg ha⁻¹ yr⁻¹)	Relative turnover rate
Ca	34	46	Low
P	4	5	Medium
Mg	6	3	High
K	13	4	Highest

Source: Sharpe et al. (1980).

render the humus fraction resistant to further decomposition by microbes. Clays in contact with humus are easily aggregated and become essentially immobilized, inhibiting translocation (Sanborn and Pawluk 1983).

Grasses and some species of trees, e.g., sugar maple, yellow birch, and aspen, are strong base-cyclers, whereas hemlock, many oaks and conifers are not (Alban 1982, Bockheim 1997b, Fujinuma et al. 2005). Because some trees do not readily biocycle base cations, and indeed some, including hemlock, hickory, and beech, actually cycle Al (Chenery 1951), base cations in many forest soils are more freely removed by percolating water, and soils here become acidic. Acidification, an important precursor to many pedogenic processes, is often a direct result of decalcification caused by acid dissolution of calcite and dolomite, deep percolation, and low rates of base cation cycling (Table 13.1, Fig. 13.4). In acidification, the exchange complex of the soil becomes increasingly occupied by H_3O^+ and Al^{3+} cations, as base cations are removed. Base cations are removed from the soil by (1) leaching from the soil in percolating water, vertically or laterally, or (2) incorporation into plant biomass. Cations in plant biomass may be further removed from the system as animals consume the plant and move elsewhere, or by wind or water erosion of ash derived from fires that burn the plant matter. Recarbonation (Fuller et al. 1999) occurs when carbonates are added to a leached soil, sometimes as a deep-rooted, base-cycling plant species invades a site that had previously been covered with a non-base-cycling plant (Table 13.1).

Organic acids also play a key role in acidification (de Vries and Breeuwsma 1984). Organic acids and H_2CO_3 are weak acids that are able to dissociate protons as a function of pH, thereby contributing to acidification of the soil solution.

Many plants, grasses in particular, accumulate silica-rich particles, usually as amorphous opal, in their roots and leaves. These usually silt-sized deposits are called phytoliths. Most grasses and sedges are particularly adept at biocycling silicon, much of which is deposited as opal within their leafy tissues, giving the leaves a certain amount of rigidity and sometimes forming knifelike edges (Sommer et al. 2006, Street-Perrott and Barker 2008, Opdekamp et al. 2012). Certain species of sedges and horsetails have

so much silica that Native Americans used them to scrub dishes. Phytoliths are incorporated into the soil when leaves and stems fall to the surface. Some soils can accumulate thousands of kilograms of phytoliths per hectare (Fanning and Fanning 1989). The shape, size, and surface textures of opal phytoliths are reasonably unique to each type of plant and can reside in the soil for long periods, especially in acid soils, making them a useful paleoecological indicator (see Chapter 16).

Leaching and Leucinization

Leaching is the primary way that base cations and other soluble compounds are removed from the profile. The term "leaching" is often inappropriately used to refer to all forms of translocation in percolating water. Rather, it should be used only for the complete removal of soluble constituents from the profile. Ions in solution are not, technically, leached if they are redeposited in the lower profile. Also, because colloids are, by definition, not soluble compounds, they cannot be leached; they are translocated.

Humid climate soils, in which water periodically wets the entire profile and percolates through it, can become leached of base cations and other soluble components. Some of these solutes end up in groundwater, especially in soils that lack artificial drainage. Those components that are more soluble and highly mobile are removed from the profile first and translocated the deepest. Eventually, a leached zone develops in the upper profile, from which the most soluble ions have been removed. In parent materials rich in carbonate minerals, the leached zone is easily determined; unleached materials below will effervesce upon exposure to a strong acid, usually 10% HCl. In many soils on calcareous, Pleistocene-age deposits, the leached zone is equivalent to the thickness of the solum.

The thickness of the leached zone in a soil is a function of several factors. Sites that have more water added to them, perhaps because of a wetter climate or their location in local depressions, may be more leached than are drier sites. Coarse-textured parent materials are more quickly and deeply leached, because they have less negatively charged surface area on which to retain cations (Fig. 13.6), and

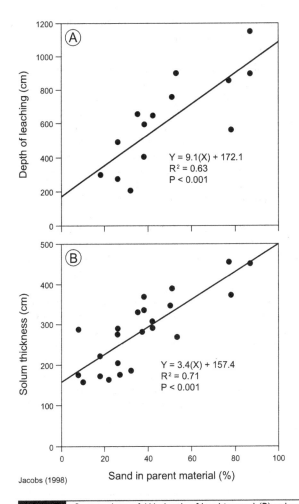

Jacobs (1998)

Fig. 13.6 Scatterplots of (A) depth of leaching and (B) solum thickness for the Sangamon Geosol in Indiana, as affected by sandiness of the parent material. The Sangamon Geosol is strongly developed in this area, having formed over a period in excess of 100,000 years.

because they usually have more quartz, which is relatively inert and does not release base cations when it weathers. Soils with less cycling of cations are also likely to be more intensively and deeply leached. Depth of leaching usually increases with time, making it – along with solum thickness – a potential relative dating tool (see Chapter 15).

Elements that are very slowly mobile in soil water, such as Zr and Ti, are considered, for the sake of communication, immobile. Through the use of mass balance pedogenesis studies, the relative mobilities of elements within soils have been established. While there are site-to-site and climate-to-climate variations, the *general* mobility sequence for semiarid climates, based on Busacca and Singer (1989) and Harden (1987), is

$$Mg \gg Na > Ca \approx Fe \approx Al > Ti \approx Si > K > Zr$$

In a humid climate, Bain *et al.* (1993) reported

$$Mg > Na > Ca > K$$

for base cations only.

The degree to which any particular cation or anion is mobile in soils is dependent on its ability to form a stable compound that would precipitate from an aqueous solutions (Paton 1978). Whether a cation persists in the soil solution depends in part on its potential for chemical reaction with hydroxyl (OH^-) anions. This potential reactivity depends on both valence and ionic size, i.e., ionic potential (Fig. 9.8). Ionic size determines how many hydroxyl anions can be accommodated around a cation, whereas its valence refers to the number of bonding electrons it has lost to become a cation.

Lessivage: Alfisols and Ultisols

One of the most common processes that follow acidification in soils is the translocation of clay-sized particles from E to Bt horizons (McKeague and St. Arnaud 1969, Buurman *et al.* 1998, Dixit 1978, Quénard *et al.* 2011, Cornu *et al.* 2014). Translocation of clay-sized particles in suspension is called *lessivage* (French *lessivé*, washed) *argilluviation* (L. *argilla*, white clay, *luv*, washed) (Table 13.1). Although most lessivage occurs from the upper profile to the lower (from A and E horizons to the B), the process does occur laterally as well. Most translocated clays are silicate clays, as oxide clays are more difficult to disperse into the soil solution. E horizons that have undergone lessivage have lost considerable amounts of clay and their associated iron-bearing minerals, thereby acquiring lighter colors and coarser textures. If this process has continued for a considerable period, the horizon may qualify as a diagnostic *albic horizon* in Soil Taxonomy (Table 8.4).

Soil components that are able to be translocated, e.g., clay and soluble substances, are collectively referred to as plasma (see Chapter 2). Gravel and sand grains are typically considered skeletal, implying that they are cannot be translocated in percolating water. Silt can be plasmic or skeletal, depending on the circumstance. Translocation of silt grains in suspension is called *pervection* and is typically a significant process only in cold climates (Table 13.1; see Chapter 11).

Associated with, and sometimes mistaken for, lessivage are two companion processes: *decomposition* and *neoformation* of clays (Table 13.1). Clays can be chemically weathered (decomposition) in the upper profile, especially in acidic soils (see Chapter 9). As clays weather, their soluble by-products are translocated in solution to the B horizon, where they can precipitate to form new minerals, via *neoformation*. In most soils, neoformation happens in the lower solum for several reasons: (1) the pH increases with depth, promoting the precipitation of hydroxyl-bearing minerals; (2) as wetting fronts continually pass through the solum, the concentrations of soluble ions at depth increase until the solubility product constant of a new mineral is exceeded, promoting neoformation (Chapter 4); and/or (3) loss of water at the wetting front concentrates ions in the soil solution, favoring precipitation and clay neoformation (Brewer and Haldane 1957, Gombeer and D'Hoore 1971, Howitt and Pawluk 1985a). Where lessivage is active, the clay mineralogy in the E and Bt horizons is usually similar, whereas under a decomposition and synthesis regime the clay mineralogy of the E and Bt horizons can be different because the minerals being synthesized in the lower profile are different from those that decomposed above.

Because lessivage is so widespread, terms that refer to illuvial clay and to horizons rich in illuvial clay permeate various soil taxonomies. Soil Taxonomy (Soil Survey Staff 1999) defines an *argillic horizon* as a Bt horizon that contains a specified increase in illuvial clay relative to its overlying eluvial horizon. In soil descriptions, B horizons are given a *t* suffix to indicate illuvial clay, for the German word for clay, *der ton* (Table 3.2). *Alfisols* must, by definition, have an argillic horizon; many Inceptisols and Entisols are developing one but do not yet contain a horizon that meets the taxonomic criteria. Many *Ultisols* also have an argillic horizon. In warm (thermic), humid environments, the Bt horizon clay mineralogy is often dominated by low-activity clay, such as kaolinite and Fe oxides, even as the clay content of the Bt horizon increases. In that instance the illuvial clay horizon is called a *kandic horizon* (Soil Survey Staff 1999). A kandic horizon can be thought of as the equivalent of an argillic horizon, but for soils dominated by low-activity (1:1 and oxide) clays. The main differences between Alfisols and Ultisols center on the degree to which they have been weathered; Alfisols are much less weathered, have more base cations present in the profile, and have more 2:1 clay minerals (Hallmark and Franzmeier 2011). Ultisol mineralogy is dominated by kaolinite, hydroxy-interlayered vermiculite, gibbsite, iron oxides, and, to a lesser extent, quartz, halloysite, and mica (Shaw *et al.* 2010, West *et al.* 2011). Some Mollisols, Aridisols, Spodosols, and Andisols also have Bt horizons, while many Oxisols and Vertisols probably have lost their Bt horizons because of weathering, argilliturbation, or bioturbation. The development of Bt horizons in Aridisols is usually thought to date back to a time when the climate was more humid.

The World Reference Base for Soil Resources defines horizons rich in illuvial clay as *argic* horizons (Food and Agriculture Organization 1998). In Canada, soils in the Luvisolic order have illuvial clay–rich Bt horizons, although the Canadian system does not have an equivalent diagnostic horizon to the argillic (Expert Committee on Soil Science 1987, Lavkulich and Arocena 2011). The French use the term *Sols Lessivés* to refer to similar soils. In Australia, soils with a clay-impoverished horizon above a clay-rich horizon are called *duplex* and *texture-contrast soils* (Gunn 1967, Koppi

and Williams 1980, Chittleborough 1992), although many of these soils have not formed entirely as a result of lessivage (Paton *et al.* 1995).

Not all clay-enriched B horizons are due to lessivage or decomposition/neoformation (Chittleborough 1992, Quénard *et al.* 2011). Formation of clay in situ is a type of neoformation called *argillation*. Relative clay enrichment in the B horizon can also be caused by (1) sand and silt destruction (weathering), in the horizons above; (2) preferential erosion of finer materials from, or additions of coarser materials to, the upper profile; (3) comminution of silt and coarse clay to fine clay in the B horizon (Oertel 1968, Smeck *et al.* 1981); and/or (4) bioturbation, which can move relatively coarser materials to the surface (Nooren *et al.* 1995). Combinations of the preceding processes are common (Phillips 2004, 2007). Bishop *et al.* (1980) concluded that texture-contrast soils in parts of Australia are due to the slow downslope movement of a clay-impoverished layer above a more sedentary, clay-rich layer, rather than to lessivage. The upper layer is assumed to be a *biomantle* that has had most of its fine particles removed by water and wind; its mobility can be confirmed by the presence of a stone line at the contact with the B or Cr horizon (see Chapter 14). This type of genetic interpretation has high explanatory ability for sloping landscapes that have been stable for long periods (Paton *et al.* 1995). On younger landscapes, such as those that date to Quaternary glaciations, the more traditional interpretation of vertical clay translocation via lessivage seems more plausible (Johnson 2000) and certainly is supported by the development of clay coatings, glossic tongues, and lamellae in the profile (see later discussion).

Landscapes: Early Views on Pedogenesis of Clay-Rich B Horizons

The presence of a Bt horizon is common in many soils. It is a key criterion in the classification of Alfisols and Ultisols and is a common characteristic in soils of several other orders, e.g., Spodosols and Aridisols. It seems simple – clay particles are translocated to the lower profile in percolating water, accumulating there as a clay-enriched B horizon. Should not this process be nearly universal in all humid climate soils? Well, not so fast, my friend.

In the early days of pedology, when general ideas about soil formation were being developed, soils were grouped into two main categories, based on assumed genesis – Pedocals and Pedalfers (Fig. 8.2). The dry grasslands, where calcium compounds accumulated in the subsoil, were the realm of Pedocals. In humid climates under forest, soils were acidic and many had reddish B horizons enriched in clay and/or iron oxides. These soils were called Pedalfers. *Calcification* was the process assigned to the Pedocals, and *podzolization* was deemed responsible for the formation of Pedalfers. However, disagreements arose over whether the reddish, clay-enriched B horizons of Pedalfers had formed mainly as a result of (1) decomposition of clays within the more acidic upper profile, accompanied by translocation of iron

Fig. 13.7 Cutans on ped faces in loamy soils. Note the differences between the smooth sheen of the cutan and the ped interiors. Photos by RJS.

oxides (byproducts of clay weathering) from the upper profile to the B horizon, or (2) translocation of clays (as particles) and iron (as dissolved ions), as discrete processes. In short, were clays being translocated to the B horizon as particles, and could this even happen in soils that are so acidic? Would not the clays simply be destroyed by the acids?

A continuous argument festered over what was involved in this podzolization process, and how to define it (Glinka 1924, Sokolovskii 1924, Rode 1937, Duchaufour 1951, Fridland 1958). Everyone agreed that podzolization involved the translocation of iron, but what about clay? Did clay *translocation* also fall under the rubric of podzolization? Or was clay *destruction* – and that alone – within the realm of podzolization? Soon, micromorphological data began to accumulate, showing that clays can and do translocate in acidic soils as discrete particles (Kubiëna 1956).

Eventually, the French pedologist Duchaufour coined the term *lessivage* to refer to the physical transport of clay particles to the B horizon. Lessivage can *accompany* podzolization, but it is a separate process that can occur in the absence of podzolization processes as well. Podzolization was found to be strongest on acid, coarse-textured soils of coniferous forest zones, while lessivage was best expressed on finer-textured parent materials under broadleaf forest (Frei and Cline 1949, Muir 1961).

What's in a name? In this case, several decades of debate.

Because Bt horizons can form in more than one way, it is best to use a combination of morphological features to verify that lessivage has been active. Obviously, in soils where lessivage has occurred, B horizons have more total clay than the E horizon above. Most classification systems require such an increase in clay concentration in the B horizon as a way to verify lessivage; in Soil Taxonomy this criterion is one of several that define an argillic or kandic horizon. Another important criterion used to infer lessivage, either in the past or ongoing, is the presence of coatings of illuvial clay (*argillans*) on ped faces and grain surfaces (Grossman

et al. 1964, Soil Survey Staff 1999). When viewed macroscopically, argillans usually appear as smooth surfaces on ped faces, and if wet they can have an almost glassy sheen (Fig. 13.7). Under the microscope, argillans appear as accumulations of fine, sometimes banded material, often within pores and at pore-grain contacts. They often have high birefringence (producing strong interference colors in cross-polarized light) when viewed with cross-polarized light (Lavkulich and Arocena 2011). Nonetheless, pedoturbation can destroy argillans (Nettleton *et al.* 1969, Reynders 1972, Howitt and Pawluk 1985a), and clay within argillans can be remobilized. Argillans observed in the field may not always be visible when examined under the microscope in thin section (McKeague *et al.* 1978).

Argillans form most readily in loamy or fine-textured soils that maintain relatively stable structure. In sandy soils, illuvial clay usually coats individual sand grains but also occurs as bridges between grains (Buol and Hole 1961, Wright and Foss 1968; Fig. 2.14). Although less well-developed argillans occupy only the walls of pores (Fig. 13.8), eventually, argillans can completely fill interped pores and channels. Gradual filling of Bt horizon pores by illuvial clay may cause the horizon to become an aquitard or even an aquiclude.

Argillans develop when layer silicates, moving as suspended particles in the soil solution, are deposited in thin layers on the surface of the ped or skeletal grain (Figs. 13.7, 13.8). The depositional process occurs as the soil water is absorbed into the peds when the wetting front slows (Buol and Hole 1961), plastering the suspended silicate clay platelets onto ped surfaces in ever-thickening layers, forming a smooth, shiny argillan. After deposition, shrink-swell activity, which adds stress to the grains within the illuvial horizon, can accentuate the orientation of the clay (Gunal and Ransom 2006).

As would also be expected, the ease with which clay can be translocated in a soil is a function of its mineralogy

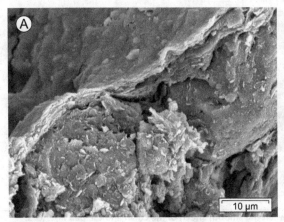

Fig. 13.8 Scanning electron photomicrographs of argillans. Images by B. Weisenborn. (A) Argillan from the Bt horizon of a Glossudalf in central Michigan. (B) Argillan on a pore wall in a Btx horizon of a Haplorthod from northern Michigan.

and size. Smectite tends to be the most readily translocated clay mineral, usually because it is one of the smallest, whereas the relatively larger kaolinite and oxide clays are more difficult to suspend and translocate (Grossman et al. 1964). Fine clay (<0.2 μm) is much more mobile than is coarse clay (0.2–2.0 μm), regardless of mineralogy (Dixit 1978, DeNovio et al. 2004, Rawilli et al. 2005, Lavkulich and Arocena 2011). Thus, the fine/coarse clay ratio is commonly used to identify whether a horizon has been undergoing lessivage. In Soil Taxonomy, fine/coarse clay ratios must be higher in argillic than in overlying eluvial horizons (Oertel 1968, Smith and Wilding 1972, Soil Survey Staff 1999). This ratio is typically lowest in E horizons. To that end, Cremeens and Mokma (1986) developed the I/E index, which reflects the degree of development of Bt horizons:

$$\text{I/E index} = (\text{fine clay}_{Bt}/\text{total clay}_{Bt})/(\text{fine clay}_{E}/\text{total clay}_{E})$$

Many Bt horizons form in freely drained, upland soils in humid (but not perhumid) climates that experience ongoing acidification (Cremeens and Mokma 1986). Frequent wet-dry cycles can also favor development of Bt horizons. Although lessivage is not an important process in arid climates, some precipitation occurs in every climate, and with each infiltration event exists the possibility for clay translocation.

How Lessivage Works

Like all eluvial-illuvial couplets, lessivage has three components: mobilization, transport, and deposition (Eswaran and Sys 1970, Lavkulich and Arocena 2011). First, clay must be dispersed and suspended in soil water. Then, the water must transport the clay, and, finally, there must be a means to deposit or attenuate the clay. In this section we discuss this threefold suite of processes.

Clay platelets exist in two states in soil. When suspended in water and acting independently of one another,

they are said to be *dispersed*. When stuck together electrochemically or physically, they are *flocculated* or coagulated. Clays that are flocculated behave as large masses that are not readily translocated. Groups of flocculated clay minerals in soil water are called floccules. In the solid phase, aggregated groups of smectite are called *quasi-crystals*. Aggregated groups of more discrete clay particles such as clay mica or kaolinite are called *domains*.

Lessivage cannot begin until clays are dispersed, usually in the eluvial zone. Clay can be dispersed both physically and chemically. If a cementing agent such as sesquioxides or silica is causing the flocculation, that agent must be chemically removed before the clay can be dispersed (Lepsch et al. 1977b). From a physical perspective, rapid wetting of a dry soil can sometimes cause dispersion of clay and a disruption of fabric (McKeague and St. Arnaud 1969, Chittleborough 1992, Soil Survey Staff 1999). For example, when heavy rain impacts a dry surface horizon, some peds are broken apart as the air pressure within the ped increases ahead of the the infiltrating water. This disaggregation is called *slaking*. Platelets of clay that are dispersed by slaking are suspended in the water and translocated down macropores. Clay that is being translocated rapidly or through large pores can move significant distances, as long as the translocating agent, i.e., percolating water, remains active (Howitt and Pawluk 1985b, Johnson et al. 2001). This fact helps explain why argillic horizons form quickly in many xeric (wet-dry) and ustic climates (Chittleborough 1981). Examples of this process include clay films deep within limestone fissures and in calcareous soils, which are not conducive, chemically, to dispersion (Gile 1970, Goss et al. 1973). However, if the pH and ionic strength of the soil solution are high, dispersion will not be favored, and the clay will soon be deposited.

Dispersion is more effective, chemically, in acidic soils. Replacement of base cations, primarily, K, Ca, and Mg, in the soil solution and on exchange sites by H (or Na) favors dispersion. According to Ciolkosz et al. (1989), dispersion is favored at base cation saturation values <40%. Lessivage is a weak or almost nonexistent process in calcareous materials, because the clay exchange sites in such material remain occupied by Ca and Mg cations. Removal of free iron oxides also assists in dispersion and translocation (Lepsch *et al.* 1977b). Thus, a period of weathering, leaching, and acidification, i.e., *decarbonation*, must usually precede lessivage (Allen and Hole 1968, Gile and Hawley 1972, Vidic and Lobnik 1997). In leaching regimes, the soil solution in the upper profile eventually becomes acidic enough and low enough in electrolytes that clay platelets can be dispersed. In the lower part of the soil, where the ionic strength if the soil solution usually increases, the dispersed platelets clump together and become too large to be mobile. Although the terms "coagulation" and "flocculation" are often used interchangeably, in *coagulation* the clumping is due to an increase in multivalent cations in solution, as explained later. The process is called *flocculation* when it is the result of bridging of clay platelets by polymers of organic matter or of sesquioxides such as iron oxides.

Recall that 2:1 layer silicate clay minerals usually have a net negative charge (see Chapter 4), which is balanced by swarms of cations in the soil solution. This cation swarm is called the the *diffuse layer* (or *outer layer*) in the clay-soil solution system (Fig. 13.9). Dispersion and coagulation are regulated by the thickness of the diffuse layer that extends away from the negatively charged outer surfaces of clay minerals. In acidic soil horizons (from which many base cations have been leached), the negative charges of layer silicates are balanced by H_3O^+ cations or by Al species such as $Al(OH)_2^+$, in addition to the remaining base cations. Large numbers of these +1 valence cations are required to satisfy the negative charges of the clay. Although the entire clay-water system is always electrostatically neutral, the large assemblage (or "swarm") of positively charged cations in the diffuse layer repel each other, causing the diffuse layer to thicken. In this situation, the clay platelets are unable to move close enough to allow the natural attraction between two uncharged objects (van der Waals forces) to take over and coagulate them. Thus, the clay platelets remain separated, or dispersed, in the soil solution. The diffuse layers repel each other. Acidic, base-poor conditions favor clay dispersion, because so many of the cations in the outer layer are hydrated, monovalent cations, and there are fewer multivalent cations such as Ca^{2+} in the soil system.

The upper (eluvial) parts of soil profiles are also favored sites for dispersal because many of the potentially coagulating divalent cations such as Ca^{2+} and Mg^{2+} can be coordinated by organic anions (Dixit 1978). In most situations, Na^+ cations also favor dispersal. Dispersion is additionally favored in the leached, upper profile because the ionic strength of the soil solution may be lower than it is in deeper horizons. Lavkulich and Arocena (2011) considered the loss of soluble ions in the upper profile as the first step in the lessivage process. Deeply flushing a soil, continually, with fresh

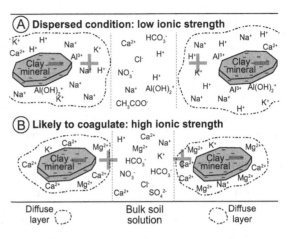

Fig. 13.9 Diagrammatic representation of the composition and thickness of the diffuse layer of cations associated with clay minerals in solutions of low and high ionic strength. Note that the diffuse layer is thinner at the higher ionic strength, allowing a closer approach of clay particles to one another and hence, a greater likelihood of coagulation.

water depletes it of electrolytes. Thus, the more water that can be flushed through a soil, the more likely clays within are to disperse, and consequently, to be translocated.

Where the ionic strength of the soil solution is high, e.g., in base cation-rich and calcareous horizons, the diffuse layer of cations near a clay surface is thin, thereby favoring coagulation. When base cation saturation values are high, e.g., >85%, the ionic strength of the soil solution is also high enough to encourage coagulation. Here, the diffuse layer is thinner because it is populated with fewer cations that have +2 and +3 valency, such as Ca^{2+} and Mg^{2+} (Fig. 13.9). The thinner diffuse layers allow clay surfaces to be close enough that either van der Waals forces can cause coagulation or divalent Ca^{2+} and Mg^{2+} ions can bridge the negative charges of two different clay layers. Deeper in the soil profile, where leaching has not been as pronounced, electrolyte concentrations are higher, and this, too, promotes flocculation by forcing the outer layer to shrink. For these reasons, in soils that form in calcareous parent materials, the base of the Bt horizon usually occurs at the top of a calcareous zone. As leaching continues to drive the depth of leaching deeper, i.e., as percolating water strips carbonate ions and divalent cations from the top of the calcareous zone, clays can be translocated deeper, allowing the Bt horizon to grow downward, and the solum to thicken.

Organic compounds may also be involved in lessivage. When they are adsorbed to clay surfaces at low concentrations, organic compounds are effective agents of dispersion because they usually contain negatively charged carboxylate groups that repel one another. Thus the clay particles cannot get close enough to one another to coagulate. As pH increases, the proportion of carboxylate groups also increases, promoting dispersion. On the other hand, at high concentrations, adsorbed organic molecules may also be attracted to one another by hydrophobic interactions, creating a polymeric glue that promotes flocculation of clay (Visser and Caillier 1988).

Lessivage requires percolating water as a vector of transport, and generally, the more frequent and rapid the percolating water, the more capable it is of translocating clay (Goddard *et al.* 1973). Rapidly percolating water is facilitated by large, highly interconnected, and vertically aligned pores, and by intense rainfall or rapid snowmelt. If these two mechanisms operate simultaneously, clay can even be translocated through calcareous material (Goss *et al.* 1973). Lessivage also appears to be promoted by rainfall and percolation events during the leaf-off season, as occurs in xeric soil moisture regions (Engel *et al.* 1997). Here, winter rains are particularly effective in translocating clay, and, as a result, soil development rates can be impressively rapid.

Three general mechanisms can lead to the deposition of clay in Bt horizons: (1) cessation of the wetting front due to (A) desiccation, (B) loss of energy (as at a water table), or (C) the presence of a lithologic discontinuity or an aquitard/aquiclude; (2) flocculation within a base cation-rich material (as mentioned earlier); and/or (3) deposition by filtration (pores are too small to allow clay to pass). Clay in suspension stops moving when and where the wetting front stops and water is pulled into peds by matric forces. Many Bt horizons are best developed, therefore, at the modal depth of wetting front penetration (Daniels *et al.* 1971a). In perhumid climates where soils are constantly wet, such as parts of southeastern Alaska, the Olympic Peninsula of Washington, and the Scottish Highlands, Bt horizons are often absent (Soil Survey Staff 1999). Here, there is no mechanism to stop clay that is migrating. Lessivage is promoted in soils that occasionally dry out at depth for three reasons: (1) Wetting a dry soil favors dispersion by slaking, mobilizing clay (Thorp *et al.* 1959, Grossman and Lynn 1967); (2) dry soils have larger pores that facilitate movement of the clay-water suspension; and/or (3) the wetting front fades as the soil dries, and with it the capillary withdrawal of water *into* dry peds and deposition of clay *onto* peds. As water penetrates these soils, clay (much of which is carried near the wetting front) is deposited at the depth where the wetting front ceases to penetrate. Clay is also deposited onto peds at the top of a water table, because the percolating water loses its energy there and stops. However, Bt horizons are usually better expressed in well-drained soils than in soils with high water tables (Cremeens and Mokma 1986), perhaps because interped pores remain closed by swelling, because water saturation leads to pH buffering that inhibits weathering, or because water table fluctuations cause illuviation to occur at varying depths, spreading the translocated materials but not concentrating them. Also, the likelihood of the dry soil being wetted by an infiltration event, a process that favors lessivage, is greater in upland soils and less in soils with high water tables.

The most common reason why clay is deposited in a preferred zone is an increase in base cations, and, as mentioned previously, this location usually coincides with the top of an unleached, calcareous layer (Applegarth and Dahms 2001, Gunal and Ransom 2006). For this reason, lessivage in calcareous parent materials occurs only after a lag period following time$_{zero}$ – time enough to form a leached zone from which clays can be translocated. Depending on the texture of the parent material and the concentration of carbonate minerals, this lag could last for tens of thousands of years. This process explains why most Bt horizons that occur on young landscapes or where the parent materials are underlain by a leached-calcareous boundary (Allan and Hole 1968, Khakural *et al.* 1993). Increased electrolyte concentrations within the calcareous materials also promote flocculation. When that depth is also the modal depth of wetting front penetration, a relatively large amount of clay can be deposited within a thin depth increment, quickly forming a Bt horizon. For soils developed in initially calcareous parent materials, formation of the Bt horizon is aided not only because the Ca and Mg derived from carbonate minerals help to coagulate illuvial clay in the lower solum, but also because calcite and dolomite dissolution in the upper solum releases clay; i.e., it is a clay source. Carbonate minerals act as cemented agents, and thus, as they dissolve, clay is freed for translocation and typically will move to the top of a calcareous zone, or deeper if the percolating water is moving rapidly. Van Ranst and DeConinck (2002) showed that Fe^{3+} oxyhydroxides are also effective at inhibiting or stopping soil clay from moving downward in suspension.

Wetting fronts also stop and deposit clay at textural or lithologic discontinuities (Schaetzl 1996, Hopkins and Franzen 2003). In particular, matric forces cause water to hang at a fine-over-coarse contact, such as a loess over gravel discontinuity (Fig. 13.10; see Chapter 10). If this discontinuity occurs below the solum, illuvial clay accumulations there may even develop into a second (deeper) clay maximum for the profile. Such horizons are called *beta horizons* because they sometimes occur below a similar (*alpha*) Bt horizon. (Note: Any second horizon, separated by genetically different material, can be considered a beta horizon; the term is not restricted to Bt horizons.) Beta Bt horizons tend to be composed of finer clay than their alpha equivalents; fine clay is more mobile than coarse clay and can be remobilized more readily as well. Beta Bt horizons are common in soils that have deep lithologic discontinuities (Bartelli and Odell 1960a, b, Ranney and Beatty 1969) and above shallow bedrock or aquicludes (Schaetzl 1992a). Clay deposited in the finer-textured material above the discontinuity enhances the textural contrast across the discontinuity, initiating a positive feedback that may even accelerate the development of the beta Bt horizon. In some soils, the beta horizon corresponds to the modal depth of

Fig. 13.10 A soil from Michigan with a pronounced concentration of illuvial clay and sesquioxides at a sand-over-gravel lithologic discontinuity. The holes have been made by a nesting bird, probably a swallow. Units on tape are 10 cm. Photo by RJS.

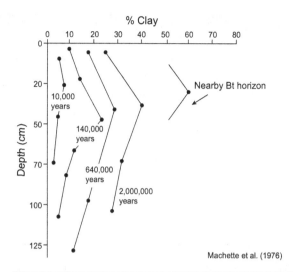

Fig. 13.11 Depth functions of clay contents in soils of different age, on the Colorado Piedmont.

wetting fronts associated with the wet season, e.g., winter rainfall, while the upper (alpha) Bt horizon corresponds to the depth of dry-season wetting fronts.

Bt horizons evolve in generally predictable ways. As clay accumulates in the horizon itself via lessivage, deposition is aided by filtering effects as the horizon grows continually finer-textured (Fig. 13.11). Generally, the upper boundary of the horizon also grows downward, as the depth of leaching and acidification increase, facilitating remobilization of clays in the upper Bt and their translocation to the lower B horizon. Remobilized clay minerals, translocated to increasingly greater depths, drive the bottom of the Bt horizon deeper. The ever-thickening and fining of the Bt horizon may be slowed as pores become blocked by illuvial clay, creating a filter effect that may explain why so many Bt horizons have their maximum clay contents near the top of the horizon, with gradually diminishing contents below. As more clay is captured by the Bt horizon, a positive-feedback loop sets up, sometimes leading to the formation of a nearly impermeable Bt horizon, which may act as an aquitard. Water may begin to perch on the Bt, leading to *ferrolysis* in the horizons above (Table 13.1; see Ferrolysis section later). Some Bt horizons do not become aquicludes mainly because pedoturbation occurs or because they have strongly expressed structure that facilitates percolation between peds. Other soils simply do not have enough clay ever to become aquitards.

Although some researchers are skeptical about lessivage as a dominant pedogenic process (Zaidelman 2007, Quénard *et al.* 2011), illuvial argillans, i.e., those with sharp boundaries relative to the underlying soil matrix, are taken by most soil scientists as strong evidence of clay migration (Brewer 1976, Bronger 1991, Chittleborough 1992; Figs. 13.7, 13.8). The lack of argillans in soils with Bt horizons, however, should not be considered evidence of lack of

lessivage (Oertel 1961, Nettleton *et al.* 1969, Chittleborough *et al.* 1984). Many argillans are destroyed by pedoturbation, especially argilliturbation and bioturbation. Because most argillans are enriched in fine clay, and fine clay has more smectite than other clay fractions, shrink-swell processes may also physically disrupt or distort argillans. For this reason, some soils retain the best expressed argillans, in the lower, not upper, Bt horizon (Buol and Hole 1961). Alternatively, argillans in the upper Bt horizon can be destroyed as the soil develops and continues to acidify, with the remobilized clay translocated deeper in the profile; this is a normal pedogenic pathway for many Alfisols and often results in *glossic horizons* (see Degradation of argillic horizons section later)

Argillic horizons can be overprinted with other illuvial materials as pedogenic pathways change. This is particularly common in soils on the humid-arid climate boundary, where small changes in climate can cause a shift from lessivage to calcification (Reheis 1987a). The different types of cutans can be used to verify such environmental changes (see Chapter 16).

Where illuviation occurs, so must eluviation. Therefore, a secondary morphologic manifestation of lessivage is the formation of clay-impoverished, often light-colored, E horizons. Lightening of a horizon is called *leucinization* (Table 13.1). The light, whitish colors are due to the predominance of quartz (and, to a lesser extent, plagioclase feldspar) grains that have been stripped of their reddish brown coats of clay or organic matter, coupled with the preferential weathering of dark-colored primary minerals in the acidic, eluvial zone. Some of the most easily weathered primary minerals are dark in color, while quartz and K-feldspar, both resistant minerals and common in many soils, are light-colored. Where eluviation is strong,

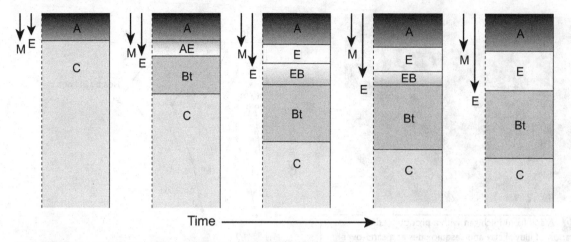

Fig. 13.12 Interplay between depth of melanization (M), a process bundle that darkens soils by the addition of humus, and eluviation (E), a process bundle that (in this case) strips soil particles of their clay coatings and promotes the loss of clay minerals to lower horizons. In the example shown, melanization forms dark A horizons while leaching/eluviation promotes leucinization, or the lightening of horizons, thereby forming E horizons. This diagram shows that both the A and E horizons are areas of eluviation, with the main difference being the intensity of melanization in the former.

E horizons can form rather quickly, perhaps in as little as a few decades.

Leucinization operates in opposition to melanization. Only when the eluvial zone becomes thick enough to out-pace melanization, or out-distance it with depth, can a distinct E horizon develop below the A (Fig. 13.12). In the uppermost part of the profile, a thin eluvial zone is often effectively counteracted by melanization. Therefore, one way that E horizon formation can be enhanced is to mini-mize the depth of melanization, e.g., in dry, sandy soils beneath acid forest (mor) litter (Fig. 13.1B). The population of soil organisms is low in such soils, and the litter decays so slowly that sometimes O horizons rest directly on E horizons; there is virtually no melanization of the mineral soil. In soils where bioturbation and additions of organic matter by root decay operate to considerable depth, e.g., grassland soils, E horizons may never form. Even if there *is* an eluvial zone, it is usually but not always masked by the deep organic matter signature. The depths at which the dark colors of the organic matter fade out are so great that the yellowish browns of the B horizon are now present; no E horizon is visible.

Lamellae

In sandy soils, evidence of lessivage often occurs as thin (<5 mm) layers of illuvial clay, associated with Fe oxides, called *lamellae* (singular: lamella). Rather than being depos-ited throughout the horizon and on ped faces, as occurs in fine-textured soils, in sandy soils the illuvial clay is depos-ited as thin, wavy, sheetlike zones of illuviation (Thorp *et al.* 1959, Ahlbrandt and Fryberger 1980, Cooper and Crellin 1996). In profile view, they appear as wavy, reddish lines, often contorted and discontinuous (Fig. 13.13). Most

lamellae are more clay- and Fe-oxide-rich than the sur-rounding matrix and are therefore redder in color (Schaetzl 1992b, Rawling 2000, Bockheim and Hartemink 2013a). When viewed in thin section, the clay within lamellae usu-ally appears as grain coatings and bridges between sand grains (Torrent *et al.* 1980a, Phillips *et al.* 2006). Lamellae are often located in the lower solum, usually overlain by an A-E-Bs, A-E-Bw, A-Bw, or comparable horizon sequence. In sandy landscapes, the presence of lamellae at depth is ecologically important, as these thin bands dramatically increase the water-holding capacity of the soil and, hence, enhance site productivity (Oliver 1978, Host *et al.* 1988, McFadden *et al.* 1994).

Although the clay in lamellae has, presumably, been eluviated from the entire profile above it, for the sake of communication, lamellae are called Bt horizons and the areas between are referred to as E horizons. Since it would be impractical to describe each lamella individually, zones of lamellae are grouped together and named E&Bt or Bt&E horizons, depending on whether the clean sand (interla-mellae areas) or the illuvial layers are dominant, respec-tively (Soil Survey Division Staff 1993). If sufficiently thick, the lamellae-rich horizon may qualify as an argillic hori-zon (Soil Survey Staff 1999).

The formation of lamellae is understood at one level, but because they can apparently form in different ways and are present in such a wide variety of soils and settings, detailed knowledge of the pedogenic processes involved in their formation continues to accrue (Rawling 2000). Pedogenic lamellae must first be differentiated from geologic lamel-lae. Many clay bands that look like lamellae, especially the thicker ones, are probably geogenic in origin (Hannah and Zahner 1970), conforming to geologic bedding planes

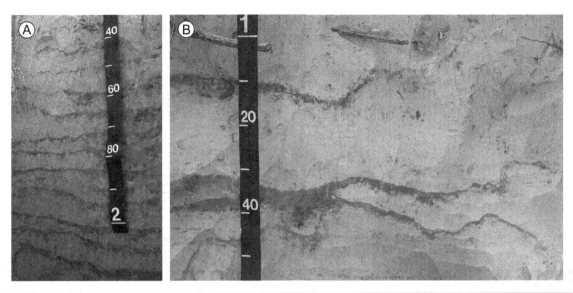

Fig. 13.13 Lamellae in sandy soils. In both photos, the lamellae are thicker than is typical, suggesting that the sand was not initially well sorted; i.e., it contained considerable amounts of silt and clay. Note the smooth upper surfaces and the more ragged lower surface of the individual lamellae, as is typical. (A) Lamellae in a sand dune from the Fair Oaks Dune Field, northern Indiana, United States. (B) Lamellae in a sandy soil in Michigan, United States. Tape increments in centimeters below the soil surface. Photos by RJS.

(Robinson and Rich 1960, Leigh 1998, Schaetzl 2001). Many lamellae of this kind occur so deep, or within calcareous substrates, that they could not possibly be pedogenic. Pedogenic lamellae are usually thinner and wavier, and they cross geologic bedding planes (Schaetzl 2001). They also are found only within the lower solum and usually have clear, sharp upper boundaries and ragged, diffuse lower boundaries (Fig. 13.13).

Pedogenic lamellae form as percolating water, carrying small amounts of clay in suspension, deposits clay in thin bands within sandy sediment (Gile 1979, Bond 1986, Kemp and McIntosh 1989). The edges of wetting fronts in sandy soils are often contorted. Then, if the percolating water was carrying most of the clay near the edge of the wetting front, when the wetting front stops, a contorted zone of deposition that resembles an incipient lamella is produced. Once deposited, this zone may act as a filter for future wetting fronts, stripping them of some clay and thickening as it does so. A related question centers on why the wetting fronts stop where they do. Perhaps the stopping point is random, but sometimes, surely the wetting fronts stop at points of textural and lithologic change. Since most sand has varying degrees of bedding and stratification, it has been assumed that lamellae begin forming at these subtle pore size discontinuities (Van Reeuwijk and de Villiers 1985, Bouabid et al. 1992, Soil Survey Staff 1999). Nonetheless, many lamellae are much too contorted to be aligned entirely with pore size discontinuities and may simply reflect random wetting front locations.

It is also possible that clay illuviation occurs while the wetting front is moving, and not at its end point. The explanation for this is as follows: As the wetting front moves through the soil, it continually picks up colloids. At some point, a maximum colloidal concentration threshold is crossed, forcing clay to be deposited even as the wetting front continues to move (Bond 1986). Flocculation, leading to the formation of a clay band, can also occur where pH or free iron oxide contents increase, perhaps across very short vertical distances (Miles and Franzmeier 1981, Berg 1984).

Sands acidify and leach quickly, enabling small amounts of clay to be easily and rapidly translocated and facilitating the formation of lamellae even in relatively young, sandy soils (Berg 1984). Most researchers assume that lamellae increase in thickness and clay content with time, although they can degrade as well, especially in the upper, more acidic, parts of the profile (Fig. 13.14). Along these lines, we paraphrase the Soil Survey Staff's view of lamellae genesis (1999: 82–83):

Lamellae nearer the soil surface have the least concentration of clay and the faintest color contrast to the overlying E horizon. These may be degrading. They are generally more wavy and discontinuous than those below. Lamellae in the midsolum appear to have the highest concentration of clay near the top of each lamella. In the lower part of these lamellae, some sand grains are devoid of clay and some have only thin clay coatings, similar to lamellae above. The deepest lamellae are commonly very thin and not very wavy. Parallel to each other, they may be, in fact, sedimentary strata or bedding planes. Clay is episodically moved from upper to lower lamellae. Clay stripped from the lower part of one lamella is redeposited in the top of the next lower lamella; lamellae thicken as clay is added to their tops. Loss of clay from its base, coupled with additions of clay to the

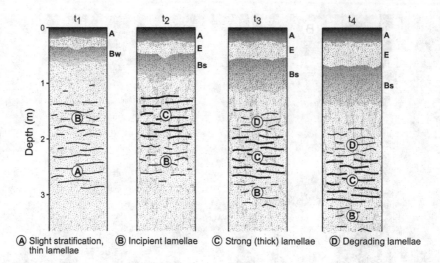

Schematic diagram showing how lamellae in a sandy soil in a udic soil moisture regime typically develop over time.

top of a lamella, cause it to move away from any bedding plane from which it may have originated. The stripping of clay from the lower part of a lamella and redeposition in the top of the next lower lamella continue this upward movement. Because the movement of each lamella upward is not uniform throughout its extent, lamellae are wavy rather than smooth. Branching in lamellae occurs where a part of a lamella has moved up more rapidly than the overlying part of the next higher lamella and they become joined in this part.

Although the preceding view of lamellae genesis is only one possible scenario, it is supported by empirical research (Van Reeuwijk and de Villiers 1985) and appears to be consistently supported by field evidence and micromorphological data (Fig. 13.14). Schaetzl (2001) has shown that an alternate explanation for lamellae genesis is also possible; they can form when thin, geogenic clay bands at bedding planes are remobilized, deformed, and moved *deeper* by percolating water. Until we know more about the genesis of lamellae, it is best to assume that they represent a good example of pedogenic *equifinality* (Rawling 2000) – the same *form* may have developed in different *ways*.

Degradation of Argillic Horizons

In the early stages of soil development, argillic horizons thicken (usually) and become more clay-rich; i.e., they continue to develop along a clear and consistent pedogenic pathway. Across the midlatitudes, most Bt horizons also continue to grow downward through time, as clay is stripped from the top of the horizon and remobilized, usually accreting onto the bottom of the horizon (Fig. 13.12). This process is accentuated by the increasing acidification of the upper solum and eluvial zone, as well as by the slow decarbonation that may occur at depth. This remobilization and translocation of clays to the lower solum, where pH and base saturation conditions are more favorable for flocculation and

deposition, are commonplace (Ranney and Beatty 1969, Bullock *et al.* 1974, Busacca and Cremaschi 1998). Every deep percolation event does a little bit to deepen the Bt horizon, and few processes operate to move clay upward in the soil profile. So, naturally, the Bt horizon, like many illuvial horizons, deepens with time, and as this happens, fine clays are preferentially translocated (Jamagne *et al.* 1984).

The stripping of clays from the top of the Bt horizon also occurs via a process called *degradation*. Degradation of the Bt horizon starts at the top, where conditions are most acidic (Cady and Daniels 1968). Acidification of the upper solum can degrade and destroy clays in some soils (DeConinck and Herbillon 1969), with some of this type of degradation at the upper Bt driven by redox processes as well (Jamagne *et al.* 1984). The increased acidity, often in conjunction with ferrolysis (see later discussion), causes the clay minerals, especially 2:1 clays, in the upper Bt horizon to become unstable. The clays and their weathering by-products are then translocated to the lower profile (DeConinck *et al.* 1968). Here, they may be deposited or they may be transformed into other clay minerals that are more stable, depending on the local soil chemistry at this depth. This type of Bt horizon degradation can be viewed as a localized change in pedogenic pathway from lessivage to clay destruction, because some of the clay that is lost from the upper Bt is destroyed by chemical weathering before it can be translocated deeper. Under this scenario, the soil might be seen as maintaining its preexisting pedogenic pathway, through which E horizons thicken and Bt horizons grow downward and thicken.

Degradation of Bt horizons is similar to the degradation of fragipans and natric horizons (see later discussion). All three of these horizons occasionally function as aquitards, forcing water to perch on and flow across the top of the horizon, eventually percolating through it, along preferred pathways (Fletcher and McDermott 1957,

Langohr and Vermeire 1982, Ransom *et al.* 1987b). Indeed, according to Jamagne *et al.* (1984), Bt horizon degradation *starts* with a decrease in hydraulic conductivity at the E-Bt boundary, followed by desaturation of the exchange complex and lowering of the pH. As Chartres (1987) noted, degradation features are most common in soils that have a marked textural contrast between the E and B horizons and that have slowly permeable B horizons. Lindbo *et al.* (2000) observed that Fe-Mn nodules, indicative of perched water and redox conditions, are common in the E horizons of soils with degraded Bt horizons. Because perched water is probably involved in Bt horizon degradation, processes related to ferrolysis (Brinkman 1970, 1977b) and oxidation-reduction (Lindbo *et al.* 2000) must be involved, too, as these conditions cause reduction of Fe and Mn in minerals in the upper Bt horizon. The Fe-poor clay that results is then more mobile and can be translocated downward or destroyed entirely (DeConinck *et al.* 1976).

Bt (argillic), Btx (fragipans), and Btn (natric) horizons all degrade from the top down, as percolating water strips away base cations, destroys clays, and translocates fine clays. As a result, argillans change into *skeletans* or *siltans* – light-colored, acid, siliceous, skeletal materials composed mainly of quartz, muscovite, and free silica, on ped faces (Bullock *et al.* 1974, Chartres 1987, Payton 1993b). These albic bodies, which are almost pure sand and silt near the top of the B horizon but become more clay-rich with depth, have also been referred to as eluvial bodies, grainy gray ped coatings, albans, albic neoskeletans, and silica powder (Daniels *et al.* 1968, Vepraskas and Wilding 1983a, Spiridonova *et al.* 1999). Even if some clay minerals remain present in them, the albic bodies are usually coarse clays with low cation exchange capacity (CEC), such as kaolinite and chloritelike hydroxy-interlayered minerals.

With time, the thickness of the albic zone (E horizon) increases, as the degradational processes continue farther into the ped centers and proceed deeper into the Bt horizon. Bt horizon degradation and clay stripping follow the most permeable conduits in the upper B horizon, i.e., along ped faces and macropores (Bouma and Schuylenborgh 1969). Roots also preferentially exploit these sites; their secretions help to degrade the clays by acidification, and their oxygen consumption also occasionally creates reducing conditions. Decaying roots within these areas further accentuate reducing conditions, as the microbes facilitating the decay use up free oxygen (Van Ranst and DeConinck 2002). For these reasons, any percolating water that moves preferentially along these ped faces (prism faces in fragipans) encounters an acidic zone of intense chemical weathering, increasing the likelihood of clay degradation and/or translocation. Leaching of perched water on top of the B horizon helps to force translocation and degradation through these preferred macroflowpaths (Bullock *et al.* 1974, Jamagne *et al.* 1984, Payton 1993b, Westergaard and Hansen 1997). Along these conduits, reducing conditions can quickly be established so that Fe^{3+} can be reduced and translocated

as Fe^{2+} (see later discussion). Indeed, many of the whitish areas on ped faces are simply redox depletions, from which Fe and Mn have been removed (Bouma and Schuylenborgh 1969, Caillier *et al.* 1985). Thus, the degrading conduit areas preferentially lose Fe and thereby become enriched in Si, as quartz is left behind (Payton 1993b). Reducing conditions allow Fe and Mn to migrate from the conduit (tongues) to ped interiors, where the oxidizing conditions facilitate precipitation as Fe-Mn nodules or neocutans (Bouma and Schuylenborgh 1969), with Mn near the edge and Fe farther inside. With time, even these nodules and cutans are not safe, as degradation continues inward.

Normally, a degrading Bt horizon has an irregular upper boundary that is marked by penetrations (tongues, fingers) of E (albic) material into the upper Bt horizon (Jha and Cline 1963, Langohr and Vermeire 1982; Fig. 13.15). This tonguing, degrading horizon, designated E/Bt or Bt/E, is called a *glossic horizon* in Soil Taxonomy (Soil Survey Staff 1999, Bockheim 2012). Glossudalfs (in Soil Taxonomy) or Albeluvisols (in the World Reference Base) are defined as having glossic horizons.

Glossic tongues form first as albic bodies – thin, whitish ped coats composed of quartz-rich silt and fine sands that are too large to translocate but have been stripped of Fe and clay by percolating, often anoxic and acidic, soil water (Fig. 13.16). Organic acids and Al cations may assist in the weathering and translocation of these materials, as the tongues are mainly in the most acidic parts of the horizon (McKeague *et al.* 1967, Duchaufour 1982). As the peds degrade from the outside in, they become like a house of cards as the glues that hold them together, i.e., clay and Fe oxides, are slowly removed. They retain their overall structure, because their skeletal grains remain in place, but the condition is tenuous. The degradation of the clays and cementing agents releases silica and other cations to the soil solution, which may precipitate to form allophane or other amorphous materials and can act as a binding agent for these albic materials, partially offsetting the loss of Fe and clay. Peds so formed tend to retain this original fabric but, when dry, will shatter easily under only finger pressure. Near the end of this process, small Fe- and Mn-rich nodules and clay-rich remnants of the Bt horizon exist amid a sea of thickening, sandy, albic materials. Fragments of argillans and small, nodular remnants of the argillic horizon remain as isolated fragments in the lower part of the E horizon, at sites that were farthest from the original albic tongues. In advanced degradational stages, all argillans and Bt horizon remnants become destroyed.

Although the mechanisms of Bt horizon degradation are reasonably understood, the causative agents for glossic horizon formation have intrigued scientists for years (Bockheim 2012). What are the external drivers, if any, of this process? Cline (1949b) assumed that this process was a natural pedogenic progression associated with long-term depletion of base cations and acidification. But was this

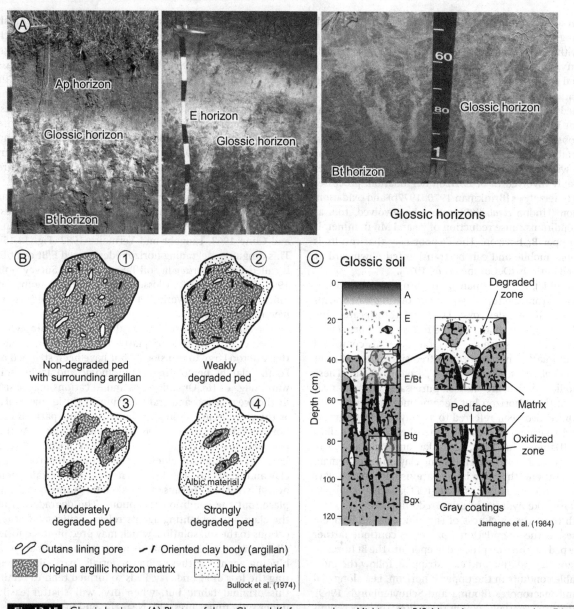

Fig. 13.15 Glossic horizons. (A) Photos of three Glossudalfs from northern Michigan, by RJS. Note the prominent white E horizon tongues extending down into the Bt horizon. Tape increments are 10 cm. (B) Diagrammatic degradation of a ped from a Bt horizon. (C) Diagrammatic representation of a glossic soil and how the glossic tongues relate to reduced and oxidized zones.

acidification forced (or reinforced) by a vegetation change, from base cation-cycling deciduous forest vegetation to mixed forest or conifers, with their acidic litter? Perhaps. Bartelli (1973) concluded that increased water flow through soils, as occurs in wetter climates and at lower slope positions, leads to increased ferrolysis and drives Bt horizon degradation (see later discussion). In Europe, a theory emerged that the degradation was induced by land clearing, deforestation, and the subsequent loss of base cations from the system (Tavernier and Louis 1971). Langohr and

Vermiere (1982) suggested that Bt horizons degrade when a permafrost-derived, dense layer exists at depth; soils without such a layer tend not to develop glossic features. Some data suggest that thin additions of loess can increase the whitish tonguing by contributing quartz-rich silts to the upper profile (Jalalian and Southard 1986, Bockheim 2012).

As Bt horizons degrade and E horizons thicken, the acid, sandy eluvial zone formed can become the parent material for another pedogenic process – podzolization. As a result, the soil can develop a second sequum. *Bisequal soils*

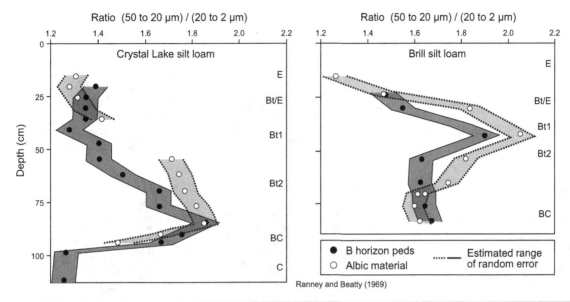

Ranney and Beatty (1969)

Fig. 13.16 Coarse:fine silt ratios of albic (glossic) tongues and Bt horizon peds from comparable depths, from two Glossudalfs in Wisconsin.

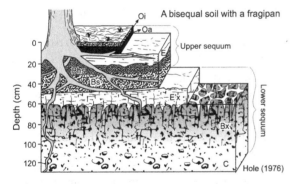

Fig. 13.17 Typical expression of a bisequal soil, probably an Oxyaquic Fragiorthod, with a fragipan in the lower sequum.

(Gardner and Whiteside 1952, Allen and Whiteside 1954, DeConinck et al. 1968, Schaetzl 1996, Bockheim 2003, Bonifacio et al. 2009) typically have an E-Bs sequum that is developing in the E part of the thicker, older E-Bt (or E-Bx) sequum (Fig. 13.17). The Bs horizon that forms is reflective of podzolization, which is best expressed in coarser-textured, acidic materials. Thus, lessivage in the upper solum is a type of preconditioning agent for podzolization. In some coarse-textured soils, however, lessivage can operate simultaneously with podzolization (Guillet et al. 1975). Nonetheless, in finer-textured soils, a period of lessivage usually precedes podzolization in the eluvial part of the profile, but only after it has become sufficiently acidic and coarse-textured.

Fragipans

Characteristics

A fragipan (Latin *fragilis*, brittle) is a dense, brittle subsurface horizon (Soil Survey Staff 1999, Bockheim and Hartemink 2013b). It is a diagnostic horizon in Soil Taxonomy (Soil Survey Staff 1999). Fragipan horizons are given an *x* suffix in soil descriptions, e.g., Bx, Btx, Ex, or Egx.

Their high bulk density (commonly >1.5 Mg m^{-3}) and nonconnective porosity make fragipans water- and root-restricting, i.e., aquitards. Many fragipans perch water at some time of the year and, as a result, exhibit redoximorphic features. With few roots, fragipans also have low organic matter contents, relative to the horizons above them. Most fragipans are in the lower solum, sometimes below a spodic (E-Bs) upper sequum (Fig. 13.17), and because they are formed in part by illuviation, they are commonly associated with Bt horizons. In other words, fragipans are usually overprinted with illuvial clay. Because other types of horizons can also be dense, root-restricting, and low in organic matter, the tendency for moist fragipan peds to be *brittle* is also a distinguishing characteristic. Brittleness is defined as the tendency for a ped to rupture suddenly when pressure is applied (Fig. 13.18), as opposed to the slow, plastic deformation that many other peds in loamy soils undergo. Similarly, air-dry fragments of fragipans *slake* when placed in water, just as some other types of noncemented peds do (Lindbo and Rhoton 1996, Szymański et al. 2012). Disaggregation by slaking can occur as a slow "crumbling like a pile of table sugar," or it can be fast, almost as if the soil

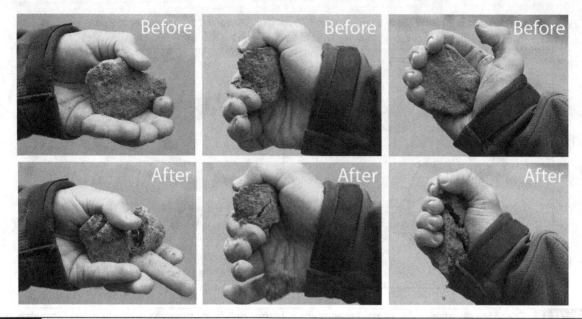

Fig. 13.18 Examples of peds removed from a fragipan, before and after pressure is applied. Note the tendency for these moist (but brittle) peds to rupture, rather than to deform slowly.

is popping apart from within, as incoming water compresses the air in the ped.

Fragipans are typically loamy or silty in texture and usually form in udic soil moisture regimes, beneath forest vegetation (Bockheim and Hartemink 2013b). Well-developed fragipans have coarse prismatic structure; an abrupt, but tongued upper boundary; and a gradual or diffuse lower boundary (Fig. 13.17). The exteriors of the large (5–50 cm diameter) prisms are typically light-colored and sandier than are the interiors, and they form a polygonal pattern in plan view (Fig. 13.17). Because they are the most permeable parts of the fragipan, the interprism tongues have been stripped of clay and iron oxides; the latter have been made mobile by redox processes. In this sense, fragipan degradation is comparable to argillic horizon degradation. And just as in argillic horizons, the E (albic) seams within are also the primary locations where (the few) roots are found.

Genesis and Evolution

Fragipans have been the focus of several major reviews (Grossman and Carlisle 1969, Smalley and Davin 1982, Smeck and Ciolkosz 1989, Ciolkosz *et al.* 1995, Weisenborn and Schaetzl 2005b, Bockheim and Hartemink 2013b). Dozens of empirical studies of fragipans also dot the literature. The result of all this work can be summed as follows: (1) Fragipan genesis is still being debated and studied, and (2) Bx horizons may form in different ways (Buntley *et al.* 1977). We examine fragipan genesis by highlighting the main processes that may be involved and the various points of debate, and we end with a model developed for

fragipans in Michigan that incorporates much of the current genetic thinking about fragipans.

Two main issues must be addressed to understand fragipan and fragic property genesis:

1. How does the pan become so dense? Most studies attribute the high bulk densities of fragipans to dense packing of individual grains, combined with intergrain bridging by clay. Many fragipans have high concentrations of silt and very fine sand, and these particle sizes are inherently prone to close packing. They take on a brick and mortar analogy, where sand and silt particles are the brick, and clay or some other cementing agent is the mortar (Knox 1957, Hutcheson and Bailey 1964).

2. What is the cause of brittleness and apparent cementation in the pan? Slaking of fragipan clods in water attests to the fact that the cementation is weak enough that isolated peds are not able to resist the compression of air ahead of a rapid wetting front (Szymański *et al.* 2012). Most studies attribute the brittleness to pedogenesis, particularly to illuviated cementation agents. Any of a number of these agents may be invoked: aluminosilicates, amorphous compounds, hydrous oxides, and gels of Si and Al, clay minerals, and sesquioxides. These materials may occur in the horizon as coatings/cutans, void in-fillings, or grain-to-grain bridges.

Because they parallel overlying soil horizons and display varying degrees of expression as a function of topography and age, fragipans are assumed to be pedogenic features. Dense subsurface zones, sometimes informally called *densipans*, occur in some till-derived soils as a result of the weight of glacial ice during deposition of the lodgment till. These zones have sometimes been identified as fragipans (Antoine 1970, Lyford and Troedsson 1973), although they lack the characteristic pedogenic features of fragipans (Miller et al. 1993, Lindbo and Veneman 1993b). Obviously, this hypothesis does not apply to fragipans formed in loess.

In other studies, the coincidence between a buried paleosol or weathering zone and the fragipan horizon was strong enough to suggest that fragipans are polygenetic features, with many of their properties inherited from a preexisting period of pedogenesis (Buntley et al. 1977, Bruce 1996).

Other scholars have suggested that fragipans (1) developed many of their characteristics while under the influence of permafrost or (2) have formed as a result of freeze-thaw processes (FitzPatrick 1956, Collins and O'Dubhain 1980, Van Vliet and Langhor 1981, Bridges and Bull 1983, Habecker et al. 1990, Payton 1992, 1993a, French et al. 2009). This family of hypotheses keys in on several morphologic properties of fragipans: (1) the depth and abruptness of the upper boundary of the fragipan and its correlation to a paleopermafrost table; (2) its polygonal/prismatic structure, which is remarkably similar, though on a smaller scale, to patterns developed by ice segregation features (Fig. 11.21); (3) the platy structure of the upper zones of some fragipans, which resembles the platy structure formed by ice lenses; and (4) the ability of freeze-thaw processes to cause consolidation and high bulk densities in soils and sediments (Van Vliet and Langohr 1981). Other morphological features of fragipans, such as silt concentrations or caps on peds and stones, clay flows, and vesicular porosity, are also associated with frost activity. To be sure, freezing conditions *can* create dense fabric as the freezing front advances downward toward a permafrost layer below (see Chapter 11).

Instead of the weight of a glacier or the pressures exerted by ice lenses, the self-weight collapse of sediment that has been appropriately preconditioned by a high water content may result in dense fabric, according to some researchers (Thompson and Smeck 1983). The main proponent of this notion – the hydroconsolidation model – is Ray Bryant (1989), and his work has been generally supported by subsequent studies (Weisenborn and Schaetzl 2005a, 2005b, Aide et al. 2006, Smalley and Marković 2013). Bryant suggested that certain types of sediment can undergo physical ripening and/or self-weight collapse, creating a dense condition that may persist, and then be pedogenically imprinted upon, to form a fragipan. Others had suggested that collapse of this sort may occur after a period of weathering (Yassoglou and Whiteside 1960, Nettleton et al. 1968), but Bryant was the first to propose that this process could occur in the early stages of soil development. Assalay et al. (1998) used experimental data to support the hydroconsolidation theory and added to it, suggesting that it works best when (1) some clay exists in the sediment,

but usually not more than 30%; (2) stress is produced via overburden pressure; and (3) the required water content is realized. Desiccation alone may also be enough to initiate a type of collapse or densification (Parfitt et al. 1984).

Because many fragipans occur at or near a lithologic discontinuity, researchers have investigated how the interruption of percolating water that occurs there, or how changes in mineralogy across the discontinuity, might have affected their genesis (Harlan and Franzmeier 1977, Habecker et al. 1990, Tremocoldi et al. 1994, McDaniel et al. 2001, Bockheim 2003, Aide et al. 2006, Wilson et al. 2010). Of particular importance is the possibility that percolating water might deposit dissolved or suspended substances at the discontinuity, and that these might act as cementing agents (Karathanasis 1987b, Smeck et al. 1989). In some studies, the discontinuity occurs in conjunction with the top of a buried paleosol (Harlan et al. 1977, Thompson et al. 1981, Thompson and Smeck 1983, Tremocoldi et al. 1994). Roots that proliferate at the discontinuity or within the paleosol might cause that zone to experience more wet-dry cycles, enhancing precipitation of solutes by absorbing soil water.

A genesis question that is necessary to resolve before fragipan formation can be fully understood centers on the nature of the brittleness and any cementing agents that may be responsible for it. This question has been difficult to answer for all fragipans, as the cementing agent is not overtly visible, unlike $CaCO_3$ or organic matter, and it varies among soils. Some have suggested that no cementing agent is necessary. Instead, they note that the organization, dense packing, and internal arrangement of the clay and silt particles within a skeleton of sand grains are adequate to explain the brittleness (Hutcheson and Bailey 1964). Most researchers, however, feel that a cementing agent *is* a part of fragipan genesis, and they fall into one (or both) of two camps:

1. Cementation is due to amorphous materials, such as silica or iron/aluminum compounds, that have been either illuviated or released by in situ weathering (Bridges and Bull 1983). Deposition mainly occurs when the soil solution becomes saturated via desiccation (De Kimpe 1970, Harlan et al. 1977, Steinhardt et al. 1982, Ajmone Marsan and Torrent 1989, Karathanasis 1987a, b, Norfleet and Karathanasis 1996). And this cementation functions best if the soil has been preconditioned by some physical mechanism that creates *close packing* (Fig. 13.20A).

2. Silicate clay minerals form bridges that hold the brittle matrix together (Grossman and Cline 1957, Knox 1957, Wang et al. 1974, Lindbo and Veneman 1993a, Szymański et al. 2012; Fig. 13.20B–D). This theory has merit because many fragipans are Btx horizons. Perhaps close packing with clay mineral bridging *and* the correct architecture are enough to create brittleness and high bulk density, especially in Ex horizons that may have been stripped of amorphous materials (Miller et al. 1993).

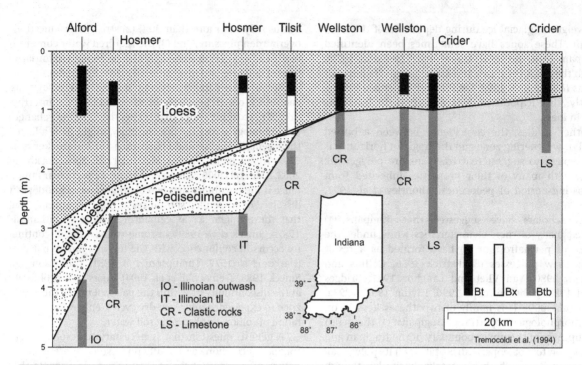

Fig. 13.19 Parent materials and major soil horizons, including fragipans, along a transect of decreasing loess thickness in southwestern Indiana.

The former argument – cementation by an amorphous silica compound or a Si-rich aluminosilicate – seems to be the favored explanation for fragipan brittleness and confirmed by data (Bridges and Bull 1983, Ajmone and Torrent 1989, Tremocoldi *et al.* 1994, Norfleet and Karathanasis 1996, Duncan and Franzmeier 1999; Fig. 13.21). Steinhardt and Franzmeier (1979) found that natural fragipan clods broke down after silica was extracted from them, although extraction of Fe and Al in the same way had little effect. SEM and EDXRA analyses have identified Si-rich bridges between silt grains in fragipans (Norton *et al.* 1984). The silica is thought to coat and form bridges between grains and plug pores. Because grasses biocycle silica, less is available as a bonding agent, perhaps explaining why fragipans are not found in grassland soils (Bockheim and Hartemink 2013b). Some of the most compelling evidence for a silica or aluminosilicate cementing agent in fragipans has been found in southern Indiana, where extractable silica normally increases in fragipans (Harlan and Franzmeier 1977, Harlan *et al.* 1977). The soils here are formed in silty loess over sandy loess, above a buried paleosol. Fragipans tend to form in the lower part of the silty loess. Harlan *et al.* (1977) hypothesized that silica-enriched soil water, generated in the weathered, upper solum, percolates and tends to hang in the silty loess because of its greater matric tension, and not move into the sandy loess below. When the loess is the optimal thickness, such that the base of the silty loess is within the rooting zone, it desiccates during the summer, precipitating the silica and creating conditions that can lead to fragipan formation. In thicker loess, the base of the silty loess is below the rooting zone, where it does not dry out as often, and less silica is precipitated; in these settings the soils lack fragipans. Because some fragipans do not exhibit high amounts of extractable silica (Yassoglou and Whiteside 1960, De Kimpe *et al.* 1972), it is likely that silica is an important, but perhaps not the only, factor contributing to fragipan brittleness.

With this background in mind, we turn to the conditions that must exist for a soil to develop a fragipan, the processes that must occur to precondition the soil for its formation, and the processes that interact to form it. Most fragipans form in acidic parent materials, or at least, in parent materials that are easily leached of carbonates (Yassoglou and Whiteside 1960, Miller *et al.* 1993). Percolating water is necessary to translocate the weathering by-products to the lower solum, where they can precipitate as cementing agents. Illuviation of silicate clay to this depth is probably also a plus. Deposition of illuvial clay and soluble materials is enhanced at a particular depth if a lithologic or weathering discontinuity occurs there. Smeck *et al.* (1989) defined a *weathering discontinuity* as the lower limit of the weathering zone. In calcareous soils it may coincide with the lower limit of leaching; in other soils it may coincide with the top of a buried paleosol. Weathering by-products, particularly hydrous oxides

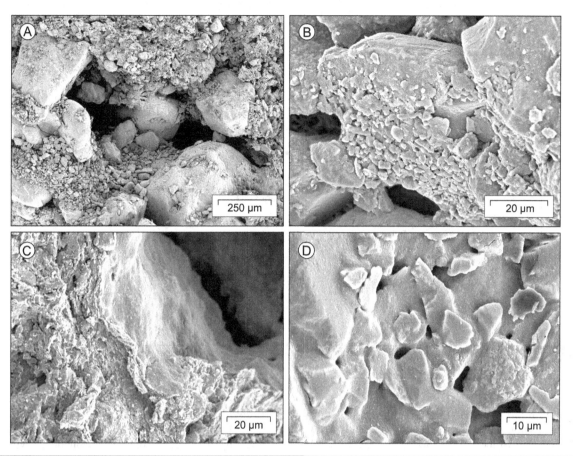

Fig. 13.20 Scanning electron micrographs of properties characteristic of fragipans, all of which are in northern Michigan. Images by B. Weisenborn. (A) Close packing of clay and silt grains around an architecture of tightly packed sand and silt particles. The image was deliberately selected to show the large void; the majority of this Btx horizon is densely packed as shown around the periphery. (B) Clay bridging between skeletal sand grains. (C) A thick, laminated ferriargillan, including some amorphous material, on a ped face in a Bt horizon. (D) Close-up of an amorphous coating draping (and cementing) some silt particles.

of Si and Al, are likely to accumulate at weathering discontinuities – they are generated *above* it, translocated *to* it, and preferentially deposited *at* it because of changes in soil chemistry. A dense, closely packed parent material, rich in very fine sand and silt, also enhances the likelihood of fragipan formation, since the effectiveness of a bonding agent is partially dependent upon a conducive architecture that has many grain-to-grain contacts and few large voids. Few, if any, fragipans form in sands or clays. Clays have so much surface area that any bonding agent is spread too thinly, and the shrink-swell propensities of clayey soils tend to be stronger than the bonding agents (Ciolkosz *et al.* 1979, Smeck *et al.* 1989). In sands, the voids may be too large for the horizon ever to achieve the necessary cementation.

In his hydroconsolidation model, Bryant (1989) illustrated how otherwise porous parent materials can become dense by processes he called physical ripening and self-weight collapse, individually or in coordination.

Once saturated, parent materials of the proper texture will, Bryant suggested, develop close packing among particles, prismatic structure, and high bulk densities; all it takes is an initial desiccation event (Jha and Cline 1963). Subsequent desiccation events may further ripen/densify the material only if they are more intense, i.e., if the soil desiccates further. Additional wet-dry cycles will not disrupt dense, ripened material unless it is expansible and has a high COLE value. In short, a slurried sediment of the proper texture will densify (ripen) upon desiccation. These desiccation events may normally occur at the very earliest stages of pedogenesis, as wet sediment dewaters (dries out). Physical ripening is most efficient in subsoil horizons that undergo fewer subsequent pedoturbation and wet-dry cycles. Thus, it works well as a prerequisite for the development of many fragipan soils.

Working in combination with physical ripening is the concept of self-weight collapse, in which fabric rearrangement and densification occur as sediment is wetted to

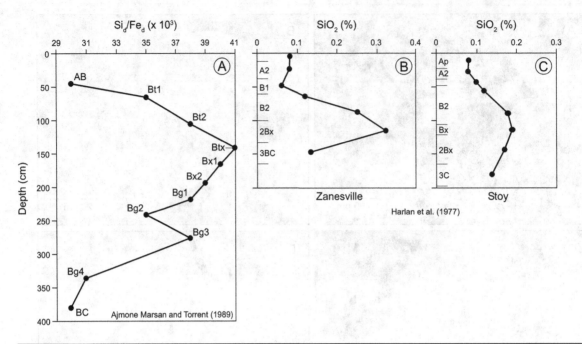

Fig. 13.21 Depth functions illustrating the coincidence of high silica content and fragipan expression. (A) Free silica/free iron ratio in a Fragiudalf from Italy. (B) and (C) Total SiO₂ in Fragiudalfs from Indiana.

saturation (Bryant 1989). Eolian sediments are especially prone to self-weight collapse; not coincidentally many fragipans occur in loess. Thus, Bryant has set up two ways in which open, porous sediment can become denser and more closely packed. Both involve saturation. In physical ripening, a saturation event is followed by desiccation, but the concept of self-weight collapse illustrates that desiccation may not even be necessary in certain types of sediments. The two processes can act independently or together.

Using the preceding theoretical guidance and some data from some fragipan soils in Michigan, Weisenborn and Schaetzl (2005b) developed a comprehensive model of fragipan evolution (Fig. 13.22). Although no one model may ever be able to explain fragipan formation for all soils, this model appears to work for Fragiorthods and Fragiudalfs in the Great Lakes region, as it addresses the three main components of fragipan genesis. It *assumes* that the parent material is *collapsible* or of the correct texture, structure, COLE, thickness, and wetness. It *accommodates* and explains how the requisite fragipan physical properties can be attained by invoking *physical ripening* and/or *self-weight collapse*. And it *accounts for* fragipan binding and brittleness by invoking the *precipitation of amorphous materials* (bonding agents) at a pedogenic *weathering front* or *lithologic discontinuity*. Importantly, the model does not exclude other processes from having an important influence; we note particularly the potential for frost action in fragipan genesis (Van Vliet and Langhor 1981).

Degradation

Some fragipans appear to be degrading, a process manifested by the formation of bleached tongues growing downward and deepening along vertical cracks. Degradation over time is observed in the northeastern United States, where soils on young glacial tills have fragipans while those on older tills do not.

As with argillic horizons, fragipan degradation stems mainly from the fact that they are aquitards and occasionally form perched zones of saturation (Payton 1993b, McDaniel *et al.* 2001, Weisenborn and Schaetzl 2005b). Some of the acidic, base-poor, perched water flows across the top of the pan or percolates vertically through it, along cracks. Seasonal drying and shrinking can greatly assist in the formation of cracks in the fragipan (Szymański *et al.* 2011). For this reason, the presence of at least some swelling clay minerals in the fragipan assists in fragipan cracking and, eventually, its degradation. Redox processes become increasingly operative in the upper fragipan, especially in the cracks, because of the anoxic conditions formed by the perched water, assisted by the fact that roots in the cracks remove oxygen (Ajmone-Marsan *et al.* 1994, Lindbo *et al.* 2000; Fig. 13.23). The water strips clay and reduces the Fe and Mn minerals there, forming light-colored, leached zones (streaks), as the pan degrades from the cracks outward, into the prism centers. This form of degradation is much like the grainy gray ped coats (skele-tans) that form in degrading argillic horizons – except that the prism streaks are much thicker in fragipans (James

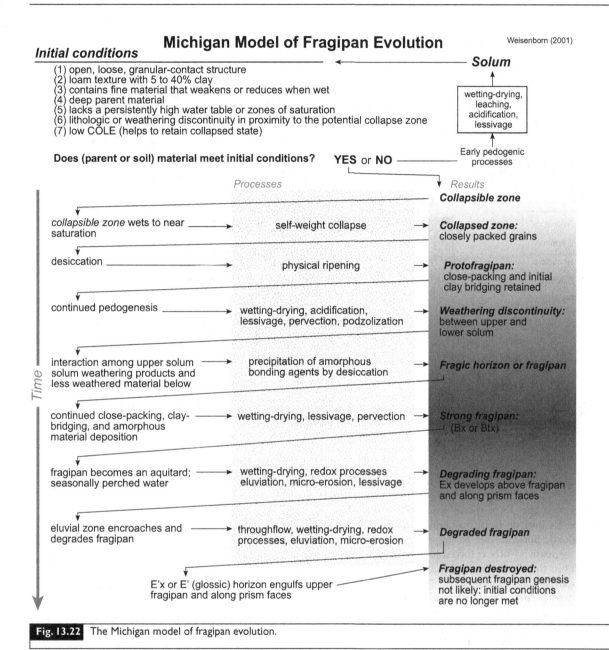

Fig. 13.22 The Michigan model of fragipan evolution.

et al. 1995). Lindbo *et al.* (2000) called these areas "vertical gray seams." Eluviation in the cracks is indicated by lower concentrations of clay minerals and Fe oxides in material collected from the vertical tongues, as compared with material collected from prismatic clods and the Btx horizon (Szymański *et al.* 2011). Some of the reduced Fe and Mn ions then migrate into the prisms of the Bx horizon, where the oxidizing conditions cause them to precipitate as nodules or Fe oxide stains, forming a reddish rim on the prism edges – common in fragipans. As these light-colored zones or seams develop in the uppermost pan, it degrades from the top down and from the edges in.

Once the bleached zones become continuous across the top of the pan, an Ex horizon is described above the Bx horizon. It is labeled an Ex as long as it retains some amount of brittleness (Figs. 13.18, 13.23). Such Ex horizons have varying amounts of Fe-Mn nodules and mottles – as fragipan relicts – within them (Lindbo *et al.* 2000, Fig. 13.23). With continued leaching and degradation, the Ex horizon may become stripped of its bonding/cementing agents, including clay and Fe, and develop into a nonbrittle, continuous E horizon (Steele *et al.* 1969). As the cracks and prism streaks in the fragipan widen, small amounts of materials from the E (or Ex) horizon above can

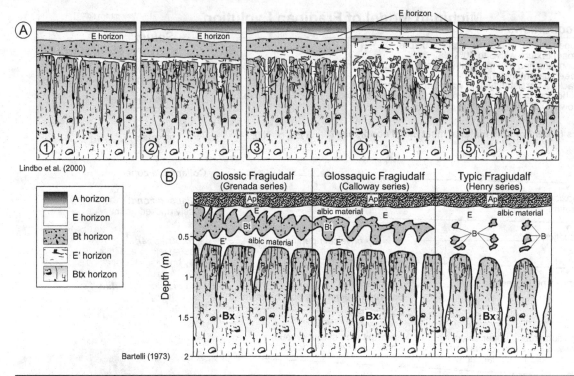

Lindbo et al. (2000)

Bartelli (1973)

Fig. 13.23 Schematic representations of the degradation of fragipans, as modified by the authors. Although each study examined different types of degradation (Bt vs. Btx), many of the same principles apply.

be translocated into them (Grossman and Carlisle 1969, Ranney *et al.* 1975).

The processes that form the albic streaks within the pan and the Ex horizon itself are quite similar to those that form glossic horizons (see earlier discussion). Acidic, chemically aggressive water is forced to pass through the few conduits in the slowly permeable horizon, removing and translocating clays in suspension (Miller *et al.* 1993), chemically weathering some clays, and reducing Fe and Mn minerals there. Payton (1993b) linked this form of microerosion of clay coatings and void walls to the destabilization of the clay-Fe system after the Fe is reduced and eluviated. Weathering by-products, many of which are potential amorphous bonding materials, are remobilized from the pan and moved downward, leaving the fragic architecture but removing the glue (Szymański *et al.* 2011). For this reason, many Ex horizons are more porous, of lower density, and considerably more fragile than the Bx horizons below.

Fragipan degradation is slightly different, and perhaps more intense, than Bt horizon degradation because it almost always involves oxidation-reduction processes. Reduced conditions tend to be more common in fragipan soils because fragipans are better aquitards, leading to increased Fe and Mn mobility under reducing conditions. Thus, many Ex horizons are actually Egx horizons.

Oxidation-Reduction and Gleization

In soils, the terms *oxidized* and *reduced* are used, respectively, to describe whether the soil system has oxygen available to microbes or whether it does not. Reduction-oxidation (redox) reactions refer to the suite of chemical give-and-take that centers on these two conditions, which can change quickly within soils and can vary greatly across even small distances.

Specifically, *redox* reactions refer to the transfer of electrons (e⁻) among atoms. Because electrons cannot exist freely with the soil solution, a reducing agent (an e⁻ donor) must be coupled with an oxidizing agent (an e⁻ acceptor) in order to complete the redox reaction. Redox processes in soils are mediated largely by several oxidizing agents, including oxygen, nitrogen, and iron, and by organic matter bioavailability. We use the following series of equations, presented by Bedard-Haughn (2011), to illustrate the interactions.

Heterotrophic microbes use soil organic matter as a carbon source for energy. A typical organic compound that is used for energy is glucose, $C_6H_{12}O_6$, and the oxidation of carbon in glucose can be represented as

$$C_6H_{12}O_6 + 6H_2O \rightarrow 6CO_2 + 24e^- + 24H^+$$

This reaction is catalyzed by microbial enzymes. In an oxygenated soil, aerobic microorganisms that produce those enzymes use O_2 as the terminal e^- acceptor, as shown here:

$$6O_2 + 24e^- + 24H^+ \rightarrow 12H_2O$$

As long as O_2 is present, other possible electron acceptors, such as N, Fe, or Mn, are not needed to oxidize carbon. However, when the soil becomes saturated and the oxygen supply is limited, anaerobic conditions will develop. Then, organic matter decomposition will slow because the aerobic microorganisms cannot be active. However, some microbes, adapted to anaerobic conditions, are able to use elements other than O, e.g., N, Fe, Mn, or S, as e^- acceptors. The reduction of Fe, for example, can be expressed as

$$Fe^{3+} + e^- \rightarrow Fe^{2+}$$

The molecular-scale mechanisms by which Fe atoms in a solid phase mineral are reduced are not well understood. It is likely that organic compounds, secreted by microorganisms, act as "electron shuttles" to carry electrons to mineral surfaces. In any case, Fe^{3+} in minerals can be reduced to Fe^{2+} in a reaction that also consumes protons, as shown in the following reactions:

$$Fe(OH)_3 + e^- + 3H^+ \rightarrow Fe^{2+} + 3H_2O$$

$$FeOOH + e^- + 3H^+ \rightarrow Fe^{2+} + 2H_2O$$

In this situation, decomposition of organic matter can continue under anaerobic conditions, as long as an alternate electron acceptor (Fe, in this case) exists. Compounds that contain Fe are destabilized, and organic matter is oxidized. This is one example of redox processes in soils.

The open atmosphere is the main *source* of oxygen in soils, into which it moves by diffusion (Table 13.3). Microbes, along with living roots, are the main oxygen *sinks*. They utilize O_2, the most energetically favorable e^- acceptor, first, where it exists as entrapped soil air (Table 13.4). When that source is no longer available, as in saturated soils, they utilize the dissolved oxygen in soil water. When *that* O_2 supply is gone, anaerobic (or anoxic) conditions develop and a series of biochemical reactions, associated with anaerobic microbes, occurs (Ponnamperuma *et al.* 1967, Patrick and Mahapatra 1968, Patrick and Henderson 1981, van Breeman 1987). The next step of this process is the gain of an electron, i.e., *reduction*, for certain other vulnerable elements (electron acceptors). Nitrogen in nitrate or nitrite is commonly the first of a series of elements to be reduced in the anaerobic soil system, as it undergoes denitrification (Table 13.4). After the nitrate is used up, the sequence continues as Mn is reduced from Mn^{3+} (or Mn^{4+}) to Mn^{2+}. Next, ferric iron (Fe^{3+}) is reduced to ferrous iron (Fe^{2+}), and finally S in sulfate is reduced (Vepraskas 1999). Most soils have so much Fe and so little sulfate that reduction of S is rare. Excellent reviews of redox chemistry in soils are available elsewhere (Bartlett and James 1993, Vepraskas and Faulkner 2001, Fiedler *et al.* 2007).

Table 13.3	Factors that regulate the supply of versus demand for oxygen in soils

Supply-side regulators	Demand-side regulators
Water table depth	Root density
Oxygen content of soil water	Microbial activity
Soil texture	Soil temperature
Soil porosity	Soil texture
Soil permeability	
Soil thickness	

Table 13.4	Order of utilization of principal electron acceptors in soils and associated reactions

Process	Reaction
O_2 utilization	$6O_2 + 24e^- + 24H^+ \rightarrow 12H_2O$
N reduction (denitrification)	$1/5NO_3^- + 6/5H^+ + e^- = 1/10N_2(g) + 3/5H_2O$
Mn^{4+} reduction	$MnO_2 + 2e^- + 4H^+ = Mn^{2+} + 2H_2O$
Fe^{3+} reduction	$FeOOH + e^- + 3H^+ = Fe^{2+} + 2H_2O$
S^{6+} reduction	$SO_4^{2-} + 6e^- + 9H^+ = HS^- + 4H_2O$

Source: Mausbach and Richardson (1994).

As soil conditions change from oxidizing to reducing and back again, cations change states as well, and the suite of secondary minerals associated with them also changes. Under oxidizing conditions, Fe^{3+} and Mn^{4+} cations are immobilized by precipitation as oxide and oxyhydroxide minerals. The ferric oxide and oxyhydroxide minerals have yellowish brown, orange, or reddish colors; that is why dry upland soils that allow free entry of oxygen into them tend to display various shades of yellow, brown, or red (Table 2.4). Reducing conditions force a new suite of soil minerals to form. Reduced iron, which often exists in the soil solution as soluble complexes with anions like HCO_3^- and dissociated organic acids, is mobile and capable of diffusing within the soil solution; oxidized iron compounds are not mobile. When destabilized by the lower redox potential of an anaerobic soil solution, ferric oxides are dissolved and stripped from grain surfaces. The result is a change in the soil matrix to dull colors – shades of gray, blue-green, or other pale, muted hues. This bears repeating: Reduction leads to gray and muted soils colors because these are the colors of most uncoated soil particles. Reduced Fe and Mn can migrate out of microsites, as we saw with fragipans, or eventually migrate completely out of gleyed horizons, rendering those microsites or horizons gray and iron-poor. Under strongly reducing conditions where water is stagnant over a long period ferrous Fe oxides called "green rusts" may form. The lack of oxidized Fe and Mn oxides or the presence of reduced Fe oxides both result in the same morphologic expression: dull, gray, muted soil colors.

The condition represented by long-term reduction is referred to as *gleying*, and the processes associated with long-term gleying are called *gleization* (Table 13.1). Wet, gleyed soils are so common and important in Canada that they are singled out at the order level – Gleysolic soils (Expert Committee on Soil Science 1987, Bedard-Haughn 2011). Although they are rare in dry climates, gleyed soils are not unique to any particular climate; rather, they are usually found in association with lowlands and wetlands where water tables are high.

In older literature, reduction associated with high water tables, i.e., groundwater, was called *groundwater gleying*. Alternatively, the term *pseudogley* has been used to refer to brief periods of reduction that occur in association with a temporarily *perched* water table, as occurs in many soils with slowly permeable subsoil horizons, e.g., fragipans (McDaniel *et al.* 2001, 2008) or shallow bedrock (Zhang and Karathanasis 1997, Hseu *et al.* 1999, Carter *et al.* 2011, Breuning-Madsen *et al.* 2012). Regardless of the type of water table, gleization occurs in soils or soil horizons that have been saturated long enough to render at least a portion of the horizon anoxic. Assisting in this process are relatively high organic matter contents – the microbial food source that fuels the demand for oxidizing agents in the soil. That is, saturated soils will become reduced and gleyed much faster if they are rich in organic matter, because organic matter is the primary e⁻ donor required for reduction to occur. Reducing conditions are most likely to occur in seasons of the year when the soil temperature is high enough for microbial activity to occur, especially in spring and summer.

Gleyed soils tell a story. They are, or have been over a long period, wet or waterlogged soils, because water saturation is the primary mechanism by which atmospheric oxygen is excluded and depleted from soils. Saturated, fine-textured soils are readily gleyed, because (1) the diffusion rate of oxygen in water is considerably slower than in air, and (2) they often have higher organic matter contents and microbial populations. Sandy soils can theoretically become gleyed, i.e., develop gray colors, more rapidly than can fine-textured soils, because the sand has much less surface area area that might retain secondary Fe oxides in ferric form. Gleyed horizons can occur within the profile or well below it, although deeper horizons are less likely to be gleyed, even if they are saturated, because the demand for oxygen in soils, associated with organic matter and microbial activity, decreases with depth. For that reason, some seasonally waterlogged soils may be gleyed within the profile but become subtly browner and brighter below about 2–3 m. Gleyed horizons are typically given the suffix g, as in Bg or Cg horizons (Table 3.2).

In many soils, alternating periods of oxidation and reduction occur, often referred to as *fluctuating redox conditions* (Vepraskas 1999, Bedard-Haughn 2011). Morphologic features associated with fluctuating redox conditions take many forms – nodules, concentrations, masses, and pore/ped linings, as well as zones from which Fe oxides have been removed – and are referred to as *redoximorphic features* (Schwertmann and Fanning 1976, Vepraskas 1999; see Chapter 14). While the soil is saturated and in a generally reduced state, soluble Fe and Mn species may move along concentration gradients or slowly with currents in the soil solution; i.e., they are relatively mobile within the soil. Later, under oxidizing conditions, the reduced Fe will be oxidized and precipitate to form insoluble minerals such as ferrihydrite, lepidocrocite, or poorly crystalline ferric hydroxide, forming reddish brown or orange Fe oxide concentrations, or blackish Fe-Mn concentrations (Table 2.4). Over time, continued oxidation-reduction cycles cause Fe and Mn to become concentrated in these reddish *redox* or *Fe concentrations*. Surrounding areas with lower chromas and grayer hues are referred to as *redox* or *Fe depletions* (Richardson and Hole 1979, Schoeneberger *et al.* 2012). Gray areas associated with Fe depletions generally have low Fe and Mn contents; the gray colors reflect uncoated sand and silt grains.

Many soil horizons are *mottled*; i.e., they have variegated patterns of soil colors due to redox depletions (gray mottles) and concentrations (orangish or reddish mottles) (Vepraskas 1999; see Chapter 14). Like gleying, the redox processes that cause soil horizons to be mottled are primarily biogeochemical. Mottling is also a useful indicator of water table fluctuations. It can provide information on the duration and depth of saturation in a soil and the dissolved oxygen status of the water (Simonson and Boersma 1972, Veneman and Bodine 1982, Khan and Fenton 1994, Mokma 1997b, Hayes and Vepraskas 2000). Mottle patterns can also be used to identify ephemeral features, such as a perched water table (Guthrie and Hajek 1979, Ransom and Smeck 1986, Megonigal *et al.* 1993).

Ferrolysis

In 1970, Robert Brinkman coined the term *ferrolysis* to explain the formation of the texture contrast conditions found in wet, acidic soils, such as the paddy soils of Southeastern Asia (Brinkman 1970, 1977a, Zi-Tong 1983), as well as similar soils worldwide (Dijkerman and Miedema 1988, Feijtel *et al.* 1988, Chittleborough 1992, Hardy *et al.* 1999). The process requires cycles of oxidation and reduction in a wet-dry climate. A related process – *xerolysis* – appears to be operative in semiarid climates (Ducloux *et al.* 2002).

Ferrolysis involves decomposition of clay, due in part to redox processes, in acid, seasonally wet soils (Van Ranst and DeConinck 2002, Barbiero *et al.* 2010, Van Ranst 2011). Ferrolysis leads to lowering of the soil cation exchange capacity, because clay minerals in the soil are destroyed

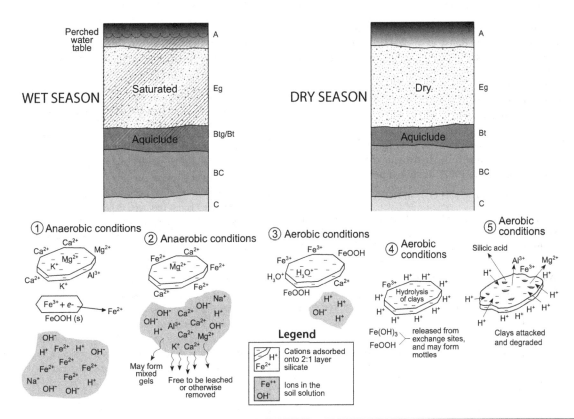

Fig. 13.24 Diagrammatic representation of the annual cycle associated with ferrolysis. (1) Under anaerobic conditions, the redox potential of the soil solution declines; Fe in Fe oxides is reduced from Fe^{3+} to Fe^{2+}, destabilizing the crystals, increasing the abundance of redox depletion features, and releasing Fe^{2+} to the soil solution. The reduction of Fe in an aqueous system consumes protons, as illustrated in the reaction $FeOOH(s) + 3H^+(aq) + e^- \rightarrow Fe^{2+}(aq) + 2H_2O(l)$. As a result, the soil solution pH rises. (2) In this anaerobic condition, Fe^{2+} ions displace adsorbed cations from clay mineral surfaces; these cations are free to translocate in the soil solution. (3) When aerobic conditions resume, the redox potential of the solution increases; Fe^{2+} on exchange sites is then oxidized to Fe^{3+}. The Fe^{3+} may react with water to precipitate as redox concentrations: $Fe^{3+}(aq) + 2H_2O(l) \rightarrow FeOOH(s) + 3H^+(aq)$. The result – protons are released, solution pH declines, and FeOOH mottles may form. (4) The increased acidity leads to acid attack at silicate mineral surfaces, destabilizing and dissolving the mineral crystals and releasing Al^{3+}, SiO_4^{4-}, and more Fe to the soil solution. Hydrolysis of the released Al^{3+} and Fe^{3+} produces more protons, and pH declines further. The released silica may form an amorphous silica precipitate. (5) The Al^{3+} precipitates as hydroxy-Al interlayers within 2:1 clay mineral structures. Thus the cyclic fluctuation in redox potential promotes a cyclic fluctuation of solution pH, leaching of base cations, and partial dissolution and interlayering of silicate minerals.

(Table 13.1). The process is best expressed in soils that are alternately saturated and dry, usually because of a perched water table on a slowly permeable B horizon during a wet season (allowing the water to become stagnant), followed by a dry season (Bartelli 1973). The B horizon, an aquiclude, is usually a strong Bt, Btn, or Btx horizon (Fig. 13.24). In the rice paddies of southeastern Asia, which have been under cultivation for centuries, the aquiclude can also be a plowpan. Long-term ferrolysis will cause the soil horizons in the upper solum to be sandier and more acidic because clay decomposition by ferrolysis is, according to Brinkman (1970), 10 times more efficient at weathering clays than are organic acids alone. The sandy upper solum of soils undergoing ferrolysis has gray matrix colors, indicating generally reducing conditions, sometimes with yellowish brown

goethite or reddish brown ferric hydroxide mottles within the matrix, but mostly along cracks and ped faces. The brightly colored mottles form during the dry season when conditions become oxidizing. The upper part of the profile is not saturated long enough, however, for true gleying to occur.

In ferrolysis, most of the action occurs in the upper part of the profile, which undergoes the most extreme wet-dry cyclicity. Here, during the *anaerobic stage*, soils are saturated and organic matter is oxidized as Fe^{3+} is reduced to Fe^{2+},, consuming protons and enriching the soil solution in OH^- anions (Brinkman 1977b). Because saturation occurs in the *upper* solum, which is near to a large source of organic matter (the A horizon), microbial activity is high and reduction can occur quickly. The resulting Fe^{2+} ions displace +2 and

+1 base cations from clay mineral exchange sites (Ransom and Smeck 1986, Van Ranst *et al.* 2011). However, these base cations (and Al) are then free to be removed from the soil system with soil water. Removal of the cations is often accomplished as lateral flow, with soil water slowly flowing along the top of the aquiclude (Fig. 13.24).

During the dry season's *aerobic stage*, the Fe^{2+} cations adsorbed to the clay surfaces are oxidized to Fe^{3+} and some precipitate as yellowish brown mottles of goethite (α-FeOOH) and/or poorly crystalline ferric hydroxide ($Fe(OH)_3$) (Bartelli 1973, Eaqub and Blume 1982), as shown here:

$$4Fe^{2+} + O_2 + 10H_2O \rightarrow 4Fe(OH)_3 + 8H^+$$

As a result, the soil solution becomes more acidic, because the oxidation of ferrous to ferric Fe and the formation of the ferric hydroxide releases H^+ cations, and indirectly because base cations are lost via leaching. Next, via hydrolysis, the remaining H^+ cations attack clay mineral particles from their edges inward, liberating Si, Al^{3+}, and some Mg^{2+}. Clays with higher CEC values are attacked most effectively; 1:1 clays are less susceptible to acidic attack. Some of the Si that is released from the weathered clays is reprecipitated as amorphous silica, showing up as gray silans or skeletans (Brinkman 1977b). Mg^{2+} cations will eventually be leached, whereas the Al^{3+} cations stay in the soil to acidify it further (Fig. 13.24).

During every wet-dry cycle, base cations and some Al are (potentially) leached as soluble polymers, some clays are destroyed, and the upper profile becomes sandier and more silica-rich (Van Ranst *et al.* 2011). Weathering by-products are removed laterally in soil water, flowing on top of the Bt horizon. Some of the solubilized Al may remain, however, to precipitate in Al hydroxide interlayers in existing 2:1 clays; such modified minerals are termed hydroxy-interlayered vermiculite or smectite (Chapter 4). Brinkman (1970) called this process *chloritization* (Dijkerman and Miedema 1988), although evidence for chlorite's existence is debatable (Van Ranst and DeConinck 2002). (Because we now refer to these minerals as hydroxy-interlayered minerals (HIMs) or HIV/ HISs, and not chloritized smectite/vermiculite or pedogenic chlorite, the term "chloritization" is no longer used.) Release of Al^{3+} cations and their precipitation in the interlayer space of expandable minerals are, therefore, other ways that the soil's CEC is reduced by weathering (Reuter 1965).

Other Al cations may combine with silica to form amorphous compounds. The release of silicic acid, coupled with the destruction of phyllosilicate clays, can cause soils undergoing ferrolysis to have an abundance of quartz (probably neoformed from the silicic acid) in the clay fraction of the upper solum (Hardy 1993). Hydrolysis of readily weatherable minerals such as biotite or augite in the sand and silt fractions will replenish the supply of base cations and slow, but not stop, the acidification process. The pH continues to drop and clay minerals continue to weather, resulting in an acid, sandy horizon, usually an Eg or E horizon, above a mottled aquiclude (B horizon). Eventually, ferrolysis may work its way down into the aquiclude and degrade it, creating windows that will lessen the duration of saturation and cause the process to slow, perhaps even to stop.

Landscapes: Pathways to Planosols

Planosols (a term used in the 1938 U.S. classification system as well as in the World Reference Base system; usually Mollic or Vertic Albaqualfs or Albaquults in Soil Taxonomy) are sometimes considered to be the classic example of ferrolized soils (Feijtel *et al.* 1988). These soils have an abrupt texture change from a coarser E horizon to the dense, clay-rich Bt horizon (see Figure). They tend to be found on older, flat landscapes and in slight depressions (Culver and Gray 1968a, b). They perch water on a dense claypan (Jarvis *et al.* 1959). In the chemically weathered, silty/ sandy E or Eg horizon are abundant Fe-Mn nodules and mottles, indicative of the periodically saturated, reducing conditions. Soils like these led Brinkman (1970) to develop his theory of ferrolysis.

For some pedologists, Planosols represent lessivage, and then ferrolysis, taken to the *n*th degree. It seems simple: Bt horizons form via lessivage and develop into aquicludes, setting the ferrolysis process in motion. Could this be a natural developmental cycle for fine-textured parent materials in wet-dry climates? Yes, we believe it could. But let us also explore alternate pedogenic pathways that can lead to Planosol formation. Like so many "things pedogenic," Planosols may represent equifinality.

Pathway 1. Aquolls can evolve to Albaqualfs (Planosols) through a suite of processes that take considerable time to gestate. Initially, melanization and base cation cycling are strong, as they are in most Mollisols. Because climate or vegetation changes occur or because the soil is located in a groundwater recharge depression with a permeable substratum, the soil becomes more intensively leached/acidified over time. Lessivage begins, and the soil develops an E-Bt sequum. On flat landscapes, water moves primarily downward, rather than laterally, facilitating lessivage. Organic matter also is translocated, leading to a profile with a secondary organic carbon maximum in the B horizon (Culver and Gray 1968a; see Figure). Eventually, the Bt horizon becomes so clay-rich that it passes an intrinsic threshold and begins to accumulate clay by sieving, as attested by clay contents that are highest at the top of the horizon. Ferrolysis may then kick in. The A and E horizons become leached of most of their base cations, whereas the lower solum stays base cation–rich. The Bt horizon becomes less and less permeable, perching water in spring and fall, promoting more intense ferrolysis. Acidification and destruction of the clays in the upper solum result, forming coarser, acidic A and Eg horizons over a clay-rich Btg horizon. Some of the silt in the upper solum is translocated to the Bt, especially during dry periods, when deep cracks open, forming siltans on ped faces (Culver and Gray 1968a). Vegetation becomes more impoverished as ferrolysis continues, and the overall productivity of the landscape declines. The direction of pedogenesis in these soils only changes if (1) landscape dissection leads to steeper slopes, deeper water tables, and better internal drainage, or (2) a change, back to base cation-cycling vegetation, occurs.

Pathway 2. Vertisols normally form in clayey, smectite-rich parent materials, under a wet-dry climate (see Chapter 11). These soils typically develop on flat landscapes, and their morphology is self-sustaining.

However, Barbiero et al. (2010) have argued that some Vertisols may evolve into Planosol-like soils via ferrolysis under certain circumstances. The wet-dry climate of the Vertisols already favors ferrolysis; in order to change the pathway, the landscape needs to be incised by rivers. As shown in India, when approaching incised river valleys, Vertisols can grade into texture-contrast soils that resemble Planosols. Here, incision allows for lateral subsurface flow, creating a pathway for the export of base cations and Al liberated during the wet season. Ferrolysis begins, and the Vertisol-argilliturbation pathway changes toward a Planosol-ferrolysis one. Ferrolysis causes protonation of clay in the upper solum, weathering 2:1 clays to 1:1 clay minerals. Surface soil horizons become sandier and more acidic. Also, amorphous silica precipitates in the upper profile during the dry season, forming sands and silts. Wedge-shaped aggregates, typical of Vertisols, remain as ghosts in the sandy loam-textured upper solum, attesting to the fact that these soils recently underwent argilliturbation, but are now deep into ferrolysis.

This change in pedogenic pathway was initiated by the crossing of an extrinsic threshold. The fluvial incision that changed the internal soil hydrology from a closed to an open system allowed evacuation of soluble materials with the soil solution during the wet season. Not only is this process intriguing, but it appears to have run its course quite quickly – in less than 4,200 years!

Pathway 3. Planosols can also develop as salt-affected soils are leached of sodium in a process called *solodization* (White 1961, 1964). This topic will be explored later in this chapter.

Figure: Planosol genesis and data. (A) Model of Planosol development in Oklahoma. (B) Some depth functions for the Nardin (end-member) soil (Mollic Albaqualf).

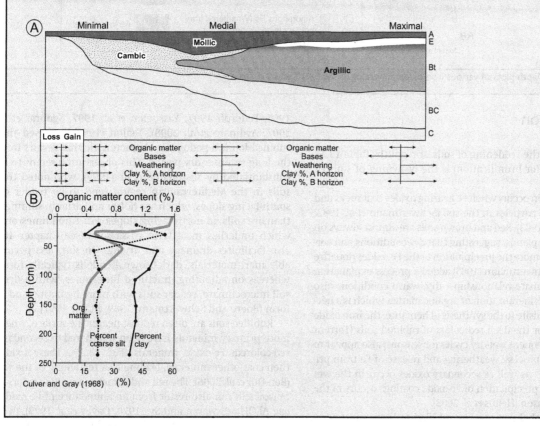

For years, ferrolysis has been invoked as a key agent in the formation of textural contrast soils in wet-dry climates, where bleaching and mottling are predominant morphological features of the upper solum, and where the lower solum has low permeability (Chartres 1987, Dijkerman and Miedema 1988). Nonetheless, soils with similar morphologies can form without invoking ferrolysis (Morrás 1979). Eaqub and Blume (1982) described two soils in Bangladesh that had morphologies typical of soils undergoing ferrolysis. They argued that weathering and lessivage alone can produce the clay-poor epipedons; ferrolysis is not required. Van Ranst and DeConinck (2002) concluded that some texture-contrast soils in Belgium and France are formed simply by lessivage, not ferrolysis, and that the fine quartz and chlorite in the upper profile, long assumed to have been secondary minerals formed by ferrolysis, are more likely formed by the physical disintegration of preexisting quartz and chlorite minerals. Subsequent studies have drawn similar conclusions: ferrolysis is an important pedogenic process, but it may be more geographically restricted than previously thought, and it may be subsidiary to more widespread processes such as lessivage (Payton 1993b, Boivin et al. 2004).

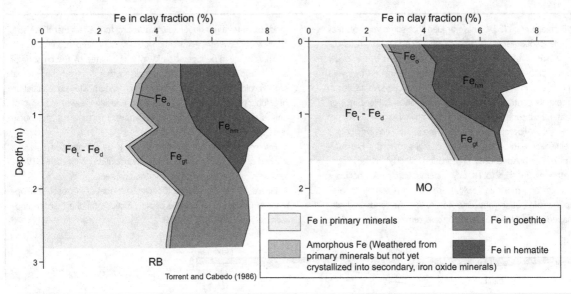

Fig. 13.25 Depth plots of various iron oxide mineralogies for two Xeralfs in Spain.

Rubification

Rubification is the reddening of soils or regolith (Table 13.1). *Braunification* (or brunification) is the browning of soils or regolith.

Rubification occurs when Fe-bearing oxides coat rocks and other skeletal particles in the soil (Schwertmann *et al.* 1982, Bronger *et al.* 1983). Red and brown soils are almost always on well-drained uplands, suggesting that dry conditions and aeration help promote the precipitation of the Fe oxides (Jouaffre *et al.* 1991). Schwertmann (1985) added a process explanation to this soil-climate relationship – dry, warm conditions also facilitate the mineralization of organic matter, which is a necessary prerequisite to the synthesis of hematite, the iron oxide responsible for the deep red colors of rubified soils (Torrent *et al.* 1980). Frequent wet-dry cycles (or seasons) also appear to facilitate this process; weathering and release of Fe from primary minerals as well as secondary oxides occur in the wet season, while precipitation of Fe-oxide coatings occurs in the latter (dry) season (Helms *et al.* 2003).

From a global perspective, rubified soils are most common in xeric soil moisture regimes and the humid or wet-dry tropics or subtropics. Factors important to rubification include good aeration, warm temperatures, parent materials rich in Fe-bearing minerals, and a wet season sufficient to permit microbial activity (specifically *iron-oxidizing bacteria*, which derive the energy they need to live by oxidizing dissolved (Fe^{2+}) iron). The oxidation of Fe^{2+} to Fe^{3+} facilitates weathering of primary minerals and turnover of organic matter (Glazovskaya and Parfenova 1974, Boero and Schwertmann 1989, Gehring *et al.* 1994). Fractured limestone uplands in Mediterranean climates fit many of these conditions; that may be why so many soils there – Rhodoxeralfs in particular – are so deeply red (Torrent

1995, Fedoroff 1997, Yassoglou *et al.* 1997, Sgouras *et al.* 2007, Aydinalp *et al.* 2008). Yaalon (1997b) stressed that rubification is a pedogenic characteristic that results from the long summer dry period. This observation seems to be substantiated by Yassouglou *et al.* (1997), who noted that soils in the Mediterranean region tend to be redder on south-facing slopes, which are hotter and drier in summers than are soils on north-facing slopes. Fractured limestone, which underlies most terra rossa soils (see Chapter 10), also facilitates drainage and drying. On soft, less permeable marl materials, dark-brown Rendolls typically form, whereas on adjoining, fractured limestones, with a drier soil microclimate, redder soils with more hematite tend to form (Boero and Schwertmann 1989, Atalay 1997).

Rubified soils are often reddest nearer the surface, where more primary minerals have been weathered to secondary, red-colored, Fe-oxide minerals (Fig. 13.25). These oxides then coat other minerals, leading to a reddening of the soil (Ben-Dor *et al.* 2006). The red and sometimes yellow colors in *tropical soils* can also result from an abundance of Fe oxides and $Al(OH)_3$ (Eswaran and Sys 1970, Davey *et al.* 1975). Even though gibbsite ($Al(OH)_3$) is colorless, it is also correlated with red colors, because Fe oxide and $Al(OH)_3$ concentrations are often positively correlated in highly weathered soils.

In most soils, the degree of redness or yellowness in soils is most strongly correlated to the amount of each particular iron mineral, for each has a unique hue and intensity of coloration. Therefore, the hue of rubified soils is actually more a function of *which* ferric pigment (mineral) predominates (Hurst 1977) than it is of the *total* concentration of iron oxides in the soil (Plice 1942, Waegemans and Henry 1954, Buol and Cook 1998; Table 2.4). For example, Soileau and McCracken (1967) found that the redness of Ultisols was not as much due to the *thickness of iron oxide coatings* on soil particles as it

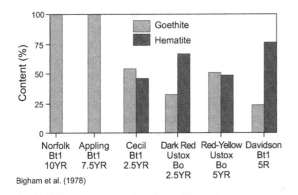

Bigham et al. (1978)

Fig. 13.26 Mineralogical composition of the iron oxide fractions of selected soil horizons from some Ultisols in North Carolina and some Oxisols from Brazil.

was to the *kind* of iron oxides. The yellowish brown soils of the midlatitudes tend to have high amounts of lepidocrocite and goethite. In tropical soils, goethite is more common in brownish/yellowish horizons, while kaolinite is the most abundant mineral in the white and yellow horizons. Soils with 7.5–10YR hues have at least some goethite and little or no hematite (Fig. 13.26; Table 2.4). Deep, almost beet-red colors are associated with submicron-size hematite (Hurst 1977, Michalet *et al.* 1993, Azzali *et al.* 2011). Hues between 2.5YR and 10YR generally indicate that hematite content exceeds goethite content (Eswaran and Sys 1970, Bigham *et al.* 1978). Because color is such a good indicator of iron oxide mineralogy, it is used at high categorical levels in several tropical soil classification systems (Soil Survey Staff 1999).

Red tropical and subtropical soils derive most of that color from ferric (oxidized) iron in the form of fine-grained hematite. Several studies have therefore attempted to develop quantitative relationships between Fe oxide content and color (see Chapter 15). Early attempts (Soileau and McCracken 1967) found no consistent relationship between hue and total Fe oxide content. As discussed previously, it is the *mix* of different oxide minerals that is most related to color. Nonetheless, Hurst's (1977) redness rating, although derived for use in saprolite, has been successfully applied to soils and has been shown to describe accurately the relationship between *concentrations* of specific iron oxides (particularly hematite) and soil color (Liebens and Schaetzl 1997). Torrent *et al.* (1983) later modified the Hurst Redness Rating (RR; see Chapter 15). They, and other workers, have reported strong relationships between RR and hematite content, both for tropical Oxisols as well as for a variety of other soil types. This work points to the ability of hematite to impart red colors to soils (Fig. 13.27 A–C). In some areas, the strong relationship between redness and iron content can be used to map soils using remote sensing (Rossel *et al.* 2008, Minasny and Hartemink 2011).

At different landscape positions, water table relations and soil wetness conditions affect the types of minerals that are likely to form, and hence soil colors. Redder colors are more common in aerated soils, usually on dry uplands (Curi

and Franzmeier 1984) and in dry climates. Wetter soils with higher water tables will generally have yellower hues, and the clay mineralogy will lean toward goethite (Macedo and Bryant 1987). Even short periods of reduction, for example, around decaying roots, will lead to the removal of hematite, leaving yellow-brown goethite as the dominant mineral (Buol and Eswaran 2000). Under prolonged wetness, gleying can and will occur in Oxisols as in other soils, leading to gray colors and a lack of oxidized iron (see Chapter 14).

One final point about rubification. Release of Fe oxides from primary minerals by weathering and their neoformation as secondary iron oxide minerals will clearly cause changes in soil color as soil particles are coated with the reds and browns of these minerals. But these colors will only be present if the minerals *persist* and Fe remains *oxidized* in the soil. In wet soils, Fe may be reduced, and the ferric oxides may be dissolved, causing the soil colors to become muted. In other soils, these mineral coatings can be stripped and the Fe translocated and re-precipitated, forming reddish-colored B horizons below less-red E horizons. The factors and conditions that regulate the dynamics of red Fe oxide minerals are complex. Nonetheless, it appears that soils with more phyllosilicate clays, less organic matter, finer overall textures, and higher pH values tend to retain secondary oxide minerals in situ and do not favor their translocation. Particularly, silicate clays can form strong bonds with Fe oxides (Duchaufour and Souchier 1978, Pedro *et al.* 1978, Kaiser and Wilcke 1996). Plus, in soils with large amounts of iron oxide minerals, and in finer-textured soils, translocation processes will be unable to form eluvial zones because iron oxide cutans (ferrans) are too numerous and too thick to strip, thereby preventing formation of an E horizon (Fig. 13.28). As a result, the soils are red-brown throughout; in older literature these soils were called *Braunerde*. This topic is discussed more thoroughly in the Podzolization section of this chapter.

Pedogenesis in the Humid Tropics: Oxisols

Highly weathered soils of the humid tropics often classify as Oxisols (Committee on Tropical Soils 1972, Beinroth *et al.* 2011). Oxisols are most common in the tropical forests of central South America and Africa. By definition, Oxisols must usually have an *oxic horizon* – a subsurface mineral horizon, usually designated Bo, of sandy loam or finer texture, with low cation exchange capacity (CEC) and few weatherable minerals (Soil Survey Staff 1999). In some cases, the presence of a kandic horizon alone is acceptable to classify a soil as an Oxisol. The genetic concept of the oxic horizon involves extreme weathering and a long period of pedogenesis under a hot, humid climate. Intense weathering and deep leaching are everywhere evident in this climate. Values of CEC, effective cation exchange capacity (ECEC), and extractable P; amounts of weatherable minerals, silica, and exchangeable base cations; as well as

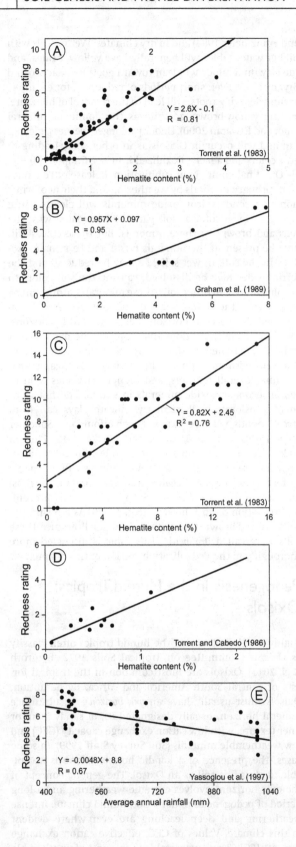

divalent/monovalent cation ratios are all very low (Tanaka *et al.* 1989), whereas the amounts of oxide clays – the most weathered kind – are high (Fig. 13.29). Silt contents are also low, because of the strong weathering of primary minerals and an associated process – desilication (van Breeman and Buurman 1998). In readily weathered parent materials and when climatic conditions are favorable, oxic horizons can form in soils on young surfaces and over a relatively short period (Soil Survey Staff 1999), but generally they require considerable lengths of time to develop. Thus, most Oxisols are highly polygenetic (Pfisterer *et al.* 1996, Muggler and Buurman 2000, Muggler *et al.* 2007).

Because of the intense weathering and leaching, Oxisols are acidic, and the clay mineralogy of oxic horizons is dominated by low activity clays, including Fe and Al oxides (Table 13.5). The clay mineralogy (see Chapter 4) is typically dominated by gibbsite, hematite, goethite, maghemite, and Ti minerals (mainly ilmenite and anatase). Goethite (α-FeOOH), indicated by yellowish colors (2.5Y–10YR), when hematite is not present to mask it, and hematite (α-Fe$_2$O$_3$), which imbues deep reddish colors (2.5YR–5R) even when present in minor amounts, are the most common clay minerals (Schmidt-Lorenz 1977, Schaefer *et al.* 2008). Goethite is considered the most stable mineral form and is widespread. Kaolinite, a 1:1 clay, is usually the only phyllosilicate mineral of any abundance (Ferreira *et al.* 2010). Gibbsite (γ-Al(OH)$_3$) is the most common aluminum oxide clay mineral. Its abundance is extremely variable, from a complete absence to >50% in some red Oxisols weathered from mafic rocks. Smectites are only observed in waterlogged sites where the potential to remove weathering by-products is low. Maghemite (γ-Fe$_2$O$_3$) is present in soils derived from basic or ultrabasic rocks (Kämpf and Schwertmann 1983a, Buol and Eswaran 2000). The silt and sand fractions are generally dominated by quartz, along with small amounts of other resistant minerals such as anatase, zircon, and rutile, depending on the parent material. Most coarse fragments are coated with sesquioxides.

Fig. 13.27 Scatterplots employing the redness rating of Torrent *et al.* (1983). In most studies, only the B horizon color is reported because of the masking effects of organic matter in the A horizon. (A) Relationship between hematite content and the redness rating for some Alfisols, Inceptisols, and Ultisols from Europe. (B) Relationship between hematite content and the redness rating for some Ultisols and Inceptisols from North Carolina. (C) Relationship between hematite content and the redness rating for some Oxisols and Ultisols from Brazil. (D) Relationship between hematite content and the redness rating for some Xeralfs and Xerepts in Spain. (E) Relationships between redness rating and mean annual precipitation for some soils in Greece.

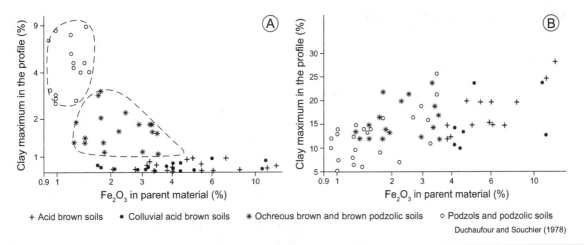

+ Acid brown soils   • Colluvial acid brown soils   ∗ Ochreous brown and brown podzolic soils   ○ Podzols and podzolic soils

Duchaufour and Souchier (1978)

Fig. 13.28 Scatterplots showing relationships between soil characteristics and Fe content of the parent material. (A) Index of podzolization intensity (which is normally low in brown-red, rubified soils) versus Fe_2O_3 contents of the parent material. (B) Clay content at the Bt_{max} versus Fe_2O_3 contents of the parent material. Note that, in this figure, Fe_2O_3 does not refer to hematite, but to the total Fe concentration reported as an oxide equivalent.

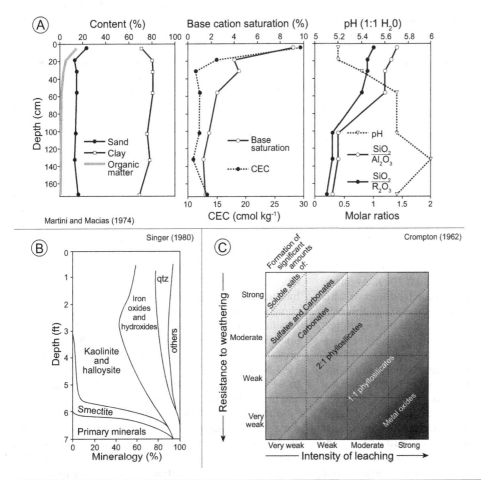

Fig. 13.29 Chemical and mineralogical data for some Oxisols. (A) Depth functions of various solid-phase components for a Typic Hapludox from Costa Rica, formed in volcanic materials on an old, alluvial terrace. The current vegetation is savanna and the mean annual rainfall is ≈260 cm, with a 2 to 3 month dry season. (B) Mineralogical changes, with depth, in an Oxisol formed on volcanic rocks. (C) General relationships among weathering resistance and leaching intensity for the various secondary mineral families found in soils (see Chapters 4 and 9).

Table 13.5	Common iron- and aluminum-rich compounds and minerals in soils
Mineral or compound	Chemical formula
Goethite	α-FeOOH
Hematite	α-Fe$_2$O$_3$
Ferrihydrite	5Fe$_2$O$_3$ • 9H$_2$O
Maghemite	γ-Fe$_2$O$_3$
Lepidocrocite	γ-FeOOH
Gibbsite	γ-Al(OH)$_3$
Amorphous ferric iron hydroxide	Fe(OH)$_3$

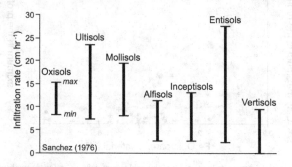

Fig. 13.30 Ranges of infiltration rates for the soil orders found in Puerto Rico.

Oxisols – *Latosols* in older terminology (Schaefer *et al.* 2008) and Ferralsols in the World Reference Base (FAO-UNESCO 1998) – normally have thick, deep profiles. They are reddish, or infrequently yellowish, usually depending on the type of Fe oxide mineral that is present. Horizon boundaries are gradual and of low contrast, especially in the subsurface, where the only differentiating characteristics may be subtle changes in structure or consistence. Surficial (A) horizons are strongly acidic and tend to be low in organic matter (Bennema 1974). Deep horizons have even less organic matter than the A horizons, making it difficult to differentiate them (Martini and Macias 1974). The thin character of Oxisol A horizons is mainly due to the rapid faunal consumption or microbial decomposition of organic materials in the hot environments.

Another important attribute is that, in many soils with oxic horizons, the clay content is relatively constant with depth, indicating little or no clay mobility. Lessivage is not a dominant process in many humid tropical soils, and where it does occur the translocated clay does not move into any one preferential depth zone, as it does in other soils (Eswaran and Bin 1978a, Soil Survey Staff 1999). Argillans are almost impossible to identify in the field in soils with kaolinitic or oxic mineralogy (Beinroth 1982). If observed in thin section, they tend to be thin and degrading. In the lab, oxide clays are not readily dispersed, probably because of cementation by sesquioxides, and thus in situ they tend to form flocculated aggregates that are relatively immobile within the profile. This tendency makes lab technicians who work on Oxisols steaming mad, and for the field scientist, it makes the soils feel and appear to be much less clayey than they really are. Some actually feel quite sandy. The oxide clay mineralogy often gives rise to a "uniquely tropical" type of friable and porous soil structure, which Buol (1973) called fine granular oxic structure (Moberg and Mmikonga 1977, Soil Survey Staff 1999). The fine crumbs, <2 mm in size, are composed of quartz sand grains held together by clays and sesquioxides (Escolar and López 1968, Schaefer 2001, Schaefer *et al.* 2004, 2008). The crumbs can adhere to each other and form stable conglomeratelike

masses that approach 5 mm in diameter, facilitating high infiltration capacities. This tendency minimizes runoff and reduces erosion on these soils (Van Wambeke 1962; Fig. 13.30). Unlike the granular structures of temperate climate soils, this type of structure is common in the B horizon, as well as the A. Bulk densities are generally low, commonly near 1 g cm^{-3}.

Another artifact of this unique clay mineralogy and structure centers on plant-available water capacity. Oxisols often have low water-holding capacities because most of their pores are either very large *between* the granules (and thus do not retain much water against the forces of gravity) and/or very small *within* the granules (and thus retain water at too great a tension to be plant-available). On the other hand, microaggregates often play an important role in some Oxisols, because they retard leaching of base cations from the inner aggregate (Moura Filho and Buol 1976, Schaefer *et al.* 2008).

Despite their widespread occurrence (Fig. 13.31), when compared to almost all other soils (Gelisols excepted), comparatively little is known about tropical soils and tropical soil landscapes. Soils of tropical landscapes are spatially diverse and quite variable (Sanchez 1976, Richter and Babbar 1991). Indeed, perhaps their main commonality is not pedologic but climatic – they lack seasonal temperature cycles (Buol 1973). Another unifying attribute of tropical soil landscapes, at least upland landscapes, is their red colors, which are due to the presence of hematite (Ardiuno *et al.* 1984, Rebertus and Buol 1985, Bech *et al.* 1997, Buol and Cook 1998; Tables 2.4, 13.5). In wet, low-lying areas, the iron oxide mineralogy changes and the soils take on yellow-orange colors.

Although tropical soil landscapes are just as varied as temperate soil landscapes, most of the Oxisols on them are acidic and infertile (Buol 1973). The nutritive status of many tropical soils can be characterized as one in which relatively large concentrations of organically bound nutrients exist in the A horizon, while low concentrations occur in the subsoil and on clay mineral exchange sites (Stark and Jordan 1978, Buol and Cook 1998). Especially problematic for crop productivity in these soils are the low levels of exchangeable P, not only because they have

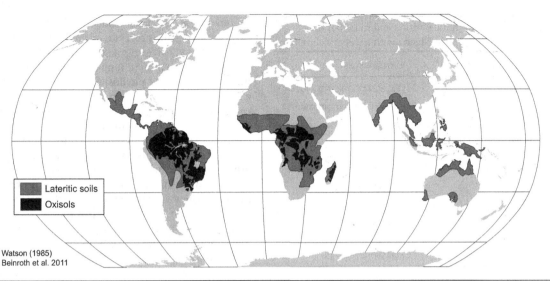

Fig. 13.31 The global distributions of Oxisols and Lateritic soils.

low P concentrations naturally but also because the sesquioxides present are capable of strongly fixing P, making it insoluble and unavailable for plant uptake (Kamprath 1973, Bigham *et al.* 1978, Ferreira *et al.* 2010). Additions of fertilizer or organic matter are only short-term solutions for crop production, for any added P is soon irreversibly adsorbed, organic matter is mineralized, and the soils are again infertile. Nonetheless, where additions of fresh parent material, such as alluvium, colluvium, loess, or volcanic ash, are present, fertility in Oxisols can be high. Indeed, atmospheric inputs have been found to be very important to many tropical soils. This type of "natural" fertilizer varies seasonally and over long-term cycles (Swap *et al.* 1992, Chadwick *et al.* 1999, Derry and Chadwick 2007, Pett-Ridge *et al.* 2009).

Pedogenic processes in the humid and subhumid tropics are influenced by several factors. Abundant rainfall (at least at some time in the year) facilitates vigorous plant growth, strong *chemical weathering*, and deep percolation and leaching of weathering products, leaving behind residual concentrations of kaolinite and sesquioxides. High temperatures speed up the rates of both inorganic chemical and biological reactions. On these landscapes, ample *time* has passed for many persistent but slow processes to have run to completion; many soils exist in a near-steady-state condition (Shuster *et al.* 2012). Many Oxisols have developed in old parent materials that were weathered and then transported from other locations before the current cyle of pedogenesis (Tavernier and Eswaran 1972, Beinroth 1982). Oxisols on in situ parent materials usually rest on thick sequences of saprolite (Lepsch and Buol 1974, Lepsch *et al.* 1977b), although this is certainly not always the case. On stable surfaces, weathering rates exceed erosion rates, effectively ruling out the exposure of new bedrock and nutrients. Pedogenesis in these kinds of environments also commonly includes

active *biomechanical agents* such as termites and ants (van Wambeke 1992; see Chapter 11). Such fauna not only mix the soil, blurring horizons, but also mine the saprolite that often occurs below and transport parts of it into the profile; Ferreira *et al.* (2010) referred to them as excavating fauna. Together, these factors combine to create a situation involving rapid mineralization of organic matter, coupled with intense and nearly complete weathering of primary minerals (De Villiers 1965, Beinroth *et al.* 2011). Al^{3+} and Fe^{3+} cations are everywhere in these soils; they are relatively insoluble and tend to remain in situ, while soluble weathering products are leached from the profile. Indeed, because of the high rainfall in the humid tropics, the potential to translocate all but the most insoluble cations from the profile is large. Even silica may be leached from the profile: a process called *desilication* (described further later) (Scholten and Andriesse 1986, Buol and Eswaran 2000, Oh and Richter 2005, Muggler *et al.* 2007, Yaro *et al.* 2008). Nonetheless, Kleber *et al.* (2007) provide evidence that silica can also be biocycled in some Oxisols.

Buol and Eswaran (2000) described three main types of landscapes where Oxisols dominate: (1) On stable slopes, where acidic, igneous rocks are slowly weathered into soils rich in low-activity clays. As expected, thick saprolite underlies these soils and their mineralogy largely reflects the bedrock. (2) On younger landscapes formed from readily weatherable, basic rocks. Here, Oxisols are (again) on the most stable surfaces, or sometimes developed in weathered alluvium. Andisols and Inceptisols are associated with volcanic materials, if present. (3) On late Tertiary high plateaus with deep weathering profiles (Du Preez 1949, Wright 1963). Oxisols on these plateaus typically have a thick, resistant plinthite cap (described furtherlater) (Thomas 1974; Table 13.6). Nearer the dry edges of the plateaus, the cap is even thicker and harder, because the edge experiences

Table 13.6	Geomorphic position and soil types on a landscape in Nigeria with a wet-dry climate
Landscape position	Dominant soils
Plateau top	Typic and Rhodic Kandiudox
Sandstone escarpment, plateau edge	Ustoxic Quartzipsamments
Lower, eroded hills and uplands (off the plateau)	Ustic Kandihumults, Typic Kanhaplustults, Typic
	Plinthaquults and Plinthustults, Typic Plinthohumults
Floodplains	Aeric Endoaquents
River's edge	Oxyaquic Quartzipsamments

Source: Gobin et al. (2000).

drying most often and ferrous iron may migrate laterally to the edge, oxidize, and precipitate there (Hunter 1961). Because the plinthite cap is rock-hard, i.e., *petroplinthite*, soils at the plateau edge are shallow and rubbly (Alexander and Cady 1962). As forests colonize the sideslopes, organic anions complex the ferric iron, allowing much of it to be translocated to footslope positions, where it may recement the rubble into plinthite. Likewise, iron-rich groundwater may penetrate the rubble on footslopes and assist in its recementation (Sivarajasingham *et al.* 1962).

In most Oxisols, the residual Fe and Al can precipitate to form oxide, oxyhydroxide, or hydroxide clays that, if dried, can become cemented into a bricklike substance of various forms and with various names: laterite, plinthite, and ferruginous nodules, and concretions (Livens 1949). And that is where this discussion now turns.

Laterization and Latosolization

Clearly, tropical soils remain poorly understood (Sanchez 1976, Hartemink 2004). Confounding this problem is a plethora of confusing and inaccurate terminology that has been perpetuated for years, much based on inference and generalizations rather than on systematic observations and science. Many older textbooks and publications on tropical soils simply refer to them as red soils, red loams, or red earths (Greene 1945). Early twentieth century agronomists would speak of "rubber soils" or "coffee soils," reflecting a bias toward agronomy rather than pedology. Later, terms such as "laterite" and "latosols" emerged, based on the travels of a few scientists to tropical regions (Sherman and Alexander 1959, Martini and Macias 1974, Cline 1975; Fig. 13.31). These terms, however, meant different things to different audiences. In this section, we will define all these terms, while also moving the discussion toward the terminology of modern pedology.

Laterization and *latosolization* are older terms used to describe pedogenic processes of the humid tropics, usually beneath broadleaf evergreen forest (Kellogg 1936). Both concepts revolve around long-term weathering of primary minerals, as well as phyllosilicate clays. Silica is leached, e.g., Figure 13.32, leaving mainly Fe and Al cations, which combine with oxygen to form iron- and aluminum-rich clay minerals such as hematite, goethite, and gibbsite. In *laterization*, Fe compounds *migrate* into and out of the soil, leading to concentrations at some sites and depletions at others (Fanning and Fanning 1989). Much of the Fe probably originates in sites outside the local profile, e.g., the saprolite below, or upslope locations. Therefore, processes that lead to Fe mobility are paramount in laterization, with the movement occurring primarily within wet, water-logged soils as ferrous Fe (Fe^{2+}), and in all soils via complexation with soluble organic acid anions (Wood and Perkins 1976). In such soils, aluminum is relatively immobile; that is why some tropical soil classification systems stress Fe contents and distributions. Evidence for translocation of Fe is taken from many sources, not the least of which is the great divergence in Fe contents across even short distances in plinthite and soils undergoing laterization (Fig. 13.33). Alexander and Cady (1962) pointed to iron-cemented quartz in a weathered sandstone as evidence that the Fe had to have been carried to the soil in solution, as Fe^{2+}.

Although redox processes undoubtedly play a role in laterization, details of the processes by which Fe accumulates in the soil profile are still unclear. Preferred sites of Fe accumulation may occur in certain parts of the profile where, for instance, oxidizing conditions exist. Iron oxide concretions may be found in association with quartz gravels, implying that the ferrous Fe in the groundwater became oxidized and precipitated when it encountered gravel layers that were more oxygenated (Livens 1949). Some red, Fe-rich zones tend to be sandier, again suggestive of increased oxidation (Wood and Perkins 1976) and a redox mechanism of formation.

In contrast to laterization, *latosolization* refers to only the residual accumulation of Fe oxides and other weathering products. It includes processes that lead to the formation of oxic (Bo) horizons, common to Oxisols and soils formerly known as Latosols (Martini and Macias 1974, Fanning and Fanning 1989). The term *ferralitization* is a synonym (van Breeman and Buurman 1998; Table 13.1). In latosolization, *residual* sesquioxides accrue, as base cations and silica are leached from the profile, via long-term weathering in a hot, humid climate. Over time, the soils become more mineralogically mature, as weatherable minerals are lost. Translocation of Fe from other parts of the landscape, as occurs in laterization, is assumed to be minimal. Although some of the base cations released by weathering are bio-cycled, leading to slightly higher concentrations in the A horizon than at depth, most are leached and leave the soil system, leading to acidic conditions. The residual material that remains is rich in sesquioxides of Fe, Al, Ti, and Mn

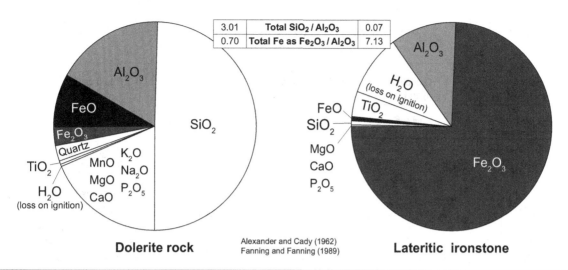

Fig. 13.32 Chemical composition of a dolerite rock parent material and the lateritic ironstone that has formed from it, via laterization, in Guyana. All values are in mass percentages. In the ironstone, the increase in water lost on ignition occurs because of the hydroxyls in the mineral structures of gibbsite, goethite, and kaolinite. These are converted to water during heating, which is necessary for the chemical analysis: Two structural OH groups → one structural O + H_2O.

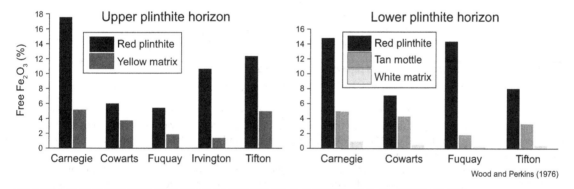

Fig. 13.33 Iron oxide contents of the various zones within the reticulate mottles that are typical of plinthite, for five different soil series in Georgia, United States. Four of the soils are Paleudults and one is a Fragiudult. Here Fe_2O_3 refers to total oxide-normalized Fe, not to hematite.

oxides; oxyhydroxides and hydroxide minerals; as well as heavy elements like Ni, Zr, and Cr (Beinroth 1982). Clay minerals such as kaolinite can readily neoform where Si concentrations are sufficiently high.

An important side effect of latosolization, with its loss of soluble materials, is collapse, or as McFarlane and Bowden (1992) put it, "the land shrinks like a rotten apple." Many Oxisols have developed on saprolite or transported sediment that has collapsed from its original parent material, and as a result, insoluble metals or resistant minerals within the saprolite tend to become enriched and concentrated in the soil and subsoil (Costa 1997, Retallack 2010). As a result, an important subsidiary process to latosolization is the mineralization of metals. Oxisols in general, and lateritic soils in particular, are often looked at as sources of valuable and

minable materials such as iron, aluminum (bauxite), kaolin, manganese, gold, nickel, copper, and phosphate (Freyssinet *et al.* 1989. Parisot *et al.* 1989, Butt 1990, Hale and Porto 1994, Porto and Hale 1995, Costa *et al.* 1999, Chisonga *et al.* 2012). Mineral neoformation in tropical swamp environments long after the original rocks were exposed to weathering leads to the formation of high-grade kaolin and refractory bauxite ores (Costa 1997). In locations where the soil cover is thick, analysis of metal contents in termitaria has been useful as a way of identifying metal enrichment in the underlying subsoil and saprolite (Arhin and Nude 2010).

Evidence for purely residual Fe accumulations in these kinds of soils in West Africa is given by Alexander and Cady (1962). Here, upland soils that were developed on basalt residuum had much thicker laterite accumulations

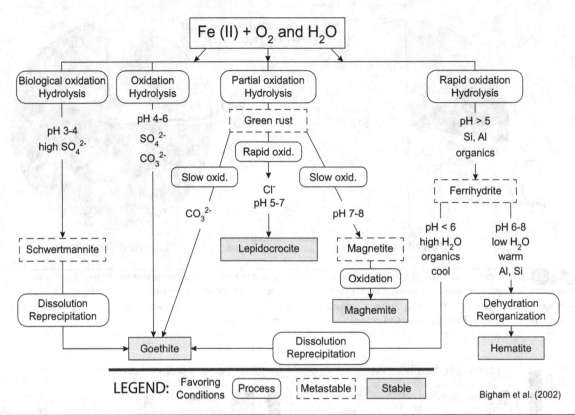

Fig. 13.34 Schematic diagram of the mineral synthesis pathways for iron oxides in soils.

than soils developed on granite, because granite has fewer Fe-bearing minerals. A significant modification to this genetic model was provided by Brimhall *et al.* (1988, 1991). Their data suggested that many upland tropical soils are enriched in mineralogically mature sediment from *eolian sources*, suggesting that only in the lowermost part of many Oxisol profiles is most of the Fe truly residual. Faure and Volkoff (1998) reported similar findings and suggested that the neutral pH values of some upland tropical soils may be due to inputs of Ca-rich dust from nearby deserts. Thus, it appears likely that not all of the Fe and Al in many Oxisols is truly residual, especially if they are downwind of eolian source areas.

Associated with laterization and latosolization is the almost complete mineralization of SOM in the B horizons. Oxisols have long been reported to contain low amounts of OM, although many contain more than their colors suggest (Gobin *et al.* 2000). The red colors, brought about by the presence of hematite, mask the organic matter in these soils. Buol (1973) even suggested that the organic matter of some Oxisols may be almost colorless. Additions of litter from vegetation does little to add organic matter to the soil, because most of what falls is quickly decomposed, and the constituents are either taken up by flora, washed from the soil, or leached. Organic matter does not form strong

complexes with clay because mineral surface area is low. In addition, the low CEC values of the oxide clays preclude significant bridging between organic anions and clay by polyvalent metals like Ca.

As mentioned previously, latosolization often leads to the neoformation of oxide clays. Perhaps because of the large amounts of residual materials present, or because of the long periods involved, the Fe and Al compounds in these soils are typically highly crystalline. That is, they have had ample time to neoform into minerals with discrete ordered structures, and these minerals are difficult to weather. Oxide clays are the dominant pigmenting agents in saprolites, producing the browns and reds typical of tropical soils and saprolites, forming where free Fe^{3+} is abundant. Iron is released by weathering and initially precipitates as poorly crystalline Fe oxides (in oxidizing situations) (Hurst 1977). Eventually in moist soils, and immediately in the dry ones, the Fe oxides dehydrate and crystallize as goethite, lepidocrocite, maghemite, or hematite (Fig. 13.34). Hematite formation tends to be favored in environments that are warmer and drier, and in soils with lower organic matter contents, because organic acids chemically complex with Fe, discouraging inorganic crystallization to hematite (Schwertmann *et al.* 1982; Fig. 13.35). Hematite is more likely to form in warm, dry conditions because dehydration is a necessary

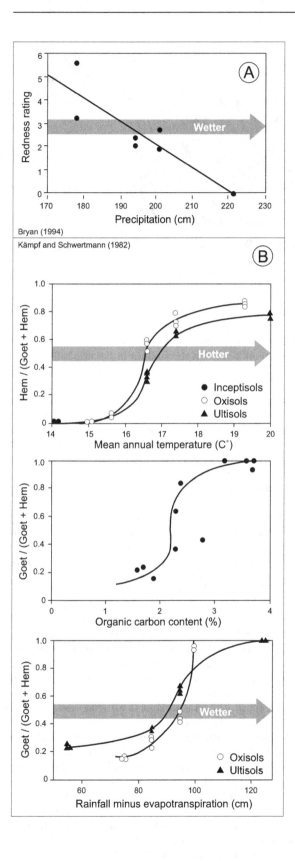

Bryan (1994)

Kämpf and Schwertmann (1982)

process in its formation, and because at warmer sites soil organic matter contents are more likely to be low. Goethite is more stable in humid, cooler environments; in organic-rich soils; and in acidic soils where organic acids are prevalent (Kämpf and Schwertmann 1983a).

Desilication

Central to the concept of latosolization, and necessary to the process of laterization, is the loss of silica from the profile of tropical soils. *Desilication* occurs when water that is undersaturated with silica percolates through a soil at the correct pH; it is facilitated by freely draining conditions in hot, humid environments (Oh and Richter 2005, Yaro *et al.* 2008). Often desilication occurs concomitantly with the loss of base cations (Curi and Franzmeier 1984). Because the solubility of silica increases in warmer temperatures, soil solutions in the tropics are more likely to be silica-undersaturated than are those elsewhere. As silica is lost from the profile, residual Al and Fe remain behind, where they can crystallize into oxide clays. The amount of silica removed from the soil depends on residence time of the water around the silicate mineral, and on mineralogy (Buol and Eswaran 2000). Therefore, like most translocation processes, desilication decreases in intensity with depth, as percolating water picks up silica and becomes less aggressive.

An important effect of desilication is a lowering of the Si:Al ratio in soils, with a concomitant enrichment of 1:1 and oxide clays (Buol and Eswaran 2000). Where the silica content is low, synthesis of 2:1 clay minerals is virtually impossible (Soil Survey Staff 1999). Given the strong desilication that occurs in well-drained soils, however, even kaolinite may be unstable and can weather. Desilication also results in collapse of saprolite and soils on uplands. Alternatively, in waterlogged bottomlands where silica-rich waters accumulate, secondary silicate minerals can precipitate (McFarlane and Bowden 1992). Evidence of desilication in humid tropical soils is provided by $SiO_2:R_2O_3$ ratios (Martini and Marcias 1974). In Figure 13.29, it is evident that the $SiO_2:R_2O_3$ ratio of the Udox is extremely low, whereas the related soils all had $SiO_2:R_2O_3$ ratios less than 3.0, indicating that they contain high amounts of kaolinite and gibbsite.

Desilication and long-term weathering can lead to high amounts of Al-rich minerals, although in some tropical

Fig. 13.35 Relationships between iron oxide mineralogy and various climate or soil climate factors. (A) Redness rating (Torrent *et al.* 1983) of soils versus precipitation in the Blue Ridge Mountains of western North Carolina. (B) Hematite or goethite content (as a percentage of total hematite + goethite content), as affected by various climatic or pedogenic variables, for some Brazilian soils.

soils even aluminum is soluble (McFarlane and Bowden 1992). Aluminum in Oxisols can lead to low pH values and, if taken to extremes, biotoxicity. Liming is not a long-term solution to the low pH values, because the Al is such a strong pH buffer (Martini and Marcias 1974). Soil and subsoil residues that are especially high in Al are often mined as bauxite (Ahmad *et al.* 1966, Scholten and Andriesse 1986).

Latosolization converts rock to saprolite (see Chapter 10). Unlike saprolite, the soil profile has *soil structure* and *soil horizons*, which in tropical regions are formed through desilication and accumulation of residual oxides, but also through the chemical *and* biomechanical actions of plant roots and soil fauna. Soil fauna disrupt and mix the weathered rock, moving some of it to the surface. Termites, in particular, mine parts of the subsurface and move those components to the surface, creating a thick biomantle, often with an underlying stone line composed of rocks or secondary concretions and ironstone nodules that are difficult to weather (see Chapter 11). Additionally, lessivage and melanization are at work to form clay-rich kandic horizons and organic matter–enriched A horizons. Where bioturbation is strong, existing kandic horizons that are rich in illuvial clay may be destroyed and converted to red oxic horizons.

Many tropical soils, especially those that had already developed or were considered to have the potential to develop laterite, would have been formerly classified as Latosols. Latosols were defined as zonal soils in the 1938 system of classification (as amended in the 1950s) with these characteristics: (1) low $SiO_2:R_2O_3$ ratios in the clay fraction, (2) low cation exchange capacities, (3) an abundance of low activity clays, (4) low contents of weatherable minerals, (5) low amounts of soluble constituents, (6) a high degree of aggregate stability, (7) no essential horizons of accumulation or illuviation, (8) thin A horizons, (9) low silt contents, and usually (10) red colors (Kellogg 1949, 1950). Latosols could contain laterite. The Latosol concept has been largely subsumed into the Oxisol order and the oxic horizon. "Latosol" is, however, still a popular and useful taxonomic term in some parts of the world, especially Brazil (Tanaka *et al.* 1989).

Laterite and Plinthite

The origin of the term "laterite" can be traced to an 1807 report by Francis Buchanan, a Scottish scholar (Beinroth *et al.* 2000). In his travels to India, Buchanan described soft, red, Fe-rich soil material that hardened upon drying, rendering it useful as a construction material (Fig. 13.36). He called this material brickstone or laterite, from the Latin *later*, brick (Sivarajasingham *et al.* 1962). Buchanan's original definition referred to "an iron-oxide-rich, indurated, quarryable slag-like or pisolitic illuvial horizon, developed in the soil profile" (van der Merwe 1949). The laterite observed by Buchanan was soft enough to be cut into blocks with a knife, but upon exposure to air it quickly became hard. Presumably, this hardening is due to the dehydration of hydrous Fe oxides and small amounts of

Fig. 13.36 A laterite brick quarry near Kano, Nigeria. Photo by J. Olson.

amorphous silica that exist in the plinthite (van Breeman and Buurman 1998). This laterite stone was found mainly as a sheetlike cap on the summit of basaltic hills and plateaus in the Kerala state of India, where even today it is exploited for construction materials. Eventually, other materials were lumped into the category of laterite, too. For example, hard gravel pisoliths, blocks, or ferralitic concretions that stand out as low humps in the terrain or lie buried a few meters below the surface (Horbe and da Costa 2005, Adzmi *et al.* 2010) have been called laterite. These materials consist of gravel-sized concretionary nodules set within a matrix of silt and clay, with thicknesses between 1 and 5 m (Alexander *et al.* 1956).

Some points were, however, undisputed. Laterite soils often had thick sequences of saprolite below them. Thicknesses greater than 200 m have been recorded (Kroonenberg and Melitz 1983). Many tropical soil landscapes are underlain by thick, weathered residual soils on stable uplands, often usually associated with erosion and redeposition cycles over long periods of geologic time (Scholten and Andriesse 1986, Bitom *et al.* 2004, Burak *et al.* 2010). Soils on these landscapes also were observed to have one or more stone lines at depth (Ruhe 1956a, Dijkerman and Miedema 1988). Laterite is common here, often acting as a resistant cap on upland surfaces (Shuster *et al.* 2012). In many dissected tropical landscapes, a laterite upland is all that remains of an old, often Tertiary-aged surface (Kroonenberg and Melitz 1983, Dijkerman and Miedema 1988). Much of the geologic literature simply refers to these caps as *duricrusts* – a more generic name for a hardened, near-surface crust, whatever the cementing agent may be. Little lateral transport of material occurs, and if Bt horizons do occur, the soils are Kandiudults or Kandiudalfs. But on sideslopes, where slope processes and bioturbation are more influential, soils can be considerably different, as laterite is replaced by creeping biomantles above stone lines.

(Stone line formation and its role in tropical soil geomorphology are discussed in more detail in Chapter 14.)

A debate arose as to whether laterite was a type of *soil* or a type of *material* (Kellogg 1950). In the original meaning of the term, laterite soils contained a laterite horizon as a *part of the soil profile* (van der Merwe 1949). Thus, laterite soils, or at least parts thereof, came to be characterized as forming hard, impenetrable, and often irreversible pans when dried. Adding to the confusion, a variety of *materials* with various compositions and origins *also* had this lateritic tendency, and they had been called laterites, too (Kellogg 1949, Carter and Pendleton 1956). These materials range from the pedogenic iron cappings on the plateaus of southern India to whole soils in the humid tropics to iron-rich breccias (rocks) and slopewash accumulations (colluvium). So what, exactly, *was* laterite?

Another source of confusion was the abundance of names for the same cemented, iron-rich material; most called it laterite, but others have referred to this material as *ironstone, duricrust,* or *ferricrete*. Because laterite is so highly variable in hardness, where should the line be drawn? Using a hardness criterion, unfortunately, made it nearly impossible for soils to be classified as lateritic, at the time of exposure, i.e., in a freshly dug soil pit. Field soil scientists and engineers were required to make a field call about whether or not the soil contained laterite based on their experience, not on the morphology exposed in the pit. To prevent this problem, some relied on definitions based on chemical criteria. Laterites contain mixtures of hydrated oxides of aluminum and iron, often with a small percentage of other oxides, chief among them Mn and Ti oxides (Varghese and Byju 1993; Fig. 13.32). Van der Merwe (1949) suggested that the term laterite should be further restricted to materials with free alumina, whereas most definitions stressed the ratio ($SiO_2/[Fe_2O_3 + Al_2O_3]$), also written as SiO_2/R_2O_3. Soils with SiO_2/R_2O_3 ratios between 1.33 and 2.0 were taken as indicative of lateritic soils, while those with ratios >2.0 were indicative of nonlateritic soils. (Care must be taken to exclude data on quartz silica from the SiO_2 part of the ratio.)

For a long time (McNeil 1964), it was thought that all tropical lateritic soils would, if cleared of vegetation, harden irreversibly and quickly become worthless brick pavement – essentially the definition of laterite. And so, for considerable time, the term "tropical soils" was taken to mean red soils high in iron that can harden irreversibly. (Later, this notion was shown to be inaccurate, at least for some tropical soils.) Part of this confusion stemmed from the fact that soil science began in the midlatitudes; information about tropical soils was minimal. Upon visiting the tropics, midlatitude soil scientists were shown extreme and unique examples of soils, e.g., laterites, and carried these ideas back to their countries (Sanchez 1976). Misguided ideas that laterites and red, bricklike soils dominated the tropics were disseminated, while vast areas of the tropics that were covered with luxuriant forests and friable soils, similar to those found in midlatitudes, were largely ignored. For example, in the 1938 system of soil classification (Baldwin *et al.* 1938), only one zonal soil order was defined for the forested tropical regions: lateritic soils of forested warm-temperate and tropical regions. By 1955, a new name, Latosols, had been proposed for soils that are developing or can develop lateritic characteristics (Kellogg 1949, Cline 1955, Sherman and Alexander 1959). *Laterization* became known as a suite of processes whereby soils developed laterite. Finally, as more and more soil scientists conducted research in the humid tropics, they reported that laterite was relatively uncommon (Hardy 1933). In 1949, Charles Kellogg advocated abandoning the term "laterite" in soil classification, and today, it is an obsolete term. Most pedologists now use other terms to describe soil materials that are cemented or enriched in Fe oxides. In Soil Taxonomy, the term *plinthite* (Greek *plinthos*, brick) refers to materials that previously would have been called laterite (Sivarajasingham *et al.* 1962, Daniels *et al.* 1978b, Soil Survey Staff 1999). Plinthite is an iron-rich, humus-poor mixture of kaolinitic clay with quartz and other minerals. It commonly displays dark red redox concentrations in platy, polygonal, or reticulate patterns (Wood and Perkins 1976). Plinthite changes irreversibly to an ironstone hardpan or to irregular aggregates when exposed to wet-dry cycles and heat from the sun (Aide *et al.* 2004). The World Reference Base system uses the term *petroplinthic horizon* or *petroplinthite* to describe plinthite that has become irreversibly hardened (Food and Agriculture Organization 1998). Moist plinthite is soft enough to be cut with a spade; after irreversible hardening, it is no longer considered plinthite but is called ironstone. Horizons with plinthite, usually B horizons, are given the suffix *v* (Table 3.2).

Plinthite is common in the lower sola of many Oxisols and in some Paleudults on old (Miocene to Pliocene) surfaces (Cady and Daniels 1968). It is usually dominated by goethite and hematite, with some kaolinite. Athough bauxitic laterites have abundant gibbsite and boehmite, many laterites contain virtually no free aluminum minerals. Geomorphically, plinthite is found in two settings: (1) as a continuous crust on flat uplands, where it acts as a hard cap that resists erosion (Alexander and Cady 1962), and (2) at footslopes and in seepage areas where reduced Fe in soil solutions encounters oxidizing conditions and precipitates.

Plinthite forms by *segregation* of iron species, much of which had probably been initially translocated into the site/horizon in soil water, and often in association with illuvial clay (Aide *et al.* 2004). The term segregation implies that iron has locally migrated within the soil, forming red, iron-rich, and white or light gray, iron-poor patches or splotches. In a fresh exposure, iron is clearly segregated in the form of soft, more or less clayey, reddish redox concentrations (see Chapter 14), but these concentrations are not considered plinthite unless enough segregation of iron has occurred to permit their irreversible hardening.

Some Fe may move also by capillarity, as water and dissolved ions in wet soil below is wicked into drier soil above. Plinthite formed in this manner was previously called groundwater laterite. It may *form in lowlands* such as alluvial plains and swamps, but later, as the landscape is dissected and the lowlands become plateaus, it may be widespread across *upland surfaces*, forming a resistant caprock (Fig. 10.8). Where the Fe in plinthite has a clearly reticulate pattern and the overlying horizons are eroded, the hardened material may take the form of nodules or *pisoliths*, as described long ago by Buchanan (Beinroth *et al.* 2000).

The deep, almost beet-red color of some plinthite is attributed to hematite (Schmidt-Lorenz 1977). Often, areas of intermediate composition, brownish yellow in color, also occur, intimately intermixed into the plinthite (Fig. 13.33). The reticulate nature of this segregation is a telltale indicator of plinthite. Generally, plinthite is thought to form in horizons that are saturated for some time during the year, further suggesting that redox processes are involved in its formation (Aide *et al.* 2004). Fe segregation in plinthite is often aided by microbial activity that regulates the redox potential during these (paleo)wet or waterlogged conditions (Gehring *et al.* 1994).

The potential for hardening of plinthite to ironstone is a key characteristic. The irreversible hardening of plinthitic materials is important to millions of people in the humid tropics who depend on the laterite bricks as roadbed material, for home construction, and for many other uses (Fig. 13.36). Obviously, it is important that a soil's capacity to harden be known before it is cultivated. Thus, people in the tropics will frequently remove subsoil material and force it to undergo several wet-dry cycles, to see whether it is potential roadbed or construction material, or whether it has agricultural potential (Buol 1973).

At a minimum, two mechanisms occur when soft plinthite hardens: crystallization of the Fe minerals, especially in the red mottled areas, and dehydration of the soil mass. Drying of the iron-rich soil matrix may cause crystallization of previously poorly crystalline iron oxides. Alexander and Cady (1962) noted that residual iron accumulations are usually not adequate to develop hard laterite; additions of Fe in solution are usually necessary to provide enough Fe to form hard petroplinthite. That is, Fe enrichment *beyond the residual enrichment* derived from in situ weathering appears to be prerequisite to the formation of ironstone; this process is now known as *plinthization* (Yaro *et al.* 2008).

Landscapes: The Three Phases of Warm-Climate Pedogenesis

A widely accepted paradigm for soil genesis in warm-hot climates was proposed in 1982 by Philippe Duchaufour. His three-phase

system of pedogenesis, although not technically a development series, can be viewed as such. Perhaps better, it could be seen as a climosequence of pedogenic processes from midlatitude to tropical climates (Table 13.1, see Figure on the following page). This three-phase pedoweathering package nicely ties together many of the concepts that have been discussed earlier in this book– weathering, clay mineralogy, lessivage, base cycling, latosolization, and desilication, among others. From Phase 1 to 3, soils show increased (1) weathering of primary minerals, (2) loss of silica, and (3) neoformation of clay minerals, formed from weathering by-products (Singh *et al.* 1998).

Phase 1, typical of Mediterranean regions and hot, wet-dry climates, involves *fersiallitization*. This process involves the inheritance and neoformation of smectitic clays, and the immobilization of iron oxides due to alkaline soil conditions (Ortiz *et al.* 2002). Because of the high pH conditions, the soils are not strongly weathered. Although this process suite can occur on any parent material, it is best expressed on porous, calcareous substrates, giving rise to reddish brown, terra rossa soils (see Chapter 10). The red colors indicate that these soils contain significant amounts of iron oxides. Little silica has been lost, base cation saturation remains high, and lessivage forms Bt horizons enriched in 2:1 clays. Many of these soils would classify as Udalfs or Eutrudepts. As the name implies, these soils retain high amounts of Fe, Si, and Al.

Phase 2 is *ferrugination* or *braunification* under a warm, humid climate. In this transitional phase, acidification and weathering increase in intensity, but some primary, weatherable minerals still persist. Desilication increases in intensity, and, as a result, some 1:1 clays, such as kaolinite, can neoform. Conditions are not yet right, however, for the formation of gibbsite. Base saturation is lower, but lessivage is still active. Plinthite and kandic horizons can form. These soils, typically Ultisols, are common on subtropical savanna and forest landscapes. Strong rubification – the reddening of the soil – occurs as Fe oxides, precipitated after weathering of primary minerals, are adsorbed to the phyllosilicate clays, particularly in the dry season.

Phase 3 is *ferrallitization* or *desilication*, typical of the humid tropics. At this point, the soils have become acidic and strongly weathered (Latham 1980). Of the primary minerals inherited from the parent material, only quartz remains. The soil has become enriched in Fe and Al oxides, because of desilication. As a result, gibbsite, kaolinite, and oxide clays neoform. Lessivage is not strong, as the clays have become increasingly resistant to dispersion. Many of these soils are Oxisols on thick sequences of saprolite.

The sequence shows that, in warm climates where water is ample for leaching, weathering and pedogenesis will run a predictable course. Along the way, soils will redden and the clay mineral mix will change, as silica is lost to the system.

Figure: Diagrammatic conceptualization of Duchaufour's (1982) three-phase weathering/pedogenesis sequence for humid and warm/hot climates, and the relative strengths of some key pedogenic processes for each stage. Although typical profiles are shown, many other morphologies are possible.

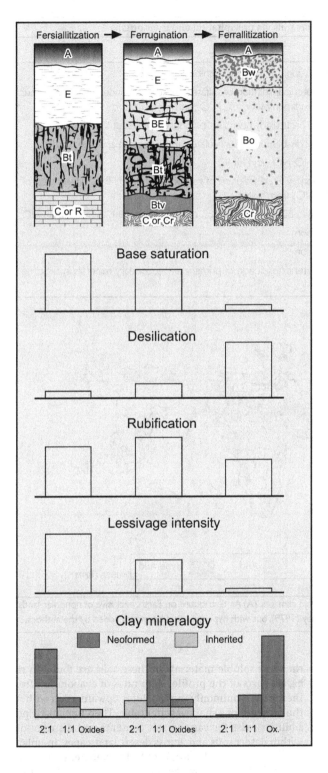

Pedogenesis in Dry Climates: Aridisols

In this section, we focus on pedogenesis in landscapes characterized by a marked deficiency of soil water in some or most of the year, i.e., deserts and grasslands (Fig. 13.37). Dry climate regions need not be hot; many cold deserts exist, and generally similar pedogenic processes occur there and in hot deserts.

The major pedogenic processes in dry climates are the accumulation of soluble materials in the profile, often enhanced by additions of dust (Ugolini et al. 2008). Blowing dust is common in most deserts (Bach et al. 1996). In contrast, the major processes in humid climates revolve around weathering and translocation of materials derived in situ. Leaching of soluble materials from the soil profile occurs in humid climate soils, but in dryland soils, translocation of soluble materials is restricted to the uppermost portions of the profile. Because of the overall deficiency in water, accumulation of those leached constituents occurs in the subsurface – or even at the surface, if percolating water is later wicked upward (Buol 1965, Khresat and Qudah 2006, Monger et al. 2011). In drylands, the extent and intensity of many pedogenic processes are less than in humid climates. Thus, weathering is slow or essentially nonexistent, and A horizons are thin and have low organic matter contents. Although chemical weathering and leaching do occur in dry climates, they do not occur in the whole soil profile.

It rains in all but the coldest and driest of desert climates. The main difference, again, is in the degree to which this rainfall and percolation affect soils. In desert soils, translocation of soluble materials (and even clays) through the profile does occur, but these substances may not necessarily be translocated *out of* the profile. After each rainfall event, some of the more readily soluble compounds are drawn into solution and translocated downward. But at some depth, the wetting front stops and the compounds are precipitated. Because many dryland soils only infrequently become flushed by deep wetting events, their B and C horizons continue to accumulate soluble compounds. As with illuvial clay in more humid climates, continued illuviation of soluble substances to a preferential depth in dryland soils can eventually result in the formation of a plugged hardpan. For this reason, taxonomic systems usually differentiate dryland soils on the basis of B horizon character.

Most translocation in dryland soils involves soluble materials, as they are most readily translocated during the short and infrequent wetting events that typify these settings. The exception to this statement are phyllosilicate clays, which may be translocated in suspension by "mud rains," especially when dispersed as a result of high Na salt concentrations. Thus, Bt horizon formation is different from the traditional manner (lessivage) in which they form in humid climates.

Many kinds of soluble materials are available in dryland areas (Table 13.7). Thus, the type of soluble material (salts, mainly) that are translocated in these soils is, first, a function of availability. How *deep* they are translocated is then a function of solubility, precipitation, and permeability; i.e., how deep the wetting fronts typically move.

Table 13.7.	Common soluble, chemical species carried to desert soils via precipitation and their origins

Constituents	Major origins
Chloride (Cl^-)	Sea water; dissolution of particulate chlorides
Sulfates SO_4^{2-}, HSO_4^-	Sea and lake waters; dissolution of sulfate particulates; condensation of gaseous H_2SO_4; sulfite oxidation by O_2, O_3, H_2O_2, and metal catalysts
Sulfites (H_2SO_3, HSO_3^-, SO_3^{2-})	SO_2 dissolution, particularly near sites of industrial emissions
Carbonates (CO_3^{2-}, HCO_3^-, H_2CO_3)	CO_2 dissolution; dissolution of particulate carbonates; sea and lake waters
Nitrites (HNO_2, NO_2^-)	Dissolution of NO, NO_2, and HNO_2
Nitrates (NO_3^-)	Nitrate oxidation by O_3; dissolution of gaseous NO^{2-} and HNO_3; dissolution of particulate $NaNO_3$ from soils
Ammonium (NH_4OH, NH_4^+)	Dissolution of gaseous NH_3; dissolution of particulate $(NH_4)_2SO_4$ and NH_4NO_3
Hydroxyls (OH^-)	Water dissociation
Active hydrogen (hydronium ion) (H_3O^+)	Water dissociation
Metal cations (Na^+, K^+, Ca^{2+}, Mg^{2+}, Fe^{3+})	Sea and lake waters; dissolution of primary and secondary minerals in airborne particulates

Source: Pye and Tsoar (1987).

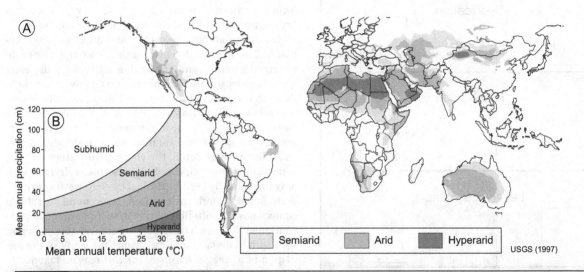

Fig. 13.37 Generalized limits of hyperarid, arid, semiarid, and subhumid climates. (A) As distributed on Earth, exclusive of nonpolar lands. (B) As based on mean annual temperature and precipitation. After Bailey (1979), but with hyperarid-arid boundary added by the authors.

Generally speaking, salts and soluble compounds have different mobilities and solubilities in soils:

$$Cl^- > SO_4^{2-} > HCO_3^- > CaSO_4 \bullet 2H_2O > CaCO_3$$

(chlorides > sulfates > bicarbonate > gypsum > carbonates).

In upland soils everywhere, the most soluble materials are translocated the deepest (Amit *et al.* 2010). In low-lying desert soils where the water table may be shallow, soluble materials in the groundwater are wicked upward by capillarity and deposited in the soil profile, meaning that the most soluble materials in these soils are found in the *highest* parts of the profile. High rates of evaporation from the surface commonly move solutes upward, even without the presence of a shallow water table. Therefore, many possibilities exist for varying salt contents and distributions within desert soils and across desert landscapes. In upland soils on the humid margin of a dry climate, only $CaCO_3$ may remain in the profile; other materials will have been leached. In soils that are a bit drier, carbonates and gypsum may remain within the profile. Here, the maximum gypsum concentration will typically be deeper, as gypsum is more soluble. Chloride salts are found in upland soils only in the driest deserts, and then only if there is a nearby source, e.g., sea water or saline groundwater. The rule of thumb is, as the climate grows drier, salts and soluble materials in upland desert soils become more common and closer to the surface.

Aridisols

Most dryland soils have aridic (or torric) soil moisture regimes, defined by the number of days a soil is dry, i.e., when water is held at tensions less than –1,500 kPa (Monger et al. 2011). Soils in an aridic soil moisture region do not have water available for plant growth for more than half the growing (warm) season, and during this same season plant-available water exists in the soil for <90 consecutive days. In other words, prolonged drought stress occurs during the growing season. Smith (1986) described the original concept of Aridisols as soils that were so dry that, if left unirrigated, they could only be grazed; they could not be successfully cultivated without irrigation.

The concept of the *Aridisol* order is centered on the aridic soil moisture regime. However, only about half of the soils in arid climates are Aridisols (Soil Survey Staff 1999, Monger et al. 2011). Entisols are particularly common there, on young sediments, e.g., dunes and recent alluvium, and on shallow, rocky sites. Many deserts have vast expanses of thin, rocky Entisols in mountainous areas. In order to qualify as an Aridisol, a soil must be within an aridic soil moisture region *and* have a diagnostic subsurface horizon, e.g., calcic, natric, argillic, cambic, and gypsic. Often, these horizons are relicts from a period in the geologic past when the climate was wetter and during which pedogenesis was relatively faster. In summary, Aridisols are soils of dry areas that show some signs of soil development and do not otherwise classify as Oxisols, Vertisols, or Andisols.

Calcification

Calcification refers to the translocation and accumulation of secondary calcium ($CaCO_3$) and magnesium ($MgCO_3$) carbonates in soils (Buol 1965; Fig. 13.38). It is a dominant process in many dryland soils, because (1) aridity, coupled with the low solubility of carbonate minerals, makes them difficult to leach (Southard 2000), and (2) carbonate sources, both external (dust) and internal (carbonate-rich parent materials), are widespread. Calcification is especially common and ongoing in most subhumid grassland soils (Sobecki and Wilding 1983, Gunal and Ransom 2006; Fig. 13.39). It is also found in humid climates where parent materials are rich in carbonates, and sola are clay-rich and only slowly permeable (Wenner et al. 1961, Schaetzl et al. 1996, Tornes et al. 2000). Eventually, as the climate becomes too humid and the soils too moist, carbonates are translocated completely out of the profile, leaving the soil system, and entering groundwater. No zone of carbonate accumulation exists in such soils. In short, wherever a source of calcium exists without adequate water and

Fig. 13.38 A Pachic Argiustoll, near Beaver, Colorado, showing accumulations of secondary carbonates in the subsoil. The abrupt boundary of the B horizon probably reflects a former erosional contact. Units on the tape are in centimeters. Photo by P. Schoeneberger.

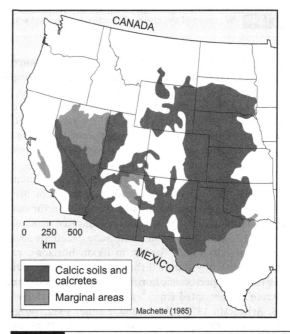

Calcic soils and calcretes

Marginal areas

Machette (1985)

Fig. 13.39 Distribution of calcic soils in the western United States. Marginal areas have discontinuous or poorly preserved calcic soils.

energy to translocate it out of the profile, evidence of calcification also exists (Stuart and Dixon 1973).

Because the dominant form of secondary carbonates in soils is $CaCO_3$, soils with abundant subsurface carbonates are referred to as *calcic soils*. In older terminology, these soils were called "Pedocals" (Marbut 1935, Jenny 1941a; Fig. 8.2). If the $CaCO_3$-enriched B horizon meets certain criteria, it can become a diagnostic *calcic horizon*

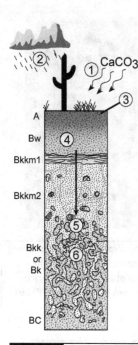

Calcification

① Carbonate ($CaCO_3$) rich dust blows onto soil surface

② Water in precipitation combines with atmospheric and soil carbon dioxide (CO_2) to form carbonic acid (H_2CO_3)

$$H_2O(l) + CO_2(g) \rightleftharpoons H_2CO_3(aq)$$

③ Calcium carbonate ($CaCO_3$) in (and on top of) the soil reacts with carbonic acid, rendering

$$H_2CO_3 + CaCO_3 \rightleftharpoons Ca^{2+}(aq) + 2HCO_3^-(aq)$$

④ Calcium (Ca^{++}) and bicarbonate (HCO_3^-) ions are translocated in the soil with percolating water

⑤ Dry conditions at depth lead to precipitation of secondary carbonates

$$Ca^{2+}(aq) + CO_3^{2-}(aq) \longrightarrow CaCO_3(s)$$

⑥ Deep wetting events may cause dissolution of secondary carbonates and translocation/precipitation at still greater depths

Fig. 13.40 Sequence of the chemical reactions associated with calcification.

(Soil Survey Staff 1999). Because the degree of carbonate accumulation is so important to interpreting dryland soil genesis, the NCSS has developed three different types of horizon nomenclature, which describe different levels of carbonate accumulation (Table 3.2). In general, calcic horizons are characterized by carbonate-coated pebbles or thin filaments of carbonate. These horizons are referred to as Bk horizons, if the carbonates occupy <50% of the horizon volume. As even more carbonates accumulate, the surrounding material commonly becomes cemented, forming a Bkk horizon. Eventually, the pores in the horizon fill and plug, forming a massive, indurated *petrocalcic* (Bkkm) horizon that resembles limestone (Machette 1985, Soil Survey Staff 1999). That is, in Bkkm horizons, carbonate has literally engulfed the preexisting soil matrix. Older terms for petrocalcic horizons include "caliche" and "calcrete" (abbreviated from "caliche concrete") (Sidwell 1943, Aristarain 1970, Watts 1980, Blümel 1982, Wright and Tucker 1991, Dixon 1994b). Birkeland (1999) and other pedologists working in the western United States used the term *K horizon* (German *Kalk*) for these horizons. The K horizon was first proposed by Gile *et al.* (1965) because no horizon designation had been established for horizons cemented with $CaCO_3$. We acknowledge the long-established use of the K horizon term but use the more accepted Bk, Bkm, Bkk, and Bkkm nomenclature of the Soil Survey Staff (2010) because (1) a term currently *exists* for such horizons (petrocalcic), (2) the K horizon is not officially recognized by the Soil Survey Staff (1999), and (3) it can be confused with Butler's (1959) K-cycle terminology (Chapter 14).

Calcification: The Process

$CaCO_3$ mediates or controls most of the major chemical and physical properties of desert soils (Schoeneberger *et al.* 2012), and hence, most research on desert soils has focused on the progressive accumulation of carbonates in soils. Salomons *et al.* (1978) and Monger (2002) reviewed the main models by which carbonates are deposited in the vadose zone: (1) in situ dissolution, followed by repricipitation (Blank and Tynes 1965, Treadwell-Steitz and McFadden 2000); (2) upward capillary flow from shallow groundwater or wet subsoils, i.e., the *per ascensum* model (Nikiforoff 1937, Espejo *et al.* 2008); (3) various biogenic models; and (4) the *per descensum* model, which involves carbonate-rich solutions descending in percolating water, to be deposited as the wetting front stops (Gile *et al.* 1966). These models involve at least these three stages: (1) provision of a carbonate-rich solution in the soil, (2) movement of that solution within the soil or near-surface environment, and (3) precipitation of carbonate minerals from solution. The *per descensum* model has garnered the most attention and is our focus (Durand 1963, Reeves 1970, Blümel 1982; Fig. 13.40).

Carbonates are rendered soluble through a process called *carbonation*, which acts on $CaCO_3$ to dissolve it into soluble components (see Chapter 9). Carbonic acid (H_2CO_3) forms as CO_2 and water combine, reacting with $CaCO_3$ and mobilizing Ca^{2+} and HCO_3^- ions:

$$CO_2(g) + H_2O \rightleftharpoons H_2CO_3 (aq)$$

$$H_2CO_3 + CaCO_3 \rightleftharpoons Ca^{2+} + 2HCO_3^-$$

Conditions that drive the reaction to the right, initiating carbonate dissolution, include a moister soil environment (as long as the water is not saturated with $CaCO_3$ and its by-products), lower pH values, higher CO_2 contents in the soil air, and cooler temperatures (Brook *et al.* 1983; Fig. 13.41). Although temperature affects carbonate chemistry to a lesser extent than do the other factors and conditions, cold water *is* able to dissolve more carbonate than warm water (Fig. 13.41). The temperature effect is only important when comparing calcification between regions, not locally.

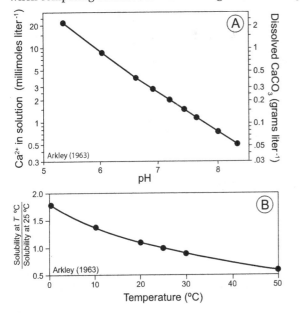

Fig. 13.41 Solubility of $CaCO_3$ as a function of (A) pH of the equilibrium solution and (B) temperature of the solution.

This discussion assumes that the permeability of the soil exceeds the rate at which wetting fronts normally move through it. If, for example, a soil has a clayey, slowly permeable layer, water with dissolved carbonates may tend to perch there, and as roots take up the water, carbonates will be precipitated (Schaetzl *et al.* 1996).

Ca^{2+} ions can move in soil water. If, at some point, the preceding reaction is driven to the left, $CaCO_3$ is precipitated as a secondary carbonate. Carbonate precipitation is driven by decreases in soil moisture and CO_2 partial pressure (pCO_2), and by increases in pH (Salomons *et al.* 1978). Desiccation (due to cessation of the wetting front and evaporation from the soil surface) is one of the main ways that secondary carbonate is precipitated in soils, as rainfall from many precipitation events in aridic soil moisture regimes enters initially dry soils and cannot wet the soil to great depths (Fig. 13.42). Wetting fronts stop because of lack of energy; i.e., the amount of water to continue the downward flow, and the matric forces required to pull them downward, are both lacking. Wetting fronts are also "pulled" back, upward, as a result of evaporation from the surface; the humidity at the surface can change very quickly, to very low levels, after a rain event in arid or hyperarid areas. Uptake by plant roots also may force the wetting front to stop. Wetting fronts may be slowed or stopped when they encounter textural changes, as at hardpans such as a petrocalcic horizon. For many desert soils, this action will force additional precipitation of carbonates on top of the pan, further thickening it and setting up a positive feedback mechanism in which a petrocalcic horizon grows upward through time. Wetting fronts can also be slowed by shallow bedrock, causing petrocalcic horizons to

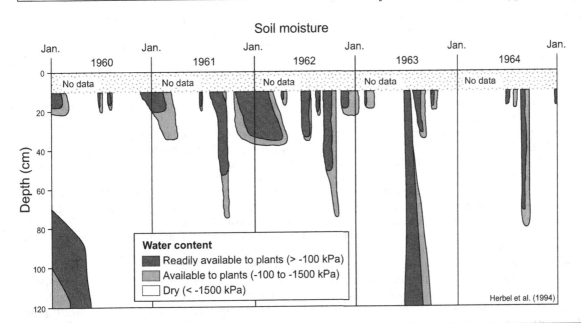

Fig. 13.42 Wetting front depth and duration, 1960–1964, in a Ustic Calciargid in New Mexico, United States. According to Monger *et al.* (2011), the abnormally wet year (1962) would not have met the aridic soil moisture regime criteria.

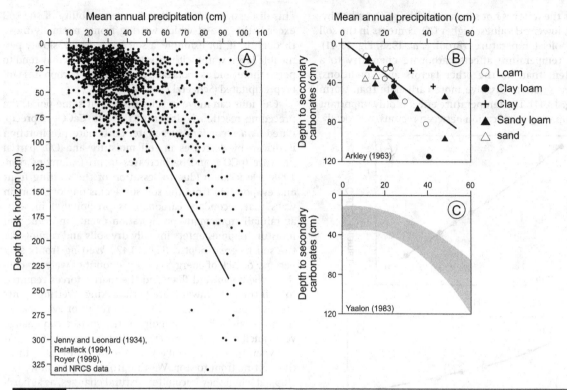

Fig. 13.43 Scatterplots showing the relationship between the depth to the top of the calcic horizon and mean annual precipitation.

form directly at the contact, e.g., Buol and Yesilsoy (1964), Stuart and Dixon (1973).

The presence of gypsum ($CaSO_4 \cdot 2H_2O$) also depresses the solubility of $CaCO_3$ (Reheis 1987b). In theory, rain and soil water acidified by carbonic acid will dissolve both gypsum and calcite when they are present separately at the soil surface (from dustfall). However, when both are present, the situation changes. Gypsum is more soluble than calcite. Therefore, as gypsum dissolves, the Ca ion concentration in the soil solution increases until equilibrium is reached. As long as the Ca ion concentration is maintained by the dissolution of gypsum, the dissolution of calcite will not be favored because the solubility product constant of calcite will always be exceeded (Reheis 1987b). In short, if the soil solution is saturated with gypsum, then carbonate will no longer dissolve, and therefore it will no longer become part of the pedogenic process. Thus, the more gypsum in a soil, the less soluble $CaCO_3$ will become, providing a feedback mechanism that prevents calcic soil development at depths where gypsum dominates.

As shown in Equation 13.1, calcification is largely governed by the production of CO_2 in the soil, which produces H_2CO_3, a weak acid capable of dissolving carbonates (Breecker *et al.* 2010). CO_2 in soils is produced by microbially mediated reactions, decomposition of organic matter, root respiration, and the dissolution of carbonates (Egli

and Fitze 2001). The main sources of CO_2 in soil air are respiration from plant roots and decaying organic material, in addition to that which diffuses in from the open atmosphere. As a result, the partial pressure of CO_2 in soil air is much greater than atmospheric levels, driving the chemical reactions shown earlier toward dissolution of calcite (Rabenhorst *et al.* 1984a). However, most researchers now agree that the *main* reason that carbonates are deposited at depth is desiccation (which can include desiccation by freezing) (Cox and Lawrence 1983), rather than changes in pCO_2 content.

Because carbonates are translocated more deeply in wet than in dry periods, the depth to the top of the secondary carbonate horizon in many soils generally reflects regional mean annual precipitation (Arkley 1963, Sehgal and Stoops 1972, Royer 1999; Fig. 13.43). Egli and Fitze (2001) showed that the amount of carbonates that *leave* the profile is also correlated to precipitation (and to percolation) (Fig. 13.44). Royer (1999) observed a significant correlation ($p < 0.05$) between the presence of carbonate-rich B horizons and annual precipitation values below 760 mm, perhaps defining the Pedocal-Pedalfer boundary in this way. Jenny (1941b) and Jenny and Leonard (1934) found that the depth to the top of the Bk horizon (D), in centimeters, is directly proportional to mean annual precipitation (P), in centimeters:

$$D = 6.35(P - 30.5)$$

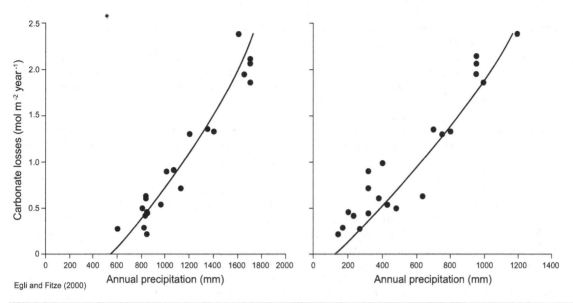

Egli and Fitze (2000)

Fig. 13.44 Relationships between (A) annual precipitation and (B) modeled annual percolation, versus loss of carbonates from several soil profiles in Switzerland and Germany. All soils are in a humid climate and have formed in calcareous parent materials. Although the loss of carbonate from the profiles is also affected by the carbonate content of the parent material, it appears that the cumulative quantity of infiltrating water passing through the soils remains the most explanatory factor.

Extrapolating this relationship to $D = 0$ would, however, imply that the Bk horizon should be at the surface, i.e., no leaching, in climates where $P \leq 30$ cm (Fig. 13.43A). However, this does not occur because the actual depth to the calcic horizon also depends on local factors such as slope, position on the landscape, permeability, presence or absence of a surface crust, vegetation cover, and the temporal distribution and intensity of precipitation. For example, runoff is common in deserts because precipitation intensities are high and because the soil surface often has a slowly permeable crust or is covered with gravel and rocks. Runoff from steeper slopes leads to pedogenically drier conditions there, whereas sites at the base of slopes are pedogenically wetter because of run-on (Yair *et al.* 1978). Thus, thickness of, and depth to, B horizons may be greater at the bases of slopes (or on level sites) than on level or steep surfaces (Hirmas and Allen 2007). Additionally, in arid and hyperarid climates, rainfall is often intense and concentrated in only a few, large storms, which almost always result in *some* translocation, further explaining why calcic horizons do not form at the surface. The depth to carbonates may be more closely related to these few, large precipitation events, rather than to some sort of annual precipitation mean value (Amundson *et al.* 1997). Plus, in some soils there is no calcic development because the presence of gypsum or other soluble Ca salts prevents carbonate dissolution.

Because calcic horizons form slowly and can persist for long intervals, the depth to the calcic horizon is also, at least partly, a function of paleoclimate, making interpreting the relationships in Figure 13.43 even more problematic

(Netterberg 1969a, McDonald *et al.* 1996, Khokhlova *et al.* 2001). Carbonates that had been precipitated at depth in a previously dry climatic interval may be redissolved and driven deeper in the profile. During wet climatic intervals, e.g., the Late Pleistocene in the deserts of the southwestern United States (Phillips 1994), deep Bk horizons may form. If that is followed by a drier period, e.g., the Holocene, soils may develop a Bk horizon higher in the profile and thereby exhibit two Bk maxima (McDonald *et al.* 1996). A modeling study by McDonald *et al.* (1996) suggested that processes operative in wet years in the Mojave Desert today are good analogs to normal years during Pleistocene pluvial climatic intervals.

Computer models point to the complexity of the calcification process. They show that the depth of Bk horizons in dryland soils is primarily a function of climate, but texture, coarse clast content, dust influx, soil CO_2 contents, soil age, presence/absence of gypsum, and type and density of vegetation cover are also very important (Marion *et al.* 1985, Mayer *et al.* 1988, McFadden *et al.* 1991). Most importantly, the depth of carbonates is dramatically influenced by eolian inputs of carbonate-rich dust (see later discussion). Dust not only provides a source of carbonate but dramatically affects the way in which water moves through the profile (Treadwell-Steitz and McFadden 2000). In the pages that follow, we will try to sort through this material and get a better handle on the widespread process of calcification.

Sources of Carbonate

Sources of pedogenic carbonate include (1) dust, (2) rainfall, (3) soil parent material, (4) CO_2 from plant respiration, and (5) groundwater (Gile *et al.* 1966, Rech *et al.* 2003, Robins *et al.* 2012). Although plants are capable of biocycling Ca^{2+} (Goudie 1996), only in rare situations do they actually *add* Ca to the profile. Machette (1985) pointed out that in most dry environments groundwater is too deep and the overlying sediment too coarse for capillary action to draw carbonate-rich water into the profile and into the reach of plant roots. Where this process does occur, e.g., in lowlands and on floodplains, carbonate-enriched horizons will form by capillary rise (Machette 1985).

Although chemical weathering is slow in dry environments, carbonaceous rocks like limestone and other rocks rich in Ca^{2+}, e.g., basalt and granites rich in Ca-feldspars, can release Ca^{2+} ions via weathering (Blank and Tynes 1965, Kahle 1977, Rabenhorst and Wilding 1986b, Boettinger and Southard 1991). However, the release of cations in this way is exceedingly slow in dry environments and thus is not considered a *major* source of carbonate (Lattman 1973). If rock weathering *were* a major carbonate source, trace elements such as Al and Ti that are found in the rock

would also be present in the secondary carbonate; they usually are not (Aristarain 1970). Indeed, many Aridisols, especially those on surfaces of Pleistocene age or older, have accumulated pedogenic carbonate within parent materials that are essentially carbonate-free (Gile 1979, Ugolini *et al.* 2008).

Landscapes: Southern Arizona Basin and Range

Centered on the state of Nevada and extending from southern Oregon to western Texas, the Basin and Range is an immense physiographic region of north-south-trending faulted mountains separated by wide, dry bolsons (basins of interior drainage). In southern Arizona this landscape extends into the Sonoran Desert with its trademark cactus, the stately Saguaro (*Carnegiea gigantea*), with its branched, tree-like form (see Figure). What little rainfall falls here does so mostly in late summer.

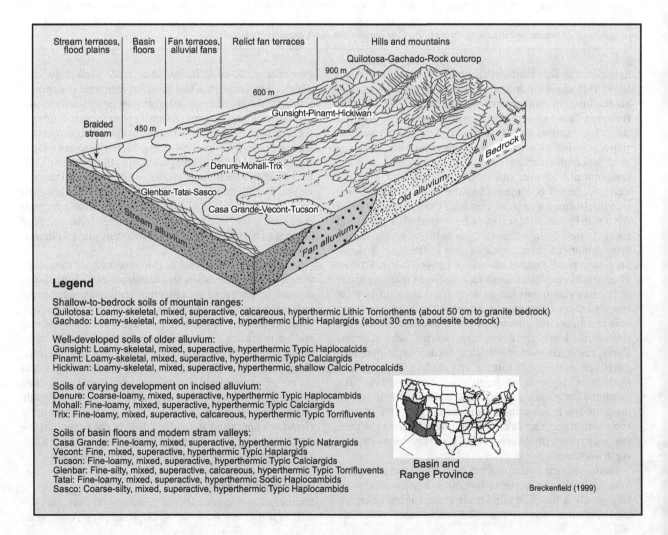

Legend

Shallow-to-bedrock soils of mountain ranges:
Quilotosa: Loamy-skeletal, mixed, superactive, calcareous, hyperthermic Lithic Torriorthents (about 50 cm to granite bedrock)
Gachado: Loamy-skeletal, mixed, superactive, hyperthermic Lithic Haplargids (about 30 cm to andesite bedrock)

Well-developed soils of older alluvium:
Gunsight: Loamy-skeletal, mixed, superactive, hyperthermic Typic Haplocalcids
Pinamt: Loamy-skeletal, mixed, superactive, hyperthermic Typic Calciargids
Hickiwan: Loamy-skeletal, mixed, superactive, hyperthermic, shallow Calcic Petrocalcids

Soils of varying development on incised alluvium:
Denure: Coarse-loamy, mixed, superactive, hyperthermic Typic Haplocambids
Mohall: Fine-loamy, mixed, superactive, hyperthermic Typic Calciargids
Trix: Fine-loamy, mixed, superactive, calcareous, hyperthermic Typic Torrifluvents

Soils of basin floors and modern stram valleys:
Casa Grande: Fine-loamy, mixed, superactive, hyperthermic Typic Natrargids
Vecont: Fine, mixed, superactive, hyperthermic Typic Haplargids
Tucson: Fine-loamy, mixed, superactive, hyperthermic Typic Calciargids
Glenbar: Fine-silty, mixed, superactive, calcareous, hyperthermic Typic Torrifluvents
Tatai: Fine-loamy, mixed, superactive, hyperthermic Sodic Haplocambids
Sasco: Coarse-silty, mixed, superactive, hyperthermic Typic Haplocambids

Basin and Range Province

Breckenfield (1999)

The Basin and Range Province was formed about 20 million years ago as the Earth's crust stretched, thinned, and then broke into more than 400 mountain blocks that partly rotated from their originally horizontal positions. To add to the complexity, Miocene volcanoes in what is now Arizona and Mexico emitted silicic lava and ash across the region. Since the Pliocene, mass wasting and erosion of adjacent mountain ranges have gradually filled the basins with thousands of meters of sediment. Massive, coarse-textured alluvial fans have formed at the contact between the basins and ranges (McAuliffe 1994).

In this dry climate, Lithic Entisols and Aridisols dominate the weathered bedrock uplands, where soils are thin. Soil development in the basins depends on the age of the geomorphic surface and sediment type. Many of the older surfaces consist of alluvial fans and pediment surfaces covered with alluvium. Here, soil development ranges from very strong, e.g., the Hickiwan series with its thick petrocalcic horizon at 35 cm, to Calcids and Argids of moderate development (see Figure). Downslope, these landforms may be incised, with inset, younger surfaces that have less developed soils. Arroyo valleys incised into the gravelly and sandy alluvium have Torriorthents. Some soils, such as the Casa Grande series with its Btknz horizons, have accumulated a variety of soluble materials.

In short, this is a landscape with a wide variety of soils typical of warm deserts, and one in which the degree of soil development is generally coincident with surface age. This desert, like most, is not all sand!

Figure: Block diagram of the soils and landscapes of a part of the Basin and Range province of the Tohono O'odham nation, in the Sonoran Desert of south-central Arizona.

Most researchers today agree with the early conclusions of Brown (1956) and Ruhe (1967) that the main source of carbonate in many Aridisols is external – carbonate-rich

eolian dust (Quade *et al.* 1995, Offer and Goossens 2001, Bockheim and Douglass 2006, Hirmas and Graham 2011). Machette (1985) outlined four lines of evidence to support the hypothesis that airborne materials are the primary source of pedogenic carbonate: (1) Most calcic soils are well above any groundwater influence, (2) relationships between soil age and development would not be present if the carbonate were from groundwater, (3) abundant sources of eolian carbonate exist in (and upwind of) many dry areas, and (4) rainfall has been shown to contain high amounts of Ca (Junge and Werby 1958). Dust is blown onto sites, and any carbonates on (or in) it are later dissolved and translocated into the soil by infiltrating water. In coastal areas, sea spray can contribute significant amounts of calcium as well. Some carbonate is carried in, already dissolved, with rain or snow. Nonetheless, most carbonate minerals enter the soil when coatings on silt- and clay-sized dust particles – formerly deposited on the soil surface – dissolve and the ions are translocated into the profile. Dust can also be simply fine fragments of limestone. Often, carbonate-rich dust is deposited on the surface and blown away before it can be dissolved. In this regard, the desert soil system is very dynamic, with dust of varying compositions blowing about, unceasingly; the carbonates on some of this dust have found a resting place in a soil because they were translocated in during a storm event. Other dust particles may be reentrained and move to yet another location (Gile *et al.* 1966).

In dry climates, dust commonly blows out of dry lake beds (playas), alluvial fans, and other alluvial flats (Young and Evans 1986, Gunatilaka and Mwango 1987, Chadwick and Davis 1990) and off eroding soils and weathering bedrock escarpments (Reheis and Kihl 1995). For soils in the

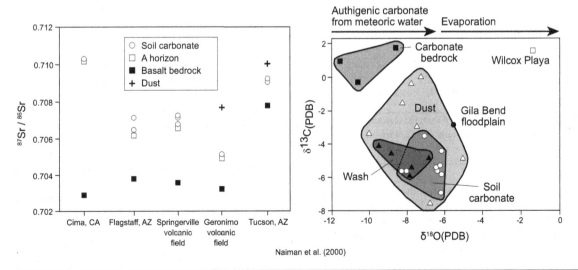

Naiman et al. (2000)

Fig. 13.45 Isotopic composition of soil carbonate, carbonate bedrock, dust, wash (fluvial material), floodplain, and playa materials from southern Arizona landscapes, illustrating that the main source material for dust falling on soils is eroded soil carbonate.

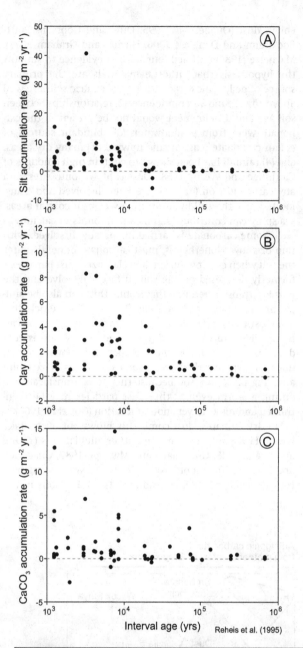

Fig. 13.46 Accumulation rates of (A) silt, (B) clay, and (C) CaCO$_3$, in soils from southern Nevada and California. Negative numbers imply losses.

immediately downwind from dust source areas tend to be better developed than soils farther away (Lattman 1973).

Determining long-term rates of carbonate accumulation is complicated by the fact that dust influx rates and climate vary considerably over geologic timescales. Moist conditions in the southwestern United States during Pleistocene glacial periods, i.e., pluvial conditions, greatly reduced influx rates, and hence, rates of calcification (Chadwick and Davis 1990). Research is continuing on the rates at which dust accumulates on, and assimilates into, desert soils, and the factors that influence these rates. Topography certainly enters into the equation: Dust that falls onto a knob or steep slope could be blown or washed away. Climate also matters. For example, dust falling onto a hyperarid site (no matter what the rate of influx) is likely to blow away without ever affecting the soil; this situation is described as *moisture-limited* (Machette 1985). The surfaces of moisture-limited sites have greater rates of Ca^{2+} influx than rates of Ca^{2+} translocation by infiltrating local rainfall, run-on, or snowmelt. Conversely, *influx-limited* sites are, theoretically, wet enough to translocate and accumulate more carbonate than they currently do, if only more dust fell onto the site. Some influx-limited sites lie in desert areas that have high amounts of snowmelt infiltration (Machette 1985). Thus, the amount of carbonate that a soil accumulates is a function of not only the influx rates and carbonate content of clastic dust (Fig. 13.46), but also climate (primarily precipitation amount and character) and surface age (Machette 1985, Harden *et al.* 1991a, Bockheim and Douglass 2006). Slate *et al.* (1991) observed that soils near major dust sources develop slowly and have carbonates evenly distributed through the profile, presumably because they accumulate eolian sediment faster than translocation of carbonates can occur. Sites farther from the dust source, with slower dust influx rates, have carbonates concentrated within a few horizons. They interpreted this relationship to mean that soils with slower rates of dust input have time to allow for carbonate translocation into the subsoil.

Influxes of carbonates from external sources are often indicated when concentrations of carbonate, gypsum, other salts, and exchangeable ions occur in soils, over and above that of the parent material. Then, if the amount of secondary carbonate that has accumulated in the soil and the surface age are known, *rates* of calcification can be estimated. For example, Machette (1985) reported CaCO$_3$ accumulation rates of 2.2–5.1 g m^{-2} yr^{-1} for Late Pleistocene soils in New Mexico. Accumulation rates were less for older soils, possibly because erosion episodes may have removed carbonates from the soils, or because during wet climatic intervals some of the carbonates were leached from the profile, or simply because influx rates were variable. Scott *et al.* (1983) reported carbonate accumulation rates of 5 g m^{-2} yr^{-1} in Utah. For the Mojave Desert, Schlesinger (1985) reported 1–3 g m^{-2} yr^{-1}.

It has been noted that gypsum (CaSO$_4$ • 2H$_2$O) can have an influence on carbonate accumulation, because it depresses

desert Southwest of the United States, dust has been confirmed as the source of carbonate minerals, using C and Sr isotopes (Naiman *et al.* 2000). In this case, the main source of dust was previously existing soil carbonate that had been eroded, not weathered bedrock (Fig. 13.45). Dust downwind from carbonate-rich areas is much higher in carbonate minerals and gypsum than dust from other sources, including playas (Reheis and Kihl 1995), and soils the solubility of $CaCO_3$. However, it is also a Ca source and is commonly associated with calcic horizons (Lattman and Lauffenburger 1974). Many Bkm and Ckm horizons smell of H_2S gas when crushed, suggesting that sulfur is present.

Physicochemical Models of Carbonate Accumulation

Starting with the work of Hawker (1927), various pedogenic models have been developed to explain the accumulation of *secondary* carbonate in soils. Differentiating the amount of secondary carbonate from primary (inherited) carbonate is a difficult but necessary first step (Fig. 13.47). With these data, one also needs an estimate of soil (or surface) age. Once those two characteristics are known, it is possible to determine *rates* of carbonate accumulation, and then, perhaps, dust influx rates (Fig. 13.48). Both of these rates are best calculated on sites where the slopes are stable and where the bedrock does not contribute significant amounts of Ca to the soil. Slate *et al.* (1991) described such a setting in New Mexico. Their data illustrate the slow, steady

increase in carbonates and clay that is thought to occur in soils on stable surfaces, with a reliable, upwind source of carbonate-rich dust.

The Desert Project Model
In association with a USDA-sponsored soil-geomorphology project on desert soils, led first by Robert Ruhe but most prominently by Leland Gile, Gile and Grossman (1979) developed the first pedogenic model of carbonate accumulation in soils. The model established not only how long soils took to develop certain kinds of carbonate morphologies, but also explained why those morphologies occurred and how calcification is manifested in the soil's morphology. Finally! A pedogenic model now existed that field personnel could use to estimate the relative age of the various surfaces they were mapping. Because of its association with the USDA-sponsored Desert Project, we will refer to this model as the Desert Project model. Like most of the traditional, time-tested models for carbonate accumulation, the Desert Project model assumes that carbonates accumulate in soils mainly by physicochemical processes as opposed to biological processes (Abtahi 1980).

The gold standard of the physicochemical models for describing secondary carbonate accumulation is that of Gile *et al.* (1966); however, others exist as well (Rabenhorst and Wilding 1986b, Alonso-Zarza *et al.* 1998). The Desert Project model relies on known physicochemical processes for the accumulation of carbonates and describes a sequence of carbonate accumulation leading ultimately to carbonate-plugged Bkkm (petrocalcic) horizons (Reeves 1970, Gile 1993, 1995, Brock and Buck 2009). It is a type of *per descensum* model because it assumes that carbonate accumulates in soils primarily by descending via percolating water. In the model, pedogenic carbonate accumulation follows a clear and predictable morphogenetic sequence. This model, developed for the southwestern United States, has unquestionably withstood the test of time.

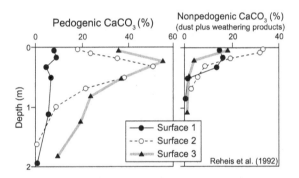

Fig. 13.47 Depth plots of pedogenic versus nonpedogenic carbonate in three soils on the Kyle Canyon fan, Nevada.

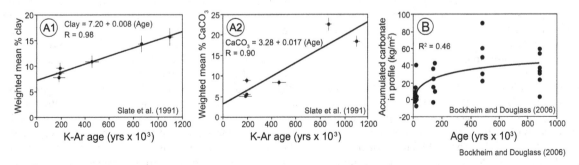

Fig. 13.48 Relationships between surface age and soil properties in areas undergoing calcification. (A) Relationship between soil age and weighted mean (A1) clay and (A2) $CaCO_3$ contents for soils in the Pinacate volcanic field, New Mexico. K-Ar ages on the underlying lava flows should be interpreted as maximum ages. The rate of carbonate accumulation at this site is slower than at many other sites in the western United States. (B) Relationship between soil age and profile carbonate content for soils on moraines in southwestern Patagonia. Negative quantities imply a loss of carbonate.

Table 13.8 | Morphologic stages[a] of $CaCO_3$ accumulation for soils developed in noncalcareous parent materials

Stage	Gravelly materials	Nongravelly (fine-grained) materials
I	Thin discontinuous coatings on pebbles, usually on undersides; some filaments; matrix can be calcareous beside stones; ≈4% $CaCO_3$	Few filaments, masses, threads, and patchy coatings in B horizon and on sand grains; <10% $CaCO_3$
I+	Thin and continuous coatings on many or all pebbles	Filaments and masses common
II	Continuous coatings on all sides of pebbles, especially on bottoms as pendants; local cementation of few to several clasts; matrix loose and calcareous enough to give whitened appearance	Few to common carbonate nodules, concretions, and masses of 5–40 mm diameter; matrix between nodules slightly whitened; carbonates occur in veinlets and as filaments; parts of matrix can be noncalcareous; ≈10–15% $CaCO_3$ in whole sample, but 15–75% in nodules
II+	Same as Stage II, except carbonate in matrix more pervasive, sometimes forming continuous fillings between gravels	Carbonate nodules common; 50–90% of matrix whitened; ≈15% $CaCO_3$ in whole sample
III	Most or all sides of clasts in the B horizon coated; 50–90% continuously filled with carbonate (K-fabric) and may be cemented; color mostly white; carbonate-rich layers more common in upper part; ≈20–25% $CaCO_3$	Nodules coalesce; so many nodules, internodule fillings, and carbonate coats on grains that >90% of horizon is white; matrix may be cemented; carbonate-rich layers more common in upper part; ≈20% $CaCO_3$
III+	Most pebbles have thick carbonate coats; matrix particles continuously coated with carbonate, or pores plugged by carbonate; nearly continuous cementation; >40% $CaCO_3$	Most grains coated with carbonate; most pores plugged; >40% $CaCO_3$
IV	Upper part of B horizon nearly pure, cemented carbonate (75–90% $CaCO_3$) consisting of laminar carbonate layerse <1 cm thick in total; lower part of horizon plugged with carbonate (50–75% $CaCO_3$)	
V	Platy, laminar layer (>1 cm thick) strongly expressed, may contain pisoliths (roughly spherical masses with concentric layers, which may have parent material clasts or broken fragments of petrocalcic material in center)	
VI	Common brecciation and recementation, with multiple generations of brecciated laminae, pisoliths, and petrocalcic fragments, many of them recemented	

a Developed for soils in the southwestern United States; see also Figures 13.49 and 13.50.
Sources: Gile *et al.* (1966), Bachmann and Machette (1977), Machette (1985) and Birkeland (1999). Recognition should be made, however, of the early contribution of Hawker (1927).

For this reason it is a powerful tool in the correlation of various deposits and geomorphic surfaces (Wells *et al.* 1985, Vincent *et al.* 1994). The original four stages of the model depict the various widely observed morphologies of $CaCO_3$ accumulation; many intermediate stages also occur (Table 13.8, Fig. 13.49). Monger *et al.* (1991a) provided micromorphological data for soils in the various stages. Machette (1985), working with data from an earlier paper (Bachman and Machette 1977), proposed two additional stages of carbonate accumulation. Brock and Buck (2009) added detail to the pedogenesis of the later stages, and Forman and Miller (1984) suggested subdivisions of Stages I and II to characterize more clearly the progressive accumulation of $CaCO_3$ over time.

The carbonate accumulation patterns and morphology of the development stages are affected by both age and parent material. Two distinct sequences of accumulation were modeled: for gravelly (>50% gravel) and for less gravelly parent materials (Fig. 13.49). Materials with intermediate amounts of gravel have intermediate morphologies. The time required to achieve each of the morphology stages depends on a number of factors, chief among them being the rate of carbonate influx, but also including porosity, climate, and various disturbance factors, e.g., bioturbation and erosion. Therefore, the time for each morphologic stage to develop can vary greatly with location. Fine-grained soils have more surface area and porosity than do gravelly soils; therefore, fine-grained soils require more carbonates to fill voids and form equivalent carbonate stages (Gile *et al.* 1966. 1981). Be aware that a given soil may contain horizons – each with different stages – due to shifting climate conditions or deposition or erosion of sediment, causing partial burial or exhumation of preexisting horizons. Please consult Figure 13.49 as you read this section.

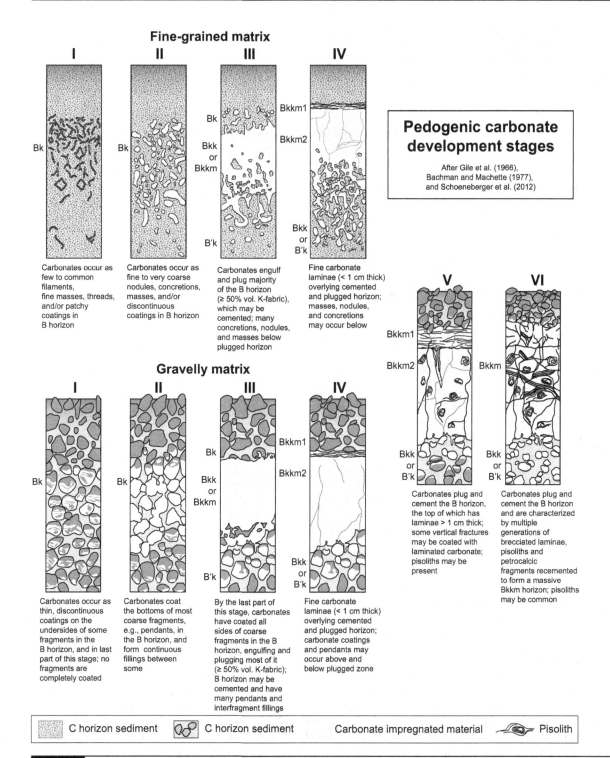

Pedogenic carbonate development stages

After Gile et al. (1966),
Bachman and Machette (1977),
and Schoeneberger et al. (2012)

Fine-grained matrix

I — Carbonates occur as few to common filaments, fine masses, threads, and/or patchy coatings in B horizon

II — Carbonates occur as fine to very coarse nodules, concretions, masses, and/or discontinuous coatings in B horizon

III — Carbonates engulf and plug majority of the B horizon (≥ 50% vol. K-fabric), which may be cemented; many concretions, nodules, and masses below plugged horizon

IV — Fine carbonate laminae (< 1 cm thick) overlying cemented and plugged horizon; masses, nodules, and concretions may occur below

Gravelly matrix

I — Carbonates occur as thin, discontinuous coatings on the undersides of some fragments in the B horizon, and in last part of this stage; no fragments are completely coated

II — Carbonates coat the bottoms of most coarse fragments, e.g., pendants, in the B horizon, and form continuous fillings between some

III — By the last part of this stage, carbonates have coated all sides of coarse fragments in the B horizon, engulfing and plugging most of it (≥ 50% vol. K-fabric); B horizon may be cemented and have many pendants and interfragment fillings

IV — Fine carbonate laminae (< 1 cm thick) overlying cemented and plugged horizon; carbonate coatings and pendants may occur above and below plugged zone

V — Carbonates plug and cement the B horizon, the top of which has laminae > 1 cm thick; some vertical fractures may be coated with laminated carbonate; pisoliths may be present

VI — Carbonates plug and cement the B horizon and are characterized by multiple generations of brecciated laminae, pisoliths and petrocalcic fragments recemented to form a massive Bkkm horizon; pisoliths may be common

Legend: C horizon sediment | C horizon sediment | Carbonate impregnated material | Pisolith

Fig. 13.49 The six-stage model of carbonate accumulation in dryland soils, developed primarily from observations of soils in the southwestern United States. As implied in the figure, Stages V and VI are associated with both gravelly and fine-grained matrix materials.

Fig. 13.50 Soil morphologies associated with calcification. (A) A series of photos showing the various carbonate stage morphologies. (1) A Typic Torrifluvent with weak Stage I morphology. Photo by C. Monger. (2) A Stage III Petrocalcid from southern New Mexico, with typical Stage III morphology. This soil has an indurated Bkkm horizon but lacks the laminar cap that defines Stage IV development. Photo by C. Monger. (3) A Typic Petroargid from southern Nevada, with Stage IV morphology. Photo by B. Buck. (4) A Stage V or VI Petrocalcid near Las Cruces, New Mexico. The soil is estimated to be 1.6 million years old. Photo by C. Monger. (B) Close-up of the upper profile of a Stage VI soil, showing pisoliths in and above the Bkkm horizon. Photo by B. Buck. (C) Effects of calcification on a landscape scale. This is the edge of Mormon Mesa, near Las Vegas, Nevada. The entire mesa is held up by Stage VI soils, much as bedrock might do. Photo by B. Buck.

In fine-grained soils, Stage I is marked with carbonates (but mainly calcite) that occur mainly as thin, white filaments. Filaments often are calcified fungal hyphae. In gravelly Stage I soils, secondary carbonates accumulate as thin, discontinuous carbonate coatings, primarily on the undersides of pebbles and gravel. They are generally difficult to see, unless the soil is being examined closely, e.g., with a hand lens. Carbonates tend to accumulate on the bottoms of clasts because percolating, carbonate-rich water will tend to accumulate there, and when it later evaporates or is taken up by roots, carbonates are precipitated (Treadwell-Steitz and McFadden 2000). The larger the rock, the more carbonate-rich water it can intercept, and thus the faster its coating will grow. A unique situation was reported by Amundson *et al.* (1997), however, for soils in southern Baja California. Here, carbonate coatings are found on the *tops* of large clasts. The explanation for this geographically isolated phenomenon may be connected with soil thermal and hydraulic gradients. In Stage I, carbonates are most common at about 15–60 cm depth. Horizons that have secondary carbonate minerals are not yet cemented are designated Bk horizons.

By Stage II, pebbles show more continuous and thicker coatings of carbonate, and some voids between pebbles have become filled with white, nonindurated carbonate (Fig. 13.50A2). The stage II coatings underneath pebbles are called *pendants*, but they do not necessarily grow thicker by younger coats being added to the bottom. Brock and Buck (2005) found that a crack can form at the contact between the clast and the pendant, allowing younger fluids to enter and precipitate younger laminae of carbonate adjacent to the clast. If the soil is fine-grained, the carbonates begin to accumulate as small nodules. Some nodules take on cylindrical form or may resemble baby rattles, perhaps because they cement in-filled cicada burrows or krotovinas (Hugie and Passey 1963, Suprychev 1963; Fig. 11.5B).

Fig. 13.51 A trench through a Petrocalcid in southern New Mexico. Note the pipes (lacunae) in the Bkkm horizon, and the possible krotovinas below and within, probably attributable to badgers. Photo by D. Cremeens.

Gravelly soils with Stage III carbonate morphology have many interpebble fillings, and all or nearly all of the skeletal grains in the upper Bk horizon are coated with carbonate (Fig. 13.49). Carbonates now coat the tops and sides of clasts in these soils both because increasing carbonate accumulations fill the pore spaces and because dust additions will, by now, have caused the soil to be much finer-textured. The fine dust materials, most common in the upper profile, fill macropores and allow percolating water to be retained in close contact with all sides of large clasts. When this water dries, carbonates are deposited on all sides of the clasts (Treadwell-Steitz and McFadden 2000), instead of only hanging from the bottoms. Soils in dust-poor areas may take considerably longer to reach Stage III. Late in Stage III, whether in gravelly or nongravelly materials, carbonates fully impregnate the entire Bk horizon and voids become plugged. At this time, it is designated a Bkk or Bkkm horizon. Part or all of it is indurated. Earlier researchers and many geologists referred to Stage III or greater B horizons as K horizons (Gile *et al.* 1965).

By Stage IV, some convergence occurs between gravelly and nongravelly soils (Gile *et al.* 1966). Because they have less void space and surface area, gravelly soils reach Stage IV more rapidly than do fine-grained soils (Gile *et al.* 1966, 1981, Marion *et al.* 1985); the process is even more rapid in soils with larger gravels (Treadwell-Steitz and McFadden 2000). The Bkkm horizon of Stage IV perches water, forcing subsequent carbonate deposition to be focused on the top of the plugged horizon (Reeves 1970). Eventually, a laminar, indurated horizon forms, displacing and engulfing surrounding materials and growing upward through time (Alonso-Zarza *et al.* 1998, Jolley *et al.* 1998, Pfeiffer *et al.* 2012; Fig. 13.50A3). The Bkkm horizon exhibits a succession of thin (≈1 mm), low porosity, $CaCO_3$-enriched layers and consists of almost pure $CaCO_3$, with little other allogenic clastic material. Upper laminae are younger than lower laminae. Although it has been happening in all previous stages as well, volume expansion of the soil is occurring at this point, as a result of silicate mineral (dust) and carbonate accumulation. Irregularities in the top of the Bkm horizon are in-filled more rapidly, as percolating water preferentially ponds there. In time, the laminar horizon attains smoothness and horizontality (Aristarain 1970). Nonetheless, the petrocalcic horizon is not totally impermeable; it has windows – sites of preferential flow. These pipes, or windows, are kept free of carbonate accumulation as percolating water is funneled rapidly through them after heavy rainstorms (Fig. 13.51). Many of these pipes are thought to be formed by mammal bioturbation (Hirmas and Allen 2007), or by any preferential flow of water that takes advantage of openings created by tap roots, surface topography, or expansion fractures in the petrocalcic horizon. In short, processes of dissolution and (re) precipitation can occur simultaneously and can vary in space and time across the landscape.

Machette (1985), working with data from an earlier paper (Bachman and Machette 1977), proposed two additional – later – stages of carbonate accumulation (Table 13.8, Fig. 13.49). He suggested that Stage IV be restricted to soils with laminar horizons thinner than 1 cm. Stage V was then defined to include soils with laminar horizons >1 cm thick. The bulk densities of Stage V and VI petrocalcic horizons are similar to that of limestone.

Stage VI is distinguished by evidence of brecciation, which is recognized by multiple zones of randomly oriented broken laminae, or other fragmented pieces of the petrocalcic horizon that are now cemented and strongly indurated. It is formed from repeated exhumation and reburial of the petrocalcic horizon, causing it to be broken into fragments, and then recemented in different orientations (Fig. 13.50A4, C). Because the laminar cap occurs at the top of Stage V petrocalcic horizons, it is this part of the horizon that is most likely to become fragmented and broken. The result is an uppermost, brecciated but strongly recemented zone, overlying the less disturbed petrocalcic horizon (Brock and Buck 2009). However, in highly eroded landforms, the uppermost, brecciated portions of the petrocalcic may be in the process of exhumation and fragmentation, resulting in a layer of broken rubble overlying a more consolidated and cemented zone. The evidence that the petrocalcic horizon has reached Stage VI may exist only in the petrocalcic rubble at the surface, which upon close examination can show evidence of previous fragmentation and recementation (House *et al.* 2010).

The multiple episodes of brecciation found in Stage VI petrocalcic horizons indicate extreme age and polygenesis (Bryan and Albritton 1943, Brock and Buck 2009, Melendez *et al.* 2011, Pfeiffer *et al.* 2012); some may date to the early Pliocene (Fig. 13.50C). Stage VI petrocalcic horizons form as a geomorphic response to climate change over significant periods (Brock and Buck 2009). An increase in aridity can decrease vegetation density (or change the type), resulting in increased erosion of the overlying, unconsolidated soil horizons. Shallow or exposed petrocalcic horizons become broken and fragmented as a result of a combination of processes that can include ponding or lateral flow of water during large rain events (Aristarain 1970), bioturbation, or freeze-thaw events (Brock and Buck 2009). Later climate and vegetation changes can increase the capture of dust and eolian sediment, thus increasing the thickness of overlying horizons, which then encourages recementation of the brecciated petrocalcic horizon. Stage VI morphologies illustrate that they are the product of multiple climatic cycles (Sanz and Wright 1994).

Stage VI and some Stage V soils have concentrically banded *pisoliths* (0.2–5 cm diameter) with interiors composed of pebbles from the original parent materials, or fragments of the petrocalcic horizon (Fig. 13.50B). The concentric laminae around these clasts (pebbles or petrocalcic fragments) are actually layers of carbonate that originally precipitated as a pendant. Thus, pisoliths are simply cross-sectional views of pendants and supply unequivocal evidence that coarse-grained clasts (parent materials, or more often, fragments of petrocalcic horizons) existed at one time in the soil (Brock and Buck 2009). Some pisoliths/pendants can extend vertically for 5 cm or more, whereas others have multiple directions of laminae formation, indicating that they were moved or rotated during formation, mostly likely by bioturbation (Brock and Buck 2009).

Stage V soils often have pisoliths – small, rounded fragments of secondary, pedogenic carbonate with enveloping concentric laminae. They may also contain interiors of broken or disrupted laminae. Presumably, they grow by slow carbonate accretion around a core fragment, which can be, e.g., a pebble or a previously broken fragment of Bkkm material. The pisoloths commonly have pendant-like laminae of secondary carbonates, sometimes as long as 5 cm, extending vertically from their undersides (Brock and Buck 2009). Some pisoliths have pendants that extend outward in different directions from the core, suggesting that after they had started forming, the pisolith was moved or rotated, probably by bioturbation or by root growth and expansion.

Throughout the formation of calcic and petrocalcic horizons, the precipitation of carbonate acts *displacively*, literally moving skeletal grains out of the way as a result of crystallization pressures. Voids and fractures become filled with secondary carbonate, which then exceeds the original pore volume, forcing expansion and causing clastic grains seemingly to float within a carbonate matrix (Reheis 1988, Reheis *et al.* 1992). Machette (1985) estimated that expansion could reach 400–700%. Secondary carbonate can also act *replacively*, wherein $CaCO_3$ replaces primary silicate grains (Millot *et al.* 1977, Watts 1980, Reheis 1988, Reheis *et al.* 1992).

Using radiocarbon and other dating methods (see Chapter 15), it is sometimes possible to establish the age of secondary carbonates and put some age constraints on this model. Because some of the CO_2 involved in the formation of the carbonic acid is derived from plant (root) and microbial respiration, a portion of the carbon in the reprecipitated $CaCO_3$ is biologic and can be dated (Amundson *et al.* 1994). Likewise, the stable isotope contents of soil carbonate can provide information about the paleoenvironment during which they were formed (Quade *et al.* 1989, Quade and Cerling 1990; see Chapter 16). [14]C dating of soil carbonate has established that the youngest carbonate occurs nearest the top of the Bkkm horizon, and that ^{14}C ages increase with depth (Buol and Yesilsoy 1964, Reeves 1970). Applying this dating information, rates of carbonate accumulation have been established, allowing soil development to be correlated to surface age. Because climate and dust influx rates vary through space and time, rates of carbonate accumulation will also vary in space and time (Machette 1985, Sehgal and Stoops 1972). Figure 13.52 illustrates that the attainment of carbonate Stages I through VI takes different amounts of time at different locations across the southwestern United States. In favored locations, Stage VI morphologies can be attained on surfaces dating back to the Middle Pleistocene, while others of presumably similar age remain in Stage III. Holocene soils in this region have formed in a dry, interglacial period and thus exhibit rapid rates of carbonate accumulation, although they are universally still in Stage I. Older, Pleistocene-aged, soils would have been exposed to at least one wetter, cooler glacial climatic interval, in which carbonates may not have accumulated rapidly, or perhaps even been partially dissolved (Fig. 13.53). Sehgal and Stoops (1972) examined a sequence of soils from dry to subhumid climates and found that the forms of secondary carbonates also changed along a climatic transect. As would be expected, the amounts of secondary carbonates in the soils also decreased from dry to subhumid climates. Rates of carbonate accumulation also appear to vary, depending on the types of rock that dominate the gravel fraction (Lattman 1973). The extent and development of secondary carbonates are greatest where carbonates and basic igneous rocks dominate, intermediate in soils with large amounts of siliceous sedimentary detritus, and least where acid igneous gravels are common.

Other Physicochemical Models

Rabenhorst *et al.* (1991) and Rabenhorst and Wilding (1986b) presented another physicochemical model of petrocalcic horizon formation, one developed for soils underlain shallowly by limestone, in dry West Texas (see also Blank and Tynes 1965). They envisioned that the carbonate horizons in these soils developed from in situ dissolution of limestone, followed by reprecipitation of soil carbonate, often at a lithologic discontinuity within the limestone (West *et al.* 1988b). In Stage II of this model, percolating water containing carbonic acid and organic compounds dissolves some of the porous limestone, enlarging some of the pores. Limestone dissolution is thought to be enhanced by the acidic secretions of algae and fungi, eventually leading to precipitation of micrite crystals in pores, a process called *sparmicritization* (Kahle 1977, Kaplan *et al.* 2013). (*Micrite* is a term used for calcite crystals < 4

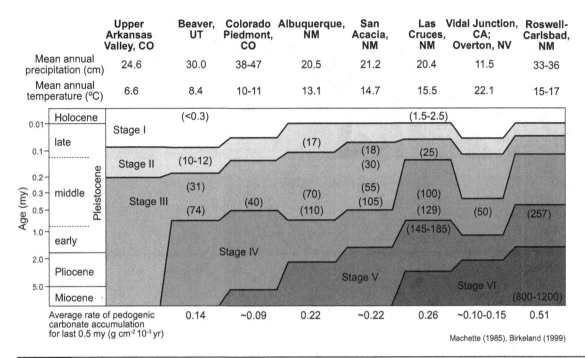

	Upper Arkansas Valley, CO	Beaver, UT	Colorado Piedmont, CO	Albuquerque, NM	San Acacia, NM	Las Cruces, NM	Vidal Junction, CA; Overton, NV	Roswell-Carlsbad, NM
Mean annual precipitation (cm)	24.6	30.0	38-47	20.5	21.2	20.4	11.5	33-36
Mean annual temperature (°C)	6.6	8.4	10-11	13.1	14.7	15.5	22.1	15-17

Machette (1985), Birkeland (1999)

Fig. 13.52 Maximum stages of carbonate morphology in gravelly alluvium in the southwestern United States. Numbers in parentheses represent mean grams of $CaCO_3$ cm^{-2} of soil column.

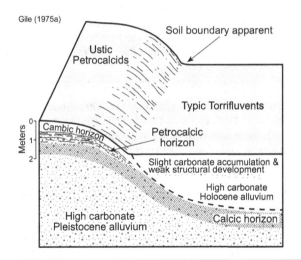

Fig. 13.53 Block diagram illustrating the difference in calcic horizon development in high-carbonate parent materials of different ages.

μm in size.) As pores continue to enlarge with sustained limestone dissolution, micrite linings continue to coalesce with persistent carbonate precipitation, eventually leading to a plugged condition, similar to Stage III in Figure 13.49. In Stage V of this model, a laminar cap of secondary carbonates forms above the plugged horizon, similar to Stage IV in Figure 13.49. Evidence in support of the Rabenhorst-

Wilding model includes (1) the low noncarbonate residue in the Bkkm horizon, i.e., horizons that were nearly pure calcite, and (2) the presence of limestone fragments within the Bkkm horizon (West *et al.* 1988a, 1988b). Grain displacement by growing calcite crystals could not explain the low (≈2%) noncarbonate residue levels. In contrast, soils in New Mexico, where Gile and his colleagues developed their model, often contain 25–54% noncarbonate residue in the Bkkm horizon. The Rabenhorst-Wilding model also explains the somewhat anomalous presence of fluorite (CaF_2) in Bkm horizons in West Texas Calcids, since this mineral is found in the bedrock as well and would not have been leached from the soil system.

A related model by West *et al.* (1988b) incorporates many of the same components but relies more heavily on the formation of the petrocalcic horizon in conjunction with a dense, weathering-resistant layer in the limestone. The resistant limestone bed remains after softer layers above and below have weathered away, forming the locus of secondary carbonate deposition. Bkkm horizons in Texas often embed fragments of limestone. Eventually, geologic erosion lowers the soil surface, placing the Bkkm horizon within the leached zone and causing it to break down.

Biogenic Models of Carbonate Accumulation

Recent research has repeatedly pointed to the influence, be it subtle or dominant, of biological agents in the calcification process (Verrecchia 1994, Monger 2002, Melendez *et al.* 2011). Although some biogenic calcification models

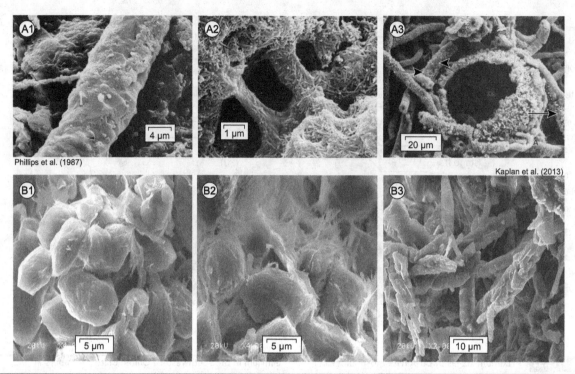

Fig. 13.54 Scanning electron photomicrographs of secondary carbonates in calcic horizons. (A) Biogenic, calcified filaments from soils in South Australia. (A1) Calcified filament with a relatively smooth surface, with calcite rhombohedra beneath a possible organic sheath. (A2) Submicron-size calcified rods of possible bacterial origin. (A3) Images of fruiting bodies associated with filaments. The large hollow sphere may be an oogonium. The small curved filament in the lower right (at A) may be an antheridium attached to the oogonium. The pustular structures at the arrows may be bacteria. (B) Nonbiogenically deposited carbonates from soils in southern Turkey. (B1 and B2) Calcite crystals having euhedral and subhedral forms, including micropores between them, and palygorskite fiber bundles that extend from the calcite crystals. (B3) Irregular networks of calcite needles and calcified filaments in a Bkm horizon.

appear to work only for instances of subaerial exposure, i.e., explaining the formation of surficial, calcium-enriched soil crusts by microbial activity (Verrecchia *et al* 1995), others stress the importance of microbial mediation of calcification within the subsoil. Most of these *biogenic models* also include a physicochemical component of Ca deposition. Organisms may act as *catalysts* for carbonate precipitation, either passively or actively, and may function as *sources* of carbonate materials (Goudie 1996). Obviously, by removing CO_2 from soil air, organisms also mediate the carbonate precipitation process (Krumbein and Giele 1979; see the reactions presented on page 380). And Goudie (1996) stressed that soil biota may play a key role in carbonate precipitation; e.g., fungi may trigger carbonate precipitation by dumping their excess Ca^{2+} (Verrecchia 1990).

Secondary $CaCO_3$ has long been known to form as precipitates on or within biological substrates, e.g., root hairs, fungi, and actinomycetes (Calvet *et al.* 1975, Klappa 1979, Phillips and Self 1987, Vaniman *et al.* 1994, Shankar and Achyuthan 2007). Calcium and carbonate ions may precipitate onto an existing surfaces in soils – in this case the surface are biological. If the substrate is a root hair, the carbonate feature produced can be referred to as a rhizolith,

root tubule, or root cast (Jaillard *et al.* 1991, Goudie 1996, Wright *et al.* 1995, Brock and Buck 2009, Pfeiffer *et al.* 2012). Indeed, secondary carbonates are commonly found as precipitates in and along root traces of all kinds, living and fossil. Roots assist in the precipitation process by removing water, causing supersaturation of the remaining soil water. Where they are cemented and adjacent to former tree roots, Alonso-Zarza and Jones (2007) referred to such features as rootcretes. They formed as roots and associated microorganisms generated cavities that acted as preferred zones of carbonate precipitation. In an earlier paper, Alonso-Zarza (1999) concluded that even the laminar, Stage IV petrocalcic horizons of the Desert Project model can have biogenically mediated origins – where calcium carbonate has precipitated as calcified root mats.

Monger *et al.* (1991b), however, documented a more *active* role for microorganisms, suggesting that they can *directly and actively promote precipitation* of calcite in soils. Cited as evidence are fossilized remains of calcified fungal hyphae, among other biogenic features, in Bk horizons (Kahle 1977, Phillips and Self 1987, Wright *et al.* 1995, Phillips *et al.* 1987; Fig. 13.54). In an interesting experiment in which soil columns were irrigated with calcium-rich water, calcite formed only in the soils that contained microorganisms; none formed in sterile soils (Phillips *et al.* 1987).

Amit and Harrison (1995) proposed a carbonate accumulation model that incorporates physicochemical and biogenic processes and, thus, may be the most holistic and widely applicable model to date. It was developed in a sand dune landscape in a hyperarid region of Israel – but one with surprisingly high biological activity. Here, carbonate accumulations show distinct biogenic origins; many are calcified fungal hyphae. All of the secondary, micritic carbonate in these young (1,425 ^{14}C years old) soils occurred in conjunction with roots, bacteria, or fungi. Even the nodules of calcite contain high amounts of biogenic material. Over time, the accumulation of secondary carbonates, along with dust additions, decreases the permeability of the soil. At some point, the soil crosses an intrinsic threshold and physicochemical processes of carbonate accumulation begin. Thus, biogenic processes initiate calcification and at some later point are joined by physicochemical processes. Addition of dust is an important component of this model because it reduces moisture loss between precipitation events, thereby enhancing carbonate precipitation. This model is well suited to highly permeable parent materials with a high degree of biogenic activity (Amit and Harrison 1995).

The calcification model of Alonso-Zarza *et al.* (1998) also involves biogenic carbonate accumulation, along with surface evolution and stability. This model highlights the complex interaction between pedogenesis and surface stability in arid regions, integrates biophysical processes, and points out that disruptions to the development sequence of the Gile *et al.* (1966) model are not only possible, but likely (Verrecchia 1987, Sancho and Meléndez 1992).

Secondary carbonates in soils have multiple pedogenic origins. Certainly, as carbonate-rich water in a soil is lost to root uptake or evaporation, the dissolved carbonate must precipitate. This is a physicochemical reaction. However, the reaction could be facilitated or even initiated biochemically. Fungal hyphae and fruiting bodies, algae, bacteria, plant roots, and even pupal cases might provide optimal sites for initial precipitation and continued growth of calcite and micrite crystals (Phillips *et al.* 1987). We still have much to learn about the intricacies of the calcification process, even though the general sequence has been studied for decades.

Gypsification

Desert soils accumulate many different kinds of soluble compounds, for similar genetic reasons – the water necessary to translocate them out of the profile is lacking (Veenenbos and Ghaith 1964, Dan and Yaalon 1982). Many desert regions are essentially large evaporation basins, and, hence, salts and evaporite minerals of various kinds can be found widely, e.g., in soils and ephemeral lake basins (Meijer 2002). Accumulations of evaporite minerals may become available for entrainment and transport by wind

at some time or another (Reheis and Kihl 1995). Some of these substances are transported downwind, deposited on soil surfaces, and possibly translocated *into* those same soils by a subsequent rainfall event. We have already discussed carbonates, which are perhaps the most common type of soluble material that blows around in deserts. Gypsum and other salts are also available to the wind and are discussed later. Generally, similar pedogenic processes of mobilization and precipitation occur with evaporites; their pedogenesis varies mainly in intensity and detail.

The process whereby soils accumulate secondary *gypsum* (or calcium sulfate: $CaSO_4 \cdot 2H_2O$) is called *gypsification*. *Gypsic* (By and Cy) horizons contain significant amounts of secondary gypsum. If cemented (or nearly so), they are referred to as *petrogypsic* (Byy for accumulations >50%, or Byym for accumulations >90%) horizons, gypsum crusts, or gypcrete (Watson 1985, Dixon 1994b, Herrero and Porta 2000, Herrero 2004, Soil Survey Staff 2010).

Whether soils accumulate gypsum and how it is distributed within the profile, depend as much on a *source* as on the leaching regime. Many dry areas of the world lack sources of gypsum and are devoid of gypsic soils (Retallack and Huang 2010; Fig. 13.55). The primary source of gypsum for most gypsic soils is dust derived from gypsum-bearing rocks (Reheis 1987b, Dixon 1994b). Gypsic soils are relatively common near ocean coasts, because dry deposition of evaporated sea spray contains sulfate that can, in the presence of Ca, precipitate as gypsum (Fig. 13.56). Hydrogen sulfide gas (H_2S), an important source of oceanic gypsum, is derived from oceanic upwelling of gases produced by sea floor bacteria. When dissolved in fog, the sulfur is oxidized to $SO_2(g)$, which reacts with water to form sulfuric acid, H_2SO_4. The sulphate anion, SO_4^{2-}, enters the soil with precipitation.

At inland locations, gypsic soils occur near parent materials that are rich in gypsum, such as gypsiferous shale (Eswaran and Zi-Tong 1991). Some of the most common sources of gypsum in dry climates are playas (*aka* chotts or sabkhas) underlain by gypsum-rich groundwater (Nash *et al.* 1994). When groundwater rises to the soil surface by capillarity, gypsum precipitates, often as a surface crust, which can subsequently be deflated and blown downwind (Busson and Perthuisot 1977, Schwenk 1977, Watson 1979, 1985, 1988). Where capillary flow leads to surficial gypsum deposits, the crust is often referred to as *croûte de nappe*. If gypsiferous soils are eroded and the By or Byym horizon is exposed, it can deflate as well, becoming a gypsum source for other soils. When present in large amounts, gypsum resembles secondary calcium carbonate, – i.e., white-colored deposits in the profile or in a surface crust (Watson 1979, 1985).

Of all the common soluble materials that accumulate in dryland soils, only $CaCO_3$ is less soluble than gypsum, meaning that the carbonate maximum (depth of maximum illuvial carbonates) is usually shallower than the gypsum max, in freely draining desert soils. That is, By horizons are usually below Bk horizons (Fig. 13.57). Gypsum-rich horizons that *overlie* carbonate-rich horizons are a telltale sign

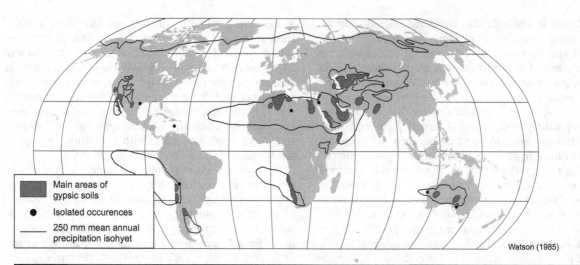

Fig. 13.55 Generalized global distribution of gypsic soils, including some playas and sinks.

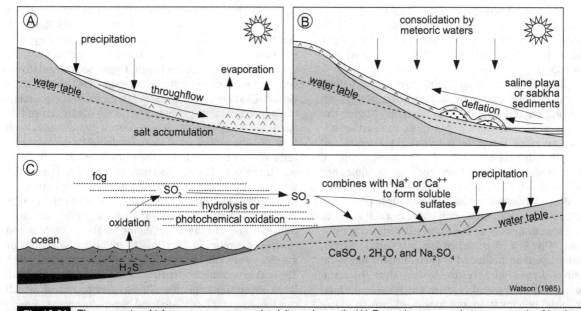

Fig. 13.56 Three ways in which gypsum sources can be delivered to soils. (A) Downslope accumulation as a result of leaching of hillslope sediments rich in gypsum, coupled with throughflow. Gypsum accumulates in lowlands via evaporation, often as a surface crust. (B) Deposition of gypsiferous dust directly onto soils; the dust is later translocated into the soil by precipitation. (C) Gypsum transported to coastal areas by fog or as dry deposition from sea spray.

that the gypsum has accumulated via capillary rise, or it may suggest a discontinuity.

In many respects, gypsification is similar to calcification. Depth to the maximum gypsum accumulation is related to mean annual precipitation (Dan and Yaalon 1982; Fig. 13.58), just as it is with carbonates (Fig. 13.43). Indeed, in many desert and semiarid soils, illuvial gypsum and carbonates are intermingled. The differences between secondary gypsum and secondary carbonate lie in gypsum's greater solubility (Carter and Inskeep 1988) and in the fact that it depresses the solubility of $CaCO_3$ (Reheis 1987b).

Because calcite and gypsum often co-occur in dryland soils, differentiating their respective illuvial accumulation horizons is critical to the accurate identification and naming of the horizon, as well as to the correct interpretation of the genesis of the soil. In the field, one can use dilute HCl to identify carbonate minerals; secondary gypsum will not effervesce when exposed to weak HCl, but carbonate

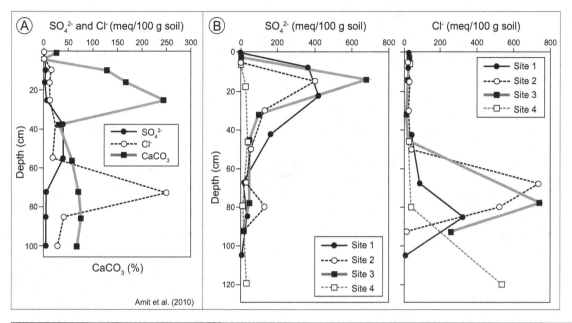

Fig. 13.57 Depth plots of soluble substances in some desert soils. (A) Chloride (reflecting NaCl and similar salts), sulfate (reflecting gypsum), and carbonate contents of a soil from the Namib Desert of Namibia, southwestern Africa. (B) Chloride and sulfate contents in some soils of the hyperarid Negev Desert of southern Israel (Sinai Peninsula).

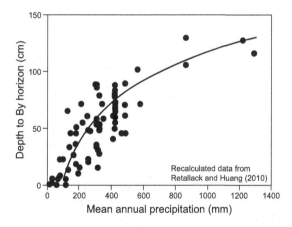

Fig. 13.58 Scatterplot showing the relationship between the depth to the gypsic horizon and mean annual precipitation, for soils of postglacial age, around the world.

minerals will. A useful test for pedogenic sulfate minerals is the identification of euhedral crystals with a hand lens (Carter and Inskeep 1988, Amit and Yaalon 1996). Indeed, secondary gypsum takes on a variety of forms in soils: powdery mottles, small to hard nodules, "snowballs," crystal clusters, pendants, interlocking plates, and sand-sized crystals ranging up to as much as ~28% by volume of the soil (Dan *et al.* 1973, Pankova and Yamnova 1987, Buck and Van Hoesen 2002). When viewed microscopically, gypsum crystals take on various shapes: single and radiating fibrously

shaped particles and random lenticular and granular crystals along channels and planar voids (Khademi and Mermut 2003). Gypsum may accumulate uniformly throughout sandy soils, while in finer-textured or gravelly material it may be more concentrated in masses or clusters. In gravelly or stony material, it also may accumulate in pendants below the rock fragments, as do carbonates. Unlike secondary gypsum, carbonates seldom occur as surface evaporite deposits, because of their lower solubility (Watson 1979). Like calcic horizons (Fig. 13.49), gypsic horizons appear to have discrete developmental stages; however, far less work has been done on this topic than for carbonates.

Silicification

Accumulation of secondary silica in soils is called *silicification*. Silicon is abundant in essentially all soils – in silicate minerals and silica-rich volcanic materials, e.g., tephra. Silicification of soils occurs in much the same way as the *per descensum* models for calcification or gypsification. Less commonly, silica-enriched horizons can also form *per ascensum*, as silica-rich groundwater moves upward via capillarity or evapotranspiration (Summerfield 1982).

Silica sources are omnipresent, but varied. Most soils have at least some silicate minerals such as feldspar and quartz, which can serve as silica sources (Boettinger and Southard 1991). Evidence for this is found, in part, in southern California, where primary silica in soils decreases with increasing duration of weathering, particularly in the more weathered upper profile (Kendrick and Graham 2004). That is, the upper profile loses silica over time, through weathering and translocation. Although quartz – a significant silica source in most soils – is difficult to weather and minimally soluble, Summerfield (1982, 1983) argued that quartz *dust* is characterized by disordered surfaces and smaller particle sizes, both of which render it more soluble (Siever 1962). Other, more weatherable, silica sources include volcanic ash and pyroclastic materials, as well as amorphous forms of silica, e.g., opal phytoliths and diatoms (Jones and Beavers 1963, Jones and Handreck 1967, Scurfield *et al.* 1974, Summerfield 1982; see Chapter 16). The geographic association of strong silica-cemented horizons called *duripans* with areas affected by volcanism lends support to the notion that volcanic materials such as tuffs, ignimbrites, and volcanic glass are excellent pedogenic silica sources (Soil Survey Staff 1999). Glass tends to weather readily and liberate a great deal of silica (Chadwick *et al.* 1989a).

Duripans

Soils that accumulate sufficient amounts of soluble silica may eventually develop Bq or Bqm horizons. The latter is a type of silica hardpan, or *duripan* (Flach *et al.* 1969, Southard *et al.* 1990, Soil Survey Staff 1999). Silica accumulation in soils is, taxonomically, set apart from that of carbonates and gypsum. Soil Taxonomy assigns names to *uncemented* horizons with illuvial carbonates (calcic) and gypsum (gypsic), but not to horizons with illuvial silica. Only when that silica becomes a *cementing agent* does the horizon become diagnostic, i.e., a duripan. Duripans, also called *silcretes* or *siliceous duricrusts*, are firm and brittle, even when wet (Stephens 1964, Summerfield 1982, Twidale and Milnes 1983, Dixon 1994b). Many duripans contain >90% silica, and often other illuvial materials, especially $CaCO_3$ (Watts 1977). Unlike carbonates or gypsum, illuvial silica does not impart lighter colors to soils. Therefore, cementation of soil fabric by silica is usually verified when fragments do *not* slake in water or after prolonged soaking in acid (HCl). Fragipans, which form in humid climates under forest vegetation (see previous discussion), appear to be partially cemented by silica as well. Fragipan and duripans differ in at least two ways: (1) Duripans are much more completely cemented, and (2) duripans do not slake in water.

Duripans (and gypcretes) have many morphological features in common with petrocalcic horizons (Callen 1983, Harden *et al.* 1991b, Reheis *et al.* 1992, Vaniman *et al.* 1994). Like other hardpans, duripans stand up as resistant layers, even when the overlying horizons have been eroded, reflecting the age of the geomorphic surface (Milnes and Twidale 1983, Moody and Graham 1997; Fig. 13.59). Most Australian duripans (silcretes) are extremely old – Tertiary age (Callen 1983, Milnes and Twidale 1983).

Most duripans form in soils of Mediterranean climates, which have xeric moisture regimes with a winter rainfall peak but a dry summer season. Weak duripans can occur in humid climates, however, where the soils have usually formed in volcanic materials.

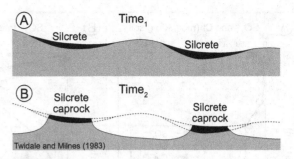

Fig. 13.59 Inversion of topography in arid parts of Australia, as induced by hard, siliceous duricrusts.

Duripans commonly break into very coarse prisms, commonly with coatings of opaline silica (amorphous silica material) lining the prism faces and large pores (Norton 1994). The prisms presumably form by slight volume changes that result from wetting and drying; they are absent in duripans of arid regions (Soil Survey Staff 1999). Like petrocalcic horizons, indurated duripans have an abrupt upper boundary, sometimes with a laminar top consisting of a nearly continuous layer of secondary silica (Boettinger and Southard 1991). Water often perches on top of the pan during the rainy winter season.

Strong duripans are dominated by spherical and ellipsoidal nodules of microcrystalline and opaline silica, with primary silicate minerals observable in partially weathered states (Boettinger and Southard 1991). These nodules can agglomerate into microscopic glaebules or microagglomerates (Chadwick *et al.* 1989a). Composed of silt and clay held together by poorly crystalline silica (or carbonate) cement, microagglomerates can eventually grow to become durinodes (Latin *durus*, hard, *nodus*, knot) – weakly cemented to indurated silica nodules ≥1 cm in diameter.

Blank and Fosberg (1991) proposed a two-step model for the formation of silica agglomerates: (1) *encapsulation* of quartz grains by silica, followed by (2) its *alteration* to opaline silica. They suggested that agglomerates can form when loess, rich in quartz and volcanic glass, is deposited on the surface of a soil with a duripan. Silt-sized loess particles are translocated, intact, to the top of the pan, where they become coated (encapsulated) with calcium carbonate and silica. Encapsulation may occur when perched water is lost to evaporation, forcing precipitation of calcite and silica. Then, in response to high pH conditions, the encapsulated loess agglomerates are altered to opal and other forms of amorphous silica. This model suggests that not all silica in duripans must be translocated in solution.

Silicification: The Process

All volcanic and most nonvolcanic soils have adequate amounts of silica-rich primary and secondary minerals, such that, via weathering, considerable amounts of silica can be released to the soil solution. And so perhaps the

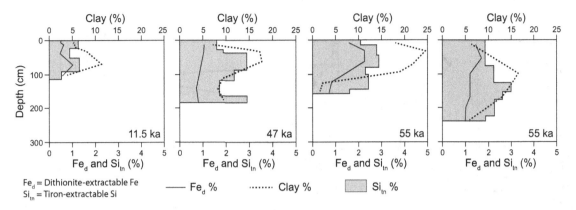

Fig. 13.60 Depth plots for some soils in southern California, showing the association of illuvial silica with clay and free Fe oxides. After Kendrick and Graham (2004); some data from Kendrick and McFadden (1996).

better question to ask is, Why does this silica remain in some soils, eventually to accumulate in illuvial horizons? Silica is translocated in soils either as monosilicic acid (H_4SiO_4) or as an aqueous sol (Beckwith and Reeve 1964). Why is it not leached? Let us consider this source-sink approach to silicification in more detail.

In arid climates, less silica enters the soil system (via weathering) than do carbonates and gypsum (via eolian influx). Thus, some Aridisols lack duripans entirely, and in those that have duripans, they are often masked and over-printed by strong (petro)calcic or (petro)gypsic horizons (Reheis 1987b, Boettinger and Southard 1991, Harden *et al.* 1991b, Eghbal and Southard 1993a, 1993b, Francis *et al.* 2007), making the duripan itself almost indistinguishable. In humid climates, amorphous silica may also enter the soil system, but it is commonly leached or biocycled. So then, what is it about the xeric moisture regime, with wet winters and hot, dry summers, that leads to the *production* of amorphous silica but not its *translocation from* the profile? Why do duripans form here?

The answer centers on weathering (production) versus silica mobility (translocation), and much of this balance is climatic. Just as in soils with illuvial carbonates, salts, and gypsum, soils with duripans occur where the climate is just wet enough to weather minerals and translocate dissolved silica, but not wet enough to translocate it out of the profile (Callen 1983, Southard *et al.* 1990, Thiry and Milnes 1991, Yaalon 1997b). Working in a Mediterranean-like climate, but with higher amounts of precipitation, both Merritts *et al.* (1992) and Langley-Turnbaugh and Bockheim (1998) documented large silica losses from soil profiles. Their data clearly show that dry summers and wet winters work in unison to liberate silica in soils. In drier, more traditional Mediterranean climates, silica is mobile but much of it does not leave the profile; it accumulates in Bq and Bqm horizons. These horizons are often overlain by horizons that have been variously depleted in silica (Torrent *et al.* 1980a; Figs. 13.60, 13.61). Conditions that enhance Si solubility

(dissolution) prevail particularly in the upper horizons of soils in xeric moisture regimes: (1) the presence of organic material, (2) low ionic concentrations in the soil water in winters, and (3) frequent wetting-and-drying cycles (Baker and Schrivner 1985). The role of plants, grasses in particular, in biocycling the free silica that exists in the upper profile may also be important to silicification in soils (White *et al.* 2012).

According to Kendrick and Graham (2004), the optimal conditions for silica *precipitation* in soils are (1) noncalcareous pH values, (2) fine textures (so as to maximize surface area), and (3) high ionic strengths (saturation or super-saturation) of the soil solution (Chadwick *et al.* 1987). In Mediterranean climates, soils usually lack $CaCO_3$ because of wintertime leaching. Therefore, for fine-textured soils with argillic horizons in xeric moisture regions, two of these three conditions are already met. These types of soils provide conditions that favor silica precipitation, especially in association with a Bt horizon – the depth at which most duripans form. Figure 13.60 illustrates the strong association of illuvial silica for clay and Fe oxides, both of which dramatically increase the surface area in soils and thereby provide a template for silica adsorption (Marsan and Torrent 1989). In the dry summer season, high solute Si concentrations are achieved in these xeric soils (White *et al.* 2012) and may promote precipitation of illuvial silica. And so it appears that the conditions and factors that promote the production of amorphous silica (in the upper profile) and its deposition (in B horizons) are all best met in the Mediterranean climates of the world.

In soils that are acidic in the upper solum, silica is released by weathering, into solution, and is precipitated in deeper horizons, where the pH is higher. Silica precipitation in a pH-friendly soil environment is primarily due to supersaturation of the soil solution. Thus, if present in the soil solution, silica tends to be deposited in horizons that have near-neutral pH values, either as amorphous opal or microcrystalline forms such as quartz and chalcedony;

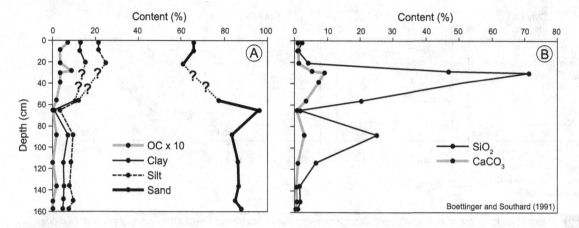

Fig. 13.61 Depth plots for a Xeric Haplodurid in southern California's Mojave Desert. Data from a thin saprolite (Crk horizon) seam at 47–75 cm have been omitted for clarity of presentation.

opal is most common (Summerfield 1983). The amorphous forms can later be transformed to quartz (Siever 1962). Repeated dissolution and reprecipitation of these silica compounds are commonplace, as evidenced by etched and embayed borders of quartz crystals and opal fragments in soils.

One important point relating to sources of silica centers on the differences between silica and carbonate solubilities. Calcite and quartz have inverse solubility relationships in alkaline environments (Summerfield 1982). The presence of carbonate in soils may itself be responsible for raising the pH to a point where forms of silica are more soluble and mobile. If the parent material contains abundant calcium, a calcic or petrocalcic horizon tends to cooccur with the duripan (Fig. 13.61), or it may occur immediately below the duripan, attesting to the slightly greater solubility of $CaCO_3$. Because of its greater solubility and abundance within dry regions, calcic horizons often form more rapidly than do horizons of illuvial silica (Reheis et al. 1992). Secondary carbonates preferentially precipitate in large voids on existing calcite crystals, plugging the soil faster than does illuvial silica. Indeed, many duripans require several tens of thousands of years to form (Othberg et al. 1997, Kendrick and Graham 2004). Also, secondary calcite is more readily *redissolved* than silica, facilitating the translocation of carbonates to lower positions within the profile (Chadwick et al. 1987).

In the initial stages of duripan formation, silica is released by weathering and reprecipitated as a cementing agent; it is translocated only short distances. Any adsorbed silica can provide a template for further precipitation. The silica deposition process, therefore, tends to predominate in small voids and at grain-grain contacts, cementing the grains without completely plugging the voids between them (Chadwick et al. 1987). Although active *biogenic processes* involved in duripan formation have not yet been

identified (Callen 1983), passive silicification of plants (replacement of plant parts with secondary silica) has been reported (Ambrose and Flint 1981).

Salinization

In this discussion of soils of dry climates we have, so far, focused on materials that are not readily soluble. But many salts *are* relatively soluble and can be translocated readily in soils by moving passively in percolating water.

Whether by natural processes or accelerated by human activity, the accumulation of water-soluble salts in soils is called *salinization* (McClelland et al. 1959, Parakshin 1974, Peterson 1980; Table 13.1). Horizons enriched in salts more soluble than gypsum are given the z suffix, e.g., Bz or Cz. Salts exist in soils as various combinations of the cations Ca^{2+}, Mg^{2+}, Na^+, K^+, and H^+, with the anions CO_3^{2-}, SO_4^{2-}, Cl^-, and NO_3^-, and water (Dan et al. 1982, van Breemen and Buurman 1998, Soil Survey Staff 1999, Meijer 2002, Amit et al. 2010). The large number of possible anion-cation combinations gives rise to a varied cocktail of salts. In this section we focus on the most soluble salts, e.g., sodium chloride (halite), soluble sulfates (thernadite, hexahydrite, epsomite, and mirabilite), and bicarbonates (trona and natron).

Salt sources are present in all soils. Ultimately, the main salt source can be traced back to the weathering of primary minerals, because weathering by-products usually contain small amounts of salt compounds (Gunn 1967). Nonetheless, salts in arid regions may have three *additional* sources: (1) dust blown onto the soil surface, (2) ions or soluble complexes in rainwater, and (3) salty groundwater (Fehrenbacher et al. 1963, Duchaufour 1982, Munn and Boehm 1983, Walthal et al. 1992, Reid et al. 1993, Bockheim 1997a; Table 13.7; Figs. 13.56, 13.62A). Eolian processes contribute salts to soil as dried sea spray, as salt coatings on clastic particles (dust), and in pyroclastic materials (Yaalon and

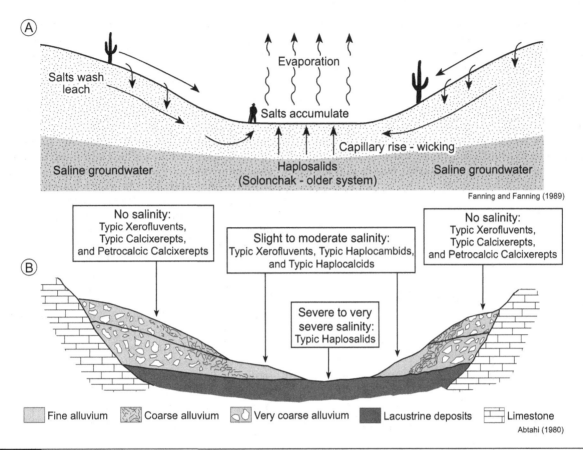

Fig. 13.62 Salt sources and processes of salt accumulation in dryland soils. (A) Idealized diagram showing how salts can accumulate in lowlands in dry climates. (B) Real-world example of the landscape depicted: a cross section of a valley in Iran, showing the relationships among parent materials, topography, and soil salinity/development. Note that the soils in the valley bottom are influenced by saline groundwater, while upland soils have mainly accumulated carbonates.

Lomas 1970, Dan and Yaalon 1982, Berger and Cooke 1997). Indeed, sea spray and fog are major sources of airborne salts (Watson 1985, Pye and Tsoar 1987, Schemenauer and Cereceda 1992, Langley-Turnbaugh and Bockheim 1998). In upland soils, the sources of salts are mainly atmospheric, especially via dry deposition from oceans and salty lakes (Gerson and Amit 1987), and – never to be ignored – from weathering bedrock. Groundwater as a salt source is, obviously, more common in low-lying areas.

Soils that contain soluble salts within the profile are most common in dry regions (for the exceptions, see later discussion), where evaporation demand exceeds precipitation, implying that salts cannot be completely leached from the soil (Szabolcs 1989; Fig. 13.63). Nonetheless, salts can accumulate in humid climate soils as well. Almost 8% of soils on ice-free lands of Earth are salt-affected (Middleton and Thomas 1997; Fig. 13.64). Pariente (2001) found that a threshold exists at about 200 mm of annual precipitation; soils receiving less than that amount often have high soluble salt contents, while soils receiving more than 200 mm typically have very few salts.

Salt content is indexed in the *laboratory* by the electrical conductivity of a saturated soil paste (Pariente 2001). Perhaps the only *field test* for salts (and gypsum) in soils is taste, because at low concentrations they impart no color to the soil. When present in high quantities, however, salts, carbonates, and gypsum readily impart white colors to soils (Fig. 13.65).

Saline (Salty) Soils

Primarily on the basis of their chemical properties, salt-affected soils are classified as either *saline* (or solonchak) or *sodic* (or alkalai) (Schofield *et al.* 2001; Table 13.9). Let us first discuss saline soils. Saline soils have high concentrations of soluble salts, often including not only sodium chloride (NaCl) but also gypsum, sodium sulfate (Na_2SO_4), and various other sulfates, bicarbonates, and chlorides. Saline soils are not normally dominated by sodium salts, and most have pH values <8.5. If present in high enough amounts, saline soils often develop white, salty surface crusts. Generally, a soil is considered saline if its salt content is high enough to inhibit plant growth. For some plants, this may occur when the electrical conductivity of a saturated paste extract is as low as 1 dS m^{-1}, but most plants are negatively affected when EC values are >4 dS m^{-1}.

Soils can become saline by the natural process of salinization or by human agency. Human-caused salinization is a serious resource concern for soil managers, because excess salts hinder the growth of crops by limiting their ability to take up water (Rengasamy 2006). Irrigated soils in desert areas often become progressively more salty over time, as small amounts of dissolved salts are added during irrigation (all natural waters contain *some* salts), but the salts are left behind as pure H_2O is evapotranspired. In many instances, insufficient irrigation water is added to the soil to leach the remaining salts thoroughly from the profile, and as a result, soils become increasingly salty (Jacobsen and Adams 1958, Hillel 1991, Qadir *et al.* 2000, Pisinaras *et al.* 2010). In other instances, poorly managed irrigation causes the local water table to rise, placing salty groundwater in contact with the rooting zone. Successful irrigation systems, like those in California's Imperial Valley (Fig. 13.66B), overirrigate the soils, thereby leaching salts from the profile, but also capture this salty water in subsurface drains and flush it from the system (Tait 2012). In the following, we focus on salinization as a *natural* pedogenic process.

In dryland settings, salts accumulate in two distinctly different landscape positions. The soil may have a zone of salt accumulation at the surface, which precipitated there as salty groundwater evaporated (Fig. 13.65A), or the salts may be scattered throughout the profile (Fig. 13.65B, C), depending on where the soil is positioned on the landscape. In lowlands, salts from salty groundwater can rise into the soil – or even to the soil surface – via capillarity (Rhoades *et al.* 1990; Figs. 13.62A, 13.65A, 13.67). These types of salty soils are often found in alluvial valleys and lowlands, especially if they have been irrigated for long periods. In upland settings, soils accumulate salts when vertical translocation of salts from the profile is minimal and deep percolation is rare. Often these soils have low permeabilities, limiting how much salt can be leached, or are immediately downwind from a salt source.

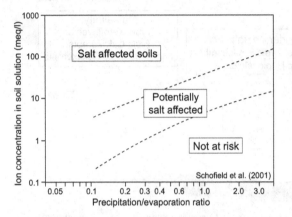

Fig. 13.63 Generalized distribution of salinization risk in soils, based on the precipitation/evaporation ratio and concentration of ions in the soil solution.

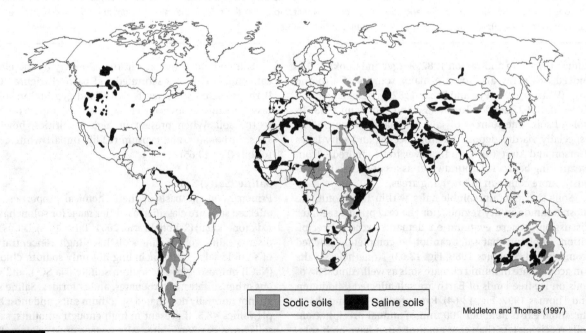

Fig. 13.64 Global distribution of saline and sodic soils.

Table 13.9 | Terms, chemical conditions, and morphologies associated with salt-affected soils

Name	Chemistry	Morphology	Soil Taxonomy information
Solonchaks, saline soils	Abundant amounts of neutral salts; pH <8.5	Some have surface salt crusts or accumulations; common in closed depressions, wetlands, and playas	Salids, with salic horizon
Solonetz, alkalai soils, sodic soils	High proportion of Na^+ (and/or Mg^{2+}) cations on exchange sites (at least 10x more than Ca^{2+}); pH >8.5; need not be saline (salty)	Shallow eluvial zone above dense, nearly impermeable, strongly columnar Btn (natric) horizon; dark, sometimes thick, A horizon; commonly fine-textured and on low-relief grasslands	Natrustalfs, Natrudalfs, Natrustolls, Natrargids
Solodized Solonetz	Acidic upper solum, but retaining high Na saturation at depth	Clay-poor, bleached E horizon; retains some columnar structure in the B horizon; structure is beginning to break down	Natrustalfs, Natrudalfs
Solod	Acidic profile, more Na at depth	Thin (or absent) E horizon over dark B horizon rich in illuvial humus	Albaqualfs, Haplustolls, Haplargids

Fig. 13.65 Salts in soils located in semiarid landscapes. (A) An evaporation-induced salt crust on the surface in a low swale, where salty groundwater is near to the surface. (B) Salts scattered throughout the profile of a Salid. (C) Salts (mainly between 18 and 35 cm depth) in a Torriorthent in the hyperarid Anza-Borrego Desert, southern California. Tape increments in centimeters below the soil surface. Photos by RJS.

Among Earth's dry climates, those that have a distinct wet season tend to favor salt accumulation most (Lewis and Drew 1973). During wet seasons, salts can weather out of minerals or be washed onto low-lying areas. Once there, salts can be translocated by percolating water to a constant depth. The dry season ensures that the salts are not completely flushed. Not only do regional and local conditions affect the movement of salts within the profile, but what you see today may be very different in just a few years. Salts are *that* mobile.

In upland, leaching regimes, salts with the highest mobilities will generally be found in the deepest horizons of the profile (Berger and Cooke 1997; Figs. 13.57, 13.68). Lowland areas are more complicated. In a lowland setting where soils are shallow to salty groundwater, the most mobile salts will move to the *highest* part of the profile by capillarity (Abtahi 1980, Young and Evans 1986; Figs. 13.62A, 13.65A). On a landscape-scale, soils at the bases of some slopes may have salts translocated *deeply* in the profile because of additional run-on of nonsalty water. Or they may have *more* salts overall because the run-on water carries additional salts with it. In sum, to understand how salts become concentrated in soils we must know where the sources are, what the forces acting on soil and groundwater are, and how both of these vary spatially and temporally (Arshad and Pawluk 1966, Seelig *et al.* 1990).

Soils undergoing salinization may develop a diagnostic subsurface horizon, the *salic* horizon, if they accumulate

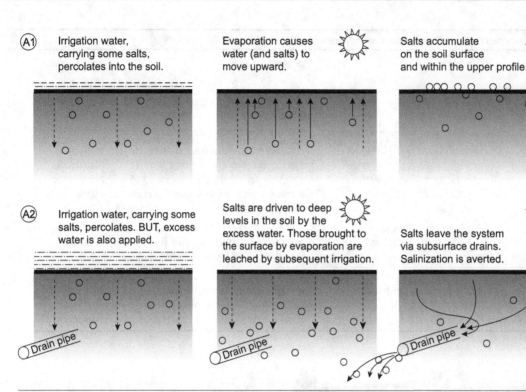

Fig. 13.66 Salinization and irrigation. (A) Schematic diagram showing (1) how salts accumulate in irrigated soils, because of evaporation, and (2) how the problem can be alleviated by overirrigation and drainage of the net percolation via subsurface drains. (B) Crops, irrigated by drip irrigation in California's Imperial Valley. In (1) is broccoli; in (2) are dates. Note the irrigated date palm grove in the background of (A), and in the far background of both pictures, the mountains of the California Desert. Salinization is not a serious issue in the Imperial Valley because of subsurface drains. Photo by RJS.

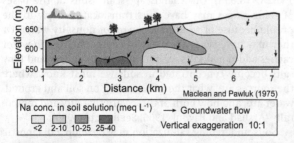

Fig. 13.67 Groundwater flow paths and Na content of groundwater in a rolling, semiarid, steppe in Alberta, Canada.

enough salts (Soil Survey Staff 1999). If cemented, they are *petrosalic* horizons (Amit and Yaalon 1996). In the FAO's World Reference Base, arid-region soils with these properties are called *Solonchaks*, from the Russian terms *sol* (salt) and *chak* (salty area). In Soil Taxonomy they usually classify as Salids. The terms *Calcium Carbonate Solonchak* and *Calcic Solonchak* are used for such soils if Ca and Mg salts are dominant (Redmond and McClelland 1959). Gypsum is also common in Solonchaks. Naturally occurring saline soils are common in saline seeps and river valleys in dry areas, on recently reclaimed marine sediments, and on parent

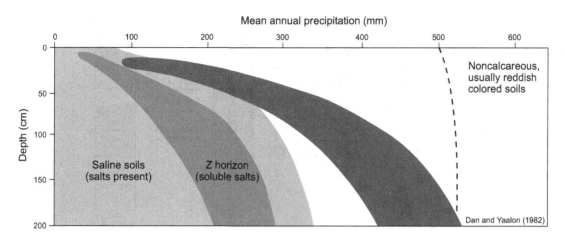

Mean annual precipitation (mm)

Fig. 13.68 Schematic diagram illustrating the depth of translocation of soluble salts, gypsum, and carbonates along a climate sequence in Israel.

materials that were originally salty. Nonetheless, this is key: Saline soils tend to be well flocculated, because of the presence of at least some divalent (Ca, Mg) cations and, hence, can retain high permeabilities (van Breeman and Buurman 1998). As we shall see (later), sodic/natric soils are much different!

Sodic (Natric) Soils

Many different types of salts exist in soils; some of the most common are compounds of Na, and to a lesser extent, Mg. Of the two, more emphasis is placed on Na$^+$ and its salts, because even a small amount of sodium can cause dramatic morphological changes in soils. Soils may be exposed to sodium salts from initially Na-rich parent materials, slightly saline groundwater, or salty run-on water (Kellogg 1934, Henry *et al.* 1985, Miller *et al.* 1985). Secondary sources may also be atmospheric, as rainout of aerosols derived from oceanic sources and playas (Downes 1954, Wilding *et al.* 1963, Gunn 1967, Dimmock *et al.* 1974, Ballantyne 1978, Gunn and Richardson 1979, Peterson 1980, Bockheim 1997). The amount of exchangeable sodium in soils is referred to as *sodicity*. Both the extent of sodic soils (Fig. 13.64) and the dramatic effects that sodium has on soil properties and behavior warrant a discussion of this topic.

In the World Reference Base, sodic, or alkaline, soils are called Solonetz (Table 13.9). The name is derived from the Russian terms *sol* (salt) and *etz* (strongly expressed). These soils are often *not* marked by high salt contents, but rather, by high *proportions* of Na (or sometimes, Mg) ions in the soil solution and adsorbed onto exchange sites (Miller and Brierley 2011). Solonetz soils usually contain enough of the alkaline salt NaCO$_3$ to cause clays to disperse, resulting in severe deterioration of the soil structure, particularly in the B horizon. Usually, sodic soils are fine-textured. Soils in which sodic salts are

dominant have pH values >8.5, reducing nutrient availabilities for plants (Duchaufour 1982). In addition to the obvious chemical differences, sodic and saline soils display different physical and biological properties, and they differ in their methods of reclamation and agricultural utilization (Schofield *et al.* 2001). They are also different because sodic soils are usually found in semiarid grasslands, not in true deserts, as is more common for saline soils (Fig. 13.64).

Sodic, or natric, soils are unique. As mentioned, Na$^+$ is very effective at deflocculating (dispersing) clay colloids. For this reason, when the Na saturation percentage in a Bt horizon exceeds 15%, it is classified as a *natric* (Btn) horizon (Soil Survey Staff 1999). Several pathways may lead a horizon to develop high levels of Na saturation, as discussed later. Nonetheless, with even as little as 15% Na saturation, clays will disperse and be free to migrate downward in the profile, forming (or enhancing) a Bt or Btn horizon below a coarser-textured, poorly structured E horizon (White 1978). Organic matter may also be dispersed through the Na-rich soil matrix, giving the soil a very dark color, hence the name "black alkali soils" (Byers *et al.* 1938). The excess Na, coupled with the dispersed organic matter, causes a decrease in structural (ped) stability, often transforming the Bt horizon from a permeable, well-structured horizon into a sticky, jellylike mass that, upon drying, is structureless and nearly impervious to water (Reid *et al.* 1993). This type of sticky, muddy condition is referred to as puddled. Air and water movement through puddled soils is extremely slow. Thus, Na salts can be considered a type of backdoor cementing agent, for they create a situation whereby soils become swollen, hard (when dry), and nearly impermeable. In this way, Na salts, such as carbonates, gypsum, and silica, can form pans (Bzm horizons). All of these pans are nearly impermeable, whether wet or dry. One notable difference: Bzm horizons are soft when wet;

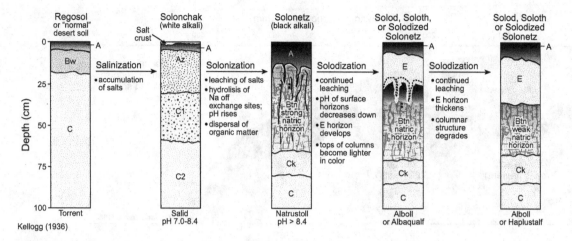

Kellogg (1936)

Fig. 13.69 Diagrammatic illustration of the classic salination-solonization-solodization sequence.

horizons cemented by carbonates, gypsum, or silica are hard even when wet. Gypsum in the lower profile is common in sodic soils (Bowser *et al.* 1962). Sodium saturation is harmful to plants in two ways: (1) Presence of too much Na⁺ in the soil is *directly toxic* to sodium-sensitive plants and disturbs uptake of essential plant nutrients, and (2) excess Na⁺ affects plant growth *indirectly* because the dense natric horizon obstructs downward percolation of water and the growth of roots.

The factors that lead to high Na contents in soils – *sodication* – are varied and usually different from those that lead to salinization. For example, the worldwide distribution of sodic soils, i.e., those high in Na, is often disjunct from that of saline soils (Fig. 13.64). Because they are normally quite soluble, the accumulation of sodium salts in these soils must, therefore, be due to some factor that either impedes the loss of sodium from the profile or adds it to the profile faster than it can be removed. Fine textures help – most sodic soils are fine-textured, keeping permeabilities low and facilitating capillary uptake of sodium in solution from below. In saline soils on old landscapes, selective removal of Ca from the soil by plants, followed by erosion of soil surface materials and plant debris, will concentrate Na on the exchange complex (Beadle 1962). It is common to see sodic soils in swales or at the bases of slopes, as a result of shallow, Na-rich groundwater (Sandoval and Shoesmith 1961, Miller *et al.* 1985) while upslope catena members are leached of sodium (Westin 1953, Lewis and Drew 1973; Fig. 13.67). An impervious subsoil, buried paleosol, or hardpan layer – especially if it also is a Na source – also facilitates sodication (Arshad and Pawluk 1966). The solubility of common Na and Mg compounds in soil, e.g., $Na_2SO_4 \cdot 10H_2O$, $Na_2CO_3 \cdot 10H_2O$, and $MgSO_4 \cdot 7H_2O$, increases sharply over the temperature range from 0°C to 30°C. Leaching of the sodium salts is much slower during wet but cold winters, leading to their

slow but steady accumulation in the profile. Salts of Ca and Mg are less soluble and, in this situation, would be precipitated above sodium salts (Munn and Boehm 1983, Sobecki and Wilding 1983, Reid *et al.* 1993).

The literature is rich with terms that describe sodium-affected soils (Byers *et al.* 1938, Johnson *et al.* 1985; Table 13.9). We review some of these terms here. *Salinization* can occur in any soil; this term refers only to the accumulation of *soluble salts*. Most soils that are saline do not necessarily become sodic, but they can if they are in the correct landscape position and exposed to internal or external sources of Na (Fig. 13.68). In order for a saline soil to become sodic, increasingly large percentages of the exchange sites on the clay minerals must become occupied by Na cations. To this end, the literature describes a process called *solonization* (or *desalinization*), which is the removal/leaching of soluble salts from the upper solum, possibly due to a change in climate or a change in the surface or subsurface hydrology. During this stage, some *salts* are leached, but high percentages of *sodium ions* may remain on the exchange complex. The term for this gradual process is *alkalization* or *sodication*, in which more of the exchange complex becomes dominated by sodium and the pH rises. Alkalization occurs because sufficiently large quantities of Na⁺ cations are added to the exchange complex, so that Ca and Mg ions are gradually replaced (White 1978, Munn and Boehm 1983). Solonization and alkalization result in the formation of sodic, or *Solonetz*, soils (Figs. 13.69, 13.70). We note, however, that if the source of Na ceases and bivalent Ca and Mg cations can compete effectively for the exchange sites (sometimes as aided by leaching), saline soils may evolve into Mollisol (Ustoll) morphologies (Byers *et al.* 1938). In fact, this is the mechanism envisioned for restoration of soils that have been enriched in Na by application of Na-rich irrigation water. Leaching the soil with low-Na water is often recommended as a remediation strategy.

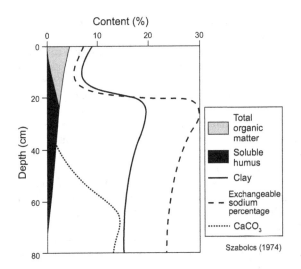

Fig. 13.70 Generalized depth plot of some physical and chemical properties of a typical Solonetz soil.

Fig. 13.71 A Natrudoll from Foster County, North Dakota, illustrating the columnar, biscuit top structure of the nitric horizon, and a light-colored calcic horizon at depth. Tape increments in inches below the soil surface. NRCS file photo.

Sodic soils may develop directly from nonsaline materials (Wilding *et al.* 1963) or may follow a stepwise evolutionary sequence from Solonchaks, as their Na contents change through time. This sequence proceeds from *solonization* or *sodication* (the accumulation of sodium salts) to *solodization* (the loss of sodium salts) (Miller and Brierley 2011; Fig. 13.69). Along this pathway, three kinds of sodium-affected soils are commonly recognized: the *Solonetz*, the *Solodized Solonetz*, and the *Solod*. These old Russian terms refer to profile morphologies associated with the accumulation of sodium, followed by its gradual loss. Although this salt cycle concept has long been accepted in Russia and the United States (Kellogg 1934, Byers *et al.* 1938), details are still emerging.

The hallmark of Solonetz soils is their coarse columnar structure within the upper part of a dense Btn (nitric) horizon (Figs. 2.11F, 13.69, 13.71). No other soil type has this kind of structure. Columnar structure exhibits light-colored, rounded *biscuit tops*, so named because of the white "powder" on the top of each column. This whitish material is *not* calcite, salts, or gypsum; it is essentially a thin zone of clean sand and silt, left behind by eluvial processes (White 1964).

Columnar structure is thought to form as follows. After the soil becomes a Solonetz, translocation pathways for soil water into the lower profile are few, mainly along desiccation cracks, as the Btn horizon has become nearly impermeable. By virtue of this focused leaching, Na in these areas is slowly replaced by H⁺ (see later discussion of solodization), allowing clay and organic matter to be translocated downward. Hence, the white biscuit tops, formed of uncoated sand and silt grains, can develop. The tops of the columns become rounded with time, as percolating water strips material more rapidly from their edges than from

the centers. Sediment in these cracks and along the tops of the columns eventually becomes increasingly stripped of organic coatings and clay, as percolating water is forced to flow repeatedly along these same, narrow pathways – not unlike in a degrading fragipan (Fig. 13.23). Under these high pH conditions, organic matter becomes more soluble and can translocate freely in soil water, further facilitating the development of the E horizon and white column tops. Indeed, black flakes of translocated organic matter can often be seen on top of the natric horizon, immediately alongside whitish, bleached mineral particles. Thus, over time, a thin, leached, slightly acidic E horizon can develop above the natric horizon. This horizon is composed of clean sand grains – the flour atop the biscuits.

This model of columnar structure formation is not universally accepted. Arshad and Pawluk (1966) suggested that the E horizon character in these soils may be due to degradation of organic matter in the lower A horizon, rather than to eluviation. White (1964) suggested another mechanism that can lead to the rounded biscuit tops: As the soil in the columns expands upon wetting, the centers of the columns swell upward.

Thorough leaching of the soil occurs in a third stage: *solodization* or *dealkalization* (Fig. 13.69). This stage sometimes is initiated as regional drainage improves, furthering deep leaching and inhibiting the capillary rise of water into the profile; hence, replenishment of the B horizon with Na is stopped. In short, the groundwater network becomes disconnected from the soil per se (Miller and Brierley 2011). As a result, Na is gradually *removed* from the profile, replaced by Ca. The calcium facilitates flocculation of the clays, thereby forming or reestablishing good structure and enhancing leaching (White and Papendick 1961, Bowser *et al.* 1962). The columnar structure breaks down, and the increasing numbers of cracks in the upper B horizon become stripped of clay and whitish colored, taking on a strong glossic look. The upper profile may become sandier and more acidic, while the unleached lower profile retains Na and has a pH near 9. The terms *Solodized Solonetz* and *Degraded Alkali* are often applied to soils like this, which were once saturated with sodium, but have since been partially leached (Mogen *et al.* 1959, White 1961). If enough calcium is retained within the profile, the soil may resemble a Planosol (White 1961).

Solodization is a long-term leaching process, and therefore, it is inhibited in extremely dry climates or where shallow bedrock impedes water flow (Arshad and Pawluk 1966). Nonetheless, in humid climates, with continued solodization, more Na is removed from the profile by leaching with fresh water (Byers *et al.* 1938). The columnar structure of the B horizon completely degrades into a crude, blocky form, resulting in a soil called a *Solod, Soloth,* or *Solidi* (White 1964; Fig. 13.69). Solods are not common. In wetter landscape positions, they have a thick E horizon over a dark B horizon rich in illuvial humus (Anderson *et al.* 1979). In older classification systems, these soils were called Planosols; many now classify as Albaqualfs (White 1964). On uplands, the resultant soil resembles an Ustoll; the E horizon is often absent. White (1971) showed that grasses, when growing on solodized soils, do not selectively biocycle sodium; if they did, it would represent a mechanism by which solodization could be offset. Invading grasses do, however, cycle Ca, further assisting in the flocculation of clay and the formation of pedogenic structure. Eventually, Solods may be so well leached of salts that they lose all outward signs of ever having been sodium-affected. As one might imagine, many soils exist that are intergrades among all of the types discussed, and often, they are difficult to compartmentalize or assign to a category sensu stricto.

Landscapes with Sodic Soils

Sodic soils seldom cover the entire landscape (Arshad and Pawluk 1966). Rather, they exist at preferred spots where the balance is shifted toward Na accumulation rather than leaching (MacLean and Pawluk 1975). The Soil Survey Staff (1951) referred to these kinds of landscapes, with their intricate complexes of sodic patches or spots, as "smallpox of the face of the steppe." Understanding the processes that flush Na from the soil versus those that replenish it, and understanding how – and why – this balance varies across the landscape are tantamount to explaining the genesis and distribution of sodic soils. This balance changes with time, season, and landscape position. Many sodic soils develop where sodium-rich bedrock is near the surface, as on summit and shoulder slopes, or where groundwater is

near to the surface (Parakshin 1974). But many others have different origins.

In some low-relief landscapes, sodium-rich soils occur in slight depressions called *slick spots* or *panspots* (Norton and Smith 1930, Westin 1953). The dispersed clays of the exposed B horizon make these areas slick or slippery when wet. Slick spots range in diameter from ≈1 m to 15 m. They are often devoid of vegetation or, at the most, are vegetated with halophytes because of their high Na contents, puddled structure, and low infiltration capacities (Jordan *et al.* 1958, Munn and Boehm 1983, Hopkins *et al.* 1991, Reid *et al.* 1993). In California, soils near slick spots had infiltration rates four to ninety-seven times greater than did soils beneath slick spots proper (Reid *et al.* 1993). Because of their minimal vegetative cover and dispersed clays (in the presence of Na), soils at slick spots are susceptible to erosion, by both wind and water. Johnson *et al.* (1985) reported on a slick spot in South Dakota with an erosional scarp on its upslope side that had retreated 7 cm in less than three years.

Slick spots represent interesting outcomes of the patchy distribution of Na in the landscape. But how do these spots form? Slick spot patterns may form and persist in at least four situations (Westin 1953, Anderson 1987): (1) An underlying source of sodium, such as saline shale, which also limits infiltration (Munn and Boehm 1983, Johnson *et al.* 1985. (2) movement of water from a source of Na in soils of microtopographic highs to those of microtopographic lows (spots). Better infiltration on microhighs fosters infiltration and loss of sodium, whereas sodium-rich water accumulates in the less permeable lows. Water movement is primarily driven by matric potential gradients, wherein moisture is preferentially pulled into clay-rich low-lying positions. As water (and clay) continue to move into low areas, they become increasingly clay-rich, further enhancing matric suction and creating a positive feedback situation (Munn and Boehm 1983). (3) Deflation of A horizon material from the spots, which moves any salty subsurface water into even closer contact with the surface. (4) An increase in bulk density as the soil structure collapses as a result of sodication, helping to maintain slick spots at low points on the landscape.

In the following we provide some examples that illustrate how soils and landscapes interact to form slick spots, as an early introduction to Chapter 14, which focuses on soils and landscapes.

Example 1. Panspots occur in semiarid Montana, where soils have deep water tables (Munn and Boehm 1983). Percolating water, carrying salts, gypsum, and carbonates, encounters Na-rich shale bedrock at shallow depths. At high points in the bedrock surface, Na replaces other base cations on the exchange complex, dispersing the subsoil (Fig. 13.72). Feedback processes set in, as the dispersed layer becomes drier than adjacent subsoil, because of enhanced runoff. Groundwater, carrying silica and Na,

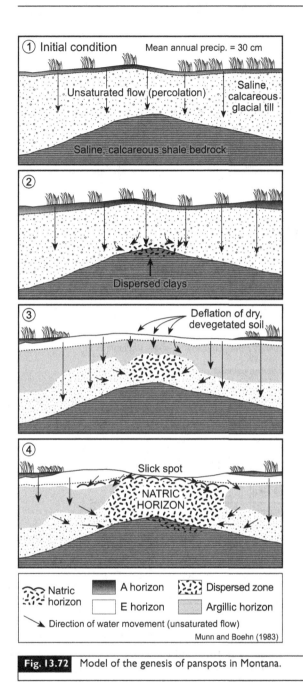

Fig. 13.72 Model of the genesis of panspots in Montana.

moves in response to matric potential toward the dispersed zone, further enriching it in sodium. The dispersed zone grows upward and outward. Vegetation on the panspot thins, facilitating erosion. Low infiltration rates in the panspot limit how much rainwater can enter, keeping it dry and further enhancing mainly unsaturated, lateral flow into it (Jordan et al. 1958). Eroded A horizon materials from the panspot expose the Btn horizon; concurrently, vegetated soils adjacent to the panspot trap eolian sediment and thicken. Although the panspots are sites of

preferred run-on, low permeabilities slow any leaching of Na from them. This model of sodic soil genesis is one of the few that do not involve a shallow water table.

Example 2. In humid southern Illinois, where one would expect soils to be leached of Na, slick spots are common on the flat, loess-covered Illinoian till plain. Here, 50–250 cm of loess overlies a paleosol in Illinoian drift, usually till (Fehrenbacher et al. 1963, Runge et al. 2008; Fig. 13.73A, B). The Na source for these soils is weathering of Na feldspars in the loess (Wilding et al. 1963, Indorante 1998). Natraqualfs and Natralbolls occur on this landscape, but most frequently where the loess cap is intermediate in thickness. Where the loess is thick, the soils are younger and, therefore, too little Na has been released via weathering to form natric horizons. In thin loess areas where soils are older, weathering and release of Na should be maximal, but mixing of loess with the weathered (leached) till below dilutes the Na-bearing mineral content.

Why the panspot pattern? The intricate pattern of Na-affected soils on these nearly level landscapes points to localized redistribution of Na ions, driven by a combination of several factors, as explained by Wilding et al. (1963) and displayed in Figure 13.73C. They assumed highly variable permeabilities in the paleosol that has formed in the till below. During wet, saturated flow conditions, wetting fronts carrying Na (weathered from Na feldspars in the loess) are diverted toward sites where the paleosol is more permeable. As a result, Na accumulates at the juncture of the flow lines, which represent the discharge of flow waters toward the more permeable zones in the underlying paleosol. However, before reaching the till paleosol, the discharge of soluble weathering products reaches the terminus of the wetting fronts. Here, Ca and Mg carbonates precipitate in the loess, aided by summer dryness. These carbonate precipitates act as nuclei for the subsequent formation of larger carbonate nodules. Then, as unsaturated conditions take over, more Ca, Mg, Na, and K salts differentially precipitate from solution, according to their own solubilities. The most likely salt(s) to precipitate first would be Ca and/or Mg carbonates (mostly $CaCO_3$) from supersaturation of Ca-bicarbonate solution upon soil drying and decreasing partial pressures of CO_2. Next would be Ca and Mg sulfates, e.g., gypsum. Precipitation of these less soluble salts increases the concentration of Na left in solution, because Na salts are more soluble than Ca and Mg salts. Eventually, Na ions begin to exchange for Ca and Mg on the soil cation exchange complex. Even when as little as 5–10% of the cations on the exchange complex has been replaced by Na ions, clays in lower subsoil B horizons begin to disperse, the soil begins to become physically unstable, and hydraulic permeabilities decrease. This situation is what Wilding et al. (1963) termed an Incipient Solonetzic condition, developing above the more permeable zones in the Illinoian till paleosols.

Water movement through the Na-rich (incipient Solonetzic) areas becomes restricted to unsaturated flow, as

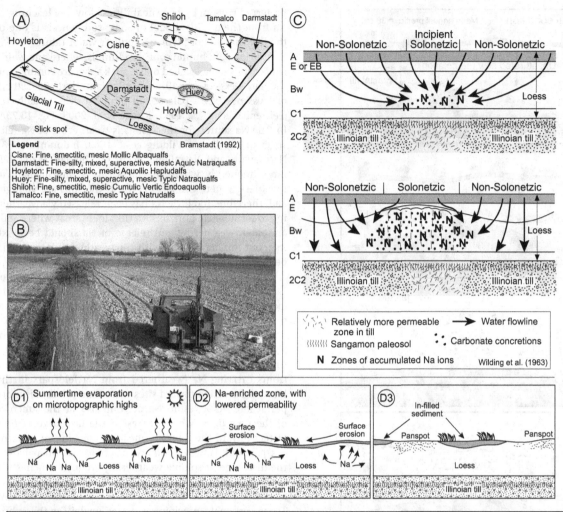

Fig. 13.73 The slick spot landscape of southern Illinois. (A) Block diagram of the Cisne-Hoyleton-Darmstadt soil association of southern Illinois, where thin loess overlies Illinoian till. (B) The flat Illinoian till plain of southern Illinois. The dark area in the distance is about 1 m lower than at the drill rig. Photo by L. Follmer. (C) A model of panspot genesis for southern Illinois. (D) Indorante's (1998) alternative model of the genesis of panspots on the southern Illinois landscape.

the dispersed B horizon in the loess soil becomes less and less permeable. Just as in the Montana landscape described earlier, Na continues to move into the drier, dispersed soil via unsaturated flow, setting up a feedback mechanism for preservation of the Na already there, and for its continued enrichment. The Na-enriched, slowly permeable B horizon grows laterally and upward, ever closer to the surface. From this point on, a kind of inversion of drainage characteristics takes place – what were once the most permeable pedons (at depth) are now Na-saturated and almost impermeable. The soils adjacent to the Solonetzic soils receive water that has moved from the slowly permeable soils, furthering their development and leaching and providing a self-limiting mechanism to the lateral expansion of the Na-enriched soils.

Indorante (1998) suggested an alternate genesis model for panspots in this landscape (Fig. 13.73D). On the most level sites on the Illinoian till plain, Na can become enriched on microtopographic highs by dry-season evaporation of soil water. Because these Na-enriched areas are less permeable, they become eroded over time, resulting in an extremely flat landscape with a seemingly random pattern of sodium-enriched panspots. Evidence of past erosion on panspot areas appears to be indicated by the shallow Bt horizons there.

Example 3. Reid *et al.* (1993) proposed a bioclimatic origin for slick spots in California, suggesting that they begin as patches are laid bare by grazing or burrowing (Fig. 13.74). The bare areas are subjected to wind and water erosion, exposing the Bt horizon. Surface crusting and dispersion of clays follow. Infiltration capacities are lower on the bare areas, and, thus, these spots become drier and less leached than adjacent soils. Matric gradients facilitate water movement in the form of unsaturated flow into the bare spots and upward, toward the dry surface (see also Ballantyne 1978). The least soluble substances (calcite) precipitate first, followed by gypsum and sodium salts (nearest the slick spot surface). In the end, solutes in soil water continue to enrich the slick spot (Munn and Boehm 1983). Silica and base cations in the soil water can even cause clays to neoform within the slick spot, further decreasing its permeability (Munn and Boehm 1983). Eventually, the vegetation

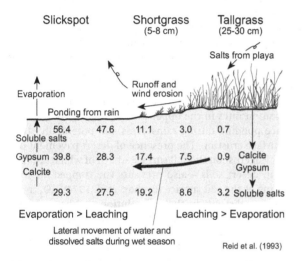

Fig. 13.74 Proposed genetic pathway for slickspot formation in a California grassland. Numbers within the figure represent electrical conductivity values (dS m⁻¹) of saturation extracts. Larger numbers imply more saline conditions.

on the slick spot dies, facilitating erosion, drying, and continued growth of the spot at the expense of adjacent vegetated pedons. Reid et al. (1993) used ^{14}C ages from relict roots in the slick spot to show that this type of salinization can occur in less than three hundred years. We conclude by noting that Hopkins et al. (1991) showed that panspots can eventually become revegetated via normal pedogenic and geomorphic processes, effectively ending the cycle.

Near-Surface Processes in Desert Soils

In this section, we transition from pedogenesis of dryland soils to some associated near-surface pedogeomorphic processes. The soil surface in deserts lies at the intersection of a wide variety of pedogenic and geomorphic processes. In some deserts, soils exhibit a surficial concentration of stones and gravel, often termed *desert pavement* (Sharon 1962, Ollier 1966). The stones in most pavements almost appear to interlock, such that little bare soil is exposed at the surface (Fig. 13.75). Often, the pavement is only one layer in thickness. Beneath many desert pavements is a porous, stone-poor horizon that typically has considerably more clay and silt than the underlying parent material (Springer 1958, Cooke 1970, McFadden et al. 1987, Dixon 1994a, Williams and Zimbelman 1994, Amit and Gerson 1986; Fig. 13.75A, B). The V horizon has well-formed vesicular porosity with many bubblelike pores. The V horizon should not be confused with horizons with the *v* suffix, which is used for plinthite (Table 3.2). When one walks on soils with V horizons, the gravel that the desert pavement comprises sinks a few millimeters into the soft V horizon below (Nikiforoff 1937).

Widespread across many different kinds of desert environments, the pavement and V horizon perform key roles in the ecology, hydrology, and dust emission potential of desert soils (Bunting 1977, Eckert et al. 1979, Young et al. 2004, Wood et al. 2005, Meadows et al. 2008, Goossens and

Fig. 13.75 Desert pavement on Torriorthents in the hyperarid Anza-Borrego Desert, southern California. The pavement stones in (B) and (C) have a strong coating of rock varnish. Tape increments in centimeters below the soil surface. Photos by RJS.

Fig. 13.76 | Vesicular (V) horizons. Photos by J. Turk.

Buck 2009). Because the genesis of V horizons and of desert pavements are interconnected (McFadden *et al.* 1998), we discuss them concurrently.

Vesicular Horizons

Originally described as having "foam" structure (Volk and Geyger 1970), V horizons are dominated by innumerable isolated, spherical pores called *vesicles*. V horizons are typically 5–20 cm thick, but they can be much thicker on old, stable surfaces (Wells *et al.* 1985, Anderson *et al.* 2002). Although the vesicles are stable and can persist for long periods, the overall soil structure in V horizons is usually weak because of lack of strong cementing agents, e.g., organic matter, clay, and carbonate minerals. Jessup (1960) referred to these horizons as apedal. These horizons are dominated by silts and fine sands, and they are generally gravel-poor (McFadden *et al.* 1998, Turk and Graham 2011). The pores tend to be ovoid, ranging in size from several hundred micrometers to a few millimeters in diameter, and not well interconnected (Fig. 13.76).

Although it is assumed that the vesicular nature of the V horizon can form very rapidly, perhaps in just a few months (Peterson 1980), until recently its actual genesis has been elusive. Various theories have been suggested, all of which center on a gas source. Most researchers have assumed that the vesicles form as gases, produced in the soil, cannot readily escape. A surface seal or trap of some sort, therefore, is often seen as a necessary prerequisite for vesicle formation. An early theory of gas and vesicle formation involved CO_2 release due to calcium carbonate crystallization when the temperature of a water-saturated surface crust increased rapidly (Paletskaya *et al.* 1958). Diurnal temperature changes in the soil have also been seen as an agent that can help form and stabilize the pores (Evenari *et al.* 1974, Brown and Dunkerley 1996). Nonetheless, recent research continues to support the early work of Springer (1958), who demonstrated that vesicular pores and tubules will form simply by wetting of dry soil, as the wetting front effectively traps air bubbles. During and after a rainfall event, the surface of the soil seals by puddling, and then, as the wetting front advances downward, gas pressures are elevated within the sediment matrix and form bubbles (Dietze *et al.* 2012). The vesicles then persist, as repeated advances of wetting fronts enhance surface sealing. Each wetting event causes gas pressures to develop, reduces intergrain connections, and, thus, forms even more of the isolated, spherical voids (Dietz *et al.* 2012). This process may be self-reinforcing, because formation of a V horizon with its noninterconnected vesicular structure lowers infiltration rates in the upper profile, leading to increased surface ponding during rainstorms and potentially stronger surface crusting. The presence of desert pavement or a biological crust on the soil surface – both of which are common in desert soils – also prevents the trapped air from escaping at the soil surface (Evenari *et al.* 1974). CO_2 in the pores may dissolve in the soil solution and be converted to carbonate ions, which combine with Ca^{2+} ions to form thin $CaCO_3$ coatings on the insides of the pores, making them more stable than they would otherwise be. Dietz *et al.* (2012) suggested that the air escaping from the soil repeatedly lifts rock fragments and, thus, may also contribute to the maintenance of desert pavements.

Most now agree that V horizons develop in what was predominantly eolian dust (Ugolini *et al.* 2008). For this reason, they are often best expressed on geomorphic surfaces that trap dust (Turk and Graham 2011). Initially rough microtopography, e.g., gravelly surfaces or recent lava flows, appears to favor dust entrapment and, hence, promote both desert pavement and V horizon formation (McFadden *et al.* 1992). This dust is later translocated into the soil – hence the high silt + clay contents, which also increase with soil age (McFadden *et al.* 1987, 1998, Anderson *et al.* 2002; Fig. 13.77). For example, V and upper B horizons in the soils of Cajon Pass, California, routinely contain 40% more silt than do the C horizons (McFadden and Weldon 1987). Likewise, *modern* dust contains significant amounts of salt and $CaCO_3$, which many desert parent materials lack, suggesting that the silty V horizon materials did not originate from the disintegration of the pavement rocks (Reheis *et al.* 1995). The eolian origin of the V horizon sediment has also been verified mineralogically and texturally; the sediment in these horizons is unlike that of the C horizon and strongly resembles dust caught in mechanical traps (Reheis *et al.* 1995, Ugolini *et al.* 2008). Dietz *et al.*

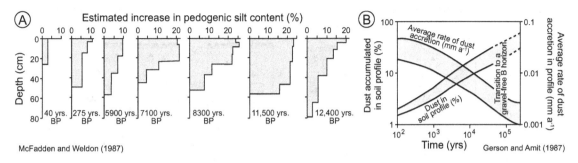

Fig. 13.77 Increases in silt content in desert soils. (A) Increased silt contents, above that of the parent material, in soils on alluvial terraces in southern California. The authors considered the minimum increase in silt content due to pedogenesis to be 3%; the remainder is from eolian dust. (B) Accumulated eolian dust in soils – and rates of accumulation– in the Negev Desert, Israel, as a function of time.

(2012) showed that the shape of vesicles may be related to dust influx intensity. Horizons with more spherical vesicles may indicate constant low dust influx on poorly vegetated surfaces and, therefore, may be used as proxies of paleoenvironmental conditions.

V horizons can form very quickly. Thin V horizons have been observed on geomorphic surfaces less than 100 years old (Gile and Hawley 1968). Chronosequence studies in the Mojave Desert have suggested that the vesicular horizons are the some of the first indicators of soil development (McFadden et al. 1986, McFadden 1988). McFadden et al. (1998) dated the sediment in V horizons by thermoluminescence, a technique that determines the last time that the sediment was exposed to sunlight (see Chapter 15). They found that most of the material was about five thousand years old, while Bw material below was last exposed to light about thirteen thousand years ago. [14]C of the organic matter within V horizons yielded comparable ages (Anderson et al. 2002). Gerson and Amit (1987) suggested that only one thousand to five thousand years is required for a continuous V horizon to form in the Negev of Israel (Fig. 13.78). These data add support to the hypothesis that the fine material in V horizons was originally eolian and explain why soils older than only a few thousand years already have developed V-Bw horizon sequences.

Desert Pavement

Armored surfaces of intricately intertwined rock mosaics that are usually one or two layers thick and overlying thin but stone-poor soil horizons are often referred to as desert pavement. Pavements are prominent in deserts worldwide, commonly being found on gently sloping sites where vegetation is minimal and precipitation is very scarce, i.e., in the driest deserts (Dan et al. 1982, Eppes et al. 2010). In Australia, they are called *gibber plains* and *stony mantles*; *hamada, reg*, and *serir* are Arabic terms for the same, whereas in parts of Asia the term *gobi* is used (Sharon 1962, Cooke 1970). Once formed, desert pavements are extremely stable. Matmon et al. (2009) reported on pavements in the Negev Desert that

may approach 2 million years in age. Extreme hyperaridity, tectonic stability, minimal relief, and armoring by clasts with highly resistant lithologies have all contributed to this extremely old pavement. As a result of their potential long-term stability, pavements may preserve and maintain any archaeological artifacts that may have been deposited on the surface (Adelsberger and Smith 2009).

Rocks in desert pavements are often aligned such that they have a flat surface parallel to the soil surface; i.e., individual rocks usually do not overtly protrude above the general level of the pavement or downward into it (Cooke 1970). They seldom overlap. The rocks are all about the same size and close-packed, often appearing pieced together like a jigsaw puzzle mosaic (Fig. 13.75C). On slopes, the rocks are often aligned preferentially (Abrahams et al. 1990, Dietze and Kleber 2012). Once formed, pavements are an important stabilizing factor for both the slope and the soil, acting as a protective armor (Sharon 1962). The lack of vegetation on desert pavements is probably a positive feedback mechanism, as the pavement itself may inhibit vegetation from becoming established, and the paucity of plants minimizes the risk of the pavement's being destroyed. Cool, high-elevation areas in deserts tend not to have pavements, probably because of increased disturbances from roots and animals (Fig. 13.79). Quade's (2001) work in the Mojave Desert of California showed that pavements form continuously only at elevations *below 400 m*, which remained hot and dry even during glacial periods. Pavements formed on sites above that elevation only during dry, warm interglacial climates. During glacial periods these areas were wetter and densely vegetated, inhibiting pavement formation.

Early work on the origin of desert pavement attributed its development to *erosion* of fines by wind and/or water, leaving behind a surficial *lag concentrate* (Lowdermilk and Sundling 1950, Symmons and Hemming 1968, Parsons et al. 1992, Khresat and Qudah 2006; Fig. 13.80). Erosion of fines by water has certainly been shown to be important in some desert settings (Parsons et al. 1992, Williams and Zimbelman 1994, Rostagno and Degorgue 2011), and

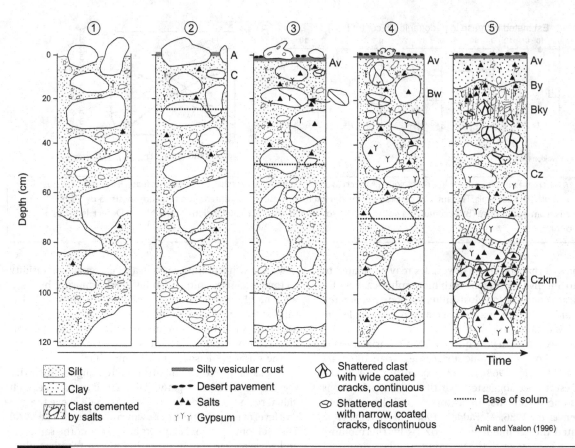

Fig. 13.78 Impact of dust on the morphology of a reg soil in the Negev Desert, Israel, showing the various developmental stages associated with pedogenesis and dust inputs to gravelly parent material. Most, but not all, horizons are named. (1) Recently deposited gravelly alluvium with low salinities. (2) Dust accumulates in the porous alluvium, forming a thin A horizon. Salts form a thin surface crust, partly biogenic in origin, and occur as thin coatings on the undersides of clasts. (3) Continued dust accumulation leads to the formation of a V horizon, concurrently with the formation of a desert pavement (early stages). Salts (mostly gypsum) in near-surface horizons cause clast shattering. Gypsum begins to accumulate in the subsurface. (4) Dust continues to accumulate, as a cambic Bw horizon forms below the V horizon. A nearly continuous desert pavement has formed. High amounts of salts accumulate, especially in the C horizon, although salt weathering of clasts is especially intense in the upper profile. Some gypsum nodules begin to form. (5) The soil now has a strong desert pavement, V, and Bky horizons. Halite dominates the petrosalic (Czkm) horizon, composed mostly of halite and shattered gravel. Thick, continuous salt cutans have led to intense rock shattering in the upper 60 cm. Gypsum peaks high in the profile.

it does appear that rain splash and water do play *some* sort of role in the reestablishment, if not in the original formation, of pavements (Poesen *et al.* 1998, Wainwright *et al.* 1999). The lag-erosion model is especially pertinent in arid areas that have some vegetative cover. For example, in arid and semiarid Patagonia, Rostagno and Degorgue (2011) provided data that support an erosional model for desert pavements. Here, grass-covered patches with soils that conserve an intact A horizon exist adjacent to soils with desert pavements. The pavements are commonly found in shrub interspaces, where processes of surface erosion are, presumably, more effective.

Today, however, the erosional model for desert pavement formation is not viewed as being widely applicable,

for several reasons. First, surface crusts, which are common in desert soils, would inhibit deflationary processes, limiting wind erosion. Erosion by water cannot explain the apparent interconnectiveness of the rocks, or the planar surface nature of the pavement (Fig. 13.75C); running water should form at least a few rills and gullies. An erosional model also cannot explain the presence of a V horizon below the pavement (Cooke 1970, Williams and Zimbelman 1994). Even if deflation (wind erosion) of fines can help *maintain* an existing pavement (Jessup 1960), it probably cannot solely *create* one.

Other models of pavement formation, e.g., Springer (1958), Jessup (1960), Ollier (1966), have suggested an alternative explanation. Could desert pavements have formed

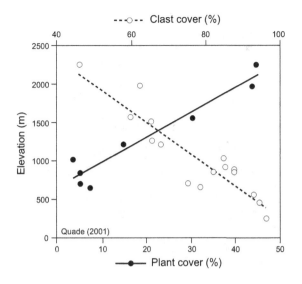

Fig. 13.79 Plant (trees and shrubs) and desert pavement (clast) coverage versus elevation for some sites in the Mojave Desert, California. These pavements formed during the last interglacial (MIS 5), when the climate was warmer and drier, and plant cover was less.

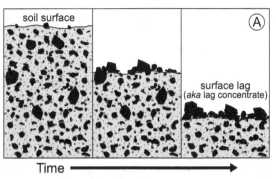

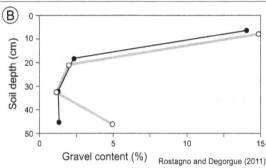

Fig. 13.80 Surface lag gravels. (A) Theoretical formation of a surface lag by selective erosion of the finer materials. Erosion can occur by wind or water. (B) Depth plots of gravel content for two soils in semiarid Patagonia that have developed lag concentrates by surface erosion.

by the upward migration of stones through clayey soil horizons, via argilliturbation (Fig. 11.47)? Cooke (1970) has extended this hypothesis to include shrink-swell processes associated with thawing and freezing or with salt crystallization. This possibility seems logical in smectite-rich soils, in wet-dry climates, but it is not likely to be applicable in deserts where the underlying B horizons are silty, and certainly not in hyperarid settings.

A more recent model, proposed by McFadden *et al.* (1987) and well supported by the work of several others (Wells *et al.* 1995, Haff and Werner 1996, Ugolini *et al.* 2008, Adelsberger and Smith 2009, Dietze and Kleber 2012), is now generally accepted as the best explanation for the concurrent formation of desert pavement *and* vesicular horizons. This model attributes desert pavement formation to (1) the accumulation of clasts on the surfaces of sloping and low-lying areas via mass movement and slope processes, in conjunction with (2) the detachment and slow uplifting of clasts as eolian fines infiltrate into and below the clast-supported matrix. The model requires that there is a source of small rocks (presumably formed by physical weathering) *as well as* inputs of eolian dust (Anderson *et al.* 2002). Rough surface topography and large interstices between rocks and gravel facilitate dust trapping and accumulation (Fig. 13.81A). As the V horizon forms below it, the pavement is actually lifted, as dust moves into fractures between rocks (Fig. 13.81A). Dust that lands on the surface but is not immediately incorporated is susceptible to deflation and water erosion. Thus, the desert pavement acts to protect the underlying, accumulated dust from further

erosion, while the rocks help to trap some dust that is later incorporated into the profile. McFadden *et al.* (1987) stressed that these pavements "are born and maintained at the surface" (Wells *et al.* 1995). That is, they need not evolve from rocks scattered at depth – a key component of any erosional model. One study, using cosmogenic ³He dating of a desert pavement consisting of basalt clasts from a lava flow, confirmed that the rocks on the pavement had been *exposed at the surface* as long as the basalt on nearby bedrock highs had been exposed; i.e., they *were never been buried* and so could not have been exhumed by erosion (Wells *et al.* 1995).

Concurrent with desert pavement formation, V horizons evolve as dust is trapped and incorporated into the soil (McFadden *et al.* 1987, 1998). As Pelletier *et al.* (2007) put it, both features form by an inflationary process. During times of low eolian input, infiltration events lead to the formation of weak, shallow Btk and Bky horizons below the V horizon. Salts can accumulate too. Gerson and Amit (1987) reported that gravelly surfaces may trap ≈0.1 mm/yr of dust initially, decreasing to several micrometers per year, as a result of plugging of the surface horizons with dust and salts. Small amounts of clay in the V horizon can lead to shrink-swell processes, small vertical cracks, and weak structure (Cooke 1970). Additionally,

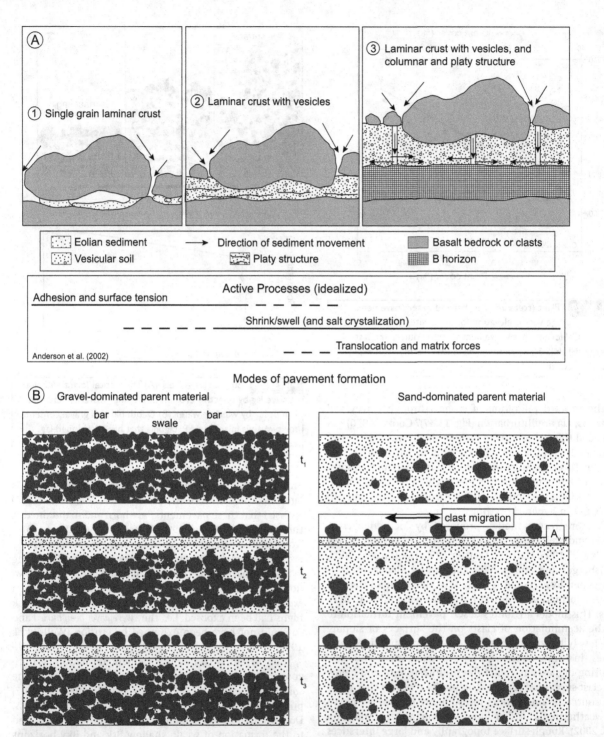

Fig. 13.81 Models of desert pavement development by aggradational processes. (A) Formation of desert pavement and the underlying V horizon by the slow accumulation of eolian dust. Emphasis is placed in the influence of eolian and pedogenic processes on the lifting of surficial clasts, which results in the formation of an aggradational desert pavement. (B) An elaboration of the preceding model by Pelletier *et al.* (2007), in which two different types of sediment (gravelly and sandy) are envisioned at time$_{zero}$, which also shows how pavement rocks are forced to move laterally by subtle processes of freeze-thaw, wetting-drying, and bioturbation.

if the dust contains some salts, precipitated crystals will expand upon drying, forcing rocks upward and outward even more. Together, these processes allow the fine materials to settle below the pavement rocks, further separating the clasts from the subsurface and subtly moving them laterally, thus smoothing out any preexisting surface microtopography and forming the typical interlocking pattern (Fig. 13.81B). Although small, the cracks facilitate translocation of silt and fine sand, allowing the uppermost V horizon to become finer and more clay-rich through time (Gerson and Amit 1987). Prism faces, therefore, are silt-coated and noncalcareous, whereas ped interiors are reddened and carbonate-enriched. As these processes continue, the soil surface grows continually upward by eolian accretion, and columns/prisms (once within the V horizon proper) are gradually buried. Coalescence of these subsurface peds forms a weak, essentially continuous Bk or By horizon. This model may apply to stone pavements in Australia that were explained incompletely by other (Mabbutt 1965) hypotheses.

Pavements can form quickly (Haff and Werner 1996, Marchetti and Cerling 2005), and after disturbance, they can also heal quickly. The fact that pavements develop and change predictably over time makes certain aspects of their characteristics useful – within limits – as relative dating indicators in some circumstances (Wells *et al.* 1985, Amit and Gerson 1986, McFadden *et al.* 1989, Al-Farraj and Harvey 2000, Helms *et al.* 2003, Valentine and Harrington 2006; see Chapter 15). For example, over time, physical weathering leads to more and more cracks in the rocks of desert pavements (Eppes *et al.* 2010), and some have suggested that older pavements have progressively smaller rocks (Al-Farraj and Harvey 2000). Older pavements also tend to be smoother and have a more interlocking pavement surface (Al-Farraj and Harvey 2000), as well as thicker and less gravelly B horizons (Amit *et al.* 1993).

Thickening of the V horizon profoundly affects the hydrology of the profile (Yair 1987, McFadden *et al.* 1998). Thicker V horizons lead to shallower penetration of wetting fronts, because the silty material has much more surface area and poorly interconnected porosity than did the gravelly materials that predated it. This, in turn, affects the depth of carbonate accumulation within the profile, such that (older) soils with V horizons accumulate carbonates and salts at shallower depths than do (younger) gravelly soils. This concept can be extended to other desert soils that have accumulated eolian materials over time: as their eolian mantles thicken, wetting fronts are unable to penetrate as deeply, maintaining dry conditions in the solum for longer periods (Yair 1987, Eppes and Harrison 1999). For this reason, in gravelly and rocky soils, other factors being equal, B horizons are deeper and soils are more deeply leached than in soils on the same geomorphic surface that have accumulated a silty mantle (Yair 1995).

Because of the pedogenetic relationships between the formation of desert pavement and the V horizon, disturbance of the pavement can very quickly lead to erosion of the underlying material (Prose and Wilshire 2000). Once its protective armor is lost, the V horizon becomes vulnerable to wind and water erosion. Disturbance by foot traffic of all kinds, as well as off-road vehicles in deserts, can destroy the interlocking pavement and make the surface susceptible to deflation (Caldwell *et al.* 2008, McLaurin *et al.* 2011). Fortunately, some of the same processes that lead to its formation also facilitate rapid healing (Sharon 1962, Haff and Werner 1996, Wainwright *et al.* 1999). Although we have made great strides in understanding the origins of desert pavements, we still have much to learn (Amundson *et al.* 1997, McFadden *et al.* 1998).

Biocrusts

Many Aridisols, with or without a desert pavement, have a thin surface crust. These *biocrusts* are readily recognized in Aridisols that lack desert pavement. But they are present in areas with pavement as well – either on top of or below the rocks in the pavement. The crusts are particularly common on bare surfaces that are occasionally wet or even flooded, such as playas and interdune areas. Although only a few millimeters thick, biocrusts can dramatically reduce infiltration rates and increase runoff after even small storms, in part because they are slightly water-repellent (Jungerius and van der Meulen 1988, Yair 1990). Such crusts have long been assumed to be rainfall-induced, because they form after rain events on apparently bare surfaces. Physical reorganization of particles in response to raindrop impact is responsible for the close-packed structure, but the source of the binding agent has been elusive. Small amounts of salts may provide some cohesion (Sharon 1962).

Most of these crusts are now known to have biogenic origins (Fletcher and Martin 1948, Macgregor and Johnson 1971, Dor and Danin 1996, Danin *et al.* 1998, Büdel 2001, Kidron *et al.* 2012, Williams *et al.* 2012). In the driest deserts, biocrusts are dominated by cyanobacteria (*Microcoleus* spp. in particular), but they also can have origins related to chlorophyte algae, heterotrophic bacteria, fungi, mosses, and lichens (Mazor *et al.* 1996, Williams *et al.* 2013). As these organisms grow in the upper few millimeters of the soil – typically after rain events – their roots, mycelia, and filaments bind the soil particles together, forming biocrusts enriched in C, N, clay, and silt (Fletcher and Martin 1948).

Soil biocrusts are an extremely important part of the dryland soil system (Pointing and Belnap 2012, Williams *et al.* 2013). For example, they facilitate the accumulation, morphology, and ecosystem function of dust, and dust is a key component of calcic horizon formation (Williams *et al.* 2012). Most of the biological activities and, therefore, the ecosystem processes, in desert soils occur in this top layer. Pointing and Belnap (2012) even went so far as to suggest that the critical zone in deserts should be restricted

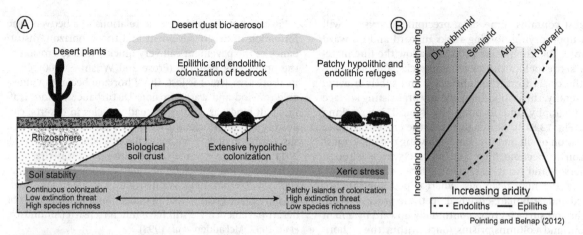

Fig. 13.82 Distribution and character of biocrusts in desert soils, across an aridity gradient. (A) Distribution and types of biocrusts. (B) Relative abundance of endolithic and epilithic organisms.

to the top few centimeters of rock or soil surfaces, rather than the more traditional "depth of rooting." It is in this shallow critical zone that biological communities, particularly those that inhabit biocrusts, operate. Processes operating in this surface layer mediate and affect almost every ecosystem process that occurs in deserts, including weathering, C and N cycles (Evans and Lange 2001), and dust capture. Biocrusts ensure the bioavailability of P and other soil nutrients, regulate decomposition, and influence soil stability (McKenna Neuman et al. 1996). They mediate almost all inputs and outputs (gases, nutrients, and water) to and from the zones above and below the surface. Unquestionably, they play important roles in surface horizon hydrology (Kidron and Yair 1997, Belnap 2006) and seedling germination (Prasse and Bornkamm 2000). As a result, any change or disturbance in their cover or biomass may have severe negative consequences on many ecosystem processes.

Considerable work has been done on the role of biocrusts in weathering (Danin and Garty 1983, Schwartzmann and Volk 1989, Viles 1995, Gorbushina 2007). Microorganisms that colonize the undersides of (usually translucent) rocks, such as quartz, are called *hypoliths* (Schlesinger et al. 2003, Warren-Rhodes et al. 2007, Azúa-Bustos et al. 2011). These lifeforms are particularly important in hyperarid environments (Tracy et al. 2010; Fig. 13.82). *Epiliths* colonize the upper surfaces of desert pavement stones and assist in the formation of rock varnish (see later discussion). *Endoliths* are found on and within exposed bedrock and thrive especially where the bedrock is porous, e.g., sandstone, or cracked, e.g., limestone (Friedmann 1980). This type of refuge provides significant protection for the microorganisms and may explain why endoliths are common across all desert types (Pointing and Belnap 2012). Bioweathering of exposed bedrock is often significantly enhanced by epilithic and endolithic microorganisms.

Rock Varnish

Over time, rocks in desert environments acquire a variety of coatings (Dorn 2009a). Many rocks on relatively old geomorphic surfaces in arid climates display a glossy, dark brownish black or gray patina called *rock varnish, desert varnish,* or *desert lacquer* (Fig. 13.75C).). The name "rock varnish" is preferred to "desert varnish," because such coatings can be found on rocks in every terrestrial environment (Krinsley et al. 2012). It is commonly also found on archaeological artifacts and petroglyphs (Loendorf 1991, Zerboni 2008). The varnish is a thin coating rich in clay minerals and hydroxides of Mn and Fe (Broecker and Liu 2001, Dorn 2004; Fig. 13.83). According to Dorn (2009a), clay minerals compose roughly two-thirds of a typical varnish, Mn and Fe oxides a quarter, with the remainder being oxides of trace elements. When compared to Fe, Mn contents in rock varnish are often enriched by more than fifty times, as compared to potential source materials such as dust, soils, water, and the underlying rock (Dorn 1998, 2009a). The Mn and Fe oxyhydroxides are effective at affixing clay particles to the host rock, presumably transported there by wind. Varnish is typically only found on the upper surfaces of rocks that have been unmoved for long periods. The undersides of the rocks (or parts that are buried) lack this varnish and take on an orange or yellow hue; some retain their original surface color (Dorn and Oberlander 1981b).

Rock varnishes are usually composed of alternating, micron-scale layers of detrital clay minerals, which cumulatively can range from 2 to ≈200 μm thick (Potter and Rossman 1977, Perry and Adams 1978, Taylor-George et al. 1983, Liu and Broecker 2008; Fig. 13.83). Across any given rock surface, its thickness can vary considerably, usually being thickest in microbasins or depressions on the surface. It is, as noted by Liu and Broecker (2000), the slowest known accumulating terrestrial sedimentary deposit, growing at rates ranging from <1 to 40 μm/1,000 years

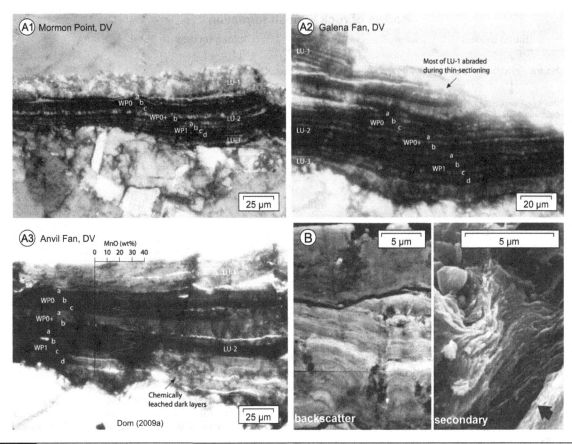

Fig. 13.83 Magnified images of rock varnish. (A) Cross-section photomicrographs of microlaminations in rock varnish from various sites in the southwestern United States. Ages shown are maximum-limiting ages for varnish formation, (cal yr BP). WH = Wet event in Holocene, WP = Wet event in Pleistocene, LU = Layering unit. Image from http://www.vmldating.com, with permission of T. Liu. (B) SEM images of rock varnish, showing its layered nature.

on subaerially exposed rock surfaces. Dorn (2004) likened rock varnish to a brick wall in which clay minerals are like the bricks (Potter and Rossman 1977), and Fe and Mn hydroxide minerals act as mortar and provide the color. Dark brown to black colors occur when elemental concentrations of Mn and Fe in the varnish are about equal; the color turns more orange when Mn concentrations drop (Dorn 2009b). The smoothness or sheen of the varnish also appears to increase as a function of Mn content. Rock varnish coatings usually become thicker, darker, and more continuous over time, as patches of varnish coalesce (Perry and Adams 1978). The oldest varnishes can be essentially black. Color, Mn content, and thickness/coverage of the coatings have all been used as relative age or surface exposure indicators (Reneau 1993). Because the varnish acquires different chemical compositions as a function of climate, it is a highly useful tool for paleoclimatic interpretation. In the drylands of the western United States, dry periods have been correlated with Mn- and Ba-poor layers,

while wetter periods favor Mn- and Ba-rich microlaminae (Liu and Broecker 2008).

Rock varnish tends to form only in *periodically* dry regions, because the microbial flora that form it are best suited to short growth periods after rain events, or during cooler and wetter climatic intervals. In humid regions, soil formation and rock dissolution are so rapid that any varnishes that might form are quickly stripped (Krumbein and Jens 1981, Dorn 2009a). Similarly, rock varnish usually forms only on hard, resistant rocks; any varnish that forms on soft limestones and granites is quickly lost as the rock disintegrates.

Many aspects of rock varnish formation are now fairly well understood, although some debate and uncertainty remain (Dorn 2009a, b). Most researchers agree that the clastic constituents of the varnish, e.g., clay minerals, are externally derived and transported to the rock by wind (Perry and Adams 1978, Dorn and Oberlander 1982). Clay minerals abound in dustfall, and the fixation of clays

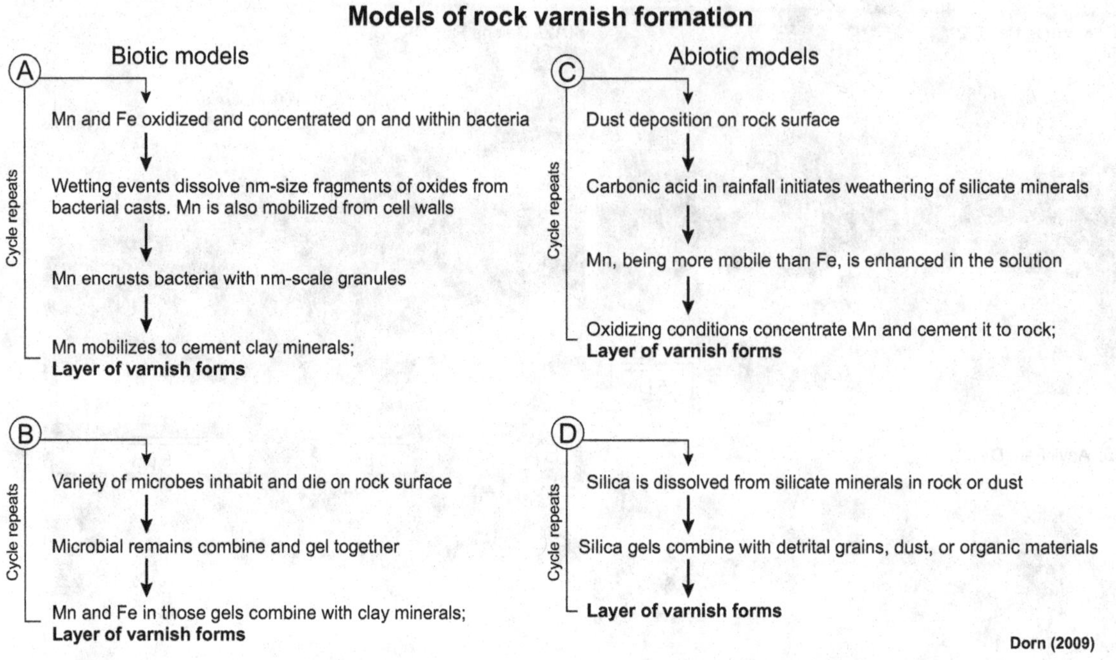

Models of rock varnish formation

(A) Biotic models

Mn and Fe oxidized and concentrated on and within bacteria

↓

Wetting events dissolve nm-size fragments of oxides from bacterial casts. Mn is also mobilized from cell walls

↓

Mn encrusts bacteria with nm-scale granules

↓

Mn mobilizes to cement clay minerals; **Layer of varnish forms**

Cycle repeats

(B)

Variety of microbes inhabit and die on rock surface

↓

Microbial remains combine and gel together

↓

Mn and Fe in those gels combine with clay minerals; **Layer of varnish forms**

Cycle repeats

(C) Abiotic models

Dust deposition on rock surface

↓

Carbonic acid in rainfall initiates weathering of silicate minerals

↓

Mn, being more mobile than Fe, is enhanced in the solution

↓

Oxidizing conditions concentrate Mn and cement it to rock; **Layer of varnish forms**

Cycle repeats

(D)

Silica is dissolved from silicate minerals in rock or dust

↓

Silica gels combine with detrital grains, dust, or organic materials

↓

Layer of varnish forms

Cycle repeats

Dorn (2009)

Fig. 13.84 Four general models of rock varnish formation. Models (A) and (B) explain varnish formation by biotic means, whereas the models shown in (C) and (D) provide abiotic mechanisms for varnish formation.

by Mn oxyhydroxides (Potter and Rossman 1977, 1979) explains why the dust remains cemented onto rock surfaces. Explanations of rock varnish formation must then focus on processes that release and precipitate Mn and Fe. So, how does this precipitation happen, and why is it in such neat layers?

Two different *abiotic* models exist to explain rock varnish formation (Fig. 13.84C, D). In the first model, wind-blown dust combines with Fe and Mn oxyhydroxides to form varnish (Smith and Whalley 1988). If dew or other water with dissolved CO_2 wets a rock, carbonic acid drives the attack of silicate minerals and releases Mn and Fe to the solution. Mn is considerably more soluble than Fe under slightly acidic surface conditions (Jones 1991), so it remains soluble as Fe oxides precipitate. Then, as the water evaporates, the Mn is fixed to eolian clay particles deposited on the rock surface. These clays become cemented there, buffer the solution pH, and help to precipitate the Mn and Fe out as varnish (Fig. 13.84A). This model may account for rock varnishes in some environments, but it fails to explain the lack of varnish in many environments where the processes mentioned earlier should be active (Dorn 2009a). Another abiotic model involves the binding of detrital grains, organics, and aerosols by a silica precipitate (Perry *et al.* 2006; Fig. 13.84D). This model could account for some kinds of rick varnish. However, it cannot lead to the formation of rock varnish that is enriched in Mn (Potter and Rossman 1979). Dorn (2009a) reviewed other shortcomings of this model.

Recent research has repeatedly documented that lithobionts – mainly Mn-fixing bacillococci bacteria and microcolonial fungi – play a clear and *active* role in most kinds of rock varnish formation (Dorn and Oberlander 1981b, Krumbein and Jens 1981, Staley *et al.* 1982, Taylor-George *et al.* 1983, Palmer *et al.* 1986, Krinsley *et al.* 2012). Dorn and Oberlander (1981a) and Dorn (2009a) reviewed the evidence for this model. The source of the Mn and Fe in the varnish is the rocks themselves, but water and dust also must be present on the rock surface (Allen 1978, Krumbein and Jens 1981). The step that distinguishes the biotic from the abiotic models is that microorganisms concentrate Mn and some Fe, which may then combine with dust to form varnish.

To that end, two slightly different biotic models have been proposed for the formation of rock varnish (Fig. 13.84A, B). The first – more polygenetic – model begins with slow-growing bacteria that oxidize Mn and Fe. Bacteria concentrate Mn and Fe in roughly equal proportions during less alkaline times; less Mn is fixed under conditions of greater alkalinity (Dorn 2009a). These substances are then abiotically cemented by clay minerals. Infrequent oxidation and concentration of Mn (and Fe) then occur during wetting events (Dorn and Oberlander 1981a, b, Grote and Krumbein 1992, Jones 1991). Infrequent wetting events dissolve nanometer-sized fragments of oxide from the bacterial casts; these fragments move a few nanometers and then recoalesce to cement the dust particles to the rock surface (Krinsley 1998, McKeown and Post 2001). As the

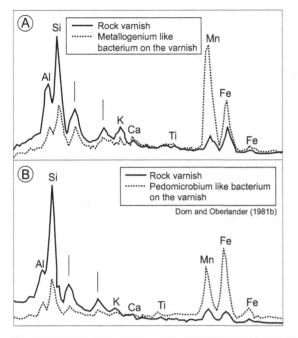

Fig. 13.85 Energy dispersive X-ray analysis spectra of rock varnish and the bacteria present on that varnish: (A) metallogeniumlike bacteria, (B) pedomicrobiumlike bacteria. Peaks show relative abundances of various elements. Peaks designated by vertical lines are emissions from the gold and palladium coating on the samples.

rock surface wets and dries, microbial activity, clay adsorption, and mobilization-immobilization of Mn in particular are enhanced. Layers of rock varnish form and thicken. Because the Mn:Fe ratio of varnishes usually exceeds 1:1, the question remains as to how Mn is enriched in the varnish relative to Fe (Jones 1991).

The other biotic model involves the rich variety of organisms that grow on or near rock surfaces. Microcolonial fungi establish themselves on rocks and help to trap eolian silt and clay on the rock surface. Under favorable conditions, the fungi grow into clusters. These lithobionts, or their organic remains, concentrate Mn and Fe (Fig. 13.85) and/or bind dust particles in the varnish after they decompose (Taylor-George *et al.* 1983). The dead fungi decompose, possibly by bacterial action or UV oxidation. Dissolution and reprecipitation of the inorganic materials associated with fungal residues occur, aided by rain and dew. In this manner, Mn becomes concentrated either directly by the microorganisms or inorganically during dissolution/reprecipitation. Patches of varnish enlarge and coalesce as fungi growing at the edges of the rock are incorporated. With time, layer upon layer of fungi and precipitated solutes develop, making the varnish thicker and more extensive.

High UV radiation levels, extremely dry conditions, high diurnal temperatures, and humidity changes, combined with virtually no chance for displacement upward or downward in/on the rock, make life on a rock surface in a desert a difficult existence! And yet, bacteria and fungi live and die there, leaving behind layer upon layer of coatings. Some evidence suggests that the clays and particulate matter in the varnish help shield the "solution front community" of microbes from the intense heat and radiation of the desert environment (Krumbein and Jens 1981). Mn oxide is a strong absorber of UV light and may thus provide shielding (Taylor-George *et al.* 1983). If so, varnish could be viewed as an ecological adaptation to protect the microbial communities from intense UV radiation and desiccation. Inadvertently perhaps, the rock surface is also protected from weathering.

Rates of rock varnish formation have been sufficiently constrained, mainly using cation-ratio and lead-profile dating (Dorn *et al.* 1990, Dorn and Krinsley 1991, Fleisher *et al.* 1999, Dorn 2004, Spilde *et al.* 2013), as well as a technique called oxidizable carbon ratio (OCR) dating (Frink 1994), that varnish coverage and chemical characteristics can be effectively used as a relative dating tool (Dorn 2009a; see Chapter 15). Nonetheless, one must bear in mind that the rate of varnish formation varies as a function of location, especially as influenced by climate (Hayden 1976). Estimates for recognizable varnish formation range from 25 (Engel and Sharp 1958) to 300,000 years (Knauss and Ku 1980), placing some varnishes well beyond the range of traditional radiocarbon dating. The chemical characteristics of rock varnish also have great utility as a paleoenvironmental indicator (Dorn 1988, 1990, Broecker and Liu 2001, Liu and Broecker 2008). This topic will be examined in more detail in Chapter 15.

Podzolization: Spodosols

And now we switch gears – moving from dry climate soils to those formed in cool, humid climates, where acidic conditions and an abundance of organic matter drive the soil system. *Podzolization* is the name given to the suite of processes by which organic carbon, Fe^{3+}, and Al^{3+}, in some combination, are translocated from the upper profile to an illuvial horizon (Petersen 1976, DeConinck 1980, Buurman and van Reeuwijk 1984, Courchesne and Hendershot 1997, Lundström *et al.* 2000a, Schaetzl and Harris 2011; Table 13.1). In some classification systems, podzolization leads to the development of soils that are classically called podzols or podzollike (Fig. 13.86). The name *podzol* is derived from the Russian *pod* (under) and *zola* (ash), referring to the gray-white, ashlike eluvial horizon that overlies variously brown-red-black illuvial subsoil horizons (Sauer *et al.* 2007b). In Soil Taxonomy, the Spodosol order is essentially synonymous with Podzols.

Podzolization is prominent in many areas where precipitation exceeds evapotranspiration, such that water frequently moves completely through the profile (McKeague *et al.* 1983). However, it is best expressed in cool, humid climates, under vegetation that produces acidic litter, such as coniferous forest and heather, and on coarse-textured

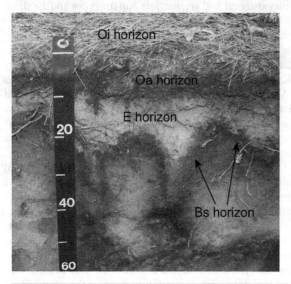

Fig. 13.86 The upper profile of a Typic Haplorthod formed under pine forest in Michigan, United States. This soil would be a Podzol in many classification systems. Photo by RJS.

parent materials where grain coatings can be quickly and readily stripped. It is especially strong in the coniferous forest zones of the high midlatitudes and weakens in intensity to the north and south (Fig. 13.87). Cool temperatures keep evapotranspiration rates low and inhibit decomposition of the acidic litter.

Many Spodosols have lamellae at depth or have a spodic horizon above an argillic horizon, a fragipan, or a gleyed zone. These morphologies illustrate that podzolization is a suite of processes associated with translocation of certain compounds in acidic soils, which can also be accompanied by other pedogenic processes (DeConinck and Herbillon 1969, Guillet *et al.* 1975).

As explained in the Landscapes box earlier in this chapter, in some literature podzolization is taken to include not only translocation of metal cations and organic colloids, but also the associated processes of (1) chemical weathering by organic and carbonic acids and (2) clay destruction and/or translocation (Sokolovskii 1924, Rode 1937, Fridland 1958, Alekseyev 1983). Under this broader definition, the focus is on mineral decomposition by strong chemical and biochemical agents in the upper profile. Some of these weathered minerals are decomposed and some of their components are lost from the profile, but the majority are simply redistributed to subsurface horizons in suspension or in solution (Guillet *et al.* 1975; Fig. 13.86).

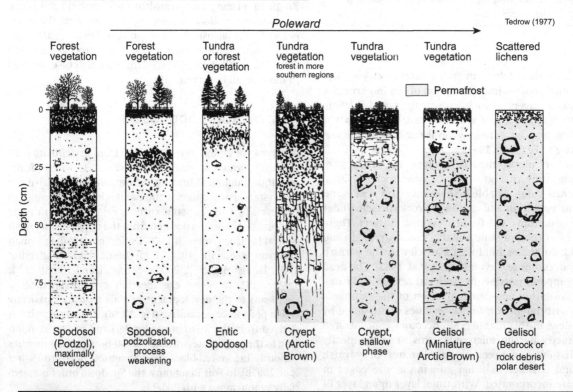

Fig. 13.87 Idealized profile morphologies along a sequence from the boreal forest of central Canada, where podzolization is strong, to the polar deserts.

The narrower, more traditional definition of podzolization – translocation of metal cations and organic matter in coarse-textured soils – is our focus. It involves mainly the translocation of organic carbon, Al^{3+}, and/or Fe^{3+} from an eluvial to an illuvial zone, mostly in solution. That is, translocation of mineral particles is not included in our definition of podzolization, although it does happen (Ugolini *et al.* 1977a). Also, in our definition, mineral weathering is deemphasized. Nonetheless, in both "types" of podzolization, pH values are low, and weathering and translocation are sufficiently intense in the upper part of the profile that readily weathered minerals *are* dissolved and removed. The dissolved ions may be completely leached from the profile, or they may reprecipitate as amorphous materials in the B horizon. Degradation of primary minerals, usually facilitated by abundant organic acids, is intense (Guillet *et al.* 1975). In either scenario, an acidic E horizon is formed; in the more narrow definition that E horizon need only be depleted of Fe, Al, and humus, whereas in the broader definition it may also have lost clay.

The central concept of soils undergoing podzolization, leading to the formation of a spodic B horizon rich in illuvial sesquixodes and humus, is embodied in Spodosols. Podzolization cannot be initiated, however, until parent materials weather enough to become acidic. During the acidification process, primary minerals begin to weather, releasing cations to the soil solution. The fate of these cations depends, in part, on their solubility, which in acidic soils is as follows (Pedro *et al.* 1978):

$$\text{base cations} > Fe^{3+} > Al^{3+} > Si^{4+}$$

As base cations are removed from a soil, it becomes more acidic. Base cations must first be depleted from the profile before podzolization per se can begin, because Al and Fe cations will not be mobile in nonacidic soils. Instead, they will precipitate to form relatively immobile and poorly crystalline oxyhydroxides. Most scholars believe that soluble organic acids, produced during litter decomposition, play an important role in podzolization. But base cations, if present in sufficient quantity, will compete with Al and Fe for complexation with the organic acid anions, further underscoring the requisite that soils be acidic before podzolization can begin.

In their book, Van Breeman and Buurman (1998) provided an excellent discourse titled "Why Podzols?" It outlines four conditions that lead to strong podzolization – the pathway to Spodosol morphology. We paraphrase their text (p. 246) here:

Production of abundant organic acids is related to (1) slow decomposition of plant litter and (2) exudation of organic acids by plants and fungi. This situation is best promoted by

1. Sandy parent materials. Sands have low buffering capacity because of their low contents of weatherable minerals; few base cations will be released by weathering. Thus, sandy soils can acidify rapidly. Coarse-textured soils also can be readily stripped of their particle coatings, allowing them to display the morphologic effects of podzolization more quickly. Sand particles retain little P and N, decreasing litter decomposition rates, because nutrient-stressed plants produce poorly decomposable substances that are rich in tannins and other phenolic compounds. Decomposers are also nutrient-limited. Nutrient-stressed plants allocate more C to mycorrhizal fungi, which in turn exude organic acids (Jongmans *et al.* 1997).

2. A cool, moist climate. Here, the litter decay period is short, and biological activity is low. The high rainfall (and large amounts of snowmelt, if applicable) promotes weathering, but also leaching of nutrients and soluble compounds from the soils. (Schaetzl and Isard (1996) suggested that podzolization is best expressed where mean summer soil temperatures are <17°C.)

3. Unpalatable litter. The litter of plants that typically inhabit sandy, nutrient-poor soils, e.g., heather, conifers, and ericaceous plants, has high contents of polyphenols, aliphatics, waxes, and lignins. These types of litter are comparatively difficult for decomposers to break down, making these types of decomposers uncommon. This situation allows the litter layer to thicken, thereby providing a year-round ready source of organic acids.

4. Minimal pedoturbation. Nutrient-poor soils in cold climates and/or wet landscape positions have few burrowing soil fauna, slowing the decay of litter and its mixing into the mineral soil. Characteristic traits of podzolic soils are their distinct horizons and abrupt horizon boundaries. As mentioned, this morphology is mainly due to low biotic activity, especially with regard to faunalturbators (Moore 1974; Fig. 13.86). In many soils, pedoturbation blurs horizon boundaries and mixes them, but such mixing is uncommon in these sandy, acidic soils (Schaetzl 1986b).

The Spodosol Profile

Spodosol profiles take on a variety of morphologies but have certain commonalities as well (Rourke *et al.* 1988). The typical podzolized soil profile has a thick O horizon, due to low rates of decomposition more than to high rates of litter production. The thickest O horizons are found under coniferous forests, where low decomposition rates are promoted by the cool climate, short growing season, and acidic litter that has high contents of slowly degradable compounds such as lignins and waxes. The O horizon varies in its degree of decomposition. Oi and Oe horizons, are sometimes referred to as *mor* horizons (see Chapter 3). Worms are almost nonexistent in these soils, and the effects of the occasional burrowing mammal are impressive, but isolated to a few locations. Most decomposition, slow as it is, is performed by fungi and is little assisted by shredders. Mycelial mats and mycorrhizae are common in the decomposing O horizon, as well as elsewhere in the upper profile (van Breeman *et al.* 2000).

The A horizon is typically thin, especially in the coarsest-textured soils, where macrofauna are rare. Bioturbation of organics into the mineral soil, normally important to A horizon formation, is minimal; thus, the primary way that the A horizon can form is through translocation of dissolved organic substances by percolating water. *Small* organic molecules often move entirely *through* the A horizon, ending up in the middle and lower profile.

The E horizon is a primary morphological indicator of podzolization. Its coarse texture is partially inherited but is also due to continued eluviation and chemical weathering of clays and fine silts. The E horizon is typically much lighter-colored (high Munsell values) than other horizons because (1) coatings of iron oxides and organic matter have been stripped from the quartz grains that dominate the mineralogy and (2) dark ferromagnesian minerals have been chemically weathered, leaving mainly quartz behind (Fig. 13.86). In many profiles the E immediately underlies the O horizon. The E is typically the most acidic horizon, having been leached of most of its base cations (Lundström *et al.* 2000b). Well-developed E horizons usually qualify as *albic* horizons (Table 8.4).

The reddish brown to black illuvial B horizon is usually given the genetic designation Bs (*s* for sesquioxide compounds of Fe and Al), Bh (*h* for humus), or Bhs (Mokma and Evans 2000). In soils with a high water table (Aquods), the B horizon may be low in Fe compounds, having lost them to the groundwater in ferrous (Fe^{2+}) forms. If the B horizon meets certain chemical and morphological criteria, it qualifies as a *spodic* horizon (Soil Survey Staff 1999). Many B horizons in strongly developed Spodosols are cemented into ortstein (see later discussion).

The boundary between the E and B horizons is typically abrupt and wavy, sometimes even deeply tongued, reflecting the preferential flow pathways in which water moves through coarse-textured soils sometimes following former root conduits (Bogner *et al.* 2012; Fig. 13.88). Tongued morphology is also typical of the lower B horizon boundary. Schaetzl (1986b, 1990) determined that at least some of these tongues are due to preferential infiltration into microtopographic lows or pits in the forest floor. Water often flows laterally within the O horizon, especially if it is mainly composed of fresh leaves. In pits and low areas, this water can leave the O horizon and flow vertically, down into the mineral soil. In these locations, the percolating water has the opportunity to be spatially focused *and* to be loaded with organic acids, because pits are sites of preferential accumulation of litter. As we shall see, organic acids in soil water are major drivers of the podzolization process.

Placic Horizons

Some Spodosols contain a thin cemented horizon called a *placic* horizon, although placic horizons are not unique to

Fig. 13.88 Deep (and often narrow) E and B horizon tongues in a Haplorthod from northern Michigan, United States. Tape increments are 10 cm. Photo by RJS.

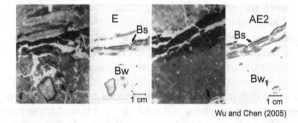

Fig. 13.89 Thin section photomicrographs of a placic horizon and surrounding soil, and drawings of the same.

Spodosols. Cementation in a placic horizon is by Fe, organic matter, and/or Mn, in the form of a thin (2–10 mm) black to dark reddish pan with sharp upper but diffuse lower boundary (Fig. 13.89, Table 8.4). The pan is often wavy and may bifurcate (Clayden *et al.* 1990, Weindorf *et al.* 2010). It is root-restricting and can perch water. The placic horizon often occurs immediately below the E horizon, confirming that its formation is due, at least in part, to illuviation.

Research on placic horizon genesis suggests that this type of horizon may form when reduced iron is mobilized in surface horizons, translocated downward in the profile, and oxidized and precipitated in the B horizon. Here, Fe oxides can adsorb soluble organic matter, but they do not necessarily form the organometallic complexes typical of Spodosols (Hseu *et al.* 1999, Soil Survey Staff 1999; see later discussion). Iron minerals in placic horizons are usually present as ferrihydrite and poorly crystalline goethite. Unlike for many other Spodosols, mobilization of Fe in the upper profiles of soils with placic horizons may be assisted by reducing conditions associated with water saturation (Clayden *et al.* 1990, Conry *et al.* 1996, Lapen and Wang 1999, Jien *et al.* 2010). Wu and Chen (2005) suggested that such saturation can be promoted by a lithologic discontinuity (Weindorf *et al.* 2010). If the discontinuity leads to occasional episodes of perched water, Fe in the upper profile may be reduced and solubilized. Then, if the discontinuity occurs at the base of the E horizon, the mobilized Fe from above can become reoxidized and precipitate as a placic horizon at the E-B horizon boundary. The presence of significant amounts of Mn in placic horizons also is strongly suggestive of a redox genesis – at least in part. By enhancing precipitation of illuvial compounds, higher pH values in the lower profile may also assist in placic horizon formation (Lapen and Wang 1999, Wu and Chen 2005). Thus, discontinuities that promote placic horizon formation may be physical (texture, coarse fragments) or chemical (pH), naturally occurring or man-made (Righi et al. 1982, Breuning-Madsen et al. 2000, Weindorf et al. 2010).

Because redox processes and discontinuities can occur in any kind of soil, placic horizons are not restricted to Spodosols. They are also commonly found in Andisols and Inceptisols, and even in Histosols (Bockheim 2011). That is, they are not necessarily formed by podzolization but may occur in conjunction with soils undergoing podzolization. Placic horizons often form in depressional landscape positions, and at sites where conditions periodically change from reducing to oxidizing (Lapen and Wang 1999, Pinheiro et al. 2004, Bockheim 2011). Wet, reducing (or epiaquic) conditions necessary for placic horizon genesis are usually accentuated by (1) high rainfall and cool temperatures, typical of perhumid climates, and often associated with Histosols, and (2) a slowly permeable subsurface layer or lithologic discontinuity (Bockheim 2011).

Ortstein

Cemented spodic (Bsm, Bhsm) horizon material that is $\geq$25 mm thick is termed *ortstein* (Wang et al. 1978, Mokma 1997a, Soil Survey Staff 1999, Bockheim 2011). Ortstein is commonly so dense and well cemented that it restricts root penetration, but when sandy it still maintains good permeability (Lambert and Hole 1971). In Aquods, ortstein is generally planar, several centimeters thick, and associated with a water table (Lee et al. 1988b, Mokma 1997a). In better-drained soils, ortstein is more spatially variable, coming and going in a seemingly random pattern of large blocks (Mokma et al. 1994). It is often more vertically oriented than planar, perhaps reflecting tongues of percolating water (Fig. 13.88). Ortstein does not slake in water, but it will break apart in reagents that dissolve Al and Fe compounds.

Although cemented like a placic horizon, ortstein has a very different genesis. Instead of forming in association with redox conditions, as do placic horizons, ortstein forms via traditional podzolization processes, as explained later. It represents a type of end point, or maximal expression of the spodic horizon, as illuvial substances accrue in such high concentrations that they cause cementation. Because ortstein commonly occurs in the *lower* B horizon, and because Al typically is translocated deeper in Spodosols

than is Fe, the cementing agent is generally assumed to be an amorphous, Al-dominated material (Moore 1976, Lee et al. 1988b, Mokma 1997a, Lapen and Wang 1999). Fe and Si may also play a role. In his review of ortstein genesis, Bockheim (2011) concluded that ortstein likely is cemented by Al-fulvic acid gels or more complex, short-range-order compounds, and that all of these compounds cement the ortstein by the bridging of sand grains (McKeague and Wang 1980, Farmer et al. 1983, Lee et al. 1988b, Freeland and Evans 1993, Kaczorek et al. 2004).

Podzolization: the Process

Models of podzolization must explain the mobilization, translocation, and eventual immobilization of oxidized metal cations and organic compounds – the central concept of podzolization. Models must also explain the intense weathering that is also typical of podzolized soils (van Hees et al. 2000). Two families of models exist regarding podzolization in well-drained soils (Lundström et al. 2000a, Sauer et al. 2007b). Both models assume that the parent material has been preconditioned (acidified) or exists in such a state that it is conducive to acidification, e.g., with coarse-textures. In addition, these models assume that the vegetation produces litter rich in low-molecular-weight organic acids and fulvic acid, and that the climate is cool and humid (Fig. 13.1).

Proto-imogolite Model

The *proto-imogolite*[1] or *inorganic sol model* was prompted by the observation that Al can exist in humus-poor Spodosols as amorphous, *inorganic* compounds such as imogolite and allophane (Farmer et al. 1980, Anderson et al. 1982, Farmer 1982, Childs et al. 1983, Gustafsson et al. 1995, Farmer and Lumsdon 2001). Lundström et al. (2000a) called this general group of compounds *imogolite type materials*, or ITM. The development of this model was prompted by the repeated identification of imogolite and ITM in illuvial horizons of Spodosols by Victor Farmer and his coworkers (Farmer et al. 1980, 1985, Farmer 1982). They believed that imogolite cannot precipitate from (out of) organic complexes, but instead forms exclusively from a hydroxy-Al-orthosilicate precursor, proto-imogolite (PI).

In this model, simple, readily soluble organic acids, e.g., formic, oxalic, and citric acids, as well as inorganic acids such as nitric acid, promote the release of Al and Fe from primary minerals in A, E, and upper B horizons. This bears repeating – the model proposes that minerals are weathered primarily by inorganic acids and small, readily biodegradable, complexing organic acids (Lundström et al. 2000b). As the organic anions are biodegraded, soluble hydroxy complexes of Al and Fe are formed. In this way, the Al that is released by mineral weathering is able to combine with Si to form Al-Si hydroxy sols in the soil

---

[1] This is not its formal name. It is used here simply for the sake of discussion.

solution. Thus Al is thought to be translocated in soil water primarily as a positively charged hydroxy-aluminum-silicate complex (Anderson *et al.* 1982). The hydroxy-Al-Si sols precipitate as ITM in the B horizon, at least in part because of its higher (>4.2) pH. Fe-Al hydroxide phases may also precipitate under these conditions.

The next step in the process involves negatively charged, colloidal organic matter, which migrates out of the upper profile and into the B horizon. There, it is immobilized by adsorption onto the solid, positively charged Al-Si ITM and Fe-Al phases. This part of the model is supported by thin-section observations of B horizons that show dark organic coatings (organs) surrounding allophane-rich coatings (allans) (Freeland and Evans 1993) or Al- and Fe-rich cutans overprinted onto Si-rich cutans (Jakobsen 1989). Dissolution/weathering processes continue to act on ITM in the B horizon. Because ITM are more easily weathered than crystalline oxides, they (ITM) will continue to dissolve and move through the soil, leaving behind an iron oxide–rich B horizon. The lower B horizon then becomes enriched in Al by its continued dissolution and remobilization from above. Because organic matter is immobilized mainly in the upper B horizons, from which Al and Fe have already been partially remobilized and translocated to the lower B horizon, a Bh horizon forms there, overlying a Bs horizon. This model acknowledges that organometallic complexes are formed in E horizons, but it also suggests that they play a role only in recycling back to the B horizon Al and Fe that is moved up to the O horizons by biological processes (Farmer and Lumsdon 2001).

Considerable research has cast doubt on the universal applicability of this model, from both chemical and mineralogical perspectives (Inoue and Huang 1990, Lundström *et al.* 2000b). Buurman and van Reeuwijk (1984) argued against this model because they could find no profiles in which the B horizons show accumulation of ITM, but not yet an accumulation of organic matter. Nonetheless, morphologic and mineralogical data do support this model, in certain circumstances (Wang *et al.* 1986a, Freeland and Evans 1993). In a hybrid model of sorts, Ugolini and Dahlgren (1987) proposed that ITM are formed as organometallic complexes migrate into the B horizon and interact with an Al-rich residue or proto-imogolite. Other models adopt some of the ideas from the imogolite model and some from the chelate-complex model (discussed later), adding key refinements to the pedogenic mechanisms (Gustafsson *et al.* 2001, Lundström *et al.* 2000a). As is often the case, no single pedogenic process model is consistent with the all the morphological expressions of major genetic horizons in soils.

In summary, this model presupposes that the Fe and Al move to the B horizon largely as inorganic sols, and that they move there *before* most of the organic matter does. Metals are translocated first, then the organics. In the next model that we shall discuss – the chelate-complex model – the metals and the organic materials are translocated *together*.

Chelate-Complex Model

The traditional and most accepted model of podzolization is the *chelate-complex* or *fulvate model*. In the chelate-complex model, dissolved organic acids form complexes with Fe^{3+} and Al^{3+}, allowing these cations to be translocated (Petersen 1976, DeConinck 1980, Mokma and Buurman 1982, Buurman and van Reeuwijk 1984, Schaetzl and Harris 2011). Normally, Fe^{3+} and Al^{3+} are not soluble in soils, but when organically *complexed*, they are readily translocated in percolating water. Research showing that 80–85% of the soluble Al in E horizons of some Spodosols is bound in organic complexes strongly supports this model (Petersen 1976, Lundström 1993). In short, the mobilization process is thought to be driven by *organic* acids, as opposed to the dominantly *inorganic* proto-imogolite pathway outlined previously.

Many low-molecular-weight (LMW) organic acids and phenolic compounds, e.g., oxalic, malic, succinic, vanillic, cinnamic, formic, benzoic, acetic, protocatechuic, *p*-hydroxybenzoic, and *p*-coumaric acids, have been identified in soils and in leachate from litter (Schnitzer and Desjardins 1969, Vance *et al.* 1986, Krzyszowska *et al.* 1996). They are also produced as root and fungal exudates. LMW compounds are typically quite soluble and easily dissolved, are derived mainly from metabolic pathways, and are mineralized in soil within a few days. As noted in Chapter 5, when a metal ion is coordinated by more than one ligand from the same molecule, the complex is called a *chelate*. Hence, the traditional name – the chelate-complex model. Not all organometallic complexes in soils are chelates, but some are. Oxalic and malic acids in the previous list could potentially form chelated complexes with Fe or Al, depending on pH. A chelation complex is stable and soluble, so translocation in mobile soil water is promoted by chelation. For this reason, the model for metal mobility in podzols is often called the *chelation model*, even though chelation per se is not required and applies to only some of the organometallic complexes that can form in such soils.

Larger, polymeric organic acids, sometimes identified as fulvic and humic acid, also occur in soils (Bravard and Righi 1991, Lundström *et al.* 2000b). These substances are defined by the technique used to extract them from soil. The terms "fulvic" and "humic" acid refer to humified organic materials that are soluble when a soil sample is treated with a strong base such as KOH at pH 11. When the subsequent extract is acidifed by the addition of HCl to lower the pH to 1.0, the organic material that is no longer soluble and coagulates is called humic acid, and the material that remains in solution is called fulvic acid. These organic acids have a large range of colloid sizes and "molecular weight." Yet there is no such thing as a molecule of humic acid or fulvic acid, so the molecular weight concept is problematic. Organic materials that can be extracted as humic acid may persist in soils for many years because they can form strong associations with minerals (van Hees *et al.* 2005). Fulvic acid consists of weakly polymeric compounds that are larger than LMW acids but smaller than humic acid colloids. Fulvic acid compounds have high metal complexation capacities and are relatively mobile in soil. In the discussion that follows, we shall see that the solubility of the components of fulvic acid is paramount to their participation in podzolization.

Organic acids are excellent at weathering primary minerals. Indeed, it has been long known from experiments on aqueous extracts from plant litter that LMW organic acids can dissolve ferric and aluminum oxides and directly weather primary minerals (Bloomfield 1953, Schnitzer and Kodama 1976, Kodama et al. 1983, Lundström et al. 1995). Because they release LMW acids, fungal hyphae are known to be effective at weathering of minerals, especially in the more acidic E horizons, thereby releasing Fe and Al to the soil solution (van Breeman et al. 2000, Bennett et al. 2001, Dorn et al. 2013). Jongmans et al. (1997) showed that ectomycorrhizal fungi, commonly associated with certain plants such as pines, exude LMW acids from their hyphal tops, and that these acids are effective at weathering minerals.

After weathering, most of the silica and the mono- and divalent cations that are released from minerals are readily removed (leached) from the profile, because of percolation of water in these coarse-textured soils. Trivalent cations, e.g., Fe^{3+} and Al^{3+}, would be insoluble except that they are complexed (sometimes chelated) by organic acids to become soluble. LMW acids, which include simple phenolic and aliphatic acids, are particularly effective at promoting the dissolution of trivalent cations from crystal structures (van Breeman and Buurman 1998). Graustein et al. (1977) showed that oxalate – excreted by fungi – forms soluble Fe and Al complexes when in contact with mineral grains.

The components of fulvic acid are also effective at forming organometallic *complexes*, and some complexes are likely to be particularly stable chelation complexes. Although fulvic acid components have a high metal complexation capacity, they become insoluble when all the anionic sites become saturated with metal ions (see later discussion). In summary, the chelate-complexation model of podzolization accounts for both weathering of primary minerals as well as mobilization and translocation of metals. Organic acids chemically complex with the otherwise insoluble weathering by-products (Al^{3+} and Fe^{3+} cations) and render them soluble in the pH range of slightly acid to acid soils (Riise et al. 2000). Swindale and Jackson (1956) coined the term *cheluviation* to refer to the combination of eluviation of chelated metal complexes (Table 13.1).

As organometallic complexes (some of which are chelated complexes) are translocated, they continue to coordinate more metal cations. The complexes remain soluble until some level of saturation is achieved (DeConinck 1980, McKeague et al. 1971, Petersen 1976, Mossin et al. 2002). At this point, the complexes reach their *zero charge point* (the negative charge on the organic molecules is sufficiently compensated by the positive charge of the bound Al and Fe cations), and they are effectively rendered immobile. The immobilization process is more likely to occur with the polymeric components of fulvic acid; the mobility of LMW complexes is less affected by metal saturation. Higher pH values in the lower part of the profile may also facilitate immobilization (Gustafsson et al. 1995), as does microbial decomposition of the organometallic complex (Lundström et al. 1995). These complexes can precipitate for other reasons as well: (1) loss of momentum in the moving water when it reaches a water table, lithologic discontinuity, or aquitard (DeConinck 1980, Buurman and van Reeuwijk 1984) or (2) adsorption onto soil particles (Sauer et al. 2007b). Immobilized, complexed metal compounds precipitate on ped faces, roots, and mineral surfaces in the B horizon.

With time, organometallic coatings on particles in the B horizon tend to thicken. They are not crystalline, and hence, these coatings tend to shrink and crack upon drying; cracked grain coatings are diagnostic of illuvial organometallic compounds (DeConinck 1980, Stanley and Ciolkosz 1981, Van Herreweghe et al. 2003; Fig. 2.14A).

Metals in such complexes may be released by subsequent complexation by other organic anions and further translocated. Alternatively, as the organic compounds decompose, the metals may precipitate as (hydr)oxides, or together with silicic acid, as amorphous silicates (van Breeman and Buurman (1998). Some of these compounds, especially if Al- and Si-rich, could be considered ITM (Buurman and van Reeuwijk 1984).

The net effect of podzolization is not only to weather primary minerals via the strong organic acids produced in the litter, but also to translocate some of the weathering by-products to the lower profile. In the initial stages of podzolization, the release of Fe and Al from primary minerals (driven by acidification) is so rapid that the chelate complexes are quickly saturated. As a result, the Fe and Al cations are rarely translocated more than a few muillimeters, and evidence of eluviation – the E horizon – is not yet manifested (Fig. 13.90). In many soils, the E horizon may be chemically present but not visible until it deepens past the depth of mixing or melanization (if one exists) (Fig. 13.12). In sandy, upland soils, where faunal mixing is minimal, E horizons can be visible within a few centimeters of the surface (Fig. 13.86). Eventually, a thickening eluvial zone of depleted Fe and Al forms – the incipient E horizon. Any new organic molecules that infiltrate past the soil surface can penetrate this eluvial zone rather quickly, without becoming saturated with metal cations. They are then free to move to the upper Bs horizon and acquire metal cations that had previously been translocated there but that had been released from their organic complexes. Duchaufour and Souchier (1978) found that podzolization is much more apparent in parent materials low in Fe, perhaps because E horizons can form more quickly in them – fewer iron-rich coatings on quartz grains to strip. Plus, fewer iron-bearing minerals exist to supply Fe to the soil solution, and, therefore, weathering by-products are more

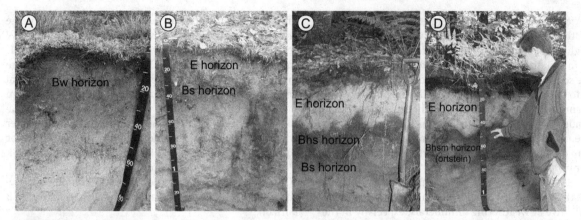

Fig. 13.90 A sequence of soil profiles from the Great Lakes region, United States, showing increasing amounts of podzolization. Increments on the tape 10 cm. Photos by RJS. (A) A Typic Udipsamment that has an A-C profile, with slight reddening in the upper solum. (B) A Spodic Udipsamment, with its thin E horizon. (C) A Typic Haplorthod with the traditional Oe-A-E-Bs-C horizonation. (D) A Typic Durorthod with an ortstein (Bsm) horizon. In this soil, the E horizon has continued to thicken while the B horizon has developed cementation.

quickly and efficiently cleared from the upper profile. Soils on carbonaceous parent materials often have a podzolic sequum over an argillic sequum, indicating that base cation depletion occurs first, followed by lessivage and then, finally, podzolization (Guillet *et al.* 1975, Schaetzl 1996). Podzolization is also inhibited in fine-textured soils, often because the organic compounds that are the major driver of the process are arrested first by clay minerals. For this reason, in soils of intermediate texture, lessivage must first remove some of the clay from the upper profile before podzolization can start in earnest.

Over time, the E horizon becomes whiter, increasingly iron-poor, and thicker, as metal cations are continually stripped from the top of the B horizon, remobilized, and redeposited lower in the solum (Figs. 13.90, 13.91). Organic acids that reach the top of the B horizon are increasingly unsaturated, leading to competition with existing organometallic complexes that have been illuviated there, followed by remobilization and transport deeper in the profile. The E horizon thickens at the expense of the B horizon. Strong weathering in the E horizon results in a white, quartz-dominated albic horizon. The B horizon – especially at the top – continues to gain organic matter, Fe, and Al as grain coatings, but the majority of the illuvial sesquioxides and humus reside near the top of the B horizon (Fig. 13.91). Indeed, many spodic B horizons have a Bh or Bhs at the top and grade downward through a series of Bs horizons, each of which is less red and has less illuvial spodic material (Figs. 13.90, 13.91).

Pros, Cons, and Contemplations of the Models

We agree with the conclusions of Jansen *et al.* (2005), that different podzolization mechanisms can dominate under different conditions. Neither of the preceding models should be considered universal, and both have merit. Nonetheless,

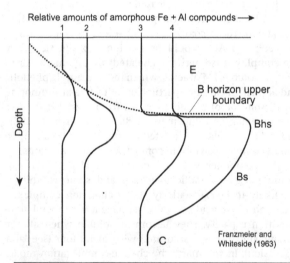

Relative amounts of amorphous Fe + Al compounds ⟶

Franzmeier and Whiteside (1963)

Fig. 13.91 Typical, theoretical development of a well-drained soil undergoing podzolization. The numbers 1, 2, 3, and 4 refer to sequential points in time.

the arguments favoring one or the other approach to podzolization are interesting and worth discussing.

One point made by ITM proponents is that Fe and Al cannot be sufficiently released by the microbial decomposition of chelate complexes to account for their high concentrations in Bs horizons, because the breakdown process is too slow. Their point is simple: Too many free (not bound to organic molecules) sesquioxides occur in some Bs horizons for the only mode of translocation to have been via organic complexes. The counterargument points out that the mean residence time (MRT) (see Chapter 15) of carbon in Bhs and Bs horizons is fairly short, sometimes less than 300 years and almost always <1,000–2,000 years (Guillet and Robin 1972, Schaetzl 1992a), implying that natural decay and turnover of organometallic complexes are rapid. Many waves of chelate complexes could have migrated into the B horizon and degraded to free, inorganic forms over a few millennia. Many metal cations are also transported by short-lived, simple organic acids, which turn over even more rapidly than do colloidal organic components.

In the proto-imogolite model, migration of ITM and iron oxides is assumed to be the first stage of podzolization; later, organic materials are deposited within the Bs horizon to form a Bhs horizon. If this sequence were to occur, podzolic profiles should exist that have E horizons depleted in Al and Fe, and Bs horizons enriched in ITM and iron compounds, but with little or no illuvial humus. Such profiles are rare, and where they do occur (Wang et al. 1986a, Freeland and Evans 1993), they might also be explained by rapid turnover of organometallic complexes in the B horizon (Buurman and van Reeuwijk 1984). Also, if organic matter were to migrate after ITM, grain coatings in the B horizons would reflect the sequence. Skeletal grains would have cutans with inner layers of ITM and outer skins of organic materials. Such coatings have been documented (Freeland and Evans 1993), but they are also rare.

Farmer et al. (1980) and Anderson et al. (1982) considered the presence of ITM in B horizons strong evidence that Al is transported as an Al-orthosilicate complex. The mere presence of a compound in a B horizon, however, does not imply that it was illuviated as such (Buurman and van Reeuwijk 1984). And while many spodic Bs horizons contain ITM (Farmer 1982, Wang and Kodama 1986, Kodama and Wang 1989, Arocena and Pawluk 1991), many do not (Mokma and Buurman 1982). The conclusion: Podzolization may include ITM, but it does not necessarily *have to involve it*. Buurman and van Reeuwijk (1984) argued that the *absence* of allophane from E horizons can, conversely, be used to support the chelate-complex model. Al-organic complexes are more stable than are Al complexes with orthosilicate ions; thus, any Al that is released from primary minerals is more likely to form organic complexes than allophane. For the same reason, allophane will not be found in B horizons that are high in organics. However, in B horizons low in humus and in the lower B horizon, allophane and ITM are likely to be found (Wang et al. 1986a); from this perspective, their presence there supports, rather than negates, the chelate-complex model. Last, the proto-imogolite model, based on solid-phase soil studies, does not explain why soil solution studies (Dahlgren and Ugolini 1989) have found that most of the Al in E horizons is organically bound.

Determining the Details of the Podzolization Process

For many pedogenic processes, we can only view and examine the chemical and morphological results of the processes, i.e., the solid phase materials left behind – the profile itself. For podzolization, however, in which so much of the translocation occurs rapidly and as dissolved substances, there is another option. We can monitor the *process itself* (Schnitzer and Desjardins 1969, Bockheim and Gennadiyev 2009). Recall that solid-phase data from the soil profile are only an indication of the accumulative pedogenic processes from time$_{zero}$. Data from soil water collected in situ provide information on processes that are *currently* occurring.

By extracting soil solutions in situ, usually using zero-tension or suction lysimeters implanted within the otherwise undisturbed soils, it is possible to examine the types and amounts of dissolved substances leaving that horizon in the soil water and entering the next lower horizon (Singer et al. 1978, Mossin et al. 2002; Fig. 13.92A). Alternatively, water can simply be extracted from moist soil samples by centrifugation (Riise et al. 2000, van Hees et al. 2000, Vestin et al. 2008). Yet another option is to implant cation-exchange resin in porous bags (Ranger et al. 1991, Barrett and Schaetzl 1998). Metals that have contact with these bags are retained and can be chemically extracted in the laboratory. Such in situ data not only provide an indication the processes that are actively occurring, but also can be used to assess the strength of the processes at various sites, in different horizons (Ugolini et al. 1987, 1988, Barrett and Schaetzl 1998), and at different times of the year (Schaetzl 1990). In soils undergoing podzolization, one normally finds that dissolved organic carbon (DOC) values are high in soil solutions exiting the O horizon, but are lower below; little DOC leaves the profile (Vestin et al. 2008; Fig. 13.92B). Similarly, higher amounts of Fe and Al in soil solutions leaving the E (but not the B) horizon confirm that active podzolization is occurring; i.e., metals are being arrested in the B horizon.

Ugolini et al. (1977b) used soil solution data to document the existence of two connected yet discrete pedogenic compartments in some Spodosols in the central Cascade Mountains of Washington State. The upper, biopedological compartment contains the O, A, E, and upper B horizons. Ionic movement here is governed by soluble organics that acidify the soil solution and depress bicarbonate concentrations. It was in these horizons that van Breeman et al. (2000) documented evidence for intense weathering by fungal hyphae. Podzolization per se is limited to the upper compartment. Most of the organic acids in the soil solution are attenuated in the upper B horizon. In the lower (geochemical) compartment, higher pH values form a weaker weathering environment dominated by the dissociation of carbonic acid (H_2CO_3).

Because podzolization is expressed morphologically as coatings (or the lack thereof) on skeletal grains, it is of interest to know the *types* of Fe and Al compounds, and their amounts, in such coatings. Soil chemists have developed a number of chemical extractants that can be used

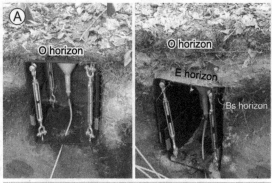

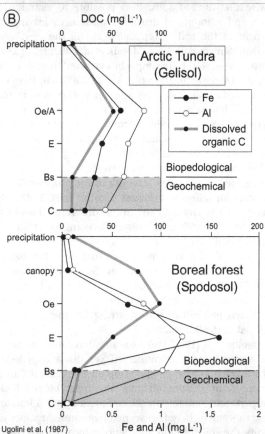

Ugolini et al. (1987)

Fig. 13.92 Methods and data resulting from the in situ study of podzolization. (A) Funnel-type, zero-tension lysimeters installed in a Spodosol. One lysimeter captures soil water leaving the O horizon, while the other captures water leaving the E horizon. A third lysimeter (not shown) captures water leaving the B horizon. The tube leading from the lysimeter drains into a lower receptacle (not shown), which can be emptied as needed. Pits are backfilled after the lysimeters are installed. Photos by RJS. (B) Solution data for a Spodosol and Gelisol in northern Alaska. Data represent dissolved organic carbon (DOC), Fe, and Al concentrations in the precipitation, in water that has passed through the vegetation canopy, and from soil solutions *exiting* each horizon.

to generate such data (McKeague and Day 1966, McKeague 1967, McKeague *et al.* 1971, Higashi *et al.* 1981, Parfitt and Childs 1988). In this process, soil samples are exposed to a liquid extractant, which is then analyzed for Fe, Al, and/or Si content. Each extractant removes a different *kind* of Fe and Al from the illuvial materials that coat the soil particles. Data from these extractants can be highly valuable in interpreting podzolization strength and pathways (Schnitzer *et al.* 1958, McKeague and Day 1966, McKeague *et al.* 1971, Mokma and Buurman 1982, Olsson and Melkerud 1989, Burt and Alexander 1996; Fig. 13.93). It should be noted that the relationships between extractants and the various forms of metal cations that they extract appear to be better established for Fe than for Al. Some forms of extractable Al are ill-defined and problematic; i.e., we are not sure that the extractant is really removing the exact form of Al that the literature suggests.

Three major types or forms of metal cations are extracted by the various chemicals. Amorphous, poorly crystalline, and crystalline (commonly referred to as "free") forms of Fe and Al oxides are extracted using a sodium citrate-dithionite solution (Mehra and Jackson 1960, Holmgren 1967, Schwertmann and Taylor 1989). These forms of Fe and Al are not tied up in the structures of primary minerals. Rather, they have been weathered and released to the soil solution and have taken on various forms subsequently. Organically bound forms of Fe and Al, consisting of organometallic complexes, are targeted by the sodium pyrophosphate extraction technique (McKeague 1967, Bascomb 1968, Higashi *et al.* 1981, Stanley and Ciolkosz 1981, Vance *et al.* 1985). This extractant, although widely used, is not without problems, especially with regard to Fe (Parfitt and Childs 1988, Skjemstad *et al.* 1992). A third extractant, acidified ammonium oxalate, is used to extract Fe and Al that is organically complexed as well as that in poorly crystalline minerals such as ferrihydrite, allophane, and imogolite (ITM) (Schwertmann 1973, Daly 1982, Kodama and Wang 1989, Mokma 1993, Wilson *et al.* 1996). In principle, ammonium-oxalate-extractable Fe and Al in spodic horizons represent what could have been transported by organic complexes and/or as ITM. Thus ratios and differences of extractable forms of Fe and Al are useful for interpreting some mechanisms of pedogenesis (Barrett 1997). For example, high Fe_p/Fe_o and Al_p/Al_o ratios suggest an important role for organic complexation in podzolization (Eger *et al.* 2011). The relative crystallinity of Fe oxides is indicated by the oxalate-/dithionite-iron activity ratio: Fe_o/Fe_d (McKeague and Day 1966). The amount of inorganic, amorphous material, often viewed as an indicator of ITM, is given by the arithmetic difference ($Fe_o - Fe_p$), even though Fe is not a required constituent of imogolite (Fig. 13.93B). Inorganic, amorphous Al ($Al_o - Al_p$) is also taken to represent poorly crystalline aluminosilicates like ITM (Jersak *et al.* 1995), as is $[(Al_o - Al_p)/Si_o]$ (Jakobsen 1991, Gustafsson *et al.* 1995). When the latter ratio exceeds 2.0, allophanelike materials are probably present. Values for

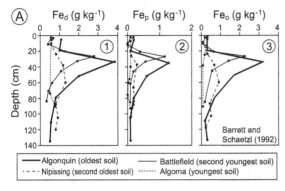

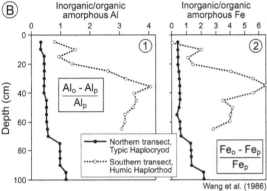

Fig. 13.93 Typical Spodosol depth functions. (A) Fe and Al depth functions for four sandy, upland soils in Michigan, as indicated by three different chemical extractants. The two youngest soils are Entisols and the two oldest are Spodosols. Note that the x-axis scales are not uniform among the three subfigures. 1. Sodium citrate-dithionite. 2. Sodium pyrophosphate. 3. Acid ammonium oxalate. (B) Depth functions of inorganic to organic, amorphous Al and Fe in two Canadian Podzols. Organic, amorphous compounds were interpreted as pyrophosphate values. Inorganic, amorphous compounds were interpreted as oxalate minus pyrophosphate values. The profile from the northern transect with the high inorganic/organic ratio contained ITM.

$Fe_o - Fe_p$ and $Al_o - Al_p$ provide some indication of the relative amounts of inorganically versus organically bound metals in the soil. For those interested in podzolization, there are lots of numbers to play with!

Factors Affecting Podzolization

Vegetation is an important, active factor in the podzolization process, because it provides the essential types of organic compounds for the process. Podzolization is best expressed under coniferous vegetation and ericaceous shrubs, whose litter decays slowly to compounds rich in fulvic acid and LMW compounds. Two trees with very acidic litter, the northern hemlock (*Tsuga canadensis*) and the kauri (*Agathis australis*), are particularly effective in

this regard, as is heather (Bloomfield 1953, Mackney 1961, Bockheim 1997b, Dijkstra and Fitzhugh 2003, Jongkind *et al.* 2007). Soil beneath these species can be much more intensively developed than are soils farther from the plant's influence.

When the vegetation changes, podzolization and related pedogenic processes change – in both strength *and* kind (Mossin *et al.* 2001, Schaetzl 2002, Falsone *et al.* 2012). Barrett and Schaetzl (1998) coined the term *depodzolization* for situations in which a vegetation (or climate) change alters the pedogenic pathway of soils that are undergoing podzolization (Miles 1985, Nielsen *et al.* 1987, Almendinger 1990). Podzolic morphology slowly degrades or becomes static under the new pedogenic regime. The effects of depodzolization are often chemically manifested long before a morphological change is apparent (Nørnberg *et al.* 1993). Hole (1975) examined the effects of logging hemlock trees from areas with Spodosols and concluded that the half-life of the spodic Bs horizon was about a century; in one hundred years these soils will have lost half their accumulated organic matter.

Podzolization occurs only where soils are thoroughly wetted during the year. It is also a dominant process in poorly drained and somewhat poorly drained soils in thermic soil temperature regimes, such as the southeastern United States, and in the humid subtropics and tropics, where it is restricted to sandy parent materials and/or sites with a high water table (Brandon *et al.* 1977, Bravard and Righi 1989, 1990, Sauer *et al.* 2007b, Schaetzl and Harris 2011). The process tends to increase in intensity as leaching increases (Jauhiainen 1973b). Translocation of organic and inorganic metal complexes cannot be accomplished without at least an occasional deep wetting event. High amounts of precipitation also facilitate weathering. In the mid- and high latitudes, podzolization is stronger on cooler sites, probably because the breakdown of organics in the O horizon is slower (Stanley and Ciolkosz 1981, Hunckler and Schaetzl 1997). Thicker, more acidic O horizons can release small amounts of organic acids at every infiltration event; in warmer macro- and microclimates the O horizon is thinner and its ability to provide organic acids to the podzolization process is diminished.

Schaetzl and Isard (1990, 1991, 1996) studied the effects of regional climate on podzolization. They noted that, on sandy sites in Michigan and Wisconsin where podzolization is a dominant process, soils were best developed in areas where the climate is coolest and snowfall thickest (Fig. 13.94). This pattern occurs despite the fact that the soils in the snowy areas are a few thousand years younger. Their work verified that snowmelt infiltration is important to the podzolization process in this region, by showing that the areas with the best-developed Spodosols had the deepest snowpacks (Schaetzl 2002). Not only is there more total infiltration in areas of thick snowpacks, there are also more *large* (>13 mm) infiltration events and more *continuous*

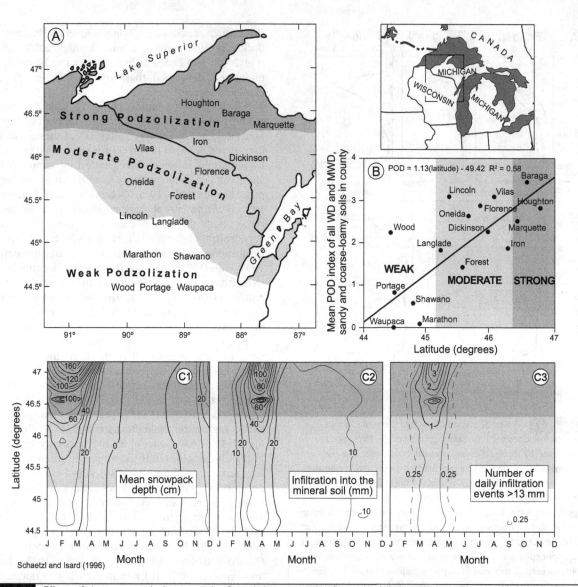

Fig. 13.94 Effects of climate on podzolization in the Great Lakes region. (A) Strength of podzolization in the western Great Lakes region. (B) Strength of podzolization, as indicated by the POD index (see Chapter 15), along a north-south transect through the region. (C) Time-space diagrams for the period 1951–1992.

infiltration events (Fig. 13.94). They suggested that large, continuous infiltration events are more effective at translocation than are smaller events, because they wet the entire profile deeply. Deep snowpacks also tend to coincide with lack of soil frost (Isard and Schaetzl 1995, 1998), facilitating uninterrupted infiltration of snowmelt water through fresh litter that retains abundant organic acids. Soil solution data taken during snowmelt and summer illustrate that translocation and hence, podzolization, are accentuated during snowmelt (Schaetzl and Isard 1990).

Until now, most of our discussion has been devoted to podzolization on upland, freely draining sites. However, many of the world's Spodosols, and most of those in subtropical and tropical regions, occur on wet sites, where the influence of a shallow water table is central to their pedogenesis (Tan et al. 1970, Dubroeucq and Volkoff 1998, Harris 2000, Watts and Collins 2008). These Spodosols commonly classify as Alaquods, reflecting their wetness and the abundance of Al in the (usually Bh) spodic horizon, which is dark (sometimes black) and rich in humus (Daniels et al. 1975, Holzhey et al. 1975, Farmer et al. 1983, Stone et al. 1993, Harris and Hollien 1999, 2000). In cooler climates, these soils are often Haplaquods, and they do not have the thick Bh horizon (Mokma et al. 1990). Much of the Fe in Aquods leaves the system as Fe^{2+}. In Alaquods, the thickness and morphological expression of the E and

Bh horizons weaken with diminishing water table influence, e.g., shorter periods of saturation or greater depths to saturation (Schaetzl and Harris 2011). B horizons in these soils tend to be more planar, whereas those in well-drained Spodosols tend to be more undulating (Mokma *et al.* 1994).

The E-Bh boundary in Alaquods represents the break between zones of uncoated (E horizon) and coated (Bh horizon) sand grains. The Bh horizon lightens in color and its upper boundary actually becomes shallower as the water table lowers. This divergence between Bh horizon and water table suggests that Bh horizon formation is not simply a consequence of precipitation of illuvial carbon at the water table. Fluctuating water tables appear to predispose these soils to podzolization, but the actual triggering mechanism for their morphologic development is still uncertain (Tan *et al.* 1999). It may relate to the effects of the frequency and duration of water saturation; e.g., anaerobic conditions could inhibit microbial degradation of complexing organic anions, such that these anions reach sufficient activity to mobilize metals. Aquods are interesting and widespread soils, but their genesis remains intriguing.

Braunification versus Podzolization

Braunification (or brunification) involves the release of iron from primary minerals followed by the formation of goethite and other iron-rich minerals, giving the soil brownish or reddish brown colors (Table 13.1). If this were the entire story, however, we would have just described something akin to *rubification*, which we discussed previously. The difference between the two is that, in braunification, the Fe oxides are quickly tied up with organic matter, leading to brown, not red colors (Cointepas 1967). Braunification is more likely to occur in midlatitude soils that have higher contents of organic matter and goethite, whereas rubification is more typical of tropical climates where organic matter contents are low and hematite is more common (Schwertmann *et al.* 1982, Vodyanitskii *et al.* 2005). Humus plays a much lesser role in rubification; that is why the Fe oxides released by weathering in tropical upland soils more often form hematite or ferrihydrite.

Braunification is typical of young, minimally weathered soils; finer-textured soils; and those that have not been fully leached. Many soils with cambic (Bw) horizons (Inceptisols, Cambisols) have a *color B* and may be considered braunified (Chen *et al.* 2001). In such soils, with finer textures and reasonably fertile status, organic matter production and turnover are rapid; much of it is not conducive to podzolization because it is sorbed to clay particles. Cation exchange capacity values are higher in braunified soils, slowing the rate at which base cations are leached. Eventually, as the braunified soils become more weathered and acidic, braunification yields to lessivage. Then, if acidification and lessivage lead to coarser textures, podzolization may begin in the upper profile (Schaetzl 1996).

Let us follow up on that concept. Braunification differs from podzolization because of iron mobility. Braunification usually occurs in soils that are more base cation– and clay-rich than does podzolization. The abundant clay minerals adsorb any free Fe cations that enter the soil solution from weathering. In podzolization, however, Fe released from primary minerals is quickly complexed with organic matter because there is little clay with which to compete. In braunification, clay and organics are competing for the Fe cations. In braunification, iron cations released by weathering precipitate to form thin coatings adjacent to clay particles or bridges between clay and organic molecules. The iron oxides, clay, and humus bind *together* in the upper solum, forming essentially *immobile* complexes. These complexes favor the development of crumb structure, which is lacking in sandy Spodosols that do not have adequate amounts of clay. When biological degradation frees the iron from the clay-humus-Fe complexes, the Fe quickly precipitates to form yellowish brown goethite, which is stable in most soils.

Yet another reason that soils become braunified instead of podzolized centers on pH; braunification operates under less acidic conditions, commonly because there is more base cation–rich litter in the O horizon. Plants growing on the more fertile soils cycle more base cations, leading to a less acidic system and different litter types, generally with lower amounts of LMW organic compounds, which are important in podzolization (Toutain and Vedy 1975). Aluminum is mobile in these only slightly acidic soils, but Fe is not. Instead, it forms stable Fe oxide-humus-clay complexes. Thus, Fe-poor E horizons do not form under braunification; since Al is colorless, Al-poor E horizons may develop, but they have brownish colors. The primary advantage that coarse-textured materials have in podzolization is that they release so few base cations or weathering products, and at such slow rates, that they do not *overwhelm* the system's ability to translocate them, allowing the soil to acidify and E horizons to develop (Alexander *et al.* 1994, van Breeman and Buurman 1998). Most coarse-textured materials are rich in quartz, which not only is slowly weatherable but also does not release base cations upon weathering.

In an interesting study of podzolization versus braunification, Duchaufour and Souchier (1978) determined that these two processes are affected as much by iron content of the parent material as by clay content (Fig. 13.95). Clay inhibits podzolization by releasing iron to the soil solution as it weathers and by immobilizing the Fe cations. Thus, E horizons do not form. Finer-textured soils also have more biological activity, fueling proisotropic pedogenic pathways and slowing the development of anisotropic, podzolic profiles (see Chapter 11). In sum, under similar climate and vegetation, podzolization tends to be favored in iron-, base cation-, and clay-poor parent materials; *braunification* is favored where iron, base cation, and clay contents are higher (Duchaufour and Souchier 1978).

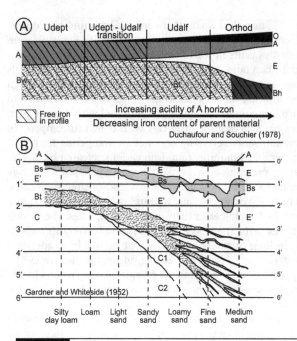

Fig. 13.95 Lithosequences of soils undergoing podzolization. (A) Effects of iron and clay content on podzolization, as illustrated by soil morphologies along a lithosequence of soils. (B) Effects of increasingly finer textures on the morphologies of some soils in Michigan.

Landscapes: The Pygmy Forest Podzol Ecosystem

Along the Mendocino coast of northern California lies a high staircase of marine terraces, formed as the land slowly rose out of the Pacific Ocean (see Figure). Known as the pygmy forest ecological staircase, the soils on the terraces represent a long chronosequence of podzolization and ecosystem succession; it is a very special place (Jenny 1960, Jenny et al. 1969).

On each terrace are beach deposits and, in some places, dunes. The lowest terrace became exposed about 100,000 years ago, and the older terraces may exceed 1 million years in age. In all cases, the dunes are younger than the terraces, and the dune soils are less weathered than are the soils formed in terrace deposits, as indicated by the more majestic and productive forest ecosystems on the dunes. The beach deposits are shallow to bedrock, generating a fluctuating water table during the rainy winter season.

The soils here follow a unique and predictable development sequence. On the first terrace, base cation–rich Mollisols have evolved beneath grassland vegetation. Acid, wet Alfisols and Ultisols occur on the much older, second terrace, where a forest of Bishop pine, hemlock, and redwood is found. Matters get interesting on the third and fourth terraces, where a globally unique pygmy forest of stunted Bishop and Bolander pines exists on a highly acid Spodosol.

This Typic Epiaquod is represented by the Blacklock soil series. It has a 36-cm-thick E horizon with a pH of 2.8! Feldspar:quartz ratios, which are about 0.2 in the parent material, are 0.0 in the Blacklock E horizon. The Bsm horizon below the E forms a hardpan that restricts rooting and perches water. This is an inhospitable soil! The trees that grow here have so few nutrients (their roots cannot penetrate into the hardpan) that they are permanently stunted. Most of the twisted and stunted "pygmy" trees are only 1.5 to 3 m tall – full grown! Large parts of the surface have no trees at all, only patches of lichen.

This chronosequence is an example of extreme long-term weathering and pedogenesis with no significant rejuvenation or pedoturbation. The bedrock below has been so compartmentalized from the soil by ortstein (thicker than a placic horizon, but functioning similarly) that fresh influxes of nutrients are rare. Nonetheless, even where bedrock is not limiting, podzolization taken to extremes will often result in loss of nutrients, and this will be reflected in the vegetation.

Figure: The pygmy forest ecological staircase.

Rates of Podzolization

Because, at its minimum, podzolization involves only the stripping of particle coatings and the translocation of soluble materials a short distance, the morphological expression of podzolization can be manifested rapidly (VandenBygaart and Protz 1995). In the most extreme example, Paton et al. (1976) found that thin E horizons had formed in mined dune sand materials in as little as 4.5 years! (Full disclosure: This material had been preconditioned to the process because it was an amalgamation of previously mined A and E horizon materials; thus, few coatings remained that actually had to be removed.) Their study illustrates an important point – podzolization rates are largely dependent upon parent material. In base cation–rich, fine-textured materials, podzolic morphologies may take many millennia to form, despite conducive environmental factors, e.g., climate and vegetation (Ragg and Ball 1964). Or they may never form, because organic acids bind with clays, and metal cations are released to the soil solution and precipitated far faster than they could ever be translocated.

Chronosequence studies provide data on the time necessary for Spodosols to form. As in many other types of soils, rates of podzolization (in the proper, preconditioned materials) start out fast, but slow with time (VandenBygaart and Protz 1995, Barrett 2001). Generally, in cool, humid areas where Spodosols are common on uplands, 1,000 to 5,000 years are typically necessary for Spodosol morphology to form (Franzmeier and Whiteside 1963, Ellis 1980b, Protz et al. 1984, Barrett and Schaetzl 1992, Petäja-Ronkainen et al. 1992, Nielsen et al. 2010). In wet-to-very wet, but still cool, areas, where leaching and acidification are more rapid, podzolic profiles can form in < 1,000 years (Chandler 1942, Jauhiainen 1973a, Singleton and Lavkulich

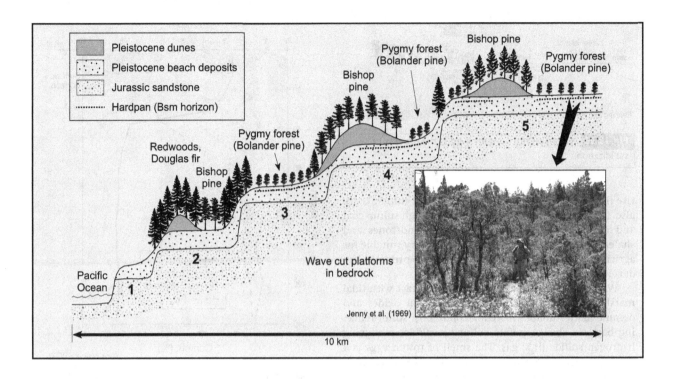

Legend:
- Pleistocene dunes
- Pleistocene beach deposits
- Jurassic sandstone
- Hardpan (Bsm horizon)

Redwoods, Douglas fir
Bishop pine
Pygmy forest (Bolander pine)
Bishop pine
Pygmy forest (Bolander pine)
Bishop pine
Pygmy forest (Bolander pine)

Pacific Ocean
Wave cut platforms in bedrock
Jenny et al. (1969)
10 km

1987, Tonkin and Basher 2001, Eger *et al.* 2011). Spodosols forming in the presence of a high water table develop at varying rates, dependent mainly upon water table fluctuations and chemistry (Tan *et al.* 1999).

Sulfidization and Sulfuricization

The two related process bundles of sulfidization and sulfuricization center on the reactions and movement of sulfur-bearing minerals within soils, which can sometimes lead to the formation of *acid sulfate soils* (Rickard 1973, Kittrick *et al.* 1982, Fanning and Fanning 1989, Dent 1991; Table 13.1). They generate sulfuric acid and typically have pH values < 3.5, sometimes as low as 2.0. Aluminum and other toxic elements associated with these extremely acid soils kill vegetation and aquatic life or render them stunted and sickly. Strong P deficiencies accompany the low pH values. Generations of people living with these soils have been impoverished. Is it any wonder that Dent and Pons (1995) describe acid sulfate soils as the "nastiest soils in the world"?

Historically, the problems started in the 1800s, in the Dutch polders. When these very wet and slightly brackish soils were drained, they quickly became highly acidic. They were some of the world's first acid sulfate soils. We quote Dent and Pons (1995, 264), "The evil reputation of acid sulphate soils stems from a combination of extraordinary characteristics: their strange colours, bad odour, the sparse and stunted vegetation contrasting with its former luxuriance, and the unusual speed of the transformation." These are quite the soils! Now let us look at their genesis.

Sulfidization or *pyritization* refers to the accumulation of sulfides and pyrite in soils, usually by the biomineralization of water containing sulfates (SO_4^{2-}) (Rabenhorst and James 1992; Fig. 13.96). It is best exemplified in anaerobic, humus-rich tidal marshes and mud flats. Sulfur is not abundant in most soils, but in the reducing environments of tidal marshes, coastal flats, and some brackish lakes, sulfur is present (Nyberg *et al.* 2012). These types of landscapes are particularly commonplace along seacoasts today, because of rising sea levels during the past ten thousand years (Brinson *et al.* 1995, Roy *et al.* 2010). Therefore, the time of permanent submergence can be regarded as time$_{zero}$ for many acid sulfate soils and the initiation of sulfidization. As marsh plants colonize the slowly submerging land, they trap sediments entering with the tides and add organic matter (Gardner *et al.* 1992). The organic matter and sediment additions facilitate the vertical accretion and lateral expansion of the coastal marsh, permitting the marsh surface to keep pace with sea level

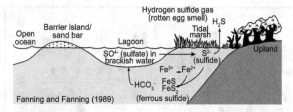

Fig. 13.96 Chemical reactions involved in the process of sulfidization.

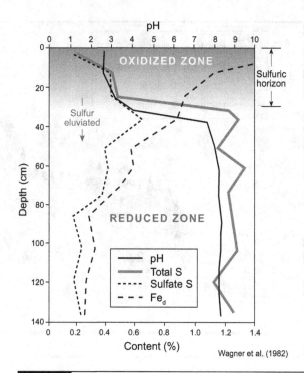

Fig. 13.97 Depth plot of pH, sulfate S, and total S in an acid-sulfate soil undergoing sulfidization. This soil was developed on sulfur-rich sediments on a scalped land surface near the Washington, D.C., beltway. The loss of sulfur from the oxidized upper solum is presumably due to leaching/eluviation. Fe_d, dithionite-extractable Fe.

rise (Hussein and Rabenhorst 1999). Sulfide minerals are also found in areas recently mined for high-sulfur coal and in soils formed on pyrite-bearing sandstones and shales (Fitzpatrick *et al.* 1996). Otherwise very suitable for agriculture, many coastal areas are highly prone to the development of acid-sulfate soils.

When sulfate-bearing sea water has contact with tidal marsh soils and sediments containing iron oxides and organic matter under low redox conditions, sulfur-reducing bacteria change sulfate sulfur to sulfide, producing hydrogen sulfide (H_2S) gas. The smell of rotten eggs is a dead giveaway for hydrogen sulfide. The H_2S reacts with dissolved ferrous iron (Fe^{2+}), forming very fine-grained and highly reactive crystals of iron sulfides, e.g., pyrite (FeS_2) or ferrous sulfide (FeS). The essential ingredients for pyrite formation are sulfate, iron-bearing minerals, organic matter (the energy source for the reduction of sulfate), sulfate-reducing bacteria, and anaerobic conditions (van Breeman and Buurman 1998). This explains why sulfur contents tend to increase as the organic matter contents of tidal marsh soils increase (Darmody *et al.* 1977). Pyrite can form and increase in content quickly (a few decades) in such soils, although the overall content often stabilizes with time (Dent 1993, Hussein and Rabenhorst 1999). The rate of pyrite formation – pyritization – is greater in areas where tidal fluctuations cause periodic flushing, which increases the supply of O_2 in the soils and facilitates the removal of bicarbonate formed during sulfate reduction. If enough sulfidic materials are present, the soils may classify as Sulfaquents or Sulfihemists (Soil Survey Staff 1999).

In *sulfuricization*, the potential acidity generated in the sulfidization process is realized when the soils are drained for agriculture, seaside development, or dredging of water channels, all of which lower the local water table and lead to oxidation. Both sulfuric acid and Fe are quickly released in these soils, as sulfide-bearing minerals, which were formed via sulfidization, are exposed to oxidizing conditions. The oxidized sulfur produces yellow jarosite ($KFe_3(SO_4)_2(OH)_6$) mottles and/or sulfuric acid (H_2SO_4). Although pyrite oxidation can proceed abiotically, it is usually performed by chemoautotrophic sulfur bacteria (*Thiobacillus* spp.). The acids created by this process

produce extremely low pH values in the soil and act to intensively weather soil minerals (Kittrick *et al.* 1982; Fig. 13.97). Weathering is so intense that the term *acid sulfate weathering* has been developed to describe just this type of process. The jarosite minerals may undergo slow hydrolysis, leading to additional production of sulfuric acid (Soil Survey Staff 1999). Jarosite can be a host for arsenic, and as jarosite dissolves at low pH, arsenic sulfide minerals may precipitate (Johnston *et al.* 2012). Fitzpatrick *et al.* (1996) described a third phase of sulfur-related pedogenesis – *ferritization* – in which the iron from the oxidized ferrous sulfide or pyrite forms a reddish precipitate, commonly ferrihydrite, which later may crystallize as maghemite, goethite, or hematite. Acidic drainage waters, colored red by precipitated Fe oxide colloids, often flow from these soils (Dent and Pons 1995).

Sulfuric horizons and acid sulfate soils (Sulfaquepts and Sulfaquents) are defined by the presence of jarosite, which usually accumulates in a distinct horizon above a pyrite-rich layer (Fig. 13.98). But what is their morphology? Unlike many other soils with a morphological record of weathering and top-down translocations, acid-sulfate soil profiles record a time sequence of soil

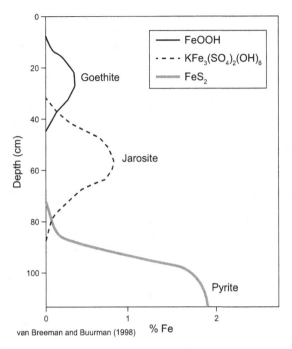

van Breeman and Buurman (1998)   % Fe

Fig. 13.98 Depth plot of goethite, jarosite, and pyrite contents in an acid-sulfate soil from Thailand that has undergone sulfuricization.

development and chemical change that is best read from the bottom up (Dent 1993, Dent and Pons 1995). At depth, where the soil remains waterlogged and fully reduced, a gleyed horizon that still contains pyrite exists. Conditions are not oxidizing enough here for jarosite formation. Above that is the actual sulfuric horizon, with its jarosite and goethite mottles. Jarosite has a straw-yellow color and frequently lines soil pores, or it occurs as pale yellow mottles. If ample calcium is present in the soil system, gypsum ($CaSO_4 \cdot 2H_2O$) can also form. Above that is a horizon with red hematite and brown goethite mottles, but still very acid (pH ≈ 4) and with high amounts of exchangeable Al. Finally, on top of all this is a very dark, humus-rich A horizon.

With increasing soil age, the jarosite-rich sulfuric horizon grows thicker, as Fe in pyrite is oxidized in the upper parts of the horizon, and hydrolysis converts the Fe to goethite. If the soil remains drained and oxidized, all the jarosite is eventually hydrolyzed. This "ripening" may ultimately produce an Al-saturated soil with a pH between 4 and 4.5. Because they have been drained relatively recently, acid-sulfate soils are usually young soils. Complete removal of the jarosite and loss of most of the acidity may take centuries (van Breeman and Buurman 1998). But these "nasty" soils may not stay that way forever.

Surface Additions and Losses

In the context of Simonson's (1959) generalized model of pedogenic processes, most of the previous discussion has centered on *internal* pedogenic processes – translocations and transformations that occur *within* the soil profile. Losses from, and additions to, the soil profile are, however, also important. Leaching is one type of profile net loss. However, many forms of whole-profile losses and additions, i.e., mass fluxes involving mineral materials, occur in soils. In particular, surficial fluxes involve sediments that are added to the soil surface or removed from it. These processes affect the thickness of the profile and the mineralogy, because many of the mass flux additions and losses are clastic mineral materials.

As discussed conceptually in Chapter 12, *upbuilding* refers to allochthonous surficial additions of mineral and organic materials to the top of the soil, e.g., eolian dust, plant matter, or slopewash sediment. When this happens, the soil thickens (Fig. 13.99). According to Johnson (1985), upbuilding can be either *developmental* or *retardant* (see Chapter 12). In *developmental upbuilding*, surface additions are slow enough that pedogenesis can keep pace and effectively incorporate the added material into the contemporary surface horizon (McDonald and Busacca 1990, Almond and Tonkin 1999, Lowe 2000, Eger *et al.* 2012; Fig. 13.99A, B). Shortly after its deposition, the new sediment is indistinguishable from what was at the surface before the addition began. This is cumulization (see later discussion). Developmental upbuilding results from *slow* additions of loess, dust, or alluvium. In *retardant upbuilding*, thick, often rapid additions of material effectively outpace the processes of horizon differentiation (Alexandrovskiy 2007; Fig. 13.99C). The soil, temporarily at least, becomes buried (Schaetzl and Sorenson 1987). In this case, the original soil per se becomes no thicker, only buried. In both cases, fresh mineral matter has entered the pedogenic system, and with it ions that can affect chemical, physical, and biological processes. Previous soil processes are altered or even stopped, as may happen if an influx of base cation–rich alluvium flocculates many of the clay colloids and causes lessivage to cease.

Sediment is *removed* from the soil surface through erosion and mass wasting. Subsurface removals by throughflow, leaching, and biomechanical processes are also important. Oxidation of organic matter may also be considered as a removal of mass (Simonson 1959). One overall pedogenic effect of removals is that, by thinning the solum, the weathering front at the base of the profile is moved closer to the surface. This may increase the rate of primary mineral weathering and foster the more rapid release of base cations. It can also foster an increase or a change in biocycling, because roots can now reach base cations and cations that they previously could not reach.

Cumulization and Burial of Soil Surfaces

How a soil and/or a surface reacts to upbuilding depends on whether the aggradation occurs at a constant or variable rate, whether periods of nondeposition or erosion are interspersed, and how the rate of aggradation compares to the rate of soil formation (McDonald and Busacca 1990, Huang *et al.* 2009, Sprafke *et al.* 2013). Rapid influx rates will bury the soil, while slower rates may lead to *cumulization*, which is the slow, upward growth of the uppermost soil horizons due to additions of sediment and their pedogenic incorporation (Fig. 13.99B). Cumulization is

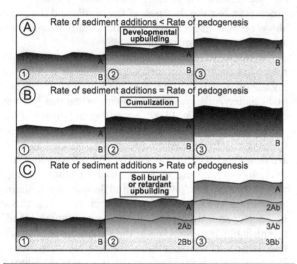

Fig. 13.99 Diagrammatic model of three different soil burial situations over time. The ultimate fate of the soil profile is dependent upon the interplay between the rate of burial and the rate of pedogenesis, i.e., assimilation of the surface additions into the profile.

driven by eolian, hydrologic, or human-induced additions of mineral particles to the surface of a soil, commonly at the base of a slope or on a floodplain (Alexandrovskiy 2007, Mandel 2008). In dust-prone areas, cumulization can occur on upland soils (Jacobs and Mason 2005, Huang *et al.* 2009). Hall and Anderson (2000) proposed three principal mechanisms for the development of cumulative soils: (1) loess deposition, (2) downslope deposition of slopewash sediment, or (3) deposition of overbank sediments on a floodplain. In all cases, gradual additions are implied, as the surface aggrades and the profile thickens (Riecken and Poetsch 1960). Intermediate and rapid rates of upbuilding may lead to soil profiles that are individually distinct yet occur atop one another. These types of soils have been variously referred to as compound soils (Morrison 1967), polymorphic soils (Simonson 1978), complex soils (Bos and Sevink 1975), welded soils (Ruhe and Olson 1980), and superimposed soils (Busacca 1989; Fig. 13.100).

How does cumulization work, pedogenically? As sediment is slowly added to the surface, it is incorporated into the profile, usually by bioturbation or as it is washed into open pores and cracks. The A horizon gains mass and thickens, resulting in an overthickened, dynamically changing A horizon (Fig. 13.99B). In some aggrading soils, the lower part of the A gradually loses more organic matter than it obtains through root additions and decay. As a result, it transforms to a B or E horizon, which also grows upward with the upbuilding surface (Wang and Follmer 1998). B horizon "upgrowth" occurs partly because the lower A horizon cannot maintain its organic matter stocks. Eventually, many characteristics of the horizons below fade completely away or are blurred by the acquisition of new characteristics; e.g., the former A acquires B horizon morphology (Hole and Nielsen 1970). Almond and Tonkin (1999) observed that the lower profiles of soils that have undergone slow upbuilding may be more weathered than

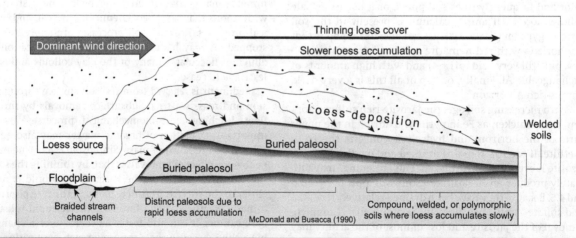

Fig. 13.100 Diagrammatic model of soil burial by loess, showing how buried and welded soils far from a loess source can bifurcate and become distinct soils at sites nearer the source.

normal, because these zones were once nearer the surface, where weathering is more intense. With this in mind, three possible scenarios can occur on an aggrading surface:

1. *Slow upward migration of the profile into the new sediment, without cumulization* (Fig. 13.99A). If additions to the soil surface are slower than the rate at which pedogenesis can assimilate them, they can be effectively incorporated into the A horizon, and the soil does not become cumulic. Because of root growth and bioturbation, the A horizon can respond faster to the changing (upbuilding) conditions than can the B, because its morphology can develop in the new sediment more rapidly. Likewise, as biological activity diminishes at its base, A horizon morphology can be obliterated from the old, lower A horizon more rapidly than can B horizon morphology. Thus, the A horizon may not become cumulic under this scenario, but the B may overthicken. For this reason, thick B horizons are commonly observed in buried soils that were (at first) very slowly buried (Follmer 1982).

2. *Cumulization* (Fig. 13.99B). If sedimentation rates generally match (or closely so) the rate of pedogenic assimilation/overprinting of the new material, the A horizon becomes overthickened (cumulic). Grass roots quickly spread into the new, thin layer of sediment, and as they decay in it, the new sediment is melanized. Equally important, bioturbation is rapid enough to incorporate the new sediment into the upper profile, even as melanization is occurring on it. The base of the A horizon may gradually grow upward but not as rapidly as does its upper boundary, so the horizon thickens. In the same way discussed earlier, in some soils, the lower part of the A horizon may be transformed to a B or E horizon. In general, B horizons cannot grow upward at this faster pace, and they may remain weakly developed, especially near the gradual boundary with the cumulic A horizon. A Cumulic Hapludoll is an example.

3. *Burial* (Fig. 13.99C). If sediment additions outpace soil-development processes, pedogenesis cannot keep pace with upbuilding. An organic matter–rich A horizon may begin to develop in the new sediment, but only when the surface is stable does the organic matter enrichment due to biological activity really begin. The A horizon thickens, but its upper part (in the new sediment) is not as dark as the lower, since melanization in the new material cannot keep pace with sedimentation. For a short period, the soil appears cumulic. The previous A and B horizons are buried and are rendered relict. The newly formed A horizon in the fresh sediment may be thinner and lighter or thicker and darker than the one that is buried. Only if the new deposit is thick enough and the surface is stable enough can strong B horizons form (Smith *et al.* 1950). Evidence of rapid sedimentation can be found as a distinct boundary between the old, now buried soil surface and the new deposits (Follmer 1982). In aggradational sediments like alluvium, one often finds buried surfaces,

where the sedimentation system stalled and the surface stabilized for a short period (Fig. 13.100).

One can argue that cumulization is more a geologic/sedimentologic process than a pedogenic one. But because it affects soils directly and because the soil predictably reacts to the new sediment, cumulization is often a focus of soil geomorphology. To quote McDonald and Busacca (1990: 449):

Soil formation on an aggrading surface is a competition between pedologic and sedimentologic processes, a competition that can lead to very complex soils and soil-stratigraphic relationships. Complexity results from the compression or dilution of soil profile features with variations in aggradation rate and from the sometimes extreme overlap of features from different episodes of soil development. Although the complexity can be great, … the opportunities for insight into soil and geomorphic processes can be equally great.

Cumulization is common on floodplains, footslopes, and toeslopes, where soils may be classified in Soil Taxonomy in cumulic subgroups. It is also common in some grasslands and deserts, as an result of additions of loess or dust, and by alluvial sedimentation (McDonald and Busacca 1990, Feng *et al.* 1994b, Wang and Follmer 1998). In humid areas, cumulization occurs in low-lying landscape positions, where sediments intermittently wash down from upslope. In Mollisols, where the major pedogenic process (melanization) is rapid and where grasses above are adept at capturing sediment, cumulization can occur slowly and continually.

Mass Balance Analysis of Pedogenesis

So much is happening during soil genesis! Would not an accurate measurement of those processes and changes help us understand the system? Perhaps. To that end, Stockmann *et al.* (2011) emphasized that to understand the complex soil system it is important to investigate pedogenic processes *quantitatively*. Quantification of pedogenesis informs answers to questions such as the following:

1. How does soil form?
2. At what rate do soils form, change, and evolve?
3. How fast is parent material converted to soil, and what influence does this rate have on pedogenesis?

We can quantify pedogenesis in a number of ways. Concentrations of minerals, elements, and other soil properties only show us what the soil is like *now*, but C horizon data suggest what soil properties were at time$_{zero}$, and by comparison we can infer the changes that must have taken place during pedogenesis. Chronosequences and chronofunctions can inform us about morphological, physical, and chemical changes in soils over time (see Chapter 15). In this section, we focus on a different kind of highly quantitative and rapidly expanding method for exploring pedogenesis: *mass balance analysis*. As an approach to quantitative pedology, mass balance analysis is a way to calculate the absolute (and relative) gains and losses of mass for a horizon or for the whole profile since time$_{zero}$. The determination of *volumetric changes* in soils and soil horizons when estimating mass fluxes is fundamental to the mass balance method.

First, let us see what is in a name. The mass balance approach is based on the idea that any changes in the mass of a soil or soil horizon over time must be due to either additions of mass from outside the soil or horizon or losses of mass from the soil or horizon. If we let M_0 = the mass of any soil constituent at time$_{zero}$ and M_t = the mass of that constituent at some subsequent time, and if we define M_{gain} as a gain of that constituent and M_{loss} as a loss, then we can write the following *mass balance* equation:

$$M_t - M_0 = M_{gain} - M_{loss} = \Delta M,$$

where ΔM = change in mass of the constituent.

Any net change in mass (ΔM) must be due to additions or losses that have taken place across the soil's or horizon's boundaries during the period under consideration. Note that there is no assumption that $\Delta M = 0$; in fact, this situation (a "balance" of sorts) would be unusual if that were the case. Indeed, the term "mass balance" is a bit of a misnomer. It implies that a gain of something in a soil or a horizon is accompanied by an equal amount of loss, i.e., a balance. That is certainly not the case for soils. A *balance* need not result.

Relative to their parent material, or even on a horizon-by-horizon basis, soils experience many kinds of additions and losses of materials via weathering, translocation, and erosion and deposition. For example, leaching of cations from a horizon is, by definition, a loss. Soils also experience gains due to surface additions, but also by transformations; e.g., the soil may gain clay as silt minerals are weathered. Gains also occur in a *relative* sense; one element or mineral is gained, relatively, as others are proportionally lost.

Mass balance analyses can be performed most confidently on soils formed in uniform parent materials, and the more uniform, the better. Thus, before undertaking a mass balance analysis, the uniformity of the parent material must be confirmed (Haseman and Marshall 1945, Evans and Adams 1975a, b, Rostad *et al.* 1976, Jersak *et al.* 1995, Egli *et al.* 2001). In some contexts, this prerequisite limits the application of the method to soils that lack discernible lithologic discontinuities (Schaetzl 1998). Alternatively, one can use the mass balance approach to infer the presence of multiple parent materials, by identifying net gains that could only have occurred when new materials were deposited, e.g., eolian additions to soils (Mason and Jacobs 1998, Heckman and Rasmussen 2011).

In mass balance analysis and soil reconstructions, the gains and losses are calculated relative to the mass of one or more stable and "unweatherable" constituents (Egli and Fitze 2000, Egli *et al.* 2001). The assumption is that the

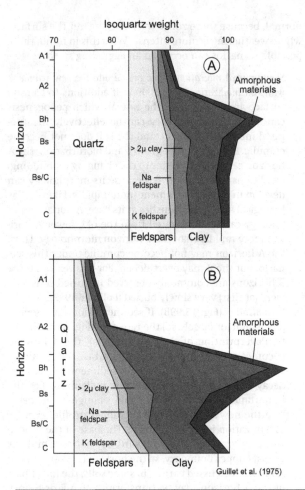

Isoquartz weight

Fig. 13.101 Diagrammatic representation of conservation of mass of an immobile element and strain (here, collapse) for a soil at two periods in its development. The mass of immobile element *i* in the soil before weathering is equal to its mass in the weathered product.

mass of a relatively immobile element is conserved during soil development, while others may be lost. Equations incorporating the amounts of stable constituents (in the presumably uniform parent material) are therefore used to quantify the gains and losses of less stable materials, e.g., those that have been translocated or dissolved (Marshall and Haseman 1942, Chadwick *et al.* 1990, Brimhall *et al.* 1991, 1992, Stolt *et al.* 1993 Fig. 13.101). Bulk density data are critical to the calculations because the masses of both stable constituents and those that might be mobile are determined from the values of bulk density and mass concentration.

Mass of the constituent (per unit of volume) = bulk density × mass concentration

Mass balance analysis uses minerals or elements that are immobile and resistant to weathering as indices to assess the degree of soil development and mass flux. Because weatherable minerals are lost at greater rates than are index minerals, the proportion of weatherable to resistant minerals will change with soil development (Barshad 1964, Haseman and Marshall 1945). Uniformity in the distribution of index minerals throughout the profile is taken as evidence that the parent material was homogeneous and that any differences in morphological features are the result of in situ pedologenesis. Zircon and tourmaline have traditionally filled the role of index minerals in mass balance studies, although quartz has also been used (Cann and Whiteside 1955, Souchier 1971, Guillet et al. 1975, Sohet et al. 1988, Tejan-Kella et al. 1991; Fig. 13.101). Quartz has an advantage over the less common heavy minerals (Sudom and St. Arnaud 1971, Guillet et al. 1975), because its measured values are not as dramatically affected by sampling technique.

Souchier (1971) outlined a mass balance formula that relies on quartz as an immobile and resistant mineral. The *isoquartz method* compares the distributions of other minerals within the profile to quartz, as a way of determining the extent to which they have been weathered and either lost from the profile or redistributed within it (Fig. 13.101). The isoquartz formula for a given horizon is

$$\Delta Xi = h_i \rho_i \cdot (Q_i/Q_o) \cdot (X_i - X_o)$$

in which ΔX_i is the gain (positive) or loss (negative) of a component X from a horizon, h_i is horizon thickness, ρ_i is the bulk density, X_i is the concentration of a specified mineral component divided by the quartz concentration of the horizon, X_o is the concentration of the same component referred to the quartz concentration in the parent material, and Q_i/Q_o is the ratio of the quartz contents of the given horizon and the parent material. The reader is referred to the original source for additional detail.

Index minerals have been superseded largely by index elements, because of great time savings in the lab (Egli and Fitze 2000). Index elements are useful only if they are uniquely present in their respective index minerals. For example, Zr is usually assumed to be present exclusively in zircon; Ti occurs in rutile, ilmenite, and titanite; and Y is used as an index element for xenotime. Niobium, thorium, and vanadium have all been used as proxies for inert minerals in mass balance analysis (Lichter 1998, Mathé et al. 1999, Egli and Fitze 2000, Stiles et al. 2003). Brown et al. (2003) advocated the use of rare earth elements hafnium, scandium, and thorium as index elements in certain settings. Combinations of index elements have been used, e.g., yttrium, Zr, and Ti (Chittleborough et al. 1984). Zr and Ti are the most commonly used index elements (Chapman and Horn 1968, Sudom and St. Arnaud 1971, Evans and Adams 1975a, b, Smeck and Wilding 1980, Rabenhorst and Wilding 1986a, Santos et al. 1986, Busacca and Singer 1989, Egli and Fitze 2000, Eger et al. 2011). For example, the mass of Zr in a soil horizon is compared with its mass in the

parent material, as a way of determining relative gain or loss in soil volume, i.e., net dilation or collapse.

A word of caution – questions exist regarding whether any mineral or element can be truly stable or immobile in soils (Sudom and St. Arnaud 1971, Evans and Adams 1975a, b, Colin et al. 1993, Mathé et al. 1999, Stiles et al. 2003). Titanium, in particular, has come under scrutiny in this regard (Sudom and St. Arnaud 1971, Brinkman 1977b, Smeck and Wilding 1980, Busacca and Singer 1989, Cornu et al. 1999, Taboada et al. 2006, Anda et al. 2009).

Mass balance analysis can be done for the entire profile, or on a horizon-by-horizon basis. For example, mass balance analysis can determine how much clay has been lost from an E horizon and how much clay has been gained in the Bt horizon, relative to what was inherited from the parent material (Brewer 1976, Stolt and Baker 1994, 2000). Likewise, it can establish how much clay the entire *profile* has gained since time$_{zero}$.

Some rock-to-soil processes are isovolumetric, e.g., the formation of saprolite from bedrock (Mathé et al. 1999). In such cases, mass balance analysis may indicate a *loss of mass but not of volume* (Stolt and Baker 2000). However, soil genesis is usually not an isovolumetric process (Jersak et al. 1995). When parent material is converted to soil, volume is usually lost as well, as the material collapses (Fig. 13.102). *Collapse* is mainly caused by solutional losses from the profile, but it is facilitated by pedoturbation. Hole and Nielsen (1970) called this process noncumulative soil genesis, which refers to the literal collapse of soil materials as soluble materials, e.g., carbonates, are leached. By definition, noncumulative soil genesis processes do not include surface additions or erosion. *Dilation* refers to the expansion of soil volume over time – the opposite of collapse – and it is also facilitated by pedoturbation.

The term *strain* is used for the deformational, volumetric changes that soils undergo over time, as a result of either collapse or dilation. Strain is a unitless ratio, but for convenience, it can be thought of as a relative change in profile thickness, i.e., a one-dimensional vertical change in thickness divided by the original thickness (Fig. 13.102). Strain is a manifestation of time-integrated mass fluxes from the profile, as facilitated by weathering and leaching, but it also is affected by root growth and decay, bioturbation, and all other pedogenic processes (Stolt and Baker 2000). Strain is calculated by subtracting 1.0 from the ratio of the mass of an immobile element in a unit volume of the parent material to its mass in a unit volume of the weathered material (Fig. 13.102). The volumetric masses are determined by multiplying the mass concentration by the bulk density in each of the two zones (Fig. 13.102). Where strain is positive, the soil has increased in volume; i.e., it has dilated. Negative strain implies collapse (Brimhall and Dietrich 1987, Brimhall et al. 1988). Collapse, for example, is indicated when an increase in the concentration of the immobile element (caused by the loss of mobile constituents) is detected

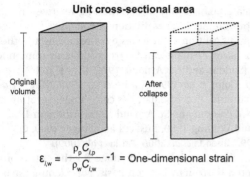

Unit cross-sectional area

Original volume

After collapse

$$\varepsilon_{i,w} = \frac{\rho_p C_{i,p}}{\rho_w C_{i,w}} - 1 = \text{One-dimensional strain}$$

ρ_p = bulk density of parent material

$C_{i,p}$ = concentration of immobile element in parent material

ρ_w = bulk density of weathered horizon

$C_{i,w}$ = concentration of immobile element in weathered horizon

Brimhall and Dietrich (1987)

Fig. 13.102 Depth plots of weathering, as indicated by mineralogy and based on the *isoquartz method* of mass balance reconstruction, for two soils undergoing podzolization in eastern France. Soil (A) is finer-textured and shows strong redistribution of clays and amorphous materials, whereas soil (B), which is coarser-textured, shows a greater amount of dissolution (subtractive pedogenesis) in the upper profile, along with redistribution of plasma components.

that is not exactly compensated by a proportional decrease in bulk density. Many soils undergo slow collapse in eluvial horizons and dilation in illuvial horizons (Fig. 13.103A). Strain may be calculated on a profile or a horizon basis, and in chronosequence studies, strain can be calculated for soil over various periods (Fig. 13.103). Another useful and related term is the *unit volume factor*, which is the number of unit volumes of parent material necessary to form a given unit volume of a soil horizon (Smeck and Wilding 1980). Thus, not only do soils evolve because of additions, removals, and transformations (Fig. 12.19), but their fabrics and hence their volumes evolve as well. Volume change and deformation are essential parts of pedogenesis, and they reflect the ongoing evolution of not just the soil plasma, but also the skeleton.

Documenting volume deformation during soil genesis is an important part of quantitative pedology studies. Many studies have documented slow, sustained collapse in soils and even in saprolites (Oh and Richter 2005), leading to the concentration of certain immobile elements and the associated loss of mobile elements. George Brimhall's work has been particularly important to understanding how metal ores become concentrated in some saprolites (Brimhall and Dietrich 1987, Brimhall *et al.* 1988, 1992). In these situations, the collapse is due to loss of soluble materials, thereby enriching the profile in insoluble elements such as Ni.

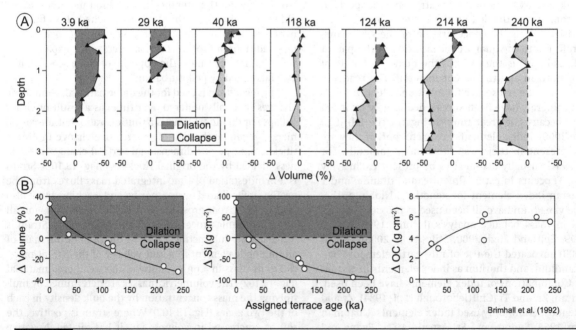

Fig. 13.103 Dilation and collapse for a series of soils on uplifted marine terraces of different age, in northern California. (A) As plotted by depth (horizon) and by time. (B) Changes in volume, Si, and organic carbon as integrated over the entire profile and by time.

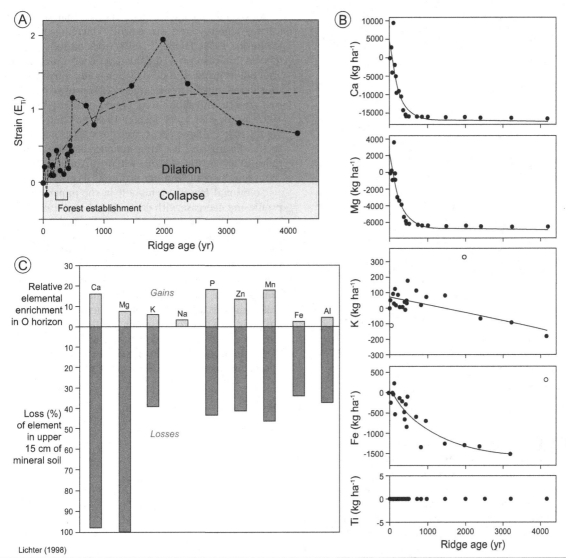

Lichter (1998)

Fig. 13.104 Mass balance data for a short chronosequence of soils forming on sand dunes in Michigan. (A) Dilation and collapse, integrated over the entire profile, over time. (B) Changes in contents of elements in the upper 15 cm of the mineral soil, shown over time. The amount of Ti is set as constant, as per the assumption of the method. As expected, no net gain was detected for any element in the upper profile. For all elements except K, large net losses are evident after only a few hundred years. (C) Net losses and gains for different elements in these same soils, integrated over 4,000 years. The lower bar represents the net percentage loss of that element in the upper 15 cm of the mineral soil, relative to what the parent material contained at time$_{zero}$. The upper bar shows relative enrichment of those same elements in the O horizon, defined as the ratio of the concentration of a particular element in the O horizon to its concentration in the upper 15 cm of mineral soil. Enrichment here is mainly due to atmospheric inputs and/or plant uptake from the mineral soil, followed by its eventual deposition within the O horizon.

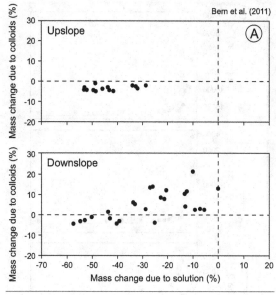

Bern et al. (2011)

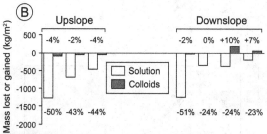

Fig. 13.105 Mass balance data for some soils on an upslope and a downslope area in South Africa, as partitioned into losses due to colloidal transfer (suspension) versus transfers in solution. (A) Percentage mass loss or gain for the <2 mm fraction. Each data point represents an individual soil horizon. (B) Calculated mass loss or gain, relative to parent material and integrated over the entire profile, via suspended solids (colloids) and in solution.

Most mass balance studies observe that soils expand (dilate) when young, as porosity increases (Cann and Whiteside 1955). Lichter's (1998) work on soils on a series of sand dunes showed that dilation occurs quickly, shortly after plants invade the site. Dilation is also much more common in the upper profile, largely because of pedoturbation, root growth, and additions of organic matter (Merritts *et al.* 1992, Jersak *et al.* 1995, Lichter 1998, Egli and Fitze 2000; Fig. 13.104). Collapse occurs in older soils as net losses due to weathering and leaching mount. Likewise, collapse is more likely in the subsoil, where mixing processes and root growth are only minor processes (Fig. 13.103).

Solutional losses are highest in the upper part of the profile and decrease with depth (Jersak *et al.* 1995, Langley-Turnbaugh and Bockheim 1998, Tonkin and Basher 2001). Only rarely does the sum of subsoil gains equal the quantity of the corresponding losses from overlying horizons. Data on profile- and horizon-based losses from soils provide important information on the varying solubilities of elements, the weathering vulnerabilities of various minerals, and the distribution of weathering across the various grain size fractions (Harden 1988a, Busacca and Singer 1989, Merritts *et al.* 1992, Oh and Richter 2005). Recent work has been able to distinguish between mass transfer in suspension and that in solution (Fig. 13.105).

Part III

Soil Geomorphology

Chapter 14

Soil Geomorphology and Hydrology

Introduction to Soil Geomorphology

Geomorphology is the study of landforms and the evolution of the Earth's surface. Because soils are so strongly linked to the landforms upon which they develop, a discipline that dealt with those relationships eventually emerged: *soil geomorphology*. Popularized by Pete Birkeland in the 1970s and 1980s, soil geomorphology came into its own with the publication of a number of books by Birkeland and other scholars during this time, all devoted to the topic and helping to define the field (Mahaney 1978, Gerrard 1981, 1992, Richards *et al.* 1985, Daniels and Hammer 1992, Paton *et al.* 1995). In 1999, Birkeland had defined soil geomorphology as the study of soils and their use in evaluating landform evolution, age, and stability, surface processes, and past climates. Wysocki *et al.* (2000) more broadly defined it as the scientific study of the origin, distribution, and evolution of soils, landscapes, and surficial deposits, and the processes that create and alter them. McFadden and Knuepfer (1990) emphasized the linkages between pedogenic and other surficial processes, in their definition of the field. Perhaps the definition we like best was presented by John Gerrard in 1992: Soil geomorphology is an assessment of the genetic relationships between soils and landforms.

Yes, soils and landforms develop together. Many times, it just makes sense to study them together. Soil geomorphology is designed to examine and elucidate the nature of that genetic dance. But it is a two-way street. Soils are affected by landforms, and through their developmental accessions and features, they in turn influence geomorphic evolution. Most importantly, pedogenic processes are variously dependent and intertwined with slope processes, e.g., erosion and sedimentation. And on top of it all, the influence of landforms on the flow of water – across the soil surface and belowground – impacts soils markedly. This is the essence of soil geomorphology – putting all these interrelationships together.

Historical Background

Soil geomorphology was first studied in its own right by the U.S. National Cooperative Soil Survey (NCSS) program in the 1930s. At that time, interests had developed among geographers, geologists, and soil scientists on the relationships between soils and landforms (Effland and Effland 1992, Holliday 2006). Acknowledging the merit in this type of approach, the NCSS adopted soil geomorphology as a paradigm for studying soil landscapes. Early soil geomorphology studies were grounded in the work of the famed geographer Carl Sauer of the University of California (Effland and Effland 1992). Many of these early studies had focused on soil erosion. Later, under the leadership of Charles Kellogg, assisted by Guy Smith, the NCSS program embarked upon a research mission to understand soil-landform relationships in the major climatic areas of the United States, in support of its soil mapping program (Grossman 2004). Smith and the NCSS established several sites where soil geomorphology was to be studied in detail: subhumid Iowa, an arid desert site in New Mexico, a humid Pacific Northwest site in Oregon, and a humid site in North Carolina. Presumably, findings and paradigms developed at these sites could be extended to other areas of the United States.

The NCSS's soil geomorphology work led to the growth of this kind of research within the university community. It had profound effects on theories of soil and landscape genesis and greatly influenced the way soils were classified (Effland and Effland 1992). Much of this effort culminated in the first textbook devoted to soil geomorphology, written by Pete Birkeland in 1974. Many other important textbooks, book chapters, and monographs have also been written on the topic (Ruhe 1969, 1975b, Daniels *et al.* 1971a, Hall 1983, Knuepfer and McFadden 1990, Daniels and Hammer 1992, Gerrard 1992, Birkeland 1984, 1999, Olson 1997). Of particular importance to the development of soil geomorphology were the establishment of a soil geomorphology committee by the Soil Science Society of America and the many field trips run under its auspices.

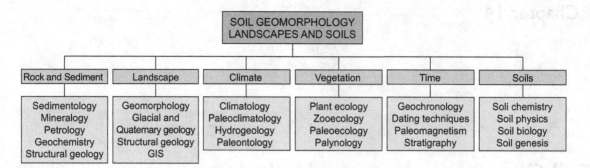

Fig. 14.1 Areas of study that contribute important concepts and background information or methods to the field of soil geomorphology. After Ruhe (1975b).

Topical Areas of Focus

Wwide-reaching and inclusive as a field, soil geomorphology embraces topics such as:

1. Soil variability and variation across landscapes, at all scales
2. Soils and paleosols as indicators of environmental/climate change
3. Soils as indicators of landscape, slope, and geomorphic stability or instability
4. Developmental studies of soils, especially in chronosequences
5. Soil-rainfall-runoff relationships, especially as they pertain to slope processes
6. Soils as indicators of past or ongoing sedimentological and depositional processes
7. Soils as indicators of Quaternary stratigraphy and parent materials
8. Hydropedology

Soil geomorphology is a geographic science. Many soil geomorphology studies are spatial, or spatiotemporal, in nature. The data generated, more often than not, assist in the interpretation of soil *patterns* on the landscape, whether on the modern landscape or the paleolandscape.

Soil geomorphology is a field-based science. Knowledge of landforms in the field is essential to obtaining a representative soil sample, for most applications. *No amount of* laboratory work, "number crunching," or library or online research can make up for poor site selection or sampling technique. Once the samples have been taken, the outcome of the research is constrained by the data derived from them. Research outcomes always depend on interpretations of the data. Thus, knowing *where* to sample on the landscape, i.e., a location that allows one to obtain a representative sample, is as important as knowing *how* to sample and knowing *what to do* with it in the lab. Never lose sight of this credo.

Soil geomorphology is an integrative science. To do it well, one must have a solid knowledge of many related fields (Fig. 14.1). Jungerius (1985) pointed out that too often

our understanding of soil geomorphology is essentially abiotic, perhaps because of the influence of William Morris Davis's cycle of erosion on the early geomorphology community. This cycle invokes little in the way of biological processes or controls on landscape evolution. Darwin's (1881) equally logical and perhaps more quantitative work had little influence on early soil scientists and geomorphologists (Johnson 2000). If Darwin's work had been more influential, soil geomorphology would today have a much stronger biophysical component (Jungerius 1985, Johnson 1999, 2002, Feller *et al.* 2003, Johnson *et al.* 2005, Meysman *et al.* 2006, Clark *et al.* 2009). One of our (many) goals in this book is to rectify the overemphasis on abiotic processes in soil geomorphology. Let us start.

Geomorphic Surfaces

Soils form on the Earth's surface. These surfaces may be flat or sloping. They may face north or south. They may be stable or eroding. They may be of all one age (having formed at the same time in the past) or time transgressive. Parent materials may change across the surface – laterally and with depth – or they may be uniform across and within it. But the *surface* defines the soil in space and time (Ruhe 1969, Hall 1983). The concept of the *geomorphic surface* captures that time-space notion.

A geomorphic surface must be definable in space and, at least in a relative sense, time. Consider it a mappable area of the Earth's surface that has a common geologic history and definite boundaries (Daniels *et al.* 1971a; Figs. 14.2, 14.3). The exposure, or formation, age of the surface is comparable or predictable across it, because it has been formed by a unifying set of processes (erosional, constructional, or both) during a similar period (Daniels *et al.* 1971a). We may not always be able to determine the age of the surface in absolute terms, but we should be able to discern its relative age vis-à-vis adjoining surfaces (Ruhe 1956b, Lepsch *et al.* 1977a, McDonald and Busacca 1990, Barrett 1993, Siame *et al.* 2001, Schaetzl *et al.* 2006).

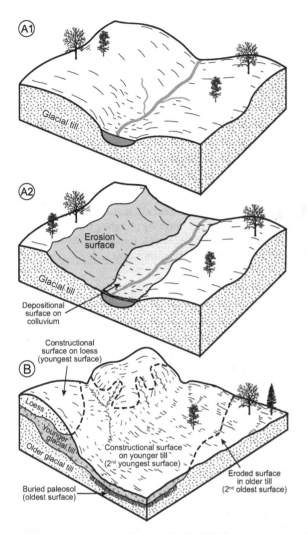

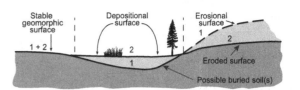

Fig. 14.3 The three main types of geomorphic surfaces, showing each surface at time$_1$ and time$_2$. After Follmer (1982).

Fig. 14.2 Hypothetical scenarios involving the formation of geomorphic surfaces of different or similar ages across landscapes. (A) A landscape formed geologically in an instant, resulting in (1) geomorphic surfaces of uniform age and (2) later, because of localized erosional and depositional events, geomorphic surfaces of differing age. (B) Deposition of parent materials at different times, some of which have subsequently been eroded, leading to surface of differing age across the landscape.

In other words, how does it relate to nearby surfaces, genetically? For example, we might know that the surface of a sand dune is younger than the surfaces on the surrounding lowlands. Ideally, its relationships to adjoining surfaces with respect to age *and* lithology should be known or discernable. Many geomorphic surfaces are also named, particularly older, buried surfaces, such as the Quaternary-aged Sangamon surface in the midwestern United States (see Chapter 16).

Types and Categories of Geomorphic Surfaces

Geomorphic surfaces can be variously subdivided and categorized. Some surfaces, inundated by water, are termed *subaqueous*. Others, when covered by allochthonous deposits, are considered *buried* (see later discussion and Chapter 16). When exposed to the atmosphere, i.e., not buried or under water, a surface is referred to as *subaerial* or *epigene* (Watchman and Twidale 2002). All subaerially exposed surfaces eventually become either buried or eroded. But while they are exposed, soils develop on them.

Geomorphic surfaces can be erosional, constructive, or a combination of both (Ruhe 1956a, b, 1975b; Figs. 14.2, 14.3). *Erosional surfaces* are those that have been formed by destructional (or degradational) processes, usually involving running water, wind, or gravitational forces. *Constructional (or depositional) surfaces* are upbuilt and buried (or aggraded) by allochthonous deposits, e.g., eolian sand, alluvium, glacial till, or loess. If the constructional event was short-lived, such as a volcanic eruption, the age of the geomorphic surface is equivalent to that of the parent material. Across a single constructional geomorphic surface (Fig. 14.2A), at time$_{zero}$ each component slope is the same age. Thus, it may be possible to determine the age of the surface by dating the sediment (see Chapter 15). Similarly, large-scale erosion of a surface may create a new surface, younger than the age of the sediment that underlies it (Fig. 14.3B). More likely, surfaces have experienced periods of erosion or aggradation since the deposition of the parent material, producing a complex pattern of geomorphic surfaces with different ages, but all relating to each other in some way (Fig. 14.2B). On the basis of this example, you may observe that it is possible to assign relative ages to these surfaces, simply by knowing how they formed.

Often, one of the first (and most important) determinations made about a geomorphic surface is whether it is erosional or depositional. Depositional surfaces are often equal in age to, or slightly younger than, the sediments in which they are formed (Daniels *et al.* 1971a; Fig. 14.2). Erosional surfaces are more complex; their age is usually determined by examining the slope or material they grade toward and merge with, since an erosional surface is the same age as the depositional surface to which it grades (Fig. 14.2A).

Often, determining the most recent erosion history of a surface can also help ascertain its relative age. For example,

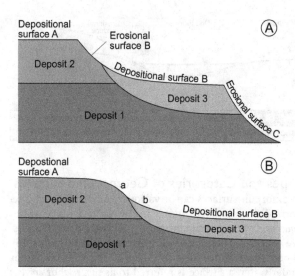

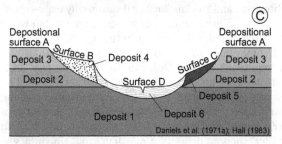

Daniels et al. (1971a); Hall (1983)

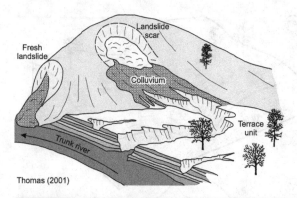

Thomas (2001)

Fig. 14.5 Block diagram model of a landscape that has had spatially isolated areas of instability (sloughing and accretion) and stability, illustrating geomorphic surfaces of different age and type (erosional, aggradational).

Fig. 14.4 Examples of relative age and stratigraphic relationships in surficial deposits. (A) Deposit 1 is older than Deposit 2 because of the principle of superposition (in a series of stratified sedimentary units, lower units are older than the ones above), and Deposit 3 is younger than 1 and 2 for the same reason. Erosional surface B is younger than Deposit 2 because it truncates that deposit. Surface B is also younger than surface A because it (B) cuts A. Erosional surface C is the youngest surface in this figure. (B) Deposit 3 rests on an erosion surface cut prior to or at the time of its emplacement. Surface A is a depositional surface and thus is younger than Deposit 1 but the same age as Deposit 2. Surface B is younger than surface A because it bevels the sediment underlying surface A, and because surface B cuts surface A. Portion *a* of surface B is the erosional element and portion *b* is the depositional element. This diagram illustrates that many surfaces have both erosional and depositional components. (C) Deposit 4 is emplaced on an erosional surface that cuts Deposits 2 and 3, and surface A. Therefore, Deposit 4 is younger than Deposit 3 and surface A. Because surface B cuts surface A, it is younger than A. Surface C is younger than surface B because it is at a lower level. Surface D is younger than surfaces B and C because it cuts them. Deposit 6 is younger than Deposits 1 and 2 because it bevels them.

erosional surfaces must, by definition, be younger than the material they are formed in, or the youngest material they cut or bevel. But, they are older than any valleys cut into them, or deposits that lie within those inset valleys. These relationships are based on the geological *principle of ascendancy and descendancy*, which clearly states that an erosion surface is younger than the youngest *deposit or surface* that it cuts across or truncates (Hallberg *et al.* 1978a, Watchman and Twidale 2002). This principle is especially useful in fluvial settings, where younger materials are inset within older sediments. Corollaries to this principle state that an erosion surface (1) is the same age as or older than other deposits lying on it and (2) is younger than erosion remnants above it. Likewise, the erosional surface of a hillslope is the same age as the alluvial fill to which it descends, but is younger than a higher surface to which it ascends (Ruhe 1975b). Many of the stratigraphic principles mentioned are illustrated in Figures 14.2, 14.3, 14.4, and 14.5 (see also Daniels *et al.* 1971a).

In the analysis of erosional and depositional surfaces, it is important not to confuse the age of the *material* (absolute or relative) with the age of the *surface* (Fig. 14.4). The *material* is the mass of bedrock, glacial till, loess, colluvium, alluvium, regolith, and so on, that the soil's parent material usually comprises. A *surface* is a two-dimensional *concept* that denotes the uppermost boundary of a material at some defined moment in time. Like soils, surfaces cannot be older than the material with which they are associated. An erosional or depositional surface either is bounded (above) by the atmosphere or was at one time bounded by the atmosphere and has been subsequently covered with younger material.

As should now be obvious, most landscapes are an assemblage of geomorphic surfaces of different kinds, ages, and histories. Many times, the boundaries of these surfaces are clear, occurring at distinct steps or breaks in the surface (Figs. 14.2, 14.5). At other times, the boundaries are more subtle. Through careful observation of the landscape – its topographic attributes mainly – these surfaces can be identified and their ages and relationships discerned. An example of this type of observation – and application – is illustrated in the K cycle concept.

The K Cycle Concept

An inherent characteristic of every geomorphic surface is its *stability*. Stability affects the length of time that is available for soil development. Soils develop best on *stable* geomorphic surfaces. Soils on unstable surfaces are being either eroded or buried (Porter *et al.* 2008; Figs. 14.3, 14.5). If the surface and the soil developed on it become eroded, a new soil may form on the surface later, after it stabilizes. If the surface is unstable because of aggradation, the soil may become buried and preserved as a paleosol, providing evidence of the past period of stability (see Chapter 16). Examples abound of stacked sequences of paleosols, buried by episodic increments of loess, dune sand, alluvium, or glacial drift that were deposited during periods of aggradation (Thorp *et al.* 1951, Arbogast and Johnson 1994, Anderton and Loope 1995, Arbogast *et al.* 2002a).

This premise – that all landscapes undergo periods of stability and instability – led Bruce Butler, an Australian soil scientist, to develop his K cycle model of soil and slope evolution (Butler 1959, 1982). Fundamental to this model are the concepts of landscape periodicity, as epitomized in the *soil cycle*. Indeed, landscape periodicity may, in fact, be the norm. Each of Butler's soil cycles includes an alternating phase of instability and stability. During the instability phase, a surface is either buried or eroded. During the stability phase, soil development proceeds on a new surface. An implicit assumption of Butler's model is that, for most landscapes, sedimentation processes overtake soil development during some intervals, leading to soil burial. Similarly, at other times and in other places, soils become eroded.

The K cycle model is built around alternating cycles of stability and instability. It contrasts with an earlier soil-landscape evolutionary model by Nikiforoff (1949), which postulated that, on many stable, gently rolling landscapes, the rate of surface erosion and deposition is generally in balance with pedogenesis. In Nikiforoff's model, the rate of sediment removal on uplands equaled the rate at which the B horizon changed into A horizon material, as the soil grew slowly downward. The result was that the A and B horizons maintained a standard thickness as the profile sank into the landscape at the same rate as the surface was lowered. Likewise, on lowlands, deposition took place at a slow enough rate that the soil could grow upward at an equal pace. A horizon material changed into B material just fast enough that the soils never became cumulic. On the landscape as a whole, this model suggests that equilibrium exists between slope erosion or deposition and soil development. Nikiforoff recognized that soil burial can occur on lower slope positions, but regarded this as a special case. His model recognized but deemphasized slope processes and made soil development into a suite of processes that seemed much more important than perhaps they are. Butler's (1959) K cycle model reemphasized slope processes, illustrating that, at times, they can overwhelm pedogenesis. The K cycle model put more emphasis on the geomorphic, i.e., slope, component of soil development.

The K cycle concept is most useful when applied as a type of *time unit*, called the *K cycle* (Greek *khronos*, time) referring to relative time, not time in an absolute sense. A full K cycle is defined as the interval of time starting with the formation (whether by erosion or deposition) of the new geomorphic surface, including the period in which soils develop on that surface, and ending when the surface becomes either buried or eroded. Butler preferred the term *groundsurface* for stable geomorphic surface. More of a continuum, one cycle leads immediately to the next, although any one cycle need not pass through all steps fully. For example, a soil could begin developing on a surface that is only ephemerally stable, only to be eroded or buried. This is a "partial" cycle because it did not experience *full* development.

Cycles are designated K_1, K_2, and so forth, with larger numbers for earlier cycles. This terminology allows the investigator to communicate about periods of past stability or instability, without having to know exact ages. The present cycle of soil formation is not the K_1 cycle; the term K_1 is reserved for the most recent cycle *previous to* the present. Further information could be presented about the paleocycles by using subscripts for the stable or unstable phases of the K cycle, e.g., K_{1s}, K_{1u}, K_{2s}, K_{2u}, K_{3s}.

The unstable phase, during which soils are eroded or buried, can be envisioned as the part of the cycle in which surfaces are *preconditioned* for pedogenesis, because new parent material becomes exposed at the surface. This phase can be spatially continuous, e.g., as when loess buries large areas. More often, erosion is focused on certain slope elements or aspects, or on surfaces underlain by more erodible soils (Pereira *et al.* 1978; Fig. 14.6). Butler envisioned that instability is usually spatially discontinuous; some parts of the landscape are undergoing erosion while others are being buried. This pattern could be easily envisioned for a periglacial landscape, where solifluction is removing material from upper slopes and burying lower slope segments and soils. Butler (1959, 1982) described these parts of the land surface as *sloughing zones* (those that erode and lose sediment). Other parts so defined were *accreting zones* (those that become buried by new sediment) and *persistent or residual zones* (those that are stable and on which the soils continue to develop). Recognizing that surface stability/instability is both time and space transgressive, Butler also referred to some areas as *alternating zones* (Fig. 14.6).

The K cycle can be a guiding model for geomorphologists who wish to interpret paleoenvironments or periods

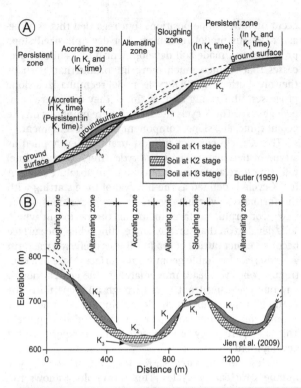

Fig. 14.6 The K cycle concept, illustrated. (A) The various zones of a slope, the erosion and burial of which are fundamental to the K cycle model. (B) Example of the K cycle, as depicted for the hillslopes of Taiwan.

of land use change (Sycheva *et al.* 2003, Sycheva 2006, Leopold and Volkel 2007, Hobbs *et al.* 2011). Applications and examples of the K cycle model can be found in a variety of landscapes around the world. Jien *et al.* (2009) described a classic example of the K cycle in the mountainous landscapes of central Taiwan. Here, because of the abundant rainfall, mass movement is common on steep slopes, burying soils in the lowlands. Working in the deserts of Israel, Wieder *et al.* (2008) describe intermittent periods of dust accumulation that bury soils in the lowlands. This type of hybrid K cycle involves only the accreting zone without an upslope sloughing zone, because burial/accretion is by eolian dustfall. Here, dry climatic intervals lead to short phases of dust accumulation. These intervals are followed by periods of soil development in a slightly wetter climate. Many other examples of application of the K cycle model to landscape evolution exist (Walker 1962, Pereira *et al.* 1978, Dijkerman and Miedema 1988, Sommer *et al.* 2008).

Butler's (1959) K cycle concept is useful for linking soil development processes with surface stability. It moved to the forefront the notion that surfaces and soils have linked histories. Both Butler's (1959) and Nikiforoff's (1949) models provide mechanisms by which pedogenesis can proceed on a landscape, despite periods of erosion and/or aggradation. A more recent model (Paton *et al.* 1995), which also incorporates slope processes into pedogenesis, is discussed later in this chapter.

Ages and Relative Dating of Geomorphic Surfaces

Establishing the age of a soil or surface is often a key component of a soil geomorphology study. Soil and geomorphic data can provide important information about not only the past stability, but also the potential age of a geomorphic surface (Mahaney 1984, Amit *et al.* 1996). So, too, can the sediment itself. For example, no geomorphic surface can be older than the *deposit on* which it is formed. It can be younger than the deposit, but not older (Fig. 14.2). Because dating is so important to soil geomorphology, we devote an entire chapter (Chapter 15) to it. For now, we will deal with this topic only in generalities.

A geomorphic surface is as old as the last period of stability. That is, if a surface has undergone several periods of instability and stability, the current soil on it probably had its time$_{zero}$ no earlier than the onset of the last period of stability. Stability and soil development go hand in hand. This pairing is one that we will continually revisit.

In 1983, the North American Commission on Stratigraphic Nomenclature published its North American Stratigraphic Code. The code defined and formalized stratigraphic nomenclature, including various units categorized by physical attributes (lithostratigraphic, biostratigraphic, and allostratigraphic units, etc.) and age (chronostratigraphic, geochronologic, and geochronometric units). These units provide a basis for the systematic ordering of the time and space relations of rock/sediment bodies and establish a time framework for the discussion of geologic history. Lithostratigraphic units are based/defined on lithology, chronostratigraphic units are based/defined on time or age, and so on. With regard to surfaces and sediments, the term *isochronous* means of equal duration, while *synchronous* means simultaneously. The term *diachronous* is used for stratigraphic units whose bounding surfaces are not synchronous. Diachronous is a synonym for *time transgressive*. Many geomorphic surfaces are time transgressive, meaning that not all parts are of the same age, i.e., the age of the surface *changes* across it, often in a systematic manner. A good example is a surface being buried by loess. As described in Chapter 13, burial occurs first and most rapidly near the loess source and only later are sites farther away buried. The buried surface is time transgressive, because its burial was not synchronous. Obviously, knowing this and knowing the span of ages on the surface are imperative to understanding soil development *on* that surface, because soils that form on unstable surfaces are soon eroded or buried, whereas soils that form on stable surfaces continue to develop. Many subaerially exposed geomorphic surfaces are undergoing erosion on some parts and burial (by slopewash or alluvium) on other areas. As the margin of the erosional or depositional surface slowly moves across the landscape, it forms a surface that is time transgressive. Using soil data to determine the length of

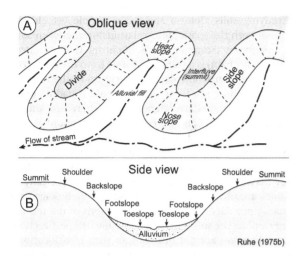

Fig. 14.7 Geomorphic components of slopes in a landscape with an open drainage system. (A) Three-dimensional components of slopes. (B) Two-dimensional components, i.e., the five elements of fully developed slopes.

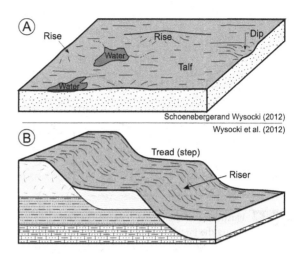

Fig. 14.8 Geomorphic components of flat landscapes and stepped landforms.

time that a geomorphic surface has been stable or whether it is currently unstable (eroding or aggrading) is a focus of soil geomorphology.

The apparent stability of a surface depends on the temporal scale one uses. Many surfaces are stable over short periods but are still undergoing erosion by natural processes, collectively referred to as *geologic erosion* or *natural erosion*. All slopes and surfaces are subject to these natural downwasting processes. On stable slopes, the rate of weathering or soil formation equals (after an initial, iterative period) the rate of geologic erosion. For example, Pavich (1989) described saprolite-mantled upland surfaces in the Appalachian Piedmont that exceed 1 million years in age. These surfaces, despite their great age and apparent stability, are not *truly* stable, in that they experience geologic erosion at rates of about 20 m per million years. Normally, geologic erosion here keeps pace with bedrock weathering and soil production (see Chapter 9). However, this normal, geologic erosion may outpace weathering and soil formation for a period, setting up a positive-feedback mechanism. The bedrock-saprolite contact grows closer to the surface, allowing weathering to accelerate, thereby deepening the bedrock-saprolite contact. During periods when geologic erosion is exceedingly slow, the weathering front will deepen faster than erosion can strip the regolith and weathering by-products. As a result, weathering and soil production will slow, because the weathering front is farther from the surface and because weathering by-products will accumulate, slowing chemical weathering reactions (see Chapter 9). Thus, over long periods, geologic erosion tends to keep pace with those processes that produce sediment/soil by a complex set of feedback mechanisms, and the slope may lower in elevation but not

change appreciably in other ways (Ahnert 1987). This interplay was envisioned by Nikiforoff (1949) in his soil-slope development essay. Cultivation or deforestation may result in *accelerated* or *anthropogenic erosion*, which may outstrip the system's ability to produce saprolite and soil horizons, thereby resulting in a thinner soil mantle.

Surface Morphometry

Surfaces can be defined genetically (as to origin) and morphometrically (as to shape and geometry). Although soil geomorphologists must always be aware of the probable *geomorphic origins* of a slope (Holmes 1955), a complicated discussion of the processes that are potentially responsible for them is beyond the scope of this book. We can, however, provide terms that can be used to *describe* a landform or landscape, regardless of origin.

Five major slope elements were derived long ago for describing rolling or hilly landscapes (Ruhe and Walker 1968, Ruhe 1975b). From the top of the slope to its base, they are the summit, shoulder, backslope, footslope, and toeslope (Fig. 14.7B). Their utility in soil geomorphology is without question (Kleiss 1970, Furley 1971, Malo *et al.* 1974, Burras and Scholtes 1987, Donald *et al.* 1993, Stolt *et al.* 1993, Korobova and Romanov 2011). Elements such as summit, shoulder, and footslope are almost universally used to describe hillslope locations.

The only places where the five-class slope element system breaks down are extremely flat landscapes and, to a lesser extent, stepped landscapes (Fig. 14.8A). On flat landscapes, which are quite common worldwide, typically the only relief is in areas of slight rises or depressions, requiring a special set of slope and landscape descriptors; Schoeneberger *et al.* (2012) describe these areas as rises and

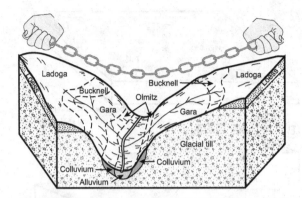

Fig. 14.9 Illustration of the catena concept. The term "catena" suggests the interlinkage of soils along a sequence, like links in a chain. The block diagram illustrates a typical catena found in Adams County, Iowa (Lensch 2008).

dips, respectively. Flat, broad summit areas are called *talfs* ("flat" spelled backward). Often these flat surfaces occur on stepped or tread-and-riser landscapes (Fig. 14.8).

The Catena Concept

Studying any one pedon in isolation holds little promise for helping to explain soil-landscape evolution. Thus, we tend to seek out the middle ground by examining small, representative, interrelated segments of the landscape, and then extrapolate what we learn from them, to larger areas; this is the same fundamental approach taken in soil survey and mapping operations. On these small segments – at the local scale – much of the variation in the soil cover is a function of slope or *relief*. The *catena*, best expressed on landscapes of moderate relief, is an embodiment of the application of Jenny's (1941b) relief factor (see Chapter 12).

We believe – and remind the reader – that the original intent of the relief factor was that of *topography*, but because it has been used and reused as *relief*, we will use that term in this book. Relief, by definition, is simply the relative *difference in elevation* between the uplands and the lowlands in a specified region. Topography refers to the relative positions, slopes, configurations, and elevations of the land surface. It is a much better expression of the lay of the land than is the term relief. We believe that both Dokuchaev and Jenny meant *topography* in their state factor equations.

Relief (topography) is essentially a passive factor, but it has the very important function of providing potential and kinetic energy to the soil system through its effect on water movement. It conditions the redistribution of matter and energy within the soil landscape system. Redistribution is a three-dimensional process (Huggett 1975), although for sake of simplicity we begin with the two-dimensional catena.

Studying soils along a slope is a simple yet elegant way to discern the spatial interrelationships between soils and topography (Sommer and Schlichting 1997). A *catena* (Latin *catena*, chain) is a transect of soils from the top to the base of a hill, perpendicular (or nearly so) to the contour lines. Milne chose the term *catena* because it had no existing usage in any modern language and, therefore, should introduce no confusion and could easily be adopted into languages other than English. Soils in a catena are often viewed as links in a hanging chain (Fig. 14.9). The concept is not unlike that of soils from hilltop to hilltop, with a valley between – each of the soils is linked genetically to the ones around it (Fig. 14.9). The pedogenic linkages are primarily provided by lateral translocations of matter and energy among the soils, mainly in the downslope direction. But the linkages are not due only to such translocations. They are also due to association. Soils closer to each other simply have more in common – because of the greater similarity of the soil forming factors – than soils farther away. Catenas include information on soils, surficial stratigraphy and hillslope hydrology, and shape. Applying the catena concept involves fitting all these pieces together into a functioning soil-geomorphic system. It provides a conceptual framework to help explain local-scale soil variability.

On the basis of work in Africa, Geoffrey Milne (1935, 1936a) originally defined a catena as the sequence of soils between the crest of a hill and the floor of the adjacent swamp. He observed that soils changed along this sequence in accordance with conditions of drainage, geomorphic history, and sediment character. Milne suggested specific mechanisms for the formation of these repeating soil-landscape patterns: (1) regular topography-geology relationships, (2) wetter drainage conditions on the lower landscape positions, and (3) soil-landscape (slope) processes, including erosion, transport, and deposition of solutes and particulates along the hillslope (Brown et al. 2004a). This catena concept does *not* exclude catenas where the soils had developed on different parent materials, i.e., lithology does not have to be homogeneous (Fig. 14.10). Milne stated that the catena was simply a unit of mapping convenience – a way of conveying repeating and predictable patterns of soils on landscapes. These soils are "repeated in the same relationships to each other wherever the same conditions are met" (Milne 1935, 197).

In 1942, Thomas Bushnell extended and at the same time constrained the application of the term catena to include all the various possible topographic, denudational, and hydrologic situations, but *on a given parent material*. He limited catenas to a single parent material, because not to do so would "spoil ... its simplicity of connotation" (Bushnell 1942: 467). In essence, Bushnell wanted a catena to differ only with regard to drainage; today we call this a *toposequence* (see Chapter 12). The toposequence concept primarily involves morphologic connotations such as changes in soil color due to changing wetness conditions (Hall 1983). Milne (1936a) would have argued that soils

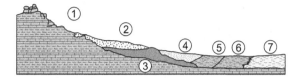

Fig. 14.10 Soils and sediments in the east African catena, as originally constructed by Milne (1936b). Note that this catena shows great variation in sediment type, i.e., its sedimentology (parent material) is not uniform. Materials 2 and 3 have most likely moved downslope, whereas materials 4–7 are alluvial deposits that may have moved both downslope and downstream. Using the principle of superposition, one can say that the ages of the materials (oldest to youngest) are most likely 3 > 2 > 4> 5 > 6 > 7. (It is possible that 2 could postdate 4–7.) All of the surfaces shown are depositional except 1, which is erosional.

along a catena, on the other hand, differ in morphology because of changes in drainage *and* because of fluxes of sediment and dissolved substances along the slope *and* because of differing original sediment or parent material composition, all included in his concept of the past history of the land surface. Ruhe (1960) stressed that the evolution of a catena of soils cannot be understood without incorporating knowledge of the geomorphic history of the landscape itself and disagreed with the notion that catenas should be restricted to one parent material. In this book we adopt that same approach.

Soils change predictably along catenas. Milne knew that, and that is why he formulated the concept. But to enhance this predictability, the two-dimensional catena must be examined in light of the three-dimensional landscape, incorporation slope curvature as seen from above. This *plan curvature* largely controls the directions of water and sediment transport on the slope (Pennock *et al.* 1987, Donald *et al.* 1993, Pennock 2003). For example, catenas on nose and head slopes are quite different, as nose slopes are water- and sediment-diffusing slopes, whereas head slopes are water- and sediment-gathering slopes (Sobecki and Karathanasis 1992; Fig. 14.7A). Perhaps the simplest versions of catenas exist on side slopes, where plan curvature effects are minimal (Fig. 14.7A). To determine whether a slope is diffusing or gathering, examine the water flow lines, which run generally perpendicular to the contour lines. In nose (or spur) slope positions, flow lines will diverge downslope. In coves or head slopes, they will converge. In two dimensions, catenas should be drawn along transects in which water will flow downslope directly along that transect, i.e., they are defined perpendicular to the contour lines.

Reasons for Soil Variation within Catenas

Topography is a highly explanatory framework within which to study soil development (Kleiss 1970, Furley 1971,

Sommer and Schlichting 1997). Early literature even went so far as to state that with time the influence of relief could dominate mature soils to the point that even parent material effects would be masked (Norton and Smith 1930). To a degree, this is correct; with time the effects of topography and geomorphology become increasingly important to soil development.

The wisdom of the catena concept highlights the fact that soils on one part of the landscape affect soils nearby, and especially those *downslope*. Materials, solutions, and suspensions move through and over landscapes, creating genetic linkages among soils on slopes (Hall 1983, Donald *et al.* 1993). Translocations and transformations, espoused by Simonson (1959) as a useful method of explaining intraprofile, i.e., horizon-related, differences, also apply to catenas. Additions to and removals from soil bodies create many of the pedogenic differences along the catena itself. As Dan and Yaalon (1964: 757) put it, "The morphology of each (catena) member is determined by its position in the landscape and related to its adjoining members."

Catenas and slopes can be classified on the basis of flow of water within and through them. In *closed catenas*, any downslope flux of water or sediment stays within the system, like water flowing into the center of a closed depression or bowl-shaped lowland. In an *open catena system*, sediment and water can exit the system, usually by way of a through-flowing river at the base of the slope. In both kinds of systems, sediment can enter and leave by eolian processes.

Soils vary along catenas for two main reasons: (1) *fluxes* of water, energy, and matter – generally but not always in the downslope direction – and (2) water table effects. Material fluxes are of two main types: debris flux (sediment and organics) and moisture flux (Malo *et al.* 1974, Schaetzl 2013). Debris flux involves erosional and depositional components. When transport is primarily by gravity-driven processes it has been referred to as *colluviation* (Goswami *et al.* 1996). When rainsplash and overland flow are the primary processes, i.e., water-driven processes, the term *slopewash* is used. In closed catena systems, sediment and water accumulate at the base of the slope (Walker and Ruhe 1968). In open drainage systems, the debris can potentially be removed from the hillslope system. Debris flux is so important that on many slopes soil properties reflect more strongly the hillslope sedimentation system than the pedologic system (Kleiss 1970). Sommer and Schlichting (1997) reviewed the importance of material fluxes to soil variation within catenas and developed an idealized scheme to represent all the possible types of material flux. Their scheme includes fluxes caused by overland flow, lateral subsurface flow, vertical infiltration/seepage, capillary rise, and return flow and shows the linkages among these fluxes (Fig. 14.11).

Although precipitation is evenly distributed along slopes, the amount of water that infiltrates versus runs off depends on a number of factors, e.g., infiltration capacity, slope steepness and plan curvature, intensity/frequency

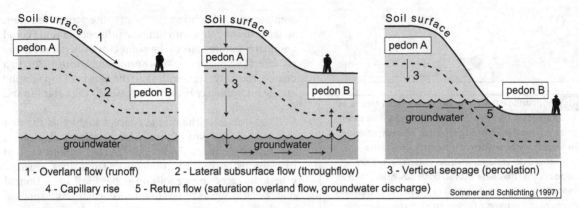

1 - Overland flow (runoff) 2 - Lateral subsurface flow (throughflow) 3 - Vertical seepage (percolation)
4 - Capillary rise 5 - Return flow (saturation overland flow, groundwater discharge)
Sommer and Schlichting (1997)

Fig. 14.11 Idealized scheme showing the various direct and indirect water-related fluxes among soils on a catena, as influenced by a water table.

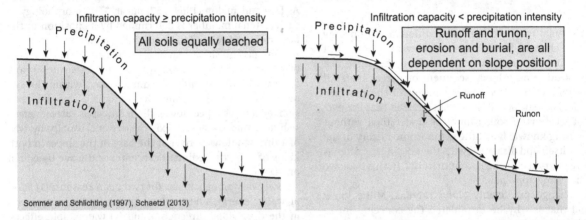

Fig. 14.12 End member scenarios for moisture movement along a catena. (A) Precipitation rate < infiltration rate, as is typical for coarse-textured soils. Little runoff occurs and soils are similarly leached/developed along the catena. (B) Precipitation rate > infiltration rate, as is typical for fine-textured soils. Runoff develops on steep slope segments, causing these soils to be drier, thinner, and more eroded. Run-on areas at the base of the slope are more deeply leached, unless inhibited by a high water table.

of rainfall events, and cover type (Fig. 14.12). Permeability is mainly affected by soil texture and structure, but vegetation also plays a role. Runoff occurs when precipitation rates exceed the infiltration capacity (permeability) of the soil and is often the driving force behind debris flux. If soil infiltration capacities exceed rates of precipitation, most of the water will enter the soil and differences due to debris and moisture flux along the catena will be small (Fig. 14.12). When runoff occurs, slope steepness and curvature affect the rate at which the water runs off the slope, and where it goes (Huggett 1975, 1976b).

On flatter summit positions, water tends to infiltrate, or only slowly run off, and as a result, soils can be deeply leached. Because they are the steepest slopes on the landscape, upper backslope and shoulder slope areas have the most potential for runoff and hence are commonly the most eroded. They also exhibit the thinnest soil profiles and are the most likely areas for rock outcrops or free faces

(Gregorich and Anderson 1985; Fig. 14.13). Slope gradient, however, is not the only factor that can influence runoff and erosion. Vegetation density and type, sediment texture, soil infiltration capacity, and biotic (especially burrowing animal) activity all influence slope erodibility by their indirect effects on permeability (Yair and Shachak 1982). The midbackslope position is dominated by transportation processes. Farther downslope, debris flux slows and deposition of sediment predominates on the areas of lower slope gradient. Many of these lower slope elements receive run-on water and are usually wetter (Fig. 14.13). High water tables here may inhibit certain pedogenic processes, although if they do not, these soils may be some of the best developed on the catena, because of the large amount of kinetic energy of infiltrating water (Runge 1973). *Cumulization* due to debris flux can occur on footslopes and toeslopes (see Chapter 13). The base of the slope is also prone to receiving inputs of sediment from outside the catena, e.g., overbank

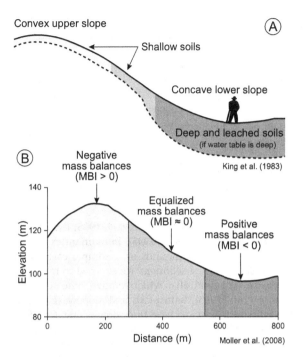

Fig. 14.13 Conceptual diagrams illustrating how soils differ along a typical catena. (A) Shallow versus deep soils. (B) Relationships among slope components and the sediment Mass Balance Index of Friedrich (1996).

deposits from streams. If all soils are equally permeable and water table effects are minimal, soils on the backslope will be the least leached, those on the toeslope will be the most leached, and summit positions will be intermediate.

During transport, the material in flux along the hillslope tends to become better sorted in the downslope direction (Fig. 14.14). In general, finer material moves farther along the slope while coarser sediment remains behind (Kleiss 1970, Malo *et al.* 1974). Thus, coarser materials are commonly left behind on shoulder and upper backslope positions (Walker and Ruhe 1968). Many soils on shoulder slopes are coarser-textured, as a result of this relationship. The fining-downward trend on shoulder slopes and backslopes is best expressed in closed basins (Walker and Ruhe 1968), because in open basins, the still finer sediment on toeslopes may be removed by streams, or sediment may be added to them by the fluvial system.

Debris Flux: Texture Contrast Soils
Most hillslopes in the midlatitudes would be considered geomorphically young, because they are usually formed in recent glacial sediment. Soils on these slopes usually have Bt horizons enriched in clay, presumably due to lessivage (Ciolkosz *et al.* 1989, 1996, Johnson 2000, Schaetzl 2000, Lavkulich and Arocena 2011). Surface horizons are coarser-

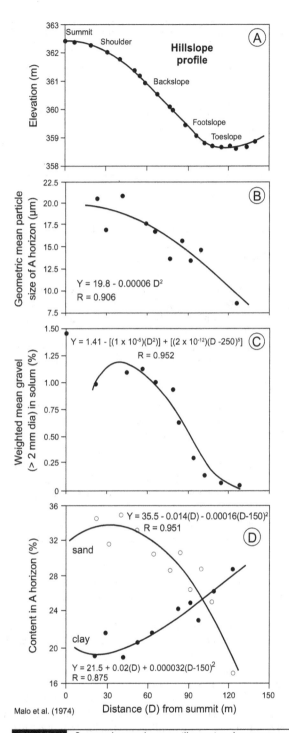

Fig. 14.14 Scatterplots and curves illustrating the strong relationship between soil textural properties and slope position in a toposequence of soils developed in till in North Dakota, United States. These relationships are largely driven by runoff and debris flux processes.

Table 14.1 Possible genetic pathways for the formation of texture contrast soils

Formation mechanism	Genetic characterization
Inheritance from parent material	Geologic/sedimentologic
Deposition of coarser materials on top of preexisting finer materials	Geologic/sedimentologic (washing/winnowing of fine materials from upper horizons by slope processes) or via bioturbation
Translocation of fine materials to subsurface	Pedogenic (lessivage, decomposition, and synthesis) and/or sedimentologic (gravitational settling) processes
Subsurface fining	Pedogenic (weathering of coarse materials from subsoil)

Source: After Phillips (2007).

textured than B horizons, presumably because of loss of clay by weathering and eluviation. Because of their relative youth and (often) dense vegetative cover, material transport processes are often only minimally operative here (Ahr *et al*. 2013). Even on the landscapes of the southeastern United States, lessivage is the explanation of choice for the Ultisols seen here. In other words, pedogenesis, particularly vertical (top-down) pedogenic processes, dominates over debris flux. Additionally, on young landscapes, the effects of bioturbation have not yet been manifested to large degree.

This is certainly not the case on many old landscapes, e.g., in interior parts of Australia, Africa, and South America. Here, many soils have loamy or coarse-textured upper (A) horizons and red-brown colored, clayey B horizons (Ollier 1959). Often, the B horizon exists in conjunction with, or above, saprolite. Because of their fine-textured B horizons, they would classify as Ultisols or Alfisols. In other classification systems they have been called Red-Brown Earths. Despite interpretations to that effect (Chittleborough and Oades 1980, Lecomte 1988, Anda *et al*. 2009), these soils clearly have not formed entirely by lessivage and top-down pedogenesis (Koppi and Williams 1980). Too often, the amount of illuvial clay in the B horizon is insufficient to account for the observed texture contrast, and few argillans are present. Rather, these *texture contrast* or *duplex* soils (Northcote 1960) are due to some combination of bioturbation, eolian processes, and debris flux, operating for millennia on old, stable hillslopes (Gunn 1967, Bishop *et al*. 1980, Paton *et al*. 1995).

The genesis of texture contrast soils – which Brown *et al*. (2004a) deemed the most ubiquitous profile feature in the world – has been elusive and long debated (Oertel 1974, Koppi and Williams 1980, Lecomte 1988, Chittleborough 1992, Paton *et al*. 1995, Phillips 2004, 2007, Ahr *et al*. 2012; Table 14.1). Here is how it presumably works, at least in Australia. For millennia, winds have redistributed sandy materials across this landscape (Walker *et al*. 1988, Krull *et al*. 2006), and bioturbation is omnipresent. If the uppermost sediment is clearly eolian, it exists as an uneven blanket of sands. Ants, termites, and other biota continually move material to the surface and churn the upper

profile (Humphreys 1981, Cowan *et al*. 1985, Debruyn and Conacher 1994, Paton *et al*. 1995). In biomantles, the sediment is deposited as mounds or as simple casts on the surface. Both types of sediment are exposed to rainsplash, slopewash, and wind, all of which remove some of the fines (Nooren *et al*. 1995). Rainsplash and slopewash presumably transport most of these fines downslope. Continued bioturbation keeps mixing this upper sediment, exposing more of it to erosional processes, and so, the upper profile becomes increasingly coarser-textured.

Several lines of evidence can be used to interpret the genesis of this upper, coarse-textured layer and its presumed slow, downslope mobility (Brückner 1955; Bishop *et al*. 1980). Notice in Figure 14.15A, for example, where bedrock and saprolite change from coarse sandstone (upslope) to igneous rock (downslope). The coarse, sandy mantle that overlies the sandstone (and sandstone saprolite) continues downslope, seamlessly, above the igneous rock. It is here, on the lower segments of the slope, that the classic texture contrast soils are found. Because the igneous rock could not have produced the sandy mantle by weathering, the sands must have been transported there by slope (or perhaps eolian) processes from upslope. The sandy mantle becomes finer downslope, suggestive of winnowing by biota, slopewash, wind, and other surficial vectors, rather than by eluvial processes (Humphreys and Mitchell 1983; Fig. 14.15B). Downslope fining is a hillslope-soil pattern that is found on landscapes like this, across the globe, including on Milne's (1936b) original catena (Brown *et al*. 2004a). The effect of the surficial winnowing penetrates to the maximum depth of the coarse upper layer, which usually coincides with the depth of bioturbation. This material also thickens downslope, perhaps as biota continually mine the upper subsoil and move that material into the upper profile, or perhaps as a result of additional eolian and slopewash additions. With time, the erosional surface wears back and lowlands in-fill, but the soils themselves do not markedly change. Rather, they evolve *with the surface*.

On other slopes in these same landscapes, similar texture contrast soils exist. But here, a stone line exists at the contact of the upper, coarse material and the finer-textured, lower material (Fig. 14.16C). Often, the stone line is

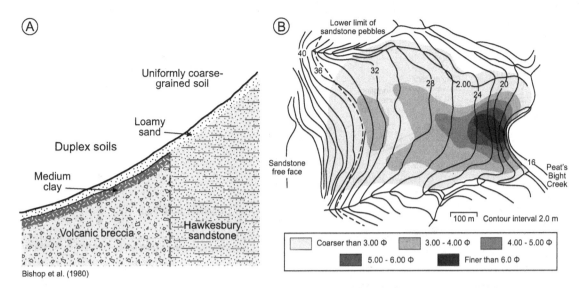

Bishop et al. (1980)

Fig. 14.15 Duplex soils on the landscape of the Sydney Basin, near the Hornsby diatreme. (A) Cross section of the soils and sediments on the slope at the edge of the diatreme. (B) Map of texture of the upper profile sediments along this slope.

present only downslope from rock outcrops or where veins of hard rock intersect the surface (Gunn 1967; Fig. 14.16C). The stone line maintains its location at the base of the sandy mantle, even in downslope areas. Imbrication of the clasts in the stone line provides further evidence of its slow transport downslope (Bishop *et al.* 1980). Clasts in the stone line also decrease in size downslope, suggesting either a sorting mechanism or downslope comminution. Last, the degree of differentiation of the stone line from the overlying A horizon increases downslope. Generally, where stone lines lie beneath loose, surficial materials, it is assumed that the stone line formed there by bioturbation and that the overlying material is a biomantle (Paton *et al.* 1995; see Chapter 11). In short, evidence for slope and biomechanical processes on this landscape is clear.

Recent work on texture contrast soils continues to document their two-layered nature and confirm that biomechanical, slope, and eolian processes – more so than vertical processes of pedogenesis – drive their genesis (Krull *et al.* 2006, Phillips 2007, Eze and Meadows 2014; Fig. 14.17). For example, radiocarbon dating of organic matter in these Australian soils confirms a marked disconnect between the A and B horizons. Bulk organic matter ages are < 800 ^{14}C years BP in the upper material, but they are considerably older below, in the B horizon (Fig. 14.17B). This striking depth distribution may reflect any of a number of surficial processes, with bioturbation of the upper sediment (but with much less bioturbation below) chief among them. Similarly, optical dating has shown that most of the sand grains in the A horizon had been exposed to light, probably by bioturbation, within the last 3000 years (Fig. 14.17C). By comparison, luminescence ages of sands in the B horizon and below (see Chapter 16) are much older, indicating that

these grains are seldom moved to the surface. Last, opal phytoliths – formed in roots and leaves (see Chapter 16) – are found only in the upper material. Together, this overwhelming array of evidence now suggests that texture contrast soils are formed as a slowly migrating (creeping) bio- or eolian mantle develops above a fine-textured B horizon that has formed and that remains largely in situ (Koppi and Williams 1980, Walsh and Humphreys 2010). In short, the A and B horizons of these soils are genetically different materials (Krull *et al.* 2006). These soils on old, tropical landscapes teach us that slope and biomechanical processes must be always considered when evaluating pedogenesis (Mitchell and Humphreys 1987, Sobecki and Karathanasis 1992).

In 1995, Tom Paton, Geoff Humphreys, and Peter Mitchell won the G. K. Gilbert Award for Excellence in Geomorphic Research, awarded by the Association of American Geographers, for their book *Soils: A New Global View.* In it, these Australian soil geomorphologists outline a view of soil and landscape evolution that deemphasizes traditional top-down processes and, instead, leans on slope processes and pedoturbation to explain soil development and patterns. Although it has come under a bit of scrutiny, as all models do (and should) (Beatty 2000, Johnson 2000), it draws together many excellent points. The model was developed mainly for, and is best applied to, old, tropical landscapes, such as in Australia, where long-term weathering and bioturbation are dominant processes. Ant and termite mounds are common here, and these mounds eventually deteriorate into stone-poor biomantles. If the underlying parent material has coarse fragments, a stone line or stone zone will form below a biomantle (see Chapter 11). And as discussed, in many residual soils the biomantle and

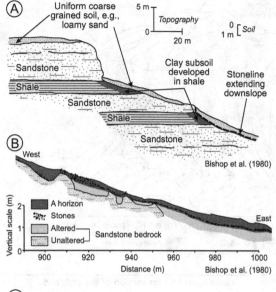

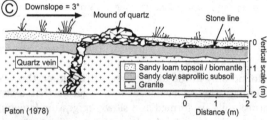

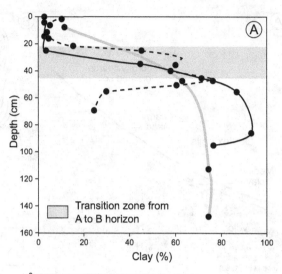

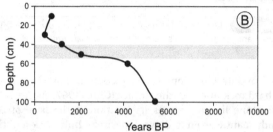

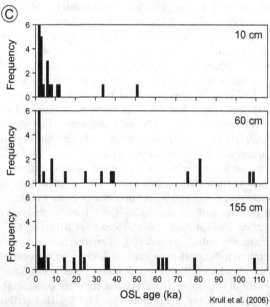

Fig. 14.16 Examples of slope-soil-sediment relationships on tropical and subtropical landscapes of Australia, where texture contrast soils and stone lines are widespread. (A) Soils and sediments across the Sydney Basin. (B) Soils and sediments along Warrah Road, near Patonga, New South Wales. Note the stone line that continues downslope from areas of sandstone outcrop in both situations, suggestive of a slowly creeping biomantle. (C) Detail of a duplex soil and stone line formed over granite saprolite in the Brisbane Valley of southern Queensland. The downslope movement of the broken quartz vein fragments provides irrefutable proof of the downslope movement of the upper mantle over the saprolite.

Fig. 14.17 Data for some texture contrast soils in southern Australia. (A) Depth functions of clay content, clearly illustrating the texture contrast in three representative soils. (B) Radiocarbon ages of organic matter in one of the texture contrast soils shown in A. (C) Distribution of luminescence ages from single quartz grains taken from three depth intervals of a duplex soil, pointing to similar interpretations.

stone line rest on a more clay-rich substrate that grades into saprolite.

One of the main purposes of the Paton *et al.* (1995) book is to highlight the formation of texture-contrast soils and to stress that many of the processes that form them are present worldwide. Let us put the texture contrast/debris flux model in an historical context. An earlier soil-landscape evolution model by Nikiforoff (1949) had deemphasized slope processes by rationalizing a way that soil development could keep pace with slope processes, to form "normal soils." Because buried soils are so common, Butler (1959) knew that Nikiforoff's ideas and model were not universally applicable. Rather than

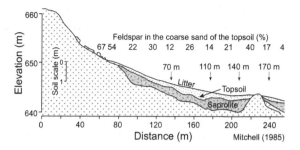

Fig. 14.18 Relationships along a slope between mineralogy and stratigraphy, for some texture contrast soils near Killonbutta, Australia.

Fig. 14.19 Photo of a soil with a pronounced stone line, near the city of Aristóbulo del Valle, Province of Misiones, Argentina. In this case, the siliceous horizon is a relict of hydrothermal veins of quartz that have weathered out of the host basalt rock. Photo by H. Morrás.

most soils and surfaces being in equilibrium, Butler saw landscapes as being subject to episodic stability and instability, leading to the development of his K cycle concept (see earlier discussion). Nonetheless, like Nikiforoff, Butler saw erosional and depositional processes on slopes as *opposing* pedogenic processes, or at least separate types of processes. Nikiforoff and Butler treated slope (erosional and depositional) processes differently, but in each model, *pedogenesis* was restricted to vertically operating processes on a stable substrate. In this respect, the model of Paton *et al.* (1995) differs – texture contrast soils form because of *interacting* slope, biomechanical, and pedogenic processes. Slope and biomechanical processes are fully *integrated into* pedogenic processes in the Paton *et al.* model, helping to explain the distribution and genesis of soils on slopes. Although lessivage is certainly operative in many soils, it is not as important, Paton *et al.* (1995) argued, to the formation of texture contrast soils as are slope processes. Watson (1961) argued long ago that vertical soil processes in the humid tropics may not be as effective as they are in the midlatitudes, because the intense tropical rainfall leads to more runoff. In the tropics the focus is on water running *parallel* to the surface. He noted that tropical soils, therefore, are not derived so much from the rock below as they are from the rock upslope.

We extend this discussion – originally begun in Chapter 11 – by defining *epimorphic processes*, which include weathering, leaching, new mineral formation, and inheritance. These processes can be distinguished from *surface processes*, which include pedoturbation, slopewash, eolian processes, and soil creep. Far too often, pedogenic explanations ignore the latter set of processes. Paton *et al.* (1995) argued that where soils are developed in bedrock, texture-contrast soils form first as weathering produces soil material from rock, via saprolite. Next, biomechanical processes begin, as biota mine the saprolite and transport some parts of it to the surface, as a biomantle (Watson 1962). Then, surficial processes winnow and slowly translocate this material downslope. Eventually, an ever-so-slowly

downslope-moving blanket of soil material, essentially a type of altered biomantle, develops above saprolite. Weathering-resistant clasts within the saprolite sink as a result of bioturbation and form a stone line at the contact between the biomantle and the saprolite, or at least above the more dense horizons below. Figure 14.16 illustrates a variety of morphologic and stratigraphic evidence that can be used to confirm that the upper, coarse-textured horizons are creeping downslope (Aleva 1987). If this model holds true, materials father downslope should be finer textured (they are; see Fig. 14.15B) and more weathered (they are; see Fig. 14.18). Weatherable minerals are, over time, depleted, leaving behind a quartz-rich, sandy layer in the downslope positions.

Stone lines on landscapes such as these are due to a combination of weathering, slope processes, *and* biomechanical processes. They document that slopes and soils are linked by a variety of dynamic processes, leading to the slow denudation of the uplands. Formed at the base of the coarse biomantle, stone lines take many forms. They may simply be an intermittent layer of stones, a continuous layer about one clast thick, or they may form a stone zone several meters thick (Fig. 14.19). The stones are usually not consolidated (Brückner 1955), and their upper contact is usually fairly abrupt.

On slopes where coarse materials are lacking, because sandstones or other coarse clastic rocks are absent, similar processes may operate but different soil morphologies evolve. Rather than a texture-contrast morphology, a *fabric contrast soil* forms (Hallsworth *et al.* 1952, Paton 1978). These soils lack stone lines. Hart (1988) described the formation of fabric contrast soils by invoking downslope movement of a bioturbated surface layer over a clay, derived in situ

Table 14.2 | Terms and concepts applicable to soil and forest hydrology

Term	Abbreviation used in this book	Definition/description
Gross precipitation	P	Incident precipitation falling from the sky
Canopy interception	I	Incident precipitation intercepted by plants and not permitted to fall directly to the soil surface
Canopy interception loss	C	Water intercepted by plants and lost to the atmosphere by passive evaporation
Net precipitation	N	Water remaining after gross precipitation has been reduced by canopy interception loss
Throughfall	T	Water that falls through, or drips off, the vegetative canopy, onto the soil surface
Stemflow	S	Water that runs down stems and onto the soil surface
Runoff	R	Water that contacts the soil surface and runs off, downslope
Litter interception loss	L	Water that contacts the O horizon and (1) is lost to the atmosphere via evaporation and/or (2) is taken up by plants and transpired
Infiltration (or pedogenic precipitation)	I or PP	Water that enters the mineral soil and is thus capable of participating in pedogenic processes

from bedrock. The contrast is not textural, but structural, i.e., from strongly granular or fine blocky in the A horizon, to a much coarser prismatic or blocky structure in the B horizon. Hart (1988) concluded that a combination of bioturbation and downslope flux produces the distinctive A horizon fabric in these soils. Different parent materials, same slope processes.

In their book, Paton *et al.* (1995) argued that many traditional pedogenic paradigms are flawed and proposed to replace them with one involving more geomorphology, in which surficial and biomechanical processes act upon an inherently mobile mantle of sediment (Schaetzl 2000). Milne's (1936b) original catena concept incorporated a view of soil formation that included concurrent lateral movement of soil material *and* traditional pedogenesis, and the model espoused by Paton *et al.* (1995) is a fitting extension. The point: Vertically percolating water, stressed in midlatitudes of the Northern Hemisphere as an overriding vector in soil genesis processes, may be less important on old soil landscapes.

These examples point to the necessity for the integration of slope and biomechanical processes into models of soil formation and landscape evolution. Epimorphism and bioturbation act as vehicles by which materials are prepared for hillslope transport (Humphreys and Mitchell 1983), and then, perhaps, while they operate, vertical pedogenesis is also ongoing (Phillips 2007). As cautioned by Phillips (2007), when it comes to texture contrast soils, it is sometimes difficult to generalize. Nonetheless, the suite of potential explanations for the genesis of texture contrast soils is limited (Table 14.1). Indeed, it is these kinds of soils

that continue to keep us thinking and pondering exactly *how* the soil-landscape system works.

Moisture Flux and Soil Hydrology

Water influences pedogenesis along any of several pathways. To that end, Lin (2006, 2010, Lin *et al.* 2005) has led the call for a new branch of soil science – one that is intertwined with hydrology – *hydropedology*. He defined hydropedology as the integration of classical pedology with soil physics and hydrology, for the purposes of studying the pathways, fluxes, storages, residence times, and spatiotemporal organization of water in the soil and subsoil. Hydropedology puts the focus of pedology on water, and how it flows with soils and affects them and the entire vadose zone. As such, it provides pedology with closer linkages with the bio- and geosciences communities (Wilding and Lin 2006). The section that follows lists and discusses many of the basic principles used in hydropedology research.

In order to facilitate this discussion, however, we must first define a few basic hydrological terms (Table 14.2). Standard rain gauges report their data as P (gross precipitation). Some P is intercepted (I) by plants as surface detention on leaves and stems (Grah and Wilson 1944). This water can then take one of two pathways: It can evaporate, or it can move from the plant to the soil surface. Water lost to the atmosphere by evaporation from plant surfaces, termed canopy interception loss (C), is a significant component of forest hydrology, sometimes exceeding 15–20% of gross precipitation (Voigt 1960, Helvey and Patric 1965, Mahendrappa and Kingston 1982, Alcock and Morton 1985, Freedman and Prager 1986). The precipitation that interacts with the soil surface is called net precipitation (*N*):

$$N = P - C$$

Net precipitation can reach the soil surface directly or by dripping off plants (called throughfall, *T*) or directly from precipitation that is not intercepted. Net precipitation can also reach the surface by running down stems as stemflow (*S*). Thus,

$$N = T + S$$

and

$$P = C + T + S$$

In vegetated areas, water potentially available for pedogenesis includes throughfall (*T*) and stemflow (*S*); they are the main water inputs to the soil surface. However, not all of this water infiltrates into the soil. Depending on slope gradient, surface cover, and infiltration capacity, some of the water will run off (*R*), some of it will be absorbed by the litter (O horizon), and some of it will be evaporated. Other water within the litter will be removed by plant roots and transpired, resulting in a cumulative litter interception loss (*L*). Litter interception losses are not inconsequential; they are usually between 2% and 4% of *P* (Blow 1955, Rowe 1955, Helvey 1964, Swank *et al*. 1972). Water lost as *L* does little to affect pedogenesis directly, although it does facilitate decomposition of the O horizon and foster biotic activity there. The remaining water enters the soil as infiltration (*I*) (or pedogenic precipitation) and fuels pedogenesis. Thus,

$$I = T + S - L$$

and

$$I = P - C - L$$

Water entering the mineral soil is called *infiltration*. Water that *has* entered the soil and is moving through it is called *percolation*. Percolating water normally moves along gradients established by the interplay between gravity and matric tensions within the soil. Laterally percolating water is referred to as *throughflow* or *interflow*. On most slopes, this type of percolation is not insignificant (Fig. 14.11A). Water may flow along the top of aquitards, e.g., a Bt or Bkkm horizon. Lateral translocation of clay through an eluvial zone, or along the surface, is sometimes referred to as *alluviation* (Jackson 1965, Brown *et al*. 2004a; Table 13.1). Water percolating vertically can, obviously, function within any number of the more traditional pedogenic pathways (see Chapter 13). Slope gradient, soil heterogeneity, and texture and the rates at which water enters the soil all affect the pathways of water through the soil. And so, we next discuss the various ways that slopes can be described and characterized.

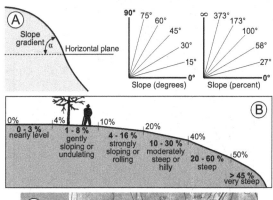

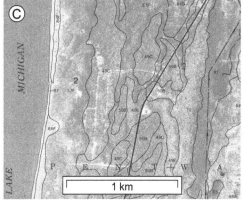

Fig. 14.20 The concept of slope gradient. (A) How slope gradient is measured, and the two ways (degrees and percent) in which it can be expressed. (B) Typical limits of slope classes used by the Natural Resources Conservation Service. After the Soil Survey Division Staff (1993). (C) An example of a map sheet from a standard NRCS county soil survey publication, illustrating how map unit slopes are depicted with letters. Note that, for areas of A slope, the A is often omitted from the map unit symbol. From the Oceana, Michigan, county soil survey by Calus (1996).

Slope Description and Catenas

Slope description must be a part of the soil geomorphologist's bag of tricks. Slopes can be described in a number of ways, both quantitatively and qualitatively. The nature and geometry of slopes can be used not only to define and describe the slope itself, but to predict soil characteristics. Terms that are frequently used to describe a slope or part of a slope include its (1) gradient or steepness, (2) length, (3) aspect, (4) curvature/shape, and (5) elevation (Aandahl 1948).

Slope Gradient

Gradient refers to the steepness or inclination of a slope from a horizontal plane (Fig. 14.20). It is commonly referred to simply as *slope*. If expressed in degrees, slopes can range from 0° to 90°, with a 90° slope being vertical. However,

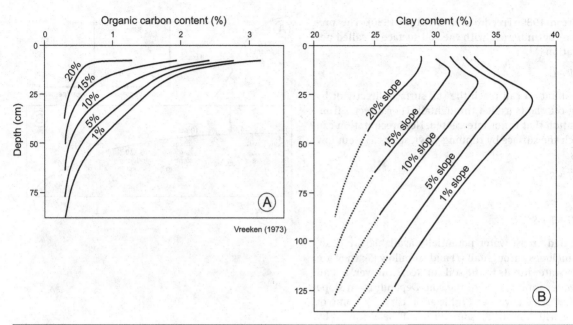

Fig. 14.21 Depth functions of (A) organic carbon and (B) clay contents of soils on an Iowa hillside of (generally) Mollisols, grouped according to slope gradient (shown as numbers near each line).

most soil scientists express slope as a percentage. A 100% slope has one unit of rise for every unit of run; hence, it is a 45° slope. Because slope gradient is a good predictor of soil patterns, and because gradient affects use and management of the soil, soil map units usually contain information on slope. In the U.S. Soil Survey system, slopes on soil maps are usually indicated by letters, after the map unit number, e.g., 143D (Fig. 14.20C).

Perhaps no other slope attribute is as closely related to soil development as is gradient, largely because slope steepness so dramatically affects the rates of water and sediment movement across the slope (Norton and Smith 1930, Acton 1965, Furley 1971). Slope gradient changes within most catenas (Ahnert 1970). For many catenas, the flowline with the maximum gradient – also commonly the shortest flowline – is of most interest. Slope gradient is a proxy for potential energy that the slope possesses. Steep slopes impose a great deal of kinetic energy onto the water and debris on slopes. Water on the soil surface has two main options: It can run off or infiltrate (Fig. 14.11). If it infiltrates, soil development is generally promoted, but if it runs off, soil development is not promoted, and in fact it may be regressed if runoff causes erosion. (However, for an exception, consider the texture contrast model discussed earlier.) For these reasons, steeper slopes tend to have thinner and less developed soils (Carter and Ciolkosz 1991; Figs. 14.13, 14.21). Vreeken (1973) found strong statistical relationships between slope gradient and various soil developmental properties, all of which could be explained with the assumption that less water moves into and through

the soil as slope gradient increases (see also Daniels *et al.* 1971a).

Many catenas have two or more gradient inflection points along the flowline; the first occurs on the shoulder position and the second at the toeslope (Fig. 14.7B). Inflection points may mark the location of lithologic discontinuities, from one parent material to another, and at the very least mark natural boundaries between soil bodies. They also may mark the edge of a different geomorphic surface, with a different geomorphic history (Fig. 14.4). Soil mappers, therefore, use inflection points to guide their estimation of soil map unit boundaries, because across them, drainage class, parent material, and surface age often change (see Chapter 8).

Slope Length

Slope length is directly correlated to erosion potential and therefore correlates with soil development (Musgrave 1935, Gard and Van Doren 1949, Liu *et al.* 2000, Rejman and Brodowski 2005). On longer slopes, more runoff can be expected, and at the base of long slopes, more slopewash or colluvium might accumulate. Along longer slopes, sediment can become increasingly sorted, so that finer materials accumulate at the toeslope.

Although theoretically a simple concept, slope length is difficult to quantify in a meaningful fashion (Aandahl 1948, Rodriguez and Suarez 2010). Because slopes change gradient and curvature in a complex manner, it is difficult to know where to begin and end measuring slope length. On complex slopes, should the entire length be used, or

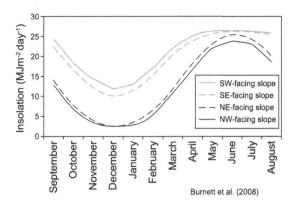

Fig. 14.22 Daily insolation (solar energy receipt) totals, tallied by month, for four sites of differing aspect in the canyons of semiarid northeastern Arizona. Note that the differences are least when solar angles are highest – in summer.

only a subset? If the summit is broad and gently sloping, should slope length information begin on the shoulder (edge)? On highly permeable slopes the length may not be as meaningful, at least from a soil erosion perspective. Therefore, the length of each slope *element* (see later discussion) is often determined. Slope length for a slope element can be expressed absolutely, in a unit of measurement, or relatively, in terms of a percentage of the total slope length.

Slope Aspect

Aspect refers to the compass direction toward which a slope faces, when looking downslope. Slope aspect matters to soil development because (1) the incidence angle of the sun varies among slopes that face different compass directions, and (2) in areas of high relief, upland areas can partially shade lowlands and other sloping areas (Lee and Baumgartner 1966, Beaudette and O'Geen 2009, Geroy *et al.* 2011). Therefore, aspect affects soils directly through its effect on microclimate, but also indirectly because of its influence on vegetation (Lee and Baumgartner 1966, Franzmeier *et al.* 1969, Small 1973, Nullet *et al.* 1990, Kutiel 1992, Hunckler and Schaetzl 1997). Differences due to aspect also tend to increase as slope gradient increases.

Exactly how the microclimate effect is manifested in soils varies greatly from landscape to landscape. For example, equatorial landscapes are virtually unaffected by aspect, because solar radiation is received from angles both north and south as the celestial zenith moves between the Tropics of Cancer (23.5° N latitude) and Capricorn (23.5° S latitude) over the course of the year. Here, the sun is seldom low in the sky. In areas of high relief located between 30° and 60° latitude, however, aspect becomes a major factor in the amount of solar radiation received. Here, the sun appears lower in the sky and is always in one part (southern or northern) of the sky. Last, in very high latitudes,

e.g., near the poles, the summer sun circumscribes a near-circle in the sky each day, potentially negating the effect of aspect. However, because[1] many poleward-facing slopes retain snowpacks for much longer into the spring, north-south slope differences in soil temperature and growing season length can be significant here too.

Most studies involving site exposure – aspect – agree that the processes responsible for the observed differences in soil chemical or morphological properties are driven by differences in receipt of solar radiation (Beaudette and O'Geen 2009). Generally, in the Northern Hemisphere, north- and northeast-facing slopes receive less direct sunlight, and thus, are cooler and moister than those of other aspects, whereas the driest slopes face south and southwest. North of the Tropic of Cancer, the sun is always in the southern part of the sky. As a result, south-facing slopes get more direct sunlight, i.e., insolation, than do north-facing slopes (Fig. 14.22), and north-facing slopes are more likely to be shaded (Lee and Baumgartner 1966). As a result, soil temperatures tend to be higher on south-facing slopes (Cantlon 1953, Whittaker *et al.* 1968, Franzmeier *et al.* 1969, Hutchins *et al.* 1976, Macyk *et al.* 1978, Hairston and Grigal 1994). Soils on south-facing slopes tend to be drier because of higher evaporation rates, coupled with increased transpiration from the vegetation (Finney *et al.* 1962, Carter and Ciolkosz 1991, Liu *et al.* 2012).

In theory, slopes that face *due south* should be the warmest and driest, although this is rarely the case. The warmest and driest slopes usually face south and southwest, while the coolest and most mesic slopes face north or northeast. A major reason for this apparent offset centers on early morning sunlight, which affects mainly east-facing slopes; much of this energy is used to evaporate water (dew) and thus has less influence as a source of heat. Late afternoon sun, which affects the southwest-facing slopes, is used more for direct heating, making this the warmest and driest aspect. North- and northeast-facing slopes are also kept cool because they remain in shadow during the warm late afternoon period.

An important aspect-related component centers on *shadowing*. Like aspect, shadowing is not really an issue in equatorial latitudes, where the sun is nearly always directly overhead. But in high latitudes, where the sun stays low in the sky, and in areas of deep, narrow valleys, many north-facing slopes stay in nearly perpetual shadow. Here, snowpacks can persist long into spring. Lower parts of south-facing slopes may also be in nearly perpetual shadow. Soils here may also be considerably cooler because of nighttime cold-air drainage (Franzmeier *et al.* 1969). In such cases, only the soils on the upper parts of the slopes will exhibit the traditional, aspect-related differences.

The pedogenic effect of slope aspect is not always intuitive. Some studies have indicated that soil

---

[1] Our discussion will be confined to Northern Hemisphere examples; the opposite situation holds in the Southern Hemisphere.

Table 14.3 | Data from soils on north-northeast and south-southeast slope aspects in northern Michigan

Variable (units)	Soils on N-NE slopes (Mean [SD])	Soils on S-SW slopes (Mean [SD])	Indicates stronger podzolization on which slope?	Statistical level of significance[a]
Data indicative of eluviation				
Depth to top of E (cm)	8.0 (3.1)	7.4 (3.5)	Neither	ns[3]
Depth to top of B (cm)	24.6 (7.2)	17.1 (5.2)	N-NE	0.02
E thickness (cm)	16.6 (6.1)	9.7 (4.0)	N-NE	0.02
E value (Munsell units)	4.6 (0.5)	4.8 (0.4)	Neither	ns
Fe_o in E (g kg^{-1})	0.2 (0.2)	0.4 (0.2)	N-NE	0.05
Al_o in E (g kg^{-1})	0.2 (0.1)	0.3 (0.1)	N-NE	0.01
Fe_p in E (g kg^{-1})	0.1 (0.0)	0.2 (0.1)	N-NE	0.05
Fe_o-Fe_p[b] in E (g kg^{-1})	0.1 (0.1)	0.3 (0.2)	Neither	ns
Data indicative of illuviation				
B thickness (cm)	34.4 (10.0)	30.6 (4.1)	Neither	ns
Uppermost B hue (Munsell YR)	4.5 (1.1)	6.0 (1.3)	N-NE	0.02
Uppermost B value (Munsell units)	3.1 (0.4)	3.9 (0.3)	N-NE	0.01
Uppermost B chroma (Munsell units)	4.3 (1.0)	5.6 (0.8)	N-NE	0.02
Fe_o in uppermost B (g kg^{-1})	5.1 (2.0)	3.5 (1.1)	N-NE	0.04
Al_o in uppermost B (g kg^{-1})	4.7 (2.2)	3.5 (1.1)	Neither	ns
Al_o in second B (g kg^{-1})	3.5 (1.7)	2.1 (0.7)	N-NE	0.01
Fe_p in uppermost B (g kg^{-1})	2.5 (1.5)	1.1 (0.5)	N-NE	0.01
Fe_p in second B (g kg^{-1})	1.1 (0.7)	0.6 (0.5)	N-NE	0.03
Fe_o-Fe_p in uppermost B (g kg^{-1})	2.4 (1.5)	2.4 (1.2)	Neither	ns
Fe_o-Fe_p in second B (g kg^{-1})	1.3 (0.7)	1.3 (0.6)	Neither	ns
ODOE[c] of uppermost B	0.3 (0.2)	0.1 (0.1)	N-NE	0.01

[a] One-tailed test using Wilcoxon test; ns: not significant at $\alpha = 0.05$.
[b] Subscripts o and p refer to oxalate-extractable and pyrophosphate-extractable elements.
[c] Optical density of the oxalate extract, an indicator of organic acid abundance.
Source: Hunckler and Schaetzl (1997).

development is greater on north- and northeast-facing slopes (Marron and Popenoe 1986, Hunckler and Schaetzl 1997; Table 14.3). But others have found the opposite (Finney et al. 1962, Franzmeier et al. 1969, Losche et al. 1970, Macyk et al. 1978, Birkeland et al. 2003). In many instances, a mix of soil development expressions may be observed on both slopes. For example, certain energy-related components of soil development, such as weathering, clay formation, acidification, and illuviation, may be better expressed on south-facing slopes (Finney et al. 1962, Small 1973), whereas other, moisture-related components, such as organic matter accumulation, melanization, eluviation, and leaching, are often better expressed under the cooler and moister conditions of north-facing slopes (Krause et al. 1959, Franzmeier et al. 1969, Zech et al. 1990, Carter and Ciolkosz 1991, Geroy et al. 2011,

Ebel 2012). Leaching-related measures, such as solum and E horizon thickness, may be better expressed on the cooler, wetter northeast-facing slopes (Finney et al. 1962, Hunckler and Schaetzl 1997). Weathering-related soil properties such as clay content, clay mineralogy, and rubification tend to be better expressed on south-facing slopes (Losche et al. 1970).

Nonetheless, even when armed with this theoretical bag of tricks, the effects of aspect on soils are not always straightforward and as expected. In southeastern Michigan, for example, Cooper (1960) found shallow, intensely developed sola on south-facing slopes, and deeper, less intensely developed sola on north-facing slopes. Marron and Popenoe (1986) found that a greater degree of soil development existed on north-facing slopes in California, as indicated by redder and more clay-rich B horizons.

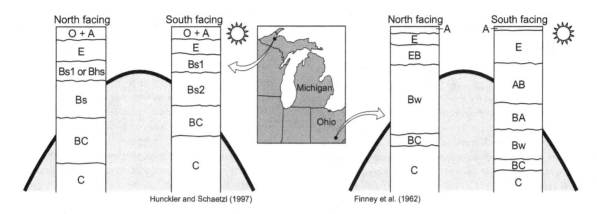

Fig. 14.23 Effects of slope aspect on soil morphology at two midlatitude sites in the United States.

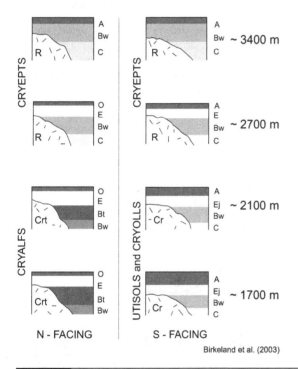

Fig. 14.24 Generalized diagram of profile morphologies on north- versus south-facing slopes in the canyons of the Colorado Front Range, near Boulder.

To resolve apparent contradictions like this, Hunckler and Schaetzl (1997) proposed that aspect must first be viewed in the context of the dominant pedogenic processes for the area. Are the pedogenic processes in the area – based on macroclimate – energy- or moisture-limited? The answer may vary, depending on which processes are being considered. And both answers may apply. For example, at cool, mesic sites with an abundance of moisture, energy may be limiting for certain processes such as weathering and humification; soils on such sites may exhibit increased development on the warmer, south-facing slopes. In these same areas, pedogenic processes that are not as dependent upon energy but upon moisture, such as eluviation, podzolization, and melanization, may be better expressed on northern aspects (Egli et al. 2006; Table 14.3). Conversely, in areas that are dry with respect to the dominant pedogenic process, moisture may be limiting. Here, the moister (poleward-facing) slope may contain better developed soils (Rusanov and Milyakova 2005). Figure 14.24 provides an excellent example of this situation. On the east slope of the Rocky Mountains, the warm, dry climate of the lower elevations would suggest that soils are moisture-limited, and in fact, the soils *are* better developed on the moister, north-facing slopes. At higher elevations, however, this aspect difference is much less apparent, as moisture limitations diminish in the wetter alpine regions. So, what about cold, dry regions? Here, energy may be more pedogenically limiting than moisture, such that, in these areas, southern aspects may have the best-developed soils. In summary, the answer to which aspect has the better developed soils is "It depends."

Another complication with regard to aspect centers on paleoclimate. Many soils have undergone more than one climatic cycle during the Pleistocene. Soils on these sites may reflect the interaction of paleoclimate and aspect as much as (or more than) they do the interactions of contemporary climate and aspect (Carter and Ciolkosz 1991). For example, soils on cinder cones in arid and semiarid eastern Arizona should be best developed on north- and northeast-facing slopes, since this is clearly a moisture-limiting environment. However, Rech et al. (2001) found a mix of aspect-related soil development. Weathering and solum thicknesses were greater on south-facing slopes, presumably because they continued to develop during cold glacial climates, while soils on north-facing slopes were often frozen.

Landscapes: The Palouse Hills, Washington State

Parts of eastern Washington State – the Palouse region – are covered with thick loess deposits (see figure). The Xerolls developed in this loess are well known for their productivity. Winter wheat is the main crop. This region, in the rainshadow of the Cascade Mountains, receives only about 50 cm of annual precipitation, most of which falls during the cool winters. Summers are particularly dry. The silty soils can retain a great deal of water, but little of it remains beyond one growing season because so much of the wintertime precipitation runs off the frozen ground.

The loess hills of the Palouse resemble ocean waves. The steep, rolling topography here has local relief >60 m. The loess hills are distinctly asymmetrical, almost crescentic in shape. Most ridges are oriented in a southeast-northwest direction, and the cooler, northeast-facing slopes are the steepest (Lotspeich and Smith 1953). Because the dominant wind direction is from the southwest, the northeast (leeward) slopes accumulate thick snow drifts, while less snow is retained on the gentle southeast-facing slopes (Rockie 1934). Snow drifts are important, because snow is the primary source of soil moisture during spring.

According to Lotspeich and Smith (1953), soil development in this landscape is mainly a function of the amount of infiltration. Thatuna soils, which have argillic horizons, are found on north-facing slopes with thicker snowdrifts, and on other slopes that receive runoff from higher landscape positions. In addition, snow that the northeast slopes receive leads to headward erosion and continual steepening of the slopes (Kirkham et al. 1931). In effect, this is a positive-feedback mechanism that keeps these slopes sheltered enough, so as to continue to be snow-receiving sites. Athena soils – those on ridge crests with Bk horizons – are the driest soils in this catena, mainly because they receive very little water from snowmelt; the snow blows off the ridges. Soils within the Palouse series, on gentle windward slopes, are intermediate in morphology and development. In sum, the driest sites are on the ridges, not so much because of runoff but because snow blows off these sites in winter. Lee sites, where large snowdrifts develop, are the most thoroughly leached and soils here are the best developed. Thus, within a few meters, one can go from soils with Bk horizons to leached soils with Bt horizons.

The Palouse catena, nontraditional in that it contains only well-drained soils, illustrates the importance of topography and landscape position to soil development. In this case, the role of topography is to redistribute the pedogenic energy provided by snowmelt and, in so doing, affect pedogenic pathways.

Figure: Soils and landscapes of the Palouse Hills of Washington State. (A) Oblique aerial image of the landscape. Photo by L. Suckow via Wikimedia. (B) Block diagram of the soils and landforms in Whitman County, Washington. (C) Typical soil pattern across one of the loess ridges.

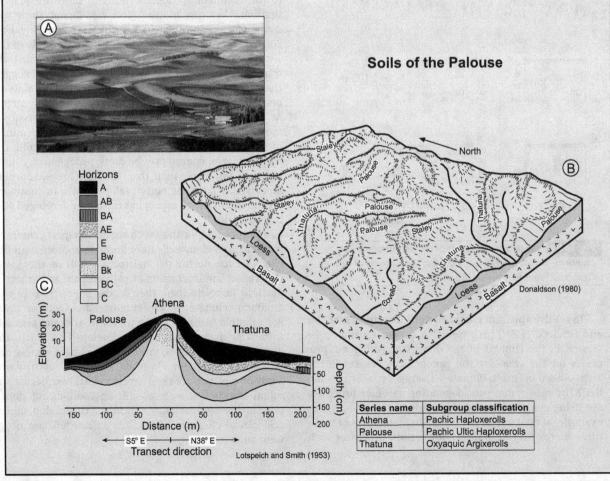

Soils of the Palouse

Series name	Subgroup classification
Athena	Pachic Haploxerolls
Palouse	Pachic Ultic Haploxerolls
Thatuna	Oxyaquic Argixerolls

Donaldson (1980)

Lotspeich and Smith (1953)

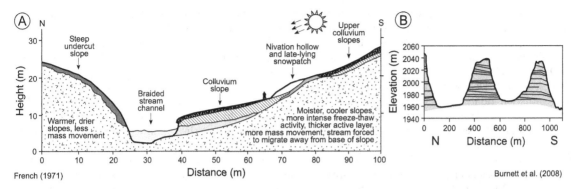

French (1971)   Burnett et al. (2008)

Fig. 14.25 Effects of slope aspect on the formation of slope asymmetry. (A) On the Beaufort Plain, northern Canada, the moister, northeast-facing slopes accumulate more snow in winter. Solifluction and nivation processes are stronger on these slopes, transporting more sediment to the base of the slopes as colluvium. The colluvium forces the river to shift position and undercut and oversteepen the southwest-facing slopes, whose soils are still drier because of its steepness. Shallow Cryorthents are the norm on these steep slopes. (B) Distinct slope asymmetry in the canyons of semiarid northeastern Arizona, where cliffs occupy 29% of the south-facing slopes, but only 2.5% of north-facing slopes. Excluding cliffs, south-facing slopes are 1°–3° steeper than are north-facing slopes and have less weathered bedrock.

Aspect also affects soil development through its control of geomorphic processes. The effects of aspect on weathering and mass wasting processes, in particular, can sometimes lead to marked slope asymmetry, as discussed previously (Small 1973, Burnett *et al.* 2008, Istanbulluoglu *et al.* 2008). For example, in high latitudes the incidence of freeze-thaw activity may be greater on north-facing slopes, whereas the opposite situation may set up in the midlatitudes. Greater freezing and thawing may lead to more mass movement and gentler slopes, creating the slope asymmetry (Fig. 14.25). Additional snow accumulations, usually on lee or north-facing slopes, also affect runoff and other slope processes (French 1971, Redding and Devito 2011), and soil development. In some regions, the increased snow accumulation on east and northeast slopes leads to slope instability, but this effect may be offset by increased infiltration and biomass production. Aspect-induced slope instability also affects soils directly; they are less developed on unstable (and younger) slopes.

In sum, slope aspect is a dynamic and complex factor that affects soils on a number of levels, all of which revolve around microclimate. Insolation receipt, including shadowing, and winds influence other processes and characteristics, e.g., snow accumulation and soil moisture and temperature. These, in turn, affect slope and pedogenic processes.

Slope Curvature or Shape

Slope curvature refers to the change in aspect along the slope face and is normally best demonstrated by the manner in which contour lines (lines of equal elevation)

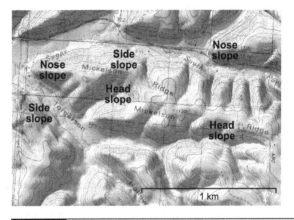

Fig. 14.26 Examples of slope curvature, as seen in plan view, in Crawford County, Wisconsin, United States. Contour interval = 20 feet.

bend or curve (Fig. 14.26). When slope curvature is minimal, contours are linear in plan view, water runs directly down the slope, and the two-dimensional catena is able to capture and explain most of the variation on that slope. Where slopes are more complex, however, they must be considered more holistically – as a three-dimensional system (Park and Burt 2002, Williamson *et al.* 2006). The formation of gullies, the deposition of sediment, and to a certain extent the depth to the water table are all affected by water flowlines, which are a function of, among other variables, slope curvature, and orientation.

On head slopes, i.e., in coves with concave plan curvature (Figs. 14.7A, 14.26), flowlines converge on the lower parts of the slope (Aandahl 1948). These types of slopes are called *convergent slopes*. Less precipitation is required to wet the soils on the bases of convergent slopes, as debris and water will preferentially accumulate there (Huggett 1976b, Pennock 2003), and soils will generally be thicker or even cumulic (King *et al.* 1983). On nose slopes, knobs, and spurs, the opposite situation holds. Flowlines diverge and the bases of the slopes are less subject to deposition and run-on; these slopes are called *divergent slopes*. Here, soil profiles are thinner (King *et al.* 1983).

Knowing this, Richard Huggett (1975), in a landmark paper, urged that soil development be examined within the three-dimensional slope-soil system, rather than just up and down the slope. He argued that, instead of a catena, the *valley basin concept* should be employed to understand soil-topography relationships, for it includes flowlines. In this regard, he equated the basic organizational unit of the soil system to that of the fluvial system – he single-order drainage basin (one with a stream that is so small as to have no tributaries). The valley basin is bounded by the soil surface on top and the weathering front at the bottom and functions as an open system. Fluxes of sediment, plasma, and water move through it, accumulating in hollows and disseminating away from nose slopes.

Subsequent work has supported Huggett's approach to the study of soil landscapes. King *et al.* (1983) found that soils on a rolling landscape were much more strongly correlated to depressions and knolls, i.e., three-dimensional slope parameters, than to two-dimensional slope positions like shoulders or backslopes (Nizeyimana and Bicki 1992). The reason for this is that any one pedon rarely occupies only one distinct slope position, as might be indicated along a transect. Rather, it may be influenced by flowlines from other directions. For example, a pedon on a backslope might be sediment and water shedding if that backslope is also a nose slope, while a nearby backslope with a much thicker and better developed soil might be located within a sediment-receiving site, such as a cove. Thus, no one slope parameter is adequate to explain the complexity of soil-landscape relationships. All the different slope parameters influence soil development, and they should be examined in that context. Think of the *whole* picture of soils and slopes.

Elevation

When describing a field site, some mention of its elevation is normally given. As discussed more fully in Chapter 12, elevation affects the temperature of a site. The normal environmental lapse rate is a 6.4°C decrease in air temperature per kilometer increase in elevation (3.5° F/1,000 feet). This type of temperature change, although potentially important to pedogenesis, is normally considered important only in terrain with high relief (Egli *et al.* 2008).

Elevation can be important locally, when examined in the context of *local relief*. Cold air drains into low areas on calm nights, making them colder than surrounding high points or slopes. This tendency may persist during the day, if the lowland is also an area that remains shaded for long periods. Although cold air drainage is of most interest to agriculturalists and fruit growers, it also affects soils. Closed basins, e.g., glacial kettles in the Great Lakes region, often have a different vegetation assemblage in them, which is reflected in the soils (Hobbs *et al.* 2011). Our experience has shown that soils on the adjoining uplands are warmer, and soils there have less organic carbon. Some deep kettles in the Great Lakes region experience frost in all months of the year and have only stunted trees and shrubs in them, whereas upland areas have mature forest, again illustrating that even a few meters of local relief can have pronounced effects on soils and biota.

Slope Elements

In many landscapes, soil properties vary consistently and predictably as a function of landscape position (Malo *et al.* 1974). Previously, we have stressed that soil landscapes must be viewed three-dimensionally, because that is how we can best discern water and sediment fluxes across landscapes (Huggett 1975). However, at times, this approach can be overwhelming, and often one first has to discern the straight up-and-down slope, catenary relationships, before proceeding on to a 3-D analysis of soil variation. Because soil geomorphology is very much concerned with the *codevelopment* of landforms, slopes, and soils, we discuss the matter in more detail in the following.

Figure 14.27 illustrates some of the schemes that have been developed to categorize the various hillslope elements or components (Wood 1942, Schoeneberger and Wysocki 2001). In the simplest, two-dimensional or catenary, view, all hillslopes can be seen as having an erosional component (near the top), a transportational component (in the middle), and a depositional component (at the base). From another perspective, a slope can be thought of as having three main components: a rounded upper edge (aka a convex, waxing slope), a constant slope of varying length (aka a pivotal point or junction point), and a concave, waning slope at the base, where sediment and debris accumulate (Wood 1942, Ahnert 1970, Ruhe 1975a). Friedrich (1996) referred to these slope components as areas of negative, equalized, and positive mass balance, respectively (Fig. 14.13B). Thus, the simplest downslope pattern is convex, straight, and concave (Fig 14.28). Compound and complex slopes vary from this pattern but retain the convex-straight-concave pattern, in a downslope direction, over at least some of the slope (Savigear 1965; Fig. 14.28). Profile breaks between slope elements have considerable diagnostic value in geomorphology, since a break may indicate a change in slope processes, lithology, and/or depositional/erosional geomorphic history (Qin *et al.* 2009; Figs. 8.16, 14.5). Most fully developed slopes are described using Ruhe's (1960) system of five slope elements: summit or crest, shoulder,

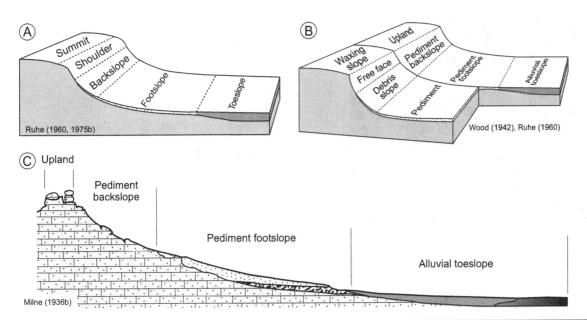

Fig. 14.27 Various terms used to describe slope elements, based on different models. The original slope element diagram – Fig. 14.7 – can be used as a basis of comparison and is generally reproduced in Part A.

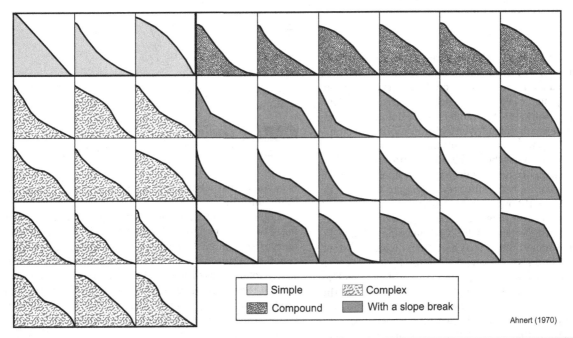

Fig. 14.28 Examples of simple, compound, and complex hillslope profiles, and those with slope breaks, as viewed in cross section.

backslope, footslope, and toeslope (Figs. 14.7, 14.27A). These elements do not need to occur consecutively down the slope and all five need not be present.

Soils respond to, and affect, slope development processes. On different parts of the slope, soils may thicken or even be buried, others may erode, yet others may maintain a balance between pedogenic and slope development. Thus, along a given slope or catena, the actual age of the soil may change, as some areas may be undergoing long-term, gradual erosion and rejuvenation, whereas others

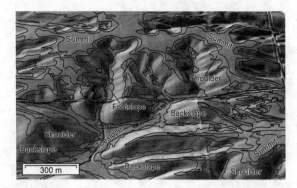

Fig. 14.29 The five major slope elements, as modeled for a landscape in eastern Iowa, using detailed topographic (LiDAR) data. Note that, for some slopes, all five slope elements do not occur or span such a small area that they cannot be shown at this map scale.

are on more stable slope elements and are, therefore, older and better developed.

Lower slope elements, where slow but progressive burial is occurring, may be time transgressive. The introduction of differing soil ages and time$_{zero}$ considerations complicates the catena concept but is important to keep in mind.

SUMMIT POSITIONS

Summits are the tops of the landscape. They can be flat or convex. Broad, flat summits are generally quite stable, with minimal sediment erosion or accretion. In this situation, soil development can proceed nearly uninterruptedly. And, according to King (1957), summits may be dominated more by chemical weathering (driven by infiltration of precipitation) than by physical weathering and erosion.

Wide summits may be the oldest and most stable of the five slope elements, because little water is likely to run off. Most precipitation here infiltrates, leading to more deeply leached soils with more organic matter, thicker sola, and better horizonation than slope elements that are immediately downslope (Fig. 14.29). Clay maxima tend to be deeper and more pronounced here than on steeper slope elements (Fig. 14.30). However, if the soil is slowly permeable or has a slowly permeable horizon at depth, such as a fragipan or a thick Bt horizon, water may perch on the soil surface or within the profile, making the summit soils much wetter than would otherwise be expected (see the Landscapes box that follows). Similar situations can occur on upland summits on bedrock, where flat benches may extend for some distance. Water may perch above the bedrock, making the upland soils quite wet. In such a setting, the soil-bedrock interface becomes a locus for illuviation (Schaetzl 1992a). On gently sloping summits, water may flow laterally at the top of such a slowly permeable layer, eventually reaching the shoulder, where it may continue as subsurface flow or emerge as a seep. Because water often

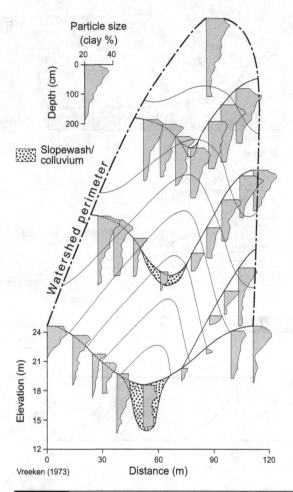

Fig. 14.30 Depth functions of clay content at various locations within a first-order watershed in Iowa.

Vreeken (1973)

behaves similarly across the summit, soils are commonly quite uniform across it.

Convex summits are more erosional in nature, approaching shoulder slopes in that regard. Similar conditions exist on narrow summits; runoff can be considerable. Likewise, runoff from summits can occur on landscapes where precipitation totals are high or where the soil itself is slowly permeable. In these cases, pedogenesis is inhibited and soils may erode as fast as they form (Birkeland *et al.* 1991). Exceptions occur on sharp-crested and undulating summits.

Landscapes: The Illinoian Till Plain of Southern Illinois

The flat landscapes of southern Illinois are underlain by Illinoian-age till. In the >100 millennia of generally humid conditions since the deposition of this parent material, a strongly developed soil with a thick, clay-rich Bt horizon has formed – the Sangamon soil, now a buried paleosol (see Chapter 16). Much of the inherited relief has been lost from the landscape. Later, the Wisconsinan ice sheet advanced to positions nearby, but never covered this landscape. Meltwater from the ice, however, did flow in the nearby Mississippi River valley, which then became a source of calcareous loess. About a meter of loess was deposited on this low-relief landscape, burying the Sangamon soil. Between the loess and the Sangamon paleosol is a layer of gritty-silty erosional sediment that likely formed as the loess was beginning to be deposited. The Sangamon paleosol, the gritty sediment, and the overlying loess then welded, or grew, together to form one thick, complex soil. The paleosol became the Bt horizon of this complex soil, which today is mapped in the Cisne series (fine, smectitic, mesic Mollic Albaqualfs). The paleosol is an aquitard, perching water in the loess above, trapping illuvial clay at its upper boundary, and slowly growing upward by clay accretion into the overlying gritty sediment and loess. Countering this process is clay destruction by ferrolysis, which occurs in the eluvial zone of the surface soil when water seasonally perches in the upper solum (see Chapter 13). The presence of an acidic paleosol at depth has helped to acidify the Cisne soils, which originally had formed in calcareous loess.

Despite being in a summit position, Cisne soils are poorly drained, with gleying in the upper B horizon (see figure). The shallow aquitard, combined with the extremely low slopes of the uplands, leads to perched water and wet conditions within the profile. This landscape clearly illustrates that summits need not be the best drained slope position. Here, the soils with the deepest water tables (relative to the surface) are on the steepest slopes, either on small ridges or knolls that rise above the flat uplands or on the sides of small drainageways, i.e., on convex shoulder slopes. The moderately well-drained Ava soils (fine-silty, mixed, active, mesic Oxyaquic Fragiudalfs), which are found on valley-side slopes, would probably be well drained if they did not have a fragipan.

It is true – uplands are not always the driest places.

Figure: Block diagram of the Hoyleton-Cisne soil association, Franklin and Jefferson Counties, Illinois.

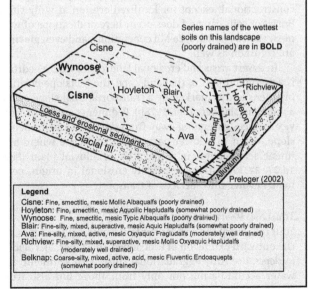

Series names of the wettest soils on this landscape (poorly drained) are in **BOLD**

Preloger (2002)

Legend
Cisne: Fine, smectitic, mesic Mollic Albaqualfs (poorly drained)
Hoyleton: Fine, smectitic, mesic Aquollic Hapludalfs (somewhat poorly drained)
Wynoose: Fine, smectitic, mesic Typic Albaqualfs (poorly drained)
Blair: Fine-silty, mixed, superactive, mesic Aquic Hapludalfs (somewhat poorly drained)
Ava: Fine-silty, mixed, active, mesic Oxyaquic Fragiudalfs (moderately well drained)
Richview: Fine-silty, mixed, superactive, mesic Mollic Oxyaquic Hapludalfs
        (moderately well drained)
Belknap: Coarse-silty, mixed, active, acid, mesic Fluventic Endoaquepts
        (somewhat poorly drained)

SHOULDER AND FREE FACE POSITIONS

Convexity is the operative concept associated with shoulder slopes (Fig. 14.7B). Ruhe (1975b) called these slope positions waxing slopes. Shoulders are usually the youngest and least stable of the five slope elements (Furley 1971). Here, runoff and erosion are maximal (Walker and Ruhe 1968, Pennock 2003), and surface instability and surficial erosion are the norm. The steepest shoulder slopes can be so nearly vertical as to be *free faces* (Fig. 14.27B). On shoulders and free faces, runoff dominates to the point that erosion outstrips pedogenesis, and thin or nonexistent soils are the result; bare bedrock is not uncommon (Roy *et al.* 1967; Fig. 14.16A). Detritus may accumulate in rock crevices and shallow pockets within the free face, as well as at the base, in what is referred to as a debris slope (Fig. 14.27B).

Soils on shoulders tend to be comparatively thin and low in organic matter (Aandahl 1948, Vreeken 1973; Figs. 14.31, 14.32, 14.33). This is especially true for shoulder slopes that are also on nose slope positions. Because erosion preferentially strips the finer material from the shoulders, soils here may also be coarser-textured than elsewhere (Fig. 14.14). On steep shoulder slopes, mass movements may even be commonplace. This instability is initiated by surface runoff, which is quite pronounced on shoulder slopes, as well as lateral flow of water in the subsurface. Subsurface flow is often concentrated along preferred flowlines, leading to gullies and seeps at a few spots, rather than uniformly across the slope. Most of the time, shoulders have the driest soils on the landscape.

BACKSLOPE POSITIONS

Backslopes are steep, transportational slopes that lie between upslope areas dominated by erosion and lower slopes that accumulate sediment (Furley 1971). Normally, they are straight slopes that lack significant curvature in profile view, although any and all types of curvature can be associated with backslopes (Figs. 14.7, 14.34). As a result, soils can change markedly along backslopes (King *et al.* 1983). Debris and water move most rapidly over or through backslopes, sometimes on top of subsurface aquitards (Gile 1958, Young 1969, Huggett 1976b, Schlichting and Schweikle 1980). The pathways along which the water flows will depend on the *curvature* of the slope (Fig. 14.34). Backslopes that lie below head slopes are especially likely to receive water from upslope locations. Slope length also determines how much material moves through and along the backslope, as does the stratification of materials within it, including sediments below the soil profile. Mass movements such as creep, slump, and solifluction are common here, sometimes producing hummocky topography (Hall 1983). In general, one should think of backslopes as transitional areas and transportational slopes.

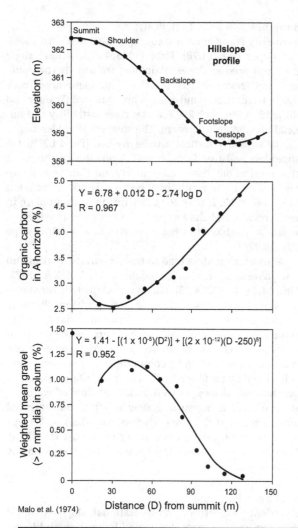

Malo et al. (1974)

Fig. 14.31 Scatterplots and curves illustrating the strong relationships– largely driven by runoff and debris flux processes – between A horizon thickness and slope position for a North Dakota hillslope.

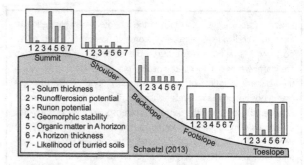

1 - Solum thickness
2 - Runoff/erosion potential
3 - Runon potential
4 - Geomorphic stability
5 - Organic matter in A horizon
6 - A horizon thickness
7 - Likelihood of buried soils

Schaetzl (2013)

Fig. 14.32 Estimates of relative variation in soil characteristics and slope process intensity along a typical catena in a humid climate.

position (Ruhe and Walker 1968, Walker *et al.* 1968, Malo *et al.* 1974; Figs. 14.32, 14.33). The increasing wetness of the lower slope positions also accentuates net primary productivity of the plants, which in turn provides more litter to the soils on these sites. Also, the relatively cooler and wetter conditions at the base of the slope may slow the decomposition of these organics. All of these factors combine to make soils in the foot- and toeslope positions high in organic matter (Kleiss 1970; Figs. 14.31, 14.33).

Not all materials that accumulate in footslopes are transported from upslope via surface wash. Throughflow along subsurface aquitards is one means by which water and materials are transported to footslopes (Buda *et al.* 2009). Dissolved material and clastic sediment can be forced to the surface from groundwater that wells up at spring sapping sites (Fig. 14.12), where gullies can form and work their way upslope. As they do, the backslope will become increasingly the focus of erosion, and small fans will form on the footslope. Footslope surfaces are therefore constructional, except for localized erosion at gully sites. Burial of soils can and does occur here. Indications of wetness, such as mottles, Fe-Mn concretions, and even gleying, are common as well.

In desert areas and erosional landscapes where bedrock is a dominant part of the landscape, the footslope position broadens and is marked by a transportational slope called a *pediment* (Hallberg *et al.* 1978b). Pediments are broad, convex surfaces extending away from the backslope or debris slope, down to a lower part of the landscape where sediment accumulates in a playa or an alluvial plain (Figs. 14.27, 14.35). Although initially erosional in origin, pediments evolve into surfaces of transport.

Toeslope Positions

Your toes are at the end of your foot; similarly, toeslopes are the outward extension of footslopes. Toeslopes, aka *alluvial toeslopes*, are fully aggradational sites, with sediment and water accumulating not only from above but also from streams that flood and deposit overbank alluvium. The

Footslope Positions

Footslopes, the most concave parts of the slope, are sediment- and water-receiving positions. Material carried downslope, in solution and in suspension, within throughflow and overland flow, begins to be deposited here, because of the lessening of the hillslope gradient (Pennock 2003). Perhaps no other slope position is more influenced by slope curvature than is the footslope. Water that follows flowlines down the slope is dramatically affected by them, and the influence of topography on flowlines is greatly diminished as waters slow at the footslope (Fig. 14.34).

The most likely sediment to be deposited here originated from the upper profile (A horizon) of soils upslope, leading to overthickened A horizons and sola in the footslope

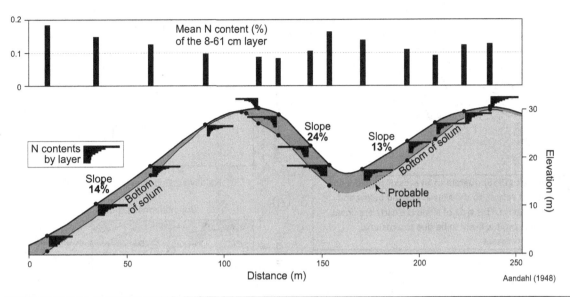

Fig. 14.33 N contents and solum thicknesses (to calcareous loess) along three hillslopes in northwestern Iowa.

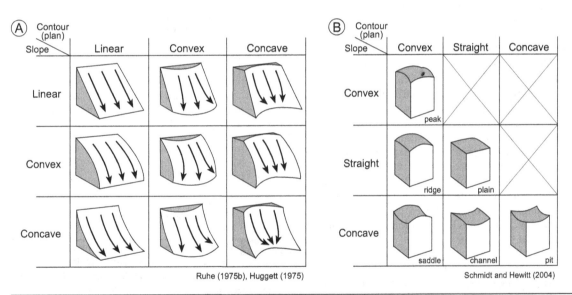

Fig. 14.34 Geometric forms of hillslopes. (A) Forms of sideslopes, with flowlines illustrating how water and debris move across them. (B) Forms of uplands.

latter type of sediment accumulation was especially common in the United States after European settlers cleared and cultivated the forest and prairies. Sediment washed onto toeslopes and plugged river channels, forcing the rivers to flood more frequently; each time they did, more sediment was deposited on the alluvial toeslopes. In parts of Wisconsin, this postsettlement alluvium approaches a meter in thickness (Lecce 1997, Faulkner 1998). Thus, cumulization and burial are dominant processes on toeslopes.

Sediment on toe- and footslopes tends to be finer-textured and more uniform than material upslope because slopewash processes transport the finer material farther (Nizeyimana and Bicki 1992; Fig. 14.14). Likewise, slope-derived sediment becomes thicker, farther out onto the toeslope (Walker and Ruhe 1968).

Although sediment is likely to accumulate gradually on all toeslopes (Vreeken 1973), the accumulation of sediment at the base of slopes is especially important in basins of closed drainage. Here, no mechanisms operate to *remove*

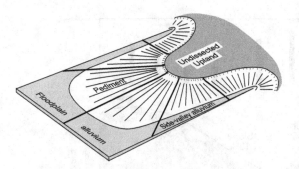

Fig. 14.35 Ruhe's (1958) diagram of an erosional (pediment) surface and how it relates to the alluvial lowland below, and the relict upland above. This type of slope is clearly erosional, and any stone line on it is likely to be due to erosional (pedimentation) processes.

the sediment (except wind and solutional losses following weathering), and thus, sediment accumulations may be particularly thick (Walker and Ruhe 1968).

Soil development on toeslopes reflects the natural wetness and sediment aggradational character of the site (Fig. 14.36). A horizons tend to be thicker than anywhere else on the slope (Gregorich and Anderson 1985; Figs. 14.32, 14.37). Indicators of wetness will generally be stronger and more prominent on toeslopes than in any other slope position, except perhaps wide, flat summits underlain by aquitards. This is particularly true for toeslopes at the base of headslopes.

The associations among slope position, wetness, and the aggradational nature of specific sites have been successfully modeled, using digital terrain data, in a number of instances. For example, Moore *et al.* (1993) and Gessler *et al.* (1995) developed an index that quantifies landscape position, using digital terrain data. The Compound Topographic Index (CTI), often referred to as the steady-state wetness index, integrates both landform position and context through this equation:

$$CTI = \ln (A_s/\tan ß)$$

where A_s is the upslope catchment area and ß is the slope angle. Small values of CTI generally occur at upper slope positions, whereas large CTI values depict lower catenary positions where the potential accumulation of water and sediment is high (Zheng *et al.* 1996, Gessler *et al.* 2000). In this regard, the CTI is one of a family of indexes and equations that quantify the relationships among landscape position (however defined) and various soil variables, typically wetness, horizon thickness, or degree of development. These indices rely on terrain data and often focus on improving our ability to visualize current, or predict future, soil wetness conditions (Beven and Kirkby 1979, O'Loughlin 1986, 1990, Thompson *et al.* 1997, Chaplot *et al.* 2000; Fig. 14.38).

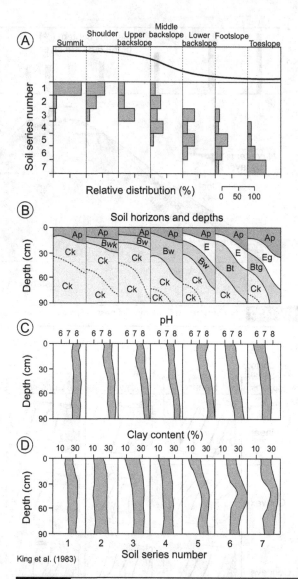

King et al. (1983)

Fig. 14.36 Soil characteristics, for each of the seven soil series along a catena of the Weyburn association in Saskatchewan, Canada. (A) Relative distributions of soil series, classified according to slope position. (B) Soil horizonation, portrayed roughly according to slope position. (C) Depth distributions of pH. (D) Depth distributions of clay.

The Nine-Unit Landsurface Model

Recognizing the utility but also the simplicity of the catena concept, and realizing that many pedogenic processes operate not only in the vertical dimension but also parallel to the slope, Conacher and Dalrymple (1977) developed their nine-unit landsurface model (Fig. 14.39). Largely based on work in Australia and England, this model stressed that catenas can be composed of up to nine interrelated *landsurface*

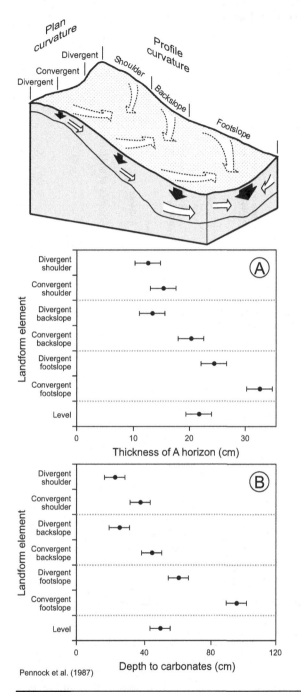

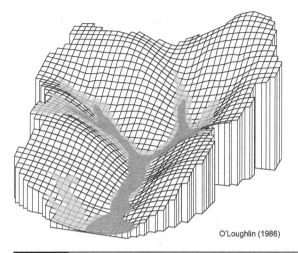

O'Loughlin (1986)

Fig. 14.38 A mesh-net depiction of a watershed, showing areas (modeled) that would develop perched water tables under different conditions of an input wetness parameter developed by O'Loughlin (1986). Darker areas imply a higher likelihood of saturation.

Pennock et al. (1987)

Fig. 14.37 Block diagram illustrating water flow pathways on a slope, how those pathways are affected by slope curvature, and how they affect soil development. (A) Thickness of the A horizon on various types of slope positions. (B) Depth to carbonates (a measure of leaching effectiveness) on various slope positions.

units. Each unit is affected by interactions between water- and gravity-based processes; by translocation and reposition of soil materials; by overland flow, throughflow, and streamflow; and by creep and mass movements. Essentially an extension of the five slope units in Ruhe's (1975b) slope model, the nine-unit landsurface model also includes processes of fluvial erosion and deposition at the base of the slope. It is a more universal soil-geomorphic model that has wide application in older, more incised and developed landscapes, and that, unlike many pedology-based catenary studies, includes the effects of slope processes (creep, mass movement, throughflow, alluviation) on the development of soil patterns.

The model begins at a stable summit with well-drained soils and an adjoining, wetter unit with shallow water tables due to the development of a subsurface aquitard. A free face, dominated by slope processes of mass movement, is also a central component. Downslope of the free face (unit 4) are slope positions that are primarily sediment-losing (5) and sediment-receiving (6) units, illustrating that runoff and downslope transport are given great weight in this model. The model is also more areally inclusive than the traditional hillslope model. For example, the lower two landsurface units are associated with fluvial landforms and sediments; unit 8 is essentially a riverine cut bank and unit 9 is the streambed itself (Fig. 14.39).

Because the model was developed on older landscapes, many of the landsurface units may not be evident on low relief or on young landscapes, and especially on the constructional, Pleistocene-age landscapes of northern Europe and North America. Some of the units will not be observed

Predominant and/or distinguishing pedological criteria

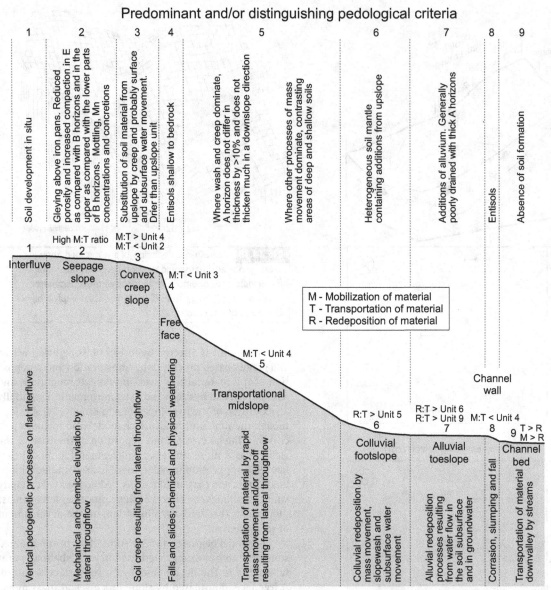

Fig. 14.39 The nine-unit landsurface model, slightly modified from Conacher and Dalrymple (1977).

on landscapes where bedrock is deep. Units 8 and 9 will not be present in catenas that end in closed depressions. However, these caveats are to be expected for a model that is essentially universal – parts of it should fit all landscapes, but few landscapes will use all the landsurface units. As landscapes age, local relief increases, and stream dissection proceeds, more of the landsurface units will apply.

The landsurface model of Conacher and Dalrymple (1977) reinforced Milne's (1935, 1936a, b) original intention – that the catena concept should include soil differences that result from variations in drainage, *as well as* from differential transport of eroded materials and chemical elements (but also see Brown *et al.* 2004a, Khomo *et al.* 2011). Through the years, the latter concepts had been variously lost or deemphasized. Conacher and Dalrymple stressed that the catena concept should not exclude soils formed in materials redeposited by slopewash and other transportational processes. As a result, the landsurface model employs more geomorphology and slope processes than do most sensu stricto pedogenic catenas. Their model is a response to the rather sterile catena concept that had dominated the mid-twentieth century – one that focused strongly on pedogenic processes and drainage, while deemphasizing hillslope processes.

Soil Wetness and Hydric Soils

Knowing the water status of a soil is crucial to understanding its genesis and for determining its proper management. One of the most important components of the soil water regime involves the depth to the zone of saturation, including how this depth and the oxygen status of the groundwater below change through time. A *water table* is the upper surface of a zone of saturation, where the hydrostatic pressure is zero (Chapter 6). Soil horizons are considered saturated when all pores are water-filled, i.e., the soil water pressure is zero or positive (Soil Survey Staff 1999). Soil water held under negative pressure will generally not flow as saturated flow out of a soil pore. In saturated situations, however, free water will flow out of the soil and into an auger hole or pit, eventually stabilizing at the level of the water table (Vepraskas 1999). Water that is wicked upward into the soil, a few centimeters or more above the water table, is referred to as the *capillary fringe* (Gillham 1984). Soluble materials in the groundwater can be moved upward in solution, into the capillary fringe, and precipitated there.

We have seen that slopes and topography affect soils through the translocation of sediment and water. Also important, however, is their ability to affect the depth to the water table, which in and of itself has a tremendous effect on soil development. Indeed, because of the strong relationship between topography and the zone of water saturation in the soil, the term *hydrosequence* is used to refer to series of soils with differing wetness, usually along a catena (Zobeck and Ritchie 1984, Cremeens and Mokma 1986, Smeck *et al.* 2002). Understanding the water regime of a soil is vital to interpreting its development, as well as its proper management. Drainage is an important component of the water regime. It refers to the rapidity and completeness with which water added to a soil is removed, as well as the frequency and duration of periods when the soil is not saturated. Drainage has internal and external drivers. Internally, drainage is controlled by permeability and water table relations. Externally, it is largely a function of slope configuration (Simonson and Boersma 1972, Crabtree and Burt 1983).

Whether a soil is normally wet or dry is a function of many factors, among which topography and local soil climate are foremost. Regional climate notwithstanding, soils are wet if they are affected by a high water table, or when they are so slowly permeable that they retain large amounts of water after precipitation events. In dry climates most soils are dry, except for some along stream courses or on playas, i.e., in lowlands. In moist climates, topography is perhaps less important, for upland soils can be either comparatively dry, provided the soil is permeable, or wet, if the soil is slowly permeable or underlain by an aquitard (Mausbach and Richardson 1994). So much of soil wetness revolves around not just the precipitation variable but what happens to the precipitation after it infiltrates.

Wetlands and hydric soils are strictly defined entities (Megonigal *et al.* 1993). By definition, wetlands and their soils are saturated with water, either permanently or seasonally, thereby acquiring the characteristics of a distinct ecosystem. They are usually delimited on the basis of characteristic vegetation, which is adapted to its unique soil conditions, i.e., hydric soils support aquatic plants. By definition, *hydric soils* have formed under conditions of saturation, flooding, or ponding, for periods long enough during the growing season to develop anaerobic conditions in the upper profile. In the United States various government agencies are responsible for defining the geologic, hydrologic, and biotic criteria used to determine whether wet soils meet the definitions of hydric soils, or whether parts of the landscape can be classified as wetland.

Both wetlands and hydric soils exist due to unique combinations of geology and climate (Mausbach and Richardson 1994). High rainfall and cool conditions generally favor the formation of wetlands. Flat topography minimizes runoff, favoring the development of broad areas of hydric soils, whereas rolling topography promotes the development of wetlands only within localized depressions and valleys. In humid areas, most wetland depressions are groundwater *discharge* areas where groundwater emerges to become surface water (Fig. 14.40). In humid regions, the water table is usually mounded under topographic highs and is lowest (in elevation) but closest to the surface beneath depressions. Water is thus moved from recharge (upland) areas to the discharge wetlands, keeping the latter wet even during climatically dry intervals. Capillary flow of water and dissolved substances from the shallow water table can, in such settings, lead to the formation of saline, sodic, or carbonaceous soils in these low spots (Richardson and Bigler 1984, Miller *et al.* 1985, Mausbach and Richardson 1994). In cool, humid climates, such depressions may develop into Histosols, whereas upland mineral soils are leached.

Conversely, those depressions on uplands where the water table is deep may be sites of focused infiltration, perhaps as recharge wetlands. Soils there may exhibit increased leaching, more organic matter production, and stronger soil development (Runge 1973, Miller *et al.* 1985; Fig. 14.40). In both cases, topography facilitates the additional influx of water, and the depth to the water table determines the fate of the additional, site-focused infiltration.

Landscapes: The Pocosins of North Carolina

In parts of humid eastern North Carolina, flat upland soils are often very wet. Here, where long-term precipitation exceeds evapotranspiration, excess water must either infiltrate or run off (Daniels *et al.* 1977). Runoff is minimal and lateral subsurface flow is slow, because the landscape is so flat. Under such conditions, organic soils as much as 2 m thick develop on the wettest, flattest uplands (Richardson 1983, Daniels *et al.* 1999). These organic accumulations are, locally, called *pocosins* ("swamp on a hill") (see figure). Pocosins are rainfall driven and lack a well-defined stream surface-flow connection to major rivers on the landscape (Richardson 2003). The accumulation of organic materials in the pocosins is aided by slow decomposition under waterlogged conditions, due to the low relief and the great distances between streams. Under the thermic soil temperature regime of North Carolina, where evapotranspiration levels are high, the development of organic soils on uplands is remarkable.

Pocosin peat covers the broad, flat interfluves of this landscape; they are not in depressions as are so many other Histosols worldwide. They develop into large, low domes of organic material. The

sapric organic materials generally are younger than 10,000 years old, suggesting that they started to develop as sea levels rose at the end of the last glaciation (Whitehead 1972, Daniels *et al.* 1977). Sediments beneath the pocosins may range from sand to clay; sandy beach ridges occasionally poke through. The pocosins are not unlike the blanket peats found in parts of northern Europe, which are formed in part because of the cool climate. During dry periods, occasional wildfires will burn parts of the pocosin, leaving behind a depression that fills with water. The smoldering histic materials cause smoke and other lingering concerns (Reardon *et al.* 2009). Some of North Carolina's largest lakes have originated in this way and are today rimmed by the remains of pocosin Histosols.

Similar peat accumulations occur in southeastern Wisconsin, for slightly different reasons. Here, the glacial stratigraphy has created flowing springs and seeps on flat but otherwise well-drained plains. Where wet enough, mounds of peat have formed above these seep areas. The peat and muck mounds can be more than 1 ha in area and 4 m thick (Ciolkosz 1965).

Figure: Cross section through a typical North Carolina pocosin. Source unknown.

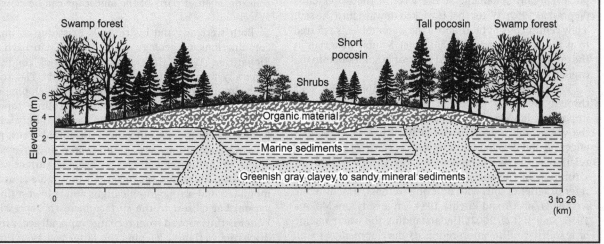

Depth to the water table is a function of several factors: (1) most importantly, landscape position, which affects runoff versus run-on tendencies (Fig. 14.40); (2) temporal patterns and amounts of precipitation, vis-à-vis evapotranspiration and evaporation demands; and (3) permeability in the surface and subsurface, especially with regard to subsoil horizons (Gile 1958, Fritton and Olson 1972, Khan and Fenton 1994). In highly permeable soils, the water table can quickly respond to precipitation (Hyde and Ford 1989), whereas in less permeable soils more water will run off. Response of the water table may significantly lag. Upslope areas tend to be recharge areas where meteoric water percolates through the *vadose* (unsaturated) zone and into the saturated zone below (Fig. 14.40). In lowland areas, where the water table intercepts the surface, soil and surface water is discharged

as water flows upward, out of the soil (Fig. 14.40). Discharge can also occur on sideslopes where a perched zone of saturation intercepts the surface. Such areas can be determined by direct observation, but also by sampling the soil for soluble products deposited from the groundwater. Khan and Fenton (1994), for example, noted that calcite/dolomite ratios were higher in discharge areas near the bases of slopes, because of deposition of secondary carbonates from shallow groundwater (see also Knuteson *et al.* 1989).

Water tables fluctuate through time. Annual fluctuations are usually greatest on higher landscape positions and least in lowlands or wetlands (Khan and Fenton 1994). Intraannually, water tables are usually highest after a cool and wet season, and lowest after a warm, dry period (Khan and Fenton 1994). For this reason, determining the depth

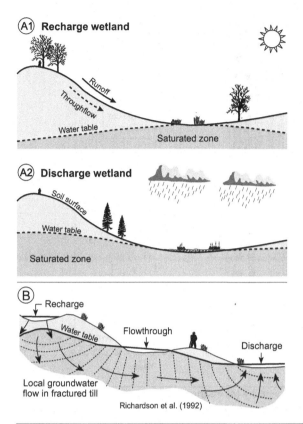

(A1) Recharge wetland

Runoff

Throughflow

Water table

Saturated zone

(A2) Discharge wetland

Soil surface

Water table

Saturated zone

(B)

Recharge

Water table

Flowthrough

Discharge

Local groundwater
flow in fractured till

Richardson et al. (1992)

Fig. 14.40 Recharge versus discharge wetlands. (A) Simplified diagram of the relationships among landscape and water table for a recharge and a discharge wetland. (B) Generalized groundwater flow net for a prairie pothole landscape underlain by fractured till. Equipotential lines represent lines of equal head. Note the recharge wetland on the upland site, and the discharge wetland in the lowland.

to the water table at any one time provides only a partial picture of the soil water regime. A better way to ascertain the characteristics of the water table is to examine its status over time and then plot its change in depth over time. Many people refer to this as "depth and duration" data (Fig. 14.41). Additional, highly useful information about the water table includes the oxygen status of the groundwater and whether the water table is perched or apparent (see later discussion; Fig. 14.42).

Measuring the depth to the water table can be accomplished by a variety of means. The low-tech way involves coring into the soil and allowing water to flow into a shallow well (Guthrie and Hajek 1979, Mokma and Cremeens 1991, Stolt et al. 2001, Callahan et al. 2012, Calzolari and Ungaro 2012). Usually the well is lined with porous tubing to prevent the walls from caving in, while still allowing water to pass through (Zobeck and Ritchie 1984). This type of well is called a *piezometer* (Blavet et al. 2000). An alternative method of water table monitoring involves neutron

probe technology (Khan and Fenton 1994, Biswas and Si 2011). Several other methods, however, are also available for determining depth to saturation and/or redox conditions in soils, e.g., platinum (redox) electrodes (Megonigal et al. 1993, Austin and Huddleston 1999, Jenkinson et al. 2002, Reuter and Bell 2003, Rabenhorst et al. 2009), ferrihydrite coated tubes (Jenkinson and Franzmeier 2006), and the chemical α,α dipyridyl, which turns pink when exposed to Fe^{2+} in soils (Childs 1981). Several excellent reviews have been published on the various methods available for monitoring water table levels and soil water contents in real time (Vorenhout et al. 2004, Fiedler et al. 2007, Rabenhorst et al. 2009, Bedard-Haughn 2011).

Conditions of Saturation in Soils

Two different types of water tables are distinguished by the soils and hydrology communities. One kind of water table is the longer-term, regional water table that normally marks the top of a thick saturated zone in the subsurface. To some, this is the "real" water table. But soil water may also form perched zones of saturation, which normally are short-lived. These *perched* water tables mark the top of a comparatively thin zone of saturation and form when the percolation rate of water exceeds the infiltration capacity of one or more zones (aquitards). The percolating water then accumulates above this slowly permeable zone, leading to a thin saturated zone and, hence, a perched water table. An *apparent water table* is defined by the depth to which water will rise in a borehole that is unlined or is lined by a perforated pipe (Fritton and Olson 1972, Mackintosh and van der Hulst 1978, Hyde and Ford 1989). It is called "apparent" because one has no way of knowing from this single observation whether one is looking at a regional water table or a temporarily perched water table. The *apparent* water table can be considered the *effective* water table for most soil processes.

Soil materials below a water table are saturated, regardless of whether it is a perched or an apparent water table. Such soils usually exhibit *aquic conditions*, i.e., they undergo continuous or periodic saturation and often, the accompanying reduction of Fe (Soil Survey Staff 1999, Vepraskas 1999). The presence of these conditions is indicated by gray (gleyed) colors, or by patches of red, gray, blue-gray, brown, and other colors, variously called mottles or *redoximorphic features* (Vepraskas 1999, see Chapter 13). The Soil Survey Staff (1999) defines three types of aquic conditions:

1. *Endosaturation*. Here, the soil is saturated in all layers from the upper boundary of the water table to >2 m depth. The surface of this zone is typically an apparent water table. Endosaturation is a slow-to-change feature because some water tables respond only slowly to recharge or drawdown (Jacobs et al. 2002). Some water tables are perched zones of saturation (see later discussion). These are more ephemeral features and respond

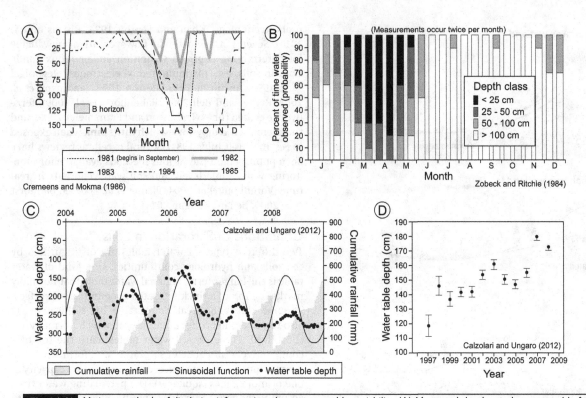

Fig. 14.41 Various methods of displaying information about water table variability. (A) Measured depths to the water table for a Typic Argiaquoll in Michigan, over a 5-year span of time. (B) Water table probability diagram for an Aeric Haplaquept in Ohio. (C) Yearly cumulative rainfall and water table depths for an area of xeric soil moisture regime in northern Italy. (D) Annual water table depths displayed as mean values with standard deviation whiskers.

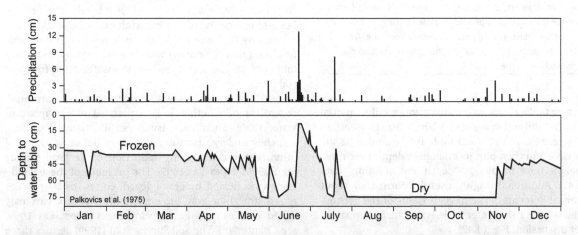

Fig. 14.42 An annual cycle of perched water in an Aquic Fragiudalf, as affected by precipitation. Each incident of water table presence, i.e., times when the chart does not show dry, indicates the presence of a perched water table above the fragipan.

to wet and dry periods more quickly, so endosaturation does not occur. An example of a soil with endosaturation is a Mollic Endoaqualf.

2. *Episaturation.* Here, the soil is saturated in one or more layers, but it *also* has one or more *unsaturated* layers within 2 m. The water table is *perched* on top of a slowly

permeable layer – an aquitard or aquiclude – below which the soil is unsaturated. Episaturation is more correctly described as a *perched zone of saturation*, rather than a perched water table. In many parts of Europe this condition is referred to as surface-water gley, or pseudogley (Chisolm *et al.* 1984, Bedard-Haughn 2011).

An example of a soil with periodic episaturation is a Mollic Epiaqualf.

Perched water can occur on top of soil (Bx or Bt) horizons or on inherited sedimentary layers like a clay lens in till or dense basal till (Fletcher and McDermott 1957, Gile 1958, Simonson and Boersma 1972, Palkovics *et al.* 1975, McDaniel and Falen 1994, McDaniel *et al.* 2001, Stolt *et al.* 2001, Jien *et al.* 2010). Water may perch on dense C horizons, whereas the solum, with good structure, is more permeable and allows water to percolate more freely (Harlan and Franzmeier 1974, King and Franzmeier 1981). Perched water tends to develop for short periods after precipitation or snowmelt events, illustrating the quick response time that episaturation requires to establish itself or be eliminated (McDaniel *et al.* 2001; Fig. 14.42). On sloping surfaces, episaturated water may flow laterally as throughflow (Gile 1958, Evans and Franzmeier 1986). Both endosaturation and episaturation are conducive to the formation of redox features. However, because the perched water condition may be short-lived, the soil horizons involved may be saturated with oxygenated water, and low (≤2) chroma mottles may not form (Evans and Franzmeier 1986). This condition is especially likely if the horizons are saturated when the soil temperature is <5°C, e.g., after snowmelt.

Clothier *et al.* (1978) described a situation that is analogous to, but slightly different from, perched water – when infiltrating water hangs at a lithologic discontinuity, as where finer-textured materials overlie coarse sediments (see Chapter 13). Mottles may then form in the overlying fine material, even though it may never have been technically saturated and so was never perched.

3. *Anthric saturation.* This special type of aquic condition occurs in soils that are cultivated and irrigated, especially by flood irrigation. Examples of anthric saturation would be rice paddies, cranberry bogs, and treatment wetlands (Eghbal *et al.* 2012).

Natural Drainage Classes
In order to assist land managers better with use and management decisions, the NCSS developed the concept of natural soil drainage classes, which refer to the frequency, depth, and duration of wet and/or saturated conditions in soils. Drainage classes usually vary predictably with topography and hillslope position (Mackintosh and van der Hulst 1978, Reuter and Bell 2001). Exceptions occur where perched water may cause a soil to be in a wetter drainage class than would otherwise be indicated by topography, or where very fine-textured parent materials allow water to move so slowly that increased wetness results. But, in general, water tables are nearer the surface in lower landscape positions, and natural drainage classes reflect this.

Drainage classes generally correlate to the water table depths, which are inferred from morphologic indicators such as mottles and gleying (Table 14.4, Fig. 14.43). The use of natural drainage classes in describing soils is tailored more to land and soil managers than to pedologists, as the classes are rather loosely defined and difficult to quantify. Drainage classes are defined for soils under relatively undisturbed conditions similar to those under which the soil developed (Soil Survey Division Staff 1993). Alteration of the water regime by humans, through either drainage or irrigation, is not usually a consideration. Nonetheless, long-term drainage or irrigation can change the drainage class of a soil and be manifested in its morphology.

Indirectly, drainage classes are reflected in the taxonomic classification of soils, thereby allowing the user to estimate the soil's natural drainage class from its subgroup classification (Chisolm *et al.* 1984, Soil Survey Division Staff 1993, Schaetzl *et al.* 2009). Table 14.5 illustrates the logic by which one can determine – with a few exceptions – the natural drainage class of a soil, solely from its subgroup classification. For example, Histosols are always very poorly drained. Soils in Aquic suborders, e.g., Haplaquods, are poorly drained, unless they are an Aeric extragrade. Soils that neither are Histosols nor have -*aqu*- anywhere in their subgroup name are usually well drained or drier. The "rules" depicted in Table 14.5 can quickly enable most users who are reasonably proficient at Soil Taxonomy to determine the drainage class of a soil, with no need for rote memorization.

Morphologic Expressions of Wetness in Soils

Drainage classes and water tables go hand in hand, for the most part. With water tables, however, we have but one observation – the one we are seeing (or not) at the moment! Because it is expensive and time-consuming to monitor water table depths, only rarely do we have actual, long-term data on water tables. Short-term water table data – the type that usually exist – may not be meaningful, as they may not represent long-term conditions (Zobeck and Ritchie 1984, Hyde and Ford 1989). Studies that report on water table fluctuations over a significant period, e.g., Pickering and Veneman (1984), Zobeck and Ritchie (1984), Hyde and Ford (1989), Mokma and Cremeens (1991), and Khan and Fenton (1994), are of the most value in determining the long-term wetness status or drainage class of a soil. But these kinds of data are few.

Therefore, our best alternative is to use *morphologic indicators* as proxy data for water table depth, duration, and fluctuation (Boersma *et al.* 1972) and as a means of ascertaining the drainage class of a soil. The best such indicators are *redoximorphic features*, which are formed by redox processes (see Chapter 13). They develop when Fe and Mn oxides are chemically reduced, oxidized, and/or translocated. Gray, olive, or pale colors (chroma ≤2) suggest that most of the iron in the soil is reduced, usually indicating prolonged wetness or saturation (although low chromas can also be caused by organic matter). Because solid-phase ferrous (Fe^{2+}) iron is bound in primary minerals and soluble Fe^{2+} is colorless, soils dominated by prolonged wetness tend to take on the gray-white color of quartz. Red or brown hues suggest that Fe has been

Table 14.4	Descriptions of the NRCS's natural soil drainage classes		
Natural drainage class	Characteristics[a]	Normal water table depth (cm)	Typical locations of mottles (redox features) and/or gleying
Excessively drained	Water is removed from the soil very rapidly; soil is commonly very coarse textured	>100	None in profile
Somewhat excessively drained	Similar to excessively drained, but the water table may not be as deep and the soil may be slightly finer-textured	>100	None in profile
Well drained	Water is removed from the soil readily but not rapidly; water is available to plants throughout most of the growing season in humid regions, and wetness does not inhibit growth of roots for most or all of the growing season	~100	Mottles in C, or BC, horizon
Moderately well drained	Water is removed from the soil somewhat slowly; soil is wet for only a short time within the rooting zone during the growing season, but long enough that most mesophytic crops are affected; soils commonly have a slowly pervious layer within the upper meter, periodically receive high rainfall, or both	75–100	Mottles in lower or middle B, and C, horizon
Somewhat poorly drained[b]	Soil is wet at shallow depths for significant periods during the growing season, and wetness restricts the growth of mesophytic crops; soils commonly have (1) a slowly pervious layer, (2) a high water table, (3) additional water from seepage, and/or (4) nearly continuous rainfall	30–75	Mottles in upper B horizon; gleyed in C and lower B horizons
Poorly drained	Water removed so slowly that the soil is wet at shallow depths, sometimes for long periods; water table is persistently shallow, such that most mesophytic crops cannot be grown; shallow water table is commonly the result of a slowly pervious layer, nearly continuous rainfall, or a combination	<30	Mottles throughout profile; gleyed in the upper B horizon and below
Very poorly drained	Similar to poorly drained soils except that the soils are commonly level or in a depression and frequently ponded; thick O horizons are typical	At surface or <15	Entire profile has mottles; gleying can extend to the surface, but is usually unobservable because of O horizon materials

[a] conditions in this table apply to soils that have **not** been artificially drained
[b] Comparable to Imperfectly drained in older and some non-U.S. publications
Source: Soil Survey Division Staff (1993).

released from iron-bearing minerals and has been oxidized to Fe^{3+} (Table 2.4). Oxidized iron occurs in soils where oxygen is not limiting, usually because the soils are unsaturated (Pickering and Veneman 1984). Mixtures of colors, often referred to as mottles or redoximorphic features, point to localized, alternating conditions of wetness and dryness. Our focus is on the morphological manifestations of oxidation-reduction processes in soils, and how these features can inform us about water table depths and fluctuations.

Before we enter into a discussion of how soil morphology can be used to infer soil hydrology, a few words of caution must be mentioned. Pedomorphologic features may reflect current processes and conditions, or they may be relict, in which case they provide information on a *former* water regime (Vepraskas and Wilding 1983a, b). Also, in wet but cold soils or in those low in organic matter, reduction may not occur because of low biological activity (Couto *et al.* 1985). Soils formed from red parent materials, however, may appear more oxidized (and hence, drier) than they actually are, as a result of the inherited color (Sprecher and Mokma 1989). Similarly, redox features that are used to determine wetness may be masked in soils formed in dark- or red-colored parent

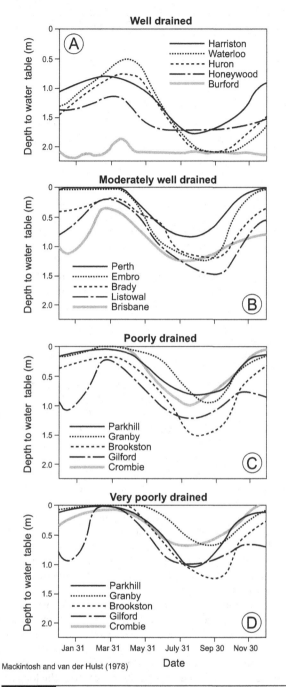

Mackintosh and van der Hulst (1978)

Fig. 14.43 Water table depths in four groups of soils, all belonging to different drainage classes. Note the annual water table cycle, as well as the variability within and between drainage classes. On the basis of these data, Mackintosh and van der Hulst (1978) advocated lumping the poorly and very poorly drained sites together; only three drainage classes appear to be justified in this landscape.

materials, or that have high concentrations of organic matter (Sprecher and Mokma 1989, Stolt *et al.* 2001). Mottle patterns also may not be readily observable in sandy Spodosols, because the colors imparted by eluviation-illuviation processes overwhelm them (Hyde and Ford 1989, Mokma and Sprecher 1994). Indeed, in many coarse-textured soils, redox patterns are difficult to discern and can be quite ephemeral. Nonetheless, despite these potential pitfalls, it is not difficult to ascertain soil hydrology and water table depths accurately on the basis of morphologic indicators, mainly revolving around soil color (Evans and Franzmeier 1986, Mausbach and Richardson 1994, Hayes and Vepraskas 2000).

Small, gray areas in soils that show evidence of reductions have long been known as gray or low-chroma (≤2) mottles. These areas have recently come to be referred to as *redox depletions*, reflective of their lower iron contents (Vepraskas 1999). Redox depletions are localized areas of decreased pigmentation that are grayer, lighter, or less red than the adjacent matrix (Schoeneberger *et al.* 1998). The term "mottles" is still used to indicate variegated patterns of soil colors due to redox depletions and concentrations, while acknowledging that "redoximorphic features" is now the preferred term (Schoeneberger *et al.* 1998, Vepraskas 1999). If a soil horizon has >50% low-chroma colors, i.e., if its matrix is gray, it is referred to as *gleyed*. Hues of 2.5Y or 5Y are also good indicators of saturated conditions (Simonson and Boersma 1972), although not necessarily of the presence of ferrous iron (Daniels *et al.* 1961).

In contrast to redox depletions, areas of concentrations of oxidized Fe and Mn are referred to as *redox* or *iron concentrations*. In older literature, iron oxide concentrations were called red mottles or high chroma mottles. Redox concentrations are indicators of saturated conditions and poor drainage, either long-term or temporary (Simonson and Boersma 1972). They are most abundant near the depth of maximum length of saturation, where wide oxidation-reduction potential fluctuations also occur. Wetter soils will have more and larger concentrations throughout more of the profile than will drier soils (Simonson and Boersma 1972). Their formation is clear; reduced Fe moves with the soil solution until it encounters free oxygen, at which point it is reoxidized and rendered immobile, usually converting to ferrihydrite, lepidocrocite, or goethite.

The color of redox concentrations can be related, in a very general sense, to the types of oxide minerals present (Table 2.4). For example, ferrihydrite has reddish brown colors, lepidocrocite has orange colors, and goethite has yellowish brown colors. Fe-Mn concentrations tend to be black or very dark red, and the Mn mineral is usually birnessite (Schwertmann and Fanning 1976). Concretions of Mn will effervesce upon exposure to weak H_2O_2, while pure iron oxyhydroxide masses will not.

Soil Color Patterns versus Soil Hydrology

Patterns of soil color, mottles, and redox features provide a great deal of information about long-term soil hydrology (Harlan and Franzmeier 1974, Vepraskas and Bouma 1976).

| Table 14.5 | A quick-and-easy way to predict the natural drainage class of a soil, using Soil Taxonomy |

Drainage class	Subgroup modifier	Great Group	Examples
Well drained or drier	-----	---------------	Typic Hapludult Lithic Eutrudept
Moderately well drained	Aquic	---------------	Aquic Haplorthod Aquic Udipsamment
Somewhat poorly drained	Aeric	-----aqu-----	Aeric Haplaquoll Aeric Haplaquept
Poorly drained	-----	-----aqu-----	Typic Haplaquod Pachic Melanaquand
Very poorly drained	-----	-----------ist	Typic Haplosaprist Terric Borohemist

Note: General rules and associations only; many exceptions to this categorization do occur.

Indeed, knowing how to interpret these patterns is vital to the correct interpretation of hydric soils. The most common pedogenic features used to infer wetness are (1) low (<2) chroma mottles (redox depletions) and matrix colors, (2) gray ped and channel coatings (albans or albic neoskeletans), (3) sesquioxide nodules or concretions, and (4) high-chroma mottles (redox concentrations) in ped interiors. We discuss these more fully in the following.

FULLY OXIDIZING CONDITIONS

In porous soils that have a deep water table, colors indicative of oxidizing conditions prevail, especially in subsurface horizons. The entire soil matrix in these horizons is red, brown, or yellow, with hues of 10YR (brown), 7.5YR (brownish-red), 5YR (red) or nearly so, depending mainly on the type and concentrations of the various iron minerals (Table 2.4; Fig. 14.44). These soils are well drained or excessively drained (Table 14.4). If they have a high water table for brief periods, either it does not affect soil color or its effect quickly fades. High water tables that also coincide with a cold season do little to impart reducing colors to a soil, because of the low oxygen demand (see Chapter 13). Such soils generally may remain red, brown, or yellow-colored throughout.

FULLY REDUCING CONDITIONS

Wet soils with high water tables (endosaturation) and ample amounts of organic matter are usually gleyed, acquiring a gray, olive-gray, or blue-gray soil matrix, with chromas ≤2 (Franzmeier *et al.* 1983; Fig. 14.44). Many of these soils are poorly drained or wetter (Table 14.4) and contain Fe in ferrous forms (Fig. 14.44). Matrix hues are typically 2.5Y or 5Y, or even GY (Fig. 14.45). In poorly drained, fully reduced soils, gleying is apparent throughout the profile. Any evidence of periodic oxidation in the upper profile is often masked by organic matter. These types of soils may or may not have a thick O horizon, depending on whether water ponds at the surface. At great depth, well below the solum, gray colors indicative of gleying change to browner hues, despite their saturated status, because, with low organic

matter concentrations, the demand for terminal electron acceptors such as oxygen or Fe there is low (Franzmeier *et al.* 1983).

FLUCTUATING WATER TABLE

In humid regions, all but the most well-drained soils are affected by a water table at some time during the year, and many experience fluctuating water table depths throughout the year (Figs. 14.41, 14.43). Most soils with water tables that fluctuate within the profile are moderately well drained or somewhat poorly drained (Table 14.4). In these soils, the horizons within the zone of the fluctuating water table periodically cross the reducing-oxidizing threshold (Fig. 14.46). The onset of Fe reduction has been commonly set at an Eh value of 170 or less (Fiedler and Sommer 2004, Fiedler *et al.* 2007). In this on-again off-again redox situation, redox depletions and concentrations can quickly develop (Lohr *et al.* 2010). For such profiles, a typical sequence has gleyed horizons at depth that change to mottled horizons nearer the surface and, depending on water table depth, to yellowish brown horizons in the upper B horizon (Fig. 14.44).

It should now be apparent that, in soils with redoximorphic features, much information about the location of the water table – and its behavior through time –can be gleaned from the patterns of the redox depletions and concentrations versus the soil matrix. Matrix colors are the colors within the centers of peds; they reflect the long-term oxidation status of the soil. Gray ped interiors result from long-term reducing conditions (Veneman *et al.* 1976), but in those horizons where saturated conditions are not permanent, iron can move from the ped interiors to the exteriors and oxidize when the water table drops during brief periods (Veneman and Bodine 1982). Redox concentrations – in the form of ferrans (see later discussion) – then can form on ped faces and within channels, giving them a brown or red color. Ferrans can persist through the ensuing period of saturation. This type of ped/soil morphology reflects long periods of saturation, with brief interludes of oxidation (Fig. 14.44).

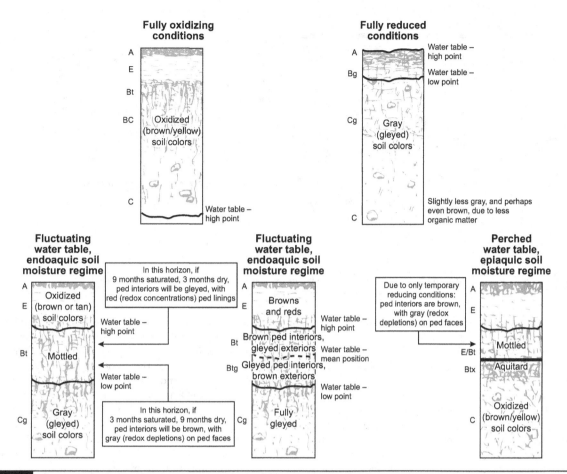

Fig. 14.44 Generalized relationships among water table depth(s), soil colors, and redox patterns.

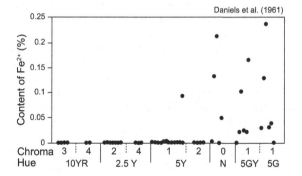

Fig. 14.45 Relationship between Munsell hue and chroma and the content of Fe^{2+} in sediments from Iowa.

Similarly, in the surface-water gley or pseudogley model, water in the upper profile is saturated for only brief periods (Fig. 14.44). As a result, most peds have brown/red interiors, reflecting their general, long-term oxidized status. They may develop bleached or gleyed faces during the brief periods of saturation because, during the saturation phase, free oxygen is first consumed by microbes and roots in the voids between peds. Thus, the soil solution in the interped spaces quickly becomes anaerobic, forcing Fe and Mn into the reduced state. Water carrying the reduced Fe and Mn may penetrate the ped, wherein it encounters some oxygen and oxidizes as a red halo or ring within the ped interior, while the ped face stays gray or has gray mottles (Richardson and Hole 1979). This situation is typical of soil horizons that perch water after snowmelt or heavy rain, or where a water table rises upward, into a soil horizon, but for only a brief period.

Types of Redoximorphic Features

Redoximorphic features are a blueprint to predict the long-term hydrological status of a soil (see Chapter 13). They can take on several forms – nodules, concretions, masses, and pore/ped linings. Ped and pore linings are most readily observed in a freshly exposed soil, because masses and concretions are commonly within peds (Schwertmann and Fanning 1976, Vepraskas 1999; Table 14.6).

Redox concentrations can form on ped exteriors in several ways. For example, in a normally saturated soil horizon, water containing Fe^{2+} can diffuse out of a ped, possibly having contact with oxidizing conditions, as can result from a

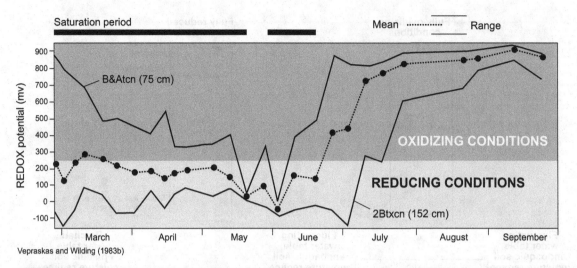

Veraskas and Wilding (1983b)

Fig. 14.46 Redox potentials during part of 1979 for two horizons of an Aeric Glossaqualf in southern Texas.

brief period of drainage. The aerated condition of the pore or ped face can develop if it has a direct, open, and unsaturated pathway to the surface (Fig. 14.44). Then, the Fe can oxidize on contact with the surface (Fig. 14.47), forming a reddish brown coating called a *ferran*. Ferrans that coat channels, such as those formed by roots, are *channel ferrans* (Veneman *et al.* 1976, Veneman and Bodine 1982). *Ferromangans* and *ferriargillans* are also possible in this type of microenvironment, if both Fe and Mn are in solution. When this happens, Fe tends to precipitate first. Modifiers such as *quasi-* and *neo-* are sometimes added to indicate the degree of development or the location of the ferran or mangan (Table 14.6).

Albans or *neoalbans* are whitish or gray ped coats that form where Fe and Mn compounds on ped or channel surfaces have been *reduced* and *removed*, forming a redox depletion zone (Veneman *et al.* 1976) (Fig. 14.47). They form in horizons that are only periodically saturated, as in pseudogley or perched water situations (Fig. 14.44). They are best expressed in finer-textured soils (Vepraskas and Wilding 1983a) and their outer boundary may be quite sharp.

The loss of Fe and Mn from ped exteriors may cause clay present there to disperse and become translocated to lower horizons, giving these types of Fe depletions a whitish appearance and sandy feel. Redox cycles can also lead to loss of clay from ped exteriors by ferrolysis (see Chapter 13). For these reasons, these types of coarse-textured eluvial features have been variously called *skeletans*, *neoskeletans*, *albic neoskeletans*, *albans*, and *grainy gray ped coatings* (Arnold and Riecken 1964, Vepraskas and Wilding 1983a, Ransom *et al.* 1987a; Table 14.6). They are common in degrading argillic horizons and fragipans, which commonly have perched water and temporarily reducing conditions. Formation of albans occurs as interped areas become saturated by infiltrating water and, because of warm temperatures and high biological activity, the areas

near pores and channels become locally anaerobic. Roots supply the soluble organic matter needed for Fe reduction and actively remove oxygen. Additionally, any rapidly infiltrating water has the ability to translocate the Fe and clay (Vepraskas and Wilding 1983a). Albans form slowly (Vepraskas 1999) and are reflective of stable pores and ped faces that have been used by many different generations of roots. In fact, they are so closely tied to reducing conditions that their abundance can be related to the length of time the soil is reduced (Vepraskas and Wilding 1983a).

Meanwhile, the ped matrix may remain oxidizing and accumulate more Fe, as it diffuses inward (Richardson and Hole 1979). In extreme cases, the amount of oxidized iron inside the ped can increase to the point where it becomes nodulelike, a feature that Veneman and Bodine (1982) referred to as a *sesquioxidic nodule*. They are thought to form when oxidation occurs rapidly (Vepraskas 1999) or when the periods of saturation are short (Veneman *et al.* 1976).

Sometimes gray argillans can be observed in lower horizons, indicating that clays and reduced Fe compounds have been translocated there, from albans above. Here, the reduced Fe diffuses, is deposited onto peds, and diffuses into them, where it forms *neopedferrans*, or Fe-Mn masses, within the ped, as a result of oxidation by air trapped within the peds (Pickering and Veneman 1984; Fig. 14.47). Thus, these types of albans or Fe depletions grow from the ped surface inward, as a result of eluviation of Fe and Mn (Vepraskas and Wilding 1983a).

In soils that are at least periodically saturated and that have Fe and Mn contents above some minimum threshold, poorly crystalline Fe-Mn *nodules* can form in zones of water table fluctuation (Richardson and Hole 1979, Vepraskas and Wilding 1983a). Nodule size and abundance generally increase with increasing duration of saturation.

Table 14.6 | Location and genesis of some common redoximorphic features in soils

Feature name(s)	Location within soil	Mechanisms of formation
Ferrans, neoferrans, quasiferrans, mangans	On and adjacent to ped faces; channel exteriors	Fe^{2+} moves in soil water, out of pore interior onto ped or channel surface, where it encounters an oxidizing environment; Fe^{2+} oxidizes and precipitates as a Fe^{3+} compound (ferran) on ped surface
Ferriargillans	Ped faces; channel exteriors	Similar process, but Fe^{3+} compounds coat ped faces that originally contained only argillans
Albans, neoalbans, gray mottles	Ped faces; channel exteriors	Fe and Mn compounds on ped or channel surfaces are reduced and removed by leaching, sometimes after heavy rains; matrix may be oxidizing
Albic neoskeletans, silt coatings, skeletans, grainy gray ped coatings	On and adjacent to ped faces; channel exteriors	Similar process, but Mn^{2+}, Fe^{2+}, and clay are eluviated from the surface, leaving silt grains behind
Neopedferrans	Inside peds	As ped faces and channels are reduced upon wetting, forming Albans, reduced iron moves into ped interiors, oxidizes, and precipitates as neopedferrans, forming a ring inside the albans

Note: Exclusive of matrix features indicative of redox processes.

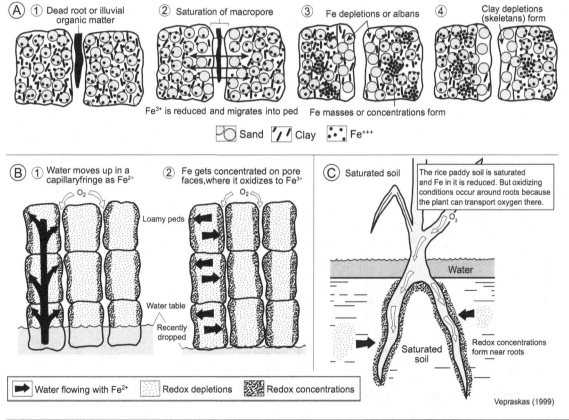

Fig. 14.47 Examples of the conditions that can lead to the formation of redox depletions and concentrations.

Landscapes: The Edge Effect

On many low-relief slightly dissected landscapes, shoulder slopes are the driest microsites (Daniels et al. 1971a), especially on stepped landscapes and flat landscapes that drop down onto a steep escarpment (Fig. 14.8). Daniels and Gamble (1967) referred to this type of site as the landscape edge. If the escarpment drops off to a drier landscape, it is called a dry edge or red edge (see figure). Escarpments that fall off into a wet depression are wet edges.

On the basis of their work on Paleudults in North Carolina, Daniels and Gamble (1967) described how and why soils on edges differ from other soils. They pointed out that soils change the most, pedon to pedon, at edges, and that these changes in soil color can occur over as little as a few meters. Much of this rapid change is related to the way the water table is influenced by the edge (Daniels et al. 1971b). At a dry edge water tables are generally deeper than farther inland, and the amount of water table variation throughout the year is greater. Dry edge soils are redder, especially the B horizons (Daniels et al. 1971b). Farther from the edge, soils are grayer, with mottles and redoximorphic features higher in the profile. They are also lower in Fe, presumably because they have lost reduced Fe in solution. Soils at a dry edge also have more clay and exhibit a greater textural contrast between E and Bt horizons. Translocation of clay is clearly a more dominant process in edge soils, perhaps because the dry edge experiences more wet-dry cycles per unit time than do other parts of the landscape. Pedons away from the edge have only thin E horizons, a glossiclike transitional horizon and a shallow, sandier B horizon.

At wet edges, soils predictably change in opposite ways. Wet soils immediately off (downslope from) the wet edge are gray, are low in iron, and have lost much of their clay to weathering and translocation (Daniels and Gamble 1967). Sandy textures are common. Strong fragipans have developed at some wet edges in North Carolina.

Figure: Changes in soils and water tables near and far from dry edges in North Carolina. (A) Idealized landscape, showing the locations of dry and wet edges. (B) Changes in color and clay content of the upper B horizons as affected by a dry edge. (C) Isolines showing the likelihood that the water table is at a given depth, at various distances from dry edges. (D) Variation in water table depth throughout the year, at various locations near dry edges.

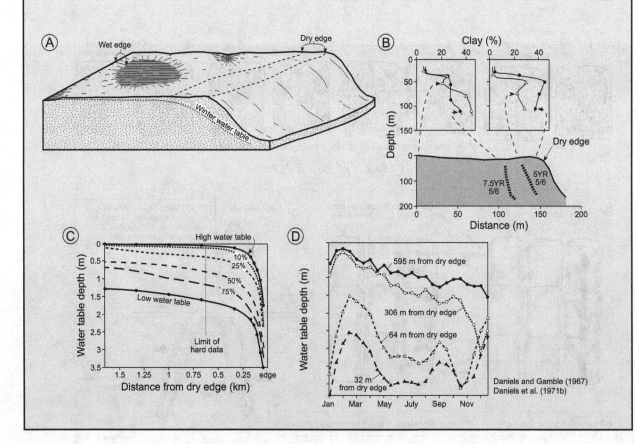

Daniels and Gamble (1967)
Daniels et al. (1971b)

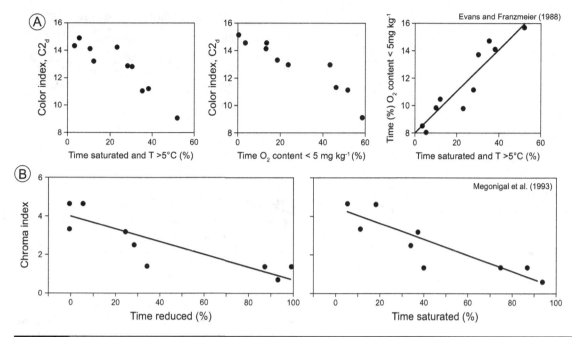

Fig. 14.48 Scatterplots showing the relationships between the Chroma Index of Evans and Franzmeier (1988) and various soil redox and morphologic indicators. (A) Relationships between the color index and various temperature and oxygen status for some Aquolls, Aqualfs, and Udalfs in Indiana. (B) Relationships between the color index and the length of time a soil was saturated or reduced (<300 mV, pH 5), for a variety of soils in South Carolina.

Quantification of Wetness

As explained previously, redox patterns can provide a great deal of information about soil hydrology. Interpreting redox patterns is critical to soil classification, for land management and especially for on-site waste disposal questions, or when delineating jurisdictional wetlands (Megonigal et al. 1993, Vepraskas et al. 2004, Dewey et al. 2006). Because there is so much variability in the expression of aquic soil conditions, Soil Taxonomy intentionally does not specify how long a soil must be saturated and free of dissolved oxygen for aquic conditions to be recognized (Soil Survey Staff 1999). Although color patterns related to wetness do appear to change after drainage, e.g., by ditches (Hayes and Vepraskas 2000), most soils retain their color patterns for some time. Sandy soils, with lower surface areas, change most quickly in response to altered hydrological status (Jacobs et al. 2002).

Quantification of soil wetness, water table characteristics, and redox conditions in soils can inform soil management decisions and enlighten studies of soil genesis (Thompson and Bell 1996, Dudley and Rochette 2005; Fig. 14.44). In Indiana, for example, Franzmeier et al. (1983) were able to quantify the following soil morphology–soil water relationships: (1) Gleyed horizons with matrix or argillan colors that are dominantly gray (chroma < 2) are saturated most of the year; (2) horizons that have a dominantly brown matrix but also have gray mottles are saturated a few

months of the year when they occur above a gleyed horizon, but are saturated most of the time if they lie below a gleyed horizon; (3) horizons with chromas of 5-6 that lack mottles with chromas ≤3 are seldom saturated. Franzmeier et al. (1983) reiterated the point that soils with a matrix chroma ≥5 are never saturated during the growing season unless they are near gleyed horizons. Jacobs et al. (2002) were able to quantify water table relationships in some Georgia Ultisols, although they cautioned that the relationships are not valid in soils that have reticulately mottled plinthite. Here, soil horizons with a gray (≤2 chroma) matrix were saturated more than half of the time. Low chroma Fe depletions could form in horizons that are saturated as little as 18% of the time. Likewise, high chroma Fe concentrations (red mottles) formed in horizons saturated about 25% of the time, but Jacobs et al. (2002) pointed out that these types of features could develop in the capillary fringe, which technically is not saturated. Taken one step further, Evans and Franzmeier (1988) were able to define a color index that has been very useful in ascertaining the relationships between redox conditions and soil morphology. Examples are provided in Figure 14.48.

The Natural Soil Drainage Index (DI) of Schaetzl et al. (2009) illustrates how well taxonomic classification can inform users about soil wetness. Using NRCS map data, which provide a taxonomic class for each map unit, and slope data, Schaetzl et al. (2009) quantified all soils on a

Fig. 14.49 Photo of pit-mound topography on a formerly forested hillslope, now used as a pasture. Photo by RJS.

Fig. 14.50 Pit-mound microtopography, formed by tree uprooting, on a moderately well-drained soil in northern Michigan. Water entering the system via snowmelt has raised the water table so high that the treethrow pits become filled with water. Photo by RJS.

scale of 0–99 (Fig. 8.24). When a digital soil map is coded by DI in a GIS, the map produced is an excellent visualization of soil wetness on landscape scales. The use of the DI in landscape and environmental modeling studies has yet to be tapped.

Microrelief

Until now, we have been focusing on the effects of topography on soil development and soil wetness, but on hillslope scales. Often overlooked, however, are the small-scale forms of topography, which, although local, are also very important to soil development. Features with vertical or horizontal dimensions of less than a few meters could be considered microtopographic in scale. Mesotopographic features are 10 to a few tens of meters in size.

The origins of microtopographic highs and lows, swells and swales, humps and hollows, are myriad. Tree uprooting is perhaps the most common origin of microtopography in forested regions, creating pit-and-mound or cradle knoll microtopography (Lyford and MacLean 1966, Hamann 1984, Schaetzl et al. 1989, 1990, Samonil et al. 2010b; Fig. 14.49). Argilliturbation forms gilgai microrelief (see Chapter 11). Tree death, followed by wood decay, forms stump holes – surface depressions that become foci for litter and water (Phillips and Marion 2006). Frost heave is responsible for various forms of microtopography in areas of permafrost (see Chapter 11). Ridge-and-swale microtopography is common on floodplains. Many glacial landscapes inherit hummocks and hillocks of various sizes (Gracanin 1971, Attig and Clayton 1993, Johnson et al. 1995, Clayton et al. 2008). Animal mounds, e.g., mainly those of arthropods, insects, and mammals, form a vast array of different microtopographic forms.

The pedogenic effect of microtopography often depends on macroclimatic and water table variables. Below microtopographic lows where the water table is high pedogenic processes associated with vertical translocation may be considerably hindered by the high water table (Fig. 14.50). Here, soils on the adjacent microtopographic highs will usually be better drained and potentially better developed. Conversely, in many humid climate soils with a deep water table, soils in pits and small depressions are almost always better developed (Låg 1951, Denny and Goodlett 1956, Veneman et al. 1984, Miller et al. 1985, Schaetzl et al. 1990, Samonil et al. 2010a, b). Pits are more leached and have thicker O horizons, better horizonation, and many other attributes associated with progressive pedogenesis (Schaetzl 1990), largely because more water percolates through these soils (Table 14.7, Fig. 14.51). This observation is predicted by the Energy Model of Runge (1973; see Chapter 12). Pits on wet sites with a high water table will not benefit from any extra potential energy and will be less developed. In some situations, a thick mat of broadleaf litter (Oi and Oe horizons) or slowly permeable soils due to clayey textures and/or frost can promote even more run-in and percolation within pits. Finally, snowpacks may be thicker in pits, leading to more meltwater inputs (Schaetzl 1990; Fig. 14.52).

A key factor that can contribute to better-developed soils beneath microtopographic lows is thicker litter layers (O horizons), especially in forests. Litter tends to collect there preferentially, and the cooler, moister conditions that prevail in pits inhibit decomposition (Armson and Fessenden 1973, Shubayeva and Karpachevskiy 1983, Schaetzl 1986b, Mueller et al. 1999; Table 14.8). Litter is the primary source of organic acids that often promote soil development in acidic forest soils (see Chapter 13). The thicker O horizons in pits also protect the soil from drying events, and the extra moisture may facilitate weathering. Microtopography also affects soil temperature, which again influences pedogenic processes and the biotic communities that inhabit the microsites (Troedsson and Lyford 1973; Tables 14.7, 14.8).

Microtopography is also important to pedogenesis in dry climates and in nonforested settings (Sharma et al. 1998).

Table 14.7 | Comparisons of soil characteristics at microsites (pits and mounds) formed by tree uprooting

Characteristic[a]	Mound	Undisturbed areas	Pit	References
Soil development or profile differentiation	Low		High	Moore 1974, Schaetzl 1990, Veneman et al. 1984, Samonil et al. 2010b
	Low	High		Denny and Goodlett 1956, Goodlett 1954
Frost action	High		Low	Beatty 1984, Denny and Goodlett 1956, Goodlett 1954, Hart et al. 1962, Lutz 1940, Schaetzl 1990
Winter temperatures	Cool		Warm	Federer 1973, Schaetzl 1990, Samonil et al. 2010a
Spring temperatures	Cool		Warm	Beatty 1984, Beatty and Stone 1986, Federer 1973
Summer temperatures	Warm		Cool	Beatty 1984, Beatty and Stone 1986, Federer 1973
H_2O content	Low		High	Beatty 1984, Beatty and Sholes 1988, Beatty and Stone 1986, Lyford and MacLean 1966, Schaetzl 1990, Shubayeva and Karpachevskiy 1983, Clinton and Baker 2000, Samonil et al. 2010a
Saturated infiltration capacity	High	Low		Goodlett 1954, Lutz 1940
Porosity	High	Low		Lutz 1940
pH	High	Low		Lutz 1940
	Low		High	Beatty and Sholes 1988, Shubayeva and Karpachevskiy 1983
CEC	Low		High	Beatty 1984
Available N content	Low		High	Beatty 1984
Organic matter content	Low		High	Beatty 1984, Beatty and Sholes 1988, Beatty and Stone 1986, Schaetzl 1990, Stone 1975
Ca content	High		Low	Stone 1975
	Low		High	Beatty 1984, Beatty and Stone 1986
Mg content	High		Low	Stone 1975
Heavy mineral content	High	Low		Lutz 1940
Leaf litter accumulation	Low		High	Beatty and Sholes 1988, Beatty and Stone 1986, Hart et al. 1962, Schaetzl 1990, Stone 1975
O horizon thickness	Thin		Thick	Beatty 1984, Goodlett 1954, Hart et al. 1962, Lyford and MacLean 1966, Moore 1974, Schaetzl 1990, Shubayeva and Karpachevskiy 1983, Veneman et al. 1984
A horizon thickness	Thin		Thick	Beatty 1984, Beatty and Sholes 1988, Beatty and Stone 1986
Snow depth	Thin		Thick	Beatty 1984, Beatty and Stone 1986, Federer 1973, Schaetzl 1990
Likelihood of being snow-free during the snowmelt period	High		Low	Beatty 1984, Schaetzl 1990

[a] "High," "Low," "Thick," "Thin," etc., indicate direction, not absolute magnitude, of variability.

Source: Schaetzl et al. (1990), and the personal observations of the authors.

Table 14.8 | Properties of mounds, pits, and undisturbed sites formed by uprooting in a beech-maple forest in New York State

Microsite	Soil moisture (%)	Summer soil temperatures (°C at 10 cm)	Spring and fall soil temperatures (°C at 10 cm)	Organic matter in upper profile (%)	O horizon thickness (cm)	A horizon thickness (cm)	Snowpack thickness (cm)
Mound	20–35	10–14	3–4	5.7	1.2	2.7	35
Undisturbed	30–45	9–13	4–5	10.0	3.5	5.2	41
Pit	40–60	8–11	5–6	17.8	5.7	9.4	47

Source: Beatty and Stone (1986).

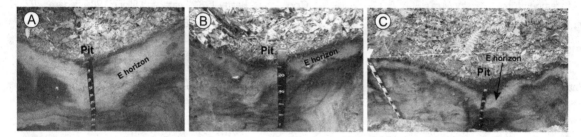

Fig. 14.51 Spodic soil development in and below some pits (microtopographic lows) formed by tree uprooting. The E horizon thickens considerably as it approaches the pit, often developing a deep tongue immediately below the pit center. Note also how the spodic B horizon is better developed in the pit. Ages of the uprooting event were determined by ^{14}C dating of buried wood and charcoal (Samonil et al. 2013) and are estimated at: A: 4,480 ± 35 BP, B: 4,550 ± 50 BP, C: 3,600 ± 30 BP. Tape increments in centimeters. Photos by P. Samonil.

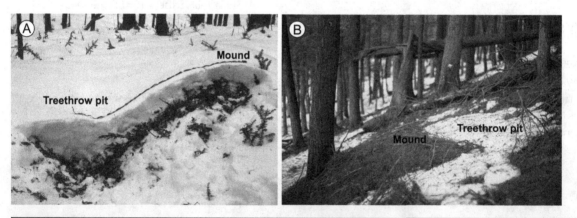

Fig. 14.52 Relationship between pit-mound microtopography and snowpack thickness. The snowpack is thicker in pits than on mounds, as shown in (A), and persists longer in pits, as shown in (B). Photos by RJS.

Small variations in microclimate may lead to significant differences in soil moisture, vegetation, and soil development in grasslands and deserts, especially with respect to pedogenic properties that can be changed readily by small amounts of water. For example, White (1964) pointed out the differences between Na-affected soils in depressions and mounds (see Chapter 13). On the Coast Prairie of Texas, calcic horizons are restricted to microhighs, as a result of capillary rise of carbonate-rich water into the microhighs, whereas subtle depressions lack calcic horizons (Sobecki and Wilding 1982; Fig. 14.53). Sobecki and Wilding (1983) proposed that carbonates are leached from microlows and redistributed to microhighs via lateral flow on top of a slowly permeable (3C) horizon. Thus, the flow of moisture is driven by a hydraulic gradient that sets up between the dry knolls and the wetter depressions.

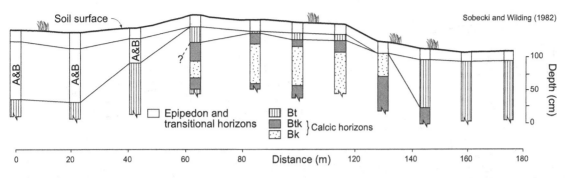

Sobecki and Wilding (1982)

Fig. 14.53 Soil horizonation through a microtopographic high on the Texas Coast Prairie.

Many of the deepest Vertisols, with the strongest expression of argilliturbation, are in microlows, which act as settling basins for clays eroded nearby, and where water can pond during the wet season (see Chapter 11). Ponding of water on these landscapes for extended periods further assists in smectite neoformation and virtually assures complete wetting of the profile, accentuating the wet-dry seasonality of the site and fostering additional argilliturbation.

Microtopography variously affects almost all landscapes. Consider it a type of microcatena. As such, microtopography accelerates or decelerates soil development by redirecting dissolved and suspended substances in water toward or away from certain sites (Beatty and Stone 1986). It can also affect soils indirectly through its interaction with vegetation establishment patterns and productivity (Beatty 1984, Beatty and Sholes 1988). For these reasons, microtopography is vital to maintaining spatial heterogeneity in soil landscapes, which in turn is important to plant and animal biodiversity.

Examples of Catenas

In this section, we provide examples of catenas that illustrate many of the points made previously, as well as providing applications of the models that we have discussed previously and will develop in the following.

Wisconsin Till Plain, Iowa

In north-central Iowa and southern Minnesota, soils have developed in calcareous, loam-textured glacial till and other surficial sediments. Although the landscape was deglaciated about 14,000 years ago (Ruhe and Scholtes 1959), the Mollisols here developed their characteristics primarily over the past 3000 years, after grasslands invaded the area (Van Zant 1979, Steinwand and Fenton 1995). The landscape is very hummocky, with many basins of interior drainage. Much of this landscape is included within the Clarion-Nicollet-Webster soil

association (CNW) which occupies >31,000 km² of the Des Moines glacial lobe (Steinwand and Fenton 1995, Reuter and Bell 2003; Fig. 14.54). Soil variation across this landscape is due to drainage (water table relations), carbonate status, and parent material texture. This landscape has innumerable examples of catenas within closed depressions. Sediment eroded from the hillslopes is deposited within the lowlands, because through-flowing streams do not exist to carry it away. Let us examine the details.

Upland Hapludolls (Clarion series) have developed in oxidized, leached loam till (Fig. 14.54). Oxidizing conditions occur in the upland soils, but in the lowlands and at depth, soils are reduced. In swales, postglacial sediments have accumulated above the till (Ruhe 1969). This sediment, which exhibits a fining upward sequence, typical of fluvially deposited materials, is thickest in the centers of swales and basins (Burras and Scholtes 1987). The lowermost material is sandy and may have been deposited during the waning stages of ice retreat (Steinwand and Fenton 1995). Collectively, it may be best described as colluvium or slopewash. Many rolling, glacial landscapes have similar sediments in swales, dating back to a former period of landscape instability (Walker and Ruhe 1968, Pennock and Vreeken 1986). Finer sediments that lack coarse fragments overlie the sandy slope alluvium and form the uppermost parent material in the swales. Stone lines, in this case indicative of an erosional contact, may be located at the discontinuity between the two materials (Burras and Scholtes 1987). Sediments become finer and thicker in the swales, suggestive of sorting during downslope transport; finer sediments were transported into the swales while coarser sands remained on the slopes. The finer sediments are believed to reflect a late Holocene period of landscape instability (Burras and Scholtes (1987).

Both the Nicollet and Webster soils have developed in various thicknesses of slopewash material (Fig. 14.54). In the wettest parts of the landscape, Harps and Canisteo soils with Bkg horizons have formed where calcareous

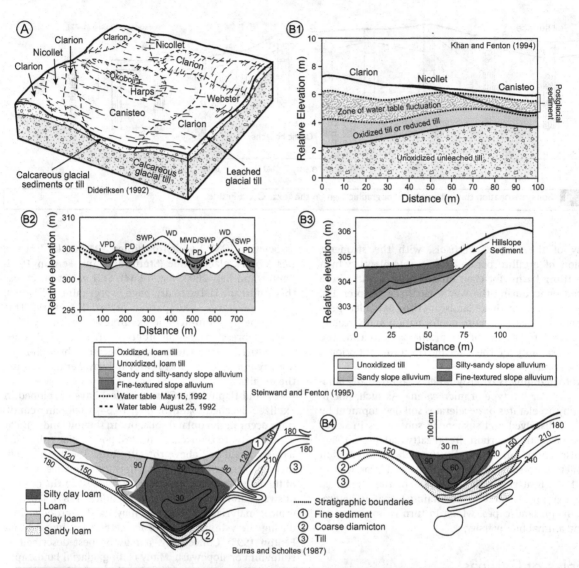

Fig. 14.54 Soil and sediment data for the Clarion-Nicollet-Webster soil association of north-central Iowa. (A) Block diagram of the landscape of the Des Moines lobe in north-central Iowa, on which the Clarion-Nicollet-Webster-Canisteo catena dominates. (B) Detailed stratigraphy of the landscape, which is floored with till but has accumulated various thicknesses of slopewash or colluvium since deglaciation. (1) Generalized slopes, stratigraphy, and hydrology of the catena. (2) Stratigraphic cross section, showing the sediment relationships to the land surface, water table, and drainage classes. (3) Detailed cross section of a hillslope showing the facies relationships along the catena. (4) Cross section of a typical basin, showing isolines of geometric mean particle size (micrometers), as well as texture classes and stratigraphy.

groundwater is discharged (Khan and Fenton 1994). The cumulic Okoboji soils form in wet, depressional areas where the colluvium/slopewash has been leached of these carbonates.

Clearly, the influence of slope processes and the effects of groundwater on the soils of this catena are considerable. In that respect, they represent many young landscapes in humid climates, in which periods of slope instability have existed in the recent past.

Negev Desert, Israel

In this desert, as in all deserts, soil development is slow. The intermittent character of the rainfall determines, in large part, the pathways of pedogenesis. Annual precipitation in the Negev is less than 250 mm. It is so dry here that inputs of dew are an important parts of the hydrology (Kidron *et al.* 2000). Inputs of Saharan dust are present but are quickly redistributed across the landscape. Much of this dust is redistributed into lower slope

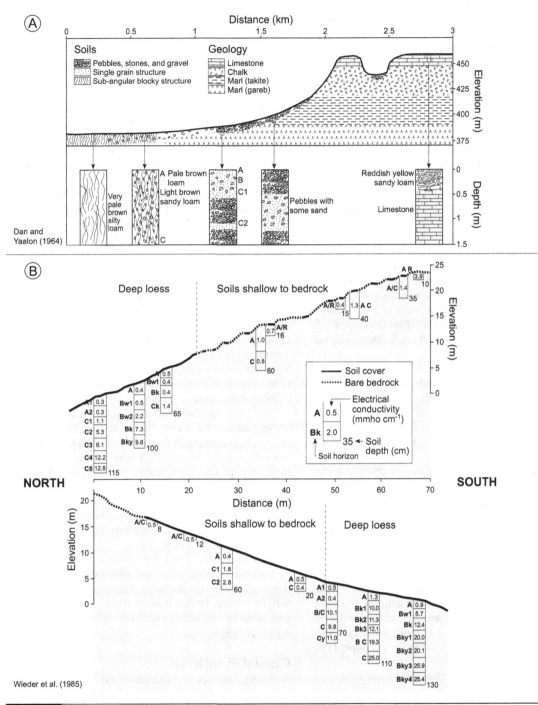

Fig. 14.55 Diagrams of, and data for, typical catenas in the hyperarid Negev Desert, Israel. (A) Diagram of the soils and stratigraphy of the catena, as developed on carbonate sediments in the southern Negev. (B) Soil morphology and electrical conductivity data for both north and south topographic aspects of a catena in the Negev.

positions, whereas the upper portions of slopes remain rocky and have thin soils, except for some rock interstices, which trap loess (Kadmon *et al.* 1989, Verrecchia and LeCoustumer 1996). In high-relief bedrock terrain, free faces in limestone and chalk occur on shoulder slopes (Fig. 14.55A). Soils on these surfaces are thin and stony for two reasons: (1) The high Ca^{2+} contents inhibit weathering, keeping pH values high, and (2) the steep slopes are unstable and erode. In bedrock pockets, soils may be more leached than almost anywhere else, because they are sites of focused run-in and percolation. On similar sites and landscapes in a Mediterranean climate, one might find terra rossa soils (Dan *et al.* 1972).

Despite the low precipitation amounts, downslope transport of sediment is a dominant factor in determining the soil and vegetation associations here (Yair and Danin 1980). Leaching intensity and percolation depth are highly dependent on the amount of soil pore space and run-on, both of which are largely a function of surface bedrock exposures. Water availability is greatest where the ratio of hard bedrock to soil is high, because water runs off the soils and into the limited amount of soil between (Yair and Berkowicz 1989).

Soils in the footslope and toeslope positions are developed in deep silts and sands that have washed down from upslope. Water infiltrates more shallowly in the colluvium/alluvium here than it does on bedrock uplands, because the entire surface is permeable, pore space is more abundant than in rocky soils, and essentially no run-on enters from upslope. Much of what infiltrates is evaporated later from the surface, leaving salts behind and making the soils biologically drier than they might otherwise be (Fig. 14.55B). Soils in this deep sediment have therefore developed calcic (Bk) horizons overlying gypsic and calcic (Bky) horizons (Fig. 14.55).

Salts and carbonates are translocated more deeply and soils remain wetter for longer periods on the rocky upper slopes than on the alluvial flats at the bases of the slopes, because on the uplands more of the limited infiltration is directed deeply into a few small fissures in the rock (Fig. 14.55). The most deeply leached areas of the slope are at the base of the rocky upper segment, where runoff is maximal, and yet the amount of land surface that can accept water is still low, because rocks are exposed at the surface. Thus, depths to gypsic and calcic horizons become progressively shallower downslope, in the colluvial materials (Yair and Berkowicz 1989; Fig. 14.55B). Likewise, soils in colluvium have calcic horizons only near the rocky footslope. Farther downslope, out on the colluvium, they have calcic *and* gypsic horizons, whereas those farthest out have only gypsic horizons, reflecting the increasing aridity with distance from the rocky, runoff-generating slopes (Wieder *et al.* 1985, Verrecchia and LeCoustumer 1996). Vegetation reflects this pattern; it is more dense and lush on the rocky, upper slopes than on the silty toeslopes (Yair and Danin 1980).

This catena illustrates that across the desert landscape the degree of soil aridity, soil development, and leaching is highly variable but predictable, primarily as a function of location *and* substrate, which are important because of their effects on runoff versus. run-in (Buis and Veldkamp 2008).

Front Range, Colorado

Soils are often linked geochemically along a catena, as materials in upslope positions are translocated downslope, where they either precipitate or make their way into a stream, lake, or underground aquifer (Glazovskaya 1968). Many soil attributes along catenas are due to *lateral* losses or gains of *soluble* materials, which are directly affected by the slope itself (Huggett 1976b, Hillel and Talpaz 1977, Knuteson *et al.* 1989, Reuter *et al.* 1998, Sommer *et al.* 2000).

An application of this concept is provided by Litaor's (1992) study of soils along a catena in the Colorado Front Range, which focuses on the movement of Al in solution. Litaor sampled the Typic Cryumbrepts and the solution that moved through soil macropores at summit, backslope, and toeslope positions (Fig. 14.56). The site is an alpine meadow with about 3–4 m of glacial till above gneiss bedrock. This catena is an excellent location to examine lateral transport of soluble materials because (1) little mass movement occurs, (2) surface runoff is minimal because of thick vegetation and numerous cracks and fissures in the surface (formed by cryoturbation), and (3) subsurface water flow, mainly fed by spring snowmelt and summer thunderstorms, is accentuated by frozen subsoil. Much of the meltwater flows laterally within the subsurface, taking ions and some clay with it. Last, vertical translocation of clay is insignificant, and silt is translocated vertically only in summit positions. Vertical translocation of aluminum, to Bw horizons, does occur in summer, however. Thus, most of the variation along the catena is due to lateral transport of soluble materials.

The catenary distribution of some materials, such as organic carbon (OC), may be due to lower decomposition rates in the wetter soils at the base of the slope (Fig. 14.56). Nonetheless, the soil solution contains more dissolved OC in downslope locations, suggesting that subsurface lateral flow of carbon is an important ongoing process. Increases in Al-organic complexes at downslope locations also point to lateral flow of soil solutions containing these soluble products. This process is probably most active at snowmelt, when the subsoil is frozen.

This study illustrates the importance of slope on *subsurface* translocation processes, which can occur whenever water inputs exceed the vertical infiltration capacity of the soil. Of course, lateral translocation processes are always promoted by subsurface aquitards, many of which are pedogenic in origin.

Coastal Plain, Israel

Normally, catenas on sand dunes are texturally uniform, and soils vary across them mainly as a function of depth to the water table. Wet, sandy soils in swales between dunes are subject to different processes, e.g., oxidation-reduction, but otherwise the soils on dune landscapes are often relatively similar. Dan *et al.* (1968) described a very different situation for a catena on a dune, less than 1 km from the Mediterranean Sea, in Israel (Fig. 14.57). The xeric climate here is humid enough (the moisture surplus is 150–200 mm in winter) that soils in the interdune swales are saturated in winter, while soils on the dune crest are dry and leached.

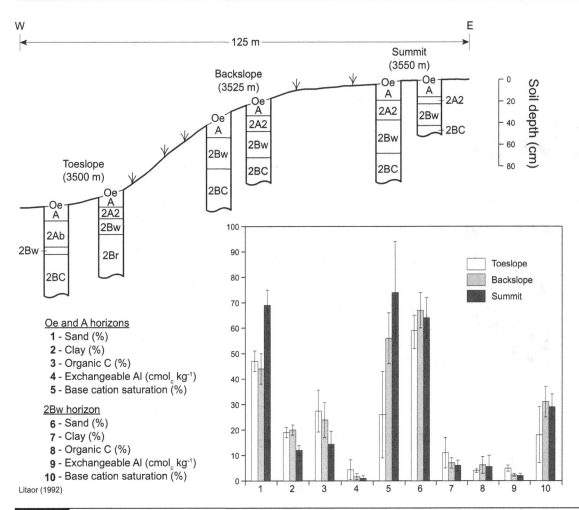

Fig. 14.56 Morphology and data for some soils along an alpine catena in the Front Range, Colorado.

The swale retains so much runoff in winter that it supports marsh vegetation. In the dry summers all soils experience severe soil moisture deficits.

The textures of the soils on this catena range from sand to clay. The clay originated as eolian dust, carried in from surrounding deserts (Yaalon and Ganor 1975). Although carbonate-rich dust blankets the landscape evenly, the plant debris and litter upon which it is deposited are preferentially washed and blown into swales, and, thus, the fine materials accumulate there. Soils in the swale (Haploxererts) become clay-rich but are underlain by dune sand. Because not enough leaching occurs in the swale to remove bases, the pH has remained high, allowing smectites to form and cracking to occur in the dry summers.

Meanwhile, on the much sandier dune crests (but not the side slopes), some of the eolian dust has been translocated into the soil, forming a Bt horizon (Fig. 14.57). Sandy eluvial zones and acidic conditions in these soils attest to the strong leaching environment. All the soils are free of

soluble salts and carbonates, and kaolinite, typical of acidic soils, is a prominent clay mineral. Soils on the steepest slopes of the dune are the least developed, because this is a runoff-generating site. Even here, however, enough vertical percolation takes place to form a weak Bt horizon. During winter, throughflow moves along the top of these Bt horizons, translocating still more clay and bases to the swales.

Footslope soils display columnar structure in the Bt horizon. Normally, this type of structure is associated with natric conditions, but these soils lack Na. The genesis of this structure is explained by Dan *et al.* (1968) as follows: The Bt horizon breaks into prisms in the dry summer, as clays slightly shrink and contract. At the onset of the winter rains, clay, sand, and other coarse materials are translocated into and through the gaps between the prisms, preserving their gross structure. Rewetted, the prisms expand, but the primary avenue for expansion is the centers of the prisms, which are forced upward, creating the round tops.

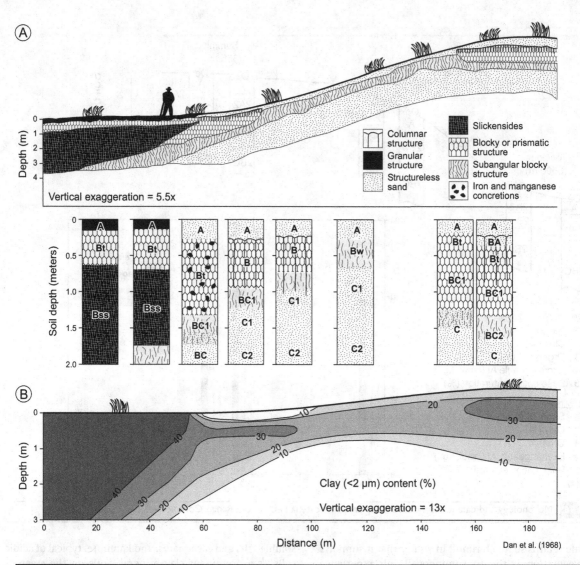

Fig. 14.57 Distribution, morphology, and clay contents of the soils of the Netanya catena, formed on a coastal dune in Israel.

This catena illustrates the importance of dust and the local intensity of leaching processes in xeric climates. Although a moisture surplus of only 150–200 mm occurs here, it is concentrated in a few months when the vegetation is dormant and, thus, leaching, translocation, and weathering can be intense.

The Humid Tropics

Perhaps the most challenging catenas, but also potentially most enlightening, exist in the humid tropics. Here, not only are the usual suspects in play – slope processes, microclimate, and traditional top-down pedogenesis – but the extreme age of some of these landscapes and the intensity of weathering and leaching must also be factored in. Several brief examples are provided here, to illustrate the complexity and extreme – but predictable – variation across these old tropical landscapes.

Many uplands in the humid tropics consist of low, rounded hills with a generally continuous mantle of Oxisols and Ultisols that are forming in saprolite (Dubroeucq and Volkoff 1998). Ironstone, plinthite, and other materials that might form a solid caprock are more typical of the semiarid or subhumid tropics; they are usually absent here (Fig. 14.58). In contrast with western Africa, Fe accumulations in the Amazon Basin, for example, produce only minimal formation of iron pans (ferricretes or cuirasses), because of the low elevations of the uplands (Fritsch *et al.* 2011). On the hillslopes, soils change to Ultisol-like morphologies, and then, in the wetter, footslope and toeslope locations, wet Spodosols are common, and Entisols or

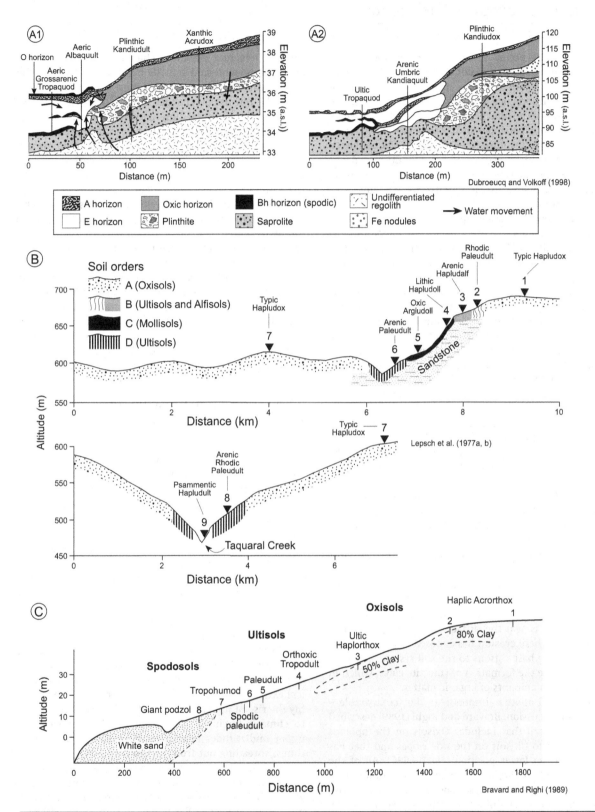

Fig. 14.58 Examples of catenas form the humid tropics. (A) Soils and parent materials on various landscape positions in the Rio Negro basin, Amazonia. (B) Soil-landscape relationships in a part of the Occidental Plateau, near São Paulo, Brazil, showing geomorphic surfaces and their relationships to soils. (C) Soils and clay contents along a tropical catena near Manaus, Brazil, showing the very gradual but predictable changes in elemental composition along the catena.

Histosols can also occur. At these lower positions on the slope, the texture changes to sands, which, combined with the abundance of organic acids in this ecosystem, favor podzolization. Dubroeucq and Volkoff (1998) hypothesized that the source of the sand is the upland soils themselves – transported to the footslopes and toeslopes by surficial processes. On the very wettest sites, Histosols (Tropofibrists) have developed overlying the sand, but Aquults have developed where saprolite is nearer to the surface. The Histosols develop as the quartz in the E horizons is fragmented, holding up water and making the soil even wetter. One possible end point of landscape evolution in these hot, humid environments may be a flat landscape dominated by wet, generally sandy Spodosols and Ultisols, underlain by white, kaolinite-rich saprolite (Dubroeucq and Volkoff 1998).

Research on soil catenas has shown that parts of the tropical landscape are old and deeply weathered, but many other parts continue to be geomorphically rejuvenated by exposure to less weathered parent materials. These materials are usually in downslope positions and have been transported there by erosion of weathered topsoil upslope, or by additions of fresh materials (colluvium, slopewash, alluvium, or ash). In rejuvenated areas, any of a number of soils might be found: Andisols in ash, Entisols or Inceptisols in alluvium, or Alfisols or Ultisols on eroding sideslopes.

This type of situation – not much different from the one described by Dubroeucq and Volkoff (1998) – was studied by Lepsch et al. (1977a, b) for Brazil (Fig. 14.58B). The uplands here are held up not by an ironstone layer, but by sandstone. Oxisols have developed on this old, stable surface. Colluvium and alluvium are common on the dissected sideslopes; in these materials, Alfisols, Inceptisols, and Ultisols have developed. Where this material is sandy and water tables are high, Spodosols and Histosols can form (Richards 1941, Andriesse 1969, Tan et al. 1970, Schwartz et al. 1986). Lateral water movement on these slopes has, presumably, initiated removal of some Fe, facilitating lessivage and Bt horizon development. As described earlier, Ultisols form where saprolite is near the surface. Mollisols have formed on sideslopes where erosion has exposed calcareous sandstone, providing base cations to the soil system. The base cations facilitate the formation of smectite clays, which in turn retain high amounts of organic matter.

Last, we add another impressive – but comparable – example for discussion. Bravard and Righi (1989) described a catena in Brazil that includes Oxisols on the uplands that transition to Ultisols on the side slopes and that has Spodosols in the lower position (Fig. 14.58C). All of the soils are strongly acidic, with very low base cation saturation. Like many Oxisols, these have experienced intense desilication and are kaolinite-dominated, as exemplified by their elemental mobilities: Si > Al > Fe = Ti. Clay contents decrease downslope, corresponding to increases in quartz. These changes have resulted from a combination of weathering and eluviation/erosion of clay minerals. In the lower catena, elemental mobilities change to Al > Fe = Ti > Si, as quartz is stable and actually neoforms in the wet, footslope areas. This catena also exhibits some dramatic – but gradual – downslope changes in mineralogy (Fig. 14.58C). It also shows the incredibly marked differences in pedogenesis that can occur from the top to the base of a slope. As explained by Fritsch et al. (2011), this catena exemplifies the two end-member, geochemical pathways in the lateritic landscapes of the humid tropical regions. And the resultant downslope changes in morphology are just as striking: Oxisols with 80–90% clay, most of which is kaolinitic, and Spodosols that are 90% quartz sand. The intense leaching of alkali and alkaline-earth elements and silica on the uplands has led to residual accumulations of Fe and Al in the freely drained parts of the landscape, but also to their remobilization under waterlogged, reducing, and acidic conditions downslope (Bryant and Macedo 1990, Fritsch et al. 2011). This and related studies of tropical catenas (Klinge 1965, Chauvel et al. 1987, Bravard and Righi 1990, 1991) are truly mind-boggling examples of the great pedologic diversity that is possible along a catena, given enough time and pedogenic energy.

The Geologic Timescale and Quaternary Paleoclimates

In order to understand fully the soil geomorphology studies that are to follow (in this chapter and in later ones), we must first delve into the geologic timescale. This digression is necessary – soils are always forming within the constraints of time. Time is the one factor that affects all soils equally.

The past 65 million years lie within the Cenozoic Era (Fig. 14.59). However, few, if any, soils currently at the Earth's surface are this old, because the original Cenozoic surfaces by now have been eroded. The oldest surface soils on Earth probably date back to the Pliocene, and most soils have their $time_{zero}$ somewhere in the Quaternary Period – the last 2.4–2.6 Ma (Fig. 14.59).

The Quaternary Period – within which we still live – has been a time of great geomorphic dynamics, instability, and change. Repeated climatic shifts, triggering waves of biological extinctions and evolutionary adaptations – most notably the rise of humankind – typify this period. In response to climatic forcings, continental ice sheets invaded the upper midlatitudes of the Northern Hemisphere several times, spreading out from accumulation zones in central Canada and northern Eurasia (Bowen 1978, Andrews 1987). These climatic forcings were almost certainly associated with cyclicity in the amounts and timing of insolation (solar radiation) received on Earth, which were in turn driven by oscillations in orbital geometry (Hays et al. 1976, Martinson et al. 1987, Ruddiman and Wright 1987, Berger and Loutre 1991, Muller et al. 1997, Wunsch 2004; Fig. 14.60). This

Global chronostratigraphical correlations for the last 2.0 Ma

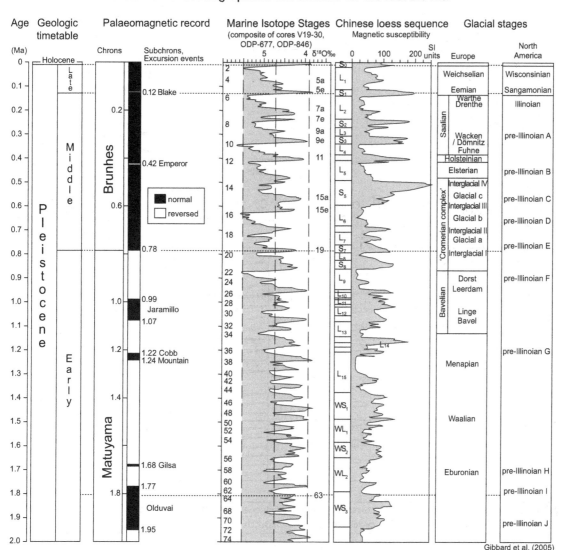

Fig. 14.59 Some of the major chronostratigraphical correlations for the last 2.0 million years of Earth history.

cyclicity is referred to as Milankovitch cycles, after the Serbian geophysicist and astronomer Milutin Milankovitch, who first described them in the 1930s.

Milankovitch cycles have three components (Berger 1988; Fig. 14.60). *Orbital eccentricity* is a measure of the ellipticity of the Earth's orbit, or how much the Earth's orbit deviates from a perfect circle, toward an ellipse. The elongation and shortening of this ellipse, i.e., the orbital eccentricity, changes on a complex ~100,000-year cycle. *Orbital obliquity* refers to the tilt of the Earth's axis with respect to its orbital plane. Today the tilt, or obliquity, is 23.5°, but it varies between about 22.5° and 24.5° on a 41,000-year cycle. Greater tilt generally leads to greater seasonality

(hotter summers and colder winters), especially in high latitudes. Cooler summers generally favor the onset of glaciation, because less of the previous winter's snow is melted at high latitude locations. Therefore, lower obliquity favors glacial periods for two reasons: (1) reductions in overall summer insolation, coupled with (2) the additional reduction in mean insolation at high latitudes. *Precession of the equinoxes* is in reference to the position in Earth's orbit when the seasons occur. Currently, the Earth is farthest from the sun in the Northern Hemisphere winter, but this gradually changes because the direction that the Earth's axis points in space slowly rotates 360° over the course of years, like a wobbling top. The precession cycle usually

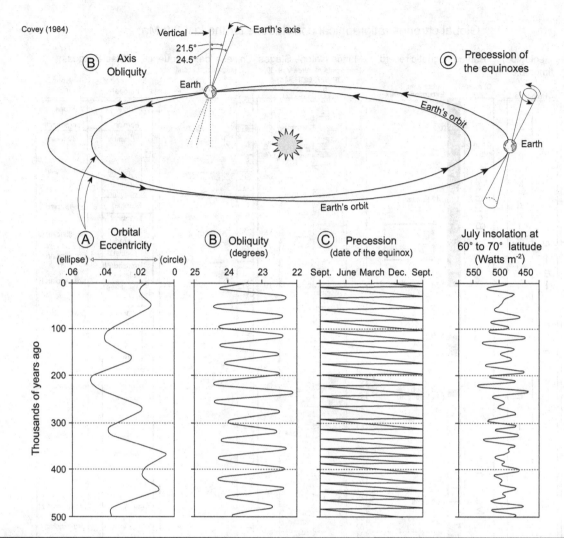

Fig. 14.60 The cyclicity of the three astronomic factors (eccentricity, obliquity, and precession of the equinoxes) that have created the insolation cycles of the Quaternary period. The cycles are called Milankovitch cycles.

spans 21,000–23,000 years. In the end, the intricate combinations of the three cycles produce great (and predictable, backward and forward in time) cyclical variations in the amount of insolation received at the Earth's surface and at different latitudes, all of which affect climate and drive the glacial cycles (Fig. 14.60). Obviously, other inputs and factors such as volcanic activity, continental alignments, changes in sea ice coverage, and albedo of the land surface also affect global climate cycles (Ruddiman 2008), but the main reason for the major climatic shifts of the Quaternary Period is tied to Milankovitch cycles.

Major glacial advances are called *glaciations* or glacials, whereas the periods of warmer climate between, when the ice sheets retreat (melt), are called *interglaciations* or interglacials. Smaller retreats within a larger glacial advance are called *interstadials*, while small readvances are

called *stadials* (Table 14.9). During the Pleistocene glaciations, each glacial advance introduced new parent materials, as it eroded and/or buried the preexisting ones. Glacial sediments were derived largely as the result of physical comminution and erosion of soils and rocks farther up-ice. Their deposition occurred as the ice ablated (melted). Sea levels fell and rose in response to the accumulation of glacier ice, which was ultimately sourced from oceanic evaporation (Fairbanks 1989, Lambeck *et al.* 2002). In mountainous areas, alpine glaciers advanced onto the piedmont lowlands and retreated back, in step with the cycles of the larger continental ice sheets. In the tropics and subtropics, the climatic cycles were often manifested as wet or dry periods, with the wet periods coinciding with cold, glacial periods in the high and midlatitudes. In many deserts, the change to a cooler, more humid climate is called a *pluvial*

Table 14.9 | Correlations between marine isotope stages and the major glacial advances or retreats

Marine isotope stage (V28–238)	Glacier characteristics	Midwestern United States	Rocky Mountain	Sierra Nevada	Alps	Northern Europe
1	Interglacial	Holocene	Holocene	Holocene	Holocene	Holocene
2–4	Glacial	Wisconsin	Pinedale	Tioga, Tenaya, and Tahoe	Würm	Weischel
5	Interglacial	Sangamon	(Varies)		Riss–Würm	Eem
6	Glacial	Illinoian	Bull Lake	Mono Basin	Riss	Saale
7–10[a]	Interglacial, but deteriorating climate	Post-Yarmouth	?	?	?	?
11[a]	Interglacial	Main Yarmouth	?	?	Mindel–Riss?	Holstein?
12+	Glacials and Interglacials	Pre-Illinoian (Kansan, Nebraskan, etc.)	Sacagawea Ridge	Sherwin	Mindel, Gunz, etc.	Elster, Menap, etc.

[a] After Sharp and Birman (1963), Richmond (1986), and Chadwick *et al.* (1997), with assistance from L. Follmer

period, because of the lakes that often formed in conjunction with it (Benson *et al.* 1990, McFadden *et al.* 1992, Adams and Wesnousky 1999).

The Deep Sea Record

Until about the 1950s, only four major glacial advances were recognized, on the basis of a suite of terraces in some meltwater valleys of the Alps and nested moraines found in Germany (Daly 1963, Cooke 1973). The assumption was that each glacial cycle filled the valleys with outwash, forming a high terrace, while during interglacial periods the terraces were incised. This cyclicity supposedly produced a suite of four terraces that reflected four major glaciations. However, by the mid-twentieth century, data from the seafloors had begun to emerge, revealing a detailed pattern of glacial cycles that was far more complex (Emiliani 1955, 1966, Shackleton 1968, 1977).

Some of the very best data on past climates and ice volumes have been obtained from the seafloors, where shells of one-celled plankton (mostly *foraminifera* (forams) but also coccoliths and diatoms) have accumulated in relatively undisturbed layers. These plankton secrete silica- and $CaCO_3$-rich shells, using sea water as the oxygen source. Most oxygen atoms are ^{16}O, but about 0.2% are a heavier isotope, ^{18}O. During glacial periods, the oceans became depleted of the lighter ^{16}O, because it is preferentially evaporated and tied up in ice sheets. As past oceans became enriched in ^{18}O during glacials, the isotopic composition of the foram shells reflected this change. During interglacials, foram shells became proportionately reenriched in ^{16}O. Therefore, preserved in the deep sea sediments – rich in forams – is a record of the isotopic composition of sea water, which is a direct proxy for the amount of water tied

up in ice sheets (Shackleton and Opdyke 1976). In other words, the oxygen isotope composition of the deep sea sediment is a proxy for global ice volume but does not indicate how this volume is distributed across the landmasses. It has become clear, through analysis of deep-sea and ice cap cores, that many more than four main glacial cycles have occurred during the Quaternary Period, and that the Quaternary climate record is complex (Shackleton and Opdyke 1973, 1976, Martinson *et al.* 1987, Bond *et al.* 1993, Grootes *et al.* 1993, Petit *et al.* 1999; Fig. 14.59).

Marine isotope stages (MIS) interpreted from these cores are numbered from top to bottom (Fig. 14.59). MIS 1 is the current interglacial (Holocene). The last major glaciation – the Wisconsinan in North America– occupies Stages 2–4, with a warm but short interstadial in the MIS 3 position. As is obvious, odd-numbered stages are assigned to interglacials or interstadials, while even-numbered stages correspond to glacials or stadials.

Because of the validity (when correlated to Milankovitch cycles) and global character of this record, MIS numbers have become the conventional way of referring to climatic intervals during the Quaternary. When working regionally, the advance or retreat associated with a particular oxygen isotope stage is often given a name unique to that region (Table 14.9). Data from other paleoclimatic indicators are often deliberately fit to the marine isotope record, as a way of placing these events in time and relating them to ice volume (and climate) fluctuations during the Quaternary (Bond and Lotti 1995, Heslop *et al.* 2000, Rasmussen *et al.* 2008, Giuliani *et al.* 2011, Oba and Irino 2012). Similarly, ice core data from the Greenland and Antarctic ice sheets have also shown the validity and correlation advantages of the deep sea data, while providing considerable additional insight into past climates (Dansgaard *et al.* 1993, Andersen *et al.* 2004, Narcisi *et al.* 2005, Lambert *et al.* 2008, Wolff 2008, Orombelli *et al.* 2010).

Scant evidence exists on land for stages prior to MIS 10, and their ages are primarily based on correlations of their paleomagnetism to the paleomagnetic timescale (Fig. 14.59). With few exceptions (Balco *et al.* 2005, Balco and Rovey 2010), little is known about the earliest dozen or more glacial advances. Details about the last few glacial advances, i.e., MIS 6 and later, derive primarily from the stratigraphic record contained within the glacial sediments and the loess deposits that indirectly resulted from them, as well as any intercalated paleosols. Thus, it is primarily a land-based record (Table 14.9). Our understanding of soil formation, burial, and erosion is, necessarily, placed within this climatic record. And now, armed with this background, let us go to Iowa to examine some of that terrestrial record.

Ruhe's Work in Iowa

This section focuses on work performed in one of the soil geomorphology study areas of the U.S. National Cooperative Soil Survey (NCSS) program – Iowa. Here, the work of Robert Ruhe (Fig. 14.61) and many others "ushered in an era of landscape evolution and soil formation research and established the importance of paleosol studies" (Olson 1989: 133–134).

Ruhe was a geologist who spent much of his career studying the stratigraphy and soil geomorphology of Iowa. A meticulous researcher and a tireless field man, his work refocused and energized many in soil geomorphology, soil stratigraphy, and paleopedology (Effland and Effland 1992). His two books, *Quaternary Landscapes in Iowa* (1969) and *Geomorphology* (1975b), are classics. Ruhe's work stressed the important linkages among soil development, landscape (slope) evolution, and Quaternary stratigraphy. Most importantly, his work quantified soil and paleosol development to enhance understanding of landscape evolution.

The Stratigraphy and Constructional Surfaces of Southern Iowa

Iowa's landform regions are largely delineated by their Quaternary stratigraphy (Fig. 14.62). At the end of the nineteenth century and near the turn of the twentieth century, glacial deposits in Iowa and nearby midwestern states were recognized and roughly correlated to the then-accepted fourfold glacial succession, introduced from Europe. Eventually, the glacial chronosequence for the midwestern United States became known as (from oldest to youngest) Nebraskan, Kansan, Illinoian, and Wisconsinan (Table 14.9). In northeastern Iowa, glacial sediment has, for the most part, been eroded, and topography there is controlled by bedrock. A late advance of the Wisconsinan glacier formed

the hummocky landscapes of the Des Moines lobe, in north-central Iowa. Southern Iowa was not glaciated during the Wisconsinan advance, and the uppermost sediments there are tills and loess deposited before the Illinoian glaciation, as well as late Wisconsinan loesses.

Early studies named two of these older tills, and their presumed ice advances, after the states of Nebraska and Kansas (Chamberlain 1895, Shimek 1909). Later, as more information was generated, multiple tills were identified, and they were not always correlated to the type Nebraskan and Kansan till sites. For example, several tills are now known to predate the type Nebraskan till (Hallberg *et al.* 1978a, Hallberg 1986). Thus, today the terms Nebraskan and Kansan are obsolete, and the entire series of older tills are lumped into an informal *Pre-Illinoian* category until better stratigraphic and dating information becomes available (Guccione 1983, Richmond and Fullerton 1986, Aber 1991, 1999, Rovey and Kean 1996). One or more strongly developed paleosols are often developed in these tills, marking the contact between them and helping to tease out their age and stratigraphic correlations. Indeed, most of Iowa is underlain by Pre-Illinoian tills, but only in the Southern Iowa Drift Plain do they constitute major soil parent materials (Fig. 14.62). Ruhe (1969) named this landscape the Kansan Drift region – for the surficial glacial sediment then recognized. Some research indicates that the youngest of the Pre-Illinoian (formerly Kansan) tills may be 780–620 ka old (Colgan 1999).

Above the youngest till and its paleosols in southern Iowa are various loess deposits of Late Pleistocene age (Figs.

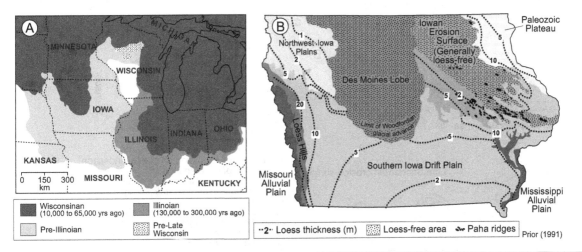

Fig. 14.62 Background information about Iowa's geomorphology. (A) Limits of the Late Pleistocene glaciations in the upper Midwest. (B) Landform regions of Iowa, showing the thickness and distribution of loess.

14.62, 14.63). The lower unit – Loveland Loess – is Illinoian (MIS 6) in age (Leighton and Willman 1950, Colgan 1999, Forman et al. 1992). Both the Mississippi and the Missouri River valleys were sources for this loess (Shimek 1909, Ruhe 1956b, Follmer 1982), which is >7 m thick near the source areas but thins to <2 m inland (Ruhe 1956b). In south-central Iowa, therefore, far from either of these rivers, the Loveland Loess is thin and difficult to recognize. Much of it has been incorporated into the underlying paleosols, developed in the till below (Woida and Thompson 1993). During the Wisconsinan (mainly MIS 2) glaciation, additional loess deposits – derived from the same source regions – were deposited across the entire Southern Iowa Drift Plain, burying the soils developed in the older loess and till (Fig. 14.62).

In southern Iowa, where landscape dissection has been deep enough to expose the Pre-Illinoian tills, buried paleosols are often observed at the top of buried till units.[1] The oldest of these soils could be as old as Early Pleistocene (Rovey and Kean 1996, Balco et al. 2005). The paleosol that was buried by the "type Kansan" drift[2] is well developed and highly transformed. It was initially named the *Afton paleosol*, identifying

an interglacial period between what was then called the Nebraskan and the Kansan glaciations (Table 14.9). Early researchers thought that the high clay contents of this (and similar) paleosols were due to long-term weathering and coined the term *gumbotil* for them (Kay 1916, Kay and Pierce 1920, Kay and Apfel 1929). They did not initially recognize it as a buried soil but thought it was simply a strongly weathered zone in the till, possibly with a soil at its top (Alden and Leighton 1917). Kay (1916) defined gumbotil (or gumbo) as a gray to dark-colored thoroughly leached, nonlaminated, reduced clay, having very sticky consistence, but with very hard consistence when dry. Simonson (1941) and Scholtes et al. (1951) later argued that gumbotil was the B horizon of a buried soil. Pedostratigraphic work by Ruhe (1956, 1969, Ruhe et al. 1967) identified similar thick, clayey paleosols on the uppermost Pre-Illinoian (Kansan) surface in Iowa. These studies and those of Frye in Illinois (Frye et al. 1960a, b), showed that gumbotil was found typically in swales. Thus, gumbotil was actually the lower part of a paleocatena, and the large amounts of clay in it had originated as a result of long-term mineral weathering and slopewash, accumulating to great thicknesses in paleodepressions. Subsequent research (Woida and Thompson 1993, Rovey 1997) suggested that these paleosols are typically gleyed throughout the paleocatena and that their unusual thickness is due to a more complex process than accumulation of slopewash alone. These clay-rich soils were later termed *accretion gleys* (Frye et al. 1960a, b, Willman 1979). The gray, clayey paleosols on older Pre-Illinoian buried surfaces (e.g., Nebraskan) must have formed during similarly long periods of subaerial exposure. The long exposure necessary for their formation was available because the ice did not advance back over the landscape until much later.

In southern Iowa, Afton paleosols are buried by younger Pre-Illinoian drift (Kansan), and exposures are limited

---

[1] We encourage the use of current terminology, but for the sake of discussion in this text, we will refer to some older terms as well to help the reader better understand the original literature.

[2] The stratigraphic story of this region is, in actuality, much more complex than is explained here, as determined from recent advances in tephrostratigraphy, magnetostratigraphy, cosmogenic dating, and till stratigraphy (Hallberg 1986, Balco and Rovey 2010). For example, several distinct till sheets clearly occur in southern Iowa and northern Missouri. However, the point of this discussion is not to elucidate the precise stratigraphic column as we know it today, for that is certain to change as more knowledge emerges. Rather, we focus on the logic behind the early discoveries, the methods that led to them, and how these lessons exemplify solid soil geomorphology research.

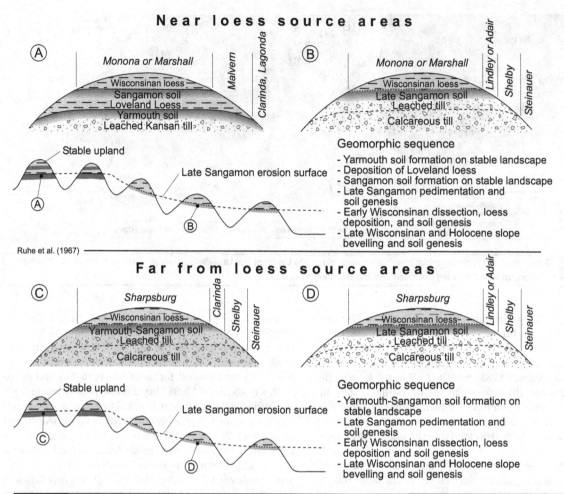

Fig. 14.63 Major types of stratigraphic sequences, geomorphic history, and associated surface soils in southwestern Iowa. (A) Sites near the Missouri and Mississippi Rivers that have deposits of Loveland (Illinoian) *and* Wisconsinan (Peoria) loess. (B) Similar sites, which only have deposits of Wisconsinan loess, being located on the Late Sangamon erosion surface. (C) Sites far from the Missouri and Mississippi Rivers, i.e., south-central Iowa, having only a cover of Wisconsinan loess. (D) Similar sites, located on the Late Sangamon erosion surface.

to some upper hillslope positions. However, gumbotillike paleosols are *also* present on top of the low-relief surface of southern Iowa (Woida and Thompson 1993). These soils are referred to as Yarmouth-Sangamon paleosols, for they formed more or less continuously from the Yarmouthian interglaciation (between the last Pre-Illinoian glaciation – Kansan – and the Illinoian glaciation), through the time of Illinoian glaciation (which did not cover this part of Iowa), and continued to form through the Sangamonian interglacial period (MIS 5; Table 14.9) and much of the Wisconsinan glaciation prior to deposition of Late Wisconsinan loess. They represent some of the thickest paleosols in North America, strongly transformed from the original parent material and polygenetic. During the latter part of the Illinoian glaciation, these soils were covered by Loveland

Loess, which was thickest near the river bluffs. However, this loess was not thick enough in most places to bury the soil formed in till below completely. Thus, across much of the landscape a polygenetic, *welded paleosol* formed (see Chapter 16). Soil welding is the process whereby a surface soil develops downward and pedogenically merges with the solum of a buried soil (Ruhe and Olson 1980, Schaetzl and Sorenson 1987, Tremocoldi *et al.* 1994). Thus, in south-central Iowa, Ruhe and his colleagues described a thick, highly weathered, welded soil that had developed in the Pre-Illinoian till and then was thinly covered by Loveland Loess during the Sangamonian interglacial period, i.e., the Yarmouth-Sangamon paleosol (Woida and Thompson 1993; Fig. 14.63C). In thicker loess areas near the Missouri River, the Yarmouth paleosol had not welded with the surface soil

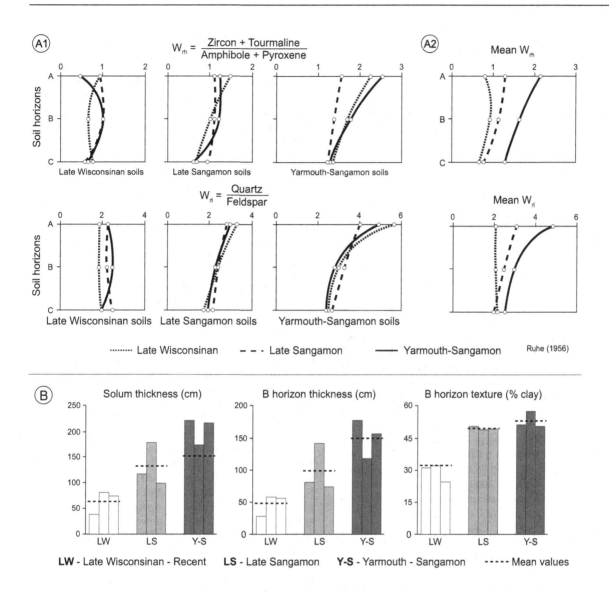

Fig. 14.64 Comparative data on soil development for various surfaces in southern Iowa. (A) Depth plots of weathering ratios for soils formed in Kansan till in southern Iowa. The soils have formed on surfaces of three different ages: Yarmouth-Sangamon, Late Sangamon, and Late Wisconsin. (1) Ratios of heavy and light minerals to weathering-resistant minerals, with depth, for three soils on each surface. (2) Mean values of the same data, summarized for each soil type. (B) Thicknesses of sola and B horizons, and comparative data on B horizon clay contents.

in the Loveland Loess because the loess was too thick, and, thus, two distinct paleosols can be seen (Fig. 14.63A). The lower Yarmouth paleosol had developed in till, while the Sangamon paleosol above had developed in Loveland Loess. Below both of these paleosols was the Afton paleosol, which formed in an older till yet. This example shows the importance that Ruhe placed on soil stratigraphy in unraveling the Quaternary history of Iowa. In this mainly constructional area of southern Iowa, this approach works well.

Both the Yarmouth-Sangamon and the Sangamon paleosols in southern Iowa are strongly developed. This

development is best indicated by weathering data and depth functions of, particularly, clay content (Fig. 14.64). In swales, the Sangamon soil is gray – usually an accretion gley. On flat uplands it is often an in situ gleyed profile. Figure 14.64 illustrates that the Yarmouth-Sangamon soil is even more weathered than the Sangamon soil – as expected, because of its longer period of subaerial exposure. This example shows how weathering and stratigraphic data can provide insight into landscape evolution, the length of soil-forming intervals, and the ages of geomorphic surfaces. One last note: The Sangamon paleosol should not be confused with

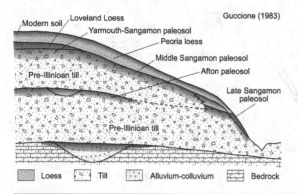

Guccione (1983)

Fig. 14.65 Idealized Quaternary stratigraphy of north-central Missouri and south-central Iowa.

one we will soon discuss – the Late Sangamon soil, or its erosion surface.

And if this story is not enough, there is more. Subsequent depositional events in southern Iowa involved loess that was deposited in association with the melting Wisconsin (MIS 2) glacier. Ruhe (1956) placed the age of the base of this loess at 16.5–29 ka. The wide age range illustrates that the surface was covered by loess in a time transgressive manner (see Chapter 10). The age of the base of the loess decreases as one moves farther from the source areas, i.e., the rivers. An early Wisconsin (MIS 4) loess, known as the Roxana silt (in Illinois and Indiana), the Pisgah Formation (in Iowa), or the Gilman Canyon Formation (in Nebraska and Kansas), was deposited first (Johnson and Follmer 1989, Leigh and Knox 1993, Leigh 1994, Grimley 2000). The Pisgah Formation is relatively thin and was sometimes incorporated into the Yarmouth-Sangamon paleosol as it aggraded upward via developmental upbuilding. Commonly, a paleosol called the Farmdale Soil formed in the Pisgah Formation. Finally, the late Wisconsinan (Peoria) Loess was deposited rapidly enough and eventually became thick enough to bury all of the preexisting surfaces; of course it, too, was thickest near the main river source areas (Fig. 14.62).

The Pisgah Formation and Peoria Loess were deposited on a low-relief, swell-and-swale surface with thick, highly transformed soils (Ruhe 1956b). All these surfaces contained either a Yarmouth and a Sangamon soil, or a polygenetic Yarmouth-Sangamon soil, depending on whether Loveland Loess deposits were thick enough to separate the two sola (Fig. 14.63). Perhaps it is easy to see why Ruhe's use of Quaternary stratigraphy, coupled with paleopedology, was a highly useful approach to unraveling the Quaternary history of Iowa.

The Erosional Surfaces and Landscape Evolution of Southern Iowa

Southern Iowa is not just a simple, layer-cake constructional landscape. Erosional events also come into play. Eventually, Ruhe ascertained the chronology of these various constructional and erosional events, some of which were beyond the range of radiocarbon dating. These events helped to shape the southern Iowa landscape. In fact, among Ruhe's major contributions were the identification and genetic interpretation of stepped/beveled erosion surfaces on this landscape. Through careful examination of cores, roadcuts, and railroad cuts, he identified what he called the *Late Sangamon erosion surface*. It became clear to Ruhe that the geomorphic history of the swell-and-swale Sangamon surface – the one formed in Loveland Loess or Pre-Illinoian till – was not simple. Many slopes/catenas were not long and continuous from uplands to lowlands. Rather, they had discrete steps, risers, and slope breaks that represented contacts between different geomorphic surfaces, some of which were erosional, but others that were constructional (Figs. 14.65, 14.66, 14.67). These stepped surfaces had a history to tell and merited further study. As mentioned earlier in this chapter, breaks in slope often indicate a change in geomorphic history, and that relationship is nowhere better shown than in southern Iowa.

Soil development on the various geomorphic surfaces, particularly the classic study of the Turkey Creek watershed (Ruhe *et al.* 1967; Fig. 14.66), filled in many important details about landscape evolution in southern Iowa. The flat uplands at the interfluves of this watershed are remnants of the Pre-Illinoian (Kansan) drift plain and have Wisconsinan aged (Peoria) loess over a Yarmouth-Sangamon paleosol with accretion gleys in the swales (Fig. 14.63C). This old, constructional surface has a relief of about 2–3 m, between swales that are about 200 m apart (Ruhe 1956b). Near the major rivers, where the Loveland Loess is thicker, a separate Sangamon soil has developed within the stratigraphic column (Fig. 14.63A). However, Turkey Creek is far enough from the Missouri River that Loveland Loess is not recognizable (as a C horizon) between the Peoria Loess and the Yarmouth-Sangamon paleosol (Fig. 14.66). The Peoria Loess here *is* thick enough, however, that the modern soil is not welded to the Yarmouth-Sangamon paleosol below. In other words, the stratigraphy at Turkey Creek is like that shown in Fig. 14.63C.

Prior to the MIS 2 (Wisconsin) glaciation, the Yarmouth-Sangamon surface began experiencing erosion. An erosion surface formed and, in places, completely removed the Yarmouth-Sangamon paleosol (Fig. 14.67). Ruhe surmised that the age of this surface was Late Sangamon because the soils on it were not as strongly developed as the Sangamon soils developed in Loveland Loess or Illinoian till in nearby areas. That is, he used relative soil development data to guide his interpretations of surface age (Fig. 14.64). Why this erosion episode occurred is unknown, but it probably had some relation to the rapid landscape and climate changes that occurred during MIS 5 and 4. That such an erosion event *did* occur actually seems more plausible envisioning long periods of stability. Nonetheless, evidence suggests that the Yarmouth-Sangamon soil on the stable

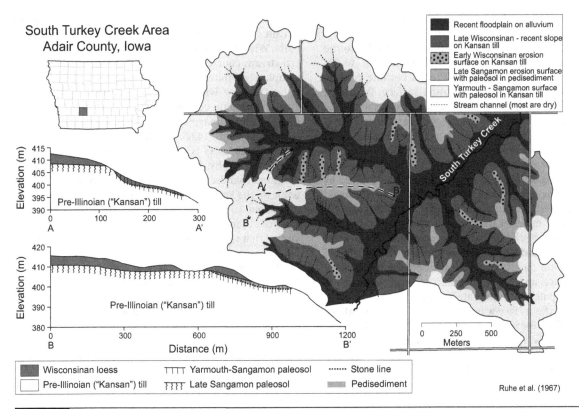

Fig. 14.66 Geomorphic surfaces in the South Turkey Creek watershed in southwestern Iowa.

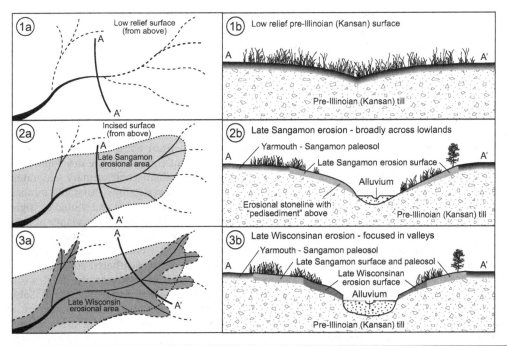

Fig. 14.67 Diagrammatic representation of the erosional surfaces in southern Iowa, in both cross section and plan view, showing their evolution through time.

uplands of southern Iowa withstood episodic periods of erosion (Woida and Thompson 1993; Fig. 14.65).

The Late Sangamon erosion surface is incised into the latest Pre-Illinoian (Kansan) till and older tills (Fig. 14.67). Ruhe believed that stones from the till became concentrated as a lag deposit on the erosion surface, forming a prominent *stone line* that often mantles this surface (Ruhe 1956b). Along with, and generally above, the stones is coarse-textured sediment that had presumably been transported downslope as slopewash and colluvium. Ruhe termed the process that produced these surfaces *pedimentation* – a type of backwasting in which slope gradients were lowered by erosion of the upper slope elements and deposition on lower slope elements. Ruhe (1956) also referred to the sandy/gravelly/stony material that mantles the erosion surfaces as *pedisediment*. Pedisediment is a type of colluvium. After the erosional period had ended, a Late Sangamon paleosol developed into and through the pedisediment, the stone line at its base, and underlying till. The degree of development of the paleosol on the Late Sangamon erosion surface implies that the surface was cut and then stabilized, allowing the soil to form. Soil formation proceeded for some time before it was buried by Wisconsinan loess, about 14 ka ago. Estimates indicate that the Late Sangamon surface may have been stable and soils may have been developing in it for as long as 40,000 years.

Soil and slope data provide evidence for yet another erosion event – this one occurring during the Early Wisconsinan glacial period (Figs. 14.66, 14.67). This erosion surface, unlike the Late Sangamon surface, has no paleosol developed on it, suggesting that it was continually undergoing erosion until it became buried by loess. This surface, called the Early Wisconsinan erosion surface (Fig. 14.66), also contains a stone line and pedisediment. The Early Wisconsinan erosion event was, geomorphically, different from that of the Late Sangamon erosional period. The Late Sangamon erosion surface is relatively gently sloping and widespread, suggesting that it may have been driven mostly by slope processes such as solifluction, creep, and other mass movements. It is not confined to areas near stream valleys (Figs. 14.66, 14.67). The Early Wisconsin erosion event, however, was quite different. Incision is deeper and more confined to stream valleys, suggesting that it was driven more by fluvial processes. Later, during the Holocene, streams widened their valley bottoms by lateral cutting, developing floodplains that are in some places erosional and in others filled with Holocene alluvium. More recently, agriculturally driven erosion on uplands has caused valleys to fill with sediment.

Early Wisconsinan incision in south-central Iowa was most pronounced along stream courses and less noticeable along nose slopes, leading to the development of two types of landscape profiles (Fig. 14.67). Some landscapes have all three surfaces: the low-relief Yarmouth-Sangamon (or Sangamon) upland surface, the Late Sangamon erosion surface, and the Early Wisconsinan erosion surface. This tripartite suite of landscapes is found between modern stream courses, on nose slopes that escaped some of the Early Wisconsinan erosion (Fig. 14.66). Farther up modern stream courses, the Early Wisconsinan erosion event completely removed the Late Sangamon surface. Here, the upland surface with Yarmouth-Sangamon soils was beveled and directly connected to the Early Wisconsin pediment as the drainage system deepened and extended into the uplands (Figs. 14.66, 14.67).

Loess thickness is an important part of the southern Iowa stratigraphy and evolution. On the flat Yarmouth-Sangamon landscape, Peoria Loess is about 4.5 m thick, while on the Early Wisconsinan erosion surface the loess is only about 2.3 m thick. The Pisgah Formation and Farmdale soil are absent. Using this information, Ruhe *et al.* (1967) argued that the erosion surface was cut during the early to middle period of Wisconsinan loess deposition.

We delve into such detail about the Iowa story because it also shows how stratigraphy and paleopedology can assist soil mappers. Ruhe's work helped soil mappers make sense of the landscape and enabled them to predict the soil-landscape relationships better by giving them a clear conceptual model to use in the field (Fig. 14.63). It also stimulated more detailed and exhaustive work on stratigraphy and paleopedology in the Midwest.

Perhaps the most academically important contributions of Ruhe's work in southern Iowa stemmed from his 1956 paper, in which he compared soil development in three soils, all developed in Kansan till. Yarmouth-Sangamon and Late Sangamon soils, also formed in till, are widespread and easily located. The latest cycle of erosion exposed some till areas for soil development as well, by stripping off the loess cover; these areas stabilized much later, and perhaps no earlier than 6.8 ka ago (Ruhe 1956b). Ruhe estimated that the Late Sangamon soils were surficially exposed for at least 13,000 years, although a longer period of formation seems likely. And of course, the Yarmouth-Sangamon soils could have been forming for hundreds of thousands of years. Because the chemistry of buried soils is so easily altered by solutions percolating into them from above, and vice versa (see Chapter 16), one way of comparing the development of these three soils is to evaluate their primary mineralogy. In principle, the ratio of resistant/weatherable minerals should increase as the soil develops and should be higher in near-surface horizons. Data on the abundance of heavy and light minerals of varying resistance to weathering exhibited these predicted trends – with age and with depth (Fig. 14.64). The most weathered soils are on the Yarmouth-Sangamon surface, whereas the least weathered soils are Late Wisconsinan in age (Fig. 14.64). Solum and B horizon thickness, as well as B horizon texture, all change consistently along this development sequence. In this regard, Ruhe's work paved the way for many other studies of relative soil development and quantification of soil characteristics (see Chapter 15).

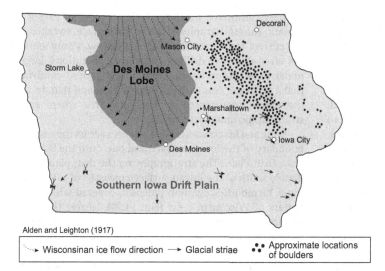

Fig. 14.68 Distribution of large granite boulders in Iowa. Inset photo: large boulders in a field on the Iowan erosion surface, Chickasaw County. Photo by D. Johnson.

The Iowan Erosion Surface

Historically, one of the most controversial soil geomorphic problems in the midwestern United States has been the Iowan erosion surface (Olson 1989; Fig. 14.62). Once thought to be a landscape formed during a separate ice advance, i.e., the Iowan, it is now known to have formed by widespread long-term erosion. Most of this low relief surface is covered with a stone line and pedisediment. Many of the stones are large and concentrated in farm fields (Fig. 14.68); this is stone country! The Iowan surface is a regional window through which we can view the great reduction of topography that was imposed during the early and mid-Wisconsinan periods and via which we can learn some excellent lessons about soil geomorphic research.

The Iowan Erosion Surface covers almost 23,000 km² in northeastern Iowa (Fig. 14.62). On its northwestern edge it abuts and passes underneath Wisconsinan aged deposits of the Des Moines Lobe, suggesting that erosion on the adjoining Iowan surface began well before ~14 ka, i.e., the age of Des Moines Lobe sediment. The low-relief Iowan erosion surface lacks many of the short slopes and kettles typical of recent, constructional landscapes like those formed by the Des Moines Lobe, or the steeper slopes of the fluvially incised Southern Iowa Drift Plain (Fig. 14.69). However, the integrated drainage networks on the Iowan surface point to an erosional origin for this landscape. The erosion event cut well into the Pre-Illinoian (Kansan) drift, exposing it at the surface. Nonetheless, scattered remnants of the Pre-Illinoian paleosurface do exist in landforms called *paha* (described further later), where the entire preerosional stratigraphy is preserved (Alden and Leighton 1917). Many of the sediments eroded from the uplands of the Iowan surface were deposited in alluvial lowlands; stones and

Fig. 14.69 Landscape views of (A) the Iowan surface in Black Hawk County, Iowa, and (B) the dissected Kansan (Pre-Illinoian) drift plain in Jones County, Iowa. From plate V of Alden and Leighton (1917). For comparison (C), a modern photo of the hummocky Des Moines Lobe landscape is also provided. Photo by B. Miller.

large boulders, left behind, dot the uplands (Fig. 14.68). The Iowan surface contains comparatively thin loess deposits, presumably because any loess that would have been deposited here would have been quickly eroded.

Erosion of the soils and uppermost sediment from the Iowan surface was probably a Late Wisconsin event, circa 29–18 ka ago (Ruhe 1969). Presumably, at this time, cold, windy periglacial conditions existed here, giving rise to slope instability and erosion. Ice wedge casts and polygons, unequivocal indicators of a periglacial climate and permafrost, are common across the surface. Faunal remains buried in lowland sediments indicate a cold, almost tundralike climate, but slightly wetter than today (Prior 1991). Walters (1994) placed the timing of periglacial conditions here at ~21–16.5 ka ago (Mason and Knox 1997). Such a climate would have facilitated extensive freeze-thaw action, widespread mass movements, and slopewash. Existing soils would have been stripped from the frozen landscape, with the exception of paleosols buried beneath relict hills. Stone lines, ubiquitous across the Iowan surface and associated with the erosion event, are slightly disrupted and slump down when they cross the ice wedge casts, implying that the periglacial conditions coincided with slope degradational processes (Walters 1994). During the erosional period that formed the Iowan surface, soils on the Southern Iowa Drift Plain were only slightly affected

The net effect of this erosion episode was to decrease the relief, but to preserve the moderately integrated nature of the preexisting drainage network (Fig. 14.69). Because of this landscape morphology – so different from that in other parts of Iowa – many early researchers saw the Iowan landscape as one that was clearly *older* than the kettled Des Moines lobe landscape, with its deranged drainage pattern, but one that was younger than the high-relief, dissected drift plains of southern Iowa. The Iowan surface was, therefore, thought to have been formed by a glaciation that deposited a thin drift onto a previously deeply dissected Pre-Illinoian surface. On the basis of geomorphology alone, these were reasonable first assumptions. However, soon it became apparent that a number of topographic/geomorphic features associated with the Iowan surface suggested that it was *not* a separate drift sheet: (1) It lacked an end moraine, (2) it extended farther south on interstream divides than in valleys, and (3) the landscape was topographically lower than the Pre-Illinoian drift plain (Hallberg *et al.* 1978b).

The events that led to the current interpretation of the genesis of the Iowan erosion surface illustrate the value of Ruhe's pedostratigraphic approach to landscape evolution. Iowan drift was thought to have been deposited between the Illinoian and Wisconsinan glacial stages (McGee 1891, Alden and Leighton 1917, Kay and Apfel 1929, Leverett 1939, Kay and Graham 1941), or perhaps during the early Wisconsinan (Leighton 1933, Kay and Graham 1941, Ruhe *et al.* 1957, Ruhe and Scholtes 1959). But because this area has such low relief, few exposures could be found at which

to study subsurface stratigraphy (Hallberg *et al.* 1978b). As a result, the stratigraphy of the Iowan surface, so critical to the correct interpretation of its genesis, was only worked out after extensive drilling and coring operations, largely under Ruhe's supervision. These investigations confirmed that a separate "Iowan drift" did not exist and that, in fact, the region represented an erosional area where entire stratigraphic units were *missing*.

Ruhe and his coworkers drilled cores across the southern boundary of the Iowan surface and out, onto the Southern Iowa Drift Plain. The stratigraphy on the drift plain immediately south of the Iowan surface consisted of till containing a Yarmouth-Sangamon paleosol, covered with several meters of Wisconsin loess (Fig. 14.70). Nearer the Iowan surface, drill core data showed increasing amounts of truncation of this paleosol, and thinning loess overlying it. Eventually, on the Iowan surface proper, the paleosol is missing, and loess overlies either leached or calcareous – but pedogenically unaltered – till. Because the Pre-Illinoian till can be traced to the erosion surface, there is no Iowan drift per se on the Iowan surface. Instead, Wisconsinan age loess lies directly above eroded Pre-Illinoian till. Ruhe's keen stratigraphic work had shown that the Iowan surface was cut into "Kansan" till (as then conceived), below the level of the Yarmouth-Sangamon paleosol (Ruhe 1969). A prominent stone line occurs, stratigraphically, between the loess and the underlying till, marking the contact and clearly indicating that an erosion event was the reason for the missing stratigraphy. The sandy zone in the thick Wisconsinan loess (Fig. 14.70) probably formed as sand blew off the eroding Iowan surface, onto the adjoining, higher and more stable, landscapes. Ventifacts – wind abraded rocks – on the Iowan erosion surface support this assumption (Walters 1994). After erosion ceased, sand deposition also stopped; however, loess deposition continued on the area, burying the sandy zone. Thus, the erosion interval is neatly placed between 29 ka (based on a ^{14}C date on organics in the Farmdale paleosol, which would later be buried by loess) and 18.3 ka ago (based on a ^{14}C date on organic matter from the base of the loess that immediately overlies the erosion surface). The earliest loess increment (below the sand zone) equates with the hiatus represented by the stone line in the till. The last increment of loess covered much of the area. Thus, Wisconsinan age loess deposition started before the erosion event and ended after it.

A key geomorphic/stratigraphic assemblage used to show the origin of the Iowan surface occurs at its many *paha*. Paha are loess-capped hills – erosional relicts that preserve the underlying stratigraphy (Fig. 14.71). They are essentially outliers of the Southern Iowa Drift Plain, preserved on the Iowan surface. Paha retain the full complement of loess, in contrast to the erosion surface, which does not (Ruhe 1969). The discovery and correct interpretation of the buried soils and stratigraphy allowed Ruhe to interpret the soil geomorphic history of the Iowan erosion surface correctly.

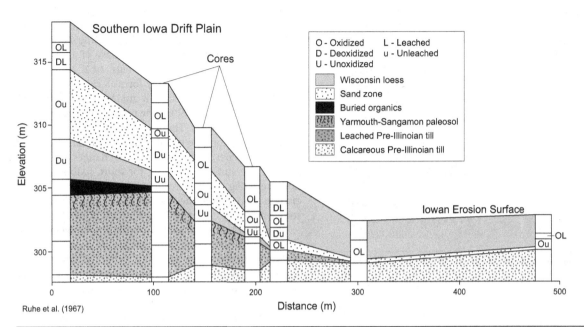

Ruhe et al. (1967)

Fig. 14.70 Stratigraphic data taken from a transect of drill cores across the Southern Iowa Drift Plain, onto the southern edge of the Iowan erosion surface.

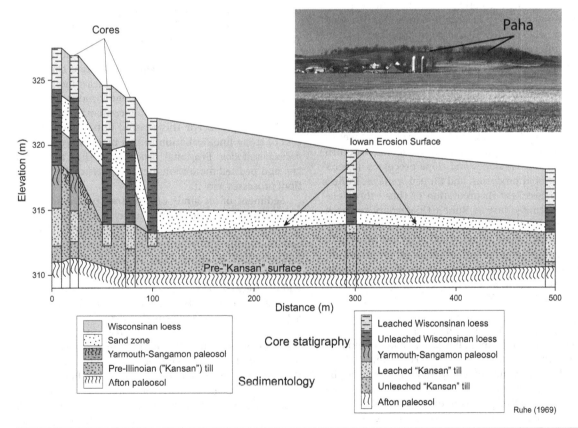

Fig. 14.71 The importance of paha and their stratigraphy to uncovering the geomorphic history of the Iowan erosion surface. (A) Stratigraphic data taken from a transect of drill cores from the Iowan erosion surface onto the 4-Mile Creek Paha. (B) Paha near Mount Vernon, Iowa. Photo by T. Kemmis.

Fig. 14.72 *Chuquiraga avellanedae* plant surrounded by an erosional desert pavement, in a desert in Patagonia. Photo by C. M. Rostagno.

Ruhe and his associates made good use of radiocarbon dating and stratigraphy in their work, and they had access to coring equipment that previous researchers did not. They worked along newly constructed railroad rights-of-way as well as during the period when interstate expressways were being constructed. Both opportunities provided very large, fresh exposures that allowed them to see laterally continuous sequences of sediments and paleosols very clearly. Data derived from these methods and exposures opened up key insights into landscape evolution in the Midwest. Ruhe's work emphasized the importance of slope stability to soil formation and showed that by determining whether a geomorphic surface is erosional or depositional the evolution of entire landscapes can be unraveled. In the end, soil geomorphic work always seems to return to three elements: (1) slopes and slope processes, (2) erosional/depositional processes, and (3) pedogenesis. Ruhe's work was an excellent incorporation of those three elements. Interestingly, just as Ruhe's Quaternary research proved to be a boon for NRCS soil mappers in Iowa, today we have come full circle – county soil maps are proving to be excellent surrogates for surficial geologic maps and are becoming invaluable data sources for Quaternary mapping endeavors (Schaetzl *et al.* 2000, 2013, Millar 2004, Miller *et al.* 2008, Quade *et al.* 1998).

Stone Lines and Erosion Surfaces

As surfaces erode, certain indicators remain behind as clues to the types of processes involved. For example, Ruhe (1969) used the presence of sandy deposits produced by downslope transport under the influence of gravity and water, which he called pedisediment, and stone lines to interpret erosional events on the Iowan surface. The correct interpretation of stratigraphic and sedimentologic information is vital to understanding landscape evolution.

Landscape-scale erosional events here are revealed by the consistent presence of broad, slightly concave erosional surfaces, usually *cut into* the base of a larger slope (Ruhe 1958, Mammerickx 1964; Figs. 14.2, 14.5). Erosion of a slope into another usually forms a riser (or a break in slope) at its upper end, where it grades into a higher, older surface (Figs. 14.4A, 14.67). Ruhe referred to these erosional surfaces as pediments. These erosional surfaces have a linear or slightly concave profile, because rapidly descending runoff loses energy where the slope changes at the lower parts of the surface, allowing for more sediment deposition there (Fig. 14.35). Profile concavity often supports an erosional origin for the surface, and a lag origin for any stones that may exist within or above it (Ruhe 1958).

Rocks and stones, many of which have been slightly rounded during erosion, are common to erosion surfaces. A layer with this type of origin, where coarse materials and rocks have been left behind on the surface after erosion, is called a *lag concentrate* (Wallace and Handy 1961, Rostagno and Degorgue 2011; Fig. 13.80). These deposits occur on both contemporary and paleoerosion surfaces and are often composed of flat or tabular stones lying on the surface (Fig. 14.72; see Chapter 13). When viewed in cross section, i.e., within a soil profile, such an erosional lag, or even a buried layer of stones, appears as a *stone line* or *stone zone*. Similar terms are *carpetolith, carpedolite,* and *carpedolith,* literally translated as "carpet of stones" (Parizek and Woodruff 1957, Aleva 1983).

In this model of formation, the stone line forms at the same time as the erosion events that create the erosion surface, and thus the stones overlie, at least momentarily, a truncated soil profile. Any sediment found in association with the stone line is presumably contemporaneous with the erosion event. For this reason, Ruhe called the process of stone line formation in this manner *pedimentation-pedisedimentation.* Erosional processes *produce* the surface lag, and pedisedimentation (perhaps assisted by bioturbation) processes *bury* it.

Sediment of all kinds may be transported across erosion surfaces, making them as much a slope of transportation as a slope of removal (Ruhe 1969, 1975b; Conacher and Dalrymple 1977). Ruhe's term *pedisediment* refers to sediment that is, episodically, in transport across an erosion surface; similar terms include *coalluvium, pedoluvium,* and the German *Decklehm.* Some of this material is usually present on some parts of an erosion surface (Ruhe and Scholtes 1956, Ruhe and Daniels 1958).

When exposed in a soil profile (Fig. 14.73), stone lines formed (and buried) in this manner mark a clear discontinuity, even if there is no mineralogical difference between the material above and below the stone line. Erosional stone lines provide evidence of a dramatic change in surficial processes in the geologic past. They are the morphological expression of an erosion pavement, cut by running water and subsequently covered with surficial sediment (Ruhe 1975b: 130). For example, Lichte and Behling (1999)

Fig. 14.73 Subsurface stone lines. Photos by Don and Diana Johnson. (A) Near Big Sur, California. (B) Near Iron Gate Reservoir, Siskiyou County, California. (C) Near Dubbo, New South Wales, Australia. (D) Along county Highway 9, near Cripple Creek, Colorado.

concluded that the stone lines in southeastern Brazil have developed on an ancient soil surface that was subsequently covered by eolian sediment. Quartz pebbles in the stone line were derived from quartz veins in the Precambrian crystalline bedrock. They were then redistributed along the slopes by heavy rain and subsequently covered by fine eolian sediment. Stratified deposits above stone lines argue strongly for burial of the stone line by slopewash or colluvium. This type of stone line represents not only a lithologic discontinuity and a change/break in sedimentation, but also a temporal discontinuity within the stratigraphic section (Parizek and Woodruff 1957). Stone lines of this sort are often interpreted to mark many erosion surfaces cut into Quaternary drift in the midwestern United States (Ruhe 1969).

Most mid-twentieth century studies ascribe stone line formation – especially in the tropics – to erosional processes, often followed by sedimentation and burial (Parizek and Woodruff 1957, Fairbridge and Finkl 1984). Even the Soil Survey Manual (Soil Survey Staff 1951) presupposes that stone lines have an erosional genesis. The historical reason for this bias is worth exploring (Johnson 2002). Curtis Marbut's successor at the U.S. Department of Agriculture Division of the NCSS, Charles Kellogg, interpreted subsurface stone lines he saw in the Belgian Congo as erosional lags that had later become buried (Johnson and Balek 1991). Kellogg later arranged for a detailed study of this area by Robert Ruhe, whose findings confirmed Kellogg's erosional interpretation of stone line genesis, despite his knowledge of other work in the region that described stone lines formed by processes of bioturbation and creep (Ruhe and Cady 1954, Ruhe 1958). The genetic interpretations of stone lines by prominent scientists like Kellogg and Ruhe have had far-reaching and long-lasting consequences. Many stone lines, even today and especially in tropical regions (Iriondo and Kröhling 1997, Beinroth et al. 2011), have since been interpreted as erosional features (Johnson 2000).

To be sure, the popularity of the erosion-lag model of stone line formation is largely due to its main proponent, Robert Ruhe, whose strong stance on the origin of stone lines in Africa was then carried to Iowa. Ruhe's academic stature and strong personality inhibited those who felt that stone lies could form in other ways, e.g., by bioturbation. His position on stone line formation has persisted for decades, despite (1) Darwin's (1881) work that had documented how soil biota could lower stones and human artifacts by bioturbation (Johnson 2002, Van Nest 2002) and (2) Ruhe's knowledge that nonerosional landscapes of the Des Moines Lobe (in Iowa) also had weak stone lines (Wallace and Handy 1961). Thus, the erosional-lag theory of stone line formation became the preferred view, while another, proposed decades earlier by Darwin (1881), languished.

Other Theories of Stone Line Development

Despite early research on stone lines that had interpreted them to be erosional relicts, other models of origin also exist, and the debate continues even today (Kröhling and Iriondo 2010, Morrás et al. 2010). Stone lines in soils were first defined by Sharpe (1938) as lines of angular to subangular fragments that parallel the surface (Fig. 14.73). Besides the erosional lag model, three other mechanisms can lead to the formation of subsurface stone lines: (1) detachment of weathering-resistant clasts (usually quartz) from dikes in weathered igneous rocks, along with slope processes to move them downslope in the subsurface; (2) lowering and concentration of stones by bioturbation; and (3) in situ chemical formation as (usually Fe-rich) concretions and nodules, directly in the subsurface (Bremer and Späth 1989, Johnson and Balek 1991, Ségalen 1994).

Often, the first step in determining the origin of the stones in the stone line is to ascertain whether the stones are autochthonous (formed in place) or allochthonous (transported in) (Morrás et al. 2009). The latter characteristic would suggest that they have formed – at least partially – via transport and pedimentation. In the tropics, for example, stone lines are often composed of Fe-rich nodules and concretions and have, therefore, likely formed in situ (Müller et al. 1981).

Perhaps the earliest theory of subsurface stone line genesis (not associated with erosion) is that the stones have become detached from dikes or other resistant layers and have been drawn along at the base of a creeping soil layer (Sharpe 1938). In this model, stones work their way downward because of the more rapid movement of the surface layers, or they become concentrated at depth because of their more rapid disintegration (weathering) near the soil surface. Sharpe was essentially describing a situation like the texture-contrast model of Paton et al. (1995) described

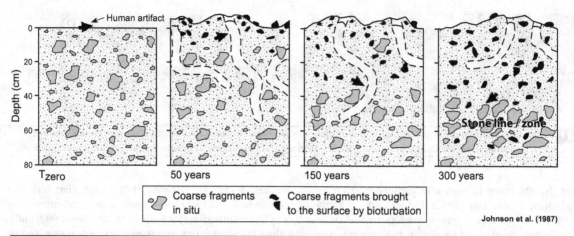

Johnson et al. (1987)

Fig. 14.74 Diagram showing how a stone zone can be formed by intense burrowing by animals – in this case, mammals. Any coarse fragments in the soil that are too large for the fauna to move to the surface settle to the depth of burrowing, forming a stone zone or stone line at that depth, and a biomantle above.

earlier in this chapter. However, the Paton *et al.* model relies more heavily on bioturbation for lowering the stones, ultimately forming a stone line at the maximum depth of burrowing (Fig. 14.74; see Chapter 11). The bioturbation model of stone line formation was proposed by Darwin (1881: 228–229) for the lowering of stones and coins and the burial of ancient Roman foundations and walls:

worms have played a considerable part in the burial and concealment of several Roman and other old buildings in England; but no doubt the washing down of soil from the neighbouring higher lands, and the deposition of dust, have together aided largely the work of concealment. Dust would be apt to accumulate wherever old broken-down walls projected a little above the then existing surface and thus afforded some shelter. The floors of old rooms, halls and passages have generally sunk, partly from the settling of the ground, but chiefly from having been undermined by worms.... The walls themselves, whenever their foundations do not lie at a great depth, have been penetrated and undermined by worms, and have subsequently subsided.

And again (Darwin 1881: 308–309):

Archaeologists ought to be grateful to worms, as they protect and preserve for an indefinitely long period every object, not liable to decay, which is dropped on the surface of the land, by burying it beneath their castings.

Unlike erosional-origin stone lines, stone lines formed in this manner will always be in the subsurface (Fig. 14.73). Also, they will commonly be parallel to the modern surface, and they will be buried by a uniform thickness of overlying material.

We suggest that stone lines formed by bioturbation – a *biomechanical* soil process further described in Chapter 8 – are more common than most people assume. Or at the very

least, bioturbation has been more involved in stone line formation – at least partially – than is commonly assumed. Biogenic stone line formation has been advocated by many, especially those working in the tropics (Tricart 1957, Ollier 1959, Webster 1965a, Williams 1968, Babalola and Lal 1977, Eswaran *et al.* 1992, Humphreys and Adamson 2000, Mercader *et al.* 2002). Stone lines can form in any soil that contains coarse fragments and burrowing soil biota. Even if the local biota are not mounders per se,– i.e., they do not move material to the surface – stones will still sink to the depth of burrowing, via subsurface mixing and settling. Fine sediment mining is best done by soil invertebrates such as termites, ants, and worms, although stone lines can be formed by any type of burrower (see Chapter 11). The relatively stone-poor material transported to the surface is called a *biomantle* of autochthonous material.

Because a biomantle takes some time to form, some degree of slope and surface stability, and/or age, is assumed, and usually accompanies biomechanical stone line formation. That is, the biomechanical model of stone line formation (Johnson 1990, 1993a, b, 2002, Johnson and Balek 1991, Johnson and Schaetzl 2014) does not normally include erosion or pedimentation. Instead, stone lines form contemporaneously with soils, on stable slopes, as subsurface deposits become covered by a biomantle (Johnson and Balek 1991). Johnson and Balek (1991) referred to this as the *dynamic denudation–soil evolution–biomantle model*, because the sediment above the stone line is, in essence, a stone-poor biomantle (Johnson 1993a, b, 1994; see later discussion). A type of hybrid model, discussed by Brückner (1955), involves erosion of gravelly soils to form a surface gravel lag (stone line), which is then buried by a biomantle.

Although the preceding models of stone line formation are plausible and have utility in some contexts, they are *not* mutually exclusive. Biogenic stone lines are always in the subsurface. They tend to be small and are often intermittent. Erosional stone lines are formed at the surface and are, whether later buried or not, often associated with sandy colluvial sediments. Different types of stone lines exist in nature, and it is up to soil geomorphologists to sort out their genesis and the resulting geomorphic implications. Stone lines are an excellent example of pedogenic equifinality – illustrating that many physical features in soils can have more than one possible origin. We end this discussion by emphasizing two points about stone lines: (1) We know too little about them – both their distribution and their origins, and (2) much can be gained from a more thorough understanding of stone line character and genesis.

Dynamic Denudation

It should now be apparent, from the many examples discussed, that soils develop in concert with slopes. On some slopes, erosion outpaces soil formation and soils become eroded. On others, soil development keeps pace with slope processes, and soil formation proceeds normally. We have seen examples of the former – Butler's (1959, 1982) K cycle and Ruhe's (1969) Iowan erosion surface – and of the latter: texture contrast soils (Bishop et al. 1980, Paton et al. 1995).

Donald Johnson, a soil geographer formerly at the University of Illinois, is credited with developing a model of soil-landscape evolution that incorporates many of the ideas put forth by Paton et al. (1995). Some of Johnson's work (Johnson 1993a, b, 1994) preceded the publication of *Soils: A New Global View* (Paton et al. 1995), and so it is not so much a follow-up as a subtly different way of viewing soil-slope development. Johnson's *dynamic denudation model* incorporates geomorphologic *and* pedologic processes; it is highly integrative.

Biomechanical Soil Processes

Johnson's dynamic denudation model makes heavy use of *biomechanical soil processes*, which alter the *physical* character and framework of soils by moving objects – sand, rocks, silt – throughout and within them. Johnson (1993a), who originally coined the term, stressed that biomechanical processes actually have had a long history in the literature, beginning with giants like Darwin (1881) and Shaler (1890), and continuing with Thorp (1949) and Hole (1961, 1981). Nonetheless, few soils researchers in the mid-twentieth century had utilized these concepts in their pedogenic bags of tricks. Clearly, the early work of Darwin and others had fallen by the wayside. Johnson, almost single-handedly, resurrected the importance of biomechanical soil processes by promoting their role in his papers and talks.

Strictly defined, biomechanical processes are those processes and effects that are *physically imparted* by biota to the landscape (see Chapter 11). Biomechanical processes include disturbance processes such as bioturbation;

loosening activities of hooves, claws, and feet; and anchoring and soil expansion by roots. Together with *biochemical* processes, they constitute the influence of all biotic processes and are captured in the *o* soil-forming factor (Jenny 1941b). Too often, however, the *o* factor is viewed as the biochemical part only, e.g., acidification by organic acids, base cycling, and melanization.

Although it was one of the earliest suites of pedogenic processes to be discovered and discussed (Darwin 1881), the idea that biomechanical processes are important to pedogenesis and slope development had to wait a century to be recognized and ranked equally with other forms of pedogenic processes (Johnson 1993a, 2002, Paton et al. 1995, Feller et al. 2003, Meysman et al. 2006). Why were *biochemical* processes stressed for so long? Part of the answer lies in the dominance of the plant component of the *o* factor in the state-factor model, particularly the amount and quality of organic matter produced by *plants*, relative to biochemical and nutrient cycling. Also, within the *o* factor, animals were never given equal status with plants (see Chapter 7); they were most often viewed as affecting soils biochemically, through their excreta. Plus, much of soil science is rooted in agronomic applications, for which base cation cycling, nutrient dynamics, and the growth of plants are paramount.

Elements of Dynamic Denudation

Johnson took key concepts from the literature and wove them together into a holistic model that has wide utility for explaining soil and slope development (Table 14.10). The model (Fig. 14.75) incorporates three pedogeomorphic agents (gravity, water, and biota) to explain the formation of three-tiered soils with stone lines (and their equivalents), which are so common on relatively stable but actively evolving surfaces on every continent but Antarctica (Johnson 1993b). In many respects, dynamic denudation is similar to the model of Paton et al. (1995) and differs only subtly from earlier work by Watson (1961) and Nye (1954). The results of dynamic denudation are best seen on older landscapes – those that have not undergone recent rejuvenation, such as recently glaciated terrains, on mountains and high hills with unstable slopes or in areas with periodic additions of volcanic ejecta. Thus, it is most applicable to soils and slopes in stable interiors of continents such as Africa, Australia, and South America. In the model, soils are assumed to have been bioturbated for a long period. This is a reasonable assumption for older, and especially tropical, landscapes. As a result, a stone line has formed at the lower limit of bioturbation (although this is not a necessary component of the model; Fig. 14.75D). Material above the stone line is seen as a biomantle, moving downslope by a variety of slow, but persistent, processes. Weathering proceeds at depth, producing saprolite from hard bedrock, and soil from saprolite (Fig. 14.75). Nonetheless, if shallow bedrock is not present, the model retains its explanatory power (Fig. 14.75C, D).

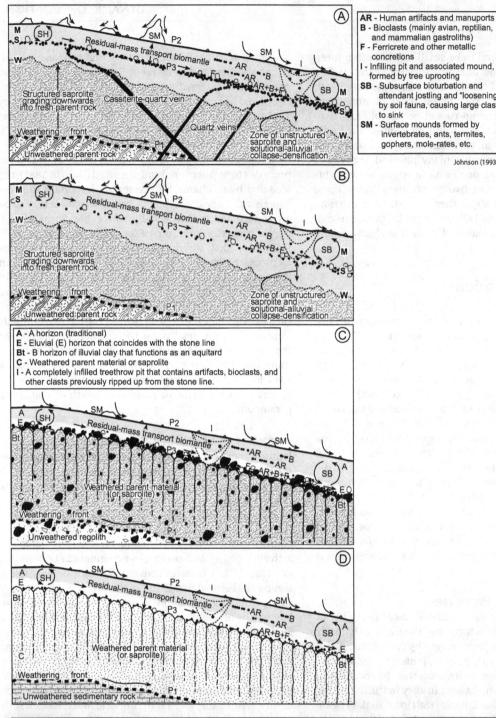

Fig. 14.75 Examples of a dynamic denudation model to understand soil-geomorphic processes on various types of substrates. (A) Dynamic denudation in landscapes underlain by dike- and vein-bearing intrusive crystalline rocks. The dashed arrow indicates the first planation (P_1) level, P_2 is the second planation (wash) surface (curved arrows), and P_3 is the third planation (throughflow) level (downslope-directed straight arrows). M is the mineral zone, S is the stone zone/line, and W is the weathered zone (saprolite). The base of the S zone is the top of the W zone. The residual-mass transport biomantle, which extends from the soil surface to the base of the S zone, thickens downslope. A Bt horizon may or may not be present below the stone line. The upper, unstructured saprolite is denser than either the structured saprolite below or the S zone above and thus functions as an aquitard. As a result, the S zone functions as an aquifer for lateral throughflow. (B) Dynamic denudation in landscapes underlain by crystalline rock that lacks veins and dikes. (C) Dynamic denudation in landscapes underlain by stony and gravelly sediments. (D) Dynamic denudation in landscapes underlain by nonstony sedimentary rock. In nonstony terrains, residual clasts are exclusively artifacts, bioclasts, and metallic nodules.

Dynamic denudation is partly built upon the ideas of others. For example, Aleva's (1983, 1987) triple planation model is a key part of the model. Triple planation refers to the three surfaces, or fronts, that are often found in bioturbated tropical soils (Büdel 1957). P_1 is the weathering front, at the rock-saprolite interface, which gradually migrates downward through time (Fig. 14.75). Many of the weathering by-products are removed laterally by groundwater, along the P_1 surface. The P_2 surface – the soil surface itself – is conceived to be a wash surface, on which sediment and soil materials are transported downslope by water, wind, and biota, including tree uprooting (Brückner 1955, Schaetzl and Follmer 1990, Norman et al. 1995, Richards et al. 2011, Eldridge et al. 2012). Finer-textured materials tend to move farther downslope, such that soils on the toeslopes are finer-textured and thicker than those upslope (Watson 1961; Fig. 14.75). In essence, the P_2 surface is the top of an actively moving biomantle. The P_3 surface is often associated with a stone line or a coarse-textured zone at the base of the biomantle. Beneath it is a denser zone that sometimes functions as an aquitard: either a Bt or C horizon, or sometimes saprolite (Morrás et al. 2009). Sometimes, the P_3 surface is a zone of lateral throughflow, within which soluble materials leached from the overlying soil can leave the soil-slope system. It is often convoluted, rather than planar (Johnson 1993b).

Over time, weathering (epimorphism) proceeds on the underlying bedrock, and as it does, the rock-turned-saprolite collapses as a result of solutional losses. Weathering is concentrated on the rock divides and outcrops, whereas the broad, concave valleys, so typical of large areas of humid tropical landscapes, are mainly areas transporting weathering by-products. The evidence for collapse and downslope transport is unequivocal – dikes and quartz veins that transgress the saprolite bend and curve downslope in a manner that could only happen if the weathering rock were collapsing and slowly moving downslope (Aleva 1987, Bremer and Späth 1989; Figs. 14.16C, 14.75A). Because the clasts in the stone line are often composed of resistant, economically important minerals, sampling the stone line may be an ingenious way to establish the presence or absence of key minerals in the rocks upslope (Aleva 1987).

Dynamic denudation helps explain the formation of stone lines by bioturbation and slope processes and can easily be fit into many prevailing theories of stone line development. It is especially important in tropical soils, where stone line genesis often involves a prior erosion event, or the formation of nodules in situ, by weathering-related processes. For example, Morrás et al. (2009) explained that some stone lines in the soils of Argentina are dominated by ferruginous nodular aggregates, formed by in situ basalt weathering. However, stone lines with dominantly siliceous clasts appear to be related to resistant quartz veins in the basalt, supporting a slopewash and bioturbation origin.

Horizon Nomenclature within the Constraints of Dynamic Denudation

We now pause to take a brief digression into the history of soil horizonation, which will help clarify the nomenclature of dynamic denudation. The traditional A-B-C horizon nomenclature system, developed by Dokuchaev in the late nineteenth century, has long been the standard throughout the midlatitudes (Tandarich et al. 2002). The A-B-C system is, according to Johnson (1994), primarily a descriptive one, which has become linked to various genetic processes. For example, the A horizon became associated with certain processes such as the accumulation of organic matter, whereas the B horizon concept evolved into one of illuviation and weathering. The E horizon is even named for a pedogenic process – eluviation (Guthrie and Witty 1982). By 1951, the U.S. Department of Agriculture encoded the A-B-C nomenclature by publishing it in their Soil Survey Manual (Soil Survey Staff 1951). The A and B horizons came to be known as the solum while everything below was considered C horizon material or "not soil" (Marbut 1935). As we have discussed in Chapter 3, the logic behind this decision was linked to practical mapping concerns – limiting the solum to about 2–3 m in thickness allowed soil mappers to focus on near-surface soil properties and saved many hours of deep augering and excavation into materials that geologists cared more about than did soil scientists. By the 1980s, the basic A-B-C horizon scheme had expanded to O-A-E-B-C (Guthrie and Witty 1982, Soil Survey Division Staff 1993). Recall that all of this nomenclatural tinkering had taken place in the United States, i.e., the midlatitudes. Fitting the A-B-C system to soils on older, tropical landscapes was more difficult because they are *much* thicker, their horizon boundaries are more blurred, and slope processes are a more important part of their genesis (Johnson 1994). For many of the younger soils of the midlatitudes, slope processes have not had nearly as dramatic an impact on soils. Rather, they are more dominated by top-down processes associated with percolating water. Last, stone lines, so common in tropical soils, are difficult to fit into the A-B-C horizon system. In short, the sedimentary and layered nature of the thick, old, tropical soils is distinctly different from the genetic, A-B-C horizonation found in many young, midlatitude soils.

Let us examine one example from West Africa. Here, the upper profiles are sandy, porous, and often heavily influenced by soil biota. Nye (1954, 1955a, b, c) differentiated these soils into two general layers – the unconsolidated material near the surface, and the more sedimentary material below (Fig. 14.76). Speculating that the uppermost layer was strongly influenced by creep processes (as did Paton et al. 1995), he named it "Cr" (De Villiers 1965). The uppermost, allochthonous, creeping zone was called Crw, for the dominant activities of worms, and the lower one Crt, for termites (Fig. 14.76). Nye (1955b) observed that the Crw horizon was only a few centimeters thick and contained almost no coarse sand, which worms do not

Table 14.10 | Fundamental elements of dynamic denudation theory

Major element(s)	Components
Process theories in geomorphology and pedology	Triple planation processes, etchplanation, chemical denudation, and leaching processes; biomantle, soil evolution, and soil thickness processes; mass transport; and soil creep processes
Soil horizon conventions	O-A-E-B-C and M-S-W horizon designations
Hydrologic principles	Throughflow, interflow, and eluviation-illuviation processes
Key pedogeomorphic agents	Biological (biota), chemical (biota, air, water), and physical (biota, gravity, water)
General definition of soil	*Soil* is rock material or sediment at the surface of planets and similar bodies altered by biological, chemical or physical agents, or a combination of them

Source: Johnson (1993a, 1993b, 2002).

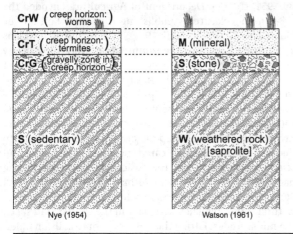

Fig. 14.76 Comparison of two early horizon schemes for tropical soils. After Johnson (1994).

ingest and hence do not carry to the surface. Conversely, the Crt horizon does contain coarse sand, which ants and termites can move. Strongly influenced by Nye's work, Watson (1961) later combined the Crw and Crt layers into a master zone (M), for the mineral soil (Fig. 14.76). It is safe to assume that the Crw, Crt, and M layers are simply biomantles.[3] Aleva (1983) also chose to differentiate the Cr layer from one above it that was more strongly influenced by biotic activity and biomechanical processes. In 1987, he went on record as saying that all necessary downslope movement of the M zone can occur by gravity movement – "there is no need to involve boring animals, such as termites" (Aleva 1987: 200). Both Nye and Watson routinely observed a gravelly/stony zones below the Cr or M zones. Nye (1954) called this stone zone CrG, while Watson (1961, 1962) labeled it S for stones (Fig. 14.76). Subsurface stone lines in pedogenic systems such as these are often a smoking gun for long-term bioturbation, and they attest to the

importance of biomechanical processes in these soils. The bottommost layer, autochthonous saprolite, Nye (1954) believed to be sedentary (not mobile) and thus he labeled it S. Watson (1961) labeled saprolite W, for weathered rock. The W zone grades downward into structured saprolite, and then to hard bedrock.

To summarize, the M, S, and W zones of old, tropical landscapes correspond to layers of mineral soil (M), stones or a stone line (S), and weathered rock (W) (Williams 1968, Johnson 1994). In most cases, the W zone corresponds to the BC, C, or upper Cr horizons of standard pedogenic terminiology; the M zone corresponds to the A horizon or the upper solum. In some soils, the S zone represents an area of enhanced lateral throughflow and leaching, above the saprolite. Hence, it has been seen as (loosely) equivalent to an eluvial zone, possibly the E (Johnson 1994). Watson's (1961) M-S-W terminology, so useful for many upland tropical sites, was nonetheless not adopted by midlatitude soil scientists, perhaps because they so rarely encountered comparably developed soils (Table 14.10).

M horizons are maintained on slopes by a dynamic interplay among processes such as slopewash, bioturbation, and other forms of mass transport. Fine particles are carried to the surface by termites, ants, worms, and other biota as a biomantle. The maximum grain size in the upper, M, zone is the same as observed in termite mounds, and in the Gold Coast the thickness of the M horizon on summits and upper slope locations rarely exceeds the normal depth of termite burrowing (Brückner 1955). Some, although not a lot, of organic matter is added to the M zone, forming a traditional A horizon. That the soil has a distinct A horizon amid all this dynamism implies that organic matter accumulation is progressing more rapidly, at least in the uppermost few centimeters, than are mixing and disturbance processes.

The coarse-textured, eluvial S zone lies at the base of the biomantle and corresponds to Aleva's (1983) P_3 planation surface. Much of the soluble material that exits the soil-slope system is hypothesized to leave through the S (E) zone (Johnson 1993b). On sloping surfaces, the stones in

---

[3] They did not actually use this term, because it had not yet been coined. But it does fit their intent.

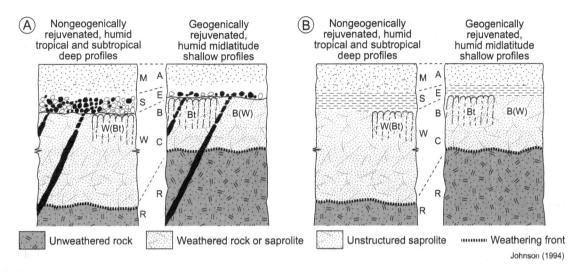

Fig. 14.77 Comparison of master horizon nomenclature in the two main horizonation schemes – for tropical soils and for midlatitude soils. Designation of the east horizon assumes that lateral throughflow has at least contributed some of its eluvial character. For the record, Watson (1961) did not equate the south horizon to the east, as Johnson (1994) did. (A) Morphologies associated with soils forming on saprolite that contains dikes and veins of resistant rock. (B) Morphologies associated with parent materials that lack coarse fragments.

the stone line are also creeping downslope (Figs. 14.16C, 14.75, 14.77). Watson (1961) argued that stones in the S horizon may be essentially stationary on gentle slopes. Often, the coarse fragments are relics from the saprolite below, commonly with resistant lithologies of quartz, cassiterite, tourmaline, topaz, anatase, corundum, or rutile (Aleva 1983, Dijkerman and Miedema 1988). Clasts are fed into the S layer from dikes and veins that are weathering out from below, or they form chemically (see later discussion). Continued bioturbation keeps most of them below the M zone. At locations where the saprolite does not contain hard relics, the stone line may be absent. However, Fe-rich, plinthitelike nodules can and do form at the saprolite-soil contact, especially when infiltrating water perches there (Dijkerman and Miedema 1988, Faure and Volkoff 1998). Nye (1955a, b) noted that the concretions will form only when soils wet and dry, and that they can weather if carried by creep to the wetter, lower slopes. Archaeological artifacts, if present, can also be lowered to the stone line by bioturbation, or they can be found above it, depending on how long they have been subjected to biomechanical processes (Brückner 1955). Bioclasts, e.g., gizzard stones and gastroliths, can also accumulate at the stone line. After large birds ingest stones (to aid in digestion), they pass them out as excreta or cough them up. Bioclasts are fairly frequent components of soils and stone lines but are only noted in soils where coarse fragments would otherwise be absent, such as on loess uplands (Cox 1998).

W zones reflect a complex interplay between weathering and pedogenic/translocation processes (Johnson 1994).

Commonly, a Bt horizon exists in the upper part of the W zone, especially if the degree of slope downwasting is slow. At least some of the illuvial clay in the Bt horizon is hypothesized to have entered the horizon laterally, along the stone line, as well as vertically from above (Nye 1955a). If the slope is downwasting rapidly, any Bt horizon that might form is stripped and degraded, as it has contact with the biomantle and the S zone, explaining perhaps why not all three-tiered soils have a Bt horizon. Under this scenario, Bt horizons would be best developed on slope elements with the lowest gradient. Certainly, the presence or absence of a Bt horizon also has a great deal to do with climate. In dry climates Bt horizons will not form, even on stable slopes.

In sum, the dynamic denudation model provides the conceptual linkage between the midlatitudes, where top-down pedogenesis is strong, and the tropics, where slope processes, great age, and bioturbation are more prominent. The model is the intellectual middle ground that brings many different elements together to explain soil formation on slopes, under the watchful eye of the omnipresent bioturbators.

Soil Geomorphic Applications in Geoarchaeology

Much of the field of geoarchaeology revolves around inferring past human behavior from the archaeological record. This work can be done only by examining the contextual association of artifacts and other traces of past human behavior, and by evaluating the factors that have produced that record (Rapp and Hill 1998). But how, exactly, is this done, and how can the study of soils help?

Geoarchaeology: Inferring Past Human Behaviors from the Physical Landscape

Geoarchaeologists address both the behavioral and nonbehavioral aspects of archaeological context from a *landscape point of view* (Cremeens and Hart 1995). The current physical landscape and the artifacts and features contained within it constitute the record from which past processes must be inferred. Geoarchaeological investigations use methods, theories, and concepts from the fields of geomorphology, sedimentology, stratigraphy, and pedology to evaluate site formation processes and to reconstruct past landscapes. In so doing, they ask two general questions (Cremeens and Hart 1995), often answerable, in part, by study of soils:

1. Why did past peoples pick a specific location for their activities? This question is largely behavioral and involves concepts of location choice. For example, what made a particular location attractive, and what influence did that particular location have on the activities there?
2. What has happened to the record of these people since they abandoned the site? This question is where the natural and behavioral contexts begin to overlap. The answer deals with more site-specific aspects of natural and human-caused processes that have affected the record or patterning of artifacts.

Because so much of the behavioral inference is derived from the patterns of artifact distributions at a site, it is crucial that we understand the specific site formation processes that have influenced it, since its abandonment. Specific site formation processes also influence the interpretation of the ages generated from site samples (Waters 1992). Without the precise stratigraphic, sedimentological, and postdepositional context, age estimates of archaeological remains can become skewed by mixing and contamination. Here, a thorough knowledge of pedogenic processes, particularly bioturbation, can be helpful.

How Soils Can Help

Geoarchaeologists have to work backward in time from the present site condition and take into account all that has happened to the landscape since site abandonment, including recent events. Recent advancements in analysis techniques have greatly increased the potential for geoarchaeologists to study soils at archaeological sites (Walkington 2010). For example, at many locations in the northeastern United States, forest has invaded abandoned agricultural fields from early settlers. To understand the original pattern of artifacts, we have to consider the effects of intensive hillslope agriculture, as practiced at the time. Agriculture introduces both order and disorder into the

landscape. Order takes the form of straight lines of field boundaries and roads. Disorder results from the erosion and sedimentation associated with cultivation. After considering the effects of agriculture and the invading forest, the natural pedogenic processes associated with it must also be evaluated. In several areas, more than 200 years of postabandonment processes have occurred at such archaeological sites – long enough to introduce a new set of processes and the resultant patterns.

The contemporary distribution and visibility of archaeological sites are a function of several factors (Cremeens and Lothrop 2001). In part, they result from human interaction with the environment existing at the time of occupation (Matthews 1965). The landscape setting directly influenced site selection by people and the nature of the activities that occurred there (Rapp and Hill 1998). After abandonment, surficial processes (pedologic and geomorphic) determined whether or not the archaeological record at that site was *preserved, modified,* or *destroyed* (Bettis 1988, Waters 1992). Thus, the archaeological record is a result of both (1) human behavior and (2) geologic and pedologic phenomena. Therefore, artifacts found at a site have both a behavioral and a nonbehavioral context (Schiffer 1983).

Contemporary archaeological investigations require that site formation processes be evaluated because variability, error, and noise can be introduced into the archaeological record. Preexisting materials may have been modified or destroyed, materials may have been added, and patterning may have been modified, destroyed, or added. Thus, the dynamic changes that operate on every surface, and the archaeological record contained in the soil, must be deciphered to make meaningful inferences about past human behavior (Cremeens and Hart 1995). The record of landscape stability, as interpreted from the soils, is significant to archaeological site investigations because the cultural remains (artifacts and features) often are associated with stable landscapes (Mandel and Bettis 2001). This association reflects the probable human preference for locating residential camps or settlements on stable surfaces and the resulting likelihood of the concentration and preservation of artifacts and features on generally stable surfaces (Holliday 2004).

The pedogenic principles discussed in this and previous chapters are therefore applicable in a wide variety of geoarchaeological investigations. Most geoarchaeologists must understand the geologic, sedimentologic, and pedologic history of a site, in order to explain the effects of human habitation on the sediments, and vice versa. The literature is rich with examples of how a thorough understanding of geomorphic principles is vital to geoarchaeology (Butzer 1982, Holliday 1992, 2004). Much of soil geomorphology involves reconstruction of paleoenvironments and paleolandscapes, and this, obviously, has clear archaeological applications. Indeed, one of the most significant aspects of geoarchaeology is the analysis of landscapes, especially

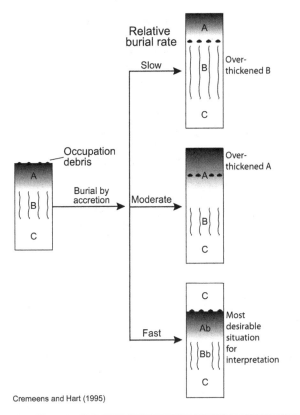

Relative
burial rate

Slow → Over-thickened B

Occupation
debris

Burial by
accretion → Moderate → Over-thickened A

Fast → Most desirable situation for interpretation

Cremeens and Hart (1995)

Fig. 14.78 Potential differences in stratigraphy and soil morphology due to varying rates of burial of the soil, which in this case has an artifact layer at the surface.

in terms of the changing options they presented to their former human habitants (Butzer 1982). Few aspects of the environment are as intimately linked to the physical landscape as are soils. Therefore, knowledge and appreciation of soils should be an integral component of every geoarchaeologist's bag of tricks (Holliday 2004). We are not able to delve into soils and geoarchaeology in great detail here, but in this section we point out several ways (in no particular order) that soil information and context can and do assist in geoarchaeological investigations.

Geoarchaeologists often use soil data (phosphorus enrichment, in particular) to suggest former human habitation, e.g., Lauer et al. (2013). Then, they use soil morphology to provide even more information about the sedimentologic and geomorphic history of a site (Andertson 1999, Grave and Kealhofer 1999). Soils and stratigraphic data can answer questions about the site's erosional or aggradational history (Fig. 14.78). For example, resolving the issue of parent material origin and its uniformity is almost always of great importance (Frederick et al. 2002, Ahr et al. 2012, Mandel 2008). At sites that are, or have been, aggrading, human artifacts may have become buried, whereas at sites that are eroding or stable, they may potentially be

at the surface or may have been transported away from the site and lost (Fig. 14.79). Many archaeological sites were or have been unoccupied for many centuries or millennia, and it is paramount to the correct interpretation of the site to know what has been happening – geologically and sedimentologically – since the former habitants left (Matthews 1965). Some kinds of soil data, if correctly interpreted, can provide that information.

This point, however, immediately raises an important corollary. Not only does sedimentation, e.g., by loess or alluvium, cause artifacts and evidence of occupations to become buried, but so can bioturbation. We have learned in earlier chapters that pedoturbation of all kinds can (and does) disrupt (or preserve) artifact stratigraphy, and geoarchaeologists are becoming increasingly aware of its existence and importance (Flegenheimer and Zárate 1993, Davidson 2002, Peacock and Fant 2002). Normally, bioturbation causes artifacts to sink through time, eventually becoming buried beneath biomantles (Leigh 1998, 2001, Alexandrovskiy 2007; Figs. 11.2, 14.74). As we noted, Darwin had proposed this concept in 1881, when he observed that the floors of old rooms, coins, and other human artifacts had sunk over time, as a result of worm bioturbation. A large literature has since shown this concept to be valid, and geoarchaeologists have been using this knowledge successfully in the field to explain buried artifact layers (Johnson 1989, 1990, Frolking and Lepper 2001, Balek 2002, Peacock and Fant 2002, Van Nest 2002).

The informed use of soil maps and soil surveys can be a useful skill for geoarchaeologists searching the landscape for former habitation sites (Almy 1978). Although not commonly, some NRCS soil surveys actually note and report on archaeological site locations. The advent of Cultural Resource Management (CRM) work in archaeology has created a demand for readily available information about regional resources such as soils, wetlands, and hydrology, much of which can be gleaned from a careful reading of large-scale county soil survey reports (Holliday 2004). Certain kinds of soils, e.g., better drained series, may have been preferred for habitation, whereas others may have been preferred for other uses, such as butchering, traveling, or food storage (Lovis 1976, Voight and O'Brien 1981, Warren et al. 1981). An excellent example is provided by Artz (1985). He assumed that any soils associated with the late Archaic period (~5–2 ka ago) in northeastern Oklahoma would have developed Bt horizons. Using the county soil survey, he noted that where sites could have been, three main soils were mapped in the valley: Hapludolls, Haplaquolls, and Arguidolls. Only the Argiudolls were dry enough and developed enough to have been the habitation sites of late Archaic peoples. The subsequent survey of these soils produced artifacts, and the ¹⁴C ages verified that these were late Archaic sites. Similar examples, using both soils and landforms, are reported by Muhs et al. (1985), Bettis (1992), Mandel (1994, 1995), and West (2012).

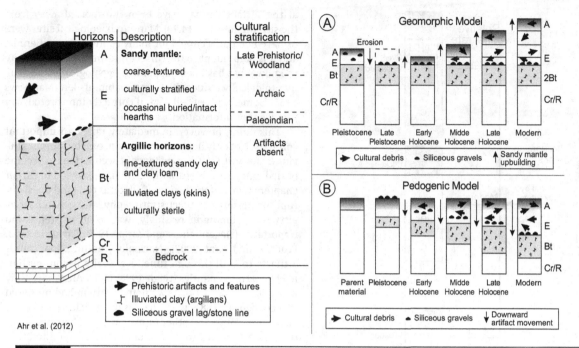

Ahr et al. (2012)

Fig. 14.79 Conceptual diagram showing two scenarios for the formation of a sandy mantle overlying buried artifacts, in eastern Texas. (A) Artifact lowering and burial by pedoturbation processes, implying minimal sedimentation subsequent to the human habitation. (B) Artifact burial by sedimentation subsequent to human habitation.

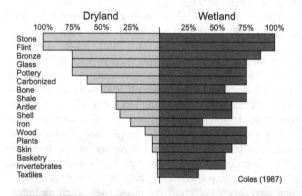

Coles (1987)

Fig. 14.80 Generalized comparisons of the preservation potential of archeological materials for dryland and wetland sites/soils, as developed for Europe. For example, the percentage of stones likely to be preserved in drylands is 100%, whereas only about 75% of glass is preserved in the same types of soils.

Another example, this one from the Ohio River valley, illustrates that soils can also guide the geoarchaeologist *away from* areas where sites not likely to be (Stafford and Creasman 2002). Here, landforms with Alfisols are underlain by Pleistocene age alluvium; these areas contain few archaeological sites. Most of the archaeology represents Woodland and later occupations, which are buried in late Holocene alluvial fills, in which Entisols, Inceptisols, and Mollisols have developed.

In Norway, Simpson *et al.* (1998) used soils to identify relict arable soils. After considerable research, it was discovered that the uncultivated soils of the area are sands or peats overlying sand. In localized areas, loamy sands were identified as formerly cultivated fields, with scattered bone, charcoal, and pottery shards. The cultural components were added as fertilizer amendments.

Another useful skill that soil geomorphology provides to geoarchaeology involves the interpretation and dating of buried soils – paleosols (see Chapter 16). Buried soils are very useful in interpreting the archaeological record, but are not always common. Their correct interpretation can be a boon to the study of an archaeological site (Mitusov *et al.* 2009; Fig. 14.78). Buried soils and the sediments above them can inform the geoarchaeologist in at least three ways by providing (1) unmistakable evidence of the paleosurface upon which former humans lived, with the potential for good preservation of artifacts associated with that occupation and that surface; (2) information about the environment at the time of soil formation (and human occupation) from the soil itself; and (3) insight into the causes of the burying event, which in and of itself may help interpret the site history. Many archaeological sites are associated with sedimentation and burial, e.g., Engovatova and Golyeva (2012), Dreibrodt *et al.* (2013). This association makes sense, because the soil surfaces associated with erosion are damaged or no longer exist! Indeed, many sites of former human habitation are on slowly aggrading surfaces, such as floodplains, alluvial fans, and coastlines (Hoyer 1980, Holliday 1985, 1987, Chatters and Hoover 1986, Ferring 1986, Hajic 1990, Mandel and Bettis 2001, Mandel 2008), and that explains why paleosols and archaeological sites commonly go hand in hand. Also, by understanding the characteristics of the soil, pre- and post burial, one can begin to estimate the likelihood of artifact preservation on and in that soil (Fig. 14.80).

Last, we note that soil data have often assisted geoarchaeological investigations in the interpretation of burial mounds (Olson *et al.* 2003). Soils on and beneath mounds can provide keys as to the age of the mound (Saunders *et al.* 2005, Alekseeva *et al.* 2007), as can soils within the mound (Parsons *et al.* 1962, Cremeens 1995, Alexandrovskiy 2000).

This brief introduction to the utility of soils in geoarchaeological research does not do justice to the wide variety of pedologic applications to the study of human habitation. But we hope that it does underscore the commonalities that exist between the two disciplines.

Chapter 15

Soil Development and Surface Exposure Dating

Basic Principles and Concepts

The age of a soil, sediment, surface, or geomorphic feature can inform us about the (1) period of initiation (time$_{zero}$) of the soil or surface, (2) time interval over which the soil or surface has developed, and (3) geomorphic and pedogenic processes that have operated during that interval. Establishing the age of features, soils, or sediments is a key component of soil geomorphology. Various dating techniques are now at our disposal. These are continually being refined, and more methods are always being developed.

Because so many soil geomorphology studies now employ some type of dating, we stress the point made by Colman *et al.* (1987) when they argued for consistency in terminology among those using dating techniques. For example, a *date* (as a noun) is a specific point in time, e.g., AD 530 or July 14, 2013. An *age* is an interval of time measured back from the present, e.g., 11.7 ka. Colman *et al.* (1987) argued that the use of *date* should generally be minimized, in favor of *age* or *age estimate*. The terms ka and Ma are to be used for ages (thousands and millions of years as measured from the present, respectively) (Table 15.1). In addition, only radiocarbon ages should be expressed in terms of years before present (yr. BP) (Colman *et al.* 1987). All others should be *calibrated* to calendar years (see later discussion). For example, all other types of ages, including numerical and calibrated ages, should be referred to as ka or k.y., and so forth.

Ages can be determined for surfaces and for bodies of sediment. A *chronostratigraphic unit* is a body of rock that was formed during a specified interval of geologic time. It serves as the material reference for all rocks formed during the same span of time, and its boundaries are synchronous. A *geochronometric unit* is a division of time, based on the rock record. It corresponds to the time span of an established chronostratigraphic unit. A chronostratigraphic unit is bounded by synchronous horizons, i.e., its beginning and end coincide with the base and top of the referent chronostratigraphic unit. Types of geochronometric units include eons, eras, periods, epochs, and ages.

Of special importance to us is the *pedostratigraphic unit* – a body of rock or sediment that consists of one (or more) pedogenic horizon(s) that is overlain by one or more formally defined stratigraphic units. Pedostratigraphic units must be buried; surface soils do not count! Soil horizons, recognized by their morphology or micromorphology, are used to distinguish pedostratigraphic units from other stratigraphic units. The boundaries of pedostratigraphic units are almost always *time transgressive*. As in surface soils, it is often much more difficult to identify the base of a pedostratigraphic unit than it is to identify the top. The fundamental pedostratigraphic unit is the *geosol*, which is similar to a paleosol except that a geosol formally fits within the guidelines of the North American Stratigraphic Code (see Chapter 16). The term geosol refers to the entire soil; it is not formally divisible into horizons. Geosols are named for a location where they were first formalized, studied, or described, e.g., the Sangamon Geosol and the Farmdale Geosol.

Numerical Dating

In Quaternary studies, sediments and surfaces can be dated in a number of different ways, each belonging to one of four different families of methods (Table 15.2). These methods can, in turn, be boiled down to two basic dating techniques: *numerical* and *relative* (Colman *et al.* 1987, Watchman and Twidale 2002; Fig. 15.1). In *numerical* (formerly called *absolute*) dating, an established radiometric method is used to read the natural chronometer that is built into the sedimentary system (Colman *et al.* 1987). Numerical methods allow for quantification of absolute differences in age among sediments or units.

Often we have the option to date the *sediment* or the *surface*. For example, we may seek to date a *sediment* by dating something buried within it that is age-contemporaneous. A piece of wood that lies within alluvium can be dated by measuring the amount of ^{14}C remaining in it (see the description of radiocarbon dating techniques that

Table 15.1 | Geochronological terminology

Common or former usage	Suggested usage	Examples
Date (verb)	Date	Radiocarbon dating; to date a sample; OSL-dated sediment
Date (noun)	Age[a] or age estimate	^{14}C age of 11.3 ka; ^{14}C age estimate of 11.3 ka; the luminescence age is 12.3 ka
Radiocarbon age	^{14}C yr BP, yr BP	11,300 ^{14}C yr BP; 11,300 yr BP
Calibrated age (kya, Mya)	ka; Ma; cal yr BP	11.3 ka; 11.3 cal yr ka
Absolute age or date (kya, Mya)	Numerical age; calibrated age; correlated age	11.3 ka
Radiometric dating	Radioisotopic dating or cosmogenic nuclide dating	^{14}C; ^{10}Be; ^{26}Al; etc.

Note: The exception: *Date* can be used as a noun if it refers to a specific date in sidereal time, e.g., The Mt. St. Helen's ash has a date of May 18, 1980.
Source: After Colman *et al.* (1987) and Bringham-Grette (1996).

Table 15.2 | Types of dating techniques typically used in Quaternary investigations

← Numeric ages →

← Calibrated ages →

← Relative ages →

← Correlated ages →

Sidereal	Isotopic	Radiogenic	Chemical and Biological	Geomorphic	Correlations
Historical records	^{14}C	Fission track	Amino acid racemization	Soil development	Tephrochronology
Dendrochronology	K-Ar and $^{39}Ar/^{40}Ar$	Luminescence	Obsidian hydration	Rock and mineral weathering	Paleomagnetism
Varve chronology	U-series		Lichenometry	Surface modification methods	Stable isotopes
	Cosmogenic isotopes			Geomorphic position	
	Uranium trend				

Source: After Bringham-Grette (1996).

follows). That date tells us how many years have passed since the woody plant died when it was buried by the sediment. We can also then say that the age of the sediment *surface* is no older than the numerical age obtained on the sediment proper. The sediment, in this case, provides a *maximum-limiting age* for the surface, i.e., the surface can be no older than that age. If an erosion surface has cut into the sedimentary deposit, the age of the erosion surface can be considerably younger than the numerical age of the sediment in which it has formed. Numerical dating of geomorphic *surfaces* can be done only if datable materials exist on that surface that are contemporaneous and associated with it, i.e., they are strictly surficial phenomena. Some of the best types of material in this regard are archaeological, e.g., hearths, storage pits, post molds, since human occupation is always associated with a surface and not the sediment below. Numerical dating of both sediments and surfaces is obviously limited by the availability of datable materials, but also by cost. Many numerical dating methods are expensive, and for that reason, researchers may rely on relative dating techniques (Hall 1983).

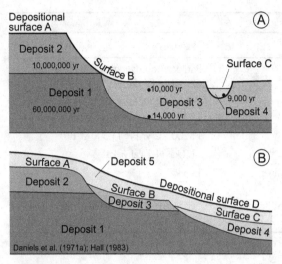

Daniels et al. (1971a); Hall (1983)

Fig. 15.1 Hypothetical examples of numerical and relative dating of sediments and surfaces. (A) Use of the principles of superposition and cross-cutting relationships to establish the relative ages of surfaces and sediments. The deposits are numbered in their relative age sequence. Deposit 3 must be younger than Deposit 2 because it buries Deposit 2. Deposit 5 is the youngest surface, formed in the topmost sediment; it covers the entire landscape as a relatively uniform blanket. Although the surficial expression of each of the buried surfaces is still present, Surface D has a single (synchronous) age. Surface B must be younger than Deposit 2 because it cross-cuts Deposit 2. (B) Use of the principles of superposition and cross-cutting relationships – combined with numerical dating – to establish more precisely the ages of surfaces and sediments. The numerical ages of Deposits 1 and 2 are 60 Ma and 10 Ma, respectively. The erosional part of surface B must therefore be younger than 10 Ma, because it cuts that sediment. A radiocarbon age of 14 ka from the base of Deposit 3 establishes that the deposit is younger than 14 ka. A 10 ka radiocarbon age from material just below surface B implies that the depositional phase associated with Deposit 3 took less than 4,000 years. Dating of material in Deposit 4, which fills a surface cut into and below surface B, establishes that surface B erosion and deposition occurred between 10 and 9 ka. Surface C is less than 9 ka; from the information given a more precise age cannot be established.

Relative Dating

In contrast to numerical dating, *relative dating* methods can only tell us whether a sediment, surface, or soil feature is older or younger than another feature – no numerical ages are determined. Relative dating cannot tell us exactly how long ago a sediment was deposited or how long ago a geomorphic surface formed, but only that it followed or preceded the deposition of another sediment (by principle of superposition), or perhaps the erosion of a geomorphic

surface (by cross-cutting relationships). Relative assessment of age is still very useful, however, especially when coupled with available numerical dates of sediments above or below the units of interest.

Before the advent of numerical dating, sediments and rocks could only be dated by relative means. In relative dating, rocks and sediments are placed in their proper *sequence of formation*, i.e., which was first, second, etc., or which is younger or older. In other words, we can only establish that A is older than B, which is older than C. Theoretically, relative dating has no temporal limits (Watchman and Twidale 2002). Nonetheless, the older the surface, the less likely it is that the materials necessary for relative dating are either preserved or exposed.

Although national and international stratigraphic codes formalize units of known age, or at least help to correlate among them, geologists have developed general laws to assist in this endeavor. They may seem obvious today, but these laws were important breakthroughs when geology was a young science. Nicolaus Steno (1638–1686), a Danish geologist and priest, developed a very simple rule – the *principle of original horizontality* – which states that layers of sediment are generally deposited in a horizontal position. Thus, rock layers that are flat have not been disturbed and maintain their original horizontality. Those not currently flat-lying must have been disturbed at a subsequent time. Steno also recognized the *principle of superposition*. It states that in an undeformed sequence of sedimentary rocks or other similar sediments, each bed is older than the one above and younger than the one below. In other words, older rocks lie deeper in the geologic column. Other laws that are useful in interpreting the relative age of the rock column include the *principle of cross-cutting relationships* – when a fault or igneous intrusion, e.g., a dike, cuts across another body of rock, it must be younger than the host rock (Fig. 15.2). A similar relative age relationship holds for rock inclusions. Material that is included within a host, such as fragments of basalt within a mud flow deposit, must be older than the host.

With some skill and a little luck, relative dating can be taken one step further – we can determine the general *magnitude* of the age differences between two or more deposits. For example, we may use the degree of soil development to determine that not only is Surface A older than Surface B, but that it is at least *three times older*, or perhaps that it is only *slightly older*. This degree of difference may allow for discrimination among the three alpine moraines, for example, and determination of whether they resulted from three distinct advances (MIS Stages 2, 4, and 6) or from one advance with three short pulses/surges (all MIS Stage 2).

Taken still further, relative dating can utilize a group of dating techniques called *correlated age methods* (Colman *et al.* 1987). Correlated age assessment requires that numerical ages have been determined for *some* of the surfaces that have *also* been dated using relative dating techniques.

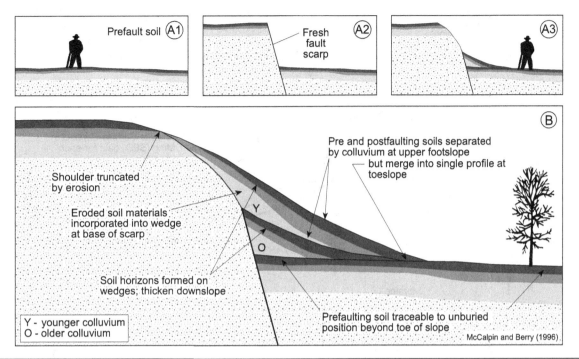

Fig. 15.2 Examples of how soil development, cross-cutting relationships, and buried soils (surfaces) can be used to infer past periods of surface stability or instability, and the relative ages of surfaces. In (A), the fresh fault scarp in (A2) has no soil development, as it is the youngest surface. The surface formed by the faulting cross-cuts both of the soils, indicating that it must be younger than they are. The buried soils in (B) illustrate at least two subsequent periods of stability in this landscape. Even within geomorphic surfaces, varying degrees of stability and soil development are expressed, as indicated by the shallow soils on the shoulder slope.

This group of methods, therefore, involves informal correlation of data from soils or rocks among surfaces of known and unknown age (Fig. 15.3). For example, weathering rind data from rocks might be generated for moraines of known ages to arrive at a statistical function, which can then be extrapolated (within limits) to nearby moraines of unknown age. Thus, correlated age assessment can provide semiquantitative estimates of the magnitude of age differences among sites, surfaces, or materials.

Relative dating techniques have varying degrees of accuracy and applicability. Some, such as lichenometry, cannot be used on surfaces older than a few millennia. Others, such as obsidian hydration dating, can provide age estimates for rocks and surfaces potentially as old as 1 Ma (Table 15.3). Additionally, the types and quality of relative age data vary. Interval- and ratio-level data that can be generated for some methods, e.g., solum thickness in centimeters, clay content as a percentage, and weathering rind thickness in millimeters, contrast with other methods that produce only grouped ordinal data, e.g., weathered rock categories high, medium, and low. Because each relative dating method has its own strengths and weaknesses, using a variety of methods is always recommended (Hall and Michaud 1988, Mills 2005; Table 15.3, Fig. 15.4). It should also be apparent that data from any

relative (or correlated) dating study can be extended only a small distance spatially. In many instances, the data may be applicable only within the region from which they originated.

Surface Exposure Dating

Soil geomorphologists make good use of both relative *and* numerical dating information. With these two methods in hand, one can date the *sediment* within which a soil has developed, and/or the *geomorphic surface* on which it is developing. The latter – an estimate of the length of time that sediment has been exposed at or near Earth's surface – is referred to as *surface exposure dating* (SED) (Colman *et al.* 1987, Dorn and Phillips 1991). The age of the surface can never be greater than the age of the sediment, and it is often much less. For sites of rapid sedimentation and for rapidly formed geomorphic surfaces such as moraines, the two ages (sediment and surface exposure) are so close in time that they are essentially the same. For erosional surfaces and for some deposits that have accrued slowly, the two ages may be widely disparate (Knuepfer 1988). Bear in mind that most surfaces exhibit a wide range of ages, i.e., they are time transgressive.

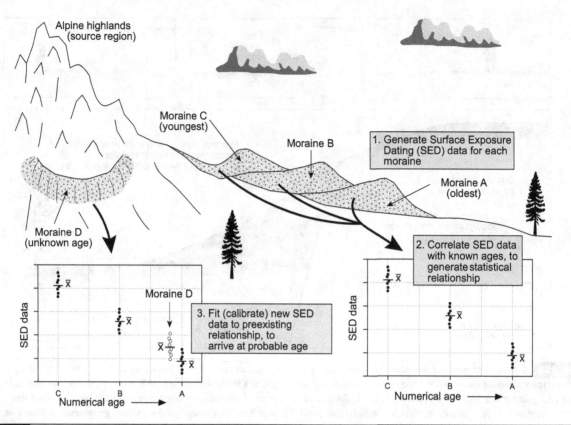

In surface dating, we determine the time when the surface first became subaerially exposed *and* geomorphically stable. The age estimate is based on data from materials on (and to a lesser extent, within) the surface that change, predictably and quantifiably, through time (Birkeland 1982). Occasionally, we may wish to date the exposure interval of a *buried* surface or soil; this would be equivalent to the time from initial surface stabilization to the time of burial.

Knowing the exposure age of a surface has many applications within soil geomorphology. First, this information will assist in the interpretation of the origin of the surface. Second, knowing the age of a surface permits estimates of the time span over which various surficial processes may have operated, e.g., pedogenic or slope processes. Last, one might use surface exposure data, when developed within a known climatic context, to add to paleogeographic interpretations and chronologies of climatic change (Watchman and Twidale 2002).

The Iowan surface (see Chapter 14) is a good example that illustrates surface exposure versus sediment age. It was formed by widespread erosion of Pre-Illinoian till between about 21 and 16.5 ka ago, during the coldest part of late Wisconsinan time (Ruhe 1969). The preexisting landsurface was reshaped by frost action, solifluction, sheetwash, and strong winds that accompanied the advancing continental ice sheet. A layer of this eroded and redeposited material remains on top of the Pre-Illinoian till, forming a loamy mantle in which soils are developed. In addition, across the southern portion of this landform region, a thin (60–100 cm) blanket of Wisconsinan loess overlies the reworked loamy material. Time$_{zero}$ for soils on the Iowan surface occurred when the erosion-deposition period stopped. Thus, while the till into which the surface was cut dates back to a Pre-Illinoian glaciation, the Iowan surface itself is much younger. Thus, soil development on the Iowan surface reflects the age of the *surface*, not the *sediment*. The surface age equals the age of the redeposited sediments because soil formation started only after the surface stabilized. Although small inputs of (younger) dust can be added to a depositional surface or to an erosion surface, thin loess deposits postdate the establishment of the surface and do not usually require one to assume a new time$_{zero}$ for it.

Each SED method has its own precision, accuracy, and time limit (Table 15.3). In general, accuracy and precision are sacrificed in methods that measure very long time

Table 15.3 | Some of the major surface exposure dating methods

Method	Common methods and/or assumptions	Maximum time span utility (years)
Geology- and *biology-based methods*	Assumes that rocks or minerals were not pre-weathered, and climatic variation over time has been minimal, or at least comparable, among sites	
Rock weathering, general	Degree of clast weathering is determined by striking the rock with a hammer and estimating the degree and ease with which it breaks apart; rocks are then categorized	10^4
Clast sound velocity	Compressional (P) waves are made to travel through clasts; the speed with which the waves move through the rock is proportional to their degree of weathering	$>10^5$
Rock angularity	Rates the degree of angularity of rock edges, assuming they become more rounded with time	10^4
Hornblende etching	Degree of etching of hornblende (or perhaps other) minerals in sand fraction is determined under a microscope and placed into categories	10^5
Rock surface oxidation, weathering rind formation, and obsidian hydration	Correlates the depth (mean, max or mode), color, hardness, or mineralogy of a weathering rind on the outsides of rocks to exposure age; obsidian hydration dating has the advantage in that obsidian can be numerically dated by the K–Ar method	10^5 (10^7 for obsidian)
Rock surface pitting	Pits are deeper, wider, and more common on older rocks	10^5
Vein height	Uses the height of veins on weathered rocks, assuming that the veins stand out higher as the softer matrix weathers away over time; also assumes that the veins have not been eroded	10^4
Lichenometry	Assumes that the size of lichen thalli on exposed rocks, usually in cold climates, increases with time; lichen growth rate is, however, affected by other factors such as climate and substrate	4.5×10^3
Rock varnish	Assumes that rock varnish thickens, darkens, and accumulates more Mn with time; indices of surface exposure correlate best to Mn content in varnish, varnish color, and areal coverage	10^4
Cation ratios of rock varnish	Assumes that the ratio of (K + Ca)/Ti in rock varnish decreases with time as the more stable Ti becomes relatively enriched	10^6
Soil-based methods[a]	Assumes that the main reasons for soil variation on the surface are a function of time, i.e., the remaining state factors have been held as constant as possible; also assumes that soil development has been, for the most part, progressive and generally linear	
Horizon thickness or depth to a particular horizon type	Depth (thickness) is determined and compared among sites, assuming that thickness increases with time	10^5
Solum thickness, sometimes equivalent to depth of leaching or oxidation, or depth of carbonate leaching	Solum thickness is most accurate as a relative age indicator in young soils, assuming that solum thickness increases with time	10^5
Content of a mobile constituent (clay, $CaCO_3$, Fe)	After controlling for parent material, content in the soil is compared among surfaces; some constituents will assumedly increase with time; others decrease; content values are usually weighted (see note)	10^5
Soil chemistry	Examples include CEC, electrical conductivity, and pH; measures changes in soil properties that reflect clay mineral weathering, leaching, and base cycling, which are at least partially time-dependent	10^4
Horizon color, especially rubification	Expected color changes are based on pedogenesis theory	10^4

(cont.)

Table 15.3 (cont.)

Method	Common methods and/or assumptions	Maximum time span utility (years)
Clay mineralogy	Changes in clay mineral abundance and type are related to weathering sequences; ratios of the abundance of one clay mineral to another are commonly used	10^5
Soil morphology and micromorphology	Expected morphologic change, e.g., structure, cutans, texture, silt coatings, are based on pedogenesis theory	5×10^4
Sand mineralogy (light or heavy fraction)	Resistant/weatherable mineral ratios are assumed to increase with time	10^5
Formation of desert pavement	Related to morphology on surfaces of known age and based on an understanding of its genesis	10^5

[a] Most quantitative, soil-based criteria can be evaluated on a mass-basis, weighted by horizon or solum thickness, or on a volumetric basis (weighted by horizon thickness and bulk density); see text for details.

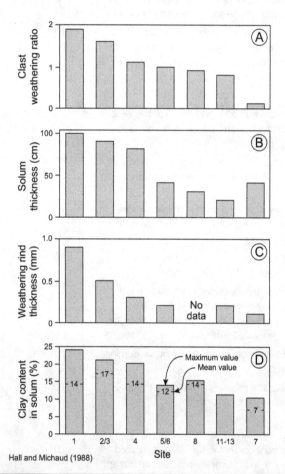

Hall and Michaud (1988)

Fig. 15.4 Comparison of four (relative) surface exposure dating methods for surfaces of different age in the Tobacco Root Range, Montana. (A) Weathering ratio. Higher ratios imply that a higher percentage of the rocks are considered weathered, as indicated by a hammer impact. (B) Mean weathering rind thicknesses. (C) Mean solum thicknesses. (D) Clay contents of soil horizons.

spans. Soil-based methods can provide high-resolution SED information for young surfaces but may not be useful for older surfaces on which the soils are more homogeneous. On older surfaces, slow-to-change parameters like rock weathering rinds may be better SED tools. Thus, a SED method that may be *possible* may not necessarily be *useful*, given the resolution required or the age range of the surfaces. Careful selection of the proper SED tool is paramount.

Theoretically, SED methods work because they use *post-depositional modifications* (PDMs) on surfaces. PDMs involve aspects of landform morphology, or physical and chemical changes in the rocks and soils on surfaces, that change predictably with surface exposure (Kiernan 1990). SED methods assume that PDMs are directly or indirectly correlated to surface age or exposure. However, complicating factors can and do influence PDMs, regardless of the dating application. Climate is a particularly important one; often its effects can overwhelm that of time (Locke 1979). Every researcher using relative or numerical dating tools must be aware of the potential complications introduced to the data by agents and factors besides age. For these reasons, the maximal number of dating techniques possible should be used to assess the age of surfaces and sediments (Birkeland 1973, Miller 1979, Kiernan 1990). Using multiple approaches is always better. When the results are in general agreement, they give us confidence in the overall conclusions. When results are not in agreement, they spur us to reexamine our assumptions (Burke and Birkeland 1979, McFadden *et al.* 1989; Figs. 15.4, 15.5). Combining relative *and* numerical age estimations is optimal, although not always possible.

Although a wide variety of SED techniques exist, we focus on those that are of most use in soil geomorphology (Table 15.3) and point the reader to reviews of other techniques for further information, e.g., Kiernan (1990) and Dorn and Phillips (1991).

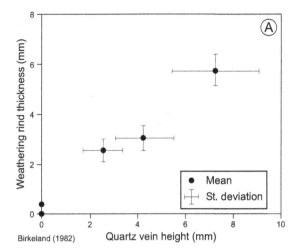

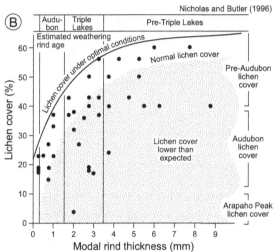

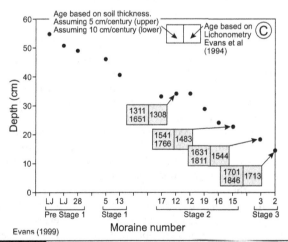

Fig. 15.5 Comparisons of relative dating methods, showing that they often correlate well to each other, and illustrating that using more than one relative dating technique is always a good approach.

Geomorphology and Stratigraphy

Leaning on the principles of cross-cutting relationships and superposition, Dorn and Phillips (1991) observed that relative position is often a simple, but elegant and highly useful SED method. Constructional landforms, such as moraines and alluvial fills, often overlap, providing unequivocal information about relative ages (Gibbons *et al.* 1984; Figs. 15.1, 15.2). Geomorphologists also use the *form* or *shape* of a landform to estimate its age (Coates 1984). Many landforms originate with sharp edges and slope breaks but become more rounded with time (Sharp and Birman 1963, Miller 1979, Nash 1984, Bursik 1991, Nelson and Shroba 1998), although Mills (2005) reported a clear exception to this rule.

Considerable advances have been made using this type of relative dating in establishing the ages of alpine moraines. Older moraines tend to be broader and lower in surface elevation (Bursik 1991). The mere presence or absence of rocks on a surface, i.e., their frequency, also decreases with time, as they weather. Thus, rock density alone, called *surface boulder frequency*, can also be used as an SED tool (Sharp 1969, Scott 1977, Burke and Birkeland 1979, Miller 1979, Nelson *et al.* 1979). Older surfaces tend to have fewer surface boulders per unit area (Sharp and Birman 1963, Birman 1964, Berry 1994, Pinter *et al.* 1994; Fig. 15.6). Another SED option involves data on the *sizes* of rocks and boulders on surfaces of different age (Fig. 15.7). The assumption is that rocks become smaller with time, as a result of weathering. Various types of data on slope inclination, microrelief, and surface roughness can also be used to estimate surface exposure (Fig. 15.8).

Obviously, all these methods assume that (1) the surfaces being compared have all been derived from similar materials, and (2) they have undergone equivalent processes of weathering and erosion, such that PDMs have developed uniformly with time (Sampson and Smith 2006).

Desert Pavement

Desert pavement is a PDM that follows a predictable development sequence and is, therefore, a good SED method for some desert areas (see Chapter 13). Despite the numerous forces that can destroy a pavement and reset the geomorphic clock to zero (Quade 2001), its use as a relative age indicator is well established (McFadden *et al.* 1987, 1989, 1998, Amit *et al.* 1993, Wells *et al.* 1995, Haff and Werner 1996, Helms *et al.* 2003, Valentine and Harrington 2006, Adelsberger and Smith 2009, Dietze and Kleber 2012). Under optimal conditions, desert pavements can form in a few thousand years. And they can form in cold and hot deserts, and those in between (Li *et al.* 2003). On older surfaces, rocks in the pavement tend to be better and more closely packed, and the vesicular (Av) horizon below is usually thicker. Pavement clasts also become increasingly

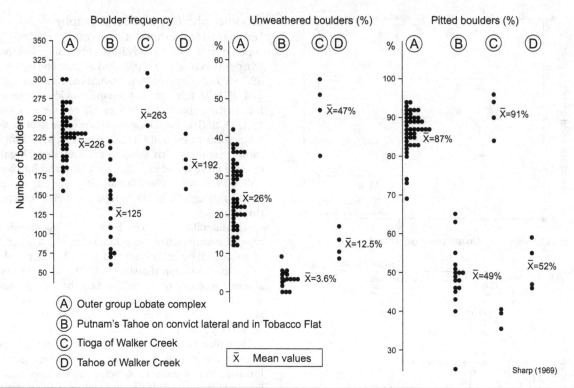

Fig. 15.6 Application of a relative dating technique – frequency of surface boulders – to determine the relative ages of four moraine complexes in the Sierra Nevada Mountains, California.

shattered, making their angularity and size, not to mention any rock varnish on them, additional SED tools (Dorn 2009a). With this knowledge in mind, Al-Farraj and Harvey (2000) devised a desert pavement development index, scaled from 0 to 4, for alluvial fan surfaces in the Middle East. These data correlate well with the relative degree of soil formation, because the pavements and the soils developed concurrently. Quade's (2001) work, however, cautioned that relative dating using desert pavement has limitations. He found that pavements on high deserts are no older than the Holocene, for these areas were cool and moist during the Pleistocene and therefore did not develop pavements. Only in the lowest, hottest deserts do pavements date back more than ~10 ka. Thus, PDMs on high-altitude surfaces in some deserts appear to reflect the effects of climate more than time.

Rock Weathering

A time-tested group of SED methods centers on the degree of alteration of rocks that are exposed on stable geomorphic surfaces (Tables 15.3, 15.4). Rocks that lie on a surface or are shallowly buried beneath it undergo slow, predictable changes with time. The PDMs that rocks undergo are usually associated with the development of weathering rinds or the accumulation of coatings of various kinds – both biotic and abiotic (Oguchi 2013). Use of rock

weathering information as a SED tool is most useful in deposits that have many large rocks of similar lithologies (Kiernan 1990), preferably with many of them exposed at the surface (Shiraiwa and Watanabe 1991). Age-weathering relationships for rocks are most commonly applied, and therefore presumably most dependable, in dry, semiarid, and alpine regions (Colman and Pierce 1981).

Most studies of rock weathering have shown a clear relationship between weathering and time (Colman 1981). Analysis of rock-related features has an advantage over soil development as a SED tool because many more rocks than pedons can be sampled. Plus, rock characteristics can often be determined quickly in the field.

Before we proceed with a discussion of rock weathering as a SED tool, some precautionary statements are in order. Lowe and Walker (1997) listed four concerns with using rock weathering as a relative age dating tool: (1) The degree of weathering is often difficult to measure objectively; (2) rates of weathering intensity are nonlinear, decreasing with time; (3) calibration of the weathering rates to numeric ages, using isotopic methods, is often problematic or not possible; and (4) weathering intensity varies with climate and other local conditions. Indeed, research into the controls on weathering processes, e.g., Goudie and Watson (1984), Jenkins and Smith (1990), and McGreevy (1985), illustrates that many weathering processes are

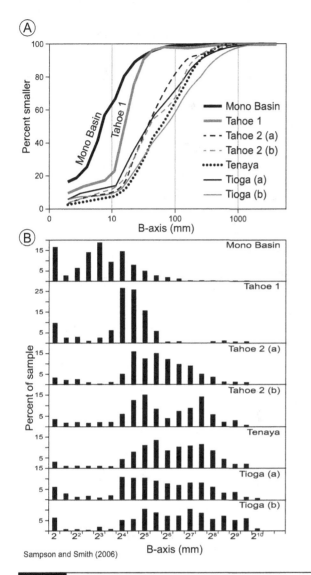

Sampson and Smith (2006)

Fig. 15.7 Relative dating comparisons using clast size distributions for moraines in the Sierra Nevada Mountains, California. The two oldest moraines, Mono Basin and Tahoe 1, are clearly separable from the younger ones, and from each other. (A) Cumulative clast size distributions. (B) Clast size histograms (grouped by bin).

Nicholas and Butler (1996) argued that the surfaces of coarse-textured rocks become rougher (less smooth) with time. As rocks weather they may develop weathering rinds or acquire coatings, e.g., rock varnish (Krinsley *et al.* 1990, Dixon *et al.* 2002, Dorn 2009a), silica, and carbonate coatings (Unger-Hamilton 1984, Curtiss *et al.* 1985; Table 15.5); coats of weathering by-products such as clay; or coverings of lichens. For rock varnish in desert environments, two relative dating methods are available to researchers: (1) the amount of Mn that has accumulated in the varnish and (2) the proportional amount of the surface that has become covered by varnish (McFadden *et al.* 1989, Reneau 1993; see Chapter 13). Dorn and Oberlander (1981a), however, cautioned that varnish forms irregularly in time and space, rendering it a potentially *supportive* SED tool, but one that should not be used as the *sole criterion* for age assessment. For example, problems due to saltating sand, which can abrade away some of the accumulated varnish, can arise.

A relatively simple method used to assess rock weathering involves determining the relative hardness of the rocks, by assuming that their integrity/hardness is directly correlated with degree of weathering. Often, this is evaluated by striking exposed rocks with a hammer and rating the ease with which the rocks break apart. Ordinal classes of hardness such as unweathered, hard, moderately hard, and friable/weak may be used. Several additional, more quantifiable methods for estimating rock hardness also exist. For example, Crook (1986) studied compressional (P) waves sent through rocks; the velocity with which the waves propagate through the rock is a function of its degree of weathering. P wave assessments are common in engineering studies of rock stability and strength (Ceryan *et al.* 2008, Yavuz 2011, Undul and Tugrul 2012). Fractures and chemical changes in the rocks, caused by weathering, cause the waves to move more slowly (Fig. 15.10). This technique, termed the clast-sound velocity (CSV) method, has advantages over others: It is nondestructive, meaning that one can obtain multiple measurements on a single specimen, and, as a result, the precision of that measurement is increased.

Another roughly equivalent method involves the Schmidt hammer – a device used to test the *hardness of a surface* (McCarroll 1987, 1989, 1991, Goudie 2006, 2013; Fig. 15.11). The hammer operates by propelling a weight against a plunger that is in contact with the rock being tested. After the impact, the weight rebounds to a distance proportional to the hardness of the rock (Nicholas and Butler 1996). Initially designed to test concrete and building stone, the method has found applications in rock weathering and geomorphology (Monroe 1966, Day and Goudie 1977, Matthews and Shakesby 1984, White *et al.* 1998, Winkler 2005, Cerna and Engel 2011, Viles *et al.* 2011), as well as on various types of soil pans, e.g., duripans, gypcrete, and caliche. In theory, the Schmidt hammer works because weathering processes cause microfractures to develop, and those fractures increase rock compressibility

site-specific, and thus, relationships between weathering and time cannot always be extrapolated to other locations. Nonetheless, the successful application of rock weathering in relative dating studies has such a long history that it merits a detailed discussion here.

As with soils, rocks on exposed surfaces change with time, and these changes can take many different, but measureable, forms (Fig. 15.9). They may become more fragmented or abraded. They may become more rounded, as weathering processes preferentially attack sharp edges.

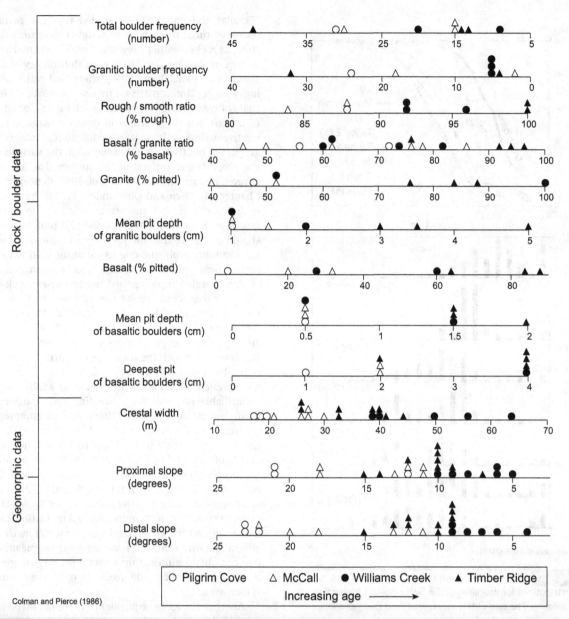

Colman and Pierce (1986)

Fig. 15.8 Detailed rock weathering and moraine morphology data, as applied to an SED study on some morainic landforms in Idaho.

(Crook 1986, Crook and Gillespie 1986). Results from Stahl *et al.* (2013) suggest that the hammer is useful for surfaces that are $<10^5$ years old, and that the age uncertainties of the hammer are consistent with those of weathering rind studies.

Rock angularity has also been used to assess weathering, on the assumption that rocks become more rounded through time (Birkeland 1973, 1982). Other researchers have measured the degree to which the rock surfaces are pitted or stained along fractures or split along cracks (Berry 1994). The height of weathering-resistant veins or posts

(usually quartz) increases through time as well, as the surrounding matrix weathers away (Birkeland 1982, Rodbell 1993; Fig. 15.5A). Quartz vein heights (and rind thicknesses) should always be viewed as *minimal relative ages* in SED studies, because parts of them could have broken off at some time in the past. Additionally, fires can cause rinds and veins to spall and fall off. Dyke (1979) determined degree of weathering by examining how many rocks had lost evidence of glacial striae.

Although some of the criteria discussed permit a quantitative assessment of weathering, many allow for only a

Table 15.4	Weathering classes for surface-exposed boulders
Weathering class	Characteristics and diagnostic weathering features
1	Completely fresh and unweathered
2	Surface staining
3	Surface is rough due to crystal relief but crystals not recoverable by hand
4	Crystals removable with fingernail
5	Crystals removable by rubbing
6	Micropitted (<1 cm), with exfoliation shells or weathering relief >1 cm
7	Macropitted (>1 cm) or inclusions protruding
8	Surface disintegrated
9	Deeply weathered or completely disintegrated

Source: Dyke (1979).

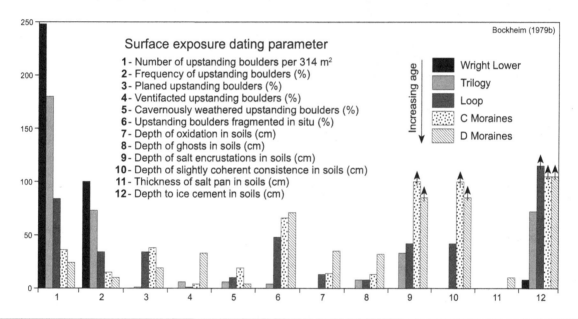

Fig. 15.9 Rock weathering and soil developmental properties on a series of moraine surfaces of increasing age in Antarctica. Arrows at the tops of the histogram bars indicate that the data included a "greater than" category.

subjective, class-based ranking. Indeed, many kinds of rock weathering data are, at best, ordinally scaled, forming classes such as highly weathered, slightly weathered, and unweathered (Sharp and Birman 1963, Rahn 1971, Hall and Michaud 1988). Along these lines, several ordinal weathering scales have been devised (Miller 1973, Brookes 1982, Whitehouse *et al.* 1986; Table 15.4).

Weathering Rinds

As rocks weather, almost always from the outside in, their outer parts oxidize and develop a discolored layer, called a *weathering rind* (Fig. 15.12). Rinds are discolored and more permeable, typically enriched in immobile oxides, e.g., Fe_2O_3, TiO_2, and Al_2O_3, and with altered primary and secondary mineralogy (Fig. 15.13; see Chapter 9). On the whole, weathering rinds are dominated by *dissolution* of minerals and materials, rather than the accumulation of weathering by-products, usually involving losses of alkali and alkaline earth metals. They also usually involve the oxidation of Fe and the development of secondary porosity (Dixon *et al.* 2002).

Rind *thickness* is one of the best and widely used SED methods (Chinn 1981, Gellatly 1984, Shiraiwa and Watanabe 1991, Douglass and Mickelson 2007), although one could also use rind hardness, color, specific gravity, or mineralogy. Use of weathering rinds as a SED tool is especially common in alpine settings, on moraines and rock glaciers, and has also been applied frequently to debris flows

Table 15.5 | Types of naturally occurring rock coatings

Coating name	Composition	Related terms
Carbonate skin	Coating of primarily $CaCO_3$, but sometimes $MgCO_3$	Calcrete, travertine
Case hardening	Any cementing agent to a rock matrix; composition widely varies	Depends of composition
Dust film	Light coating of clay- and silt-sized particles, attached to rough surfaces and in rock fractures	Clay skins, clay films, soiling
Heavy metal skins	Coating of iron, manganese, copper, zinc, nickel, mercury, lead, and other heavy metals, in natural or human-altered settings	Depends on composition
Iron film	Coating of iron oxides and oxyhydroxides	Ferric oxide, iron staining
Lithobiontic coatings	Organisms forming coatings, e.g., lichens, moss, fungi, cyanobacteria, and algae	Organic mat, biofilm, biotic crust
Nitrate crust	Coating of potassium and calcium nitrates, often in caves and rock shelters	Saltpeter, niter, icing
Oxalate crust	Coating of mainly calcium oxalate and silica, with other impurities; often near or in association with lichens	Patina, oxalate patina, lichen-produced crust, scialbatura
Phosphate skin	Coating of various phosphate minerals, e.g., iron phosphate or apatite, sometimes mixed with clay or manganese	Organophosphate film, epilithic biofilm
Rock varnish	Coatings of clay minerals, manganese and iron oxides, and minor trace elements, ranging from orange to black in color	Desert varnish, patina, Wüstenlack
Salt crust	Coating of chloride-rich precipitates	Halite crust, efflorescence
Silica glaze	Coating of usually clear white to orange luster, but can be darker, composed primarily of amorphous silica and aluminum, and commonly with iron	Desert glaze, turtle-skin patina, siliceous crust, silica-alumina coating, silica skin
Sulfate crust	Coating of sulfates, e.g., barite, gypsum	Sulfate skin

Source: Dorn (2013).

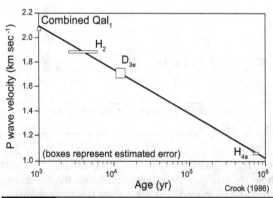

Fig. 15.10 Curve of *P* wave velocity versus age for rocks on geomorphic surfaces in the San Gabriel Valley, California.

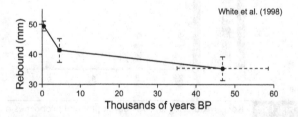

Fig. 15.11 Rebound of the Schmidt hammer for rocks on three alluvial fan segments of different age.

(Ricker *et al.* 1993). Caution should be used when applying rind data alone in SED studies, because weathering rates are rapid at first and increase more slowly with time. This trend occurs because (1) weathering by-products accumulate in the rind and (2) as the rind thickens, the unweathered core becomes increasingly protected (Brookes 1982; Fig. 15.14). Assessment of rind data and weathering rates must always keep this fact in mind.

Assessing the degree of rock weathering on a geomorphic surface is usually accomplished by examining rind thicknesses on ~30–50 rocks. Rocks on the surface are preferred, although successful work has been accomplished on shallowly buried rocks (Yoshida *et al.* 2011). If subsurface clasts are used, this should be indicated. As would be expected, weathering rind data for subsurface clasts are more variable, as the weathering intensity within the soil is variable. Buried clasts also tend to weather more slowly than do surface-exposed clasts, depending upon climate and soil conditions. In dry climates or in salty soils, however, the reverse may be true. Mean, maximal, or modal rind thicknesses are then calculated and used to characterize surface exposure, on the assumption that rind thickness is positively correlated to surface age (Birkeland 1973, Chinn 1981; Fig. 15.15). Measurement precision is essential, because errors of even <1 mm can dramatically affect the age estimate. Thus, the rind/no rind interface must be precisely defined; this is often not possible in some types of rocks (Fig. 15.16).

Weathering rind thicknesses are a function of many factors in addition to exposure age. Climate, lithology, and the effects of slow burial or exposure all affect weathering rind thicknesses (Chinn 1981, Knuepfer 1988, Yoshida *et al.* 2011). Rind thicknesses also tend to differ as a function of clast shape and curvature, i.e., the core is often more spherical than is the clast exterior, indicating that rinds are thicker (and weathering rates faster) near protuberances on the rock surface (Sak *et al.* 2010). Care must also be taken with any study of weathering rinds, because as rinds form,

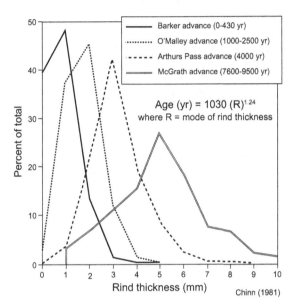

Age (yr) = 1030 (R)$^{1.24}$
where R = mode of rind thickness

Fig. 15.12 Changes in weathering rind thickness on moraines of increasingly greater age, Southern Alps, New Zealand.

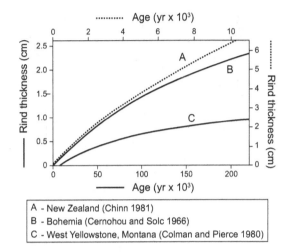

A - New Zealand (Chinn 1981)
B - Bohemia (Cernohou and Solc 1966)
C - West Yellowstone, Montana (Colman and Pierce 1980)

Fig. 15.14 Weathering rind thickness versus time for three representative SED studies, based on Colman's (1981) compilation.

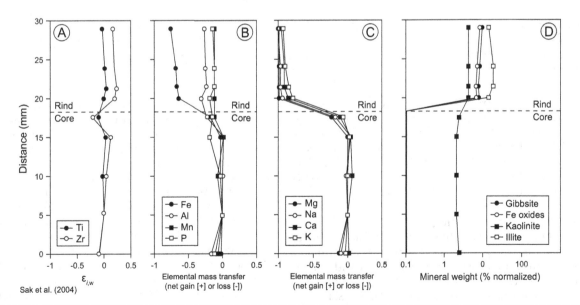

Fig. 15.13 Weathering characteristics of a rock and its weathering rind, as indicated by elemental and mineralogical data, shown by distance from the center of the rock from a fluvial terrace in Costa Rica.

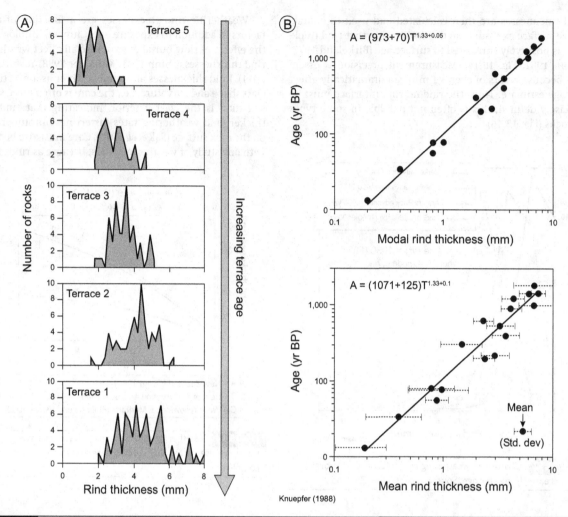

$A = (973+70)T^{1.33+0.05}$

$A = (1071+125)T^{1.33+0.1}$

Knuepfer (1988)

Fig. 15.15 Weathering rind thickness data for rocks on some fluvial terraces, South Island, New Zealand. (A) Variability of rind thickness for the rocks on the Saxton River terraces. (B) Calibration curves for rind thickness versus terrace surface age for a number of rivers in the region. Most of the ages used to calibrate the terraces are on wood from the underlying alluvium.

they may spall or erode. For this reason, weathering rind analysis has a limited span of utility. Coarse-grained rocks, such as granite, develop rinds rapidly, but they quickly spall from the rock, making them of little use (Brookes 1982, Gordon and Dorn 2005, Mahaney *et al.* 2012). Fine-textured, extrusive igneous rocks are preferred, as rinds form more slowly and persist longer than on granites.

Obsidian Hydration Dating

A very fine-textured rock, obsidian, has a special place in the world of rind dating. Obsidian absorbs water and hydrates slowly through time, developing a whitish rind with a sharp inner front (Friedman and Smith 1960, Friedman and Long 1976). *Obsidian hydration dating*, a special form of SED using weathering rinds on obsidian clasts, is the most quantitative of all rind analyses. Rind development on glasslike

obsidian occurs extremely slowly and persists well; most obsidian rinds are so thin that they must be measured under a high-magnification microscope. Obsidian has a distinct utility in dating archaeological sites, because it is a raw material used in the manufacture of stone tools such as projectile points, knives, or other cutting tools (Pierce *et al.* 1976, Meighan 1983, Ridings 1996). In this application, after the exposed surfaces of ancient obsidian artifacts had absorbed water, a hydration rim formed, whose width is, theoretically, dependent on time (exposure), chemical composition (obsidian source), and temperature.

For years, obsidian hydration was viewed as a powerful SED tool, and one that had many applications in archaeology (Michels 1967, Meighan *et al.* 1968, Friedman and Trembour 1978, Stevenson *et al.* 1989, Tripcevich *et al.* 2012). Its potential utility derives not only from the fact that

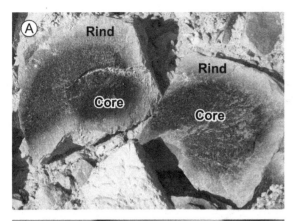

Fig. 15.16 Weathering rinds. Photos by RJS. (A) Diffuse but thick weathering rinds on quartzite clasts. (B) Sharp edge to a weathering rind in a shaly sandstone.

obsidian is numerically datable by K-Ar, but also because (1) the rate of hydration under standardized conditions is known and has been modeled (Friedman and Long 1976), and (2) rinds appear to develop at similar rates in both shallowly buried and subaerial rocks. Recent work and critical analyses, however, have cast doubt on obsidian hydration dating (Ridings 1996). Anovitz et al. (1999) argued that optical techniques cannot provide the needed precision for measuring rind thicknesses, and that the theoretical basis on which the weathering rates have been determined is incorrect. Others have noted that the approach is sensitive to temperature and humidity under surficial conditions (Anovitz et al. 2006). In short, this method – in its traditional form – "has been long on promise but short on results" (Anovitz et al. 1999: 735). Nonetheless, in a recent review of obsidian hydration dating, Liritzis and Laskaris (2011) argued that new technologies may soon make this method more appealing and applicable to archaeological and geomorphic studies in the future. Perhaps, obsidian hydration data may become a better paleothermometer than a dating tool (Anovitz et al. 2006).

Hornblende Etching

The weathering characteristics of individual, usually silt- or sand-sized, grains within sediment is also a potentially valuable SED tool (Hall and Michaud 1988). Hornblende is an especially useful mineral for this type of analysis (Locke 1979, Hall and Horn 1993, Mikesell et al. 2004), as are apatite (Lång 2000) and feldspar (Read et al. 1996). Hornblende develops cockscomblike terminations and etch pits in such a predictable manner that the depth of the etch pits can be quantified and correlated to weathering intensity, and these features are often a function of age (Mikesell et al. 2004; Velbel and Losiak 2010, Velbel 2011, Howard et al. 2012; Fig. 15.17). Like many age-related functions, hornblende etching is initially rapid, but the rate of etching decreases with time. And like many other weathering-related SED methods, it varies in intensity with depth in the soil and as a function of climate. Strong etching is also associated with increased amounts of effective precipitation (Locke 1979). Similar etching/weathering data could be generated for minerals such as garnet, feldspars, and amphiboles, but little work has been done in this area (Read 1998). A word of caution – when applying this type of method and sampling for sand- and silt-sized grains, one must always consider that they can be easily moved within soil by pedoturbation processes (see Chapter 11).

Rock Coatings

Over time, rock surfaces can weather to form rinds, or they can acquire various kinds of coatings. Dorn (2013) listed 14 different types of coatings that can cover rock surfaces (Table 15.5). We are just beginning to understand the details of the genesis of many of these coatings, and we must better understand the conditions under which they are maintained or eroded. Only after that can we begin to use their physical and chemical characteristics as a SED tool. Thus, the potential for developing new SED tools using rock coatings is enormous.

For coatings to form, bare rock faces must first exist. For these coatings to be used in SED context, the exposure of the bare rock surface must coincide with time$_{zero}$ of surface exposure. Because most bare rock surfaces occur in deserts and high alpine settings, most work on rock coatings, e.g., rock varnish, has been in these environments (Dorn 1988, 1998, 2009a, 2013, Dorn and Oberlander 1981b). Coatings can occur on the tops or undersides of rocks, or within cracks, each with different life spans and implications. Erosion of the coating – always a consideration – is much more likely on the top of a rock. Many rock coatings occur on top of a previous one. When the outer coating is eroded, a lower one may be exposed (Fig. 15.18). Thus, the story can become complicated, but also can hold great promise for understanding past environments.

Dating rock varnish in deserts has been the focus of a great deal of work. Dorn (2013) reviews the many ways (eleven!) that rock varnish has been chronometrically assessed. These methods range from relative dating

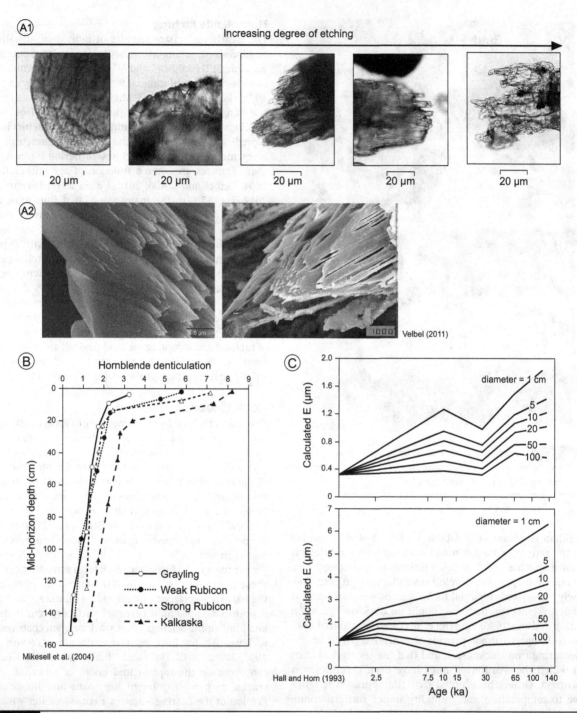

Fig. 15.17 Hornblende etching as a relative dating tool. (A) Images of fine sand-sized hornblende grains. (1) A series of photomicrographs that show increasing degrees of etching on the margins of grains. Images by M. Velbel and L. Mikesell. (2) SEM images of the same type of etching. (B) Mean etching category versus depth within four soils on an outwash plain in Michigan, illustrating that factors other than age, e.g., depth and pedogenesis, also affect etching. (C) Etching versus age and depth. Note that the x-axis is logarithmic, indicating that the increase in etching rates slows with time.

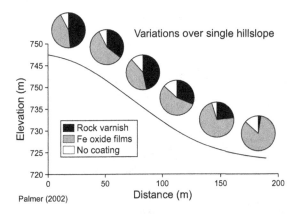

Palmer (2002)

Fig. 15.18 Rock coating characteristics along a catena on basalt, in the Mojave Desert. Less erosion on the summit allows more rock varnish to form. The increased amounts of Fe oxide films downslope reflect greater amounts of soil erosion, which has exposed Fe oxide films that were originally formed when the rocks were buried.

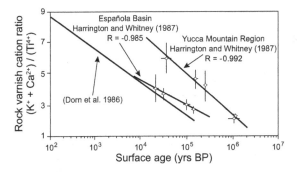

Fig. 15.19 Examples of rock varnish cation ratios as a relative dating tool, using cation leaching curves, for three lava flows dated with K-Ar and U-series in the Mojave Desert, California.

approaches to numeric dating techniques. From a relative dating perspective, simple appearance seems to be a plausible SED method. For example, the surface darkens and the varnish thickens and increases in coverage over time. However, because color and thickness are affected by inherited coatings and the nature of the underlying weathering rinds, Dorn (2013) cautioned that this method does not permit highly accurate or precise age assignments. Another approach that is commonly used for assessing rock varnish involves the accumulated Fe and Mn in the varnish, on the assumption that as more varnish accumulates, the amount of Fe and Mn gradually increases. Nonetheless, this method was demonstrated to be inaccurate (Dorn 2001). Other methods that can be used to inform us about rock varnish, and hence, exposure ages, include (1) ^{14}C dating of carbonate or organic matter found in association with rock varnish, (2) U-series dating, and (3) lead profiles (Dorn

2013). Lead profile dating operates on the knowledge that Fe and Mn in the rock varnish scavenge lead and other metals from twentieth century pollutants, resulting in a spike in the surface varnish layer. Finally, a technique that holds great promise is the use of microlaminations in the varnish (Liu and Broecker 2007, 2008a, 2008b; Fig. 13.83). Past climate fluctuations change the pattern and chemistry of varnish microlaminations, enabling a rough correlation to known intervals of past climate. Confidence in the potential of the microlamination method is high (Marston 2003).

Cation Ratio Dating of Rock Varnish

First proposed by Dorn and Oberlander (1981b), cation ratio dating is a calibrated relative dating method for use on desert varnish. In theory, the ratio of soluble to insoluble cations in the varnish decreases with time because soluble cations are replaced or depleted, relative to less mobile cations (Dorn *et al.* 1986, 1990, Dorn 1989, Dorn and Krinsley 1991, Dorn 2001). Choosing the correct type of varnish for this application appears to be important (Dorn 2013). Although a number of ratios could be used with this application, the most popular one is [(Ca + K)/Ti], hence the term cation ratio dating.

This ratio assumes that the amount of time the varnish has been exposed to cation leaching, and the initial ratio, was constant for the rock varnish within the region of interest. Within thick varnishes, the uppermost layers should be most leached (Fig. 15.19).

Cation ratio ages are generally assumed to provide minimum ages and obviously are applicable only to dry regions (Harry 1995). The method has potential for dating human artifacts and petroglyphs in desert areas (Nobbs and Dorn 1988, Loendorf 1991) and has been applied in China (Zhang *et al.* 1990) and Israel (Patyk-Kara *et al.* 1997), among other places. Although some researchers have questioned the accuracy and validity of cation ratio dating (Dragovich 1988, Krinsley *et al.* 1990, Reneau and Raymond 1991, Bierman and Gillespie 1994, Harry 1995, Watchman 2000), it has performed well in blind tests (Loendorf 1991).

Lichenometry

Biological organisms can also inhabit rock surfaces; these organisms are called *lithobionts*. One of the most common lithobionts used in SED is the lowly crustose lichen. *Lichenometry* is the branch of relative dating that uses lichen sizes as a SED tool.

In 1950, Roland Beschel established that the maximum diameter of the largest thallus of a crustose lichen on a rock surface is proportional to time since colonization, or surface exposure. As such, he ushered in a new SED tool – lichenometry (Fig. 7.3). A *thallus* (plural, *thalli*) is the body of a lichen, which for epipetric lichens usually exists as a circular form on a hard substrate, such as a rock. Crustose species are the slowest growing of all lichens; their slow

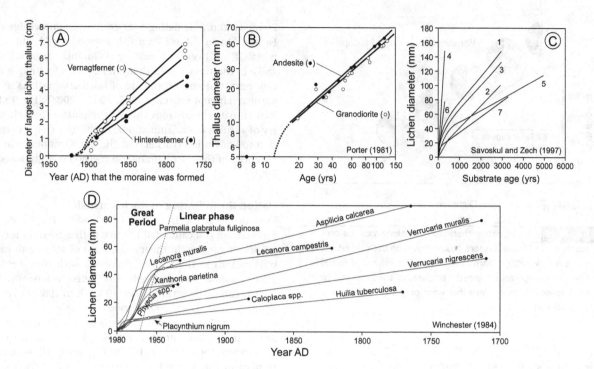

Fig. 15.20 Examples of lichen growth curves. (A) Roland Beschel's (1950) original lichen growth curves, for moraines in the Austrian Alps. (B) Growth curves for lichens on andesite and granodiorite boulders on Mt. Rainier, Washington. (C) *Rhizocarpon geographicum* growth curves, established at various alpine sites around the world. (D) Growth curves for eleven different lichen species, from tombstones in southern England.

growth and longevity made them the group of choice for many lichenometric studies (Armstrong and Bradwell 2010). The method assumes that lichens begin colonizing rocks shortly after the rocks are subaerially exposed and expand in area predictably over time. Theoretically, lichenometry involves the measurement of lichen sizes on surfaces of different, but known, ages, and using these data to calibrate a lichen radial *growth curve* (Benedict 1967, 2009, Lock *et al.* 1979). Growth curves can also be determined by directly monitoring the increase in size of individual thalli over a period of years. Unique to each lichen species, a growth curve can then be used to estimate the exposure ages of other surfaces for which numerical age control has *not* been established.

Lichenometry is especially applicable to cold, harsh alpine regions, where rocky surfaces are common, lichen growth is slow, and vascular plant competition is low (Webber and Andrews 1973, Bradwell 2009, Harvey and Smith 2013). Plus, the absence of trees in the alpine tundra eliminates dendrochronology (see later discussion) as an SED tool. The utility of lichenometry is further accentuated by the fact that soil development, another possible SED tool, is minimally useful in these cold, dry environments, where pedogenesis is slow, and where cryoturbation is common. Lichenometry is most useful on young (late Holocene) surfaces that are <1,000 years

old (Birkeland 1973, Porter 1981). Although Benedict (1967) was able to construct a growth curve that could be used to establish ages of deposition to perhaps 3.5 ka, Nicholas and Butler (1996) argued that extrapolation of lichenometry curves beyond ca. 2.5 ka involves increasing uncertainty. Benedict (2009) suggested that the practical limit of the method approaches 5 ka. In general, as the environment becomes more mild and moist, lichens will grow more rapidly, increasing dating resolution but shortening the useful time span of the method (Ten Brink 1973, Benedict 2009). The method has good precision on surfaces of very young age, as indicated by the fact that lichens on gravestones are commonly used to provide age control for the early stages of lichen growth curves (Beschel 1958, Carrara and Andrews 1973; Fig. 15.20). Crustose lichens grow especially slowly in cold climates (Armstrong and Bradwell 2010). Thus, their thalli may not exceed the diameters of the rocks they are growing on for thousands of years. Beschel (1950) described an initial period of rapid growth, called a *great period*, that lasts only a few decades (Fig. 15.20C). During this time, radial growth rates are about 15–50 mm/century (Webber and Andrews 1973). Some believe that the great period is followed by a longer period of much slower but steady growth – the linear phase (Porter 1981, McCarthy and Smith 1995). Linear phase growth rates are 2–4 mm/century, or less. According

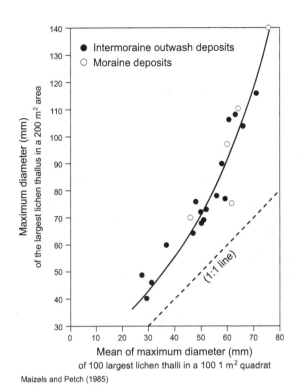

Maizels and Petch (1985)

Fig. 15.21 Relationship between the maximum and mean diameters of lichen thalli, from a study in Norway.

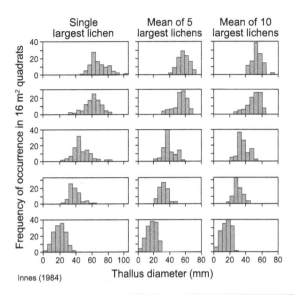

Innes (1984)

Fig. 15.22 Histograms of *Rhizocarpon* lichen thalli diameters, as recorded for moraines of different age in Norway. Note that the variation in sizes decreases when sample sizes increase from 1 to 10.

to Benedict (2009), the linear phase should be regarded as a working hypothesis, subject to testing.

Sampling of lichens for lichenometry requires considerable forethought (Benedict 2009). If one chooses to use a lichen growth curve for an individual species, this will add precision to age estimates (Fig. 15.20C), but identification of lichens to species is difficult. A way around this is simply to measure percentage cover of lichens on a deposit, regardless of species (Nicholas and Butler 1996). This method – which can be accomplished digitally (McCarthy and Zaniewski 2001, Bowker *et al.* 2008) – precludes the problems of species identification and establishment of a growth curve, by working on the underlying assumption that older deposits should exhibit heavier lichen cover than more recent deposits.

Lichens of all sizes, shapes, and ages occur on rocks. However, the small (young) ones probably reveal little about surface age, as they began growing long after the surface stabilized. Thus, in theory, we want to find the *one lichen* that began growing when the surface was first stabilized and has continued to grow up to the present day. This theory explains why the size of the *single largest lichen* is most often correlated to surface age (Webber and Andrews 1973, Calkin and Ellis 1980, Orombelli and Porter 1983, Harvey and Smith 2013). However, the one largest lichen could be an anomaly (Matthews and Trenbirth 2011). For

this reason, some researchers do not sample abnormally large thalli (Calkin and Ellis 1980, Rapp and Nyberg 1981). To quote Matthews (1975: 104):

There is no reason why the single largest lichens should be preferred to the use of means (*averages*) of more than one largest lichen.... Indeed there are grounds for preferring means in that a single largest lichen is more likely to be an anomaly.

This discussion raises a valid question about lichenometry, which also applies to other SED methods: What is the sampling strategy? How many lichens should be sampled per surface? Should lichen *size* or simply lichen *coverage* (the relative amounts of the rock surfaces that have lichens) be the focus? Many researchers believe it more statistically valid to use the mean or modal size of a *sample* of large lichens, rather than relying on data from one large lichen thallus (Matthews 1974, Innes 1985). Maizels and Petch (1985) found a strong relationship between these two types of data, perhaps making the distinction less critical (Fig. 15.21). Nonetheless, by sampling larger numbers of lichens, the variation in lichen size on each surface decreases (Fig. 15.22), as does the likelihood of an erroneous result due to one anomalously large thallus. However, as sample size increases, it becomes increasingly likely that greater numbers of sampled lichens began growing long *after* the surface was subaerially exposed. Thus, a clear-cut answer to the question of how many lichens should be sampled remains elusive.

A related question asks, "How large an area should be canvassed for lichens?" As the area sampled becomes

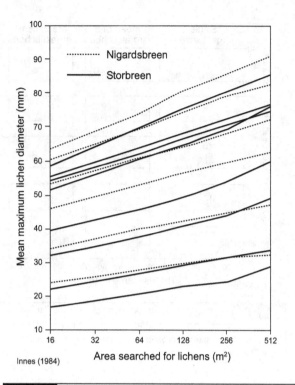

Innes (1984)

Fig. 15.23 Increase in the mean maximum lichen diameter with increasing size of the area sampled.

larger, more large lichens are found (Fig. 15.23). Probably the best rule of thumb to keep in mind with regard to SED sampling strategies is to remain *consistent*, both with the literature and within one's own study.

All relative dating methods have assumptions that are commonly (and sometimes, necessarily) violated; lichenometry is particularly enigmatic in this regard (Jochimsen 1973). We have discussed some sampling concerns, but also consider these issues:

1. *Species effects.* Lichen species grow at different rates, necessitating that sampled thalli be from the same species. However, it is often difficult to identify each thallus to species. Conversely, Winchester (1984) suggested that the accuracy of the method is, in fact, enhanced if several lichen species are used and the data agglomerated (Fig. 15.20C).

2. *Thallus shape.* Thalli vary in shape as well as size. Seldom are they perfectly circular. What should be measured? Maximum or minimum diameter? Area? Typically, diameter is used as a size metric, although evidence suggests that perimeter or area may correlate better to age (McCarthy and Henry 2012). Furthermore, when lichens grow together and overlap, interaction may affect growth rates.

3. *Microclimate effects.* Although this factor can be controlled, or taken into account, microclimate is a particularly important factor in affecting lichen growth rates (Beschel 1958, Webber and Andrews 1973). Lichens growing on the top versus sides of rocks vary in growth rates because of differing microclimates (Birkeland

1973, Matthews and Trenbirth 2011). Quartz veins that stand in relief cast shadows that can affect growth rates and restrict the spread of the thalli (Birkeland 1982). Also, on microclimatically cold sites or sites where snow patches persist well into summer, lichens can be much smaller than at more favorable sites (Benedict 1990). Last, in sites near cities, air pollution can affect lichen growth rates.

4. *Substrate effects.* In lichenometry, one always tries to keep the rock type constant, as growth rates do vary as a function of lithology (Innes 1984). Some rocks are not conducive to lichen growth (Carrara and Andrews 1973), while others may be problematic because they weather so rapidly that the spall and erode, forcing loss of lichen data. Colonization of rock surfaces by lichens is also controlled in part by surface texture; lichens colonize rough surfaces faster than smooth surfaces (Beschel 1973, Topham 1977). Thus, older deposits, with their more weathered clasts, would be likely to show greater lichen cover, other factors being equal.

5. *Pathogen effects.* Porter's (1981) work illustrated how a widespread lichen kill can seriously limit the technique.

Nonetheless, despite all the potential pitfalls, lichenometry has been repeatedly shown to be a reliable SED tool for surfaces younger than about 4,500 years. It is especially reliable when used in conjunction with other methods, such as weathering rind thickness. And for the environments in which it is normally applied, i.e., alpine and young surfaces, it is often the best option for exposure dating.

Dendrochronology and Dendrogeomorphology

Dendrochronology (tree ring dating) is a related and useful tool for determining the paleoenvironmental conditions for a site (Cook and Kairiukstis 1990, Leavitt and Bannister 2009, Speer 2010). Commonly, the width of a tree ring is correlated to a climatic variable, usually temperature or moisture availability, even in the nongrowing season (Opała and Mendecki 2013). In cold areas, where temperature limits growth, e.g., at alpine treeline, the width of the rings is best correlated to temperature variables. In dry areas, tree ring width might respond best to precipitation variables. Salzer and Hughes (2007) found a signal in the tree-ring record that is related to volcanic eruptions.

Dendrochronologists use the technique of cross-dating (Fig. 15.24) to take the tree ring sequence, and hence the climate, record back much further than can the record of current living trees (Douglas 1941, Fritts 1976). Cross-dating can also add to the accuracy of tree ring records. A simple annual ring count, without cross-dating, can produce errors in analysis of true age of geomorphic episodes due to locally absent, or false, rings (Speer 2010). By applying cross-dating techniques in the very old bristlecone pine (*Pinus longaeva*) forests of California, dendrochronologists have extended the tree ring record back more than 8,300 years (Ferguson and Graybill 1983, Suess and Linick 1990). However, because most sites have trees that are less than a few hundred years old,

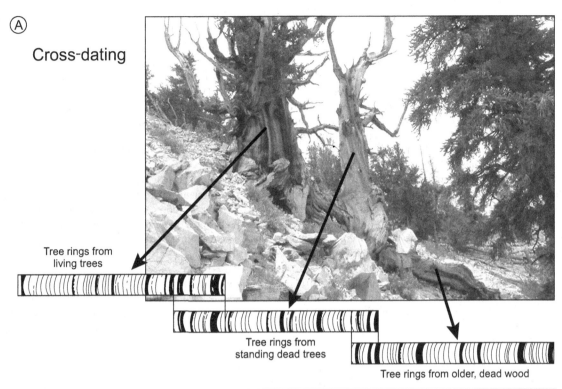

Ⓐ

Cross-dating

Tree rings from
living trees

Tree rings from
standing dead trees

Tree rings from older, dead wood

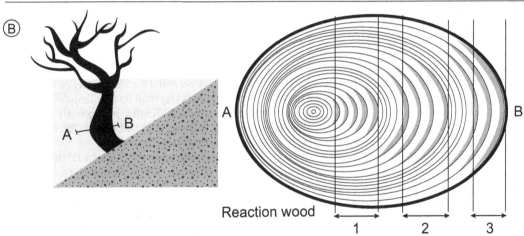

Ⓑ

A

B

A ⊢ B

Reaction wood

1   2   3

Fig. 15.24 Cross-dating and dendrogeomorphology. (A) The principle of cross-dating, using tree rings. From a U.S. Forest Service interpretive guide, Inyo National Forest. Original artwork by J. Janish. The tree pictured is a bristlecone pine. Image by RJS. (B) An example of the use of tree rings to address a geomorphic question. Tree rings respond after three distinct periods of slope instability, each of which has caused the tree to lean. Wider rings (tension wood) on the upslope side of the deciduous tree are its response to this leaning, as it attempts to right itself. Figure provided by P. Owczarek.

dendrochronology usually sacrifices longevity for this high (annual) level of SED (and paleoclimatic proxy) accuracy.

Dendrogeomorphology is the study of geomorphic processes and surface exposure through the use of tree ring analysis (Shroder 1978, 1980, Stoffel *et al.* 2010a). The technique is primarily useful on surfaces younger than a few centuries. By counting the annual rings of the oldest trees growing on a surface, one can establish the minimum-limiting age of that surface, because one can assume that the surface had to exist before the tree could have started growing on it (Bryan and Hupp 1984, Hupp and Carey 1990). That is, the surface cannot be younger than the tree.

Because trees respond to geomorphic stimuli, the tree ring record is often a good proxy for the geomorphic record. To that end, Shroder (1980) identified the various responses of trees to geomorphic events; they can develop *reaction wood* or callous growth, or exhibit abrupt growth changes. A change in growth position of a tree, e.g., by tilting during landslide movement, causes the trunk to develop bent and eccentric shapes, as well as the production of reaction wood. This type of reaction wood, composed of gelatinous cells with thickened walls, can be used to determine the age of events that caused the tree to tilt (Speer 2010), e.g., landslides (Corominas and Moya 1999). Deneller and Schweingruber (1993) and Pawlik *et al.* (2012b) also used this signal to examine soil creep on unstable mountain slopes.

Other dendrogeomorphological signals are provided by events that wound or injure the tree, as, for example, by high-energy geomorphic processes like rock falls or debris flows (Stoffel *et al.* 2010b). Scars on the tree are overgrown by callus tissue around the wound. A count of the number of growth rings outside the scar enables the researcher to determine the age of the event (Hupp *et al.* 1987, Stoffel and Bollschweiler 2008). Not only tree trunks but also their roots can provide information about geomorphic events. Gärtner (2007) used such data to determine soil erosion, because abrupt changes in wood anatomy appear after roots become exposed by erosion (Ballesteros-Cánovas *et al.* 2013). Mass movement activity and other geomorphic phenomena can also be dated dendrochronologically, not only by using trees but also by using dwarf shrubs and shrubs in which measurable annual rings, reaction wood, and scars are clearly visible. Thus, this method can be useful in the areas where trees are absent, e.g., above alpine treeline and in polar regions. Owczarek *et al.* (2013) demonstrated new data about growth ring structures in High Arctic dwarf shrubs and, on the basis of this record, reconstructed debris flow activity over the last 70 years. Recently, the potential of these plants in the dendrogeomorphological research has increased because of the rapid environmental changes occurring in Arctic and high alpine areas.

SED Methods Based on Soil Development

It is truly a challenge and potentially even a pitfall to attempt to quantify soil development, so as to place soils on some sort of development sequence.

In 1968, Francis Hole penned an article entitled "What is a well-developed soil?" In it, he asked, "Which is better developed, a Spodosol or a Vertisol? This question is in a way as useless as asking which is better developed, a chipmunk or a giraffe?" (p. 12). Hole's wisdom does not escape us. There is clearly a challenge in determining what a well-developed soil is, or which of several soils is better

developed. But it is a challenge with great opportunity as well.

Daniels and Hammer (1992: 196) warned that "probably very little if anything" can be used from soil profile data to place a geomorphic surface within a relative age sequence. Despite this warning, many soil properties have been successfully used in SED applications (Markewich *et al.* 1989, Nelson and Shroba 1998). Perhaps Daniels and Hammer were so cautious because, for surfaces of different ages, it is difficult to hold the other four state factors (climate, organisms, relief, and parent material) constant. Additionally, soils do not develop and soil properties are not usually attained and lost *linearly* over long periods, further complicating matters. However, because many chronosequence studies have successfully related soil development to time, the use of soil properties as SED tools need not be entirely discarded – just employed judiciously (Table 15.3).

To use soil-based data for SED, one must understand not only pedogenesis, but also the paleoenvironmental history of a site. Knowledge of soil development pathways and what properties could have been acquired in a predictable manner for the site is essential. For example, one would not use carbonate content of the B horizon as an SED tool for soils in which carbonates do not accumulate, or in which they accumulate only for a short period and then are leached from the profile. Knowledge of paleoenvironments is also essential because soil processes may have been different in the past; a soil that is accumulating carbonates today may have only begun doing so after a recent climatic change toward aridity (see Chapter 16). Thus, although soil development has potential for SED, it may be limited in its application by one's less-than-perfect knowledge of pedogenesis and paleoclimate. Finally, pedogenesis is not always predictable. Soils regress and change because of thresholds and accessions (see Chapter 13).

In the application of soils as a relative dating tool, we attempt to relate soil development to surface age, or the state factor time. To use this approach, soil development must be evaluated in at least a semiquantitative manner – we have to put a number on it. Purely qualitative terms such as infantile, juvenile, immature, mature, and senile, as well as terms like weakly developed and well developed, have application and are widely used. They are better than chipmunk or giraffe! However, in an ideal SED situation, soil development should be quantifiable.

Individual Properties or Attributes

Soil properties that change regularly and predictably with time – and are quantifiable – are the most applicable in SED studies (Table 15.3, Fig. 15.25). The list of potential soil properties is a long one, e.g., horizon or profile thickness, accumulation or loss of mobile constituents from specific horizons, or some integrative index of soil development (Mills 2005). One of the better SED tools is mineralogy,

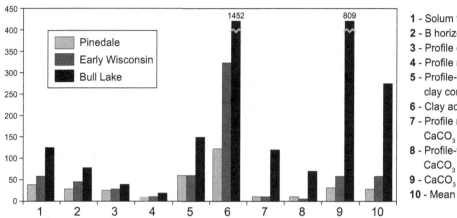

Fig. 15.25 Comparative data for soils on moraines of three different ages in Wyoming. For each soil property, the data represent mean values for a number of soils.

both clay mineralogy and sand or silt mineralogy (Ruhe 1956b, Soller and Owens 1991). As an example, iron concentration and the abundance of Fe-bearing minerals were shown to be good indicators of relative soil age in Italy (Ardiuno *et al.* 1986).

We start this discussion of potential soil properties with the easy ones and move to properties that are more difficult to obtain, but also may yield additional insight. Solum or profile thickness is an easy and useful parameter with which to assess development in young soils, especially those in which it is equivalent to depth of leaching, e.g., humid climate soils developing on calcareous parent materials (Allen and Whiteside 1954, Wang *et al.* 1986b; Fig. 15.26). Thicknesses of individual horizons are also useful parameters (Fig. 15.26). For soils that develop illuvial and eluvial zones, various measures of illuviation, as related to clay, Fe, Al, organic carbon, and $CaCO_3$ content, are appropriate measures of soil development that may relate nicely to surface age. Argillic horizons, for example, form relatively slowly. On many landscapes, Holocene age soils lack argillic horizons, whereas those of Pleistocene age have well-developed argillic horizons. Thus, an argillic horizon indicates that the geomorphic surface has been stable for a significant amount of time, usually at least 3–5 k years (Bilzi and Ciolkosz 1977, Ciolkosz *et al.* 1989, Cremeens 1995). A unique soil property used in relative dating studies is phosphorus retention, presumably because a soil's ability to retain more phosphorus from the soil solution increases as primary minerals weather and more secondary Fe oxides are precipitated (Smeck 1973; Fig. 15.27). Micromorphological features are also useful as soil development indicators (Berry 1994; Fig. 15.28). Weathering-related parameters in soil development often include either (1) ratios of resistant to weatherable minerals or (2) total content of a mineral or element (Dorronsoro and Alonso 1994).

Ideally, when applied to a SED study, soil properties should change systematically and, preferably, unidirectionally through time. Conversely, soil properties that are short-lived or reversible, or that vary in rate of development, are less useful or even problematic. Yaalon (1971) provided much of the theory that is used to judge whether a soil property will be useful or not. In the context of determining which soil attributes were most useful for examining *buried* soils, his focus was on their relative *persistence*. Properties that persist in a soil after burial would be useful for ascertaining what the soil was like at the time of burial. Those that do not persist and are quickly altered after burial, such as pH, would tell us little about the original soil. His threefold categorization has great utility for SED of surface (unburied) soils as well. Soil properties that tend to persist are often the most useful as SED tools, especially on older surfaces. Yaalon's categories, which provide a sound theoretical basis for deciding which soil properties to choose for an SED study (Table 15.3), are the following:

1. *Reversible, self-regulating processes and properties.* These properties rapidly attain a state of dynamic equilibrium and are therefore subject to rapid alteration or reversal when environmental conditions change. Nonetheless, they might be useful for determining relative ages on young surfaces, because they change rapidly over short periods. Examples include pH, base saturation, organic matter, contents of soluble salts and gypsum, mottles, gilgai, slickensides, and most kinds of soil structure.
2. *Processes and properties that slowly achieve a state of equilibrium.* These properties or horizons are slow to develop and relatively resistant to change, e.g., spodic, natric, argillic, gyspic, and calcic horizons; histic epipedons; fragipans; and some types of redox features.
3. *Irreversible, self-terminating processes and properties.* Once formed, these pedogenic features or properties are very

Legend for Fig. 15.25:
1 - Solum thickness (cm)
2 - B horizon thickness (cm)
3 - Profile development index
4 - Profile max clay (%)
5 - Profile-weighted mean clay content (% x 10)
6 - Clay accumulation index
7 - Profile maximum $CaCO_3$ content (% x 10)
8 - Profile-weighted mean $CaCO_3$ content (% x 10)
9 - $CaCO_3$ accumulation index
10 - Mean of all variables

Hall and Shroba (1995)

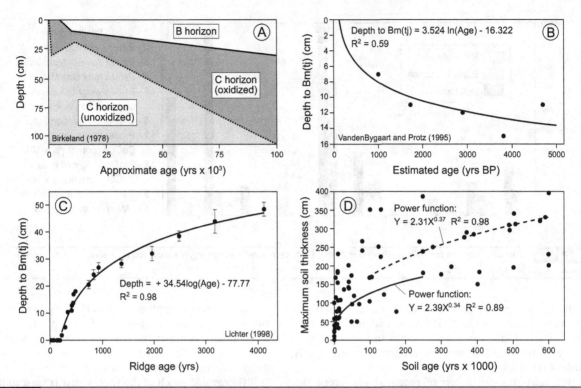

Fig. 15.26 Solum, profile, and B horizon depths, as indicators of their utility as SED tools. (A) Data (maximum values) for some soils on Baffin Island. (B) Regression data for some soils on a series of dated sand dunes in Ontario. (C) Single-log chronofunction describing the depth to the B horizon in some soil on beach ridges in Michigan. (D) Maximum soil thicknesses, as reported from 15 Mediterranean soil chronosequences, versus estimated soil age. Consult Sauer (2010) for details.

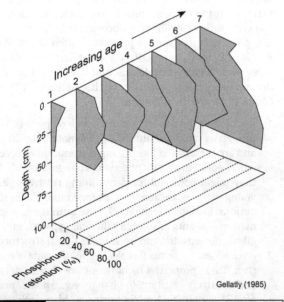

Fig. 15.27 Phosphorus retention values for soils on a series of moraines in New Zealand.

difficult to obliterate. Many form extremely slowly, such as oxic, petrocalcic, placic, and petrogypsic horizons; plinthite; clay mineralogy; and certain strongly developed argillic and natric horizons. Embodied within the group is the concept of an irreversible *loss* of certain properties, such as weatherable primary minerals; after these minerals are gone they do not neoform. These properties all are potentially useful on the oldest surfaces.

Weighted Soil Properties

Many soil characteristics, if they are ratio- or interval-scale data and evaluated on a horizon basis, can be mathematically weighted by multiplying them by *horizon thickness* (Sauer *et al.* 2007a; Fig. 15.29). For example, imagine two Alfisols in which the amount of illuvial clay in the Bt horizon was being used as an SED tool and assume that both had formed in the same loamy sand parent material and that the bulk densities of the illuvial horizons of the soils are not different from one another. Soil A has 25% clay in a 10-cm-thick Bt horizon, whereas Soil B has 20% clay

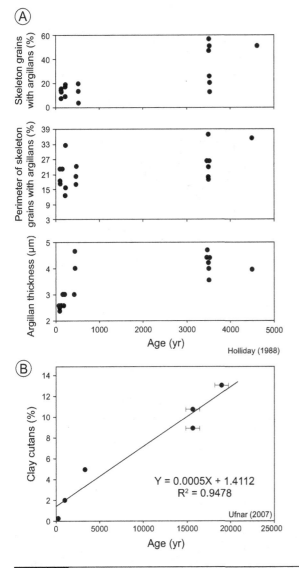

Use of argillans as a relative dating tool. (A) Comparisons of various properties of argillans viewed in thin section, with soil age. The soils are Ustepts, Ustalfs, and Ustolls. (B) Modal percentages of argillans (quantified via point counting of thin sections in soils of different ages from southeastern Mississippi).

subhorizons, each can be weighted in turn by its thickness and summed to arrive at a weighting for the entire horizon. Comparisons of thickness-weighted values for horizons in different soils assume that both horizons have the same bulk density.

Another possible way to examine the degree to which soil constituents have been translocated in a soil is to express their concentrations in illuvial/eluvial ratios. This kind of mathematical manipulation may be applied to the study of soils in which translocation is an important and time-dependent process. For example, one could calculate ratios of clay concentration or Fe concentration between B and E horizons and use these as indices of soil development. Such data could also be weighted by horizon thicknesses if the bulk densities of the horizons were similar.

Soil properties can be weighted in a number of other ways (Hall and Shroba 1995). One can determine *horizon-weighted* values, i.e., property value × horizon thickness, as discussed previously. Similarly, one can sum the values weighted by horizon thickness to arrive at a *profile-weighted* value (Douglass and Mickelson 2007; Fig. 15.29). Thicker sola would be advantaged in such a scheme, and this advantage is often justified, because a thicker soil normally indicates more extensive pedogenic development. To compare thick to thin profiles, the summed profile-weighted data can be divided by overall profile thickness, thereby producing weighted mean concentration data that are not biased by profile thickness (Goodman *et al.* 2001). Birkeland and Burke (1988) called this the profile-weighted mean (PWM) value. The possibilities are many for quantification of soils data.

Birkeland *et al.* (1991) took this analysis one step further and calculated a mean, profile-weighted value for the soils along an entire catena, because they assumed that these data were a better representation of overall soil development on landscapes of uniform age. Bottom line: More data, and more integrated data, may be better for quantifying soil development, especially because we know that soils are spatially variable, and any one pedon or profile may not be truly representative (Birkeland and Burke 1988). As noted before, such calculations provide only rough guides for comparisons of soils or catenas because they ignore any volume changes that have occurred during soil development or make assumptions about such data.

Another type of quantification is the accumulation index (AI). To calculate the AI, the value of a soil property in each horizon is compared to that of the unweathered C horizon. For most properties that are assumed to increase with time, e.g., clay, iron, or organic matter concentrations, the C horizon value is subtracted *from the values of the other horizons.* Negative numbers are ignored. For soil properties that are assumed to decrease with time, such as feldspar content, horizon values should *be subtracted from that of the C horizon.* The difference obtained for each horizon is then multiplied by horizon thickness and summed for the solum (Fig. 15.25). The solum's thickness-weighted AI value

in a 20-cm-thick Bt horizon. Prior to weighting the data, it might be assumed that soil A was better developed, because of its increased content of illuvial clay. But that would be a concentration comparison, not a comparison of the mass of clay in each horizon. In actuality, soil B's Bt horizon is better developed, because it has accumulated more overall clay, i.e., in the thicker horizon. This difference is reflected in the horizon-thickness-weighted clay contents (250 versus 400), which in this case would have units of percent-centimeters. If a master horizon has

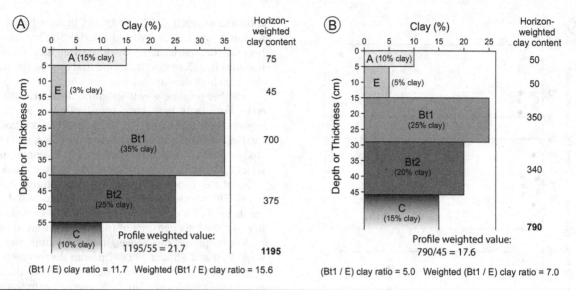

Fig. 15.29 Graphical illustration of horizon-weighted, profile-weighted, and profile-summed data can be used for comparison in a SED study.

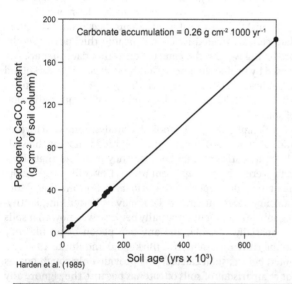

Harden et al. (1985)

Fig. 15.30 Accumulation of secondary carbonates in some soils of a chronosequence. The data are reported as mass per unit (1 cm²) column of soil – a useful means of expressing soil data.

can also be adjusted by dividing it by solum thickness. As before, this calculation is quantitatively defensible only if it is assumed that bulk density values are the same in the horizons being compared and that the bulk density of the solum has not changed from that of the C horizon during the period of soil development.

Another variation of the approach to weighting soil properties is to multiply their mass concentrations by bulk

density to arrive at a *volume concentration*. Most soil properties, e.g., clay, quartz, and Fe content, are determined as mass concentrations, i.e., on gravimetric bases. That is, the units of mass concentration are the mass of a soil constituent in a given mass of soil, sometimes expressed as a percentage [(mass/mass) × 100]. However, the total volume of a soil or soil horizon includes the volume of both solids and pores. So a dense horizon with the same mass-based concentration of a soil property as a less dense horizon would actually have more of the soil constituent per unit volume of soil. Thus to attribute quantitative differences between the properties of one horizon or soil and another to pedogenesis, we cannot always compare mass concentrations. Bulk density data, which add in the volume dimension, help to alleviate this problem. Bulk densities of soils change naturally over time, commonly becoming less dense in the upper profile and denser in the subsoil (Lichter 1998). If the bulk density has declined with time and no net gain or loss of a constituent's mass occurs, that mass will be relatively more diffuse in the horizon's volume. Volumetric concentrations may be weighted by multiplying by horizon thickness. When the thickness-weighted values are summed and the sum is divided by the total solum thickness, an areal mass for the constituent may be reported (Smiles 2009; Fig. 15.30). These data have units of mass per unit area, e.g., clay mass is the mass of clay per unit area (usually per m² or cm²) integrated through a selected horizon or profile thickness (Markewich and Pavich 1991, Liebens and Schaetzl 1997). Thickness-weighted volumetric mass data often provide a meaningful way to compare soils on a profile-to-profile or horizon-to-horizon basis. Where soils contain abundant coarse fragments, it is necessary

Table 15.6 | An ordinal scale for ranking the degree of soil and paleosol development

Morphological expression	Key character	Horizons[a]	Defining characteristics	Degree of mineral alteration
0 None	Geologic	D or R	Unaltered sediment	None
1 "Not soil" ("protosoil")	"Geologic"	C or Cr D R	Evidence of an "altered" horizon (C) over unaltered material	Detectable
2 Weak ("band")	Weak solum	A C D/R	Evidence of an A/C profile	Slight
3 Weak	Weak B	A Bw C D/R	Evidence of an A/Bw/C profile	Weak
4 Moderate	Weak E or Bt	A E Bt C D R	First evidence of Bt or E/Bt horizons	Weak
5 Strong	"Normal" E and Bt	A E Bt BCt C D/R	"Normal" Bt or E/Bt horizons	Weak
6 Very strong	Thick E and Bt	A E Bt Bt Ct D/R	Thick Bt or E/Bt horizons	Moderate
7 Very strong	Maximum Bt	A E Btt Bt Ct D/R	Occurrence of a Bt horizon with >50% clay (Btt)	Moderate
8 Strong	Thick horizons	AE Bt Bto Ct Cr R	Occurrence of a Bto horizon	Strong
9 Moderate	"?"	EA Bto Bo Ct Cr R	Transitional	Very strong
10 Weak	Poor horizonation	EA Bo Bo Ct Cr R	Strongly developed Bo horizon	Complete

[a] The use of the D and Btt horizon designations is informal.
Source: Follmer (1998a).

to correct the bulk density values for the impact of coarse fragments before calculating the volume-based concentrations and performing further comparisons.

Whenever possible, changes in bulk density that occur over time should be incorporated into studies of soil genesis by calculating profile *strain*. As discussed in Chapter 13, strain is the deformational, volumetric change that soil materials undergo over time, due to either collapse or dilation (see Chapter 13). Where strain is positive, the material has gained volume; this condition is termed *dilation*. Negative strain implies *collapse* (Brimhall and Dietrich 1987, Brimhall *et al.* 1988). Strain is determined using measurements of bulk density of the horizons to be compared, along with the concentration of a stable element or mineral, i.e., one whose mass is not likely to change over time. Mass data that are corrected for changes in bulk density are then compared on a profile-by-profile or horizon-by-horizon basis to determine the overall mass balances of elements or constituents (Jersak *et al.* 1995, Ozaytekin and Ozcan 2013). An easy-to-follow development of mass balance calculations for soil genesis studies was presented by Brewer in his 1976 book *Fabric and Mineral Analysis of Soils*. Data calculated in this manner are more meaningful for comparison of changes in soil constituents than are values that ignore bulk density, because strain is a more time-integrated assessment of volumetric changes in the soil, having been derived from immobile, slowly weatherable components (Fig. 13.101).

There are many ways to manipulate soil data; each has its advantages and disadvantages. However, soil data are only as good as the sampling technique used to acquire

them, reinforcing the important point that representative site selection or unbiased sampling approaches are as important to soil interpretation as are any mathematical or statistical manipulations of the resultant data.

Indices of Soil Development

Researchers have long known that it is difficult to compare a variety of soil data in a meaningful and easily comprehensible way. Rather than compare a number of different data sets and lines of evidence among soils of differing development, soil scientists have devised schemes and mathematical indices, designed to incorporate all the various and related data into *one* value – an *index*. Such indices provide a simple and more conceptually integrative way of comparing soils (Goodman *et al.* 2001). No index is universal; each has limited utility and should only be used on the types of soils *for which it was developed*.

Some of the early indices were primarily morphological, facilitating comparisons on an *ordinal scale*, e.g., 1, 2, 3, 4 (Gile *et al.* 1966; Table 15.6). Later, these indices became more quantitative, providing *interval scale* data on soil development, thereby allowing researchers not only to rank soils from strongest to most weakly developed, but to assign placeholders along that scale and to compare the data statistically. For example, an index that increases in value as soil development increases could assign one soil a value of 1, another a value of 3.5, and a third might be 6.7. Although the same sort of comparisons could be made for individual soil properties, an index usually incorporates more than one variable or characteristic in its formulation.

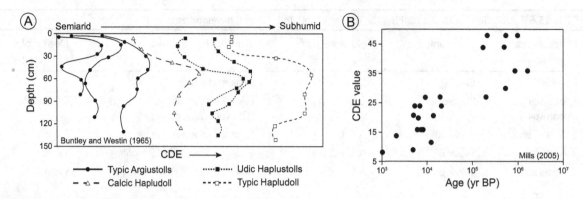

Fig. 15.31 The Buntley-Westin (1965) color index. (A) Color development equivalent (CDE) curves (CDEs) for a transect of soils across the ustic-udic boundary in the northern Great Plains. The CDE is the same as the Buntley-Westin index. (B) CDE values for alluvial terraces in the unglaciated Appalachians and southeastern Coastal Plain of the United States.

Color Indices

Some of the simpler yet highly useful indices of soil development center on only one morphological aspect: color. As discussed in Chapters 2 and 13, color is often highly correlated with various soil properties (Fernandez *et al.* 1988, Mokma 1993). Indices that incorporate hue, value, and/ or chroma provide an integration of soil color that can be even more insightful.

One of the earliest color indices was the Buntley-Westin (B-W) Index (Buntley and Westin 1965). Developed for Mollisols, the index converts hue to a single number from 0 to 7, generally with each Munsell page separated from another by a whole number, e.g., 10YR = 3, 7.5YR = 4, 5YR = 5. Redder hues are assigned higher numbers because they often signify increased weathering and development. The number for hue is then multiplied by chroma to arrive at a color development equivalent, or CDE. The CDE is a classic example of an index that provides more information, and is more useful, than its component parts, e.g., color hue or chroma, individually could provide (Leigh 1996). It was not so much designed to be used as an SED tool as it was a discriminatory tool, to separate soils on the basis of depth distributions of their horizon colors (Fig. 15.31). Nonetheless, it has found application as an SED tool in many types of environments, particularly those with young soils, and in cold environments (Bockheim 1979a, 2008, Leigh 1996, Evans 1999, Munroe 2007, McLeod *et al.* 2009).

Because many different kinds of soils redden with age (Arduino *et al.* 1984, McFadden and Hendricks 1985, Markewich and Pavich 1991, Howard *et al.* 1993, Lindeburg *et al.* 2013), quantification of *rubification* is an important SED tool. Rubification is generally ascribed to increased amounts of Fe oxide minerals, especially hematite (Costantini and Damiani 2004, Azzali *et al.* 2011). But because determining Fe_2O_3 content in the laboratory is time-consuming, indices have been developed to estimate its content solely on the basis of color. Early attempts (Soileau and McCracken

1967) found no consistent relationship between hue and iron oxide content. The first successful attempt that incorporated hue, value, and chroma into one index value was developed by a geologist, for use in saprolite (Hurst 1977). Like the CDE, the Hurst Index converts hue to a single numerical value (5R = 5, 10R = 10, 5YR = 15, 10YR = 20, etc.) and multiplies this number by the Munsell value/chroma quotient. The Hurst Index decreases as iron content and redness increase. Although developed for saprolite, it has been successfully applied to many different kinds of soils and appears to correlate especially well to degree of development in red Mediterranean soils (Shiraiwa and Watanabe 1991, Leigh 1996, Liebens and Schaetzl 1997, Sgouras *et al.* 2007; Fig. 13.35) and mine waste deposits (Shum and Lavkulich 1999, Azzali *et al.* 2011).

The widespread utility of the Hurst Index prompted others to suggest subtle improvements. Alexander's (1985) suggested changes to the Hurst Index formulation are one example (Ajmone Marsan *et al.* 1988). Perhaps the most widely utilized modified index is the Redness Rating (RR) of Torrent *et al.* (1980b, 1983; Fig. 13.35), defined thus:

$$RR = (10 - H) \times C/V$$

where *C* and *V* are numerical values of Munsell chroma and value, respectively, and *H* (hue) is the number preceding YR in the Munsell hue. The RR behaves inversely to the Hurst Index; it increases as iron oxide and hematite content of soils increases. The RR has shown good correlation with hematite/goethite ratios and hematite contents in soils (Kämpf and Schwertmann 1983a, Fontes and Weed 1996, Fontes and Carvalho 2005). Additional modifications to the Hurst Index and RR continue to be suggested (Harden 1982, Santana 1984, Gobin *et al.* 2000).

Thompson and Bell (1996) developed a Profile Darkness Index (PDI), which recognized that darker colors in Mollisols imply not only more organic carbon, but also wetter conditions:

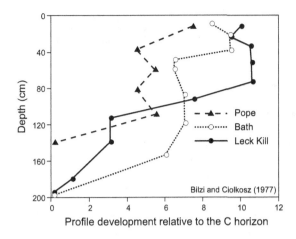

Fig. 15.32 RPD (Index of Relative Profile Development) for some soils in Pennsylvania. Larger values indicate greater differentiation of that horizon, with respect to its parent material. Pope soils are presumably younger than are the other two, and hence have lower B-C index values.

$$PDI = \sum \frac{A \text{ horizon thickness}}{(V \cdot C) + 1}$$

The PDI is calculated for each A subhorizon and then summed; V and C are Munsell value and chroma. As A horizons get thicker and/or darker, PDI values increase. In a catena of Mollisols in Minnesota, PDI consistently increased downslope and was highly correlated with A horizon organic carbon content (Thompson and Bell 1996).

Another entire suite of field morphology and color indices designed to assess soil wetness also exists; some were discussed in Chapter 14. Most soil wetness studies center on the direct measurement of redox feature colors in the soil and correlate them to water table conditions or redox potentials, e.g., Jacobs et al. (2002), He et al. (2003). Other indices of wetness that derive from a formulaic integration of soil colors, from both redox features and the soil matrix (Evans and Franzmeier 1988, Megonigal et al. 1993, Gobin et al. 2000), have been proposed, and the potential for growth in this arena is large (Jien et al. 2004).

Field/Morphology Indices

The success and simplicity of deriving various soil-based indices using color – a morphological attribute – have spurred work on other soil development indices that are based on field morphology. These kinds of indices have the advantage of not requiring laboratory data for their application, and they typically take into consideration more morphological indicators than simply color. Most assume a uniform parent material at time$_{zero}$.

Perhaps the simplest and most elegant index is the one developed by Follmer (1998a) (Table 15.6). Originally developed for buried soils (paleosols) in which gross morphology

is preserved but potentially little else, the 1–10 ordinal scheme is applicable to surface soils as well (Grimley et al. 2003). It will perhaps be applied most often in situations in which soils of wide ranges of development are being compared, and for buried soils whose chemical properties have been altered since burial.

The Relative Profile Development (RPD) Index of Bilzi and Ciolkosz (1977) was developed for leached soils of humid regions but can potentially be applied worldwide. Although not developed as an SED tool, it nonetheless has applicability in this context. The index is used to compare the morphologies of adjacent horizons to each other and/or to the C horizon, hence its alternative name, the Relative Horizon Distinctness (RHD) Index. With this index, points are arbitrarily given for differences between horizons. One point is assigned for each "unit difference" in color, texture, structure, consistence, mottles, horizon boundary, and argillans (the index was first applied to Alfisols). In short, most major morphological criteria are determined for each horizon in the field, and the *magnitudes* of the combined differences are then ascertained. When used to compare the morphologies of adjacent horizons to each other, the method is useful for evaluating profile anisotropy, which for many soils is an indicator of development and age (Duchaufour and Souchier 1978, Meixner and Singer 1981, Ajmone Marsan et al. 1988). When comparing soil horizons to the C horizon, points are assigned on the basis of differences between them and the C horizon, implying that the RPD index value for the C horizon is always zero (Fig. 15.32). Used in this way, the index provides information on roughly how far soil development has proceeded beyond conditions at time$_{zero}$, i.e., the C horizon state. For a particular horizon, the larger the rating scale the greater its pedological development. The method is not applicable to soils in which the C horizon is in a different parent material. Meixner and Singer (1981) modified the RPD to evaluate soil development in Mediterranean climates, e.g., the San Joaquin Valley, California. They observed that the maximum RPD values for A horizons were obtained within 12,000–14,000 years, whereas the RPD values for B horizons showed continuous increases over the entire 25 k year chronosequence. Thus, the RPD has shown good utility as a relative dating tool.

Perhaps the most used field morphology index is that of Jennifer Harden (1982). The Harden index, commonly called the Profile Development Index (PDI), is a modification of the RPD index. In both schemes, points are assigned to each horizon as particular properties are developed or increase in magnitude. In the RPD index, the points for each horizon are compared to that of the C horizon; in the Harden index, the horizon values are compared to the assumed parent material. If the parent material is not available, it must be sought out from a nearby location. The PDI examines any or all of the following soil properties: argillans, texture plus wet consistence, rubification, structure, dry consistence, moist consistence, color value, and pH. First

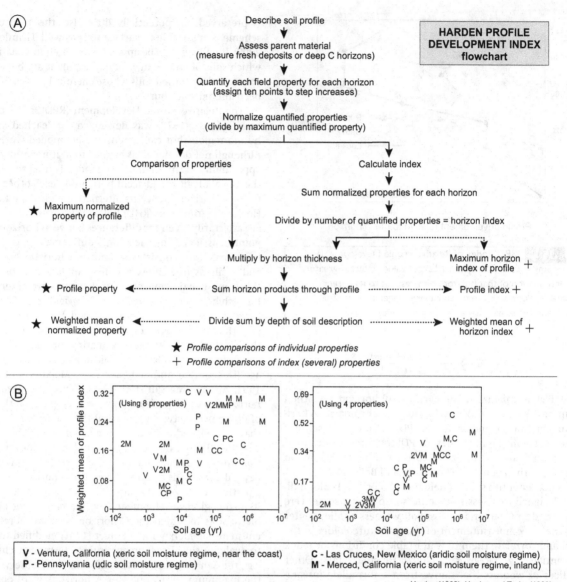

Fig. 15.33 The profile development index (PDI). (A) Flowchart to assist in the derivation of the PDI from morphological data. (B) PDI data for soils of four chronosequences from different climates.

applied to soils in central California, the index was modified by Harden and Taylor (1983) to make it more applicable to arid climate soils by adding two properties typical of desert soils: color paling (increases in hue and chroma) and color lightening (increases in value). The index design is open-ended – researchers can add properties to this list or delete those that are not applicable (Knuepfer 1988).

The PDI is mathematically involved, but flexible (Fig. 15.33). Like the RPD index, PDI values can be obtained and reported on a horizon basis (Figs. 15.32, 15.34). When choosing how to use these data, one also has to decide

whether soil thickness should be included in the PDI calculation; a soil development index that includes thickness may increase in a soil chronosequence (because soil thickness increases with soil age), even if other development-related parameters do not (Sauer 2010). Increasing soil thickness might quantitatively mask other soil development trends (Harden 1982).

Birkeland (1999) described several options for merging horizon-based values of the PDI to obtain one value for a profile, thereby accounting for soil thickness (Fig. 15.33). First, the horizon PDI values are multiplied by horizon

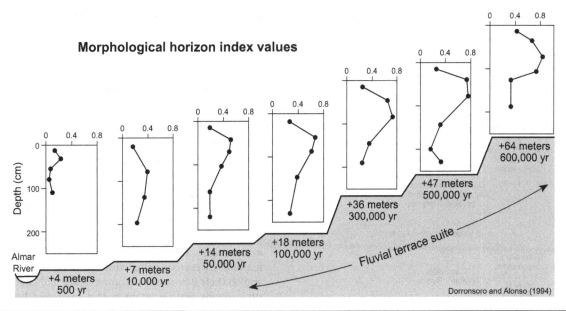

Morphological horizon index values

Dorronsoro and Alonso (1994)

Fig. 15.34 Distribution of horizon-based PDI values for a chronosequence of soils on fluvial terraces in western Spain.

thicknesses. Then, the horizon-weighted PDI data can be (1) summed; (2) summed and divided by the thickness of the profile, resulting in the weighted mean PDI; or (3) summed to the thickness of the thickest profile in the data set, after artificially adjusting all profiles to this same thickness, and then dividing the result by the largest thickness value. Option 3 attributes to each profile (except the thickest one) an index of zero for the part between its lowermost soil depth and the soil depth of the thickest profile. Essentially, for thinner profiles, the C horizon thickness is increased so that the data for all profiles are based on an equivalent thickness of soil plus regolith.

PDI data can also be normalized, to facilitate more robust interpedon comparisons. To do this, the individual horizon values are divided by the maximum value for the profile, yielding a rating between 0 and 1. All the normalized values are then summed, the total is divided by the number of properties used, and the latter is multiplied by horizon thicknesses. A final sum of these values yields a normalized PDI value for the profile. Higher PDI values correlate with increased soil development. Profile-summed and normalized PDI values usually increase logarithmically with time.

The PDI is flexible and widely applicable, e.g., Amit *et al.* (1996), Vidic (1998), Harrison *et al.* (1990), Dorronsoro and Alonso (1994), Evans (1999), Treadwell-Steitz and McFadden (2000), Al-Farraj and Harvey (2000), Dahms (2002), Zielhofer *et al.* (2009), Sauer (2010).

An earlier and perhaps simplified version of the PDI and the RPD, the Index of Profile Anisotropy (IPA) is similar to other whole-profile indices in that it is applicable only to

soils that lack lithologic discontinuities (Walker and Green 1976). The IPA assumes that profiles are isotropic at time$_{zero}$, at which time the IPA also equals zero. As the soil develops, anisotropy, taken as a surrogate for development, also increases (Walker *et al.* 1996). The IPA is defined as:

$$IPA = \sum D(100/M)$$

where D is the mean deviation of a horizon from the overall weighted mean value for the profile (M). Horizon deviation values are summed for each horizon to arrive at the IPA. As with the PDI, the investigator has the option to choose which soil property or properties to use when calculating the IPA.

Birkeland (1999) suggested a modified formulation:

$$mIPA = \frac{\sum [(t \cdot D) / PM]}{T}$$

where D is the numerical deviation of a soil property from that of the parent material (PM). Deviations are determined for each horizon, normalized to the parent material value, and then multiplied by horizon thickness (t). The sum of these values is then divided by the profile thickness (T).

Recognizing that the PDI requires parent material (C horizon) data in its calculation, Langley-Turnbaugh and Evans (1994) developed a modified index that does not require parent material data. It weights soil structure more heavily than does the PDI and minimizes the quantitative influence of horizon thickness. Its primary use is in

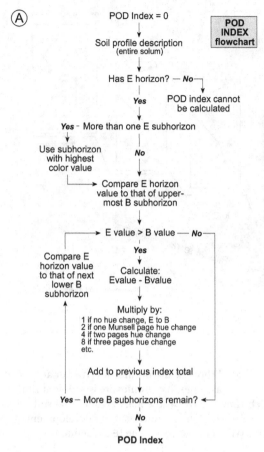

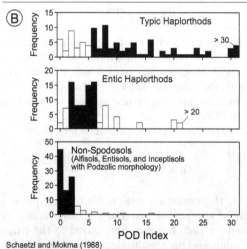

Schaetzl and Mokma (1988)

Fig. 15.35 The POD index, a numerical index of soil development for soils developing spodic morphologies. (A) Flowchart used in the derivation of POD index values from field (morphological) data. (B) POD index values for a sample of Typic Haplorthods, Entic Haplorthods, and sandy non-Spodosols in the northeastern United States.

distinguishing soil from not-soil sediment and, therefore, has utility in paleopedology.

Schaetzl and Mokma (1988) developed a field-based index specifically for Spodosols and soils developing toward podzolic morphologies. Like the IPA and mIPA, it assumes that soils become more anisotropic as they develop, especially with respect to color. The POD Index uses field data on horizon color and number as inputs to the following equation:

$$\text{POD index} = \sum \Delta V \cdot 2^{\Delta H}$$

where ΔV is the Munsell color value difference between the E and B subhorizons, and ΔH is the difference in the number of Munsell pages between the horizons (Fig. 15.35). The data are summed over all the B subhorizons. The index assumes that, as soils develop toward Spodosols, (1) their E horizons increase in color value and become less red (hue), and (2) their B horizons become thicker and develop more subhorizons and (3) attain redder hues and lower color values. POD Index values increase as soils show increased evidence of podzolization (see Chapter 13). Most soils with POD Indices ≤ 2 are Entisols, whereas those with POD indices ≥ 6 are within Typic subgroups of Spodosols, e.g., Typic Haplorthods. Entic subgroups of Spodosols commonly have POD values between 2 and 6. Application of the POD Index in different environments has confirmed its utility as an index of spodic development for both surface soils (Goldin and Edmonds 1996, Arbogast and Jameson 1998, Wilson 2001, Schaetzl *et al.* 2006, Falsone *et al.* 2012) and buried soils developing toward Spodosols (Anderton and Loope 1995, Arbogast *et al.* 2004).

Laboratory Indices

Soil development indices that incorporate laboratory data, e.g., mineralogy, clay content, pH, and contents of ions, have some advantages over field-based indices. Having more data is, generally, better, and thus lab data combined with field indices tend to be better indicators of soil development. However, laboratory data have a cost, because any index that requires laboratory data cannot be generated in the field and is therefore more expensive with regard to time and money. As with field-based indices, it is important that the compared soils have formed in similar (and preferably, uniform) parent materials and that little (or at least comparable amounts of) material has been added to the soil since time$_{zero}$, e.g., as dust.

Textural data are some of the most commonly used laboratory data for soil development indices. As soils become better developed (to a point), most tend to become more clay-rich, and the sand and silt fractions weather away. In very old soils, however, clay begins to be destroyed; in others, the potential for argillan formation decreases as pedoturbation continues to mix the soil.

Relying on these assumptions, Van Wambeke (1962) argued that clay/silt ratios were good indicators of soil age and weathering status in the humid tropics. He rationalized this as a developmental index by assuming that the fine silt content is a good indicator of the soil's remaining weatherable mineral storehouse; as these minerals are weathered and lost from the silt fraction, they become a part of the clay fraction. Pécsi and Richter (1996) developed their weathering index (K_d), with applications to paleosols (Terhorst et al. 2011). The K_d index (not to be confused with the distribution coefficient, K_d, described in Chapter 5) estimates the weathering intensity of paleosols by dividing the contents of coarse and medium silt by the content of fine silt and clay. Martini (1970) also developed a weathering index – for *red* tropical soils – that takes into account that (1) soils become more clay-rich through time, (2) the CEC is derived from clay and the organic matter in these soils, and (3) as the soils weather, clay mineral suites change from amorphous materials to 2:1 clays to 1:1 clays and finally to oxide clays. The index assumes that CEC values decrease as soils become more weathered. Contrary to this trend, however, is the tendency for older, more weathered soils to have more clay. Martini suggested that the lowering of CEC through time was the more important factor and therefore developed his weathering index (Iw) as

$$Iw = CEC \text{ (cmol (–) kg}^{-1} \text{ of clay)/clay concentration (\%)}$$

The index appears to reflect tropical weathering in soils quite well; older soils have lower values. Generally, Oxisols have Iw values <1.0.

An index of desilication, formalized by Singh et al. (1998) but used for decades is the molar ratio of silica to resistant oxides:

$$\text{molar ratio} = SiO_2/R_2O_3$$

where $R_2O_3 = Fe_2O_3 + Al_2O_3 + TiO_2$, with each component being expressed as a molar concentration. These values are calculated by horizon, weighted by horizon, or summed for the entire profile. This index is most applicable for humid tropical soils (see Chapter 13).

A soil texture index that has been used effectively is Levine and Ciolkosz's (1983) clay accumulation index (CAI). The index essentially measures the difference in clay contents between the Bt and C horizons, weighted by horizon thickness:

$$CAI = \sum [(Clay_{Bt} - Clay_C) \cdot T]$$

where $Clay_{Bt}$ is the clay content (as a percentage) of the Bt or other clay-enriched B horizons, $Clay_C$ is the clay content of the parent material, and T is the thickness (in cm) of the B horizon used in the equation. The values are summed for as many Bt or Bw subhorizons as apply. In essence, this equation represents a measure of illuvial clay. It works well in soils where lessivage is a primary pedogenic process, i.e., Alfisols and Ultisols. The index has been applied in only a few studies (Singh et al. 1998, Ortiz et al. 2002) but has great potential.

Modification of the index for use with other illuvial materials, e.g., iron oxides, has some utility (Ortiz et al. 2002). It could also be modified to include bulk density data, making it a volumetrically based index. Perhaps even better, although not as widely applicable, would be a formulaic modification by examining E-Bt differences, rather than C-Bt differences. A fruitful avenue for future research involves extending the CAI to other illuvial materials, e.g., Fe in Spodosols, Ca in Calcids, or gypsum in Gypsids, such as Machette's (1985) *cS* index.

The *cS* index is a profile-based index of secondary carbonate accumulation that uses a mass balance approach (Machette 1985). It is calculated as the sum of the differences between the mass of carbonate in each soil horizon and that in an equivalent thickness of the parent material. *cS* (mass per unit area) is calculated as

$$cS = \sum (c_h \bullet \rho_h \bullet d_h) - (c_p \bullet \rho_p \bullet d_p)$$

where the c = mass concentration of $CaCO_3$, ρ = bulk density, d = thickness, and the subscripts h and p refer to horizon-based and parent material-based data, respectively. In short, the carbonate present in each horizon is determined by multiplying its mass concentration by its bulk density and thickness. The difference between this value and an estimate of the carbonate of that horizon at time$_{zero}$ (derived from the parent material data) represents the secondary carbonate that has accumulated in that horizon via calcification. The sum of these differences over all horizons is the profile index of carbonate accumulation, or *cS*. The concentration of carbonate in the parent material and the bulk density of the parent material must be estimated for soils in which they cannot be easily measured. In addition, the thickness of the each horizon before carbonate accumulation starts, i.e., d_p, may be estimated as

$$d_p = d_h \bullet [\rho_p / \rho_h]$$

Generically, the *cS* index formulation provides a measure of soil development that is based on the change in mass of carbonate relative to the C horizon, but without reference to an immobile constituent. Machette (1985) reviewed the strengths and weaknesses of the *cS* index as a tool for dating geomorphic surfaces; surface age correlates to the amount of accumulated $CaCO_3$ (Fig. 15.36). Although it may not be as accurate as other, more traditional mass balance approaches (Chapter 13), the same formulation could be used for any other type of soil constituent that accumulates in soils over time.

Duchaufour and Souchier (1978) developed a podzolization index based on laboratory data. Knowing that soils undergoing podzolization lose Al from their eluvial zones and gain it in their B horizons, they developed an index to reflect this trend. The K_{Al} index is best envisioned graphically (Fig. 15.37). To generate it, the amount of total aluminum in the A and C horizons is determined and plotted as a depth function. A line connecting these two points is drawn, as is a horizontal line at the depth of maximum Al content. The ratio of the measured value to the interpolated value, as illustrated in Fig. 15.37, is the K_{Al} index, which represents the ratio of translocated to inherited aluminum in the maximally developed B horizon. Although the original index uses total aluminum, other podzolization-related products can also be used to generate this index, e.g., total Fe, Fe_d, Fe_p, Fe_o, $(Al + Fe)$, and so on (see Chapter 13).

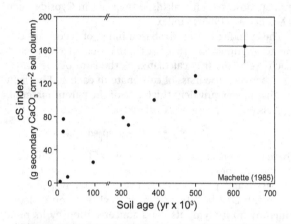

Fig. 15.36 Scatterplot showing how the ages of calcic soils in New Mexico compare to their cS indexes. Note the break in the x-axis between 100 and 300 ka.

Resistant/Weatherable Ratios of Primary Minerals

As early as 1956, Ruhe had successfully applied ratios of resistant/weatherable (R/W) minerals in soils as a relative dating and comparative soil development tool. Over time, the R/W mineral ratio increases because the concentrations of weatherable minerals decrease, while resistant minerals, because they are less affected by weathering, increase proportionately (Soller and Owens 1991, Howard *et al.* 1993, Dorronsoro and Alonso 1994; Figs. 14.64, 15.39). In well-developed soils, these ratios also decrease with depth, indicating the degree to which the upper profile has been altered by weathering (Brophy 1959). A number of generally accepted mineral weathering sequences exist for primary minerals, many of which vary as a function of pH (Fig. 9.1).

Although ratios of clay minerals can be used as indices of weathering or relative soil development, this practice is less precise because it is affected by many variables that can introduce noise. With primary mineral weathering, minerals weather and are lost (or not), whereas with clay minerals, many other outcomes exist. They can weather and be lost, as with primary minerals, but they can also (1) neoform in place, (2) be translocated into the horizon from an overlying horizon, or (3) enter the soil from dust additions. Thus, most studies that use mineral weathering as a relative dating tool use primary minerals, not secondary clay minerals (Read *et al.* 1996, Read 1998).

To study primary mineral weathering patterns, grains are isolated in the laboratory and a minimum of 300 grains

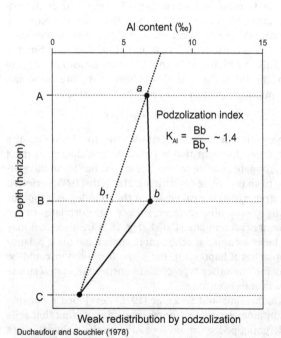

Duchaufour and Souchier (1978)

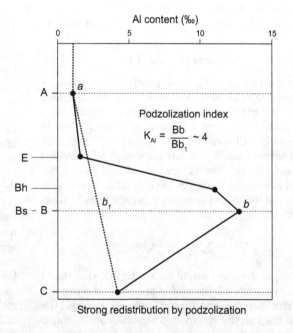

Fig. 15.37 Graphical representation of the K_{Al} index.

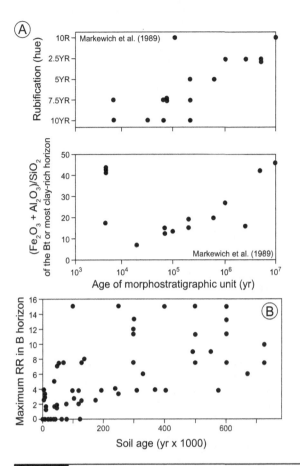

Fig. 15.38 Scatterplots showing the relationship between soil redness and age. (A) Rubification for some soils on the Coastal Plain of the southeastern United States. Rubification was defined as the Munsell hue of the Bt horizon or the reddest B horizon. Both rubification and sesquioxide content correlate well with age, despite the wide variety of soils included in the data set. (B) Maximum Redness Rating (Torrent et al. 1983) for B horizons in soils in 15 Mediterranean soil chronosequences versus estimated soil age. Consult Sauer (2010) for details.

of a single size class are identified and counted, using a petrographic microscope. Fine or very fine sand is often the grain size of choice for this type of analysis, because (1) in larger size fractions, the weatherable minerals are almost completely depleted, especially in older soils, and (2) although the silt size fractions may retain more of the relatively rare, resistant minerals (Chapman and Horn 1968), they are more difficult to work with under the microscope. Also, because they are skeleton grains, sands are assumed to have been immobile during soil formation, pedoturbation notwithstanding.

Resistant minerals are often referred to as *index minerals*. Because index minerals are (presumably immobile and) resistant to weathering, they provide a stable concentration against which other, mobile and weatherable

minerals, can be compared. The choice of minerals to use is a function of their abundance (rare minerals are generally avoided because they will constitute a small fraction of the total grains counted and statistical confidence in the results will be low) and resistance to weathering. The latter has generally been well established by geochemists and mineralogists on the basis of thermodynamics (Table 15.7). A consideration, and sometimes a problem, associated with this analysis centers on the relative paucity of some types of resistant minerals (Chittleborough *et al.* 1984). For example, zircon and tourmaline can be rare in some soils. Extreme care must be taken to ensure that large, *representative* samples are taken from the soil under study. These must be carefully split into subsamples to assure that they are fully representative of the horizon or profile. Additionally, some minerals, e.g., certain feldspars or garnets, are easier to identify under the petrographic microscope and may be preferred for that reason alone. Ratios of resistant/weatherable minerals can utilize light minerals, heavy minerals, or a combination of the two (Brophy 1959; Fig. 15.39). (Heavy minerals have specific gravity values that exceed 2.9 g cm^{-3}.) Dyes and stains that are selective for certain minerals can assist in the identification of feldspars and other minerals (Reeder and McAllister 1957, Norman 1974, Houghton 1980).

Before a given weathering sequence is assumed for interpreting the results of grain counts, the specific conditions under which the expected sequence was developed should be ascertained. For example, was it a sequence based on weathering studies in the field or weathering that was simulated in the laboratory? What were the pH values of the soils in the study? Howard *et al.* (1993) reported the following general stability suite for heavy minerals in soils:

tourmaline > zircon > rutile > silimanite ≥ kyanite > hornblende/amphiboles > augite/pyroxenes

This weathering sequence was published by Howard *et al.* in 2012:

zircon, rutile, magnetite, ilmenite > tourmaline > chloritoid > epidote

Compare the preceding lists to one determined in the laboratory, under moderately acid (pH < 5.6) conditions, by Nickel (1973):

zircon > epidote > amphiboles > garnet > apatite

Because grain counting under the microscope is a time-consuming process, and one that has some user bias (not all grains are easy to identify!), most researchers have now turned to an alternate approach – geochemical weathering indices using elemental data (Ding *et al.* 2001, Price and Velbel 2003, Schellenberger and Veit 2006, Souri *et al.* 2006, Buggle *et al.* 2011, Araki and Kyuma 2012). In this approach, which has its beginnings in the rock weathering literature, chemical elements, not mineral grains, are evaluated

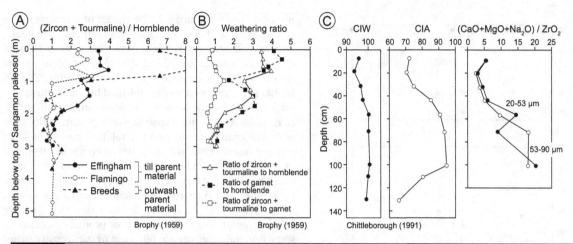

Fig. 15.39 Depth plots of weathering profiles. (A) Weathering profiles, based on resistant/weatherable heavy mineral suites in the fine sand fraction, in the buried Sangamon paleosol in Illinois. These data show the variation in weathering ratios as a function of parent material. (B) Variation in weathering ratios in a Sangamon paleosol formed in till, as a function of minerals chosen for the ratios. (C) Variation among three different indices of weathering for Xeralf in Australia.

Table 15.7 | Examples of resistant/weatherable mineral ratios that have been used to assess weathering and soil development

Quartz/feldspar

(Zircon + tourmaline)/(amphibole + pyroxene)

(Zircon + tourmaline)/hornblende

Garnet/hornblende

(Zircon + tourmaline)/garnet

Tourmaline/(kyanite + staurolite)[a]

Tourmaline/biotite

Tourmaline/zoisite

[a] Kyanite and staurolite are only of moderate resistance; use of other minerals with less resistance to weathering is recommended.

(Rabenhorst and Wilding 1986a, Santos *et al.* 1986). The theory behind the method is similar to weathering ratios using mineral grains – soluble and mobile *elements* are depleted, whereas less soluble and immobile *elements* are enriched. Fortunately, the most useful resistant minerals are often sole sources of certain elements. Zircon is the main source of zirconium; anatase, rutile, and tourmaline are the main sources of titanium; and yttrium occurs in the resistant mineral xenotime (Murad 1978, Chittleborough *et al.* 1984). The use of elements has an additional advantage: The silt fraction, which may have higher concentrations of these minerals than sand fractions, can be included. In loess-derived soils, silt dominates the suite of particle sizes, so one can evaluate weathering on the dominant, not minor fraction of the soil.

Here, grains of a given size fraction (or the whole soil is used – a practice that is generally discouraged) are isolated and then dissolved chemically. The solution may then be analyzed by using inductively coupled plasma (ICP) mass spectrometry, and the results are reported as elemental concentration in the sample mass. In short, index minerals have often been superseded by index elements, which are useful if they are present only in their respective index minerals (see Chapter 13).

Selection of elements for analysis must be based on what mineral(s) they mainly derive from, along a weatherability sequence. That is the first consideration. If the elemental analysis included the whole soil, i.e., secondary minerals and weathering by-products, the fate (mobile or not, soluble or not) of each element must also be considered. After they have been released from primary minerals, the relative solubility/mobility of elements in soils revolves around their ionic potential (IP), i.e., the ratio between ionic charge and ionic radius (Buggle *et al.* 2011; Fig. 9.8). Cations having IP values <3 form only weak bonds with oxygen. When in solution, they exist as soluble cations, e.g., Ca, K, and Na. If the IP of the element is between 3 and 8, the high density of the positive charge enables strong bonds with oxygen, and thus, these elements form weathering-resistant oxides, e.g., Fe, Ti, and Al. Elements with an IP of between 3 and 10–12 can form insoluble hydroxides or oxyhydroxides. Elements of this category, when released during weathering, precipitate quickly as insoluble and immobile hydroxides (Buggle *et al.* 2011). One last point: Solubility does not always mean mobility. Cations with low IP values are still likely to be retained by electrostatic attraction to clay mineral surfaces, reducing their mobility in the short term (Chapter 5).

The postweathering fate of the elements in the soil must always be considered when choosing elements for a

Table 15.8 Examples of named elemental ratios that have been used to assess rock weathering and soil development

Ratio	Commonly applied name	Source
$(Al_2O_3 + K_2O)/(MgO + CaO + Na_2O)$	Vogt's Residual Index	Vogt (1927)
$[SiO_2/(TiO_2 + Fe_2O_3 + SiO_2 + Al_2O_3)] \times 100$	Product Index	Reiche (1950)
SiO_2/Al_2O_3	Ruxton Ratio	Ruxton (1968)
$([(2Na)_a/0.35] + [(Mg)_a / 0.9] +$ $[(K_a) / 0.25] + [(Ca)_a / 0.7]) \times 100$ a = percent mass / atomic mass	Parker Index; Weathering Index of Parker	Parker (1970)
$[(Na_2O + K_2O + CaO + MgO) /$ $(Na_2O + K_2O + CaO + MgO + SiO_2 + Al_2O_3 +$ $Fe_2O_3)] \times 100$	Weathering Index	Vogel (1975), after Reiche (1943)
$(CaO + Na_2O + K_2O)/(Al_2O_3 + CaO + Na_2O + K_2O)$	Index B	Kronberg and Nesbitt (1981)
$[Al_2O_3/(Al_2O_3 + Na_2O + CaO + K_2O)] \times 100$	Chemical Index of Alteration	Nesbitt and Young (1982)
$[Al_2O_3/(Al_2O_3 + Na_2O + CaO)] \times 100$	Chemical Index of Weathering	Harnois (1988)
$(CaO + MgO + Na_2O)/(ZrO_2)$	Weathering Ratio	Chittleborough (1991)
$[(Al_2O_3 - K_2O)/(Al_2O_3 + CaO + Na_2O - K_2O)] \times 100$	Plagioclase Index of Alteration	Fedo et al. (1995)
$(SiO_2 + CaO + K_2O + Na_2O) /$ $(Al_2O_3 + SiO_2 + CaO + K_2O + Na_2O)$	A Index	Guggenberger et al. (1998)
$\ln (K/Hf)$	--	Brown et al. (2003)
$2 \arcsin\{[(K/Na)/(K/Na_f)]^{1/2}\}$ f = conc. in feldspars	--	Brown et al. (2003)
$(Si + Ca) / (Fe + Ti)$	WI 1	Darmody et al. (2005b)
$(Si + Ca) / (Fe + Ti + Al)$	WI 2	Darmody et al. (2005b)
$(CaO + K_2O + Na_2O) /$ $(Al_2O_3 + CaO + K_2O + Na_2O)$	B Index	Guggenberger et al. (1998)
$(Ca + K + Mg + Na)/Al$	Σ Base cations / Al ratio	Colman (1982), Sheldon and Tabor (2009)
$[Al_2O_3/(Al_2O_3 + Na_2O)] \times 100$	Chemical Proxy of Alteration	Buggle et al. (2011)

geochemical weathering study. Other factors that can influence the solubility of ions, e.g., soil solution chemistry, pH, redox conditions, and temperature, also must be considered (Buggle *et al.* 2011). Under near-neutral and oxidizing conditions, however, the classification according to the IP successfully predicts the behavior of the most common elements of interest for geochemical weathering indices.

Generally, elements such as Li, Cs, Rb, Ba, Na, Ca, and Mg are used as indicators of weatherable minerals. Because they have relatively high intrinsic solubilities (Harriss and Adams 1966, Nesbitt *et al.* 1980, Chittleborough 1991, Liu *et al.* 1993, Muhs *et al.* 2001), their concentrations in the soil should decrease particularly well as soil development proceeds under leaching conditions. In particular, Ca, Mg, and Sr are common elements in many weathering-susceptible silicate minerals (Nesbitt *et al.* 1980), and because they are also mobile in the weathering environment, they appear in several commonly used geochemical weathering indices, e.g., Ba/Sr, Rb/Sr, Sr/K, Sr/Zr, Mg/K, Mg/Ti, Ca/K, Ca/Zr, and Ca/Ti. Of these, Ba/Sr may have the most potential as

a proxy for weathering intensity and soil development (Bokhorst *et al.* 2009).

Elements that are used to indicate resistant minerals include ions of intermediate ionic potential, i.e., ions that tend to form insoluble hydroxides. Additionally, Rb, Ba, and K, i.e., ions that can be immobilized by fixation in interlayer spaces of 2:1 clay minerals, may be used as immobile references, e.g., Rb/Sr, Ba/Sr, or Na/K (Liu *et al.* 1993, Gallet *et al.* 1996, Chen *et al.* 1999). However, under intense weathering conditions, significant losses of these elements can nonetheless occur during the transformation of micas, feldspars, and other host minerals into secondary clay minerals (Buggle *et al.* 2011). Thus, most researchers have chosen to focus on elements of the insoluble hydroxide category, e.g., Al, Si, Ti, and Zr. Of these, Zr and Ti may be the most reliable, because Al and Si are mobile in certain pedogenic environments (Beavers *et al.* 1963, Chittleborough 1991). To add to these more common ratios, Table 15.8 provides a listing of the major "named" weathering indices that use ratios of elements in their formulations. Many of

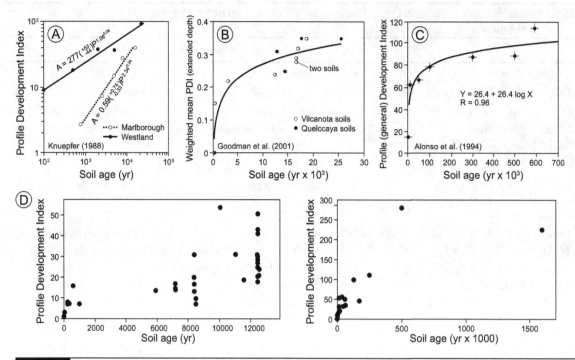

Fig. 15.40 Examples of application of Harden's (1982) Profile Development Index. Note that the PDI increases logarithmically with time; data plot as a straight line on a logarithmic axis. (A) Stream terraces, South Island, New Zealand. (B) Alpine moraines, southern Peru. (C) Stream terraces, Spain. (D) Various Mediterranean soil chronosequences, plotted versus estimated soil age. Consult Sauer (2010) for details.

the studies that focus on weathering of paleosols also make rich use of rare earth elements in their analysis. Obviously, it is possible to choose from many options of geochemical weathering or pedogenesis indexes. It is paramount that investigators choose indices that fit their local weathering, leaching, and pedogenic situation. Elemental data should be acquired from dissolution of an immobile size fraction, e.g., silt, very fine sand (Chittleborough 1991), rather than whole, fine-earth samples. Ultimately, the choice of elements is dependent upon the expected pedogenic processes, e.g., in areas of podzolization, Fe is mobile while in dry climates it is not.

A wide variety of applications exist for data on soil development and weathering using geochemical approaches (Liu *et al.* 1993). Over time, resistant minerals increase in abundance as less resistant ones (and their by-products) are removed from the profile. Thus, these data are useful in assessing relative soil age (Fig. 15.40). The data are highly useful in discerning the weathered nature of buried soils, in which other pedogenic tests of development, especially those associated with soluble components of the soil system such as organic matter and some minerals, do not work (Brophy 1959). For example, Muhs *et al.* (2001) reported a number of elemental data and ratios for a stratigraphic sequence that extended from the modern soil formed in loess, through unaltered loess, and into two different paleosols below

(Fig. 15.41). Assuming that the paleosols are more weathered than the overlying loess, these data showed that some indices tend to be better indicators of weathering than others. Many of these indices correlate nicely with climate or paleoclimate data and have been used to provide information on the relationships among soil development, weathering, and (paleo)climate (Chen *et al.* 1999, Jeong *et al.* 2008, Nordt *et al.* 2010, Adams *et al.* 2011, Buggle *et al.* 2011). R/W ratios and geochemical data are particularly useful in chronosequence formulations (Bockheim *et al.* 1996; Fig. 15.42).

Pedogenic Mass Balance

A perhaps more quantitatively robust technique for assessing relative amounts of soil development centers on creating a balance sheet of elemental or mineralogical gains and losses for the soil profile (Bourne and Whiteside 1962). This type of study – called pedogenic mass balance analysis – is discussed earlier in this chapter and in more detail in Chapter 13. The focus in pedogenic mass balance is on the minerals and elements that are lost versus. the minerals or elements that are gained, both determined relative to stable minerals. The gains and losses may be calculated on a relative *or* a total basis. Losses or gains can be examined over the entire profile, for the eluvial zone only, or by horizon. This type of study can be performed only on soils developing in uniform parent materials.

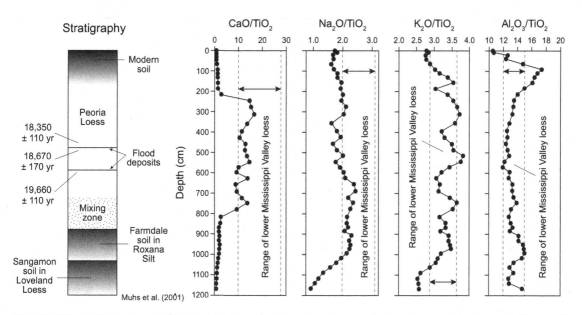

Fig. 15.41 Stratigraphy and elemental ratios – reflective of weathering – for a loess section in Illinois. Ranges of major element ratios in deep, unleached Peoria Loess from the southern Mississippi River valley are from Pye and Johnson (1988).

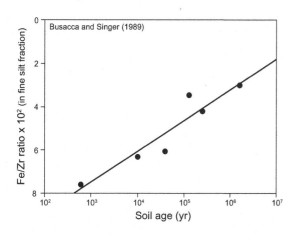

Fig. 15.42 Changes in content Zr/Fe content ratio in soils of varying age in the Sacramento Valley, California.

Applying pedogenic mass balance principles is much like using R/W mineral (or geochemical) ratios, in that it assumes that certain minerals, usually zircon, rutile, anatase, tourmaline, ilmenite, and monzanite, are resistant to weathering and thereby provide indices of chemical stability. If sand-sized minerals are of interest, mass balance analysis assumes that the grains have been pedogenically immobile. This is not true in strongly pedoturbated soils. Nonetheless, depth distributions of stable and slowly weatherable sand-sized minerals are one of the best ways to assess parent material uniformity – a critical first step in any mass balance study (Evans and Adams 1975a, Chittleborough et al. 1984; Figs. 10.51, 15.43).

Landscapes: The ~500 ka Loess-Paleosol Sequence at Thebes, Southern Illinois

Only a few places worldwide have such a well-preserved sequence of loess deposits and paleosols as at Thebes, in southern Illinois. (But see, for example, Wang et al. [2009].) And seldom are the sediments and soils studied in such detail as has been done by Grimley et al. (2003). The Thebes site contains an unusually well-preserved loess-paleosol sequence, which formed as a result of an apparently continuous record of eolian deposition and soil development since the middle Quaternary. Thus, it is an ideal site to compare soil development, because all the soils and paleosols have developed in loess parent materials of generally similar composition. Where such conditions exist, comparative soils data can be used to improve understanding of the character and duration of pedogenesis in the past, because parent material factors are negated. Let us explore the wide variety of data these authors used to determine the relative amounts of soil development for the surface soil and buried paleosols at Thebes.

The 20-m-thick Thebes section lies outside the glacial limit in southern Illinois, and close to the junction of two major meltwater rivers – the Mississippi and Ohio (see figure). Thus, the section has received thick increments of loess during glacial periods; during interglacial periods, soils formed in the loess deposits. A modern soil has formed at the surface. From bottom to top, the stratigraphic sequence (and likely ages) is as follows: (1) residuum, (2) Crowley's Ridge Silt (MIS 12, ~450 ka) with the interglacial Yarmouth geosol formed in it, (3) Loveland Loess (MIS 6, or ~190–130 ka) with the interglacial Sangamon Geosol (MIS 4–5, ~135–55 ka) formed in it, (4) Roxana Silt (MIS 4) with the interstadial Farmdale Geosol (MIS 3) formed in it, and (5) Peoria Loess (MIS 2, ~25–12 ka) with the modern soil formed in it.

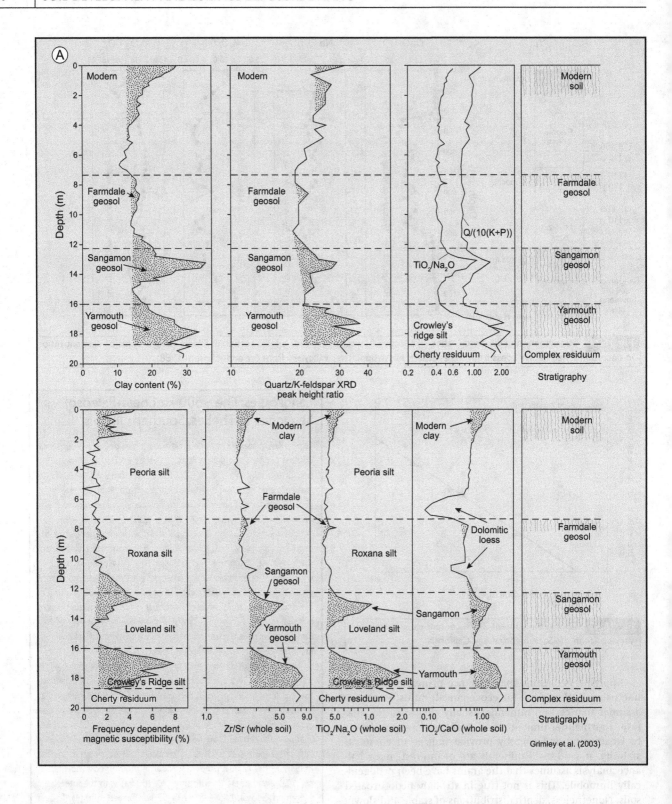

Grimley et al. (2003)

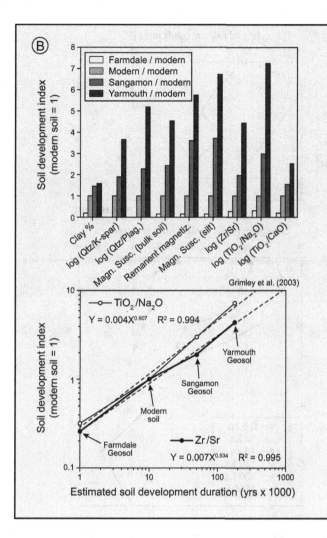

Grimley et al. (2003)

Several types of data were determined for the loesses and intercalated paleosols at Thebes, all of which were used by Grimley et al. (2003) to estimate the relative intensity and duration of soil development for the four soil-forming intervals (Yarmouth, Sangamon, Farmdale, and modern) represented in the section (see Figures). Many of these soil development indicators have been discussed or will be discussed in Chapter 16: clay content, quartz/feldspar ratios, various elemental ratios, magnetic susceptibility, and remanent magnetization. When these data are compared against the modern soil by giving it an index value of 1.0, the degrees of soil development become clear. Because, with few exceptions, four of the five soil-forming factors are comparable for the four soils (all except time), the data in Figure B neatly show the effects of time on soil development. Using these data, Grimley et al. (2003) were also able to estimate the length of the soil-forming interval for each of these four soils – what they called the effective soil development duration. The durations were estimated at 1,000 years for the Farmdale Geosol, 10,000 years for the modern soil, 50,000 years for the Sangamon Geosol, and 180,000 years for the Yarmouth Geosol. These data seem fairly robust, because we know that the modern soil has been forming here since the end of Peoria Loess deposition, ca. 12–16 ka ago, and that the Farmdale interstadial was considerably shorter. These data illustrate, if nothing else, how massive the Yarmouth Geosol is and suggest that it may have developed continuously during the period encompassing MIS 7, 9, and 11. That is a long time!

Figure: Depth plots and comparative soil data for the Thebes section. (A) Depth plots of data chosen to indicate the relative degree of soil development. (B) Compiled data that suggest the relative amount of soil development for the four soils at Thebes.

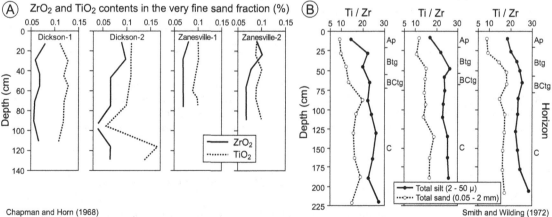

Chapman and Horn (1968)

Smith and Wilding (1972)

Fig. 15.43 Depth plots of Zr and Ti, used to ascertain parent material uniformity, a necessary step in mass balance studies. (A) ZrO_2 and TiO_2 in the very fine sand fractions of six loess soils in Arkansas. The ZrO_2 and TiO_2 contents were determined by chemically fusing (melting) the very fine sand fraction, allowing it to cool into a glass, and analyzing it in an X-ray spectrograph. (B) Ti/Zr ratios for some soils developed in till in Ohio. The low ratio values for near-surface horizons may be due to the greater mobility of Ti versus Zr, or a subtle lithologic discontinuity.

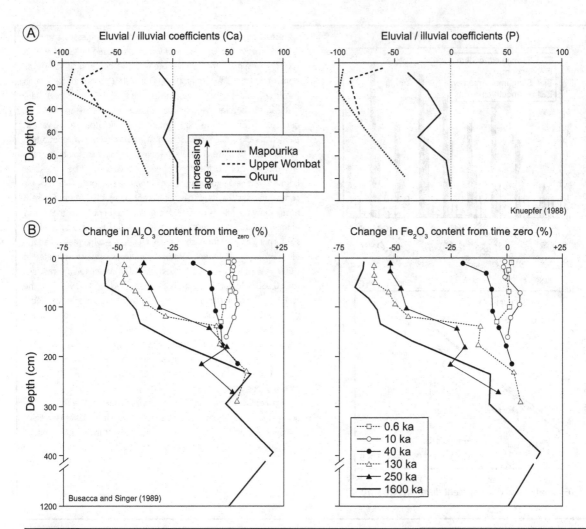

Fig. 15.44 Mass balance data showing relative gains or losses of elements in soils, determined by comparing them to amounts of a resistant and immobile element. (A) Eluvial/illuvial coefficient (EIC) values for soils on a series of Quaternary stream terraces in New Zealand. Negative EIC values indicate a loss of the element relative to the parent material. (B) Percentage gains and losses, relative to the parent material, for soils on a series of stream terraces in the Sacramento Valley, California.

The main difference between mass balance calculations and R/W mineral (or elemental) ratios is that in mass balance analysis the *actual amount* of loss or gain, relative to the parent material, of a certain mineral or element is quantitatively determined (Figs. 13.103, 13.104, 13.105). For example, Muir and Logan (1982) determined the eluvial/illuvial coefficient, or EIC:

$$EIC = \{[(S_h/R_h)/(S_p/R_p)] - 1\} \cdot 100$$

where S_h and S_p are the concentrations of element S (not sulfur) in the horizon and parent material, respectively, and R_h and R_p are the concentrations of a resistant element such as Zr or Ti in the horizon and parent material, respectively (Knuepfer 1988). A positive EIC value means that the element has been enriched relative to the parent material (Fig. 15.44). This method could also be used to determine

the relative solubility and mobilities of elements, and hence the intensities of various pedogenic processes (Fig. 15.45). These data can then inform researchers who are using elemental data in weathering or pedogenic studies. We have come full circle!

Chronosequences

As originally defined within the functional–factorial model of Jenny (1941b), a *chronosequence* is a series of soils of known age. In a chronosequence, time (soil age) is allowed to vary while, assumedly, all other soil-forming factors are held constant (Stevens and Walker 1970, Yaalon 1975, Huggett 1998b, Schaetzl *et al.* 1994). Although the latter condition is never fully realized, in most chronosequences

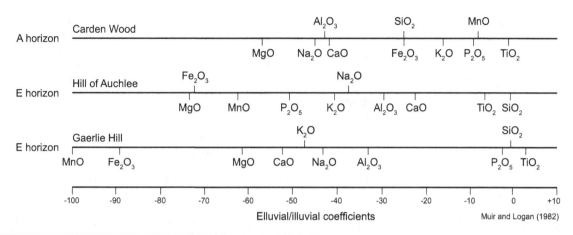

Fig. 15.45 Eluvial/illuvial coefficients for the A and E horizons of three soils undergoing podzolization in Scotland. The low values for Fe and Mn for the latter two soils are due to wet, reducing conditions in the E horizon.

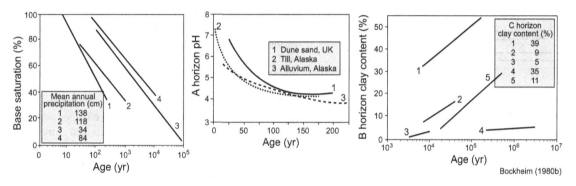

Fig. 15.46 Various chronofunction summaries.

the impact of time on soil development so outweighs the other state factors that the effect of time on soil development can be generally determined. Bockheim's (1980b) review of chronofunctions illustrated the value, from a process standpoint, that such studies can contribute to soil geomorphology and pedology (Fig. 15.46), and Huggett's (1998b) study reaffirmed it.

If soil data from a chronosequence are plotted against age, with age as the independent variable, the resultant equation is called a *chronofunction* (Fig. 12.3):

$$S = f_t(\text{time})_{cl,o,r,p} \dots$$

Bain *et al.* (1993: 276) referred to chronofunctions as rate equations of soil formation. Indeed, careful preparation and evaluation of a chronofunction can provide much insightful information about rates of development for the soils as a whole and for individual horizons.

Theoretical Considerations

Soils form on a geomorphic surface after the surface becomes stable and when environmental conditions are

suitable. This period of time – its duration and characteristics – is often referred to as a *soil-forming interval* or pedogenic interval (Morrison 1967, Vreeken 1984a; see Chapter 16). For surface soils, the soil-forming interval began at some time in the past and continues today, whereas the soil-forming interval for buried soils ceased upon burial. Many soil-forming intervals have gradational beginnings and endings. A major goal of chronofunction studies is to determine the pedogenic outcomes of the soil-forming interval: What type of soil formed? What degree of development did it reach? At what rate did soil formation occur during that period? How long did soil property X or Y take to form?

The theoretical underpinnings of chronosequences involve the ergodic hypothesis, aka the comparative geographical approach, in which space is substituted for time (Huggett 1998b). For example, since we cannot remain in one *place* and examine soil development over long periods of *time*, we substitute space for time by examining a number of soils *at the same moment in time but at different places*. For example, a series of raised beach ridges, all of different ages but otherwise similar, are allowed to substitute for

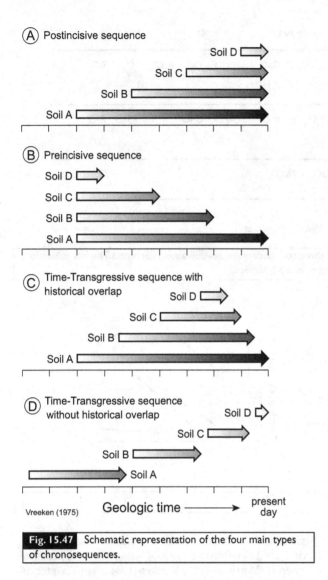

(A) Postincisive sequence

(B) Preincisive sequence

(C) Time-Transgressive sequence with historical overlap

(D) Time-Transgressive sequence without historical overlap

Vreeken (1975)   Geologic time ⟶ present day

Fig. 15.47 Schematic representation of the four main types of chronosequences.

development or steady-state conditions (Gile *et al.* 1966, Reheis 1987b, Holliday 1988), it can be assumed that the progressive pedogenic pathway in many soils is at least as strong as the regressive one (see Chapter 12). On the other hand, Hall (1999b) explained chronofunctions that did not have good age-time trends as indicative of soil regression, cryoturbation, and erosion, and changes in external climatic forcings and pedogenic pathways. In perhaps one of the longest chronosequences on alluvial terraces in Virginia, Howard *et al.* (1993: 201) made the point that

not all soil properties show unidirectional development, nor is a steady state of pedon development observed even after approximately 10^7 yr of chemical weathering. Soil development … is episodic. The transition from one phase to the next is marked by a change in rate, and sometimes a reversal in the direction, of development of one or more soil properties.

The progressive-regressive-steady-state discussion, as it pertains to soil evolution, is not over, and chronosequence data will continue to inform this discussion.

Chronosequences have theoretical issues that must be seriously considered in their construction and interpretation. For example, rarely can all the soil-forming factors except time be held constant over the duration of the chronosequence. Climate is almost certain to have changed, and often vegetation evolves in conjunction with climate and soils. Topography also evolves and changes over time, although less rapidly than do vegetation and climate, and probably over much longer timescales. Most chronosequences have a limited time span within which they can be applied, because numerical dating of surfaces has limits, i.e., we may not be able to ascertain the ages of the oldest soils of the chronosequence. And even if we could, soils evolve so dramatically over long timescales that the predictive power of the chronofunction often diminishes. Pedogenic thresholds and accessions also add complexity, as they dramatically change the rate and direction of pedogenesis. Huggett (1998b) also pointed out that not all pedogenic events are recorded in the soil's morphology or chemistry, rendering the chronosequence only a *partial* record of the past. Because of random variations in soil-forming processes and deterministic uncertainty, all geomorphic surfaces are spatially variable, prompting questions as to which soil on a surface is most representative of the soil-forming processes on that surface (Sondheim and Standish 1983, Harrison *et al.* 1990, Vidic 1998, Eppes and Harrison 1999). To mitigate this question, Barrett and Schaetzl (1993) sampled a number of soils on each surface and only used data from the modal profile in their chronofunction.

Types of Chronosequences
Vreeken (1975) defined four different kinds of chronosequences, based on the time of initiation and/or termination (Fig. 15.47). Soils in a chronosequence may all begin developing at the same time but cease development at different times. Or, like most soils sampled in chronosequences,

time and thereby provide the experimental construct for the chronosequence (Fig. 15.34).

Chronosequences assume that (1) the soil sequence represents successive stages of one or several pedogenic processes and (2) the soils all pass through the stages and in the same order (Vreeken 1984a, Huggett 1998b). Both assumptions invoke progressive pedogenic development, which, although commonly observed, does not always occur. Huggett (1998b: 155) actually attributed the popularity and widespread applicability of chronofunctions to the fact that many researchers support the notion of a developmental view of pedogenesis. A purely progressive/developmental viewpoint is counter to the notions of regressive pedogenesis and soil evolution that we have discussed in earlier chapters (Johnson and Watson-Stegner 1987). Nonetheless, much can be learned about both progressive and regressive soil genesis from the study of chronosequences. For example, because most chronosequences report progressive soil

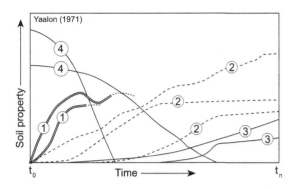

Fig. 15.48 Schematic representation of the three types of soil properties and how they theoretically change from time$_{zero}$ to some period in the future (t_n). (1) Rapidly forming but reversible, self-regulating processes, e.g., pH, salt content, and soil structure. (2) Processes and properties that more slowly achieve a state of equilibrium or steady-state condition. (3) Irreversible, self-terminating processes and properties. Many of these properties, e.g., chlorite minerals, are exhausted from the soil after a period of time (3a). Others increase with little short-term likelihood of achieving a steady-state condition.

they began development at different times and have ceased development (or have been sampled) at the same time. Additionally, the beginning and ending times for the soils may be highly variable among the group and may be time transgressive. In all cases, the length of development (or period of development) differs among the soils, making it a chronosequence. Soil development can be ended by erosion or burial, or, technically, if we sample the soil for our study, its development has just "ended."

The simplest and most common type of chronosequence is postincisive (Muhs 1982, Huggett 1998b, Scarciglia et al. 2006, Nielsen 2008, Huang et al. 2010, Sauer et al. 2010; Fig. 15.47A). Soils in a *postincisive chronosequence* all began developing at different times in the past, i.e., each had a different time$_{zero}$. These soils may still be developing now, or burial may have forced pedogenesis to stop, but in all cases their end points are the same. Soils in *preincisive chronosequences* started forming at the same time, but their development was ended, usually by burial, at different times in the past (Khokhlova et al. 2001, Alekseeva et al. 2007; Fig. 15.47B). If neither the starting nor the ending times of soil development are coincident, the chronosequence falls into Vreeken's third category: *time transgressive with historical overlap* (Fig. 15.47C). This type of chronosequence often forms when landscapes become progressively buried, but the burial is space and time transgressive. Soils in these three types of chronosequences always have some degree of historical overlap, i.e., at some point in the past, at least two of them were concurrently undergoing pedogenesis. However, many surfaces are exposed to soil development and later buried, but the soil-forming intervals for these

soils do not overlap. Vreeken (1975) called this situation *time transgressive without historical overlap* (Fig. 15.47D). This type of chronosequence usually occurs in a stacked series of buried soils, as in a till or loess column with intercalated paleosols (Karlstrom and Osborn 1992). Clear pedogenic interpretations drawn from this type of chronosequence, in which no two soils were ever forming at the same time in the past, are difficult (Stevens and Walker 1970).

Most chronosequences are postincisive (James 1988, Barrett and Schaetzl 1992, Scalenghe et al. 2000). A series of moraines or stream terraces of different age provide a possible postincisive chronosequence – soils on these surfaces are still forming but they have different time$_{zeroes}$. The analogy in biology would be a forest with numerous individual trees that started growing at different points in the past. By examining their morphologies today, we could learn about growth rates and how the trees' morphologies change over time. Preincisive chronosequences, which Vreeken (1975) favored on theoretical grounds but which are fairly rare, are equivalent to a stand where all the trees started growing after a disturbance event, but later, parts of the stand were cut down at different times (Gardiner and Walsh 1966). In soils, the event that stops the pedogenic clock is usually burial, and like trees that run the risk of decomposition after they are cut, buried soils are variously altered after burial (Mausbach et al. 1982, Olson and Nettleton 1998). Another problem associated with preincisive chronosequences centers on burial itself – one can never determine the degree to which burial is time transgressive.

Steady State

Chronofunctions are developed primarily to ascertain rates and directions of soil development (Reheis et al., 1989). Another application centers on the steady-state condition that many soils assumedly develop, or develop toward. What is this state, does it exist, and is this the theoretical end point of soil development? How long does it take to get there? If the rate of soil development slows and appears to approach an asymptote, one might assume that the soil system either is approaching steady state or may already be there. The chronofunction can then be used to determine (1) *whether* a steady state has been achieved, (2) *what* the steady-state value is, and (3) *how long* the system took to reach it. Certainly not all soils or soil properties achieve steady-state conditions (Bockheim 1980b, Dorronsoro and Alonso 1994). In these cases, pedogenic theories such as deterministic uncertainty, chaos or catastrophe theory, or soil evolution principles may help explain the nonlinear and perhaps multidimensional aspects of pedogenesis. Often, some soil *properties* achieve more or less steady-state conditions, yet others continue to change with time (Fig. 15.48). Soils or soil properties that *do* achieve steady-state conditions cannot help determine soil age for those periods after they have reached that state (Catt 1990). Those properties that take a long time to achieve steady-state conditions are most useful for dating old soils, whereas rapidly adjusting

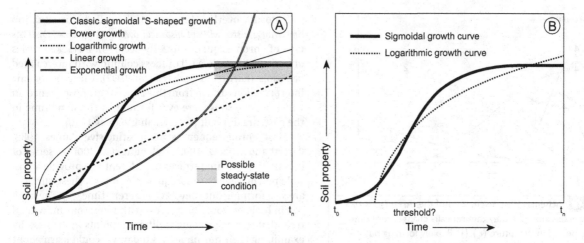

Fig. 15.49 Theoretical soil development curves. Both axes are scaled linearly. (A) The five main types of curves. In each case, the inverse of the curve would represent decay or regression, as might happen with an eroding soil, or one in which the y-axis was a metric for weatherable minerals remaining in the soil. (B) Illustration of how a delayed logarithmic growth function could be mistaken for sigmoidal growth; the latter is common to biological populations.

properties such as organic matter content are most useful in shorter chronofunctions.

Many soil properties are assumed to develop in a way that is analogous to the classic, S-shaped growth curve (Crocker and Major 1955, Yaalon 1975, Sondheim et al. 1981, Birkeland 1999, Darmody et al. 2005a). In this model, growth (development) is slow initially and then increases rapidly, only to slow as it approaches a steady state (Fig. 15.49). However, many studies that point to logarithmic changes in soils suggest that the S-shaped, sigmoidal growth curve does not fit most soil properties. Even soils that do not show rapid pedogenic gains in their early stages of development, as would have been predicted by a logarithmic curve, do not necessarily support the sigmoidal growth curve. Rather, their growth or development could simply be delayed and later, after a threshold has been passed, development proceeds along a different pathway (James 1988, Schaetzl 1994; Fig. 15.49).

Statistical Circularity

To develop a chronofunction, one must have numerical estimates of surface ages, usually obtained via numerical (as opposed to relative) surface exposure dating. These ages are then correlated to a soil property or properties.

Chronofunctions not only lean on SED in their application, but can also provide information for future SED studies, regarding the likely minimum or maximum age of a surface (Vincent et al. 1994). Once established, chronofunctions can be used to develop and enhance pedogenic theory and foster a better understanding of pedogenic processes. In a type of circular logic, chronofunctions therefore provide much of the theory that is used in SED – they tell us how much time soils require

to develop property X or to lose property Y or to thicken to depth Z (Dethier et al. 2012). For example, a chronosequence of soils in northern Michigan established that at least 4000 years are required to form a spodic (Bs) horizon in this region (Barrett and Schaetzl 1992). Future SED studies can now lean on this finding and use it as a key baseline datum in this region. This example highlights the differences between chronofunctions and SED studies involving soils. A SED study can only be as accurate as the known library of chronofunction data allow. In short, chronofunctions provide the *theory* and *age estimation* for pedogenic features that are then *applied* in SED. Soils data, however, should not be used as the *primary* age determinant for geomorphic surfaces. In other words, using SED data to understand soil development better is much more accepted and common than is using soil development to estimate the age of a surface. We caution against using chronofunction data circularly, i.e., using time-soil relationships to infer the age of surfaces and then using soils information on those surfaces to generate additional chronofunctions, and so on (Vidic 1998). One must be aware of the limits of the data and not overextend their applicability.

Statistical Considerations

Although strictly evaluating soil development, or a part of it, as a statistical (chrono)function has been questioned (Yaalon 1975, James 1988, Harrison et al. 1990), the potential benefits outweigh the potential pitfalls. Not only do chronofunctions allow us to understand the soil system today and in the past better, but many can also provide a measure of prediction. How we analyze and provide order and explanation to the array of these soil data not only

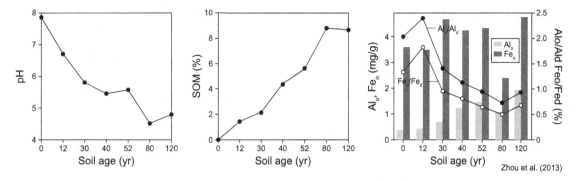

Fig. 15.50 Data for a sequence of soils developing on young glacial moraines in southwestern China. Note that no attempt has been made to regress the data statistically to any type of best-fit line. Rather, the data are simply connected by short line segments.

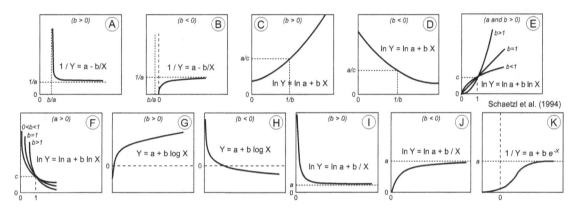

Fig. 15.51 The main types of statistical functions that are fitted to chronofunction data. For all chronofunctions, time increases along the x-axis. All are linearizable; the linear form of the curves is provided within each subfigure. (A) and (B) are hyperbolic functions. (C) and (D) are exponential functions. (E) and (F) are power functions. (G) and (H) are logarithmic functions. (I) and (J) are other functions. (K) is an S-shaped or sigmoidal curve.

are challenging, but will dramatically affect the interpretations we make.

Bockheim (1980b), Schaetzl *et al.* (1994), Huggett (1998b), and Holliday (2004) have provided summaries of the many types of dependent soil data that have been applied in chronofunctions. In a chronofunction, the dependent, i.e., soil, data are normally regressed against surface age, usually using least-squares methods (Dorronsoro and Alonso 1994). In some instances, however, a best-fit line is not determined, but rather, the data are simply shown and connected with line segments (Fig. 15.50). Although simple concentration data may be used as the dependent variable, calculated indices of soil development or scores derived from factor analysis and principal component analysis are attractive alternatives (Sondheim *et al.* 1981, Scalenghe *et al.* 2000). In chronosequences in which parent material cannot be held constant, the use of *ratios* of multiple dependent variables also holds promise (Mellor 1985).

Early attempts at the development of chronofunctions involved simply hand-fitting a line to a scatter of points

(Crocker 1952, Wilson 1960). Although this technique is still used with some success (Schaetzl and Mokma 1988, Amit *et al.* 1993, Vincent *et al.* 1994), most approaches today employ mathematical models. An inherent dilemma in chronofunctions is that the data may fit more than one model. The model that is chosen should be *based on the theoretical understanding of pedogenesis* for the site (Schaetzl *et al.* 1994). Soil systems function at a variety of rates and move along different pedogenic pathways, but chronofunctions are often fit to any of several mathematical models (Levine and Ciolkosz 1983, Mellor 1985; Figs. 15.51, 15.52).

In these chronofunction equations, Y is the soil property being examined, t is time, a is the y-axis intercept, and b is the slope of the regression line. Although other models exist (Schaetzl *et al.* 1994), including polynomial models ($Y = a + bt + ct^2$), the four listed here are the most common. And of these, the first two usually provide the best fits to chronosequence data (Little and Ward 1981, Muhs 1982, Dorronsoro and Alonso 1994). In most cases, the most

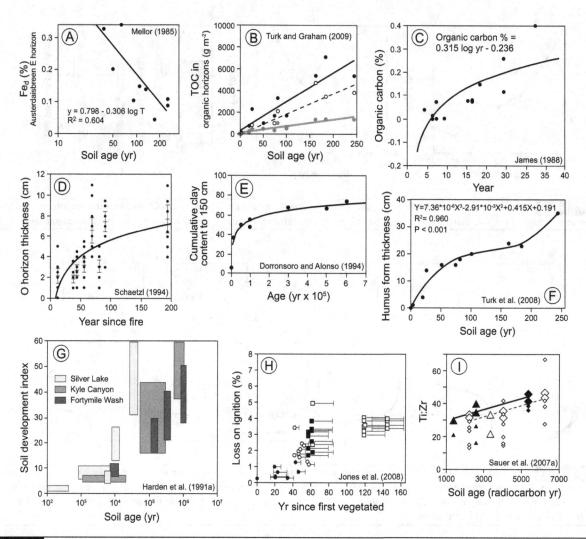

Fig. 15.52   Examples of different kinds of chronofunctions, and the various graphical ways that they can be protrayed.

successful mathematical fit will have the highest coefficient of determination, R^2.

Sometimes chronosequence data fit a linear model, implying that values of the soil property change at a constant rate (or nearly so) with respect to time. In that instance, a simple interpretation is that the slope of the fitted line (b, in the linear model) is equal to the rate of change and the intercept (a) represents the value of the property at time$_{zero}$.

More often, however, the rate of change in a soil property is not constant over time, and a scatterplot of the property value versus time indicates a nonlinear relationship between the variable and time. Rather than iteratively seeking an equation that would fit the nonlinear data well, we might hypothesize that they fit one of the three common nonlinear models: the logarithmic, exponential, and power models. For example, logarithmic

functions plot as curvilinear lines when both x- and y-axes are linearly scaled. But if the same chronofunction data are plotted on linear-log axes (i.e., the log of time), a straight line with a constant slope will confirm the presence of a logarithmic relationship (Fig. 15.53). To test our hypotheses, we use a logarithmic transformation of the values of the property, the values of time, or both, and then use a linear regression model to determine whether the log-transformed values fit a straight line. Linearizing the data in this way allows us to calculate fitting parameters more easily.

After a model is chosen and a regression line is calculated, the slope of the line can be used to infer rates of pedogenesis over specific time spans of the chronofunction (Harden *et al.* 1991a; Fig. 15.30). If the best-fit model is linear, it is justifiable to infer that pedogenesis has been proceeding at a generally constant rate for the period of

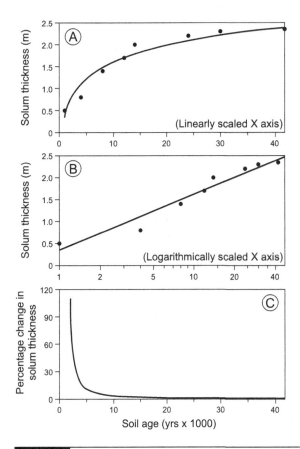

Fig. 15.53 Theoretical chronofunction showing a soil property that increases logarithmically through time. The same chronofunction is plotted in (A) and (B), but on different axes. If it is plotted on linear axes (A), one might (correctly) conclude that the soil is approaching a steady state. Plot (C) verifies this conclusion by showing that the percentage change with time rapidly approaches zero. If the same regression equation is plotted on log-linear axes (B), however, one might (erroneously) conclude that the soil property is not approaching steady state. This figure illustrates how the conclusions drawn from chronofunctions can be impacted by the graphical method of presentation.

study, and even that it may continue to do so in the near-future. Although linear models are frequently successfully fit to chronofunction data (Mellor 1986, Koutaniemi et al. 1988, Merritts et al. 1991), nonlinear models are also very commonly used (Huang et al. 2010; Fig. 15.52). Nonlinear (mainly sigmoidal and logarithmic) chronofunctions imply that the soil system has had differing rates of pedogenesis at different times (Jones et al. 2008; Figs. 15.49B, 15.52H, and several examples in Figure 15.51). In particular, the *interpretation* of *logarithmic* models is important in this context. To interpret this type of modeled relationship correctly, we need to reevaluate the original data set before it was transformed.

Importantly, logarithmic functions plot as curvilinear lines when both x- and y-axes are linearly scaled (Figs. 15.46 [center], 15.51, 15.53). If the same equation is plotted on log-linear axes, however, the regression line is straight, and with a constant slope (Fig. 15.53B). If one does not examine the x-axis of such a chronofunction, one might assume that the soil or soil property is *not* approaching a steady-state condition. When plotted on linear axes, however, the data *do* appear to show gradually slowing pedogenic development, perhaps indicative of an approaching steady-state condition.

Some (Muhs 1982, Busacca 1987, Stockman et al. 2011) have claimed that unless the slope of such a chronofunction goes to zero, steady state has not been achieved. Because *time is linear*, one must interpret the slope of the regression line in the chronofunction on the basis of *linearly scaled axes*. In short, chronofunctions are best interpreted as if the data were presented on *linearly scaled* axes (Fig. 15.53). Although the rate of change will never completely level off in a logarithmic model, visual inspection of the original data and the modeled line together *on linearly scaled axes* and in the context of measurement uncertainty may reveal that an interpretation of steady-state conditions would be reasonable.

Chronofunction data often have a great deal of statistical uncertainty, especially with regard to age, but also for dependent data, due to sampling constraints (Harden et al. 1991a). To address this problem, Switzer et al. (1988) developed a Monte Carlo approach of refitting the regression line to various data combinations. Their method is iterative. The standard deviation of the various rate constants derived from the combinations was interpreted to represent uncertainty in the rate of soil development (Fig. 15.54).

Chronofunction formulation is almost always affected by the paucity of data. Many chronofunctions have only three or four data points. Because of this, use of exponential or polynomial functions is not encouraged (Bockheim 1980b), although they have been used with some success (Bockheim 1990, Harden et al. 1991a). Figure 15.55a illustrates the risk of working with too few data. Even though the chronofunction in the figure displays five measured points for profile-weighted organic carbon concentration, with soil ages ranging from 4,000 to ~11,000 years, visual inspection of the graph reveals that the five data points occur in two clusters that effectively define the slope and intercept of the modeled line. Even though the R^2 value for this linear regression model is quite high, it is an artifact of the two clusters of data. It would be unwise to conclude that a linear model is appropriate because there are no data between the two end points of the line. The researcher cannot know whether the real relationship is linear or nonlinear.

Many soils undergo changes in pedogenic rates, due perhaps to the crossing of a threshold or the acquisition of some type of accession, e.g., development of a fragipan.

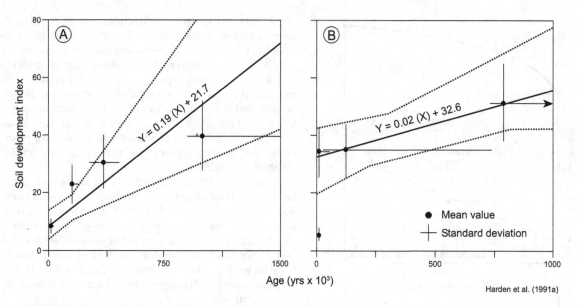

Fig. 15.54 Chronofunctions for soils at (A) Fortymile Wash and (B) Kyle Canyon, Nevada, showing the uncertainty in the statistical function, using maximum likelihood estimation.

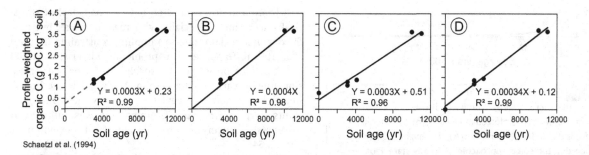

Schaetzl et al. (1994)

Fig. 15.55 An illustration of how treatment of data near $time_{zero}$ can affect a chronofunction. (A) The original chronofunction, with its five data points, showing the relationship between profile-weighted organic carbon and time. (B) The same five data points, but with the chronofunction statistically forced to go through the origin. The assumption here is that the soil had no organic carbon at $time_{zero}$. (C) The same five data points, but with a sixth point added, reflecting the condition of the soil at $time_{zero}$. The data for the sixth point were derived through analysis of a sample from the contemporary C horizon. Its use was based on the assumption that the profile-weighted organic carbon content of the soil system at $time_{zero}$ was the same as that of the C horizon. (D) The same five data points, but with a sixth point (0, 0) added, assuming that at $time_{zero}$ the soil had no organic carbon.

Two-phase regression may be used to model this type of pedogenic pathway. Two-phase regression, similar to a step function, is a method by which a scatter of points is fitted to two separate linear chronofunctions (Turk and Graham 2009). This method has a great deal of as-yet-untapped potential, especially in soils where pedogenic pathways have changed (Bacon and Watts 1971). For example, if the slope of the regression equation is significantly different for the periods before and after the pedogenic accession, two-phase regression might be an unbiased way to determine the existence of the threshold. Again, one must be careful because the paucity of data that plagues most chronofunctions rarely lends itself to two-phase regression.

For many chronosequences, conditions and rates of change near $time_{zero}$ are of special concern (Turk and Graham 2009, Caner et al. 2010, Zhou et al. 2013). For example, the remediation of disturbed soils, such as mine soils or urban soils, is a branch of pedology that necessarily focuses on incipient pedogenic processes (Leisman 1957). The processes and pathways in the early phases of pedogenesis are often different from those later in time. To address the need for information near $time_{zero}$, aka the *boundary condition*, it is tempting to extrapolate the chronofunction regression line beyond the range of the data, either back to zero or forward in time, as a potential predictive tool. Almost all researchers warn against the latter

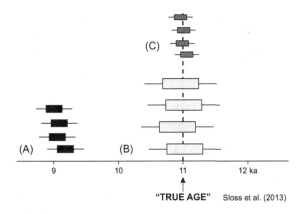

"TRUE AGE" Sloss et al. (2013)

Fig. 15.56 An illustration of the difference between dating precision and accuracy. (A) Age estimates are precise but inaccurate – they are ~2 ka too young. (B) Age estimates are accurate, because they are close to the true age, but imprecise. (C) Age estimates are accurate and precise.

practice (James 1988, Yaalon 1992, Schaetzl *et al.* 1994), especially for chronofunctions where the confidence limits on the regression equation are broadest near the ends of the regression (Fig. 15.54).

Obtaining information about boundary conditions from chronofunctions can, however, be accomplished in several ways. One way involves the use of the origin (0, 0) as a data point in the chronofunction. In linear chronofunctions where it can be assumed that the value of Y (i.e., the soil property) is zero at time$_{zero}$, insertion of the 0, 0 point into a data set may be warranted (Fig. 15.55D). Examples might be solum or horizon thicknesses, horizon-weighted or solum-weighted data, various pedogenic indices, and pedogenically *acquired properties* such as organic carbon or illuvial clay. If this option is not feasible, Schaetzl *et al.* (1994) suggest one of three *additional options* for chronofunctions in which no data are available for time$_{zero}$:

1. Retain the chronofunction in its original form, even if it does not pass through the origin and the researcher knows that at time$_{zero}$ Y = 0 (Fig. 15.55A).
2. Statistically force the regression line through the origin by fixing the intercept at 0. This option should only be used for soils where there is independent evidence that the value of Y was zero at time$_{zero}$ (Fig. 15.55B).
3. Estimate the state of the soil system at time$_{zero}$ from deep C horizon data and use that value as a data point in the chronofunction. This is generally an acceptable option, although in older soils it is difficult to obtain samples of unaltered parent material (Fig. 15.55C).

Chronofunction research is only in its infancy. As more chronofunction data accrue, better dating methods are developed, and pedogenic theory advances, researchers will continue to develop more and better chronofunctions, for a wider variety of areas. It is vital, however, that both statistical and pedogenic theory be considered when chronofunctions are developed and interpreted.

Numerical Dating Techniques in Soil Geomorphology

A wide variety of numerical dating techniques exist that have applicability in soil geomorphology and related disciplines, and the number of such techniques continues to expand as new technologies are developed (Wintle 1996, Watchman and Twidale 2002, Cornu *et al.* 2009, Sloss *et al.* 2013). These techniques are employed to provide a numerical estimate of surface, sediment, or material age. In this section, we discuss some of the major methods of numerical, or geochronometric, dating that are applicable to soil geomorphology.

Each dating method has its own age range limitations and inherent uncertainties. A key concept in any numerical dating exercise centers on the accuracy and precision of the ages (Sloss *et al.* 2013; Fig. 15.56). *Precision* refers to the statistical uncertainty that is associated with the analytical process of the dating technique (Lowe and Walker 1997, Noller *et al.* 2000), i.e., How reproducible is the age estimate? Precision is usually represented as uncertainties and is often shown as a range within which the likely true age exists. Once the numerical age has been derived (precision), the researcher must assess the *accuracy* of the age determination, in relation to the landscape, event, or depositional environment that is being dated, i.e., How close is the age estimate to the true age of the material being dated? Although we often do not know a great deal about the accuracy of an age estimate, we should always keep this parameter in mind when we report and discuss the age.

Paleomagnetism

The intensity and orientation of the Earth's magnetic field, as preserved in the orientation of ferromagnetic minerals (particularly magnetite) in rocks and sediments, are called *paleomagnetism*. When initially deposited in a loosely packed body of sediment, or as they grow from a melt, many minerals acquire *remanant magnetism*, i.e., they become aligned to the Earth's magnetic field (Barendregt 1984). Until they are disturbed, this orientation preserves a record of the Earth's magnetic field at the time of deposition or mineral formation. The best types of unconsolidated sediments for this method contain grains of silt and fine sand that can be strongly magnetized and that are free from secondary mineralization, weathering, or pedoturbation. Igneous rocks and volcanic deposits are particularly good at preserving paleomagnetic information, although other fine-grained sediments such as loess, marine, fluvial, and lacustrine deposits – and even glacial tills – are also applicable (Barendregt 1981, DukRodkin *et al.* 1996, Karlstrom

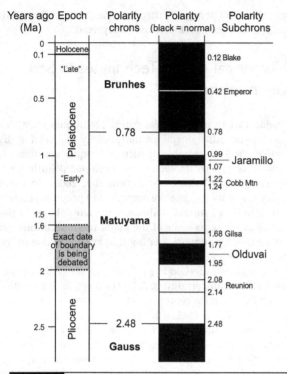

Fig. 15.57 The paleomagnetic timescale for the past 2.8 Ma. Compiled from various sources.

procedures rather than representing real magnetic changes (Barendregt 1981). Finally, within the polarity record are still shorter (10–100 ka) periods of polarity reversal called polarity *subchronozones*.

Geophysicists who study *magnetostratigraphy* have been able to determine the timing of these paleomagnetic changes from the rock record (Cox *et al.* 1963). About 171 paleomagnetic reversals have occurred in the past 76 million years. The current period of normal polarity, called the Brunhes epoch, began ~770 ka ago (Sarna-Wojcicki *et al.* 2000, Suganuma *et al.* 2010; Fig. 15.57). Previous to that, the Matuyama reversed polarity event spanned about 2,480–770 ka. The Gauss normal epoch preceded the Matuyama. Each of these events contained short-lived geomagnetic excursions. Most soil geomorphology studies are not concerned with paleomagnetic reversals prior to the Pleistocene. One exception is Early Pleistocene glacial deposits and surfaces, with reversed polarity (Karlstrom 2000, Rovey and Balco 2009).

Scientists use paleomagnetic data in Pleistocene stratigraphic studies for correlations as well as for relative and numerical dating of deposits (Barendregt 1981, Hambach *et al.* 2008). Although paleomagnetism seldom allows for great precision in dating the *entire* stratigraphic column, it does provide two key baselines at ~770 ka and ~2,500 ka, and several smaller ones (Harden *et al.* 1985, Busacca 1989, Jacobs and Knox 1994). The Brunhes-Matuyama split is a key one in thick loess sections that extend back beyond it, as well as in Early Pleistocene lacustrine sediments. Loess preserves paleomagnetism fairly well (Nabel 1993). The plasma portions of soils, e.g., illuvial clay, sesquioxides, and carbonates, also acquire paleomagnetic signatures, making paleosols (if they are suspected of being ≥700 ka old) potential applications of this method as well (Cioppa *et al.* 1995, Yang *et al.* 2007, Rovey and Balco 2009).

Tephrochronology

Tephra (Greek *tephra*, ashes) refers to volcanic materials, usually ash and pyroclastics, that have been transported aerially. Tephrochronology is a stratigraphic method for linking, dating, and synchronizing geological, paleoenvironmental, or archaeological sequences or events, by using dated tephra layers (Lowe 2011, Pyne-O'Donnell *et al.* 2012). Sediments above ash marker beds are younger than the ash, thereby establishing a maximum age for soil development within the overlying sediments (Nettleton and Chadwick 1991, Kemp *et al.* 1998). Likewise, sediments (including paleosols) below tephra beds provide clues about climate and environmental conditions before the ashfall (Ward and Carter 1998). Tephrochronology has two strong advantages as a numerical dating technique: (1) Ash layers often span large areas, providing a single chronostratigraphic unit that cuts across (potentially) many landscapes, and (2) on a geologic timescale, ash is deposited in an instant. Ashfall beds are, therefore, isochronous horizons.

and Barendregt 2001, Wang *et al.* 2005, Deng 2008, Rovey and Balco 2009, Sun *et al.* 2011, Jin *et al.* 2012).

Paleomagnetism works as a geochronometer because the Earth's magnetic field has changed dramatically over geologic time. Our current polarity is considered normal, i.e., the north-seeking end of the compass needle points toward the north magnetic pole. Periods of *normal polarity* have, however, alternated with periods of *reversed polarity*, when the north-seeking end of the compass needle pointed to the south magnetic pole. Essentially, at many times in the geologic past, the magnetic field of the Earth has done a complete flip-flop: South becomes north (Fig. 15.57). The cause of these magnetic reversals is not clearly understood, but because the transitional (changeover) periods are usually quite short (<10 ka), the long periods of normal or reversed polarity are useful stratigraphic markers for worldwide correlation and dating (Fig. 14.59). Within these long *epochs* of reversed or normal polarity are short (100–1,000 years) periods in which the magnetic pole moves sharply toward the equator for 100–1,000 years and then returns to a more stable, polar position (Mankinen and Dalrymple 1979). These geomagnetic or polarity *excursions* or *events*, of which there have been several, are equally important magnetostratigraphic markers (Harrison and Ramirez 1975, Barendregt 1984). Some of these excursions are considered questionable, however, and may actually reflect problems with the sediment or the analytical

Fig. 15.58 A thick ash bed – probably Lava Creek A – exposed in a lowland landscape position. Note the thin strata of silt that have washed into this site, along with the ash. Photo by A. Arbogast.

Many ashfall beds are tens of meters in thickness near the source and, yet, remain identifiable hundreds of kilometers away (Fitzsimmons *et al.* 2013). Although ash is deposited across the entire landscape, soon it is modified by wind and water. Thin ashfalls are mixed into the existing soils. When they are no longer visible to the naked eye in the field, these ashes are called *cryptotephras*. Thicker ash deposits are washed or blown into lowlands and other accumulative landscape positions (Fig. 15.58). When viewed in exposures, many ashes are found only in low-lying parts of the paleolandscape, as a result of redeposition. Indeed, ash beds often thicken in a downslope direction and may be absent on uplands (Carter *et al.* 1990). Thus, even though an ashfall bed may have initially covered a landscape, it is absent from most exposures.

Glass shards are the primary constituent of volcanic ash, rendering them amenable to dating by K-Ar, fission track, or neutron-activation dating, or by thermoluminescence methods (Dalrymple *et al.* 1965, Westgate and Briggs 1980). Young ashfall layers can be dated by historical means or by radiocarbon dating of organic materials within the layers or immediately above or below them (Hallett *et al.* 1997, Kaufman *et al.* 2012). Tephrochronology provides additional utility when a numerical age obtained for a tephra layer is transferred from one site to another using stratigraphy and by comparing and matching, aka *fingerprinting*, compositional and morphological features of these deposits, many of which exist as cryptotephras (Davis 1985, Westgate *et al.* 1987, Lowe 2011). Used this

way, tephrochronology is an age-equivalent dating method that provides exceptionally precise ages. In other words, the ash marker bed or the cryptotephra does not always need to be "dated." Rather, it can often be identified on the basis of shard size and shape, elemental composition, mineralogy, hydration, weathering, or other intrinsic characteristics, and its age correlated to equivalent beds elsewhere that *have* been dated (Kuehn and Foit 2006, Kuehn *et al.* 2009, Pyne-O'Donnell *et al.* 2012; Fig. 15.59). As a result, ash can often be utilized as geostratigraphic marker beds across wide areas (Izett 1981, McDonald and Busacca 1988, Pyne-O'Donnell *et al.* 2012). Although analysis of cryptotephras in soils far from their source volcano is a challenge, the net gain has been the extension of tephra isochrons hundreds or several thousands of kilometers from their sources, and even onto distant ice caps.

Tephrochronology can be used in a macro-sense to date sediments and stratigraphic layers and in a microsense to date pedogenic processes. For example, Kemp *et al.* (1998) identified argillans coating known ash shards and tephra pockets, thereby providing evidence of contemporary lessivage as well as quantitative information on the amount of lessivage that has occurred since the ashfall event.

Most of the major ashfalls and tuff beds of Quaternary age are named and their extents mapped (Fig. 15.60, Table 15.9). In older literature, and still in informal discussions, the name Pearlette Ash is mentioned (Swineford and Frye 1946). Pearlette Ash refers to any of a number of ash beds on the Great Plains, which have since been recorrelated and renamed (Izett 1981). Many more Pliocene and older beds are also known (Smith *et al.* 1999).

Tephrochronology is a highly useful dating tool in volcanically active areas of the world, e.g., New Zealand (Vucetich 1968) and Iceland. Nonetheless, hundreds of distinct ash beds have also been identified in the United States (Izett 1981, Bacon 1983, Ward *et al.* 1993). Tephrochronology has particular utility in western North America, where most of the major ashfalls are from the Cascade and Rocky Mountains (Bacon and Lanphere 2006, Kuehn *et al.* 2009); these ash beds are seldom discernible east of the Mississippi River (Fig. 15.60). Most originate in three major regions: the Yellowstone Park area, the Long Valley area of California, and the northern Cascades (Izett *et al.* 1970, Porter 1978). Ashes from these different centers are readily distinguishable from each other, but differentaged eruptions from the same center are more difficult to differentiate.

Radioisotopic Dating

Radioisotopic dating is a large family of numerical dating techniques that has gained increasing application; it is centered on the use of various isotopes – both stable and some unstable. Formerly termed radiometric dating, this term is now disfavored because radiometry is the measurement of electromagnetic wavelengths, which has nothing to do

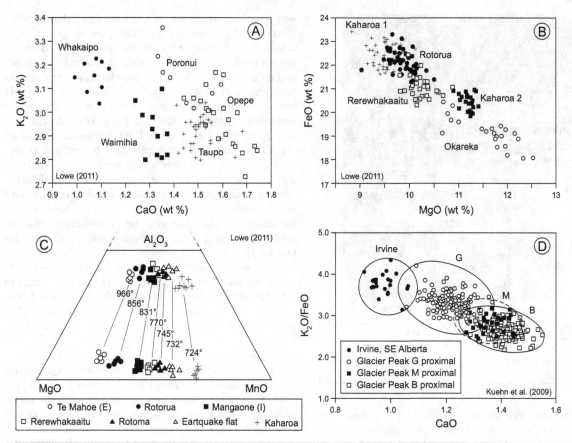

Fig. 15.59 Use of elemental data to distinguish (*fingerprint*) different tephras. (A) Binary plot of K_2O versus CaO contents for rhyolitic glass from five Holocene age tephras in New Zealand. The data show that some of the tephras (Taupo, Whakaipo, Waimihia) are generally distinguishable from one another, but others (Poronui, Opepe, Taupo) partly overlap and, hence, are not clearly distinguishable using these oxides. (B) Binary plot of FeO versus MgO contents of biotites from four late Quaternary tephras in New Zealand, showing that Okareka, Rerewhakaaitu, and Rotorua tephras are distinguishable from one another. (C) Ternary plot of minor elements in titanomagnetites and illmenites from seven late Quaternary tephras, showing their fingerprinting utility. Also shown are tie lines with estimated eruption temperatures. (D) Elemental data for some ash deposits in North America.

with radioisotopic dating. Many of the isotopes used in radioisotopic dating have found great use in SED and geomorphologic dating applications (Nishiizumi *et al.* 1986, Chiverrell *et al.* 2009). Unstable isotopes are termed radioactive. Related radioisotopic methods, for example, K-Ar and U-Th, utilize the buildup of daughter products from primordial radionuclides; because they are most useful in geology and not soil geomorphology, we do not dwell on them here. Our focus here is on the most popular form of radioisotopic dating, i.e., radiocarbon dating.

Radiocarbon

Compounds containing carbon occur in many forms that cycle among their major reservoirs, including CO_2 in the atmosphere, biomass in the biosphere, and humus, peat, carbonate, marl, and other substances in the pedosphere.

Radiocarbon dating determines the ages of carbon-bearing materials on the basis of their content of the radioisotope ^{14}C.

THEORY AND MEASUREMENT

About 98.9% of all carbon atoms are ^{12}C, with an atomic number of 12. Another 1.1% are ^{13}C. An even smaller proportion are ^{14}C, an unstable isotope called *radiocarbon*. With two extra neutrons in its nucleus, ^{14}C is produced when neutrons emitted from cosmic rays impact in various ways (directly and indirectly) nitrogen ($^{14}N_2$) gas, which composes more than 78% of the atmosphere (Vogel 1969; Fig. 15.61). Both ^{12}C and ^{13}C are stable, but ^{14}C is not, instead decaying at a known rate, relative to ^{12}C. The decay process produces ß-particles (beta particles, or electrons) (Taylor 1987). Eventually, ^{14}C decays back to ^{14}N.

Volumes of
eruptive material:

1 Huckleberry Ridge tuff, 2.0 Ma, 2500 km³
2 Lava Creek tuff, 0.6 Ma, 1000 km³
3 Mesa Falls tuff, 1.2 Ma, ~280 km³
4 Tambora, 1815, 150 km³
5 Mazama (Crater Lake), 7.6 Ka, 75 km³
6 Krakatoa, 1883, 18 km³
7 Pinatubo, 1991, 10 km³
8 Mount St Helens, 1980, ~1-2 km³

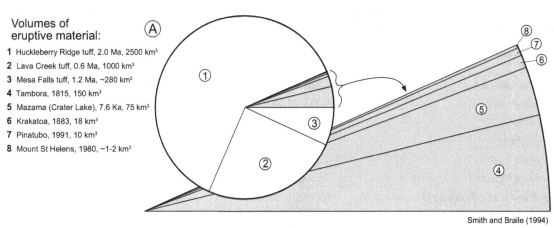

Smith and Braile (1994)

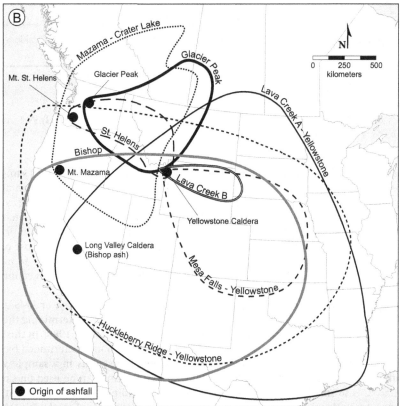

Fig. 15.60 The major ashfall marker beds of central North America. (A) Volumes of some Pleistocene and recent volcanic eruptions. (B) Distribution of the major ashfall beds (tephrostratigraphic units) in the western United States. From many sources, but primarily after Izett and Wilcox (1982), Smith and Braile (1994), and Kuehn et al. (2009).

The radiocarbon cycle is well understood (Guilderson et al. 2005, Hajdas 2008, Ramsey 2008, Olsson 2009; Fig. 15.61). Radiocarbon is present in different quantities in different global reservoirs because of a dynamic equilibrium between its production in the atmosphere and its loss through radioactive decay (Ramsey 2008). Radiocarbon produced in the upper atmosphere is quickly converted to $^{14}CO_2$ and mixed throughout. Much of it is dissolved in the oceans as bicarbonate, but some is also taken up by plants during photosynthesis and incorporated as biomass. The amount of radiocarbon in plant biomass quickly equilibrates to that in the atmosphere. (Actually, for most plants, the amount of radiocarbon in their biomass is 3–4% below that of the atmosphere, but this difference can be taken into account.) As long as the plant is alive, it is absorbing radiocarbon from the atmosphere. Living plants form the

Table 15.9 | Approximate ages and origins of major ashfall beds in central North America

Tephrochronological unit[a]	Age (years BP)	Area of origin
White River ash	1,250[b]	Mount Bona, Alaska
Mazama (Crater Lake) ash	7,630[b]	Mount Mazama, Oregon
Mount St. Helens J ash	8,000–12,000[b]	Mount St. Helens, Oregon
Glacier Peak B ash	11,250[b]	Glacier Peak, Washington
Glacier Peak G, M ash	12,750[b]	Glacier Peak, Washington
Mount St. Helens S ash	12,120–13,650[b]	Mount St. Helens, Oregon
Mount St. Helens M tephra	18,560–20,350[b]	Mount St. Helens, Oregon
Lava Creek A and B ash[c] (Pearlette O)	580,000–850,000 (620,000 most likely)	Yellowstone caldera, Wyoming–Idaho
Bishop ash (and overlying ash flow Bishop tuff)	759,000	Long Valley caldera, California
Mesa Falls ash (Pearlette S)	1,270,000–1,290,000	Yellowstone caldera, Wyoming–Idaho
Huckleberry Ridge ash (Pearlette B)	1,800,000–2,180,000 (1,970,000 most likely)	Yellowstone caldera, Wyoming–Idaho

[a] Most ashfall units consist of a number of smaller units that cluster together; for simplicity, we refer to them here as one unit.
[b] In ^{14}C yrs BP; all other ages have been determined by other means.
[c] Lava Creek A is the older of the two ashes.
Sources: Westgate and Briggs (1980), Izett *et al.* (1981), Davis (1985), Porter *et al.* (1983), Ward *et al.* (1993), Williams (1994), Ward and Carter (1998) and Sarna-Wojcicki *et al.* (2000).

base of the food chain, such that higher organisms ultimately feed on plant material and thus have radiocarbon in their biomass in proportion to that of the plant and the atmosphere (Burleigh 1974). However, after plant or animal metabolism ceases – i.e., the organism dies – uptake of radiocarbon stops, and the store of radiocarbon in the dead plant (seed, log, leaf) or animal (bone, hair, tooth) is governed by radioactive decay processes. In other words, the amount of radiocarbon in dead tissue is a function of the time elapsed since death. The ^{14}C concentration in the biomass diminishes logarithmically, having a half-life of 5,730 years (Linick *et al.* 1989). Thus, in 5,730 years, half of the radiocarbon has decayed to ^{14}N, and after another 5,730 years another half is gone (leaving only one-quarter of the original store of radiocarbon). After about 10 half-lives, the amount of residual ^{14}C is so small that it cannot be determined with current technology, meaning that samples older than ~50 ka are not normally datable.

Most radiocarbon ages provide an estimate of the time since the death of a plant or animal. A radiocarbon age is based on a measurement of a sample's residual ^{14}C content, a comparison to modern standards, and knowledge of the half-life decay constant (Taylor 1987). Determining the amount of residual ^{14}C in a sample can be done in either of two ways. In the conventional method, the sample is isolated in a chamber and the number of ß-particles emitted over a period of time, usually at least 24 hours, is determined. Usually ~2–200 g of carbonaceous sample is required, depending on its age and type, to obtain a ^{14}C age

using this method. Old and/or small samples will not yield enough beta particles to determine an age accurately. For example, in a gram of relatively young carbonaceous material, only 14 ß-particles are produced each minute, whereas in an older sample there might only be 14 particles produced each hour. Eventually, the cost of laboratory time and the increased size of the error term become prohibitive, explaining why older samples cannot be dated in this manner.

Accelerator mass spectrometry, or AMS dating, is a newer, alternative technique for determining the radiocarbon content of samples (Linick *et al.* 1989). In this method, the total amount of radiocarbon is determined by directly counting the numbers of ^{14}C atoms in a sample with a mass spectrometer. AMS dating can be used on samples as small as 2–30 mg (depending on carbon content). Although more expensive than traditional methods, AMS dating is faster, is more accurate, and requires a significantly smaller sample, with a weight 1,000 times less than that of material dated using the conventional method. Using the AMS method, the required sample size is so small that one has the option to date only the tiniest, most pristine fragments of carbon from a sample locale, reducing the potential for contamination (Hormes *et al.* 2004) but also potentially increasing the likelihood that the sample is not representative. It is also possible to obtain datable samples from sediments that typically have only finely comminuted fragments of carbonaceous material, further widening the possible applications (Pohl 1995). Consequently, AMS dating

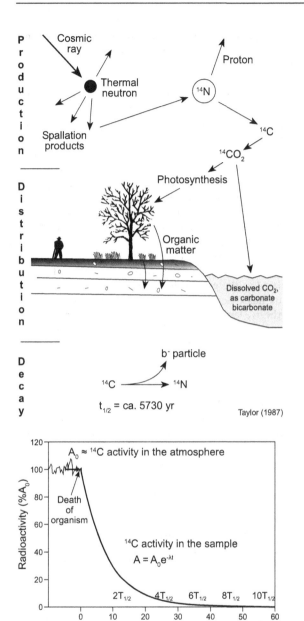

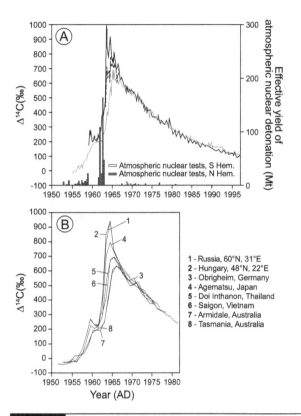

Fig. 15.62 Effects of nuclear bomb detonations of the mid-twentieth century on ^{14}C contents. (A) Atmospheric ^{14}C contents from 1950 to 1997, and the magnitude of atmospheric nuclear detonations. Data are from sites in Austria, Germany, Ethiopia, Australia, and New Zealand. (B) ^{14}C contents in tree rings. See Hua and Barbetti (2004) for data sources.

Fig. 15.61 The cycle of radiocarbon production, distribution, and decay.

has extended the radiocarbon dating method to 50 ka or even 60 ka BP.

Because the amount of ^{14}C detected (in either method) is based on probabilities, one must always consider the *statistical uncertainty* in the reported age. The amount of radiocarbon in a sample can be determined only within a certain accuracy range, yielding age estimates that are usually reported with an envelope of error, expressed as a standard deviation, or sigma (σ), e.g., 7,560 ± 120 BP. The radiocarbon age is only a *range* ± *of years* within which the *true age* lies, according to certain probability rules. Always, the most probable age is the number obtained by the measurement. If the reported error range is one standard deviation, the true age has a 68% probability of falling within that envelope. Although it would be advisable to report ages to two standard deviations, widening the range but increasing the likelihood of that range containing the actual age to 95%, conventional and AMS ^{14}C dating facilities do not report ages as such. Instead, ages are reported to one standard deviation.

Radiocarbon ages are reported in years *Before Present*, or BP. Because the method was devised in the year 1949 by Willard Libby (Libby *et al.* 1949), the BP refers to years before AD 1950. This date has stood as the standard, largely because after 1950 the open-air detonation of nuclear devices worldwide caused the ^{14}C content of the atmosphere and plants to spike (Fig. 15.62). As a result, radiocarbon ages on postbomb materials are erroneously enriched in ^{14}C and are, thus, not reported.

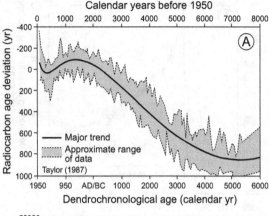

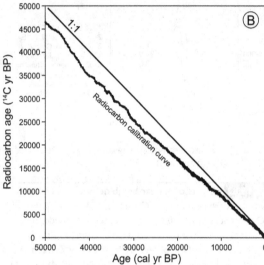

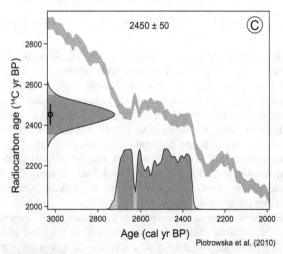

Fig. 15.63 Radiocarbon-calendar year corrections. (A) Relationship between the ^{14}C age (in radiocarbon years BP) and the calendar age based on dendrochronology, showing the deviation of radiocarbon years from calendar years for the past 8,000 years. (B) A recent calibration curve for

Although ^{14}C dating is a good measure of sample age, it is not an exact correlation to astronomical (calendar) time. This occurs because ^{14}C production in the atmosphere has not been uniform over geologic time, but varies as a result of cosmic ray fluctuations related to sunspot activity. To correct for this problem, and thus make radiocarbon ages comparable to astronomical time, ^{14}C ages are now *calibrated* to calendar years before AD 1950 (Ramsey 2009). Indeed, most radiocarbon laboratories report both uncalibrated (raw) and calibrated ^{14}C ages. In fact, it is now suggested that only calibrated ^{14}C ages be reported on everything except soil carbon mean residence time (MRT) ages (see later discussion). The parameters used for this calibration curve were originally derived from the tree ring record, which provided detailed information to about 10 ka. Use of information from varves (carbonate-rich lake sediment that accumulates in annual layers and thus can be dated to the exact calendar year of formation) and sea corals has extended the calibration curve to ~50 ka for Northern Hemisphere atmospheric and marine samples (Reimer *et al.* 2009, Bronk Ramsey *et al.* 2012, Reimer 2012; Fig. 15.63). Corals can be dated by ^{14}C and U-series methods, making them good for calibration (Bard *et al.* 1998, Cutler *et al.* 2004, Fairbanks *et al.* 2005, Chiu *et al.* 2007); forams deposited in ocean-bottom sediments can also be calibrated to other chronologies (Bard *et al.* 1998, Hughen *et al.* 2004). In general, the divergence between ^{14}C ages and the calibration curve is not significant over the last 3,500 years, but before this time the calibration uncertainties become progressively larger. One of the stickiest problems associated with calibration centers on a few select periods, i.e., ca. 10, 12.3, 14.4, and 18.3 ka, and a few other places where the calibration curve is not linear. At these points in time, called "radiocarbon plateaus," calibrations are less accurate, and more intricate age estimation models are sometimes required (Guilderson *et al.* 2005, Hajdas 2008, Blauuw 2010, Piotrowska *et al.* 2010).

Fig. 15.63 caption (cont.)

converting radiocarbon years to calendar years. Primarily after Reimer *et al.* (2004, 2009), but redrawn from Piotrowska *et al.* (2010). (C) An example of how calibration of a ^{14}C sample can be affected by plateaus on the calibration curve. The calibration for a sample of radiocarbon age 2,450 ± 50 BP is shown as an open circle with 1 sigma range, and with the full probability distribution plotted on the vertical axis. The INTCAL09 calibration curve and the probability distribution of the sample's calendar age are plotted on the horizontal axis. Probabilities are plotted arbitrarily, although darker areas represent 95.4% ranges of age. This example shows the wide possible calendar age range delivered for a sample with a radiocarbon age on a plateau in the radiocarbon calibration curve.

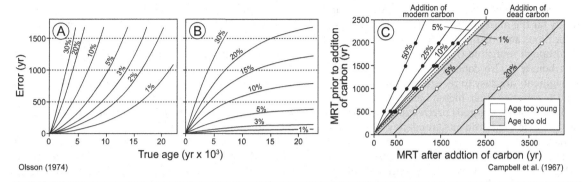

Fig. 15.64 Potential errors due to contamination of ^{14}C samples. (A) Effects of the introduction of dead carbon to a sample. Values on the y-axis indicate the difference, in years, between the true and reported sample ages. (B) Effects of the introduction of modern carbon (with high ^{14}C activities) to a sample. Values on the y-axis indicate the difference, in years, between the true and reported sample ages. (C) Effects of dead and modern carbon on mean residence time (MRT) ages on soil organic matter. Because MRT ages are a statistical average of many different organic fractions, the effects of this type of contamination are different.

Assumptions

The radiocarbon method has several assumptions, some of which we know are violated and, hence, we have adjusted our interpretations accordingly. First, we assume that the amount of ^{14}C in each C reservoir (particularly the atmosphere) has remained constant over the time span of the method. For reasons discussed previously and confirmed by detailed ^{14}C analysis of the wood in tree rings, from long-lived trees, e.g., bristlecone pine, redwood, sequoia, and European oak, this assumption (for the atmosphere) does not hold (Willis et al. 1960, Ferguson and Graybill 1983, Kovaltsov et al. 2012), validating earlier research (Anderson and Libbey 1951). Dendrochronology is ideal for use in radiocarbon calibration, because tree rings are near-perfect archives of atmospheric ^{14}C levels, and the tree ring timescale can be built with minimal error and with high replication (Kromer 2009). Tree ring and coral records extend back thousands of years, and, in principle, they provide a growing season average of the atmospheric and marine radiocarbon concentrations, on decadal and perhaps even annual timescales (McCormac et al. 2004, Reimer et al. 2009).

Differences in the amount of radiocarbon in the various reservoirs are due to (1) variation in production, which is commonly traced back to changes in the intensity of the Earth's magnetic field, and (2) for modern or near-surface samples, large-scale burning of fossil fuels and open-air testing of nuclear devices (Stuiver and Quay 1980. Ramsey 2008). Thus, radiocarbon ages vary somewhat from the standard astronomical calendar, necessitating that they be reported in radiocarbon years BP and then converted to calendar years using the IntCal09 calibration curve (Reimer et al. 2009; Fig. 15.63). An uncalibrated radiocarbon age is usually reported as ^{14}C yr BP. A calibrated, or calendar, age is abbreviated as cal yr BP or cal BP, which is interpretable as calibrated years before present, or calendar years before present.

A second basic assumption applied to ^{14}C dating is that all plants accumulate radiocarbon in equal proportions to each other; this assumption has been shown to be incorrect (Olsson and Osadebe 1974). Most plants take their carbon as CO_2 directly from the air, or from the water in the case of submerged aquatic plants. In more complicated environments, most notably peat, the sources of carbon are more complex and may vary in ^{14}C content (Piotrowska et al. 201). However, the small variation among plant species is usually not a large source of error.

Another assumption centers on the lack of sample contamination, either by old (dead) carbon such as coal, black shale, or any carbonaceous material that no longer retains ^{14}C (>60 ka), or by young (modern) carbon with high amounts of ^{14}C. The most common contaminants in soils are roots, especially fine root hairs, humic substances, and carbonates from hard (carbonate-rich) water and those deposited by pedogenesis (Olsson 1974). Hard water contamination should be seriously evaluated when dating material from soils with high water tables. Contamination by young carbon has the greater potential to skew the age, however, especially if it is old (Fig. 15.64). Even so, contamination is not a large problem, statistically, for samples that are less than a few thousand years old (Vogel 1969). Fortunately, many radiocarbon laboratories are able to pretreat and leach samples of possible secondary carbon contaminants, especially if using the AMS method. Nonetheless, great care must be taken during sampling and subsequent storage to prevent the sample from touching skin or hair and/or the growth of mold or other organisms on the sample, by always keeping the samples refrigerated or frozen. Samples with large numbers of roots in and among them should be carefully treated to remove as many roots as possible.

Applications

In principle, any material containing carbon that at some time has been in exchange with atmospheric CO_2 can be radiocarbon dated (Vogel 1969, Ramsey 2008). Olsson (2009) provided an excellent historical review of the many different types of radiocarbon dating applications and materials. By far the most common type of materials used in ^{14}C applications are organic materials such as wood and charcoal, although inorganic C such as cave speleothems and soil carbonate is also datable (Pohl 1995). Wood and charcoal fragments are easy to recognize in soils and sediments, and because of their high-molecular-weight components, they allow for rigorous pretreatment procedures to extract possible contaminants, rendering them quite "safe" to date (Geyh et al. 1971, Taylor 1987). Charcoal is particularly attractive because it is more commonly preserved. However, it does have the potential to absorb humic substances – a source of contamination – from soils. Dating charcoal, as with any other substance, should not be regarded as reliable until evidence is found, usually via field observations, that supports the reliability of the radiocarbon age (Watchman and Twidale 2002). Nonwoody plant parts such as leaves, seeds, reed, roots, and their derivatives (papyrus, paper) are also candidates for dating. Thorn et al. (2009) even described a procedure whereby arbuscular mycorrhizae fungal spores in buried soils were dated by ^{14}C. The spores are produced by mycorrhizal fungi associated with plant roots and are particularly attractive as a dating resource because they may not have interacted with other soil constituents after burial.

In carbonate-rich shells, ^{14}C determinations are based on the inorganic C fraction, as $CaCO_3$. Organisms construct their shells using ^{14}C from water or atmospheric sources. Thin shells, such as those of some kinds of snails, are easily contaminated by dead C (from groundwater) or other radioactive sources (Goodfriend and Hood 1983, Wilson and Farrington 1989, Innocent et al. 2005, Broecker et al. 2006, Olsson 2009). With proper pretreatment, however, ^{14}C ages on shell material, especially thick shells, can be reliable. Pigati et al. (2010) concluded that shells of land snails (gastropods) can be successfully radiocarbon dated if they meet two criteria: (1) While living, the ^{14}C activity of their shells must have been in equilibrium with that of the atmosphere and (2) after burial, their shells must behave as closed systems with respect to carbon. Land snails are common in loess and other Quaternary age deposits; they have long been important in establishing loess chronologies (Forman et al. 1992, Pearce et al. 2010, Xu et al. 2011, Rech et al. 2012, Pigati et al. 2013; Fig. 15.65) and can also be dated by other methods (Goodfriend 1992). Bone is also commonly dated, especially within archaeological and paleontological contexts. Because bone is highly susceptible to contamination, particularly by humic acids, proper chemical pretreatment is essential (Taylor 1982, Olsson 2009).

In soil geomorphology, ^{14}C ages usually provide a minimum- or maximum-limiting age for either a *sediment* or *geomorphic surface* (Nelson et al. 1979, Pohl 1995). Maximum-limiting ages imply that the sediment or surface can be *no older than* the ^{14}C (or other type of) age. Conversely, minimum-limiting ages indicate that a surface or sediment can be *no younger than* that age. These constraints provide numerical age control on stratigraphy, providing a temporal framework for the timing of erosional and depositional events. For example, envision wood within an alluvial deposit dated at 11,300 ^{14}C yr BP[1] (Fig. 15.66). This age suggests that the alluvium is *no older than* 11,300 ^{14}C yr BP, because the tree died at approximately that time (if the date was from an outer ring, as is preferred). If it was deposited within the alluvium that same year it died, then the alluvium was emplaced ca. 11,300 ^{14}C yr BP. More likely, the wood was intermittently carried to the site and deposited there some time later. Thus, the ^{14}C age provides a *maximum-limiting age estimate* for the alluvium and the geomorphic surface that overlies it; neither the alluvium nor the geomorphic surface above can be older than 11,300 ^{14}C yr BP. Likewise, a geomorphic surface or sediment that lies stratigraphically below this alluvium can be no younger than 11,300 ^{14}C yr BP. The wood, therefore, provides a *minimum-limiting age* for that underlying surface or sediment.

Another example of a minimum-limiting age would be an in situ tree stump, dated to 1,560 ^{14}C yr BP, in an alluvial terrace (Fig. 15.66). In this case, the terrace surface can be *no younger than* 1,560 ^{14}C yr BP. That is, the surface had to exist for a tree to grow on it. Therefore, the stump provides a minimum-limiting age for the surface.

DATING SOIL ORGANIC MATTER

Despite being somewhat problematic, dating the organic matter fraction of soils, both modern and buried, is frequently performed with success. Interpreting the results must, however, be done with care (Perrin et al. 1964, Scharpenseel and Schiffmann 1977, Gilet-Blein et al. 1980, Martin and Johnson 1995). Nonetheless, if the assumptions and limitations of the method are understood beforehand, dating soil organic matter can yield important information about buried soil and surface ages, e.g., Wolfe et al. (2000).

Soils primarily accumulate organic carbon from the excreta of living animals and via the decomposed parts of plants and animals that inhabit the soil. They also accumulate inorganic forms of carbon such as $CaCO_3$. Soils are a veritable storehouse of organic materials in various stages of decomposition that originated at various times in the past. If one were able to extract each particle of soil organic matter from an A horizon, it would be evident that the horizon contains fragments of carbon-rich materials of a wide variety of ages (Tipping et al. 2010; Fig. 15.67). Some originated from plants that grew on the soil thousands of years ago, yet others are only weeks old microbial residues. And of course, there i everything in between! How can we make sense of this mishmash of carbon ages?

Early researchers tried to answer this question by focusing on the component parts of the carbon. They extracted and dated humic acids – the larger organic particles – from soils (Felgenhauser et al. 1959). Then, in 1964, Paul et al.

---

[1] In this book we will typically omit the standard deviation error terms on radiocarbon dates, for the sake of brevity.

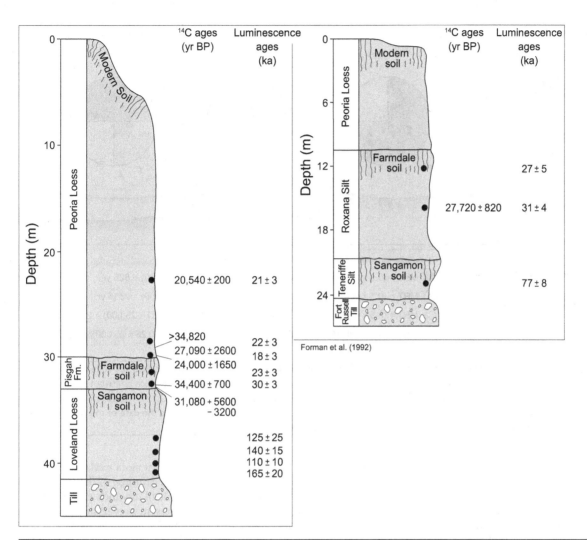

Forman et al. (1992)

Fig. 15.65 Typical loess and paleopedological stratigraphy for the midcontinent United States, showing the major loess units and paleosols, along with ^{14}C ages (on shells) and luminescence ages (on the loess itself). (A) At the Loveland loess type locality, Pottawattamie County, Iowa. (B) At the Pleasant Grove School section, Madison County, Illinois.

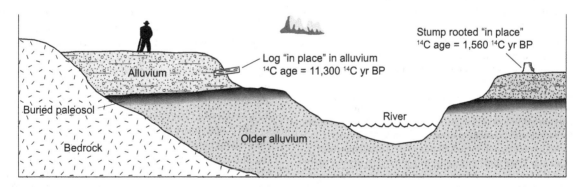

Fig. 15.66 Hypothetical valley cross section used to illustrate the concepts of minimum- and maximum-limiting ages. The log provides a maximum-limiting age for the alluvium surrounding it, and the geomorphic surface that overlies it. The stump provides a minimum-limiting age for the lower terrace surface in which it is rooted.

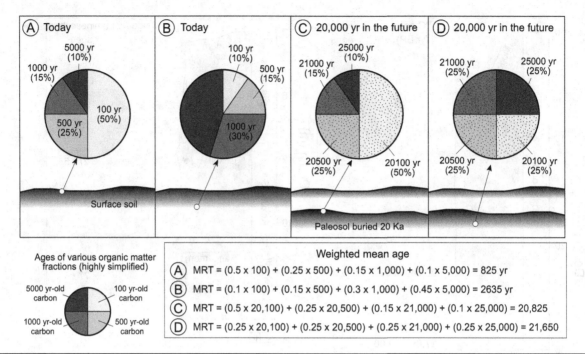

Fig. 15.67 Simplified and conceptual diagram showing how MRT ages on soil organic matter from a surface soil reflect, roughly, the ages of the various organic fractions in the soil. Although innumerable age fractions exist in soils, all of which contribute to the MRT age, only four are shown here for simplicity. For the purposes of complete disclosure, in parts C and D, the ^{14}C ages obtained 20 ka in the future would have been set to AD 1950. Thus, if you date this buried soil 20,000 years in the future the date of the soil will still be 5000 ^{14}C yr BP! We are using this example notwithstanding that knowledge/assumption.

reported on a new way to study the age of soil carbon – dating the entire suite of soil humic compounds and arriving at an overall mean age, reflecting the time that the carbon had *resided* in the soil – all of it. This *mean residence time* (MRT) method drew attention to the dynamics of soil organic matter and is still in use. Let us examine what an MRT age actually means, and how to interpret it.

Organic matter begins to accumulate as humus and as macroscopic forms almost from time$_{zero}$ and this accumulation continues to the present, or until the soil becomes deeply buried (Schaetzl and Sorenson 1987). Geyh *et al.* (1983: 409) elaborated on the inherent problems:

Soil dating is a questionable attempt to date only a small part of the total humic matter of a soil horizon and to interpret the result as representative of the whole sample. The discrepancy between ^{14}C soil dates and true ages results from the complexity of soil genesis, which is a continuous process of accumulation and decomposition of organic substances. Penetration of rootlets, bioturbation, and percolation of soluble humic substances ... cause rejuvenation, and the admixture of allochthonous plant residues may cause apparent aging. As a result, the organic matter in a soil is *a mixture of an unknown number of compounds of unknown chemical composition, concentration, and age* [emphasis is ours].

MRT ages are based on the entire soil organic matter fraction, usually obtained from alkaline extracts (Campbell

et al. 1967). Also called *apparent mean residence time* age, an MRT age reflects the *weighted mean ages* of the many organic components within the soil, all of which exist in different amounts. Older carbon is present in ever-diminishing quantities, as new carbon is continually added to the soil. Thus, the "age mixture" of the soil carbon changes over time. MRT ages reflect the fact that younger carbon is continually added to the soil – usually near the surface – whereas older carbon is continually lost through decay and mineralization (Anderson *et al.* 2002). Thus, any age estimate on bulk soil carbon will be younger if taken nearer the surface, where little old carbon remains and much of the carbon is young (Fig. 15.68). The degree to which each component is represented in a soil is a function of its *residence time* and the amount and type of carbon cycling within the soil (Geyh *et al.* 1971, Stout and Goh 1980). Organic materials within the soil that are slow to decompose, i.e., have long residence times, will be overrepresented.

MRT ages must be interpreted correctly. For example, they always provide a *minimum age* for surface soils, i.e., the soil can be no younger than the MRT age (Polach and Costin 1971, Gilet-Blein *et al.* 1980, Cherkinsky and Brovkin 1991, 1993, Wang *et al.* 1996a, Mayer *et al.* 2008). Visualize this by realizing that the oldest organic carbon in the soil can be no older than time$_{zero}$, unless it had been introduced from outside the soil system. If we were to date the one

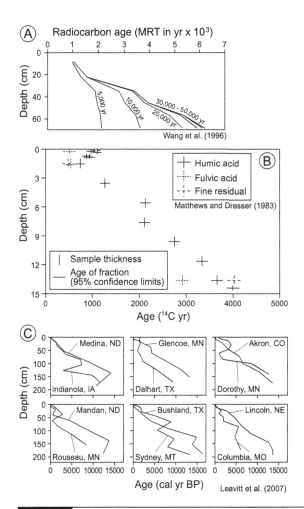

Fig. 15.68 MRT ages obtained by ¹⁴C for soil organic matter, as a function of depth, assuming minimal amounts of bioturbation. (A) Theoretically derived MRT ages of soil organic matter, with depth, for a prairie soil at different stages in its development. Numbers along the lines indicate the estimated actual age of the soil. (B) Depth plot of radiocarbon ages on various soil organic matter fractions, for a soil buried beneath a Neoglacial moraine in Norway. (C) Depth plots of ¹⁴C ages (MRT) for some undisturbed grassland soils of the Great Plains, United States.

humus of that age, assuming none was inherited in the parent material. If we assume the soil is 5,000 years old, then the MRT age of 825 years is only 16.5% of the actual age. This young age occurs in the upper profile, where much of the organic matter is composed of recently added organic materials (Geyh *et al.* 1971; Fig. 15.68). Note that the proportions of older carbon increase deeper in the profile, producing an older MRT age (Fig. 15.67B). In this case, the MRT age is closer to the actual age (52.5%), but it is still a *minimum* age estimate. Figure 15.67B illustrates the point that when sampling surface soils for MRT purposes, the deepest carbon-rich horizon should be sampled to obtain the oldest possible age and, hence, the one closest to the time$_{zero}$.

MRT ages in soils increase with depth and, of course, as the soil age itself increases (Catt 1990, Wang *et al.* 1996a, Leavitt *et al.* 2007; Fig. 15.68). MRT dates are usually youngest for A horizons (Herrera and Tamers 1970, Ganzhara 1974, Stout and Goh 1980). Although this depth-age relationship holds for soils in which organic matter is not generally mobile within the profile, e.g., Mollisols and Aridisols (Table 15.10), in soils where soluble organic matter is readily translocated as part of their genesis (Spodosols and some Andisols), MRT ages in the B horizon may be similar to those of the A (Tamm and Holmen 1967). Likewise, soils that are strongly pedoturbated, e.g., Gelisols and Vertisols, may show irregular depth-age trends (Kovda *et al.* 2001). MRT data can also provide information about the stability, turnover rates, and overall longevity of OC in different microsites of soils, e.g., in small versus large soil aggregates, or in different types of soil profiles (Balesdent 1987, Bruun *et al.* 2005, Rabbi *et al.* 2013).

The effect of contamination – both young and old – on MRT ages is different than it is on whole fragments of carbonaceous material that are assumed to have one age, because MRT ages represent a weighted mean (Fig. 15.67A). Additions of old or dead carbon, therefore, have a dramatic effect on MRT ages. Organic matter with long MRTs can include recent additions of aged C or compounds at an advanced stage of decay (Trumbore 2009, Kleber and Johnson 2010). Senescence of stored C molecules photosynthesized centuries ago in long-lived plants, microbial recycling, OM transport through physical translocation or leaching can all inflate the MRT age on a soil sample. Also, atmospheric deposition of fossil fuel–derived C can inflate radiocarbon-based ages of recent inputs of OM in soils and aquatic bodies (Stubbins *et al.* 2012).

When dating the organic fraction of surface horizons, one must always consider the effect of bomb carbon from twentieth century open-air testing of nuclear devices (Broecker and Olson 1960, Wang *et al.* 1996a, Hua and Barbetti 2004, Rabbi *et al.* 2013). Since about 1955, thermonuclear tests have added considerably to the amount of ¹⁴C in the atmosphere (Fig. 15.62), almost doubling the quantity of ¹⁴C activity in terrestrial carbon-bearing materials (Hajdas 2008). Contamination by bomb carbon has

microfragment of humus that dates to time$_{zero}$, we could obtain the exact age of the soil, i.e., when time$_{zero}$ occurred. But since we, in fact, date a *mixture* of carbon compounds, even if they did contain that one old fragment, the mean age will always be statistically weighted toward a younger age. The mean age obtained on the whole organic matter fraction of a surface horizon, therefore, must be younger than the age of the soil (Fig. 15.67A).

The soil in Figure 15.67 illustrates some of these points. The soil must be at least, 5000 years old, because it contains

Table 15.10 | Age of humus fractions in Chernozems (Ustolls) on the Russian Plain

Soil	Depth (cm)	Organic matter content (%)	^{14}C age of humic acids bound to Ca (yr BP)	^{14}C age of humin (yr BP)
Kursk	10–20	7.7	1,680 ± 80	1,110 ± 70
Chernozem	50–60	4.7	2,970 ± 110	1,230 ± 180
	70–80	2.3	4,020 ± 90	2,970 ± 90
	120–130	0.7	6,100 ± 200	–
Voronezh	20–30	8.1	660 ± 80	1,590 ± 100
Chernozem	40–50	5.6	2,200 ± 100	–
Tambov	60–70	3.1	–	3,000 ± 160
Chernozem	0–22	7.8	1,330 ± 80	2,120 ± 130
	30–40	7.5	2,800 ± 120	2,380 ± 150
	70–80	5.4	2,460 ± 120	2,970 ± 130

Source: After Ganzhara (1974).

made MRT ages on modern surface soils much younger than they normally would have been (Taylor 1987). Work by Campbell *et al.* (1967), however, indicated that bomb carbon may not be a major influence on MRT ages in the long term, because of the rapid turnover of fresh carbon residues in soils.

The most common application of MRT ages occurs in paleopedology, where the goal could be either (1) determining $time_{zero}$ of soil formation for the buried soil or (2) ascertaining the time since burial (Kusumgar *et al.* 1980, Matthews and Dresser 1983, Nesje *et al.* 1989, Orlova and Zykina 2002, Demkina *et al.* 2008, Mason *et al.* 2008; see Chapter 16). The MRT age on a buried soil reflects the MRT age of the soil at the time of burial, plus the period during which it has been buried. Thus, two clear applications exist for MRT dating of buried soils: (1) establishing $time_{zero}$ for the buried soil and (2) establishing the age of burial. For both cases, we turn to Figure 15.67. When examining the MRT ages from buried paleosols, one must always keep in mind the two possible contaminants – both of which make the age younger – modern rootlet penetration and decomposition in situ, and soluble humic acids translocated into the soil by percolating water.

How can an MRT age on a buried soil determine the $time_{zero}$ for that soil, before it was buried? In Figure 15.67C, we hypothetically bury a 5,000 year old soil with an MRT age of 825 years in its A horizon for 20 ka and then reevaluate its MRT age. Now, because all the organic fractions are 20 ka older, the analysis yields an MRT age of 20,825 ^{14}C yr BP. This MRT age is a *maximum estimate* of the time since burial, i.e., the soil could not have been buried any longer than the age reported in the ^{14}C age (Nesje *et al.* 1989, Mayer *et al.* 2008). In some soils, however, the MRT age obtained from the upper A horizon is close to the actual date of burial (Haas *et al.* 1986, Rawling *et al.* 2003). These types of ages represent a *minimum estimate* for $time_{zero}$, in some

cases being as much as several thousand years younger than the initiation of pedogenesis. Recall that the $time_{zero}$ for the soil in Figure 15.67C is actually 25 ka (it formed for 5,000 years before being buried for 20,000). The MRT age on the buried soil is actually now 83.3% of the actual age (20,825/25,000), which is a far better estimate of its $time_{zero}$ than when it was dated as a surface soil (825/5,000 = 16.5%). This example illustrates why MRT ages on buried soils have an advantage over those on surface soils, for estimating $time_{zero}$. Now, factor in the assumption that, after burial, the younger carbon fractions will decay more rapidly than will the older fractions, yielding a more even distribution, as shown in Figure 15.67D. Now, the MRT age on the paleosol is an even closer reflection of reality, because these data yield an MRT age of 21,650 ^{14}C yr BP, or 86.6% of the actual $time_{zero}$ age. Thus, to obtain the overall age of a buried soil, from its $time_{zero}$ to present, sample the lowermost part of the A horizon. One problem with sampling deeper horizons is that they may contain less carbon.

A second application of MRT dating of buried soils allows us to evaluate the age of burial. MRT ages obtained from the upper profile of a buried soil will always provide a maximum-limiting estimate for the period of burial – the soil could not have been buried for longer than that. Note that the MRT age from the upper profile (Fig. 15.67D) is closer to the actual burial period (20,000 years) than is the age from the lower profile, where the carbon is older. Thus, to obtain the best estimate of the length of time that a soil has been buried, one must always sample the part of the soil with the youngest carbon, i.e., the uppermost part of the A horizon (Haas *et al.* 1986, Nesje *et al.* 1989). Matthews and Dresser (1983) used an application of this theory. By regressing ^{14}C ages with depth in a buried soil they obtained the expected age of the organic matter at the very surface of the buried soil and assumed that this age was the best maximum estimate of the time of burial (Fig. 15.68C).

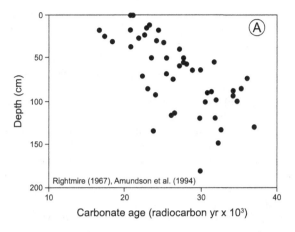
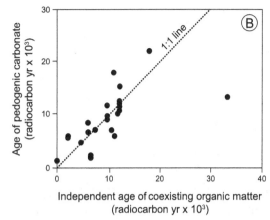

Fig. 15.69 Radiocarbon ages of soil carbonate. (A) Depth plot of published ages of soil carbonate, for some soils in Texas. The younger ages nearer the surface could reflect the true age of the carbonate or may be due to the drier Holocene climate, which caused carbonates to be deposited at shallower depths than they were during the Pleistocene. (B) Radiocarbon ages of pedogenic carbonate as compared to other, established ages for the same sediment or surface. See Amundson *et al.* (1994) for references to the ^{14}C ages.

Several other factors must be considered when dating the carbon in buried soils. All buried soils are modified by the burial process itself, as well as postburial processes (see Chapter 16). Thus, one potential problem centers on erosion prior to burial. In such circumstances, MRT ages from the upper profile may be slightly older, possibly resulting in an overestimate of the time since burial (Matthews 1980). Similar problems can result if older carbon is mixed into the soil prior to burial (Geyh *et al.* 1983). From a dating perspective, postburial alteration of the organic carbon in buried soils is so common as to be expected in all but the very driest environments. Oxidation and mineralization processes, after burial, reduce the amount of carbon in the buried soil and increase the error in the estimate of time since burial, if we assume that the younger organic fractions are disproportionately lost (but this assumption is not always true; see Kleber 2010, Kleber *et al.* 2011). Translocation of dissolved organic substances from overlying sediments can seriously affect the MRT age of the buried soil; this is so problematic in cool, humid environments as to make dating of all but the most deeply buried soils impractical (Matthews 1980, Geyh *et al.* 1983). Haas *et al.* (1986) also cautioned that MRT ages on buried soils that have been exposed for several years, e.g., in a trench wall, are less reliable than those taken from freshly exposed faces.

Because of the inherent problems in using MRT ages, efforts have continually been made to isolate and date a *fraction* of the soil organic matter, through chemical or physical means, that is (ideally) the oldest and presumably most biologically recalcitrant (protected), so as to estimate time$_{zero}$ of soil formation better (Scharpenseel 1972a, b, Martin and Johnson 1995). The fractionation is done on the assumption that one fraction will have a greater age than another and, hence, provide a better approximation of time$_{zero}$ for surface horizons. Most laboratories break the organic matter into the NaOH-soluble fraction (humic acid, aka humate) and the NaOH-insoluble fraction (residue, or humin). Others have extracted and dated the lipid and fulvic acid fractions (Gilet-Blein *et al.* 1980). The humic acid fraction is probably more mobile and usually yields older ages, possibly because humin is an earlier intermediary than is humic acid in the humification process (Polach and Costin 1971, Dalsgaard and Odgaard 2001). Whichever fraction yields the oldest ages continues to be, nonetheless, debatable and remains under study (Gilet-Blein *et al.* 1980, Haas *et al.* 1986, Martin and Johnson 1995, Kleber *et al.* 2011).

DATING SOIL CARBONATE

The results of ^{14}C dating soil carbonates have generally been consistent with those of other dating methods (Fig. 15.69). Like MRT ages on soil organic matter, ^{14}C ages on carbonate are not highly precise but often can provide stratigraphic age restrictions or minimal estimates of soil age, both of which may be helpful in certain contexts (Kusumgar *et al.* 1980).

Pedogenic carbonate is datable because $CaCO_3$ incorporates ^{14}C into its structure, as $Ca^{14}CO_3$ (Leamy 1974). Pedogenic carbonate acquires ^{14}C from the soil air and from dissolved gases in soil water. The carbonate forms are assumed to be in isotopic equilibrium with soil CO_2 (Cerling 1984, Quade *et al.* 1989), which is derived mostly from biological processes – roots, respiring fauna, and microbes that are consuming organic matter (Dörr and Münnich 1986). The first two of these are generally in isotopic equilibrium

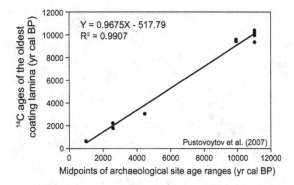

Fig. 15.70 Correlation of the midpoints of age-constrained archaeological sites in the eastern Mediterranean and southern Siberia with the mean weighted ^{14}C ages of the oldest carbonate coatings on clasts in the soil.

with the atmosphere, but depending upon the age of the organic matter being decomposed, soil microbes could release old $^{14}CO_2$ to the soil air (Wang et al. 1994). Therefore, the radiocarbon content of soil CO_2 is assumed to be only *generally* in equilibrium with the atmosphere.

Another potential problem lies in the effects of contamination from preexisting, older caliche or even limestone that is dissolved and reprecipitated in the new secondary carbonate (Williams and Polach 1969, Amundson et al. 1989b, Monger et al. 1998). This is especially problematic in highly polygenetic soils that have undergone humid and dry climatic intervals. During humid periods soil carbonate is dissolved and translocated to greater depths, where it can reprecipitate. During dry periods, carbonate-rich soils are eroded and the dust derived from them, rich with old carbonate, affects broad areas (Chen and Polach 1986). Still another source of contamination is dead C from groundwater.

An important source of carbon in soil CO_2 derives from the respiration/mineralization of soil organic matter. In young soils with low organic matter contents, little of the soil CO_2 content is derived from this source. Knowing the degree of contamination from this source is very difficult without the use of stable isotopes, such as $\delta^{13}C$ (see Chapter 16). Most SOM has a ^{14}C activity that is less than that of the atmosphere because it is in various stages of decay from plants that died long ago. However, this carbon source is always younger than the surface age, since essentially none of it formed prior to time$_{zero}$. Because some soil CO_2 has a modern age while other CO_2, derived from the decay of old organic matter, can be thousands of years old, dates on pedogenic carbonate *in very young soils* may overestimate the soil age, i.e., they are maximum ages. Soils with high respiration rates due to high organic matter contents also may have low $^{14}CO_2$ contents and thereby yield disproportionately *old* ^{14}C carbonate ages (Amundson et al. 1994).

Despite these complications, ^{14}C ages on soil carbonate from all but the youngest soils are usually *minimum ages* because (1) most of the carbonate is from soil air that is in equilibrium with the atmosphere and (2) the age on the carbonate is essentially a mean age that includes considerable amounts of young C (Callen et al. 1983).

Dating soil carbonate has the advantage of being able to utilize laminae on carbonate nodules and concretions. Nodules of carbonate grow, and preexisting clasts become coated with carbonate, in roughly concentric rings; inner rings are older than outer ones. Ages obtained from ^{14}C on carbonate nodules confirm this and thus provide a means of establishing the length of time necessary for these types of pedogenic features to form, and a way of isolating older carbonate (Chen and Polach 1986; Fig. 15.70). Mixing of old and young carbonate is usually minimal because the older rings protect the inner ones from contamination. Although inclusion of detrital carbonates, e.g., limestone gravel, within a carbonate mass would obviously cause an overestimation of the age (Callen et al. 1983), the likelihood of this occurrence is minimal in laminar Bkkm horizons or concretions.

Like MRT ages on organic matter, ^{14}C ages on soil carbonate increase with depth and time (Gile and Grossman 1968; Fig. 15.69A). This trend occurs because (1) carbonates continue to build upward in the profile (younger carbonates form nearer the surface), (2) near-surface carbonates are affected by younger C sources, and/or (3) drier conditions in the Holocene have led to carbonate deposition at shallower depths while Pleistocene carbonate occurs deeper. The last line of reasoning should remind us that desert soils are generally polygenetic; most date back to the wetter, cooler Pleistocene.

Dating pedogenic carbonate and silica is also possible using the U-series method (Ku et al. 1979, Schwarcz and Gascoyne 1984, Ludwig and Paces 2002). Care must be taken to date only pedogenic carbonates that are free of old, detrital limestone fragments (Bischoff et al. 1981, Rabenhorst et al. 1984b, Radtke et al. 1988). U-series ages of soil carbonate are minimum ages (Slate et al. 1991).

Other Cosmogenic Isotopes

Cosmogenic isotopes are produced as solar radiation interacts with atoms (Lal 1988, Cerling and Craig 1994), either in situ, e.g., within rocks or soils, or in the atmosphere (called meteoric or fallout nuclides). Some of these isotopes are stable, e.g., ^{21}Ne and ^{3}He, but others decay over time, e.g., ^{36}Cl. Within the family of cosmogenic isotope dating methods, one group uses the decay of meteoric, cosmogenic isotopes, such as ^{14}C and meteoric ^{10}Be, which are *produced in the atmosphere* and then accumulated by fallout onto terrestrial reservoirs. After these nuclides reach the surface, they may or may not be leached, taken up by plants, eroded, or remain in place. The other group of cosmogenic isotope dating methods utilizes isotopes that form directly in rocks because of exposure to cosmic rays, e.g., in situ ^{10}Be (Table 15.11).

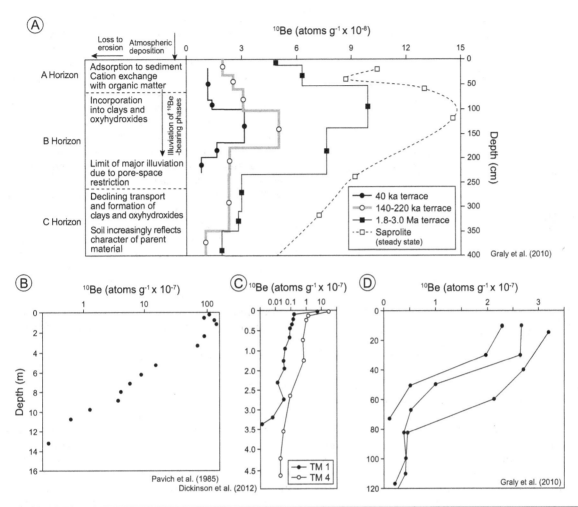

Fig. 15.71 Distribution and behavior of meteoric ^{10}Be. (A) Conceptual model of the control of soil processes on meteoric ^{10}Be distribution. (B) Depth plot of ^{10}Be in a soil-saprolite sequence in Virginia. (B) Depth plot of ^{10}Be in a soil located in a dry Antarctic valley. (D) Depth plot of ^{10}Be from hillslope profiles near Contra Costa, California.

To determine the exposure time of rocks on the surface, one must know or estimate (1) the rate of nuclide production, (2) its half-life, (3) the amount of inherited nuclides in the rock or reservoir, and (4) the burial/erosion history of the surface. Many parent materials, e.g., alluvium, loess, colluvium, and residuum, have a history of nuclide inheritance from previous sites that must be taken into consideration before an accurate age can be determined (Phillips *et al.* 1998). Burial/erosion history and the amount of nuclide inheritance, both of which are particularly important, are also sometimes quite difficult to determine (Anderson *et al.* 1996, Hancock *et al.* 1999).

The applications of cosmogenic dating are multifaceted and growing (Cerling and Craig 1994, Ivy-Ochs and Kober 2008). These methods have been used to date moraines, debris flow deposits, fluvial terraces, and glacial depositional landforms (Brown *et al.* 1991, Gosse *et al.* 1995, Zreda and Phillips 1995, Akcar *et al.* 2012, Goehring *et al.* 2012, Mentlik *et al.* 2013; Fig. 15.71) and are rapidly being used to date surface exposures in bedrock terrain. Isotopic methods have the advantage of providing information not only about surface exposure, but also about erosional or burial histories of surfaces, which can be used to inform us about both soil erosion and production (Lal 1988, 1991, Portenga and Bierman 2011).

FALLOUT FROM THE ATMOSPHERE

The primary fallout nuclide used in soil geomorphology research is beryllium-10 (^{10}Be), a cosmogenic nuclide that forms in the atmosphere, is delivered to surfaces, and can accumulate in soils. It has been widely used as an SED tool (McHargue and Damon 1991, Morris 1991, Graly *et al.* 2010).

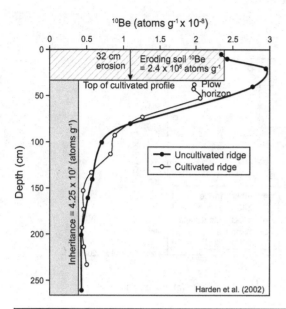

Fig. 15.72 Meteoric ^{10}Be depth profiles from uncultivated and cultivated ridges in western Iowa, with the depth of the cultivated profile data corrected for erosional loss.

This form of ^{10}Be is referred to as meteoric ^{10}Be, as opposed to that produced in situ in rocks (Nishiizumi et al. 1986, Monaghan and Elmore 1994).

Meteoric ^{10}Be is produced in the upper atmosphere as cosmic radiation interacts with O and N atoms. Some also is recycled back to the atmosphere, from the oceans, as hygroscopic nuclei. ^{10}Be attaches to aerosols, which are carried to the Earth's surface mainly by wet precipitation, where it is either fixed in soils or carried away by water or wind; some is locked in ice sheets (Dickinson et al. 2012). ^{10}Be is also transported to sites by dust – a form of recycled ^{10}Be (Baumgartner et al. 1997, Shen et al. 2010). Deposition of ^{10}Be onto the surface is primarily dependent upon precipitation type and amount, as most of it is carried down with rain and snow, and latitude (Graly et al. 2011). The theoretical global average rate of ^{10}Be fallout is 2.15 to 4.0 × 10^{-2} atom cm^{-2} s^{-1} (O'Brien 1979, Pavich et al. 1984), or about 1.7 to 1.8 × 10^{24} atoms yr^{-1} (McHargue and Damon 1991).

The assumption that most ^{10}Be is *retained* in soils is critical to its application in SED studies (You et al. 1988, Morris 1991, Willenbring and von Blanckenburg 2010), i.e., soils can potentially be long-term dosimeters for beryllium. After ^{10}Be reaches the surface, that which is not removed by runoff or wind is free to infiltrate. Virtually all natural species of Be are insoluble at pH values >4 (Takahashi et al. 1999) and are, therefore, capable of being retained (chemically) in most soils. Whether the atoms can be physically retained in soils is key to their use as a dosimeter. Assuming some degree of surface stability and minimal amounts of ^{10}Be leaching and inheritance (from the parent material),

one need only know the rate of ^{10}Be influx and its decay constant to estimate the elapsed surface exposure time.

As a cation, ^{10}Be is subject to the same translocation, erosion, and mixing processes typical of most soils. In coarse-textured soils where deep leaching is common and CEC values low, accumulation of ^{10}Be is not as effective an SED method as in more clay-rich, slowly permeable soils (Pavich et al. 1984). Nonetheless, in all but the oldest and sandiest soils, ^{10}Be is found in highest concentrations at the soil surface, and its concentrations decreases with depth, suggesting that it is reasonably well retained and pedogenically stable (Fig. 15.71). Many researchers have assumed that ^{10}Be is quickly adsorbed onto clay minerals because it is found in only very low concentrations in surface waters. Strong correlations between clay and ^{10}Be contents (Pavich et al. 1986, Graly et al. 2010) support the assumption that the fate of meteoric ^{10}Be in most soils depends largely on the mobility of clays and other fine particles to which it is adsorbed (Monaghan et al. 1983). We do note, however, that in older, slowly eroding soils, the highest ^{10}Be concentrations are found at depth, usually at 50–200 cm (Graly et al. 2010), presumably because of the additional time for translocation in these older soils.

Most soils also inherit small amounts of ^{10}Be from parent material. Two potential methods exist for estimating inheritance: (1) Modern sediment can be analyzed and these concentrations applied to older sediments (Pavich et al. 1986) or (2) sediment unaffected by ^{10}Be delivery can be found and analyzed. Typically the latter material is identified by a consistent minimum ^{10}Be concentration deep in the soil profile (Reusser et al. 2010, Graly et al. 2011). In all soils, however, because of surficial erosion and leaching, some ^{10}Be is lost, rendering SED data derived from ^{10}Be *minimum-limiting exposure ages* (Pavich et al. 1985). For example, Monaghan et al. (1983) found that ages of marine terraces determined with meteoric ^{10}Be were 16–380 times too young. Also, because ^{10}Be is being produced in the soil as cosmic rays interact with rocks within (see later discussion), any surface exposure estimate based on accumulated meteoric ^{10}Be must be a minimum age. Additionally, the MRT of the isotope, probably ca. 10^4–10^5 years, puts an upper limit on the method (Pavich et al. 1985, 1986). Meteoric ^{10}Be data from soils are also sensitive to long-term and even human-influenced erosion (Fig. 15.72), which in many cases must be estimated – adding uncertainty to the exposure age.

Nonetheless, despite the potential pitfalls, meteoric ^{10}Be has been successfully used in at least three different types of applications, in the determination of (1) duration of surface exposure (Maejima et al. 2005), (2) local- to regional-scale erosion or denudation rates (Valette-Silver et al. 1986, Brown 1987, Brown et al. 1988, Harden et al. 2002), and (3) rock-to-soil conversion rates (Pavich 1989, Monaghan et al. 1992, Monaghan and Elmore 1994). In a unique application of another sort, Dickinson et al. (2012) used ^{10}Be data from soils in Antarctica to confirm that these soils were

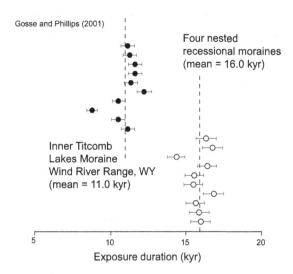

Gosse and Phillips (2001)

Four nested
recessional moraines
(mean = 16.0 kyr)

Inner Titcomb
Lakes Moraine
Wind River Range, WY
(mean = 11.0 kyr)

Exposure duration (kyr)

Fig. 15.73 In situ cosmogenic nuclide (^{10}Be) ages (and their precision) from fresh, granitic boulders with that were at least 1.2 m aboveground.

formerly unfrozen. Depth plots of ^{10}Be and the presence of illuvial clay suggested that vertical translocation of ^{10}Be and clays had been occurring in these soils, which are now completely frozen (Fig. 15.71B). They used these data not only to infer a warmer than present climate in the past, but also to model denudation rates of 1–6 cm Myr^{-1} since freezing occurred.

In sum, although meteoric ^{10}Be inventory data will not typically provide a highly precise exposure age, they often can be used to bracket the age and are a useful proxy for a soil's erosion and deposition history (Graly et al. 2010). As future work informs the scientific community about uncertainties in the method, and as we acquire more data from different soils, the utility of meteoric ^{10}Be analysis in soils will become an increasingly robust monitor of surficial processes on 10^3–10^6 year timescales.

FORMATION IN ROCKS AT THE SURFACE

Nuclides also accumulate in situ within rocks at the surface, as cosmic rays interact with atoms in mineral lattices (Lal 1988, 1991, McHargue and Damon 1991). The five most widely used nuclides used in these applications, ^{3}He, ^{10}Be, ^{21}Ne, ^{26}Al, and ^{36}Cl, enable surface exposure dating of rock surfaces of virtually any lithology, at any latitude and altitude, for exposures ranging from 10^2 to 10^7 years (Zreda et al. 1991, Gosse and Phillips 2001, Ivy-Ochs et al. 2006). Data from single or multiple nuclides, obtained from bedrock or boulders, can be used to estimate a variety of useful information, e.g., erosion rates on boulder and bedrock surfaces, fluvial incision rates, denudation rates of individual surfaces or entire drainage basins, burial histories of rock surfaces and sediment, rock-to-soil conversion rates, scarp

retreat, fault slip rates, paleoseismology, and paleoaltimetry (Gosse and Phillips 2001, Dehnert and Schlüchter 2008). In situ nuclide data have been used to reconstruct glacial fluctuations and erosion rates from continental and mountain settings, using data from rocks in moraines and from bedrock, respectively (Gosse et al. 1995, Jackson et al. 1997, Putkonen and Swanson 2003, Balco et al. 2005; Fig. 15.73). Because the concentrations of these nuclides are sensitive to surface erosion and vary with depth below the surface, many advances have been made in determining local and large-scale erosion rates and soil development rates and, in general, informing the academic community about landscape evolution. The future is indeed bright for this family of SED methods. The following reviews provide additional details about theory and methods (Lal 1988, 1991, Morris 1991, Bierman 1994, Gosse and Phillips 2001).

Gosse and Phillips (2001) used the analogy of a person's skin and tanning as akin to in situ nuclide production. The degree of redness on a person's skin is mainly proportional to the *duration* of sunlight exposure. But also important are sunscreen applications, cloudiness, and latitude. Rocks are similar; they accumulate nuclides mainly as a function of exposure duration, but they are also affected by latitude and shielding, for example, by regolith or snow. People tan at different rates, just as in situ nuclide production rates vary in different minerals. The change in the color of a tan after peeling may result in an overestimation or underestimation of the total duration of skin exposure, just as erosion may result in an overestimation or underestimation of cosmic ray exposure duration in rocks. People who return for a second day of tanning will begin the process partially tanned from the previous exposure, just as cosmogenic isotopes can be inherited. Most importantly, a tan will gradually wear away, just as (most, but not all) cosmogenic nuclides decay over time. All in all, an excellent analogy!

Each type of nuclide preferentially accumulates in certain target minerals (Nishiizumi et al. 1986; Table 15.11). And as mentioned, the total amount of accumulation is dependent upon a number of site factors: duration (mainly), but also altitude, latitude, rock chemistry, and density; geometry of the exposed rock; depth of burial or shielding (for example, by deep snow or glacier ice); exposure history; and cosmic ray flux (Dorn and Phillips 1991). Because the cosmic flux has varied with time, data on the accumulation of nuclides primarily reflect long-term mean rates. Most cosmic rays can penetrate to about 50–60 cm in rock, meaning that below that depth the rock will accumulate almost no nuclides, i.e., it is effectively shielded (Nishiizumi et al. 1993). In soils, because of porosity, a depth of 3 m is equivalent for complete shielding. Rocks buried more deeply by regolith than 3 m, therefore, accumulate few or no nuclides.

For stable isotopes, such as ^{3}He and ^{21}Ne, accumulation rates in rocks are linear, making them useful dating tools on stable surfaces that have had no prior exposure (Cerling

Table 15.11 | Major cosmogenic isotopes used in surface exposure dating, and their half-lives

Isotope[a]	Useful dating range (years ×10³)	Half-life (years)	Comments
^{36}Cl	1,000	301,000	Can be used for both erosion and burial investigations; useful for a wide range of lithologies
^{26}Al	2,500	716,000	Restricted to quartz; simple stoichiometric chemistry
^{10}Be	5,000–6,000v	1,510,000	Two approaches: in situ accumulation in quartz in rocks versus accumulation in soils by meteoric fallout Useful for long exposures due to long half-life, but difficult to use on young surfaces; meteoric ^{10}Be is a potential source of contamination for in situ work
^{21}Ne	Unlimited	Stable	Can be used for extremely long exposures, because it is stable; preexposure can be problematic
^{3}He	Unlimited	Stable	Useful for both long and short exposures; preexposure can be problematic

[a] Only those isotopes that are most applicable to soil geomorphology and SED are listed here. For a more complete list, see Lal (1988).
Source: In part, after Gosse and Phillips (2001) and Ivy-Ochs and Kober (2008).

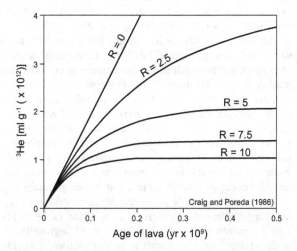

Fig. 15.74 Accumulation of ^{3}He in surface-exposed rocks, as a function of time and erosion rate. *R* represents the erosion rate of the rock in meters per million years (m Ma⁻¹).

1990, Ammon *et al.* 2009, Gillen *et al.* 2010, Goehring *et al.* 2010). If a rock is eroded, loss of accumulated isotopes occurs (Fig. 15.74), but with correct interpretation, stable isotopes can be quite useful in the determination of surface erosion rates and history (Lal 1988, Foeken *et al.* 2009). Use of ^{3}He is also advantageous because it has the highest in situ production rate of any cosmogenic nuclide (Kurz *et al.* 1990).

For radioactive nuclides, net accumulation over time is more complicated, but potentially promising as well. Their data reflect gains by exposure versus loss by radioactive decay. Half-lives of the radionuclides, and hence, effective time spans of the method, vary (Table 15.11). Saturation of nuclides is eventually reached after about four times the half-life, effectively determining the time span of the method (Fig. 15.75). ^{10}Be has the longest potential usefulness of all the major isotopes (2.2 Ma).

In situ production of cosmogenic nuclides can be used as a chronometer for both *surface exposure* and *length of burial* (Granger 2006, Dehnert and Schlüchter 2008, Li and Harbor 2009). By using a number of different isotopes with widely varying half-lives, detailed information on surface exposure and erosion history can be obtained (Nishiizumi *et al.* 1986, Gosse and Phillips 2001).

Determining the duration of *surface exposure* is one of the main applications of cosmogenic dating (Fig. 15.76). Few surfaces have simple exposure histories (Dockhorn *et al.* 1991, Liu *et al.* 1996, Phillips *et al.* 1998), but therein lies the utility of cosmogenic isotopes. They can be used to unravel the exposure history of a geomorphic surface, even if it has had periods of stability, erosion, and/or burial (Nishiizumi *et al.* 1993). The simplest scenario involves rock that has had no prior exposure history, usually because it has always been deeply buried. Then, it is instantaneously exposed by some geomorphic process, e.g., an erosion event, or because it was plucked from bedrock by glacial ice and deposited on the surface of a moraine. In the latter case, the implantation of the rock coincides with the formation of a surface and with time$_{zero}$ of soil development. A similar alternative involves rock that has just *formed* at the surface, e.g., a lava flow (Phillips *et al.* 1986). In these cases, the exposure of the rock to cosmogenic radiation coincides exactly with the formation of a subaerial geomorphic surface; isotopes begin to accumulate in the rock as soon as

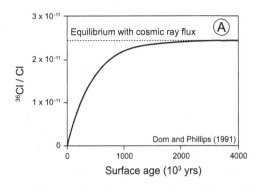

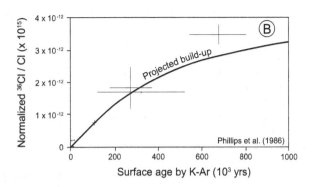

Fig. 15.75 Accumulations of ^{36}Cl with time. (A) Calculated ^{36}Cl accumulation in a lava exposed at 3,500 m elevation on Mauna Kea, Hawaii. (B) $^{36}Cl/Cl$ ratios for rocks from lava flows in the western United States.

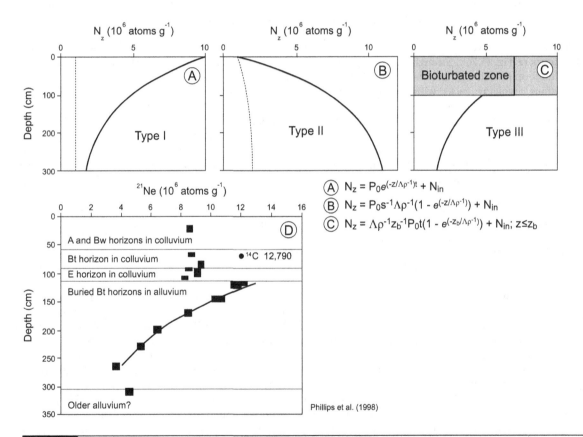

$$\text{Ⓐ} \quad N_z = P_0 e^{(-z/\Lambda\rho^{-1})t} + N_{in}$$

$$\text{Ⓑ} \quad N_z = P_0 s^{-1}\Lambda\rho^{-1}(1 - e^{(-z/\Lambda\rho^{-1})}) + N_{in}$$

$$\text{Ⓒ} \quad N_z = \Lambda\rho^{-1}z_b^{-1}P_0 t(1 - e^{(-z_b/\Lambda\rho^{-1})}) + N_{in}; \ z \leq z_b$$

Fig. 15.76 Theoretical and actual cosmogenic nuclide depth profiles for different types of surfaces. Dotted lines represent (base level) inherited nuclide contents. Solid lines represent nuclide contents at the time of sampling. All surfaces are assumed to have had uniform nuclide contents at time$_{zero}$. (A) A stable surface with minimal burial or erosion. (B) A surface experiencing a constant rate of burial. (C) A stable surface with minimal burial or erosion, but with a bioturbated zone at the surface. (D) Actual ^{21}Ne depth distributions for an uneroded Pleistocene surface on a fluvial terrace that has been buried by colluvium. The buried soil was eroded during burial, and the ^{21}Ne depth function clearly shows the discontinuity.

it is exposed. The signal given by radionuclides within the rock then will reflect the age of the surface, being greatest in surface layers (Fig. 15.76A). Interpreting the age of these kinds of surfaces is easiest when using stable isotopes, for the accumulation rate is linear, although it is also possible with radioactive isotopes such as ^{10}Be and ^{26}Al.

Efforts have been made to distinguish the effect of nuclide inheritance (initial nuclides taken to the site in the sediment) from that of accumulation after surface exposure. Anderson et al. (1996) showed that inheritance can be corrected by examining depth profiles of stable nuclides. Phillips et al. (1998) extended their work by examining depth profiles of stable isotopes with respect to conditions of burial and bioturbation. Similar inferences can be used for radio-isotopes if the depth data are adjusted for decay. The theory is as follows. Beneath (erosional or constructional) surfaces that form quickly, nuclide contents will initially be zero but will eventually (assuming surface stability) build up at the surface, decreasing exponentially with depth (Fig. 15.76A). On slowly aggrading surfaces, nuclide contents will, therefore, increase with depth and then decrease if preexposed materials occur there (Fig. 15.76B). Strongly bioturbated layers will show uniform nuclide concentrations (Fig. 15.76C).

Interpretations involving rocks with multiple exposures and burials are more complicated, requiring the use of more than one isotope (Lal 1991, Bierman and Turner 1995, Phillips et al. 1998; Fig. 15.76C).

In situ cosmogenic isotopes are commonly used to determine rates of bedrock erosion and, hence, landscape denudation (Dockhorn et al. 1991, Granger et al. 1996, Cockburn and Summerfield 2004, Granger 2006). For surfaces of known age, one way to determine the effect and magnitude of surface erosion or burial, or whether either of these two processes has been operative, is to compare the amounts of nuclides in the rock to the amounts expected. An example of this approach, by Craig and Poreda (1986), used cosmogenic nuclides to calculate erosion rates of lavas. The age of the lavas was known by other means, and because the content of ^{3}He in the lavas was below what would have been expected for lavas of this age, an erosion rate could be generated (Fig. 15.74). In situ produced ^{10}Be and ^{26}Al can also be used to quantify the extent (as well as the rates) of saprolite formation and weathering, thereby informing the soil production equation (Burke et al. 2007).

Cosmogenic-nuclide burial dating relies on comparisons between a pair of nuclides that are produced in the same rock or mineral target at a fixed ratio but have different half-lives (Dehnert and Schlüchter 2008). Burials can be by sediment or by ice. The nuclides most commonly used for this purpose, ^{26}Al and ^{10}Be, are produced in quartz at a ratio of 6.75:1 (Balco and Shuster 2009, Rovey and Balco 2008, 2010). Quartz in rocks that experiences a single period of surface exposure has ^{26}Al and ^{10}Be concentrations governed by this ratio. If a rock containing quartz is exposed long enough that nuclide concentrations reach a balance between production and loss (by radioactive decay and

physical erosion) and is then buried and shielded from the cosmic ray flux, nuclide production stops and inventories of both nuclides decrease by radioactive decay constants (Balco et al. 2005). Because the half-life of ^{26}Al is shorter than that of ^{10}Be (Table 15.11), the ^{26}Al/^{10}Be ratio decreases exponentially after burial, and this ratio can therefore be used to determine the age of the burial event (Balco and Rovey 2008, Balco and Shuster 2009). This type of dating requires in situ quartz that has been exposed at the surface for a time and then buried. Because quartz is so common in rocks, this method has wide application potential. For example, Balco et al. (2005) used this method to date the expansion of early ice sheets in North America. Knowing that the method is most accurate when sediments are exposed for long periods (>10^5 years) and then buried rapidly to at least several meters depth, they studied tills and soils in Missouri that met these criteria. Here, soils had developed during long periods of landscape stability and were then buried deeply by till during subsequent ice sheet advances. ^{26}Al and ^{10}Be concentrations in the quartz rocks from within the buried soil provided the age of the overlying till, i.e., the burial event. Another example: Wells et al. (1995) compared ^{3}He contents in desert pavement clasts to uneroded basalt nearby. Similar values for both areas indicated that the rocks on the pavement had been exposed at the surface continually and had never been buried.

Although most studies of burial dating use ^{26}Al and ^{10}Be, largely because they are relatively easy to measure and their production in quartz is well established, Balco and Shuster (2009) argued that ^{26}Al-^{21}Ne and ^{10}Be-^{21}Ne approaches to burial dating offer several advantages over ^{26}Al-^{10}Be burial dating. These alternate isotopes have an increased useful age range and improved accuracy. Although the useful range of ^{26}Al-^{10}Be burial dating is ~0.5–6 Ma under ideal conditions, the useful range of the ^{10}Be-^{21}Ne pair is potentially more than double that, potentially facilitating burial dating into the Miocene.

Lastly, we caution that all surface exposure dating is done on landscapes, which have variously stable and unstable segments. Soils on erosional versus aggradational landscape positions will have vastly different histories and isotopic signatures, which can be problematic if not interpreted correctly, but also potentially enlightening if interpreted correctly (Fig. 15.77). For this reason, it is important to be mindful of slope position when sampling for SED analyses.

Although potentially complicated, the possibilities for generating additional information on surface exposure dating, through the use of in situ and meteoric cosmogenic nuclides, are excellent.

Luminescence Dating

Luminescence dating, a suite of methods used to date sediment, takes advantage of a natural chronometer that accrues in quartz and feldspar grains after burial, in order to estimate the duration of burial. This chronometer takes the form of energy stored from ionizing radiation present

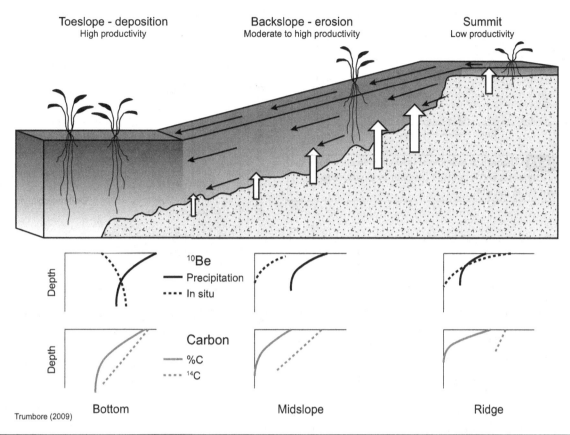

Fig. 15.77 Influence of slope and landscape position on soil properties and cosmogenic isotopies within, all of which are routinely studied and used for surface exposure dating. Depth plots show predicted profiles for meteoric and in situ ^{10}Be (the latter content is exaggerated to show detail), soil OC, and ^{14}C.

in the surrounding environment. In the lab, the sediment emits photons of light (*luminescence*) in proportion to the amount of accrued energy – hence the name. In essence, luminescence dating estimates the time since *burial* of sediment that *had been* exposed to sunlight, usually during a prior transportation event. Luminescence dating relies on four steps or events in the sediment history, to obtain an estimate of the interval since it was last exposed to light: (1) removal of a trapped charge within mineral grains by natural daylight exposure that occurred during a previous transportation event; (2) the gradual buildup of a charge or signal as the buried grains are exposed to ambient, environmental radiation; (3) sample collection and measurement of the trapped charge concentration in the grains; and (4) determination of the environmental dose rate experienced by grains at each sample location (Rhodes 2011). We will explain these steps in the following.

In general, luminescence refers to the photons of light resulting from energy recombination following exposure to radiation, emitted from mineral grains when heated or exposed to light. Thermoluminescence (TL) refers to resetting of the luminescence signal by heat, whereas the suite of *optical* dating methods exploits the resetting

of the luminescence signal by particular wavelengths of light. Within optical dating, optically stimulated luminescence (OSL) generally refers to signals reset by blue-green wavelengths, and infrared stimulated luminescence (IRSL) refers to signals reset by infrared wavelengths. The similarities of OSL, IRSL, and TL justify discussing them together. For more detail, the reader is referred to any of several excellent reviews of luminescence dating (Huntley *et al.* 1985, Aitken 1994, Prescott and Robertson 1997, Wallinga 2002, Duller 2004, 2006, Lian and Roberts 2006, Fuchs and Owen 2008, Preusser *et al.* 2008, Wintle 2008, Rhodes 2011).

Luminescence dating is possible because most common silicate minerals (mainly quartz and feldspars) contain crystal lattice defects/imperfections that are potential sites of electron storage. The source of the electrons, which accrue in these crystal *traps*, is the decay of isotopes in the surrounding (within a few decimeters) sediment. Electrons caught in these traps remain stored there, making the minerals essentially long-term dosimeters. The amount of stored electrons is called the *equivalent dose* and the rate at which they are produced by the ambient sediment is the *dose rate*. Exposure to light releases the electrons,

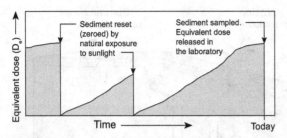

Fig. 15.78 Theoretical example of how luminescence dose (D_e) accumulates and is released (zeroed) by exposure to sunlight (or in the lab).

emptying the traps and *rezeroing* the grain. This release of the stored electrons (in the lab) is detected as luminescence. Performed in the lab under controlled light conditions, this procedure zeroes the dosimeter clock by emptying the traps (Fig. 15.78). The age of a sample is determined by dividing the amount of ionizing radiation the sample absorbed during burial, i.e., the equivalent dose, by the dose rate derived from the sediment surrounding the sample (Rittenour 2008), as follows.

Luminescence age (years) = (equivalent dose (D_e, in grays) / dose rate (in grays yr^{-1})

In essence, the luminescence obtained in the lab is equivalent to the radiation energy deposited within the crystal since its last sunlight (or heat) exposure during transport through the environment. Ages reported by luminescence dating are in calendar years, and, thus, calibration is unnecessary. Traditionally, luminescence ages refer to years before present, whereas ^{14}C ages refer to years before AD 1950.

Luminescence dating is best applied to sandy-silty sediment that was fully exposed to sunlight, either by transportation in the atmosphere or in clear water, for as little as a minute or two, and then buried. This process is easily visualized by saltating sand that is zeroed by sunlight and then buried in a sand dune. The luminescence age then establishes when the dune formed – the last moment of instability/transportation, or the first moment of the dune's existence. Similar examples could be used for loess, alluvium, or lacustrine sediment that was transported in clear water, or for beach sand. In summary, luminescence methods directly date *sediment* and establish the ages of sedimentation events; that is why these methods are so valuable in the understanding of geomorphic systems (Stokes 1999). If the sediment age is the same as the surface age, these age estimates can have excellent utility in soil applications (Nielsen *et al.* 2010).

PARTIAL BLEACHING AND GRAIN SENSITIVITY
Two main issues affect the likelihood of success of luminescence dating – complete zeroing of the luminescence signal within the sediment prior to burial and the sensitivity

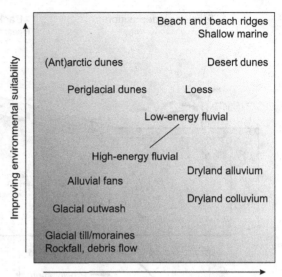

Rhodes (2011)

Fig. 15.79 Relative rankings of different sediments/environments and their appropriateness for luminescence dating, based on *likelihood of bleaching* achieved by quartz grains during transport and *sample sensitivity*. Environments characterized by repeated and/or lengthy light exposure, e.g., beaches and dunes, contrast with those characterized by little light exposure, e.g., debris flows. Grains recently eroded from bedrock commonly have lower sensitivities and bleach more slowly than those that have had an extended residence at the surface. Areas of darker shading correspond to less effective bleaching.

of quartz grains themselves. Sensitivity is the luminescence generated (luminescence signal size) in response to one unit of dose. It is often limiting in grains that have only recently been released from the parent rock and have developed relatively few and shallow traps (Rhodes 2011).

The more critical of these two constraints is that the sample must have no residual luminescence signal from the period before the event being dated, i.e., the sampled sediment must have been fully bleached or zeroed prior to burial (Singarayer *et al.* 2005, Preusser *et al.* 2008). In samples that are incompletely zeroed, a small amount of *remnant dose* remains stored in the grains (Li 1994). Remnant dose error is important for young samples but is negligible for old samples, simply because of the proportion of total luminescence signal it represents. Usually, eolian sand is completely reset while in transport, but sediment such as glacial outwash or alluvium often carries with it a remnant dose (Singarayer *et al.* 2005). Grains transported within the many different kinds of sedimentary environments experience diverse amounts of light exposure, ranging from subglacial transport or, in deep karstic systems, where grains are not exposed to light, to repeated exposure in a beach swash zone (Rhodes 2011; Fig. 15.79). The phenomenon

Fig. 15.80 A series of photos illustrating the acquisition of a sample for OSL dating. (A) An opaque sample tube is driven into a freshly exposed face in a sand dune. (B) The tube and sample are removed intact from the face and then (C) covered with opaque tape and labeled. Together with information about the site such as elevation and sample depth and a dose rate sample, the sample shown in (C) is ready to be submitted to an OSL lab for dating. Photos by RJS.

whereby some grains are zeroed but others are not is termed partial bleaching, and it is particularly common in many fluvial and glaciofluvial environments (Hutt and Jungner 1992, Klasen et al. 2007, Rittenour 2008). Some of the grains may be transported at night and receive no light exposure. Important additional considerations are the total transport time for the grains and the number of repeated bursts of movement they undergo, and whether they are interspersed with temporary halts or shallow burial. The last transport event before deposition is not necessarily the most important, as previous movement can often provide sufficient light exposure for bleaching.

Several methods have been proposed to combat the influence of partial bleaching. One uses the multiple components of the quartz luminescence signal to isolate and date only the most light-sensitive, thermally, and temporally stable OSL traps (Tsukamoto et al. 2003, Jain et al. 2005). As a result, it is now possible to investigate whether a significant portion of the quartz grains was likely to have been zeroed by sunlight exposure (Wintle 2008). This method has been successfully applied to glacial (Duller 2006) and fluvial (Arnold et al. 2007, Rustomji and Pietsch 2007) sediment. A second group of methods uses recent advances in the measurement of increasingly smaller aliquot sizes (Olley et al. 1998), culminating in the development of single-grain dating techniques (see later discussion). Single-grain OSL procedures have been finding wide application, e.g., on archaeological sediments from cave sites where mixing of grains may have occurred after deposition (Bateman et al. 2003), or where grains may have been added by in situ weathering of the host rock (Jacobs and Roberts 2007).

Improving the techniques necessary to detect, and compensate for, incomplete bleaching is an area of much current research (Rhodes 2011). Researchers are continuing to push the envelope of luminescence dating, as they attempt to date sediment from depositional environments where bleaching may not have been complete, with

varying degrees of success (Rodnight et al. 2006, Schaetzl and Forman 2008, Lukas et al. 2007). Will it be "technology to the rescue" for partial bleaching? Perhaps. But we also paraphrase Fuchs and Owen (2008), who stressed that a geomorphic understanding of the sedimentary environment and its deposits is essential to identify the proper sediments for sampling, so that samples will have had the maximum likelihood of daylight exposure.

TAKING A LUMINESCENCE SAMPLE

Samples for dating are removed from sediment below the soil profile (pedogenesis, in particular bioturbation, can affect the signal and mix grains of different ages), usually by driving an opaque tube into a freshly exposed face (Preusser et al. 2008; Fig. 15.80). A sample from the zone immediately surrounding the tube is also collected for dose rate analysis. Each end of the tube is sealed immediately after removal. At the lab, an analyst (operating in near-darkness) will discard the sample near the ends of the tube, using only that which is in the center and has not been exposed to light during the sampling process. Measurements of the gamma component of the dose rate are often made in situ using portable gamma spectrometers. In the laboratory, once the sample has been processed to isolate the particular mineral (quartz or feldspar) of the target grain size desired for measurement, the sample in the tube is exposed to intense light (in OSL) or heat (in TL) to release the charge from the appropriate traps, and the photons emitted are counted. Generally, the more photons emitted, the more electrons had been trapped in the minerals and the longer the sample had presumably been buried.

Dose rate is a function of several factors. It is primarily the radioactivity of the sediment surrounding the sample, which can readily be determined by obtaining a sample of the sediment surrounding the luminescence sample. Dose rate is also affected by long-term water content, because water attenuates the radiation within the sediment. In essence, wetter soils – long term – have lower dose rates.

In a luminescence age calculation, a mean value for soil water content for the entire period of burial is needed. This value remains the greatest source of uncertainty in a luminescence age (Preusser *et al.* 2008). Often the water content of the luminescence sample is simply used as a guideline for long-term water content, but it is crucial that potential changes in soil hydrology be considered when using this value. That is, the water content of the sample is only representative of the water content at that instant in time and must be evaluated as an estimate of longer-term soil water content, which is used in the estimation of a luminescence age (Schaetzl and Forman 2008). Uncertainties related to long-term water content are most important for water-lain sediments, but must be considered for all sediments.

Last, latitude, elevation, and depth also affect dose rate as a result of the exposure to ionizing radiation from cosmic rays (Prescott and Hutton 1994); that is why laboratories always request this information. Most OSL samples must be taken from at least 1 m depth, in order to minimize the exponentially decreasing effects of cosmic radiation penetrating into the soil and affecting the sample, i.e., a *cosmic dose*. Samples also must be taken at least 2 m deep to minimize the effects of pedoturbation, which can carry grains to the surface, where they can become rezeroed (Wilkinson and Humphreys 2005, Hansen *et al.* in press). If these grains are later cycled into the subsurface, they may be sampled along with the grains that were buried much longer ago, producing a too-young age. For these reasons, most near-surface and pedoturbated samples from soils return abnormally young luminescence ages (Bateman *et al.* 2003, 2007a, 2007b, Ahr *et al.* 2013, Hansen *et al.* in press). Optical dating of samples from within the soil profile is further complicated by the effects of weathering, which can alter ^{40}K, U, and Th trajectories within the profile and thus affect the environmental dose rate (Ahr *et al.* 2013).

Perspective, Opportunities, and Overview

Luminescence dating was slow to catch on with Quaternary scientists (Dreimanis *et al.* 1978). Many viewed it as simply a relative dating tool. As technological advances were made during the 1970s, 1980s, and 1990s, the accuracy and precision of the method improved markedly, and sampling methods and protocols also improved (Fig. 15.81). Luminescence dating is now critical to studies of Quaternary sediments and soils because it provides a surface exposure and sediment dating tool with acceptable accuracy that not only extends well beyond the range of radiocarbon, but can date sediments that are post bomb (Dreimanis *et al.* 1978, Nielsen *et al.* 2006). Rhodes (2011) argued that luminescence dating can be applied to sediment as young as one year, to as old as several hundred thousand years (Madsen and Murray 2009). The age limitations of luminescence dating are a function of the dose rate and the capacity of the sample grains to store electrons. Young sediment, recently released from rock by physical weathering, has low sensitivities and, thus, cannot retain

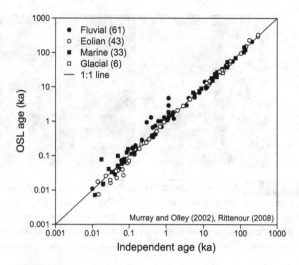

Fig. 15.81 Comparisons between OSL ages and those generated independently, for fluvial, eolian, marine, and glacial samples, as shown by the solid line.

a large amount of photons (Olley *et al.* 1998). Generally, K-feldspars have more capacity than quartz, rendering their effective age range nearly three times greater. The precision of OSL and TL ages is generally around ±10% of the actual age (when expressed as one standard deviation), but this varies widely, depending on the character of the sediment and the certainty that it was fully bleached at the time of deposition (Fig. 15.81).

An exciting new development in OSL dating is the application of single-grain techniques. In this method, luminescence is examined grain by grain, instead of in larger aliquots (Murray and Roberts 1997, Jacobs *et al.* 2003, Duller 2006, 2008). Recall that limited exposure to light results in grains' retaining a part of any prior trapped charge, and, if unaccounted for, this remnant dose causes overestimation of the age. Traditionally, many grains are measured simultaneously (as an aliquot) and the luminescence signal is averaged. As a result, any variability in resetting between grains is obscured. Single-grain dating facilitates the determination of the true age of a sample by allowing the population of grains not bleached at deposition to be identified. Single-grain results commonly show positively skewed D_e distributions, with the youngest population representing the grains fully bleached at deposition (Fig. 15.82). The advantages of the single-grain method are myriad, especially in samples that have experienced incomplete bleaching or a degree of mixing, e.g., in cave or fluvial deposits (Galbraith *et al.* 1999, Henshilwood *et al.* 2002, Wallinga 2002, Olley *et al.* 2004, Rhodes *et al.* 2010). The method also has application in situations in which grains have been variously rezeroed after deposition – perhaps by pedoturbation. The researcher can separate those grains from ones that have a clearer signal. In short, single-grain methods open up a wider range of potential environments and

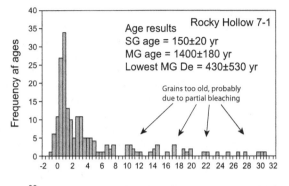

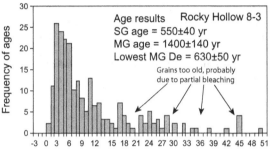

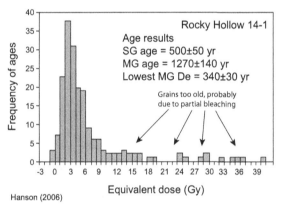

Hanson (2006)

Fig. 15.82 Single-grain (SG) and multiple-grain (MG) OSL data from three young fluvial samples from an ephemeral stream in western Nebraska.

new possibilities for understanding postdepositional grain movement, although we should note that this method is ineffective and time-consuming where grain sensitivity is low, because few grains provide sufficient signal for measurement (Duller 2006, Rhodes 2011).

Applications of luminescence dating are myriad and growing (Berger 1984, 1988, Rees-Jones and Tite 1997). They range from soils to geomorphology (Stokes 1999) to glaciology and even archaeology (Roberts *et al.* 1998). Eolian sediment (loess, dune sand) is optimal for the method, because the grains are usually fully bleached during transport, and because they contain large numbers of grains of the optimal size for dating (coarse silt through medium sand).

For this reason, dune sand and loess have been successfully dated by luminescence for decades (Dreimanis *et al.* 1978, Pye and Johnson 1988, Wintle and Catt 1985, Wintle 1993, Duller and Augustinus 1997, Arbogast 2000, Murray and Clemmensen 2001, Arbogast *et al.* 2002b, Reiman *et al.* 2011, Blumer *et al.* 2012, Loope *et al.* 2012). Other applicable sediment types include colluvium, glacial outwash, and fluvial sediment, providing that the water below which they were deposited was clear enough to allow for solar resetting of the inherited luminescence signal (Forman 1989, Schaetzl and Forman 2008). Alluvium has been successfully dated (Fuchs and Lang 2001, Wallinga *et al.* 2001, Rittenour 2008, Hansen *et al.* 2010, Hanson *et al.* 2010), as has coastal sediment (Ollerhead *et al.* 1994, Bateman and Catt 1996, Forman 1999, Nielsen *et al.* 2006, Clemmensen and Murray 2010, Fuchs *et al.* 2012, Reiman *et al.* 2012, Tamura *et al.* 2012), lacustrine sediment (Attig *et al.* 2011, Carson *et al.* 2012, Moller *et al.* 2013), and ice wedge casts (Bateman 2008, Guhl *et al.* 2013, Liu and Lai 2013), among others. Because most pedogenic carbonates are carried in as dust, they are also datable by OSL (Singhvi and Krbetschek 1996). Luminescence dating can provide age constraints on erosion of surface soils, if the sediment eroded and deposited elsewhere can be dated (Eriksson *et al.* 2000). The method can also be used to establish the age of soil burial, providing that the sediment above the soil can be dated (Berger and Mahaney 1990, Hutt *et al.* 1993). Because heat, in addition to light, can also reset the clock, TL can be used to determine the last time a sample was heated. Therefore, it is especially applicable to dating of pottery, hearths, and baked flints (Zimmerman 1971, Wintle 1996).

Like any dating method, luminescence dating has some issues and assumptions that must be taken into consideration. Samples that have not been sufficiently zeroed before being buried will report ages that are too old, whereas in older or less-sensitive samples, saturation of the electron traps can cause ages to be anomalously young. *Anomalous fading* – the loss of electrons from the traps on a timescale that is short compared to the lifetime predicted on the basis of trap size (Wintle 1973, Lamothe 1984, Huntley and Lamothe 2001, Strickertsson *et al.* 2001, Lamothe *et al.* 2003) – is primarily a problem with feldspars, resulting in an age estimate that is too young. However, continuing investigations promise new solutions to overcome this problem (Buylaert *et al.* 2012). One should also be aware that sediments that have been affected by pedogenesis are, in some circumstances, unreliable for luminescence dating. One of the fundamental assumptions made in luminescence is that the dose rate has been consistent through time, yet pedogenesis may result in a variable dose rate that cannot be measured or modeled. Illuvial materials such as carbonate, Fe, or silica typically have higher concentrations of radioisotopes. In such a case, the laboratory would measure an anomalously high (current) dose rate, which would be quite different from the *long-term* dose rate. Thus, samples from the profile or even the upper C horizon may yield anomalously *young ages*. Nonetheless, Nielsen and Murray (2008) found that podzolization had no effect on dose rates in young sandy soils. One should also be aware that near-surface samples are more likely to have been affected by pedoturbation and postdepositional rezeroing.

Another potential worry is disequilibrium in the U-series decay chain, which can also change the dose rate with time (Olley *et al.* 1996). This can happen when water moves through the sediment, removing daughter products from the decay chain. The cosmic ray component of dose rate can also be a concern in dynamic environments such as dunes, where the depth of the sample might have varied considerably. Because sand in dunes typically has very low dose rates, the cosmic component can be a high proportion, and the incorrect depth assumption can affect the OSL age estimate by several thousand years.

Despite the few concerns, scientists of all kinds are turning to luminescence dating as a sound alternative when more traditional methods, like radiocarbon, are not an option. Recent technological advances, particularly the development of new lab procedures for determining D_e, have resulted in more widespread and successful applications of OSL for dating (Nielsen *et al.* 2006, Roberts and Plater 2007, Wintle 2008). Technological advances are making it possible to test explicitly whether all the grains have had been reset at deposition for a specific sample, allowing the reliability of the luminescence age to be assessed and improved. Researchers are using smaller samples, i.e., single-grain OSL, and the age limits for luminescence dating are expanding to include the past few years, to 100 k or more, with the potential for even more advances (Wintle 2008). It is clear that luminescence dating – optical dating in particular – is here to stay and will be applied to an increasing variety of environments in the future.

Chapter 16

Soils, Paleosols, and Paleoenvironmental Reconstruction

Paleosols as Palimpsests

Commonly, the first step in a soil geomorphic study is to ascertain the age of the soil or surface and then use information obtained from the soil to better understand the character of the geologic timespan during which the soil has formed. That is, we try to determine the starting and ending points of the *soil-forming interval* – when the surface was stable and the soil was forming. Soils are, however, an integration of the many surficial processes that occur over these time intervals. No one soil-forming interval is a monotony of similar situations, factors, and inputs. Climate ebbs and flows, vegetation changes, and surfaces gradually aggrade or erode. For this reason, Don Johnson often argued that all soils are polygenetic (Johnson 2002, Johnson *et al.* 2005; see Chapter 12). The older the soil, the more likely it is to have experienced a wider variety of inputs, and the more polygenetic it will be – the more like a *palimpsest* it will be.

A palimpsest (Latin *palimpsestus*, scraped again) is a parchment used one or more times, with earlier writings having been incompletely erased from it. Recently erased passages are not very difficult to make out, but older passages are much more difficult, yet not impossible, to read and interpret. It simply takes skill and patience. In this sense, soils are palimpsests, and in this chapter we provide information useful to their interpretation. Interpreting such a palimpsest is an exciting challenge and an important application of soil geomorphology (Catt 1990). The information gleaned from pedopalimpsests is indicative of past soil-forming intervals and the changes that they may have undergone with respect to *climate*, *vegetation*, or *geomorphology* (Rutter 2009).

In this chapter we discuss how a careful reading of soil palimpsests, coupled with a knowledge of how soil development is related to *contemporary* soil-forming processes, can often (for older soils) provide a wealth of information about conditions under which old or buried soils formed (Sheldon and Tabor 2009). That is, soils – because they are

palimpsests – can be highly useful keys to conditions of the geologic past (Catt 1991). This chapter must by necessity follow chapters on pedogenesis, weathering, pedoturbation, parent materials, and soil geomorphology, because *paleo*pedologic, *paleo*climatic, and *paleo*geomorphic interpretations require a thorough knowledge of *contemporary* processes. Our ability to use soils to help understand the past is only as good as our ability to understand soils on the surface today.

If we dropped out of the sky into a soil pit, with the intent of determining all that we could about the evolution of the landscape and its soils, of what should we be mindful? Included in the plethora of possible answers to this question are data related to soil mineralogy, horizonation, morphology, sedimentology, various isotopes, and biogenic materials.

A focus of this chapter is *paleopedology* – the science of old soils, or paleosols. The paleosols that are buried can provide important information about the environment at a time when their geomorphic surfaces were subaerially exposed.

One must always be mindful of the fact that soils respond to more than external forcings and inputs (see Chapter 12). Many soils react to intrinsic thresholds and develop accessions (Muhs 1984, Johnson and Watson-Stegner 1987, Phillips 2001c). The pedogenic pathway of a soil may be altered after it reaches and passes a pedogenic threshold or develops an accession; this change may have little or nothing to do with external climatic or geomorphic forcings. Soil processes and properties may develop, ebb, or flow in ways no one could predict (Phillips *et al.* 1996). Various kinds of feedbacks are also an important part of the soil geomorphic story. Soil development affects landscape change and evolution, e.g., when soils develop Bkkm horizons that act as caprocks, or when well-developed Btx horizons become aquicludes and force water to move laterally across them, producing surface seepage and springs at downslope areas, thereby initiating gully formation. Indeed, understanding soil history is a complex but exciting challenge.

Paleosols and Paleopedology

Nowhere is the concept of soils as palimpsests better exemplified than when we peer into the past through the window of *paleosols*, i.e., fossil soils. Ruhe (1965) defined a paleosol as a soil that formed on a landscape of the geologic past.

When used in formal *stratigraphic* applications, the term *geosol* is appropriate as the fundamental pedostratigraphic unit, based on the North American Stratigraphic Code (Morrison 1967, 1998, Catt 1998). When capitalized, a Geosol is a fundamental, formally recognized geosol unit. Formally recognized geosols are given a geographic name, e.g., the Farmdale Geosol or the Sangamon Geosol. Formally named geosols are buried soils that have formed in a consistent stratigraphic position and, like all stratigraphic units, have a type locality. Because soils are diachronous units that begin forming and may be buried at different times in different parts of the landscape (Hall 1999a), they should not be assigned strict chronostratigraphic status. When considered in its less formal, "lowercase" form, the term *geosol* is equivalent to the term *paleosol*, which is defined by pedogenic characteristics in addition to stratigraphic characteristics and need not be buried. However, an important difference separates geosols and paleosols: The term *geosol* communicates that the observation of a soil has stratigraphic and chronologic attributes, whereas the term *paleosol* is more ambiguous.

Paleopedology: Why Study Paleosols?

Paleopedology is the study of paleosols, their genesis, morphology, stratigraphy, and classification. Paleopedology also encompasses the study of the environment in which the soil formed. Much of the early work in this field was done in the midwestern (Olson 1989) and southwestern United States (Holliday 1998). Although there is a growing interest in pre-Quaternary paleosols (Retallack 1990, Nesbitt 1992, Dahms 1998, Costantini and Priori 2007, Metzger and Retallack 2010), our focus here is on those formed in the Quaternary Period.

So, why study paleosols? Constantini *et al.* (2009) listed several reasons, all of which center on the ability of paleosols to inform us about (1) paleoclimate and paleoenvironmental conditions, for they serve as palimpsests and proxies of times past; (2) pedogenic processes and rates, not unlike those of chronosequences but in a slightly different way; (3) mobilization and sedimentation episodes in the past, e.g., of dust, loess, and periglacial sediments, which were strongly controlled by climatic factors and are often interrupted by periods of stability and soil formation; and (4) the relationships among climate, environmental processes, and human adaptation and reactions, i.e., the effects of environmental changes on past civilizations (Riehl *et al.* 2009). Additionally, Constantini *et al.* (2009) argued that paleosols are important from the standpoint of historical documentation (Constantini *et al.* 2007, Constantini

and L'Abate 2009). Because they store information about paleoenvironmental conditions, paleosols reflect the natural and cultural heritage of the landscape, a condition that Targulian and Goryachkin (2004) referred to as *soil memory*, which makes them particularly valuable.

We have devoted large portions of this book to the understanding of *surficial* pedogenic processes. But we have also noted that, when shallowly buried, paleosols dramatically influence pathways of soil genesis at the surface. Indeed, when soil scientists first tried to explain the patterns of soils on landscapes such as southern Iowa (see Chapter 14), they were confounded. Only when they understood the stratigraphy and paleopedology of these landscapes did the surface patterns make sense. Our point: Where buried soils are present, knowledge of those paleosols is necessary to understand the surficial pedogenic system. And of course, vice versa.

Probably the main utility of paleopedology, however, is the way it can inform us about paleoenvironments (Dahms 1998, Kemp 1999). Like most surface soils, paleosols and catenas of paleosols are palimpsests from a time when the surface was stable and soil development outpaced erosional or burial processes (Nesje *et al.* 1989). Maher (1998, 26) referred to them as "natural archives of paleoclimatic information." Much of this chapter is devoted to the application of paleosols to paleoenvironmental interpretation.

It should be kept in mind, however, that the information provided by paleosols is *indirect* and their interpretation is often *difficult*; only rarely are they physically complete and/or monogenetic. Maher (1998) stressed that the integrity, accuracy, and retrieval of this information depend both on the degree of *development* and of *preservation* of pedogenic properties in paleosols, and whether or not these properties *can be interpreted* via modern analogues.

Paleosols change across space, just as surface soils do, prompting many investigators to observe that the best way to interpret (and, indeed, to identify) paleosols is to examine them on a landscape basis, as we do for surface soils (Follmer 1982). Paleosols on the same geomorphic surface existed in catenas, even though only remnants of those catenas may be preserved today (Valentine and Dalrymple 1976). Likewise, since paleosols are stratigraphic entities, their spatial variation should be fully known if they are to be used accurately as stratigraphic markers, just as with other stratigraphic units (Ruhe 1965).

Classes of Paleosols

Some have attempted to classify paleosols taxonomically, as we do for surface soils (Retallack 1990, Mack *et al.* 1993, Buurman 1998, Nettleton *et al.* 1998, 2000, Krasilmkov and Calderon 2006, Iriondo 2009). Taxonomic names carry connotations, and so, the thinking goes, taxonomic naming of paleosols will help elucidate linkages between paleosols and certain aspects of the formative environment. However, no systematic paleosol classification scheme has been widely adopted. If paleosols are classified at all, the assigned class names are usually analogues to those of Soil Taxonomy, e.g., a buried Udalf.

The simplest and most widely accepted classification of paleosols is that of Ruhe (1956, 1965, 1970). His early definitions of three main types of paleosols – buried, relict, and exhumed – (recognizing that Catt (1998) has recently updated them) have stood the test of time (Rutter 2009). *Relict* paleosols are soils that have formed on preexisting landscapes, under previous and presumably different paleoenvironmental conditions than those of the present, and remain at the surface today (Ruellan 1971, Hall and Goble 2012). They have never been buried. Most relict paleosols are polygenetic, and the oldest ones have endured major climatic shifts during the Pleistocene (Heinz 2002). Evidence for polygenesis in soils is discussed in later sections of this chapter, but we agree with the statement of Chadwick *et al.* (1995, 2) in reference to them: "Optimal conditions for *interpreting* paleoclimates from polygenetic soils occur when precipitation and/or temperature changes are great enough to produce new soil properties without obliterating existing properties" [Emphasis ours].

How long a soil must have remained at the surface to be unquestionably considered a relict paleosol is a topic of considerable debate. Nettleton *et al.* (1989) argued that a relict soil is one that has its time$_{zero}$ in the Pleistocene, has remained at the surface since that time, and exhibits horizons or features that formed in environments that were different from today's. By this definition, soils formed in the Holocene cannot be relict paleosols. Catt (1990) made the point that should a surface soil become buried by a debris flow, the soil beneath it would unquestionably be defined as a buried soil and a paleosol. However, this raises the issue of what to call the *unburied* soil that is pedostratigraphically continuous with the buried soil. The portion of the soil that becomes a paleosol because of burial is perhaps better considered a special kind or variant of the surface soil, rather than the other way around.

If one argues that *all soils* are polygenetic (Johnson and Watson-Stegner 1987, Johnson 2002), then either (1) all surface soils, even the youngest ones, must be considered relict paleosols or (2) the term relict paleosol must be abandoned, because it lacks discriminating power. In this book, we recognize the legitimacy of this debate but take the position that, just as there are degrees of polygenesis, there are relict paleosols and surface soils of all kinds that have been variously influenced by paleoenvironmental conditions. To be sure, excellent examples of relict paleosols, as originally defined, certainly exist in the world. Many occur on surfaces that date back to the Middle Pleistocene or earlier (Daniels *et al.* 1978a, Machette 1985, Nettleton *et al.* 1989, Brock and Buck 2009). Many might also argue that many Holocene age soils with clear evidence of polygenesis, e.g., the Mollic Hapludalfs of the midwestern United States (White and Riecken 1955), might also be considered minimally relict

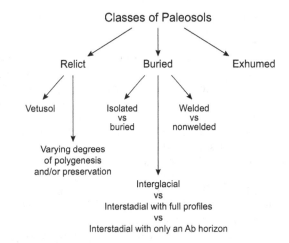

Classes of Paleosols

Relict — Buried — Exhumed

Vetusol — Isolated vs buried — Welded vs nonwelded

Varying degrees of polygenesis and/or preservation

Interglacial vs Interstadial with full profiles vs Interstadial with only an Ab horizon

Fig. 16.1 Some general classes of paleosols, arranged schematically.

paleosols. But many soils in some middle ground may have intermediate characteristics, which might or might not fit with the definition. Ruhe did not concern himself with these soils, and neither do we. In short, the term "relict paleosol" has value and should be retained, although when possible, the degree of its "relictness" (read: *age*) should be noted and the effects of age should be assessed.

A term that has been slow to catch on, *Vetusol*, refers to a soil, surface or buried, that has evolved under the same or similar processes from its time$_{zero}$ to the present (Cremaschi 1987, Busacca and Cremaschi 1998) and exhibits few or no signs of polygenesis (Fig. 16.1). Vetusols have developed "under the influence of a single set of major pedogenic processes throughout their history" (Busacca and Cremaschi 1998: 96). They exhibit little evidence of environmental *change* during the period of their development.

Buried Paleosols

No one denies that buried soils are paleosols (Johnson 1998b). *Buried* paleosols have formed on landscapes of the past but have since been buried by younger sediment such as loess, till, colluvium, or alluvium (Catt 1998). For this reason, they are the easiest kind of paleosol to recognize (Ruhe 1965, Ruellan 1971, Nettleton *et al.* 1989). The Soil Survey Staff (1999) defined buried soils as having a surface mantle material of new, pedogenically unaltered material that is generally ≥50 cm thick. Bos and Sevink (1975) suggested that even thicker accumulations are necessary, whereas Schaetzl and Sorenson (1987) felt that putting a specific limit on depth of burial should be less important to the definition than assessing how the thickness of the overlying sediment actually affects the paleosol. They suggested that any soil buried by material that has not yet been pedogenically incorporated into the profile is, effectively, buried. Thus, any sediment, of any thickness, can be viewed as burying a soil. The soil *stays* buried unless the sediment above has been pedogenically incorporated into the buried profile (Johnson 1985). The latter process is termed *soil welding* (Ruhe and Olson 1980, Dahms 1994, Kemp *et al.* 1998, Olson and Nettleton 1998). After the overlying sediment has been mixed into and pedologically incorporated, the soil is no longer buried, although it may be cumulic. Because any amount of sediment can bury a soil (at least for a while), Schaetzl and Sorenson (1987) also defined a subtype of buried paleosol – an *isolated paleosol* – one that is buried so deeply that it is essentially cut off from surficial pedogenic processes (Figs. 16.1, 16.2). Isolated paleosols are affected only by postburial modifications due to diagenesis, not due to pedogenesis (Ruellan 1971; see later discussion).

Buried paleosols may be modified by (1) the burial process, (2) the weight of overlying sediments, and (3) the fluids that have migrated through them. Use of paleosols to interpret paleoenvironmental conditions requires that one be able to identify what, if any, postburial modification the paleosol has undergone (see later discussion). Isolated paleosols, by definition, are not currently being modified by surficial processes.

When the sediment that overlies a buried paleosol is removed, the soil is said to be exhumed (Ruhe 1965). *Exhumed paleosols* are once again subject to the entire suite of surficial pedogenic and geomorphic processes (Bushue *et al.* 1974). Most have been truncated, either during burial or during exhumation, or sediment from the current surface has been mixed into their upper sola (Nettleton *et al.* 2000). Thus, the B horizon is often the main criterion used to recognize exhumed paleosols. Many have been so changed that they can only be recognized as an exhumed paleosol by tracing them laterally to the point where they remain buried.

Recognition and Alteration of Paleosols

Recognizing that a layer in a stratigraphic sequence is a buried soil, or a nonpedogenic stratum, for that matter, is a highly important observation; it will affect every interpretation thereafter. If it is a soil, the paleosol represents a formerly stable surface and thus could have been occupied by humans – information that is important to paleopedologists, archaeologists (Mandel and Bettis 2001), geomorphologists, and paleoenvironmental scientists alike. Recognition of buried paleosols ranges from the very obvious to the extremely difficult (Ruhe 1965, Ruellan 1971).

Problems of alteration must be addressed when trying to recognize and interpret paleosols. First, postburial modifications (diagenesis) sometimes make recognition of buried paleosols problematic (Simonson 1954a, Catt 1990). Second, in relict and welded paleosols it is difficult to differentiate relict features from those associated with contemporary pedogenesis. Therefore, the use of multiple criteria, usually assessed in the field and followed up with

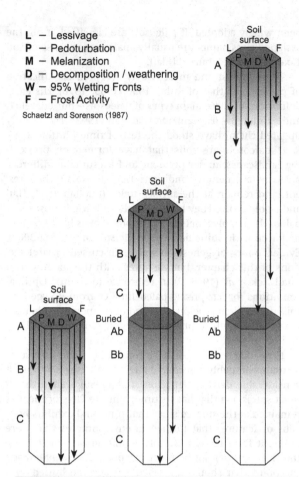

L – Lessivage
P – Pedoturbation
M – Melanization
D – Decomposition / weathering
W – 95% Wetting Fronts
F – Frost Activity

Schaetzl and Sorenson (1987)

Fig. 16.2 Conceptualization of paleosols buried at different depths, showing how they are affected by surficial processes. The model assumes no influence of a water table. The degree to which the buried soil is affected by surficial processes determines whether a paleosol is considered to be pedogenically isolated.

laboratory data, is required to assist in the identification, recognition, and interpretation of paleosols, and their differentiation from unmodified sediment (Hall and Anderson 2000, Grimley *et al.* 2003).

Burial of a geomorphic surface and its associated soils is a time and space transgressive process. Like an erosion event, it represents an unstable period in the pedogeomorphic system. During burial, vegetative cover may have been changing; fires, floods, or drought may have been occurring; or glacial ice may have been advancing onto the landscape. Any of these processes, among others, could also have (indirectly or directly) caused widespread erosion of all or parts of the surface, prior to burial. For this reason, many buried soils are partially eroded, often displaying an abrupt upper boundary and lacking O and A horizons (Catt 1990). Conversely, an abrupt upper boundary in a buried

Fig. 16.3 Photo of the Late Sangamon (MIS 5) paleosol near Clinton, in eastern Iowa. Here, the soil has a thick profile with a red Bt horizon and well-developed A and E horizons. Note the gradual upper boundary of the Ab horizon, indicative of slow burial by the Peoria (MIS 2) loess. Photo by D. Johnson.

paleosol may be due only to rapid burial of an otherwise stable surface, such as by eolian sand, glacial outwash, or volcanic ash. Some buried soils have been *interpreted* as having been truncated during burial simply because the postburial loss (mineralization) of organic matter has made it appear as though the soil lacks an A horizon (Busacca 1989). In such cases, the micromorphology of the humus-poor A horizon, specifically its granular structure, is all that remains to indicate the existence of an Ab horizon.

Buried paleosols may have been, and likely were, modified *during* burial. Although many soils are eroded during periods of geomorphic instability, others are rapidly buried and well preserved. Some may even develop progressively during slow and/or shallow burial. If buried slowly, e.g., by eolian (loess), alluvial (alluvium), or volcanic deposits (ash), the soil may grow upward into the burying sediment and become overthickened, before pedogenic processes finally become ineffective (see Chapter 13). These types of cumulic or upbuilt paleosols usually have thick Ab horizons that gradually merge into unaltered sediment above (Fig. 16.3). Such paleosols are also typically mixed zones, in which materials from the upper paleosolum and overlying, burying sediment are intermingled (Ransom *et al.* 1987b).

Buried paleosols span the range between those that are unmistakable and easily recognizable to those that are weak and extremely difficult to discern. Thus, many have called for the establishment of a set of criteria to use when describing paleosols. These criteria would eliminate layers in stratigraphic sequences that are not soils from those that clearly are paleosols. To recognize, describe, and interpret a buried paleosol, standard pedologic techniques and assumptions are used to document pedogenic features (Woida and Thompson 1993). A key point in the discrimination of buried (and surface) soils is that, unlike many other sedimentary units, soils have distinct patterns in their depth distributions of properties (see Landscapes). Clay content may decrease and then increase with depth, as do many other pedogenic attributes; the changes may occur gradually and without abrupt breaks. Organic matter, meteoric isotope, and opal phytolith contents are commonly maximal at the top of the profile and decrease with depth, sometimes with a secondary peak in the upper B horizon. These trends are uniquely pedogenic; geologic strata frequently have much more abrupt upper or lower boundaries. Retallack (1990) listed three main features that differentiate paleosols from sediment: (1) root traces, which sometimes become more prominent *after* burial (although using *only* traces can be problematic; see Berry and Staub [1998]); (2) soil horizons that can be correlated with diagnostic horizons of contemporary soils; and (3) pedogenic structure and other pedogenically acquired features such as cutans and nodules. Indeed, cutans and peds are some of the most characteristic pedogenic signatures observed in stratigraphic sections (Rutter and Ding 1993, Horiuchi *et al.* 2009), because they are commonly preserved when paleosols become part of the rock record (Retallack 2004). Both micro- and macromorphological observations, therefore, may be critically important to differentiate pedogenic from sedimentary fabrics (Dalrymple 1958, Botha and Fedoroff 1995, Kühn *et al.* 2013), and to determine previous pedogenic pathways (Fedoroff and Goldberg 1982).

Mineralogical data, e.g., clay mineralogy and weathering data of all kinds (including indices), provide some of the best criteria for the identification of buried paleosols (Fig. 16.4). Skeletal grains are useful because of their persistence and immobile nature. Mineralogical methods also have the advantage of providing information about the relative degree of paleosol development, prior to burial (Ruhe 1956b; Fig. 14.64). Mineral signatures in paleosols are also unlikely to change markedly; they frequently become even more stable after burial (Droste and Tharin 1958, Mausbach *et al.* 1982, Markewich *et al.* 1998, Nettleton *et al.* 2000). For this reason, weathering ratios of sand- or silt-sized grains are some of the most useful indicators of paleosol development (Ruhe 1956b, 1965), as are certain Fe and Al species (Evans 1982). Use of mineralogical measures for comparison of development between buried and surface soils,

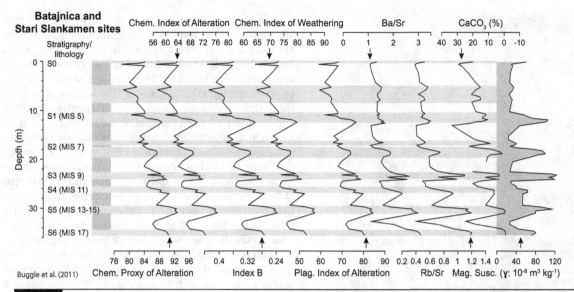

Fig. 16.4 Depth plots of various indices and indicators of weathering (CPA, Rb/Sr, and CaCO₃) (see Chapter 15 and Table 15.8), in a thick, composite loess-paleosol sequence at Batajnica and Stari Slankamen, Serbia, representing the last 17 Marine Isotope Stages (Buggle *et al.* 2009, Markovich *et al.* 2009). The two sites, located ~40 km apart, are near the banks of the Danube River. The Chemical Proxy of Alteration is defined as CPA = $[Al_2O_3/(Al_2O_3 + Na_2O)] \times 100$.

however, generally necessitates that they both be formed in similar materials (Grimley *et al.* 2003, Buggle *et al.* 2011).

Postburial modification of buried paleosols occurs by *pedodiagenesis*, which is a combination of (1) pedogenesis (effects of surficial pedogenic processes trickling down into a buried soil) and (2) diagenesis (processes associated with compaction, heat, and circulation of fluids) (Ruhe and Olson 1980, Olson and Nettleton 1998). The longer a soil is buried the more likely pedodiagenesis has operated and the greater effect it will have had on the buried soil. Soils buried deeply are more affected by diagenesis, whereas shallowly buried soils are more affected by pedodiagenesis. Some pedogenic processes that occur at the surface may also affect buried paleosols, particularly the translocation of particles and ions and pedoturbation processes of different kinds.

Processes associated with pedodiagenesis include, first and foremost, compaction (Sheldon and Retallack 2001). Peds may become welded together to the extent that their discrete nature is lost or muted. The entire buried profile is compacted – a process that is especially pronounced in Histosols and other humus-rich soils. Despite compaction, many buried paleosols remain more permeable than the surrounding sediment, readily conducting fluids. When the fluid flow ceases, precipitation of secondary minerals may occur.

Also included in the pedodiagenesis bundle are additions of base cations from overlying calcareous material by percolating soil solutions or from groundwater, losses of organic carbon by oxidation/mineralization, changes in redox patterns, additions or losses of Fe and Mn (depending

on redox conditions), soil color changes, and an overall blurring of horizon boundaries (Simonson 1941, Stevenson 1969, Ruhe *et al.* 1974, Valentine and Dalrymple 1976, Mausbach *et al.* 1982, Ransom *et al.* 1987b). Deep burial environments are usually mildly alkaline and reducing, favoring carbonate and silica precipitation (Catt 1990). As a result, diagenesis also involves formation of new minerals (sulfides, carbonates, clay minerals), dissolution and recrystallization of existing minerals, and the termination of much of the preexisting microbial and macrobiotic activity (Catt 1990). Also associated with compaction are subsidiary processes such as loss of interstitial pore water.

To interpret altered, buried paleosols, we must understand which soil properties are most prone to postburial modification and which are persistent. To this end, Yaalon (1971) placed soil properties into a series of categories based on the degree of persistence after burial. Properties that persist in a soil after burial are most useful for ascertaining what the soil was like before it was buried (Valentine and Dalrymple 1976, Karlstrom and Osborn 1992). Those that do not persist and are quickly altered after burial, such as pH, are less likely to provide useful information about the original soil. Yaalon's threefold categorization is detailed in Chapter 15. Catt (1990), however, noted that organic matter content, a property that is routinely altered by oxidation after burial and according to Yaalon is reversible, can be preserved if the soil is rapidly buried in a stagnant, anaerobic environment. We might also add that organic carbon can be preserved in fine-textured soils of semiarid and subhumid settings, e.g., in the loess plains of eastern Europe and the Great Plains of North America (Arbogast and Johnson 1994, Martin and Johnson 1995, Marković *et al.* 2009). Whether a soil property is preserved depends not only on its morphology, but also on the physicochemical conditions at the location where it is buried.

Under most conditions, it can be broadly assumed that morphological properties, e.g., horizonation, structure, root casts, concretions, krotovinas, are the most resistant to modification, whereas some chemical properties (unless manifested as morphological or mineralogical features) are more ephemeral (Busacca 1989). For example, many paleosols buried by calcareous loess have been resaturated with base cations, i.e., their chemistry has changed, but their Bt horizons – formed under more acidic conditions – remain. Although postburial enrichment in base cations does erase most of the chemical signatures that the paleosol may have had, it also effectively stops lessivage into or out of the buried soil and limits further weathering. Such changes allow the investigator to use textural and weathering-related depth plots to interpret pedogenic development (Figs. 14.64, 16.4). For this reason, clay and primary mineral signatures – unlikely to change markedly after burial – are some of the most useful indicators of paleosol development (Ruhe 1956b, 1965, Droste and Tharin 1958, Mausbach *et al.* 1982).

Shallowly buried paleosols are prone to modification by postburial pedogenesis from above, and their presence has dramatic effects on surface soil genesis as well. Particularly noteworthy are situations in which an acid paleosol is overlain by a thin layer of sediment. The acidity of the underlying soil may enhance acidification, weathering, and pedogenesis in the surface soil (Kleiss 1973, James *et al.* 1995). Kleiss (1973) studied a trend in the expression of *surface* soils developed in Peoria loess in Illinois, a trend first observed by Guy Smith (1942); he called this the Illinois soil development sequence. Soils along this sequence become better developed as the loess cover thins, and the paleosol below becomes progressively closer to the surface. In many parts of the midwestern United States, strongly developed paleosols buried by only 1–2 m of loess are believed to have influenced the formation of fragipans in the surface soil, perhaps through perching of silica-rich water on top of the dense, clay-rich paleosol (Harlan and Franzmeier 1977, Ruhe and Olson 1980, Ransom *et al.* 1987b, Wilson *et al.* 2010). In southern Illinois, variable thicknesses of loess above a shallowly buried paleosol have led to a spotty pattern of Na-rich slick spots (Wilding *et al.* 1963, Indorante 1998).

Ruhe's (1969) work on the Sangamon paleosol in Iowa, where it is buried by Peoria loess, argued that the paleosol is better developed, i.e., as indexed by stronger Bt horizon, at sites that are progressively farther from the loess source (Fig. 16.5). The amount of clay in the Sangamon Bt horizon increases mathematically as

Clay (%) = 19.7 + 1.23 Distance (miles)

The increase in Bt clay concentration of the Sangamon paleosol at sites farther from the river is coincident with

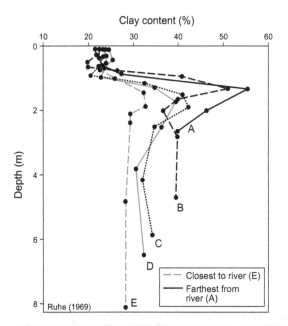

Fig. 16.5 Clay content of the Bt horizons of the Sangamon (MIS 5) paleosol, formed in Loveland (MIS 6) loess, in southwestern Iowa, along an east-west traverse away from the Missouri River.

three distance-related trends: (1) The Yarmouth paleosol below it becomes increasingly closer to the surface as the loess thins, (2) the loess becomes finer-textured, and (3) the period within which the Sangamon soil has had to form, i.e., its soil-forming interval, increases. It is difficult to know exactly which of the three trends most affected the development of the Sangamon soil in Iowa, but some influence of the Yarmouth paleosol below has to be considered.

Shallowly buried paleosols may become overprinted by the downward-developing profile of the surface soil. When these two soil profiles pedogenically connect, such that C horizon material is absent between them, they are welded (Ruhe and Olson 1980) or superimposed (Gerrard 1992). In a sense, a welded paleosol is intermediate between a buried and a relict paleosol, as it is influenced by surficial processes, yet it retains some characteristics of the buried soil (Fig. 16.6). Welded paleosols are affected by contemporary pedogenesis but have incorporated a cover of sediment that did not exist when they initially formed. Soil welding complicates recognition and interpretation of buried soils and makes detailed interpretations of their pedostratigraphy more difficult (Kemp *et al.* 1998; Fig. 16.7).

Buried Paleosols and Quaternary Stratigraphy

Paleosols provide distinct stratigraphic markers, representing a period when a geomorphic surface was both temporarily stable and subaerial (Ruhe *et al.* 1971). They are some of the best records of episodes when little or no erosion or

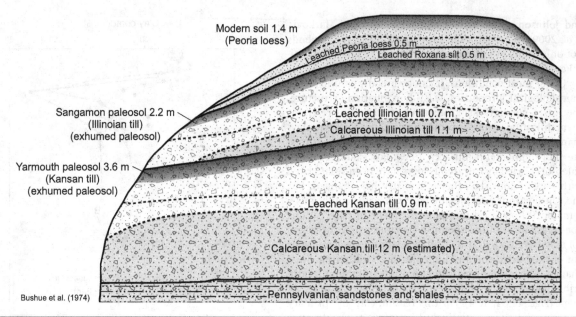

Fig. 16.6 Quaternary stratigraphy in a part of western Illinois, illustrating the concepts of buried, exhumed, and welded paleosols.

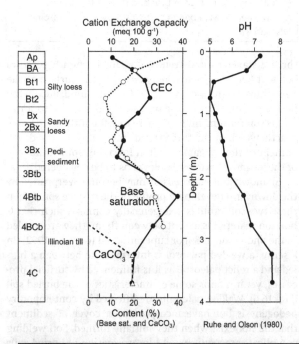

Fig. 16.7 Depth functions for a surface soil welded to a Sangamon (MIS 5) paleosol in southern Indiana. Note the Bt horizon clay maxima for both the surface soil and the paleosol. Although the paleosol has retained some signature of its preburial, acidic status (low base cation saturation), pH, a more ephemeral property, fails to provide any paleopedological discriminating information.

sediment deposition was occurring on a landscape. Indeed, much of the early interest in buried paleosols stemmed from their use as stratigraphic marker beds, rather than as indicators of paleoenvironmental conditions (Leverett 1898, Kay and Apfel 1929, Leighton and MacClintock 1930, Frye and Leonard 1949). Stratigraphically, a buried paleosol can be used (1) to discriminate between deposits of different ages, (2) to correlate deposits across landscapes, and (3) to provide information on the length of depositional hiatuses, i.e., length of the soil-forming interval, or erosional episodes within the stratigraphic column (Ely *et al.* 1996, Olson *et al.* 1997, Marković *et al.* 2009; Fig. 16.8). The most common application of buried paleosols in this context occurs in areas of constructional topography like the midwestern United States and the loess plains of China, Russia, and eastern Europe, where loess and glacigenic sediments of different types that were deposited during periods of regional instability are intercalated with paleosols (Wascher *et al.* 1947, Feng *et al.* 1994a, Dlusky 2007, Marković *et al.* 2009; Figs. 16.6, 16.8).

The loess-paleosol record, which in places can be traced back to the Miocene (Deng *et al.* 2005, Hao *et al.* 2008), is perhaps the best preserved and most studied –and can be found at a variety of locations across the Earth (Porter 2001, Buggle *et al.* 2009, 2011, Marković *et al.* 2009, Pierce *et al.* 2011). The paleosols and the paleoenvironmental record they contain can be evaluated in a number of ways, and new paleoenvironmental proxies are continually being developed from this record, e.g., magnetic susceptibility and other magnetic properties; Fe^{2+}/Fe^{3+} ratios; OC and $\delta^{13}C$ contents; pollen and macrofossil remains; carbonate contents; clay mineralogy; and grain size, among several

Surduk site

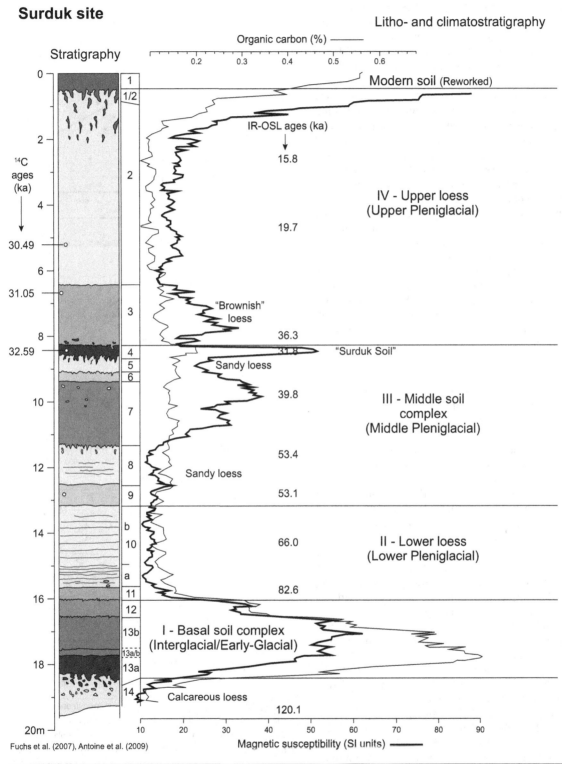

Fig. 16.8 Stratigraphy, dating, magnetic susceptibility, and OC data for the Surduk loess/paleosol sequence in Serbia.

others (Farnham *et al.* 1964, Feng and Johnson 1995, Botha and Fedoroff 1995, Grimley *et al.* 2003). These topics are discussed in more detail in the following.

The thickness of the China loess column and the number of buried paleosols within it make this paleoenvironmental record second to none (Liu 1985, Pye 1987, Kemp and Derbyshire 1998, Kemp 2001, Ding *et al.* 2002). The loess forms a huge plateau southwest of Beijing (Fig. 10.35). Although eolian deposition began here almost 7 Ma ago, loess deposition did not begin in earnest until about 2.6–2.4 Ma ago. The main loess source region was the Takla Makan and Gobi Deserts to the northwest and north (see Chapter 9). Loess was blown in from these deserts by strong, cold winter monsoon winds. Hence, loess deposition occurred during glacial periods, when the cold landscapes to the northwest destabilized; soil formation occurred as surfaces stabilized during the warmer interglacial periods, which were dominated by the wet summer monsoon (Hovan *et al.* 1989, Zhongli *et al.* 1993, Zhang *et al.* 1994). For this reason, the loess and its paleosols carry a strong paleoenvironmental signal, and their study has provided a wealth of information on paleoclimate (Verosub *et al.* 1993, Xiao *et al.* 1999, Panaiotu *et al.* 2001, Tang *et al.* 2003; Fig. 16.9).

The Sangamon Geosol

The most extensively studied and, one could argue, impressive paleosol in the United States is the Sangamon Geosol (Fig. 16.3). It has long been known to represent the last major interglacial period, i.e., MIS 5. A MIS 5 soil is widely developed worldwide (Tarnocai 1990, Rousseau *et al.* 2001), but is particularly well developed and studied in the central United States, where it is developed on Illinoian drift and its correlative loess, Loveland loess (Ruhe 1974b, Follmer 1978, 1979a, 1982, Schaetzl 1986c, Jacobs 1998, Hall and Anderson 2000, Rutter *et al.* 2006, Markewich *et al.* 2011; Fig. 15.65). In places where these deposits are absent, it occurs directly on bedrock or residuum. In almost every instance, the buried Sangamon soil is considerably better developed than its counterpart on the surface, with respect to illuvial clay in the Bt horizon, clay mineralogy, redness (if on a well-drained part of the paleolandscape), weathered character, and solum thickness (Ruhe *et al.* 1974, Grimley *et al.* 2003). Within the midcontinent, the Sangamon Geosol formed over a minimum period of 60,000–80,000 years, within at least two soil-forming intervals: ca. 130–90 ka and 74–58 ka (Wang *et al.* 2009, Markewich *et al.* 2011). Meteoric [10]Be data confirm the long period of soil formation represented in this MIS 5 soil (Fig. 16.10). At many places throughout the midcontinent region, the soil was effectively buried by loess between 55 and 25 ka ago (Hall and Anderson 2000). The Sangamon soil is commonly developed to 3 m or more in thickness when in coarse-textured materials, but is usually ~2 m thick when developed in till or loess. Where formed on sloping surfaces, it

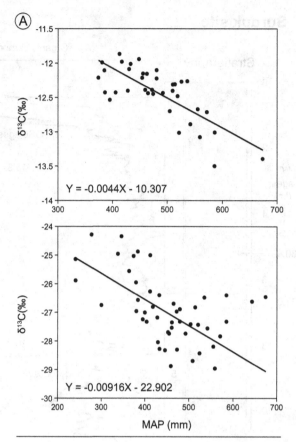

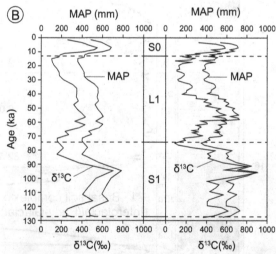

Fig. 16.9 An example of the type of paleoenvironmental data that can be developed from stacked loess-paleosol sequences. Using the correlation between the $\delta^{13}C$ values of plants and mean annual precipitation (MAP) shown in (A), as well as the relation between the C_4/C_3 ratios and annual precipitation published by An *et al.* (2005), Ning *et al.* (2008) were able to estimate the variation in MAP for two loess-paleosol sections in China, for the past 130 ka.

Phillips Bayou, Phillips County, Arkansas

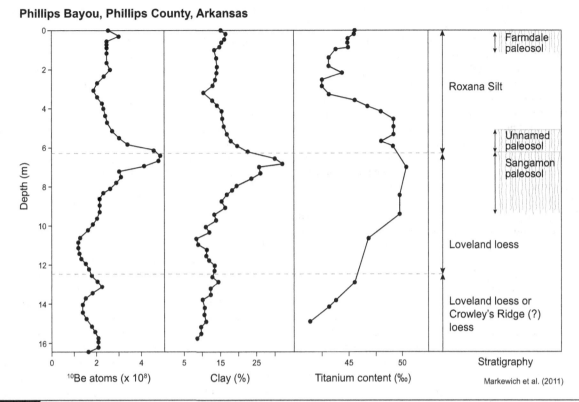

Fig. 16.10 Stratigraphy and depth plots of soil data, indicative of the strong development of the Sangamon Geosol in the midcontinent, United States. The data are from a site at Phillips Bayou, Arkansas.

typically has a stone line, ascribed either to pedimentation processes immediately preceding burial or to bioturbation (see Chapter 14). Where developed on deposits older than Illinoian age, the Sangamon soil is the highly polygenetic Yarmouth-Sangamon soil (see Chapter 14). In parts of the midwestern United States, soils that developed only during the latter part of Sangamonian time can be distinguished and are called Late Sangamon paleosols.

The morphological variation in the Sangamon Geosol is reflective of the variety (and length) of soil-forming environments that it encountered, as well as the geomorphology of the surface it developed on (Follmer 1979a; Fig. 14.67). Depending upon landscape position, the Sangamon Geosol can be a deep, reddish, clay-rich Ultisol on uplands or a thick, gleyed, grayish black Aquoll or Aqualt in swales. On paleosummits the soil commonly developed in situ in Illinoian till or Loveland loess. It commonly has a strongly developed Aquoll- or Aqualf-like profile in many places in Illinois, whereas on the drier, western end of its range within the central United States upland Sangamon soils are redder, resembling a Paleustoll. Trace the red Sangamon soil from under the glacial deposits of the midcontinent and it surfaces as a red, clayey, relict paleosol at locations south of the glacial border, e.g., Kentucky, Missouri, and Oklahoma. In Oklahoma, it is mapped as a red Ustoll. In

Kentucky and Missouri, it is a Udult. As with all paleosols, recognition of its full variation is essential if it is to be properly correlated as a stratigraphic marker bed. The Sangamon Geosol – something for everyone!

The Soil-Forming Interval

Soils form on stable geomorphic surfaces and continue to evolve with them, as long as they remain relatively stable. Eventually, all surfaces and their soils will become (or have become) buried or destroyed by erosion. The period during which the soil formed, before interruption, is called its *soil-forming interval* (Yaalon 1971, Chadwick and Davis 1990). For surface soils, the soil-forming interval starts at the cessation of an erosion or depositional episode, i.e., when the surface stabilizes at time$_{zero}$, and continues to the present. For soils on surfaces that have long been stable, soil-forming intervals might be associated with a period or periods of particularly strong pedogenesis – punctuated by periods of slowed or regressive development (Morrison 1967, Chadwick and Davis 1990, Hall and Anderson 2000).

For a paleosol buried deeply enough, the soil-forming interval ends at the time of burial. Thus, in a stratigraphic context, the soil-forming interval is bracketed by the ages of adjacent depositional units. The presence of a buried paleosol implies that a stable period/interval once existed on that landscape. The fact that the soil is now buried also implies that the pedogeomorphic processes have, at some time since, changed. Thus, as suggested by Butler (1959), buried soils provide excellent proxy information about landscape stability (Nesje *et al.* 1989, Anderton and Loope 1995, Olson *et al.* 1997), which when dated can be related to the drivers of that instability, e.g., climatic change or human activity.

Soil properties can reveal information about paleoenvironmental conditions during the soil-forming interval, as well as the length of that interval. Often, however, the relative importance of *length versus intensity* of the soil-forming interval is unclear. For example, the Sangamon soil was long assumed to have attained its strong morphology as a result of an intense (warm) soil-forming climate during MIS 5, particularly the hottest initial part, 5e. However, the counterargument is that pedogenesis was not abnormally strong or intense during MIS 5, but rather, the great *length* of the soil-forming interval facilitated the strong development (Ruhe 1965, Boardman 1985, Hall and Anderson 2000). Length or intensity? We will revisit this topic later.

Soils, paleosols, and soil-forming intervals are the essence of paleopedology. They provide a suitable lead-in to the next section, which focuses on pedogenic features that are particularly useful for interpreting the duration, intensity, and characteristics of the soil-forming interval.

Dating of Buried Paleosols

Paleopedological studies are often concerned with one or more of the following questions: (1) What was the time$_{zero}$ for the buried soil? (2) How long was the soil-forming interval, and (3) When was the soil effectively buried? Answers to these questions can usually be obtained using various dating and pedological techniques (see Chapter 15).

Some of the best materials that can be used, in a radiocarbon context, for estimating the age of burial (question (3) above) are plant and animal remains, e.g., charcoal, seeds, wood, and shells, that were on the soil surface immediately prior to burial (Catt 1990; Fig. 15.66). However, because such material is not always available, one usually has to turn to mean residence time (MRT) data (see Chapter 15) on soil organic matter to obtain ages for the soil-forming interval. MRT data provide an estimate of either the time since burial (question (3) above) or the time$_{zero}$ of soil formation (question (1) above) (Scharpenseel and Schiffmann 1977). When dating soil organic matter in buried paleosols, the length of time that the soil has been buried (if it is not highly polygenetic) is best determined by sampling the uppermost part of the A horizon, because it will have the youngest MRT age (Haas *et al.* 1986, Nesje *et al.* 1989; see Chapter 15). And even then, the age obtained will be a maximum age of the time since burial. Put another way, the soil could not have been buried any longer than the MRT age from the A horizon of the buried paleosol.

To establish the age of time$_{zero}$ of the buried soil from an MRT age, it is advisable to sample the lowermost part of the A horizon, because it is most likely to contain the oldest carbon (Leavitt *et al.* 2007). Sampling lower-profile horizons could theoretically result in older ages that better approximate time$_{zero}$, but unfortunately, often they do not contain adequate amounts of carbon. The MRT age obtained from the lower profile will be a minimum estimate of the time$_{zero}$ of the buried soil, i.e., the soil could not have begun forming any later than that time. Soils that could contain ancient or dead carbon, e.g., coal, must be dated with caution, because MRT age estimates on these soils could significantly predate time$_{zero}$ (Catt 1990).

Estimating the length of the soil-forming interval on the basis of MRT ages from the youngest and oldest organic carbon in the soil is theoretically possible, but usually not advisable, given the error terms and uncertainties of the method. If a more reliable age estimate of the time of burial were to be obtained, e.g., on charcoal that was on the surface of the buried soil, then one could compare that age to an MRT age from the lowermost A horizon to arrive at a minimum estimate of the soil-forming interval of the buried paleosol. Another potentially useful method to estimate the length of the soil-forming interval involves meteoric ^{10}Be, which accumulates in surface soils as a function of exposure time (see Chapter 15). Other surface exposure dating (SED) techniques, such as mineral weathering indices, might also provide information on the *relative* length of soil-forming intervals of buried paleosols versus in correlation to others in which the length of interval is known with more certainty.

For shallowly buried paleosols, contamination of the soil organic matter with modern carbon from the surface soil is likely, eliminating the option to use MRT dating. Likewise, roots may penetrate the buried soil in search of nutrients and proliferate there.

Environmental Pedosignatures Associated with Soils

Landforms and soils provide many clues about paleoenvironments; this is one of the main reasons why we study them. Most geomorphology texts discuss how certain *landforms* provide indications of paleogeomorphic processes. Examples include stone circles and frost wedges as indicators of permafrost conditions and periglacial activity (Kozarski 1974, Washburn 1980b, Ballantyne and Matthews 1982, Mader and Ciolkosz 1997); colluvium as indicators of slope instability, forced by high-magnitude rainfall events (Liebens 1999, Leopold and Volkel 2007); and innumerable others. Our focus is on *soils* and *pedogenic features/attributes* as paleoenvironmental indicators (Gao and Chen 1983, Chadwick *et al.* 1995, Han *et al.* 1996). Certain soil properties form in, and therefore are associated with, a particular type of environmental situation or condition. These pedofeatures may carry a unique environmental signature.

Sedimentological characteristics, including the presence or absence of lithologic discontinuities, and the

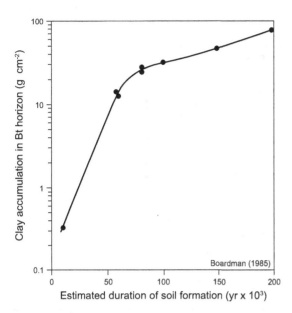

Fig. 16.11 Generalized diagram showing the relationship between Bt horizon clay accumulation and duration of soil formation and suggesting that greater clay accumulations in Pleistocene paleosols are due to longer soil-forming intervals, not a more favorable climate. The data set includes modern (Holocene age) soils, as well as the major paleosols of the midwestern United States.

paleotemperature, paleoprecipitation, paleowinds, and paleoecology (Ruhe 1974a, b). Examples abound of buried paleosols with characteristics that imply that they formed in a different paleoenvironment than that of the surface soil (e.g., Bryson *et al.* 1965, Follmer 1978, Reider *et al.* 1988). In such cases, an environmental change has occurred since the paleosol's soil-forming interval. The task of the paleoenvironmental scientist is to determine *what* changes have taken place, and if possible, *when* those changes occurred.

In this chapter, our goal is to provide examples of soil properties that have a usable signature or provide a discernable environmental, and paleoenvironmental, signal. The main soil properties that apply in this context are associated with *mineralogy* and *morphology* (Barshad 1966, Dormaar and Lutwick 1983, Karlstrom and Osborn 1992). We also note that many aspects of soil development that were discussed in Chapter 15, especially with regard to elemental and other chemical signatures, are also commonly applied as paleoenvironmental proxies, e.g., Bokhorst *et al.* (2009).

Horizonation and Morphology

The *horizonation* of soils is usually the first characteristic that one observes, and it generally sets the tone for the overall pedogenic interpretation. For example, soils with carbonate-rich horizons provide definite evidence of a climatically dry period of formation in which carbonate-rich dust affected the soil surface (Bryan and Albritton 1943). When such soils are present in areas that, today, are humid, this morphology alone usually indicates a dry-to-wet climate change (Reider 1983). Similarly, soils with sesquioxide-rich horizons or with accumulations of salts or gypsum clearly indicate a distinct type of pedogenesis. Additional quantification can be accomplished as well; depth to carbonates has been correlated to annual precipitation (Royer 1999), and this relationship may also apply to paleoprecipitation values (for paleosols) (Retallack 1994). Ruhe and Scholtes (1956) used the morphology of a buried Alfisol in Iowa to infer that it formed in a forested paleoenvironment. Ruhe *et al.* (1974) suggested that the buried Sangamon soil in southern Indiana was morphologically like the forested soils that are forming today, several hundred kilometers to the south. Van Ryswyk and Okazaki (1979) observed that soil profiles in the alpine zone of the Cascade Range had spodiclike morphologies, and Rovey and Balco (2009) interpreted a 2.47 Ma buried paleosol in Missouri as a Gelisol, on the basis of morphologic evidence for cryoturbation. Because this type of soil genesis usually occurs under forest, they suggested that the treeline was once higher. In sum, the ability to deduce bioclimatic signatures from profile morphologies, which is gained by studies of modern soil genesis, is a key part of paleoenvironmental reconstruction.

Before an assessment of pedogenic pathways can be undertaken, the parent material(s) that pedogenic

texture and mineralogy of the parent material, as well as water table effects and slope/aspect, all set the environmental stage for the more active pedogenic vectors to operate within. Pedogenic processes are driven by these active vectors, e.g., precipitation, temperature, vegetation, snowfall, and aerosol inputs. Together, parent material characteristics and active sets of vectors form unique pedosignatures within all soils, within the constraints of time (Fig. 16.11). Every soil *has* these pedosignatures, although they may be weak and difficult to ascertain. For example, because A horizons are common to all soils, one might assume that the A horizon can provide little or no information about the soil vectors or processes that have affected it. In a general sense, this is correct; the mere presence of an A horizon tells little about paleoenvironmental conditions. However, a more careful examination might reveal information about the A horizon that may be useful in this regard, e.g., thickness and organic matter content, isotopic composition of the organic matter, types and density of krotovinas, or lower horizon boundary characteristics. Thus, pedosignature information *is present*, and as our knowledge of pedogenesis increases, we will sharpen our abilities to decipher and understand it. That is the challenge, and it entails great potential rewards.

Pedosignatures from paleosols or older surface soils, when coupled with geomorphic data, can provide a variety of paleoenvironmental information on, for example,

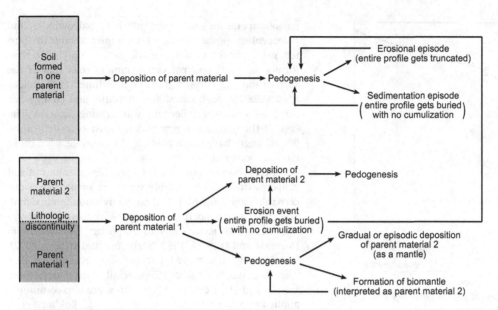

Fig. 16.12 Pathways associated with soils that have developed in one versus two parent materials. Knowledge of the lithologic nature of the soil parent material enables one to ascertain the sedimentological history of a soil, which then has utility in determining the paleoenvironmental history of that site/soil.

processes have been acting upon must be discerned and evaluated (see Chapter 10). In short, knowing the sedimentological and geomorphic history of the site is often as important as knowing its paleoclimate or paleovegetation (Amba *et al.* 1990). Soils provide many clues about geomorphic history in their parent materials, although strong pedogenic overprinting may have occurred on these sediments (Evans 1978, Meixner and Singer 1981, Schaetzl 1996, 1998). Because many parent materials accumulate vertically over time via surficial aggradation, another aspect of soil morphology that can be used to unravel paleoenvironmental conditions centers on parent material stratification/sedimentology (Benedict 1970, Reider 1983, Rindfleisch and Schaetzl 2001). For example, a period of dust accumulation is often used to infer dry conditions, whereas lack of dust could be associated with more humid conditions when surfaces are better vegetated.

Regarding parent material, if a buried geomorphic surface/soil can be identified, this observation has important implications for climatic change, landscape stability, and archaeology, among many other factors (Holliday 1988, Cremeens and Hart 1995, Curry and Pavich 1996, Hobbs *et al.* 2011). One can ascertain whether periods of erosion and deposition occurred, what agents were responsible for those depositional or erosion episodes, or whether a biomantle existed (Johnson 1990, Humphreys 1994, Nooren *et al.* 1995). In other words, one of the first observations that should be made about a soil revolves around the parent material – one or more? Is a lithologic discontinuity present? Does a buried soil exist within the profile? And so forth (Fig. 16.12).

For soils developed in a single parent material, only a single period of deposition is assumed. Another option for soils with one parent material, however, involves a period of deposition, then an intense erosion episode, in which most of the preexisting soil is eroded, followed by a period of stability (Fig. 16.12). In any event, at some point in the past a surface became stable and subaerial, the material beneath that surface was relatively uniform and probably developed in association with a single geomorphic event, and nothing significant has been added to the top of that surface since time$_{zero}$.

Soils that have developed in two or more parent materials, as indicated by one or more lithologic discontinuities, have more complicated depositional histories (Amba *et al.* 1990). For these soils, at some time in the past, two or more periods of deposition must have occurred (unless the upper material is determined to be a biomantle; see Chapters 11 and 14). These periods may also have been separated by an erosion event. A period of stability and pedogenesis may have intervened between deposition events, or it may not have, i.e., the two depositional events followed closely on the heels of each other. If one envisions no erosion event, it follows that pedogenesis probably began acting on the lower material, followed by additions of a second material at a later time, and this second material was pedogenically incorporated into the profile, i.e., no buried paleosol is present. Dixon (1991) used sedimentologic data of this kind for some alpine soils to infer a period of eolian influx (during the Holocene) onto a soil developed in till. Dixon's hypothesis was strengthened because the upper material was less weathered than the material below, suggesting

Table 16.1 | Taxonomic groups, horizons, and materials that sometimes provide useful climatic pedosignatures

Climatic signal/type	Taxonomic classes indicative of that climate[a,b]	Horizon types[a] or materials indicative of that climate
Cold climate with permafrost (tundra)	Gelisols, Cryids	Gelic materials
Cold desert	Gelisols, Cryids	Gelic materials, salic horizon, desert pavement, and varnish
Hot desert	Calcids, Gypsids, Durids, Salids	Calcic, gypsic, and salic horizons (and petroversions thereof), duripan, desert pavement, and varnish
Subhumid and semiarid grassland and savanna	Mollisols, Vertisols	Mollic epipedon (possibly a calcic or natric horizon), gilgai, slickensides
Cool, humid (conifer or mixed conifer-hardwood forest)	Spodosols	Spodic materials, umbric and spodic horizons (possibly a fragipan)
Temperate humid (deciduous forest)	Udalfs, Udolls, Udults	Argillic and kandic horizons, plinthite (possibly a fragipan)
Humid tropics	Oxisols	Oxic horizons, plinthite
Mediterranean climate	Xeralfs, Xerolls, Xerults	Duripans

[a] Based on Soil Taxonomy (Soil Survey Staff 1999). This table is our interpretation and is not meant to imply endorsement by the NRCS.

[b] Classes with weak development, i.e., Entisols, are common on many landscapes and provide little climatic information. They are not included here.

that it was younger. His study is an example of how a lithologic discontinuity can be used to understand changes in pedogenic/depositional systems over the life span of a soil or geomorphic surface. Soil geomorphologists *must* perform detailed analysis of parent material, looking for subtle lithologic breaks or changes down-profile, because geologists and sedimentologists may examine only the sediments below the soil profile per se. Careful examination of the last depositional event in many cases reveals it to be the most important one (Allan and Hole 1968, Dahms 1993, Schaetzl *et al.* 2000, Schaetzl 2008, Luehmann *et al.* 2013; Fig. 10.39). By interpreting near-surface discontinuities, pedologists can maximize their understanding of soil-geomorphic history.

Because certain soil morphologies are associated with particular climates, soil classification – a reflection of morphology – can often provide information on soil-forming climate. Then, this information can be used to make broad interpretations of paleoenvironmental conditions (Arkley 1967, Follmer 1978, Tarnocai 1990, Baker *et al.* 1991, Arbogast *et al.* 2004, Markewich *et al.* 2011; Table 16.1). Be aware, however, that several of the major taxonomic groups, e.g., Entisols, Inceptisols, Andisols, Fluvents, and Anthrepts, can form in a variety of climates. Although classes of soil temperature and moisture regime may be assigned to soils at the present landsurface, these groupings reflect contemporary soil conditions and are not intended to correspond to paleoclimate (Soil Survey Staff 1999). Many taxonomic groups are also influenced most by local circumstances, rather than macroclimate, e.g., a wet site that leads to the formation of a Histosol, or a swale that is shallow to saline shale that leads to the formation of a Natrustoll. Still other classifications provide virtually no pedoclimatic signature because they were designed to indicate incipient development. Some taxonomic classes or materials are so ephemeral that they can be used only to understand recent conditions. Some groups or horizons are reflective primarily of parent material, e.g., Sulfaquents, Vitrands, while others reflect human impact (Plaggepts). Nonetheless, despite this long list of caveats, the classification of a soil or a paleosol can be a useful first step in its use as a paleoenvironmental proxy (Retallack 1990).

Organic Matter Content

The amount of organic matter in a soil and its distribution through the profile are potentially useful indicators of paleoclimate and paleovegetation. Generally, organic matter contents increase in soils as the climate becomes cooler and moister (Jenny 1941b). Obviously, this relationship does not hold for the entire range of climate types on Earth, because in the coldest climates the decrease in organic matter production outpaces the decrease in mineralization rate. Indeed, the organic matter content of Gelisols varies widely; despite this caveat, organic matter contents in Gelisols are generally associated with environmental variables. For example, Gelisols of cold, dry deserts have very low organic matter contents (Beyer *et al.* 2000).

Many studies have examined the rates of organic matter production, decomposition, and accumulation in soils across the many global ecosystems and taxonomic

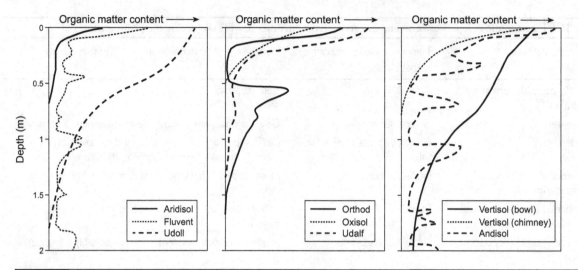

Fig. 16.13 Generalized depth distributions of organic matter in a few representative soil types.

categories. The purpose of these studies is often to determine an inventory of soil carbon reservoirs and, in so doing, inform our understanding of the soil carbon cycle. Histosols contain the most organic matter of all soils, because the decomposition rates of the fresh plant material are inhibited by cold conditions and/or a high water table that keeps the soil system anaerobic (see Chapter 13). Many wet mineral soils have high carbon contents for the same reasons, but their high organic carbon (OC) contents primarily reflect local conditions, not regional climate. Most studies of upland mineral soils (excluding Entisols) tend to show that cool grassland Udolls and high alpine Gelisols and Cryepts soils store the most organic carbon, whereas hyperarid and ultraxerous Aridisols have the least (Burke *et al.* 1989). Thus, with a few exceptions, soil organic carbon contents appear to be broadly related to climatic parameters (soil temperature and moisture), as influenced by parent material, especially clay content (Nichols 1984).

The distribution of OC with depth is also an indicator of pedogenesis and, hence, paleoclimate (Fig. 16.13). Soils in dry climates tend to have little OC, and they retain it shallowly, in near-surface horizons. This depth distribution occurs for two main reasons: (1) High base cation saturation in these soils facilitates Ca-humus bonds with minerals, limiting the mobility of OC, and (2) the generally low precipitation values limit how much organic matter is produced and limit how deeply it can be translocated. Mollisols, which form under grasslands, have high amounts of OC in thick, dark A horizons, whereas most forest soils have thin A horizons. In some soils, the depth distribution of OC shows a second peak in the B horizon. Many Spodosols, especially Humods, and Andisols display such trends because of the soluble nature of organic acids. Fluvents are defined as having irregular decreases in the concentration of OC with depth, due to frequent burial of

A horizons by alluvium. In sum, both the total amount of OC and its distribution with depth provide gross climatic signatures, although in many cases the OC signature does not always persist fully after burial (Holliday 1988, Dahms 1998).

Clay Mineralogy

Clay mineralogy is a persistent and sometimes almost irreversible pedogenic property that has, in all but the youngest soils, strong ties to bioclimate (Folkoff and Meentemeyer 1985, Wilson 1999). Parent material sets the starting point for the clay mineral suite in a soil. As a result, the clay mineralogy of most Entisols and Inceptisols mainly reflects parent material, and not paleoclimate. In such soils, the clay mineral suite has been so recently inherited that it is generally of little value as a paleoenvironmental indicator. However, in older and better developed soils, clay mineralogy may be a good indicator of not only climate, but also age and degree of development (Pal *et al.* 1989). As time progresses and chemical weathering has more time to affect the soil, the clay mineral suite becomes increasingly in equilibrium with the chemistry and leaching regime of the soil, i.e., the *soil* climate (Fig. 16.14). Thus, clay mineralogy is often reflective of climate, as affected and modified by biota (van der Merwe and Weber 1963, Barshad 1966, Arkley 1967, Folkoff and Meentemeyer 1985, 1987, Velde 2001). We include the caveat "as affected by the biota" because for many soils, the soil climate is different from the macroclimate. The pedogenic climate of an upland soil might be wetter than would be predicted by macroclimatic information alone because of an aquiclude that perches water. Soils in aquic soil moisture regimes are always wetter than macroclimatic data might indicate. Wet, clayey soils often show redox features even though they may not frequently be saturated, because water percolates so slowly

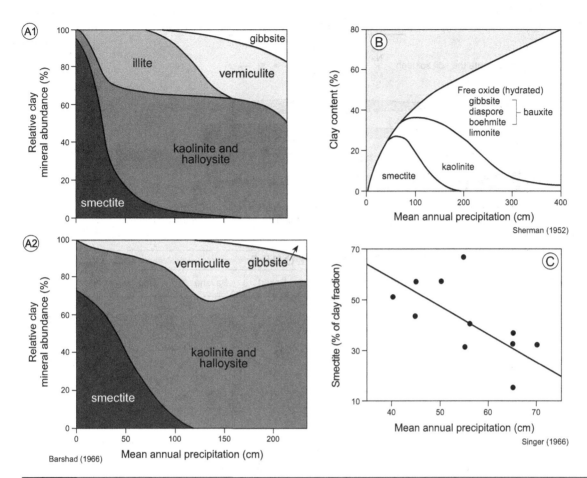

Fig. 16.14 Relationships between clay mineral suites in soils and climatic parameters. (A) Effect of precipitation on the frequency distribution of clay minerals in residual soils in California. The climate here is xeric, i.e., it has a winter rainfall maximum and mean annual temperatures between 10°C and 15°C. A1: Acidic igneous rocks. A2: Basic igneous rocks. (B) Generalized relationship between precipitation and clay mineralogy in a continuously wet climate (Hawaii). (C) Relationship between precipitation and smectite content in soils near Galilee (Israel).

through them. Depletion of base cations may be minimal in a poorly drained soil while well-drained counterparts on uplands are thoroughly leached and acidified. Conversely, some soils are drier than the macroclimate would suggest because they are coarse-textured or on a convex part of the landscape. Also, vegetation influences clay mineralogy through base cycling and acidification, or lack thereof. Thus, in sum, clay mineralogy, in older soils, is a reasonable reflection of *soil* climate.

A scale factor is at work here as well; On local scales clay mineralogy is reflective of parent material or topography, whereas across large regions the clay mineralogy may be more reflective of soil climate. Time also matters; after longer periods, the effect of soil climate will come to dominate the clay mineral suite across larger segments of the landscape. The clay mineral suites of soils that have had several thousand or more years to acclimate confirm this bioclimate–clay mineralogy relationship.

What is it about soil climate that matters most, when it comes to clay mineralogy? The primary clay mineral–environment relationship is centered on leaching, weathering regimes, pH, and precipitation. Folkoff and Meentemeyer (1985) argued that the two climatic factors that relate most to clay mineralogy are the degree of seasonal leaching and, if present, drying (Chadwick *et al.* 1995; Fig. 16.14). Along the leaching-weathering sequence, various suites of clay minerals exist, according to their geochemical stabilities. Increased leaching tends to decrease silica activity and increase weathering, resulting in the formation of more stable, weathering-resistant minerals, such as oxide clays, and an increase in Al activities. Where leaching is minimal, weathering by-products and cations, especially Si but also K, remain in the soil solution and can precipitate to form new clay minerals that are stable in, and therefore reflective of, that environment. Temperature affects all this primarily through its influence on the *rate* of the

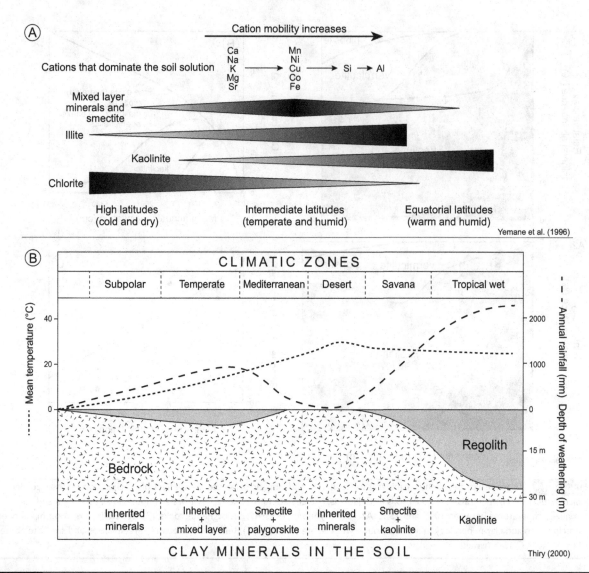

Fig. 16.15 Predictable changes in clay mineralogy by latitude, which enable these data to be used as a proxy for soil climate. (A) Generalized relationships among cation mobility, clay mineral suites, and the latitudinal process/chemical drivers that influence them. (B) Like the latitudinal weathering diagram shown in Figure 9.11, this figure shows the main clay minerals, as arranged by latitude and organized by climate types.

weathering reactions, not the *direction* of the weathering pathways.

Both theoretical and geochemical models, as well as field observations and lab data, point to the strong linkage between clay mineralogy and environmental conditions, specifically related to leaching and weathering intensity (Beaven and Dumbleton 1966, Rai and Lindsay 1975; Figs. 16.14, 16.15; see Chapter 4). Smectites tend to form, and are stable, in nonleaching environments – generally dry climates or areas of shallow groundwater where deep leaching is inhibited (Beaven and Dumbleton 1966, Birkeland 1969, Ojanuga 1979). Smectites are associated

with abundant Ca- and Mg-base cations that would normally be leached from other soils. They are especially common in soils with distinct wet and dry seasons (Singer and Navrot 1977; see Chapter 13). Worldwide, many soils on base-rich parent materials or with calcic horizons are rich in smectite (Fig. 16.16A). The smectite example illustrates the point that *soil climate*, not *regional atmospheric climate*, directly affects clay mineralogy by influencing soil solution chemistry.

Additional linkages between clay mineralogy and environment have been determined to improve our understanding of the relationships between climate and clay

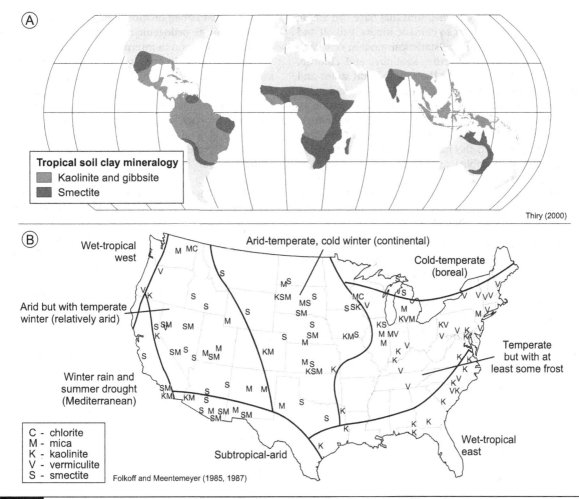

Fig. 16.16 Examples of the geographic distributions of clay mineral types in soils. (A) Map of the major smectitic versus oxidic soil mineralogies, with emphasis on the tropics. (B) The dominant clay minerals in the A horizons of 99 soils in the contiguous United States. The map is divided into climatic regions to show the strong relationship between climate and clay mineralogy.

mineralogy, and they may be useful for paleoenvironmental reconstruction as well. For example, allophane forms in humid or at least semihumid climates; in climates with a dry season where desiccation prevents allophane formation, halloysite forms instead. Vermiculite is an alteration product of biotite, forming where annual precipitation is >100 cm (Barshad 1966). In dry climates where base cation saturation values are high, illite weathers to smectite. In intense leaching environments, silica is depleted from the soil system and oxide clays come to dominate. Oxide minerals, e.g., goethite, gibbsite, and hematite, are associated with strong weathering regimes with ample precipitation and strong desilication (van der Merwe and Weber 1963, Schaefer *et al.* 2008, Thiry 2000, Fig. 16.15). The association of hematite-rich, deep red soils with hot, dry conditions was discussed in Chapter 13 (Fig. 13.35). In slightly drier, but still intensively leached environments, the loss of base cations and silica leads to the formation of 1:1 clays such as kaolinite and

halloysite (van der Merwe and Weber 1963, Ojanuga 1979, Wagner *et al.* 2007). Palygorskite and sepiolite are associated with aridic climates, alkaline pH values, and relatively high concentrations of Mg and Si (Bachmann and Machette 1977, Khormali and Abtahi 2003). And the list goes on.

Clay mineralogy, however, has some notable limitations as an indicator of paleoenvironments, even for mature soils. For example, granitic rocks usually contain the primary mineral muscovite and give rise to illite in the clay fraction, regardless of climate (van der Merwe and Weber 1963, Singer 1980). Likewise, in areas where the parent material is slowly permeable, leaching is inhibited, and the clay mineralogy may be reflective of a drier climatic regime; the same relationship holds on wet sites as a result of topography and a high water table (Kantor and Schwertmann 1974).

In a classic study, Folkoff and Meentemeyer (1987) identified the dominant clay minerals in 99 soils in the United

States (Fig. 16.16B). They studied soils that have had ample time to develop and adjust to climatic inputs. Folkoff and Meentemeyer developed five statistical models, one each for mica, vermiculite, smectite, kaolinite, and chlorite, based on correlations among the clay mineral suites and various soil (texture, drainage class, pH, etc.), parent material (acid rocks, basic rocks, till, loess, alluvium, carbonate rocks, etc.), and climatic variables (annual temperature, precipitation, water surplus and deficit variables, leaching indexes, etc.). For each clay mineral model listed in the following, a climate variable was the first one selected by the regression equation, illustrating the strong link between clay mineralogy and climate. With the exception of mica, a climatic variable was *also* the second variable selected! Their models suggested the following:

Mica model Mica, particularly biotite, dominates in cold, arid climates, on calcareous parent materials. Mica abundance is negatively correlated with annual precipitation. This model suggests that as the climate grows colder and more arid, weathering is increasingly inhibited, allowing more of the weathering-susceptible micaceous clays to remain in the soil.

Vermiculite model Water surplus, an index of leaching intensity, was the variable most strongly correlated to vermiculite content. After oxidation decreases the layer charge of biotite, leaching removes the interlayer K^+ cations, forming vermiculite.

Smectite model Smectite formation is favored in low leaching environments, caused by local-scale wetness or by intense, long dry seasons, explaining why smectite abundance is most strongly (but negatively) correlated to water surplus. Unlike that of mica, smectite formation does require a minimal level of leaching, tending to make it most common in dry climates with a short wet season and in some types of warm and humid, continental climates. Not coincidentally, these environments are also areas where Vertisols are common (see Chapter 11).

Kaolinite/halloysite model These 1:1 minerals optimally form in warm, humid climates with a prolonged leaching season. The first variable to enter this model was mean annual temperature, whereas the second and third variables were leaching intensity and water availability.

Chlorite model Chlorite is generally an unstable and easily weathered clay mineral, making its relationship to climate the weakest of the five clay minerals discussed here. It is limited to areas with weakly developed soils derived from chlorite-bearing parent materials, commonly in cold climates (Gao and Chen 1983, Yemane *et al.* 1996).

The linkages between clay mineralogy and environment of formation are useful, but they are complex. As a result, they continue to be the focus of much study. It is not our intent to summarize this literature in detail, but simply to point out that strong relationships exist between clay mineralogy and the pedogenic environment; such knowledge can be highly useful in interpreting soil evolution.

One last point: In order to assess the *paleo*environmental significance of clay mineralogy, one must know not only the environmental conditions under which clays have formed (above), but also their *resistance to alteration* as the environment changes, or after the soil is buried (Birkeland 1969). The clay mineral suite in old *surface* soils commonly reflects only the most long-lasting and intense paleoclimatic interval. In general, the low-silica minerals (kaolinite, gibbsite) formed in humid conditions will persist even if a change to a drier or cooler environment occurs. Conversely, the more siliceous minerals, perhaps inherited from an earlier cool or dry period, are likely to weather in a wetter and warmer climate (Catt 1990). Thus, clay mineralogy is often a one-way palimpsest, reflecting only the most intense soil-forming interval that the soil has undergone.

Nonetheless, if this caveat is taken into consideration, the clay mineral suite may be a paleoenvironmental indicator for both surface and buried soils (Singer 1980). The drawbacks we note are (1) that clay minerals do not provide a great deal of *detail* about the formative pedoenvironment (Fernández Sanjurjo *et al.* 2001) and (2) that they are mobile and can be translocated into or out of paleosols. Clay minerals in *buried* soils have an advantage, however, over other paleoenvironmental proxies in that they are slow to change and undergo little postburial alteration (Yaalon 1971).

Cutans

Cutans of various kinds typically form in response to a particular pedogenic regime (Table 2.6). They are frequently the consequence of illuvial processes that have a distinct climatic signature, e.g., lessivage, calcification, gypsification, podzolization (Jakobsen 1989). For example, not only do argillans carry the climatic signature associated with a humid (but not perhumid) leaching regime with a short dry season, but their mineralogy can be used to fine-tune paleoenvironmental interpretations further (Chadwick *et al.* 1995). Cutans – whether they are on ped faces or on the surface of clasts (Pustovoytov 1998, Pustovoytov and Targulian 1996, Dlusky 2007) – have a powerful advantage as a paleoenvironmental palimpsest because their signature is often not erased, but simply *written over*, building up as thick, multilayer ped coatings and potentially retaining multiple paleoenvironmental signatures (Pustovoytov 2002).

In some cases, however, cutans can be ephemeral features. Pedoturbation by shrinking and swelling, for example, can destroy argillans (Gile and Grossman 1968, Nettleton *et al.* 1969), limiting their utility as a pedogenic markers.

Like some other paleoenvironmental indicators, cutans and other illuvial coatings have the most potential to be useful in ecotonal areas (Busacca 1989, Eppes *et al.* 2008). An ecotone is a transition zone between bioclimatic regimes, e.g., grassland to forest. Subtle bioclimatic shifts may be recorded in soils at the ecotone as it shifts position, whereas soils in the central part of the region may not undergo noticeable change. Thus, soils within the ecotone may register a record of the change, but those in the core will not. With this in mind, Reheis (1987a) studied soils in an area of Montana that experienced dramatic climatic shifts over the past 2 Ma. During warm interglacials, calcification occurred, and the soils accumulated calcans in the upper profile. During the relatively cool glacial periods, acidification occurred and lessivage was initiated. Argillans were then overprinted onto the calcans. In the oldest soils, nine distinct layers (five calcan layers alternating with four argillan layers) point to the highly dynamic nature of the climate in this region. A similar, although less complicated, polygenetic soil is described by Reider (1983) for an arid lowland to humid upland climosequence in Wyoming. Here, carbonates were shown to engulf argillans in the lower B horizon. The current climate is humid, leading one to suspect that the carbonate-forming episode dates to a mid-Holocene warm period. Radiocarbon ages obtained on soil carbonates (5,230 and 4,735 ^{14}C yr BP) confirmed this assumption.

These examples highlight the fact that perhaps the most usable and well-preserved cutans are argillans (in humid climates) and calcans (in dry climates) (Kleber 2000). They can form quickly, attain reasonable thicknesses, are readily preserved, and are easily observed and measured. As discussed in Chapter 13, the depth to carbonates is a rough indicator of paleoprecipitation. Even upon a change to a more humid climate, some of the calcans may remain, especially on the undersides of clasts. The presence of pedogenic carbonates in buried paleosols provides a reliable indicator of a subhumid or drier environment of formation (Feng et al. 1994) and the amount of secondary carbonates in them might be related to the length or intensity of the soil-forming interval.

The paleoenvironmental signatures contained *within* calcans are actually quite multifaceted (Courty et al. 1994, Chadwick et al. 1989b), as demonstrated by the fact that the crystal form of calcite in carbonate coatings can reveal a climatic signature. In desert areas, the micrite of the calcans takes the form of equant or parallel prismatic crystals whereas in more humid areas the crystals are randomly oriented, euhedral, prismatic, and fibrous. Pustovoytov (2002) suggested other ways in which the microlaminae of calcans can provide paleoenvironmental information. Laminae of pure $CaCO_3$ with large calcite crystals are suggestive of the driest climates with low biological activity, whereas laminae with admixtures of organic matter and other minerals, containing poorly formed calcite crystals, are indicative of more humid paleoclimates (see also Pustovoytov and Targulian 1996).

Magnetic Susceptibility

A highly useful paleoenvironmental proxy that has been widely applied in stacked loess-paleosol sequences is *magnetic susceptibility* (MS). Proportional to the amount of strongly magnetic minerals (ferrimagnets) in the sediment or soil, MS is easily measured in the laboratory and the field, enabling large numbers of samples to be analyzed rapidly and inexpensively (Mullins 1977). Although the factors that influence MS in soils are manifold and not fully understood (Blundell et al. 2009), for most paleopedological studies it is assumed that MS increases proportionately to soil development, age, or weathering intensity (Bokhorst et al. 2009). Thus, it has become widely used as a proxy for comparative soil development in situations in which most of the other soil-forming factors are held constant. And nowhere is that situation better exemplified than in thick, stacked loess-paleosol sequences (Fig. 16.8). But what *actually* is the MS indicating, i.e., what is causing and influencing the signal? Let us discuss that aspect first.

High MS values in a soil or sediment are generally due to large amounts (and large sizes) of *ferrimagnets*, particularly magnetite and maghemite but also titanomagnetites and iron sulfides (Mullins 1977, Stanjek et al. 1994). High numbers of ferrimagnets in a soil or sediment are usually due to either (1) dilution of surrounding sediments with materials low in ferrimagnets or (2) in situ alteration of the material, thereby forming more ferrimagnets (Reynolds and King 1995). In the former case, it is assumed that magnetic enhancement exists in paleosols buried by loess partly because dilution of the magnetic signal occurs in the overlying loess (Kukla et al. 1988). The second case ascribes increased MS to weathering and pedogenesis (Torrent et al. 1980b, Singer and Fine 1989, Zhou et al. 1990, Singer et al. 1992, Fine et al. 1993, Heller et al. 1993, Verosub et al. 1993, Geiss and Zanner 2007), but as influenced by drainage. Wetter soils can lose ferrimagnets as a result of weathering and subsequent loss of Fe^{2+} by reduction and translocation. This observation explains why MS interpretations are more straightforward in freely draining, drier soils. Nonetheless, in sediment-paleosol columns, increased levels of MS are most commonly associated with paleosols, which represent the more weathered and pedogenically altered zones (Fig. 16.17). Because intensity of the MS signal is often correlated with other indicators of soil development, it is a highly useful tool for recognizing even the weakest paleosols within a stratigraphic column. These soils have lost most of their OC to mineralization and as a result may not otherwise even be visible. MS is also a type of surface exposure dating (SED) tool, as shown by Grimley (1998) for a loess column with intercalated paleosols of varying strength in Illinois (Fig. 16.17). The MS signal appeared to become stronger as soil development strengthened. Last, MS is also useful for correlating stratigraphic sequences where paleosols are the primary type of stratigraphic marker bed (Maher and Thompson 1991, Han et al. 1996).

Paleosols, especially those formed in loess, usually have higher MS values than their intervening parent material. However, in certain settings, where chemical reduction, weathering, or complexation of iron has been a dominant process, e.g., Spodosols or poorly drained soils, a buried

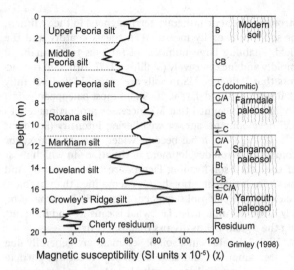

Fig. 16.17 Magnetic susceptibility and paleopedology of a loess-paleosol sequence from southern Illinois. Low susceptibility values for the Bt horizon on the Sangamon soil have been ascribed to weathering of ferrimagnets under acidic or reducing conditions.

paleosol or a surface soil may actually exhibit *lower* MS values than the overlying C horizon (Hayward and Lowell 1993, Grimley 1998). A buried Spodosol, for example, may have high quantities of Fe, but because it exists in organic complexes and not in magnetic minerals, the soil will have low MS values. Thus, little direct correlation can often be found between iron content and MS (Fine *et al.* 1995, Maher 1998). Think of the relationship between iron content and MS in this way: Fe contents are rarely a limiting factor for magnetic enhancement; instead, *opportunities for ferrimagnetic production* and *preservation* are the key constraint on MS values. It follows, then, that we need to know what pedogenesis actually *does* to create ferrimagnets, and in what kinds of soil environments MS is particularly enhanced, if we are to interpret MS correctly as a paleoenvironmental indicator (Verosub *et al.* 1993).

The formation of ferrimagnets by pedogenesis and how these data can be used as a paleoclimate indicator has been the topic of considerable study (Mullins 1977, de Jong *et al.* 2000, Fialova *et al.* 2006, Alekseeva *et al.* 2007, Blundell *et al.* 2009). Parent material provides the magnetic and nonmagnetic minerals that, through weathering and pedogenesis, are responsible for the increased MS of soils. Magnetite may be inherited from the parent material (Schwertmann and Taylor 1989), it may be formed by magnetotactic bacteria in the soil, or it may precipitate where Fe^{3+}-reducing bacteria are active (Maher 1998). Because magnetite is susceptible to reduction, dissolution, and removal from the profile, the *persistence* of the MS signal is dependent on a mechanism to convert magnetite to a more stable form – maghemite. Maghemite formation is favored by oxidizing

conditions, via repeated redox cycles (Mullins 1977) or excessive heating of the soil, as with fires. Maghemite is not stable in wet soils where conditions are reducing (de Jong *et al.* 2000). The optimal conditions for the formation of ferrimagnets, therefore, involves a warm, wet-dry climate in which organic matter production is high; here, magnetite can be formed by weathering during periods of intermittent soil wetness, and is transformed to maghemite in the dry season (Maher 1998). Maher (1998) listed several reasons why MS is increased by pedogenesis, the most important being "fermentation," in which iron-reducing bacteria in soils reduce ferric iron during anaerobic respiration. Under near-neutral pH conditions, the Fe may precipitate as ultrafine-grained magnetite, which then may or may not be subsequently transformed to maghemite (Grimley 1998, Maher 1998). In soils where ferrimagnet production and preservation are favored, a correlation is apparent between the maximum MS value and annual rainfall, at least for situations in which the annual precipitation is >140–200 cm (Vidic and Verosub 1999, Balsam *et al.* 2011). Ferrimagnet depletion (or nonformation) occurs in permanently wet soils, acid Spodosols, hyperarid soils, and those forming in the humid tropics where annual precipitation values are >2 m.

In unaltered sediments, the amount of ferrimagnets is dependent upon the source lithology and, therefore, the source area (Grimley 1998). For this reason, any comparisons of MS among various soils or paleosols must strive for parent material uniformity; fortunately, this is common in sequences of stacked paleosols in loess. Thus, it should come as no surprise that the MS signal in the 150-m-thick Chinese loess column, with its many intercalated paleosols, has been highly enlightening. The MS signals of the soils in the sequence not only are highly correlated to pedogenesis, but also have a strong climatic signal (Kukla *et al.* 1988, Zhou *et al.* 1990, Verosub *et al.* 1993, Chlachula *et al.* 1998). Generally, paleosols in the loess-paleosol sequence correspond to interglacial periods in the marine isotope record, indicating a strong climatic coupling of marine and continental climatic processes (Liu 1985, Hovan *et al.* 1989). The MS signal here correlates so well to pedogenesis because many properties have joined to create a situation in which highly contrasting amounts of ferrimagnets were formed (or not) in the column, as the glacial-interglacial cycles switched on and off. The loess parent material is uniform, so that the MS signal can be ascribed to pedogenesis and not to depositional anomalies. Most of the Chinese paleosols have also formed under well-drained conditions, which have favored preservation of pedogenically formed ferrimagnets. Vegetation has covaried with climate, making the MS signal a paleovegetation signal as well (Maher 1998). The climate of formation was wet-dry (monsoonal), leading to near-neutral pH values in the loess soils. Thus, soils within the loess column record intensity of pedogenesis, which in this area reflects paleorainfall (Fig. 16.18). The amount of paleorainfall, in turn, correlates to the intensity of the summer monsoon, which changed in concert with the glacial cycles recorded in the marine isotope record.

The paleoclimatic information we have gleaned from the Chinese loess record, which spans 2.4 Ma, is truly amazing. Its use has been particularly important to our understanding of the regional climate-geomorphic shifts; it also has reinforced confidence in the marine isotope record as a climatic proxy. The record shows changes in paleoclimate not only through time (expressed vertically in the loess column) but also through space (Maher *et al.* 1994; see also Busacca 1989). Data gleaned from this record include not only MS and morphological-paleopedological information from the many paleosols within, but also isotopic data from carbonates contained within the buried soils with information about the paleofloristic composition of past landscapes (Ding and Yang 2000). Study and correlation of the China loess and marine isotope records have shown that cool, glacial periods in the Late Pleistocene loess were times of loess deposition and worldwide dust transport, whereas during interglacials the landscape stabilized and soil formation occurred (Hovan *et al.* 1989). These soils would later be buried and preserved within them a record of pedogenesis and regional geomorphic stability.

The role of duration versus intensity of pedogenesis applies to studies of magnetic susceptibility. Are high values of MS due to a *longer* period of pedogenesis (and hence, a longer period of surface stability) or to a more *intense* period of pedogenesis (perhaps a warmer and wetter soil-forming interval)? For MS, or any other parameter, to reflect short-term changes in paleoclimate, it helps if the pedogenic regime favors rapid development, with soils reaching a steady state with regard to climate in a few thousand years or less. This may or may not be the case over parts of the Chinese loess plateau, where pedogenic properties reflect *intensity* of pedogenic drivers, in this case, climate. Conversely, a steady, continually developing soil property, one that requires many thousands of years to reach a steady state, will reflect *duration* of soil development more than intensity. Thus, clay mineralogy would not be a good paleorainfall proxy within the loess column, but it might reflect length of surface exposure for individual paleosols better than MS does. This rule of thumb about intensity versus duration is, however, countermanded by studies that show that MS in soils continues to increase with time (Singer *et al.* 1992). Maher (1998) argued that, as soils weather and thicken, *cumulative*, i.e., profile-weighted, susceptibility might increase but the *maximum* susceptibility of any one horizon might attain an equilibrium value with climate.

[13]C in Soils and Paleosols

Like ^{14}C, ^{13}C is a carbon isotope that has been used to great advantage in soil geomorphology and paleoenvironmental studies. And also like ^{14}C, ^{13}C accumulates in biomass and soil carbonate. Unlike ^{14}C, ^{13}C is a stable isotope.

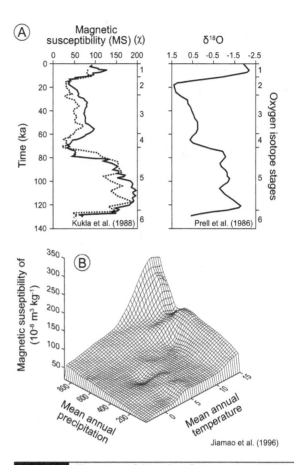

Fig. 16.18 Illustration of the utility of magnetic susceptibility in buried soils and sediments as a paleoenvironmental indicator. (A) Comparison of the magnetic susceptibility of the loess at Xifeng and Luochuan, China, and the marine isotope record. (B) Magnetic susceptibility of modern soils in the loess plateau of China, as it relates to contemporary values of precipitation and temperature.

Natural isotopic fractionations in plant tissue and soil carbonate are very small but they are measurable. The ratio of ^{13}C to ^{12}C in a sample is expressed relative to that of an internationally recognized standard carbonate – a Cretaceous belemnite formation at Peedee, South Carolina – by dividing the difference between the sample's ^{13}C/^{12}C ratio and the standard's ratio by the standard's ratio. This value is then multiplied by 1,000, and the result is reported in parts per thousand (‰):

$$\delta^{13}C\ (\text{‰}) = [^{13}C/^{12}C\ (\text{sample}) - {}^{13}C/^{12}C\ (\text{standard})]/[^{13}C/^{12}C\ (\text{standard})] \times 1,000$$

For most plants, the atmosphere, and pedogenic carbonate, the $\delta^{13}C$ value is negative, indicating that they contain less of the ^{13}C isotope than does the standard (Troughton 1972; Fig. 16.19).

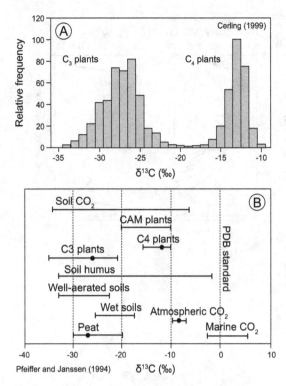

Material	Mean δ¹³C
C₄ soil carbonate	3
PDB δ¹³C standard	0
Marine HCO₃	-1
Speleothems	-9
C₃ soil carbonate	-11
Bone apatite and original carbonate	-12
Grasses in arid zone, sedges	-13
Straw, flax	-14
Marine organisms (organic)	-15
Freshwater plants (submerged)	-16
Succulents (cactus, pineapple, agave, yucca, etc.)	-17
Bone collagen (C₃ diet), wood cellulose	-20
Fossil wood, charcoal	-24
Recent wood charcoal	-25
Raw tree leaves, wheat, straw, etc.	-27

Fig. 16.19 Values of δ¹³C for some common C compounds and reservoirs.

¹³C in Soil Humus

To the extent that the δ¹³C values of humus and carbonates in soils (and hence, paleosols) reflect the δ¹³C values of vegetation growing in the soil (Tieszen *et al.* 1997; Fig. 16.20), these data represent a (paleo)biotic signature (Victoria *et al.* 1995, Boutton *et al.* 1998, Leavitt *et al.* 2007, Wittmer *et al.* 2010, Zani *et al.* 2012, Solis-Castillo *et al.* 2013). Organic matter derived from forest litter, for example, has a much more negative δ¹³C value than does humus from a short-grass prairie (Fig. 16.19). δ¹³C values can also be used to examine carbon dynamics and turnover (Balesdent *et al.*

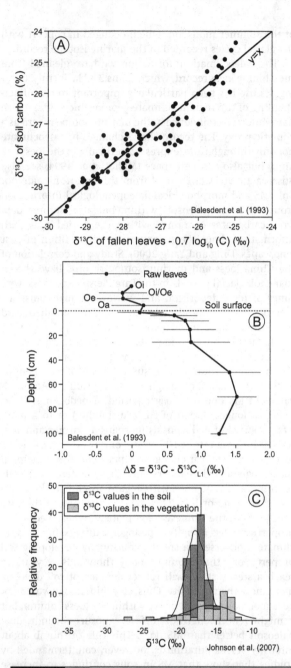

Fig. 16.20 Illustrations of the correspondence between contemporary vegetation and δ¹³C values. (A) Scatterplot showing the correlation between δ¹³C values of recently fallen leaves and the soil C for the same soil. (B) Mean isotopic (δ¹³C) enrichment or depletion, with depth, in some Alfisols, Inceptisols, and Spodosols in France, as compared to that of the O horizon and fresh leaves. (C) A comparison of isotopic compositions, as expressed by normal curves and relative frequencies, of vegetation and soil samples from a C4-dominated tallgrass prairie site in eastern Kansas, United States.

1987, Martin *et al.* 1990). It should be noted that the $\delta^{13}C$ signature in soils is easier to interpret for soil organic matter than it is for soil carbonate.

^{13}C values in soil organic matter may be good paleobotanic indicators because the ^{13}C values of plant materials vary considerably, and predictably (Fig. 16.19). Let us examine how and why.

Plants require CO_2 for photosynthesis and take it in through pores in their leaves called stomata. When plants open their stomata, however, they risk losing water vapor through those same vents. Therefore, some plants have adapted mechanisms to utilize CO_2 quite efficiently, given their particular environmental constraints. For example, under hot and dry environmental conditions the stomata close during the daytime to reduce the loss of water vapor, but this also results in a greatly diminished intake of CO_2.

Plants can be grouped on the basis of how they utilize CO_2 and their pathways of carbon fixation: C3, C4, and the CAM (Smith and Epstein 1971, O'Leary 1988, Ehleringer and Monson 1993, Ehleringer and Vogel 1993). Most humid climate plants use the Calvin-Benson (C3) cycle to fix CO_2 – it is the default pathway for trees, most shrubs, and herbs and many grasses. Indeed, most plants on Earth are C3 plants. The stereotypical photosynthetic plant is called a C3 plant because the first stable compound formed from CO_2 is a three-carbon compound at the beginning of the Calvin cycle. C3 plants are typical of humid climates, and some plants adapted to disturbance, including herbaceous annuals and some shrubs, are also C3 (Grimm 2001, Feggestad 2004). Values of $\delta^{13}C$ for C3 plants (and the humus they generate) range from –21‰ to –35‰, with a mean value between –26‰ and –27‰ (Deines 1980, Mikhailova *et al.* 2008, Stinchcomb *et al.* 2013; Fig. 16.19).

The C4 (*aka* Hatch-Slack) pathway is more CO_2 efficient than is the C3 pathway. Plants that use this pathway are known as C4 plants because the initial carboxylation reaction in photosynthesis produces a four-carbon compound. C4 plants fix CO_2 so efficiently that they do not need to have their stomata open as much as plants operating by other pathways, enabling them to be more water-efficient. The C4 pathway is mostly found among tropical grasses and some sedges and herbs growing in warm, sunny environments, because it allows for efficient C fixation in dry, warm environments. Feggestad *et al.* (2004) pointed out, however, that the C3-C4 mix of plants is not always a direct function of annual precipitation; C4 grass species dominate areas where summer moisture and/or nutrients are limiting, because they have higher water-use efficiencies. C3 plants often outcompete C4 plants in moist, colder, and less sunny environments. Although <3% of all plant species use the C4 photosynthetic pathway, these species are important in temperate prairies, tropical savannas, and arid grasslands. Values of $\delta^{13}C$ for C4 plants range from –10‰ to –16‰ with a mean value between –11‰ and –13‰ (Deines 1980, Cerling 1999, Johnson *et al.* 2007, Stinchcomb *et al.* 2013). Geoarchaeologists have measured the isotopic signature of bone collagen and other tissues to determine whether the principal diet of a human or other animal consisted primarily of C3 plants (rice, wheat, soybeans, potatoes) or C4 plants (corn, or corn-fed animals). Leavitt *et al.* (2007) assumed that a $\delta^{13}C$ value of –20‰ could be used as a midpoint between C3 and C4 carbon isotope compositions, i.e., it represents an equal mixture of C3 and C4 carbon, but for a variety of reasons outlined later, that value in soil organic matter does not necessarily mean that the soil's vegetation is an even mixture of C3 and C4 plants.

The third and least common pathway, CAM, is found in plants that live in dry, desertlike conditions, e.g., cacti and succulents, as well as some tropical succulents. The CAM name points to the fact that this pathway occurs mainly in succulent plants of the Crassulaceae and Cactaceae families, i.e., Crassulacean acid metabolism. CAM plants have adapted to the dry conditions by opening their stomata only at night, which is when they store new CO_2 in their tissues. During the day, when the stomata are closed (preventing unnecessary loss of water vapor), the CO_2 is removed from storage and enters into photosynthetic reactions, which are fueled by light energy from the sun. Values of $\delta^{13}C$ for CAM plants range from –10‰ to –20‰ (O'Leary 1988). Thus, their $\delta^{13}C$ values serve to distinguish them easily from C3 plants. Differentiating CAM from C4 plants is more difficult.

With time, the isotopic composition of SOC enters into rough equilibrium with that of the vegetation, as decomposition of plant matter adds C residues representative of the original flora (Boutton *et al.* 1998; Fig. 16.20A). Carbon residues enter the profile via litter decomposition and are transported into the mineral soil; root exudates and root decomposition at depth also contribute carbon. Because of this, it is possible to use the $\delta^{13}C$ values of a paleosol to assess the general paleovegetation assemblages present during its formation; we assume that the storehouse of organic matter in the paleosol *isotopically reflects* the vegetation that existed when the soil was forming (Guillet *et al.* 1988, Nordt *et al.* 1994, Fredlund and Tieszen 1997, Ehleringer and Cerling 2002, Feggestad *et al.* 2004; Fig. 16.21).

As discussed, most plants in humid and/or cool climates and most shade-tolerant plants follow the C3 pathway while most C4 plants are in warm, semiarid, or subhumid climates (Ehleringer and Monson 1993). A good rule of thumb for grassland soils is that the warmer the climate is, the more C4 plants are likely to be there, with the temperature optimum for C4 grass being ca. 24–40°C (Leavitt *et al.* 2007). The $\delta^{13}C$ values derived from soil organic matter, along a climatic gradient, generally follow this rule (Koch 1998). For example, Quade and Cerling (1990) noted that, along an altitudinal transect in a desert climate near Yucca Mountain in Nevada, United States, the vegetation changed from CAM and C4 plants at low altitudes to C3 plants at higher altitudes. The $\delta^{13}C$ values of pedogenic carbonates within the soils varied as well, from about

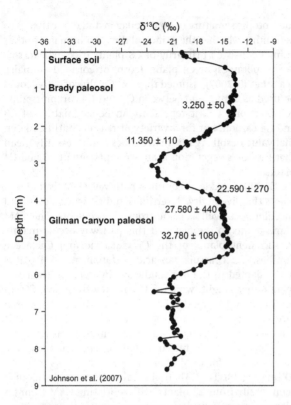

Fig. 16.21 Isotopic compositions of the modern (but welded) soil, loess, and the Gilman Canyon (MIS 3) paleosol, at a prairie site in central Kansas. ¹⁴C MRT ages on soil organic matter (reported in thousands of years) are also provided.

zero in the creosote bush–desert holly zone at the base of the mountain, to ~ –7‰ in the pinyon-juniper shrubland upslope, and to –9‰ in the ponderosa pine forests still farther upslope. It should also be noted that, within soils, there is a tendency for the $\delta^{13}C$ value to vary slightly with depth (Fig. 16.20B).

Finally, we point out several complicating factors that can influence $\delta^{13}C$ values of soil organic carbon and that must be considered when evaluating and interpreting $\delta^{13}C$ values in soils. First, the burning of fossil fuels, which are preferentially depleted in ¹³C, has decreased atmospheric $\delta^{13}C$ values by –1.3‰ in the last 250 years (Ehleringer et al. 2000). This phenomenon is called the Suess effect. Second, a shift in $\delta^{13}C$ values may reflect a change in litter input values, i.e., a shift from C3 to C4 plants, which will lead to lighter (less negative) ¹³C values (Ehleringer and Cerling 2002). Third, the components of plant tissues, e.g., lignin and cellulose, have somewhat different $\delta^{13}C$ values to begin with, and they may decay at different rates (Balesdent et al. 1993, Ehleringer et al. 2000, Wynn and Bird 2007). Fourth, microbial decomposition results in further isotope fractionation, as plant organic matter is gradually replaced by less negative ¹³C microbial carbon (Nadelhoffer and Fry

1988, Balesdent et al. 1993). Fifth, pedoturbation can mix older, heavier (less negative) carbon at depth with younger, lighter carbon at the surface (Nadelhoffer and Fry 1988). In the absence of vegetation change, it is likely that a combination of these effects, especially increased microbial carbon and the Seuss effect, work in tandem to change $\delta^{13}C$ values of SOC (Ehleringer et al. 2000).

The work of Johnson et al. (2007) on the Great Plains, United States, provides an excellent example of using isotopic signatures in soils and sediments to interpret past environments. They examined the $\delta^{13}C$ values from a core, which included the surface soil (a modern soil welded to a shallow paleosol [the Brady paleosol]), the loess below, the Gilman Canyon paleosol below that, and, finally, deeper loess (Fig. 16.21). Starting from the base of the core, the carbon in the loess of the Gilman Canyon Formation has a mean isotopic signal of ca. –21‰, reflecting a strong contribution from the C3 plants that characterized the central Great Plains during the last glaciation, when this loess was deposited. The Gilman Canyon paleosol, formed in the upper part of this loess, developed during the Farmdalian Interstade of the last glaciation (MIS 3) (Reed and Dreeszen 1965). It is a Mollisol and formed mainly under C4 grasses (isotopic values –15.7‰ to –13.8‰) that invaded this region as a result of warming temperatures. In the overlying Peoria loess, $\delta^{13}C$ values fall, because of climatic cooling that occurred at the end of the interstade; this loess (MIS 2 in age) was being deposited in a prevailing C3 plant environment (isotopic values –23.4‰ to 22.7‰). As the climate warmed at the end of MIS 2, a soil (the Brady paleosol) developed within the uppermost Peoria loess (Schultz and Stout 1945). This soil formed under a C4-dominated plant community (isotopic values –15‰ to –13.5‰) (Feggestad et al. 2004). The Brady paleosol was subsequently buried by a thin increment of Holocene loess, the thickness of which was insufficient to separate it from the overlying modern soil, resulting in a welded surface soil-paleosol complex. The $\delta^{13}C$ values above about 40 cm probably reflect the wheat (a C3 plant) that has been cultivated recently at the site. This example highlights the point that $\delta^{13}C$ values can be useful paleoenvironment indicators not only for buried paleosols but also for loess, which while it was accumulating was also under the influence of the local plant community, even if a paleosol did not develop in it.

Knowing the types of plants that typify the C3 and C4 pathways is also useful in conjunction with ¹⁴C analysis (Guillet et al. 1988). Most ¹⁴C laboratories routinely report the $\delta^{13}C$ value of samples, allowing the investigator the opportunity to know both sample age *and* likely floristic composition of the site at a particular period in the past. We caution that bulk soil organic carbon can only yield long-term, integrated $\delta^{13}C$ values, which may be difficult to interpret in polygenetic soils.

Like cutans, the $\delta^{13}C$ method is particularly advantageous in areas where environmental shifts force a change in biota that have one photosynthetic pathway to biota that have another. For example, glacial-interglacial climatic shifts on Great Plains grasslands may have forced a biotic change from C3-dominated ecosystems to C4 grasses (Mandel 2008). This method is particularly useful in ecotonal areas, where small paleoenvironmental changes are easily registered, e.g., where grassland has invaded forest, or vice versa (Steuter *et al.* 1990, Ambrose and Sikes 1991). Paleosols formed under one botanical association often become buried when the climate changes, because periods of climatic change are also periods of geomorphic instability. The end result might be a buried soil with carbon isotope ratios indicative of the preburial paleoenvironment, and the modern soil (or a soil higher in the stratigraphic column) that is reflective of the environment at a later time (Fig. 16.21). Such paleosols would carry carbon isotopes containing information about the paleovegetation that occupied the site while they formed and, by proxy, information about paleoclimate (Khokhlova *et al.* 2001).

Applications of $\delta^{13}C$ values of organic matter in surface soils are also interesting. Because the isotopic composition of soil organic matter remains independent of the vegetative cover for some time after a vegetative change, the method can be used to track the extent of recent, or even ongoing, floristic shifts (Martin *et al.* 1990). Steuter *et al.* (1990) did precisely this on a landscape where C3 trees were presumably invading a C4 grassland. They were able to show that, in sites that had been invaded by forest in the recent past, the isotopic composition of roots was significantly more negative than that of the soil organic matter. On sites that had been invaded less recently, the isotopic difference was less.

13C in Soil Carbonate

$\delta^{13}C$ values on soil carbonate can also provide a useful paleoenvironmental signature – one that is especially applicable in dry climates. The $\delta^{13}C$ values of pedogenic carbonate are mainly controlled by the $\delta^{13}C$ values of soil CO_2, which are a function of mixing from two isotopically distinct sources: atmospheric CO_2 that mixes and diffuses into the soil and biologically respired CO_2. The $\delta^{13}C$ value of atmospheric CO_2 is usually around –6‰ to –11‰. Soil CO_2 derived from plant root respiration and decay of organic matter is usually correlated to the type and density of the overlying flora, and the fractions of C3 and C4 present in the flora (Amundson *et al.* 1988, Koch 1998, Ding and Yang 2000, Achyuthan *et al.* 2007). It is also, however, correlated to soil respiration rates, which are low in deserts and cold climates, in which case much of the soil CO_2 could have originated in the atmosphere and is therefore less reflective of the flora (Dörr and Münnich 1986). In short, at high respiration rates, the $\delta^{13}C$ value of soil CO_2 will reflect that of the vegetation, whereas at lower rates, it will be more influenced by atmospheric inputs. Therefore, the $\delta^{13}C$ values of soil carbonate can range from that of the atmosphere to that of the existing, much more isotopically negative, vegetation (Emrich *et al.* 1970; Fig. 16.22). In humid and

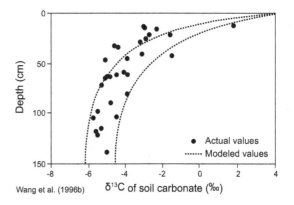

Wang et al. (1996b) $\delta^{13}C$ of soil carbonate (‰)

Fig. 16.22 Variation in $\delta^{13}C$ values for soil carbonate with depth, in some desert soils in southern California. Predicted values, based on a diffusion model, are also shown.

subhumid areas, where C3 and C4 plants dominate, soil respiration is high and thus little atmospheric CO_2 mixes with that of the soil; here, the isotopic composition of soil carbonate (if any actually forms) is more reflective of the flora (Cerling 1984). In arid climates, however, where soil respiration rates are low, the $\delta^{13}C$ values of soil carbonate may be related to the density of vegetation and, therefore, rates of soil respiration (Amundson *et al.* 1988, Wang *et al.* 1996b). Along these lines, Cerling *et al.* (1989) pointed out that the relationship between the isotopic composition of soil carbonate and biota is less useful in true desert soils; they have respiration rates that are so low that diffusion of atmospheric CO_2 into the soil raises the isotopic values above what would be expected from the flora alone.

As Kelly *et al.* (1991) pointed out, interpreting results from isotopic studies of soils is a difficult task, because the circumstance outlined previously presents a potential problem in the use of soil carbonate isotopes[1] for paleoenvironmental reconstruction. Nonetheless, data from Wang *et al.* (1996b) demonstrate that, despite the potential problems in the method, the isotopic composition of soil carbonates can be used to provide information on paleoenvironments

---

[1] This discussion focuses on the $\delta^{13}C$ values of soil organic matter and carbonate as paleoenvironmental proxies. The reader should also be aware of ^{18}O isotopes, which in carbonates are primarily controlled by the $\delta^{18}O$ values of soil water. On a general level, therefore, the $\delta^{18}O$ values of precipitation are usually related to those of soil carbonate (Cerling 1984), making this isotope a potentially usable indicator of paleoclimate (Sheldon and Tabor 2009). In calcite, ^{13}C and ^{18}O isotopes may be clumped together with ^{16}O in a small fraction of CO_3^{2-} units of the mineral. The degree of this clumping can be related to the temperature at which the mineral precipitated from a homogeneous solution (Eiler 2007). Clumping can be assessed by mass spectrometry of the CO_2 molecules that are released by thermal decomposition of the calcite and that have a molecular mass of 47 (13 + 18 + 16). The correlation of clumped isotope values with climatic parameters may provide insight about the temperature and paleoclimate in which pedogenic carbonates formed (Quade *et al.* 2013).

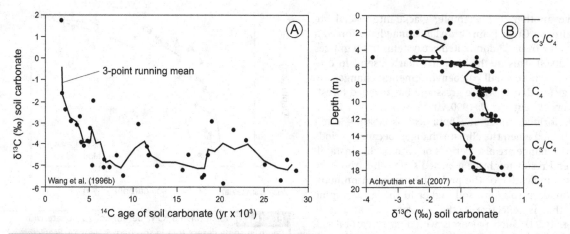

Fig. 16.23 Use of $\delta^{13}C$ values on soil carbonate as a paleoenvironmental proxy. (A) $\delta^{13}C$ values and ^{14}C ages of soil carbonates for some desert soils in southern California. Note the shift to less negative $\delta^{13}C$ values shortly after the beginning of the Holocene, i.e., the last 10 ka. (B) $\delta^{13}C$ values with depth in an aggrading, almost 400,000 year old dune profile in the Thar Desert, India. Note the changing dominance of C3 and C4 plants during this period.

(Figs. 16.23, 16.24). They documented a change in isotopic composition of soil carbonates after about 9 ka, which suggests either an increase in C4 and CAM plants or a decrease in plant density. Either way, the trend is toward increasing climatic dryness. Cerling (1984) developed equations that can be used to determine the fraction of C4 plants in the local flora, using the isotopic composition of soil carbonate (Fig. 16.24). He assumed that the $\delta^{13}C$ values for the atmosphere and soil air (the sources of which are plant root respiration and decaying litter) were –6‰ and –27‰, respectively, and was then able to calculate the steady-state isotopic composition of soil carbonate under various C4-C3 floral mixes.

The $\delta^{13}C$ values of pedogenic carbonate precipitated in equilibrium with soil CO_2 are normally higher (less negative) by 14–17‰ relative to CO_2 respired from plant roots and soil microbes (Fig. 16.25). That is, isotopic values of *soil organic matter* are far more negative because of biotic fractionation than are the values of *pedogenic carbonate*, which are much less fractionated. Thus, the $\delta^{13}C$ value for soil *organic matter* is primarily a reflection of the aboveground flora, whereas *carbonate* values are shifted toward heavier (less negative) values (Cerling 1984, Quade *et al.* 1989). This shift will increase as the environment becomes drier and, to a lesser extent, colder, largely as a result of the increased effects of atmospheric CO_2. Under conditions of moisture stress, therefore, $\delta^{13}C$ values of soil carbonate of –8‰ and +4‰ indicate nearly pure C3 and C4 ecosystems, respectively (Ding and Yang 2000).

Our discussion has focused on secondary pedogenic carbonates, which should not be confused with carbonate inherited from parent material; the $\delta^{13}C$ values of soil carbonate can actually help distinguish between the two. Carbonate derived from dissolution of marine limestone

has near-zero to slightly positive $\delta^{13}C$ values, whereas purely pedogenic carbonate has $\delta^{13}C$ values that are much more negative (Magaritz and Amiel 1980, Rabenhorst *et al.* 1984, Quade and Cerling 1990; Fig. 16.19). Thus, in soils where some of the pedogenic carbonate has resulted from the dissolution of limestone, it should be evident because the caliche will have an abnormally high $\delta^{13}C$ value. However, it is also likely that, once the marine carbonate (limestone) dissolves, it equilibrates, partially or wholly, with soil-respired CO_2. Pedogenic carbonate formed by reprecipitation of dissolved marine carbonates may therefore have the respiration isotopic signature and not the marine carbonate signature. Interpretations must be based on knowledge of the diffusion coefficients for $^{12}CO_2$ and $^{13}CO_2$ (Cerling *et al.* 1989).

Interpretations of isotopic values for soil carbonate have a distinct advantage over those for organic matter, since carbonates can persist in an unaltered state despite burial or slight changes in climate, whereas organic matter is quickly oxidized and lost (Amundson *et al.* 1989b). Carbonate is readily preserved in dry paleosols, giving the $\delta^{13}C$ value of carbonates from paleosols particular application to paleoenvironmental study (Cerling and Hay 1986, Amundson *et al.* 1989b, Jolley *et al.* 1998). However, several potential complications can arise. In soils with shallow Bk horizons, much of the soil CO_2 has probably been derived from the atmosphere, which has a less negative $\delta^{13}C$ value than CO_2 derived from the decay of organic matter, regardless of the type of plant (Fig. 16.19). Soil carbonate at depth will have little of this influence, with most of its carbon having been derived from plant respiration. Also problematic is "overprinting," or the imposition of later climatic information contained in soil carbonate onto information from much older carbonate-generating climatic cycles

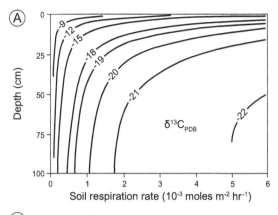

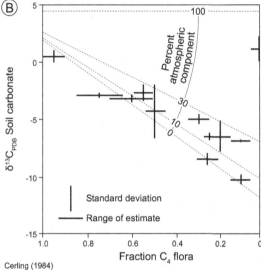

Cerling (1984)

Fig. 16.24 Isotopic composition of soil carbonate and paleoenvironments. (A) Calculated steady-state isotopic composition of soil CO₂ for different soil respiration rates, assuming that the δ¹³C values for the atmosphere and soil air are −6‰ and −27‰, respectively. (B) Isotopic composition of soil carbonate compared to the fraction of C4 plants in the local flora.

(Cerling 1984). Polygenesis is common in desert environments; it remains to be seen whether isotopic analysis of soil carbonate will ultimately help to resolve these climatic and floristic changes, or whether the method will be defeated by them.

Opal Phytoliths

Phytoliths are microscopic mineral deposits, usually silica-rich, formed in the epidermal cells of plants, especially grasses (Blinnikov 2005, Kamenik *et al.* 2013). They range in size from 2 to 1,000 μm; most are between 20 and 200 μm in length. In transmitted light, phytoliths are usually translucent, yellowish brown in color. Phytoliths are formed when plants uptake monosilicic acid through their roots, and the silica precipitates within plant cells in leaves (and in the roots of some plants) during transpiration (Cornelis *et al.* 2014). They are, essentially, copies of plant cell bodies. After the plant dies, the phytoliths are liberated by normal decay processes and enter the soil with plant litter. Because phytoliths are fairly resistant to decomposition in all but the most extreme pH conditions, they accumulate in soils (Smithson 1958, Baker 1959, Fredlund *et al.* 1985, Zhao and Pearsall 1998), mostly in A horizons where roots are most dense (Beavers and Stephen 1958) and within the coarse silt fraction. Opal (silica-rich) phytoliths have been the subject of most study, although calcium oxalate phytoliths also occur, especially in cacti (Jones and Handreck 1967, Franceschi and Horner 1980).

Like pollen grains (discussed later), opal phytoliths have unique morphologies that can be identified to plant genus or even species, providing important paleobotanical and paleoenvironmental proxy data (Fredlund and Tieszen 1997, Delhon *et al.* 2003, Calegari *et al.* 2013, Cordova 2013; Fig. 16.26). As with pollen, phytolith studies require development of a modern phytolith (specimen) database for identification purposes, before these studies can be successfully implemented (Runge 1999, Peto 2013). Phytoliths are also very important to many archaeological investigations. For example, phytoliths of maize, wild rice, and other plants can be preserved in food residues on ceramic

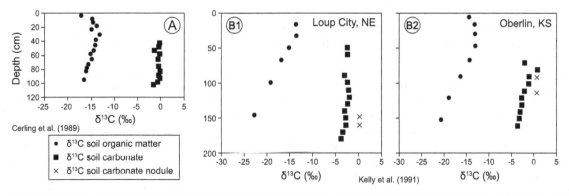

Fig. 16.25 Depth plots of the isotopic composition of soil organic matter and pedogenic carbonate. (A) For an Argiustoll in Kansas. (B) For some soils in Nebraska (B1) and Kansas (B2).

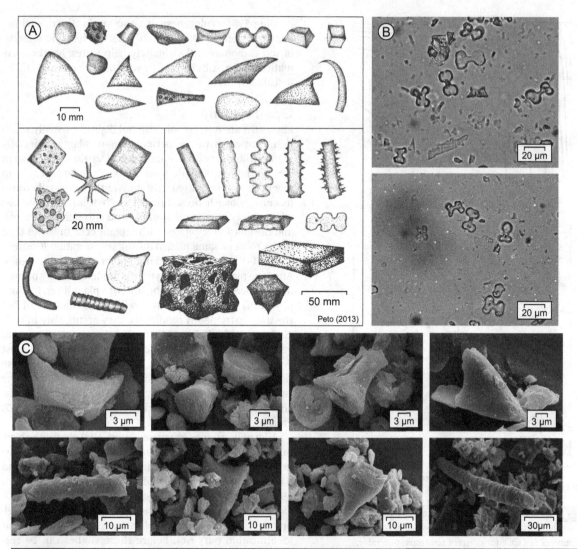

Fig. 16.26 Examples of opal phytoliths. (A) Drawings, (B) light micrographs, and (C) SEM photomicrographs of some common phytolith morphotypes recovered from soils. Drawings and photos in (A) and (B) by Á. Pető, S. Baklanov, and A. Golyeva. Photos in (C) by T. Messner.

pot fragments in archaeological contexts. Grass forms have been recovered from the cusps of mammalian teeth, providing important information of the diets of humans and animals, respectively (Pearsall 2000, Piperno 2006). In essence, examination of soil phytoliths provides a signature of the flora that once inhabited the soil.

Phytoliths have an advantage over pollen because they do not require lakes or swampy sediments for preservation – some landscapes, such as grassland and desert habitats, have few such wetland deposits (Lewis 1981, Piperno and Becker 1996). Phytoliths are also more likely than pollen grains to represent the local vegetation, as long-distance wind transport of pollen grains is more common than is transport of phytoliths (Fredlund and Tieszen

1997). Phytoliths may contain traces of organic matter that is potentially datable by radiocarbon techniques, although the analysis requires a considerable amount of sample pretreatment (Santos et al. 2010). However, phytoliths rarely occupy more than about 3% of a horizon by volume, making isolation of individual phytoliths a laborious and painstaking task (Wilding 1967).

Fairly widespread production coupled with resistance to decomposition gives phytoliths good potential as paleoenvironmental indicators (Rovner 1971, 1988, Stinchcomb et al. 2013). In the first instance, their abundance can be an indication of long-term stability or change in the dominance of grasses versus forest on a site (Beavers and Stephen 1958, Witty and Knox 1964, Reider et al. 1988, Fuller and Anderson

1993). Additionally, phytoliths provide information on the *types* of grasses that have been at the site, as they can be taxonomically identified to various grass genera or types. Specially, phyoliths can be distinguished into the following classes: Festucoid, which are C3 (cool-season) grasses, and two types of C4 (warm-season) grasses: the Chloridoid (short grasses of the western Plains) and the Panicoid (tall grasses of the eastern Plains and forests). For some, they can be further identified to genus and species (Twiss *et al.* 1969, Fredlund and Tieszen 1997, Cordova 2013). Typically, phytolith and stable carbon isotope analyses are performed on buried soils in grassland locales. The former type of analyses (see earlier discussion) can identify shifts in the dominance of C3 versus C4 vegetation over time, whereas phytolith data provide more accurate information about the taxonomic composition of the paleoflora. Phytolith analysis can help identify whether the C3 vegetation was primarily composed of grasses, forbs or trees, and shrubs and can distinguish between the C4 short and tall grasses (Bozarth 1995, Fredlund and Tieszen 1997). Obviously, times of greater precipitation in the past would be inferred from higher amounts of tree, shrub, and Panicoid (tall grass) phytoliths, which would impact soil formation processes. Besides grasses (Poaceae or Gramineae), Mulholland and Rapp (1992) reported that the following families are consistent accumulators of opal phytoliths: Cyperaceae (sedges), Ulmaceae (elm trees), Fabaceae or Leguminosae (beans and peas), Cucurbitaceae (squash), and Asteraceae or Compositae (asters and sunflowers).

Content of opal phytoliths can also be used in the identification of buried paleosols and can even help locate the very top of a buried soil, because they are most concentrated in the A horizon (Beavers and Stephen 1958, Dormaar and Lutwick 1969). They can also be used to determine whether a horizon rich in organic matter is a buried A horizon or simply a Bh horizon (Anderson *et al.* 1979). An Ab horizon, especially if it had been formed under grass, would have high numbers of phytoliths, with decreases in phytolith occurrence below. Alternatively, a Bh horizon would contain fewer phytoliths than the horizons above it, because root numbers (and hence, phytolith production) would have been higher in overlying horizons and because phytoliths are not normally translocated in soils.

Pollen and Macrofossils

Like opal phytoliths, pollen (microscopic grains released from the male part of plants) and plant macrofossils (seeds, fragments of stems and roots, and leaf buds or scales) have great potential to assist in the interpretation of paleoenvironments. Pollen is particularly resistant to decomposition in acidic environments such as peat bogs and may sometimes even be preserved in mineral soils (Bryant and Holloway 1985, Faegri and Iverson 1989). Like other organic compounds, pollen can be dated using radiocarbon methods (Zhou *et al.* 1997, Gavin and Brubaker 1999). *Palynology* is the study of pollen.

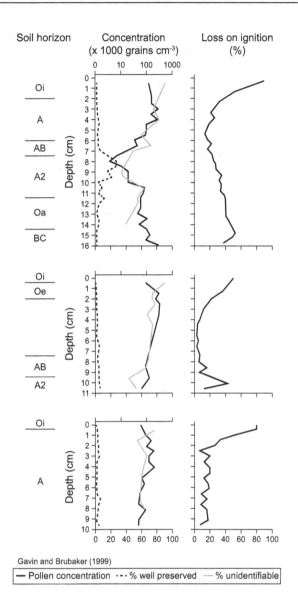

Gavin and Brubaker (1999)

— Pollen concentration ··· % well preserved — % unidentifiable

Fig. 16.27 Pollen concentrations and preservation in two mineral soil profiles on Meadow Ridge, in the Olympic Mountains, Washington, showing that even when organic matter concentrations are high, the proportion of identifiable pollen grains in an oxidized soil may be low.

Preferential preservation of pollen in bogs (and in water-saturated lacustrine and palustrine sediments) makes pollen analysis perhaps more a sedimentological than pedological study. The low pH and higher water level in bogs inhibit fungal and bacterial destruction of pollen grains. The decay of pollen grains is greatest in oxidized, neutral to high pH mineral soils (Hall 1981, Bryant and Hall 1993; Fig. 16.27). Consequently, only occasionally have pollen grains been extracted from upland mineral soils, and these studies have shown mixed but potentially promising results (Dimbleby 1957, Havinga 1974, Schaetzl and Johnson 1983, Gavin and Brubaker 1999). Pollen analysis of mineral soils shows more tenable results when it is performed on buried soils, where accurate depth functions of pollen spectra are less important than simple acquisition of a snapshot of the vegetation at the time of burial, e.g., Caseldine (1984). Nonetheless, palynological research has been successfully applied to thick O horizons in some well-drained mineral soils (Dijkstra and VanMourik 1996).

The initial stratification of sedimentary deposits in mineral soils is soon lost through pedoturbation and translocation processes (Ray 1959, Walch et al. 1970, Kelso 1974). Pollen grains originate at the soil surface and have the potential for translocation, which is due to their small sizes (20–120 μm diameter) (Havinga 1974, Dimbleby 1985, Russell 1993). Only in soils with pH values <5.5 is significant preservation of pollen found, and even then one must be wary of preferential preservation, i.e., taphonomy, that will distort the palynological signal in peat bog levels above the water table.

Besides site conditions, the chemical structure of the pollen grains determines preservation potential. Specifically, the amount of sporopollenin, a tough protein polymer, in the exine (external structure) of pollen grains largely controls this potential, favoring certain types over others (Faegri and Iverson 1989). For example, pollen of vesiculate conifers (pine, spruce, and fir) and Cheno-Ams (weedy plants of the Chenopodiaceae and Amaranthaceae families) contains the highest amounts of sporopollenin and hence are the last to deteriorate in soils, whereas the most fragile are pollen grains of *Populus* (poplar/aspen) and *Cupressaceae* (juniper or cedar); these grains are typically not preserved (Hall 1981).

Faunal *macrofossils* in mineral soils are also useful as paleoenvironmental indicators. For example, sponge spicules are animal microfossils preserved in peat bogs, and taxonomic distinctions can indicate changes in water levels and hence be used as a proxy for paleoprecipitation (Amesbury et al. 2012).

The primary way that pollen studies contribute to soil genesis/geomorphology is by providing paleoecological information for segments of the geologic past. In a typical palynological study, a core of sediment is removed from stratified, acidic peat or sediment. Radiocarbon ages of in situ organic materials from the core provide a chronostratigraphic framework for the pollen and/or macrofossils extracted from it. Charcoal and plant macrofossils (wood, seeds, etc.) are especially useful for dating, because these relatively larger fragments are typically deposited in situ. In their absence, pollen grains can be dated, but there is a greater likelihood of contamination as a pollen concentrate sample is submitted for dating that will include some microscopic algae and other materials (Grimm et al. 2009).

The palynologist operates on the assumption that the pollen and macrofossils for each layer within the core are representative of the regional or local vegetation at that moment in the past; the palynologist statistically determines the relationship between modern pollen rain and modern plant communities and then uses those data to backcast pollen-flora relationships (Liu and Lam 1985, Overpeck et al. 1985). Obviously, this assumption is violated somewhat by differential preservation and the characteristic that some plants, particularly conifer trees, produce pollen in amounts that far exceed their relative abundance in the local flora (Hall 1981). Hence, cutoff values have been established to indicate the local presence of trees, on the basis of modern pollen deposition studies. For example, a minimum *Picea* pollen value of 20% is required to indicate the local presence of spruce, because of its long-distance transport potential (Ritchie 1987). The other assumption in bog-based pollen research is that the pollen in each layer represents that layer only, i.e., no vertical translocation and little interlayer mixing have occurred. This is a reasonable assumption in most acidic, bog environments.

Pollen rains from the air onto lakes or bogs and settles there, accumulating in the sediment that underlies the water. Thus, it has derived from the regional pollen rain and represents the vegetation from within a few kilometers of the site. Plant macrofossils are larger and, thus, enter the site mainly via runoff from the soil surface or are carried only short distances by small tributary streams. As a result, the macrofossil signature represents a much more localized picture of the paleovegetation (Lacroix et al. 2011, Galka and Sznel 2013, Jimenez-Moreno and Anderson 2013).

Most pollen grains are identified to the genus level, with a limited number to species (or type), and a few others only to the family level, providing a coarser taxonomic precision than typically provided by phytolith and plant macrofossil analyses. The morphological characters used to identify them are based on the sculpturing patterns of the exine, the number and type of apertures (openings), and the size and shape of the grains (Faegri and Iverson 1989). A series of different acid and base treatments, outlined by Faegri and Iverson (1989), are used to remove most of the sediment and wood fragments and concentrate the fossil pollen grains, which are usually identified and counted under 400X magnification.

The final step in the process is the development of a depth plot – a pollen (or macrofossil) diagram. These types of diagrams often provide a reasonably clear picture of the temporal changes in vegetation that have occurred at a site (Maher 1972). Vegetation changes often show clearly in such a profile, as pollen spectra change with depth, which is a proxy for time. And, as is typically done, ^{14}C ages on organic materials from levels within the core (and other ages extrapolated between dated layers) place these paleobotanical changes within a chronologic context. The data can then be correlated to climate, prehistoric human activities, natural plant succession, soil development, or any of a number of other factors (Baker et al. 1989). Figure 16.28

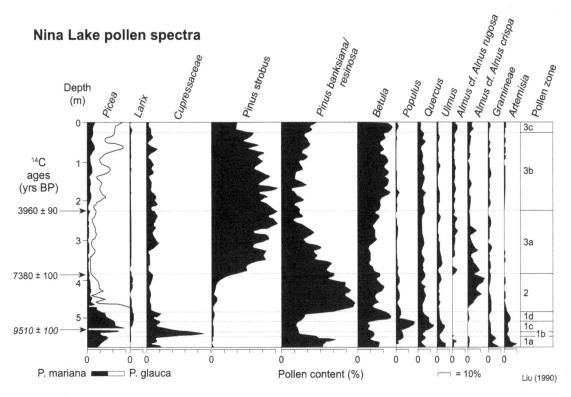

Fig. 16.28 Pollen percentage diagram for Nina Lake in central Ontario, just north of Lake Huron. Not all the pollen spectra originally reported are shown.

provides an example of a typical pollen diagram – this one from Ontario. The diagram, with its pollen assemblage biozones, illustrates the climate-induced changes in Holocene vegetation rather nicely – from a spruce-dominated boreal forest assemblage to a jack pine–red pine community to the current white pine–white birch forest. Similar examples abound in the literature (e.g., Watts and Wright 1966, Davis 1969, Webb 1974, Davis *et al.* 1975, Birks 1976, Bartlein *et al.* 1984, Tonkov and Marinova 2005, Grimley *et al.* 2009, Schaetzl *et al.* 2013).

Chapter 17

Conclusions

The Importance of Soils

With the exception of ice-covered surfaces and areas of bare bedrock, soils of one kind or another exist on almost every part of Earth's landsurface. The wide variety and range of expression that soils have are truly remarkable, especially when one considers that this plethora of colors, mineral assemblages, and horizon types and sequences is due to the interaction of only a few (five) soil-forming factors. In mathematics, 5 is a small number; 5^5, which (minimally) reflects the complex interactions among those five factors, is considerably larger (3,125). So it is in soils, as those five soil-forming factors can team together in myriad ways to form a world of soils that is complex, spatially diverse, and manifaceted. Unraveling and explaining that world, or at least some of the better-understood parts of it, have been our goals. Indeed, in the context of its complexity, when we think about it, it should be clear that we still know relatively little about soil and how it functions as a natural system. Yaalon (2000) paraphrased Leonardo da Vinci as stating, "Why do we know more about distant celestial objects than we do about the ground beneath our feet?"

Almost all of our food and sustenance are from the soil. As the saying goes, "If you eat, you are involved in agriculture," and, we would add, you are also dependent upon the soil. Most of the oxygen that we breathe originates from plants rooted in the soil. Soil is also one of the best natural filters we have. What would our world be like without soil?

The name Adam is from the Hebrew *adama*, which means earth, or soil. Adam's name, in the Bible's book of Genesis, is meant to capture humankind's intimate link with the soil, to which we are tied while we live, and to which we return upon our death (Hillel 1991). Remember Francis Hole's expression, TNS? *Temporarily Not Soil* – that is what we are. And what all terrestrial life on this planet is. Clearly, soils have been significant to humankind for as long as we have existed and will be here long after we are gone. The importance of soils cannot be overestimated. Their value cannot be overstated.

And despite this, soils can be treated with indifference, disdain, or lack of respect (Hillel 1991). We call it "dirt." Our clothes get "soiled." Our reservoirs become silted in. Paving over prime agricultural land is sometimes accepted in the name of short-term economic "progress" without a clear appreciation for the long-term consequences. Indeed, and perhaps instead, the Earth needs people who really *care* about soils and see soils as *essential* to our existence. The Earth needs people who consider these natural systems *fascinating* and their colors *beautiful*, who like to smell and taste and peer at soils, who wonder what is under their feet as they walk in the woods or even mow the front yard.

We hope that, after reading this book, you will add yourself to that list! We hope that this book has made you a land and soil lover (Hole 1997). We hope that you now think of soil as does William Logan (2007) – as the *ecstatic skin* of the Earth! We hope we have in some way enlightened you, as the giants of the field have enlightened us in even larger ways. We mention especially Charles Darwin, Vasili Dokuchaev, Curtis Marbut, Guy Smith, Hans Jenny, Roy Simonson, James Thorp, Kirk Bryan, John Frye, Gerry Richmond, Ray Daniels, Bob Ruhe, Roger Morrison, Lee Gile, Philippe Duchaufour, Francis Hole, Dan Yaalon, Geoff Humphreys, Don Johnson, and Pete Birkeland, among many, many others, far too numerous to mention here.

With few exceptions, most soils have horizons. The near-omnipresence of genetic soil horizons immediately below the surface implies that soils represent a degree of stability and a certain amount of equifinality. To be sure, some types of horizons form rather quickly, but in most cases the typical soil, with its distinct O, A, and B horizons, should send the signal that pedogenesis has been ongoing and that a soil has developed because of its relatively stable location, from a geomorphic standpoint. The morphological characteristics of the soil and the relationship of these characteristics to nearby soils provide additional evidence that each soil evolves on, within, and concurrently with its landscape. That is a major point of this book: Soils form on and along with landscapes, and it is best to study the two systems together. That is, indeed, the essence of soil geomorphology, which forms the core of this book. Soil genesis can only be explained if we know about the landscape upon which the soil is forming. And each landscape has its own unique geography, pointing to the undeniable conclusion that knowing the *geography* of soils is vital to understanding the *essence* of the physical landscape.

What of the Future?

Where might the discipline of soil geomorphology and the study of soil genesis go from here? What are the challenges and wherein lie the greatest potentials for advancing our understanding of the soil geomorphic system?

First, we emphasize dating. Establishing the time$_{zero}$ for soils is essential to understanding many aspects of their genesis, including the rates of many pedogenic processes. Knowing time$_{zero}$ with some degree of accuracy is also a highly useful piece of information from a geomorphic standpoint. Much progress has recently been made in the arena of surface and sediment dating, and vast potentials remain.

Mapping: Soils are so critically important that we must make better soil maps at scales that match our needs to understand and predict ecosystem processes or the impacts of future soil management. Although soil map coverages exist for all of the world, they are at widely varying levels of accuracy and scales. Soil maps are vital to environmental management and food production efforts, especially in this era of precision agriculture and global positioning systems (Pierzynski et al. 2000, Lal 2004, Foley et al. 2005, Brevik et al. 2003). For that reason, we must continue to improve soil maps, no matter how good we may think they are (Miller 2012). The digital revolution has led us to believe that if we have reasonable terrain and land cover data, we can map soils adequately, or even more accurately and efficiently, from a computer screen than in the field. To be sure, early efforts to this end have shown promise (McBratney et al. 2000, Hash and Noller 2009, Shi et al. 2009), but only after being informed by data from field mapping. We believe that remotely sensed data on terrain and land cover will never be more than a good start toward the making of a great soil map. Better base data will certainly help the mappers – who have their boots in the field and augers in the ground – make a better soil map. However, to map soils using only remotely sensed data, as is the goal of digital soil mapping, is akin to trying to check someone's tonsils with a camera. There is only so much you can do. Eventually, you must get inside the animal. To that end, we argue that there is no substitute for mapping done in the field. Soils exist *under*gound. We must be inside them to see them truly and, therefore, map them. This effort must not fail.

Biologists commonly lament the rapid rate at which the human species, by its actions, is wiping out other species across the globe. Extinctions occur daily. But what of soils? Unique soil types are destroyed as well (Amundson et al. 2003), and information from *them* is lost forever. Our position is not necessarily one of soil preservation, as one might advocate for for buffalo, red-legged frogs, or Peregrine falcons. Instead we must realize that there are many soils across the globe that have never been studied, some of which we do not even know *exist*! Consider this a call for soil scientists everywhere to beat the corners of the globe for unique and interesting soils to study. Often, the study of unique, almost freaklike soils can lead to major advances in pedogenic theory, e.g., Jenny et al. (1969). Travel, travel, travel, to the ends of the Earth, especially to and through the tropics, and glean all that you can! There is much to learn, opportunities to be had, and no time to lose.

The definition of soil should be broadened, certainly with regard to soil depth (Cremeens et al. 1994, Richter and Markewitz 1995, Johnson 1998) but also regarding the definition of soil horizons. Most, if not all, C horizons have been altered by pedogenic processes to the extent that they really no longer fit the concept of "unaltered parent material." Are stone lines soil horizons, because most are pedogenically formed? Stone lines in tropical soils are almost always pedogenic features, although the notion that they are purely geological entities continues to be promoted (Beinroth et al. 2011). To draw more practitioners into the fold we should rethink the definition of *soil* and broaden it (Richter and Yaalon 2012). More geologists and geomorphologists would and could use soil data and contribute to soil knowledge if pedogenic processes could be recognized where they operate at depth. The great area of overlap, and the arena of immense promise, lies in the deep subsoil, for here it is that pedologists, geomorphologists, hydrologists, geologists, and scientists of weathering can find common ground. Yes, it *is* a critical zone.

For decades, soil fauna and biomechanical soil processes have been largely ignored or relegated to a low spot on the list of important pedogenic processes (Johnson 2002). As a result, we have fallen far behind in our understanding of the interactions among soil biota and soil development (Johnson et al. 2005). It is time to study soils more holistically by connecting with our animal ecology colleagues and including faunal processes in every discussion of pedogenesis.

Let us always allow soil research and knowledge to drive taxonomy, not vice versa. Taxonomic systems should work *for* soil users, researchers, and theorists. The pendulum is swinging in this direction already, and this trend should continue. Taxonomy is necessary and it is good. But it is the *tail* on the horse; it should never *be* the horse.

Basic research in soil genesis and geomorphology must be encouraged, and funding for it must become a priority. In the modern world, applied soils research is laudable and necessary, but truly significant advances in applications that benefit humanity are also founded on concomitant significant advances in fundamental research. The long-term applications of basic research may not always be perceived at the time they accrue, particularly in a discipline like soils, in which the system and its complex linkages with other systems are still only minimally understood. The classic example of the utility of basic soils research are the soil geomorphology studies of the 1950s through the 1970s. Started by the USDA Soil Conservation Service as a way to facilitate soil mapping, they have become, in hindsight, so much more (Effland and Effland 1992, Holliday 1998, 2006). Many classic and highly influential papers have resulted from these studies, e.g., Ruhe (1956b, 1969), Daniels and Gamble (1967), Ruhe and Daniels (1968), Daniels *et al.* (1971a), Gile et al. (1965), Gile (1966, 1979), Parsons *et al.* (1970). And who would have guessed that these studies – these field studies – could have had such wide and long-lasting influence?

Contemplation and Reflection

Last, we encourage all scholars of soils and geomorphology to be firmly educated in the literature of soil science – large as it is (Hartemink 2012)! Be field scientists where necessary, but above all free and open-minded thinkers. "Thinking outside the box" has become such a common platitude that we tend to ignore it or downplay it as something we either all do anyway or a useless dead end. We disagree. This book was inspired by, and is dedicated to, freethinkers – Francis Hole, Donald Johnson, and Pete Birkeland – all of whom thought outside the box before it was popular. Many giants of soil science – Darwin especially, but also Dokuchaev – operated at least partly outside the box. Unfortunately, after the deaths of those who see the furthest it may be generations before they are recognized. We exhort all soil geomorphologists to be free thinkers, to think of new and exciting ways to study, explore, and understand the soil system. The rewards will be many – for you, for science, and for our progeny.

Soils are important. We have made that point. Soils are interesting, exciting, and complex. You know that by now. Soils are genetically intertwined with the landscapes on which they have formed. That is our mantra. The soil system is a highly complex one, and we are optimistic that our simple, ground-based treatise on soils has helped explain some of these complexities to you, the reader. To that end, we conclude this book as we started it, by reiterating the words of St. John Vianney, the curé of Ars, who said, "I have been privileged to give great gifts from my empty hands."

Thank you for your trust and attention.

References

The following list provides journal name abbreviations used in this References section.

Aeolian Research *AE*
American Journal of Science *AJS*
American Midlands Naturalist *AMN*
Annals of the Association of American Geographers *AAAG*
Arctic and Alpine Research *AAR*
Arctic, Antarctic and Alpine Research *AAAR*
Australian Journal of Botany *AJB*
Australian Journal of Soil Research *AJSR*
Canadian Journal of Earth Sciences *CJES*
Canadian Journal of Forest Research *CJFR*
Canadian Journal of Soil Science *CJSS*
Earth and Planetary Science Letters *EPSL*
Earth Surface Processes and Landforms *ESPL*
European Journal of Soil Science *EJSS*
Geological Society of America Bulletin *GSA Bull*
Geological Society of America Special Paper *GSA Spec. Paper*
Geophysical Research Letters *GRL*
Global Change Biology *GCB*
Journal of Archaeological Science *JAS*
Journal of Ecology *JE*
Journal of Quaternary Science *JQS*
Journal of Soil Science *JSS*
Journal of Soil Science and Plant Nutrition *JSSPN*
Palaeogeography Palaeoclimatology Palaeoecology *PPP*
Physical Geography *PG*
Proceedings of the National Academy of Sciences of the United States of America *PNAS*
Progress in Physical Geography *PPG*
Quaternary Geochronology *QG*
Quaternary International *QI*
Quaternary Research *QR*
Quaternary Science Reviews *QSR*
Soil Science *SS*
Soil Science Society of America Journal *SSSAJ*
Soil Science Society of America Proceedings *SSSAP*
Soil Science Society of America Special Publication *SSSASP*
Soil Survey Horizons *SSH*
Soviet Soil Science *SSS*
Transactions of the American Geophysical Union *TAGU*
United States Forest Service *USFS*
United States Geological Survey Professional Paper *USGS* Prof. Paper
Zeitschrift fur Geomorphologie *ZG*

Aandahl, AR. 1948. The characterization of slope positions and their influence on the total nitrogen content of a few virgin soils of western Iowa. *SSSAP* 13:449–454.

Abe, SS, Watanabe, Y, Onishi, T, Kotegawa, T and T Wakatsuki. 2011. Nutrient storage in termite (*Macrotermes bellicosus*) mounds and the implications for nutrient dynamics in a tropical savanna Ultisol. *JSSPN* 57:786–795.

Aber, JS. 1991. The glaciation of northeastern Kansas. *Boreas* 20:297–314.
   1999. *Pre-Illinoian glacial geomorphology and dynamics in the central United States, west of the Mississippi.* GSA Spec. Paper 337:113–119.

Abrahams, AD, Parsons, AJ, Cooke, RU and RW Reeves. 1984. Stone movement on hillslopes in the Mojave Desert, California: A 16-year record. *ESPL* 9:365–370.

Abrahams, AD, Soltyka, N, Parsons, AJ and PJ Hirsch. 1990. Fabric analysis of a desert debris slope: Bell Mountain, California. *J. Geology* 98:264–272.

Abrahams, PW and JA Parsons. 1996. Geophagy in the tropics: A literature review. *Geog. J.* 162:63–72.
   1997. Geophagy in the tropics: An appraisal of three geophagical materials. Env. Geochem. *Health* 19:19–22.

Abtahi, A. 1980. Soil genesis as affected by topography and time in highly calcareous parent materials under semiarid conditions in Iran. *SSSAJ* 44:329–336.

Achyuthan, H, Quade, J, Roe, L and C Placzek. 2007. Stable isotopic composition of pedogenic carbonates from the eastern margin of the Thar Desert, Rajasthan, India. *QI* 162/163:50–60.

Ackerman, IL, Teixeira, WG, Riha, SJ, Lehmann, J and ECM Fernandes. 2007. The impact of mound-building termites on surface soil properties in a secondary forest of Central Amazonia. *Appl. Soil Ecol.* 37:267–276.

Ackermann, O, Maeir, AM and HJ Bruins. 2004. Unique human-made catenary changes and their effect on soil and vegetation in the semi-arid Mediterranean zone: A case study on *Sarcopoterium spinosum* distribution near Tell es-Safi/Gath, Israel. *Catena* 57:309–330.

Acquaye, DK, Dowuona, GN, Mermut, AR and RJ St. Arnaud. 1992. Micromorphology and mineralogy of cracking soils from the Accra Plains of Ghana. *SSSAJ* 56:193–201.

Acton, DF. 1965. The relationship of pattern and gradient of slopes to soil type. *CJSS* 45:96–101.

Acworth, RI. 1987. The development of crystalline basement aquifers in a tropical environment. *Q. J. Eng. Geol.* 20:265–272.

Adams, KD, Locke, WW and R Rossi. 1992. Obsidian-hydration dating of fluvially reworked sediments in the West Yellowstone region, Montana. *QR* 38:180–195.

Adams, JS, Kraus, MJ and SL Wing. 2011. Evaluating the use of weathering indices for determining mean annual precipitation in the ancient stratigraphic record. *PPP* 309:358–366.

Adams, KD and SG Wesnousky. 1999. The Lake Lahontan highstand: Age, surficial characteristics, soil development, and regional shoreline correlation. *Geomorphology* 30:357–392.

Adams, WA. 1973. The effect of organic matter on the bulk and tree densities of some uncultivated podzolic soils. *JSS* 24:10–17.

Adelsberger, KA and JR Smith. 2009. Desert pavement development and landscape stability on the Eastern Libyan Plateau, Egypt. *Geomorphology* 107:178–194.

Adzmi, Y, Suhaimi, WC, Husni, MSA, Ghazali, HM, Amir, SK and I Baillie. 2010. Heterogeneity of soil morphology and hydrology on the 50 ha long-term ecological research plot at Pasoh, Peninsular Malaysia. *J. Trop. For. Sci.* 22:21–35.

Aerts, R. 1997. Climate, leaf litter chemistry and leaf litter decomposition in terrestrial ecosystems: A triangular relationship. *Oikos* 79:439–449.

Afanasiev, JN. 1927. *The Classification Problem in Russian Soil Science, Russian Pedological Invest.* #5. USSR Academy of Science, Moscow.

Afanas'yeva, YeA. 1966. Thick Chernozems under grass and tree coenoses. *SSS* 615–625.

Aghamiri, R and DW Schwartzman. 2002. Weathering rates of bedrock by lichens: A mini watershed study. *Chem. Geol.* 188:249–259.

Ahlbrandt, TS and SG Fryberger. 1980. Eolian Deposits in the Nebraska Sand Hills. USGS Prof. Paper 1120.

Ahmad, N. 1983. Vertisols. In: LP Wilding, NE Smeck and GF Hall (eds). *Pedogenesis and Soil Taxonomy*. Elsevier, New York. pp. 91–123.

Ahmad, N and RL Jones. 1969. Genesis, chemical properties and mineralogy of Caribbean grumusols. *SS* 107:166–174.

Ahmad, N, Jones, RL and AH Beavers. 1966. Genesis, mineralogy and related properties of West Indian soils. I. Bauxite soils of Jamaica. *SSSAP* 30:719–722.

Ahmadjian, V. 1993. *The Lichen Symbiosis*. Wiley, New York.

Ahn, JH and DR Peacor. 1987. Kaolinitization of biotite: TEM data and implications for an alteration mechanism. *Am. Mineral.* 72:353–356.

Ahnert, F. 1970. An approach towards a descriptive classification of slopes. *ZG* 9:71–84.

1987. Process-response models for denudation at different spatial scales. *Catena Suppl.* 10:31–50.

1996. *Introduction to Geomorphology*. Edward Arnold, London.

Ahr, SW, Nordt, LC and SG Driese. 2012. Assessing lithologic discontinuities and parent material uniformity within the Texas sandy mantle and implications for archaeological burial and preservation potential in upland settings. *QR* 78:60–71

Ahr, SW, Nordt, LC and SL Forman. 2013. Soil genesis, optical dating, and geoarchaeological evaluation of two upland Alfisol pedons within the Tertiary Gulf Coastal Plain. *Geoderma* 192:211–226.

Ahrens, RJ and RW Arnold. 2011. Soil taxonomy. In: PM Huang, Y Li and ME Sumner (eds). *Handbook of Soil Sciences*, 2nd ed. CRC Press, New York. pp. 31-1–31-13.

Aide, M, Pavich, Z, Lilly, ME, Thornton, R and W Kingery. 2004. Plinthite formation in the coastal plain region of Mississippi. *SS* 169:613–623.

Aide, M and C Smith-Aide. 2003. Assessing soil genesis by rare-earth elemental analysis. *SSSAJ* 67:1470–1476.

Aide, MT, Dunn, D, and G Stevens. 2006. Fragiudults genesis involving multiple parent materials in the eastern ozarks of Missouri. *SS* 171:483–491.

Aitken, MJ. 1994. Optical dating: A non-specialist review. *QG* 13:503–508.

Ajmone-Marsan, F, Pagliai, M, and R Pini. 1994. Identification and properties of fragipan soils in the Piemonte region of Italy. *SSSAJ* 58:891–900.

Ajmone-Marsan, F and J Torrent. 1989. Fragipan bonding by silica and iron oxides in a soil from northwestern Italy. *SSSAJ* 53:1140–1145.

Ajmone-Marsan, F, Barberis, E, and E Ardiuno. 1988. A soil chronosequence in northwestern Italy: Morphological, physical and chemical characteristics. *Geoderma* 42:51–64.

Akamigbo, F. 1984. The role of the nasute termites in the genesis and fertility of Nigerian soils. *Pédologie* 36:179–189.

Akcar, N, Deline, P, Ivy-Ochs, S, Alfimov, V, Hajdas, I, Kubik, PW, Christl, M and C Schluchter. 2012. The AD 1717 rock avalanche deposits in the upper Ferret Valley (Italy): A dating approach with cosmogenic Be-10. *JQS* 27:383–392.

Alban, DH. 1974. Soil Variation and Sampling Intensity under Red Pine and Aspen in Minnesota, USFS Res. Paper NC-106.

2000. Desert pavement characteristics on wadi terrace and alluvial fan surfaces: Wadi Al-Bih, U.A.E and Oman. *Geomorphology* 35:279–297.

Alban, DH, and E. Berry. 1994. Effects of earthworm invasion on morphology, carbon, and nitrogen of a forest soil. *Appl. Soil Ecol.* 1:246–249.

Albert, DA and BV Barnes. 1987. Effects of clearcutting on the vegetation and soil of a sugar maple-dominated ecosystem, western upper Michigan. *For. Ecol. Mgt.* 18:283–298.

Albrecht, CF, Joubert, JJ and PH de Rycke. 2001. Origin of the enigmatic, circular barren patches ("Fairy Rings") of the pro-Namib. *S. Afr. J. Sci.* 97:23–27.

Alcock, MR and AJ Morton. 1985. Nutrient content of throughfall and stem-flow in woodland recently established on heathland. *JE* 73:625–632.

Alden, WC and MM Leighton. 1917. The Iowan drift, a review of the evidences of the Iowan stage of glaciation. Iowa Geol. *Survey Ann. Rept.* 26:49–212.

Aleinikoff, JN, Muhs, DR, Bettis, EA, Johnson, WC, Fanning, CM and R Benton. 1999. Late Quaternary loess in northeastern Colorado. II. Pb isotopic evidence for the variability of loess sources. *GSA Bull.* 111:1876–1883.

2008. Isotopic evidence for the diversity of late Quaternary loess in Nebraska: Glaciogenic and nonglaciogenic sources. *GSA Bull.* 120:1362–1377.

Alekseeva, T, Alekseev, A, Maher, BA and V Demkin. 2007. Late Holocene climate reconstructions for the Russian steppe, based on mineralogical and magnetic properties of buried palaeosols. *PPP* 249:103–127.

Alekseyev, VYe. 1983. Mineralogical analysis for the determination of podzolization, lessivage, and argillation. *SSS* 15:21–28.

Aleva, GJJ. 1965. The buried bauxite deposits of Onverdacht, Suriname, South America. *Geol. Mijn.* 44:45–58.

1983. On weathering and denudation of humid tropical interfluves and their triple planation surfaces. *Geol. Mijn.* 62:383–388.

1987. Occurrence of stone-lines in tin-bearing areas in Belitung, Indonesia, and Rondônia, Brazil. *Intl. J. Trop. Ecol. Geog.* 11:197–203.

Alexander, EB. 1980. Bulk densities of California soils in relation to other soil properties. *SSSAJ* 44:689–692.

1982. Volume estimates of coarse fragments in soils: A combination of visual and weighing techniques. *J. Soil Water Cons.* 37:62–63.

1985. Estimating relative ages from iron-oxide/total-iron ratios of soils in the western Po Valley, Italy: A discussion. *Geoderma* 35:257–259.

1986. Stones: An Earth scientist's conception. *SSH* 27:15–17.

1990. Influences of inorganic substrata on chemical properties of Lithic Cryofolists in southeast Alaska. *SSH* 31:92–99.

1991. Soil temperatures in forest and muskeg on Douglas Island, southeast Alaska. *SSH* 32:108–116.

Alexander, EB and J DuShey. 2011. Topographic and soil differences from peridotite to serpentinite. *Geomorphology* 135:271–276.

Alexander, EB, Mallory, JI, and WL Colwell. 1993. Soil-elevation relationships on a volcanic plateau in the southern Cascade Range, Northern California, USA. *Catena* 20:113–128.

Alexander, EB, Ping, CL, and P Krosse. 1994. Podzolisation in ultramafic materials in southeast Alaska. *SS* 157:46–52.

Alexander, JB, Beavers, AH, and PR Johnson. 1962. Zirconium content of coarse silt in loess and till of Wisconsin age in northern Illinois. *SSSAP* 26:189–191.

Alexander, LT and JG Cady. 1962. Genesis and Hardening of Laterite in Soils. USDA Tech. Bull. 1282.

Alexander, LT, Cady, JG, Whittig, LD, and RF Dever. 1956. Mineralogical and chemical changes in the hardening of laterite. *Proc. 6th Intl. Congr. Soil Sci.* 11:67–72.

Alexandrovskaya, EI and AL Alexandrovskaya. 2000. History of the cultural layer in Moscow and accumulation of anthropogenic substances in it. *Catena* 41:249–259.

Alexandrovskiy, AL. 2000. Holocene development of soils in response to environmental changes: The Novosvobodnaya archaeological site, North Caucasus. *Catena* 41:237–248.

2007. Rates of soil-forming processes in three main models of pedogenesis. *Rev. Mexic. Ciencias Geol.* 24:283–292.

Al-Farraj, A and AM Harvey. 1982. Effects of nutrient accumulation by aspen, spruce, and pine on soil properties. *SSSAJ* 46:853–861.

2000. Desert pavement characteristics on Wadi terrace and aluvial fan surfaces: Wadi Al-Bih, UAE and Oman. *Geomorphology* 35:279–297.

Allan, RJ, Brown, J and S Rieger. 1969. Poorly drained soils with permafrost in interior Alaska. *SSSAP* 33:599–605.

Allan, RJ and FD Hole. 1968. Clay accumulation in some Hapludalfs as related to calcareous till and incorporated loess on drumlins in Wisconsin. *SSSAP* 32:403–408.

Allen, BL and BF Hajek. 1989. Mineral occurrence in soil environments. In: JB Dixon and SB Weed (eds). *Minerals in Soil Environments*, 2nd ed., Soil Sci. Soc. Am. Madison, WI. pp. 331–378.

Allen, BL and EP Whiteside. 1954. The characteristics of some soils on tills of Cary and Mankato age in Michigan. *SSSAP* 18:203–206.

Allen, CC. 1978. Desert varnish of the Sonoran Desert: Optical and electron probe microanalysis. *J. Geol.* 86:743–752.

Allen, JRL. 1970. A quantitative model of grain size and sedimentary structures in lateral deposits. *Geol. J.* 7:129–146.

Allison, FE. 1968. Soil aggregation-some facts and fallacies as seen by a microbiologist. *SS* 106:136–143.

Allison, RJ and GE Bristow. 1999. The effects of fire on rock weathering: Some further considerations of laboratory experimental simulation. *ESPL* 24:707–713.

Allison, RJ and AS Goudie. 1994. The effects of fire on rock weathering: An experimental study. In: DA Robinson and RBG Williams (eds). *Rock Weathering and Landform Evolution*. Wiley, Chichester. pp. 41–56.

Almendinger, JC. 1990. The decline of soil organic matter, total-N, and available water capacity following the late-Holocene establishment of jack pine on sandy Mollisols, north-central Minnesota. *SS* 150:680–694.

Almond, P, Roering, J and TC Hales. 2007. Using soil residence time to delineate spatial and temporal patterns of transient landscape response. *J. Geophys. Res. Earth Surf.* 112:1–19.

Almond, PC and PJ Tonkin. 1999. Pedogenesis by upbuilding in an extreme leaching and weathering environment, and slow loess accretion, south Westland, New Zealand. *Geoderma* 92:1–36.

Almy, MM. 1978. The archeological potential of soil survey reports. *Florida Anthropol.* 31:75–91.

Aloni, K and J Soyer. 1987. Cycle des matériaux de construction des termitières d'humivores en savane au Shaba méridional (Zaïe). *Rev. Zool. Africa* 101:329–358.

Aloni, K, Malaisse, F and I Kapinga. 1983. Rôles des termites dans la décomposition du bois et la transfert de terre dans une forêt claire zambézienne. In: P Lebrun, H André, A de Medts, C Gregoire-Wibo and G Wanthy (eds). *New Trends in Soil Biology*. Louvain-la-Neuve, Dieu-Brichart. pp. 600–602.

Alonso, P, Sierra, C, Ortega, E and C Dorronsoro. 1994. Soil development indices of soils developed on fluvial terraces (Peñaranda de Bracamonte, Salamanca, Spain). *Catena* 23:295–308.

Alonso-Zarza, AM. 1999. Initial stages of laminar calcrete formation by roots: Examples from the Neogene of central Spain. *Sed. Geol.* 126:177–191.

Alonso-Zarza, AM and B Jones. 2007. Root calcrete formation on Quaternary karstic surfaces of Grand Cayman. *Geol. Acta* 5:77–88.

Alonso-Zarza, AM, Silva, PG, Goy, JL and C Zazo. 1998. Fan-surface dynamics and biogenic calcrete development: Interactions during ultimate phases of fan evolution in the semiarid SE Spain (Murcia). *Geomorphology* 24:147–167.

Amba, EA, Smeck, NE, Hall, GF and JM Bigham. 1990. Geomorphic and pedogenic processes operative in soils of a hill slope in the unglaciated region of Ohio. *Ohio J. Sci.* 90:4–12.

Ambrose, GJ and RB Flint. 1981. A regressive Miocene lake system and silicified strandlines, in northern Australia: Implications for regional stratigraphy and silcrete genesis. *J. Geol. Soc. Austral.* 28:81–94.

Ambrose, SH and NE Sikes. 1991. Soil carbon isotope evidence for Holocene habitat change in the Kenya Rift Valley. *Science* 253:1402–1405.

Amesbury, MJ, Mallon, G, Charman, DJ, Hughes, PDM, Booth, RK, Daley, TJ and M Garneau. 2012. Statistical testing of a new testate amoeba based transfer function for water-table depth reconstruction on ombrotrophic peatlands in north-eastern Canada and Maine, United States. *JQS* 28:27–39.

Amit, R, Enzel, Y, Grodek, T, Crouvi, O, Porat, N and A Ayalon. 2010. The role of rare rainstorms in the formation of calcic soil horizons on alluvial surfaces in extreme deserts. *QR* 74:177–187.

Amit, R and R Gerson. 1986. The evolution of Holocene reg (gravelly) soils in deserts – an example from the Dead Sea region. *Catena* 13:59–79.

Amit, R, Gerson, R and DH Yaalon. 1993. Stages and rate of the gravel shattering process by salts in desert Reg soils. *Geoderma* 57:295–324.

Amit, R and JBJ Harrison. 1995. Biogenic calcic horizon development under extremely arid conditions, Nizzana Sand Dunes, Israel. *Adv. GeoEcol.* 28:65–88.

Amit, R, Harrison, JBJ and Y Enzel. 1995. Use of soils and colluvial deposits in analyzing tectonic events: The southern Arava Rift, Israel. *Geomorphology* 12:91–107.

Amit, R, Harrison, JBJ, Enzel, Y and N Porat. 1996. Soils as a tool for estimating ages of Quaternary fault scarps in a hyperarid environment: The southern Arava Valley, the Dead Sea Rift, Israel. *Catena* 28:21–45.

Amit, R and DH Yaalon. 1996. The micromorphology of gypsum and halite in reg soils: The Negev Desert, Israel. *ESPL* 21:1127–1143.

Ammer, S, Weber, K, Abs, C, Ammer, C and J Prietzel. 2006. Factors influencing the distribution and abundance of earthworm communities in pure and converted Scots pine stands. *Appl. Soil Ecol.* 33:10–21.

Ammon, K, Dunai, TJ, Stuart, FM, Meriaux, AS and E Gayer. 2009. Cosmogenic He-3 exposure ages and geochemistry of basalts from Ascension Island, Atlantic Ocean. *QG* 4:525–532.

Amoozegar, A and AW Warrick. 1986. Hydraulic conductivity of saturated soils. In: A Klute (ed). *Methods of Soil Analysis*. Part 1. *Physical and Mineralogical Methods*. Agronomy Mon. 9, 2nd ed. Soil Sci. Soc. Am., Madison, WI. pp. 735–770.

Amos, DF and EP Whiteside. 1975. Mapping accuracy of a contemporary soil survey in an urbanizing area. *SSSAP* 39:937–942.

Amundson, R. 1998a. Discussion of the paper by J. D. Phillips. *Geoderma* 86:25–27.

1998b. Do soils need our protection? *Geotimes* 43:16–20.

Amundson, R, Guo, Y and P Gong. 2003. Soil diversity and land use in the United States. *Ecosystems* 6:470–482.

Amundson, RG, Chadwick, OA, Kendall, C, Wang, Y and M DeNiro. 1996. Isotopic evidence for shifts in atmospheric circulation patterns during the late Quaternary in mid-North America. *Geology* 24:23–26.

Amundson, RG, Chadwick, OA, Sowers, JM and HE Doner. 1988. Relationship between climate and vegetation and the stable carbon isotope chemistry of soils in the eastern Mojave Desert, Nevada. *QR* 29:245–254.

1989a. Soil evolution along an altitudinal transect in the eastern Mojave Desert of Nevada, *USA Geoderma* 43:349–371.

1989b. The stable isotope chemistry of pedogenic carbonates at Kyle Canyon, Nevada. *SSSAJ* 53:201–210.

Amundson, RG, Graham, RC and E Franco-Vizcaino. 1997. Orientation of carbonate laminations in gravelly soils along a winter/summer precipitation gradient in Baja California, Mexico. *SS* 162:940–952.

Amundson, RG, Wang, Y, Chadwick, OA, Trumbore, S, McFadden, L, McDonald, E, Wells, S and M DeNiro. 1994. Factors and processes governing the 14C content of carbonate in desert soils. *EPSL* 125:385–405.

An, Z, Huang, Y, Liu, W, Guo, Z, Clemens, S, Li, L, Prell, P, Ning, Y, Cai, Y, Zhou, W, Lin, B, Zhang, Q, Cao, Y, Qiang, X, Chang, H and Z Wu. 2005. Expansion of C_4 vegetation in Loess Plateau and stages uplifts of Qing-Tibet Plateau. *Geology* 33:705–708.

An, ZS, Kukla, G, Porter, SC and JL Xiao. 1991. Late Quaternary dust flow on the Chinese loess plateau. *Catena* 18:125–132.

Anda, M, Chittleborough, DJ and RW Fitzpatrick. 2009. Assessing parent material uniformity of a red and black soil complex in the landscapes. *Catena* 78:142–153.

Ande, OT and B Senjobi. 2010. Lithologic discontinuity and pedogenetic characterization on an aberrant toposequence associated with a rock hill in South Western Nigeria. *Intl. J. Phys. Sci.* 5:596–604.

Andersen, KK, Azuma, N, Barnola, JM, Bigler, M, Biscaye, P, Caillon, N, Chappellaz, J, Clausen, HB, DahlJensen, D, et al. 2004. High-resolution record of Northern Hemisphere climate extending into the last interglacial period. *Nature* 431:147–151.

Anderson, DW. 1979. Processes of humus formation and transformation in soils of the Canadian Great Plains. *JSS* 30:77–84.

1987. Pedogenesis in the grassland and adjacent forests of the Great Plains. *Adv. Soil Sci.* 7:53–93.

Anderson, DW, De Jong, E and DS McDonald. 1979. The pedogenetic origin and characteristics of organic matter of Solod soils. *CJSS* 59:357–362.

Anderson, DW, Paul, EA and RJ St. Arnaud. 1974. Extraction and characterization of humus with reference to clay-associated humus. *CJSS* 54:317–323.

Anderson, EC and WF Libby. 1951. World-wide distribution of natural radiocarbon. *Phys. Rev.* 81:64–69.

Anderson, HA, Berrow, ML, Farmer, VC, Hepburn, A, Russell, JD and AD Walker. 1982. A reassessment of Podzol formation processes. *JSS* 33:125–136.

Anderson, JU, Bailey, OF and D Rai. 1975. Effects of parent materials on the genesis of Borolls and Boralfs in south-central New Mexico mountains. *SSSAP* 39:901–904.

Anderson, JU, Fadul, KE and GA O'Connor. 1973. Factors affecting the coefficient of linear extensibility in Vertisols. *SSSAP* 37:296–299.

Anderson, K, Wells, S and R Graham. 2002. Pedogenesis of vesicular horizons, Cima Volcanic Field, Mojave Desert, California. *SSSAJ* 66:878–887.

Anderson, MA and GM Browning. 1949. Some physical and chemical properties of six virgin and six cultivated Iowa soils. *SSSAP* 14:370–374.

Anderson, MG and TP Burt. 1978. The role of topography in controlling throughflow generation. *ESPL* 3:331–334.

Anderson, RS, Repka, JL and GS Dick. 1996. Explicit treatment of inheritance in dating depositional surfaces using in situ [10]Be and 26Al. *Geology* 24:47–51.

Anderson, SP. 1988. The upfreezing process: Experiments with a single clast. *GSA Bull.* 100:609–621.

Anderton, JB. 1999. The soil-artifact context model: A geoarchaeological approach to paleoshoreline site dating in the Upper Peinsula of Michigan, USA. *Geoarchaeol.* 14:265–288.

Anderton, JB and WL Loope. 1995. Buried soils in a perched dunefield as indicators of Late Holocene lake-level change in the Lake Superior basin. *QR* 44:190–199.

Andrade, G, Mihara, KL, Linderman, RG and GJ Bethlenfalvay. 1998. Soil aggregation status and rhizobacteria in the mycorrhizosphere. *Plant Soil* 202:89–96.

Andren, T, Bjorck, J and S Johnsen. 1999. Correlation of Swedish glacial varves with the Greenland (GRIP) oxygen isotope record. *JQS* 14:361–371.

Andrews, JT. 1987. The Late Wisconsin glaciation and deglaciation of the Laurentide ice sheet. In: WF Ruddiman and HE Wright, Jr (eds). *North America and Adjacent Oceans during the Last Deglaciation.* Geol. Soc. Am. Vol. K-3. pp. 13–37. Boulder, CO.

Andriesse, JP. 1969. A study of the environment and characteristics of tropical podzols in Sarawak (East-Malaysia). *Geoderma* 2:201–226.

Anduaga, S and G Halffter. 1991. Beetles associated with rodent burrows (Coleoptera Scarabaedae, Scararaeinae). *Folia Entomol. Mexicana* 81:185–197.

Anonymous. 2004. Soil and trouble. *Science* 304:1614–1615.

Anovitz, LM, Elam, JM, Riciputi, LR and DR Cole. 1999. The failure of obsidian hydration dating: Sources, implications, and new directions. *JAS* 26:735–752.

Anovitz, LM, Riciputi, LR, Cole, DR, Fayek, M and JM Elam. 2006. Obsidian hydration: A new paleothermometer. *Geology* 34:517–520.

Antoine, P, Rousseau, D-D, Fuchs, M, Hatté, C, Gauthier, C, Marković, SB, Jovanović, M, Gaudenyi, T, Moine, O and J Rossignol. 2009. High-resolution record of the last climatic cycle in the southern Carpathian Basin (Surduk, Vojvodina, Serbia). *QI* 198:19–36.

Antoine, PP. 1970. Mineralogical, chemical and physical studies on the genesis and morphology of a Rockwood sand loam (Typic Fragiboralf). PhD dissertation, Univ. of Minnesota.

Anton, D and F Ince. 1986. A study of sand color and maturity in Saudi Arabia. *ZG* 30:339–356.

Applegarth, MT and DE Dahms. 2001. Soil catenas of calcareous tills, Whiskey Basin, Wyoming, USA. *Catena* 42:17–38.

April, R, Newton, R and LT Coles. 1986. Chemical weathering in two Adirondack watersheds: Past and present-day rates. *GSA Bull.* 97:1232–1238.

Araki, S and K Kyuma. 2012. Characterization of red and/or yellow colored soil materials in the southwestern part of Japan in terms of lithology and degree of weathering. *Soil Sci. Plant Nutr.* 31:403–410.

Aran, D, Gury, M and E Jeanroy. 2001. Organometallic complexes in an Andisol: A comparative study with a Cambisol and a Podzol. *Geoderma* 99:65–79.

Arbogast, AF. 1996. Stratigraphic evidence for Late-Holocene aeolian sand mobilization and soil formation in south-central Kansas, USA. *J. Arid Envs* 34:403–414.

2000. Estimating the time since final stabilization of a perched dune field along Lake Superior. *Prof. Geog.* 52:594–606.

Arbogast, AF, Hansen, EC and MD Van Oort. 2002a. Reconstructing the geomorphic evolution of large coastal dunes along the southeastern shore of Lake Michigan. *Geomorphology* 46:241–255.

Arbogast, AF and TP Jameson. 1998. Age estimates of inland dunes in East-Central Lower Michigan using soils data. *PG* 19:485–501.

Arbogast, AF and WC Johnson. 1994. Climatic implications of the Late Quaternary alluvial record of a small drainage basin in the central Great Plains. *QR* 41:298–305.

Arbogast, AF and WL Loope. 1999. Maximum-limiting ages of Lake Michigan coastal dunes: Their correlation with Holocene lake level history. *J. Great Lakes Res.* 25:372–382.

Arbogast, AF and DR Muhs. 2000. Geochemical and mineralogical evidence from eolian sediments for northwesterly mid-Holocene paleowinds, central Kansas, USA. *QI* 67:107–118.

Arbogast, AF, Schaetzl, RJ, Hupy, JP and EC Hansen. 2004. The Holland Paleosol: An informal pedostratigraphic unit in the coastal dunes of southeastern Lake Michigan. *CJES* 41:1385–1400.

Arbogast, AF, Scull, P, Schaetzl, RJ, Harrison, J, Jameson, TP and S Crozier. 1997. Concurrent stabilization of some interior dune fields in Michigan. *PG* 18:63–79.

Arbogast, AF, Wintle, AG and SC Packman. 2002b. Widespread middle Holocene dune formation in the eastern Upper Peninsula of Michigan and the relationship to climate and outlet-controlled lake level. *Geology* 30:55–58.

Arduino, E, Barberis, E, Carraro, F and MG Forno. 1984. Estimating relative ages from iron-oxide/total-iron ratios of soils in the western Po Valley, Italy. *Geoderma* 33:39–52.

Arduino, E, Barberis, E and F Ajmone Marsan. 1986. Iron oxides and clay minerals within profiles as indicators of soil age in northern Italy. *Geoderma* 37:45–55.

Arhin, E and PM Nude. 2010. Use of termitaria in surficial geochemical surveys: Evidence for >125-mµ m size fractions as the appropriate media for gold exploration in northern Ghana. *Geochem.-Explor. Environ. Analysis* 10:401–406.

Arino, X, Ortega-Calvo, JJ, Gomezbolea, A and C Saizjimenez. 1995. Lichen colonization of the Roman pavement at Baelo-Claudia (Cadiz, Spain) – biodeterioration vs bioprotection. *Sci. Total Env.* 167:353–363.

Aristarain, LF. 1970. Chemical analyses of caliche profiles from the High Plains, New Mexico. *J. Geol.* 78:201–212.

Arkley, RJ. 1963. Calculation of carbonate and water movement in soil from climatic data. *SS* 96:239–248.

1967. Climates of some Great Soil Groups of the western United States. *SS* 103:389–400.

Armson, KA and RJ Fessenden. 1973. Forest windthrows and their influence on soil morphology. *SSSAP* 37:781–783.

Armstrong, R and T Bradwell. 2010. Growth of crustose lichens: A review. *Geog. Annaler* 92A:3–17.

Arnaulds, O. 2004. Volcanic soils of Iceland. *Catena* 56:3–20.

Arnalds, O and J Kimble. 2001. Andisols of deserts inIceland. *SSSAJ* 65:1778–1786.

Arnalds, O, Hallmark, CT, and LP Wilding. 1995. Andisols from four different regions of Iceland. *SSSAJ* 59:161–169.

Arno, SF. 1979. Forest Regions of Montana. US For. Serv. Res. Paper INT-218.

Arnold, LJ, Bailey, RM and GE Tucker. 2007. Statistical treatment of fluvial dose distributions from southern Colorado arroyo deposits. *QG* 2:162–167.

Arnold, RW. 1968. Pedological significance of lithologic discontinuities. *Trans. 9th Intl. Congr. Soil Sci.* 4:595–603.

1983. Concepts of Soils and Pedology. In: LP Wilding, NE Smeck and GF Hall (eds). *Pedogenesis and Soil Taxonomy.* Elsevier, New York. pp. 1–21.

1992. Becoming a pedologist. *SSH* 33:33–36.

1994. Soil geography and factor functionality: Interacting concepts. In: R Amundson, JW Harden and MJ Singer (eds). Factors of Soil Formation: A Fiftieth Anniversary Retrospective. Soil Sci. Soc. Am. Spec. Publication 33, pp. 99–109.

2005. Pedological significance of discontinuities: A sequel. *SSH* 46:48–58.

Arnold, RW and FF Riecken. 1964. Grainy gray ped coatings in Brunizem soils. *Proc. Iowa Acad. Sci.* 71:350–360.

Arocena, JM and S Pawluk. 1991. The nature and origin of nodules in Podzolic soils from Alberta. *CJSS* 71:411–426.

Arriaga, FJ, Lowery, B and MD Mays. 2006. A fast method for determining soil particle size distribution using a laser instrument. *SS* 171:663–674.

Arshad, MA and S Pawluk. 1966. Characteristics of some solonetzic soils in the Glacial Lake Edmonton basin of Alberta. I. Physical and chemical. *JSS* 17:36–47.

Arshad, MA, Schnitzer, M and CM Preston. 1988. Characterization of humic acids from the termite mounds and surrounding soils, Kenya. *Geoderma* 42:213–225.

Artz, JA. 1985. A soil-geomorphic approach to locating buried late Archaic sites in northeast Oklahoma. *Am. Archeol.* 5:142–150.

Arveti, N, Reginald, S, Kumar, KS, Harinath, V and Y Sreedhar. 2012. Biogeochemical study of termite mounds: A case study from Tummalapalle area of Andhra Pradesh, India. *Env. Monitor. Assess.* 184:2295–2306.

Arzhanova, VS. 2003. Soils of alpine geosystems of the Sikhote Alin Ridge and some aspects of their genesis. *Euras. Soil Sci.* 36:809–817.

Asadu, CLA and FOR Akamigbo. 1987. The use of abrupt changes in selected soil properties to assess lithological discontinuities in soils of eastern Nigeria. *Pédologie* 37:43–56.

Asady, GH and EP Whiteside. 1982. Composition of a Conover-Brookston map unit in southeastern Michigan. *SSSAJ* 46:1043–1047.

Asamoa, GK and R Protz. 1972. Influence of discontinuities in particle size on the genesis of two soils of the Honeywood catena. *CJSS* 52:497–511.

Asawalam, DO and S Johnson. 2007. Physical and chemical characteristics of soils modified by earthworms and termites. *Comm. Soil Plant Anal.* 38:513–521.

Ascaso, C and J Galvan. 1976. Studies on the pedogenetic action of lichen acids. *Pedobiologia* 16:321–331.

Ascaso, C, Galvan, J and C Rodriguezpascual. 1982. The weathering of calcareous rocks by lichens. *Pedobiologia* 24:219–229.

Ashley, GM, Shaw, J and ND Smith (eds). 1985. *Glacial Sedimentary Environments*, Society for Sedimentary Geology Short Course #16. Publ. by SEPM, Tulsa, OK.

Assallay, AM, Jefferson, I, Rogers, CDF and IJ Smalley. 1998. Fragipan formation in loess soils: Development of the Bryant hydroconsolidation hypothesis. *Geoderma* 83:1–16.

Assallay, AM, Rogers, CDF, Smalley, IJ and IF Jefferson. 1998. Silt: 2–62 um, 9–4. *Earth-Sci. Revs.* 45:61–88.

Atalay, I. 1997. Red Mediterranean soils in some karstic regions of Taurus mountains, Turkey. *Catena* 28:247–260.

Atkinson, HJ and JR Wright. 1957. Chelation and the vertical movement of soil constituents. *SS* 84:1–11.

Atkinson, RJC. 1957. Worms and weathering. *Antiquity* 31:219–233.

Attig, JW and L Clayton. 1993. Stratigraphy and origin of an area of hummocky glacial topography, northern Wisconsin, USA. *QI* 18:61–67.

Attig, JW, Hanson, PR, Rawling, JE, Young, AR and EC Carson. 2011. Optical ages indicate the southwestern margin of the Green Bay Lobe in Wisconsin, USA, was at its maximum extent until about 18,500 years ago. *Geomorphology* 130:384–390.

Austin, WE and JH Huddleston. 1999. Viability of permanently installed platinum redox electrodes. *SSSAJ* 63:1757–1762.

Avakyan, ZA, Karavaiko, GI, Melnikova, Krutsko, VS and YI Ostroushyo. 1981. Role of microscopic fungi in weathering of rocks and minerals from a pegmatite deposit. *Mikrobiologiya* 50:156–162.

Aydinalp, C and MS Cresser. 2008. Red soils under Mediterranean type of climate: Their properties use and productivity. *Bulg. J. Agric. Sci.* 14:576–582.

Ayres, E, Steltzer, H, Berg, S and DH Wall. 2009. Soil biota accelerate decomposition in high-elevation forests by specializing in the breakdown of litter produced by the plant species above them. *JE* 97:901–912.

Azmatch, TF, Sego, DC, Arenson, LU and KW Biggar. 2012. New ice lens initiation condition for frost heave in fine-grained soils. *Cold Regions Sci. Tech.* 82:8–13.

Azúa-Bustos, A, González-Silva, C, Mancilla, RA, Salas, L, Gómez-Silva, B, McKay, CP and R Vicuña. 2011. Hypolithic cyanobacteria supported mainly by fog in the coastal range of the Atacama Desert. *Microb. Ecol.* 61:568–581.

Azzali, E, Carbone, C, Marescotti, P and G Lucchetti. 2011. Relationships between soil colour and mineralogical composition: Application for the study of waste-rock dumps in abandoned mines. *Neues Jahr. fur Mineral. Abhandl.* 188:75–85.

Babalola, O and R Lal. 1977. Subsoil gravel horizon and maize root growth. I. Gravel concentration and bulk density. *Plant Soil* 46:337–346.

Bach, AJ, Brazel, AJ and N Lancaster. 1996. Temporal and spatial aspects of blowing dust in the Mojave and Colorado Deserts of Southern California (1973–1994). *PG* 17:329–353.

Bachmann, GO and MN Machette. 1977. Calcic soils and calcretes in the southwestern United States. USGS Open File Rept. 77–794.

Bacon, CR. 1983. Eruptive history of Mount Mazama and Crater Lake caldera Cascade Range, USA. *J. Volcan. Geotherm. Res.* 18:57–115.

Bacon, CR and MA Lanphere. 2006. Eruptive history and geochronology of Mount Mazama and the Crater Lake region, Oregon. *GSA Bull.* 118:1331–1359.

Bacon, DW and DG Watts. 1971. Estimating the transition between two intersecting straight lines. *Biometrika* 58:525–534.

Bagine, RKN. 1984. Soil translocation by termites of the genus *Odontotermes* (Holgren) (*Isoptera: Macrotermitinae*) in an arid area of northern Kenya. *Oecologia* 64:263–266.

Bailey, HP. 1979. Semi-Arid climates: Their definition and distribution. In: AE Hall, GH Cannell and HW Lawton (eds). *Agriculture in Semi-arid Environments*. Springer-Verlag, Berlin. pp. 73–97.

Bailey, LW, Odell, RT and WR Boggess. 1964. Properties of selected soils developed near the forest-prairie border in east-central Illinois. *SSSAP* 28:257–263.

Bailey, SW. 1980. Structures of layer silicates. In: GW Brindley and G Brown (eds). *Crystal Structures of Clay Minerals and their X-Ray Identification*. Mon. 5. Mineralogical Society, London. pp. 1–123.

Bain, DC, Mellor, A, Robertson-Rintoul, MSE and ST Buckland. 1993. Variations in weathering processes and rates with time in a chronosequence of soils from Glen Feshie, Scotland. *Geoderma* 57:275–293.

Baker, DG. 1971. Snow Cover and Winter Soil Temperatures at St. Paul, Minnesota, US Dept. of Interior. Bull. #37. St. Paul, MN, Water Resources Research Center, Univ. of Minnesota.

Baker, G. 1959. Opal phytoliths in some Victorian soils and "red rain" residues. *AJB* 7:64–87.

Baker, JC and CL Schrivner. 1985. Simulated movement of silicon in a Typic Hapludalf. *SS* 139:255–261.

Baker, RG, Schwert, DP, Bettis, EA III and CA Chumbley. 1991. Mid-Wisconsinan stratigraphy and paleoenvironments at the St. Charles site in south-central Iowa. *GSA Bull.* 103:210–220.

Baker, RG, Sullivan, AE, Hallberg, GR and DG Horton. 1989. Vegetational changes in western Illinois during the onset of Late Wisconsinan glaciation. *Ecology* 70:1363–1376.

Bakker, L, Lowe, DJ and AG Jongmans. 1996. A micromorphological study of pedogenic processes in an evolutionary soil sequence formed on Late Quaternary rhyolitic tephra deposits, North Island, New Zealand. *QI* 34–36:249–261.

Balco, G and CW Rovey II. 2008. An isochron method for cosmogenic-nuclide dating of buried soils and sediments. *AJS* 308:1083–1114.

——— 2010. Absolute chronology for major Pleistocene advances of the Laurentide Ice Sheet. *Geology* 38:795–798.

Balco, G, Rovey, CW II, and JOH Stone. 2005. The first glacial maximum in North America. *Science* 307:222.

Balco, G and DL Shuster. 2009. ^{26}Al-^{10}Be-^{21}Ne burial dating. *EPSL* 286:570–575.

Baldock, JA. 2001. Interactions of organic materials and microorganisms with minerals in the stabilization of soil structure. In PM Huang, JM Bollag and N Senesi (eds). *Interactions between Soil Particles and Microorganisms: Impact on the Terrestrial Ecosystem*. Wiley, New York. pp. 85–131.

Baldock, JA and JO Skjemstad. 2000. Role of the soil matrix and minerals in protecting natural organic materials against biological attack. *Org. Geochem.* 31:697–710.

Baldwin, M, Kellogg, CE and J Thorp. 1938. Soil classification. In: *Soils and Men: Yearbook of Agriculture*. US Dept. of Agric., Washington, DC, US Govt. Printing Off. pp. 979–1001.

Balek, CL 2002. Buried artifacts in stable upland sites and the role of bioturbation: A review. *Geoarchaeol.* 17:41–51.

Balesdent, J. 1987. The turnover of soil organic fractions estimated by radiocarbon dating. *Sci. Total Environ.* 62:405–408.

Balesdent, J, Girardin, C and A Mariotti. 1993. Site-related 13C of tree leaves and soil organic matter in a temperate forest. *Ecology* 74:1713–1721.

Balesdent, J, Mariotti, A and B Guillet. 1987. Natural 13C abundance as a tracer for soil organic matter dynamics studies. *Soil Biol. Biochem.* 19:25–30.

Ball, DF and WM Williams. 1968. Variability of soil chemical properties in two uncultivated Brown Earths. *JSS* 19:379–391.

Ballagh, TM and ECA Runge. 1970. Clay-rich horizons over limestone: Illuvial or residual? *SSSAP* 34:534–536.

Ballantyne, AK. 1978. Saline soils in Saskatchewan due to wind deposition. *CJSS* 58:107–108.

Ballantyne, CK and JA Matthews. 1982. The development of sorted circles on recently deglaciated terrain, Jotunheimen, Norway. *AAR* 14:341–354.

Ballesteros-Cánovas JA, Bodoque JM, Lucía A, Martín-Duque JF, Díez-Herrero A, Ruiz-Villanueva V, Rubiales JM and M Genova. 2013. Dendrogeomorphology in badlands: Methods, case studies and prospects. *Catena* 106:113–122.

Balsam, WL, Ellwood, BB, Ji, JF, Williams, ER, Long, XY and A El Hassani. 2011. Magnetic susceptibility as a proxy for rainfall: Worldwide data from tropical and temperate climate. *QSR* 30:2732–2744.

Banak, A, Pavelić, D, Kovačić, M and O Mandic. 2013. Sedimentary characteristics and source of loess in Baranja (Eastern Croatia). *AR* 11:129–139.

Barbiero, L, Kumard, MS, Violette, A, Oliva, P, Braun, JJ, Kumar, C, Furian, S, Babic, M, Riotte, J and V Valles. 2010. Ferrolysis induced soil transformation by natural drainage in Vertisols of sub-humid South India. *Geoderma* 156:173–188.

Bard, E, Arnold, M, Hamelin, B, Tisnerat-Laborde, N and G Cabioch. 1998. Radiocarbon calibration by means of mass spectrometric Th-230/U-234 and C-14 ages of corals: An updated database including samples from Barbados, Mururoa and Tahiti. *Radiocarbon* 40:1085–1092.

Barendregt, RW. 1981. Dating methods of Pleistocene deposits and their problems. *VI. Paleomagnetism. Geosci. Can.* 8:56–64.

1984. Using paleomagnetic remanence and magnetic susceptibility data for the differentiation, relative correlation and absolute dating of Quaternary sediments. In: WC Mahaney (ed). *Quaternary Dating Methods.* Elsevier, New York. pp. 101–122.

Barnes, CJ and GB Allison. 1983. The distribution of deuterium and 18O in dry soils. *I. Theory. J. Hydrol.* 60:141–156.

Barnes, GW. 1879. The hillocks or mound formations of San Diego, California. *Am. Nat.* 13:565–571.

Barnhisel, RI, Bailey, HH and S Matondang. 1971. Loess distribution in central and eastern Kentucky. *SSSAP* 35:483–487.

Barnhisel, RI and PM Bertsch. 1989. Chlorites and hydroxy-interlayered vermiculite and smectite. In: JB Dixon and SB Weed (eds). *Minerals in Soil Environments*, 2nd ed. Soil Sci. Soc. Am., Madison, WI. pp. 729–788.

Barnhisel, RI and CI Rich. 1967. Clay mineral formation in different rock types of a weathering boulder conglomerate. *SSSAP* 31:627–631.

Barrett, LR. 1993. Soil development and spatial variability on geomorphic surfaces of different age. *PG* 14:39–55.

1997. Podzolization under forest and stump prairie vegetation in northern Michigan. *Geoderma* 78:37–58.

1998. Regressive pedogenesis following a century of deforestation: Evidence for depodzolization. *SS* 163:482–497.

2001. A strand plain soil development sequence in Northern Michigan, USA. *Catena* 44:163–186.

Barrett, LR and RJ Schaetzl. 1992. An examination of podzolization near Lake Michigan using chronofunctions. *CJSS* 72:527–541.

1992. An examination of podzolization near Lake Michigan using chronofunctions. *CJSS* 72:527–541.

1993. Soil development and spatial variability on geomorphic surfaces of different age. *PG* 14:39–55.

Barshad, I. 1964. Chemistry of soil development. In: FE Bear (ed). *Chemistry of the Soil.* Chapman and Hall, London. pp. 1–71.

1966. The effect of a variation in precipitation on the nature of clay mineral formation in soils from basic and acid igneous rocks. *Proc. Intl. Clay Conf. (Jerusalem)* 1:167–173.

Barshad, I, Halevy, E, Gold, HA and I Hagin. 1956. Clay minerals in some limestone soils from Israel. *SS* 81:423–437.

Barshad, I and FM Kishk. 1969. Chemical composition of soil vermiculite clays as related to their genesis. *Contrib. Mineral. Petrol.* 24:136–155.

Bartelli, LJ. 1973. Soil development in loess in the southern Mississippi Valley. *SS* 115:254–260.

Bartelli, LJ and RT Odell. 1960a. Field studies of a clay-enriched horizon in the lowest part of the solum of some Brunizem and grey-brown podzolic soils in Illinois. *SSSAP* 24:388–390.

1960b. Laboratory studies and genesis of a clay-enriched horizon in the lowest part of the solum of some Brunizem and grey-brown podzolic soils in Illinois. *SSSAP* 24:390–395.

Bartgis, RL and GE Lang. 1984. Marl wetlands in eastern West Virginia: Distribution, rare plant species, and recent history. *Castanea* 49:17–25.

Bartlein, PJ, Webb, T III and EC Fleri. 1984. Holocene climatic change in the northern Midwest: Pollen-derived estimates. *QR* 22:361–374.

Bartlett, M and K Ritz (eds). 2011. *The Zoological Generation of Soil Structure.* CABI Publ., Wallingford, England. p. 244.

Bartlett, RJ and BR James. 1993. Redox chemistry of soils. In: *Advances in Agronomy*, Academic Press, San Diego. 50:151–208.

Barton, BJ, Kirschbaum, CD and CE Bach. 2009. The impacts of ant mounds on sedge meadow and shrub carr vegetation in a prairie fen. *Natural Areas J.* 29:293–300.

Barton, RN. 1987. Vertical distribution of artefacts and some post-depositional factors affecting site deformation. In: P Rowley-Conwy, M Zvelebil and HP Blankholm (eds). *Mesolithic in Northwest Europe: Recent Trends.* Sheffield, UK, Dept. of Archaeology and Prehistory, Univ. of Sheffield. pp. 55–62.

Bascomb, CL. 1968. Distribution of pyrophosphate-extractable iron and organic carbon in soils of various groups. *JSS* 19:251–268.

Basher, LR. 1997. Is pedology dead and buried? *AJSR* 35:979–994.

Batjes, NH. 2011. *Overview of soil phosphorus data from a large international database.* Report 2011/01, Plant Research Intl., Wageningen UR, and ISRIC – World Soil Information, Wageningen.

Bateman, MD. 2008. Luminescence dating of periglacial sediments and structures. *Boreas* 37:574–588.

Bateman, MD, Boulter, CH, Carr, AS, Frederick, CD, Peter, D and M Wilder. 2007a. Detecting post-depositional sediment disturbance in sandy deposits using luminescence. *QG* 2:57–64.

2007b. Preserving the palaeoenvironmental record in Drylands: Bioturbation and its significance for luminescence-derived chronologies. *Sed. Geol.* 195:5–19.

Bateman, MD and JA Catt. 1996. An absolute chronology for the raised beach and associated deposits at Sewerby, East Yorkshire, England. *JQS* 11:389–395.

Bateman, MD, Frederick, CD, Jaiswal, MK and AK Singhvi. 2003. Investigations into the potential effects of pedoturbation on luminescence dating. *QSR* 22:1169–1176.

Baumgartner, S, Beer, J, Wagner, G, Kubik, P, Suter, M, Raisbeck, GM and F Yiou. 1997. [10]Be and dust. Nucl. Inst. *Methods Physics Res.* 123B:296–301.

Baveye, P. 2006. A future for soil science. *J. Soil Water Cons.* 61:148–151.

Baveye, P, Jacobson, AR, Allaire, SE, Tandarich, J and R Bryant. 2006. Whither goes soil science in the US and Canada? Survey results and analysis. *SS* 171:501–518.

Baxter, FP and FD Hole. 1967. Ant (Formica cinerea) pedoturbation in a prairie soil. *SSSAP* 31:425–428.

Beadle, NCW. 1962. An alternative hypothesis to account for the generally low phosphate content of Australian soils. *Austr. J. Agric. Res.* 13:434–42.

Beattie, JA. 1970. Peculiar features of soil development in parna deposits in the eastern Riverina, N. S.W. *AJSR* 8:145–156.

Beatty, SW. 1984. Influence of microtopography and canopy species on spatial patterns of forest understory plants. *Ecology* 65:1406–1419.

2000. On the rocks: Shaken, not stirred. *AAAG* 90:785–787.

Beatty, SW and ODV Sholes. 1988. Leaf litter effect on plant species composition of deciduous forest treefall pits. *CJFR* 18:553–559.

Beatty, SW and EL Stone. 1986. The variety of soil microsites created by tree falls. *CJFR* 16:539–548.

Beaudette, DE and AT O'Geen. 2009. Quantifying the aspect effect: An application of solar radiation modeling for soil survey. *SSSAJ* 73:1345–1352.

Beauvais, A and F Colin. 1993. Formation and transformation processes of iron duricrust systems in tropical humid environment. *Chem. Geol.* 106:77–101.

Beaven, PJ and MJ Dumbleton. 1966. Clay minerals and geomorphology in four Caribbean islands. *Clay Mins.* 6:371–382.

Beavers, AH, Fehrenbacher, JB, Johnson, PR and RL Jones. 1963. CaO-ZrO2 molar ratios as an index of weathering. *SSSAP* 27:408–412.

Bech, J, Rustullet, J, Garrigó Tobías, FJ and R Martínez. 1997. The iron content of some Mediterranean soils from northeast Spain and its pedogenic significance. *Catena* 28:211–229.

Beavers, AH and I Stephen. 1958. Some features of the distribution of plant-opal in Illinois soils. *SS* 86:1–5.

Becher, M, Olid, C and J Klaminder. 2013. Buried soil organic inclusions in non-sorted circles fields in northern Sweden: Age and paleo-climatic context. *J. Geophys. Res. Biogeosci.* 118:1–8.

Beckel, DKB. 1957. Studies on seasonal changes in the temperature gradient of the active layer of soil at Fort Churchill, Manitoba. *Arctic* 10:151–183.

Becker, T and S Getzin. 2000. The fairy circles of Kaokoland (North-West Namibia) – origin, distribution, and characteristics. *Basic Appl. Ecol.* 1:149–159.

Beckett, PHT and R Webster. 1971. Soil variability: A review. *Soils and Fert.* 34:1–15.

Beckmann, GG. 1984. The place of "genesis" in the classification of soils. *AJSR* 22:1–14.

Beckmann, GG, Thompson, CH and BR Richards. 1984. Relationships of soil layers below gilgai in black earths. In: JW McGarity, EH Hoult and HB So (eds). *The Properties and Utilization of Cracking Clay Soils.* Reviews in Rural Science #5. Univ. of New England, Armidale, NSW. pp. 64–72.

Beckwith, RS and R Reeve. 1964. Studies on soluble silica in soils. II. The release of monosilicic acid from soils. *AJSR* 2:33–45.

Bedard-Haughn, A. 2011. Gleysolic soils of Canada: Genesis, distribution, and classification. *CJSS* 91:763–779.

Beer, J, Shen, CD, Heller, F, Liu, TS, Bonani, G, Dittrich, B and M Suter. 1993. Be and magnetic susceptibility in Chinese loess. *GRL* 20:57–60.

Behrens, T, Schmidt, K, Zhu, AX and T Scholten. 2010. The ConMap approach for terrain-based digital soil mapping. *EJSS* 61:133–143.

Beinroth, FH. 1982. Some highly weathered soils of Puerto Rico. I. Morphology, formation and classification. *Geoderma* 27:1–73.

Beinroth, FH, Eswaran, H, Uehara, G and PF Reich. 2000. Oxisols. In: ME Sumner (ed). *Handbook of Soil Science.* CRC Press, Boca Raton, FL. pp. E-373–E-392.

Beinroth, FH, Eswaran, H, Uehara, G, Smith, CW and PF Reich. 2011. Oxisols. In: PM Huang, Y Li and ME Sumner (eds). *Handbook of Soil Sciences,* 2nd ed. CRC Press, New York. pp. 33–177–33–190.

Beinroth, FH, Uehara, G and H Ikawa. 1974. Geomorphic relationships of Oxisols and Ultisols on Kauai, Hawaii. *SSSAP* 38:128–131.

Bell, JC, Cunningham, RL and MW Havens. 1992. Calibration and validation of a soil-landscape model for predicting soil drainage class. *SSSAJ* 56:1860–1866.

Belnap, J. 2006. The potential roles of biological soil crusts in dryland hydrologic cycles. *Hydrol. Process.* 20:3159–3178.

Benac, C and G Durn. 1997. Terra rossa in the Kvarner area: Geomorphological conditions of formation. *Acta Geog. Croatica* 32:7–17.

Ben-Dor, E, Levin, N, Singer, A, Karnieli, A, Braun, O and GJ Kidron. 2006. Quantitative mapping of the soil rubification process on sand dunes using an airborne hyperspectral sensor. *Geoderma* 131:1–21.

Benedict, JB. 1967. Recent glacial history of an alpine area in the Colorado Front Range, USA. I. Establishing a lichen-growth curve. *J. Glac.* 6:817–832.

1970. Downslope soil movement in a Colorado alpine region: Rates, processes and climatic significance. *AAR* 2:165–226.

1976. Frost creep and gelifluction features: A review. *QR* 6:55–76.

1990. Lichen mortality due to late-lying snow: Results of a transplant study. *AAR* 22:81–89.

2009. A review of lichenometric dating and its applications to archaeology. *Am. Antiq.* 74:143–172.

Bennema, J. 1974. Organic carbon profiles in Oxisols. *Pédologie* 24:119–146.

Bennett, DR and T Entz. 1989. Moisture-retention parameters for coarse-textured soils in southern Alberta. *CJSS* 69:263–272.

Bennett, PC, Rogers, JR, Choi, WJ and FK Hiebert. 2001. Silicates, silicate weathering, and microbial ecology. *Geomicrobiol. J.* 18:3–19.

Benson, LV, Currey, DR, Dorn, RI, Lajoie, KR, Oviatt, CG, Robinson, SW, Smith, GI and S Stine. 1990. Chronology of expansion and contraction of four Great Basin lake systems during the past 35 000 years. *PPP* 78:241–286.

Berg, AW. 1990. Formation of Mima mounds: A seismic hypothesis. *Geology* 18:281–284.

Berg, RC. 1984. The origin and early genesis of clay bands in youthful sandy soils along Lake Michigan, USA. *Geoderma* 32:45–62.

Berger, A. 1988. Milankovitch theory and climate. *Rev. Geophys.* 26:624–657.

Berger, A and MF Loutre. 1991. Insolation values for the climate of the last 10000000 years. *QSR* 10:297–317.

Berger, GW. 1984. Thermoluminescence dating studies of glacial silts from Ontario. *CJES* 21:1393–1399.

1988. TL dating studies of tephra, loess and lacustrine sediments. *QSR* 7:295–303.

2012. Effectiveness of natural zeroing of the thermoluminescence in sediments. *J. Geophys. Res. Solid Earth Planets* 95:12375–12397.

Berger, GW and W Mahaney. 1990. Test of thermoluminescence dating of buried soils from Mt. Kenya, Kenya. *Sed. Geol.* 66:45–56.

Berger, IA and RU Cooke. 1997. The origin and distribution of salts on alluvial fans in the Atacama Desert, Northern Chile. *ESPL* 22:581–600.

Bern, CR, Chadwick, OA, Hartshorn, AS, Khomo, LM and J Chorover. 2011. A mass-balance model to separate and quantify colloidal and solute redistributions in soil. *Chem. Geol.* 282:113–119.

Bernard, MJ, Neatrour, MA and TS McCay. 2009. Influence of soil buffering capacity on earthworm growth, survival, and community composition in the Western Adirondacks and Central New York. *Northeastern Natural.* 16:269–284.

Bernier, N. 1998. Earthworm feeding activity and the development of the humus profile. *Biol. Fert. Soils* 26:215–223.

Bernoux, M, Arrouays, D, Cerri, C, Volkoff, B and C Jolivet. 1998. Bulk densities of Brazilian Amazon soils related to other soil properties. *SSSAJ* 62:743–749.

Berry, EC and JK Radke. 1995. Biological processes: Relationships between earthworms and soil temperatures. *J. Minnesota Acad. Sci.* 59:6–8.

Berry, L and BP Ruxton. 1959. Notes on weathering zones and soils on granitic rocks in two tropical regions. *JSS* 10:54–63.

Berry, ME. 1987. Morphological and chemical characteristics of soil catenas on Pinedale and Bull Lake moraine slopes in the Salmon River Mountains, Idaho. *QR* 28:210–225.

1994. Soil-geomorphic analysis of Late-Pleistocene glacial sequences in the McGee, Pine, and Bishop Creek Drainages, East-Central Sierra Nevada, California. *QR* 41:160–175.

Berry, ME and JR Staub. 1998. Root traces and the identification of paleosols. *QI* 51/52:9–10.

Bertran, P, Allenet, G, Ge, T, Naughton, F, Poirier, P and MFS Goni. 2009. Coversand and Pleistocene palaeosols in the Landes region, south-western France. *JQS* 24:259–269.

Beschel, RE. 1950. Flechten als Altersmasstab rezenter Moränen. *Gletscherkd. Glazialgeol.* 1:152–161.

1958. Lichenometrical studies in West Greenland. *Arctic* 11:254.

1973. Lichens as a measure of the age of recent moraines. *AAR* 5:303–309.

Beshay, NF and AS Sallam. 1995. Evaluation of some methods for establishing uniformity of profile parent materials. *Arid Soil Res. Rehabil.* 9:63–72.

Betard, F. 2012. Spatial variations of soil weathering processes in a tropical mountain environment: The Baturite massif and its piedmont (Ceara, NE Brazil). *Catena* 93:18–28.

Bettis, EA III. 1988. Pedogenesis in late prehistoric Indian mounds, upper Mississippi Valley. *PG* 263–279.

——— 1992. Soil morphologic properties and weathering zone characteristics as age indicators in Holocene alluvium in the Upper Midwest. In: VT Holliday (ed). *Soils and Landscape Evolution*. Smithsonian Institution Press, Washington, DC. pp. 119–144.

——— 1998. Subsolum weathering profile characteristics as indicators of the relative rank of stratigraphic breaks in till sequences. *QI* 51/52:72–73.

——— 2003. Patterns in Holocene colluvium and alluvial fans across the prairie-forest transition in the midcontinent USA. *Geoarch.* 18:779–797.

Bettis, EA III, Muhs, DR, Roberts, HM and AG Wintle. 2003. Last glacial loess in the conterminous USA. *QSR* 22:1907–1946.

Beuselinck, L, Govers, G, Poesen, J, Degraer, G and L Froyen. 1998. Grain-size analysis by laser diffractometry: Comparison with the sieve-pipette method. *Catena* 32:193–208.

Beven, K and P Germann. 1982. Macropores and water flow in soils. *Water Resour. Res.* 18:1311–1325.

Beven, KJ and MJ Kirkby. 1979. A physically based, variable contributing area model of basin hydrology. *Hydrol. Sci. Bull.* 24:43–69

Beyer, L, Bockheim, JG, Campbell, IB and GGC Claridge. 1999. Genesis, properties and sensitivity of Antarctic Gelisols. *Antarctic Sci.* 11:387–398.

Beyer, L, Knicker, H, Blume, H-P, Bölter, M, Vogt, B, and D Schneider. 1997. Soil organic matter of suggested spodic horizons in relic ornithogenic soils of coastal continental Antarctica (Casey Station, Wilkes Land) in comparison with that of spodic soil horizons in Germany. *SS* 162:518–527.

Beyer, L, Pingpank, K, Wriedt, G and M Bölter. 2000. Soil formation in coastal Antarctica (Wilkes Land). *Geoderma* 95:283–304.

Bhattacharjee, JC, Landey, RJ and AR Kalbande. 1977. A new approach in the study of Vertisol morphology. *J. Ind. Soc. Soil Sci.* 25:221–232.

Bhattacharyya, T, Pal, DK and M Velayutham. 1999. A mathematical equation to calculate linear distance of cyclic horizons in Vertisols. *SSH* 40:127–133.

Bidwell, OW and FD Hole. 1965. Man as a factor of soil formation. *SS* 99:65–72.

Bierman, PR. 1994. Using in situ produced cosmogenic isotopes to estimate rates of landscape evolution: A review from the geomorphic perspective. *J. Geophys. Res.* 99:13885–13896.

Bierman, PR and AR Gillespie. 1994. Evidence suggesting that methods of rock-varnish cation-ratio dating are neither comparable nor consistently reliable. *QR* 42:82–90.

Bierman, P and J Turner. 1995. ^{10}Be and ^{26}Al evidence for exceptionally low rates of Australian bedrock erosion and the likely existence of pre-Pleistocene landscapes. *QR* 44:378–382.

Bigham, JM, Fitzpatrick RW and DG Schulze. 2002. Iron oxides. In: JB Dixon and DG Schulze (eds). *Soil Mineralogy with Environmental Applications*. Soil Sci. Soc. Am., Madison, WI. pp. 323–366.

Bigham, JM, Golden, DC, Buol, SW, Weed, SB and LH Bowen. 1978. Iron oxide mineralogy of well-drained Ultisols and Oxisols. II. Influence on color, surface area, and phosphate retention. *SSSAJ* 42:825–830.

Bigham, JM, Smeck, NE, Norton, LD, Hall, GF and ML Thompson. 1991. Lithology and general stratigraphy of Quaternary sediments in a section of the Teays River valley of southern Ohio. In: WN Melhorn and JP Kempton (eds). *Geology and Hydrogeology of the Teays-Mahomet Bedrock Valley System*. Spec. Paper #258. Geological Society of America, Boulder, CO. pp. 19–27.

Billings, WD and KM Peterson. 1980. Vegetational change and ice-wedge polygons through the thaw-lake cycle in Arctic Alaska. *AAR* 12:413–432.

Bilzi, AF and EJ Ciolkosz. 1977. A field morphology rating scale for evaluating pedological development. *SS* 124:45–48.

Birkeland, PW. 1968. Correlation of Quaternary stratigraphy of the Sierra Nevada with that of the Lake Lahontan area. In: RB Morrison and HE Wright, Jr (eds). Means of Correlation of Quaternary Successions. Proc. Intl. Assoc. Quat. Res., VII Congress, Proc. 8:469–500.

——— 1969. Quaternary paleoclimatic implications of soil clay mineral distribution in a Sierra Nevada-Great Basin transect. *J. Geol.* 77:289–302.

——— 1973. Use of relative age-dating methods in a stratigraphic study of rock glacier deposits, Mt. Sopris, Colorado. *AAR* 5:401–416.

——— 1974. *Pedology, Weathering and Geomorphological Research*. Oxford, New York.

——— 1978. Soil development as an indication of relative age of Quaternary deposits, Baffin Island, N. W.T., Canada. *AAR* 10:733–747.

——— 1982. Subdivision of Holocene glacial deposits, Ben Ohau Range, New Zealand, using relative dating methods. *GSA Bull.* 93:433–449.

——— 1984. *Soils and Geomorphology*, 2nd ed. Oxford, New York.

——— 1999. *Soils and Geomorphology*, 3rd ed. Oxford, New York.

Birkeland, PW, Berry, ME and DK Swanson. 1991. Use of soil catena field data for estimating relative ages of moraines. *Geology* 19:281–283.

Birkeland, PW and RM Burke. 1988. Soil catena chronosequences on eastern Sierra Nevada moraines, California, USA. *AAR* 20:473–484.

Birkeland, PW, Shroba, RR, Burns, SF, Price, AB and PJ Tonkin. 2003. Integrating soils and geomorphology in mountains – an example from the Front Range of Colorado. *Geomorphology* 55:329–344.

Birks, HJB. 1976. Late-Wisconsinan vegetation history at Wolf Creek, central Minnesota. *Ecol. Monogr.* 46:395–429.

Birman, JH. 1964. *Glacial geology across the crest of the Sierra Nevada*, California. GSA Spec. Paper 75.

Biscaye, PE, Chesselet, R and JM Prospero. 1974. Rb-Sr, ^{87}Sr/^{86}Sr isotope system as an index of provenance of continental dusts in the open Atlantic Ocean. *J. Rech. Atmos.* 8:819–829.

Bischoff, JL, Shlemon, RJ, Ku, TL, Simpson, RD, Rosenbauer, RJ and FE Budinger Jr. 1981. Uranium-series and soil-geomorphic dating of the Calico archaeological site, California. *Geology* 9:576–582.

Bisdom, EBA, Nauta, R and B Volbert. 1983. STEM-EDXRA and SEM-EDXRA investigations of iron-coated organic matter in thin sections with transmitted, secondary and backscattered electrons. *Geoderma* 30:77–92.

Bishop, PM, Mitchell, PB and TR Paton. 1980. The formation of duplex soils on hillslopes in the Sydney Basin, Australia. *Geoderma* 23:175–189.

Bishop, TFA, McBratney, AB and GM Laslett. 1999. Modelling soil attribute depth functions with equal-area quadratic smoothing splines. *Geoderma* 91:27–45.

Biswas, A and BC Si. 2011. Revealing the controls of soil water storage at different scales in a hummocky landscape. *SSSAJ* 75:1295–1306.

Bitom, D, Volkoff, B, Beauvais, A, Seyler, F and PD Ndjigui. 2004. Landscape and soil evolution control by lateritic heritages and hydrostatic base level in the intertropical rain forest zone. *Compt. Rendus Geosci.* 336:1161–1170.

Bjelland, T and IH Thorseth. 2002. Comparative studies of the lichen-rock interface of four lichens in Vingen, western Norway. *Chem. Geol.* 192:81–98.

Black, RF. 1969. Climatically significant fossil periglacial phenomena in northcentral United States. *Biul. Peryglac.* 20:227–238.

1976. Periglacial features indicative of permafrost: Ice and soil wedges. *QR* 6:3–26.

Blackwelder, E. 1926. Fire as an agent in rock weathering. *J. Geol.* 35:134–140.

1931. Pleistocene glaciation in the Sierra Nevada and basin ranges. *GSA Bull.* 42:865–922.

Blake, GR and KH Hartage. 1986. Bulk density. In: A Klute (ed). *Methods of Soil Analysis*, 2nd ed. Agronomy Mon. 9. American Society Agron., Madison, WI. pp. 363–375.

Blanchart, E. 1992. Restoration by earthworms (Megascolecidae) of the macroaggregate structure of a destructed savanna soil under field conditions. *Soil Biol. Biochem.* 24:1587–1594.

Blanchart, E, Albrecht, A, Alegre, J, Duboisset, A, Pashanasi, B, Lavelle, P and L Brussaard. 1999. *Effects of earthworms on soil structure and physical properties.* In: P Lavelle, L Brussaard and P Hendrix (eds). *Earthworm Management.* CAB International, Wallingford, UK. pp. 139–162.

Bland, W and D Rolls. 1998. *Weathering: An Introduction to the Scientific Principles.* Edward Arnold, London.

Blank, HR and EW Tynes. 1965. Formation of caliche in situ. *GSA Bull.* 76:1387–1392.

Blank, RR, Cochrane, B and MA Fosberg. 1998. Duripans of southwestern Idaho: Polygenesis during the Quaternary deduced through micromorphology. *SSSAJ* 62:701–709.

Blank, RR and MA Fosberg. 1991. Duripans of Idaho, USA: In situ alteration of eolian dust (loess) to an opal-A/X-ray amorphous phase. *Geoderma* 48:131–149.

Blank, RR, Young, JA and T Lugaski. 1996. Pedogenesis on talus slopes, the Buckskin range, Nevada, USA. *Geoderma* 71:121–142.

Blaser, P, Pannatier, EG and L Walthert. 2008a. The base saturation in acidified Swiss forest soils on calcareous and noncalcareous parent material. A pH-base saturation anomaly. *J. Plant Nutr. Soil Sci.* 171:155–162.

Blaser, P, Walthert, L, Zimmermann, S, Pannatier, EG and J Luster. 2008b. Classification schemes for the acidity, base saturation, and acidification status of forest soils in Switzerland. *J. Plant Nutr. Soil Sci.* 171:163–170.

Blauuw, M. 2010. Methods and code for "classical" age-modelling of radiocarbon sequences. *QG* 5:512–518.

Blavet, D, Mathe, E and JC Leprun. 2000. Relations between soil colour and waterlogging duration in a representative hillside of the West African granito-gneissic bedrock. *Catena* 39:187–210.

Blinnikov, MS. 2005. Phytoliths in plants and soils of the interior Pacific Northwest, USA. *Rev. Palaeobot. Palynol.* 135:71–98.

Blokhuis, WA. 1982. Morphology and genesis of Vertisols. Proc. 12th Intl. Congr. Soil Sci. 23–47.

Bloomfield, C. 1953. A study of podzolization. II. The mobilization of iron and aluminum by the leaves and bark of Agathis australis (Kauri). *JSS* 4:17–23.

Blow, FE. 1955. Quantity and hydrologic characteristics of litter upon upland oak forests in eastern Tennessee. *J. For.* 53:190–195.

Blum, WE and R Ganssen. 1972. Bodenbildende prozesse der Erde, ihre erscheinungsformen und diagnostischen Merkmale in tabellarischer Darstellung. *Die Erde* 103:7–20.

Blume, H-P, Beyer, L, Bölter, M, Erlenkeuser, H, Kalk, E, Kneesch, S, Pfisterer, U and D Schneider. 1997. Pedogenic zonation in soils of the southern circum-polar region. *Adv. GeoEcol.* 30:69–90.

Blume, HP and P Leinweber. 2004. Plaggen soils: Landscape history, properties, and classification. *J. Plant Nutr. Soil Sci.* 167:319–327.

Blümel, WD. 1982. Calcretes in Namibia and SE Spain: Relations to substratum, soil formation and geomorphic factors. *Catena Suppl.* 1:67–82.

Blumer, BE, Arbogast, AF and SL Forman. 2012. The OSL chronology of eolian sand deposition in a perched dune field along the northwestern shore of Lower Michigan. *QR* 77:445–455.

Blundell, A, Dearing, JA, Boyle, JF and JA Hannam. 2009. Controlling factors for the spatial variability of soil magnetic susceptibility across England and Wales. *Earth-Sci. Revs.* 95:158–188.

Boardman, J. 1985. Comparison of soils in Midwestern United States and western Europe with the interglacial record. *QR* 23:62–75.

Bocek, B. 1986. Rodent ecology and burrowing behavior: Predicted effects on archaeological site formation. *Am. Antiq.* 51:589–603.

Bochter, R and W Zech. 1985. Organic compounds in cryofolists developed on limestone under subalpine coniferous forest, Bavaria. *Geoderma* 36:145–157.

Bockheim, JG. 1979a. Properties and relative age of soil of southwestern Cumberland peninsula, Baffin Island, NWT, Canada. *AAR* 11:289–306.

1979b. Relative age and origin of soils in eastern Wright Valley, Antarctica. *SS* 128:142–152.

1980a. Properties and classification of some desert soils in coarse-textured glacial drift in the Arctic and Antarctic. *Geoderma* 24:45–69.

1980b. Solution and use of chronofunctions in studying soil development. *Geoderma* 24:71–85.

1982. Properties of a chronosequence of ultraxerous soils in the Trans-Antarctic Mountains. *Geoderma* 28:239–255.

1990. Soil development rates in the Transantarctic Mountains. *Geoderma* 47:59–77.

1997a. Properties and classification of cold desert soils from Antarctica. *SSSAJ* 61:224–231.

1997b. Soils in a hemlock-hardwood ecosystem mosaic in the Southern Lake Superior Uplands. *CJFR* 27:1147–1153.

2003. Genesis of bisequal soils on acidic drift in the upper Great Lakes region, USA. *SSSAJ* 67:612–619.

2005. Soil endemism and its relation to soil formation theory. *Geoderma* 129:109–124.

2007. Importance of cryoturbation in redistributing organic carbon in permafrost-affected soils. *SSSAJ* 71:1335–1342.

2008. Functional diversity of soils along environmental gradients in the Ross Sea region, Antarctica. *Geoderma* 144:32–42.

2010. Soil-factorial models and earth-system science: A review. *Geoderma* 159:243–251.

2011. Distribution and genesis of ortstein and placic horizons in soils of the USA: A review. *SSSAJ* 75:994–1005.

2012. Origin of glossic horizons in Cryalfs of the eastern Rocky Mountains, USA. *Geoderma* 187:1–7.

Bockheim, JG and DC Douglass. 2006. Origin and significance of calcium carbonate in soils of Southwestern Patagonia. *Geoderma* 136:751–762.

Bockheim, JG, Everett, LR, Hinkel, KM, Nelson, FE and J Brown. 1999. Soil organic carbon storage and distribution in Arctic Tundra, Barrow, Alaska. *SSSAJ* 63:934–940.

Bockheim, JG and AN Gennadiyev. 2000. The role of soil-forming processes in the definition of taxa in Soil Taxonomy and the World Soil Reference Base. *Geoderma* 95:53–72.

2009. The value of controlled experiments in studying soil-forming processes: A review. *Geoderma* 152:208–217.

2010. Soil-factorial models and earth-system science: A review. *Geoderma* 159:243–251.

Bockheim, JG, Gennadiyev, AN, Hammer, RD and JP Tandarich. 2005. Historical development of key concepts in pedology. *Geoderma* 124:23–36.

Bockheim, JG and AE Hartmemink. 2013a. Classification and distribution of soils with lamellae in the USA. *Geoderma* 206:92–100.

2013b. Soils with fragipans in the USA. *Catena* 104:233–242.

Bockheim, JG and KM Hinkle. 2007. The importance of "deep" organic carbon in permafrost-affected soils of Arctic Alaska. *SSSAJ* 71:1889–1892.

Bockheim, JG, Marshall, JG and HM Kelsey. 1996. Soil-forming processes and rates on uplifted marine terraces in southwestern Oregon, USA. *Geoderma* 73:39–62.

Bockheim, JG, Mazhitova, G, Kimble, JM and C Tarnocai. 2006. Controversies on the genesis and classification of permafrost-affected soils. *Geoderma* 137:33–39.

Bockheim, JG and C Tarnocai. 2000. Gelisols. In: ME Sumner (ed). *Handbook of Soil Science*. CRC Press, Boca Raton, FL. pp. E-256–E-269.

2011. Gelisols. In: PM Huang, Y Li and ME Sumner (eds). *Handbook of Soil Sciences*, 2nd ed. CRC Press, New York. pp. 33-72–33-82.

Bockheim, JG, Tarnocai, C, Kimble, JM and CAS Smith. 1997. The concept of gelic materials in the new Gelisol order for permafrost-affected soils. *SS* 162:927–939.

Bockheim, JG and FC Ugolini. 1990. A review of pedogenic zonation in well-drained soils of the southern circumpolar region. *QR* 34:47–66.

Bocock, KL, Jeffers, JNR, Lindley, DK, Adamson, JK and CA Gill. 1977. Estimating woodland soil temperature from air temperature and other climatic variables. *Agric. Metr.* 18:351–372.

Boe, AG, Murray, A and SO Dahl. 2007. Resetting of sediments mobilised by the LGM ice-sheet in southern Norway. *QG* 2:222–228.

Boero, V and U Schwertmann. 1989. Iron oxide mineralogy of terra rossa and its genetic implications. *Geoderma* 4:319–327.

Boersma, L, Simonson, GH and DG Watts. 1972. Soil morphology and water table relations. I. Annual water table fluctuations. *SSSAP* 36:644–648.

Boettinger, JL and DW Ming. 2002. Zeolites. In: JB Dixon and DG Schulze (eds). *Soil Mineralogy with Environmental Applications*. SSSA Book Series no.7. Soil Sci. Soc. Am., Madison, WI. pp. 585–610.

Boettinger, JL and RJ Southard. 1991. Silica and carbonate sources for Aridisols on a granitic pediment, western Mojave Desert. *SSSAJ* 55:1057–1067.

Bogner, C, Borken, W and B Huwe. 2012. Impact of preferential flow on soil chemistry of a podzol. *Geoderma* 175:37–46.

Bohlen, PJ, Pelletier, DM, Groffman, PM, Fahey, TJ and MC Fisk. 2004. Influence of earthworm invasion on redistribution and retention of soil carbon and nitrogen in northern temperate forests. *Ecosystems* 7:13–27.

Boivin, P, Saejiew, A, Grunberger, O and S Arunin. 2004. Formation of soils with contrasting textures by translocation of clays rather than ferrolysis in flooded rice fields in Northeast Thailand. *EJSS* 55:713–724.

Bokhorst, MP, Beets, CJ, Marković, SB, Gerasimenko, NP, Matviishina, ZN and M Frechen. 2009. Pedo-chemical climate proxies in Late Pleistocene Serbian–Ukranian loess sequences. *QI* 198:113–123.

Bold, HD and MJ Wynne. 1979. *Introduction to the Algae*. Prentice-Hall, Englewood Cliffs, NJ.

Boll, J, van Rijn, RPG, Weiler, KW, Ewen, JA, Daliparthy, J, Herbert, SJ and TS Steenhuis. 1996. Using ground penetrating radar to detect layers in a sandy field soil. *Geoderma* 70:117–132.

Bonan, G and H Shugart. 1989. Environmental factors and ecological processes in boreal forests. *Ann. Rev. Ecol. Syst.* 20:1–28.

Bond, G, Broecker, W, Johnsen, S, McManus, J, Labeyrie, L, Jouzel, J and G Bonani. 1993. Correlations between climate records from North Atlantic sediments and Greenland ice. *Nature* 365:143–147.

Bond, GC and R Lotti. 1995. Iceberg discharges into the North-Atlantic on millennial time scales during the last glaciation. *Science* 267:1005–1010.

Bond, WJ. 1986. Illuvial band formation in a laboratory column of sand. *SSSAJ* 50:265–267.

Bonifacio, E, Falsone, G, Simonov, G, Sokolova, T and I Tolpeshta. 2009. Pedogenic processes and clay transformations in bisequal soils of the Southern Taiga zone. *Geoderma* 149:66–75.

Bonifacio, E, Santoni, S, Celi, L and E Zanini. 2006. Spodosol-Histosol evolution in the Krkonose National Park (CZ). *Geoderma* 131:237–250.

Borchardt, G. 1989. Smectites. In: JB Dixon and SB Weed (eds). *Minerals in Soil Environments*, 2nd ed. Soil Sci. Soc. Am., Madison, WI. pp. 675–727.

Borchardt, GA, Hole, FD and ML Jackson. 1968. Genesis of layer silicates in representative soils in a glacial landscape of southeastern Wisconsin. *SSS AP* 32:399–403.

Borgaard, OK. 1983. Effect of surface area and mineralogy of iron oxides on their surface charge and anion adsorption properties. *Clays Clay Mins.* 31:230–232.

Bormann, FH, Siccama, TG, Likens, GE and RH Whittaker. 1970. The Hubbard Brook ecosystem study: Composition and dynamics of the tree stratum. *Ecol. Monogr.* 40:373–388.

Bornyasza, MA, Graham, RC and MF Allen. 2005. Ectomycorrhizae in a soil-weathered granitic bedrock regolith: Linking matrix resources to plants. *Geoderma* 126:141–160.

Borst, G. 1968. The occurrence of crotovinas in some southern Californian soils. *Trans. 9th Intl. Congr. Soil Sci.* 2:19–27.

Bos, RHG and J Sevink. 1975. Introduction of gradational and pedomorphic features in descriptions of soils. *JSS* 26:223–233.

Bossuyt, H, Six, J and PF Hendrix. 2004. Rapid incorporation of carbon from fresh residues into newly formed stable microaggregates within earthworm casts. *EJSS* 55:393–399.

Botha, GA and N Fedoroff. 1995. Palaeosols in Late Quaternary colluvium, northern KwaZnhr-Natal, South Africa. *J. Afr. Earth-Sci.* 21:291–311.3

Bouabid, R, Nater, EA and P Barak. 1992. Measurement of pore size distribution in a lamellar Bt horizon using epifluorescence microscopy and image analysis. *Geoderma* 53:309–328.

Bouchard, M and MJ Pavich. 1989. Characteristics and significance of pre-Wisconsinan saprolites in the northern Appalachians. *ZG* 72:125–137.

Boulaine, J. 1975. *Géographie des sols*. Presses Universités de France, Paris.

Bouma, J and LW Dekker. 1978. A case study on infiltration into dry clay soil. I. Morphological observations. *Geoderma* 20:27–40.

Bouma, J, Paetzold, RF and RB Grossman. 1982. Measuring hydraulic conductivity for use in soil survey. USDA Soil Survey Investigations Rept. 39. US Gov. Printing Off. 38 pp.

Bouma, J and V van Schuylenborgh. 1969. On soil genesis in a temperate humid climate. VII. The formation of a glossaqualf in a silt-loam terrace deposit. *Neth. J. Agric. Sci.* 17:261–271.

Bourne, WC and EP Whiteside. 1962. A study of the morphology and pedogenesis of a medial Chernozem developed in loess. *SSSAP* 26:484–490.

Boutton, TW, Archer, SR, Midwood, AJ, Zitzer, SF and R Bol. 1998. δ13C values of soil organic carbon and their use in documenting vegetation change in a subtropical savanna ecosystem. *Geoderma* 82:5–41.

Bowen, DQ. 1978. *Quaternary Geology: A Stratigraphic Framework for Multidisciplinary Work*. Pergamon, New York.

Bowen, LH and SB Weed. 1981. Mössbauer spectroscopic analysis of iron oxides in soil. In: JG Stevens and GK Shenoy (eds), *Mössbauer Spectroscopy and Its Applications*. American Chemical Society, Washington, DC. pp. 247–261.

Bowker, MA, Johnson, NC, Belnap, J and GW Koch. 2008. Short-term monitoring of aridland lichen cover and biomass using photography and fatty acids. *J. Arid Environs.* 72:869–878.

Bowler, JM. 1973. Clay dunes: Their occurrence, formation and environmental significance. *Earth-Sci. Revs.* 9:315–338.

Bowser, WE, Milne, RA and RR Cairns. 1962. Characteristics of the major soil groups in an area dominated by solonetzic soils. *CJSS* 42:165–179.

Boyer, P. 1973. Actions de certains termites constructeurs sur l'évolution des soils tropicaux. *Ann. Sci. Nat. Zool. (Paris)* 15:329–498.

1975. Etude particulières de Bellioositermes et de leur action sur sols tropicaux. *Ann. Sci. Nat. Zool. (Paris)* 17:273–446.

Boynton, D and W Reuther. 1938. A way of sampling soil gas in dense subsoil and some of its advantages and limitations. *SSSAP* 3:37–42.

Bozarth, S. 1995. Analysis of fossil biosilicates from the valley fill. In: V Holliday (ed). Stratigraphy and Paleoenvironments of Late Quaternary Valley Fills on the Southern High Plains. *GSA Memoir* 186:161–171.

Bradbury, JP, Dean, WE and RY Anderson. 1993. Holocene climatic and limnologic history of the north-central United States as recorded in the varved sediments of Elk Lake, Minnesota: A synthesis. In: JP Bradbury, and WE Dean (eds). Elk Lake, Minnesota: Evidence for Rapid Climatic Change in the North-Central United States. *GSA Spec. Paper* 276:309–328.

Bradley, WC. 1963. Large-scale exfoliation in massive sandstones of the Colorado Plateau. *GSA Bull.* 74:519–528.

Bradwell, T. 2009. Lichenometric dating: A commentary, in the light of some recent statistical studies. *Geog. Annaler* 91A:61–69.

Brady, NC. 1974. *The Nature and Properties of Soils*. MacMillian, New York.

2001. *The Nature and Properties of Soils*, 13th ed. Prentice-Hall, Upper Saddle River, NJ.

Brady, NC and RR Weil. 1999. *The Nature and Properties of Soils*, 12th ed. Prentice-Hall, Upper Saddle River, NJ.

Brakensiek, DL and WJ Rawls. 1994. Soil containing rock fragments: Effects on infiltration. *Catena* 23:99–110.

Bramstedt, MW. 1992. *Soil Survey of Jasper County, Illinois*. USDA-SCS, US Govt. Printing Off., Washington, DC.

Branch, LC. 1993. Inter- and intra-group spacing in the plains vizcacha (Lagostomus maximus). *J. Mammal.* 74:890–900.

Brandon, CE, Buol, SW, Gamble, EE and RA Pope. 1977. Spodic horizon brittleness in Leon (Aeric Haplaquod) soils. *SSSAJ* 41:951–954.

Brantley, SL. 2003. Reaction kinetics of primary rock-forming minerals under ambient conditions. In: JI Drever (ed), *Treatise on Geochemistry*. Vol. 5. Elsevier. pp. 73–117.

Brantley, SL and M Lebedeva. 2011. Learning to read the chemistry of regolith to understand the critical zone. *Ann. Rev. Earth Planet. Sci.* 39:387–416.

Brantley, SL, Buss, H, Lebedevaa, M, Fletchera, RC and L Mac. 2011. Investigating the complex interface where bedrock transforms to regolith. *Applied Geochem.* 26:S12–S15.

Bravard, S and D Righi. 1989. Geochemical differences in an Oxisol-Spodosol toposequence of Amazonia, Brazil. *Geoderma* 44:29–42.3

1990. Podzols in Amazonia. *Catena* 17:461–475.

1991. Characterization of fulvic and humic acids from an Oxisol-Spodosol toposequence of Amazonia, Brazil. *Geoderma* 48:151–162.

Bray, JR and E Gorham. 1964. Litter production in forests of the world. *Adv. Ecol. Res.* 2:101–157.

Breckenfield, DJ. 1999. *Soil Survey of Tohono O'odham Nation, Arizona: Parts of Maricopa, Pima, and Pinal Counties*. USDA-NRCS, US Govt. Printing Off., Washington, DC.

Breecker, DO, Sharp, ZD and LD McFadden. 2010. Atmospheric CO2 concentrations during ancient greenhouse climates were similar to those predicted for A.D. 2100. *PNAS* 107:576–580.

Breiman, L. 2001. Random forests. *Machine Learning* 45:5–32.

Bremer, H and H Späth. 1989. Geomorphological observations concerning stone-lines. *Intl. J. Trop. Ecol. Geog.* 11:185–195.

Brener, AGF and JF Silva. 1995. Leaf-cutting ant nests and soil fertility in a well-drained savanna in western Venezuela. *Biotropica* 27:250–254.

Breuning-Madsen, H, Awadzi, TW and HR Mount. 2004. Classification of soils modified by termite activity in tropical moist semideciduous forests of West Africa. *SSH* 45:111–120.

Breuning-Madsen, H, Holst, MK and PS Henriksen. 2012. The hydrology in huge burial mounds built of loamy tills: A case study on the genesis of perched water tables and a well in a Viking Age burial mound in Jelling, Denmark. *Geog. Tidsskrift* 112:40–51.

Breuning-Madsen, H, Ronsbo, J and MK Holst. 2000. Comparison of the composition of iron pans in Danish burial mounds with bog iron and spodic material. *Catena.* 39:1–9.

Brevik, EC. 2002. Soil classification in geology textbooks. *J. Geosci. Educ.* 50:539–543.

Brevik, EC, Fenton, TE, and DB Jaynes. 2003. Evaluation of the accuracy of a central Iowa soil survey and implications for precision soil management. *Precision Agric.* 4:331–342.

Brewer, R. 1976. *Fabric and Mineral Analysis of Soils*. Robert E. Krieger, Huntington, NY.

Brewer, R and AD Haldane. 1957. Preliminary experiments in the development of clay orientation in soils. *SS* 84:301–309.

Bridges, EM and PA Bull. 1983. The role of silica in the formation of compact and indurated horizons in the soils of South Wales. In: P Bullock and CP Murphy (eds). *Soil Micromorphology*. Academic Press, New York. pp. 605–613.

Brierley, JA, Stonehouse, HB and AR Mermut. 2011. Vertisolic soils of Canada: Genesis, distribution, and classification. *CJSS* 91:903–916.

Briese, DT. 1982. The effects of ants on the soil of a semi-arid salt bushland habitat. *Insectes Soc.* 29:375–386.

Brimecombe, MJ, DeLeij, FA and JM Lynch. 2001. The effect of root exudates on rhizosphere microbial populations. In: R Pinton, Z Varanini and P Nannipieri (eds). *The Rhizosphere: Biochemistry and Organic Substances at the Soil-Plant Interface*. Marcel Dekker, New York. pp. 95–140.

Bringham-Grette, J. 1996. Geochronology of glacial deposits. In: J Menzies (ed). *Past Glacial Environments: Sediments, Forms, and Techniques*. Vol. 2. Butterworth-Heinemann Ltd., Oxford, UK. pp. 377–410.

Brimhall, GH, Chadwick, OA, Lewis, CJ, Compston, W, Williams, IS, Danti, KJ, Dietrich, WE, Power, ME, Hendricks, D and J Bratt. 1992. Deformational mass transport and invasive processes in soil evolution. *Science* 255:695–702.

Brimhall, GH and WE Dietrich. 1987. Constitutive mass balance relations between chemical composition, volume, density, porosity, and strain in metasomatic hydrochemical systems: Results of weathering and pedogenesis. *Geochim. Cosmochim. Acta* 51:567–587.

Brimhall, GH, Lewis, CJ, Ague, JJ, Dietrich, WE, Hampel, J, Teague, T and P Rix. 1988. Metal enrichment in bauxites by deposition of chemically mature aeolian dust. *Nature* 333:819.

Brimhall, GH, Lewis, CJ, Ford, Bratt, J, Taylor, G and O Warin. 1991. Quantitative geochemical approach to pedogenesis: Importance of parent material reduction, volumetric expansion, and eolian influx in lateritization. *Geoderma* 51:51–91.

Brindley, GW. 1980. Quantitative X-ray mineral analysis of clays. In: GW Brindley and G Brown (eds). *Crystal Structures of Clay Minerals and their X-Ray Identification.* Mon. 5. Mineralogical Society, London. pp. 411–438.

Bringham-Grette, J. 1996. Geochronology of glacial deposits. In: J Menzies (ed). *Past Glacial Environments: Sediments, Forms, and Techniques.* Vol. 2. Butterworth-Heinemann Ltd., Oxford, UK. pp. 377–410.

Brinkman, R. 1970. Ferrolysis, a hydromorphic soil forming process. *Geoderma* 3:199–206.

1977a. Problem hydromorphic soils in north-east Thailand. II. Physical and chemical aspects, mineralogy and genesis. *Neth. J. Agric. Sci.* 25:170–181.

1977b. Surface-water gley soils in Bangladesh: Genesis. *Geoderma* 17:111–144.

Brinkman, R, Jongmans, AG, Miedema, R and P Masskant. 1973. Clay decomposition in seasonally wet, acid soils: Micromorphological, chemical and mineralogical evidence from individual argillans. *Geoderma* 10:259–270.

Brinson, MM, Christian, RR and LK Blum. 1995. Multiple states in the sea-level induced transition from terrestrial forest to estuary. *Estuaries* 18:648–659.

Britton, ME. 1957. Vegetation of the arctic tundra. Proc. Biol. Colloq. (Corvallis) 26–61.

Brock, AL and BJ Buck. 2005. A new formation process for calcic pendants from Pahranagat Valley, Nevada, USA, and implication for dating Quaternary landforms. *QR* 63:359–367.

2009. Polygenetic development of the Mormon Mesa, NV petrocalcic horizons: Geomorphic and paleoenvironmental interpretation. *Catena* 77:65–75.

Broecker, W, Barker, S, Clark, E, Hajdas, I and G Bonani. 2006. Anomalous radiocarbon ages from foraminifera shells. *Paleoceanography* 21:PA2008. doi:10.1029/2005PA001212

Broecker, WS and EA Olson. 1960. Radiocarbon from nuclear tests. II. *Science* 132:712–721.

Broecker, WS and T Liu. 2001. Rock varnish: Recorder of desert wetness? Geol. Soc. Am. *Today* 11:4–10.

Broersma, K and LM Lavkulich. 1980. Organic matter distribution with particle-size in surface horizons of some sombric soils in Vancouver Island. *CJSS* 60:583–586.

Bronfenbrener, L and R Bronfenbrener. 2010. Modeling frost heave in freezing soils. *Cold Regions Sci. Tech.* 61:43–64.

Bronger, A. 1991. Argillic horizons in modern loess soils in an ustic soil moisture regime: Comparative studies in forest-steppe and steppe areas from Eastern Europe and the United States. *Adv. Soil Sci.* 15:41–90.

2003. Correlation of loess-paleosol sequences in East and Central Asia with SE Central Europe: Towards a continental Quaternary pedostratigraphy and paleoclimatic history. *QI* 106:11–31.

Bronger, A, Ensling, J, Gütlich, P and H Spiering, 1983. Rubification of Terrae Rossae in Slovakia: Mossbauer effect study. *Clays Clay Mins.* 31:269–276.

Bronk Ramsey, C, Staff, RA, Bryant, CL, Brock, F, Kitagawa, H, van der Plicht, J, Schlolaut, G, Marshall, MH, Brauer, A, Lamb, HF, Payne, RL, Tarasov, PE, Haraguchi, T, Gotanda, K, Yonenobu, H, Yokoyama, Y, Tada, R and T Nakagawa. 2012. A complete terrestrial radiocarbon record for 11. 2 to 52.8 kyr B.P. *Science* 338:370–374.

Bronswijk, JJB. 1991. Relations between vertical soil movements and water-content changes in cracking clays. *SSSAJ* 55:1220–1226.

Brook, GA, Folkoff, ME and EG Box. 1983. A world model of soil carbon dioxide. *ESPL* 8:79–88.

Brookes, IA. 1982. Dating methods of Pleistocene deposits and their problems. VIII. *Weathering. Geosci. Canada.* 9:188–199.

Brophy, JA. 1959. Heavy mineral ratios of Sangamon weathering profiles in Illinois. *Ill. St. Geol. Surv. Circ.* 273:1–22.

Brossard, M, Lopez-Hernandez, D, Lepage, M and J-C Leprun. 2007. Nutrient storage in soils and nests of mound-building *Trinervitermes* termites in Central Burkina Faso: Consequences for soil fertility. *Biol. Fert. Soils* 43:437–447.

Brown, CN. 1956. The origin of caliche in the northeast Llano Estacado, Texas. *J. Geol.* 46:1–15.

Brown, DJ, Clayton, MK and K McSweeney. 2004a. Potential terrain controls on soil color, texture contrast and grain-size deposition for the original catena landscape in Uganda. *Geoderma* 122:51–72.

Brown, DJ, Helmke, PA and MK Clayton. 2003. Robust geochemical indices for redox and weathering on a granitic landscape in central Uganda. *Geochimica Cosmochim. Acta* 67:2711–2723.

Brown, DJ, McSweeney, K and PA Helmke. 2004b. Statistical, geochemical, and morphological analyses of stone line formation in Uganda. *Geomorphology* 62:217–237.

Brown, ET, Edmond, JM, Raisbeck, GM, Yiuo, F, Kurz, MD and EJ Brook. 1991. Examination of surface exposure ages of Antarctic moraines using in situ produced [10]Be and [26]Al. *Geochim. Cosmochim. Acta* 55:2269–2283.

Brown, G and GW Brindley. 1980. X-ray diffraction procedures for clay mineral identification. In: GW Brindley and G Brown (eds). *Crystal Structures of Clay Minerals and their X-Ray Identification.* Mon. 5. Mineralogical Society, London. pp. 305–359.

Brown, GG. 1995. How do earthworms affect microfloral and faunal community diversity? *Plant Soil* 170:209–231.

Brown, J. 1969. Soils of the Okpilak River region, Alaska. In: TL Péwé (ed). *The Periglacial Environment Past and Present.* McGill-Queens' Univ. Press, Montreal. pp. 93–128.

Brown, JL. 1979. Etude systématique de la variabilité d'un sol Podzolique, le long d'une tranchée dans une erablière à bouleau jaune. *CJSS* 59:131–146.

Brown, KJ and DL Dunkerley. 1996. The influence of hillslope gradient, regolith texture, stone size and stone position on the presence of a vesicular layer and related aspects of hillslope hydrologic processes: A case study from the Australian arid zone. *Catena* 26:71–84.

Brown, L. 1987. [10]Be as a tracer of erosion and sediment transport. *Chem. Geol.* 65:189–196.

Brown, L, Pavich, MJ, Hickman, RE, Klein, J and R Middleton. 1988. Erosion of the eastern United States observed with [10]Be. *ESPL* 13:441–457.

Bruce, JG. 1996. *Morphological characteristics and interpretation of some polygenetic soils in loess in southern South Island,* New Zealand. *QI* 34–36:205–211.

Bruce, JP, Frome, M, Haites, E, Janzen, H, Lal, R and K Paustian. 1999. Carbon sequestration in soils. *J. Soil Water Cons.* 54:382–389.

Brückner, W. 1955. The mantle rock ("laterite") of the Gold Coast and its origin. *Geol. Rundschau* 43:307–327.

Brugam, RB and SM Johnson. 1997. Holocene lake-level rise in the Upper Peninsula of Michigan, USA, as indicated by peatland growth. *Holocene* 7:355–359.

Bruun, S, Six, J, Jensen, LS and K Paustian. 2005. Estimating turnover of soil organic carbon fractions based on radiocarbon measurements. *Radiocarbon* 47:99–113.

Bryan, BA and CR Hupp. 1984. Dendrogeomorphic evidence for channel changes in an east Tennessee coal area stream. *TAGU* 65:891.

Bryan, D. 1994. Factors controlling the occurrence and distribution of hematite and goethite in soils and saprolites derived from schists and gneisses in western North Carolina. MS thesis, Mich. State Univ.

Bryan, K and CC Albritton Jr. 1943. Soil phenomena as evidence of climatic changes. *AJS* 241:469–490.

Bryan, K. 1946. Cryopedology: The study of frozen ground and intensive frost-action with suggestions on nomenclature. *AJS* 244:622–642.

Bryan, WH and LJH Teakle. 1949. Pedogenic intertia: A concept in soil science. *Nature* 164:969.

Bryant, RB. 1989. Physical processes of fragipan formation. In: NE Smeck and EJ Ciolkosz (eds). *Fragipans Their Occurrence, Classification, and Genesis*. Soil Sci. Soc. Am. Spec. Publication #24. Madison, WI. pp. 141–150.

Bryant, RB and J Macedo. 1990. Differential chemoreductive dissolution of iron oxides in a Brazilian Oxisol. *SSSAJ* 54:819–821.

Bryant, VM and SA Hall. 1993. Archaeological palynology in the United States: A critique. *Am. Antiq.* 58:277–286.

Bryant, VM, Jr and RG Holloway (eds) 1985. *Pollen Records of Late Quaternary North American Sediments*. American Association of Stratigraphic Palynologists Foundation. Dallas, TX.

Brye, KR. 2004. Pedogenic interpretation of a loess-covered, Pleistocene-glaciated toposequence using the energy model. *SS* 169:282–294.

Brye, KR, West, CP and EE Gbur. 2004. Soil quality differences under native tallgrass prairie across a climosequence in Arkansas. *AMN* 152:214–230.

Bryson, RA, Irving, WN and JA Larsen. 1965. Radiocarbon and soil evidence of former forest in the southern Canadian tundra. *Science* 147:46–48.

Bucher, EH and RB Zuccardi. 1967. Significación de los hormigueros de Atta vollenweideri Foreal como alteradores del suelo en la Provinces de Tucuman. *Acta Zool. Lilloana* 23:83–96.

Buck, BJ and JG Van Hoesen. 2002. Snowball morphology and SEM analysis of pedogenic gypsum, southern New Mexico, U. S.A. *J. Arid Env.* 51:469–487.

Buda, AR, Kleinman, PJA, Srinivasan, MS, Bryant, RB and GW Feyereisen. 2009. Factors influencing surface runoff generation from two agricultural hillslopes in central Pennsylvania. *Hydrol. Proc.* 23:1295–1312.

Büdel, B. 2001. Synopsis: Comparative biogeography of soil-crust biota. In: J Belnap and OL Lange (eds). *Biological Soil Crusts: Structure, Function, and Management*. Springer, Berlin. pp. 141–152.

Büdel, J. 1957. Die "Doppelten Einebnungsflächen" in den feuchten Tropen. *ZG* 2:201–228.

1982. *Climatic Geomorphology*. Princeton Univ. Press, Princeton, NJ.

Buggle, B, Glaser, B, Hambach, U, Gerasimenko, N and S Markovic. 2011. An evaluation of geochemical weathering indices in loess-paleosol studies. *QI* 240:12–21.

Buggle, B, Hambach, U, Glaser, B, Gerasimenko, N, Marković, SB, Glaser, I and L Zöller. 2009. Stratigraphy and spatial and temporal paleoclimatic trends in East European loess paleosol sequences. *QI* 196:86–106.

Buhmann, C and PLC Grubb. 1991. A kaolin-smectite interstratification sequence from a red and black complex. *Clay Mins.* 26:343–358.

Buis, E and A Veldkamp. 2008. Modelling dynamic water redistribution patterns in arid catchments in the Negev Desert of Israel. *ESPL* 33:107–122.

Buj, O, Gisbert, J, McKinley, JM and B Smith. 2011. Spatial characterization of salt accumulation in early stage limestone weathering using probe permeametry. *ESPL* 36:383–394.

Bullock, P, Federoff, N, Jongerius, A, Stoops, G, Tursina, T, Babel, U, Aguilar, J, Altemüller, HJ, FitzPatrick, EA, Kowalinski, S, Paneque, G, Rutherford, GK and EA Yarilova. 1985. *Handbook for Soil Thin Section Description*. Waine Research Publications, Wolverhampton, UK.

Bullock, P, Milford, MH and MG Cline. 1974. Degradation of argillic horizons in Udalf soils in New York State. *SSSAP* 38:621–628.

Bullock, P and CP Murphy (eds). 1983. *Soil Micromorphology*. Vol. 2. *Soil Genesis*. AB Academic Publications, Berkhamsted, UK

Bulmer, CE and LM Lavkulich. 1994. Pedogenic and geochemical processes of ultramafic soils along a climatic gradient in southwestern British Columbia. *CJSS* 74:165–177.

Bunting, BT. 1977. The occurrence of vesicular structures in arctic and subarctic soils. *ZG* 21:87–95.

Buntley, GJ and RI Papendick. 1960. Worm-worked soils of eastern South Dakota: Their morphology and classification. *SSSAP* 24:128–132.

Buntley, GJ and FC Westin. 1965. A comparative study of developmental color in a Chestnut-Chernozem-Brunizem soil climosequence. *SSSAP* 29:579–582.

Buntley, GJ, Daniels, RB, Gamble, EE and WT Brown. 1973. Soil genesis, morphology and classification. In: PA Sanchez (ed). *A Review of Soils Research on Tropical Latin America*. North Carolina Agric. Exp. Station Tech. Bull. #219. Raleigh. pp. 1–37.

1977. Fragipan horizons in soils of the Memphis-Loring-Grenada sequence in west Tennessee. *SSSAJ* 41:400–407.

Buol, SW. 1965. Present soil-forming factors and processes in arid and semiarid regions. *SS* 99:45–49.

1973. Soil genesis, morphology and classification. In: PA Sanchez (ed). A Review of Soils Research on Tropical Latin America. *N. Carolina Agric. Exp. Sta. Tech. Bull.* 219:1–37.

Buol, SW and MG Cook. 1998. Red and lateritic soils of the world: Concept, potential, constraints and challenges. In: J Sehgal, WE Blum and KS Gajbhiye (eds). *Red and Lateritic Soils: Managing Red and Lateritic Soils for Sustainable Agriculture*. Vol. 1. Oxford and IBH Publ. Co., New Delhi. pp. 49–56.

Buol, SW and H Eswaran. 2000. Oxisols. *Adv. Agron.* 68:151–195.

Buol, SW and FD Hole. 1961. Clay skin genesis in Wisconsin soils. *SSSAP* 25:377–379.

Buol, SW, Hole, FD, McCracken, RJ and RJ Southard. 1997. *Soil Genesis and Classification*, 4th ed. Iowa State Univ. Press, Ames.

Buol, SW, Southard, RJ, Graham, RC, and PA McDaniel. 2011. *Soil Genesis and Classification*, 6th ed. Wiley-Blackwell, Chichester. 543 pp.

Buol, SW and MS Yesilsoy. 1964. A genesis study of a Mohave sandy loam profile. *SSSAP* 28:254–256.

Buondonno, C, Ermice, A, Buondonno, A, Murolo, M and ML Pugliano. 1998. Human-influenced soils from an iron and steel works in Naples, Italy. *SSSAJ* 62:694–700.

Burak, DL, Fontes, MPF, Santos, NT, Monteiro, LVS, Martins, ED and T Becquer. 2010. Geochemistry and spatial distribution of heavy metals in Oxisols in a mineralized region of the Brazilian Central Plateau. *Geoderma* 160:131–142.

Burges, A. 1968. The role of soil micro-flora in the decomposition and synthesis of soil organic matter. *Trans. 9th Intl.Congr. Soil Sci. (Adelaide)* 2:29–35.

Burghardt, W. 1994. Soils in urban and industrial environments. *Z. Pflanzen. Bodenkd.* 157:205–214.

Burke, BC, Heimsath, AM and AF White. 2007. Coupling chemical weathering with soil production across soil-mantled landscapes. *ESPL* 32:853–873.

Burke, IW, Yonker, CM, Parton, WJ, Cole, CV, Flach, K and DS Schimel. 1989. Texture, climate, and cultivation effects on soil organic matter content in US grassland soils. *SSSAJ* 53:800–805.

Burke, RM and PW Birkeland. 1979. Re-evaluation of multiparameter relative dating techniques and their application to the glacial sequence along the eastern escarpment of the Sierra Nevada, California. *QR* 11:21–51.

Burleigh, R. 1974. Radiocarbon dating: Some practical considerations for the archaeologist. *JAS* 1:69–87.

Burn, CR. 2004. The thermal regime of Cryosols. In: JM Kimble (ed). *Cryosols: Permafrost-Affected Soils*. Springer-Verlag, Berlin. pp. 391–413.

Burn, CR and CAS Smith. 1988. Observations of the thermal offset in near-surface mean annual ground temperatures at several sites near Mayo, Yukon Territory, Canada. *Arctic* 41:99–104.

Burnett, BN, Meyer, GA and LD McFadden. 2008. Aspect-related microclimatic influences on slope forms and processes, northeastern Arizona. *J. Geophys. Res. Earth Surf.* 113. doi:10.1029/2007JF000789

Burns, SF and PJ Tonkin. 1982. Soil-geomorphic models and the spatial distribution and development of alpine soils. In: C Thorn (ed). *Space and Time in Geomorphology*. Allen and Unwin, London. pp. 25–43.

Burras, CL and WH Scholtes. 1987. Basin properties and postglacial erosion rates of minor moraines in Iowa. *SSSAJ* 51:1541–1547.

Bursik, M. 1991. Relative dating of moraines based on landform degradation, Lee-Vining Canyon, California. *QR* 35:451–455.

Burt, R. (ed) 2004. *Soil Survey Laboratory Methods Manual*. Soil Survey Invest. Rept. 42, ver. 4.0. USDA, Natural Resources Conservation Service, Lincoln, NE.

Burt, R and EB Alexander. 1996. Soil development on moraines of Mendenhall Glacier, southeast Alaska. 2. Chemical transformations and micromorphology. *Geoderma* 72:19–36.

Busacca, AJ. 1987. Pedogenesis of a chronosequence in the Sacramento Valley, California, USA. I. Application of a soil development index. *Geoderma* 41:123–148.

1989. Long Quaternary record in eastern Washington, U.S.A., interpreted from multiple buried paleosols in loess. *Geoderma* 45:105–122.

Busacca, AJ and M Cremaschi. 1998. The role of time versus climate in the formation of deep soils of the Apennine fringe of the Po Valley, Italy. *QI* 51/52:95–107.

Busacca, AJ and MJ Singer. 1989. Pedogenesis of a chronosequence in the Sacramento Valley, California, USA. II. Elemental chemistry of silt fractions. *Geoderma* 44:43–75.

Bushnell, TM. 1942. Some aspects of the soil catena concept. *SSSAP* 7:466–476.

1945. The catena caldron. *SSSAP* 10:335–340.

Bushue, LJ, Fehrenbacher, JB and BW Ray. 1974. Exhumed paleosols and associated modern till soils in western Illinois. *SSSAP* 34:665–669.

Busson, G and JP Perthuisot. 1977. Intérêt de laSabkha el Melah (Sud Tunisien) pour l'interpretation des séries évaporitiques anciennes. *Sed. Geol.* 19:139–164.

Butler, BE. 1956. Parna: An eolian clay. *Austral J. Sci.* 18:145–151.

1959. *Periodic Phenomena in Landscapes as a Basis for Soil Studies*. CSIRO Soil Publication #14. Melbourne, Australia.

1982. A new system for soil studies. *JSS* 33:581–595.

Butt, CRM. 1990. Genesis of supergene gold deposits in the lateritic regolith of the Yilgarn Block, Western Australia. *Econ. Geol. Monog.* 6:460–470.

Butzer, KW. 1982. *Archaeology as Human Ecology*. Cambridge Univ. Press, Cambridge.

Buurman, P. 1998. Classification of paleosols – a comment. *QI* 51/52:17–33.

Buurman, P, Jongmans, AG and MD PiPujol. 1998. Clay illuviation and mechanical clay infiltration – is there a difference? *QI* 51:66–69.

Buurman, P and LP van Reeuwijk. 1984. Proto-imogolite and the process of podzol formation: A critical note. *JSS* 35:447–452.

Buylaert, JP, Jain, M, Murray, AS, Thomsen, KJ, Thiel, C and R Sohbati. 2012. A robust feldspar luminescence dating method for Middle and Late Pleistocene sediments. *Boreas* 41:435–451.

Byers, HG, Kellogg, CE, Anderson, MS and J Thorp. 1938. Formation of soil. In: *Soils and Men: Yearbook of Agriculture*. US Dept. of Agric., US Govt. Printing Off., Washington, DC. pp. 948–978.

Cabrera-Martinez, F, Harris, WG, Carlisle, VW and ME Collins. 1989. Evidence for clay translocation in coastal plain soils with sandy/loamy boundaries. *SSSAJ* 53:1108–1114.

Cady, JG. 1950. Rock weathering and soil formation in the North Carolina Piedmont region. *SSSAP* 15:337–342.

Cady, JG and RB Daniels. 1968. Genesis of some very old soils: The Paleudults. *Trans. 9th Intl. Congr. Soil Sci.* 4:103–112.

Cady, JG, Wilding, LP and LR Drees. 1986. Petrographic microscope techniques. In: A Klute (ed). *Methods of Soil Analysis*. Part 1. *Physical and Mineralogical Methods*. Agronomy Mon. 9. American Society for Agronomy, Madison, WI. pp. 185–218.

Cai, SS and ZC Yu. 2011. Response of a warm temperate peatland to Holocene climate change in northeastern Pennsylvania. *QR* 75:531–540.

Cailleux, A. 1942. Les actiones e'oliennes pe'riglaciaires en Europe. *Mem. Soc. Geol. Fr.* 41:1–166.

1942. Les actions éoliennes periglaciaires en Europe. *Mém. Soc. Geol. France* 46:1–166.

Caillier, M, Guillet, B and M Gury. 1985. Differential alteration of phyllosilicate minerals under hydromorphic conditions. *Appl. Clay Sci.* 1:57–64.

Caldwell, RE and J Pourzad. 1974. Characterization of selected Paleudults in west Florida. *Soil Crop Sci. Soc. Florida Proc.* 33:143–147.

Caldwell, TG, McDonald, EV, Bacon, SN and G Stullenbarger. 2008. The performance and sustainability of vehicle dust courses for military testing. *J. Terramechanics* 45:213–221.

Calegari, MR, Madella, M, Vidal-Torrado, P, Pessenda, LCR and FA Marques. 2013. Combining phytoliths and delta C-13 matter in Holocene palaeoenvironmental studies of tropical soils: An example of an Oxisol in Brazil. *QI* 287:47–55.

Calhoun, FG, Smeck, NE, Slater, BL, Bigham, JM and GF Hall. 2001. Predicting bulk density of Ohio soils from morphology, genetic principles, and laboratory characterization data. *SSSAJ* 65:811–819.

Calkin, PE and JM Ellis. 1980. A lichenometric dating curve and its application to Holocene glacier studies in the Central Brooks Range, Alaska, USA. *AAR* 12:245–264.

Callahan, TJ, Vulava, VM, Passarello, MC and CG Garrett. 2012. Estimating groundwater recharge in lowland watersheds. *Hydrol. Proc.* 26:2845–2855.

Callen, RA. 1983. Late Tertiary "grey billy" and the age and origin of surficial silicifications (silcrete) in South Australia. *J. Geol. Soc. Austral.* 30:393–410.

Callen, RA, Wasson, RJ and R Gillespie. 1983. Reliability of radiocarbon dating of pedogenic carbonate in the Australian arid zone. *Sed. Geol.* 35:1–14.

Calus, JK. 1996. *Soil Survey of Oceana County, Michigan*. USDA Natural Resources Cons. Service, US Govt. Printing Off., Washington, DC.

Calvert, CS, Buol, SW and SB Weed. 1980a. Mineralogical characteristics and transformations of a vertical rock-saprolite-soil sequence in the North Carolina Piedmont. I. Profile morphology, chemical composition, and mineralogy. *SSSAJ* 44:1096–1103.

1980b. Mineralogical characteristics and transformations of a vertical rock-saprolite-soil sequence in the North Carolina Piedmont. II. Feldspar alteration products: Their transformations through the profile. *SSSAJ* 44:1104–1112.

Calvet, F, Pomar, L and M Esteban. 1975. Las Rizocreciones del Pleistoceno de Mallorca. *Inst. Invest. Geol., Univ. Barcelona* 30:36–60.

Calzolari, C and F Ungaro. 2012. Predicting shallow water table depth at regional scale from rainfall and soil data. *J. Hydrol.* 414:374–387.

Campbell, CA, Paul, EA, Rennie, DA and KJ McCallum. 1967. Factors affecting the accuracy of the carbon-dating method of soil humus studies. *SS* 104:81–85.

Campbell, IB and GCC Claridge. 1990. *Classification of cold desert soils*. In: JM Kimble and WD Nettleton (eds). *Characterization, Classification, and Utilization of Aridisols*. Proc. 4th Intl. Soil Correlation Meeting (ISCOM IV). USDA-SCS, Lincoln, NE. pp. 37–43.

Campbell, JB. 1977. Variation of selected properties across a soil boundary. *SSSAJ* 41:578–582.

1978. Spatial variation of sand content and pH within single contiguous delineations of two mapping units. *SSSAJ* 42:460–464.

1979. Spatial variability of soils. *AAAG* 69:544–556.

Campbell, JB and WJ Edmonds. 1984. The missing geographic dimension to Soil Taxonomy. *AAAG* 74:83–97.

Campos, MCC, Marques, J, de Souza, ZM, Siqueira, DS and GT Pereira. 2012. Discrimination of geomorphic surfaces with multivariate analysis of soil attributes in sandstone – basalt lithosequence. *Revista Ciencia Agron.* 43:429–438.

Caner, L, Joussein, E, Salvador-Blanes, S, Hubert, F, Schlicht, JF and N Duigou. 2010. Short-time clay-mineral evolution in a soil chronosequence in Oleron Island (France). *J. Plant Nutr. Soil Sci.* 173:591–600.

Cann, DB and EP Whiteside. 1955. A study of the genesis of a Podzol-Gray-Brown Podzolic intergrade soil profile in Michigan. *SSSAP* 19:497–501.

Canti, MG and TG Piearce. 2003. Morphology and dynamics of calcium carbonate granules produced by different earthworm species. *Pedobiologia* 47:511–521.

Cantlon, JE. 1953. Vegetation and microclimates on north and south slopes of Cushetunk Mountains, New Jersey. *Ecol. Monogr.* 23:241–270.

Cárcamo, HA., Abe, TA, Prescott, CE, Holl, FB, and CP Chanway. 2000. Influence of millipedes on litter decomposition, N mineralization, and microbial communities in a coastal forest in British Columbia, Canada. *CJFR* 30:817–826.

Carlson, DC and EM White. 1987. Effects of prairie dogs on mound soils. *SSSAJ* 51:389–393.

1988. Variations in surface-color, texture, pH, and phosphorous content across prairie dog mounds. *SSSAJ* 52:1758–1761.

Carpenter, CC. 1953. A study of hibernacula and hibernating associations of snakes and amphibians in Michigan. *Ecology* 34:445–453.

Carrara, PE and JT Andrews. 1973. Problems and application of lichenometry to geomorphic studies, San Juan Mountains, Colorado. *AAR* 5:373–384.

Carrington, EM, Hernes, PJ, Dyda, RY, Plante, AF and J Six. 2012. Biochemical changes across a carbon saturation gradient: Lignin, cutin, and suberin decomposition and stabilization in fractionated carbon pools. *Soil Biol. Biochem.* 47:179–190.

Carson, EC, Hanson, PR, Attig, JW and AR Young. 2012. Numeric control on the late-glacial chronology of the southern Laurentide Ice Sheet derived from ice-proximal lacustrine deposits. *QR* 78:583–589.

Carson, MA. 1984. The meandering-braided river threshold: A reappraisal. *J. Hydrol.* 73:315–334.

Carson, MA and MJ Kirkby. 1972. *Hillslope Form and Process*. Cambridge Univ. Press, Cambridge.

Carter, BJ and EJ Ciolkosz. 1980. Soil temperature regimes of the central Appalachians. *SSSAJ* 44:1052–1058.

1991. Slope gradient and aspect effects on soils developed from sandstone in Pennsylvania. *Geoderma* 49:199–213.

Carter, BJ and WP Inskeep. 1988. Accumulation of pedogenic gypsum in western Oklahoma soils. *SSSAJ* 52:1107–1113.

Carter, BJ, Kelley, JP, Sudbury, JB and DK Splinter. 2009. Key aspects of A horizon formation for selected buried soils in Late Holocene alluvium; Southern Prairies, USA. *SS* 174:408–416.

Carter, BJ, Ward, PA III and JT Shannon. 1990. Soil and geomorphic evolution within the Rolling Red Plains using Pleistocene volcanic ash deposits. *Geomorphology* 3:471–488.

Carter, GF and RL Pendleton. 1956. The humid soil: Process and time. *Geog. Rev.* 46:488–507.

Carter, JT, Gotkowitz, MB and MP Anderson. 2011. Field verification of stable perched groundwater in layered bedrock uplands. *Ground Water* 49:383–392.

Carter, NEA and HA Viles. 2005. Bioprotection explored: The story of a little known earth surface process. *Geomorphology* 67:273–281.

Cary, JW, Campbell, GS and RI Papendick. 1978. Is the soil frozen or not? An algorithm using weather records. *Water Resources Res.* 14:1117–1122.

Caseldine, CJ. 1984. Pollen analysis of a buried Arctic-Alpine Brown soil from Vestre Memurubreen, Jotunheimen, Norway: Evidence for postglacial high-altitude vegetation change. *AAR* 16:423–430.

Cates, KJ and FW Madison. 1993. *Soil-Attentuation-Potential Map of Trempealeau County, Wisconsin, Soil map #14*. Wisconsin Geology and Natural History Survey, Madison, WI.

Catoni, M, Falsone, G and E Bonifacio. 2012. Assessing the origin of carbonates in a complex soil with a suite of analytical methods. *Geoderma* 175:47–57.

Catt, JA. 1990. Paleopedology manual. *QI* 6:1–95.

1991. Soils as indicators of Quaternary climate change in mid-latitude regions. *Geoderma* 51:167–187.

1998. Report from working group on definitions used in paleopedology. *QI* 51/52:84.

Cerling, TE. 1984. The stable isotope composition of modern soil carbonate and its relationship to climate. *EPSL* 71:229–240.

1990. Dating geomorphologic surfaces using cosmogenic ^{3}He. *QR* 33:148–156.

1999. Paleorecords of C_4 plants and ecosystems. In: RF Sage and RK Monson (eds). *C4 Plant Biology*. Academic Press, San Diego, CA. pp. 445–471.

Cerling, TE and H Craig. 1994. Geomorphology and in-situ cosmogenic isotopes. *Ann. Rev. Earth Planet. Sci.* 22:273–317.

Cerling, TE and RL Hay. 1986. An isotopic study of paleosol carbonates from Olduvai Gorge. *QR* 25:63–78.

Cerling, TE, Quade, J, Wang, Y and JR Bowman. 1989. Carbon isotopes in soils and palaeosols as ecology and palaeoecology indicators. *Nature* 341:138–139.

Cerna, B and Z Engel. 2011. Surface and sub-surface Schmidt hammer rebound value variation for a granite outcrop. *ESPL* 36:170–179.

Černohouz, J and I Šolc. 1966. Use of sandstone wanes and weathered basalt crust in absolute chronology. *Nature* 212:806–807.

Certini, G and R Scalenghe. 2011. Anthropogenic soils are the golden spikes for the Anthropocene. *Holocene* 21:1269-1274.

Certini, G, Scalenghe, R and WI Woods. 2013. The impact of warfare on the soil environment. *Earth-Sci. Revs.* 127:1-15.

Ceryan, S, Tudes, S and N Ceryan. 2008. Influence of weathering on the engineering properties of Harsit granitic rocks (NE Turkey). *Bull. Engineer. Geol. Environ.* 67:97-104.

Chadwick, OA and J Chorover. 2001. The chemistry of pedogenic thresholds. *Geoderma* 100:321-353.

Chadwick, OA and JO Davis. 1990. Soil-forming intervals caused by eolian pulses in the Lahontan Basin, northwestern Nevada. *Geology* 18:243-246.

Chadwick, OA, Brimhall, GH and DM Hendricks. 1990. From a black box to a gray box: A mass balance interpretation of pedogenesis. *Geomorphology* 3:369-390.

Chadwick, OA, Derry, LA, Vitousek, PM, Huebert, BJ and LO Hedin. 1999. Changing sources of nutrients during four million years of ecosystem development. *Nature* 397:491-497.

Chadwick, OA, Hall, RD and FM Phillips. 1997. Chronology of Pleistocene glacial advances in the central Rocky Mountains. *GSA Bull.* 109:1443-1452.

Chadwick, OA, Hendricks, DM and WD Nettleton. 1987. Silica in duric soils. I. A depositional model. *SSSAJ* 51:975-982.

Chadwick, OA, Nettleton, WD and GJ Staidl. 1995. Soil polygenesis as a function of Quaternary climate change, northern Great Basin, USA. *Geoderma* 68:1-26.

Chadwick, O, Sowers, J and R Amundson. 1989a. Silicification of Holocene soils in northern Monitor Valley, Nevada. *SSSAJ* 53:158-164.

1989b. Morphology of calcite crystals in clast coatings from four soils in the Mojave Desert region. *SSSAJ* 53:211-219.

Chamberlain, EJ and AJ Gow. 1979. Effect of freezing and thawing on the permeability and structure of soils. *Engin. Geol.* 13:73-92.

Chamberlain, EJ, Iskandar, I and SE Hunsicker. 1990. Effect of freeze-thaw cycles on the permeability and macrostructure of soils. In: *Frozen Soil Impacts on Agricultural, Range, and Forest Lands*. US Army Cold Regions Research and Engineering Laboratory, Hanover, NH. pp. 144-155.

Chamberlain, TC. 1895. The classification of American glacial deposits. *J. Geol.* 3:270-277.

Chambers, FM. 1999. The Quaternary history of Llangorse Lake: Implications for conservation. *Aquatic Cons. Marine Fresh. Ecosyst.* 9:343-359.

Chandler, RF Jr. 1942. The time required for Podzol profile formation as evidenced by the Mendenhall glacial deposits near Juneau, Alaska. *SSSAP* 7:454-459.

Chang, CW. 1950. Effects of long-time cropping on soil properties. *SS* 69:359-368.

Chaplot, V, Walter, C and P Curmi. 2000. Improving soil hydromorphy prediction according to DEM resolution and available pedological data. *Geoderma* 97:405-422.

Chapman, SL and ME Horn. 1968. Parent material uniformity and origin of silty soils in northwest Arkansas based on zirconium-titanium contents. *SSSAP* 32:265-271.

Chappel, J. 1983. Thresholds and lags in geomorphologic changes. *Austral. Geog.* 15:357-366.

Chartres, CJ. 1987. The composition and formation of grainy void cutans in some soils with textural contrast in southeastern Australia. *Geoderma* 39:209-233.

Chatters, JC and KA Hoover. 1986. Changing late Holocene flooding frequencies on the Columbia River, Washington. *QR* 26:309-320.

Chauvel, A, Boulet, R and Y Lucas. 1987. On the genesis of the soil mantle of the region of Manaus, Central Amazonia, Brazil. *Experientia* 43:234-241.

Chauvel, A and G Pedro. 1978. Sur l'importance de l'extrême dessication des sols (ultradessication) dans l'évolution des zones tropicales à saisons contrastées. *C. R. Acad. Sci. Paris* 286:1581-1584.

Chen, J, An, Z and J Head. 1999. Variation of the Rb/Sr ratios in the loess-paleosol sequences of Central China during the last 130,000 years and their implications for monsoon paleoclimatology. *QR* 51:215-219.

Chen, J, Blume, HP and L Beyer. 2000. Weathering of rocks induced by lichen colonization: A review. *Catena* 39:121-146.

Chen, LM, Zhang, GL and WR Effland. 2011. Soil characteristic response times and pedogenic thresholds during the 1000-year evolution of a paddy soil chronosequence. *SSSAJ* 75:1807-1820.

Chen, Y and H Polach. 1986. Validity of ^{14}C ages of carbonates in sediments. *Radiocarbon* 28:464-472.

Chen, ZS, Tsou, TC, Asio, VB and CC Tsai. 2001. Genesis of Inceptisols on a volcanic landscape in Taiwan. *SS* 166:255-266

Chendev, YG, Burras, CL and TJ Sauer. 2012. Transformation of forest soils in Iowa (United States) under the impact of long-term agricultural development. *Euras. Soil Sci.* 45:357-367.

Chenery, EM. 1951. Some aspects of the aluminium cycle. *JSS* 2:97-109.

Cheng, W, Coleman, DC, Carroll, CR and CA Hoffman. 1993. In situ measurement of root respiration and soluble C concentrations in the rhizosphere. *Soil Biol. Biochem.* 25:1189-1196.

Chenu, C and D Cosentino. 2011. Microbial regulation of soil structural dynamics. In: *Architecture and Biology of Soils: Life in Inner Space*. Cabi, Publ., Oxon, England. pp. 37-70.

Cherkinsky, AE and VA Brovkin. 1991. A model of humus formation in soils based on radiocarbon data of natural ecosystems. *Radiocarbon* 33:186-187.

1993. Dynamics of radiocarbon in soils. *Radiocarbon* 35:363-367.

Chertov, OG. 1966. Description of humus-profile types for podzolic soils of Leningrad Oblast. *SSS* 266-275.

Chester, R, Baxter, GG, Behairy, AKA, Connor, K, Cross, D, Elderfield, H and RC Padgham. 1977. Soil-sized eolian dusts from the lower troposphere of the eastern Mediterranean Sea. *Marine Geol.* 24:201-217.

Chesworth, W. 1973. The parent rock effect in the genesis of soil. *Geoderma* 10:215-225.

Childs, CW. 1981. Field test for ferrous iron and ferric-organic complexes (on exchange sites in water-soluble forms) in soils. *AJSR* 19:175-180.

Childs, CW, Parfitt, RL and R Lee. 1983. Movement of aluminum as an inorganic complex in some podzolised soils, New England. *Geoderma* 29:139-155.

Chinn, TJH. 1981. Use of rock weathering-rind thickness for Holocene absolute age-dating in New Zealand. *AAR* 13:33-45.

Chisolm, PS, Irwin, RW and CJ Acton. 1984. Intepretation of soil drainage groups from soil taxonomy – southern Ontario. *CJSS* 64:383-393.

Chisonga, BC, Gutzmer, J, Beukes, NJ and JM Huizenga. 2012. Nature and origin of the protolith succession to the Paleoproterozoic Serra do Navio manganese deposit, Amapa Province, Brazil. *Ore Geol. Revs.* 47:59-76.

Chittleborough, DJ. 1981. Genesis of red-brown earths. In: JM Oades, DG Lewis and K Norrish (eds). *Red-Brown Earths of Australia*. Waite Agric. Res. Inst., Adelaide, South Australia. pp. 29-45.

1991. Indices of weathering for soils and palaeosols formed on silicate rocks. *Austr. J. Earth Sci.* 38:115-120.

1992. Formation and pedology of duplex soils. *Austral. J. Exp. Agric.* 32:815-825.

Chittleborough, DJ and JM Oades. 1980. The development of a red-brown earth. III. The degree of weathering and translocation of clay. *AJSR* 18:375-385.

Chittleborough, DJ, Walker, PH and JM Oades. 1984. Textural differentiation in chronosequences from eastern Australia. III. Evidence from elemental chemistry. *Geoderma* 32:227–248.

Chiu, TC, Fairbanks, RG, Cao, L and RA Mortlock. 2007. Analysis of the atmospheric ^{14}C record spanning the past 50,000 years derived from high-precision $^{230}Th/^{234}U/^{238}U$, $^{231}Pa/^{235}U$ and ^{14}C dates on fossil corals. *QSR* 26:18–36.

Chiverrell, RC, Foster, GC, Thomas, GSP, Marshall, P and D Hamilton. 2009. Robust chronologies for landform development. *ESPL* 34:319–328.

Chlachula, J, Evans, ME and NW Rutter. 1998. A magnetic investigation of a Late Quaternary loess/palaeosol record in Siberia. *Geophys. J. Intl.* 132:128–132.

Christopherson, RW. 2000. *Geosystems: An Introduction to Physical Geography.* Prentice-Hall, Upper Saddle River, NJ.

Church, M. 2002. Geomorphic thresholds in riverine landscapes. *Freshwater Biol.* 47:541–557.

Churchward, M. 1989. Whither field pedology. *Soils News* 37:94–95.

Ciolkosz, EJ. 1965. Peat mounds in southeastern Wisconsin. *SSH* 6:15–17.

Ciolkosz, EJ, Cronce, RC, Cunningham, RL and GW Petersen. 1985. Characteristics, genesis and classification of Pennsylvania minesoils. *SS* 139:232–238.

Ciolkosz, EJ, Petersen, GW, Cunningham, RL and RP Matelski. 1979. Soils developed from colluvium in the Ridge and Valley area of Pennsylvania. *SS* 128:153–162.

Ciolkosz, EJ, Waltman, WJ, Simpson, TW and RR Dobos. 1989. Distribution and genesis of soils of the northeastern United States. *Geomorphology* 2:285–302.

Ciolkosz, EJ, Waltman, WJ and NC Thurman. 1995. Fragipans in Pennsylvania soils. *SSH* 36:5–20.

Ciolkosz, EJ, Waltman, WJ, Thurman, NC, Cremeens, DL and MD Svoboda. 1996. Argillic horizons in Pennsylvania soils. *SSH* 37:20–44.

Cioppa, MT, Karlstrom, ET, Irving, E and RW Barendregt. 1995. Paleomagnetism of tills and associated paleosols in southwestern Alberta and northern Montana: Evidence for Late Pliocene-Early Pleistocene glaciations. *CJES* 32:555–564.

Cipra, JE, Bidwell, OW, Whitney, DA and AM Feyerherm. 1972. Variations with distance in selected fertility measurements of pedons of Western Kansas Ustolls. *SSSAP* 36:111–115.

Claridge, GGC and IB Campbell. 1968. Soils of the Shackleton Glacier region, Queen Maud Range, Antarctica. *N. Z. J. Sci.* 11:171–218.

1982. A comparison between hot and cold desert soils and soil processes. *Catena Suppl.* 1:1–28.

Clark, B, York, R and JB Foster. 2009. Darwin's worms and the skin of the Earth: An introduction to Charles Darwin's The Formation of Vegetable Mould, Through the Action of Worms, With Observations on Their Habits (Selections). *Organiz. Environ.* 22:338–350.

Clark, MM. 1972. Intensity of shaking estimated from displaced stones. *USGS Prof. Paper* 787:175–182.

Clayden, B. 1992. The role of todays pedologist. *New Zealand Soil News* 40:148–150.

Clayden, B, Daly, BK, Lee, R and G Mew. 1990. The nature, occurrence and genesis of placic horizons. In: JM Kimble and RD Yeck (eds). *Characterization, Classification, and Utilization of Spodosols.* Proc. 5th Intl. Soil Correlation Mtg (ISCOM). USDA-SCS, Lincoln, NE. pp. 88–104.

Clayton, L and JW Attig. 1989. *Glacial Lake Wisconsin.* Wisc. Geol. Nat. Hist Survey Memoir 173, Madison, WI.

Clayton, L, Attig, JW, Ham, NR, Johnson, MD, Jennings, CE and KM Syverson. 2008. Ice-walled-lake plains: Implications for the origin of hummocky glacial topography in middle North America. *Geomorphology* 97:237–248.

Cleaves, ET, Godfrey, AE and OP Bricker. 1970. Geochemical balance of a small watershed and its geomorphic implications. *GSA Bull.* 81:3015–3032.

Clements, FE. 1936. Nature and structure of the climax. *JE* 24:252–284.

Clemmensen, LB and AS Murray. 2010. Luminescence dating of Holocene spit deposits: An example from Skagen Odde, Denmark. *Boreas* 39:154–162.

Cline, AJ and DD Johnson. 1963. Threads of genesis in the Seventh Approximation. *SSSAP* 27:220–222.

Cline, MG. 1949a. Basic principles of soil classification. *SS* 67:81–91.

1949b. Profile studies of normal soils of New York. I. Soil profile sequences involving Brown Forest, Gray-Brown Podzolic, and Brown Podzolic soils. *SS* 68:259–272.

1955. *Soil Survey of the Territory of Hawaii.* USDA-SCS, US Govt. Printing Off., Washington, DC.

1961. The changing model of soil. *SSSAP* 25:442–446

1963. Logic of the new system of soil classification. *SS* 96:17–22.

1975. Origin of the term Latosol. *SSSAP* 39:162.

1977. Historical highlights in soil genesis, morphology, and classification. *SSSAJ* 41:250–254.

Clinton, BD and CR Baker. 2000. Catastrophic windthrow in the southern Appalachians: Characteristics of pits and mounds and initial vegetation responses. *For. Ecol. Mgmt.* 126:51–60.

Clothier, B and D Scotter. 2002. The soil solution phase: Unsaturated water transmission parameters obtained from infiltration. In: JH Duane and GC Topp (eds). *Methods of Soil Analysis. Part 4. Physical Methods.* SSSA Book Series no. 5. Soil Sci. Soc. Am. Madison, WI. pp. 879–898.

Clothier, BE, Pollok, JA and DR Scotter. 1978. Mottling in soil profiles containing a coarse-textured horizon. *SSSAJ* 42:761–763.

Coates, DR. 1984. *Landforms and landscapes as measures of relative time.* In: WC Mahaney (ed). Quaternary Dating Methods. Elsevier, New York. pp. 247–267.

Coates, DR and JD Vitek (eds) 1980. *Thresholds in Geomorphology.* Allen and Unwin, London.

(eds) 1980. *Perspectives on geomorphic thresholds.* In: DR Coates and JD Vitek (eds). *Thresholds in Geomorphology.* Allen and Unwin, London. pp. 3–23.

Cockburn, HAP and MA Summerfield. 2004. Geomorphological applications of cosmogenic isotope analysis. *PPG* 28:1–42.

Coester, M, Pfisterer, U and HK Siem. 1997. Soil development and heavy metal contents of a street sweepings dump. *Z. Pflanzen. Bodenk.* 160:89–92.

Coile, TS. 1953. Moisture content of small stone in soil. *SS* 75:203–207.

Cointepas, JP. 1967. *Les sols rouges et bruns meditérranéens de Tunisie.* Trans. Conf. Mediterr. Soils (Madrid). pp. 187–194.

Coleman, DC, Crossley, DA Jr and PF Hendrix. 2004. *Fundamentals of Soil Ecology,* 2nd ed. Elsevier, Amsterdam.

Coles, B. 1987. Tracks across the wetlands: Multi-disciplinary studies in the Somerset levels of England. In: JM Coles and AJ Lawson (eds). *European Wetlands in Prehistory.* Clarendon Press, Oxford. pp. 145–167.

Colgan, PM. 1999. Early middle Pleistocene glacial sediments (780 000–620 000 BP) near Kansas City, northeastern Kansas and northwestern Missouri, USA. *Boreas* 28:477–489.

Colin, F, Alarçon, C and P Vieillard. 1993. Zircon: An immobile index in soils? *Chem. Geol.* 107:273–276.

Collins, JF and T O'Dubhain. 1980. A micromorphological study of silt concentrations in some Irish podzols. *Geoderma* 24:215–224.

Collins, ME and RJ Kuehl. 2001. Organic matter accumulation and organic soils. In: MJ Vepraskas and JL Richardson (eds). *Wetland Soils: Genesis, Hydrology, Landscapes and Classification*. CRC Press, Boca Raton, FL. pp. 137–163.

Colman, SM. 1981. Rock-weathering rates as functions of time. *QR* 15:250–264.

——— 1982. Chemical weathering of basalts and andesites: Evidence from the weathering rinds. USGS Prof. Paper 1246.

Colman, SM and KL Pierce. 1981. Weathering rinds on basaltic and andesitic stones as a Quaternary age indicator, western United States. USGS Prof. Paper 1210:1–56.

——— 1986. Glacial sequence near McCall, Idaho: Weathering rinds, soil development, morphology, and other relative-age criteria. *QR* 25:25–42.

Colman, SM, Pierce, KL and PW Birkeland. 1987. Suggested terminology for Quaternary dating methods. *QR* 28:314–319.

Committee on Tropical Soils. 1972. *Soils of the Humid Tropics*. National Academy of Sciences, Washington, DC.

Conacher, AJ and JB Dalrymple. 1977. The nine unit landscape model: An approach to pedogeomorphic research. *Geoderma* 18:1–154.

Conkin, JE and BM Conkin. 1961. Fossil land snails from the loess at Vicksburg, Mississippi. *Trans. Kentucky Acad. Sci.* 22:10–15.

Connin, SL, Virginia, RA and CP Chamberlain. 1997. Isotopic study of environmental change from disseminated carbonate in polygenetic soils. *SSSAJ* 61:1710–1722.

Conry, MJ. 1971. Irish plaggen soils – their distribution, origin, and properties. *JSS* 22:401–415.

Conry, MJ, DeConinck, F and G Stoops. 1996. The properties, genesis and significance of a man-made iron pan Podzol near Castletownbere, Ireland. *EJSS* 47:279–284.

Corti, G, Ugolini, FC and A Agnelli. 1998. Classing the soil skeleton (greater than two millimeters): Proposed approach and procedure. *SSSAJ* 62:1620–1629.

Constantini, EAC and D Damiani. 2004. Clay minerals and the development of Quaternary soils in central Italy. *Revista Mex. De Ciencias Geol.* 21:144–159.

Constantini, EAC and G L'Abate. 2009. The soil cultural heritage of Italy: Geodatabase, maps, and pedodiversity evaluation. *QI* 209:142–153.

Constantini, EAC, Makeev, A and D Sauer. 2009. Recent developments and new frontiers in paleopedology. *QI* 209:1–5.

Costantini, EAC, Malucelli, F, Brenna, S and A Rocca. 2007. Using existing soil databases to consider paleosols in land planning. The case study of the Lombardy Region (Northern Italy). *QI* 162/163:166–171.

Constantini, EAC and S Priori. 2007. Pedogenesis of plinthite during early Pliocene in the Mediterranean environment – Case study of a buried paleosol at Podere Renieri, central Italy. *Catena* 71:425–443.

Cook, ER and LA Kairiukstis. 1990. *Methods of Dendrochronology: Applications in the Environmental Sciences*. Kluwer, Dordrecht.

Cooke, HBS. 1973. Pleistocene chronology: Long or short? *QR* 3:206–220.

Cooke, RU. 1970. Stone pavements in deserts. *AAAG* 60:560–577.

Cooke, RU, Warren, A and AS Goudie. 1993. *Desert Geomorphology*. UCL Press, London.

Cooper, AW. 1960. An example of the role of microclimate in soil genesis. *SS* 90:109–120.

Cooper, TH and J Crellin. 1996. Lamellae morphology in a sandy outwash soil of east-central Minnesota. *Soil Survey Hor.* 37:87–92.

Cooper, WS. 1913. The climax forest of Isle Royale, Lake Superior, and its development. *I. Bot. Gaz.* 55:1–44.

Cordova, CE. 2013. C$_3$ *Poaceae* and *Restionaceae* phytoliths as potential proxies for reconstructing winter rainfall in South Africa. *QI* 287:121–140.

Corio, K, Wolf, A, Draney, M and G Fewless. 2009. Exotic earthworms of Great Lakes forests: A search for indicator plant species in maple forests. *For. Ecol. Mgmt.* 258:1059–1066.

Cornelis, J-T, Dumon, M, Tolossa, AR, Delvaux, B, Deckers, J and E Van Ranst. 2014. The effect of pedological conditions on the sources and sinks of silicon in the Vertic Planosols in south-western Ethiopia. *Catena* 112:131–138.

Cornu, S, Lucas, Y, Lebon, E, Ambrosi, JP, Luizao, F, Rouiller, J, Bonnay, M and C Neal. 1999. Evidence of titanium mobility in soil profiles, Manaus, central Amazonia. *Geoderma* 91:281–295.

Cornu, S, Montagne, D and PM Vasconcelos. 2009. Dating constituent formation in soils to determine rates of soil processes: A review. *Geoderma* 153:293–303.

Cornu, S, Quénard, L and I Cousin. 2014. Experimental approach of lessivage: Quantification and mechanisms. *Geoderma* 213:357–370.

Corominas J and J Moya. 1999. Reconstructing recent landslide activity in relation to rainfall in the Llobregat River basin, eastern Pyrenees, Spain. *Geomorphology* 30:79–93.

Corte, AE. 1963. Particle sorting by repeated freezing and thawing. *Science* 142:499–501.

Cortes, A and DP Franzmeier. 1972. Climosequence of ash-derived soils in the central Cordillera of Columbia. *SSSAP* 36:653–659.

Corti, G, Ugolini, FC and A Agnelli. 1998. Classing the soil skeleton (greater than two millimeters): Proposed approach and procedure. *SSSAJ* 62:1620–1629.

Costa, ML. 1997. Lateritization as a major process of ore deposit formation in the Amazon region. *Explor. Mining Geol.* 6:79–104.

Costa, ML, Angelica, RS and NC Costa. 1999. The geochemical association Au-As-B-(Cu)-Sn-W in latosol, colluvium, lateritic iron crust and gossan in Carajas, Brazil: Importance for primary ore identification. *J. Geochem. Explor.* 67:33–49.

Costello, DM, Tiegs, SD and GA Lamberti. 2011. Do non-native earthworms in Southeast Alaska use streams as invasional corridors in watersheds harvested for timber? Biol. *Invasions* 13:177–187.

Costin, AB. 1955a. A note on gilgaies and frost soils. *JSS* 6:32–34.

——— 1955b. Alpine soils in Australia with reference to conditions in Europe and New Zealand. *JSS* 6:35–50.

Coulombe, CE, Wilding, LP and JB Dixon. 1996. Overview of Vertisols: Characteristics and impacts on society. *Adv. Agron.* 57:289–375.

Courchesne, F. 2000. Breaking the barrier of conceptual locks. *AAAG* 90:782–785.

Courchesne, F and GR Gobran. 1997. Mineralogical variations of bulk and rhizosphere soils from a Norway Spruce stand. *SSSAJ* 61:1245–1249.

Courchesne, F and WH Hendershot. 1997. La Genèse des Podzols. *Geog. Phys. Quat.* 51:235–250.

Courty, M-A, Goldberg, P and R Macphail. 1989. *Soils and Micromorphology in Archaeology*. Cambridge Univ. Press, New York.

Courty, M-A, Marlin, C, Dever, L, Tremblay, P and P Vachier. 1994. Morphology, geochemistry and origin of calcitic pendants from the High Arctic (Spitsbergen). *Geoderma* 61:71–102.

Coutard, JP and HJ Mücher. 1985. Deformation of laminated silt loam due to repeated freezing and thawing cycles. *ESPL* 10:309–319.

Couto, W, Sanzonowicz, C and A de O Barcellos. 1985. Factors affecting oxidation-reduction processes in an Oxisol with a seasonal water table. *SSSAJ* 49:1245–1248.

Coventry, RJ, Holt, JA and DF Sinclair. 1988. Nutrient cycling by mound-building termites in low-fertility soils of semi-arid tropical Australia. *AJSR* 26:375–390.

Covey, C. 1984. The Earth's orbit and the ice ages. *Sci. Am.* 250:58–66.

Cowan, JA, Humphreys, GS, Mitchell, PB and CL Murphy. 1985. An assessment of pedoturbation by two species of mound-building ants, *Camponotus intrepidus* (Kirby) and *Iridomyrmex purpureus* (F. Smith). *AJSR* 23:95–107.

Cox, GW. 1984. The distribution and origin of Mima mound grasslands in San Diego County, California. *Ecology* 65:1397–1405.

1990. Soil mining by pocket gophers along topographic gradients in a Mima moundfield. *Ecology* 71:837–843.

Cox, GW and DW Allen. 1987a. Soil translocation by pocket gophers in a Mima moundfield. *Oecologia* 72:207–210.

1987b. Sorted stone nets and circles of the Columbia Plateau: A hypothesis. *Northwest Sci.* 61:179–185.

Cox, A, Doell, RR and GB Dalrymple. 1963. Geomagnetic polarity epochs and Pleistocene geochronometry. *Nature* 198:1049–1051.

Cox, GW and CG Gakahu. 1986. A latitudinal test of the fossorial rodent hypothesis of Mima mound origin. *ZG* 30:485–501.

1987. Biogeographical relationships of rhizomyid mole rats with Mima mound terrain in the Kenya highlands. *Pedobiologia* 30:263–275.

Cox, GW, Gakahu, CG and DW Allen. 1987a. Small-stone content of Mima mounds of the Columbia Plateau and Rocky Mountain regions: Implications for mound origin. *Gr. Basin Natural.* 47:609–619.

Cox, GW, Gakahu, CG and JM Waithaka. 1989. The form and small stone content of large earth mounds constructed by mole rats and termites in Kenya. *Pedobiologia* 33:307–314.

Cox, GW and J Hunt. 1990. Form of Mima mounds in relation to occupancy by pocket gophers. *J Mammal.* 71:90–94.

Cox, GW and WT Lawrence. 1983. Cemented horizon in subarctic Alaskan sand dunes. *AJS* 283:369–373.

Cox, GW, Lovegrove, BG and WR Siegfried. 1987b. The small stone content of Mima-like mounds in the South African cape region: Implications for mound origin. *Catena* 14:165–176.

Cox, GW, Mills, JN and BA Ellis. 1992. Fire ants (Hymenoptera, Formicidae) as major agents of landscape development. *Environ. Entom.* 21:281–286.

Cox, GW and VG Roig. 1986. Agentinian Mima mounds occupied by Ctenomyid rodents. *J. Mamm.* 67:428–432.

Cox, GW and VB Scheffer. 1991. Pocket gophers and mima terrain in North America. *Natural Areas J.* 11:193–198.

Cox, NJ. 1980. On the relationship between bedrock lowering and regolith thickness. *Earth Surf. Proc.* 5:271–274.

Cox, T. 1998. Origin of stone concentrations in loess-derived interfluve soils. *QI* 51/52:74–75

Crabtree, RW and TP Burt. 1983. Spatial variation in solutional denudation and soil moisture over a hillslope hollow. *ESPL* 8:151–160.

Craig, H and RJ Poreda. 1986. Cosmogenic 3He in terrestrial rocks: The summit lavas of Maui. *PNAS USA* 83:1970–1974.

Crang RE, Holsen RC and JB Hitt. 1968. Calcite production in mitochondria of earthworm calciferous glands. *Bioscience* 18:299–301.

Crawford, RMM, Jeffree, CE and WG Rees. 2003. Paludification and forest retreat in northern oceanic environments. *Ann. Bot.* 91:213–226.

Crawford, TW, Jr, Whittig, LD, Begg, EL and GL Huntington. 1983. Eolian influence on development and weathering of some soils of Point Reyes peninsula, California. *SSSAJ* 47:1179–1185.

Cremaschi, M. 1987. Paleosols and Vetusols in the Central Po Plains (Northern Italy): A Study in Quaternary Geology and Soil Development, Studie ricerche sul territorio #28. Milan, Edizioni Unicopli.

Cremeens, DL. 1995. Pedogenesis of Cotiga Mound, a 2100-year-old Woodland mound in southwest West Virginia. *SSSAJ* 59:1377–1388.

2000. Pedology of the regolith-bedrock boundary: An example from the Appalachian plateau of northern West Virginia. *Southeast. Geol.* 39:329–339.

2003. Geoarchaeology of soils on stable geomorphic surfaces: Mature soil model for the glaciated northeast. In: DL Cremeens and JP Hart (eds). *Geoarchaeology of Landscapes in the Glaciated Northeast.* Bull. 497. New York State Museum, Albany. pp. 49–60.

Cremeens, DL, Brown, RB and JH Huddleston. (eds) 1994. *Whole Regolith Pedology.* SSSASP 34.

Cremeens, DL and JP Hart. 1995. On chronostratigraphy, pedostratigraphy, and archaeological context. In: ME Collins, BJ Carter, BG Gladfelter and RJ Southard (eds). Pedological Perspectives in Archaeological Research. Soil Sci. Soc. Am. Spec. Publ. 44:15–33.

Cremeens, DL and DL Mokma. 1986. Argillic horizon expression and classification in the soils of two Michigan hydrosequences. *SSSAJ* 50:1002–1007.

Cremeens, DL and JC Lothrop. 2001. Geomorphology of upland regolith in the unglaciated Appalachian Plateau. In: LP Sullivan and SC Prezzano (eds). *Archaeology of Appalachian Highlands.* Univ. of Tennessee Press, Knoxville. pp. 31–48.

Cremeens, DL, Norton, LD, Darmody, RG and IJ Jansen. 1988. Etch-pit measurements on scanning electron micrographs of weathered grain surfaces. *SSSAJ* 52:883–885.

Crocker, RL. 1952. Soil genesis and the pedogenic factors. *Q. Rev. Biol.* 27:139–168.

1960. The plant factor in soil formation. *Proc. 9th Pacific Sci. Congr.* 18:84–90.

Crocker, RL and J Major. 1955. Soil development in relation to vegetation and surface age at Glacier Bay, Alaska. *JE* 43:427–448.

Crompton, E. 1962. Soil formation. *Outlook Agric.* 3:209–218.

Crook, R. Jr. 1986. Relative dating of Quaternary deposits based on P-wave velocities in weathered granitic clasts. *QR* 25:281–292.

Jr and AR Gillespie. 1986. Weathering rates in granite boulders measured by P-wave speeds. In: SM Colman and DP Dethier (eds). *Rates of Chemical Weathering of Rocks and Minerals.* Academic Press, London. pp. 395–417.

Crosson, LS and R Protz. 1974. Quantitative comparison of two closely related soil mapping units. *CJSS* 54:7–14.

Crownover, SH, Collins, ME and DA Lietzke. 1994. Parent materials and stratigraphy of a doline in the valley and ridge province. *SSSAJ* 58:1738–1746.

Crutzen, PJ and EF Stoermer. 2000. The "Anthropocene." *Global Change Newsletter* 41:17–18.

Cullity, BD. 1978. *Elements of X-Ray Diffraction*, 2nd ed. Addison-Wesley, New York.

Culver, JR and F Gray. 1968a. Morphology and genesis of some grayish claypan soils of Oklahoma. I. Morphology, chemical and physical measurements. *SSSAP* 32:845–851.

1968b. Morphology and genesis of some grayish claypan soils of Oklahoma. II. Mineralogy and genesis. *SSSAP* 32:851–857.

Curi, N and DP Franzmeier. 1984. Toposequence of Oxisols from the Central Plateau of Brazil. *SSSAJ* 48:341–346.

Curry, BB and MJ Pavich. 1996. Absence of glaciation in Illinois during Marine Isotope Stages 3 through 5. *QR* 46:19–26.

Curtis, JT. 1959. *The Vegetation of Wisconsin.* Univ. of Wisc. Press, Madison.

Curtis, RO and BW Post. 1964. Estimating bulk density from organic-matter content in some Vermont forest soils. *SSSAP* 28:285–286.

Curtiss, B, Adams, JB and MK Ghiorso. 1985. Origin, development and chemistry of silica-alumina rock coatings from the semi-arid regions of the island of Hawaii. *Geochim. Cosmochim. Acta* 49:49–56.

Cutler, EJB. 1981. The texture profile forms of New Zealand soils. *AJSR* 19:97–102.

Cutler, KB, Gray, SC, Burr, GS, Edwards, RL, Taylor, FW, Cabioch, G, Beck, JW, Cheng, H and J Moore. 2004. Radiocarbon calibration and comparison to 50 kyr bp with paired C-14 and Th-230 dating of corals from Vanuatu and Papua New Guinea. *Radiocarbon* 46:1127–1160.

D'Amore, DV and WC Lynn. 2002. Classification of forested Histosols in southeast Alaska. *SSSAJ* 66:554–562.

Dahlgren, RA, Boettinger, JL, Huntington, GL and RG Amundson. 1997. Soil development along an elevational transect in the western Sierra Nevada, California. *Geoderma* 78:207–236.

Dahlgren, RA, Saigusa, M and FC Ugolini. 2004. The nature, properties, and management of volcanic soils. *Adv. Agron.* 82:113–182.

Dahlgren, RA and FC Ugolini. 1989. Aluminum fractionation of soil solutions from unperturbed and tephra-treated Spodosols, Cascade Range, Washington, USA. *SSSAJ* 53:559–566.

1991. Distribution and characterization of short-range-order minerals in Spodosols from the Washington Cascades. *Geoderma* 48:391–413.

Dahms, DE. 1993. Mineralogical evidence for eolian contributions to soils of Late Quaternary moraines, Wind River Mountains, Wyoming, USA. *Geoderma* 59:175–196.

1994. Mid Holocene erosion of soil catenas on moraines near the type Pinedale till, Wind River Range, Wyoming. *QR* 42:41–48.

1998. Reconstructing paleoenvironments from ancient soils: A critical review. *QI* 51/52:58–60.

2002. Glacial stratigraphy of Stough Creek Basin, Wind River Range, Wyoming. *Geomorphology* 42:59–83.

Dalrymple, GB, Cox, A and RR Doell. 1965. Potassium-argon age and paleomagnetism of the Bishop tuff, California. *GSA Bull.* 76:665–674.

Dalrymple, JB. 1958. The application of soil micromorphology to fossil soils and other deposits from archaeological sites. *JSS* 9:199–209.

Dalsgaard, K, Baastrup, E and BT Bunting. 1981. The influence of topography on the development of Alfisols on calcareous clayey till in Denmark. *Catena* 8:111–136.

Dalsgaard, K and BV Odgaard. 2001. Dating sequences of buried horizons of podzols developed in wind-blown sand at Ulfborg, Western Jutland. *QI* 78:53–60.

Daly, BK. 1982. Identification of Podzol and podzolized soils in New Zealand by relative absorbance of oxalate extracts of A and B horizons. *Geoderma* 28:29–38.

Daly, RA. 1963. *The Changing World of the Ice Age.* Hafner Publ., New York.

Dan, J. 1990. Effect of dust deposition on the land of Israel. *QI* 5:107–111.

Dan, J, Moshe, R and N Alperovich. 1973. The soils of Sede Zin. *Israel J. Earth Sci.* 22:211–227

Dan, J and DH Yaalon. 1964. The application of the catena concept in studies of pedogenesis in Mediterranean and desert fringe regions. *Trans. 8th Intl. Congr. Soil Sci.* 83:751–758.

1982. Automorphic saline soils in Israel. *Catena Suppl.* 1:103–115.

Dan, J, Yaalon, DH and H Koyumdjisky. 1968. Catenary soil relationships in Israel. I. The Netanya catena on coastal dunes of the Sharon. *Geoderma* 2:95–120.

1972. Catenary soil relationships in Israel. II. The Bet Gurvin catena on chalk and nari limestone crust in the Shefela. *Isr. J. Earth-Sci.* 21:99–114.

Dan, J, Yaalon, DH, Moshe, R and S Nissim. 1982. Evolution of reg soils in southern Israel and Sinai. *Geoderma* 28:173–202.

Dane, JH and JW Hopmans. 2002. The soil solution phase: Laboratory. In: JH Duane and GC Topp (eds). *Methods of Soil Analysis.* Part 4. Physical Methods, SSSA Book Series no. 5. Soil Sci. Soc. Am. Madison, WI. pp. 675–719.

Daniels, RB. 1988. Pedology, a field or laboratory science? *SSSAJ* 52:1518–1519.

Daniels, RB and SW Buol. 1992. *Water table dynamics and significance to soil genesis.* In: JM Kimble (ed). *Characterization, Classification, and Utilization of Wet Soils.* Proc. 8th Intl. Soil Correlation Mtg. (ISCOM). USDA-SCS, Lincoln, NE. pp. 66–74.

Daniels, RB, Buol, SW, Kleiss, HJ and CA Ditzler. 1999. *Soil Systems in North Carolina.* N. Carolina State Tech. Bull. 314, Raleigh.

Daniels, RB and EE Gamble. 1967. The edge effect in some Ultisols in the North Carolina coastal plain. *Geoderma* 1:117–124.

Daniels, RB, Gamble, EE and LJ Bartelli. 1968. Eluvial bodies in B horizons of some Ultisols. *SS* 106:200–206.

Daniels, RB, Gamble, EE and JG Cady. 1971a. The relation between geomorphology and soil morphology and genesis. *Adv. Agron.* 23:51–88.

Daniels, RB, Gamble, EE and CS Holzhey. 1975. Thick Bh horizons in the North Carolina coastal plain. I. Morphology and relation to texture and soil ground water. *SSSAP* 39:1177–1181.

Daniels, RB, Gamble, EE and LA Nelson. 1971b. Relations between soil morphology and water-table levels on a dissected North Carolina coastal plain surface. *SSSAP* 35:781–784.

Daniels, RB, Gamble, EE and WH Wheeler. 1978a. Age of soil landscapes in the coastal plain of North Carolina. *SSSAJ* 42:98–105.

Daniels, RB, Gamble, EE Wheeler, WH, and CS Holzhey. 1977. The stratigraphy and geomorphology of the Hofmann Forest pocosin. *SSSAJ* 41:1175–1180.

Daniels, RB and RD Hammer. 1992. *Soil Geomorphology.* Wiley, New York.

Daniels, RB, Handy, RL and GH Simonson. 1960. Dark-colored bands in the thick loess of western Iowa. *J. Geol.* 68:450–458.

Daniels, RB, Perkins, HF, Hajek, BF and EE Gamble. 1978b. Morphology of discontinuous phase plinthite and criteria for its field identification in the southeastern United States. *SSSAJ* 42:944–949.

Daniels, RB, Simonson, GH and RL Handy. 1961. Ferrous iron content and color of sediments. *SS* 91:378–382.

Daniels, WL, Everett, CJ and LW Zelazny. 1987. Virgin hardwood forest soils of the southern Appalachian Mountains. I. Soil morphology and geomorphology. *SSSAJ* 51:722–729.

Danin, A, Dor, I, Sandler, A and R Amit. 1998. Desert crust morphology and its relations to microbiotic succession at Mt. Sedom, Israel. *J. Arid Envs* 38:161–174.

Danin, A and J Garty. 1983. Distribution of cyanobacteria and lichens on hillsides of the Negev Highlands and their impact on biogenic weathering. *ZG* 27:423–444.

Danin, A, Gerson, R and J Garty. 1983. Weathering patterns on hard limestone and dolomite by endolithic lichens and cyanobacteria: Supporting evidence for eolian contribution to terra rossa soil. *SS* 136:213–217.

Dansgaard, W, Johnsen, SJ, Clausen, HB, Dahljensen, D, Gundestrup, NS, Hammer, CU, Hvidberg, CS, Steffensen, JP, Sveinbjornsdottir, AE, Jouzel, J and G Bond. 1993. Evidence for general instability of past climate from a 250-kyr ice-core record. *Nature* 364:218–220.

Dansgaard, W, White, JWC and SJ Johnsen. 1989. The abrupt termination of the Younger Dryas climate event. *Nature* 339:532–534.

Dare-Edwards, AJ. 1984. Aeolian clay deposits of southeastern Australia: Parna or loessic clay? *Trans. Inst. Brit. Geog.* 9:337–344.

Darmody, RG, Allen, CE and CE Thorn. 2005a. Soil topochronosequences at Storbreen, Jotunheimen, Norway. *SSSAJ* 69:1275–1287.

Darmody, RG, Fanning, DS, Drummond, WJ, Jr and JE Foss. 1977. Determination of the total sulfur in tidal marsh soils by X-ray spectroscopy. *SSSAJ* 41:761–765.

Darmody, RG and JE Foss. 1979. Soil-landscape relationships of the tidal marshes of Maryland. *SSSAJ* 43:534–541.

Darmody, RG, Thorn, CE and CE Allen. 2005b. Chemical weather and boulder mantles, Kärkevagge, Swedish Lapland. *Geomorphology* 67:159–170.

Darwin, C. 1881. *On the Formation of Vegetable Mould through the Action of Worms.* John Murray, London.

Darwish, TM and RA Zurayk. 1997. Distribution and nature of Red Mediterranean soils in Lebanon along an altitudinal sequence. *Catena* 28:191–202.

Dasog, GS, Acton, DF and AR Mermut. 1987. Genesis and classification of clay soils with vertic properties in Saskatchewan. *SSSAJ* 51:1243–1250.

Davey, BG, Russell, JD and MJ Wilson. 1975. Iron oxide and clay minerals and their relation to colours of red and yellow podzolic soils near Sydney, Australia. *Geoderma* 14:125–138.

David, MB, McIsaac, GF, Darmody, RG and RA Omonode. 2009. Long-term changes in Mollisol organic carbon and nitrogen. *J. Envir. Qual.* 38:200–211.

Davidson, DA. 2002. Bioturbation in old arable soils: Quantitative evidence from soil micromorphology. *JAS* 29:1247–1253.

Davis, JO. 1985. Correlation of Late Quaternary tephra layers in a long pluvial sequence near Summer Lake, Oregon. *QR* 23:38–53.

Davis, MB. 1969. Climatic changes in southern Connecticut recorded by pollen deposition at Rogers Lake. *Ecology* 50:409–422.

Davis, RB, Bradstreet, TE, Stuckenrath, R and HW Borns. 1975. Vegetation and associated environments during the past 14000 years near Moulton Pond, Maine. *QR* 5:435–465.

Davis, WM. 1899. The geographical cycle. *Geog. J.* 14:481–494.

Day, MJ and AS Goudie. 1977. Field assessment of rock hardness using the Schmidt test hammer. *Br. Geom. Res. Group Tech. Bull.* 18:19–29.

Dazzi, C, Lo Papa, G and V Palermo. 2009. Proposal for a new diagnostic horizon for WRB Anthrosols. *Geoderma* 151:16–21.

Dean, WRJ and SJ Milton. 1991. Disturbances in semi-arid shrubland and arid grassland in the Karoo, South Africa: Mammal diggings as germination sites. *Afr. J. Ecol.* 29:11–16.

Debruyn, LAL and AJ Conacher. 1994. The bioturbation activity of ants in agricultural and naturally vegetated habitats in semiarid environments. *AJSR* 32:555–570.

DeConinck, F. 1980. Major mechanisms in formation of spodic horizons. *Geoderma* 24:101–128.

DeConinck, F, Favrot, JC, Tavernier, R and M Jamagne. 1976. Dégradation dans les sols lessivés hydromorphes sur matériaux argilo-sableux. Exemple des sols de la nappe détritique bourbonnaise (France). *Pédologie* 26:105–151.

DeConinck, F and A Herbillon. 1969. Evolution minéralogique et chémique des fractions argileuses dans des Alfisols et des Spodosols de la Campine (Belgique). *Pédologie* 19:159–272.

DeConinck, F, Herbillon, AJ, Tavernier, R and JJ Fripiat. 1968. Weathering of clay minerals and formation of amorphous material during the degradation of a Bt horizon and podzolization in Belgium. *Trans. 9th Intl. Congr. Soil Sci.* 4:353–365.

Degens, BP, Sparling, GP and LK Abbott. 1996. Increasing the length of hyphae in a sandy soil increases the amount of water-stable aggregates. *Appl. Soil Ecol.* 3:149–159.

Dehnert, A and C Schlüchter. 2008. Sediment burial dating using terrestrial cosmogenic nuclides. *Quat. Sci. J.* 57:210–255.

Deines, P. 1980. The isotopic composition of reduced organic carbon. In: P Fritz and JC Fontes (eds). *Handbook of Environmental Isotope Geochemistry.* Vol. 1. *The Terrestrial Environment.* Elsevier, Amsterdam. pp. 329–406.

de Jong, E, Kozak, LM and HPW Rostad. 2000. Effects of parent material and climate on the magnetic susceptibility of Saskatchewan soils. *CJSS* 80:135–142.

De Jonge, LW, Kjaergaard, C and P Moldrup. 2004. Colloids and colloid-facilitated transport of contaminants in soils: An introduction. *Vadose Zone J.* 3:321–325.

De Kimpe, CR. 1970. Chemical, physical and mineralogical properties of a Podzol soil with fragipan derived from glacial till in the province of Quebec. *CJSS* 50:317–330.

De Kimpe, CR, Baril, RW and R Rivard. 1972. Characterization of a toposequence with fragipan: The Leeds-Ste. Marie-Brompton series of soils, province of Quebec. *CJSS* 52:135–150.

De Kimpe, CR and JL Morel. 2000. Urban soil management: A growing concern. *SS* 165:31–40.

Dekker, LW and CJ Ritsema. 1996. Uneven moisture patterns in water repellent soils. *Geoderma* 70:87–99.

de la Rosa, JPM, Warke, PA and BJ Smith. 2012. Microscale biopitting by the endolithic lichen Verrucaria baldensis and its proposed role in mesoscale solution basin development on limestone. *ESPL* 37:374–384.

Delgado, R, Martin-Garcia, JM, Calero, J, Casares-Porcel, M, Tito-Rojo, J and G Delgado. 2007. The historic man-made soils of the generalife garden (La Alhambra, Granada, Spain). *Europ. J. Soil Sci.* 58:215–228.

Delhon, C, Alexandre, A, Berger, J-F, Thiébault, S, Brochier, J-L, and J-D Meunier. 2003. Phytolith assemblages as a promising tool for reconstructing Mediterranean Holocene vegetation. *QR* 59:48–60.

Del Villar, EH. 1944. The tirs of Morocco. *SS* 57:313–339.

Demkina, TS, Khomutova, TE, Kashirskaya, NN, Demkina, EV, Stretovich, IV, El-Registan, GI and VA Demkin. 2008. Age and activation of microbial communities in soils under burial mounds and in recent surface soils of steppe zone. *Euras. Soil Sci.* 41:1439–1447.

Deng, CL. 2008. Paleomagnetic and mineral magnetic investigation of the Baicaoyuan loess-paleosol sequence of the western Chinese Loess Plateau over the last glacial-interglacial cycle and its geological implications. *Geochem. Geophys. Geosystems* 9. doi:10.1029/2007GC001928

Deng, CL, Vidic, NJ, Verosub, KL, Singer, MJ, Liu, QS, Shaw, J and RX Zhu. 2005. Mineral magnetic variation of the Jiaodao Chinese loess/paleosol sequence and its bearing on long-term climatic variability. *J. Geophys. Res. Solid Earth* 110. doi:10.1029/2004JB003451

Denneler, B and FH Schweingruber. 1993. Slow mass movement. A dendrogeomorphological study in Grams, Swiss Rhine Valley. *Dendrochronologia* 11:55–67.

Denny, CS. 1951. Pleistocene frost action near the border of the Wisconsin drifts in Pennsylvania. *Ohio J. Sci.* 51:116–125

Denny, CS and JC Goodlett. 1956. Microrelief resulting from fallen trees. *USGS Prof. Paper* 288:59–66.

1968. Tree-throw origin of patterned ground on beaches of the ancient Champlain Sea near Plattsburgh, New York. USGS Prof. Paper 600B:157–164.

DeNovio, NM, Saiers, JE and JN Ryan. 2004. Colloid movement in unsaturated porous media: Recent advances and future directions. *Vadose Zone J.* 3:338–351.

Dent, D, Hartemink, A and J Kimble. 2005. Soil – Earth's Living Skin. Earth Sciences for Society Foundation, Year of Planet Earth. http://yearofplanetearth.org/content/downloads/Soils.pdf.

Dent, DL. 1991. Reclamation of acid sulphate soils. *Adv. Soil Sci.* 17:79–122.

—— 1993. Bottom-up and top-down development of acid sulphate soils. *Catena* 20:4519–425.

Dent, DL and LJ. Pons. 1995. A world perspective on acid sulphate soils. *Geoderma* 67:262–276.

De Ploey, J. 1964. Nappes de gravats et couvertures argilo-sableuses au Bas-Congo: Leur genèse et l'action des termites. In: A Bouillon (ed). *Etudes sur les termites africains*. Editions de l'Université de Léopoldville, Léopoldville, Congo. pp. 399–414.

De Ploey, J, Savat, J and J Moeyersons. 1976. The differential impact of some soil loss factors on flow, runoff creep and rainwash. *ESPL* 1:151–161.

Derbyshire, E. 1984. Granulometry and fabric of the loess at Jiuzhoutai, Lanzhou, People's Republic of China. In: DN Eden and RJ Furkert (eds). *Loess Its Distribution, Geology and Soils*. Proc. Intl. Sympos. on Loess, New Zealand. AA Balkema, Rotterdam. pp. 93–103.

Derry, LA and OA Chadwick. 2007 Contributions from Earth's atmosphere to soil. *Elements* 3:333–338.

Dethier, DP, Birkeland, PW and JA McCarthy. 2012. Using the accumulation of CBD-extractable iron and clay content to estimate soil age on stable surfaces and nearby slopes, Front Range, Colorado. *Geomorphology* 173:17–29.

De Villiers, JM. 1965. Present soil-forming factors and processes in tropical and subtropical regions. *SS* 99:50–57.

De Vos, JH and KJ Virgo. 1969. Soil structure in Vertisols of the Blue Nile Clay Plains, Sudan. *JSS* 20:189–206.

de Vries, W and A Breeuwsma. 1984. Causes of soil acidification. *Neth. J. Agric. Sci.* 32:159–161.

Dewey, JC, Schoenholtz, SH, Shepard, JP and MG Messina. 2006. Issues related to wetland delineation of a Texas, USA bottomland hardwood forest. *Wetlands* 26:410–429.

Dexter, AR. 1978. Tunnelling of soil by earthworms. *Soil Biol. Biochem.* 10:447–449.

Diamond, DD and FE Smeins. 1985. Composition, classification and species response patterns of remnant tallgrass prairies in Texas. *AMN* 113:294–308.

Di-Bonaventura, MP, Del Gallo, M, Cacchio, P, Ercole, C and A Lepidi. 1999. Microbial formation of oxalate films on monument surfaces: Bioprotection or biodeterioration? *Geomicrobiology J.* 16:55–64.

Dickinson, WW, Schiller, M, Ditchburn, BG, Graham, IJ and A Zondervan. 2012. Meteoric Be-10 from Sirius Group suggests high elevation McMurdo Dry Valleys permanently frozen since 6 Ma. *EPSL* 355:13–19.

Dideriksen, RO. 1992. *Soil Survey of Wright County, Iowa*. USDA-NRCS, US Govt. Printing Off., Washington, DC.

Dietrich, WE, Reiss, R, Hsu, M and DR Montgomery. 1995. A process-based model for colluvial soil depth and shallow landsliding using digital elevation data. *Hydrol. Proc.* 9:383–400.

Dietze, M, Bartel, S, Lindner, M and A Kleber. 2012. Formation mechanisms and control factors of vesicular soil structure. *Catena* 99:83–96.

Dietze, M and A Kleber. 2012. Contribution of lateral processes to stone pavement formation in deserts inferred from clast orientation patterns. *Geomorphology* 139:172–187.

Dijkerman, JC. 1974. Pedology as a science: The role of data, models and theories in the study of natural soil systems. *Geoderma* 11:73–93.

Dijkerman, JC, Cline, MG and GW Olson. 1967. Properties and genesis of textural subsoil lamellae. *SS* 104:7–16.

Dijkerman, JC and R Miedema. 1988. An Ustult-Aquult-Tropept catena in Sierra Leone, West Africa. I. Characteristics, genesis and classification. *Geoderma* 42:1–27.

Dijkstra, EF and JM VanMourik. 1996. Reconstruction of recent forest dynamics based on pollen analysis and micromorphological studies of young acid forest soils under Scots pine plantations. *Acta Botanica Neerland.* 45:393–410.

Dijkstra, FA and RD Fitzhugh. 2003. Aluminum solubility and mobility in relation to organic carbon in surface soils affected by six tree species of the northeastern United States. *Geoderma* 114:33–47.

Dimbleby, GW. 1957. Pollen analysis of terrestrial soils. *New Phytol.* 56:12–28.

—— 1985. *The Palynology of Archaeological Sites*. Academic Press, London.

Dimmock, GM, Bettenay, E and MJ Mulcahy. 1974. Salt content of lateritic soil profiles in the Darling Range, Western Australia. *AJSR* 12:63–69.

Dimo, VN. 1965. Formation of a humic–illuvial horizon in soils on permafrost. *SSS* 9:1013–1021.

Ding, ZL, Derbyshire, E, Yang, SL, Yu, ZW, Xiong, SF and TS Liu. 2002. Stacked 2.6-Ma grain size record from the Chinese loess based on five sections and correlation with the deep-sea delta O-18 record. *Paleoceanography* 17. doi:10.1029/2001PA000725

Ding, Z, Liu, T, Rutter, NW, Yu, Z, Guo, Z and R Zhu. 1995. Ice-volume forcing of East Asia winter monsoon variations in the past 800000 years. *QR* 44:149–159.

Ding, ZL, Rutter, N, Han, JT and TS Liu. 1992. A coupled environmental system formed at about 2. 5 Ma in East Asia. *PPP* 94:223–242.

Ding, ZL, Rutter, N and L Tungsheng. 1993. Pedostratigraphy of Chinese loess deposits and climatic cycles in the last 2. 5 Myr. *Catena* 20:73–91.

Ding, ZL, Sun, JM, Rutter, NW, Rokosh, D and TS Liu. 1999. Changes in sand content of loess deposits along a north-south transect of the Chinese Loess Plateau and the implications for desert variations. *QR* 52:56–62.

Ding, ZL, Sun, JM, Yang, SL and TS Liu. 2001. Geochemistry of the Pliocene red clay formation in the Chinese Loess Plateau and implications for its origin, source provenance and paleoclimatic change. *Geochim. Cosmochim. Acta* 65:901–913.

Ding, ZL and SL Yang. 2000. C3/C4 vegetation evolution over the last 7. 0 Myr in the Chinese Loess Plateau: Evidence from pedogenic carbonate ^{13}C. *PPP* 160:291–299.

Dinis, P and H Castilho. 2012. Integrating sieving and laser data to obtain bulk grain-size distributions. *J. Sed. Res.* 82:747–754.

Dittmar, G, Guggenberger, IA, Janssens, M, Kleber, I, Kogel-Knabner, J, Lehmann, DA, Manning, C, Nannipieri, P, Rasse, DP, Weiner, S and SE Trumbore. 2011. Persistence of soil organic matter as an ecosystem property. *Nature* 478:49–56.

Ditzler, CA. 2005. Has the polypedon's time come and gone? Newsletter of the Natl. *Cooperative Soil Survey*, Lincoln, NE. Vol. 30, pp. 1–3.

Ditzler, CA and RJ Ahrens. 2006. Development of soil taxonomy in the United States of America. *Euras. Soil Sci.* 39:141–146.

Dixit, SP. 1978. Measurement of the mobility of soil colloids. *JSS* 29:557–566.

Dixon, JB. 1989. Kaolin and serpentine group minerals. In: JB Dixon and SB Weed (eds). *Minerals in Soil Environments*, 2nd ed. Soil Sci. Soc. Am., Madison, WI. pp. 467–525.

Dixon, JB and SB Weed (eds) 1989. *Minerals in Soil Environments*, 2nd ed. Soil Sci. Soc. Am., Madison, WI.

Dixon, JC and RW Young. 1981. Character and origin of deep arenaceons weathering mantles on the Bega Batholith, Southeastern Australia. *Catena* 8:97–109.

Dixon, JC. 1991. Alpine and subalpine soil properties as paleoenvironmental indicators. *PG* 12:370–384.

——— 1994a. Desert soils, patterned ground, and desert pavements. In: AD Abrahams and AJ Parsons (eds). *Geomorphology of Desert Environments*. Chapman and Hall, London. pp. 61–78.

——— 1994b. Duricrusts. In: AD Abrahams and AJ Parsons (eds). *Geomorphology of Desert Environments*. Chapman and Hall, London. pp. 82–105.

Dixon, JC, Thorn, CE, Darmody, RG and SW Campbell. 2002. Weathering rinds and rock coatings from an Arctic alpine environment, northern Scandinavia. *GSA Bull.* 114:226–238.

Djukic, I, Zehetner, F, Tatzber, M and MH Gerzabek. 2010. Soil organic-matter stocks and characteristics along an Alpine elevation gradient. *J. Plant Nutr. Soil Sci.* 173:30–38.

Dlusky, KG. 2007. Likhvin interglacial polygenetic paleosol: A reconstruction on the Russian Plain. *QI* 162/163:141–157.

Dobereiner, L and CG Porto. 1990. Considerations on the weathering of gneissic rocks. Eng. Group Meeting on the Geology of Weak Rock, Proc. 26th Annual Conf. of Brit. Geol. Soc., Leeds. pp. 228–241.

Dobrovolskii, GV. 1996. Dokuchaev and the present-day soil science. *Euras. Soil Sci.* 29:105–109.

Dockhorn, B, Neumaier, S, Hartmann, FJ, Petitjean, C, Faestermann, H and E Nolte. 1991. Determination of erosion rates with cosmic ray produced ^{36}Cl. *Zeitschrift Physik* 341A:117–119.

Dokuchaev, VV. 1883. *Russkii Chernozem*. Moscow.

——— 1886. Materials on Land Evaluation of the Nizhni Novgorod Governorate. Natural and Historical Part: Report to the Nizhni Novgorod Governorate Zemstvo. 1: Key Points in the History of Land Evaluation in the European Russia, with Classification of Russian Soils. Tipogr. Evdokimova. St. Petersburg.

——— 1893. The Russian Steppes/Study of the Soil in Russia in the Past and Present. St. Petersburg, Dept. of Agricultural Ministry of Crown Domains for the World's Columbian Exposition at Chicago.

——— 1899. *On the Theory of Natural Zones*. St. Petersburg.

Donahue, RL, Miler, RW and JC Shickluna. 1983. *Soils: An Introduction to Soils and Plant Growth*. Prentice-Hall, Englewood Cliffs, NJ.

Donald, RG, Anderson, DW and JWB Stewart. 1993. The distribution of selected soil properties in relation to landscape morphology in forested Gray Luvisol soils. *CJSS* 73:165–172.

Donaldson, NC. 1980. *Soil Survey of Whitman County*, Washington. USDA-SCS, US Govt. Printing Off., Washington, DC.

Doner, HE and PR Grossl. 2002. Carbonates and evaporites. In JB Dixon and D Schulze (eds). *Soil Mineralogy with Environmental Applications*. SSSA Book Series no. 7. Soil Sci. Soc. Am., Madison, WI. pp. 199–228.

Doner, HE and WC Lynn. 1989. Carbonate, halide, sulfate, and sulfide minerals. In: JB Dixon and SB Weed (eds). *Minerals in Soil Environments*, 2nd ed. Soil Sci. Soc. Am., Madison, WI. pp. 279–330.

Dong, H, Peacor, DR and SF Murphy. 1998. TEM study of progressive alteration of igneous biotite to kaolinite throughout a weathered soil profile. *Geochim. Cosmochim. Acta* 62:1881–1887.

Doornkamp, JC and HAM Ibrahim. 1990. Salt weathering. *PPG* 14:335–348.

Dor, I and A Danin. 1996. Cyanobacterial desert crusts in the Dead Sea Valley, Israel. *Algol. Stud.* 83:197–206.

Dormaar, JF and LE Lutwick. 1969. Infrared spectra of humic acids and opal phytoliths as indicators of paleosols. *CJSS* 49:29–37.

Dorn, RI. 1983. Extractable Fe and Al as an indicator for buried soil horizons. *Catena* 10:167–173.

——— 1988. A rock varnish interpretation of alluvial-fan development in Death Valley, California. *Natl Geog. Res.* 4:56–73.

——— 1989. Cation-ratio dating: A geographic perspective. *PPG* 13:559–596.

——— 1990. Quaternary alkalinity fluctuations recorded in rock varnish microlaminations on western USA volcanics. *PPP* 76:291–310.

——— 1998. *Rock coatings*. Elsevier, Amsterdam.

——— 2001. Chronometric techniques: Engravings. In: DS Whitley (ed). *Handbook of Rock Art Research*. Altamira Press, Walnut Creek. pp. 167–189.

——— 2003. Boulder weathering and erosion associated with a wildfire, Sierra Ancha Mountains, Arizona. *Geomorphology* 55:155–171.

——— 2004. Experimental approaches to dating petroglyphs and geoglyphs with rock varnish in the California Deserts: Current status and future directions. In: MW Allen and J Reed (eds). *The Human Journey and Ancient Life in California's Deserts*. Maturango Museum Publ. 15, Ridgecrest, CA. pp. 211–224.

——— 2009a. Desert rock coatings. In: AJ Parsons and AD Abrahams (eds). *Geomorphology of Desert Environments*, 2nd ed. Chapman and Hall, New York. pp. 153–186.

——— 2009b. The rock varnish revolution: New insights from microlaminations and the contributions of Tanzhuo Liu. *Geog. Compass* 3:1–20.

——— 2011. Revisiting dirt cracking as a physical weathering process in warm deserts. *Geomorphology* 135:129–142.

——— 2013. Rock coatings. In: JF Shroder (ed). *Treatise on Geomorphology*. Vol. 4. Academic Press, San Diego, CA. pp. 70–97.

Dorn, RI, Bamforth, DB, Cahill, TA, Dohrenwend, JC, Turrin, BD, Donahue, DJ, Jull, AJT, Long, A, Macko, ME, Weil, EB, Whitley, DS and TH Zabel. 1986. Cation-ratio and accelerator radiocarbon dating of rock varnish on Mojave artifacts and landforms. *Science* 231:830–833.

Dorn, RI, Cahill, TA, Eldred, RA, Gill, TE, Bach, AJ and DL Elliott-Fisk. 1990. Dating rock varnishes by the cation ratio method with PIXE, ICP, and the electron microprobe. *Intl. J. PIXE* 1:157–195.

Dorn, RI, Gordon, SJ, Krinsley, D and K Langworthy. 2013. Nanoscale:mineral weathering boundary. In: JF Shroder (ed). *Treatise on Geomorphology*. Vol. 4. Academic Press, San Diego, CA. pp. 44–68.

Dorn, RI and DH Krinsley. 1991. Cation-leaching sites in rock varnish. *Geology* 19:1077–1080.

Dorn, RI and FM Phillips. 1991. Surface exposure dating: Review and critical evaluation. *PG* 12:303–333.

Dorn, RI and TM Oberlander. 1981a. Microbial origin of rock varnish. *Science* 213:1245–1247.

——— 1981b. Rock varnish origin, characteristics, and usage. *ZG* 25:420–436.

——— 1982. Rock varnish. *PPG* 74:308–322.

Dörr, H and KO Münnich. 1986. Annual variations of the ^{14}C content of soil CO_2. *Radiocarbon* 28:338–406.

Dorronsoro, C and P Alonso. 1994. Chronosequence in Almar River fluvial-terrace soil. *SSSAJ* 58:910–925.

Dort, W and GAM Dreschhoff. 2002. The enigmatic and controversial Merna Crater, central Nebraska. *Meteor. Planetary Sci. Suppl.* 37:A42.

Dosseto, A, Buss, H and PO Suresh. 2011. The delicate balance between soil production and erosion, and its role on landscape evolution. *Appl. Geochem.* 26:S24–S27.

Douglas AE. 1941. Crossdating in dendrochronology. *J. For.* 39:825–831.

Douglas, CL Jr, Fehrenbacher, JB and BW Ray. 1967. The lower boundary of selected Mollisols. *SSSAP* 31:795–800.

Douglas, LA and JCF Tedrow. 1959. Organic matter decomposition rates in Arctic soils. *Soil Sci.* 88:305–312.

1979. Tundra soils of Arctic Alaska. *Trans. 7th Intl. Cong. Soil Sci.* 4:291–304.

Douglas, LA. 1989. Vermiculites. In: JB Dixon and SB Weed (eds). *Minerals in Soil Environments*, 2nd ed. Soil Sci. Soc. Am., Madison, WI. pp. 635–674.

Douglas, LA and ML Thompson (eds) 1985. *Soil Micromorphology and Soil Classification*. Soil Sci. Soc. Am. Spec. Publ. 15. Madison, WI.

Douglass, DC and JG Bockheim. 2006. Soil-forming rates and processes on Quaternary moraines near Lago Buenos Aires, Argentina. *QR* 65:293–307.

Douglass, DC and DM Mickelson. 2007. Soil development and glacial history, west fork of Beaver Creek, Uinta Mountains, Utah. *AAAR* 39:592–602.

Downes, RG. 1954. Cyclic salt as a factor in soil genesis. *AJSR* 5:448–464.

Draaijers, GPJ, van Leeuwen, EP, De Jong, PGH and JW Erisman. 1997. Base cation deposition in Europe. Part II. Acid neutralization capacity and contribution to forest nutrition. *Atmos. Environ.* 31:4159–4168.

Dragovich, D. 1988. A preliminary electron probe study of microchemical variations in desert varnish in western South Wales, Australia. *ESPL* 13:259–270.

1993. Fire-accelerated boulder weathering in the Pilbara, Western Australia. *ZG* 37:295–307.

Dredge, LA. 1992. Breakup of limestone bedrock by frost shattering and chemical-weathering, eastern Canadian Arctic. *AAR* 24:314–323.

Drees, LR and LP Wilding. 1973. Elemental variability within a sampling unit. *SSSAP* 37:82–87.

Dreibrodt, S, Jarecki, H, Lubos, C, Khamnueva, SV, Klamm, M and H-R Bork. 2013. Holocene soil formation and soil erosion at a slope beneath the Neolithic earthwork Salzmünde (Saxony-Anhalt, Germany). *Catena* 107:1–14.

Dreimanis, A. 1989. Tills: Their genetic terminology and classification. In: RP Goldthwait and CL Matsch (eds). Genetic Classification of Glacigenic Deposits. Intl. Quat. Assoc. pp. 17–83.

Dreimanis, A, Hütt, G, Raukas, A and PW Whippey. 1978. Dating methods of Pleistocene deposits and their problems. I. Thermoluminescence dating. *Geosci. Can.* 5:55–60.

Drever, JI. 1994. The effect of land plants on weathering rates of silicate minerals. *Geochim. Cosmochim. Acta* 58:2325–2332.

Drohan, PJ and TJ Farnham. 2006. Protecting life's foundation: A proposal for recognizing rare and threatened soils. *SSSAJ* 70:2086–2096.

Droste, JB and JC Tharin. 1958. Alteration of clay minerals in Illinoian till by weathering. *GSA Bull.* 69:61–68.

Drury, WH and ICT Nisbet. 1971. Interrelationships between developmental models in geomorphology, plant ecology, and animal ecology. *Gen. Syst.* 16:57–68.

Du Preez, JW. 1949. Laterite: A general discussion with a description of Nigerian occurrences. *Bull. Agric. Congo Belge* 40:53–66.

Duan, L, Hao, JM, Xie, SD, Zhou, ZP and XM Ye. 2002. Determining weathering rates of soils in China. *Geoderma* 110:205–225.

Dubroeucq, D, Geissert, D and P Quantin. 1998. Weathering and soil-forming processes under semi-arid conditions in two Mexican volcanic ash soils. *Geoderma* 86:99–122.

Dubroeucq, D and B Volkoff. 1988. Evolution des couvertures pédologiques sableuses à podzols géants d'Amazonie (Bassin du Haut rio Negro). *Cahiers ORSTOM (Série pédol.)* 24:191–214.

1998. From Oxisols to Spodosols and Histosols: Evolution of the soil mantles in the Rio Negro basin (Amazonia). *Catena* 32:245–280.

Duchaufour, P. 1951. Lessivage et podzolization. *Rev. de forestière Française* 10.

1976. Dynamics of organic matter in soils of temperate regions: Its action on pedogenesis. *Geoderma* 15:31–40.

1982. *Pedology*. Allen and Unwin, Winchester, MA.

Duchaufour, PH and B Souchier. 1978. Roles of iron and clay in genesis of acid soils under a humid, temperate climate. *Geoderma* 20:15–26.

Ducloux, J, Guero, Y and P Fallavier. 1998. Clay particle differentiation in alluvial soils of Southern Niger (West Africa). *SSSAJ* 62:212–222.

Ducloux, J, Guero, Y, Sardini, P and A Decarreau. 2002. Xerolysis: A hypothetical process of clay particles weathering under Sahelian climate. *Geoderma* 105:93–110.

Ducloux, J, Meunier, A and B Velde. 1976. Smectite, chlorite and a regular interstratified chlorite-vermiculite in soils developed on a small serpentinite body Massif Central, France. *Clay Mins.* 11:121–135.

Dudal, R. 1987. The role of pedology in meeting the increasing demands on soils. *Soil Survey Land Eval.* 7:101–110.

Dudal, R and H Eswaran. 1988. Distribution, properties and classification of Vertisols. In: LP Wilding and R Puentes (eds). *Vertisols: Their Distribution, Properties, Classification and Management*. Texas A&M Univ., College Station, TX. pp. 1–22.

Dudas, MJ and S Pawluk. 1969. Naturally occurring organo-clay complexes of Orthic Black Chernozems. *Geoderma* 3:5–17.

Dudley, K and EA Rochette. 2005. Indicators of saturation in albic horizons of New Hampshire and Maine. *SSH* 46:59–67

DukRodkin, A, Barendregt, RW, Tarnocai, C and FM Phillips. 1996. Late Tertiary to late Quaternary record in the Mackenzie Mountains, Northwest Territories, Canada: Stratigraphy, paleosols, paleomagnetism, and chlorine-36. *CJES* 33:875–895.

Duller, GAT. 2004. Luminescence dating of Quaternary sediments: Recent advances. *JQS* 19:183–192

2006. Single grain optical dating of glacigenic deposits. *QG* 1:296–304.

2008. Single-grain optical dating of Quaternary sediments: Why aliquot size matters in luminescence dating. *Boreas* 37:589–612

Duller, GAT and P Augustinus. 1997. Luminescence studies of dunes from north-eastern Tasmania. *QSR* 16:357–365.

Duncan, MM and DP Franzmeier. 1999. Role of free silicon, aluminum, and iron in fragipan formation. *SSSAJ* 63:923–929.

Durand, JH. 1963. Les croûtes calcaires et gypseuses en Algérie: Formation et âge. *Bull. Géol. Soc. France* 7:959–968.

Durn, G, Ottner, F and D Slovenec. 1999. Mineralogical and geochemical indicators of the polygenetic nature of terra rossa in Istria, Croatia. *Geoderma* 91:125–150.

Dutta, PK, Zhou, Z and PR dos Santos. 1993. A theoretical study of mineralogical maturation of eolian sand. *GSA Spec. Paper* 284:203–209.

Dyke, AS. 1979. Glacial and sea-level history, southwestern Cumberland Peninsula, Baffin Island, Canada. *AAR* 11:179–202.

Dzulinski, S. 1966. Sedimentary structures resulting from convection-like pattern of motion. *Ann. Socíktt Gkoloyique Pologne* 36:3–21.

Eaqub, M and H-P Blume. 1982. Genesis of a so-called ferrolysed soil of Bangladesh. *Z. Pflanzen. Bodenkd.* 145:470–482.

Eaton, JS, Likens, GE and FH Bormann. 1973. Throughfall and stemflow chemistry in a northern hardwood forest. *JE* 61:495–508.

Ebel, BA. 2012. Impacts of wildfire and slope aspect on soil temperature in a mountainous environment. *Vadose Zone J.* 11. doi:10.2136/vzj2012.0017

Eckert, RE, Wood, M, Blackburn, WH and FF Peterson. 1979. Impacts of off-road vehicles on infiltration and sediment production of two desert soils. *J. Range Mgmt.* 32:394–397.

Eden, DN, Qizhong, W, Hunt, JL and JS Whitton. 1994. Mineralogical and geochemical trends across the Loess Plateau, North China. *Catena* 21:73–90.

Edmonds, WJ, Campbell, JB and M Lemtner. 1985. Taxonomic variation within three soil mapping units in Virginia. *SSSAJ* 49:394–401.

Edwards, C. 1999. *Earthworms, PA-1637*. USDA, NRCS Soil Quality Institute, Washington, DC. pp. H1–H8.

Edwards, CA and PJ Bohlen. 1996. *Biology and Ecology of Earthworms*. Chapman and Hall, New York.

Edwards, WM, Norton, LD and CE Redmond. 1988. Characterizing macropores that affect infiltration into nontilled soil. *SSSAJ* 52:483–487.

Effland, ABW and WR Effland. 1992. Soil geomorphology studies in the US soil survey program. *Agric. Hist.* 66:189–212.

Effland, WR and RV Pouyat. 1997. The genesis, classification, and mapping of soils in urban areas. *Urban Ecosyst.* 1:217–228.

Eger, A, Almond, PC and LM Condron. 2011. Pedogenesis, soil mass balance, phosphorus dynamics and vegetation communities across a Holocene soil chronosequence in a super-humid climate, South Westland, New Zealand. *Geoderma* 163:185–196.

2012. Upbuilding pedogenesis under active loess deposition in a super-humid, temperate climate – quantification of deposition rates, soil chemistry and pedogenic thresholds. *Geoderma* 189–190: 491–501.

Eggleton, RA and G Taylor. 2008. Effects of some macrobiota on the Weipa Bauxite, northern Australia. *Austr. J. Earth-Sci.* 55:S71–S82.

Eghbal, MK, Givi, J, Torabi, H and M Miransari. 2012. Formation of soils with fragipan and plinthite in old beach deposits in the South of the Caspian Sea, Gilan province, Iran. *Appl. Clay Sci.* 64:44–52.

Eghbal, MK and RJ Southard. 1993a. Micromorphological evidence of polygenesis of three Aridisols, western Mojave Desert, California. *SSSAJ* 57:1041–1050.

1993b. Stratigraphy and genesis of Durorthids and Haplargids on dissected alluvial fans, western Mojave Desert, California. *SSSAJ* 59:151–174.

Egli, M, Brandová, D, Böhlert, R, Favilli, F and P Kubik. 2010. ^{10}Be inventories in Alpine soils and their potential for dating land surfaces. *Geomorphology* 119:62–73.

Egli, M and P Fitze. 2000. Formulation of pedologic mass balance based on immobile elements: A revision. *SS* 165:437–443.

2001. Quantitative aspects of carbonate leaching of soils with differing ages and climates. *Catena* 46:35–62.

Egli, M, Fitze, P and A Mirabella. 2001. Weathering and evolution of soils formed on granitic, glacial deposits: Results from chronosequences of Swiss alpine environments. *Catena* 45:19–47.

Egli, M, Merkli, C, Sartori, G, Mirabella, A and M Plötze. 2008. Weathering, mineralogical evolution and soil organic matter along a Holocene soil toposequence on carbonate-rich materials. *Geomorphology* 97:675–696.

Egli, M, Mirabella, A, Sartori, G, Zanelli, R and S Bischof. 2006. Effect of north and south exposure on weathering rates and clay mineral formation in Alpine soils. *Catena* 67:155–174.

Ehleringer, JR, Buchmann, N and LB Flanagan. 2000. Carbon isotope ratios in belowground carbon cycle processes. *Ecol. Appl.* 10:412–422.

Ehleringer, JR and TE Cerling. 2002. C3 and C4 Photosynthesis. In: HA Mooney and JG Canadell (eds). *Encyclopedia of Global Environmental Change*. Wiley, New York. pp. 186–190.

Ehleringer, JR and RK Monson. 1993. Evolutionary and ecological aspects of photosynthetic pathway variation. *Ann. Rev. Ecol. Syst.* 24:411–439.

Ehleringer, JR and JC Vogel. 1993. Historical aspects of stable isotopes. In: JR Ehleringer, AE Hall and GD Farquhar (eds). *Stable Isotopes and Plant Carbon-Water Relations*. Academic Press, New York. pp. 9–18.

Ehrlich, WA, Rice, HM and JH Ellis. 1955. Influence of the composition of parent materials on soil formation in Manitoba. *Can. J. Agric. Sci.* 35:407–421.

Eidt, RC. 1977. Detection and examination of Anthrosols by phosphate analysis. *Science* 197:1327–1333.

Eiler, JM. 2007. "Clumped-isotope" geochemistry – The study of naturally-occurring, multiply substituted isotopologues. *EPSL* 262:309–327.

Eisenhauer, N, Partsch, S, Parkinson, D and S Scheu. 2007. Invasion of a deciduous forest by earthworms: Changes in soil chemistry, microflora, microarthropods and vegetation. *Soil Biol. Biochem.* 39:1099–1110.

Eisenhauer, N, Schlaghamersky, J, Reich, PB and LE Frelich. 2011. The wave towards a new steady state: Effects of earthworm invasion on soil microbial functions. *Biol. Invasions* 13:2191–2196.

El Abedine, Z, Robinson, GH and A Commissaris. 1971. Approximate age of the Vertisols of Gezira, central clay plain, Sudan. *SS* 111:200–207.

Eldredge, N and SJ Gould. 1972. Punctuated equilibria: An alternative to phyletic gradualism. In: TJM Schopf (ed). *Models of Paleobiology*. Freeman, Cooper, San Francisco, CA. pp. 82–115.

Eldridge, DJ. 2004. Mounds of the American badger (*Taxidea taxus*): Significant geomorphic features of North American shrub-steppe ecosystems. *J. Mammalogy* 85:1060–1067.

Eldridge, DJ, Koen, TB, Killgore, A, Huang, N and WG Whitford. 2012. Animal foraging as a mechanism for sediment movement and soil nutrient development: Evidence from the semi-arid Australian woodlands and the Chihuahuan Desert. *Geomorphology* 157–158:131–141.

Eldridge, DJ and A Mensinga. 2007. Foraging pits of the short-beaked echidna (*Tachyglossus aculeatus*) as small-scale patches in a semi-arid Australian box woodland. *Soil Biol. Biochem.* 39:1055–1065.

Eldridge, DJ and D Rath. 2002. Hip holes: Kangaroo (*Macropus* spp.) resting sites modify the physical and chemical environment of woodland soils. *Austral Ecol.* 27:527–536.

Elghamry, W and M Elashkar. 1962. Simplified textural classification triangles. *SSSAP* 26:612–613.

Elliott, JA and E DeJong. 1992. Quantifying denitrification on a field scale in hummocky terrain. *CJSS* 72:21–29.

Ellis, S. 1979. The identification of some Norwegian mountain soil types. *Norsk. Geogr. Tidsskr.* 33:205–211.

1980a. An investigation of weathering in some Arctic-Alpine soils on the northeast flank of Oksskolten, north Norway. *JSS* 31:371–385.

1980b. Physical and chemical characteristics of a podzolic soil formed in neoglacial till, Okstindan, northern Norway. *AAR* 12:65–72.

Elvidge, CD and RM Iverson. 1983. Regeneration of desert pavement varnish. In: RH Webb and HG Wilshire (eds). *Environmental Effects of Off-Road Vehicles: Impacts and Management in Arid Regions*. Springer-Verlag, New York. pp. 225–243.

Ely, LL, Enzel, Y, Baker, VR, Kale, VS and S Mishra. 1996. Changes in the magnitude and frequency of late Holocene monsoon floods on the Narmada River, central India. *GSA Bull.* 108:1134–1148.

Elzenga, W, Schwan, J, Baumfalk, YA, Vandenberghe, J and L Krook. 1987. Grain surface characteristics of periglacial aeolian and fluvial sands. *Geol. En Mijn.* 65:273–286.

Emiliani, C. 1955. Pleistocene temperatures. *J Geol.* 63:538–578.

1966. Palaeotemperature analysis of Caribbean cores and a generalized temperature curve for the last 425 000 years. *J. Geol.* 74:109–126.

Emrich, KD, Ehalt, H and JC Vogel. 1970. Carbon isotope fractionation during the precipitation of calcium carbonate. *EPSL* 8:363–371.

Engel, CG and RP Sharp. 1958. Chemical data on desert varnish. *GSA Bull.* 69:487–518.

Engel, RJ, Witty, JE and H Eswaran. 1997. The classification, distribution, and extent of soils with a xeric moisture regime in the United States. *Catena* 28:203–209.

Engovatova, A and A Golyeva. 2012. Anthropogenic soils in Yaroslavl (Central Russia): History, development, base for landscape reconstruction, *QI* 265:54–62.

Enloe, HA, Graham, RC and SC Sillett. 2006. Arboreal Histosols in old-growth redwood forest canopies, Northern California. *SSSAJ* 70:408–418.

Eppes, MC, Bierma, R, Vinson, D and F Pazzaglia. 2008. A soil chronosequence study of the Reno valley, Italy: Insights into the relative role of climate versus anthropogenic forcing on hillslope processes during the mid-Holocene. *Geoderma* 147:97–107.

Eppes, MC and JBJ Harrison. 1999. Spatial variability of soils developing on basalt flows in the Potrillo volcanic field, southern New Mexico: Prelude to a chronosequence study. *ESPL* 24:1009–1024.

Eppes, MC, McFadden, LD, Wegmann, KW and LA Scuderi. 2010. Cracks in desert pavement rocks: Further insights into mechanical weathering by directional insolation. *Geomorphology* 123:97–108.

Erginal, AE and B Ozturk. 2009. Andesite weathering caused by crustose lichens Xanthoria calcicola and Diploschistes scruposus: A case study. *Fresenius Env. Bull.* 18:499–504.

Eriksson, MG, Olley, JM and RW Payton. 2000. Soil erosion history in central Tanzania based on OSL dating of colluvial and alluvial hillslope deposits. *Geomorphology* 36:107–128.

Erlandson, JM. 1984. A case study in faunalturbation: Delineating the effects of the burrowing pocket gopher on the distribution of archaeological materials. *Am. Antiq.* 49:785–790.

Eschner, AR and JH Patric. 1982. Debris avalanches in eastern upland forests. *J. For.* 80:343–347.

Escolar, RP and MAL López. 1968. Nature of aggregation in two tropical soils of Puerto Rico. *Univ. Puerto Rico J. Agric.* 52:227–232.

Eshel, G, Levy, GJ, Mingelgrin, U and MJ Singer. 2004. Critical evaluation of the use of laser diffraction for particle-size distribution analysis. *SSSAJ* 68:736–743.

Espejo, JMR, Faust, D, Granados, MAN and C Zielhofer. 2008. Accumulation of secondary carbonate evidence by ascending capillary in Mediterranean argillic horizons (Cordoba, Andalusia, Spain). *SS* 173:350–358.

Eswaran, H and WC Bin. 1978a. A study of a deep weathering profile on granite in peninsular Malaysia. I. Physico-chemical and micromorphological properties. *SSSAJ* 42:144–149.

1978b. A study of a deep weathering profile on granite in peninsular Malaysia. II. Mineralogy of the clay, silt, and sand fractions. *SSSAJ* 42:149–153.

Eswaran, H, Kimble, J, Cook, T and FH Beinroth. 1992. Soil Diversity in the tropics: Implications for agricultural development. In: R Lal and P Sanchez (eds). *Myths and Science of Soils of the Tropics.* Soil Sci. Soc. Am., Madison, WI. pp. 1–16.

Eswaran, H and C Sys. 1970. An evaluation of the free iron in tropical basaltic soils. *Pédologie* 20:62–85.

Eswaran, H, van den Berg, E, and P Reich. 1993. Organic carbon in soils of the world. *SSSAJ* 57:192–194.

Eswaran, H, van Wambeke, A and FH Beinroth. 1979. A study of some highly weathered soils of Puerto Rico: Micromorphological properties. *Pédologie* 29:139–162.

Eswaran, H and G Zi-Tong. 1991. Properties, genesis, classification, and distribution of soils with gypsum. In: WD Nettleton (ed). Occurrence, Characteristics, and Genesis of Carbonate, Gypsum, and Silica Accumulations in Soils. *SSSASP* 26:89–119.

Etienne, S and J Dupont. 2002. Fungal weathering of basaltic rocks in a cold oceanic environment (Iceland): Comparison between experimental and field observations. *ESPL* 27:737–748.

Evans, CV and WA Bothner. 1993. Genesis of altered Conway Granite (grus) in New Hampshire, USA. *Geoderma* 58:201–218.

Evans, CV and DP Franzmeier. 1986. Saturation, aeration, and color patterns in a toposequence of soils in north-central Indiana. *SSSAJ* 50:975–980.

1988. Color index values to represent wetness and aeration in some Indiana soils. *Geoderma* 41:353–368.

Evans, DJA. 1999. A soil chronosequence from neoglacial moraines in western Norway. *Geog. Annal.* 81A:47–62.

Evans, LJ. 1978. *Quantification of pedological processes.* In: WC Mahaney (ed). Quaternary Soils. Geo Abstracts, Norwich, UK. pp. 361–378.

1982. Dating methods of Pleistocene deposits and their problems. VII. Paleosols. *Geosci. Can.* 9:155–160.

Evans, LJ and WA Adams. 1975a. Quantitative pedological studies on soils derived from Silurian mudstones. IV. Uniformity of the parent material and evaluation of internal standards. *JSS* 26:319–326.

1975b. Quantitative pedological studies on soils derived from Silurian mudstones. V. Redistribution and loss of mobilized constituents. *JSS* 26:327–335.

Evans, RD and OL Lange. 2001. Biological soil crusts and ecosystem nitrogen and carbon dynamics. In: J Belnap and OL Lange (eds). *Biological Soil Crusts: Structure, Function, and Management.* Springer, Berlin. pp. 263–279.

Evenari, J, Yaalon, DH and Y Gutterman. 1974. Note on soils with vesicular structures in deserts. *ZG* 18:162–172.

Everett, KR. 1971. Composition and genesis of the organic soils of Amchitka Island, Aleutian Islands, Alaska. *AAR* 3:1–16.

1979. Evolution of the soil landscape in the sand region of the Arctic coastal plain as exemplified at Atkasook, Alaska. *Arctic* 32:207–223.

1980. Distribution and variability of soils near Atkasook, Alaska. *AAR* 12:433–446.

Expert Committee on Soil Science, Agriculture Canada Research Branch. 1987. *The Canadian System of Soil Classification,* 2nd ed, Publication #1646. Ottawa, Agriculture Canada.

Eze, PN and ME Meadows. 2014. Texture contrast profile with stone-layer in the Cape Peninsula, South Africa: Autochthony and polygenesis. *Catena* 118:103–114.

Fadl, AE. 1971. A mineralogical characterization of some Vertisols in the Gezira and the Kenana clay plains of the Sudan. *JSS* 22:129–135.

Faegri, K and J Iverson. 1989. *Textbook of Pollen Analysis.* Wiley, New York.

Fahey, BD and TH Lefebure. 1988. The freeze-thaw weathering regime at a section of the Niagara escarpment on the Bruce Peninsula, southern Ontario, Canada. *ESPL* 13:293–304.

Fairbanks, RG. 1989. A 17,000-year glacio-eustatic sea-level record – Influence of glacial meting rates on the Younger Dryas event and deep-ocean circulation. *Nature* 342:637–642.

Fairbanks, RG, Mortlock, RA, Chiu, T-C, Cao, L, Kaplan, A, Guilderson, TP, Fairbanks, TW and AL Bloom. 2005. Marine radiocarbon calibration curve spanning 0 to 50,000 years B. P. based on paired $^{230}Th/^{234}U$ and ^{14}C dates on pristine corals. *QSR* 24:1781–1796.

Fairbridge, RW and Tropical stone lines and podzolized sand plains as paleoclimatic indicators for weathered cratons. *QSR* 3:41–72.

Falsone, G, Celi, L, Caimi, A, Simonov, G and E Bonifacio. 2012. The effect of clear cutting on podzolisation and soil carbon dynamics in boreal forests (Middle Taiga zone, Russia). *Geoderma* 177:27–38.

Fanning, DS and MCB Fanning. 1989. *Soil Morphology, Genesis, and Classification*. Wiley, New York.

Fanning, DS, Keramidas, VZ and MA El-Desoky. 1989. Micas. In: JB Dixon and SB Weed (eds). *Minerals in Soil Environments*, 2nd ed. Soil Sci. Soc. Am., Madison, WI. pp. 551–634.

Fanning, DS, Rabenhorst, MC, Burch, SN, Islam, KR and SA Tangren. 2002. Sulfides and sulfates. In JB Dixon and DG Schulze (eds). *Soil Mineralogy with Environmental Applications*. SSSA Book Ser. 7. Soil Sci. Soc. Am., Madison, WI. pp. 229–260.

FAO-UNESCO. 1988. *FAO-UNESCO Soil Map of the World, revised legend*. World Soil Resources Rept. 60. FAO, Rome.

Farmer, VC. 1982. Significance of the presence of allophane and imogolite in Podzol Bs horizons for podzolization mechanisms: A review. *JSSPN* 28:571–578.

Farmer, VC and DG Lumsdon. 2001. Interactions of fulvic acid with aluminium and a proto-imogolite sol: The contribution of E-horizon eluates to podzolization. *EJSS* 52:177–188.

Farmer, VC, McHardy, WJ, Robertson, L, Walker, A and MJ Wilson. 1985. Micromorphology and sub-microscopy of allophone and imogolite in a Podzol Bs horizon: Evidence for translocation and origin. *JSS* 36:87–95.

Farmer, VC, Russell, JD and ML Berrow. 1980. Imogolite and proto-imogolite allophane in Spodic horizons: Evidence for a mobile aluminum silicate complex in Podzol formation. *JSS* 31:673–684.

Farmer, VC, Skjemstad, JO and CH Thompson. 1983. Genesis of humus B horizons in hydromorphic humus podzols. *Nature*. 304:342–344.

Farnham, RS, McAndrews, JH and HE Wright Jr. 1964. A Late-Wisconsin buried soil near Aitkin, Minnesota, and its paleobotanical setting. *AJS* 262:393–412.

Faulkner, DJ. 1998. Spatially variable historical alluviation and channel incision in west-central Wisconsin. *AAAG* 88:666–685.

Faure, P and B Volkoff. 1998. Some factors affecting regional differentiation of the soils in the Republic of Benin (West Africa). *Catena* 32:281–306.

Favre, F, Boivin, P and MCS Wopereis. 1997. Water movement and soil swelling in a dry, cracked Vertisol. *Geoderma* 78:113–123.

Feddema, J and TC Meierding. 1987. Marble weathering and air pollution in Philadelphia. *Atmos. Env.* 21:143–157.

Federer, CA. 1973. *Annual Cycles of Soil and Water Temperatures at Hubbard Brook*. USFS Experimental Station Research Note #NE-167. Newton Square, PA.

Fedo, CM, Nesbitt, HW and GM Young. 1995. Unraveling the effects of potassium metasomatism in sedimentary rocks and paleosols, with implications for paleoweathering conditions and provenance. *Geology* 23:921–924.

Fedoroff, N. 1997. Clay illuviation in Red Mediterranean soils. *Catena* 28:171–189.

Fedoroff, N and P Goldberg. 1982. Comparative micromorphology of two Late Pleistocene paleosols (in the Paris basin). *Catena* 9:227–251.

Fedorova, NN and A Yarilova. 1972. Morphology and genesis of prolonged seasonally frozen soils in Western Siberia. *Geoderma* 7:1–13.

Feggestad, AJ, Jacobs, PM, Miao, X and JA Mason. 2004. Stable carbon isotope record of Holocene environmental change in the central Great Plains. *PG* 25:170–190.

Fehrenbacher, JB, Olson, KR, and IJ Jansen. 1986. Loess thickness in Illinois. *SS* 141:423–431.

Fehrenbacher, JB, White, JL, Beavers, AH and RL Jones. 1965a. Loess composition in southeastern Illinois and southwestern Indiana. *SSSAP* 29:572–579.

Fehrenbacher, JB, White, JL, Ulrich, HP and RT Odell. 1965b. Loess distribution in southeastern Illinois and southwestern Indiana. *SSSAP* 29:566–572.

Fehrenbacher, JB, Wilding, LP, Odell, RT and SW Melsted. 1963. Characteristics of solonetzic soils in Illinois. *SSSAP* 27:421–431.

Feijtel, TC, Jongmans, AG, van Breemen, N and R Miedema. 1988. Genesis of two Planosols in the Massif Central, France. *Geoderma* 43:249–269.

Felgenhauser, F, Fink, J and H de Vries. 1959. Studien zur absoluten und relativen Chronologie der fossil Böden in Österreich. *Archaeol. Austr.* 25:35–73.

Felixhenningsen, P, Kohl, A and H Zakosek. 1983. The quantification of the leaching of calcium-ions from Holocene soils. *Zeit. fur Kultur. Flurbereinigung* 24:288–297.

Feller, C, Brown, GG, Blanchart, E, Deleporte, P and SS Chernyanskii. 2003. Charles Darwin, earthworms and the natural sciences: Various lessons from past to future. *Agric. Ecosystems Environ.* 99:29–49.

Feng, J-L, Hua, Z-G, Ju, J-T and L-P Zhu. 2011. Variations in trace element (including rare earth element) concentrations with grain sizes in loess and their implications for tracing the provenance of eolian deposits. *QI* 236:116–126.

Feng, J-L, Zhu, L-P and Z-J Cui. 2009. Quartz features constrain the origin of terra rossa over dolomite on the Yunnan-Guizhou Plateau, China. *J. Asian Earth-Sci.* 36:156–167.

Feng, Z-D and WC Johnson. 1995. Factors affecting the magnetic susceptibility of a loess-soil sequence, Barton County, Kansas, USA. *Catena* 24:25–37.

Feng, Z-D, Johnson, WC, Lu, Y-C and PA Ward III. 1994a. Climatic signals from loess-soil sequences in the central Great Plains, USA. *PPP* 110:345–358.

Feng, Z-D, Johnson, WC, Sprowl, DR, and Y-C Lu. 1994b. Loess accumulation and soil formation in central Kansas of United States during the past 400 000 years. *ESPL* 19:55–67.

Fenton, N, Lecomte, N, Legare, S and Y Bergeron. 2005. Paludification in black spruce (*Picea mariana*) forests of eastern Canada: Potential factors and management implications. *For. Ecol. Mgmt.* 213:151–159.

Fenton, TE. 1983. Mollisols. In: LP Wilding, NE Smeck and GF Hall (eds). *Pedogenesis and Soil Taxonomy*. Elsevier, New York. pp. 125–163.

Ferguson, CW and DA Graybill. 1983. Dendrochronology of bristlecone pine: A progress report. *Radiocarbon* 25:287–288.

Fernandes, EAD and FAM Bacchi. 1998. Lanthanides in the study of lithologic discontinuity in soils from the Piracicaba River basin. *J. Alloys Compounds* 275:924–928.

Fernandez, RN, Schulze, DG, Coffin, DL and GE Van Scoyoc. 1988. Color, organic matter, and pesticide adsorption relationships in a soil-landscape. *SSSA J* 52:1023–1026.

Fernández Sanjurjo, MJ, Corti, G and FC Ugolini. 2001. Chemical and mineralogical changes in a polygenetic soil of Galicia, NW Spain. *Catena* 43:251–265.

Ferreira, CA, Silva, AC, Torrado, PV and WW Rocha. 2010. Genesis and classification of Oxisols in a highland toposequence of the Upper Jequituinhonha valley (MG). *Revista Brasil. Ciência Solo* 34:195–209.

Ferrians, OJ, Jr, Kachadoorian, R and GW Greene. 1969. Permafrost and Related Engineering Problems in Alaska. USGS Prof. Paper 678.

Ferring, CR. 1986. Rates of fluvial sedimentation: Implications for archaeological variability. *Geoarchaeol.* 1:259–274.

Fey, M. 2010. *Soils of South Africa*. Cambridge Univ. Press, New York.

Fialova, H, Maier, G, Petrovský, E, Kapička, A, Boyko, T and R Scholger. 2006. Magnetic properties of soils from sites with different geological and environmental settings. *J. Appl. Geophys.* 59:273–283.

Fiedler, S and M Sommer. 2004. Water and redox conditions in wetland soils – Their influence on pedogenic oxides and morphology. *SSSAJ* 68:326–335.

Fiedler, S, Vepraskas, MJ and JL Richardson. 2007. Soil redox potential: Importance, field measurements, and observations. In: *Advances in Agronomy*, Elsevier Press, San Diego, CA. 94:1–54.

Fine, P, Singer, MJ and KL Verosub. 1992. Use of magnetic-susceptibility measurements in assessing soil uniformity in chronosequence studies. *SSSAJ* 56:1195–1199.

1993. New evidence for the origin of ferrimagnetic minerals in loess from China. *SSSAJ* 57:1537–1542.

1995. Pedogenic and lithogenic contributions to the magnetic susceptibility record of the Chinese loess/palaeosol sequence. *Geophys. J. Intl.* 122:97–107.

Finlayson, B. 1981. Field measurements of soil creep. *ESPL* 6:35–48.

Finney, HR, Holowaychuk, N and MR Heddleson. 1962. The influence of microclimate on the morphology of certain soils of the Allegheny Plateau of Ohio. *SSSAP* 26:287–292.

Finzi, AC, Van Breemen, N and CD Canham. 1998. Canopy tree – soil interactions within temperate forests: Species effects on pH and cations. *Ecol. Appl.* 8:440–446.

Fiskell, JGA and VW Carlisle. 1963. Weathering of some Florida soils. *Soil Crop Soc. Florida Proc.* 23:32–44.

FitzPatrick, EA. 1956. An indurated soil horizon formed by permafrost. *JSS* 7:248–254.

1975. Particle size distribution and stone orientation patterns in some soils of north east Scotland. In: AMD Gemmell (ed). *Quaternary Studies in North East Scotland*. Quat. Res. Assoc., Aberdeen, UK. pp. 49–60.

1993. *Soil Microscopy and Micromorphology*. Wiley, New York.

Fitzpatrick, RW, Fritsch, E and PG Self. 1996. Interpretation of soil features produced by ancient and modern processes in degraded landscapes. V. Development of saline sulfidic features in non-tidal seepage areas. *Geoderma* 69:1–29.

Fitzsimmons, KE, Hambach, U, Veres, D and R Iovita. 2013. The Campanian ignimbrite eruption: New data on volcanic ash dispersal and its potential impact on human evolution. *PLoS ONE* 8:e65839. doi:10.1371/journal.pone.0065839

Fiuza, SD, Kusdra, JF and DT Furtado. 2011. Chemical properties and microbial activity in castings of Chibui bari (*Oligochaeta*) and surrounding soil. *Revista Brasil. Ciência Solo* 35:723–728.

Flach, KW, Cady, JG and WD Nettleton. 1968. Pedogenic alteration of highly weathered parent materials. *Trans. 9th Intl. Congr. Soil Sci.* 4:343–351.

Flach, KW, Holzhey, CS, DeConinck, F and RJ Bartlett. 1980. Genesis and classification of Andepts and Spodosols. In: BKG Theng (ed). *Soils with Variable Charge*. New Zealand Soc. Soil Sci., Palmerston North, New Zealand. pp. 411–426.

Flach, KW, Nettleton, WD, Gile, LH and JG Cady. 1969. Pedocementation: Induration by silica, carbonates, and sesquioxides in the Quaternary. *SS* 107:442–453.

Flegenheimer, N and M Zárate. 1993. The archaeological record in Pampean loess deposits. *QI* 17:95–100.

Fleisher, M, Liu, T, Broecker, W and W Moore. 1999. A clue regarding the origin of rock varnish. *Geophys. Res. Letts.* 26:103–106.

Fletcher, JE and WP Martin. 1948. Some effects of algae and molds in the rain-crust of desert soils. *Ecology* 29:95–100.

Fletcher, PW and RE McDermott. 1957. Moisture depletion by forest cover on a seasonally saturated Ozark ridge soil. *SSSAP* 21:547–550.

Fletcher, RC and SL Brantley. 2010. Reduction of bedrock blocks as corestones in the weathering profile: Observations and model. *AJS* 310:131–164.

Florinsky, IV. 2012. The Dokuchaev hypothesis as a basis for predictive digital soil mapping (on the 125th anniversary of its publication). *Euras. Soil Sci.* 45:445–451.

Foeken, JPT, Day, S and FM Stuart. 2009. Cosmogenic He-3 exposure dating of the Quaternary basalts from Fogo, Cape Verdes: Implications for rift zone and magmatic reorganization. *QG* 4:37–49.

Foley, JA, DeFries, R, Asner, GP, Barford, C, Bonan, G, Carpenter, SR, Chapin, FS, Coe, MT, Daily, GC, Gibbs, HK, Helkowski, JH, Holloway, T, Howard, EA, Kucharik, CJ, Monfreda, C, Patz, JA, Prentice, IC, Ramankutty, N and PK Snyder. 2005. Global consequences of land use. *Science* 309:570–574.

Folkoff, ME and V Meentemeyer. 1985. Climatic control of the assemblages of secondary clay minerals in the A-horizon of United States soils. *ESPL* 10:621–633.

1987. Climatic control on the geography of clay minerals genesis. *AAAG* 77:635–650.

Follmer, LR. 1978. The Sangamon soil in its type area: A review. In: WC Mahaney (ed). *Quaternary Soils*. Geo Abstracts, Norwich, UK. pp. 125–165.

1979a. A historical review of the Sangamon soil. In: *Wisconsinan, Sangamonian, and Illinoian Stratigraphy in Central Illinois*. Illinois State Geological Survey Guidebook #13. Urbana-Champaign, IL, Midwest Friends of the Pleistocene Field Conference Guidebook. pp. 79–91.

1979b. Explanation of pedologic terms and concepts used in the discussion of soils for this guidebook. In: *Wisconsinan, Sangamonian, and Illinoian Stratigraphy in central Illinois*. Illinois State Geological Survey Guidebook #13. Champaign-Urbana, IL, Midwest Friends of the Pleistocene Field Conference Guidebook. pp. 129–134.

1982. The geomorphology of the Sangamon surface: Its spatial and temporal attributes. In: C Thorn (ed). *Space and Time in Geomorphology*. Allen and Unwin, Boston, MA. pp. 117–146.

1984. Soil: An uncertain medium for waste disposal. Proc. 7th Ann. Madison Waste Conf. 296–311.

1998a. A scale for judging degree of soil and paleosol development. *QI* 51/52:12–13.

1998b. Preface. *QI* 51/52:1–3.

Follmer, LR, McKay, ED, Lineback, JA and DL Gross. 1979. Wisconsinan, Sangamonian, and Illinoian stratigraphy in central Illinois, Illinois State Geological Survey Guidebook #13. Midwest Friends of the Pleistocene Field Conference Guidebook.

Fölster, H, Kalk, E and N Moshrefi. 1971. Complex pedogenesis of ferrallitic savanna soils in South Sudan. *Geoderma* 6:135–149.

Fontes, MPF and IA Carvalho. 2005. Color attributes and mineralogical characteristics, evaluated by radiometry, of highly weathered tropical soils. *SSSA J.* 69:1162–1172.

Fontes, MPF and SB Weed. 1996. Phosphate adsorption by clays from Brazilian Oxisols: Relationships with specific surface area and mineralogy. *Geoderma* 72:37–51.

Food and Agriculture Organization. 1998. *World Reference Base for Soil Resources*. Food and Agriculture Organization of the United Nations, Rome.

Fookes, G. 1997. *Tropical Residual Soils, a Geological Society Engineering Group Working Party Revised Report*. Geological Society, London.

Forman, S and GH Miller. 1984. Pedogenic processes and time-dependent soil morphologies on raised beaches, Bröggerhalvöya, Spitzbergen, Norway. *AAR* 16:381–394.

Forman, SL. 1989. Application and limitations of thermoluminescence to date Quaternary sediments. *QI* 1:47–59.

1999. Infrared and red stimulated luminescence dating of Late Quaternary near-shore sediments from Spitsbergen, Svalbard. *AAAR* 31:34–49.

Forman, SL, Bettis, EA III, Kemmis, TJ and BB Miller. 1992. Chronologic evidence for multiple periods of loess deposition during the Late Pleistocene in the Missouri and Mississippi River Valley, United States: Implications for the activity of the Laurentide Ice Sheet. *PPP* 93:71–83.

Forman, SL and GH Miller. 1984. Time-dependent soil morphologies and pedogenic processes on raised beaches, Bröggerhalvöya, Spitsbergen, Svalbard Archipelago. *Arctic Alpine Res.* 16:381–394.

Forman, SL, Nelson, AR and JP McCalpin. 1991. Thermoluminescence dating of fault-scarp-derived colluvium: Deciphering the timing of paleoearthquakes on the Weber Segment of the Wasatch fault zone, north central Utah. *J. Geophys. Res. B* 96:595–605.

Fosberg, MA. 1965. Characteristics and genesis of patterned ground in Wisconsin time in a chestnut soil zone of southern Idaho. *SS* 99:30–37.

Foss, JE, Fanning, DS, Miller, FP and DP Wagner. 1978. Loess deposits of the eastern shore of Maryland. *SSSAJ* 42:329–334.

Foss, JE and RH Rust. 1968. Soil genesis study of a lithologic discontinuity in glacial drift in western Wisconsin. *SSSAP* 32:393–398.

Foster, RC. 1988. Microenvironments of soil microorganisms. *Biol. Fert. Soils* 6:189–203.

1994. The ultramicromorphology of soil biota in situ in natural soils – A review. In: AJ Ringrose-Voase and GS Humphreys (eds). *Soil Micromorphology: Studies in Management and Genesis.* Vol. 22. Elsevier, Amsterdam. pp. 381–393.

Foster, RJ. 1971. *Physical Geology.* Merrill, Columbus, OH.

Fowler, KD, Greenfield, HJ and LO Van Schalkwyk. 2004. The effects of burrowing activity on archaeological sites: Ndondondwane, South Africa. *Geoarchaeol.* 19: 441–470.

Fox, BJ, Fox, MD and GA McKay. 1979. Litter accumulation after fire in a eucalypt forest. *AJB* 27:157–165.

Fox, CA and C Tarnocai. 2011. Organic soils of Canada. Part 2. Upland Organic soils. *CJSS* 91:823–842.

Fox, CA, Preston, CM and CA Fyfe. 1994. Micromorphological and 13C NMR characterization of a Humic, Lignic, and Histic Folisol from British Columbia. *CJSS* 74:1–15.

Fox, CA, Trowbridge, R and C Tarnocai. 1987. Classification, macromorphology and chemical characteristics of Folisols from British Columbia. *CJSS* 67:765–778.

Fox, DM, Bryan, RB and AG Price. 2004. The role of soil surface crusting in desertification and strategies to reduce crusting. *Environ. Monitor. Assess.* 99:149–159.

Frakes, LA and S Jianzhong. 1994. A carbon isotope record of the upper Chinese loess sequence: Estimates of plant types during stadials and interstadials. *PPP* 108:183–189.

Franceschi, VR and HT Horner Jr. 1980. Calcium oxalate crystals in plants. *Bot. Rev.* 46:361–427.

Francis, ML, Fey, MV, Prinsloo, HP, Ellis, F, Mills, AJ and TV Medinski. 2007. Soils of Namaqualand: Compensations for aridity. *J. Arid Environs.* 70:588–603.

Fraser, J, Teixeira, W, Falcao, N, Woods, W, Lehmann, J and AB Junqueira. 2011. Anthropogenic soils in the Central Amazon: From categories to a continuum. *Area* 43:264–273.

Franzmeier, DP, Bryant, RB and GC Steinhardt. 1985. Characteristics of Wisconsinan glacial tills in Indiana and their influence on argillic horizon development. *SSSAJ* 49:1481–1486.

Franzmeier, DP, Pedersen, EJ, Longwell, TJ, Byrne, JG and CK Losche. 1969. Properties of some soils in the Cumberland Plateau as related to slope aspect and position. *SSSAP* 33:755–761.

Franzmeier, DP and EP Whiteside. 1963. A chronosequence of Podzols in northern Michigan. II. Physical and chemical properties. Michigan. *State Univ. Agr. Exp. St. Quart. Bull.* 46:21–36.

Franzmeier, DP, Yahner, JE, Steinhardt, GC and HR Sinclair Jr. 1983. Color patterns and water table levels in some Indiana soils. *SSSAJ* 47:1196–1202.

Frazee, CJ, Fehrenbacher, JB and WC Krumbein. 1970. Loess distribution from a source. *SSSAP* 34:296–301.

Frazier, BE and GB Lee. 1971. Characteristics and classification of three Wisconsin Histosols. *SSSAP* 35:776–780.

Frazier, CS and RC Graham. 2000. Pedogenic transformation of fractured granitic bedrock, southern California. *SSSAJ* 64:2057–2069.

Frechen, M, Vanneste, K, Verbeeck, K, Paulissen, E and T Camelbeeck. 2001. The deposition history of the coversands along the Bree Fault Escarpment, NE Belgium. *Netherlands J. Geosciences* 80:171–185.

Frederick, CD, Bateman, MD and R Rogers. 2002. Evidence for eolian deposition in the sandy uplands of East Texas and the implications for archaeological site integrity. *Geoarchaeol.* 17:191–217.

Fredlund, GG, Johnson, WC and W Dort Jr. 1985. A preliminary analysis of opal phytoliths from the Eustis Ash Pit, Frontier County, Nebraska. *TERQUA Symp. Ser.* 1:147–162.

Fredlund GG and LL Tieszen. 1997. Phytolith and carbon isotope evidence for Late Quaternary vegetation and climate change in the southern Black Hills, South Dakota. *QR* 47:206–217.

Freedman, B and U Prager. 1986. Ambient bulk deposition, throughfall, and stemflow in a variety of forest stands in Nova Scotia. *CJFR* 16:854–860.

Freeland, JA and CV Evans. 1993. Genesis and profile development of Success soils, northern New Hampshire. *SSSAJ* 57:183–191.

Frei, E and MG Cline. 1949. Profile studies of normal soils of New York. II. Micromorphological studies of the Gray-Brown Podzolic-Brown Podzolic soil sequence. *SS* 68:333–344.

Freiberg, M and E Freiberg. 2000. Epiphyte diversity and biomass in the canopy of lowland and montane forest in Ecuador. *J. Trop. Ecol.* 16:673–688.

Frelich, LE, Hale, CM, Scheu, S, Holdsworth, AR, Heneghan, L, Bohlen, PJ and PB Reich. 2006. Earthworm invasion into previously earthworm-free temperate and boreal forests. *Biol. Invasions* 8:1235–1245.

French, C, Periman, R, Cummings, LS, Hall, S, Goodman-Elgar, M and J Boreham. 2009. Holocene alluvial sequences, cumulic soils and fire signatures in the middle Rio Puerco basin at Guadalupe Ruin, New Mexico. *Geoarch. Intl. J.* 24:638–676.

French, H, Demitroff, M and WL Newell. 2009. Past permafrost on the Mid-Atlantic Coastal Plain, eastern United States. *Permafrost Periglac. Proc.* 20:285–294.

French, HM. 1971. Slope asymmetry of the Beaufort Plain, Northwest Banks Island, NWT, Canada. *CJES* 8:717–731.

1993. Cold-climate processes and landforms. In: HM French and O Slaymaker (eds). *Canada's Cold Environments.* McGill-Queen's Univ. Press, Montreal. pp. 143–167.

French, MH. 1945. Geophagia in animals. *E. Afr. Med. J.* 22:103–110.

Frenot, Y, Van Vliet-Lanoë, B and J-C Gloaguen. 1995. Particle translocation and initial soil development on a glacier foreland, Kerguelen Islands, Subantarctic. *AAR* 27:107–115.

Freppaz, M, Filippa, G, Caimi, A, Buffa, G and E Zanini. 2010. Soil and plant characteristics in the alpine tundra (NW Italy). In: B Gutierrez and C Pena (eds). *Tundras: Vegetation, Wildlife and Climate Trends.* Environmental Research Advances, Nova Sci. Publ., Hauppauge, NY. pp. 81–110.

Freyssinet, P, Zeegers, H and Y Tardy. 1989. Morphology and geochemistry of gold grains in lateritic profiles of southern Mali. *J. Geochem. Explor.* 32:17–31.

Fridland, VM. 1958. Podzolization and illimerization (clay migration). *SSS* 24–32.

——— 1965. Makeup of the soil cover. *SSS* 4:343–354.

——— 1974. Structure of the soil mantle. *Geoderma* 12:35–41.

Friedman, GM. 1961. Distinction between dune, beach, and river sands from their textural characteristics. *J. Sed. Petr.* 31:514–529.

Friedman, I and W Long. 1976. Hydration rate of obsidian. *Science* 191:347–352.

Friedman, I and RL Smith. 1960. A new dating method using obsidian. Part I. The development of the method. *Am. Antiquity* 25:476–493.

Friedman, I and FW Trembour. 1978. Obsidian: The dating stone. *Am. Sci.* 66:44–52.

Friedmann, EI. 1980. Endolithic microbial life in hot and cold deserts. *Orig. Life* 10:223–235.

Friedrich, K. 1996. *Digitale reliefgliederungsverfahren zur ableitung boden-kundlich relevanter flächeneinheiten.* Frankfurter Geowissenschaftliche Arbeiten D21, Frankfurt.

Frink, DS. 1994. The oxidizable carbon ratio (OCR): A proposed solution to some of the problems encountered with radiocarbon data. *North Am. Archaeol.* 15:17–29.

Fritsch, E, Balan, E, DoNascimento, NR, Allard, T, Bardy, M, Bueno, G, Derenne, S, Melfi, AJ and G Calas. 2011. Deciphering the weathering processes using environmental mineralogy and geochemistry: Towards an integrated model of laterite and podzol genesis in the Upper Amazon Basin. *Compt. Rendus Geosci.* 343:188–198.

Fritton, DD and GW Olson. 1972. Depth to the apparent water table in 17 New York soils from 1963 to 1970. NY Food Life Sci. Bull. 13.

Fritts HC. 1976. *Tree Rings and Climate.* Academic Press, New York.

Froedge, RD. 1980. *Soil Survey of Christian County, Kentucky.* USDA-SCS, US Govt. Printing Off., Washington, DC.

Frolking, TA, Jackson, ML and JC Knox. 1983. Origin of red clay over dolomite in the loess-covered Wisconsin Driftless uplands. *SSSAJ* 47:817–820.

Frolking, TA and BT Lepper. 2001. Geomorphic and pedogenic evidence for bioturbation of artifacts at a multicomponent site in Licking County, Ohio, USA. *Geoarchaeol.* 16:243–262.

Frye, JC and AB Leonard. 1949. Pleistocene stratigraphic sequence in northeastern Kansas. *AJS* 247:883–899.

Frye, JC, Shaffer, PR, Willman, HB and GE Ekblaw. 1960a. Accretion gley and the gumbotil dilemma. *AJS* 258:185–190.

Frye, JC and HB Willman. 1960. Classification of the Wisconsin Stage in the Lake Michigan glacial lobe. Illinois Geol. Surv. Circ. 285.

Frye, JC, Willman, HB and HD Glass. 1960b. Gumbotil, accretion-gley, and the weathering profile. Ill. Geol. Survey Circ. 295.

Fuchs, M, Kreutzer, S, Fischer, M, Sauer, D and R Sorensen. 2012. OSL and IRSL dating of raised beach sand deposits along the southeastern coast of Norway. *QG* 10:195–200.

Fuchs, M and A Lang. 2001. OSL dating of coarse-grain fluvial quartz using single-aliquot protocols on sediments from NE Peloponnese, Greece. *QSR* 20:783–787.

Fuchs, M and LA Owen. 2008. Luminescence dating of glacial and associated sediments: Review, recommendations and future directions. *Boreas* 37:636–659.

Fuchs, M, Rousseau, D-D, Antoine, P, Hatté, C and C Gauthier. 2007. Chronology of the last climatic cycle (Upper Pleistocene) of the Surduk loess sequence, Vojvodina, Serbia. *Boreas* 37:66–73.

Fujinuma, R, Bockheim, J and N Balster. 2005. Base-cation cycling by individual tree species in old-growth forests of Upper Michigan, USA. *Biogeochem.* 74:357–376.

Fuller, LG and DW Anderson. 1993. Changes in soil properties following forest invasion of black soils of the aspen parkland. *CJSS* 73:613–627.

Fuller, LG, Wang, D and DW Anderson. 1999. Evidence for solum recarbonation following forest invasion of a grassland soil. *CJSS* 79:443–448.

Furbish, DJ and S Fagherazzi. 2001. Stability of creeping soil and implications for hillslope evolution. *Water Resour. Res.* 37:2607–2618.

Furian, S, Barbiero, L, Boulet, R, Curmi, P, Grimaldi, M and C Grimaldi. 2002. Distribution and dynamics of gibbsite and kaolinite in an Oxisol of Serra do Mar, southeastern Brazil. *Geoderma* 106:83–100.

Furley, PA. 1971. Relationships between slope form and soil properties developed over chalk parent materials. *Inst. Br. Geog. Spec. Pub.* 3:141–163.

Furman, T, Thompson, P and B Hatchl. 1998. Primary mineral weathering in the central Appalachians: A mass balance approach. *Geochim. Cosmochim. Acta* 62:2889–2904.

Gabet, EJ and SM Mudd. 2010. Bedrock erosion by root fracture and tree throw: A coupled biogeomorphic model to explore the humped soil production function and the persistence of hillslope soils. *J. Geophys. Res. Earth Surf.* 115. doi:10.1029/2009JF001526.

Gabet, EJ, Reichman, OJ and EW Seabloom. 2003. The effects of bioturbation on soil processes and sediment transport. *Ann. Rev. Earth Planet. Sci.* 31:249–273.

Gaikawad, ST and FD Hole. 1965. Characteristics and genesis of a gravelly Brunizemic regosol. *SSSAP* 29:725–728.

Galbraith, RF, Roberts, RG, Laslett, GM, Yoshida, H and JM Olley. 1999. Optical dating of single and multiple grains of quartz from Jinmium rock shelter, northern Australia. Part I. Experimental design and statistical models. *Archaeometry* 41:339–364.

Galka, M and M Sznel. 2013. Late Glacial and Early Holocene development of lakes in northeastern Poland in view of plant macrofossil analyses. *QI* 292:124–135.

Gallet, S, Jahn, B and M Torii. 1996. Geochemical characterization of the Luochuan loess paleosol sequence, China, and paleoclimatic implications. *Chem. Geol.* 133:67–88.

Galloway, WE and DK Hobday. 1983. *Terrigenous Clastic Depositional Systems.* Springer-Verlag, New York.

Gamble, EE, Daniels, RB and RJ McCracken. 1969. A2 horizons of coastal plain soils pedogenic or geologic origin. *Southeast. Geol.* 11:137–152.

Ganzhara, NF. 1974. Humus formation in Chernozem soils. *SSS* 6:403–407. (Translated from Pochvovedeniye 7:39–43.)

Gao, Y-X and H-Z Chen. 1983. Salient characteristics of soil-forming processes in Xizang (Tibet). *SS* 135:11–17.

Garciamiragaya, J and T Herreramarcano. 1993. Charge-distribution of selected Venezuelan soils with some pedogenic and agrotechnological implications. *Comm. Soil Sci. Plant Analysis* 24:1495–1508.

Gard, LE and CA Van Doren. 1949. Soil losses as affected by cover, rainfall and slope. *SSSAP* 14:374–378.

Gardiner, MJ and T Walsh. 1966. Comparison of soil material buried since Neolithic times with those of the present day. *Proc. Roy. Irish Acad.* 65C:29–35.

Gardner, DR and EP Whiteside. 1952. Zonal soils in the transition region between the Podzol and Gray-Brown Podzolic regions in Michigan. *SSSAP* 16:137–141.

Gardner, LR. 1972. Origin of the Mormon Mesa caliche, Clark County, Nevada. *GSA Bull.* 83:143–156.

Gardner, LR, Smith, BR and WK Michener. 1992. Soil evolution along a forest-salt marsh transect under a regime of slowly rising sea level, southeastern United States. *Geoderma* 55:141–157.

Gardner, WH. 1986. Water content. In: A Klute (ed). *Methods of Soil Analysis.* Part 1. *Physical and Mineralogical Methods.* Agronomy Mon. 9, 2nd ed., Soil Sci. Soc. Am., Madison, WI. pp. 493–544.

Gärtner, H. 2007. Tree roots – methodological review and new development in dating and quantifying erosive processes. *Geomorphology* 86:243–251.

Garzanti, E, Ando, S, Vezzoli, G, Lustrino, M, Boni, M and P Vermeesch. 2012. Petrology of the Namib Sand Sea: Long-distance transport and compositional variability in the wind-displaced Orange Delta. *Earth-Sci. Revs.* 112:173–189.

Gassama, N and S Violette. 2012. Atmospheric, weathering and biological contributions in the chemical signature of stream water: The upper Iskar Reka watershed, Bulgaria. *Hydrol. Sci. J.* 57:535–546.

Gaston, LA, Mansell, RS and HM Selim. 1992. Predicting removal of major soil cations and anions during acid infiltration: Model evaluation. *SSSAJ* 56:944–950.

Gates, FC. 1942. The bogs of northern lower Michigan. *Ecol. Monogr.* 12:216–254.

Gavin, DG and LB Brubaker. 1999. A 6000-year soil pollen record of subalpine meadow vegetation in the Olympic Mountains, Washington, USA. *JE* 87:106–122.

Geertsema, M, Clague, JJ, Schwab, JW and SG Evans. 2006. An overview of recent large catastrophic landslides in northern British Columbia, Canada. *Eng. Geol.* 83:120–143.

Gehrels, R. 2010. Sea-level changes since the Last Glacial Maximum: An appraisal of the IPCC Fourth Assessment Report. *JQS* 25:26–38.

Gehring, AU, Langer, MR and CA Gehring. 1994. Ferriferous bacterial encrustations in lateritic duricrusts from southern Mali (west-Africa). *Geoderma* 61:213–222.

Geis, JW, Boggess, WR and JD Alexander. 1970. Early effects of forest vegetation and topographic position on dark-colored, prairie-derived soils. *SSSAP* 34:105–111.

Geiss, CE and CW Zanner. 2007. Sediment magnetic signature of climate in modern loessic soils from the Great Plains. *QI* 162/163:97–110.

Gellatly, AF. 1984. The use of rock weathering-rind thickness to redate moraines in Mount Cook National Park, New Zealand. *AAR* 16:225–232.

1985. Phosphate retention: Relative dating of Holocene soil development. *Catena* 12:227–240.

Gerakis, A and B Baer. 1999. A computer program for soil textural classification. *SSSAJ* 63:807–808.

Gerasimov, IP. 1973. Chernozems, buried soils and loesses of the Russian plain their age and genesis. *SS* 116:202–210.

Geroy, IJ, Gribb, MM, Marshall, HP, Chandler, DG, Benner, SG and JP McNamara. 2011. Aspect influences on soil water retention and storage. *Hydrol. Proc.* 25:3836–3842.

Gerrard, AJ. 1981. *Soils and Landforms. An Integration of Geomorphology and Pedology*. Allen and Unwin, Boston.

1992. *Soil Geomorphology. An Integration of Pedology and Geomorphology*. Chapman and Hall, New York.

Gerson, R and R Amit. 1987. Rates and modes of dust accretion and deposition in an arid region: The Negev, Israel. In: LE Frostick and I Reid (eds). *Desert Sediments: Ancient and Modern*. Blackwell Scientific Publications, Boston, MA. pp. 157–169.

Gessler, PE, Chadwick, OA, Chamran, F, Althouse, L and K Holmes. 2000. Modeling soil-landscape and ecosystem properties using terrain attributes. *SSSAJ* 64:2046–2056.

Gessler, PE, Moore, ID, McKenzie, NJ and PJ Ryan. 1995. Soil-landscape modeling and spatial prediction of soil attributes. *Intl. J. GIS* 9:421–432.

Geyh, MA, Benzler, J-H and G Roeschmann. 1971. Problems of dating Pleistocene and Holocene soils by radiometric means. In: DH Yaalon (ed). *Paleopedology. Origin, Nature and Dating of Paleosols.* Israel Univ. Press, Jerusalem. pp. 63–75.

Geyh, MA, Roeschmann, G, Wijmstra, TA and AA Middeldorp. 1983. The unreliability of 14C dates obtained from buried sandy Podzols. *Radiocarbon* 25:409–416.

Gholz, HL, Wedin, DA, Smitherma, SM, Harmon, ME and WJ Parton. 2000. Long-term dynamics of pine and hardwood litter in contrasting environments: Toward a global model of decomposition. *GCB* 6:751–765.

Gibbard, PL, Boreham, S, Cohen, KM and A Moscariello. 2005. Global chronostratigraphical correlation table for the last 2.7 million years, v. 2005c. Subcommission on Quaternary Stratigraphy, Dept. of Geography, Univ. of Cambridge, Cambridge, England.

Gibbons, AB, Megeath, JD and KL Pierce. 1984. Probability of moraine survival in a succession of glacial advances. *Geology* 12:327–330.

Gibbs, JA and HF Perkins. 1966. Properties and genesis of the Hayesville and Cecil series of Georgia. *SSSAP* 30:256–260.

Giesler, R, Ilvesniemi, H, Nyberg, L, van Hees, P, Starr, M, Bishop, K, Kareinen, T and US Lundström. 2000. Distribution and mobilization of Al, Fe and Si in three podzolic soil profiles in relation to the humus layer. *Geoderma* 94:249–263

Gilbert, GK. 1877. *Report on the geology of the Henry Mountains*. U.S. Geog. and Geol. Survey of the Rocky Mtn. Region. Washington, DC.

Gile, LH. 1958. Fragipan and water-table relationships of some Brown Podzolic and Low Humic-Gley soils. *SSSAP* 22:560–565.

1970. Soils of the Rio Grande Valley border in southern New Mexico. *SSSAP* 34:465–472.

1975a. Causes of soil boundaries in an arid region. I. Age and parent materials. *SSSAP* 39:316–323.

1975b. Causes of soil boundaries in an arid region. II. Dissection, moisture, and faunal activity. *SSSAP* 39:324–330.

1979. Holocene soils in eolian sediments of Bailey County, Texas. *SSSAJ* 43:994–1003.

Gile, LH and RR Grossman. 1968. Morphology of the argillic horizon in desert soils of southern New Mexico. *SS* 106:6–15.

1979. *The Desert Project*. USDA-SCS, US Govt. Printing Off., Washington, DC.

Gile, LH and JW Hawley. 1968. Age and comparative development of desert soils at the Gardner Spring Radiocarbon Site, New Mexico. *SSSAP* 32:709–716.

1972. The prediction of soil occurrence in certain desert regions of the southwestern United States. *SSSAP* 36:119–123.

1993. Carbonate stages in sandy soils of the Leasburg Surface, Southern New Mexico. *SS* 156:101–110.

1995. Pedogenic carbonate in soils of the Isaack's ranch surface, southern New Mexico. *SSSAJ* 59:501–508.

Gile, LH, Hawley, JW and RB Grossman. 1981. *Soils and Geomorphology in the Basin and Range Area of Southern New Mexico: A Guidebook to the Desert Project, Memoir #39*. New Mexico Bureau of Mines and Mineral Research, Socorro, NM.

Gile, LH, Peterson, FF and RB Grossman. 1965. The K horizon: A master soil horizon of carbonate accumulation. *SS* 99:74–82.

1966. Morphological and genetic sequences of carbonate accumulation in desert soils. *SS* 101:347–360.

Gilet-Blein, N, Marien, G and J Evin. 1980. Unreliability of 14C dates from organic matter of soils. *Radiocarbon* 22:919–929.

Gilichinsky, DA, Barry, RG, Bykhovets, SS, Sorokovikov, VA, Zhang, T, Zudin, SL and DG Fedorov-Davydov. 1998. A century of temperature observations of soil climate: Methods of analysis and long-term trends. In: AG Lewkowicz and M Allard (eds). *Permafrost. Proc. 7th Intl. Conf. Université Laval, Centre d'Etudes Nordiques, Yellowknife, Canada. pp. 313–317.

Gilkes, RJ, Scholz, G and GM Dimmock. 1973. Lateritic deep weathering of granite. *JSS* 24:523–536.

Gillen, D, Honda, M, Chivas, AR, Yatsevich, I, Patterson, DB and PF Carr. 2010. Cosmogenic Ne-21 exposure dating of young basaltic lava flows from the Newer Volcanic Province, western Victoria, Australia. *QG* 5:1–9.

Gillham, RA. 1984. The capillary fringe and its effect on water table response. *J. Hydrol.* 67:307–324.

Gillman, LR, Jeffries, MK and GN Richards. 1972. Non-soil constituents of termite (*Coptotermes acinaciformes*) mounds. *Austr. J. Biol. Sci.* 25:1005–1013.

Giovannini, G and P Sequi. 1976. Iron and aluminum as cementing substances of soil aggregates: Changes in stability of soil aggregates following extraction of iron and aluminum by acetylacetone in a non polar solvent. *JSS* 27:148–153.

Giuliani, S, Capotondi, L, Maffioli, P, Langone, L, Giglio, F, Yam, R, Frignani, M and M Ravaioli. 2011. Paleoenvironmental changes in the Pacific sector of the Southern Ocean (Antarctica) during the past 2. 6 Ma. Global Planet. *Change* 77:34–48.

Glaccum, RA and JM Prospero. 1980. Saharan aerosols over the tropical North Atlantic: Mineralogy. *Marine Geol.* 37:295–321.

Glaser, B and JJ Birk. 2012. State of the scientific knowledge on properties and genesis of Anthropogenic Dark Earths in Central Amazonia (terra preta de Indio). *Geochim. Cosmochim. Acta* 82:39–51.

Glaser, B, Haumaier, L, Guggenberger, G and W Zech. 2001. The "Terra Preta" phenomenon: A model for sustainable agriculture in the humid tropics. *Naturwissenschaften* 88:37–41.

Glazovskaya, MA. 1968. Geochemical landscapes and types of geochemical soil sequences. *Trans. 9th Intl. Congr. Soil Sci.* 4:303–312.

Glazovskaya, MA and EI Parfenova. 1974. Biogeochemical factors in the formation of terra rossa in the southern Crimea. *Geoderma* 12:57–82.

Glinka, KD. 1914. *Die Typen der Bodenbildung, irhe Klassifikation und Geographische Verbreitung.* Berlin, Borntraeger. (See Marbut (1927).)

——— 1924. Degradatsiya I podzolistyi protsess (Degradation and the podzolic process). Pochvovedeniye 3–4.

Global Soil Map. 2012. Specifications, Version 1 GlobalSoilMap.net products. Release 2.2.

Glover, F, Whitworth, KL, Kappen, P, Baldwin, DS, Rees, GN, Webb, JA and E Silvester. 2011. Acidification and buffering mechanisms in acid sulfate soil wetlands of the Murray-Darling basin, Australia. *Environ. Sci. Tech.* 45:2591–2597.

Gobin, A, Campling, P, Deckers, J and J Feyen. 2000. Quantifying soil morphology in tropical environments: Methods and application in soil classification. *SSSAJ* 64:1423–1433.

Goble, RJ, Mason, JA, Loope, DB and JB Swinehart. 2004. Optical and radiocarbon ages of stacked paleosols and dune sands in the Nebraska Sand Hills, USA. *QSR* 23:1173–1182.

Gocke, M, Pustovoytov, K and Y Kuzyakov. 2012. Pedogenic carbonate formation: Recrystallization versus migration-process rates and periods assessed by C-14 labeling. Global Biogeochem. *Cycles* 26. doi:10.1029/2010GB003871

Goddard, TM, Runge, ECA and BW Ray. 1973. The relationship between rainfall frequency and amount to the formation and profile distribution of clay particles. *SSSAP* 37:299–304.

Godfrey, CL and FF Riecken. 1954. Distribution of phosphorous in some genetically related loess-derived soils. *SSSAJ* 18:80–84.

Goehring, BM, Kurz, MD, Balco, G, Schaefer, JM, Licciardi, J and N Lifton. 2010. A reevaluation of in situ cosmogenic He-3 production rates. *QG* 5:410–418.

Goehring, BM, Lohne, OS, Mangerud, J, Svendsen, JI, Gyllencreutz, R, Schaefer, J and R Finkel. 2012. Late glacial and holocene [10]Be production rates for western Norway. *JQS* 27:89–96.

Gokceoglu, C, Ulusay, R and H Sonmez. 2000. Factors affecting the durability of selected weak and clay-bearing rocks from Turkey, with particular emphasis on the influence of the number of drying and wetting cycles. *Engin. Geol.* 57:215–237.

Goldich, SS. 1938. A study of rock weathering. *J. Geol.* 46:17–58.

Goldin, A and J Edmonds. 1996. A numerical evaluation of some Florida Spodosols. *PG* 17:242–252.

Goldthwait, RP. 1976. Frost sorted patterned ground: A review. *QR* 6:27–35.

Gombeer R and J D'Hoore. 1971. Induced migration of clay and other moderately mobile constituents. III. *Critical soil/water dispersion ratio, colloid stability and electrophoretic mobility Pédologie* 21:311–342.

Goodfriend, GA. 1992. The use of land snails in paleoenvironmental reconstruction. *QSR* 11:665–685.

Goodfriend, GA and DG Hood. 1983. Carbon isotope analysis of land snail shells: Implications for carbon sources and radiocarbon dating. *Radiocarbon* 27:33–42.

Goodlett, JC. 1954. Vegetation adjacent to the border of the Wisconsin drift in Potter County, Pennsylvania. Harvard For. Bull. 25.

Goodman, AY, Rodbell, DT, Seltzer, GO and BG Mark. 2001. Subdivision of glacial deposits in southeastern Peru based on pedogenic development and radiometric ages. *QR* 56:31–50.

Goossens, D and B Buck. 2009. Dust emission by off -road driving: Experiments on 17 arid soil types, Nevada, USA. *Geomorphology* 107:118–138.

Gorbunov, NL. 1961. Movement of colloidal clay particles in soils. (Problem of leaching and podzolization.) *SSS* 712–724.

Gorbushina, AA. 2007. Life on the rocks. *Envir. Microbiol.* 9:1613–1631.

Gordon, A and CB Lipman. 1926. Why are serpentine and other magnesian soils infertile? *SS* 22:291–302.

Gordon, M, Jr, Tracey, JI, Jr and MW Ellis. 1958. Geology of the Arkansas Bauxite Region. USGS Prof. Paper 299.

Gordon, SI and RI Dorn. 2005. In situ weathering rind erosion. *Geomorphology* 67:97–113.

Gorham, E. 1957. The development of peatland. *Quart. Rev. Biol.* 32:145–166.

Goryachkin, SV, Karavaeva, NA, Targulian, VO and MV Glazov. 1999. Arctic soils: Spatial distribution, zonality and transformation due to global change. *Permafrost Periglac. Proc.* 10:235–250.

Goss, DW, Smith, SJ and BA Stewart. 1973. Movement of added clay through calcareous materials. *Geoderma* 9:97–103.

Gosse, JC, Klein, J, Evenson, EB, Lawn, B and R Middleton. 1995. Beryllium-10 dating of the duration and retreat of the last Pinedale glacial sequence. *Science* 268:1329–1333.

Gosse, JC and PM Phillips. 2001. Terrestrial in situ cosmogenic nuclides: Theory and application. *QSR* 20:1475–1560.

Goswami, A, Das, AL, Sah, KD and D Sarkar. 1996. Pedological studies in Great Nicobar Island. *Geog. Rev. Ind.* 58:162–168.

Gosz, JR, Likens, GE and FH Bormann. 1976. Organic matter and nutrient dynamics of the forest and forest floor in the Hubbard Brook Forest. *Oecologia* 22:305–320.

Goudie, AS. 1996. Organic agency in calcrete development. *J. Arid. Envs* 32:103–110.

——— 2006. The Schmidt Hammer in geomorphological research. *PPG* 6:703–718.

——— 2013. The Schmidt Hammer and related devices in geomorphological research In: JF Shroder (ed). *Treatise on Geomorphology*. Vol. 14. Academic Press, San Diego, CA. pp. 338–345.

Goudie, AS and H Viles. 1997. *Salt Weathering Hazard*. Wiley, New York.

Goudie, AS and A Watson. 1984. Rock block monitoring of rapid salt weathering in southern Tunisia. *ESPL* 9:95–98.

Gozdzik, J. 1980. Zastosowanie morfometrii i graniformametrii do badan´ osado´w w kopalni we,gla brunatnego Belchato´w. *Studia Regionalne* 4:101–114.

Gozdzik, JS and HM French. 2004. Apparent upfreezing of stones in Late-Pleistocene coversand, Belchatow Vicinity, central Poland. *Permafrost Periglac. Proc.* 15:359–366.

Gracanin, Z. 1971. Age and development of the hummocky meadow (buckelwiese) in the Lechtaler Alps (Austria). In: DH Yaalon (ed). *Paleopedology. Origin, Nature and Dating of Paleosols*. Israel Univ. Press, Jerusalem. pp. 117–127.

Graf, WL. 1978. Fluvial adjustments to the spread of tamarisk in the Colorado Plateau region. *GSA Bull.* 89:1491–1501.

Grah, RF and CC Wilson. 1944. Some components of rainfall interception. *J. For.* 42:890–898.

Graham, RC, Daniels, RB and SW Buol. 1990a. Soil-geomorphic relations on the Blue Ridge Front. I. Regolith types and slope processes. *SSSAJ* 54:1362–1367.

Graham, RC, Diallo, MM and LJ Lund. 1990b. Soils and mineral weathering on phyllite colluvium and serpentine in northwestern California. *SSSAJ* 54:1682–1690.

Graham, RC and E Franco-Vizcaino. 1992. Soils on igneous and metavolcanic rocks in the Sonoran Desert of Baja California, Mexico. *Geoderma* 54:1–21.

Graham, RC and AT O'Geen. 2010. Soil mineralogy trends in California landscapes. *Geoderma* 154:418–437.

Graham, RC and AR Southard. 1983. Genesis of a Vertisol and an associated Mollisol in northern Utah. *SSSAJ* 47:552–559.

Graham, RC, Weed, SB, Bowen, LH, Amarasiriwardena, DD and SW Buol. 1989. Weathering of iron-bearing minerals in soils and saprolite on the North Carolina Blue Ridge front. II. Clay mineralogy. *Clay Mins.* 37:29–40.

Graham, SA. 1941. Climax forests of the upper peninsula of Michigan. *Ecology* 22:355–362.

Graly, JA, Bierman, PR, Reusser, LJ and MJ Pavich. 2010. Meteoric [10]Be in soil profiles – a global meta-analysis. *Geochim. Cosmochim. Acta* 74:6814–6829.

Graly, JA, Reusser, LJ and PR Bierman. 2011. Short and long-term delivery rates of meteoric Be-10 to terrestrial soils. *EPSL* 302:329–336.

Granger, D. 2006. A review of burial dating methods using [26]Al and [10]Be. *GSA Spec. Paper* 415:1–16.

Granger, DE, Kirchner, JW and R Finkel. 1996. Spatially averaged long-term erosion rates measured from in-situ produced cosmogenic nuclides in alluvial sediment. *J. Geol.* 104:249–257.

Graustein, WC, Cromack, K, Jr and P Sollins. 1977. Calcium oxalate: Occurrence in soils and effect on nutrient and geochemical cycles. *Science* 198: 1252–1254.

Grave, P and L Kealhofer. 1999. Assessing bioturbation in archaeological sediments using soil morphology and phytolith analysis. *J. Archaeol. Sci.* 26:1239–1248.

Gray, JM, Humphreys, GS and JA Deckers. 2011. Distribution patterns of World Reference Base soil groups relative to soil forming factors. *Geoderma* 160:373–383.

Green, RD and GP Askew. 1965. Observations on the biological development of macropores in soils of Romney Marsh. *JSS* 16:342–349.

Green, RN, Trowbridge, RL and K Klinka. 1993. Towards a taxonomic classification of humus forms. *For. Sci.* 39:1–48.

Greene, D, Zasada, J, Sirois, L, Kneeshaw, D, Morin, H, Charron, I and M-J Simard. 1999. A review of the regeneration dynamics of North American boreal forest tree species. *CJFR* 29:824–839.

Greene, H. 1945. Classification and use of tropical soils. *SSSAP* 10:392–396.

Greenslade, PJN. 1974. Some relations of the meat ant, Iridomyrmex purpureus (Hymenoptera: Formicidae) with soil in South Australia. *Soil Biol. Biochem.* 6:7–14.

Gregorich, EG and DW Anderson. 1985. Effects of cultivation and erosion on soils of four toposequences in the Canadian prairies. *Geoderma* 36:343–354.

Grieve, IC. 2000. Effects of human disturbance and cryoturbation on soil iron and organic matter distributions and on carbon storage at high elevations in the Cairngorm Mountains, Scotland. *Geoderma* 95:1–14.

——— 2001. Human impacts on soil properties and their implications for the sensitivity of soil systems in Scotland. *Catena* 42:361–374.

Grieve, R, Rupert, J, Smith, J and A Therriault. 1995. The record of terrestrial impact cratering. *GSA Today* 5:189, 194–196.

Griffey, NJ and S Ellis. 1979. Three in situ paleosols buried beneath neoglacial moraine ridges, Okstindan and Jotunheimen, Norway. *AAR* 11:203–214.

Grimley, DA. 1998. Pedogenic influences on magnetic susceptibility patterns in loess-paleosol sequences of southwestern Illinois. *QI* 51/52:51.

——— 2000. Glacial and nonglacial sediment contributions to Wisconsin Episode loess in the central United States. *GSA Bull.* 112:1475–1495.

Grimley, DA, Follmer, LR, Hughes, RE and PA Solheid. 2003. Modern, Sangamon and Yarmouth soil development in loess of unglaciated southwestern Illinois. *QSR* 22:225–244.

Grimley, DA, Larsen, D, Kaplan, SW, Yansa, CH, Curry, BB and EA Oches. 2009. A multi-proxy palaeoecological and palaeoclimatic record within full glacial lacustrine deposits, western Tennessee, USA. *JQS* 24:960–981.

Grimm, E. 2001. Trends and palaeoecological problems in the vegetation and climate history of the Northern Great Plains, U.S.A. Biology and Environment, Proc. *Royal Irish Acad.* 101B:47–64.

Grimm, EC, Maher LJ, Jr and DM Nelson. 2009. The magnitude of error in conventional bulk-sediment radiocarbon dates from central North America. *QR* 72:301–308.

Groffman, PM and JM Tiedje. 1989. Denitrification in north temperate forest soils: Relationships between denitrification and environmental factors, at the landscape scale. *Soil Biol. Biochem.* 21:621–626.

Grootes, PM, Stuivor, M, White, JCW, Johnsen, S and J Houzel. 1993. Comparison of oxygen isotope records from the GISP2 and GRIP Greenland ice cores. *Nature* 366:552–554.

Grossman, RB. 2004. Note on C. E. Kellogg by a junior staff member. *SSH* 45:144–148.

Grossman, RB and FJ Carlisle. 1969. Fragipan soils of the eastern United States. *Adv. Agron.* 21:237–279.

Grossman, RB and MG Cline. 1957. Fragipan horizons in New York soils. II. Relationships between rigidity and particle size distribution. *SSSAP* 21:322–325.

Grossman, RB and WC Lynn. 1967. Gel-like films that may form at air-water interfaces in soils. *SSSAP* 31:259–262.

Grossman, RB, Odell, RT and AH Beavers. 1964. Surfaces of peds from B horizons of Illinois soils. *SSSAP* 28:792–798.

Grote, G and WE Krumbein. 1992. Microbial precipitation of manganese by bacteria and fungi from desert rock and rock varnish. *Geomicrobiol.* 10:49–57.

Grube, S. 2001. Soil modification by the harvester termite Hodotermes mossambicus (Isoptera: Hodotermitidae) in a semiarid savanna grassland of Namibia. *Sociobiology* 37:757–767.

Gubin, SV. 1994. Relic features in recent tundra soil profiles and tundra soils classification. In: JM Kimble and RJ Ahrens (eds). *Classification Correlation, and Management of Permafrost-Affected Soils*. Proc. Intl. Soil Correlation Mtg (ISCOM). USDA-SCS, Lincoln, NE. pp. 63–65.

Guccione, M. 1983. Quaternary sediments and their weathering history in northcentral Missouri. *Boreas* 12:217–226.

Gugalinskaya, LA and VM Alifanov. 2000. Hypothetical lithogenic profile of loamy soils in the center of the Russian Plain. *Euras. Soil Sci.* 33:89–98.

Guggenberger, G., Bäumler, and W. Zech. 1998. Weathering of soils developed in eolian material overlying glacial deposits in eastern Nepal. *Soil Sci.* 163:325–337.

Guggenberger, G, Elliott, ET, Frey, SD, Six, J and K Paustian. 1999. Microbial contributions to the aggregation of a cultivated grassland soil amended with starch. *Soil Biol. Biochem.* 31:407–419.

Guhl, A, Bertran, P, Zielhofer, C and KE Fitzsimmons. 2013. Optically stimulated tuminescence (OSL) dating of sand-filled wedge structures and their fine-grained host sediment from Jonzac, SW France. *Boreas* 42:317–332.

Guilderson, TP, Reimer, PJ and TA Brown. 2005. The boon and bane of radiocarbon dating. *Science* 307:364–364.

Guillet, B and AM Robin. 1972. Interprétation de datations par le ¹⁴C d'horizons Bh de deux podzols humo-ferrugineux, l'un formé sous callune, l'autre sous chênaie-hêtraie. *C. R. Acad. Sci. (Paris)* 274D:2859–2861.

Guillet, B, Faivre, P, Mariotti, A and J Khobzi. 1988. The 14C dates and 13C/12C ratios of soil organic matter as a means of studying the past vegetation in intertropical regions: Examples from Colombia (South America). *PPP* 65:51–58.

Guillet, B, Rouiller, J and B Souchier. 1975. Podzolization and clay migration in Spodosols of eastern France. *Geoderma* 14:223–245.

Gunal, H and MD Ransom. 2006. Clay illuviation and calcium carbonate accumulation along a precipitation gradient in Kansas. *Catena* 68:59–69.

Gunatilaka, A and S Mwango. 1987. Continental sabkha pans, and associated nebkhas in southrn Kuwait, Arabian Gulf. In: LE Frostick and I Reid (eds). *Desert Sediments: Ancient and Modern*. Blackwell Scientific Publications, Boston, MA. pp. 187–203.

Gundlach, HF, Campbell, JE, Huffman, TJ, Kowalski, WL, Newbury, RL and DC Roberts. 1982. *Soil Survey of Shawano County, Wisconsin*. USDA-SCS, US Govt. Printing Off., Washington, DC.

Gunn, RH. 1967. A soil catena on denuded laterite profiles in Queensland. *AJSR* 5:117–132.

Gunn, RH and DP Richardson. 1979. The nature and possible origins of soluble salts in deeply weathered landscapes of eastern Australia. *AJSR* 17:197–215.

Gunnarsson, T, Sundin, P and A Tunlid. 1988. Importance of leaf litter fragmentation for bacterial growth. *Oikos* 52:303–308.

Guo, Y, Amundson, R, Gong, P and R Ahrens. 2003. Taxonomic structure, distribution, and abundance of the soils in the USA. *SSSAJ* 67:1507–1516.

Guo, Y, Amundson, R, Gong, P and Q Yu. 2006a. Analysis of factors controlling soil carbon in the conterminous United States. *SSSAJ* 70:601–612.

2006b. Quantity and spatial variability of soil carbon in the conterminous United States. *SSSAJ* 70:590–600.

Gupta, SR, Rajvanshi, R and JS Singh. 1981. The role of the termite *Odontotermes gurdaspurensis* (Isoptera: Termitidae) in plant decomposition in a tropical grassland. *Pedobiologia* 22:254–261.

Gustafsson, JP, Berggren, D, Simonsson, M, Zysset, M and J Mulder. 2001. Aluminium solubility mechanisms in moderately acid Bs horizons of podzolized soils. *EJSS* 52:655–665.

Gustafsson, JP, Bhattacharya, P, Bain, DC, Fraser, AR and WJ McHardy. 1995. Podzolization mechanisms and the synthesis of imogolite in northern Scandinavia. *Geoderma* 66:167–184.

Guthrie, RL and BF Hajek. 1979. Morphology and water regime of a Dothan soil. *SSSAJ* 43:142–144.

Guthrie, RL and JE Witty. 1982. New designations for soil horizons and layers and the new soil survey manual. *SSSAJ* 46:443–444.

Gutterman Y. 1997. The influences of depressions made by ibex on the annual vegetation along cliffs of the Zin Valley in the Negev Desert Highlands, Israel. *Isr. J. Plant Sci.* 45:333–338.

Haantjens, HA. 1965. Morphology and origin of patterned ground in a humid tropical lowland area, New Guinea. *AJSR* 3:111–129.

Haas, H, Holliday, V and R Stuckenrath. 1986. Dating of Holocene stratigraphy with soluble and insoluble organic fractions at the Lubbock Lake archaeological site, Texas: An ideal case study. *Radiocarbon* 28:473–485.

Haase, D, Fink, J, Haase G, Ruske, R, Pecsi, M, Richter, H, Altermann, M and K-D Jaeger. 2007. Loess in Europe – its spatial distribution based on a European Loess Map, scale 1:2,500,000. *QSR* 26:1301–1312.

Habecker, MA, McSweeney, K and FW Madison. 1990. Identification and genesis of fragipans in Ochrepts of North Central Wisconsin. *SSSAJ* 54:139–146.

Haberman, GM and FD Hole. 1980. Soilscape analysis in terms of pedogeomorphic fabric: An exploratory study. *SSSAJ* 44:336–340.

Haff, PK and BT Werner. 1996. Dynamical processes on desert pavements and the healing of surficial disturbances. *QR* 45:38–46.

Haile-Mariam, S and DL Mokma. 1990. Soils with carbonate-rich zones in east central Michigan. *SSH* 31:23–29.

Haimi, J, and V Huhta. 1990. Effects of earthworms on decomposition processes in raw humus forest soil: A microcosm study. *Biol. Fertil. Soils* 10:178–183.

Hairston, AB and DF Grigal. 1994. Topographic variation in soil water and nitrogen for two forested land forms in Minnesota, USA. *Geoderma* 64:125–138.

Hajdas, I. 2008. Radiocarbon dating and its applications in Quaternary studies. *Quat. Sci. J.* 57:2–24.

Hajic, ER. 1990. *Koster Site Archeology I: Stratigraphy and Landscape Evolution*. Center for Amer. Archeol., Res. Series 8. Kampsville, IL.

Hale, CM. 2008. Perspective: Evidence for human-mediated dispersal of exotic earthworms: Support for exploring strategies to limit further spread. *Mol. Ecol.* 17:1165–1169.

Hale, CM, Frelich, LE and PB Reich. 2005. Exotic European earthworm invasion dynamics in northern hardwood forests of Minnesota, USA. *Ecol. Appl.* 15:848–860.

2006. Changes in hardwood forest understory plant communities in response to European earthworm invasions. *Ecology* 87:1637–1649.

Hale, M and CG Porto. 1994. Geomorphological evolution and supergene gold ore at Posse, Goias State, Brazil. *Catena* 21:145–157.

Hale, MG, More, LD and GJ Griffin. 1978. Root exudates and exudation. In: YR Dommergues and SV Krupa (eds). *Interactions between Non-Pathogenic Soil Microorganisms and Plants*. Elsevier, New York. pp. 163–204.

Halfen, AF, Fredlund, GG and SA Mahan. 2010. Holocene stratigraphy and chronology of the Casper Dune Field, Casper, Wyoming, USA. *Holocene* 20:773–783.

Hall, GF. 1983. Pedology and geomorphology. In: LP Wilding, NE Smeck, and GF Hall (eds). *Pedogenesis and Soil Taxonomy*. Vol. 1. *Concepts and Interactions*. Elsevier, Amsterdam. pp. 117–140.

Hall, K. 1999. The role of thermal stress fatigue in the breakdown of rock in cold regions. *Geomorphology* 31:47–63.

2007. Evidence for freeze-thaw events and their implications for rock weathering in northern Canada. II. The temperature at which water freezes in rock. *ESPL* 32:249–259.

Hall, K and MF Andre. 2001. New insights into rock weathering from high-frequency rock temperature data: An Antarctic study of weathering by thermal stress. *Geomorphology* 41:23–35.

Hall, K and A Hall. 1996. Weathering by wetting and drying: Some experimental results. *ESPL* 21:365–376.

Hall, K and N Lamont. 2003. Zoogeomorphology in the Alpine: Some observations on abiotic-biotic interactions. *Geomorphology* 55:219–234.

Hall, K, Thorn, CE, Matsuoka, N and A Prick. 2002. Weathering in cold regions: Some thoughts and perspectives. *PPG* 26:577–603.

Hall, K, Thorn, C and P Sumner. 2012. On the persistence of "weathering." *Geomorphology* 149:1–10.

Hall, RD. 1999a. A comparison of surface soils and buried soils: Factors of soil development. *SS* 164:264–287.

1999b. Effects of climate change on soils in glacial deposits, Wind River Basin, Wyoming. *QR* 51:248–261.

Hall, RD and AK Anderson. 2000. Comparative soil development of Quaternary paleosols of the central United States. *PPP* 158:109–145.

Hall, RD and LL Horn. 1993. Rates of hornblende etching in soils in glacial deposits of the northern Rocky Mountains (Wyoming-Montana, USA): Influence of climate and characteristics of the parent material. *Chem. Geol.* 105:17–29.

Hall, RD and D Michaud. 1988. The use of hornblende etching, clast weathering, and soils to date alpine glacial and periglacial deposits: A study from southwestern Montana. *GSA Bull.* 100:458–467.

Hall, RD and RR Shroba. 1995. Soil evidence for a glaciation intermediate between the Bull Lake and Pinedale glaciations at Fremont Lake, Wind River Range, Wyoming, USA. *AAR* 27:89–98.

Hall, SA. 1981. Deteriorated pollen grains and the interpretation of Quaternary pollen diagrams. *Rev. Paleobot. Palynol.* 2: 193–206.

Hall, SA and RJ Goble. 2012. Berino Paleosol, Late Pleistocene argillic soil development on the Mescalero Sand Sheet in New Mexico. *J. Geol.* 120:333–345.

Hallberg, GR. 1984. The US System of Soil Taxonomy: From the outside looking. In: RB Grossman, H Eswaran and RH Rust (eds). *Soil Taxonomy: Achievements and Challenges*. Soil Science Society of America Press, Madison, WI. pp. 45–59.

1986. Pre-Wisconsin glacial stratigraphy of the Central Plains region in Iowa, Nebraska, Kansas, and Missouri. *QSR* 5:11–15.

Hallberg, GR, Fenton, TE and GA Miller. 1978a. Part 5: Standard weathering zone terminology for the description of Quaternary sediments in Iowa. In: GR Hallberg (ed). *Standard Procedures for Evaluation of Quaternary Materials in Iowa*. Tech. Info. Series 8. Iowa Geological Survey, Iowa City, IA. pp. 75–109.

Hallberg, GR, Fenton, TE, Miller, GA and AJ Luteneggar. 1978b. The Iowan Erosion Surface: An Old Story, an Important lesson, and Some New Wrinkles, Guidebook, 42nd Annual Tri-State Geological Field Conference, Geology of East-Central Iowa. *Iowa Geological Survey, Iowa City, IA.* pp. 2-1-2-94.

Hallet, B, Anderson, SP, Stubbs, CW and EC Gregory. 1988. Surface soil displacements in sorted circles, western Spitzbergen. *Proc. 5th Intl. Permafrost Conf. (Trondheim)* 1:770–775.

Hallet, B and S Prestrud. 1986. Dynamics of periglacial sorted circles in western Spitsbergen. *QR* 26:81–99.

Hallett, DJ, Hills, LV and JJ Clague. 1997. New accelerator mass spectrometry radiocarbon ages for the Mazama tephra layer from Kootenay National Park, British Columbia, Canada. *CJES* 34:1202–1209.

Hallmark, CT and DP Franzmeier. 2011. Alfisols. In: PM Huang, Y Li and ME Sumner (eds). *Handbook of Soil Sciences*, 2nd ed. CRC Press, New York. pp. 33–153–33–167.

Hallsworth, EG and GG Beckmann. 1969. Gilgai in the Quaternary. *SS* 107:409–420.

Hallsworth, EG, Coston, AE, Gibbons, FR and GK Robertson. 1952. Studies in pedogenesis in New South Wales. II: The chocolate soils. *JSS* 3:103–124.

Hallsworth, EG, Robertson, GK and FR Gibbons. 1955. Studies in pedogenesis in New South Wales. VII. The "gilgai" soils. *JSS* 6:1–31.

Halsey, DP, Mitchell, DJ and SJ Dews. 1998. Influence of climatically induced cycles in physical weathering. *Quart. J. Engin. Geol.* 31:359–367.

Hamann, C. 1984. Windwurfals ursache der bodenbuckelung an sudrand des Tennengebirges, ein beitrag zur genese der Buckelwiesen. *Berl. Geogr. Abh.* 36:69–76.

Hambach, U, Rolf, C and E Schnepp. 2008. Magnetic dating of Quaternary sediments, volcanites and archaeological materials: An overview. *Quat. Sci. J.* 57:25–51.

Han, J, Lu, H, Wu, N and Z Guo. 1996. The magnetic susceptibility of modern soils in China and its use for paleoclimate reconstruction. *Stud. Geoph. Geodet.* 40:262–275.

Hanawalt, RB and RH Whittaker. 1976. Altitudinally coordinated patterns of soils and vegetation in the San Jacinto Mountains, California. *SS* 121:114–124.

Hancock, GS, Anderson, RS, Chadwick, OA and RC Finkel. 1999. Dating fluvial terraces with ^{10}Be and ^{26}Al profiles: Application to the Wind River, Wyoming. *Geomorphology* 27:41–60.

Hannah, PR and R Zahner. 1970. Nonpedogenetic texture bands in outwash sands of Michigan: Their origin, and influence on tree growth. *SSSAP* 34:134–136.

Hansen, E, Arbogast, AF and B Yurk. 2004. The history of dune growth and migration along the southeastern shore of Lake Michigan: A perspective from Green Mountain Beach. *Mich. Acad.* 35:455–478.

Hansen, EC, Fisher, TG, Arbogast, AF and M Bateman. 2010. Geomorphic history of low perched, transgressive dune complexes along the southeastern shore of Lake Michigan. *AR* 1:111–127.

Hanson, CT and RL Blevins. 1979. Soil water in coarse fragments. *SSSAJ* 43:819–820.

Hanson, PR. 2006. Dating ephemeral stream and alluvial fan deposits on the central Great Plains: Comparing multiple-grain OSL, single-grain OSL, and radiocarbon ages. *USGS Open File Rept.* 2006-1351. 14 pp.

Hanson, P, Arbogast, AF and WC Johnson. 2010. Megadroughts and late Holocene dune activation at the eastern margin of the Great Plains, North-Central Kansas, USA. *AR* 1:101–110.

Hao, QZ, Oldfield, F, Bloemendal, J and ZT Guo. 2008. The magnetic properties of loess and paleosol samples from the Chinese Loess Plateau spanning the last 22 million years. *PPP* 260:389–404.

Harden, DR, Biggar, NE and ML Gillam. 1985. Quaternary deposits and soils in and around Spanish Valley, Utah. *GSA Spec. Paper* 203:43–64.

Harden, JW. 1982. A quantitative index of soil development from field descriptions: Examples from a chronosequence in central California. *Geoderma* 28:1–28.

1987. *Soils developed on granitic alluvium near Merced*, California. USGS Bull. 1590-A.

1988a. Genetic interpretations of elemental and chemical differences in a soil chronosequence, California. *Geoderma* 43:179–193.

1988b. Measurements of water penetration and volume percentage water-holding capacity for undisturbed, coarse-textured soils in southwestern California. *SS* 146:374–383.

Harden JW, Fries TL and ML Pavich. 2002. Cycling of beryllium and carbon through hillslope soils in Iowa. *Biogeochem.* 60:317–336.

Harden, JW and EM Taylor. 1983. A quantitative comparison of soil development in four climatic regimes. *QR* 20:342–359.

Harden, JW, Taylor, EM and C. Hill. 1991a. Rates of soil development from four soil chronosequences in the southern Great Basin. *QR* 35:383–399.

Harden, JW, Taylor, EM, McFadden, LD and MC Reheis. 1991b. Calcic, gypsic, and siliceous soil chronosequences in arid and semiarid environments. In: WD Nettleton (ed). Occurrence, Characteristics, and Genesis of Carbonate, Gypsum, and Silica Accumulations in Soils. *Soil Science Society of Am. Spec. Publ.* 26:1–16.

Hardy, F. 1933. Cultivation properties of tropical red soils. *Emp. J. Exp. Agric.* 1:103–112.

Hardy, M. 1993. Influence of geogenesis and pedogenesis on clay mineral distribution in northern Vietnam soils. *SS* 156:336–345.

Hardy, M, Jamagne, M, Elsass, F, Robert, M and D Chesneau. 1999. Mineralogical development of the silt fractions of a Podzoluvisol on loess in the Paris Basin (France). *EJSS* 50:443–456.

Harlan, PW and DP Franzmeier. 1974. Soil-water regimes in Brookston and Crosby soils. *SSSAP* 38:638–643.

1977. Soil formation on loess in southwestern Indiana. I. Loess stratigraphy and soil morphology. *SSSAJ* 41:93–98.

Harlan, PW, Franzmeier, DP and CB Roth. 1977. Soil formation on loess in southwestern Indiana. II. Distribution of clay and free oxides and fragipan formation. *SSSAJ* 41:99–103.

Harland, WB, Cox, AV, Llewellyn, PG, Pickton, CAG, Smith, AG and R Walters. 1982. *A Geologic Time Scale.* Cambridge Univ. Press, New York.

Harnois, L. 1988. The CIW index: A new chemical index of weathering. *Sed. Geol.* 55:319–322

Harper, WG. 1957. Morphology and genesis of Calcisols. *SSSAP* 21:420–424.

Harrington, CD and JW Whitney. 1987. Scanning electron microscope method for rock-varnish dating. *Geology* 15:967–970.

Harris, C. 1982. The distribution and altitudinal zonation of periglacial landforms, Okstindan, Norway. *ZG* 26:283–304.

Harris, C and S Ellis. 1980. Micromorphology of soils in soliflucted materials, Okstindan, Northern Norway. *Geoderma* 23:11–29.

Harris, C, Murton, J and MCR Davies. 2000. Soft-sediment deformation during thawing of ice-rich frozen soils: Results of scaled centrifuge modelling experiments. *Sedimentol.* 47:687–700.

Harris, P and O Farrington. 1989. Radiocarbon dating of the Holocene evolution of Magilligan Foreland, Co, Londonberry. *Proc. Roy. Irish Acad. Sci.* 89B:1–23.

Harris, WG. 2000. Hydrologically-linked Spodosol formation in the Southeastern United States. In: JL Richardson and MJ Vepraskas (eds). *Wetland Soils: Their Genesis, Hydrology, Landscape, and Separation into Hydric and Nonhydric Soils.* CRC Press, Boca Raton, FL. pp. 331–342.

2002. Phosphate minerals. In: JB Dixon and D Schulze (eds). *Soil Mineralogy with Environmental Applications.* SSSA Book Series no. 7. Soil Sci. Soc. Am., Madison, WI. pp. 637–666.

Harris, WG and KA Hollien. 1999. Changes in quantity and composition of crystalline clay across E – Bh boundaries of Alaquods. *SS* 164:602–608.

2000. Changes across artificial E – Bh boundaries generated under simulated fluctuating water tables. *SSSAJ* 64:967–973.

Harris, WG, Crownover, SH and J Hinchee. 2005. Problems arising from fixed-depth assessment of deeply weathered sandy soils. *Geoderma* 126:161–165.

Harris, W and GN White. 2008. X-ray Diffraction Techniques for Soil Mineral Identification. In: A Ulery and R Drees (eds). *Methods of Soil Analysis.* Part 5. *Mineralogical Methods.* SSSA Book Series, no. 5. Soil Sci. Soc. Am., Madison, WI. pp. 81–115.

Harris, WG, Zelazny, LW, Baker, JC, and DC Martins. 1985. Biotite kaolinization in Virginia Piedmont soils. II. Zonation in single grains. *SSSAJ* 49:1297–1302.

Harrison, CGA and E Ramirez. 1975. Areal coverage of spurious reversals of the earth's magnetic field. *J Geomag. Geoelec.* 27:139–151.

Harrison, JBJ, McFadden, LD, and RJ Weldon. 1990. Spatial soil variability in the Cajon Pass chronosequence: Implications for the use of soils as a geochronological tool. *Geomorphology* 3:399–416.

Harriss, RC and JAS Adams. 1966. Geochemical and mineralogical studies on the weathering of granitic rocks. *AJS* 264:146–178.

Harry, DG and JS Gozdzik. 1988. Ice wedges: Growth, thaw transformation, and palaeoenvironmental significance. *JQS* 3:39–55.

Harry, KG. 1995. Cation-ratio dating of varnished artifacts: Testing the assumptions. *Am. Antiq.* 60:118–130.

Hart, DM. 1988. A fabric contrast soil on Dolerite in the Sydney Basin, Australia. *Catena* 15:27–37.

Hart, G, Leonard, RE and RS Pierce. 1962. Leaf fall, humus depth, and soil frost in a northern hardwood forest. USFS Res. Note NE-131.

Hart, G and HW Lull. 1963. Some relationships among air, snow, and soil temperatures and soil frost. USFS Res. Note NE-3.

Hartemink, AE. 2004. Soils of the tropics. *Geoderma* 123:373–375.

2012. Soil science reference books. *Catena* 95:142–144.

Hartemink, AE and AB McBratney. 2008. A soil science renaissance. *Geoderma* 148:123–129.

Hartgrove, NT, Ammons, JT, Khiel, AR and JD O'Dell. 1993. Genesis of soils on two stream terrace levels on the Tennessee River. *SSH* 34:78–88.

Hartshorn, JH. 1958. Flowtill in southeastern Massachusetts. *GSA Bull.* 69:477–481.

Harvey, JE and DJ Smith. 2013. *Lichenometric dating of Little Ice Age glacier activity in the central British Columbia Coast Mountains, Canada. Geog. Annaler* 95A:1–14.

Haseman, JF and CE Marshall. 1945. The use of heavy minerals in studies of the origin and development of soils. *Missouri Agric. Exp. Stn Res. Bull.* 387:1–75.

Hash, SJ and JS Noller. 2009. Incorporating predictive mapping to advance initial soil survey: An example from Malheur County, Oregon. *SSH* 50:111–115.

Hastie, T, Tibshirani, R and J Friedman. 2009. *The Elements of Statistical Learning: Data Mining, Inference and Prediction,* 2nd ed. Springer-Verlag, New York.

Hatcher, PG, Schnitzer, M, Vassallo, AM and MA Wilson. 1989. The chemical structure of highly aromatic humic acids in three volcanic ash soils as determined by dipolar dephasing NMR studies. *Geochim. Cosmochim. Acta* 53:125–130.

Hatzenpichler, R. 2012. Diversity, physiology, and niche differentiation of ammonia-oxidizing Archaea. *Appl. Environ. Microbiol.* 78:7501–7510.

Havinga, AJ. 1974. Problems in the interpretation of pollen diagrams of mineral soils. *Geol. Mijn.* 53:449–453.

Havlin, JL, Tisdale, SL, Nelson, WL and JD Beaton. 2004. *Soil Fertility and Fertilizers: An Introduction to Nutrient Management,* 7th ed. Prentice Hall, Upper Saddle River, NJ.

Hawker, HW. 1927. A study of the soils of Hildago County, Texas, and the stages of their soil lime accumulation. *SS* 23:475–485.

Hay, RL. 1960. Rate of clay formation and mineral alteration in a 4000-year-old volcanic ash soil on St. Vincent, B. W. I. *AJS* 258:354–368.

Hayakawa, SI and AR Hayakawa. 1990. *Language in Thought and Action*. Harcourt Inc., Orlando, FL.

Hayden, JD. 1976. Pre-altithermal archaeology in theSierra Pinacate, Sonora, Mexico. *Am. Antiq.* 41:274–289.

Hayes, MHB. 2009. Evolution of concepts of environmental natural non-living organic matter. In: N Senesi, B Xing and PM Huang (eds). *Biophysico-Chemical Processes Involving Natural Nonliving Organic Matter in Environmental Systems*. pp. 1–39. Wiley, Hoboken, NJ.

Hayes, WA, Jr and MJ Vepraskas. 2000. Morphologic changes in soils produced when hydrology is altered by ditching. *SSSAJ* 64:1893–1904.

Haynes, CV. 1982. Great Sand Sea and Selima Sand Sheet, eastern Sahara: Geochronology of desertification. *Science* 217:629–633.

Hays, JD, Imbrie, J and NJ Shackleton. 1976. Variations in the Earth's Orbit: Pacemaker of the Ice Ages. *Science* 194:1121–1132.

Hayward, RK and TV Lowell. 1993. Variations in loess accumulation rates in the mid-continent, United States, as reflected by magnetic susceptibility. *Geology* 21:821–824.

He, X, Vepraskas, MJ, Lindbo, DL and RW Skaggs. 2003. A method to predict soil saturation frequency and duration from soil color. *SSSAJ* 67:961–969.

Heath, GW. 1965. The part played by animals in soil formation. In: EG Hallsworth and DV Crawford (eds). *Experimental Pedology*. Butterworth, London. pp. 236–243.

Heckman, K and C Rasmussen. 2011. Lithologic controls on regolith weathering and mass flux in forested ecosystems of the southwestern USA. *Geoderma* 164:99–111.

Hedges, JI and RG Keil. 1999. Organic geochemical perspectives on estuarine processes: Sorption reactions and consequences. *Marine Chem.* 65:55–65.

Hedl, R, Petrik, P and K Boublik. 2011. Long-term patterns in soil acidification due to pollution in forests of the Eastern Sudetes Mountains. *Environ. Poll.* 159:2586–2593.

Heiberg, SO and RF Chandler. 1941. A revised nomenclature of forest humus layers for the northeastern United States. *SS* 52:87–99.

Heimsath, AM, Chappell, J, Dietrich, WE, Nishiizumi, K and RC Finkel. 2000. Soil production on a retreating escarpment in southeastern Australia. *Geology* 28:787–790.

Heimsath, AM, Chappell, J, Spooner, NA and DG Questiaux. 2002. Creeping soil. *Geology* 30:111–114.

Heimsath, AM, Dietrich, WE, Nishiizumi, K and RC Finkel. 1997. The soil production function and landscape equilibrium. *Nature* 388:358–388.

1999. Cosmogenic nuclides, topography, and the spatial variation of soil depth. *Geomorphology* 27:151–172.

2001. Stochastic processes of soil production and transport: Erosion rates, topographic variation and cosmogenic nuclides in the Oregon Coast Range. *ESPL* 26:531–552.

Heimsath, AM, Fink, D and GR Hancock. 2009. The "humped" soil production function: Eroding Arnhem Land, Australia. *ESPL* 34:1674–1684.

Heinonen, R. 1960. Das volumgewicht als Kennzeichen der "normalen" Bodenstruktur. *J. Sci. Soc. Finl.* 32:81–87.

Heinselman, ML. 1963. Forest sites, bog processes and peatland types in the glacial Lake Agassiz region, Minnesota. *Ecol. Monog.* 33:327–374.

1970. Landscape evolution, peatland types, and the environment in the Lake Agassiz Peatlands Natural Area, Minnesota. *Ecol. Monogr.* 40:235–261.

Heinz, V. 2002. Relict soils as paleoclimatic indicators: Examples from the Austrian Alps and the Central Andes. *Proc. 17th World Congr. Soil Sci.* 4:1520.

Heller, F, Shen, CD, Beer, J, Liu, XM, Liu, TS, Bronger, A, Suter, M and G Bonani. 1993. Quantitative estimates of pedogenic ferromagnetic mineral formation in Chinese loess and paleoclimatic implications. *EPSL* 114:385–390.

Heller, JL. 1963. The nomenclature of soils or what's in a name? *SSSAP* 27:216–220.

Helmisaari, HS. 1995. Nutrient cycling in Pinus sylvestris stands in eastern Finland. *Plant Soil* 169:327–336.

Helms, JG, McGill, SF and TK Rockwell. 2003. Calibrated, late Quaternary age indices using clast rubification and soil development on alluvial surfaces in Pilot Knob Valley, Mojave Desert, southeastern California. *QR* 60:377–393.

Helvey, JD. 1964. Rainfall interception by hardwood forest litter in the southern Appalachians. USFS Res. Paper SE-8.

Helvey, JD and JH Patric. 1965. Canopy and litter interception of rainfall by hardwoods of eastern United States. *Water Resources Res.* 1:193–206.

Henry, J and AM Kwong. 2003. Why is geophagy treated like dirt? Dev. *Behav.* 24:353–371.

Henry, JL, Bullock, PR, Hogg, TJ and LD Luba. 1985. Groundwater discharge from glacial and bedrock aquifers as a soil salinization factor in Saskatchewan. *CJSS* 65:749–768.

Henshilwood, CS, d'Errico, F, Yates, R, Jacobs, Z, Tribolo, C, Duller, GA, Mercier, N, Sealy, JC, Valladas, H, Watts, I, and AG Wintle. 2002. Emergence of modern human behavior: Middle Stone Age engravings from South Africa. *Science* 295:1278–80.

Herbel, CH, Gile, LH, Fredrickson, EL and RP Gibbens. 1994. Soil water and soils at the soil water sites, Jornada Experimental Range. In: LH Gile and RJ Ahrens (eds). *Supplement to the Desert Project Soil Monograph*. Vol. 1. Soil Survey Invest. Rept. 44. National Soil Survey Center, Lincoln, NE.

Herrera, R and MA Tamers. 1970. Radiocarbon dating of tropical soil associations in Venezuela. Symp. *Age of Parent Materials and Soils (Amsterdam)* 109–115.

Herrero, J. 2004. Revisiting the definitions of gypsic and petrogypsic horizons in Soil Taxonomy and World Reference Base for Soil Resources. *Geoderma* 120:1–5.

Herrero, J and J Porta. 2000. The terminology and the concepts of gypsum-rich soils. *Geoderma* 96:47–61.

Herwitz, SR and DR Muhs. 1995. Bermuda solution pipe soils: A geochemical evaluation of eolian parent materials. *GSA Spec. Paper* 300:311–323.

Heslop, D, Langereis, CG and MJ Dekkers. 2000. A new astronomical timescale for the loess deposits of Northern China. *EPSL* 184:125–139.

Hess, D. 2014. *Mcknight's Physical Geography*. Pearson Higher Education, Upper Saddle River, NJ.

Hesse, PR. 1955. A chemical and physical study of the soils of termite mounds in East Africa. *JE* 43:449–461.

Higashi, T, DeConinck, F and F Gelaude. 1981. Characterization of some spodic horizons of the Campine (Belgium) with dithionite-citrate, pyrophosphate, and sodium hydroxide-tetraborate. *Geoderma* 25:131–142.

Hilgard, EW. 1906. *Soils*. Macmillan, New York.

Hill, DE and BL Sawhney. 1971. Electron microprobe analysis of soils. *SS* 112:32–38.

Hillel, D. 1991. *Out of the Earth*. Univ. of California Press, Berkeley.

Hillel, D and H Talpaz. 1977. Simulation of soil water dynamics in layered soils. *SS* 123:54–62.

Hiller, DA. 2000. Properties of Urbic Anthrosols from an abandoned shunting yard in the Ruhr area, Germany. *Catena* 39:245–266.

Hintikka, V. 1972. Wind-induced root movements in trees. *Commun. Inst. For. Fenn.* 76:1–56.

Hirmas, DR and BL Allen. 2007. Degradation of pedogenic calcretes in west Texas. *SSSAJ* 71:1878–1888.

Hirmas, DR and RC Graham. 2011. Pedogenesis and soil-geomorphic relationships in an arid mountain range, Mojave Desert, California. *SSSAJ* 75:192–206.

Hirschfield, E and RS Hirschfield. 1937. Soil problems in Brigalow and Belah country. *Qld. Agric. J.* 47:586.

Hladil, J, Cejchan, P, Babek, O, Koptikova, L, Navratil, T and P Kubinova. 2010. Dust – A geology-oriented attempt to reappraise the natural components, amounts, inputs to sediment, and importance for correlation purposes. *Geol. Belgica* 13:367–383.

Hobbs, WH. 1943. The glacial anticyclone and the continental glaciers of North America. *Proc. Am. Phil. Soc.* 86:368–402.

Hobbs, T, Schaetzl, RJ and MD Luehmann. 2011. Evidence for periodic, Holocene loess deposition in kettles in a sandy interlobate landscape, Michigan, USA. *AR* 3:215–228.

Hoffland, E, Giesler, R, Jongmans, T and N van Breemen. 2002. Increasing feldspar tunneling by fungi across a north Sweden podzol chronosequence. *Ecosystems* 5:11–22.

Höfle, C, Ping, C-L and JM Kimble. 1998. Properties of permafrost soils on the Northern Seward Peninsula, Northwest Alaska. *SSSAJ* 62:1629–1639.

Holdsworth, AR, Frelich, LE and PB Reich. 2007. Effects of earthworm invasion on plant species richness in northern hardwood forests. *Cons. Biol.* 21:997–1008.

2012. Leaf litter disappearance in earthworm-invaded northern hardwood forests: Role of tree species and the chemistry and diversity of litter. *Ecosystems* 15:913–926.

Hole, FD. 1950. (reprinted, 1968). *Aeolian sand and silt deposits of Wisconsin.* Map. Wisc. Geol. Nat. Hist. Survey. Madison, WI.

1953. Suggested terminology for describing soils as three-dimensional bodies. *SSSAP* 17:131–135.

1961. A classification of pedoturbations and some other processes and factors of soil formation in relation to isotropism and anisotropism. *SS* 91:375–377.

1968. What is a well developed soil? *SSH* 9:12–13.

1975. Some relationships between forest vegetation and Podzol B horizons in soils of Menominee tribal lands, Wisconsin, USA. *SSS* 7:714–723.

1976. *Soils of Wisconsin.* Univ. of Wisconsin Press, Madison.

1978. An approach to landscape analysis with emphasis on soils. *Geoderma* 21:1–13.

1980. *Soil Guide for Wisconsin Land Lookers, Bull. #88, Soil Series #63.* Wisconsin Geological and Natural History Survey, Madison, WI.

1981. Effects of animals on soil. *Geoderma* 25:75–112.

1988. Terra vibrata: Some observations on the dynamics of soil landscapes. *PG* 9:175–185.

1997. The Earth beneath our feet: Explorations in community. *SSH* 38:40–53.

Hole, FD and JB Campbell. 1985. *Soil Landscape Analysis.* Rowman and Allanheld, Totowa, NJ.

Hole, FD and GA Nielsen. 1970. *Soil genesis under prairie.* In: Proc. Symp. Prairie and Prairie Restoration, Knox College, Galesburg, IL. pp. 28–34.

Holliday, VT. 1985. Early and middle Holocene soils at the Lubbock Lake archeological site, Texas. *Catena* 12:61–78.

1987. Geoarchaeology and late Quaternary geomorphology of the middle South Platte River, northeastern Colorado. *Geoarchaeol.* 2:317–329.

1988. Genesis of a late-Holocene soil chronosequence at the Lubbock Lake archaeological site, Texas. *AAAG* 78:594–610.

1989. Paleopedology in archaeology. In: A Bronger and J Catt (eds). Paleopedology: Nature and Application of Paleosols. *Catena Suppl.* 16:187–206.

(ed) 1992. Soils in Archeology. Landscape Evolution and Human Occupation. Proc. 1st Fryxell Symposium, Phoenix, AZ. Smithsonian Institution Press, Washington, DC.

(ed) 1994. The "state factor" approach in geoarcheology. *SSSASP* 33:65–86.

(ed) 1998. Origins of soil-stratigraphic and soil-geomorphic research in the southwestern United States. *QI* 51/52:20–21.

(ed) 2001. Stratigraphy and geochronology of upper Quaternary eolian sand on the southern High Plains of Texas and New Mexico, USA. *GSA Bull.* 112:88–108.

(ed) 2004. *Soils in Archaeological Research.* Oxford Univ. Press, New York.

(ed) 2006. *A history of soil geomorphology in the United States.* In: BP Warkentin (ed). *Footprints in the Soil. People and Ideas in Soil History.* Elsevier Press, Amsterdam. pp. 187–254.

Holmer, B. 1998. Flaking by insolation drying and salt weathering on the Swedish west coast. *ZG* 42:39–55.

Holmes, CD. 1955. Geomorphic development in humid and arid regions: A synthesis. *AJS* 253:377–390.

Holmes, WH. 1893. Vestiges of early man in Minnesota. *Am. Geol.* 11:219–240.

Holmgren, GGS. 1967. A rapid citrate–dithionite extractable iron procedure. *SSSAP* 31:210–211.

Holt, JA, Abe, T and N Kirtibutr. 1998. Microbial biomass and some chemical properties of *Macrotermes carbonarius* mounds near Korat, Thailand. *Sociobiology* 31:1–8.

Holt, JA, Coventry, RJ and DF Sinclair. 1980. Some aspects of the biology and pedological significance of mound building termites in a red and yellow earth landscape near Charters Towers, North Queensland. *AJSR* 18:97–109.

Holtmeier, F-K and G Broll. 1992. The influence of tree islands and microtopography on pedoecological conditions in the forest-alpine tundra ecotone on Niwot Ridge, Colorado Front Range, USA. *AAR* 24:216–228.

Holzhey, CS, Daniels, RB and EE Gamble. 1975. Thick Bh horizons in the North Carolina coastal plain. II. Physical and chemical properties and rates of organic additions from surface sources. *SSSAP* 39:1182–1187.

Hooda, PS, Henry, CJK, Seyoum, TA, Armstrong, LDM and MB Fowler. 2004. The potential impact of soil ingestion on human mineral nutrition. *Sci. Total Environ.* 333:75–87.

Hoosbeek, MR and RB Bryant. 1992. Towards the quantitative modeling of pedogenesis: A review. *Geoderma* 55:183–210.

1994. Developing and adapting soil process submodels for use in the pedodynamic orthod model. *SSSASP* 39:111–128.

Hoover, MD and HA Lunt. 1952. A key for the classification of forest humus types. *SSSAP* 16:368–370.

Hoover, MT and EJ Ciolkosz. 1988. Colluvial soil parent material relationships in the ridge and valley physiographic province of Pennsylvania. *SS* 145:163–172.

Hopfensperger, KN, Leighton, GM and TJ Fahey. 2011. Influence of invasive earthworms on above and belowground vegetation in a northern hardwood forest. *AMN* 166:53–62.

Hopkins, DG and DW Franzen. 2003. Argillic horizons in stratified drift: Luverne end moraine, eastern North Dakota. *SSSAJ* 67:1790–1796.

Hopkins, DG, Sweeney, MD and JL Richardson. 1991. Dispersive erosion and Entisol-panspot genesis in sodium-affected landscapes. *SSSAJ* 55:171–177.

Horbe, AMC and ML da Costa. 2005. Lateritic crusts and related soils in eastern Brazilian Amazonia. *Geoderma* 126:225–239.

Horiuchi, Y, Hisada, K and YI Lee. 2009. Paleosol profiles in the Shiohama Formation of the Lower Cretaceous Kanmon Group, Southwest Japan and implications for sediment supply frequency. *Cretaceous Res.* 30:1313–1324.

Hormes, A, Karlen, W and G Possnert. 2004. Radiocarbon dating of palaeosol components in moraines in Lapland, northern Sweden. *QSR* 23:2031–2043.

Horwath, JL and DL Johnson. 2006. Mima-type mounds in southwest Missouri: Expressions of point-centered and locally thickened biomantles. *Geomorphology* 77:308–319.

Horwath Burnam, JL and DL Johnson. 2012. Mima mounds: The case for polygenesis and bioturbation. GSA Spec. Paper 490.

Host, GE and KS Pregitzer. 1992. Geomorphic influences on ground-flora and overstory composition in upland forests of northwestern lower Michigan. *CJFR* 22:1547–1555.

Host, GE, Pregitzer, KS, Ramm, CW, Hart, JB and DT Cleland. 1988. Variation in overstory biomass among glacial landforms and ecological land units in northwestern Lower Michigan. *CJFR* 18:659–668.

Houghton, HF. 1980. Refined techniques for staining plagioclase and alkalai feldspars in thin section. *J. Sed. Petr.* 55:629–631.

House, PK, Buck, BJ and AR Ramelli. 2010. Geologic Assessment of Piedmont and Playa Flood Hazards in the Ivanpah Valley Area, Clark County, Nevada. Nevada Bureau of Mines and Geology, Rept. #20.

Hovan, SA, Rea, DK, Pisias, NG and NJ Shackleton. 1989. A direct link between the China loess and marine δ18O records: Aeolian flux to the north Pacific. *Nature* 340:296–298.

Howard, JL, Amos, DF and WL Daniels. 1993. Alluvial soil choronosequence in the Inner Coastal Plain, Central Virginia. *QR* 39:201–213.

Howard, JL, Clawson, CR and WL Daniels. 2012. A comparison of mineralogical techniques and potassium adsorption isotherm analysis for relative dating and correlation of Late Quaternary soil chronosequences. *Geoderma* 179–180:81–95

Howard, JL and D Olszewska. 2010. Pedogenesis, geochemical forms of heavy metals, and artifact weathering in an urban soil chronosequence, Detroit, Michigan. *Envir. Poll.* 159:754–761.

Howitt, RW and S Pawluk. 1985a. The genesis of a Gray Luvisol within the boreal forest region. I. Static pedology. *CJSS* 65:1–8.

1985b. The genesis of a Gray Luvisol within the boreal forest region. II. Dynamic pedology. *CJSS* 65:9–19.

Hoyer, BE. 1980. The geology of the Cherokee sewer site. In: DC Anderson and HA Semken (eds). *The Cherokee Excavations.* Academic Press, NY. pp. 21–66.

Hseu, Z-Y, Chen, Z-S and Z-D Wu. 1999. Characterization of placic horizons in two subalpine forest Inceptisols. *SSSAJ* 63:941–947.

Hsu, PH. 1989. Aluminum oxides and hydroxides. In: JB Dixon and SB Weed (eds). *Minerals in Soil Environments,* 2nd ed. Soil Sci. Soc. Am., Madison, WI. pp. 199–278.

Hua, Q and M Barbetti. 2004. Review of tropospheric C-14 data for carbon cycle modeling and age calibration purposes. *Radiocarbon* 46:1273–1298.

Huang, CC, Pang, JL, Su, HX, Wang, LJ and YZ Zhu. 2009. The Ustic Isohumisol (Chernozem) distributed over the Chinese Loess Plateau: Modern soil or palaeosol? *Geoderma* 150:344–358.

Huang, PM, Wang, MK, Kämpf, N and DG Schulze. 2002. Aluminum hydroxides. In: JB Dixon and DG Schulze (eds). *Soil Mineralogy with Environmental Applications.* SSSA Book Series no.7. Soil Sci. Soc. Am., Madison, WI. pp. 261–290.

Huang, W-S, Tsai, H, Tsai, C-C, Hseu, Z-Y and Z-S Chen. 2010. Subtropical soil chronosequence on Holocene marine terraces in eastern Taiwan. *SSSAJ* 74:1271–1283.

Huggett, RJ. 1975. Soil landscape systems: A model of soil genesis. *Geoderma* 13:1–22.

1976a. Conceptual models in pedogenesis: A discussion. *Geoderma* 16:261–262.

1976b. Lateral translocations of soil plasma through a small valley basin in the Northaw Great Wood, Hertfordshire. *Earth Surf. Proc.* 1:99–109.

1991. *Climate, Earth Processes, and Earth History.* Springer-Verlag, Berlin.

1998a. Discussion of the paper by JD Phillips. *Geoderma* 86:23–25.

1998b. Soil chronosequences, soil development, and soil evolution: A critical review. *Catena* 32:155–172.

Hughen, KA, Baillie, MGL, Bard, E, Beck, JW, Bertrand, CJH, Blackwell, PG, Buck, CE, Burr, GS, Cutler, KB, Damon, PE, Edwards, RL, Fairbanks, RG, Friedrich, M, Guilderson, TP, Kromer, B, McCormac, G, Manning, S, Bronk Ramsey, C, Reimer, PJ, Reimer, RW, Remmele, S, Southon, JR, Stuiver, M, Talamo, S, Taylor, FW, van der Plicht, J and CE Weyhenmeyer. 2004. Marine04 marine radiocarbon age calibration, 0–26 cal kyr bp. *Radiocarbon* 46:1059–1086.

Hughes, OL, Tarnocai, C and CE Schweger. 1993. Plesitocene stratigraphy, paleopedology, and paleoecology of a multiple till sequence exposed on the Little-Bear River, western District of Mackenzie, NWT, Canada. *CJES* 30:851–866.

Hugie, VK and HB Passey. 1963. Cicadas and their effect upon soil genesis in certain soils in southern Idaho, northern Utah, and northeastern Nevada. *SSSAP* 27:78–82.

Humphreys, GS. 1981. The rate of ant mounding and earthworm casting near Sydney, New South Wales. *Search* 12:129–131.

1994. Bioturbation, biofabrics and the biomantle: An example from the Sydney Basin. In: AJ Ringrose-Voase and GS Humphreys (eds). *Soil Micromorphology: Studies in Management and Genesis.* Elsevier, London. pp. 421–436.

Humphreys, GS and DA Adamson. 2000. Inadequate pedogeomorphic evidence for dry and cold climatic conditions in southeastern Brazil. *ZG* 44:529–531.

Humphreys, GS and PB Mitchell. 1983. A preliminary assessment of the role of bioturbation and rainwash on sandstone hillslopes in the Sydney Basin. In: RW Young and GC Nanson (eds). *Aspects of Australian Sandstone Landscapes.* Dept of Geography, Univ. of Wollongong, Wollongong, NSW. pp. 66–80.

Humphreys, GS and MT Wilkinson. 2007. The soil production function: A brief history and its rediscovery. *Geoderma* 139:73–78.

Hunckler, RV and RJ Schaetzl. 1997. Spodosol development as affected by geomorphic aspect, Baraga County, Michigan. *SSSAJ* 61:1105–1115.

Hunt, CB. 1972. *Geology of Soils.* WH Freeman, San Francisco.

Hunter, CR, Frazier, BE and AJ Busacca. 1987. Lytell series: A nonvolcanic Andisol. *SSSAJ* 51:376–383.

Hunter, JM. 1961. Morphology of a bauxite summit in Ghana. *Geog. J.* 77:469–476.

1973. Geophagy in Africa and in the United States: A culture-nutrition hypothesis. *Geog. Rev.* 63:170–195.

1984a. Geophagy in Central America. *Geog. Rev.* 74:157–169.

1984b. Insect geophagy in Sierra Leone. *J. Cult. Geog.* 4:2–13.

1993. Macroterme geophagy and pregnancy clays in southern Africa. *J. Cult. Geog.* 14:69–92.

Hunter, JM, Horst, OH and RN Thomas. 1989. Religious geophagy as a cottage industry: The Holy Clay Tablet of Esquipulas, Guatemala. *Natl Geog. Res.* 5:281–295.

Huntley, DJ, Godfrey-Smith, DI and MLW Thewalt. 1985. Optical dating of sediments. *Nature* 313:105–107.

Huntley, DJ and M Lamothe. 2001. Ubiquity of anomalous fading in K-feldspars, and the measurement and correction for it in optical dating. *CJES* 38:1093–1106.

Hupp, CR and WP Carey. 1990. Dendrogeomorphic approach to estimating slope retreat, Maxey Flats, Kentucky. *Geology* 18:658–661.

Hupp, CR, Osterkamp, WR and JL Thornton. 1987. Dendrogeomorphic evidence and dating of debris flows on Mount Sharta, Northern California. *USGS Prof. Paper* 1396-B. 39 pp.

Hupy, JP. 2006. A varying perspective on Military Geography: The effects of explosive munitions on the WWI battlefield landscape surface of Verdun, France. *Scottish Geog. J.* 122:167–184.

2011. Khe Sanh, Vietnam: Examining the long-term impacts of warfare on the physical landscape. In: E Palka and F Galgano (eds). *Modern Military Geography*. Routledge, New York. pp. 312–326.

Hupy, JP and T Koehler. 2012. Modern warfare as a significant form of zoogeomorphic disturbance upon the landscape. *Geomorphology* 157–158:169–182.

Hupy, JP and RJ Schaetzl. 2006. Introducing "bombturbation," a singular type of soil disturbance and mixing. *SS* 171:823–836.

2008. Soil development following disturbance due to explosive munitions on the WWI battlefield of Verdun, France. *Geoderma* 145:37–49.

Hurst, VJ. 1977. Visual estimation of iron in saprolite. *GSA Bull.* 88:174–176.

Hussain, MS, Amadi, TH and MS Sulaiman. 1984. Characteristics of soils of a toposequence in northeastern Iraq. *Geoderma* 33:63–82.

Hussein, AH and MC Rabenhorst. 1999. Modeling of sulfur sequestration in coastal marsh soils. *SSSAJ* 63:1954–1963.

2001. Tidal inundation of transgressive coastal areas: Pedogenesis of salinization and alkalinization. *SSSAJ* 65:536–544.

Hutcheson, TB, Jr and HH Bailey. 1964. Fragipan soils: Certain genetic implications. *SSSAP* 28:684–685.

Hutchins, RB, Blevins, RL, Hill, JD and EH White. 1976. The influence of soils and microclimate on vegetation of forested slopes in eastern Kentucky. *SS* 121:234–241.

Hutt, G and H Jungner. 1992. Optical and TL dating on glaciofluvial sediments. *QSR* 11:161–163.

Hutt, G, Jungner, H, Kujansuu, R, and M Saarnisto. 1993. OSL and TL dating of buried podsols and overlying sands in Ostrobothnia, western Finland. *JQS* 8:125–132.

Hutton, CE. 1947. Studies of loess-derived soils in southwestern Iowa. *SSSAP* 12:424–431.

Hyde, AG and RD Ford. 1989. Water table fluctuations in representative Immokalee and Zolfo soils of Florida. *SSSAJ* 53:1475–1478.

Ibanez, JJ, Ballexta, RJ and AG Alvarez. 1990. Soil landscapes and drainage basins in Mediterranean mountain areas. *Catena* 17:573–583.

IGBP-GAIM. 1997. The First Five Years: Setting the Stage for Synthesis. Inst. for the Study of the Earth, Oceans and Space. Univ. of New Hampshire, Durham, NH. pp. 77.

Imbellone, PA and JE Gimenez. 1998. Parent materials, buried soils and fragipans in northwestern Buenos Aires province, Argentina. *QI* 51-2:115–126.

Indorante, SJ. 1998. Introspection of natric soil genesis on the loess-covered till plain in south central Illinois. *QI* 51/52:41–42.

Indorante, SJ and IJ Jansen. 1984. Perceiving and defining soils on disturbed land. *SSSAJ* 48:1334–1337.

Ingham, ER. 1999a. Soil fungi. *USDA-NRCS Soil Qual. Inst. PA* 1637:D1–D4.

1999b. Soil nematodes. *USDA-NRCS Soil Qual. Inst. PA* 1637:F1–F4.

1999c. Soil protozoa. *USDA-NRCS Soil Qual. Inst. PA* 1637:E1–E4.

1999d. The soil food web. *USDA-NRCS Soil Qual. Inst. PA* 1637:A1–A8.

Inglis, DR. 1965. Particle sorting and stone migration by freezing and thawing. *Science* 148:1616–1617.

Innes, JL. 1984. The optimal sample size in lichenometric studies. *AAR* 16:233–244.

1985. An examination of some factors affecting the largest lichens on a substrate. *AAR* 17:99–106.

Innis, RP and DJ Pluth. 1970. Thin section preparation using an epoxy impregnation for petrographic and electron microprobe analysis. *SSSAP* 34:483–485.

Innocent, C, Flehoc, C and F Lemeille. 2005. U-Th vs. AMS C-14 dating of shells from the Achenheim loess (Rhine Graben). *Bull. Soc. Geol. France* 176:249–255.

Inoue, I and PM Huang. 1990. Perturbation of imogolite formation by humic substances. *SSSAJ* 54:1490–1497.

Inoue, K and C Satoh. 1992. Electric charge and surface characteristics of hydroxyaluminosilicate- and hydroxyaluminum-vermiculite complexes. *Clays Clay Mins.* 40:311–315.

Inoue, Y, Nagatomo, Y and H Takaki. 1997. Identification of tephra deposits in the cumulative Andisols profile in Miyakonojo basin. *Pedologist* 41:42–54.

Iriondo, M. 2009. Multisol – a proposal. *QI* 209:131–141.

Iriondo, M and D Kröhling. 1997. The tropical loess. *In: Proc. 30th Intl. Geol. Congress, Intl. Union of Geol. Sci., Beijing.* 21:61–77.

2007. Non-classical types of loess. *Sed. Geol.* 202:352–368.

Isard, SI. 1986. Factors influencing soil-moisture and plant community distribution on Niwot Ridge, Front Range, Colorado, USA. *AAR* 18:83–96.

Isard, SA and RJ Schaetzl. 1995. Estimating soil temperatures and frost in the lake effect snowbelt region, Michigan, USA. *Cold Regions Sci. Tech.* 23:317–332.

1998. Effects of winter weather conditions on soil freezing in southern Michigan. *PG* 19:71–94.

Isard, SA, Schaetzl, RJ and JA Andresen. 2007. Soils cool as climate warms in Great Lakes Region, USA. *AAAG* 97:467–476.

Isbell, RF. 1991. Australian Vertisols. In: JM Kimble (ed). *Characterization, Classification, and Utilization of Cold Aridisols and Vertisols.* Proc. 6th Intl. Soil Correlation Mtg. USDA-SCS, Lincoln, NE. pp. 73–80.

Isherwood, D and A Street. 1976. Biotite-induced grusification of the Boulder Creek Granodiorite, Boulder County, Colorado. *GSA Bull.* 87:366–370.

Isphordi, WC. 1973. Origin, mineralogy, and economic potential of terra rosa soils of the Yucatan-Peninsula. *Econ. Geol.* 68:1215.

Istanbulluoglu, E, Yetemen, O, Vivoni, ER, Gutierrez-Jurado, HA and RL Bras. 2008. Eco-geomorphic implications of hillslope aspect: Inferences from analysis of landscape morphology in central New Mexico. *Geophys. Res. Letts.* 35. doi:10.1029/2008GL034477

IUSS. 2006. *World Reference Base for Soil Resources*, 2nd ed. IUSS Working Group, World Soil Resources Rept. 103. FAO, Rome. ISBN 92-5-1-5511-4.

Ivy-Ochs, S, Kerschner, H, Reuther, A, Maisch, M, Sailer, R, Schaefer, J, Kubik, PW, Synal, HA and C Schlüchter. 2006. The timing of glacier advances in the northern European Alps based on surface exposure dating with cosmogenic ^{10}Be, ^{26}Al, ^{36}C and ^{21}Ne. *GSA Spec. Paper* 415:43–60.

Ivy-Ochs, S and F Kober. 2008. Surface exposure dating with cosmogenic nuclides. *Quat. Sci. J.* 57:179–209.

Izett, GA. 1981. Volcanic ash beds: Recorders of upper Cenozoic silicic pyroclastic volcanism in the western United States. *J. Geophys. Res.* 88:10200–10222.

Izett, GA and RE Wilcox. 1982. Map showing localities and inferred distributions of the Huckleberry Ridge, Mesa Falls, and Lava Creek ash beds (Pearlette family ash beds) of Pliocene and Pleistocene age in the western United States and southern Canada. USGS Misc. Invest. Map 1325.

Izett, GA, Wilcox, RE, Powers, HA and GA Desborough. 1970. The Bishop ash bed, a Pleistocene marker bed in the western United States. QR 1:121–132.

Jackson, LEJ, Phillips, FM, Shimamura, K and EC Little. 1997. Cosmogenic 36Cl dating of the foothills erratics train, Alberta, Canada. Geology 25:195–198.

Jackson, ML. 1965. Clay transformations in soil genesis during the Quaternary. SS 99:15–22.

Jackson, ML and TA. Frolking. 1982. Mechanism of terra rosa red coloration. Clay Res. 1:1–5.

Jackson, ML, Levelt, TWM, Syers, JK, Rex, RW, Clayton, RN, Sherman, DG and G Uehara. 1971. Geomorphological relationships of tropospherically derived quartz in the soils of the Hawaiian Islands. SSSAP 35:515–525.

Jackson, ML, Lim, CH and L Zelazny. 1986. Oxides, hydroxides, and aluminosilicates. In: A Klute (ed). Methods of Soil Analysis, Part 1. Physical and Mineralogical Methods. 2nd ed. Am. Soc. Agron., Madison, WI. pp. 101–150.

Jackson, ML, Tyler, SA, Willis, AL, Bourbeau, GA and RP Pennington. 1948. Weathering sequence of clay size minerals in soils and sediments. J. Phys. Coll. Chem. 52:1237–1260.

Jacob, JS and LC Nordt. 1991. Soil and landscape evolution: A paradigm for pedology. SSSAJ 55:1194

Jacobs, PM. 1998. Influence of parent material grain size on genesis of the Sangamon Geosol in south-central Indiana. QI 51/52:127–132.

Jacobs, PM and JC Knox. 1994. Provenance and pedology of a long-term Pleistocene depositional sequence in Wisconsin's Driftless Area. Catena 22:49–68.

Jacobs, PM and JA Mason. 2004. Paleopedology of soils in thick Holocene loess, Nebraska, USA. Revista Mexicana Ciencias Geol. 21:54–70.

2005. Impact of Holocene dust aggradation on A horizon characteristics and carbon storage in loess–derived Mollisols of the Great Plains, USA. Geoderma 125:95–106.

2007. Late Quaternary climate change, loess sedimentation, and soil profile development in the central Great Plains: A pedosedimentary model. GSA Bull. 119:462–475.

Jacobs, PM, Mason, JA and PR Hanson. 2011. Mississippi Valley regional source of loess on the southern Green Bay Lobe land surface, Wisconsin. QR 75:574–583.

2012. Loess mantle spatial variability and soil horizonation, southern Wisconsin, USA. QI 265:42–53.

Jacobs, PM, West, LT and JN Shaw. 2002. Redoximorphic features as indicators of seasonal saturation, Lowndes County, Georgia. SSSAJ 66:315–323.

Jacobs, Z, Duller, GAT and AG Wintle. 2003. Optical dating of dune sand from Blombos Cave, South Africa. II – single grain data. J. Human Evol. 44:613–25.

Jacobsen T and RM Adams. 1958. Salt and silt in ancient Mesopotamian agriculture. Science 128:1251–1258.

Jacobson, GL, Jr and HJB Birks. 1980. Soil development on recent end moraines of the Klutlan Glacier, Yukon Territory, Canada. QR 14:87–100.

Jaillard, B, Guyon, A and AF Maurin. 1991. Structure and composition of calcified roots, and their identification in calcareous soils. Geoderma 50:197–210.

Jain, M, Murray, AS, Bøtter-Jensen, L and AG Wintle. 2005. A single-aliquot regenerative-dose method based on IR (1. 49 eV) bleaching of the fast OSL component in quartz. Rad. Meas. 39:309–318.

Jakobsen, BH. 1989. Evidence for translocations into the B horizon of a subarctic Podzol in Greenland. Geoderma 45:3–17.

1991. Aspects of soil geography in South Greenland. Folia Geog. Danica 14:155–164.

Jalalian, A and AR Southard. 1986. Genesis and classification of some Paleborolls and Cryoboralfs in northern Utah. SSSAJ 50:668–672.

Jamagne, M, DeConinck, F, Robert, M and J Maucorps. 1984. Mineralogy of clay fractions of some soils on loess in northern France. Geoderma 33:319–342.

James, HR, Ransom, MD and RJ Miles. 1995. Fragipan genesis in polygenetic soils on the Springfield Plateau of Missouri. SSSAJ 59:151–160.

James, LA. 1988. Rates of organic carbon accumulation in young mineral soils near Burroughs Glacier, Glacier Bay, Alaska. PG 9:50–70.

Janeau, JL, and C Valentin. 1987. Relations entre les termitières Trinervitermes sp. et la surface du sol: Réorganisations, ruissellement et érosion. Rev. Ecol. Biol. Sol 24:637–647.

Jankowitz, WJ, van Rooyen, MW, Shaw, D, Kaumba, JS and N van Rooyen. 2008. Mysterious circles in the Namib Desert. S. Afr. J. Bot. 74:332–334.

Jansen, B, Nierop, KGJ and JM Verstraten. 2005. Mechanisms controlling the mobility of dissolved organic matter, aluminium and iron in podzol B horizons. EJSS 56:537–550.

Jardine, PM, Weber, NL and JF McCarthy. 1989. Mechanisms of dissolved organic carbon adsorption on soil. SSSAJ 53:1378–1385.

Jarvis, NL, Ellis, R, Jr and OW Bidwell. 1959. A chemical and mineralogical characterization of selected Brunizem, Reddish Prairie, Grumusol, and Planosol soils developed in Pre-Pleistocene materials. SSSAP 23:234–239.

Jarvis, P, Rey, A, Petsikos, C, Wingate, L, Rayment, M, Pereira, J, Banza, J, David, J, Miglietta, F, Borghetti, M, Manca, G and R Valentini. 2007. Drying and wetting of Mediterranean soils stimulates decomposition and carbon dioxide emission: The "Birch effect." Tree Physiol. 27:929–940.

Jauhiainen, E. 1973a. Age and degree of podzolization of sand soils on the coastal plain of northwest Finland. Comment. Biol. 68:5–32.

1973b. Effect of climate on podzolization in southwest and eastern Finland. Comment. Phys.-Math. 43:213–242.

Jayawardane, NS and EL Greacen. 1987. The nature of swelling in soils. AJSR 25:107–113.

Jenkins, DA and RP Bower. 1974. The significance of the atmospheric contribution to the trace element content of soils. Trans. 10th Intl. Congr. Soil Sci. 6:466–474.

Jenkins, KA and BJ Smith. 1990. Daytime rock surface temperature variability and its implications for mechanical rock weathering: Tenerife, Canary Islands. Catena 17:449–459.

Jenkinson, BJ and DP Franzmeier. 2006. Development and evaluation of iron-coated tubes that indicate reduction in soils. SSSAJ 70:183–191.

Jenkinson, BJ, Franzmeier, DP and WC Lynn. 2002. Soil hydrology on an end moraine and a dissected till plain in west-central Indiana. SSSAJ 66:1367–1376.

Jenny, H. 1935. The clay content of the soil as related to climatic factors, particularly temperature. SS 40:111–128.

1941a. Calcium in the soil. III. Pedologic relations.SSSAP 6:27–37.

1941b. Factors of Soil Formation. McGraw-Hill, New York.

1946. Arrangement of soil series and types according to functions of soil-forming factors. SS 61:375–391.

1958. Role of the plant factor in the pedogenic functions. *Ecology* 39:5–16.

1960. Podsols and pygmies: Special need for preservation. Sierra Club Bull. 8–9.

1961. Derivation of state factor equations of soils and ecosystems. *SSSAP* 25:385–388.

Jenny, H, Arkley, RJ and AM Schultz. 1969. The Pygmy forest-Podsol ecosystem and its dune associates of the Mendocino Coast. *Madrono* 20:60–74.

Jenny, H and CD Leonard. 1934. Functional relationships between soil properties and rainfall. *SS* 38:363–381.

Jensen, HI. 1911. The nature and origin of gilgai country. *Proc. Roy. Soc. NSW* 45:337–358.

Jensen, ME. 1984. Soil moisture regimes on some rangelands of southeastern Idaho. *SSSAJ* 48:1328–1330.

Jeong, GY. 2008. Bulk and single-particle mineralogy of Asian dust and a comparison with its source soils. *J. Geophys. Res.-Atmosph.* 113. doi:10.1029/2007JD008606

Jeong, GY, Hillier, S and RA Kemp. 2008. Quantitative bulk and single particle mineralogy of a thick Chinese loess-paleosol section: Implications for loess provenance and weathering. *QI* 27:1271–1287.

Jersak, J, Amundson, R and G Brimhall Jr. 1995. A mass balance analysis of podzolization: Examples from the northeastern United States. *Geoderma* 66:15–42.

Jessup, RW. 1960. The stony tableland soils of the southeastern portion of the Australian arid zone and their evolutionary history. *JSS* 11:188–196.

Jha, PP and MG Cline. 1963. Morphology and genesis of a Sol Brun Acide with fragipan in uniform silty material. *SSSAP* 27:339–344.

Jien, SH, Chen, TH, Chiu, CY and S Nagatsuka. 2009. Relationships between soil mass movement and relief in humid subtropical low-elevation mountains. *SS* 174:563–573.

Jien, S-H, Hseu, Z-Y, and ZS Chen. 2004. Relations between morphological color index and soil wetness condition of anthraquic soils in Taiwan. *SS* 169:871–882.

Jien, S-H, Hseu, Z-Y, Iizuka, Y, Chen, T-H and C-Y Chiu. 2010. Geochemical characterization of placic horizons in subtropical montane forest soils, northeastern Taiwan. *EJSS* 61:319–332.

Jimenez-Moreno, G and RS Anderson. 2013. Pollen and macrofossil evidence of Late Pleistocene and Holocene treeline fluctuations from an alpine lake in Colorado, USA. *Holocene* 23:68–77.

Jin, CS, Liu, QS and JC Larrasoana. 2012. A precursor to the Matuyama-Brunhes reversal in Chinese loess and its palaeomagnetic and stratigraphic significance. *Geophys. J. Intl.* 190:829–842.

Jobbagy, EG and RB Jackson. 2004. The uplift of soil nutrients by plants: Biogeochemical consequences across scales. *Ecology* 85:2380–2389.

Jochimsen, M. 1973. Does the size of lichen thalli really constitute a valid measure for dating glacial deposits? *AAR* 5:417–424.

Joffe, JS. 1936. *Pedology*. Rutgers Univ. Press, New Brunswick, NJ.

1949. *Pedology*, 2nd ed. Somerset Press, Somerville, NJ.

Johnson, DL. 1985. Soil thickness processes. *Catena Suppl.* 6:29–40.

1989. Subsurface stone lines, stone zones, artifact-manuport layers, and biomantles produced by bioturbation via pocket gophers (*Thomomys bottae*). *Am. Antiq.* 54:370–389.

1990. Biomantle evolution and the redistribution of earth materials and artifacts. *SS* 149:84–102.

1993a. Biomechanical processes and the Gaia paradigm in a unified pedo-geomorphic and pedo-archaeologic framework: Dynamic denudation. In: JE Foss, ME Timpson and MW Morris (eds). Proc.

1st Intl. Conf. *Pedo-Archaeology*, Univ. Tenn. Agric. Exp. Station Spec. Publ. 93–03, pp. 41–67. Knoxville, TN.

1993b. Dynamic denudation evolution of tropical, subtropical and temperate landscapes with three tiered soils: Toward a general theory of landscape evolution. *QI* 17:67–78.

Johnson, DL. 1994. Reassessment of early and modern soil horizon designation frameworks and associated pedogenetic processes: Are midlatitude A E B-C horizons equivalent to tropical M S W horizons? Soil Sci. *(Trends Agric. Sci.)* 2:77–91.

1998a. A universal definition of soil. *QI* 51/52:6–7.

1998b. Paleosols are buried soils. *QI* 51/52:7.

1999. Darwin the archaeologist: A lesson in unfulfilled language. *Disc. Archaeol.* 1:6–7.

2000. Soils and soil-geomorphology theories and models: The Macquarie connection. *AAAG* 90:775–782.

2002. Darwin would be proud: Bioturbation, dynamic denudation, and the power of theory in science. *Geoarchaeol.* 17:7–40.

Johnson, DL and CL Balek. 1991. The genesis of Quaternary landscapes with stone lines. *PG* 12:385–395.

Johnson, DL, Domier, JEJ and DN. Johnson. 2005. Reflections on the nature of soil and its biomantle. *AAAG* 95:11–31.

Johnson, DL and NC Hester. 1972. Origin of stone pavements on Pleistocene marine terraces in California. *Proc. Assoc. Am. Geog.* 4:50–53.

Johnson, DL and FD Hole. 1994. Soil formation theory: A summary of its principal impacts on Geography, Geomorphology, Soil-Geomorphology, Quaternary Geology and Paleopedology. *SSSASP* 33:111–126.

Johnson, DL and JL Horwath Burnham. 2012. Introduction: Overview of concepts, definitions, and principles of soil mound studies. In: JL Horwath Burnam and DL Johnson (eds). Mima Mounds: The Case for Polygenesis and Bioturbation. GSA Spec. Paper 490. pp. 1–19.

Johnson, DL, Johnson, DN and DM Moore. 2001. *Deep and actively forming illuvial clay in the regolith and on bedrock.* Proc. 12th Intl. Clay Conf., Bahia Blanca, Argentina. pp. 205–210.

Johnson, DL, Keller, EA and TK Rockwell. 1990. Dynamic pedogenesis: New views on some key soil concepts, and a model for interpreting Quaternary soils. *QR* 33:306–319.

Johnson, DL and RJ Schaetzl. 2014. Differing views of soil and pedogenesis by two masters: Darwin and Dokuchaev. *Geoderma* 237–238:176–189.

Johnson, DL and D Watson-Stegner. 1987. Evolution model of pedogenesis. *SS* 143:349–366.

Johnson, DL, Watson-Stegner, D, Johnson, DN and RJ Schaetzl. 1987. Proisotropic and proanisotropic processes of pedoturbation. *SS* 143:278–292.

Johnson, DW, Miller, WW, Susfalk, RB, Murphy, JD, Dahlgren, RA and DW Glass. 2009. Biogeochemical cycling in forest soils of the eastern Sierra Nevada Mountains, USA. *For. Ecol. Mgmt.* 258:2249–2260.

Johnson, EP. 1995. *Soil Survey of Mason County, Michigan.* Soil Cons. Service, U.S. Govt. Printing Off., Washington, DC.

Johnson, MD, Mickelson, DM, Clayton, L and JW Attig. 1995. Composition and genesis of glacial hummocks, western Wisconsin, USA. *Boreas* 24:97–116.

Johnson, WC, Willey, KL and GL Macpherson. 2007. Carbon isotope variation in modern soils of the tallgrass prairie: Analogues for the interpretation of isotopic records derived from paleosols. *QI* 162–163:3–20.

Johnson, WF, Mausbach, MJ, Gamble, EE and RE Nelson. 1985. Natric horizons on some erosional landscapes in northwestern South Dakota. *SSSAJ* 49:947–952.

Johnson, WH. 1990. Ice-wedge casts and relict patterned ground in central Illinois and their environmental significance. *QR* 33:51–72.

Johnson, WH and LR Follmer. 1989. Source and origin of Roxana silt and Middle Wisconsinan midcontinent glacial activity. *QR* 31:319–331.

Johnson, WH, Hansel, AK, Bettis, EA, Karrow, PF, Larson, GJ, Lowell, TV and AF Schneider. 1997. Late Quaternary temporal and event classifications, Great Lakes Region, North America. *QR* 47:1–12.

Johnson, WM. 1963. The pedon and the polypedon. *SSSAP* 27:212–215.

Johnson, WM, Cady, JG and MS James. 1962. Characteristics of some Brown Grumusols of Arizona. *SSSAP* 26:389–393.

Johnson-Maynard, J, Anderson, MA, Green, S and RC Graham. 1994. Physical and hydraulic properties of weathered granitic rock in southern California. *SS* 158:375–380.

Johnson-Maynard, JL, Shouse, PJ, Graham, RC, Castiglione, P and SA Quideau. 2004. Microclimate and pedogenic implications in a 50-year old chaparral and pine biosequence. *SSSAJ* 68:876–884.

Johnsson, H and L-C Lundin. 1991. Surface runoff and soil water percolation as affected by snow and soil frost. *J. Hydrol.* 122:141–159.

Johnston, CA, Lee, GB and FW Madison. 1984. The stratigraphy and composition of a lakeside wetland. *SSSAJ* 48:347–354.

Johnston, SG, Burton, ED, Keene, AF, Planer-Friedrich, B, Voegelin, A, Blackford, MG and GR Lumpkin. 2012. Arsenic mobilization and iron transformations during sulfidization of As(V)-bearing jarosite. *Chem. Geol.* 334:9–24.

Jolley, DM, Grossman, EL and NR Tilford. 1998. Dating calcic soils under marginal climatic or lithologic conditions using a soil development index: An example from the Stockton Plateau, Terrell County, Texas. *Environ. Engin. Geosci.* 4:209–223.

Jones, CE. 1991. Characteristics and origin of rock varnish from the hyperarid coastal deserts of Northern Peru. *QR* 35:116–129.

Jones, LHP and KA Handreck. 1967. Silica in soils, plants and animals. *Adv. Agron.* 19:107–149.

Jones, MJ. 1985. The weathered zone aquifers of the basement complex areas of Africa. *QJ Engin. Geol.* 18:35–46.

Jones, MLM, Sowerby, A, Williams, DL and RE Jones. 2008. Factors controlling soil development in sand dunes: Evidence from a coastal dune soil chronosequence. *Plant Soil* 307:219–234.

Jones, PE. 1970. The occurrence of *Chtonius ischnocheles* (Hermann) (*Chelonethi: Chtoniidae*) in two types of hazel coppice leaf litter. *Bull. Br. Arachnol. Soc.* 1:77–79.

Jones, RL and AH Beavers. 1963. Some mineralogical and chemical properties of plant opal. *SS* 96:375–379.

Jones, RL, Hay, WW and AH Beavers. 1963. Microfossils in Wisconsinan loess and till from western Illinois and eastern Iowa. *Science* 140:1222–1224.

Jones, RL, Ray, BW, Fehrenbacher, JB and AH Beavers. 1967. Mineralogical and chemical characteristics of soils in loess overlying shale in northwestern Illinois. *SSSAP* 31:800–804.

Jones, TA. 1959. Soil classification: A destructive criticism. *JSS* 10:196–200.

Jongkind, AG, Velthorst, E and P Buurman. 2007. Soil chemical properties under kauri (*Agathis australis*) in the Waitakere Ranges, New Zealand. *Geoderma* 141:320–331.

Jongmans, AG, Pulleman, MM, Balabane, M, van Oort, F and JCY Marinissen. 2003. Soil structure and characteristics of organic matter in two orchards differing in earthworm activity. *Appl. Soil Ecol.* 24:219–232.

Jongmans, AG, van Breeman, N, Lundström, U, van Hees, PAW, Finlay, RD, Srinivasan, T, Unestam, T, Giesler, R, Melkerud, PA and M Olsson. 1997. Rock-eating fungi. *Nature* 389:682–683.

Jordan, JV, Lewis, GC and MA Fosberg. 1958. Tracing moisture movement in slick-spot soils with radiosulfur. *I. TAGU* 39:446–450.

Joseph, KT. 1968. A toposequence on limestone parent materials in North Kedah, Malaya. *J. Trop. Geog.* 27:19–22.

Joshi, AB, Vann, DR, Johnson, AH and EK Miller. 2003. Nitrogen availability and forest productivity along a climosequence on Whiteface Mountain, New York. *CJFR* 33:1880–1891.

Joshi, S and DK Upreti. 2010. Lichenometric studies in vicinity of Pindari Glacier in the Bageshwar district of Uttarakhand, India. *Current Sci.* 99:231–235.

Jouaffre, D, Bruckert, S, Williams, AF, Herbillon, AJ and B Kubler. 1991. Post-Würmian rubification in humid Jurassian Mountainous climate – Role of pedoclimate and current processes. *Geoderma* 50:239–257.

Jouquet, P, Dauber, J, Lagerlof, J, Lavelle, P and M Lepage. 2006. Soil invertebrates as ecosystem engineers: Intended and accidental effects on soil and feedback loops. *Appl. Soil Ecol.* 32:153–164.

Jouquet, P, Mamou, L, Lepage, M and B Velde. 2002. Effect of termites on clay minerals in tropical soils: Fungus-growing termites as weathering agents. *EJSS* 53:521–527.

Jouquet, P, Traore, S, Choosai, C, Hartmann, C and D Bignell. 2011. Influence of termites on ecosystem functioning. Ecosystem services provided by termites. *Eur. J. Soil Biol.* 47:215–222.

Joussein, E, Petit, S, Churchman, J, Theng, B, Righi, D and B Delvaux. 2005. Halloysite clay minerals – A review. *Clay Mins.* 40:383–426.

Jouzel, J, Lorius, C, Petit, JR, Genthon, C, Barkov, NI, Kotlyakov, VM and VM Petrov. 1987. Vostok ice core: A continuous isotope temperature record over the last climatic cycle (160 000 years). *Nature* 329:403–408.

Juergens, N. 2013. The biological underpinnings of Namib Desert fairy circles. *Science* 339:1618–1621.

Junge, CE and RT Werby. 1958. The concentration of chloride, potassium, calcium, and sulfate in rain water over the United States. *J. Metr.* 15:417–425.

Jungerius, PD. 1985. Soils and geomorphology. *Catena Suppl.* 6:1–18.

Jungerius, PD and F van der Meulen. 1988. Erosion processes in a dune landscape along the Dutch coast. *Catena* 15:217–288.

Jury, WA and B Bellantouni. 1976. Heat and water movement under surface rocks in a field soil. II. Moisture effects. *SSSAJ* 40:509–513.

Jury, WA and R Horton. 2004. *Soil Physics*, 6th ed. Wiley, Hoboken NJ.

Kabrick, JM, Clayton, MK, McBratney, AB and K McSweeney. 1997a. Cradle-knoll patterns and characteristics on drumlins in northeastern Wisconsin. *SSSAJ* 61:595–603.

Kabrick, JM, Clayton, MK and K McSweeney. 1997b. Spatial patterns of carbon and texture on drumlins in northeastern Wisconsin. *SSSAJ* 61:541–548.

Kaczorek, D, Sommer, M, Andruschkewitsch, I, Oktaba, L, Czerwinski, Z and K Stahr. 2004. A comparative micromorphological and chemical study of "raseneisenstein" (bog iron ore) and "ortstein." *Geoderma* 121:83–94.

Kadmon, R, Yair, A and A Danin. 1989. Relationship between soil properties, soil moisture, and vegetation along loess-covered hillslopes, northern Negev, Israel. *Catena Suppl.* 14:43–57.

Kahle, CF. 1977. Origin of subaerial Holocene calcareous crusts: Role of algae, fungi and sparmicritisation. *Sedimentology* 24:413–435.

Kahle, M, Kleber, M and R Jahn. 2002. Review of XRD-based quantitative analyses of clay minerals in soils: The suitability of mineral intensity factors. *Geoderma* 109:191–205.

Kahle, P. 2000. Heavy metals in garden soils from the urban area of Rostock. *J. Plant Nutr. Soil Sci.* 163:191–196.

Kaiser, J. 2004. Wounding Earth's fragile skin. *Science* 304:1616–1618.

Kaiser, K and W Wilcke. 1996. Pedogenetic differentiation of soil properties in aggregates. *Zeit. fur Pflanz. Boden.* 159:599–603.

Kamenik, J, Mizera, J and Z Randa. 2013. Chemical composition of plant silica phytoliths. *Environ. Chem. Letts.* 11:189–195.

Kämpf, N and U Schwertmann. 1983a. Goethite and hematite in a climosequence in southern Brazil and their application in classification of kaolinitic soils. *Geoderma* 29:27–39.

1983b. Relacções entre óxidos de ferro e a cor de solos cauliníticos do Rio Grande do Sul. *Revista Brasil. Ciência Solo* 7:27–31.

Kamprath, EJ. 1973. Phosphorous. In: PA Sanchez (ed). *A Review of Soils Research on Tropical Latin America.* Tech. Bull. 219. North Carolina Agric. Exp. Station, Raleigh. pp. 138–161.

Kantor, W and U Schwertmann. 1974. Mineralogy and genesis of clays in red-black soil toposequences on basic igneous rocks in Kenya. *JSS* 25:67–78.

Kanzari, S, Hachicha, M, Bouhlila, R and J Battle-Sales. 2012. Characterization and modeling of water movement and salts transfer in a semi-arid region of Tunisia (Bou Hajla, Kairouan) – Salinization risk of soils and aquifers. *Comp. Electr. Agric.* 86:34–42.

Kaplan, MY, Eren, M, Kadir, S and S Kapur. 2013. Mineralogical, geochemical and isotopic characteristics of Quaternary calcretes in the Adana region, southern Turkey: Implications on their origin. *Catena* 101:164–177.

Kapur, S, Mermut, A and G Stoops (eds). 2008. *New Trends in Soil Micromorphology.* Springer-Verlag, Berlin.

Karathanasis, AD. 1987a. Mineral solubility relationships in Fragiudalfs of western Kentucky. *SSSAJ* 51:474–481.

1987b. Thermodynamic evaluation of amorphous aluminosilicate binding agents in fragipans of western Kentucky. *SSSAJ* 51:819–824.

2008. Thermal analysis of soil minerals. In: A Ulery and R Drees (eds). *Methods of Soil Analysis.* Part 5. *Mineralogical Methods.* SSSA Book ser. 5. Soil Sci. Soc. Am., Madison, WI. pp. 415–430.

Karathanasis, AD and PA Golrick. 1991. Soil formation on loess/sandstone toposequences in west-central Kentucky. I. Morphology and physiochemical properties. *SS* 152:14–24.

Karathanasis, AD and BR Macneal. 1994. Evaluation of parent material uniformity criteria in loess-influenced soils of west-central Kentucky. *Geoderma* 64:73–92.

Karberg, NJ and EA Lilleskov. 2008. White-tailed deer (*Odocoileus virginianus*) fecal pellet decomposition is accelerated by the invasive earthworm *Lumbricus terrestris. Biol. Invasions* 3:761–767.

Karlstrom, ET. 2000. Fabric and origin of multiple diamictons within the pre-Illinoian Kennedy Drift east of Waterton-Glacier International Peace Park, Alberta, Canada, and Montana, USA. *GSA Bull.* 112:1496–1506.

Karlstrom, ET and RW Barendregt. 2001. Fabric, paleomagnetism, and interpretation of pre-Illinoian diamictons and paleosols on Cloudy Ridge and Milk River Ridge, Alberta and Montana. *Geog. Phys. Quat.* 55:141–157.

Karlstrom, ET and G Osborn. 1992. Genesis of buried paleosols and soils in Holocene and late Pleistocene tills, Bugaboo Glacier area, British Columbia, Canada. *AAR* 24:108–123.

Kaufman, DS, Jensen, BJL, Reyes, AV, Schiff, CJ, Froese, DG and NJG Pearce. 2012. Late Quaternary tephrostratigraphy, Ahklun Mountains, SW Alaska. *JQS* 27:344–359.

Kaufman, IR and BR James. 1991. Anthropic epipedons in oyster shell middens of Maryland. *SSSAJ* 55:1191–1193.

Kay, BD and DA Angers. 2000. Soil structure. In: ME Sumner (ed). *Handbook of Soil Science.* CRC Press, Boca Raton, FL. pp. A-229-A-276.

Kay, GF. 1916. Gumbotil, a new term in Pleistocene geology. *Science* 44:637–638.

Kay, GF and ET Apfel. 1929. The Pre-Illinoian Pleistocene geology of Iowa. *Iowa Geol. Surv. Ann. Rept* 34:1–304.

Kay, GF and JB Graham. 1941. The Illinoian and Post-Illinoian Pleistocene geology of Iowa. *Iowa Geol. Survey Ann. Rept.* 38:1–262.

Kay, GF and JN Pierce. 1920. The origin of gumbotil. *J. Geol.* 28:89–125.

Keen, KL and LCK Shane. 1990. A continuous record of Holocene eolian activity and vegetation change at Lake Ann, east-central Minnesota. *GSA Bull.* 102:1646–1657.

Kellogg, CE. 1930. Preliminary study of the profiles of the principal soil types of Wisconsin. Wisc. Geol. Nat. Hist. Surv. Bull. 77A.

1934. Morphology and genesis of the Solonetz soils of western North Dakota. *SS* 38:483–500.

1936. *Development and Significance of the Great Soil Groups of the United States, Miscellaneous Publication #229.* US Dept. of Agric., US Govt. Printing Off., Washington, DC.

1941. Climate and soil. In: *Climate and Men: Yearbook of Agriculture.* US Dept. of Agric., US Govt. Printing Off., Washington, DC. pp. 276–277.

1949. Preliminary suggestions for the classification and nomenclature of great soil groups in tropical and equatorial regions. *CAB Soil Sci. Tech. Commun.* 46:76–85.

1950. Tropical soils. *Trans. 4th Intl. Congr. Soil Sci.* 1:266–276.

1974. Soil genesis, classification, and cartography: 1924–1974. *Geoderma* 12:347–362.

Kellogg, CE and IJ Nygard. 1951. *The Principal Soil Groups of Alaska,* Agric. Mon. 7. US Dept. of Agric., US Govt. Printing Off., Washington, DC.

Kelly, EF, Amundson, RG, Marino, BD and MJ DeNiro. 1991. Stable carbon isotopic composition of carbonate in Holocene grassland soils. *SSSAJ* 55:1651–1658.

Kelso, GK. 1974. Pollen percolation rates in Euroamerican-era cultural deposits in the northeastern United States. *JAS* 21:481–488.

Kemp, RA. 1999. Micromorphology of loess-paleosol sequences: A record of paleoenvironmental change. *Catena* 35:181–198.

2001. Pedogenic modification of loess: Significance for palaeoclimatic reconstructions. *Earth-Sci. Revs.* 54:145–156.

Kemp, RA and E Derbyshire. 1998. The loess soils of China as record of climatic change. *EJSS* 49:525–539.

Kemp, RA, McDaniel, PA and AJ Busacca. 1998. Genesis and relationship of macromorphology and micromorphology to contemporary hydrological conditions of a welded Argixeroll from the Palouse in Idaho. *Geoderma* 83:309–329.

Kemp, RA and PD McIntosh. 1989. Genesis of a texturally banded soil in Southland, New Zealand. *Geoderma* 45:65–81.

Kendrick, KJ and RC Graham. 2004. Pedogenic silica accumulation in chronosequence soils, southern California. *SSSAJ* 68:1295–1303.

Kendrick, KJ and LD McFadden. 1996. Comparison and contrast of processes of soil formation in the San Timoteo Badlands with chronosequences in California. *QR* 46:149–160.

Kevan, DKM. 1968. Soil fauna and humus formation. *Trans. 9th Intl. Congr. Soil Sci.* 2:1–10.

Khademi, H and AR Mermut. 2003. Micromorphology and classification of Argids and associated gypsiferous Aridisols from central Iran. *Catena* 54:439–455.

Khadjeh, M. Ghauomian, J and S Faiznia. 2004. The study of lateral variation of grain size and mineralogy in order to determine prevailing winds direction in the formation of loess sediments of Golestan Province. *Biaban* 9:293–306.

Khakural, BR, Lemme, GD and DL Mokma. 1993. Till thickness and argillic horizon development in some Michigan Hapludalfs. *SSH* 34:6–13.

Khan, FA and TE Fenton. 1994. Saturated zones and soil morphology in a Mollisol catena of central Iowa. *SSSAJ* 58:1457–1464.

Khangarot, AS, Wilding, LP and GF Hall. 1971. Composition and weathering of loess mantled Wisconsin- and Illinoian-age terraces in central Ohio. *SSSAP* 35:621–626.

Khokhlova, OS, Kovalevskaya, IS and SA Oleynik. 2001. Records of climatic changes in the carbonate profiles of Russian Chernozems. *Catena* 43:203–215.

Khomo, L, Hartshorn, AS, Rogers, KH and OA Chadwick. 2011. Impact of rainfall and topography on the distribution of clays and major cations in granitic catenas of southern Africa. *Catena* 87:119–128.

Khormali, F and A Abtahi. 2003. Origin and distribution of clay minerals in calcareous arid and semi-arid soils of Fars Province, southern Iran. *Clay Mins.* 38:511–527.

Khresat, SA and EA Qudah. 2006. Formation and properties of aridic soils of Azraq Basin in northeastern Jordan. *J. Arid Environs.* 64:116–136.

Kidron, GJ, Barinova, S and A Vonshak. 2012. The effects of heavy winter rains and rare summer rains on biological soil crusts in the Negev Desert. *Catena* 95:6–11.

Kidron, GJ and A Yair. 1997. Rainfall-runoff relationships over encrusted dune surfaces, Nizzana, western Negev, Israel. *ESPL* 22:1169–1184.

Kidron, GJ, Yair, A and A Danin. 2000. Dew variability within a small arid drainage basin in the Negev Highlands, Israel. *Quart. J. Royal Metr. Soc.* 126:63–80.

Kieffer, SW. 1981. Blast dynamics at Mount St. Helens on 18 May 1980. *Nature* 291:568–570.

Kiernan, K. 1990. Weathering as an indicator of the age of Quaternary glacial deposits in Tasmania. *Austral. Geog.* 21:1–17.

Kikuchi, R and TT Gorbacheva. 2005. Resistance of organic horizon to acidification from snowmelt in podzolic soil of the Kola Peninsula. *J. Soils Seds.* 5:143–148.

Kimble, JM, Ping, CL, Sumner, ME and LP Wilding. 2000. Andisols. In: ME Sumner (ed). *Handbook of Soil Science*. CRC Press, Boca Raton, FL. pp. E-209–E-224.

King, D, Bourennane, H, Isambert, M and JJ Macaire. 1999. Relationship of the presence of a non-calcareous clay-loam horizon to DEM attributes in a gently sloping area. *Geoderma* 89:95–111.

King, GJ, Acton, DF and RJ St. Arnaud. 1983. Soil-landscape analysis in relation to soil distribution and mapping at a site within the Weyburn association. *CJSS* 63:657–670.

King, JJ and DP Franzmeier. 1981. Estimation of saturated hydraulic conductivity from soil morphological and genetic information. *SSSAJ* 45:1153–1156.

King, LC. 1957. The uniformitarian nature of hillslopes. *Trans. Edinburgh Geol. Soc.* 17:81–102.

King, TJ. 1977. The plant ecology of ant-hills in calcareous grasslands. *JE* 65:235–316.

Kirkby, A and MJ Kirkby. 1974. Surface wash at the semi-arid break in slope. *ZG Suppl.* 21:151–176.

Kirkham, VRD, Johnson, MM and D Holm. 1931. Origin of Palouse Hills topography. *Science* 73:207–209.

Kirschner, M. 2013. A perverted view of "impact." *Science* 339:250.

Kishne, AS, Ge, Y, Morgan, CLS and WL Miller. 2012. Surface cracking of a Vertisol related to the history of available water. *SSSAJ* 76:548–557.

Kishne, AS, Morgan, CLS and WL Miller. 2009. Vertisol crack extent associated with gilgai and soil moisture in the Texas Gulf Coast Prairie. *SSSAJ* 73:1221–1230.

Kittrick, JA, Fanning, DS and LR Hossner (eds). 1982. *Acid Sulphate Weathering*. SSSASP 10. Madison, WI.

Kittrick, JA. 1973. Mica-derived vermiculites as unstable intermediates. *Clay Mins.* 21:479–488.

Kladivko, EJ and HJ Timmenga. 1990. Earthworms and agricultural management. In: JE Box and LC Hammond (eds). *Rhizosphere Dynamics*. Am. Assoc. Advancement Science, Selected Symposium, Westview Press, Boulder, CO. pp. 192–216.

Klappa, CF. 1979. Calcified filaments in Quaternary calcretes: Organo-mineral interactions in the subaerial vadose environment. *J. Sed. Petr.* 49:955–968.

Klasen, N, Fiebig, M, Preusser, F, Reitner, JM and U Radtke. 2007. Luminescence dating of proglacial sediments from the Eastern Alps. *QI* 164–165:21–32.

Kleber, A. 1997. Cover-beds as soil parent materials in midlatitude regions. *Catena* 30:197–213.

——— 2000. Compound soil horizons with mixed calcic and argillic properties – examples from the northern Great Basin, USA. *Catena* 41:111–131.

Kleber, M. 2010. What is recalcitrant soil organic matter? *Environ. Chem.* 7:320–332.

Kleber, M and MG Johnson. 2010. Advances in understanding the molecular structure of soil organic matter: Implications for interactions in the environment. *Adv. Agron.* 106:77–142.

Kleber, M, Nico, PS, Plante, AF, Filley, T, Kramer, M, Swanston, C and P Sollins. 2011. Old and stable soil organic matter is not necessarily chemically recalcitrant: Implications for modeling concepts and temperature sensitivity. *Global Change Biol.* 17:1097–1107.

Kleber, M, Schwendenmann, L, Veldkamp, E, Rossner, J and R Jahn. 2007. Halloysite versus gibbsite: Silicon cycling as a pedogenetic process in two lowland neotropical rain forest soils of La Selva, Costa Rica. *Geoderma* 138:1–11.

Kleber, M, Sollins, P and R Sutton. 2007. A conceptual model of organo-mineral interactions in soils: Self-assembly of organic molecular fragments into zonal structures on mineral surfaces. *Biogeochem.* 85:9–24.

Kleiss, HJ. 1970. Hillslope sedimentation and soil formation in northeastern Iowa. *SSSAP* 34:287–290.

——— 1973. Loess distribution along the Illinois soil-development sequence. *SS* 115:194–198.

Kline, JR. 1973. Mathematical simulation of soil-plant relationships and soil genesis. *SS* 115:240–249.

Klingbiel, AA, Horvath, EH, Moore, DG and WU Reybold. 1987. Use of slope, aspect and elevation maps derived from digital elevation model data in making soil surveys. *SSSASP* 20:77–90. In: A Klute (ed). *Methods of Soil Analysis. Part 1. Physical and Mineralogical Methods.* Agronomy Mon. 9, 2nd ed., Soil Sci. Soc. Am., Madison, WI. pp. 635–662.

Klinge, H. 1965. Podzol soils in the Amazon Basin. *JSS* 16:95–103.

Klinger, LF. 1996. The myth of the classic hydrosere model of bog succession. *AAR* 28:1–9.

Kloprogge, JT, Komarneni, S and JE Amonette. 1999. Synthesis of smectite clay minerals: A critical review. *Clays Clay Mins.* 47:529–554.

Knapp, BD. 1993. *Soil Survey of Presque Isle County, Michigan*. USDA-SCS, US Govt. Printing Off., Washington, DC.

Knauss, KG and T Ku. 1980. Desert varnish: Potential for age dating via uranium-series isotopes. *J. Geol.* 88:95–100.

Knight, MJ. 1980. Structural analysis and mechanical origins of gilgai at Boorook, Victoria, Australia. *Geoderma* 23:245–283.

Knollenberg, WG, Merritt, RW and DL Lawson. 1985. Consumption of leaf litter by Lumbricus terrestris (Oligochaeta) on a Michigan woodland floodplain. *AMN* 113:1–6.

Knox, EG. 1957. Fragipan horizons in New York soils. III. The basis of rigidity. *SSSAP* 21:326–330.

——— 1965. Soil individuals and soil classification. *SSSAP* 29:79–84.

Knuepfer, PLK. 1988. Estimating ages of late Quaternary stream terraces from analysis of weathering rinds and soils. *GSA Bull.* 100:1224–1236.

Knuepfer, PLK and LD McFadden (eds) 1990. Soils and landscape evolution. *Geomorphology* 3: issues 3 and 4.

Knuteson, JA, Richardson, JL, Patterson, DD and L Prunty. 1989. Pedogenic carbonates in a Calciaquoll associated with a recharge wetland. *SSSAJ* 53:495–499.

Kobojek, S. 1997. Zawartos´c´ skaleni w utworach neoplejstocen´ – skich okolic Cze̜stochowy. *Acta Univ. Lodz. Folia Geogr. Physica* 1:196–202.

Koch, PL. 1998. Isotopic reconstruction of past continental environments. *Ann. Rev. Earth Planet. Sci.* 26:573–613.

Kocurek, G and N Lancaster. 1999. Aeolian system sediment state: Theory and Mojave Desert Kelso dune field example. *Sedimentology* 46:505–515.

Kocurek, G and J Nielson. 1986. Conditions favorable for the formation of warm-climate aeolian sand sheets. *Sedimentology* 33:795–816.

Kodama, H, Schnitzer, M and M Jaakkimainen. 1983. Chlorite and biotite weathering by fulvic acid solutions in closed and open systems. *CJSS* 63:619–629.

Kodama, H and C Wang. 1989. Distribution and characterization of noncrystalline inorganic compounds in Spodosols and Spodosol-like soils. *SSSAJ* 53:526–534.

Kohnke, H, Stuff, RG and PA Miller. 1968. Quantitative relations between climate and soil formation. *Z. Pflanzen. Bodenkd.* 119:24–33.

Kolka, RK, Grigal, DF and EA Nater. 1996. Forest soil mineral weathering rates: Use of multiple approaches. *Geoderma* 73:1–21.

Konert, M and J Vandenberghe. 1997. Comparison of laser grain size analysis with pipette and sieve analysis: A solution for the underestimation of the clay fraction. *Sedimentology* 44:523–535.

Konrad, J-M. 1989. Physical processes during freeze-thaw cycles in clayey silts. *Cold Regions Sci. Tech.* 16:291–303.

Koppi, AJ and DJ Williams. 1980. Weathering and development of two contrasting soils formed from grandodiorite in south-east Queensland. *AJSR* 18:257–271.

Korobova, E and S Romanov. 2011. Experience of mapping spatial structure of Cs-137 in natural landscape and patterns of its distribution in soil toposequence. *J. Geochem. Explor.* 109:139–145.

Koshel'kov, SP. 1961. Formation and subdivision of forest floor in southern taiga coniferous forests. *SSS* 1065–1073.

Koutaniemi, L, Koponen, R and K Rajanen. 1988. Podzolization as studied from terraces of various ages in two river valleys, northern Finland. *Silvia Fennica* 22:113–133.

Kouwenhoven, JK and R Terpstra. 1979. Sorting action of tines and tine-like tools in the field. *J. Agric. Engin. Res.* 24:95–113.

Kovaltsov, GA, Mishev, A and IG Usoskin. 2012. A new model of cosmogenic production of radiocarbon ^{14}C in the atmosphere. *EPSL* 337/338:114–120.

Kovda, I, Lynn, W, Williams, D and O Chichagova. 2001. Radiocarbon age of Vertisols and its interpretation using data on gilgai complex in the North Caucasus. *Radiocarbon* 43:603–609.

Koven, CD, Ringeval, B, Friedlingstein, P, Ciais, P, Cadule, P, Khvorostyanov, D, Krinner, G and C Tarnocai. 2011. Permafrost carbon-climate feedbacks accelerate global warming. *PNAS* 108:14769–14774.

Kowal, VA, Schmolke, A, Kanagaraj, R and D Bruggeman. 2013. Resource selection probability functions for Gopher Tortoise: Providing a management tool applicable across the species' range. *Ecol. Mgmt.* doi:10.1007/s00267-013-0210-x

Kowalkowski, A. 1995. Lithological-pedogenic discontinuity on the slopes of the Lysogóry Massif in the Holy Cross Mountains. *Quaes. Geog.* 17/18:25–39.

Kozarski, S. 1974. Evidences of Late-Wurm permafrost occurrence in north-west Poland. *Quaest. Geog.* 1:65–86.

Kram, P, Oulehle, F, Stedra, V, Hruska, J, Shanley, JB, Minocha, R and E Traister. 2009. Geoecology of a forest watershed underlain by serpentine in Central Europe. *Northeastern Natural.* 16:309–328.

Krasilmkov, P and NEG Calderon. 2006. A WRB-based buried paleosol classification. *QI* 156:176–188.

Krause, HH, Rieger, S and SA Wilde. 1959. Soils and forest growth on different aspects in the Tanana watershed of interior Alaska. *Ecology* 40:492–495.

Krinsley, D. 1998. Models of rock varnish formation constrained by high resolution transmission electron microscopy. *Sedimentol.* 45:711–725.

Krinsley, DH, Dorn, RI and SW Anderson. 1990. Factors that interfere with the age determination of rock varnish. *PG*11:97–119.

Krinsley, DH, Dorn, RI, DiGregorio, BE, Langworthy, KA and J Ditto. 2012. Rock varnish in New York: An accelerated snapshot of accretionary processes. *Geomorphology* 138:339–351.

Krishnamani, R and WC Mahaney. 2000. Geophagy among primates: Adaptive significance and ecological consequences. *Anim. Behav.* 59:899–915.

Krist, F and RJ Schaetzl. 2001. Paleowind (11,000 BP) directions derived from lake spits in northern Michigan. *Geomorphology* 38:1–18.

Kroetsch, DJ, Geng, X, Chang, SX and DD Saurette. 2011. Organic soils of Canada. Part 1. Wetland organic soils. *CJSS* 91:807–822.

Kröhling, DM and MH Iriondo. 2010. Comment on: "Genesis of subtropical soils with stony horizons in NE Argentina: Autochthony and polygenesis." H. Morras, L. Moretti, G. Piccolo, W. Zech. Quaternary International (2009), Vol. 196 (1-2): 137-159. *QI* 227:190–192.

Kromer, B. 2009. Radiocarbon and dendrochronology. *Dendrochronol.* 27:15–19.

Kronberg, GI and HW Nesbitt. 1981. Quantification of weathering of soil chemistry and soil fertility. *JSS* 32:453–459.

Kroonenberg, SB and PJ Melitz. 1983. Summit levels, bedrock control and the etchplain concept in the basement of Suriname. *Geol. Mijn.* 62:389–399.

Krull, ES, Bestland, EA, Skjemstad, JO and JF Parr. 2006. Geochemistry ($\delta^{13}C$, $\delta^{15}N$, ^{13}C NMR) and residence times (^{14}C and OSL) of soil organic matter from red-brown earths of South Australia: Implications for soil genesis. *Geoderma* 132:344–360.

Krumbein, WC. 1941a. Measurement and geological significance of shape and roundness of sedimentary particles. *J. Sed. Petr.* 11:64–72.

——— 1941b. The effects of abrasion on the size, shape and roundness of rock fragments. *J. Geol.* 49:482–520.

Krumbein, WE and C Giele. 1979. Calcification in a coccoid-cyanbacterium associated with the formation of desert stromatolites. *Sedimentology* 26:593–604.

Krumbein, WE and K Jens. 1981. Biogenic rock varnishes of the Negev Desert (Israel): An ecological study of iron and manganese transformation by cyanobacteria and fungi. *Oecologia* 50:25–38.

Krzyszowska, AJ, Blaylock, MJ, Vance, GF and MB David. 1996. Ion chromatographic analysis of low molecular weight organic acids in spodosol forest floor solutions. *SSSAJ* 60:1565–1571.

Ku, TL, Bull, WG, Freeman, ST and KG Knauss. 1979. ^{230}Th/^{234}U dating of pedogenic carbonates in gravelly desert soils of Vidal Valley, southeastern California. *GSA Bull.* 90:1063-1073.

Kubiëna, WL. 1938. *Micropedology.* Collegiate Press, Ames, IA.

1956. Zur mikronorfologie, systematic un entwicklung der rezenten und fossilen lössboden. *Eiszetalter Gegenwart* 7:102-112.

1970. *Micromorphological Features of Soil Geography.* Rutgers Univ. Press, New Brunswick, NJ.

Kudryashova, SY, Baikov, KS, Titlyanova, AA, Dits, LY, Kosykh, NP, Makhatkov, ID, and SV Shibareva. 2011. Distributed GIS for estimation of soil carbon stock of West Siberia boreal zone. *Contemp. Prob. Ecol.* 4:475-486.

Kuehn, SC and FF Foit. 2006. Correlation of widespread Holocene and Pleistocene tephra layers from Newberry Volcano, Oregon, USA, using glass compositions and numerical analysis. *QI* 148:113-137.

Kuehn, SC, Froese, DG, Carrara, PE, Foit, FF, Pearce, NJG and P Rotheisler. 2009. Major- and trace-element characterization, expanded distribution, and a new chronology for the latest Pleistocene Glacier Peak tephras in western North America. *QR* 71:201-216.

Kühn, P, Techmer, A and M Weidenfeller. 2013. Lower to middle Weichselian pedogenesis and palaeoclimate in Central Europe using combined micromorphology and geochemistry: The loess-paleosol sequence of Alsheim (Mainz Basin, Germany). *QSR* 75:43-58.

Kukla, G. 1987. Loess stratigraphy in central China. *QSR* 6:191-219.

Kukla, G and Z An. 1989. Loess stratigraphy in central China. *PPP* 72:203-225.

Kukla, G, Heller, F, Ming, LX, Chun, XT, Sheng, LT and AZ Sheng. 1988. Pleistocene climates in China dated by magnetic susceptibility. *Geology* 16:811-814.

Künelt, W. 1961. *Soil Biology with Special Reference to the Animal Kingdom.* Faber and Faber, London.

Kunze, GW and JB Dixon. 1986. Pretreatment for mineralogical analysis. In: A Klute (ed). *Methods of Soil Analysis.* Part 1. *Physical and Mineralogical Methods.* Agronomy Mon. 9, 2nd ed. Soil Sci. Soc. Am., Madison, WI. pp. 91-100.

Kunze, GW, Oakes, H and ME Bloodworth. 1963. Grumusols of the Coast Prairie of Texas. *SSSAP* 27:412-421.

Kurz, MD, Colodner, D, Trull, TW, Moore, RB and K O'Brien. 1990. Cosmic ray exposure dating with in situ produced cosmogenic 3He: Results from young Hawaiian lava flows. *EPSL* 97:177-189.

Kusumgar, S, Agrawal, DP and RV Krishnamurthy. 1980. Studies of the loess deposits of the Kashmir valley and ^{14}C dating. *Radiocarbon* 22:757-762.

Kutiel, P. 1992. Slope aspect effect on soil and vegetation in a Mediterranean ecosystem. *Isr. J. Bot.* 41:243-250.

Kuzila, MS. 1995. Identification of multiple loess units within modern soils of Clay County, Nebraska. *Geoderma* 65:45-57.

Kuzyakov, Y and G Domanski. 2000. Carbon input by plants into the soil. *J. Plant Nutr. Soil Sci.* 163:421-431.

Lacroix, C, Lavoie, M and N Bhiry. 2011. New macrofossil evidence for early postglacial migration of jack pine (*Pinus banksiana*) in the James Bay region of northwestern Quebec. *Ecoscience* 18:273-278.

Lafleur, B, Bradley, RL, and A Francoeur. 2002. Soil modifications created by ants along a post-fire chronosequence in lichen-spruce woodland. *Ecoscience* 9:63-73.

Låg, J. 1951. Illustration of the influence of topography on depth of A2 layer in Podzol profiles. *SS* 71:125-127.

Lagacherie, P, Andrieux, P and R Bouzigues. 1996. Fuzziness and uncertainty of soil boundaries: From reality to coding in GIS. In: PA Burrough and AU Frank (eds). *Geographic Objects with Indeterminate Boundaries.* Taylor and Francis, London. pp. 275-286.

Laidlaw, MAS and GM Filippelli. 2008. Resuspension of urban soils as a persistent source of lead poisoning in children: A review and new directions. *Appl. Geochem.* 23:2021-2039.

Lair, GJ, Zehetner, F, Hrachowitz, M, Franz, N, Maringer, FJ and MH Gerzabek. 2009. Dating of soil layers in a young floodplain using iron oxide crystallinity. *QG* 4:260-266.

Laird DA and ML Thompson. 2009. The ultrastructure of clay-humic complexes in an Iowa Mollisol. In: DA Laird and J Cervini-Silva (eds). *Carbon Stabilization by Clays in the Environment: Process and Characterization Methods.* CMS Workshop Lectures. Vol. 16. Clay Minerals Society, Chantilly, VA. pp. 95-118.

Laker, MC, Hewitt, PH, Nel, A and RP Hunt. 1982. Effects of the termite *Trinervitermes trinervoides* Sjöstedt on the organic carbon and nitrogen contents and particle size distribution of the soils. *Rev. Ecol. Biol. Sol* 19:27-39.

Lal, D. 1988. In situ-produced cosmogenic isotopes in terrestrial rocks. *Ann. Rev. Earth Planet. Sci.* 16:355-388.

1991. Cosmic ray labeling of erosion surfaces: In situ nuclide production rates and erosion models. *EPSL* 104:424-439.

Lal, R. 2003. Global potential of soil carbon sequestration to mitigate the greenhouse effect. *Crit. Revs. Plant Sci.* 151-184.

2004. Soil carbon sequestration impacts on global climate change and food security. *Science* 304:1623-1627.

Lambeck, K, Yokoyama, Y and T Purcell. 2002. Into and out of the Last Glacial Maximum: Sea-level change during Oxygen Isotope Stages 3 and 2. *QSR* 21:343-360.

Lambert, F, Delmonte, B, Petit, JR, Bigler, M, Kaufmann, PR, Hutterli, MA, Stocker, TF, Ruth, U, Steffensen, JP and V Maggi. 2008. Dust-climate couplings over the past 800,000 years from the EPICA Dome C ice core. *Nature* 452:616-619.

Lambert, JL and FD Hole. 1971. Hydraulic properties of an ortstein horizon. *SSSAP* 35:785-787.

Lambert, V, Boukhari, R, Misslin-Tritsch, C and G Carles. 2013. Geophagia: Progress toward understanding its causes and consequences. *Rev. Medecine Interne* 34:94-98.

Lammers, DA and MG Johnson. 1991. Soil mapping concepts for environmental assessment. *SSSASP* 28:149-160.

Lamothe, M. 1984. Apparent thermoluminescence ages of St-Pierre sediments at Pierreville, Quebec, and the problem of anomalous fading. *CJES* 21:1406-1409.

Lamothe, M, Auclair, M, Hamzaoui, C and S Huot. 2003. Towards a prediction of long-term anomalous fading of feldspar IRSL. *Rad. Measure.* 37:493-498.

Landeweert, R, Hoffland, E, Finlay, RD, Kuyper, TW and N van Breemen. 2001. Linking plants to rocks: Ectomycorrhizal fungi mobilize nutrients from minerals. *Trends Ecol. Evol.* 16:248-254.

Lång, L-O. 2000. Heavy mineral weathering under acidic soil conditions. *Appl. Geochem.* 15:415-423.

Lang, R. 1915. Versuch einer exakten Klassifikation der Böden in klimatischer und geologischer Hinsicht. *Intl. Mitteil. Bodenkd.* 5:312-346.

Langbein, WB and SA Schumm. 1958. Yield of sediment in relation to mean annual precipitation. *TAGU* 39:1076-1084.

Langley-Turnbaugh, SJ and JG Bockheim. 1998. Mass balance of soil evolution on late Quaternary marine terraces in Coastal Oregon. *Geoderma* 84:265-288.

Langley-Turnbaugh, SJ and CV Evans. 1994. A determinitive soil development index for pedo-stratigraphic studies. *Geoderma* 61:39-59.

Langmaid, KK. 1964. Some effects of earthworm invasion in virgin podzols. *CJSS* 44:34-37.

Langohr, R, Scoppa, CO and A Van Wambeke. 1976. The use of a comparative particle size distribution index for the numerical classification of soil parent materials: Application to Mollisols of the Argentine Pampa. *Geoderma* 15:305–312.

Langohr, R and R Vermeire. 1982. Well-drained soils with a "degraded" Bt horizon in loess deposits in Belgium: Relationship with paleoperiglacial processes. *Biul. Peryglac.* 29:203–212.

Lapen, DR and C Wang. 1999. Placic and ortstein horizon genesis and peatland development, southeastern Newfoundland. *SSSAJ* 63:1472–1482.

Lark, RM and PHT Beckett. 1995. A regular pattern in the relative areas of soil profile classes and possible applications in reconnaissance soil survey. *Geoderma* 68:27–37.

La Roi, GH and RJ Hnatiuk. 1980. The Pinus contorta forests of Banff and Jasper National Parks: A study in comparative synecology and syntaxonomy. *Ecol. Monogr.* 50:1–29.

Larson, ER, Kipfmueller, KF, Hale, CM, Frelich, LE and PB Reich. 2010. Tree rings detect earthworm invasions and their effects in northern hardwood forests. *Biol. Invasions* 12:1053–1066.

Latham, M. 1980. Ferrallitization in an oceanian tropical environment. In: KT Joseph (ed). *Proc. Conf. on Classification and Management of Tropical Soils.* Malaysian Society for Soil Science, Kuala Lumpur. pp. 20–26.

Lattman, LH. 1973. Calcium carbonate cementation of alluvial fans in southern Nevada. *GSA Bull.* 84:3013–3028.

Lattman, LH and SK Lauffenburger. 1974. Proposed role of gypsum in the formation of caliche. *ZG Suppl.* 20:140–149.

Lauer, F, Pätzold, S, Gerlach, R, Protze, J, Willbold, S and W Amelung. 2013. Phosphorus status in archaeological arable topsoil relicts—Is it possible to reconstruct conditions for prehistoric agriculture in Germany? *Geoderma* 207–208:111–120.

Laundré, JW. 1989. Estimating soil bulk density with expanding polyurethane foam. *SS* 147:223–224.

Laurent, TE, Graham, RC and KR Tice. 1994. Soils of the red fir forest-barrens mosaic, Siskiyou Mountains crest, California. *SSSAJ* 58:1747–1752.

Lavkulich, LM and JM Arocena. 2011. Luvisolic soils of Canada: Genesis, distribution, and classification. *CJSS* 91:781–806

Lavoie, M, Paré, D, Fenton, N, Groot, A and K Taylor. 2005. Paludification and management of forested peatlands in Canada: A literature review. *Environ. Rev.* 13:21–50.

Lawrence, D and D Foster. 2002. Changes in forest biomass, litter dynamics and soils following shifting cultivation in southern Mexico: An overview. *Interciencia* 27:400–408.

Lawrey, JD. 1977. Elemental partitioning in Pinus resinosa leaf litter and associated fungi. *Mycologia* 69:1121–1128.

Leahy, A. 1963. The Canadian system of soil classification and the Seventh Approximation. *SSSAP* 27:224–225.

Leamy, ML. 1974. The use of pedogenic carbonate to determine the absolute age of soils and to assess rates of soil formation. *Trans. 10th Intl. Congr. Soil Sci.* 6:331–338.

Leavitt, SW and R Bannister. 2009. Dendrochronology and radiocarbon dating: The laboratory of tree-ring research connection. *Radiocarbon* 51:373–384.

Leavitt, SW, Follett, RF, Kimble, JM and EG Pruessner. 2007. Radiocarbon and δ13C depth profiles of soil organic carbon in the U.S. Great Plains: A possible spatial record of paleoenvironment and paleovegetation. *QI* 162/163:21–34.

Lecce, SA. 1997. Spatial patterns of historical overbank sedimentation and floodplain evolution, Blue River, Wisconsin. *Geomorphology* 18:265–277.

Lecomte, P. 1988. Stone line profiles: Importance in geochemical exploration. *J. Geochem. Explor.* 30:35–61.

Lee, BD, Graham, RC, Laurent, TE, Amrhein, C and RM Creasy. 2001. Spatial distributions of soil chemical conditions in a serpentinitic wetland and surrounding landscape. *SSSAJ* 65:1183–1196.

Lee, FY, Yuan, TL and VW Carlisle. 1988b. Nature of cementing materials in ortstein horizons of selected Florida Spodosols. I. Constituents of cementing materials. *SSSAJ* 52:1411–1418.

Lee, GB, Bullington, SW and FW Madison. 1988a. Characteristics of histic materials in Wisconsin as arrayed in four classes. *SSSAJ* 52:1753–1758.

Lee, KE. 1967. Microrelief features in a humid tropical lowland area, New Guinea, and their relation to earthworm activity. *AJSR* 5:263–274.

1985. *Earthworms: Their Ecology and Relationships with Soils and Land Use.* Academic Press, Orlando, FL.

Lee, KE and JHA Butler. 1977. Termites, soil organic matter decomposition and nutrient cycling. *Ecol. Bull.* 25:544–548.

1971b. *Termites and Soils.* Academic Press, New York.

Lee, KE and RC Foster. 1991. Soil fauna and soil structure. *AJSR* 29:745–775.

Lee, KE and TG Wood. 1968. Preliminary studies of the role of *Nasutitermes exitiosus* (Hill) in the cycling of organic matter in a Yellow Podzolic soil under dry sclerophyll forest in South Australia. *Trans. 9th Intl. Congr. Soil Sci.* 2:11–18.

1971a. Physical and chemical effects on soils of some Australian termites and their pedological significance. *Pedobiologia* 11:376–409.

Lee, MR, Hodson, ME and G Langworthy. 2013. Earthworms produce granules of intricately zoned calcite. *Geology* 36:943–946.

Lee, R and A Baumgartner. 1966. The topography and insolation climate of a mountainous forest area. *For. Sci.* 12:258–267.

Leigh, DL. 1994. Roxana silt of the Upper MississippiValley: Lithology, source, and paleoenvironment. *GSA Bull.* 106:430–442.

1996. Soil chronosequence of Brasstown Creek, Blue Ridge Mountains, USA. *Catena* 26:99–114.

Leigh, DL and JC Knox. 1993. AMS radiocarbon age of the Upper Mississippi Valley Roxana Silt. *QR* 39:282–289.

Leigh, DS. 1998. Evaluating artifact burial by eolian versus bioturbation processes, South Carolina sandhills, USA. *Geoarchaeol.* 13:309–330.

2001. Buried artifacts in sandy soils. Techniques for evaluating pedoturbation versus sedimentation. In: P Goldberg, VT Holliday and CR Ferring (eds). *Earth Sciences and Archaeology.* Kluwer, New York. pp. 269–293.

Leighton, MM. 1926. A notable type Pleistocene section: The Farm Creek exposure near Peoria, Illlinois. *J. Geol.* 34:167–174.

1933. The naming of the subdivisions of theWisconsin glacial age. *Science* 77:168.

1958. Principles and viewpoints in formulating the stratigraphic classifications of the Pleistocene. *J. Geol.* 66:700–709.

Leighton, MM and P MacClintock. 1930. Weathered zones of drift sheets of Illinois. *J. Geol.* 38:28–53.

1962. The weathered mantle of glacial tills beneath original surfaces in north-central United States. *J. Geol.* 70:267–293.

Leighton, MM and HB Willman. 1950. Loess formations of the Mississippi Valley. *J. Geol.* 58:599–623.

Leininger, S, Urich, T, Schloter, M, Schwark, L, Qi, J, Nicol, GW, Prosser, JI, Schuster, SC and C Schleper. 2006. Archaea predominate among ammonia-oxidizing prokaryotes in soils. *Nature* 442:806–809.

Leisman, GA. 1957. A vegetation and soil chronosequence on the Mesabi Iron Range spoil banks, Minnesota. *Ecol. Monogr.* 27:221–245.

Lensch, RA 2008. *Soil Survey of Adams County, Iowa*. USDA-NRCS, US Govt. Printing Off., Washington, DC.

Leonard, EM. 1997. The relationship between glacial activity and sediment production: Evidence from a 4450-year varve record of neoglacial sedimentation in Hector Lake, Alberta, Canada. *J. Paleolimnol.* 17:319–330.

Leonardi, G and M Miglavacca. 1999. Soil phosphorus analysis as an integrative tool for recognizing buried ancient ploughsoils. *JAS* 26:343–352.

Leopold, LB and JP Miller. 1954. *A Postglacial Chronology for some Alluvial Valleys in Wyoming, Water Supply Paper #1261. US Geological Survey*, US Govt. Printing Off., Washington, DC.

Leopold, M and J Volkel. 2007. Colluvium: Definition, differentiation, and possible suitability for reconstructing Holocene climate data. *QI* 162:133–140.

Lepage, M. 1972. Recherches écologiques sur une savane sahélienne du Ferlo Septentrional, Sénégal: Données préliminaires sur l'écologie des termites. *Terre et la Vie* 26:384–409.

1973. Recherches écologiques sur une savane sahélienne du Sénégal Septentrional: Termites – répartition, biomasse et récolte de nourriture. *Ann. Univ. Adidjan* 6:139–145.

1984. Distribution, density and evolution of *Macrotermes bellicosus* nests (*Isoptera: Macrotermitinae*) in the north-east of Ivory Coast. *J. Anim. Ecol.* 53:107–118.

Lepsch, IF and SW Buol. 1974. Investigations in an Oxisol-Ultisol toposequence in Sao Paulo state, Brazil. *SSSAP* 38:491–496.

Lepsch, IF, Buol, SW and RB Daniels. 1977a. Soil-landscape relationships in the Occidental Plateau of Sao Paulo state, Brazil. I. Geomorphic surfaces and soil mapping units. *SSSAJ* 41:104–109.

1977b. Soil-landscape relationships in the Occidental Plateau of Sao Paulo state, Brazil. II. Soil morphology, genesis, and classification. *SSSAJ* 41:109–115.

Lesovaya, SN, Goryachkin, SV and YS Polekhovskii. 2012. Soil formation and weathering on ultramafic rocks in the mountainous tundra of the Rai-Iz massif, Polar Urals. *Euras. Soil Sci.* 45:33–44.

Lev, A and RH King. 1999. Spatial variation of soil development in a high arctic soil landscape: Truelove Lowland, Devon Island, Nunavut, Canada. *PPP* 10:289–307.

Levan, MA and EL Stone. 1983. Soil modification by colonies of black meadow ants in a New York old field. *SSSAJ* 47:1192–1195.

Leverett, F. 1939. The place of the Iowan drift. *J. Geol.* 47:398–407.

1898. The weathered zone (Yarmouth) between the Illinoian and Kansan till sheets. *J. Geol.* 6:238–243.

Lévieux, J. 1976. Deux aspects de l'action des fourmis (Hymenoptera, Formicidae) sur le sol d'une savane préforestière de Côte-d'Ivoire. *Bull. Ecol.* 7:283–295.

Levine, ER and EJ Ciolkosz. 1983. Soil development in till of various ages in northeastern Pennsylvania. *QR* 19:85–99.

1986. A computer simulation model for soil genesis applications. *SSSAJ* 50:661–667.

Levine, SJ, Hendricks, DM and JF Schreiber Jr. 1989. Effect of bedrock porosity on soils formed from dolomitic limestone residuum and eolian deposition. *SSSAJ* 53:856–862.

Lewis, DT and JV Drew. 1973. Slick spots in southeastern Nebraska: Patterns and genesis. *SSSAP* 37:600–606.

Lewis, RO. 1981. Use of opal phytoliths in paleoenvironmental reconstruction. *J. Ethnobiol.* 1:175–181.

Li, C, Yang, SY and WG Zhang. 2012. Magnetic properties of sediments from major rivers, aeolian dust, loess soil and desert in China. *J. Asian Earth-Sci.* 45:190–200.

Li, SH. 1994. Optical dating – insufficiently bleached sediments. *Rad. Meas.* 23:563–567.

Li, XL, Liu, XH, Ju, YT and FX Huang. 2003. Properties of soils in Grove Mountains, East Antarctica. *Sci. China Ser. D Earth Sci.* 46:683–693.

Li, XY, Li, BG and YC Shi. 2001. A soil development index and its application to loess-paleosol sequences. *Acta Pedol. Sinica* 38:153–159.

Li, YK and J Harbor. 2009. Cosmogenic nuclides and geomorphology: Theory, limitations, and applications. In: DM Ferrari and AR Guiseppi (eds). *Geomorphology and Plate Tectonics*. Nova Science Publ., Hauppauge, New York. pp. 1–33.

Lian, OB and RG Roberts. 2006. Dating the Quaternary: Progress in luminescence dating of sediments. *QSR* 25:2449–2468.

Libby, WF, Anderson, EC and JR Arnold. 1949. Age determination by radiocarbon content: World-wide assay of natural radiocarbon. *Science* 109: 227–228.

Lichte, M and H Behling. 1999. Dry and cold climatic conditions in the formation of the present landscape in Southeastern Brazil. An interdisciplinary approach to a controversially discussed topic. *ZG* 43:341–358.

Lichter, J. 1998. Rates of weathering and chemical depletion in soils across a chronosequence of Lake Michigan sand dunes. *Geoderma* 85:255–282.

Lide, DR (ed) 2005. *CRC Handbook of Chemistry and Physics*, 86th ed. CRC Press, Boca Raton, FL.

Liebens, J. 1999. Characteristics of soils on debris aprons in the Southern Blue Ridge, North Carolina. *PG* 20:27–52.

1999. Characteristics of soils on debris aprons in the Southern Blue Ridge, North Carolina. *PG* 20:27–52.

Liebens, J and RJ Schaetzl. 1997. Relative-age relationships of debris flow deposits in the Southern Blue Ridge, North Carolina. *Geomorphology* 21:53–67.

Liegh, DS. 1998. A 12 000-year record of natural levee sedimentation along the Broad River near Columbia, South Carolina. *Southeast. Geog.* 28:95–111.

Likens, GE. 2001. Biogeochemistry, the watershed approach: Some uses and limitations. *Mar. Freshw. Res.* 52:5–12.

Likens, GE, Driscoll, CT, Buso, DC, Siccama, TG, Johnson, CE, Lovett, GM, Fahey, TJ, Reiners, WA, Ryan, DF, Martin, CW and SW Bailey. 1998. The biogeochemistry of calcium at Hubbard Brook. *Biogeochemistry* 41:89–173.

Lima, HN, Schaefer, CER, Mello, JWV, Gilkes, RJ and JC Ker. 2002. Pedogenesis and pre-Colombian land use of "Terra Preta Anthrosols" ("Indian black earth") of Western Amazonia. *Geoderma* 110:1–17.

Lin, H. 2006. Hydropedology and modern soil survey applications. *SSH* 47:18–22.

2010. Earth's Critical Zone and hydropedology: Concepts, characteristics, and advances. *Hydrology Earth System Sci.* 14:25–45.

Lin, H, Bouma, J, Wilding, L, Richardson, J, Kutilek, M and D Nielsen. 2005. Advances in hydropedology. *Adv. Agron.* 85:1–89.

Lin, HS, McInnes, KJ, Wilding, LP and CT Hallmark. 1999. Effects of soil morphology on hydraulic properties. I. Quantification of soil morphology. *SSSAJ* 63:948–954.

Lindbo, DL and FE Rhoton. 1996. Slaking in fragipan and argillic horizons. *SSSAJ* 60:552–554.

Lindbo, DL, Rhoton, FE, Hudnall, WH, Smeck, NE, Bigham, JM and DD Tyler. 2000. Fragipan degradation and nodule formation in Glossic Fragiudalfs of the Lower Mississippi Valley. *SSSAJ* 64:1713–1722.

Lindbo, DL and PLM Veneman. 1993a. Micromorphology of selected Massachusetts fragipan soils. *SSSAJ* 57:437–442.

PLM Veneman. 1993b. Morphological and physical properties of selected fragipan soils in Massachusetts. *SSSAJ* 57:429–436.

Lindeburg, KS, Almond, P, Roering, JJ and OA Chadwick. 2013. Pathways of soil genesis in the Coast Range of Oregon, USA. *Plant Soil*. doi:10.1007/s11104-012-1566-z

Lindsay, WL, Vlek, PLG and SH Chien. 1989. Phosphate minerals. In: JB Dixon and SB Weed (eds). *Minerals in Soil Environments*, 2nd ed. Soil Sci. Soc. Am., Madison, WI. pp. 1089–1130.

Linick, TW, Damon, PE, Donahue, DJ and AJT Jull. 1989. Accelerator mass spectrometry: The new revolution in radiocarbon dating. *QJ* 1:1–6.

Liptzin, D and TR Seastedt. 2010. Regional and local patterns of soil nutrients at Rocky Mountain treelines. *Geoderma* 160:208–217.

Liritzis, I and N Laskaris. 2011. Fifty years of obsidian hydration dating in archaeology. *J. Non-Cryst. Solids* 357:2011–2023.

Litaor, MI. 1987. The influence of eolian dust on the genesis of alpine soils in the Front Range, Colorado. *SSSAJ* 51:142–147.

1992. Aluminum mobility along a geochemical catena in an alpine watershed, Front Range, Colorado. *Catena* 19:1–16.

Litaor, MI, Manicelli, R and JC Halfpenny. 1996. The influence of pocket gophers on the status of nutrients in alpine soils. *Geoderma* 70:37–48.

Litaor, MI, Williams, M and TR Seastedt. 2008. Topographic controls on snow distribution, soil moisture, and species diversity of herbaceous alpine vegetation, Niwot Ridge, Colorado. *J. Geophys. Res. Biogeosci.* 113. doi:10.1029/2007JG000419

Little, IP and WT Ward. 1981. Chemical and mineralogical trends in a chronosequence developed on alluvium in eastern Victoria. Australia. *Geoderma* 25:173–188.

Liu, B, Phillips, FM, Pohl, MM and P Sharma. 1996. An alluvial surface chronology based on cosmogenic 36Cl dating, Ajo Mountains (Organ Pipe Cactus National Monument), southern Arizona. *QR* 45:30–37.

Liu, BY, Nearing, MA, Shi, PJ and ZW Jia. 2000. Slope length effects on soil loss for steep slopes. *SSSAJ* 64:1759–1763.

Liu, C-Q, Masuda, A, Okada, A, Yabuki, S, Zhang, J and Z-L Fan. 1993. A geochemical study of loess and desert sand in northern China: Implications for continental crust weathering and composition. *Chem. Geol.* 106:359–374.

Liu, HY, He, SY, Anenkhonov, OA, Hu, GZ, Sandanov, DV and NK Badmaeva. 2012. Topography-controlled soil water content and the coexistence of forest and steppe in northern China. *PG* 33:561–573.

Liu, K-B. 1990. Holocene paleoecology of the boreal forest and Great Lakes-St. Lawrence forest in northern Ontario. *Ecol. Monogr.* 60:179–212.

Liu, K-B and NSN Lam. 1985. Paleovegetational reconstruction based on modern and fossil pollen data: An application of discriminant analysis. *AAAG* 75:115–130.

Liu, TS. 1985. *Loess and the Environment*. China Ocean Press, Beijing.

Liu, TZ and WS Broecker. 2000. How fast does rock varnish grow? *Geology* 28:183–186.

Liu, T and WS Broecker. 2007. Holocene rock varnish microstratigraphy and its chronometric application in drylands of western USA. *Geomorphology* 84:1–21.

2008. Rock varnish evidence for latest Pleistocene millennial-scale wet events in the drylands of western United States. *Geology* 36:403–406.

2008. Rock varnish microlamination dating of late Quaternary geomorphic features in the drylands of the western USA. *Geomorphology* 93:501–523.

Liu, X, Monger, HC and WG Whitford. 2007. Calcium carbonate in termite galleries – Biomineralization or upward transport? *Biogeochem.* 82:241–250.

Liu, X, Rolph, T, Bloemendal, J, Shaw, J and T Liu. 1995. Quantitative estimates of palaeoprecipitation at Xifeng, in the Loess Plateau of China. *PPP* 113:243–248.

Liu, XF, Wang, QF, Deng, J, Zhang, QZ, Sun, SL and JY Meng. 2010. Mineralogical and geochemical investigations of the Dajia Salento-type bauxite deposits, western Guangxi, China. *J. Geochem. Explor.* 105:137–152.

Liu, XJ and ZP Lai. 2013. Optical dating of sand wedges and ice-wedge casts from Qinghai Lake area on the northeastern Qinghai-Tibetan Plateau and its palaeoenvironmental implications. *Boreas* 42:333–341.

Liu, Y, Steenhuis, TS, Parlange, J-Y and JS Selker. 1991. *Hysteretic finger phenomena in dry and wetted sands*. In: Preferential Flow. Am. Soc. Agric. Engineers, St. Joseph, MI. pp. 160–172.

Livens, PJ. 1949. Characteristics of some soils of the Belgian Congo. *Commonw. Bur. Soil Sci. Tech. Pub.* 46:29–35.

Lobry de Bruyn, AL and AJ Conacher. 1990. The role of termites and ants in soil modification: A review. *Austral. J. Soil Res.* 28:55–93.

Lock, WW, Andrews, PJ and JT Webber. 1979. A manual for lichenometry. *Br. Geomorph. Res. Group Tech. Bull.* 26:1–25.

Locke, WW. 1979. Etching of hornblende grains in Arctic soils: An indicator of relative age and paleoclimate. *QR* 11:197–212.

Loendorf, LL. 1991. Cation-ratio varnish dating and petroglyph chronology in southeastern Colorado. *Antiquity* 65:246–255.

Löffler, E and C Margules. 1980. Wombats detected from space. *Remote Sens. Env.* 9:47–56.

Logan, WB. 2007. *Dirt: The Ecstatic Skin of the Earth*. Riverhead Books, New York.

Lohr, SC, Grigorescu, M, Hodgkinson, JH, Cox, ME and SJ Fraser. 2010. Iron occurrence in soils and sediments of a coastal catchment: A multivariate approach using self organising maps. *Geoderma* 156:253–266.

Loizeau, JL, D Arbouille, S Santiago and JP Vernet. 1994. Evaluation of a wide range laser diffraction grain size analyzer for use with sediments. *Sedimentology.* 41:353–361.

Lokaby, BG and JC Adams. 1985. Pedoturbation of a forest soil by fire ants. *SSSAJ* 49:220–223.

Loope, WL, Loope, HM, Goble, RJ, Fisher, TG, Lytle, DE, Legg, RJ, Wysocki, DA, Hanson, PR and AR Young. 2012. Drought drove forest decline and dune building in eastern USA, as the upper Great Lakes became closed basins. *Geology* 40:315–318.

Lorz, C and JD Phillips. 2006. Pedo-ecological consequences of lithological discontinuities in soils – examples from Central Europe. *J. Plant Nutr. Soil Sci.* 169:573–581.

Losche, CK, McCracken, RJ and CB Davey. 1970. Soils of steeply sloping landscapes in the southern Appalachian Mountains. *SSSAP* 34:473–478.

Lotspeich, FB and HW Smith. 1953. Soils of thePalouse loess. I. The Palouse catena. *SS* 76:467–480.

Lousier, JD and D Parkinson. 1978. Chemical element dynamics in decomposing leaf litter. *Can. J. Bot.* 56:2795–2812.

Lovegrove, BG and WR Siegfried. 1986. Distribution and formation of Mima-like earth mounds in the western Cape Province of South Africa. *S. Afr. J.Sci.* 82:432–436.

Lovis, WA. 1976. Quarter sections and forests: An example of probability sampling in the Northeastern woodlands. *Am. Antiq.* 41:364–372.

Lowdermilk, WC and HL Sundling. 1950. Erosion pavement, its formation and significance. *TAGU* 31:96–100.

Lowe, DJ. 2000. Upbuilding pedogenesis in multisequal tephra-derived soils in the Waikato Region. In Soil 2000: New Horizons for a New Century. *Australian and New Zealand 2nd Joint Soils Conf.* 3:183–184. N.Z. Soc. Soil Science.

2011. Tephrochronology and its application: A review. *QG* 6:107–153.

Lowe, JJ and MJC Walker. 1997. *Reconstructing Quaternary Environments*, 2nd ed. Longman, London.

Lu, H and Z An. 1998. *Paleoclimatic significance of grain size of loess-palaeosol deposit in Chinese Loess Plateau*. Sci. China 41D:626–631.

Lu, H and D Sun. 2000. Pathways of dust input to the Chinese Loess Plateau during the last glacial and interglacial periods. Catena 40:251–261.

Ludwig, KR and JB Paces. 2002. Uranium-series dating of pedogenic silica and carbonate, Crater Flat, Nevada. Geochim. Cosmochim. Acta 66:487–506.

Luehmann, MD, Schaetzl, RJ, Miller, BA and M Bigsby. 2013. Thin, pedoturbated and locally sourced loess in the western Upper Peninsula of Michigan. AR 8:85–100.

Lukas, S, Spencer, JQG, Robinson, RAJ and AJ Benn. 2007. Problems associated with luminescence dating of Late Quaternary glacial sediments in the NW Scottish Highlands. QG 2:243–248.

Lundqvist, J and K Bengtsson. 1970. The red snow: A meteorological and pollen analytic study of longtransported material from snowfalls in Sweden. Geol. Foreningens i Stockholm Forhandl. 92:288–301.

Lundström, US. 1993. The role of organic acids in soil solution chemistry in a podzolized soil. JSS 44:121–133.

Lundström, US, van Breemen and N Bain. 2000b. The podzolization process: A review. Geoderma 94:91–107.

Lundström, US, van Breemen, N, Bain, DC, van Hees, PAW, Giesler, R, Gustafsson, JP, Ilvesniemi, H, Karltun, E, Melkerud, P-A, Olsson, M, Riise, G, Wahlberg, O, Bergelin, A, Bishop, K, Finlay, R, Jongmans, AG, Magnusson, T, Mannerkoski, H, Nordgren, A, Nyberg, L, Starr, M and LT Strand. 2000a. Advances in understanding the podzolization process resulting from a multidisciplinary study of three coniferous forest soils in the Nordic Countries. Geoderma 94:335–353.

Lundström, US, van Breemen, N and AG Jongmans. 1995. Evidence for microbial decomposition of organic acids during podzolization. EJSS 46:489–496.

Luque, A, Ruiz-Agudo, E, Cultrone, G, Sebastian, E and S Siegesmund. 2011. Direct observation of microcrack development in marble caused by thermal weathering. Environ. Earth-Sci. 62:1375–1386.

Lusch, DP, Stanley, KE, Schaetzl, RJ, Kendall, AD, van Dam, RL, Nielsen, A, Blumer, BE, Hobbs, TC, Archer, JK, Holmstadt, JLF and CL May. 2009. Characterization and mapping of patterned ground in the Saginaw Lowlands, Michigan: Possible evidence for Late-Wisconsin permafrost. AAAG 99:445–466.

Lutz, HJ. 1940. Disturbance of Forest Soil Resulting from the Uprooting of Trees. Yale Forestry Bull 45.

1960. Movement of rocks by uprooting of forest trees. AJS 258:752–756.

1963. Importance of Ants to Brown Podzolic Soil Genesis in New England. Harvard Forest Paper 7.

1974. Narrow soils and intricate soil patterns in southern New England. Geoderma 11:195–208.

Lyford, WH. 1938. Horizon variations of three New Hampshire Podzol profiles. SSSAP 2:242–246.

1963. Importance of ants to brown podzolic soil genesis in New England. Harvard Forest Pap. 7.

1974. Narrow soils and intricate soil patterns in southern New England. Geoderma 11:195–208.

Lyford, WH and DW MacLean. 1966. *Mound and Pit Microrelief in Relation to Soil Disturbance and Tree Distribution in New Brunswick, Canada*. Harvard Forest Paper 15.

Lyford, WH and T Troedsson. 1973. Fragipan horizons in soils on moraines near Garpenburg, Sweden. Stud. Fort. Suecica 108:1–21.

Lynn, W and D Williams. 1992. The making of a Vertisol. SSH 33:45–50.

Lyons, A and J Lyons. 1979. Cradle-knolls. Wis. Nat. Resources Bull. 3:10–11.

Lyttle, A, Yoo, K, Hale, C, Aufdenkampe, A and A Sebestyen. 2011. Carbon-mineral interactions along an earthworm invasion gradient at a sugar maple forest in Northern Minnesota. Applied Geochem. 26:S85–S88.

Mabbutt, JA. 1965. Stone distribution in a stony tableland soil. AJSR 3:131–142.

MacDougall, AH, Avis, CA and AJ Weaver. 2012. Significant contribution to climate warming from the permafrost carbon feedback. Nature Geosci. 5:719–721.

Macedo, J and RB Bryant. 1987. Morphology, mineralogy, and genesis of a hydrosequence of Oxisols in Brazil. SSSAJ 51:690–698.

Macgregor, AN and DE Johnson. 1971. Capacity of desert algal crusts to fix atmospheric nitrogen. SSSAP 35:843–844.

Machette, MN. 1985. Calcic soils of the southwestern United States. GSA Spec. Paper 203:1–21.

Machette, MN, Birkeland, PW, Markos, G and MJ Guccione. 1976. Soil development in Quaternary deposits in the Golden-Boulder portion of the Colorado Piedmont. Colo. School Mines Prof. Contrib. 8:217–259.

Mack, GH, James, WC and HC Monger. 1993. Classification of paleosols. GSA Bull. 105:129–136.

Mackay, AD, Syers, JK, Springett, JA and PEM Greg. 1982. Plant availability of phosphorous in superphosphate and a phosphate rock as influenced by earthworms. Soil Biol. Biochem. 14:281–287.

Mackay, JR. 1980. The origin of hummocks, western arctic coast, Canada. CJES 17:996–1006.

1983. Downward water movement into frozen ground, western arctic coast, Canada. CJES 20:120–134.

1984. The frost heave of stones in the active layer above permafrost with downward and upward freezing. AAR 16:439–446.

Mackay, JR and C Burrous. 1979. Uplift of objects by an upfreezing ice surface. Can. Geotech. J. 16:609–613.

Mackay, JR and DK Mackay. 1976. Cryostatic pressures in non-sorted circles (mud hummocks), Inuvik, Northwest Territories. CJES 13:889–897.

Mackay, JR and JV Matthews. 1983. Pleistocene ice and sand wedges, Hooper Island, Northwest Territories. CJES 20:1087–1097.

MacKinney, AL. 1929. Effects of forest litter on soil temperature and soil freezing in autumn and winter. Ecology 10:312–321.

Mackintosh, EE and J van der Hulst. 1978. Soil drainage classes and soil water table relations in medium and coarse textured soils in southern Ontario. CJSS 58:287–301.

Mackney, D. 1961. A podzol development sequence in oakwoods and heath in central England. JSS 12:23–40.

Maclean, AH and S Pawluk. 1975. Soil genesis in relation to groundwater and soil moisture regimes near Vegreville, Alberta. JSS 26:278–293.

MacLeod, DA. 1980. The origin of the redMediterranean soils in Epirus, Greece. JSS 31:125–136.

MacNamara, EE. 1969. Soils and geomorphic surfaces in Antarctica. Biul. Peryglac. 20:299–320.

Macyk, TM, S Pawluk and JD Lindsay. 1978. Relief and microclimate as related to soil properties. CJSS 58:421–438.

Mader, DL. 1963. Soil variability: A serious problem in soil-site studies in the Northeast. SSSAP 27:707–709.

Mader, WF and EJ Ciolkosz. 1997. The effects of periglacial processes on the genesis of soils on an unglaciated northern Appalachian Plateau landscape. SSH 38:19–30.

Madsen, AT and AS Murray. 2009. Optically stimulated luminescence dating of young sediments: A review. Geomorphology 109:3–16.

Maejima, Y, Matsuzaki, H and T Higashi. 2005. Application of cosmogenic [10]Be to dating soils on the raised coral reef terraces of Kikai Island, southwest Japan. Geoderma 126:389–399.

Magaritz, M and AJ Amiel. 1980. Calcium carbonate in a calcareous soil from the Jordan Valley, Israel: Its origin as revealed by the stable carbon isotope method. *SSSAJ* 44:1059–1062.

Mahaney, WC. (ed). 1978. *Quaternary Soils*. Geo Abstracts, Norwich, England.

(ed). (ed) 1984. *Quaternary Dating Methods*. Elsevier, New York.

Mahaney, WC, Krinsley, DH, Allen, CCR, Langworthy, K, Ditto, J and MW Milner. 2012. Weathering Rinds: Archives of Paleoenvironments on Mount Kenya, East Africa. *J. Geol.* 120:591–602.

Mahendrappa, MK and DGO Kingston. 1982. Prediction of throughfall quantities under different forest stands. *CJFR* 12:474–481.

Maher, BA. 1998. Magnetic properties of modern soils and Quaternary loessic paleosols: Paleoclimatic implications. *PPP* 137:25–54.

Maher, BA, Mutch, TJ and D Cunningham. 2009. Magnetic and geochemical characteristics of Gobi Desert surface sediments: Implications for provenance of the Chinese Loess Plateau. *Geology* 37:279–282.

Maher, BA and R Thompson. 1991. Mineral magnetic record of the Chinese loess and paleosols. *Geology* 19:3–6.

1995. Paleorainfall reconstructions from pedogenic magnetic susceptibility variations in Chinese loess and paleosols. *QR* 44:383–391.

Maher, BA, Thompson, R and LP Zhou. 1994. Spatial and temporal reconstructions of changes in the Asian palaeomonsoon: A new mineral magnetic approach. *EPSL* 125:461–471.

Maher, LJ Jr. 1972. Absolute pollen diagram of Redrock Lake, Boulder County, Colorado. *QR* 2:531–553.

Maizels, JK and JR Petch. 1985. Age determination of intermontane areas, Austerdalen, Southern Norway. *Boreas* 14:51–65.

Maldague, ME. 1964. Importance des populations de termites dans les sols equatoriaux. *Trans. 8th Intl. Congr. Soil Sci.* 3:743–751.

Malde, HE. 1964. Patterned ground on the westernSnake River plain, Idaho and its possible cold-climate origin. *GSA Bull.* 75:191–207.

Mallah, PB. 2002. Vermiculites. In: JB Dixon and D Schulze (eds). *Soil Mineralogy with Environmental Applications*. Soil Sci. Soc. Am. Book Ser. 7. Soil Sci. Soc. Am., Madison, WI. pp. 501–530.

Malo, DD, Worcester, BK, Cassel, DK and KD Matzdorf. 1974. Soil-landscape relationships in a closed drainage system. *SSSAP* 38:813–818.

Malterer, TJ, Verry, ES and J Erjavec. 1992. Fiber content and degree of decomposition in peats: Review of national methods. *SSSAJ* 56:1200–1211.

Mammerickx, J. 1964. Quantitative observations on pediments in the Mojave and Sonoran Deserts (southwestern United States). *A. J. Sci.* 262:417–435.

Mandel, RD. 1994. Holocene Landscape Evolution in the Pawnee River Valley, Southwestern Kansas. Kansas Geol. Survey Bull. 236.

1995. Geomorphic controls of the Archaic record in the Central Great Plains of the United States. *GSA Spec. Paper* 297:37–66.

2008. Buried paleoindian-age landscapes in stream valleys of the central plains, USA. *Geomorphology* 101:342–361.

Mandel, RD and EA Bettis. 2001. Use and analysis of soils by archaeologists and geoscientists. In: P Goldberg, VT Holliday, and CR Ferring (eds). *Earth Sciences and Archaeology*. Plenum Press, New York. pp. 173–204.

Mandel, RD and CJ Sorenson. 1982. The role of the Western Harvester ant (*Pogonomyrmex occidentalis*) in soil formation. *SSSAJ* 46:785–788.

Manikowska, B. 1982. Upfreezing of stones in boulder clay of central and north Poland. *Biul. Peryglac.* 29:87–115.

Mankinen, EA and GB Dalrymple. 1979. Revised geomagnetic polarity timescale for the interval 0–5 My BP. *J. Geophys. Res.* B84:615–626.

Manning, G, Fuller, LG, Eilers, RG and I Florinsky. 2001. Topographic influence on the variability of soil properties within an undulating Manitoba landscape. *CJSS* 81:439–447.

Manrique, LA and CA Jones. 1991. Bulk density of soils in relation to soil physical and chemical properties. *SSSAJ* 55:476–481.

Mansberg, L and TR Wentworth. 1984. Vegetation and soils of a serpentine barren in western North Carolina. *Bull. Torrey Bot. Club* 111:273–286.

Manzoni, S and A Porporato. 2009. Soil carbon and nitrogen mineralization: Theory and models across scales. *Soil Biol. Biochem.* 41:1355–1379.

Marbut, CF. 1923. Soils of the Great Plains. *AAAG* 13:41–66.

1927a. A scheme for soil classification. *Proc. 1st Intl. Congr. Soil Sci.* 4:1–31.

1927b. *The Great Soil Groups of the World and Their Development*. Ann Arbor, MI, Edwards Bros. (Trans. from Glinka [1914]).

1935. *Soils of the United States*. III. In: Atlas of American Agriculture. US Dept. of Agric., Washington, DC. pp. 1–98.

Marbut, CF, Bennett, HH and JE Lapham. 1913. *Soils of the United States, US Bureau of Soils Bull*. 96. US Govt. Printing Off., Washington, DC.

Marchetti DW and TE Cerling. 2005. Cosmogenic ^{3}He exposure ages of Pleistocene debris flows and desert pavements in Capitol Reef National Park, Utah. *Geomorphology* 67:423–435.

Marcotullio, PJ. 2011. Urban soils. In: DI Goode, M Houck, and R Wang (eds). *Routledge Handbook of Urban Ecology*. Routledge, London. pp. 164–186.

Marin-Spiotta, E, Chadwick, OA, Kramer, M and MS Carbone. 2011. Carbon delivery to deep mineral horizons in Hawaiian rain forest soils. *J. Geophys. Res. Biogeosci.* 116. doi:10.1029/2010JG001587

Marion, GW, Schlesinger, WH and PJ Fonteyn. 1985. CALDEP: A regional model for soil $CaCO_3$ (caliche) deposition in Southwestern deserts. *SS* 139:468–481.

Markewich, HW and MJ Pavich. 1991. Soil chronosequence studies in temperate to subtropical, low-latitude, low-relief terrain with data from the eastern United States. *Geoderma* 51:213–239.

Markewich, HW, Pavich, MJ, Mausbach, MJ, Johnson, RG and VM Gonzalez. 1989. A guide for using soil and weathering profile data in chronosequence studies of the coastal plain of the eastern United States. USGS Bull. 1589.

Markewich, HW, Wysocki, DA, Pavich, MJ and EM Rutledge. 2011. Age, genesis, and paleoclimatic interpretation of the Sangamon/Loveland complex in the Lower Mississippi Valley, U.S.A. *GSA Bull.* 123:21–39.

Markewich, HW, Wysocki, DA, Pavich, MJ, Rutledge, EM, Millard, HT Jr, Rich, FJ, Maat, PB, Rubin, M and JP McGeehin. 1998. Paleopedological plus TL, ^{10}Be and ^{14}C dating as tools in stratigraphic and paleoclimatic investigations, Mississippi River Valley, USA. *QI* 51/52:143–167.

Marković, SB, Hambach, U, Catto, N, Jovanović, M, Buggle, B, Machalett, B, Zöller, L, Glaser, B and M Frechen. 2009. Middle and Late Pleistocene loess sequences at Batajnica, Vojvodina, Serbia. *QI* 198:255–266.

Marques, EAG., Barroso, EV, Menezes, AP and ED Vargas. 2010. Weathering zones on metamorphic rocks from Rio de Janeiro-Physical, mineralogical and geomechanical characterization. *Eng. Geol.* 111:1–18.

Marron, DC and JH Popenoe. 1986. A soil catena on schist in northwestern California. *Geoderma* 37:307–324.

Marsan, FA and J Torrent. 1989. Fragipan bonding by silica and iron oxides in a soil from northwest Italy. *SSSAJ* 53:1140–1145.

Marshall, CE and JF Haseman. 1942. The quantitative evaluation of soil formation and development by heavy mineral studies: A Grundy silt loam study. *SSSAP* 7:448–453.

Marston, RA. 2003. Editorial note. *Geomorphology* 53:197.

Martel, YA and AF Mackenzie. 1980. Long-term effects of cultivation and land use on soil quality in Quebec. *CJSS* 60:411–420.

Martel, YA and EA Paul. 1974. Effects of cultivation on the organic matter of grassland soils as determined by fractionation and radiocarbon dating. *CJSS* 54:419–426.

Martin, A and JCY Marinissen. 1993. Biological and physico-chemical processes in excrement of soil animals. *Geoderma* 56:331–347.

Martin, A, Mariotti, A, Balesdent, J, Lavelle, P and R Vouattoux. 1990. Estimates of the organic matter turnover rate in a savanna soil by 13C natural abundance. *Soil Biol. Biochem.* 22:517–523.

Martin, CW and WC Johnson. 1995. Variation in radiocarbon ages of soil organic matter fractions from Late Quaternary buried soils. *QR* 43:232–237.

Martin, JP and K Haider. 1971. Microbial activity in relation to soil humus formation. *SS* 111:54–63.

Martin, JP and SA Waksman. 1940. Influence of microorganisms on soil aggregation and erosion. *SS* 50:29–47.

Martinelli, N. 2004. Climate from dendrochronology: Latest developments and results. *Global Planet. Change* 40:129–139.

Martini, JA. 1970. Allocation of cation exchange capacity to soil fractions in seven surface soils from Panama and the application of a cation exchange factor as a weathering index. *SS* 109:324–331.

Martini, JA and M Macias. 1974. A study of six "Latosols" from Costa Rica to elucidate the problems of classification, productivity and management of tropical soils. *SSSAP* 38:644–652.

Martinson, DG, Pisias, NG, Hays, JD, Imbrie, J, Moore, TC Jr and NJ Shackleton. 1987. Age dating and the orbital theory of the Ice Ages: Development of a high resolution 0–300 000 yr chronostratigraphy. *QR* 27:1–29.

Martius, C. 1994. Diversity and ecology of termites in Amazonian forests. *Pedobiologia* 38:407–428.

Marye, WB. 1955. The great Maryland barrens. *Maryland Hist. Mag.* 50:11–23, 120–142, 234–253.

Mason, JA. 2001. Transport direction of Peoria loess in Nebraska and implications for loess source areas on the central Great Plains. *QR* 56:79–86.

Mason, JA and PM Jacobs. 1998. Chemical and particle-size evidence for addition of fine dust to soils of the midwestern United States. *Geology* 26:1135–1138.

1999. High resolution particle size analysis as a tool for interpreting incipient soils in loess. *Chinese Sci. Bull.* 44:70–74

Mason, JA, Joeckel, RM and EA Bettis III. 2007. Middle to Late Pleistocene loess record in eastern Nebraska, USA, and implications for the unique nature of Oxygen Isotope Stage 2. *QSR* 26:773–792.

Mason, JA and JC Knox. 1997. Age of colluvium indicates accelerated late Wisconsinan hillslope erosion in the Upper Mississippi Valley. *Geology* 25:267–270.

Mason, JA, Miao, X, Hanson, PR, Johnson, WC, Jacobs, PM and RJ Goble. 2008. Loess record of the Pleistocene–Holocene transition on the northern and central Great Plains, USA. *QSR* 27:1772–1783.

2008. Loess record of the Pleistocene–Holocene transition on the northern and central Great Plains, USA. *QSR* 27:1772–1783.

Mason, JA and EA Nater. 1994. Soil morphology-Peoria loess grain size relationships, southeastern Minnesota. *SSSAJ* 58:432–439.

Masson, PH. 1949. Circular soil structures in northeast California. *Calif. Div. Mines Bull.* 151:61–71.

Materechera, SA, Dexter, AR and AM Alston. 1992. Formation of aggregates by plant roots in homogenised soils. *Plant Soil* 142:69–79.

Mathé, PE, Rochette, P, Vandamme, D and F Colin. 1999. Volumetric changes in weathered profiles: Iso-element mass balance method questioned by magnetic fabric. *EPSL* 167:255–267.

Matmon, A, Simhai, O, Amit, R, Haviv, I, Porat, N, McDonald, E, Benedetti, L and R Finkel. 2009. Desert pavement-coated surfaces in extreme deserts present the longest-lived landforms on Earth. *GSA Bull.* 121:688–697.

Matsumoto, T. 1976. The role of termites in an equatorial rain forest ecosystem of West Malaysia. I. Population density, biomass, carbon, nitrogen and calorific content and respiration rate. *Oecologia* 22:153–178.

Matsuoka, N. 1995. Rock weathering processes and landform development in the Sor-Rondane Mountains, Antarctica. *Geomorphology* 12:323–339.

2001. Microgelivation versus macrogelivation: Towards bridging the gap between laboratory and field frost weathering. *Perm. Periglac. Proc.* 12:299–313.

2008. Frost weathering and rockwall erosion in the southeastern Swiss Alps: Long-term (1994–2006) observations. *Geomorphology* 99:353–368.

Matsuoka, N and J Murton. 2008. Frost weathering: Recent advances and future directions. *Perm. Periglac. Proc.* 19:195–210.

Matthews, JA. 1974. Families of lichenometric dating curves from the Storbreen gletschervorfeld, Jotunheimen, Norway. *Norsk Geog. Tidsskr.* 28:215–235.

1975. Experiments on the reproducibility and reliability of lichenometric dates, Storbreen gletschervorfeld, Jotunheimen, Norway. *Norsk Geog. Tidsskr.* 29:97–109.

1980. Some problems and implications of 14C dates from a Podzol buried beneath an end moraine at Haugabreen, southern Norway. *Geog. Ann.* 62A:185–208.

Matthews, JA and PQ Dresser. 1983. Intensive 14C dating of a buried palaeosol horizon. *Geol. Foreningens i Stockholm Forhandl.* 105:59–63.

Matthews, JA and RA Shakesby. 1984. The status of the Little Ice-Age in Southern-Norway: Relative-age dating of neoglacial moraines with Schmidt hammer and lichenometry. *Boreas* 13:333–346.

Matthews, JA and HE Trenbirth. 2011. Growth rate of a very large crustose lichen (*Rhizocarpon* subgenus) and its implications for lichenometry. *Geog. Annaler* 95A:27–39.

Matthews, JM. 1965. Stratigraphic disturbance: The human element. *Antiquity* 39:295–298.

Mattson, S and H Lönnemark. 1939. The pedography of hydrologic podsol series. I. Loss on ignition, pH and amphoteric reactions. *Ann. Agric. Coll. Sweden* 7:185–227.

Matzdorf, KD. 1994. *Soil Survey of Faribault County, Minnesota*. USDA Soil Cons. Service, US Govt. Printing Off., Washington, DC.

Mausbach, MJ and JL Richardson. 1994. Biogeochemical processes in hydric soil formation. *Curr. Topics Wetland Biogeochem.* 1:68–127.

Mausbach, MJ, Wingard, RC and EE Gamble. 1982. Modification of buried soils by postburial pedogenesis, southern Indiana. *SSSAJ* 46:364–369.

Mayer, JH, Burr, GS and VT Holliday. 2008. Comparisons and interpretations of charcoal and organic matter radiocarbon ages from buried soils in north-central Colorado, USA. *Radiocarbon* 50:331–346.

Mayer, L, McFadden, LD and JW Harden. 1988. Distribution of calcium carbonate in desert soils: A model. *Geology* 16:303–306.

Mazor, G, Kidron, GJ, Vonshak, A and A Abelovich. 1996. The role of cyanobacterial exopolysaccharides in structuring desert microbial crusts. *FEMS Microbiol. Ecol.* 21:121–130.

Mbila, MO, Thompson, ML, Mbagwu, JSC and DA Laird. 2001. Distribution and movement of sludge-derived trace metals in selected Nigerian soils. *J. Environ. Qual.* 30:1667–1674.

McAuliffe, JR. 1994. Landscape evolution, soil formation, and ecological patterns and processes in Sonoran Desert bajadas. *Ecol. Mon.* 64:111–148.

McBratney, AB, Bishop, TFA and IS Teliatnikov. 2000. Two soil profile reconstruction techniques. *Geoderma* 97:209–221.

McBratney, AB, Mendonça Santos, ML and B Minasny. 2003. On digital soil mapping. *Geoderma* 117:3–52.

McBratney, AB, Odeh, IOA, Bishop, TFA, Dunbar, MS and TM Shatar. 2000. An overview of pedometric techniques for use in soil survey. *Geoderma* 97:293–327.

McBrearty, S. 1990. Consider the humble termite: Termites as agents of post-depositional disturbance at African archaeological sites. *JAS* 17:111–143.

McBride, MB. 1994. *Environmental Chemistry of Soils.* Oxford Univ. Press, New York.

McBurnett, SL and DP Franzmeier. 1997. Pedogenesis and cementation in calcareous till in Indiana. *SSSAJ* 61:1098–1104.

McCahon, TJ and LC Munn. 1991. Soils developed in Late Pleistocene till, Medicine Bow Mountains, Wyoming. *SS* 152:377–388.

McCaleb, SB. 1954. Profile studies of normal soils ofNew York. IV. Mineralogical properties of the Gray-Brown Podzolic-Brown Podzolic soil sequence. *SS* 77:319–333.

McCalpin, JP and ME Berry. 1996. Soil catenas to estimate ages of movements on normal fault scarps, with an example from the Wasatch fault zone, Utah, USA. *Catena* 27:265–286.

McCarroll, D. 1987. The Schmidt Hammer in geomorphology: Five sources of instrument error. *BGRG Technical Bulletin* 36:16–27.

1989. Schmidt hammer relative-age evaluation of a possible pre-"Little Ice Age" Neoglacial moraine. *Leirbreen, southern Norway. Norsk Geol. Tidsskrift* 69:125–130.

1991. The Schmidt hammer, weathering and rock surface-roughness. *ESPL* 16:477–480.

McCarthy, DP and N Henry. 2012. Measurement of growth in the lichen *Rhizocarpon geographicum* using a new photographic technique. *Lichenologist* 44:679–693.

McCarthy, DP and DJ Smith. 1995. Growth curves for calcium-tolerant lichens in the Canadian Rocky Mountains. *AAR* 27:290–297.

McCarthy, DP and K Zaniewski. 2001. Digital analysis of lichen cover: A technique for use in lichenometry and lichenology. *AAAR* 33:107–113.

McCarthy, JJ (ed). 2001. *Climate Change 2001: Impacts, Adaptation, and Vulnerability.* Cambridge Univ. Press, Cambridge.

McClelland, JE, Mogen, CA, Johnson, WM, Schroer, FW and JS Allen. 1959. Chernozems and associated soils of eastern North Dakota: Some properties and topographic relationships. *SSSAP* 23:51–56.

McComb, AL and WE Loomis. 1944. Subclimax prairie. *Bull. Torrey Bot. Club* 71:46–76.

McCormac, FG, Hogg, AG, Blackwell, PG, Buck, CE, Higham, TFG and PJ Reimer. 2004. SHCal04 Southern Hemisphere calibration, 0–11. 0 cal kyr bp. *Radiocarbon* 46:1087–1092.

McDaniel, PA and AL Falen. 1994. Temporal and spatial patterns of episaturation in a Fragixeralf landscape. *SSSAJ* 58:1451–1457.

McDaniel, PA, Fosberg, MA and AL Falen. 1993. Expression of andic and spodic properties in tephra-influenced soils of northern Idaho. *Geoderma* 58:79–94.

McDaniel, PA, Gabehart, RW, Falen, AL, Hammel, JE and RJ Reuter. 2001. Perched water tables on Argixeroll and Fragixeralf hillslopes. *SSSAJ* 65:805–810.

McDaniel, PA, Lowe, DJ, Arnaulds, O and C-L Ping. 2011. Andisols. In: PM Huang, Y Li and ME Sumner (eds). *Handbook of Soil Sciences*, 2nd ed. CRC Press, New York. pp. 33-29 – 33-48.

McDaniel, PA, Regan, MP, Brooks, E, Boll, J, Bamdt, S, Falen, A, Young, SK and JE Hammel. 2008. Linking fragipans, perched water tables, and catchment-scale hydrological processes. *Catena* 73:166–173.

McDole, RE and MS Fosberg. 1974a. Soil temperatures in selected Southeastern Idaho soils. I. Evaluation of sampling techniques and classification of soils. *SSSAP* 38:480–486.

1974b. Soil temperatures in selected Southeastern Idaho soils. II. Relation to soil and site characteristics. *SSSAP* 38:486–491.

McDonald, EV and AJ Busacca. 1988. Record of pre-late Wisconsin floods in the Channeled Scabland interpreted from loess deposits. *Geology* 16:728–731.

1990. Interaction between aggrading geomorphic surfaces and the formation of a Late Pleistocene paleosol in the Palouse loess of eastern Washington state. *Geomorphology* 3:449–470.

1992. Late Quaternary stratigraphy of loess in theChanneled Scabland and Palouse Regions of Washington State. *QR* 38:141–156.

McDonald, EV, Pierson, FB, Flerchinger, GN and LD McFadden. 1996. Application of a soil-water balance model to evaluate the influence of Holocene climate change on calcic soils, Mojave Desert, California, USA. *Geoderma* 74:167–192.

McDonnell, MJ and STA Pickett. 1990. The study of ecosystem structure and function along the urban-rural gradients: An unexploited opportunity. *Ecology* 71:1232–1237.

McFadden, JP, MacDonald, NW, Witter, JA and DR Zak. 1994. Fine-textured soil bands and oak forest productivity in northwestern lower Michigan, USA. *CJFR* 24:928–933.

McFadden, LD. 1988. Climatic influences on rates and processes of soil development in Quaternary deposits of southern California. *GSA Spec. Paper* 216:153–177.

McFadden, LD, Amundson, RG and OA Chadwick. 1991. Numerical modeling, chemical, and isotopic studies of carbonate accumulation in soils of arid regions. *SSSASP* 26:17–35.

McFadden, LD, Eppes, MC, Gillespie, AR and B Hallet. 2005. Physical weathering in arid landscapes due to diurnal variation in the direction of solar heating. *GSA Bull.* 117:161–173.

McFadden, LD and DM Hendricks. 1985. Changes in the content and composition of pedogenic iron oxyhydroxides in a chronosequence of soils in southern California. *QR* 23:189–204.

McFadden, LD and PLK Knuepfer. 1990. Soil geomorphology – the linkage of pedology and surficial processes. *Geomorphology* 3:197–205.

McFadden, LD, McDonald, EV, Wells, SG, Anderson, K, Quade, J and SL Forman. 1998. The vesicular layer and carbonate collars of desert soils and pavements: Formation, age and relation to climate change. *Geomorphology* 24:101–145.

McFadden, LD, Ritter, JB and SG Wells. 1989. Use of multiparameter relative-age methods for age estimation and correlation of alluvial fan surfaces on a desert piedmont, eastern Mojave Desert, California. *QR* 32:276–290.

McFadden, LD and RJ Weldon Jr. 1987. Rates and processes of soil development on Quaternary terraces in Cajon Pass, California. *GSA Bull.* 98:280–293.

McFadden, LD, Wells, SG, Brown, WJ and Y Enzel. 1992. Soil genesis on beach ridges of pluvial Lake Mojave: Implications for Holocene lacustrine and eolian events in the Mojave Desert, southern California. *Catena* 19:77–97.

McFadden, LD, Wells, SG and JC Dohrenwend. 1986. Influences of Quaternary climatic changes on processes of soil development on desert loess deposits of the Cima Volcanic Field, California. *Catena* 13:361–389.

McFadden, LD, Wells, SG and MJ Jercinovich. 1987. Influence of eolian and pedogenic processes on the origin and evolution of desert pavements. *Geology* 15:504–508.

McFarlane, MJ and DJ Bowden. 1992. Mobilization of aluminum in the weathering profiles of the African surface in Malawi. *ESPL* 17:789–805.

McFarlane, MJ and S Pollard. 1989. Some aspects of stone-lines and dissolution fronts associated with regolith and dambo profiles in parts of Malawi and Zimbabwe. *Intl. J. Trop. Ecol. Geog.* 11:23–35.

McGee, WJ. 1891. The Pleistocene history of northeastern Iowa. *USGS Report* 11:189–577.

McGreevy, JP. 1985. Thermal properties as controls on rock surface temperature maxima and possible implications for rock weathering. *ESPL* 10:125–136.

McHargue, LP and PE Damon. 1991. The global beryllium 10 cycle. *Rev. Geophys.* 29:141–158.

Mckay, CP. 2009. Snow recurrence sets the depth of dry permafrost at high elevations in the McMurdo Dry Valleys of Antarctica. *Antarctic Sci.* 21:89–94.

McKay, ED. 1979. Wisconsinan loess stratigraphy of Illinois. In: *Wisconsinan, Sangamonian, and Illinoian Stratigraphy in Central Illinois*. Illinois State Geol. Survey Guidebook #13. Midwest Friends of the Pleistocene Field Conference Guidebook. pp. 95–108. Urbana-Champaign, IL.

McKeague, JA. 1967. An evaluation of 0. 1 M pyrophosphate and pyrophosphate-dithionite in comparison with oxalate as extractants of the accumulation products in podzols and some other soils. *CJSS* 47:95–99.

McKeague, JA, Bourbeau, GA and DB Cann. 1967. Properties and genesis of a bisequa soil from Cape Breton Island. *CJSS* 47:101–110.

McKeague, JA, Brydon, JE and NM Miles. 1971. Differentiation of forms of extractable iron and aluminum in soils. *SSSAP* 35:33–38.

McKeague, JA and JH Day. 1966. Dithionite- and oxalate-extractable Fe and Al as aids in differentiating various classes of soils. *CJSS* 46:13–22.

McKeague, JA, DeConinck, F and DP Franzmeier. 1983. Spodosols. In: LP Wilding, NE Smeck and GF Hall (eds). *Pedogenesis and Soil Taxonomy*. Elsevier, New York. pp. 217–252.

McKeague, JA, Guertin, RK, Page, F and KW Valentine. 1978. Micromorphological evidence of illuvial clay in horizons designated Bt in the field. *CJSS* 58:179–186.

McKeague, JA and RJ St. Arnaud. 1969. Pedotranslocation: Eluviation-illuviation in soils during the Quaternary. *SS* 107:428–434.

McKeague, JA and C Wang. 1980. Micromorphology and energy dispersive analysis of ortstein horizons of podzolic soils from New Brunswick and Nova Scotia, Canada. *CJSS* 60:9–21.

McKean, JA, Dietrich, WE, Finkel, RC, Southon, JR and MW Caffee. 1993. Quantification of soil production and downslope creep rates from cosmogenic ^{10}Be accumulations on a hillslope profile. *Geology* 21:343–346.

McKenna Neuman, C, Maxwell, CD and JW Boulton. 1996. Wind transport of sand surfaces crusted with photoautotrophic microorganisms. *Catena* 27:229–247.

McKenzie, RM. 1989. Manganese oxides and hydroxides. In: JB Dixon and SB Weed (eds). *Minerals in Soil Environments*, 2nd ed. Soil Sci. Soc. Am., Madison, WI. pp. 439–465.

McKeown, DA and JE Post. 2001. Characterization of manganese oxide mineralogy in rock varnish and dendrites using X-ray absorption spectroscopy. *Am. Min.* 86:701–713.

McLaurin, BT, Goossens, D and BJ Buck. 2011. Combining surface mapping and process data to assess, predict, and manage dust emissions from natural and disturbed land surfaces. *Geosphere* 7:260–275.

McLeod, M, Bockheim, J, Balks, M and J Aislabie. 2009. Soils of western Wright Valley, Antarctica. *Antarctic Sci.* 21:355–365.

McManus, DA. 1991. Suggestions for authors whose manuscripts include quantitative clay mineral analysis by X-ray diffraction. *Marine Geol.* 98:1–5.

McMillan, BR, Cottam, MR and DW Kaufman. 2000. Wallowing behavior of American Bison (*Bos bison*) in tallgrass prairie: An examination of alternate explanations. *AMN* 144:159–167.

McNeil, M. 1964. Lateritic soils. *Sci. Am.* 211:96–102.

McNeill, JR and V Winiwarter. 2004. Breaking the sod: Humankind, history and soil. *Science* 304:1627–1629.

McPherson, TS and VR Timmer. 2002. Amelioration of degraded soils under red pine plantations on the Oak Ridges Moraine, Ontario. *CJSS* 82:375–388.

McSweeney, K, Leigh, DS, Knox, JC and RH Darmody. 1988. Micromorphological analysis of mixed zones associated with loess deposits of the midcontinental United States. In: DN Eden and RJ Furkert (eds). *Loess Its Distribution, Geology and Soils. Proc. Intl. Sympos. on Loess, New Zealand*. A.A. Balkema, Rotterdam. pp. 117–130.

McTainsh, G. 1987. Desert loess in northern Nigeria. *ZG N.F.* 26:417–435.

Meadows, DG, Young, MH and EV McDonald. 2008. Influence of relative surface age on hydraulic properties and infiltration on soils associated with desert pavements. *Catena* 72:169–178.

Megonigal, JP, Patrick, WH Jr and SP Faulkner. 1993. Wetland identification in seasonally flooded forest soils: Soil morphology and redox dynamics. *SSSAJ* 57:140–149.

Mehra, OP and ML Jackson. 1960. Iron oxide removal from soils and clays by a dithionite-citrate system buffered with sodium bicarbonate. *Clays Clay Mins.* 7:317–327.

Meighan, C. 1983. Obsidian dating in California. *Am. Antiq.* 48:600–609.

Meighan, CW, Foote, LJ and PV Aiello. 1968. Obsidian dating in west Mexican archeology. *Science* 160:1069–1075.

Meijer, A. 2002. Conceptual model of the controls on natural water chemistry at Yucca Mountain, Nevada. *Appl. Geochem.* 17:793–805.

Meixner, RE and MJ Singer. 1981. Use of a field morphology rating system to evaluate soil formation and discontinuities. *SS* 131:114–123.

Melendez, A, Alonso-Zarza, AM and C Sancho. 2011. Multi-storey calcrete profiles developed during the initial stages of the configuration of the Ebro Basin's exorrheic fluvial network. *Geomorphology* 134:232–248.

Melillo, JM, Aber, JD, Linkins, AE, Ricca, A, Fry, B and KJ Nadelhoffer. 1989. Carbon and nitrogen dynamics along the decay continuum: Plant litter to soil organic matter. *Plant Soil* 115:189–198.

Mellor, A. 1985. Soil chronosequences on Neoglacial moraine ridges, Jostedalsbreen and Jotunheimen, southern Norway: A quantitative pedogenic approach. In: KS Richards, RR Arnett and S Ellis (eds). *Geomorphology and Soils*. Allen and Unwin, London. pp. 289–308.

1986. A micromorphological examination of two alpine soil chronosequences, southern Norway. *Geoderma* 39:41–57.

1987. A pedogenic investigation of some soil chronosequences on neoglacial moraine ridges, southern Norway: Examination of soil chemical data using principal components analysis. *Catena* 14:369–381.

Melton, DA. 1976. The biology of aardvark (Tubulidentata-Orycteropodidae). *Mammal Rev.* 6:75–88.

Melton, FA. 1935. Vegetation and soil mounds. *Geog. Rev.* 25:430–433.

Menendez, I, Cabrera, L, Sanchez-Perez, I, Mangas, J and I Alonso. 2009a. Characterisation of two fluvio-lacustrine loessoid deposits on the island of Gran Canaria, Canary Islands. *QI* 196:36–43.

Menendez, I, Derbyshire, E, Engelbrecht, J, von Suchodoletz, H, Zoller, L, Dorta, P, Carrillo, T and FR de Castro. 2009b. Saharan dust and the aerosols on the Canary Islands: Past and present. In: M

Cheng and W Liu (eds). *Airborne Particulates*. Nova Science Publ., Hauppauge, NY. pp. 39–80.

Meng, X, Derbyshire, E and RA Kemp. 1997. Origin of the magnetic susceptibility signal in Chinese loess. *QSR* 16:833–839.

Mentlik, P, Engel, Z, Braucher, R and L Leanni. 2013. Chronology of the Late Weichselian glaciation in the Bohemian Forest in Central Europe. *QSR* 65:120–128.

Mercader, J, Marti, R, Martinez, JL and A Brooks. 2002. The nature of "stone-lines" in the African Quaternary record: Archaeological resolution at the rainforest site of Mosumu, Equatorial Guinea. *QI* 89:71–96.

Mermut, AR and H Eswaran. 2001. Some major developments in soil science since the mid-1960s. *Geoderma* 100:403–426.

Merrill, GP. 1897. *A Treatise on Rocks, Rock-Weathering and Soils*. MacMillan, New York.

1906. *Rocks, Rock Weathering and Soils*. MacMillan, London.

Merritts, DJ, Chadwick, OA and DM Hendricks. 1991. Rates and processes of soil evolution on uplifted marine terraces, northern California. *Geoderma* 51:241–275.

Merritts, DJ, Chadwick, OA, Hendricks, DM, Brimhall, GH and CJ Lewis. 1992. The mass balance of soil evolution on late Quaternary marine terraces, northern California. *GSA Bull.* 104:1456–1470.

Metzger, CA and GJ Retallack. 2010. Paleosol record of Neogene climate change in the Australian outback. *Austr. J. Earth Sci.* 57:871–885.

Mew, G and CW Ross. 1994. Soil variation on steep greywacke slopes near Reefton, Western South Island. *J. Roy. Soc. New Zealand* 24:231–242.

Meyers, NL and K McSweeney. 1995. Influence of treethrow on soil properties in Northern Wisconsin. *SSSAJ* 59:871–876.

Meysman, FJR, Middelburg, JJ and CHR Heip. 2006. Bioturbation: A fresh look at Darwin's last idea. *Trends Ecol. Evol.* 21:688–695.

Miall, AD. 1985. Architectural-element analysis: A new method of facies analysis applied to fluvial deposits. *Earth-Sci. Revs.* 22:261–308.

Michaelson, GJ, Ping, CL and JM Kimble. 1996. Carbon content and distribution in tundra soils in arctic Alaska. *AAR* 28:414–424.

Michalet, R, Guillet, B and B Souchier. 1993. Hematite identification in psuedoparticles of Moroccan rubified soils. *Clay Mins.* 28:233–242.

Michel, RFM, Schaefer, CEGR, Dias, LE, Simas, FNB, Benites, VM and E Mendonça. 2006. Ornithogenic Gelisols (Cryosols) from maritime Antarctica: Pedogenesis, vegetation and carbon studies. *SSSAJ* 70:1370–1376.

Michéli, E and OC Spaargaren. 2011. Other systems of soil classification. In: PM Huang, Y Li and ME Sumner (eds). *Handbook of Soil Sciences*, 2nd ed. CRC Press, New York. pp. 32–1 – 32–34.

Michels, JW. 1967. Archaeology and dating by obsidian hydration. *Science* 158:211–214.

Middleton, LT and MJ Kraus. 1980. Simple technique for thin-section preparation of unconsolidated materials. *J. Sed. Petr.* 50:622–623.

Middleton, NJ and DSG Thomas. 1997. *World Atlas of Desertification*. Edward Arnold, London.

Migliavacca, M, Pizzeghello, D, Busana, MS and S Nardi. 2012. Soil chemical analysis supports the identification of ancient breeding structures: The case-study of Cà Tron (Venice, Italy). *QI* 275:128–136.

Migoń, P 1997. Palaeoenvironmental significance of grus weathering profiles: A review with special reference to northern and central Europe. *Proc. Geol. Assoc.* 108:57–70.

2009. Are any granite landscapes distinctive of the humid tropics? Reconsidering multiconvex topographies. *Sing. J. Trop. Geog.* 30:327–342.

Migoń, P and MF Thomas. 2002. Grus weathering mantles – problems of interpretation. *Catena* 49:5–24.

Mikesell, LR, Schaetzl, RJ and MA Velbel. 2004. Hornblende etching and quartz/feldspar ratios as weathering and soil development indicators in some Michigan soils. *QR* 62:162–171.

Mikhailova, EA, Post, CJ and DG Bielenberg. 2008. Photosynthetic pathway of steppe vegetation and the stable carbon isotope composition of organic matter in the russian chernozem. *Comm. Soil Sci. Plant Anal.* 39:641–651.

Milan, H, Garcia-Formaris, I and M Gonzalez-Posada. 2009. Nonlinear spatial series analysis from unidirectional transects of soil physical properties. *Catena* 77:56–64.

Miles, J. 1985. The pedogenic effects of different species and vegetation types and the implications of succession. *JSS* 36:571–584.

Miles, RJ and DP Franzmeier. 1981. A lithochronosequence of soils formed in dune sand. *SSSAP* 45:362–367.

Milfred, CJ and RW Kiefer. 1976. Analysis of soil variability with repetitive aerial photography. *SSSAJ* 40:553–557.

Millar, CS. 1974. Decomposition of coniferous leaf litter. In: CH Dickinson and GF Pugh (eds). *Biology of Plant Litter Decomposition*. Vol. 1. Academic Press, London. pp. 105–128.

Millar, SWS. 2004. Identification of mapped ice-margin positions in western New York from digital terrain-analysis and soil databases. *PG* 25:347–359.

Miller, BA. 2012. The need to continue improving soil survey maps. *SSH* 53. doi:10.2136/sh12-02-0005

2013. *Incorporating tacit knowledge of soil-landscape relationships for digital soil and landscape mapping applications*. PhD Diss., Michigan State University, East Lansing, MI.

2014. Semantic calibration of digital terrain analysis scale. *Cart. and GIS* 41:166–176.

Miller, BA and RJ Schaetzl. 2012. Precision of soil particle size analysis using laser diffractometry. *SSSAJ* 76:1719–1727.

Miller, BA, Burras, CL and WG Crumpton. 2008. Using soil surveys to map Quaternary parent materials and landforms across the Des Moines lobe of Iowa and Minnesota. *Soil Survey Hor.* 49:91–95.

Miller, CD. 1979. A statistical method for relative-age dating of moraines in the Sawatch Range, Colorado. *GSA Bull.* 90:1153–1164.

Miller, DL, Mora, CI and SG Driese. 2007. Isotopic variability in large carbonate nodules in Vertisols: Implications for climate and ecosystem assessments. *Geoderma* 142:104–111.

Miller, G and JM Greenwade. 2001. *Soil Survey of McLennan County, Texas*. USDA-NRCS, US Govt. Printing Off., Washington, DC.

Miller, GH. 1973. Late Quaternary glacial and climatic history of northern Cumberland Peninsula, Baltin Island, N. W.T., Canada. *QR* 3:561–583

Miller, JJ, Acton, DF and RJ St. Arnaud. 1985. The effect of groundwater on soil formation in a morainal landscape in Saskatchewan. *CJSS* 65:293–307.

Miller, JJ and JA Brierley. 2011. Solonetzic soils of Canada: Genesis, distribution, and classification. *CJSS* 91:889–902.

Miller, LL, Hinkel, KM, Nelson, FE, Paetzold, RF and SI Outcalt. 1998. Spatial and temporal patterns of soil moisture and thaw depth at Barrow, Alaska, USA. In: AG Lewkowicz and M Allard (eds). *Permafrost. Proc. 7th Intl. Conf., Université Laval, Yellowknife*, Canada. pp. 731–737.

Miller, MB, Cooper, TH and RH Rust. 1993. Differentiation of an eluvial fragipan from dense glacial till in northern Minnesota. *SSSAJ* 57:787–796.

Miller, NG and RP Futyma. 1987. Paleohydrological implications of Holocene peatland development in northern Michigan. *QR* 27:297–311.

Miller, WL, Kishne, AS and CLS Morgan. 2010. Vertisol morphology, classification, and seasonal cracking patterns in the Texas Gulf Coast Prairie. *SSH* 51:10–16.

Millot, G, Nahon, D, Paquet, H, Ruellan, A and Y Tardy. 1977. L'Epigenie calcaire des roches silicatées dans les encroutements carbonates en pays subaride Antiatlas. *Maroc. Soc. Geol. Bull.* 30:129–152.

Mills, AJ, Milewski, A, Fey, MV, Groengroeft, A and A Petersen. 2009. Fungus culturing, nutrient mining and geophagy: A geochemical investigation of *Macrotermes* and *Trinervitermes* mounds in southern Africa. *J. Zool.* 278:24–35.

Mills, HH. 1981. Some observations on slope deposits in the vicinity of Grandfather Mountain, North Carolina, USA. *Southeastern Geol.* 22:209–222

1981. Some observations on slope deposits in the vicinity of Grandfather Mountain, North Carolina, USA. *Southeast. Geol.* 22:209–222.

2005. Relative-age dating of transported regolith and application to study of landform evolution in the Appalachians. *Geomorphology* 67:63–96.

Milne, AE, Webster, R and RM Lark. 2010. Spectral and wavelet analysis of gilgai patterns from air photography. *AJSR* 48:309–325.

Milne, G. 1935. Some suggested units for classification and mapping, particularly for East African soils. *Soil Res. Berlin* 4:183–198.

1936a. A Provisonal Soil Map of East Africa, African Agricultural Research Station, Amani Memoirs. Tanganyika Territory.

1936b. Normal erosion as a factor in soil profile development. *Nature* 138:548–549.

Milnes, AR and CR Twidale. 1983. An overview of silicification in Cainozoic landscapes of arid central and southern Australia. *AJSR* 21:387–410.

Ming, DW and FA Mumpton. 1989. Zeolites in soils. In: JB Dixon and SB Weed (eds). *Minerals in Soil Environments*, 2nd ed. Soil Sci. Soc. Am., Madison, WI. pp. 873–911.

Minasny, B and AE Hartemink. 2011. Predicting soil properties in the tropics. *Earth-Sci. Revs.* 106:52–62.

Minasny, B and AB McBratney. 2001. A rudimentary mechanistic model for soil formation and landscape development II; A two-dimensional model incorporating chemical weathering. *Geoderma* 103:161–179.

2006. Mechanistic soil–landscape modelling as an approach to developing pedogenetic classifications. *Geoderma* 133:138–149.

Minasny, B, McBratney, AB, and S Salvador-Blanes. 2008. Quantitative models for pedogenesis – A review. *Geoderma* 144:140–157.

Mitchell, MJ. 1980. *Soil Survey of Winnebago County, Wisconsin.* USDA-SCS, US Govt. Printing Off., Washington, DC.

Mitchell, PB. 1985. Some aspects of the role of bioturbation in soil formation in south-eastern Australia. PhD thesis, School of Earth Sciences, MacQuarie University.

Mitchell, PB and GS Humphreys. 1987. Litter dams and microterraces formed on hillslopes subject to rainwash in the Sydney Basin, Australia. *Geoderma* 39:331–357.

Mitusov, AV, Mitusova, OE, Pustovoytov, K, Lubos, CCM, Dreibrodt, S and HR Bork. 2009. Palaeoclimatic indicators in soils buried under archaeological monuments in the Eurasian steppe: A review. *Holocene* 19:1153–1160.

Mizota, C, Izuhara, H and M Noto. 1992. Eolian influence on oxygen isotope abundance and clay minerals in soils of Hokkaido, northern Japan. *Geoderma* 52:161–171.

Moberg, JP and AA Mmikonga. 1977. Content of stable micro-peds in tropical soils. In: KT Joseph (ed). *Proc. Conf. on Classification and Mgmt. of Tropical Soils.* Kuala Lumpur, Malaysia. Malaysian Soc. Soil Science. pp. 124–131.

Moeyersons, J. 1975. An experimental study of pluvial processes on granite gruss. *Catena* 2:289–308.

1978. The behaviour of stones and stone implements, buried in consolidating and creeping Kalahari sands. *ESPL* 3:115–128.

1989. The concentration of stones into a stone-line, as a result from subsurface movements in fine and loose soils in the tropics. *Geo-Eco-Trop.* 11:11–22.

Mogen, CA, McClelland, JE, Allen, JS and FW Schroer. 1959. Chestnut, Chernozem, and associated soils of western North Dakota. *SSSAP* 23:56–60.

Mohanty, BP, Kanwar, RS and CJ Everts. 1994. Comparison of saturated hydraulic conductivity measurement methods for a glacial till soil. *SSSAJ* 58:672–677.

Mokma, DL. 1993. Color and amorphous materials in Spodosols from Michigan. *SSSAJ* 57:125–128.

1997a. Ortstein in selected soils from Washington and Michigan. *SSH* 38:71–75.

1997b. Water tables and color patterns in sandy soils in Michigan. *SSH* 38:54–59.

Mokma, DL, Doolittle, JA and LA Tornes. 1994. Continuity of ortstein in sandy Spodosols, Michigan. *SSH* 35:6–10.

Mokma, DL and CV Evans. 2000. Spodosols. In: ME Sumner (ed). *Handbook of Soil Science.* CRC Press, Boca Raton, FL. pp. E-307–E-321.

Mokma, DL and P Buurman. 1982. *Podzols and Podzolization in Temperate Regions.* Wageningen, the Netherlands, International Soil Museum.

Mokma, DL and DL Cremeens. 1991. Relationships of saturation and B horizon colour patterns in soils of three hydrosequences in south-central Michigan, USA. *Soil Use Mgt.* 7:56–61.

Mokma, DL, Schaetzl, RJ, Doolittle, JA and EP Johnson. 1990. Ground-penetrating radar study of ortstein continuity in some Michigan Haplaquods. *SSSAJ* 54:936–938.

Mokma, DL and SW Sprecher. 1994. Water table depths and color patterns in Spodosols of two hydrosequences in northern Michigan, USA. *Catena* 22:275–286.

1995. How frigid is frigid? *SSH* 36:71–76.

Moldenke, AR. 1999. Soil arthropods. *USDA-NRCS Soil Qual. Inst. PA* 1637:G1–G8.

Moll, EJ. 1994. The origin and distribution of fairy rings in Namibia. In: JH Seyani and AC Chikuni (eds). *Proc. 13th Plenary Mtg AETFAT*, Malawi. pp. 1203–1209.

Moller, M, Volk, M, Friedrich, K and L Lymburner. 2008. Placing soil-genesis and transport processes into a landscape context: A multi-scale terrain-analysis approach. *J. Plant Nutr. Soil Sci.* 171:419–430.

Moller, P, Anjar, J and AS Murray. 2013. An OSL-dated sediment sequence at Idre, west-central Sweden, indicates ice-free conditions in MIS 3. *Boreas* 42:25–42.

Monaghan, MC and D Elmore. 1994. *Garden-variety Be-10 in soils on hill slopes.* *Nucl. Instrum. Meth.* B92:357–361.

Monaghan, MC, Krishnaswami, S and JH Thomas. 1983. ^{10}Be concentrations and the long-term fate of particle-reactive nuclides in five soil profiles from California. *EPSL* 65:51–60.

Monaghan, MC, McKean, J, Dietrich, W and J Klein. 1992. ^{10}Be chronometry of bedrock-to-soil conversion rates. *EPSL* 111:483–492.

Monger, HC. 2002. Pedogenic carbonate: Links between biotic and abiotic $CaCO_3$. *Proc. 17th World Congr. Soil Sci.* 2:796.

Monger, HC, Cole, DR, Gish, JW and TH Giordano. 1998. Stable carbon and oxygen isotopes in Quaternary soil carbonates as indicators of ecogeomorphic changes in the northern Chihuahuan Desert, USA. *Geoderma* 82:137–172.

Monger, HC, Daugherty, LA and LH Gile. 1991a. A microscopic examination of pedogenic calcite in an Aridisol of southern New Mexico. *SSSASP* 26:37–60.

Monger, HC, Daugherty, LA, Lindemann, WC and CM Liddell. 1991b. Microbial precipitation of pedogenic calcite. *Geology* 19:997–1000.

Monger, HC, Southard, RJ and JL Boettinger. 2011. Aridisols. In: PM Huang, Y Li and ME Sumner (eds). *Handbook of Soil Sciences*, 2nd ed. CRC Press, New York. pp. 33-127–33-153.

Monroe, WH. 1966. Formation of tropical karst topography by limestone solution and precipitation. *Caribb. J. Sci.* 6:1–7.

Montagne, D, Cornu, S, LeForestier, L and I Cousin. 2009. Soil drainage as an active agent of recent soil evolution: A review. *Pedosphere* 19:1–13.

Montagne, D, Cornu, S, LeForestier, L, Hardy, M, Josiere, O, Caner, L and I Cousin. 2008. Impact of drainage on soil-forming mechanisms in a French Albeluvisol: Input of mineralogical data in mass-balance modeling. *Geoderma* 145:426–438.

Moody, LE and RC Graham. 1997. Silica-cemented terrace edges, central California coast. *SSSAJ* 61:1723–1729.

Moore, DM and RC Reynolds Jr. 1989. *X-Ray Diffraction and the Identification and Analysis of Clay Minerals*. Oxford, Oxford, UK.

Moore, ID, Burch, GJ and DH Mackenzie. 1988. Topographic effects on the distribution of surface soil water and the location of ephemeral gullies. *Trans. Am. Soc. Agric. Engin.* 31:1383–1395.

Moore, ID, Gessler, PE, Nielsen, GA and GA Peterson. 1993. Soil attribute prediction using terrain analysis. *SSSAJ* 57:443–452.

Moore, TR. 1974. Pedogenesis in a subarctic environment: Cambrian Lake, Quebec. *AAR* 6:281–291.

1976. Sesquioxide-cemented soil horizons in northern Quebec: Their distribution, properties and genesis. *CJSS* 56:333–344.

Moore, TR and J Turunen. 2004. Carbon accumulation and storage in mineral subsoil beneath peat. *SSSAJ* 68:690–696.

Moores, JE, Pelletier, JD and PH Smith. 2008. Crack propagation by differential insolation on desert surface clasts. *Geomorphology* 102:472–481.

Moorhead, DL, Fisher, FM and WG Whitford. 1988. Cover of spring annuals on nitrogen-rich kangaroo rat mounds in a Chihuahuan Desert grassland. *AMN* 120:443–447.

Moresi, M and G Mongelli. 1988. The relation between the terra rossa and the carbonate-free residue of the underlying limestones and dolostones in Apulia, Italy. *Clay Mins.* 23:439–446.

Morrás, H. 1979. Quelques élements de discussion sur les mécanismes de pédogenese des Planosols et d'autres sols apparentés, *Science du Sol* 7:57–66.

Morrás, H, Moretti, L, Píccolo, G and W Zech. 2009. Genesis of subtropical soils with stony horizons in NE Argentina: Autochthony and polygenesis. *QI* 196:137–159.

2010. A reply to D. Kröhling and M. Iriondo's comment on "Genesis of subtropical soils with stony horizons in NE Argentina: Autochthony and polygenesis" by Morrás, H., Moretti, L., Píccolo, G. and Zech, W. *QI* 227:193–195.

Morris, JD. 1991. Applications of cosmogenic [10]Be to problems in the Earth Sciences. *Ann. Rev. Earth Planet. Sci.* 19:313–350.

Morris, WJ. 1985. A convenient method of acid etching. *J. Sed. Petr.* 55:600.

Morrison, RB. 1967. Principles of Quaternary soil stratigraphy. *Proc. 7th INQUA Congr.* 9:1–69.

1998. How can the treatment of pedostratigraphic units in the North American Stratigraphic Code be improved? *QI* 51/52:30–33.

Moser, M and F Hohensinn. 1983. Geotechnical aspects of soil slips in alpine regions. *Engin. Geol.* 19:185–211.

Moss, RP. 1965. Slope development and soil morphology in a part of south west Nigeria. *JSS* 16:192–209.

Mossin, L, Jensen, BT and P Nørnberg. 2001. Altered podzolization resulting from replacing heather with Sitka spruce. *SSSAJ* 65:1455–1462.

Mossin, L, Mortensen, M and P Nornberg. 2002. Imogolite related to podzolization processes in Danish podzols. *Geoderma* 109:103–116.

Mottershead, D and G Lucas. 2000. The role of lichens in inhibiting erosion of a soluble rock. *Lichenologist* 32:601–609.

Mount, HR. 1998. Global change remote soil temperature network. *SSH* 39:92.

Moura Filho, W and SW Buol. 1976. Studies of a Latosol Roxo (Eutrustox) in Brazil: Micromorphology effect on ion release. *Experientiae* 21:161–177.

Muckenhirn, RJ, Whiteside, EP, Templin, EH, Chandler, RF Jr and LT Alexander. 1949. Soil classification and the genetic factors of soil formation. *SS* 67:93–105.

Mueller, G, Broll, G and C Tarnocai. 1999. Biological activity as influenced by microtopography in a Cryosolic soil, Baffin Island, Canada. *Permafrost Periglac. Proc.* 10:279–288.

Muggler, CC and P Buurman. 2000. Erosion, sedimentation and pedogenesis in a polygenetic oxisol sequence in Minas Gerais, Brazil. *Catena* 41:3–17.

Muggler, CC, Buurman, P and JDJ van Doesburg. 2007. Weathering trends and parent material characteristics of polygenetic oxisols from Minas Gerais, Brazil: I. Mineralogy. *Geoderma* 138:39–48.

Muhs, DR. 1982. A soil chronosequence on Quaternary marine terraces, San Clemente Island, California. *Geoderma* 28:257–283.

1984. Intrinsic thresholds in soil systems. *PG* 5:99–110.

2004. Mineralogical maturity in dunefields of North America, Africa and Australia. *Geomorphology* 59:247–269.

Muhs, DR, Aleinikoff, JN, Stafford, TW, Jr, Kihl, R, Been, J, Mahan, SA and C Cowherd. 1999. Late Quaternary loess in northeastern Colorado. I. Age and paleoclimatic significance. *GSA Bull.* 111:1861–1875.

Muhs, DR and JB Benedict. 2006. Eolian additions to late Quaternary alpine soils, Indian Peaks Wilderness Area, Colorado Front Range. *AAAR* 38:120–130.

Muhs, DR and EA Bettis III. 2000. Geochemical variations in Peoria loess of western Iowa indicate paleowinds of midcontinental North America during last glaciation. *QR* 53:49–61.

Muhs, DR, Bettis, EA, Aleinikoff, JN, McGeehin, JP, Beann, J, Skipp, G, Marshall, BD, Roberts, HM, Johnson, WC and R Benton. 2008a. Origin and paleoclimatic signifi cance of late Quaternary loess in Nebraska: Evidence from stratigraphy, chronology, sedimentology, and geochemistry. *GSA Bull.* 120:1378–1407.

Muhs, DR, Bettis, EA III, Been, J and JP McGeehin. 2001. Impact of climate and parent material on chemical weathering in loess-derived soils of the Mississippi River Valley. *SSSAJ* 65:1761–1777.

Muhs, DR and JR Budahn. 2009. Geochemical evidence for African dust and volcanic ash inputs to terra rossa soils on carbonate reef terraces, northern Jamaica, West Indies. *QI* 196:13–35.

Muhs, DR, Budahn, JR, Johnson, DL, Reheis, M, Beann, J, Skipp, G, Fisher, E and JA Jones. 2008b. Geochemical evidence for airborne dust additions to soils in Channel Islands National Park, California. *GSA Bull.* 120:106–126.

Muhs, DR, Budahn, JR, Prospero, JM and SN Carey. 2007. Geochemical evidence for African dust inputs to soils of western Atlantic islands: Barbados, the Bahamas, and Florida. *J. Geophys. Res.-Earth Surface* 112. doi:10.1029/2005JF000445

Muhs, DR, Budahn, JR, Prospero, JM, Skipp, G and SR Herwitz. 2012. Soil genesis on the island of Bermuda in the Quaternary: The importance of African dust transport and deposition. *J. Geophys. Res.* 117. F03025, doi:10.1029/2012JF002366

Muhs, DR, Bush, CA, Stewart, KC, Rowland, TR and RC Crittenden. 1990. Geochemical evidence of Saharan dust parent material for soils developed on Quaternary limestones of Caribbean and western Atlantic islands. QR 33:157–177.

Muhs, DR, Crittenden, RC, Rosholt, JN, Bush, CA and KC Stewart. 1987. Genesis of marine terrace soils, Barbados, West Indies: Evidence from mineralogy and geochemistry. ESPL 12:605–618.

Muhs, DR and VT Holliday. 2001. Origin of lateQuaternary dune fields on the southern High Plains of Texas and New Mexico. GSA Bull. 113:75–87.

Muhs, DR, Kautz, RR and JJ MacKinnon. 1985. Soils and the location of cacao orchards at a Maya site in western Belize. JAS 12:121–137.

Muhs, DR, Stafford, TW, Jr, Been, J, Mahan, SA, Burdett, J, Skipp, G and ZM Rowland. 1997a. Holocene eolian activity in the Minot dune field, North Dakota. CJES 34:1442–1459.

Muhs, DR, Stafford, TW, Jr, Swinehart, JB, Cowherd, SD, Mahan, SA, Bush, CA, Madole, RF and PB Maat. 1997b. Late Holocene eolian activity in the mineralogically mature Nebraska Sand Hills. QR 48:162–176.

Muhs, DR and SA Wolfe. 1999. Sand dunes of the northern Great Plains of Canada and the United States. Geol. Surv. Can. Bull. 534:183–197.

Muhs, DR and M Zárate. 2001. Late Quaternary eolian records of the Americas and their paleoclimatic significance. In: V Markgraf (ed). Interhemispheric Climate Linkages. Academic Press, San Diego, CA. pp. 183–216.

Muir, A. 1961. The podzol and podzolic soils. Adv. Agron. 13:1–57.

Muir, JW and J Logan. 1982. Eluvial/illuvial coefficients of major elements and the corresponding losses and gains in three soil profiles. JSS 33:295–308.

Mujinya, BB, Mees, F, Boeckx, P, Bode, S, Baert, G, Erens, H, Delefortrie, S, Verdoodt, A, Ngongo, M and E Van Ranst. 2011. The origin of carbonates in termite mounds of the Lubumbashi area, D. R. Congo. Geoderma 165:95–105.

Mulholland, SC and G Rapp Jr. 1992. Phytolith systematics: An introduction. In: G Rapp, Jr and SC Mulholland (eds). Phytolith Systematics. Plenum Press, New York. pp. 1–13.

Müller, D, Bocquier, G, Nahon, D and H Pacquet. 1981. Analyse des différentiations minéralogiques et structurales d'un sol ferralitique à horizons nodulaires du Congo. Cahiers de l'ORSTOM, Sér. Pédol. 18:87–109.

Muller, PE. 1879. Studier over skovjord, som bidrag til skovdyrkningens theori. Tidsskr. Skovbr. 3:1–124.

Muller, RA, Gordon J and F MacDonald. 1997. Glacial cycles and astronomical forcing. Science 277: 215–218.

Mullins, CE. 1977. Magnetic susceptibility of the soil and its significance in soil science: A review. JSS 28:223–246.

Munday, TJ, Reilly, NS, Glover, M, Lawrie, KC, Scott, T, Chartres, CJ and WR Evans. 2000. Petrophysical characterisation of parna using ground and downhole geophysics at Marinna, central New South Wales. Explor. Geophys. 31:260–266.

Munn, LC and MM Boehm. 1983. Soil genesis in a Natrargid-Haplargid complex in northern Montana. SSSAJ 47:1186–1192.

Munn, LC. 1993. Effects of prairie dogs on physical and chemical properties of soils. US Fish and Wildlife Serv. Biol. Rept 13:11–17.

Munroe, JS. 2007. Properties of alpine soils associated with well-developed sorted polygons in the uinta mountains, Utah, USA. AAAR 39:578–591.

Murad, E. 1978. Yttrium and zirconium as geochemical guide elements in soil and stream sediment sequences. JSS 29:219–223.

Murray, AS and LB Clemmensen. 2001. Luminescence dating of Holocene aeolian sand movement, Thy, Denmark. QSR 20:751–754.

Murray AS and JM Olley. 2002. Precision and accuracy in the optically stimulated luminescence dating of sedimentary quartz: A status review. Geochronometria 21:1–16.

Murray, AS and RG Roberts. 1997. Determining the burial time of single grains of quartz using optically stimulated luminescence. EPSL 152:163–180.

Murray, DF. 1967. Gravel mounds at Rocky Flats, Colorado. Mountain Geol. 4:99–107.

Murton, JB and HM French. 1993. Thaw modification of frost-fissure wedges, Richards Island, Pleistocene Mackenzie Delta, Western Arctic, Canada. JQS 8:185–196.

Musgrave, GW. 1935. Some relationships between slope length, surface runoff and the silt load of surface runoff. TAGU 11:472–478.

Mutakyahwa, MKD, Ikingura, JR and AH Mruma. 2003. Geology and geochemistry of bauxite deposits in Lushoto District, Usambara Mountains, Tanzania. J. Afr. Earth Sci. 36:357–369.

Mycielska-Dowgiallo, E and B Woronko. 1998. Analiza obtoczenia i zmatowienia powierzchni ziarn kwarcowych frakcji piaszczystej i jej wartos´c´ interpretacyjna. Prz. Geol. 46:1275–1281.

2004. The degree of aeolization of Quaternary deposits in Poland as a tool for stratigraphic interpretation. Sed. Geol. 168:149–163.

Myrcha, A and A Tatur. 1991. Ecological role of the current and abandoned rookeries in the land environment of the maritime Antarctic. Polish Polar Res. 12:3–24.

Nabel, P. 1993. The Brunhes-Matuyama boundary in Pleistocene sediments of Buenos Aires province, Argentina. QI 17:79–85.

Nadelhoffer, KJ and B Fry. 1988. Controls on natural nitrogen-15 and carbon-13 abundances in forest soil organic matter. SSSAJ 52:1633–1640.

Nadkarni, NM and JT Longino. 1990. Macroinvertebrate communities in canopy and forest floor organic matter in a montane cloud forest, Costa Rica. Biotropica 22:286–289.

Nadkarni, NM, Schaefer, D, Matelson, TJ and R Soleno. 2002. Comparison of arboreal and terrestrial soil characteristics in a lower montane forest, Monteverde, Costa Rica. Pedobiologia 46:24–33.

Nahon, D. 1991. Introduction to the Petrology of Soils and Chemical Weathering. Wiley, New York.

Naiman, Z, Quade, J, and PJ Patchett. 2000. Isotopic evidence for eolian recycling of pedogenic carbonate and variations in carbonate dust sources throughout the southwest United States. Geochim. Cosmochim. Acta 64:3099–3109.

Narcisi, B, Petit, JR, Delmonte, B, Basile-Doelsch, I and V Maggi. 2005. Characteristics and sources of tephra layers in the EPICA-Dome C ice record (East Antarctica): Implications for past atmospheric circulation and ice core stratigraphic correlations. EPSL 239:253–265.

Nash, DB. 1984. Morphologic dating of fluvial terrace scarps and fault scarps near West Yellowstone, Montana. GSA Bull. 95:1413–1424.

Nash, MH, Daugherty, LA, Buchanan, BA and BW Hunyadi. 1994. The effect of groundwater table variation on characterization and classification of gypsiferous soils in the Tularosa Basin, New Mexico. SSH 35:102–110.

National Institute of Standards and Technology. 2011. Chemistry WebBook. http://webbook.nist.gov/cgi/cbook.cgi?Name=carbon+dioxide&Units=SI&cSO=on#Solubility.

Natural Resources Conservation Service, US Dept. Agric. National Cooperative Soil Survey Soil Characterization Data. http://soils.usda.gov/survey/nscd/. Accessed Nov. 2013.

Naudé, Y, van Rooyen, MW and ER Rohwer. 2011. Evidence for a geochemical origin of the mysterious circles in the Pro-Namib desert. J. Arid Environ. 75:446–456.

Neary, DG, Swift, LW, Jr, Manning, DM and RG Burns. 1986. Debris avalanching in the southern Appalachians: An influence on forest soil formation. *SSSAJ* 50:465–471.

Nel, JJC and EM Malan. 1974. The distribution of the mounds of *Trinervitermes trinervoides* in the central Orange Free State. *J. Entomol. Soc. S. Afr.* 37:251–256.

Nelson, AR and RR Shroba. 1998. Soil relative dating of moraine and outwash-terrace sequences in the northern part of the upper Arkansas Valley, Central Colorado, USA. *AAR* 30:349–361.

Nelson, AR, Millington, AC, Andrews, JT and H Nichols. 1979. Radiocarbon-dated upper Pleistocene glacial sequence, Fraser Valley, Colorado Front Range. *Geology* 7:410–414.

Nelson, LA, Kunze, GW and CL Godfrey. 1960. Chemical and mineralogical properties of a San Saba clay, a Grumusol. *SS* 89:122–131.

Nelson, RL. 1954. Glacial geology of the Frying Pan River drainage, Colorado. *J. Geol.* 62:325–343.

Nesbitt, HW. 1992. Diagenesis and metasomatism of weathering profiles, with emphasis on Precambrian paleosols. In: IP Martini and W Chesworth (eds). *Weathering, Soils and Paleosols.* Elsevier, Amsterdam. pp. 127–152.

Nesbitt, HW, Markovics, G and RC Price. 1980. Chemical processes affecting alkalis and alkaline earths during continental weathering. *Geochim. Cosmochim. Acta* 44:1659–1666.

Nesbitt, HW and GM Young. 1982. Early Proterozoic climates and plate motions inferred from major element chemistry of lutites. *Nature* 299:715–717.

Nesje, A, Kvamme, M and N Rye. 1989. Neoglacial gelifluction in the Jostedalsbreen region, western Norway: Evidence from dated buried palaeopodzols. *ESPL* 14:259–270.

Netterberg, F. 1969a. Ages of calcretes in southern Africa. *S. Afr. Archaeol. Bull.* 24:88–92.

1969b. The interpretation of some basic calcrete types. *S. Afr. Archaeol. Bull.* 24:117–122.

Nettleton, WD. (ed) 1991. *Occurrence, Characteristics, and Genesis of Carbonate, Gypsum, and Silica Accumulations in Soils.* Soil Sci. Soc. Am. Spec. Publication #26. Madison, WI.

Nettleton, WD, Brasher, BR, Benham, EC and RJ Ahrens. 1998. A classification system for buried paleosols. *QI* 51/52:175–183.

Nettleton, WD and OA Chadwick. 1991. Soil-landscape relationships in the Wind River Basin, Wyoming. *Mountain Geol.* 28:3–11.

Nettleton, WD, Daniels, RB and RJ McCracken. 1968. Two North Carolina coastal plain catenas. I. Morphology and fragipan development. *SSSAP* 32:577–582.

Nettleton, WD, Flach, KW and BR Brasher. 1969. Argillic horizons without clay skins. *SSSAP* 33:121–125.

Nettleton, WD, Gamble, EE, Allen, BL, Borst, G and FF Peterson. 1989. Relict soils of the subtropical regions of the United States. *Catena Suppl.* 16:59–93.

Nettleton, WD, Goldin, A and R Engel. 1986. Differentiation of Spodosols and Andepts in a western Washington soil climosequence. *SSSAJ* 50:987–992.

Nettleton, WD, Olson, CG, and DA Wysocki. 2000. Paleosol classification: Problems and solutions. *Catena* 41:61–92.

Nettleton, WD, Witty, JE, Nelson, RE and JW Holly. 1975. Genesis of argillic horizons in soils of desert areas of the southwestern United States. *SSSAP* 39:919–926.

Neustruev, SS. 1927. Genesis of Soils. Russian Pedological Investigations. Vol. III. Leningrad, Academy of Sciences of the USSR.

Neustruyev, SS. 1915. Soil combinations in plains and mountainous countries. *Pochvovedeniye* 17:62–73.

Newcomb, RC. 1952. Origin of Mima mounds, Thurston County region, Washington. *J. Geol.* 60:461–472.

Newman, AL. 1983. Vertisols in Texas: Some comments. *SSH* 24:8–20.

Nicholas, JW and DR Butler. 1996. Application of relative-age dating techniques on rock glaciers of the La Sal mountains, Utah: An interpretation of Holocene paleoclimates. *Geog. Annaler* 78A:1–18.

Nichols, JD, Brown, PL and WJ Grant (eds) 1984. *Erosion and Productivity of Soils Containing Rock Fragments.* Soil Sci. Soc. Am. Spec. Publication #13. Madison, WI.

Nichols, JD. 1984. Relation of organic carbon to soil properties and climate in the southern Great Plains. *SSSAJ* 48:1382–1384.

Nickel, E. 1973. Experimental dissolution of light and heavy minerals in comparison with weathering and intrastratal solution. *Contrib. Sedimentol.* 1:1–68.

Nielsen, AH. 2008. Soil development on a 1,500-year-old beach ridge plain, Sturgeon Bay, NW Lower Michigan. *Mich. Academ.* 38:159–174.

Nielsen, A, Murray, AS, Pejrup, M and B Elberling. 2006. Optically stimulated luminescence dating of a Holocene beach ridge plain in Northern Jutland, Denmark. *QG* 4:305–312.

Nielsen, AH, Elberling, B and M Pejrup. 2010. Soil development rates from an optically stimulated luminescence-dated beach ridge sequence in Northern Jutland, Denmark. *CJSS* 90:295–307.

Nielsen, AH and AS Murray. 2008. The effects of Holocene podzolisation on radionuclide distributions and dose rates in sandy coastal sediments. *Geochronometrica* 31:53–63.

Nielsen, GA and FD Hole. 1964. Earthworms and the development of coprogenous A1 horizons in forest soils of Wisconsin. *SSSAP* 28:426–430.

Nielsen, KE, Dalsgaard, K and P Nornberg. 1987. Effects on soils of an oak invasion of a Calluna heath, Denmark. I. Morphology and chemistry. *Geoderma* 41:79–95.

Nikièma, P, Rothstein, DE and RO Miller. 2012. Initial greenhouse gas emissions and nitrogen leaching losses associated with converting pastureland to short-rotation woody bioenergy crops in northern Michigan, USA. *Biomass Bioenergy* 39:413–426.

Nikiforoff, CC. 1937. General trends of the desert type of soil formation. *SS* 43:105–131.

1949. Weathering and soil evolution. *SS* 67:219–230.

1959. Reappraisal of the soil. *Science* 129:186–196.

Nikiforoff, CC and M Drosdoff. 1943. Genesis of a claypan soil. *II. SS* 56:43–62.

Nimlos, TJ and M Tomer. 1982. Mollisols beneath conifer forests in southwestern Montana. *SS* 134:371–375.

Ning, YF, Liu, WG and ZS An. 2008. A 130-ka reconstruction of precipitation on the Chinese Loess Plateau from organic carbon isotopes. *PPP* 270:59–63.

Nishiizumi, K, Kohl, CP, Arnold, JR, Dorn, R, Klein, J, Fink, D, Middleton, R and D Lal. 1993. Role of in situ cosmogenic nuclides ^{10}Be and ^{26}Al in the study of diverse geomorphic processes. *ESPL* 18:407–425.

Nishiizumi, K, Lal, D, Klein, J, Middleton, R and JR Arnold. 1986. Production of ^{10}Be and ^{26}Al by cosmic rays in terrestrial quartz in situ and implications for erosion rates. *Nature* 319:134–136.

Nizeyimana, E and TJ Bicki. 1992. Soil and soil-landscape relationships in the north central region of Rwanda, east-central Africa. *SS* 153:225–236.

Njiru, H, Elchalal, U and O Paltiel. 2011. Geophagy during pregnancy in Africa: A literature review. *Obstet. Gynecol. Surv.* 66:452–459.

Nobbs, MF and RI Dorn. 1988. Age determinations for rock varnish formation within petroglyphs: Cation-ratio dating of 24 motifs from the Olary region, South Australia. *Rock Art Res.* 5:108–146.

Noller, JS, Sowers, JM and WR Lettis. (eds). 2000. *Quaternary Geochronology: Methods and Applications.* American Geophysical Union, Washington, DC.

Nooren, CAM, van Breemen, N, Stoorvogel, JJ and AG Jongmans. 1995. The role of earthworms in the formation of sandy surface soils in a tropical forest in Ivory Coast. *Geoderma* 65:135–148.

Nordt, LC, Boutton, TW, Hallmark, CT and MR Waters. 1994. Late Quaternary vegetation and climate change in central Texas based on the isotopic composition of organic carbon. *QR* 41:109–120.

Nordt, LC and SD Driese. 2010. New weathering index improves paleo-rainfall estimates from Vertisols. *Geology* 38:407–410.

Nordt, LC, Wilding, LP, Lynn, WC and CC Crawford. 2004. Vertisol genesis in a humid climate of the coastal plain of Texas, *U. S.A. Geoderma* 122:83–102.

Norfleet, ML and AD Karathanasis. 1996. Some physical and chemical factors contributing to fragipan strength in Kentucky soils. *Geoderma* 71:289–301.

Norman, MB II. 1974. Improved techniques for selective staining of feldspar and other minerals using amaranth. *USGS J. Res.* 2:73–79.

Norman, SA, Schaetzl, RJ and TW Small. 1995. Effects of slope angle on mass movement by tree uprooting. *Geomorphology* 14:19–27.

Nørnberg, P, Sloth, L and KE Nielsen. 1993. Rapid changes in sandy soils caused by vegetation changes. *CJSS* 73:459–468.

North American Commission on Stratigraphic Nomeclature. 1983. North American Stratigraphic Code. *Am. Ass. Petrol. Geol. Bull.* 67:841–875.

Northcote, KN. 1960. *A Factual Key for the Recognition of Australian Soils.* CSIRO Div. of Soils. Rept. 4/60. Melboune.

Norton, EA and RS Smith. 1930. The influence of topography on soil profile character. *J. Am. Soc. Agron.* 22:251–262.

Norton, LD. 1994. Micromorphology of silica cementation in soils. *Dev. Soil Sci.* 22:811–824.

Norton, LD, Bigham, JM, Hall, GF and NE Smeck. 1983. Etched thin sections for coupled optical and electron microscopy and microanalysis. *Geoderma* 30:55–64.

Norton, LD and GF Hall. 1985. Differentiation of lithologically similar soil parent materials. *SSSAJ* 49:409–414.

Norton, LD, Hall, GF, Smeck, NE and JM Bigham. 1984. Fragipan bonding in a Late-Wisconsinan loess-derived soil in east-central Ohio. *SSSAJ* 48:1360–1366.

Noy-Meir, I. 1974. Multivariate analysis of the semiarid vegetation in south-eastern Australia. II. Vegetation catenae and environmental gradients. *AJB* 22:115–140.

Nullet, D, Ikawa, H and P Kilham. 1990. Local differences in soil temperature and soil moisture regimes on a mountain slope, Hawaii. *Geoderma* 47:171–184.

Nutting, WL, Haverty, MI and JP LaFage. 1987. Physical and chemical alteration of soil by two subterranean termite species in Sonoran Desert grassland. *J. Arid Envs* 12:233–239.

Nuzzo, VA, Maerz, JC and B Blossey. 2009. Earthworm invasion as the driving force behind plant invasion and community change in northeastern North American forests. *Cons. Biol.* 23:966–974.

Nyberg, ME, Osterholm, P and MI Nystrand. 2012. Impact of acid sulfate soils on the geochemistry of rivers in south-western Finland. *Environ. Earth Sci.* 66:157–168.

Nye, PH. 1954. Some soil-forming processes in the humid tropics. I. A field study of a catena in the west African forest. *JSS* 5:7–21.

1955a. Some soil-forming processes in the humid tropics. II. The development of the upper-slope member of the catena. *JSS* 6:51–62.

1955b. Some soil-forming processes in the humid tropics. III. Laboratory studies on the development of a typical catena over granitic gneiss. *JSS* 6:63–72.

1955c. Some soil-forming processes in the humid tropics. IV. The action of the soil fauna. *JSS* 6:73–83.

O'Brien, K. 1979. Secular variations in the production of cosmogenic isotopes in the Earth's atmosphere. *J. Geophys. Res.* 84:423–431.

O'Connell, AM. 1987. Litter dynamics in Karri (Eucalyptus diversicolor) forests of south-western Austral. *JE* 75:781–796.

Oades, JM. 1989. An introduction to organic matter in mineral soils. In: JB Dixon and SB Weed (eds). *Minerals in Soil Environments*, 2nd ed. Soil Sci. Soc. Am., Madison, WI. pp. 89–159.

Oakes, H and J Thorp. 1950. Dark-clay soils of warm regions variously called Rendzina, Black Cotton Soils, Regur, and Tirs. *SSSAP* 15:348–354.

Oba, T and T Irino. 2012. Sea level at the last glacial maximum, constrained by oxygen isotopic curves of planktonic foraminifera in the Japan Sea. *JQS* 27:941–947.

Oertel, AC. 1961. Pedogenesis of some Red-BrownEarths based on trace element profiles. *JSS* 12:242–258.

1968. Some observations incompatible with clay illuviation. *Trans. 9th Intl. Congr. Soil Sci.* 4:482–488.

1974. The development of a typical red-brown earth. *AJSR* 12:97–105.

Oertel, AC and JB Giles. 1966. Quantitative study of a layered soil. *AJSR* 4:19–28.

Offer, ZY and D Goossens. 2001. Ten years of aeolian dust dynamics in a desert region (Negev desert, Israel): Analysis of airborne dust concentration, dust accumulation and the high-magnitude dust events. *J. Arid Envs* 47:211–249.

Oganesyan, AS and NG Susekova. 1995. Parent materials of Wrangel Island. *Euras. Soil Sci.* 27:20–35.

Ogg, CM and JC Baker. 1999. Pedogenesis and origin of deeply weathered soils formed in alluvial fans of the Virginia Blue Ridge. *SSSAJ* 63:601–606.

Ogg, CM, Edmonds, WJ and JC Baker. 2000. Statistical verification of soil discontinuities in Virginia. *SS* 165:170–183.

Oguchi, CT. 2013. Weathering rinds: Formation processes and weathering rates. In: JF Shroder (ed). *Treatise on Geomorphology*. Vol. 4. Academic Press, San Diego, CA. pp. 98–110.

Oh, NH and DD Richter. 2005. Elemental translocation and loss from three highly weathered soil-bedrock profiles in the southeastern United States. *Geoderma* 126:5–25.

Ohnuki, Y, Terazono, R, Ikuzawa, H, Hirata, I, Kanna, K and H Utagawa. 1997. Distribution of colluvia and saprolites and their physical properties in a zero-order basin in Okinawa, southwestern Japan. *Geoderma* 80:75–93.

Ojanuga, AG. 1979. Clay mineralogy of soils in theNigerian tropical savanna region. *SSSAJ* 43:1237–1242.

O'Leary, MH. 1988. Carbon isotopes in photosynthesis. *Bioscience* 38:328–336.

O'Loughlin, EM. 1986. Prediction of surface saturation zones in natural catchments by topographic analysis. *Water Resources Res.* 22:794–804.

1990. Modelling soil water status in complex terrain. *Agric. For. Meteor.* 50:23–38.

Oliver, CD. 1978. Subsurface geologic formations and site variation in Upper Sand Hills of South Carolina. *J. For.* 76:352–354.

Oliver, SA, Oliver, HR, Wallace, JS and AM Roberts. 1987. Soil heat flux and temperature variation with vegetation, soil type, and climate. *Agric. For. Meteorol.* 39:257–269.

Ollerhead, J, Huntley, DJ and GW Berger. 1994. Luminescence dating of sediments from Buctouche Spit, New Brunswick. *CJES* 31:523–531.

Olley, J, Caitcheon, G and A Murray. 1998. The distribution of apparent dose as determined by optically stimulated luminescence in small aliquots of fluvial quartz: Implications for dating young sediments. *QSR* 17:1033–1040.

Olley, JM, Murray, A and RG Roberts. 1996. The effects of disequilibria in the uranium and thorium decay chains on burial dose rates in fluvial sediments. *QSR* 15:751–760.

Olley, JM, Pietsch, T and RG Roberts. 2004. Optical dating of Holocene sediments from a variety of geomorphic settings using single grains of quartz. *Geomorphology* 60:337–358.

Ollier, C. 1984. *Weathering*. Longman, Harlow.

Ollier, CD. 1959. A two-cycle theory of tropical pedology. *JSS* 10:137–148.

1966. Desert gilgai. *Nature* 212:581–583.

Ollier, CD and JE Ash. 1983. Fire and rock breakdown. *ZG* 27:363–374.

Ollier, CD, Drover, DP and M Godelier. 1971. Soil knowledge amongst the Baruya of Wonenara, New Guinea. *Oceania* 42:33–41.

Ollier, CD and C Pain. 1996. *Regolith, Soils and Landforms*. Wiley, New York.

Olson, CG. 1989. Soil geomorphic research and the importance of paleosol stratigraphy to Quaternary investigations, midwestern USA. *Catena Suppl.* 16:129–142.

1997. Systematic soil-geomorphic investigations: Contributions of R. V. Ruhe to pedologic interpretation. *Adv. Geoecol.* 29:415–438.

Olson, CG and WD Nettleton. 1998. Paleosols and the effects of alteration. *QI* 51/52:184–194.

Olson, CG, Nettleton, WD, Porter, DA and BR Brasher. 1997. Middle Holocene aeolian activity on the High Plains of west-central Kansas. *Holocene* 7:255–261.

Olson, CG, Ruhe, RV and MJ Mausbach. 1980. The terra rossa-limestone contact phenomena in karst, southern Indiana. *SSSAJ* 44:1075–1079.

Olson, KR, Jones, RL, Gennadiyev, AN, Chernyanskii, S and WI Woods. 2003. Soil development on Monks Mound at the Cahokia Archaeological site, Illinois. *SSH* 44:73–85.

Olsson, IU. 1974. Some problems in connection with the evaluation of ^{14}C dates. *Geol. Foreningens Stockholm Förhandl.* 96:311–320.

2009. Radiocarbon dating history: Early days, questions, and problems met. *Radiocarbon* 51:1–43.

Olsson, IU and FAN Osadebe. 1974. Carbon isotope variations and fractionation corrections in ^{14}C dating. *Boreas* 3:139–146.

Olsson, M and P-A Melkerud. 1989. Chemical and mineralogical changes during genesis of a Podzol from till in Southern Sweden. *Geoderma* 45:267–287.

Opala M and M Mendecki. 2013. An attempt to dendroclimatic reconstruction of winter temperature based on multispecies tree-ring widths and extreme years chronologies (example of Upper Silesia, Southern Poland). *Theor. Applied Climatol.* 112. doi:10.1007/s00704-013-0865-5

Opdekamp, W, Teuchies, J, Vrebos, D, Chormanski, J, Schoelynck, J, van Diggelen, R, Meire, P and E Struyf. 2012. Tussocks: Biogenic silica hot-spots in a riparian wetland. *Wetlands* 32:1115–1124.

Oreskes, N. 2004. The scientific consensus on climate change. *Science* 306:1686.

Orlova, LA and VS Zykina. 2002. Radiocarbon dating of buried Holocene soils in Siberia. *Radiocarbon* 44:113–122.

Orombelli, G, Maggi, V and B Delmonte. 2010. Quaternary stratigraphy and ice cores. *QI* 219:55–65.

Orombelli, G and SC Poter. 1983. Lichen growth curves for the southern flank of the Mont Blanc Massif, Western Italian Alps. *AAR* 15:193–200.

Ortiz, I, Simon, M, Dorronsoro, C, Martin, F and I Garcia. 2002. Soil evolution over the Quaternary period in a Mediterranean climate (SE Spain). *Catena* 48: 131 148.

Orton, GJ. 1996. Volcanic environments. In: HG Reading (ed). *Sedimentary Environments Processes, Facies and Stratigraphy*. Blackwell, Cambridge, MA. pp. 485–567.

Osterkamp, WR, Toy, TJ and MT Lenart. 2006. Development of partial rock veneers by root throw in a subalpine setting. *ESPL* 31:1–14.

Othberg, KL, McDaniel, PA and MA Fosberg. 1997. Soil development on a Pleistocene terrace sequence, Boise Valley, Idaho. *Northwest Sci.* 71:318–329.

Ovalles, FA and ME Collins. 1986. Soil-landscape relationships and soil variability in north-central Florida. *SSSAJ* 50:401–408.

Overpeck, JT, Webb, T III and IC Prentice. 1985. Quantitative interpretation of fossil pollen spectra: Dissimilarity coefficients and the method of modern analogs. *QR* 23:87–108.

Owczarek, P, Latocha, A, Wistuba, M and I Malik. 2013. Reconstruction of modern debris flow activity in the arctic environment with the use of dwarf shrubs (south-western Spitsbergen) – a new dendrochronological approach. *ZG Supp.* 57:75–95.

Ozaytekin, HH and S Ozcan. 2013. Mass balance of soil evolution on Mt. Erenler volcanic materials in central Anatolia – A case study. *Carpath. J. Earth Environ. Sci.* 8:5–18.

Pachepsky, Y. 1998. Discussion of the paper by JD Phillips. *Geoderma* 86:31–32.

Paetzold, RF. 1990. Soil climate definitions used in Soil Taxonomy. In: JM Kimble and WD Nettleton (eds). *Characterization, Classification, and Utilization of Aridisols*. Proc. 4th Intl. Soil Correlation Mtg. USDA-NRCS, Lincoln, NE. pp. 151–165.

Pal, BC and A Bhattacharya. 1986. Wallowing behavior and wallows used by great Indian onehorned Rhinoceros at Garumara and Jaldapara wildlife sanctuaries, West Bengal, India. *Proc. Zoologic. Soc. Calcutta* 35:79–83.

Pal, DK, Deshpande, SB, Venogupal, KR and AR Kalbande. 1989. Formation of di- and trioctahedral smectite as evidence for paleoclimatic changes in southern and central peninsular India. *Geoderma* 45:175–184.

Paletskaya, L, Lavrov, A and S Kogan. 1958. Pore formation in takyr crust. *SSS* 3:245–250.

Palkovics, WE, Petersen, GW and RP Matelski. 1975. Perched water table fluctuation compared to streamflow. *SSSAP* 39:343–348.

Palmer, E. 2002. Feasibility and implications of a rock coating catena: Analysis of a desert hillslope. MA thesis, Arizona State Univ., Tempe.

Palmer, FE, Staley, JT, Murray, RGE, Counsell, T and JB Adams. 1986. Identification of manganese-oxidizing bacteria from desert varnish. *Geomicrobiol. J.* 4:343–360.

Panaiotu, CG, Panaiotu, EC, Grama, A and C Necula. 2001. Paleoclimatic record from a loess-paleosol profile in southeastern Romania. *Phys. Chem. Earth Solid Earth Geod.* 26A:893–898.

Pankova, YI and IA Yamnova. 1987. Forms of gypsic neoformations as a controlling factor affecting the meliorative properties of gypsiferous soils. *Soviet Soil Sci.* 19:94–102.

Pape, JC. 1970. Plaggen soil in the Netherlands. *Geoderma* 4:229–255.

Paradise, TR. 1995. Sandstone weathering thresholds in Petra, Jordan. *PG* 16:205–222.

1995. Sandstone weathering thresholds in Petra, Jordan. *PG* 16:205–222.

2002. Sandstone weathering and aspect in Petra, Jordan. *ZG* 46:1–17.

Parakshin, YP. 1974. "Hill" solonetzes of the low-hill region of Kazakhstan. *SSS* 6:408–411. (Translated from Pochvovedeniye 8:137–140.)

Parfitt, RL and CW Childs. 1988. Estimation of forms of Fe and Al: A review, and analysis of contrasting soils by dissolution and Mössbauer methods. *AJSR* 26:121–144.

Parfitt, RL and B Clayden. 1991. Andisols – The development of a new order in Soil Taxonomy. *Geoderma* 49:181–198.

Parfitt, RL and M Saigusa. 1985. Allophane and humus-aluminum in Spodosols and Andepts formed from the same volcanic ash beds in New Zealand. *SS* 139:149–155.

Parfitt, RL, Saigusa, M and DN Eden. 1984. Soil development processes in an Aqualf-Ochrept sequence from loess with admixtures of tephra, New Zealand. *JSS* 35:625–640.

Pariente, S. 2001. Soluble salts dynamics in the soil under different climatic conditions. *Catena* 43:307–321.

Parisio, S. 1981. The genesis and morphology of a serpentine soil in Staten Island, New York. *Proc. Staten Island Inst. Arts Sci.* 31:2–17.

Parisot, JC, Soubies, F, Audry, P and F Espourteille. 1989. Some implications of laterite weathering on geochemical prospecting – two Brazilian examples. *J. Geochem. Explor.* 32:133–147.

Parizek, EJ and JF Woodruff. 1957. Description and origin of stone layers in soils of the southeastern states. *J. Geol.* 65:24–34.

Park, SJ and TP Burt. 2002. Identification and characterization of pedogeomorphological processes on a hillslope. *SSSAJ* 66:1897–1910.

Parker, A. 1970. An index of weathering for silicate rocks. *Geol. Mag.* 107:501–504.

Parker, GR, McFee, WW and JM Kelly. 1978. Metal distribution in forested ecosystems in urban and rural northwestern Indiana. *J. Env. Qual.* 7:337–342.

Parsons, AJ, Abrahams, AD and JR Simanton. 1992. Microtopography and soil-surface materials on semi-arid piedmont hillslopes, southern Arizona. *J. Arid Envs* 22:107–115.

Parsons, RB and CA Balster. 1966. Morphology and genesis of six "Red Hill" soils in the Oregon coast range. *SSSAP* 30:90–93.

Parsons, RB, Balster, CA and AO Ness. 1970. Soil development and geomorphic surfaces, Willamette Valley, Oregon. *SSSAP* 34:485–491.

Parsons, RB and RC Herriman. 1975. A lithosequence in the mountains of southwestern Oregon. *SSSAP* 39:943–947.

Parsons, RB, Schotes, WH and FF Riecken. 1962. Soil of Indian mounds in Northeastern Iowa as benchmarks for studies of soil genesis. *SSSAP* 26:491–496.

Parsons, RB, Simonson, GH and CA Balster. 1968. Pedogenic and geomorphic relationships of associated Aqualfs, Albolls, and Xerolls in Western Oregon. *SSSAP* 32:556–563.

Paton, TR. 1974. Origin and terminology for gilgai in Australia. *Geoderma* 11:221–242.

———. 1978. *The Formation of Soil Material*. Allen and Unwin, Boston, MA.

Paton, TR, Humphreys, GS and PB Mitchell. 1995. *Soils A New Global View*. Yale Univ. Press, New Haven, CT.

Paton, TR, Mitchell, PB, Adamson, D, Buchanon, RA, Fox, MD and G Bowman. 1976. Speed of podzolization. *Nature* 260:601–602.

Patrick, WH, Jr and IC Mahapatra. 1968. Transformation and availability of nitrogen and phosphorous in waterlogged soils. *Adv. Agron.* 20:323–359.

——— Jr and RE Henderson. 1981. Reduction and reoxidation cycles of manganese and iron in flooded soil and in water solution. *SSSAJ* 45:855–859.

Patro, BC and BK Sahu. 1974. Factor analysis of sphericity and roundness data of clastic quartz grains: Environmental significance. *Sed. Geol.* 11:59–78.

Patton, PC and SA Schumm. 1975. Gully erosion, northern Colorado: A threshold phenomenon. *Geology* 3:88–90.

Patyk-Kara, NG, Gorelikova, NV, Plakht, J, Nechelyustov, GN and IA Chizhova. 1997. Desert varnish as an indicator of the age of Quaternary formations (Makhtesh Ramon Depression, Central Negev). *Trans. Russian Acad. Sci./Earth Sci.* 353A:348–351.

Paul, EA and FE Clark (eds) 1996. *Soil Microbiology and Biochemistry*, 2nd ed. Academic Press, San Diego, CA.

Paul, EA, Campbell, CA, Rennie, DA and KJ McCallum. 1964. Investigations of the dynamics of soil humus utilizing carbon dating techniques. *Trans. 8th Intl. Congr. Soil Sci. (Bucharest)* 3:201–208.

Pavich, MJ. 1989. Regolith residence time and the concept of surface age of the Piedmont "peneplain." *Geomorphology* 2:181–196.

Pavich, MJ, Brown, L, Harden, J, Klein, J and R Middleton. 1986. [10]Be distribution in soils from Merced River terraces, California. *Geochim. Cosmochim. Acta* 50:1727–1735.

Pavich, MJ, Brown, L, Klein, J and R Middleton. 1984. [10]Be accumulation in a soil chronosequence. *EPSL* 68:198–204.

Pavich, MJ, Brown, L, Valette-Silver, JN, Klein, J and R Middleton. 1985. [10]Be analysis of a Quaternary weathering profile in the Virginia Piedmont. *Geology* 13:39–41.

Pavich, MJ, Leo, GW, Obermeier, SF and JF Estabrook. 1989. Investigations of the Characteristics, Origin, and Residence Time of the Upland Residual Mantle of the Piedmont of Fairfax County, Virginia. USGS Prof. Paper 1352.

Pavlik, HF and FD Hole. 1977. Soilscape analysis of slightly contrasting terrains in southeastern Wisconsin. *SSSAJ* 41:407–413.

Pawlik, L, Kacprzak, A, Musielok, L and J Bebak. 2013a. Distinct form of microrelief on steep slopes of the Rogowa Kopa, Stołowe Mts., SW Poland – in the light of geomorphological and pedological evidence. In: P Migoà and M Kasprzak (eds). *Sandstone Landscapes. Diversity, Ecology and Conservation. Proc. 3rd Intl. Conf. on Sandstone Landscapes*. Kudowa-Zdràj (Poland). April 25–28, 2012. pp. 132–138.

Pawlik, L, Migon P, Owczarek P and A Kacprzak. 2013b. Surface processes and interactions with forest vegetation on a steep mudstone slope, Stołowe Mountains, SW Poland. *Catena* 109:203–216.

Pawluk, S and MJ Dudas. 1982. Floralpedoturbations in black chernozemic soils of the Lake Edmonton Plain. *CJSS* 62:617–629.

Payton, RW. 1992. Fragipan formation in argillic brown earths (Fragiudalfs) of the Milfield Plain, north-east England. I. Evidence for a periglacial stage of development. *JSS* 43:621–644.

———. 1993a. Fragipan formation in argillic brown earths (Fragiudalfs) of the Milfield Plain, north-east England. II. Post-Devensian developmental processes and the origin of fragipan consistence. *JSS* 44:703–723.

———. 1993b. Fragipan formation in argillic brown earths (Fragiudalfs) of the Milfield Plain, north-east England. III. Micromorphological, SEM and EDXRA studies of fragipan degradation and the development of glossic features. *JSS* 44:725–729.

Peacock, E and DW Fant. 2002. Biomantle formation and artifact translocation in upland sandy soils: An example from the Holly Springs National Forest, north-central Mississippi, USA. *Geoarchaeol.* 17:91–114.

Pearce, F. 2006. Arctic permafrost set to disappear over next century. *New Scientist* 2537:15.

Pearce, TA, Olori, JC and KW Kemezis. 2010. Land snails from St. Elzear cave, Gsape Peninsula, Quebec: Antiquity of *Cepaea hortensis* in North America. *Ann. Carnegie Museum* 79:65–78.

Pearsall, DM. 2000. *Paleoethnobotany. A Handbook of Procedures*, 2nd ed. Academic Press, San Diego, CA.

Pecher, K. 1994. Hydrochemical analysis of spatial and temporal variations of solute composition in surface and subsurface waters of a high Arctic catchment. *Catena* 21:305–327.

Pedro, G. 1966. Caractérisation géochimique des différents processus zonaux résultant de l'altération des roches superficielles. *C.R. Acad. Sci. (Paris)* 262:1828–1831.

———. 1982. The conditions of formation of secondary constituents. In: M Bonneau and B Souchier (eds). *Constituents and Properties of Soils*. Academic Press, London. pp. 63–81.

1983. Structuring of some basic pedological processes. *Geoderma* 31:289–299.

Pedro, G, Jamagne, M and JC Begon. 1978. Two routes in genesis of strongly differentiated acid soils under humid, cool-temperate conditions. *Geoderma* 20:173–189.

Peel, RF. 1974. Insolation weathering: Some measurements of diurnal temperature changes in exposed rocks in the Tibesti region, central Sahara. *ZG Suppl.* 21:19–28.

Peet, GB. 1971. Litter accumulation in jarrah and karri forest. *Austral. For.* 35:258–262.

Pelletier, JD, Cline, M and SB DeLong. 2007. Desert pavement dynamics: Numerical modeling and field-based calibration. *ESPL* 32:1913–1927.

Peltier, LC. 1950. The geographic cycle in periglacial regions as it is related to climatic geomorphology. *AAAG* 40:214–236.

Pennisi, E. 2004. The secret life of fungi. *Science* 304:1620–1622.

Pennock, DJ. 2003. Terrain attributes, landform segmentation, and soil redistribution. *Soil Tillage Res.* 69:15–26.

Pennock, DJ and WJ Vreeken. 1986. Soil-geomorphic evolution of a Boroll catena in southwestern Alberta. *SSSAJ* 50:1520–1526.

Pennock, DJ, Zebarth, BJ and E DeJong. 1987. Landform classification and soil distribution in hummocky terrain, Saskatchewan, Canada. *Geoderma* 40:297–315.

Pereira, M., Vazquez, FM and FG Ojea. 1978. Pedological and geomorphological cycles in a catena of Galacia (NW Spain). *Catena* 5:375–387.

Pérez, FL. 1984. Striated soil in an Andean Paramo of Venezuela: Its origin and orientation. *AAR* 16:277–289.

1986. The effect of compaction on soil-disturbance by needle ice growth. *Acta Geocriogen.* 4:111–119.

1987a. Downslope stone transport by needle ice in a high Andean Area (Venezuela). *Rev. Géomorphol. Dynam.* 36:33–51.

1987b. Needle-ice activity and the distribution of stem-rosette species in a Venezuelan Paramo. *AAR* 19:135–153.

1987c. Soil surface roughness and needle ice-induced particle movement in a Venezuelan paramo. *Carib. J. Sci.* 23:454–460.

1991. Particle sorting due to off-road vehicle traffic in a high Andean paramo. *Catena* 18:239–254.

Perkins, SM, Filippelli, GM and CJ Souch. 2000. Airborne trace metal contamination of wetland sediments at Indiana Dunes National Lakeshore. *Water Air Soil Poll.* 122:231–260.

Perrin, RMS, Willis, EH and DAH Hodge. 1964. Dating of humus Podzols by residual radiocarbon activity. *Nature* 202:165–166.

Perry, RS and JB Adams. 1978. Desert varnish: Evidence for cyclic deposition of managnese. *Nature* 276:489–491.

Perry, RS, Lynne, BY, Sephton, MA, Kolb, VM, Perry, CC and JT Staley. 2006. Baking black opal in the desert sun: The importance of silica in desert varnish. *Geology* 34:537–540.

Pécsi, M and G Richter. 1996. Löss Herkunft – Gliederung – Landschaften. *Zeit. für Geomorph. Neue Folge Supplement.* 98:391.

Petäja-Ronkainen, A, Peuraniemi, V and R Aario. 1992. On podzolization in glaciofluvial material in northern Finland. *Ann. Acad. Sci. Fenn. Geol. Geogr. Ser.* A3:156.

Petersen, L. 1976. *Podzols and Podzolization.* Royal Veterinary and Agric. Univ., Copenhagen.

Peterson, FF. 1980. Holocene desert soil formation under sodium salt influence in a playa-margin environment. *QR* 13:172–186.

1981. *Landforms of the Basin and Range Province. Nevada Agric. Exp. Station Tech. Bull.* 28. 52 pp.

Petit, JR, Jouzel, J, Raynaud, D, Barkov, NI, Barnola, J-M, Basile, I, Bender, M, Chappellaz, J, Davis, M, Delaygue, G, Delmotte, M, Kotlyakov, VM, Legrand, M, Lipenkov, VY, Lorius, C, Pépin, L, Ritz, C, Saltzman, E and M Stievenard. 1999. Climate and atmospheric history of the past 420,000 years from the Vostok ice core, Antarctica. *Nature* 399:429–436.

Peto, A. 2013. Studying modern soil profiles of different landscape zones in Hungary: An attempt to establish a soil-phytolith identification key. *QI* 287:149–161.

Pett-Ridge, JC, Derry, LA and AC Kurtz. 2009. Sr isotopes as a tracer of weathering processes and dust inputs in a tropical granitoid watershead, Luquillo Mountians, Puerto Rico. *Geochim. Cosmochim. Acta* 73:25–43.

Pettapiece, WW. 1975. Soils of the subarctic in the lower Mackenzie basin. *Arctic* 28:35–53.

Pettijohn, FJ. 1941. Persistence of heavy minerals and geologic age. *J. Geol.* 49:610–625.

Péwé, TL. 1966. Paloeclimatic significance of fossil ice wedges. *Biul. Peryglac.* 5:65–73.

(ed) 1969. *The Periglacial Environment Past and Present.* McGill-Queens' Univ. Press, Montreal.

Pfeiffer, E-M and H Janssen. 1994. Characterization of organic carbon, using the ^{13}C-value of a permafrost site in the Kolyma-Indigirka-lowland, northeast Siberia. In: JM Kimble and RJ Ahrens (eds). *Classification, Correlation, and Management of Permafrost-Affected Soils.* Lincoln, NE, USDA-SCS. pp. 90–98.

Pfeiffer, M, Aburto, F, Le Roux, JP, Kemnitz, H, Sedov, S, Solleiro-Rebolledo, E and O Seguel. 2012. Development of a Pleistocene calcrete over a sequence of marine terraces at Tongoy (north-central Chile) and its paleoenvironmental implications. *Catena* 97:104–118.

Pfisterer, U, Blume, HP and M Kanig. 1996. Genesis and dynamics of an Oxic Dystrochrept and a Typic Haploperox from ultrabasic rock in the tropical rain forest climate of south-east Brazil. *Zeit. Fur Pflanz. Bodenk.* 159:41–50.

Phillips, DH, Foss, JE and AC Goodyear. 2006. Micromorphology of lamellae formed in an alluvial soil, Big Pine Tree archaeological site, South Carolina. *SSH* 47:46–50.

Phillips, FM. 1994. Environmental tracers for water movement in desert soils of the American southwest. *SSSAJ* 58:15–24.

Phillips, FM, Leavy, BD, Jannik, NO, Elmore, D and PW Kubik. 1986. The accumulation of cosmogenic chlorine-36 in rocks: A method for surface exposure dating. *Science* 231:41–43.

Phillips, FM, Zreda, MG, Gosse, JC, Klein, J, Evenson, EB, Hall, RD, Chadwick, OA and P Sharma. 1997. Cosmogenic ^{36}Cl and ^{10}Be ages of Quaternary glacial and fluvial deposits of the Wind River Range, Wyoming. *GSA Bull.* 109:1453–1463.

Phillips, JD. 1989. An evaluation of the state factor model of soil ecosystems. *Ecol. Model.* 45:165–177.

1993a. Chaotic evolution of some coastal plain soils. *PG* 14:566–580.

1993b. Progressive and regressive pedogenesis and complex soil evolution. *QR* 40:169–176.

1993c. Stability implications of the state factor model of soils as a nonlinear dynamical system. *Geoderma* 58:1–15.

1999. *Earth Surface Systems.* Blackwell, Oxford, UK.

2001a. Divergent evolution and the spatial structure of soil landscape variability. *Catena* 43:101–113.

2001b. Inherited vs acquired complexity in east Texas weathering profiles. *Geomorphology* 40:1–14.

2001c. The relative importance of intrinsic and extrinsic factors in pedodiversity. *AAAG* 91:609–621.

2004. Geogenesis, pedogenesis, and multiple causality in the formation of texture-contrast soils. *Catena* 58:275–295.

2006. Evolutionary geomorphology: Thresholds and nonlinearity in landform response to environmental change. *Hydrol. Earth System Sci.* 10:731–742.

2007. Development of texture contrast soils by a combination of bioturbation and translocation. *Catena* 70:92–104.

2008. Soil system modeling and generation of field hypotheses. *Geoderma* 145:419–425.

2010. The convenient fiction of steady-state soil thickness. *Geoderma* 156:389–398.

Phillips, JD, Golden, H, Capiella, K, Andrews, B, Middleton, T, Downer, D, Kelli, D and L Padrick. 1999. Soil redistribution and pedologic transformations in coastal plain croplands. *ESPL* 24:23–39.

Phillips, JD and C Lorz. 2008. Origins and implications of soil layering. *Earth-Sci. Revs.* 89:144–155.

Phillips, JD, Luckow, K, Marion, DA and KR Adams. 2005. Rock fragment distributions and regolith evolution in the Ouachita Mountains, Arkansas, USA. *ESPL* 30:429–442.

Phillips, JD and DA Marion. 2005. Biomechanical effects, lithological variations, and local pedodiversity in some forest soils of Arkansas. *Geoderma* 124:73–89.

2006. Biomechanical effects of trees on soil and regolith: Beyond treethrow. *AAAG* 96:233–247.

Phillips, JD, Marion, DA, Luckow, K and KR Adams. 2005. Nonequilibrium regolith thickness in the Ouachita Mountains. *J. Geology* 113:325–340.

Phillips, JD, Marion, DA and AV Turkington. 2008. Pedologic and geomorphic impacts of a tornado blowdown event in a mixed pine-hardwood forest. *Catena* 75:278–287.

Phillips, JD, Perry, D, Garbee, AR, Carey, K, Stein, D, Morde, MB and JA Sheehy. 1996. Deterministic uncertainty and complex pedogenesis in some Pleistocene dune soils. *Geoderma* 73:147–164.

Phillips, MR. 2011. *The depth distribution of organic carbon in mineral Cryosols at two sites in the Canadian Arctic.* MS Thesis, Univ. of Saskatchewan.

Phillips, SE, Milnes, AR and RC Foster. 1987. Calcified filaments: An example of biological influences in the formation of calcrete in South Australia. *AJSR* 25:405–428.

Phillips, SE and PG Self. 1987. Morphology, crystallography and origin of needle-fibre calcite in Quaternary pedogenic calcretes of South Australia. *AJSR* 25:429–444.

Phillips, WM, McDonald, EV, Reneau, SL and J Poths. 1998. Dating soils and alluvium with cosmogenic 21Ne depth profiles: Case studies from the Pajarito Plateau, New Mexico, USA. *EPSL* 160:209–223.

Picker, MD, Ross-Gillespie, V, Vlieghe, K and E Moll. 2012. Ants and the enigmatic Namibian fairy circles – cause and effect? *Ecol. Entomol.* 37:33–42.

Pickering, EW and PLM Veneman. 1984. Moisture regimes and morphological characteristics in a hyrosequence in central Massachusetts. *SSSAJ* 48:113–118.

Pierce, KL, Muhs, DR, Fosberg, MA, Mahan, SA, Rosenbaum, JG, Licciardi, JM and MJ Pavich. 2011. A loess-paleosol record of climate and glacial history over the past two glacial-interglacial cycles (similar to 150 ka), southern Jackson Hole, Wyoming. *QR* 76:119–141.

Pierce, KL, Obradovich, JD and I Friedman. 1976. Obsidian hydration dating and correlation of Bull Lake and Pinedale glaciations near West Yellowstone, Montana. *GSA Bull.* 87:703–710.

Pierce, RS, Lull, HW and HC Storey. 1958. Influence of land use and forest condition on soil freezing and snow depth. *For. Sci.* 4:246–263.

Pierzynski, GM, Sims, JT and GF Vance. 2000. *Soils and Environmental Quality.* CRC Press, New York.

Pietsch, D. 2013. Krotovinas – soil archives of steppe landscape history. *Catena* 104:257–264.

Pigati, JS, McGeehin, JP, Muhs, DR and EA Bettis III. 2013. Radiocarbon dating late Quaternary loess deposits using small terrestrial gastropod shells. *QSR* 76:114–128.

Pigati, JS, Rech, JA and JC Nekola. 2010. Radiocarbon dating of small terrestrial gastropod shells in North America. *QG* 5:519–532.

Ping, CL. 1987. Soil temperature profiles of two Alaskan soils. *SSSAJ* 51:1010–1018.

1997. Characteristics of permafrost soils along a latitudinal transect in arctic Alaska. *Agroborealis* 29:35–36.

2000. Volcanic soils. B Houghton, H Rymer, J Stix, S McNutt and H Sigurdsson (eds). *Encyclopedia of Volcanoes.* Academic Press, San Diego, CA. pp. 1259–1270.

Ping, CL, Sumner, ME and LP Wilding. 2000. Andisols. In: ME Sumner (ed). *Handbook of Soil Science.* CRC Press, Boca Raton, FL. pp. E-209-E-224.

Pinheiro, J, Tejedor Salguero, M and A Rodriguez. 2004. Genesis of placic horizons in Andisols from Terceira Island Azores—Portugal. *Catena* 56:85–94.

Pinter, N, Keller, EA and RB West. 1994. Relative dating of terraces of the Owens River, Northern Owens Valley, California, and correlation with moraines of the Sierra Nevada. *QR* 42:266–276.

Pinton, R, Varanini, Z and P Nannipieri (eds) 2001. *The Rhizosphere: Biochemistry and Organic Substances at the Soil-Plant Interface.* Marcel Dekker, New York.

Piotrowska, N, Blaauw, M, Mauquoy, D and FM Chambers. 2010. Constructing deposition chronologies for peat deposits using radiocarbon dating. *Mires and Peat* 7:1–14.

Piperno, DR. 2006. *Phytoliths: A Comprehensive Guide for Archaeologists and Paleoecologists.* Alta Mira Press, Oxford, UK.

Piperno, DR and P Becker. 1996. Vegetational history of a site in the central Amazon Basin derived from phytolith and charcoal records from natural soils. *QR* 45:202–209.

Pisinaras, V, Tsihrintzis, VA, Petalas, C and K Ouzounis. 2010. Soil salinization in the agricultural lands of Rhodope District, northeastern Greece. *Envir. Monit. Assess.* 166:79–94.

Pissart, A. 1969. Le mécanisme périglaciaire dressant les pierres dans le sol: Résultats d'expériences. *C.R. Acad. Sci. (Paris)* 268:3015–3017.

Pitty, AF. 1968. Particle size of the Saharan dust which fell in Britain in July, 1968. *Nature* 220:364–365.

Plaster, RW and WC Sherwood. 1971. Bedrock weathering and residual soil formation in central Virginia. *GSA Bull.* 82:2813–2826.

Plice, MJ. 1942. Factors affecting soil color (progress report). *Proc. Oklahoma Acad. Sci.* 23:49–51.

Plummer, LN, Wigley, TML and DL Parkhurst. 1978. The kinetics of calcite dissolution in CO_2-water systems at 5° to 60° and 0. 0 to 1.0 atm CO_2. *AJS* 278:179–216.

Podniesinski, GS and DJ Leopold. 1998. Plant community development and peat strategy in forested fens in response to ground-water flow systems. *Wetlands* 18:409–430.

Poesen, J. 1987. Transport of rock fragments by rill flow: A field study. *Catena Suppl.* 8:35–54.

Poesen, J, Ingelmo-Sanchez, F and H Mücher. 1990. The hydrological response of soil surfaces to rainfall as affected by cover and position of rock fragments in the top layer. *ESPL* 15:653–671.

Poesen, J and H Lavee. 1994. Rock fragments in top soils: Significance and processes. *Catena* 23:1–28.

Poesen, JWB, van Wesemael, B, Bunte, K and AS Benet. 1998. Variation of rock fragment cover and size along semiarid hillslopes: A case-study from southeast Spain. *Geomorphology* 23:323–335.

Pohl, MM. 1995. Radiocarbon ages on organics from piedmont alluvium, Ajo Mountains, Arizona. *PG* 16:339–353.

Pointing, SB and J Belnap. 2012. Microbial colonization and controls in dryland systems. *Nature Revs. Microbiol.* 10:551–562.

Pokorny, P and V Jankovska. 2000. Long-term vegetation dynamics and the infilling process of a former lake (Svarcenberk, Czech Republic). *Folia Geobot.* 35:433–457.

Pokras, EM and AC Mix. 1985. Eolian evidence for spatial variability of late Quaternary climates in tropical Africa. *QR* 24:137–149.

Polach, HA and AB Costin. 1971. Validity of soil organic matter radiocarbon dating: Buried soils in Snowy Mountains, Southeastern Australia as example. In: DH Yaalon (ed). *Paleopedology. Origin, Nature and Dating of Paleosols.* Israel Univ. Press, Jerusalem. pp. 89–96.

Pomeroy, DE. 1976. Some effects of mound-building termites on soils in Uganda. *JSS* 27:377–394.

Ponce-Hernandez, R, Marriott, FHC and PHT Beckett. 1986. An improved method for reconstructing a soil profile from analyses of a small number of samples. *JSS* 37:455–467.

Ponge, J-F, Chevalier, R and P Loussot. 2002. Humus Index: An integrated tool for the assessment of forest floor and topsoil properties. *SSSAJ* 66:1996–2001.

Ponge, J-F, Jabiol, B and JC Gegout. 2011. Geology and climate conditions affect more humus forms than forest canopies at large scale in temperate forests. *Geoderma* 162:187–195.

Ponnamperuma, FN, Tianco, EM and TA Loy. 1967. Redox equilibria in flooded soils. I. The iron hydroxide systems. *SS* 103:374–382.

Ponomarenko, SV. 1988. Probable mechanism of redistribution of coarse fractions in the soil profile. *SSS* 20:35–41.

Ponomareva, VV. 1974. Genesis of the humus profile of Chernozem. *SSS* 6:393–402. (Translated from Pochvovedeniye 7:27–37.)

Pope, GA, Dorn, RI and JC Dixon. 1995. A new conceptual model for understanding geographical variations in weathering. *AAAG* 85:38–64.

Pope, GA, Meierding, TC and TR Paradise. 2002. Geomorphology's role in the study of weathering of cultural stone. *Geomorphology* 47:211–225.

Poppe, LJ, Paskevich, VF, Hathaway, JC and DS Blackwood. No date. A Laboratory Manual for X-Ray Powder Diffraction. USGS Open-File Rept. 01–041. http://pubs.usgs.gov/of/2001/of01-041/index.htm.

Portenga, EW and PR Bierman. 2011. Understanding Earth's eroding surface with [10]Be. *GSA Today* 21:4–10.

Porter, SC. 1978. Glacier Peak tephra in the NorthCascade Range, Washington: Stratigraphy, distribution, and relationship to late-glacial events. *QR* 10:30–41.

    1981. Lichenometric studies in the Cascade Range of Washington: Establishment of Rhizocarpon geographicum growth curves at Mount Ranier. *AAR* 13:11–23.

    2001. Chinese loess record of monsoon climate during the last glacial-interglacial cycle. *Earth-Sci. Revs.* 54:115–128.

Porter, SC, Pierce, KL and TD Hamilton. 1983. Late Wisconsin mountain glaciation in the western United States. In: HE Wright and SC Porter (eds). *Late-Quaternary Environments of the United States.* Vol. 1. *The Late Pleistocene.* Univ. of Minnesota Press, Minneapolis. pp. 71–111.

Porter, TP, Owens, PR, Lee, BD and D Marshall. 2008. Soil and landform characteristics related to landslide actvity – a review. *SSH* 49:22–26.

Porto, CG and M Hale. 1995. Gold redistribution in the stone line lateritic profile of the Posse deposit, central Brazil. *Econ. Geol.* 90:308–321

Post, FA and FR Dreibelbis. 1942. Some influence of frost penetration and microclimate on the water relationships of woodland, pasture, and cultivated soils. *SSSAP* 7:95–104.

Potter, RM and GR Rossman. 1977. Desert varnish: The importance of clay minerals. *Science* 196:1446–1448.

    1979. The manganese- and iron-oxide mineralogy of desert varnish. *Chem. Geol.* 25:79–94.

Prasolov, IJ. 1965. Soil regions of European Russia. *SSS* 4:343–354.

Pray, WS. 2005. Diarrhea: Sweeping changes in the OTC market. U.S. *Pharmacist* 30:1.

Prasse, R and R Bornkamm. 2000. Effect of microbiotic soil surface crusts on emergence of vascular plants. *Plant Ecol.* 150:65–75.

Prell, WL, Imbrie, J, Martinson, DG, Morley, JJ, Pisias, NG, Shackleton, NJ and HF Streeter. 1986. Graphic correlation of the oxygen isotope stratigraphy application to the Late Quaternary. *Paleoceanography* 1:137–162.

Preloger, DE. 2002. *Soil Survey of Franklin and Jefferson Counties, Illinois.* USDA Natural Resources Cons. Service, US Govt. Printing Office, Washington, DC.

Prescott, CE. 2010. Litter decomposition: What controls it and how can we alter it to sequester more carbon in forest soils? *Biogeochem.* 101:133–149.

Prescott, JR and JT Hutton. 1994. Cosmic-ray contributions to dose-rates for luminescence and ESR dating – Large depths and long-term time variations. *Rad. Measur.* 23:497–500.

Preusser, F, Degering, D, Fuchs, M, Hilgers, A, Kadereit, A, Klasen, N, Krbetschek, M, Richter, D and JQG Spencer. 2008. Luminescence dating: Basics, methods and applications. *Quat. Sci. J.* 57:95–149.

Prevost, DJ and BA Lindsay. 1999. *Soil Survey of Hualapai-Havasupai Area, Arizona: Parts of Coconino, Mojave, and Yavapai Counties.* USDA-NRCS, US Govt. Printing Off., Washington, DC.

Price, AG and BO Bauer. 1984. Small-scale heterogeneity and soil-moisture variability in the unsaturated zone. *J. Hydrol.* 70:277–293.

Price, JR and MA Velbel. 2003. Chemical weathering indices applied to weathering profiles developed on heterogeneous felsic metamorphic parent rocks. *Chem. Geol.* 202:397–416.

Price, TW, Blevins, RL, Barnhisel, RI and HH Bailey. 1975. Lithologic discontinuities in loessial soils of southwestern Kentucky. *SSSAP* 39:94–98.

Price, WA. 1949. Pocket gophers as architects of Mima (pimple) mounds of the western United States. *Texas J. Sci.* 1:1–17.

Prikryl, R, Melounova, L, Varilova, Z and Z Weishauptova. 2007. Spatial relationships of salt distribution and related physical changes of underlying rocks on naturally weathered sandstone exposures (Bohemian Switzerland National Park, Czech Republic). *Env. Geol.* 52:283–294.

Prior, JC. 1991. *Landforms of Iowa.* Univ. of Iowa Press, Iowa City.

Pritchett, WL. 1979. *Properties and Management of Forest Soils.* Wiley, New York.

Proctor, J. 1971. The plant ecology of serpentine. II. Plant response to serpentine soils. *JE* 59:397–410.

Proctor, J and SRJ Woodell. 1975. The ecology of serpentine soils. *Adv. Ecol. Res.* 9:256–365.

Prodi, F and G Fea. 1979. A case of transport and deposition of Saharan dust over the Italian peninsula and southern Europe. *J. Geophys. Res.* 84:6951–6960.

Prose DV and HG Wilshire. 2000. The lasting effect of tank maneuvers on desert soils and intershrub flora. USGS Open File Rept. 00–512.

Prospero, JM and PJ Lamb. 2003. African droughts and dust transport to the Caribbean: Climate change implications. *Science* 302:1024–1027.

Protz, R, Arnold, RW and EW Presant. 1968. The approximation of the true modal profile with the use of the high speed computer and landscape control. *Trans. 9th Intl. Congr. Soil Sci.* 4:193–204.

Protz, R, Ross, GJ, Martini, IP and J Terasmae. 1984. Rate of podzolic soil formation near Hudson Bay, Ontario. *CJSS* 64:31–49.

Pullan, RA. 1979. Termite hills in Africa: Their characteristics and evolution. *Catena* 6:267–291.

Pullen, A, Kapp, P, McCallister, AT, Chang, H, Gehrels, GE, Garzione, CN, Heermance, RV and L Ding. 2011. Qaidam Basin and northern Tibetan Plateau as dust sources for the Chinese Loess Plateau and paleoclimatic implications. *Geology* 39:1031–1034.

Pustovoytov, KE. 1998. Pedogenic carbonate cutans as a record of the Holocene history of relic tundra-steppes of the Upper Kolyma Valley (North-Eastern Asia). *Catena* 34:185–195.

2002. Pedogenic carbonate cutans on clasts in soils as a record of history of grassland ecosystems. *PPP* 177:199–214.

Pustovoytov, K, Schmidt, K and H Parzinger. 2007. Radiocarbon dating of thin pedogenic carbonate laminae from Holocene archaeological sites. *Holocene* 17:835–843.

Pustovoytov, KE and VO Targulian. 1996. Illuviation coatings on rock fragments as a source of pedogenic information. *Euras. Soil Sci.* 3:335–347.

Putkonen, J and T Swanson. 2003. Accuracy of cosmogenic ages for moraines. *QR* 59:255–261.

Putman, BR, Jansen, IJ and LR Follmer. 1989. Loessial soils: Their relationship to width of the source valley in Illinois. *SS* 146:241–247.

Pye, K. 1984. Loess. *PPG* 8:176–217.

1987. *Aeolian Dust and Dust Deposits.* Academic Press, London.

1988. Bauxites gathering dust. *Nature* 333:800–801.

1995. The nature, origin and accumulation of loess. *QSR* 14:653–667.

Pye, K and D Croft. 2007. Forensic analysis of soil and sediment traces by scanning electron microscopy and energy-dispersive X-ray analysis: An experimental investigation. *Forensic Sci. Intl.* 165:52–63.

Pye, K and R Johnson. 1988. Stratigraphy, geochemistry, and thermoluminescence ages of lower Mississippi Valley loess. *ESPL* 13:103–124.

Pye, K and H Tsoar. 1987. The mechanics and geological implications of dust transport and deposition in deserts with particular reference to loess formation and dune sand diagenesis in the northern Negev, Israel. In: LE Frostick and I Reid (eds). *Desert Sediments: Ancient and Modern.* Blackwell, Boston, MA. pp. 139–156.

1990. *Aeolian Sand and Sand Dunes.* Unwin Hyman, London.

Pyne-O'Donnell, SDF, Hughes, PDM, Froese, DG, Jensen, BJL, Kuehn, SC, Mallon, G, Amesbury, MJ, Charman, DJ, Daley, TJ, Loader, NJ, Mauquoy, D, Street-Perrott, FA and J Woodman-Ralph. 2012. High-precision ultra-distal Holocene tephrochronology in North America. *QSR* 52:6–11.

Qadir, M, Ghafoor, A and G Murtaza. 2000. Amelioration strategies for saline soils: A review. *Land Degrad. Develop.* 11:501–521.

Qi, F, Zhu, AX, Harrower, M and JE Burt. 2006. Fuzzy soil mapping based on prototype category theory. *Geoderma* 136:774–787.

Qin, CZ, Zhu, AX, Shi, X, Li, BL, Pei, T and CH Zhou. 2009. Quantification of spatial gradation of slope positions. *Geomorphology* 110:152–161.

Quade, DJ. 1998. *Surficial Geologic Materials of Linn County, Iowa.* Iowa Geol. Survey, Iowa City. Open File Map Series 98-2.

Quade, J. 2001. Desert pavements and associated rock varnish in the Mojave Desert: How old can they be? *Geology* 29:855–858.

Quade, J and TE Cerling. 1990. Stable isotope evidence for a pedogenic origin of carbonates in Trench 14 near Yucca Mountain, Nevada. *Science* 250:1549–1552.

Quade, J, Cerling, TE and JR Bowman. 1989. Systematic variations in the carbon and oxygen isotopic composition of pedogenic carbonate along elevation transects in the southern Great Basin, United States. *GSA Bull.* 101:464–475.

Quade, J, Chivas, AR and MT McCulloch. 1995. Strontium and carbon isotope tracers and the origins of soil carbonate in South Australia and Victoria. *PPP* 113:103–117.

Quade, J, Eiler, J, Daëron, M, and H Achyuthan. 2013. The clumped isotope geothermometer in soil and paleosol carbonate. *Geochim Cosmochim Acta* 105:92–107.

Quénard, L, Samouelian, A, Laroche, B and S Cornu. 2011. Lessivage as a major process of soil formation: A revisitation of existing data. *Geoderma* 167-68:135–147.

Quideau, SA, Chadwick, OA, Graham, RC and HB Wood. 1996. Base cation biogeochemistry and weathering under oak and pine: A controlled long-term experiment. *Biogeochem.* 35:377–398.

Raad, AT and R Protz. 1971. A new method for the identification of sediment stratification in soils of the Blue Springs Basin, Ontario. *Geoderma* 6:23–41.

Rabbi, SMF, Hua, Q, Daniel, H, Lockwood, PV, Wilson, BR and IM Young. 2013. Mean residence time of soil organic carbon in aggregates under contrasting land uses based on radiocarbon measurements. *Radiocarbon* 55:127–139.

Rabenhorst, MC and JE Foss. 1981. Soil and geologic mapping over mafic and ultramafic parent materials in Maryland. *SSSAJ* 45:1156–1160.

Rabenhorst, MC, Foss, JE and DS Fanning. 1982. Genesis of Maryland soils formed from serpentinite. *SSSAJ* 46:607–616.

Rabenhorst, MC and BR James. 1992. Iron sulfidization in tidal marsh soils. In: RW Fitzpatrick and HCW Skinner (eds). Biomineralization Processes of Iron and Manganese in Modern and Ancient Environments. *Catena Suppl.* 21:203–217.

Rabenhorst, MC, Hively, WD and BR James. 2009. Measurements of soil redox potential. *SSSAJ* 73:668–674.

Rabenhorst, MC and D Swanson. 2000. Histosols. In: ME Sumner (ed). *Handbook of Soil Science.* CRC Press, Boca Raton, FL. pp. E-183–E-209.

Rabenhorst, MC, West, LT and LP Wilding. 1991. Genesis of calcic and petrocalcic horizons in soils over carbonate rocks. *SSSASP* 26:61–74.

Rabenhorst, MC and LP Wilding. 1984. Rapid method to obtain carbonate-free residues from limestone and petrocalcic materials. *SSSAJ* 48:216–219.

1986a. Pedogenesis on the Edwards Plateau, Texas. I. Nature and continuity of parent materials. *SSSAJ* 50:678–687.

1986b. Pedogenesis on the Edwards Plateau, Texas. III. New model for the formation of petrocalcic horizons. *SSSAJ* 50:693–699.

Rabenhorst, MC, Wilding, LP and CL Girdner. 1984a. Airborne dusts in the Edwards Plateau region of Texas. *SSSAJ* 48:621–627.

Rabenhorst, MC, Wilding, LP and LT West. 1984b. Identification of pedogenic carbonates using stable carbon isotope and microfabric analyses. *SSSAJ* 48:125–132.

Radke, JK and EC Berry. 1998. Soil water and solute movement and bulk density changes in repacked soil columns as a result of freezing and thawing under field conditions. *SS* 163:611–624.

Radtke, U, Bruckner, H, Mangini, A and R Hausmann. 1988. Problems encountered with absolute dating (U-series, ESR) of Spanish calcretes. *QSR* 7:439–445.

Ragg, JM and DF Ball. 1964. Soils of the ultra-basic rocks of the Island of Rhum. *JSS* 15:124–133.

Rahn, PH. 1971. The weathering of tombstones and its relation to the topography of New England. *J. Geol. Educ.* 19:112–118.

Rai, D and WL Lindsay. 1975. A thermodynamic model for predicting the formation, stability and weathering of common soil minerals. *SSSAP* 32:443–444.

Raison, RJ, Woods, PV and PK Khanna. 1986. Decomposition and accumulation of litter after fire in sub-alpine eucalypt forests. *Austral. J. Ecol.* 11:9–19.

Rajakaruna, N, Harris, TB and EB Alexander. 2009. Serpentine geoecology of eastern North America: A review. *Rhodora* 111:21–108.

Ramsey, CB. 2008. Radiocarbon dating: Revolutions in understanding. *Archaeometry* 50:249–275.

2009. Bayesian analysis of radiocarbon dates. *Radiocarbon* 51:337–360.

Ranger, J, Dambrine, E, Robert, M, Righi, D and C Felix. 1991. Study of current soil-forming processes using bags of vermiculite and resins placed within soil horizons. *Geoderma* 48:335–350.

Ranney, RW and MT Beatty. 1969. Clay translocation and albic tongue formation in two Glossoboralfs in west-central Wisconsin. *SSSAP* 33:768–775.

Ranney, RW, Ciolkosz, EJ, Cunningham, RL, Petersen, GW and RP Matelski. 1975. Fragipans in Pennsylvania soils: Properties of bleached prisms face materials. *SSSAP* 39:695–698.

Ransom, MD and NE Smeck. 1986. Water table characteristics and water chemistry of seasonally wet soils of southwestern Ohio. *SSSAJ* 50:1281–1290.

Ransom, MD, Smeck, NE and JM Bigham. 1987a. Micromorphology of seasonally wet soils on the Illinoian till plain, USA. *Geoderma* 40:83–99.

1987b. Stratigraphy and genesis of polygenetic soils on the Illinoian till plain of southwestern Ohio. *SSSAJ* 51:135–141.

Rapp, A. 1984. Are terra rossa soils in Europe eolian deposits from Africa? Geol. *Foreningens Stockholm Forhandl.* 105:161–168.

Rapp, A and R Nyberg. 1981. Alpine debris flows in northern Scandinavia. *Geog. Ann.* 63A:183–196.

Rapp, G Jr and CL Hill. 1998. *Geoarchaeology: The Earth-Science Approach to Archaeological Interpretation.* Yale Univ. Press, New Haven, CT.

Rasmussen, SO, Selerstad, IK, Andersen, KK, Bigler, M, Dahl-Jensen, D and SJ Johnsen. 2008. Synchronization of the NGRIP, GRIP, and GISP2 ice cores across MIS 2 and palaeoclimatic implications. *QSR* 27:18–28.

Raukas, A, Mickelson, DM and A Dreimanis. 1978. Methods of till investigation in Europe and North America. *J. Sed. Petr.* 48:285–294.

Raw, F. 1962. Studies of earthworm populations in orchards. I. Leaf burial in apple orchards. *Ann. Appl. Biol.* 50:389–404.

Rawilli, JF, Chittleborough, DJ and R Beckett. 2005. Particle-size and elemental distribution of soil colloids: Implications for colloid transport. *SSSAJ* 69:1173–1184.

Rawling, JE III. 2000. A review of lamellae. *Geomorphology* 35:1–9.

Rawling, JE III, Fredlund, GG and S Mahan. 2003. Aeolian cliff-top deposits and buried soils in the White River Badlands, South Dakota, USA. *Holocene* 13:121–129.

Rawling, JE III, Hanson, PR, Young, AR, and JW Attig. 2008. Late Pleistocene dune construction in the central sand plain of Wisconsin, USA. *Geomorphology* 100:494–505.

Rawls, WJ. 1983. Estimating bulk density from particle size analysis and organic matter content. *SS* 135:123–125.

Ray, A. 1959. The effect of earthworms in soil pollen distribution. *J. Oxford Univ. For. Soc.* 7:16–21.

Rea, DL. 1990. Aspects of atmospheric circulation: The Late Pleistocene (0–950,000 yr) record of eolian deposition in the Pacific Ocean. *PPP* 78:217–227.

Read, G. 1998. The establishment of a buried soil chronosequence with a feldspar weathering index. *QI* 51/52:39.

Read, G, Kemp, RA and J Rose. 1996. Development of a feldspar weathering index and its application to a buried soil chronosequence in southeastern England. *Geoderma* 74:267–280.

Reardon, J, Curcio, G and R Bartlette. 2009. Soil moisture dynamics and smoldering combustion limits of pocosin soils in North Carolina, USA. *Intl. J. Wildland Fire* 18:326–335.

Rebertus, RA and SW Buol. 1985. Iron distribution in a developmental sequence of soils from mica gneiss and schist. *SSSAJ* 49:713–720.

Rech, JA, Nekola, JC and JS Pigati. 2012. Radiocarbon ages of terrestrial gastropods extend duration of ice-free conditions at the Two Creeks forest bed, Wisconsin, USA. *QR* 77:289–292.

Rech, JA, Quade, J and WS Hart. 2003. Isotopic evidence for the origin of Ca and S in soils gypsum, anhydrite, and calcite in the Atacama Desert, Chile. *Geochim. Cosmochim. Acta* 67:575–586.

Rech, JA, Reeves, RW and DM Hendricks. 2001. The influence of slope aspect on soil weathering processes in the Springerville volcanic field, Arizona. *Catena* 43:49–62.

Redding, T and K Devito. 2011. Aspect and soil textural controls on snowmelt runoff on forested Boreal Plain hillslopes. *Hydrol. Res.* 42:250–267.

Redmond, CE and JE McClelland. 1959. The occurrence and distribution of lime in calcium carbonate Solonchak and associated soils of eastern North Dakota. *SSSAP* 23:61–65.

Reed, EC and VH Dreeszen. 1965. Revision of the classification of the Pleistocene deposits of Nebraska. *Nebraska Geol. Survey Bull.* 23. 65pp.

Reeder, SW and AL McAllister. 1957. A staining method for the quantitative determination of feldspars in rocks and sands from soils. *CJSS* 37:57–59.

Rees-Jones, J and MS Tite. 1997. Optical dating results for British archaeological sediments. *Archaeometry* 39:177–188.

Reeves, CC Jr. 1970. Origin, classification, and geologic history of caliche on the southern High Plains, Texas and eastern New Mexico. *J. Geol.* 78:352–362.

Reheis, MC. 1987a. Climatic implications of alternating clay and carbonate formation in semiarid soils of south-central Montana. *QR* 27:270–282.

1987b. Gypsic soils on the Kane alluvial fans, Big Horn County, Wyoming. *USGS Bull.* 1590-C.

1988. Pedogenic replacement of aluminosilicate grains by CaCO3 in Ustollic Haplargids, South-Central Montana, U. S.A. *Geoderma* 41:243–261.

Reheis, MC, Goodmacher, JC, Harden, JW, McFadden, LD, Rockwell, TK, Shroba, RR, Sowers, JM and EM Taylor. 1995. Quaternary soils and dust deposition in southern Nevada and California. *GSA Bull.* 107:1003–1022.

Reheis, MC, Harden, JW, McFadden, LD and RR Shroba. 1989. Development rates of late Quaternary soils, Silver Lake Playa, California. *SSSAJ* 53:1127–1140.

Reheis, MC and R Kihl. 1995. Dust deposition in southern Nevada and California, 1984–1989: Relations to climate, source area, and source lithology. *J. Geophys. Res.* 100:8893–8918.

Reheis, MC, Sowers, JM, Taylor, EM, McFadden, LD and JW Harden. 1992. Morphology and genesis of carbonate soils on the Kyle Canyon Fan, Nevada, USA. *Geoderma* 52:303–342.

Reiche, P. 1943. Graphic representation of chemical weathering. *J. Sed. Petr.* 13:58–68.

Reiche, R. 1950. *A survey of weathering processes and products.* Univ. of New Mexico Publ. in Geology 3. Univ. of New Mexico Press, Albuquerque.

Reichman, OJ and SC Smith. 1991. Burrows and burrowing behavior by mammals. *Curr. Mammal.* 2:197–244.

Reicosky, DC, Kemper, WD, Langdale, GW, Douglas, CL Jr and PE Rasmussen. 1995. Soil organic matter changes resulting from tillage and biomass production. *J. Soil Water Cons.* 50:253–262.

Reid, DA, Graham, RC, Southard, RJ and C Amrhein. 1993. Slickspot soil genesis in the Carrizo Plain, California. *SSSAJ* 57:162–168.

Reider, RG. 1983. A soil catena in the Medicine Bow Mountains, Wyoming, USA., with reference to paleoenvironmental influences. *AAR* 15:181–192.

Reider, RG, Huckleberry, GA and GC Frison. 1988. Soil evidence for postglacial forest-grassland fluctuation in the Absaroka Mountains of northwestern Wyoming, USA. *AAR* 20:188–198.

Reid-Soukup, DA and AL Ulery. 2002. Smectites. In: JB Dixon and D Schulze (eds). *Soil Mineralogy with Environmental Applications*. Soil Sci. Soc. Am. Book Ser. 7. Soil Sci. Soc. Am., Madison, WI. pp. 467–500.

Reimann, T, Lindhorst, S, Thomsen, KJ, Murray, AS and M Frechen. 2012. OSL dating of mixed coastal sediment (Sylt, German Bight, North Sea). *QG* 11:52–67.

Reimann, T, Tsukamoto, S, Harff, J, Osadczuk, K and M Frechen. 2011. Reconstruction of Holocene coastal foredune progradation using luminescence dating – An example from the Swina barrier (southern Baltic Sea, NW Poland). *Geomorphology* 132:1–16.

Reimer, A and CF Shaykewich. 1980. Estimation of Manitoba soil temperatures from atmospheric meteorological measurements. *CJSS* 60:299–309.

Reimer, PJ. 2012. Refining the radiocarbon time scale. *Science* 338:337–338.

Reimer, PJ, Baillie, MGL, Bard, E, Bayliss, A, Beck, JW, Bertrand, CJH, Blackwell, PG, Buck, CE, Burr, GS, Cutler, KB, Damon, PE, Edwards, RL, Fairbanks, RG, Friedrich, M, Guilderson, TP, Hogg, AG, Hughen, KA, Kromer, B, McCormac, G, Manning, S, Bronk Ramsey, C, Reimer, RW, Remmele, S, Southon, JR, Stuiver, M, Talamo, S, Taylor, FW, van der Plicht, J and CE Weyhenmeyer. 2004. IntCal04 terrestrial radiocarbon age calibration, 0–26 cal kyr bp. *Radiocarbon* 46:1029–1058.

Reimer, PJ, Baillie, MGL, Bard, E, Bayliss, A, Beck, JW, Blackwell, PG, Bronk Ramsey, C, Buck, CE, Burr, GS, Edwards, RL, Friedrich, M, Grootes, PM, Guilderson, TP, Hajdas, I, Heaton, TJ, Hogg, AG, Hughen, KA, Kaiser, KF, Kromer, B, McCormac, FG, Manning, SW, Reimer, RW, Richards, DA, Southon, JR, Talamo, S, Turney, CSM, van der Plicht, J, and CE Weyhenmeyer. 2009. IntCal09 and Marine09 radiocarbon age calibration curves, 0–50,000 years cal BP. *Radiocarbon* 51:1111–1150.

Reinhart, KG. 1961. The problem of stones in soil-moisture measurement. *SSSAP* 25:268–270.

Rejman, J and R Brodowski. 2005. Rill characteristics and sediment transport as a function of slope length during a storm event on loess soil. *ESPL* 30:231–239.

Reneau, SL. 1993. Manganese accumulation in rock varnish on a desert piedmont, Mojave Desert, California, and application to evaluating varnish development. *QR* 40:309–317.

Reneau, SL, Dietrich, WE, Rubin, M, Donahue, DJ and AJT Jull. 1989. Analysis of hillslope erosion rates using dated colluvial deposits. *J. Geol.* 97:45–63.

Reneau, SL and WE Dietrich. 1990. Depositional history of hollows on steep hillslopes, coastal Oregon and Washington, Natl. *Geogr. Res.* 6:220–230.

Reneau, SL and R Raymond Jr. 1991. Cation-ratio dating of rock varnish: Why does it work? *Geology* 19:937–940.

Rengasamy, P. 2006. World salinization with emphasis on Australia. *J. Exper. Bot.* 57:1017–1023.

Resner, K, Yoo, K, Hale, C, Aufdenkampe, -A, Blum, A and S Sebestyen. 2011. Elemental and mineralogical changes in soils due to bioturbation along an earthworm invasion chronosequence in Northern Minnesota. *Appl. Geochem.* 26:S127–S131.

Retallack, GJ. 1990. *Soils of the Past*. Unwin Hyman, Boston, MA.

1994. The environmental factor approach to the interpretation of paleosols. In: R Amundson, JW Harden and MJ Singer (eds). Factors of Soil Formation: A Fiftieth Anniversary Retrospective. *Soil Sci. Soc. Am. Spec. Publ.* 33:31–64.

2004. Late Oligocene bunch grassland and early Miocene sod grassland paleosols from central Oregon, USA. *PPP* 207:203–237.

2010. Lateritization and bauxitization events. *Econ. Geol.* 105:655–667.

Retallack, GJ and CM Huang. 2010. Depth to gypsic horizon as a proxy for paleoprecipitation in paleosols of sedimentary environments. *Geology* 38:403–406.

Retzer, JL. 1956. Alpine soils of the Rocky Mountains. *JSS* 7:22–32.

1965. Present soil-forming factors and processes in Arctic and alpine regions. *SS* 99:38–44.

Reusser, L, Graly, J, Bierman, P and D Rood. 2010. Calibrating a long-term meteoric ^{10}Be accumulation rate in soil. *Geophys. Res. Lett.* 37:L19403.

Reuter, G. 1965. Tonminerals in Pseudogleyboden. *Wiss. Z. Friedrich Schiller Univ. Jena, Math. Naturwiss. Reihe.* 14:75–78.

Reuter, RJ and JC Bell. 2001. Soils and hydrology of a wet–sandy catena in east–central Minnesota. *SSSAJ* 65:1559–1569.

2003. Hillslope hydrology and soil morphology for a wetland basin in south-central Minnesota. *SSSAJ* 67:365–372.

Reuter, RJ, McDaniel, PA, Hammel, JE and AL Falen. 1998. Solute transport in seasonal perched water tables in loess-derived soilscapes. *SSSAJ* 62:977–983.

Reynders, JJ. 1972. A study of argillic horizons in some soils in Morocco. *Geoderma* 8:267–279.

Reynolds, RL and JW King. 1995. Magnetic records of climate change. *Rev. Geophys. Suppl.* 33:101–110.

Reynolds, WD, Elrick, DE, Youngs, EG, Amoozegar, A, Booltink, HWG and J Bouma. 2002. The soil solution phase: Saturated and field-saturated water flow parameters. In: JH Duane and GC Topp (eds). *Methods of Soil Analysis*. Part 4. *Physical Methods*. SSSA Book Series no. 5. Soil Sci. Soc. Am. Madison, WI. pp. 797–878.

Rhoades, JD, Lesch, SM, Shouse, PJ and WJ Alves. 1990. Locating sampling sites for salinity mapping. *SSSAJ* 54:1799–1803.

Rhodes, EJ. 2011. Optically stimulated luminescence dating of sediments over the past 200,000 years. *Ann Rev. Earth Planet. Sci.* 39:461–488.

Rhodes, EJ, Fanning, PC and SJ Holdaway. 2010. Developments in optically stimulated luminescence age control for geoarchaeological sediments and hearths in western New South Wales, Australia. *QG* 5:348–52.

Rice, TJ, Jr, Buol, SW and SB Weed. 1985. Soil-saprolite profiles derived from mafic rocks in the North Carolina piedmont. I. Chemical, morphological, and mineralogical characteristics and transformations. *SSSAJ* 49:171–178.

Richards, KS, Arnett, RR and S Ellis. (eds) 1985. *Geomorphology and Soils*. Allen and Unwin, London.

Richards, PJ. 2009. *Aphaenogaster* ants as bioturbators: Impacts on soil and slope processes. *Earth-Sci. Revs.* 96:92–106

Richards, PJ, Hohenthal, JM and GS Humphreys. 2011. Bioturbation on a south-east Australian hillslope: Estimating contributions to soil flux. *ESPL* 36:1240–1253.

Richards, PL and LR Kump. 1997. Application of the geographical information systems approach to watershed mass balance studies. *Hydrol. Proc.* 11:671–694.

Richards, PW. 1941. Lowland tropical podzols and their vegetation. *Nature* 148:129–131.

1952. *The Tropical Rainforest*. Cambridge Univ. Press, London.

Richardson, CJ. 1983. Pocosins: Vanishing wastelands or valuable wetlands? *BioScience* 33:626–633.

2003. Pocosins: Hydrologically isolated or integrated wetlands on the landscape? *Wetlands* 23:563–576.

Richardson, JL and RJ Bigler. 1984. Principal component analysis of prairie pothole soils in North Dakota. *SSSAJ* 48:1350–1355.

Richardson, JL and WJ Edmonds. 1987. Linear regression estimations of Jenny's relative effectiveness of state factors equation. *SS* 144:203–208.

Richardson, JL and FD Hole. 1979. Mottling and iron distribution in a Glossoboralf-Haplaquoll hydrosequence on a glacial moraine in northwestern Wisconsin. *SSSAJ* 43:552–558.

Richardson, JL, Wilding, LP and RB Daniels. 1992. Recharge and discharge of groundwater in aquic conditions illustrated with flow-net analysis. *Geoderma* 53:65–78.

Richmond, GM and DS Fullerton. 1986. Summation of Quaternary glaciations in the United States of America. *QSR* 5:183–196.

Richmond, GR. 1986. Stratigraphy and chronology of glaciations in Yellowstone National Park. *QSR* 5:83–98.

Richter, DD. 2007. Humanity's transformation of Earth's soil: Pedology's new frontier. *SS* 172:957–967.

Richter, DD and LI Babbar. 1991. Soil diversity in the tropics. *Adv. Ecol. Res.* 21:315–389.

Richter, DD and D Markewitz. 1995. How deep is soil? *Bioscience* 45:600–609.

Richter, DD, Jr, and D Markewitz. 2001. *Understanding Soil Change: Soil Sustainability over Millennia, Centuries, and Decades.* Cambridge Univ. Press, Cambridge.

Richter, DD and DH Yaalon. 2012. "The Changing Model of Soil" revisited. *SSSAJ* 76:766–778.

Rickard, DT. 1973. Sedimentary iron sulphide formation. In: H Dost (ed). *Acid Sulfate Soils.* Vol. 1. ILRI. Publ. 18. Wageningen, The Netherlands. pp. 28–65.

Ricker, KE, Chinn, TJ and MJ McSaveney. 1993. A late Quaternary moraine sequence dated by rock weathering rinds, Craigieburn Range, New Zealand. *CJES* 30:1861–1869.

Ridge, JC, Evenson, EB and WD Sevon. 1992. A model of Late Quaternary Landscape development in the Delaware valley, New Jersey and Pennsylvania. *Geomorphology* 4:319–345.

Ridings, R. 1996. Where in the world does obsidian hydration dating work? *Am. Antiq.* 61:136–148.

Riecken, FF. 1945. Selection of criteria for the classification of certain soil types of Iowa into Great Soil Groups. *SSSAP* 10:319–325.

Riecken, FF and E Poetsch. 1960. Genesis and classification considerations of some prairie-formed soil profiles from local alluvium in Adair County, Iowa. *Proc. Iowa Acad. Sci.* 67:268–276.

Riehl, S, Pustovoytov, KE, Hotchkiss, S and RA Bryson. 2009. Local Holocene environmental indicators in Upper Mesopotamia: Pedogenic carbonate record vs. archaeobotanical data and archae-oclimatological models. *QI* 209:154–162.

Riestenberg, MM and S Sovonick-Dunford. 1983. The role of woody vegetation in stabilizing slopes in the Cincinnati area, Ohio. *GSA Bull.* 94:506–518.

Riggins, SG, Anderson, RS and AM Tye. 2011. Solving a conundrum of a steady-state hilltop with variable soil depths and production rates, Bodmin Moor, UK. *Geomorphology* 128:73–84.

Righi, D, Bravard, S, Chauvel, A, Ranger, J and M Robert. 1990. In situ study of soil processes in an Oxisol-Spodosol sequence of Amazonia Brazil. *SS* 150:438–445.

Righi, D, van Ranst, E, DeConinck, F and B Guillet. 1982. Microprobe study of a Placohumod in the Antwerp Campine (North Belgium). *Pédologie* 32:117–134.

Rightmire, CT. 1967. *A radiocarbon study of the age and origin of caliche deposits.* MA thesis, Univ. of Texas, Austin.

Riise, G, Van Hees, P, Lundström, U and LT Strand. 2000. Mobility of different size fractions of organic carbon, Al, Fe, Mn and Si in Podzols. *Geoderma* 94:237–247.

Rindfleisch, PR and RJ Schaetzl. 2001. Soils and geomorphic evidence for a high lake stand in a Michigan drumlin field. *PG* 22:483–501.

Ritchie, AM. 1953. The erosional origin of Mima mounds of southwest Washington. *J. Geol.* 61:41–50.

Ritchie, JC. 1987. *Postglacial Vegetation of Canada.* Cambridge Univ. Press, Cambridge, England.

Rittenour, TM. 2008. Luminescence dating of fluvial deposits: Applications to geomorphic, palaeoseismic and archaeological research. *Boreas* 4:613–635.

Roach, J, Griffith, B, Verbyla, D and J Jones. 2011. Mechanisms influencing changes in lake area in Alaskan boreal forest. *GCB* 17:2567–2583.

Robert, M. 1973. The experimental transformation of mica toward smectite: Relative importance of total charge and tetrahedral charge. *Clay Mins.* 21:168–174.

Roberts, BA. 1980. Some chemical and physical properties of serpentine soils from western Newfoundland. *CJSS* 60:231–240.

Roberts, DC, Campbell, JE and TL Kroll. 1988. *Soil survey of Oconto County, Wisconsin.* USDA Soil Cons. Service, US Govt. Printing Off., Washington, DC.

Roberts, EG. 1961. Soil settlement caused by tree roots. *J. For.* 59:25–26.

Roberts, FJ and BA Carbon. 1972. Water repellence in sandy soils of south-western Australia. *AJSR* 10:35–42.

Roberts, HM, Muhs, DR, Wintle, AG, Duller, GAT and EA Bettis III. 2003. Unprecedented last-glacial mass accumulation rates determined by luminescence dating of loess from western Nebraska. *QR* 59:411–419.

Roberts, HM and AJ Plater. 2007. Reconstruction of Holocene foreland progradation using optically stimulated luminescence (OSL) dating: An example from Dungeness, UK. *Holocene* 17:495–505.

Robertson, KM and DL Johnson. 2004. Vertical redistribution of pebbles by crayfish in Mollisol catenas of central Illinois. *SS* 169:776–786.

Robins, CR, Brock-Hon, AL and BJ Buck. 2012. Conceptual mineral genesis models for calcic pendants and petrocalcic horizons, NV. *SSSAJ* 76:1887–1903.

Robinson, GH and CI Rich. 1960. Characteristics of the multiple yellowish-red bands common to certain soils of the southeastern United States. *SSSAP* 24:226–230.

Rockie, WA. 1934. Snowdrifts and the Palouse topography. *Geog. Rev.* 24:380–385.

Rodbell, DT. 1993. Subdivision of Late Pleistocene moraines in the Cordillera Blanca, Peru, based on rock-weathering features, soils, and radiocarbon dates. *QR* 39:133–143.

Rode, AA. 1937. The process of podzolization. *Pochvovedeniye.* 32:849–862. In: JS Joffe (ed). *Pedology,* 2nd ed., 1949. *Pedology Pubs.,* New Brunswick, NJ.

———. 1964. Podzolization and lesivage. *Pochvov.* 7:9–23.

Rodnight, H, Duller, GAT, Wintle, AG and S Tooth. 2006. Assessing the reproducibility and accuracy of optical dating of fluvial deposits. *QG* 1:109–120.

Rodriguez, JLG and MC Suarez. 2010. Estimation of slope length value of RUSLE factor L Using GIS. *J. Hydrol. Engin.* 15:714–717.

Roering, JJ, Marshall, J, Booth, AM, Mort, M and QS Jin. 2010. Evidence for biotic controls on topography and soil production. *Earth Planet. Sci. Letts.* 298:183–190.

Rogers, LE. 1972. The ecological effects of the western harvester ant (Pogonomyrmex occidentalis) in the shortgrass plain ecosystem. *USA/BP Grassland Biome Tech. Rept.* 206.

Rohde, R., Muller, RA, Jacobsen, R., Muller, E., Perlmutter, S., Rosenfeld, A., Wurtele, J., Groom, D., and C. Wickham. 2012. A new estimate of the average Earth surface temperature spanning 1753 to 2011. *Geoinform. Geostat.* 1. doi.org/10.4172/gigs.1000101

Rolfsen, P. 1980. Disturbance of archaeological layers by processes in the soils. *Norw. Arch. Rev.* 13:110–118.

Romanovskij, NN. 1973. Regularities in formation of frost-fissures and development of frost-fissure polygons. *Biul. Peryglac.* 23:237–277.

Romell, LG and SO Heiberg. 1931. Types of humus layer in the forests of northeastern US. *Ecology* 12:567–608.

Roose, E. 1980. *Dynamique actuelle de sols ferrallitiques et ferrugineux tropicaux d'Afrique Occidentale: Etudeexpérimentale des transferts hydrologiques et biologiques de matières sous végétations naturelles ou cultivées.* Bordeaux, France, ORSTOM.

Rossel, RAV, Fouad, Y and C Walter. 2008. Using a digital camera to measure soil organic carbon and iron contents. *Biosyst. Engr.* 100:149–159.

Rostad, HPW, Smeck, NE and LP Wilding. 1976. Genesis of argillic horizons in soils derived from coarse-textured calcareous gravels. *SSSAJ* 40:739–744.

Rostagno, CM and G Degorgue. 2011. Desert pavements as indicators of soil erosion on aridic soils in north-east Patagonia (Argentina). *Geomorphology* 134:224–231.

Roth, K, Schulin, R, Fluhler, H and W Attinger. 1990. Calibration of time domain reflectometry for water content measurement using a composite dielectric approach. *Water Resources Res.* 26:2267–2273.

Rounsevell, MDA, Evans, SP and P Bullock. 1999. Climate change and agricultural soils: Impacts and adaptation. *Clim. Change* 43:683–709.

Rourke, RV, Brasher, BR, Yeck, RD and FT Miller. 1988. Characteristic morphology of U. S. Spodosols. *SSSAJ* 52:445–449.

Rousseau, DD, Gerasimenko, N, Matviishina, Z and G Kukla. 2001. Late Pleistocene environments of the Central Ukraine. *QR* 56:349–356.

Rousseau, D-D and G Kukla. 1994. Late Pleistocene climate record in the Eustis loess section, Nebraska, based on land snail assemblages and magnetic susceptibility. *QR* 42:176–187.

Rousseau, D-D, Wu, N and Z Guo. 2000. The terrestrial molluses as new indices of the Asian paleomonsoons in the Chinese loess Plateau. *Global Planet. Change* 26:199–206.

Rovey, CW. 1997. The nature and origin of gleyed polygenetic paleosols in the loess covered glacial drift plain of Northern Missouri, USA. *Catena* 31:153–172.

Rovey, CW II and G Balco. 2009. Periglacial climate at the 2. 5 Ma onset of Northern Hemisphere glaciation inferred from the Whippoorwill Formation, northern Missouri, USA. *QR* 73:151–161.

Rovey, CW II and WF Kean. 1996. Pre-Illinoian glacial stratigraphy in north-central Missouri. *QR* 45:17–29.

Rovner, I. 1971. Potential of opal phytoliths for use in paleoecological reconstruction. *QR* 1:343–359.

——— 1988. Macro- and micro-ecological reconstruction using plant opal phytolith data from archaeological sediments. *Geoarcheology* 3:155–163.

Rowe, PB. 1955. Effects of the forest floor on disposition of rainfall in pine stands. *J. For.* 53:342–348.

Roy, BB and NK Barde. 1962. Some characteristics of the Black Soils of India. *SS* 93:142–147.

Roy, BB, Ghose, B and S Pandey. 1967. Landscape-soil relationship in Chohtan Block in Barmer District in western Rajasthan. *J. Ind. Soc. Soil Sci.* 15:53–59.

Roy, M, Martin, JB, Cherrier, J, Cable, JE and CG Smith. 2010. Influence of sea level rise on iron diagenesis in an east Florida subterranean estuary. *Geochim. Cosmochim. Acta* 74:5560–5573.

Royer, DL. 1999. Depth to pedogenic carbonate horizon as a paleoprecipitation indicator? *Geology* 27:1123–1126.

Rozanov, BG. 1957. The bleached contact horizon in soils formed from twofold parent material. *Pochvov.* 6:16–23.

Rubilin, YV. 1962. Prairie soils of North America. *SSS* 610–617.

Rubin, J. 1967. Optimal classification into groups: An approach for solving the taxonomy problem. *J. Theor. Biol.* 15:103–144.

Rubio, A and A Escudero. 2005. Effect of climate and physiography on occurrence and intensity of decarbonation in Mediterranean forest soils of Spain. *Geoderma* 125:309–319.

Ruckamp, D, Martius, C, Bornemann, L, Kurzatkowski, D, Naval, LP and W Amelung. 2012. Soil genesis and heterogeneity of phosphorus forms and carbon below mounds inhabited by primary and secondary termites. *Geoderma* 170:239–250.

Rudberg, S. 1972. Periglacial zonation: A discussion. *Göttinger Geog. Abh.* 60:221–233.

Ruddiman, WF. 2008. *Earth's Climate: Past and Future.* WH Freeman and Co., New York.

Ruddiman, WF and HE Wright Jr. 1987. Introduction. In: WF Ruddiman and HE Wright Jr (eds). *North America and Adjacent Oceans during the Last Deglaciation.* Geol. Soc. Am., Boulder, CO. pp. 1–12.

Ruedrich, J, Kirchner, D and S Siegesmund. 2011. Physical weathering of building stones induced by freeze-thaw action: A laboratory longterm study. *Env. Earth-Sci.* 63:1573–1586.

Ruellan, A. 1971. The history of soils: Some problems of definition and interpretation. In: DH Yaalon (ed). *Paleopedology. Origin, Nature and Dating of Paleosols.* Israel Univ. Press, Jerusalem. pp. 3–13.

Ruhe, RV. 1954. Relations of the properties of Wisconsin loess to topography in western Iowa. *AJS* 252:663–672.

——— 1956a. Landscape evolution in the High Ituri, Belgian Congo. INEAC Ser. Sci. 66.

——— 1956b. Geomorphic surfaces and the nature of soils. *SS* 82:441–455.

——— 1958. Stone lines in soils. *SS* 84:223–231.

——— 1960. Elements of the soil landscape. *Trans. 7th Intl. Congr. Soil Sci.* 4:165–170.

——— 1965. Quaternary paleopedology. In: HE Wright and DG Frey (eds). *Quaternary of the United States.* Princeton Univ. Press, Princeton, NJ. pp. 755–764.

——— 1967. Geomorphic surfaces and surficial deposits in southern New Mexico. New Mexico Bur. Mines Mineral Resources Mem. 18.

——— 1969. *Quaternary Landscapes in Iowa.* Iowa State Univ. Press, Ames.

——— 1970. Soils, paleosols, and environment. In: W Dort and JK Jones (eds). *Pleistocene and Recent Environments of the Central Great Plains.* Univ. of Kansas Press, Lawrence. pp. 37–52.

——— 1973. Background of model for loess-derived soils in the upper Mississippi Valley. *SS* 115:250–253.

——— 1974a. Holocene environments and soil geomorphology in Midwestern United States. *QR* 4:487–495.

——— 1974b. Sangamon paleosols and Quaternary environments in midwestern United States. *Geog. Monogr.* 5:153–167.

——— 1975a. Climatic geomorphology and fully developed slopes. *Catena* 2:309–320.

——— 1975b. *Geomorphology.* Houghton-Mifflin, Boston, MA.

Ruhe, R and J Cady. 1954. Latosolic soils on central African interior high plateaus. *Trans. 5th Intl. Congr. Soil Sci.* 4:401–407.

Ruhe, RV and RB Daniels. 1958. Soils, paleosols, and soil-horizon nomenclature. *SSSAP* 22:66–69.

Ruhe, RV, Daniels, RB and JG Cady. 1967. *Landscape Evolution and Soil Formation in Southwestern Iowa,* Tech. Bull. 1349. USDA-SCS, US Govt. Printing Off., Washington, DC.

Ruhe, RV, Hall, RD and AP Canepa. 1974. Sangamon paleosols of southwestern Indiana, USA. *Geoderma* 12:191–200.

Ruhe, RV, Miller, GA and WJ Vreeken. 1971. Paleosols, loess sedimentation and soil stratigraphy. In: DH Yaalon (ed). *Paleopedology. Origin,*

Nature and Dating of Paleosols. Israel Univ. Press, Jerusalem. pp. 41–60.

Ruhe, RV and CG Olson. 1980. Soil welding. *SS* 130:132–139.

Ruhe, RV, Rubin, M and WH Scholtes. 1957. Late Pleistocene radiocarbon chronology in Iowa. *AJS* 255:671–689.

Ruhe, RV and WH Scholtes. 1956. Ages and development of soil landscapes in relation to climatic and vegetational changes in Iowa. *SSSAP* 20:264–273.

———. 1959. Important elements in the classification of the Wisconsinan glacial stage: A discussion. *J. Geol.* 67:585–593.

Ruhe, RV and PH Walker. 1968. Hillslope models and soil formation. I. Open systems. *Trans 9th Intl. Cong. Soil Sci.* 4:551–560.

Rumelhart, D and J McClelland. 1986. *Parallel Distributed Processing.* MIT Press, Cambridge, MA.

Rumpel, C, Kogel-Knabner, I and RF Huttl. 1999. Organic matter composition and degree of humification in lignite-rich mine soils under a chronosequence of pine. *Plant Soil* 213:161–168.

Runge, ECA. 1973. Soil development sequences and energy models. *SS* 115:183–193.

Runge, ECA, Peck, TR and RL Hoeft. 2008. Amelioration of Typic and Albic Natraqualfs using calcium chloride. *SS* 173:831–836.

Runge, F. 1999. The opal phytolith inventory of soils in Central Africa: Quantities, shapes, classification, and spectra. *Rev. Paleobot. Palynol.* 107:23–53.

Rusanov, AM and EA Milyakova. 2005. The effect of slope aspect on the properties of southern chernozems in the Cis-Ural region. *Euras. Soil Sci.* 38:569–575.

Russell, EWB. 1993. Early stages of secondary succession recorded in soil pollen on the North Carolina Piedmont. *AMN* 129:384–396.

Rustomji, P and T Pietsch. 2007. Alluvial sedimentation rates from southeastern Australia indicate post-European settlement landscape recovery. *Geomorphology* 90:73–90.

Rutledge, EM, Holowaychuk, N, Hall, GF and LP Wilding. 1975a. Loess in Ohio in relation to several possible source areas. I. Physical and chemical properties. *SSSAP* 39:1125–1132.

Rutledge, EM, Wilding, LP, Hall, GF and N Holowaychuk. 1975b. Loess in Ohio in relation to several possible source areas. II. Elemental and mineralogical composition. *SSSAP* 39:1133–1139.

Rutter, N. 2009. International Year of Planet Earth 4. Utilizing paleosols in Quaternary climate change studies. *Geosci. Canada* 36:65–77.

Rutter, N and Z Ding. 1993. Paleoclimates and monsoon variations interpreted from micromorphogenic features of the Baoji paleosols, China. *QSR* 12:853–862.

Rutter, NW, Velichko, AA, Dlussky, KG, Morozova, TD, Little, EC, Nechaev, VP and ME Evans. 2006. New insights on the loess/paleosol Quaternary stratigraphy from key sections in the US Midwest. *Catena* 67:15–34.

Ruxton, BP. 1968. Measures of the degree of chemical weathering of rocks. *J. Geol.* 76:518–527.

Ruzyla, K and DI Jezek. 1987. Staining method for recognition of pore space in thin and polished sections. *J. Sed. Petr.* 57:777–778.

Sadleir, SB and RJ Gilkes. 1976. Development of bauxite in relation to parent material near Jarrahdale, Western Australia. *J. Geol. Soc. Austral.* 23:333–334.

Sak, PB, Fisher, DM, Gardner, TW, Murphy, K and SL Brantley. 2004. Rates of weathering rind formation on Costa Rican basalt. *Geochim. Cosmochim. Acta* 68:1453–1472.

Sak, PB, Navarre-Sitchler, AK, Miller, CE, Daniel, CC, Gaillardet, J, Buss, HL, Lebedeva, MI and SL Brantley. 2010. Controls on rind thickness on basaltic andesite clasts weathering in Guadeloupe. *Chem. Geol.* 276:129–143.

Salem, MZ and FD Hole. 1968. Ant (Formica exsectoides) pedoturbation in a forest soil. *SSSAP* 32:563–567.

Salik, J, Herrera, R and CF Jordan. 1983. Termitaria: Nutrient patchiness in nutrient deficient rain forests. *Biotropica* 15:1–7.

Salleh, KO. 1994. Colluvium thickness and its relationships to vegetation cover density and slope gradient: An observation for part of Murcia Province, SE Spain. *Geog. Fis. Dinam. Quat.* 17:187–195.

Salomons, W, Goudie, A and WG Mook. 1978. Isotopic composition of calcrete deposits from Europe, Africa and India. *ESPL*:43–57.

Salter, PJ and JB Williams. 1965. The influence of texture on the moisture characteristics of soils. II. Available-water capacity and moisture release characteristics. *JSS* 16:310–317.

Salvador-Blanes, S, Minasny, B and AB McBratney. 2007. Modelling long-term in situ soil profile evolution: Application to the genesis of soil profiles containing stone layers. *EJSS* 58:1535–1548.

Salzer, MW and MK Hughes. 2007. Bristlecone pine tree rings and volcanic eruptions over the last 5000 yr. *QR* 67:57–68.

Šamonil, P, Kral, K, and L Hort. 2010a. The role of tree uprooting in soil formation: A critical literature review. *Geoderma* 157:65–79.

Šamonil, P, Schaetzl, RJ, Valtera, M, Goliáš, V, Baldrian, P, Vašíčková, I, Adam, D, Janík, D and L Hort. 2013. Crossdating of disturbances by tree uprooting: Can treethrow microtopography persist for 6,000 years? *For. Ecol. Manag.* 307:123–135.

Šamonil, P, Tejnecky, V, Boruvka, L, Sebkova, B, Janik, D and O Sebek. 2010b. The role of tree uprooting in Cambisol development. *Geoderma* 159:83–98.

Šamonil, P, Valtera, M, Bek, S, Sebkova, B, Vrska, T and J Houska. 2011. Soil variability through spatial scales in a permanently disturbed natural spruce-fir-beech forest. *Eur. J. For. Res.* 130:1075–1091.

Sampson, KM and LC Smith. 2006. Relative ages of Pleistocene moraines discerned from pebble counts: Eastern Sierra Nevada, California. *PG* 27:223–235.

Sanborn, P and S Pawluk. 1983. Process studies of a chernozemic pedon, Alberta (Canada). *Geoderma* 31:205–237.

Sanchez, PA. 1976. *Properties and Management of Soils in the Tropics.* Wiley, New York.

Sanchez, PA, Ahamed, S, Carre, F, Hartemink, AE, Hempel, J, Huising, J, Lagacherie, P, McBratney, AB, McKenzie, NJ, Mendonca-Santos, M de L, Minasny, B, Montanarella, L, Okoth, P, Palm, CA, Sachs, JD, Shepherd, KD, Vågen, T-G, Vanlauwe, B, Walsh, MG, Winowiecki, LA and G-L Zhang. 2009. Digital soil map of the world. *Science* 325:680–681.

Sancho, C and A Meléndez. 1992. Génesis y significado ambiental de los caliches Pleistocenos de la región del Cinca (Depresión del Ebro). *Rev. Soc. Geol. España* 5:81–93.

Sandler, A. 2013. Clay distribution over the landscape of Israel: From the hyper-arid to the Mediterranean climate regimes. *Catena* 110:119–132.

Sandor, JA, Gersber, PL and JW Hawley. 1986a. Soils at prehistoric terracing sites in New Mexico. I. Site placement, soil morphology, and classification. *SSSAJ* 50:166–173.

———. 1986b. Soils at prehistoric terracing sites in New Mexico. II. Organic matter and bulk density changes. *SSSAJ* 50:173–177.

———. 1986c. Soils at prehistoric terracing sites in New Mexico. III. Phosphorous, selected micronutrients, and pH. *SSSAJ* 50:177–180.

Sandoval, FM and L Shoesmith. 1961. Genetic soil relationships in a saline glacio-lacustrine area. *SSSAP* 25:316–320.

Santana, DP. 1984. *Soil formation in a toposequence of Oxisols from Patos de Minas region, Minas Gerais State, Brazil.* PhD diss., Purdue Univ., West Lafayette, IN.

Santos, GM, Alexandre, A, Coe, HG, Reyerson, PE, Southon, JR and ND Cacilda. 2010. The phytolith ¹⁴C puzzle: A tale of background determinations and accuracy tests. *Radiocarbon* 52:113–128.

Santos, MCD, St. Arnaud, RJ and DW Anderson. 1986. Quantitative evaluation of pedogenic changes in Boralfs (Gray Luvisols) of East Central Saskatchewan. *SSSAJ* 50:1013–1019.

Sanz, ME and VP Wright. 1994. Modelo alternativo para el desarrollo de calcretas: Un ejemplo del Plio-Cuaternario de la Cuenca de Madrid. *Geogaceta* 16:116–119.

Sarna-Wojcicki, AM, Pringle, MS and J Wijbrans. 2000. New Ar-40/Ar-39 age of the Bishop Tuff from multiple sites and sediment rate calibration for the Matuyama-Brunhes boundary. *J. Geophys. Res. Solid Earth* 105:21431–21443.

Saucier, RT. 1966. Soil-survey reports and archaeological investigations. *Am. Antiquity* 31:419–422.

Sauer, D. 2010. Approaches to quantify progressive soil development with time in Mediterranean climate. I. Use of field criteria. *J. Plant Nutr. Soil Sci.* 173:822–842.

Sauer, D, Schellmann, G and K Stahr. 2007a. A soil chronosequence in the semi-arid environment of Patagonia (Argentina). *Catena* 71:382–393.

Sauer, D, Sponagel, H, Sommer, M, Giani, L, Jahn, R and K Stahr. 2007b. Podzol: Soil of the year 2007. A review on its genesis, occurrence, and functions. *JSSPN* 170:581–597.

Sauer, D, Wagner, S, Bruckner, H, Scarciglia, F, Mastronuzzi, G and K Stahr. 2010. Soil development on marine terraces near Metaponto (Gulf of Taranto, southern Italy). *QI* 222:48–63.

Sauerbeck, DR. 2001. CO₂ emissions and C sequestration by agriculture – perspectives and limitations. *Nutr. Cycling Agroecosyst.* 60:253–266.

Saunders, JW, Mandel, RD, Sampson, CG, Allen, CM, Allen, ET, Bush, DA, Feathers, JK, Gremillion, KJ, Hallmark, CT, Jackson, HE, Johnson, JK, Jones, R, Saucier, RT, Stringer, GL and MF Vidrine. 2005. Watson Break, a Middle Archaic mound complex in northwest Louisiana. *Am. Antiq.* 70:631–668.

Savigear, RAG. 1965. A technique of morphological mapping. *AAAG* 55:514–538.

Savoskul, OS and W Zech. 1997. Holocene glacier advances in the Topolovaya valley, Bystrinskiy Range, Kamchatka, Russia dated by tephrochronology and lichenometry. *AAR* 29:143–155.

Sawhney, BL. 1986. Electron microprobe analysis. In: A Klute (ed). *Methods of Soil Analysis. Part 1. Physical and Mineralogical Methods.* Agronomy Mon. 9, 2nd ed., Soil Sci. Soc. Am., Madison, WI. pp. 271–290.

Scalenghe, R and FA Marsan. 2009. The anthropogenic sealing of soils in urban areas. *Landscape Urban Plann.* 90:1–10.

Scalenghe, R, Zanini, E and DR Nielsen. 2000. Modeling soil development in a post-incisive chronosequence. *SS* 165:455–462.

Scarciglia, F, Pulice, I, Robustelli, G and G Vecchio. 2006. Soil chronosequences on Quaternary marine terraces along the northwestern coast of Calabria (southern Italy). *QI* 156/157:133–155.

Schaefer CE, Gilkes, GR and RBA Fernandes. 2004. EDS/SEM study on microaggregates of Brazilian Latosols, in relation to P adsorption and clay fraction attributes. *Geoderma* 123:69–81.

Schaefer, CEGR, Fabris, JD and JC Ker. 2008. Minerals in the clay fraction of Brazilian Latosols (Oxisols): A review. *Clay Mins.* 43:137–154.

Schaefer, CEGR, Lima, HN, Gilkes, RJ and JWV Mello. 2004. Micromorphology and electron microprobe analysis of phosphorus and potassium forms of an Indian Black Earth (IBE) Anthrosol from Western Amazonia. *AJSR* 42:401–409.

Schaefer, CER. 2001. Brazilian latosols and their B horizon microstructure as long-term biotic constructs. *AJSR* 39:909–926.

Schaefer, K, Zhang, TJ, Bruhwiler, L and AP Barrett. 2011. Amount and timing of permafrost carbon release in response to climate warming. *Tellus Ser. B* 63:165–180.

Schaetzl, RJ. 1986a. A soilscape analysis of contrasting glacial terrains in Wisconsin. *AAAG* 76:414–425.

1986b. Complete soil profile inversion by tree uprooting. *PG* 7:181–189.

1986c. The Sangamon paleosol in Brown County, Kansas. *Kansas Acad. Sci. Trans.* 89:152–161.

1990. Effects of treethrow microtopography on the characteristics and genesis of Spodosols, Michigan, USA. *Catena* 17:111–126.

1991a. A lithosequence of soils in extremely gravelly, dolomitic parent materials, Bois Blanc Island, Lake Huron. *Geoderma* 48:305–320.

1991b. Factors affecting the formation of dark, thick epipedons beneath forest vegetation, Michigan, USA. *JSS* 42:501–512.

1992a. Beta Spodic horizons in podzolic soils (Lithic Haplorthods and Haplohumods). *Pédologie* 42:271–287.

1992b. Texture, mineralogy and lamellae development in sandy soils in Michigan. *SSSAJ* 56:1538–1545.

1994. Changes in O horizon mass, thickness and carbon content following fire in northern hardwood stands. *Vegetatio* 115:41–50.

1996. Spodosol-Alfisol intergrades: Bisequal soils in NE Michigan, USA. *Geoderma* 74:23–47.

1998. Lithologic discontinuities in some soils on drumlins: Theory, detection, and application. *SS* 163:570–590.

2000. Shock the world (and then some). *AAAG* 90:772–774.

2001. Morphologic evidence of lamellae forming directly from thin, clayey bedding planes in a dune. *Geoderma* 99:51–63.

2002. A Spodosol-Entisol transition in northern Michigan: Climate or vegetation? *SSSAJ* 66:1272–1284.

2008. The distribution of silty soils in the Grayling Fingers region of Michigan: Evidence for loess deposition onto frozen ground. *Geomorphology* 102:287–296.

2013. Catenas and Soils. In: JF Shroder (ed). *Treatise on Geomorphology.* Vol. 4. Academic Press, San Diego, CA. pp. 145–158.

Schaetzl, RJ and JW Attig. 2013. The loess cover of northeastern Wisconsin. *QR* 79:199–214.

Schaetzl, RJ, Barrett, LR and JA Winkler. 1994. Choosing models for soil chronofunctions and fitting them to data. *EJSS* 45:219–232.

Schaetzl, RJ, Burns, SF, Small, TW and DL Johnson. 1990. Tree uprooting: Review of types and patterns of soil disturbance. *PG* 11:277–291.

Schaetzl, RJ, Drzyzga, SA, Weisenborn, BN, Kincare, KA, Lepczyk, XC, Shein, KA, Dowd, CM and J Linker. 2002. Measurement, correlation, and mapping of Glacial Lake Algonquin shorelines in northern Michigan. *AAAG* 92:399–415.

Schaetzl, RJ, Enander, H, Luehmann, MD, Lusch, DP, Fish, C, Bigsby, M, Steigmeyer, M, Guasco, J, Forgacs, C and A Pollyea. 2013. Mapping the physiography of Michigan using GIS. *PG* 34:1–38.

Schaetzl, RJ and LR Follmer. 1990. Longevity of treethrow microtopography: Implications for mass wasting. *Geomorphology* 3:113–123.

Schaetzl, RJ and SL Forman. 2008. OSL ages on glaciofluvial sediment in northern Lower Michigan constrain expansion of the Laurentide ice sheet. *QR* 70:81–90.

Schaetzl, RJ, Frederick, WE and L Tornes. 1996. Secondary carbonates in three fine and fine-loamy Alfisols in Michigan. *SSSAJ* 60:1862–1870.

Schaetzl, RJ and J Hook. 2008. Characterizing the silty sediments of the Buckley Flats outwash plain: Evidence for loess in NW Lower Michigan. *PG* 29:1–18.

Schaetzl, RJ and SA Isard. 1990. Comparing "warm season" and "snowmelt" pedogenesis in Spodosols. In: JM Kimble and RD Yeck (eds).

Characterization, Classification, and Utilization of Spodosols. Proc. 5th Intl. Soil Corr. Mtg. USDA-SCS, Lincoln, NE. pp. 303–318.

1991. The distribution of Spodosol soils in southern Michigan: A climatic interpretation. *AAAG* 81:425–442.

1996. Regional-scale relationships between climate and strength of podzolization in the Great Lakes region, North America. *Catena* 28:47–69.

Schaetzl, RJ and W Harris. 2011. Spodosols. In: PM Huang, Y Li and ME Sumner (eds). *Handbook of Soil Sciences*, 2nd ed. CRC Press, New York. pp. 33–113 – 33–127.

Schaetzl, RJ, Johnson, DL, Burns, SF and TW Small. 1989. Tree uprooting: Review of terminology, process, and environmental implications. *CJFR* 19:1–11.

Schaetzl, RJ and WC Johnson. 1983. Pollen and spore stratigraphy of a Mollic Hapludalf (Degraded Chernozem) in northeastern Kansas. *Prof. Geog.* 35:183–191.

Schaetzl, RJ, Krist, FJ Jr and BA Miller. 2012. A taxonomically based, ordinal estimate of soil productivity for landscape-scale analyses. *SS* 177:288–299.

Schaetzl, RJ, Krist, F, Rindfleisch, P, Liebens, J and T Williams. 2000. Postglacial landscape evolution of northeastern lower Michigan, interpreted from soils and sediments. *AAAG* 90:443–466.

Schaetzl, RJ, Krist, FJ Jr, Stanley, KE and CM Hupy. 2009. The Natural Soil Drainage Index: An ordinal estimate of long-term, soil wetness. *PG* 30:383–409.

Schaetzl, RJ and WL Loope. 2008. Evidence for an eolian origin for the silt-enriched soil mantles on the glaciated uplands of eastern Upper Michigan, USA. *Geomorphology* 100:285–295.

Schaetzl, RJ and MD Luehmann. 2013. Coarse-textured basal zones in thin loess deposits: Products of sediment mixing and/or paleoenvironmental change? *Geoderma* 192:277–285.

Schaetzl, RJ, Mikesell, LR and MA Velbel. 2006. Soil characteristics related to weathering and pedogenesis across a geomorphic surface of uniform age in Michigan. *PG* 27:170–188.

Schaetzl, RJ and DL Mokma. 1988. A numerical index of Podzol and Podzolic soil development. *PG* 9:232–246.

Schaetzl, RJ and C Schwenner. 2006. An application of the Runge "Energy Model" of soil development in Michigan's Upper Peninsula. *SS* 171:152–166.

Schaetzl, RJ and CJ Sorenson. 1987. The concept of "buried" vs "isolated" paleosols: Examples from northeastern Kansas. *SS* 143:426–435.

Schaetzl, RJ and DM Tomczak. 2002. Wintertime soil temperatures in the fine-textured soils of the Saginaw Valley, Michigan. *Great Lakes Geog.* 8:87–99.

Schaetzl, RJ, Yansa, CH and MD Luehmann. 2013. Paleobotanical and environmental implications of a buried forest bed in northern Lower Michigan, USA. *CJES* 50:483–493.

Schaller, M, Blum, JD and TA Ehlers. 2009. Combining cosmogenic nuclides and major elements from moraine soil profiles to improve weathering rate estimates. *Geomorphology* 106:198–205.

Scharpenseel, HW. 1972a. Messung der natürlichen C-14 konzentration in der organischen substanz von rezenten Böden. *Z. Pflanzen. Bodenkd.* 133:241–263.

1972b. *Natural radiocarbon measurement of soil and organic matter fractions and on soil profiles of different pedogenesis*. Proc. 8th Intl. Conf. Radiocarbon Dating (Wellington, NZ) p. E1.

Scharpenseel, HW and H Schiffmann. 1977. Radiocarbon dating of soils, a review. Z. Pflanzen. *Bodenkd.* 140:159–174.

Schatz, A. 1963. Soil microorganisms and soil chelation: The pedogenic action of lichens and lichen acids. *Agric. Food Chem.* 11:112–118.

Scheffer, VB. 1947. The mystery of the mima mounds. *Sci. Monthly* 65:283–294.

Schellenberger, A and H Veit. 2006. Pedostratigraphy and pedological and geochemical characterization of Las Carreras loessepaleosol sequence, Valle de Tafi, NW Argentina. *QSR* 25:811–831.

Schemenauer, RS and P Cereceda. 1992. The quality of fog water collected for domestic and agricultural use in Chile. *J. Appl. Meteorol.* 31:275–290.

Schenck, CA. 1924. Der Waldbau des Urwalds. *Allg. Forst Jagd-Zeit.* 100:377–388.

Schenk, HJ and RB Jackson. 2005. Mapping the global distribution of deep roots in relation to climate and soil characteristics. *Geoderma* 126:129–140.

Scheu, S and V Wolters. 1991. Influence of fragmentation and bioturbation on the decomposition of carbon-14-labelled beech leaf litter. *Soil Biol. Biochem.* 23:1029–1034.

Schiffer, MB. 1983. Toward the identification of formation processes. *Am. Antiq.* 48:675–706.

Schiffman, PM. 1994. Promotion of exotic weed establishment by endangered giant kangaroo rats (Dipodomys ingens) in a California grassland. *Biodivers. Conserv.* 3:524–537.

Schimel, J and O Chadwick. 2013. What's in a name? The importance of soil taxonomy for ecology and biogeochemistry. *Frontiers Ecol. Environ.* 11:405–406.

Schlesinger, DR and BL Howes. 2000. Organic phosphorus and elemental ratios as indicators of prehistoric human occupation. *JAS* 27:479–492.

Schlesinger, WH. 1985. The formation of caliche in soils of the Mojave Desert, California. *Geochim. Cosmochim. Acta* 49:57–66.

Schlesinger, WH, Pippen, JS, Wallenstein, MD, Hofmockel, KS, Klepeis, DM and BE Mahall. 2003. Community composition and photosynthesis by photoautotophs under quartz pebbles, southern Mojave Desert. *Ecology* 84:3222–3231.

Schlezinger, DR and BL Howes. 2000. Organic phosphorus and elemental ratios as indicators of prehistoric human occupation. *J. Archaeol. Sci.* 27:479–492.

Schlichting, E and V Schweikle. 1980. Interpedon translocations and soil classification. *SS* 130:200–204.

Schlichting, E and U Schwertmann (eds). 1973. *Pseudogley and Gley: Genesis and Use of Hydromorphic Soils*. Trans. Comm. V and VI Intl. Soc. Soil Sci., Chemie Verlag, Weinheim, Germany. 71–80.

Schmidlin, TW, Peterson, FF and RO Gifford. 1983. Soil temperature regimes in Nevada. *SSSAJ* 47:977–982.

Schmidt, J and A Hewitt. 2004. Fuzzy land element classification from DTMs based on geometry and terrain position. *Geoderma* 121:243–256.

Schmidt, MWI, Torn, MS, Abiven, S, Dittmar, T, Guggenberger, G, Janssens, IA, Kleber, M, Kogel-Knabner, I, Lehmann, J, Manning, DAC, Nannipieri, P, Rasse, DP, Weiner, S and SE Trumbore. 2011. Persistence of soil organic matter as an ecosystem property. *Nature* 478:49–56.

Schmidt-Lorenz, R. 1977. Soil reddening through hematite from plinthized saprolite. In: KT Joseph (ed). *Proc. Conf. on Class. and Mgmt. of Tropical Soils*, Kuala Lumpur, Malaysian Soc. Soil Sci. pp. 101–106.

Schnitzer, M and JG Desjardins. 1969. Chemical characteristics of a natural soil leachate from a Humic Podzol. *CJSS* 49:151–158.

Schnitzer, M and H Kodama. 1976. The dissolution of micas by fulvic acid. *Geoderma* 15:381–391.

Schnitzer, M and CM Monreal. 2011. Quo vadis soil organic matter research?: A biological link to the chemistry of humification. In: DL Sparks (ed). *Advances in Agronomy*. pp. 139–213. Elsevier, San Diego.

Schnitzer, M, Wright, JR and JG Desjardins. 1958. A comparison of the effectiveness of various extractants for organic matter from two horizons of a podzol profile. *CJSS* 38:49–53.

Schoeneberger, PJ and DA Wysocki. 2001. *A Geomorphic Description System (version 3.0).* Lincoln, NE, National Soil Survey Center. US Dept. of Agriculture.

2005. Hydrology of soils and deep regolith: A nexus between soil geography, ecosystems and land management. *Geoderma* 126:117–128.

Schoeneberger, PJ, Wysocki, DA, Benham, EC and WD Broderson. 1998. *Field book for describing and sampling soils.* Lincoln, NE, US Dept. of Agric., National Soil Survey Center.

Schoeneberger, PJ, Wysocki, DA, Benham, EC and Soil Survey Staff. 2012. *Field book for describing and sampling soils, Version 3.0.* Natural Resources Cons. Service, National Soil Survey Center, Lincoln, NE.

Schofield, R, Thomas, DSG and MJ Kirkby. 2001. Causal processes of soil salinization in Tunisia, Spain and Hungary. *Land Degrad. Develop.* 12:163–181.

Scholten, JJ and W Andriesse. 1986. Morphology, genesis and classification of three soils over limestone, Jamaica. *Geoderma* 39:1–40.

Scholtes, WH, Ruhe, RV and FF Riecken. 1951. Use of the morphology of buried soil profiles in the Pleistocene of Iowa. *Proc. Iowa Acad. Sci.* 58:295–306.

Schulmann, OP and AV Tiunov. 1999. Leaf litter fragmentation by the earthworm Lumbricus terrestris L. *Pedobiologia* 43:453–458.

Schultz, CB and TM Stout. 1945. Pleistocene loess deposits of Nebraska. *AJS* 243:231–244.

Schulze, DG. 1984. The influence of aluminum on iron oxides. VI. Unit cell dimensions of Al substituted goethites and estimation of Al from them. *SSSAJ* 43:793–799.

Schumm, SA. 1967. Rates of surficial rock creep on hillslopes in western Colorado. *Science* 155:560–562.

1979. Geomorphic thresholds: The concept and its applications. *Trans. Inst. Brit. Geogr. NS* 4:485–515.

1980. Some applications on the concept of geomorphic thresholds. In: DR Coates and JD Vitek (eds). *Thresholds in Geomorphology.* Allen and Unwin, London. pp. 473–485.

Schumm, SA and RS Parker. 1973. Implications of complex response of drainage systems for Quaternary alluvial stratigraphy. *Nature* 243:99–100.

Schwarcz, H and M Gascoyne. 1984. Uranium-series dating of Quaternary deposits. In: WC Mahaney (ed). *Quaternary Dating Methods.* Elsevier, New York. pp. 33–51.

Schwartz, D, Guillet, B, Villemin, G and F Toutain. 1986. Les alios humiques des Podzols tropicaux du Congo: Constituants, micro- et ultra-structure. *Pédologie* 36:179–198.

Schwartz, D. 1996. Archéologie préhistorique et processus de formation des stone-lines en Afrique central (Congo-Brazzaville et zones peripheriques). *Geo.-Eco.-Trop.* 20:15–38.

Schwartzmann, DW and T Volk. 1989. Biotic enhancement of weathering and the habitability of Earth. *Nature* 340:457–460.

Schwenk, S. 1977. Krusten und verkrustungen in Sudtunesien. *Stuttgart. Geog. Stud.* 91:83–103.

Schwertmann, U. 1973. Use of oxalate for Fe extraction from soils. *CJSS* 53:244–246.

1984. The influence of aluminium oxides on iron oxides. IX. Dissolution of Al-goethites in 6M HCl. *Clay Mins.* 19:9–19.

1985. The effects of pedogenic environments on iron oxide minerals. *Advances Soil Sci.* 1:171–200.

Schwertmann, U and DS Fanning. 1976. Iron-manganese concretions in hydrosequences of soils in loess in Bavaria. *SSSAJ* 40:731–738.

Schwertmann, U, Murad, E and DG Schulze. 1982. Is there Holocene reddening (hematite formation) in soils of xeric temperate areas? *Geoderma* 27:209–223.

Schwertmann, U and RM Taylor. 1989. Iron oxides. In: JB Dixon and SB Weed (eds). *Minerals in Soil Environments,* 2nd ed. Soil Sci. Soc. Am., Madison, WI. pp. 379–438.

Schwintzer, CR and G Williams. 1974. Vegetation changes in a small Michigan bog from 1917 to 1972. *AMN* 92:447–459.

Scott, AD and SJ Smith. 1968. Mechanism for soil potassium release by drying. *SSSAP* 32:443–444.

Scott, WE. 1977. Quaternary glaciation and volcanism, Metolius River area, Oregon. *GSA Bull.* 88:113–124.

Scott, WE, McCoy, WD, Shroba, RR and M Rubin. 1983. Reinterpretation of the exposed record of the last two cycles of Lake Bonneville, western United States. *QR* 20:261–285.

Scull, P, Franklin, J, Chadwick, OA and D McArthur. 2003. Predictive soil mapping: A review. *PPG* 27:171–197.

Scull, P and RJ Schaetzl. 2011. Using PCA to characterize and differentiate the character of loess deposits in Wisconsin and Upper Michigan, USA. *Geomorphology* 127:143–155.

Scurfield, G, Segnit, ER and CA Anderson. 1974. Silica in woody stems. *AJB* 22:211–229.

Seelig, BD and AR Gulsvig. 1988. *Soil Survey of Kidder County, North Dakota.* USDA-SCS, US Govt. Printing Off., Washington, DC.

Seelig, BD, Richardson, JL and CJ Heidt. 1990. Sodic soil variability and classification in coarse-loamy till of central North Dakota. *SSH* 31:33–43.

Ségalen, P. 1994. *Les sols ferrallitiques et leur répartition géographique.* Office de la Recherche Scient. Tech. d'Outre-Mer. ORSTOM Éditions, Paris. Tome 1.

Sehgal, JH and JC Bhattacharjee. 1990. Typic Vertisols of India and Iraq: Their characteristics and classification. *Pédologie* 38:67–95.

Sehgal, JL and G Stoops. 1972. Pedogenic calcite accumulation in arid and semi-arid regions of the Indo-Gangetic alluvial plain of Erstwhile Punjab (India): Their morphology and origin. *Geoderma* 8:59–72.

Seibert, DR, Weaver, JB, Bush, RD, Belz, DJ, Rector, DD, Hallowich, JS and RG Grubb. 1983. *Soil Survey of Greene and Washington Counties, Pennsylvania.* USDA-SCS, US Govt. Printing Off., Washington, DC.

Seidl, DE and P Klepeis. 2011. Human dimensions of earthworm invasion in the Adirondack State Park. *Human Ecol.* 39:641–655.

Seki, T. 1932. Volcanic ash loams of Japan proper (their classification, distribution, and characteristics). *Trans. 2nd Intl. Congr. Soil Sci.* 5:141–143.

Selby, MJ. 1980. A rock mass strength classification for geomorphic purposes: With tests from Antarctica and New Zealand. *ZG* 24:31–51.

Semmel, A and B Terhorst. 2010. The concept of the Pleistocene periglacial cover beds in central Europe: A review. *QI* 222:120–128.

Sequeira Braga, MA, Paquet, H and A Begonha. 2002. Weathering of granites in a temperate climate (NW Portugal): Granitic saprolites and arenization. *Catena* 49:41–56.

Setz, EZF, Enzweiler, J, Solferini, VN, Amendola, MP and RS Berton. 1999. Geophagy in the golden-faced saki monkey (Pithecia pithecia chrysocephala) in the Central Amazon. *J. Zool.* 247:91–103.

Severson, RC and HF Arneman. 1973. Soil characteristics of the forest-prairie ecotone in northwestern Minnesota. *SSSAP* 37:593–599.

Sgouras, ID, Tsadilas, CA, Barbayiannis, N and N Danalatos. 2007. Physicochemical and mineralogical properties of red Mediterranean soils from Greece. *Comm. Soil Sci. Plant Anal.* 38:695–711.

Shackleton, NJ. 1968. Depth of pelagic Foraminifera and isotope changes in Pleistocene oceans. *Nature* 218:79–80.

1977. The oxygen isotope record of the Late Pleistocene. *Phil. Trans. Roy. Soc.* B280:169–182.

Shackleton, NJ and ND Opdyke. 1973. Oxygen isotope and palaeomagnetic stratigraphy of equatorial Pacific core V28-238: Oxygen isotope temperatures and ice volumes on a 105-year and 106-year scale. *QR* 3:39–55.

1976. Oxygen-isotope and paleomagnetic stratigraphy of Pacific core V28-239, Late Pliocene to Latest Pleistocene. *Mem. Geol. Soc. Am.* 145:449–464.

Shafer, WM. 1979. Variability of minesoils and natural soils in southeastern Montana. *SSSAJ* 43:1207–1212.

Shaler, NS. 1890. The origin and nature of soils. *USGS Ann. Rept* 12:213–345.

Shamoot, S, McDonald, L and WV Bartholomew. 1968. Rhizo-deposition of organic debris in soil. *SSSAP* 32:817–820.

Shankar, N and H Achyuthan. 2007. Genesis of calcic and petrocalcic horizons from Coimbatore, Tamil Nadu: Micromorphology and geochemical studies. *QI* 175:140–154.

Shannon, RD. 1976. Revised effective ionic radii and systematic studies of interatomic distances in halides and chalcogenides. *Acta Cryst.* A32:751–767.

Sharma, BD, Mukhopadhyay, SS and PS Sidhu. 1998. Microtopographic controls on soil formation in the Punjab region, India. *Geoderma* 81:357–367.

Sharon, D. 1962. On the nature of hamadas in Israel. *ZG* 6:129–147.

Sharp, RP. 1942. Periglacial involutions in north-eastern Illinois. *J. Geol.* 50:113–133.

1949. Pleistocene ventifacts east of the Big Horn Mountains, Wyoming. *J. Geol.* 57:175–195.

1969. Semi-quantitative differentiation of glacial moraines near Convict Lake, Sierra Nevada, California. *J. Geol.* 77:68–91.

Sharp, RP and JH Birman. 1963. Additions to the classical sequence of Pleistocene glaciations, Sierra Nevada, California. *GSA Bull.* 74:1079–1086.

Sharpe, CFS. 1938. *Landslides and Related Phenomena.* Columbia Univ. Press, New York.

Sharpe, DM, Cromack, K, Jr, Johnson, WC and BS Ausmus. 1980. A regional approach to litter dynamics in southern Appalachian forests. *CJFR* 10:395–404.

Sharratt, BS, Baker, DG, Wall, DB, Skaggs, RH and DL Ruschy. 1992. Snow depth required for near steady-state soil temperatures. *Agric. For. Meteorology.* 57:243–251.

Sharratt, BS, Saxton, KE and JK Radke. 1995. Freezing and thawing of agricultural soils: Implications for soil, water and air quality. *J. Minnesota Acad. Sci.* 59:1–5.

Shaw, CF. 1927. Report of committee on soil terminology. *Am. Soil Surv. Ass. Bull.* 8:66–98.

1930. Potent factors in soil formation. *Ecology* 11:239–245.

Shaw, J. 1994. Magnetic dating of Chinese loess. *Quat. Newslett.* 73:60–61.

Shaw, JN. 2001. Iron and aluminum oxide characterization for highly weathered Alabama Ultisols. *Comm. Soil Sci. Plant Anal.* 32:49–64.

Shaw, JN, Haek, BF and JM Beck. 2010. Highly weathered mineralogy of select soils from Southeastern U. S. Coastal Plain and Piedmont landscapes. *Geoderma* 154:447–456.

Shaw, JN, Odom, JW and BF Hajck. 2003. Soils on Quaternary terraces of the Tallapoosa River, central Alabama. *SS* 168:707–717.

Shchetenko, AY. 1968. *The Most Ancient Agrarian Cultures of the Deccan [Drevneishie zemledel'cheskie kul'tury Dekana].* Nauka, Leningrad.

Sheldon, ND and GJ Retallack. 2001. Equation for compaction of paleosols due to burial. *Geology* 29:247–250.

Sheldon, ND and NJ Tabor. 2009. Quantitative paleoenvironmental and paleoclimatic reconstruction using paleosols. *Earth Sci. Revs.* 95:1–52.

Shen, CD, Beer, J, Kubik, PW, Sun, WD, Liu, TS and KX Liu. 2010. [10]Be in desert sands, falling dust and loess in China. *Nucl. Inst. Methods Physics Res.* 268B:1050–1053.

Sherman, GD. 1952. The genesis and morphology of the alumina-rich laterite clays. In: *Problems of Clay and Laterite Genesis.* Am. Inst. Mining and Metall. Engineers Sympos., St. Louis, MO. pp. 154–161.

Sherman, GD and LT Alexander. 1959. Characteristics and genesis of Low Humic Latosols. *SSSAP* 23:168–170.

Sherman, GD, Cady, JG, Ikawa, H and NE Blomberg. 1967. Genesis of the bauxitic Haliii soils. Hawaii Agric. Exp. Stn Tech. Bull. 56.

Shi, X, Girod, L, Long, R, DeKett, R, Philippe, J and T Burke. 2012. A comparison of LiDAR-based DEMs and USGS-sourced DEMs in terrain analysis for knowledge-based digital soil mapping. *Geoderma* 170:217–226.

Shi, X, Long, R, Dekett, R and J Philippe. 2009. Integrating different types of knowledge for digital soil mapping. *SSSAJ* 73:1682–1692.

Shields, JA, Paul, EA, St. Arnaud, RJ and WK Head. 1968. Spectrophotometric measurement of soil color and its relationship to moisture and organic matter. *CJSS* 48:271–280.

Shilts, WW. 1978. Nature and genesis of mudboils, Central Keewatin, Canada. *CJES* 15:1053–1068.

Shimek, B. 1909. Aftonian sands and gravels in western Iowa. *GSA Bull.* 20:339–408.

Shipp, RF and RP Matelski. 1965. Bulk-density and coarse-fragment determinations on some Pennsylvania soils. *SS* 99:392–397.

Shiraiwa, T and T Watanabe. 1991. Late Quaternary glacial fluctuations in the Langtang Valley, Nepal Himalaya, reconstructed by relative dating methods. *AAR* 23:404–416.

Shoji, S, Ito, T, Saigusa, M and I Yamada. 1985. Properties of nonallophanic Andosols from Japan. *SS* 140:264–277.

Shoji, S, Nanzyo, M and RA Dahlgren. 1993. Volcanic ash soils – genesis, properties, and utilization. *Developments in Soil Sci.* 21:1–288.

Shoji, S and T Ono. 1978. Physical and chemical properties and clay mineralogy of Andosols from Kitakami, Japan. *SS* 126:297–312.

Shoji, S, Takahashi, T, Saigusa, M, Yamada, I and FC Ugolini. 1988. Properties of Spodosols and Andisols showing climosequential and biosequential relations in southern Hokkoda, northeastern Japan. *SS* 145:135–150.

Short, JR, Fanning, DS, Foss, JE and JC Patterson. 1986a. Soils of the Mall in Washington, DC. II. Genesis, classification and mapping. *SSSAJ* 50:705–710.

Short, JR, Fanning, DS, McIntosh, MS, Foss, JE and JC Patterson. 1986b. Soils of the Mall in Washington, DC. I. Statistical summary of properties. *SSSAJ* 50:699–705.

Short, NM. 1961. Geochemical variations in four residual soils. *J. Geol.* 69:534–571.

Shrader, WD. 1950. Differences in the clay contents of surface soils developed under prairie as compared to forest vegetation in the central United States. *SSSAP* 15:333–337.

Shroder, JF Jr. 1978. Dendrogeomorphological analysis of mass movement on Table Cliffs Plateau, Utah. *QR* 9:168–185.

1980. Dendrogeomorphology: Review and new techniques in tree-ring dating. *PPG* 4:161–188.

Shubayeva, VI and LO Karpachevskiy. 1983. Soil-windfall complexes and pedogenesis in the Siberian stone pine forests of the maritime territory. *SSS* 50–57.

Shum, M and LM Lavkulich. 1999. Use of sample color to estimate oxidized Fe content in mine waste rock. *Env. Geol.* 37:281–289.

Shuster, DL, Farley, KA, Vasconcelos, PM, Balco, G, Monteiro, HS, Waltenberg, K and JO Stone. 2012. Cosmogenic He-3 in hematite and goethite from Brazilian "canga" duricrust demonstrates the extreme stability of these surfaces. *EPSL* 329:41–50.

Siame, LL, Braucher, R, Bourles, DL, Bellier, O and M Sebrier. 2001. Cosmic ray exposure dating of geomorphic surface features using in situ-production Be-10: Tectonic and climatic implications. *Bull. Soc. Geol. France* 172:223–236.

Sideri, DI. 1938. On the formation of structure in soil. V. Granular structure. *SS* 46:267–271.

Sidwell, R. 1943. Caliche deposits on the southern High Plains, Texas. *AJS* 241:257–261.

Siever, R. 1962. Silica solubility, 0–200 C, and the diagenesis of siliceous sediments. *J. Geol.* 70:127–150.

Sileshi, GW, Arshad, MA, Konate, S and POY Nkunika. 2010. Termite-induced heterogeneity in African savanna vegetation: Mechanisms and patterns. *J. Veg. Sci.* 21:923–937.

Simard, M, Bernier, PY, Bergeron, Y, Pare, D and L Guerine. 2009. Paludification dynamics in the boreal forest of the James Bay Lowlands: Effect of time since fire and topography. *CJFR* 39:546–552.

Simard, M, Lecomte, N, Bergeron, Y, Bernier, PY and D Paré. 2007. Forest productivity decline caused by successional paludification of boreal soils. *Ecol. App.* 17:1619–1637.

Simas, FNB, Schaefer, CEGR, Melo, VF, Albuquerque-Filho, MR, Michel, RFM, Pereira, VV, Gomes, MRM and LM da Costa. 2007. Ornithogenic cryosols from Maritime Antarctica: Phosphatization as a soil forming process. *Geoderma* 138:191–203.

Simonson, GH and L Boersma. 1972. Soil morphology and water table relations II. Correlation between annual water table fluctuations and profile features. *SSSAP* 36:649–653.

Simonson, RW. 1941. Studies of buried soils formed from till in Iowa. *SSSAP* 3:73–381.

1952. Lessons from the first half century of soil survey I. Classification of soils. *SS* 74:249–257.

1954a. Identification and interpretation of buried soils. *AJS* 252:705–732.

1954b. The regur soils of India and their utilization. *SSSAP* 18:199–203.

1959. Outline of a generalized theory of soil genesis. *SSSAP* 23:152–156.

1978. A multiple-process model of soil genesis. In: WC Mahaney (ed). Quaternary Soils. Geo Abstracts, Norwich, UK. pp. 1–25.

1979. Origin of the name "Ando Soils." *Geoderma* 22:333–335.

1986. Historical aspects of soil survey and soil classification. I. 1899–1910. *SSH* 27:3–11.

1995. Airborne dust and its significance to soils. *Geoderma* 65:1–43.

Simonson, RW and DR Gardner. 1960. Concept and function of the pedon. *Trans. 8th Intl. Congr. Soil Sci.* 4:129–131.

Simonson, RW and CE Hutton. 1942. Distribution curves for loess. *AJS* 252:99–105.

Simonson, RW, Riecken, FF and GD Smith. 1952. *Great soil groups in Iowa.* In: Understanding Iowa Soils. Wm. C. Brown Co., Dubuque, IA. pp. 23–25.

Simonson, RW and S Rieger. 1967. Soils of theAndept suborder in Alaska. *SSSAP* 31:692–699.

Simpson, IA, Bryant, RG and U Tveraabak. 1998. Relict soils and early arable land management in Lofoten, Norway. *JAS* 25:1185–1198.

Sinai, G, Zaslavsky, D and P Golany. 1981. The effect of soil surface curvature on moisture and yield: Beer Sheba observations. *SS* 132:367–375.

Singer, A. 1966. The mineralogy of the clay fractions from basaltic soils in the Galilee (Israel). *JSS* 17:136–147.

1980. The paleoclimatic interpretation of clay minerals in soils and weathering profiles. *Earth-Sci. Revs.* 15:303–326.

Singer, A and J Navrot. 1977. Clay formation from basic volcanic rocks in a humid Mediterranean climate. *SSSAJ* 41:645–650.

Singer, M, Ugolini, FC and J Zachara. 1978. In situ study of podzolization on tephra and bedrock. *SSSAJ* 42:105–111.

Singer, MJ and P Fine. 1989. Pedogenic factors affecting magnetic susceptibility in northern California soils. *SSSAJ* 53:1119–1127.

Singer, MJ, Fine, P, Verosub, KL and OA Chadwick. 1992. Time dependence of magnetic susceptibility of soil chronosequences on the California coast. *QR* 37:323–332.

Singh, LP, Parkash, B and AK Singhvi. 1998. Evolution of the Lower Gangetic Plain landforms and soils in West Bengal, India. *Catena* 33:75–104.

Singhvi, AK and MR Krbetschek. 1996. Luminescence dating: A review and a perspective for arid zone sediments. *Ann. Arid Zone* 35:249–279.

Singleton, GA and LM Lavkulich. 1987. A soil chronosequence on beach sands, Vancouver Island, British Columbia. *CJSS* 67:795–810.

Sivarajasingham, S, Alexander, LT, Cady, JG and MG Cline. 1962. Laterite. *Adv. Agron.* 14:1–60.

Six, J, Bossuyt, H, Degryze, S and K Denef. 2004. A history of research on the link between (micro)aggregates, soil biota, and soil organic matter dynamics. *Soil Tillage Res.* 79:7–31.

Sjöberg, A. 1976. Phosphate analysis of Anthropic soils. *J. Field Archaeol.* 3:447–454.

Skidmore, EL, Carstenson, WA and EE Banbury. 1975. Soil changes resulting from cropping. *SSSAP* 39:964–967.

Skjemstad, JO, Fitzpatrick, RW, Zarcinas, BA and CH Thompson. 1992. Genesis of Podzols on coastal dunes in southern Queensland. 2. Geochemistry and forms of elements as deduced from various soil extraction procedures. *AJSR* 30:615–644.

Slamova, R, Trckova, M, Vondruskova, H, Zraly, Z and I Pavlik. 2011. Clay minerals in animal nutrition. *Appl. Clay Sci.* 51:395–398.

Slate, JL, Bull, WB, Ku, T-L, Shafiqullah, M, Lynch, DJ and Y-P Huang. 1991. Soil-carbonate genesis in the Pinnacate Volcanic Field, northwestern Sonora, Mexico. *QR* 35:400–416.

Sloss, CR, Westaway, KE, Hua, Q and CV Murray-Wallace. 2013. An Introduction to Dating Techniques: A Guide for Geomorphologists. In: JF Shroder (ed). *Treatise on Geomorphology.* Vol. 14. Academic Press, San Diego, CA. pp. 346–369.

Small, TW. 1973. Morphological properties of Driftless Area soils relative to slope aspect and position. *Prof. Geog.* 24:321–326.

1997. The Goodlett-Denny mound: A glimpse at 45 years of Pennsylvania treethrow mound evolution with implications for mass wasting. *Geomorphology* 18:305–313.

Small, TW, Schaetzl, RJ and JM Brixie. 1990. Redistribution and mixing of soil gravels by tree uprooting. *Prof. Geog.* 42:445–457.

Smalley, IJ. 1972. The interaction of Great Rivers and large deposits of primary loess. *Trans. New York Acad. Sci.* 34:534–542.

(ed). 1975. *Loess: Lithology and Genesis.* Benchmark Papers in Geology #26. Dowden, Hutchinson and Ross, Stroudsburg, PA.

Smalley, I, Blake-Smalley-R, O'Hara-Dhand, K, Jary, Z and Z Svircev. 2012. Sand Martins favour loess: How the properties of loess ground facilitate the nesting of sand martins/bank swallows/uferschwalben (*Riparia riparia* Linnaeus 1758). *QI* 296:216–219.

Smalley, IJ and JE Davin. 1982. *Fragipan Horizons in Soils: A Bibliographic Study and Review of Some of the Hard Layers in Loess and Other Material.* Bibliographical Report #30. New Zealand Soil Bureau, Lower Hutt, NZ.

Smalley, IJ and DH Krinsley. 1978. Loess deposits associated with deserts. *Catena* 5:53–66.

Smalley, IJ and SB Marković. 2013. Loessification and hydroconsolidation: There is a connection. *Catena* 117:94–99.

Smalley, I, O'Hara-Dhand, K, McClaren, S, Svircev, Z and H Nugenta. 2013. Loess and bee-eaters I: Ground properties affecting the nesting of European bee-eaters (*Merops apiaster* L. 1758) in loess deposits. *QI* 296:220–226.

Smalley, IJ and C Vita-Finzi. 1968. The formation of fine particles in sandy deserts and the nature of "desert" loess. *J. Sed. Petr.* 38:766–774.

Smart, P and NK Tovey. 1982. *Electron Microscopy of Soils and Sediments: Examples.* Oxford, New York.

Smeck, NE. 1973. Phosphorous: An indicator of pedogenetic weathering processes. *SS* 115:199–206.

Smeck, NE, Bigham, JM, Guertal, WF and GF Hall. 2002. Spatial distribution of lepidocrocite in a soil hydrosequence. *Clay Mins.* 37:687–697.

Smeck, NE and EJ Ciolkosz. (eds) 1989. *Fragipans: Their Occurrence, Classification, and Genesis.* Soil Sci. Soc. Am., Madison, WI.

Smeck, NE, Ritchie, A, Wilding, LP and LR Drees. 1981. Clay accumulation in sola of poorly drained soils of western Ohio. *SSSAJ* 45:95–102.

Smeck, NE and ECA Runge. 1971. Factors influencing profile development exhibited by some hydromorphic soils in Illinois. In: E Schlichting and U Schwertmann (eds). *Pseudogley and Gley: Genesis and Use of Hydromorphic Soils.* Trans. Comm. V and VI Int. Soc. Soil Sci., Chemie Verlag. pp. 169–179.

Smeck, NE, Runge, ECA and EE MacKintosh. 1983. Dynamics and genetic modelling of soil systems. In: LP Wilding, NE Smeck, and GF Hall (eds). *Pedogenesis and Soil Taxonomy.* Elsevier, New York. pp. 51–81.

Smeck, NE, Thompson, ML, Norton, LD and MJ Shipitalo. 1989. Weathering discontinuities: A key to fragipan formation. *SSSASP* 24:99–112.

Smeck, NE and LP Wilding. 1980. Quantitative evaluation of pedon formation in calcareous glacial deposits in Ohio. *Geoderma* 24:1–16.

Smeck, NE, Wilding, LP and N Holowaychuk. 1968. Genesis of argillic horizons in Celina and Morley soils of western Ohio. *SSSAP* 32:550–556.

Smiles, DE. 2009. Quantifying carbon and sulphate loss in drained acid sulphate soils. *EJSS* 60:64–70.

Smith, BJ, Warke, PA, McGreevy, JP and HL Kane. 2005. Salt-weathering simulations under hot desert conditions: Agents of enlightenment or perpetuators of preconceptions? *Geomorphology* 67:211–227.

Smith, BJ, Warke, PA and CA Moses. 2000. Limestone weathering in contemporary arid environments: A case study from southern Tunisia. *ESPL* 25:1343–1354.

Smith, BJ and WB Whalley. 1988. A note on the characteristics and possible origins of desert varnishes from southeast Morocco. *ESPL* 13:251–258.

Smith, BN and S Epstein. 1971. Two categories of 13C/12C ratios of higher plants. *Plant Physiol.* 47:380–384.

Smith, CAS, Ping, CL, Fox, CA and H Kodama. 1999. Weathering characteristics of some soils formed from White River Tephra, Yukon Territory, Canada. *CJSS* 79:603–613.

Smith, CAS, Sanborn, PT, Bond, JD and G. Frank. 2009. Genesis of Turbic Cryosols on north-facing slopes in a dissected, unglaciated landscape, west-central Yukon Territory. *CJSS* 89:611–622.

Smith, GD. 1942. Illinois loess: Variations in its properties and distribution. Univ. Illinois Agric. Exp. Stn Bull. 490.

1979. Conversations in taxonomy. *NZ Soil News* 27:43–47.

1983. Historical development of Soil Taxonomy: Background. In: LP Wilding, NE Smeck and GF Hall (eds). *Pedogenesis and Soil Taxonomy.* Elsevier, New York. pp. 23–49.

1986. *The Guy Smith Interviews: Rationale for Concepts in Soil Taxonomy*, Soil Management Support Services Tech. Mon. 11., Cornell Univ., Ithaca, NY.

Smith, GD, Allaway, WH and FF Riecken. 1950. Prairie soils of the upper Mississippi valley. *Adv. Agron.* 2:157–205.

Smith, GD, Newhall, DF and LH Robbins. 1964. *Soil-Temperature Regimes, their Characteristics and Predictability.* USDA-SCS, US Govt. Printing Off., Washington, DC.

Smith, H and LP Wilding. 1972. Genesis of argillic horizons in Ochraqualfs derived from fine textured till deposits of northwestern Ohio and southeastern Michigan. *SSSAP* 36:808–815.

Smith, HTU. 1962. Periglacial frost features and releate phenomena in the United States. *Biul. Perygl.* 11:325–342.

Smith, J. 1956. Some moving soils in Spitzbergen. *JSS* 7:10–21.

Smith, ME and TJ Price. 1994. Aztec-period agricultural terraces in Morelos, Mexico: Evidence for household-level agricultural intensification. *J. Field Archaeol.* 21:169–179.

Smith, MW. 1985. Observations of soil freezing and frost heave at Inuvik, Northwest Territories, Canada. *CJES* 22:283–290.

Smith, RB and LW Braile. 1994. The Yellowstone hotspot. *J. Vulcanal. Geotherm. Res.* 61:121–187.

Smithson, F. 1958. Grass opal in British soils. *JSS* 9:148–154.

Snead, RE. 1972. *Atlas of World Physical Features.* Wiley, New York.

Sobecki, TM and AD Karathanasis. 1992. Soils on hillslopes in acid gray and black shales. *SSSAJ* 56:1218–1226.

Sobecki, TM and LP Wilding. 1982. Calcic horizon distribution and soil classification in selected soils of the Texas Coast Prairie. *SSSAJ* 46:1222–1227.

1983. Formation of calcic and argillic horizons in selected soils of the Texas coast prairie. *SSSAJ* 47:707–715.

Sohet, K, Herbauts, J and W Gruber. 1988. Changes caused by Norway spruce in an ochreous brown earth, assessed by the isoquartz method. *JSS* 39:549–561.

Soil Classification Working Group. 1998. *The Canadian System of Soil Classification.* Publ. 1646., Agric. and Agri-Food Canada, Ottawa, ON.

Soil Science Society of America. 1997. *Glossary of Soil Science Terms.* Madison, WI.

Soil Survey Staff. 2014. *Keys to Soil Taxonomy*, 12th ed. USDA Natural Resources Cons. Service, Washington, DC.

Soil Survey Division Staff. 1993. *Soil Survey Manual, US Dept. of Agric.* Handbook #18. US Govt. Printing Off., Washington, DC.

Soil Survey Staff. 1951. *Soil Survey Manual, US Dept. of Agric.* Handbook #18. US Govt. Printing Off., Washington, DC.

1960. *Soil Classification: A Comprehensive System – 7th Approximation.* US Govt. Printing Off., Washington, DC.

1975. *Soil Taxonomy.* US Dept. of Agric. Handbook #436. US Govt. Printing Off., Washington, DC.

1992. *Keys to Soil Taxonomy*, 5th ed, Soil Management Support Services Tech. Mon. 19. Pocahontas Press, Blacksburg, VA.

1999. *Soil Taxonomy.* US Dept. of Agric. Handbook #436. US Govt. Printing Off., Washington, DC.

2010. *Keys to Soil Taxonomy*, 11th ed. USDA Natural Resources Cons. Service, Washington, DC.

Soileau, JM and RJ McCracken. 1967. Free iron coloration in certain well-drained coastal plain soils in relation to their other properties and classification. *SSSAP* 31:248–255.

Sokolov, IA and DE Konyushkov. 1998. Soils and the soil mantle of the northern circumpolar region. *Euras. Soil Sci.* 31:1179–1193.

Sokolovskii, NA. 1924. K pozaniyu svoistv kolloidal'noi chaste pochvy (Some properties of the colloidal portion of the soil). Pochvovedeniye 1-2.

Solis-Castillo, B, Solleiro-Rebolledo, E, Sedov, S, Liendo, R, Ortiz-Perez, M and S Lopez-Rivera. 2013. Paleoenvironment and Human Occupation in the Maya Lowlands of the Usumacinta River, Southern Mexico. Geoarchaeol. 28:268–288.

Soller, DR and JP Owens. 1991. The use of mineralogic techniques as relative age indicators for weathering profiles on the Atlantic Coastal Plain, USA. Geoderma 51:111–131.

Sommer, M, Gerke, HH and D Deumlich. 2008. Modelling soil landscape genesis – A "time split" approach for hummocky agricultural landscapes. Geoderma 145:480–493.

Sommer, M, Halm, D, Weller, U, Zarei, M and K Stahr. 2000. Lateral podzolization in a granite landscape. SSSAJ 64:1434–1442.

Sommer, M, Kaczorek, D, Kuzyakov, Y and J Breuer. 2006. Silicon pools and fluxes in soils and landscapes – A review. J. Plant Nutr. Soil Sci. 169:310–329.

Sommer, M and E Schlichting. 1997. Archetypes of catenas in respect to matter: A concept for structuring and grouping catenas. Geoderma 76:1–33.

Sondheim, MW, Singleton, GA and LM Lavkulich. 1981. Numerical analysis of a chronosequence, including the development of a chronofunction. SSSAJ 45:558–563.

Sondheim, MW and JT Standish. 1983. Numerical analysis of a chronosequence including an assessment of variability. CJSS 63:501–517.

Sorenson, CJ. 1977. Reconstructed Holocene bioclimates. AAAG 67:215–222.

Souchier, B. 1971. Evolution des soils sur roches cristallines à l'étagemontagnard (Vosges). Mem. Serv. Carte Geol. Alsace Lorraine 33.

Souirji, A. 1991. Classification of aridic soils, past and present. In: JM Kimble (ed). Characterization, Classification and Utilization of Cold Aridisols and Vertisols. Proc. 6th Intl. Soil Correlation Mtg. USDA-SCS, Lincoln, NE. pp. 175–184.

Soumya, BS, Sekhar, M, Riotte, J, Audry, S, Lagane, C and JJ Braun. 2011. Inverse models to analyze the spatiotemporal variations of chemical weathering fluxes in a granito-gneissic watershed: Mule Hole, South India. Geoderma 165:12–24.

Souri, B, Watanabe, M and K Sakagami. 2006. Contribution of Parker and Product indexes to evaluate weathering condition of Yellow Brown Forest soils in Japan. Geoderma 130:346–355.

Southard, RJ. 2000. Aridisols. In: ME Sumner (ed). Handbook of Soil Science. CRC Press, Boca Raton, FL. pp. E-321–E-338.

Southard, RJ, Boettinger, JL and OA Chadwick. 1990. Identification, genesis, and classification of duripans. In: JM Kimble and WD Nettleton (eds). Characterization, Classification, and Utilization of Aridisols. Proc. 4th Intl. Soil Correlation Mtg. USDA-SCS, Lincoln, NE. pp. 45–60.

Southard, RJ, Driese, SG and LC Nordt. 2011. Vertisols. In: PM Huang, Y Li and ME Sumner (eds). Handbook of Soil Sciences, 2nd ed. CRC Press, New York. pp. 33–82 – 33–97.

Southard, RJ and RC Graham. 1992. Cesium-137 distribution in a California Pelloxerert: Evidence of pedoturbation. SSSAJ 56:202–207.

Spain, AV and JG McIvor. 1988. The nature of herbaceous vegetation associated with termitaria in north-eastern Australia. JE 76:181–191.

Spain, AV, Okello-Oloya, T and AJ Brown. 1983. Abundance, aboveground masses and basal areas of termite mounds at six locations in tropical north-eastern Australia. Rev. Ecol. Biol. Sol 20:547–566.

Speer JH. 2010: Fundamentals of Tree Ring Research. Univ. Arizona Press, Tucson.

Spence, DHN. 1957. Studies on the vegetation of Shetland. I. The serpentine debris vegetation in Unst. JE 45:917–945.

Spilde, MN, Melim, LA, Northup, DE and PJ Boston. 2013. Anthropogenic lead as a tracer of rock varnish growth: Implications for rates of formation. Geology 41:263–266.

Spiridonova, IA, Sedov, SN, Bronnikova, MA and VO Targulian. 1999. Arrangement, composition, and genesis of bleached components of loamy soddy-podzolic soils. Eurasian Soil Sci. 32:507–513.

Sprafke, T, Thiel, C and B Terhorst. 2013. From micromorphology to palaeoenvironment: The MIS 10 to MIS 5 record in Paudorf (Lower Austria). Catena 117:60–72.

Sprecher, SW and DL Mokma. 1989. Refining the color waiver for Aqualfs and Aquepts. SSH 30:89–91.

Springer, ME. 1958. Desert pavement and vesicular layer of some soils of the desert of the Lahontan Basin, Nevada. SSSAP 22:63–66.

Środoń, J, Drits, VA, McCarty, DK, Hsieh, JCC, and DD Eberl. 2001. Quantitative x-ray diffraction analysis of clay-bearing rocks from random preparations. Clays Clay Miner. 49:514–528.

St. Arnaud, RJ and EP Whiteside. 1964. Morphology and genesis of a chernozemic to podzolic sequence of soil profiles in Saskatchewan. CJSS 44:88–99.

Stace, HCT. 1956. Chemical characteristics of terra rossas and rendzinas of South Australia. J. Soil Sci. 7:280–293.

Stafford, CR and SD Creasman. 2002. The hidden record: Late Holocene landscapes and settlement archaeology in the Lower Ohio valley. Geoarch. 17:117–140.

Stahl, T, Winkler, S, Quigley, M, Bebbington, M, Duffy, B and D Duke. 2013. Schmidt hammer exposure-age dating (SHD) of late Quaternary fluvial terraces in New Zealand. ESPL. doi:10.1002/esp.3427

Staley, JT, Palmer, F and JB Adams. 1982. Microcolonial fungi: Common inhabitants of desert rocks. Science 213:1093–1094.

Stanjek, H, Fassbinder, JWE, Vali, H, Wägele, H and W Graf. 1994. Evidence of biogenic greigite (ferromagnetic Fe_3S_4) in soil. Eur. J. Soil Sci. 45:97–103.

Stanley, KE and RJ Schaetzl. 2011. Characteristics and paleoenvironmental significance of a thin, dual-sourced loess sheet, North–Central Wisconsin. AR 2:241–251.

Stanley, SR and EJ Ciolkosz. 1981. Classification and genesis of Spodosols in the Central Appalachians. SSSAJ 45:912–917.

Stark, NM and CF Jordan. 1978. Nutrient retention by the root mat of an Amazonian rain forest. Ecology 59:434–437.

Steele, F, Daniels, RB, Gamble, EE and LA Nelson. 1969. Fragipan horizons and Be masses in the middle coastal plain of north central North Carolina. SSSAP 33:752–755.

Stein, JK. 1983. Earthworm activity: A source of potential disturbance of archaeological sediments. Am. Antiq. 48:277–289.

Steinhardt, GC and DP Franzmeier. 1979. Chemical and mineralogical properties of the fragipans of the Cincinnati catena. SSSAJ 43:1008–1013.

Steinhardt, GC, Franzmeier, DP and LD Norton. 1982. Silica associated with fragipan and non-fragipan horizons. SSSAJ 46:656–657.

Steinwand, AL and TE Fenton. 1995. Landscape evolution and shallow groundwater hydrology of a till landscape in central Iowa. SSSAJ 59:1370–1377.

Stepanov, IN. 1962. Snow cover and soil formation in high mountains. SSS 270–276.

Stephen, I. 1952. A study of rock weathering with reference to the soils of the Malvern Hills. I. Weathering of biotite and granite. JSS 3:20–33.

Stephen, I, Bellis, E and A Muir. 1956. Gilgai phenonmena in tropical black clays of Kenya. JSS 1–9.

Stephens, CG. 1947. Functional synthesis in pedogenesis. *Trans. Roy. Soc. S. Austral.* 71:168–181.

1964. Silcretes of central Australia. *Nature* 203:1407.

Stephens, EP. 1956. The uprooting of trees: A forest process. *SSSAP* 20:113–116.

Steuter, AA, Jasch, B, Ihnen, J and LT Tieszen. 1990. Woodland/grassland boundary changes in the middle Niobrara Valley of Nebraska identified by ^{13}C values of soil organic matter. *AMN* 124:301–308.

Stevens, PR and TW Walker. 1970. The chronosequence concept and soil formation. *Quart. Rev. Biol.* 45:333:350.

Stevenson, CM, Carpenter, J and BE Scheetz. 1989. Obsidian dating: Recent advances in the experimental determination and application of hydration rates. *Archaeometry* 31:193–206.

Stevenson, FJ. 1969. Pedohumus accumulation and diagenesis during the Quaternary. *SS* 107:470–479.

1994. *Humus Chemistry: Genesis, Composition, Reactions*, 2nd ed. Wiley, New York.

Stewart, AD, Anand, RR and J Balkau. 2012. Source of anomalous gold concentrations in termite nests, Moolart Well, Western Australia: Implications for exploration. *Geochem.-Explor. Env. Analysis* 12:327–337.

Stewart, VI, Adams, WA and HH Abdullah. 1970. Quantitative pedological studies on soils derived from Silurian mudstones. II. The relationship between stone content and the apparent density of the fine earth. *JSS* 21:2484–255.

Stiles, CA, Mora, CI and SG Driese. 2003. Pedogenic processes and domain boundaries in a Vertisol climosequence: Evidence from titanium and zirconium distribution and morphology. *Geoderma* 116:279–299.

Stinchcomb, GE, Messner, TC, Williamson, FC, Driese, SG and LC Nordt. 2013. Climatic and human controls on Holocene floodplain vegetation changes in eastern Pennsylvania based on the isotopic composition of soil organic matter. *QR* 79:377–390.

Stockdill, SMJ. 1982. Effects of introduced earthworms on the productivity of New Zealand pastures. *Pedobiology* 24:29–35.

Stockmann, U, Minasny, B and AB McBratney. 2011. Quantifying processes of pedogenesis. *Adv. Agron.* 113:1–71.

Stoffel, M and M Bollschweiler. 2008. Tree-ring analysis in natural hazards research – an overview. *Nat. Hazards Earth Syst. Sci.* 8:187–202.

Stoffel, M, Bollschweiler, M, Butler, DR and BH Luckman. 2010a. Whither dendrogeomorphology? In: M Stoffel, DR Butler and BH Luckman (eds). *Tree Rings and Natural Hazards: A State-of-the-Art. Advances in Global Change Research, Book Series.* Springer, Dordrecht. pp. 495–502.

Stoffel, M, Butler, DR and BH Luckman (eds). 2010b. *Tree Rings and Natural Hazards: A State-of-the-Art. Advances in Global Change Research, Book Series.* Springer, Dordrecht.

Stokes, S, Bray, HE and MD Blum. 2001. Optical resetting in large drainage basins: Tests of zeroing assumptions using single-aliquot procedures. *QSR* 20:879–885.

Stokstad, E. 2004. Defrosting the carbon freezer of the north. *Science* 304:1618–1620.

Stolt, MH and JC Baker. 1994. Strategies for studying saprolite and saprolite genesis. *SSSASP* 34:1–19.

2000. Quantitative comparison of soil and saprolite genesis: Examples from the Virginia Blue Ridge and Piedmont. *Southeast. Geol.* 39:129–150.

Stolt, MH, Baker, JC and TW Simpson. 1991. Micromorphology of the soil-saprolite transition zone in Hapludults of Virginia. *SSSAJ* 55:1067–1075.

1992. Characterization and genesis of saprolite derived from gneissic rocks of Virginia. *SSSAJ* 56:531–539.

1993. Soil-landscape relationships in Virginia. II. Reconstruction analysis and soil genesis. *SSSAJ* 57:422–428.

Stolt, MH, Lesinski, BC and W Wright. 2001. Micromorphology of seasonally saturated soils in carboniferous glacial till. *SS* 166:406–414.

Stone, EL. 1975. Windthrow influences on spatial heterogeneity in a forest soil. *Eidgen. Anstalt Forstl. Versuch. Swes. Mitt.* 51:77–87.

1977. Abrasion of tree roots by rock during wind stress. *For Sci.* 23:333–336.

1993. Soil burrowing and mixing by a crayfish. *SSSAJ* 57:1096–1099.

Stone, EL, Harris, WG, Brown, RB and RJ Kuehl. 1993. Carbon storage in Florida Spodosols. *SSSAJ* 57:179–182.

Stoops, G. 2003. Guidelines for Analysis and Description of Soil and Regolith Thin Sections. MJ Vepraskas (ed). Soil Sci. Soc. Am., Madison.

Stoops, G, Marcelino, V and F Mees (eds). 2010. *Interpretation of Micromorphological Features of Soils and Regoliths.* Elsevier.

Stottlemyer, R. 1987. Evaluation of anthropogenic atmospheric inputs on United States national park ecosystems. *Env. Mgt.* 11:91–97.

Stout, JD and KM Goh. 1980. The use of radiocarbon to measure the effects of earthworms on soil development. *Radiocarbon* 22:892–896.

Strahler, AH and AN Strahler. 1992. *Modern Physical Geography.* Wiley, New York.

Strain, MR and CV Evans. 1994. Map unit development for sand- and gravel-pit soils in New Hampshire. *SSSAJ* 58:147–155.

Strakhov, NM. 1967. *Principles of Lithogenesis.* Oliver and Boyd, Edinburgh, UK.

Straube, D, Johnson, EA, Parkinson, D, Scheu, S and N Eisenhauer. 2009. Nonlinearity of effects of invasive ecosystem engineers on abiotic soil properties and soil biota. *Oikos* 118:885–896.

Street-Perrott, FA and PA Barker. 2008. Biogenic silica: A neglected component of the coupled global continental biogeochemical cycles of carbon and silicon. *ESPL* 33:1436–1457.

Strickertsson, K, Murray, AS and H Lykke-Andersen. 2001. Optically stimulated luminescence dates for Late Pleistocene sediments from Stensnæs, Northern Jutland, Denmark. *QSR* 20:755–759.

Stroganova, MN and MG Agarkova. 1993. Urban soils: Experimental study and classification (exemplified by the soils of southwestern Moscow). *Euras. Soil Sci.* 25:59–69.

Stromsoe, JR and O Paasche. 2011. Weathering patterns in high-latitude regolith. *J. Geophys. Res. Earth Sfc.* 116. doi:10.1029/2010JF001954

Stuart, DM and RM Dixon. 1973. Water movement and caliche formation in layered arid and semiarid soils. *SSSAP* 37:323–324.

Stubbins, A, Hood, E, Raymond, PA, Aiken, GR, Sleighter, RL, Hernes, PJ, Butman, D, Hatcher, PG, Striegl, RG, Schuster, P, Abdulla, HAN, Vermilyea, AW, Scott, DT and RGM Spencer. 2012. Anthropogenic aerosols as a source of ancient dissolved organic matter in glaciers. *Nature Geosci.* 5:198–201.

Stuiver, M, Kromer, B, Becker, B and CW Ferguson. 1986. Radiocarbon age calibration back to 13300 years BP and the 14C age matching of the German oak and US bristlecone pine chronologies. *Radiocarbon* 28:969–979.

Stuiver, M and PD Quay. 1980. Changes in atmospheric carbon-14 attributed to a variable sun. *Science* 207:11–19.

Stuiver, M, Reimer, PJ, Bard, E, Beck, JW, Burr, GS, Hughen, KA, Kromer, B, McCormac, G, van der Plicht, J and M Spurk. 1998. INTCAL98 radiocarbon age calibration, 24000–0 cal BP. *Radiocarbon* 40:1041–1083.

Stumm, W and JJ Morgan. 1996. *Aquatic Chemistry. Chemical Equilibria and Rates in Natural Waters*. Wiley, New York.

Stützer, A. 1999. Podzolisation as a soil forming process in the alpine belt of Rondane, Norway. *Geoderma* 91:237–248.

Suarez, ER, Fahey, TJ, Groffman, PM, Yavitt, JB and PJ Bohlen. 2006a. Spatial and temporal dynamics of exotic earthworm communities along invasion fronts in a temperate hardwood forest in south-central New York (USA). *Biol. Invasions* 8:553–564.

Suarez, ER, Fahey, TJ, Yavitt, JB, Groffman, PM and PJ Bohlen. 2006b. Patterns of litter disappearance in a northern hardwood forest invaded by exotic earthworms. *Ecol. App.* 16:154–165.

Sudd, JH. 1969. The excavation of soil by ants. *Z. Tierpsychal.* 26:257–276.

Sudom, MD and RJ St. Arnaud. 1971. Use of quartz, zirconium and titanium as indices in pedological studies. *CJSS* 51:385–396.

Suess, HE and TW Linick. 1990. The ^{14}C record in bristlecone pine wood of the past 8000 years based on the dendrochronology of the late C.W. Ferguson. *Phil. Trans. Royal Soc. London.* A330:403–412.

Suganuma, Y, Yokoyama, Y, Yamazaki, T, Kawamura, K, Horng, CS and H Matsuzaki. 2010. Be-10 evidence for delayed acquisition of remanent magnetization in marine sediments: Implication for a new age for the Matuyama-Brunhes boundary. *EPSL* 296:443–450.

Sugden, A, Stone, R and C Ash. 2004. Ecology in the underworld. *Science* 304:1613.

Summerfield, MA. 1982. Distribution, nature and probable genesis of silicrete in arid and semi-arid southern Africa. *Catena Suppl.* 1:37–65.

1983. Petrology and diagenesis of silicrete from theKalahari Basin and Cape coastal zone, southern Africa. *J. Sed. Petr.* 53:895–909.

Sumner, PD, Hedding, DW and KI Meiklejohn. 2007. Rock surface temperatures in southern Namibia and implications for thermally-driven physical weathering. *ZG* 51:133–147.

Sun, DH, Zhang, YB, Han, F, Zhang, Y, Yi, ZY, Li, ZJ, Wang, F, Wu, S and BF Li. 2011. Magnetostratigraphy and palaeoenvironmental records for a Late Cenozoic sedimentary sequence from Lanzhou, Northeastern margin of the Tibetan Plateau. *Global Planet. Change* 76:106–116.

Sun, JM, Li, SH, Muhs, DR and B Li. 2007. Loess sedimentation in Tibet: Provenance, processes, and link with Quaternary glaciations. *QSR* 26:2265–2280.

Sunico, A, Bouza, P and H DelValle. 1996. Erosion of subsurface horizons in northeastern Patagonia, Argentina. *Arid Soil Res. Rehab.* 10:359–378.

Superson, J, Gebica, P and T Brzezinska-Wojcik. 2010. The origin of deformation structures in periglacial fluvial sediments of the Wislok Valley, Southeast Poland. *Perm. Periglac. Proc.* 21:301–314.

Suprychev, VA. 1963. New calcareous formations in the soil-forming parent material in the Sivash region. *SSS* 383–386.

Sutton, JC and BR Sheppard. 1976. Aggregation of sand dune soil by endomycorrhizal fungi. *Can. J. Bot.* 54:326–333.

Suzuki, Y, Matsubara, T and M Hoshino. 2003. Breakdown of mineral grains by earthworms and beetle larvae. *Geoderma* 112:131–142.

Sveistrup, TE, Haraldsen, TK and F Englestad. 1998. Earthworm channels in cultivated clayey and loamy Norwegian soils. *Soil Tillage Res.* 43:253–264.

Sveistrup, TE, Haraldsen, TK, Langohr, R, Marcelino, V and J Kvaerner. 2005. Impact of land use and seasonal freezing on morphological and physical properties of silty Norwegian soils. *Soil Till. Res.* 81:39–56.

Svoray, T and S Shoshany. 2004. Multi-scale analysis of intrinsic soil factors from SAR-based mapping of drying rates. *Rem. Sens. Environ.* 92:233–246.

Swank, WT, Goebel, NB and JD Helvey. 1972. Interception loss in loblolly pine stands of the South Carolina piedmont. *J. Soil Water Conserv.* 27:160–164.

Swanson, DK. 1996. Soil geomorphology on bedrock and colluvial terrain with permafrost in central Alaska, USA. *Geoderma* 71:157–172.

Swanson, DK, Ping, CL and GJ Michaelson. 1999. Diapirism in soils due to thaw of ice-rich material near the permafrost table. *Perm. Periglac. Proc.* 10:349–367.

Swap, R, Garstang, M, Greco, S, Talbot, R and P Kållberg. 1992. Saharan dust in the Amazon Basin. *Tellus* 44B:133–149.

Swindale, LD and ML Jackson. 1956. Genetic processes in some residual podzolized soils of New Zealand. *Trans. 6th Intl. Congress Soil Sci.* E:233–239.

Swineford, A and JC Frye. 1946. Petrographic comparison of Pliocene and Pleistocene volcanic ash from western Kansas. *Kansas St. Geol. Surv. Bull.* 64:1–32.

Swinehart, AL and GR Parker. 2000. Palaeoecology and development of peatlands in Indiana. *AMN* 143:267–297.

Switzer, P, Harden, JW and RK Mark. 1988. A statistical method for estimating rates of soil development and ages of geologic deposits: A design for soil-chronosequence studies. *Math. Geol.* 20:49–61.

Sycheva, SA. 2006. Long-term pedolithogenic rhythms in the Holocene. *QI* 152:181–191.

Sycheva, S, Glasko, M and O Chichagova. 2003. Holocene rhythms of soil formation and sedimentation in the Central Russian upland. *QI* 106:203–213.

Symmons, PM and CF Hemming. 1968. A note on wind-stable stone-mantles in the southern Sahara. *Geog. J.* 134:60–64.

Syverson, KM. 2007. Pleistocene Geology of Chippewa County, Wisconsin. Wisc. Geol. and Nat. Hist. Survey Bull. 103.

Szabolcs, I. 1974. *Salt Affected Soils in Europe*. Martinus Nijhoff, The Hague.

1989. *Salt Affected Soils*. CRC Press, Boca Raton, FL.

Szyma ski, W, Skiba, M and S Skiba. 2011. Fragipan horizon degradation and bleached tongues formation in Albeluvisols of the Carpathian Foothills, Poland. *Geoderma* 167–168:340–350.

2012. Origin of reversible cementation and brittleness of the fragipan horizon in Albeluvisols of the Carpathian Foothills, Poland. *Catena* 99:66–74.

Taber, S. 1943. Perennially frozen ground in Alaska: Its origin and history. *GSA Bull.* 54:1433–1458.

Taboada, T, Cortizas, AM, Garcia, C and EG Rodeja. 2006. Particle-size fractionation of titanium and zirconium during weathering and pedogenesis of granitic rocks in NW Spain. *Geoderma* 131:218–236.

Tait, CE. 2012. *Irrigation in Imperial Valley*, California. Its Problems and Possibilities. General Books, LLC.

Takahashi, T and S Shoji. 2002. Distribution and classification of volcanic ash soils. *Global Environ. Res.* 6:83–97.

Takahashi, Y, Minai, Y, Ambe, S, Makide, Y and F Ambe. 1999. Comparison of adsorption behavior of multiple inorganic ions on kaolinite and silica in the presence of humic acid using the multi-tracer technique. *Geochim. Cosmochim. Acta* 63:815–836.

Talbot, M. 1953. Ants of an old field community on the Edwin S. George Reserve, Livingston County, Michigan. *Contrib. Lab. Vertebr. Biol. Univ. Michigan* 69:1–9.

Tamm, CO and H Holmen. 1967. Some remarks on soil organic matter turn-over in Swedish Podzol profiles. *Meddel. Norske Skogforsoks.* 85:69–88.

Tamura, T, Saito, Y, Bateman, MD, Nguyen, VL, Ta, TKO and D Matsumoto. 2012. Luminescence dating of beach ridges for characterizing multi-decadal to centennial deltaic shoreline changes during Late Holocene, Mekong River delta. *Marine Geol.* 326:140–153.

Tan, KH. 1965. The Andosols of Indonesia. *SS* 99:375–378.

(ed) 1984. *Andosols.* Van Nostrand Reinhold, New York.

Tan, KH, Perkins, HF and RA McCreeery. 1970. The characteristics, classification and genesis of some tropical Spodosols. *SSSAP* 34:775–779.

Tan, KH and J Van Schuylenborgh. 1961. On the classification and genesis of soils developed over acid volcanic materials under humid tropical conditions. II. *Neth. J. Agric. Sci.* 9:41–54.

Tan, Z, Harris, WG and RS Mansell. 1999. Watertable dynamics across an Aquod-Udult transition in Florida flatwoods. *SS* 164:10–17.

Tanaka, A, Sakuma, T, Okagawa, N, Imai, H, Sato, T, Yamaguchi, J and K Fujita. 1989. Agro-Ecological Condition of the Oxisol-Ultisol Area of the Amazon River System. Sapporo, Faculty of Agriculture, Hokkaido University.

Tandarich, JP. 2001. Wisconsin agricultural geologists: Ahead of their time. *Geoscience Wisc.* 18:21–26.

Tandarich, JP, Darmody, RG and LR Follmer. 1988. The development of pedological thought: Some people involved. *PG* 9:162–174.

1994. The pedo-weathering profile: A paradigm for whole-regolith pedology from the glaciated midcontinental United States of America. *SSSASP* 34:97–117.

Tandarich, JP, Darmody, RG, Follmer, LR and DL Johnson. 2002. The historical development of soil and weathering profile concepts. *SSSAJ* 66:1–14.

Tandarich, JP and SW Sprecher. 1994. The intellectual background for the Factors of Soil Formation. *SSSASP* 33:1–13.

Tang, YJ, Jia, JY and XD Xie. 2003. Records of magnetic properties in Quaternary loess and its paleoclimatic significance: A brief review. *QI* 108:33–50.

Targulian, VO and SV Goryachkin. 2004. Soil memory: Types of record, carriers, hierarchy and diversity. *Revista Mex. Ciencias Geol.* 21:1–9.

Tarnocai, C. 1972. Some characteristics of cryic organic soils in northern Manitoba. *CJSS* 52:485–496.

1990. Paleosols of the interglacial climates of Canada. *Geog. Phys. Quat.* 44:363–374.

1994. Genesis of permafrost-affected soils. In: JM Kimble and RJ Ahrens (eds). *Classification, Correlation, and Management of Permafrost-Affected Soils.* USDA-SCS, Lincoln, NE. pp. 143–154.

Tarnocai, C and JG Bockheim. 2011. Cryosolic soils of Canada: Genesis, distribution, and classification. *CJSS* 91:749–762.

Tarnocai, C, Canadell, JG, Schuur, EAG, Kuhry, P, Mazhitova, G and S. Zimov. 2009. Soil organic carbon pools in the northern circumpolar permafrost region. *Global Biogeochem. Cycles* 23. doi:10.1029/2008GB003327

Tarnocai, C and SC Zoltai. 1978. Earth hummocks of the Canadian Arctic and Subarctic. *AAR* 10:581–594.

Tate, RL. 1991. Microbial biomass measurements in acidic soils: Effect of fungal:bacterial activity ratios and soil amendment. *SS* 152:220–225.

Tatur, A. 1989. Ornithogenic soils of the maritime Antarctica. *Pol. Polar Res.* 10:481–532.

Tatur, A, and A Barczuk. 1985. Ornithogenic phosphates on King George Island, Maritime Antarctic. In: WR Siegfried, PR Condy and RM Laws (eds). *Antarctic Nutrient Cycles and Food Webs.* Springer-Verlag, Berlin. pp. 163–169.

Tavernier, R and H Eswaran. 1972. Basic concepts of weathering and soil genesis in the humid tropics. *Proc. 2nd ASEAN Soils Conf., Jakarta, Indonesia.* 1:383–392.

Tavernier, R and A Louis. 1971. La dégradation des sols limoneaux sous monoculture de hêtres de la forêt de Soignes (Belgique). *Ann. Intl. St. Cerc. Pedolog.* 38:165–191.

Tavernier, R and GD Smith. 1957. The concept ofBraunerde (Brown Forest soils) in Europe and the United States. *Adv. Agron.* 9:217–289.

Taylor, RE. 1982. Problems in the radiocarbon dating of bone. In: LA Currie (ed). *Nuclear and Chemical Dating Techniques: Interpreting the Environmental Record.* American Chemistry Society, Washington, DC. pp. 453–473.

1987. *Radiocarbon Dating: An Archaeological Perspective.* Academic Press, New York.

Taylor-George, S, Palmer, F, Staley, JT, Borns, DJ, Curtiss, B and JB Adams. 1983. Fungi and bacteria involved in desert varnish formation. *Microb. Ecol.* 9:227–245.

Tedrow, JCF. 1962. Morphological evidence of frost action in Arctic soils. *Biul. Peryglac.* 11:343–352.

1968. Pedogenic gradients of the Polar regions. *JSS* 19:197–204.

1977. *Soils of the Polar Landscapes.* Rutgers Univ. Press, New Brunswick, NJ.

Tedrow, JCF and J Brown. 1962. Soils of the Northern Brooks Range, Alaska: Weakening of the soil-forming potential at high Arctic altitudes. *SS* 93:254–261.

Tejan-Kella, MS, Fitzpatrick, RW and DJ Chittleborough. 1991. Scanning electron microscope study of zircons and rutiles from podzol chronosequences at Cooloola, Queensland, Australia. *Catena* 18:11–30.

Templin, EH, Mowery, IC and GW Kinze. 1956. Houston Black Clay, the type Grumusol. I. Field morphology and geography. *SSSAP* 20:88–90.

Ten Brink, NW. 1973. Lichen growth rates in West Greenland. *AAR* 5:323–331.

Theoulakis, P and A Moropoulou. 1999. Salt crystal growth as weathering mechanism of porous stone on historic masonry. *J. Porous Matls.* 6:345–358.

Theron, GK. 1979. Die verskynsel van kaal kolle in Kaokoland, SuidWes-Afrika. *J. South Afr. Biol. Soc.* 20:43–53.

Thiry, M. 2000. Palaeoclimatic interpretation of clay minerals in marine deposits: An outlook from the continental origin. *Earth-Sci. Revs.* 49:201–221.

Thiry, M and AR Milnes. 1991. Pedogenic and groundwater silcretes at Stuart Creek Opal Field, South Australia. *J. Sed. Petr.* 61:111–127.

Thomas, DJ. 1996. *Soil Survey of Macon County, North Carolina.* USDA-NRCS, US Govt. Printing Off., Washington, DC.

Thomas, MF. 1974. *Tropical Geomorphology: A Study of Weathering and Landform Development in Warm Climates.* Wiley, New York.

2001. Landscape sensitivity in time and space – An introduction. *Catena* 42:83–98

Thompson, JA and JC Bell. 1996. Color index for identifying hydric conditions for seasonally saturated Mollisols in Minnesota. *SSSAJ* 60:1979–1988.

Thompson, JA, Bell, JC and CA Butler. 1997. Quantitative soil-landscape modeling for estimating the areal extent of hydromorphic soils. *SSSAJ* 61:971–980.

Thompson, ML, Fedoroff, N, and B Fournier. 1990. Morphological features related to agriculture and faunal activity in three loess-derived soils in France. *Geoderma* 46:329–349.

Thompson, JA, Roecker, S, Grunwald, S and PR Owens. 2012. Digital soil mapping: Interactions with and Applications for Hydropedology. In: H Lin (ed). *Hydropedology. Synergistic Integration of Soil Science and Hydrology.* Elsevier, Amsterdam. pp. 665–709.

Thompson, ML and NE Smeck. 1983. Micromorphology of polygenetic soils in the Teays River valley, Ohio. *SSSAJ* 47:734–742.

Thompson, ML, Smeck, NE and JM Bigham. 1981. Parent materials and paleosols in the Teays River Valley, Ohio. *SSSAJ* 45:918–925.

Thompson, ML and L Ukrainczyk. 2002. Micas. In: JB Dixon and D Schulze (eds), *Soil Mineralogy with Environmental Applications*. Soil Sci. Soc. Am. Book Ser. 7. Madison, WI. pp. 431–466.

Thompson, R and BA Maher. 1995. Age models, sediment fluxes and palaeoclimatic reconstructions for the Chinese loess and palaeosol sequences. *Geophys. J. Intl.* 123:611–622.

Thorn, CE. 1976. A model of stony earth circle development, Schefferville, Quebec. *Proc. Ass. Am. Geog.* 8:19–23.

1979. Bedrock freeze-thaw weathering regime in an alpine environment, Colorado Front Range. *ESPL* 4:211–228.

Thorn, CE and RG Darmody. 1980. Contemporary eclian sediments in the alpine zone, Colorado Front Range. *PG* 1:162–171.

Thorn, CE, Darmody, RG, Dixon, JC and P Schlyter. 2001. The chemical weathering regime of Karkevagge, arctic-alpine Sweden. *Geomorphology* 41:37–52.

Thorn, CE, Darmody, RD, Holmqvist, J, Jull, AJT, Dixon, JC and P Schlyter. 2009. Comparison of radiocarbon dating of buried paleosols using arbuscular mycorrhizae spores and bulk soil samples. *Holocene* 19:1031–1037.

Thornthwaite, CW. 1931. The climates of North America according to a new classification. *Geog. Rev.* 21:633–655.

Thorp, J. 1947. Practical problems in Soil Taxonomy and soil mapping in Great Plains states. *SSSAP* 12:445–448.

1948. *How soils develop under grass*. In: Yearbook of Agriculture. US Dept. of Agric., Washington, DC. pp. 55–66.

1949. Effects of certain animals that live in soils. *Sci. Monogr.* 68:180–191.

Thorp, J, Cady, JG and EE Gamble. 1959. Genesis of Miami silt loam. *SSSAP* 23:156–161.

Thorp, J, Johnson, WM and EC Reed. 1951. Some post-Pliocene buried soils of central United States. *JSS* 2:1–19.

Thorp, J and GD Smith. 1949. Higher categories of soil classification. *SS* 67:117–126.

Thorson, RM and CA Schile. 1995. Deglacial eolian regimes in New England. *GSA Bull.* 107:751–761.

Tiessen, H and JWB Stewart. 1983. Particle-size fractions and their use in studies of soil organic matter. II. Cultivation effects on organic matter composition in size fractions. *SSSAJ* 47:509–514.

Tieszen, LL, Reed, BC, Bliss, NB, Wylie, BK and DD Dejong. 1997. NDVI, C-3 and C-4 production and distributions in Great Plains grassland land cover classes. *Ecol. Appl.* 7:59–78.

Tilahun, A, Kebede, F, Yamoah, C, Erens, H, Mujinya, BB, Verdoodt, A and E Van Ranst. 2012. Quantifying the masses of *Macrotermes subhyalinus* mounds and evaluating their use as a soil amendment. *Agric. Ecosystems Environ.* 157:54–59.

Tipping, E, Chamberlain, PM, Bryant, CL and S Buckingham. 2010. Soil organic matter turnover in British deciduous woodlands, quantified with radiocarbon. *Geoderma* 155:10–18.

Tisdall, JM, Smith, SE and P Rengasamy. 1997. Aggregation of soil by fungal hyphae. *AJSR* 35:55–60.

Titeux, H and B Delvaux. 2010. Properties of successive horizons in a thick forest floor (mor) reflect a sequence of soil acidification. *Geoderma* 158:298–302.

Tiunov, AV, Hale, CM, Holdsworth, AR and TS Vsevolodova-Perel. 2006. Invasion patterns of Lumbricidae into the previously earthworm-free areas of northeastern Europe and the western Great Lakes region of North America. *Biol. Invasions* 8:1223–1234.

Todhunter, PE. 2001. A hydroclimatological analysis of the Red River of the north snowmelt flood catastrophe of 1997. *J. Am. Water Resources Assoc.* 37:1263–1278.

Tombácz, E and M Szekeres. 2004. Colloidal behavior of aqueous montmorillonite suspensions: The specific role of pH in the presence of indifferent electrolytes. *Appl. Clay Sci.* 27:75–94.

2006. Surface charge heterogeneity of kaolinite in aqueous suspension in comparison with montmorillonite. *Appl. Clay Sci.* 34:105–124.

Tonkin, PJ and LR Basher. 2001. Soil chronosequences in subalpine superhumid Cropp Basin, western Southern Alps, New Zealand. *New Zeal. J. Geol. Geophys.* 44:37–45.

Tonkov, S and E Marinova. 2005. Pollen and plant macrofossil analyses of radiocarbon dated mid-Holocene profiles from two subalpine lakes in the Rila Mountains, Bulgaria. *Holocene* 15:663–671.

Toomanian, N, Jalalian, A, Khamedi, H, Eghbal, MK and A Papritz. 2006. Pedodiversity and pedogenesis in Zayandeh-red Valley, central Iran. *Geomorphology* 81:376–393.

Topham, PB. 1977. *Colonization, growth, succession and competition*. In: MRD Seaward (ed.), Lichen Ecology. Academic Press, London. pp. 31–68.

Topp, GC and JL Davis. 1985. Time-domain reflectometry (TDR) and its application to irrigation scheduling. *Adv. Irrig.* 3:107–127.

Topp, GC, Davis, JL and AP Anan. 1980. Electromagnetic determination of soil water content: Measurement in coaxial transmission lines. *Water Resources Res.* 16:574–582.

Torn, MS, Swanston, CW, Castanha, C and SE Trumbore. 2009. Storage and turnover of organic matter in soil. In: N Senesi, B Xing and PM Huang (eds). *Biophysico-Chemical Processes involving Natural Nonliving Organic Matter in Environmental Systems*. Wiley, New York. pp. 219–272.

Tornes, LA, Miller, KE, Gerken, JC and NE Smeck. 2000. Distribution of soils in Ohio that are described with fractured substratums in unconsolidated materials. *Ohio J. Sci.* 100:56–62.

Torok, A and N Rozgonyi. 2004. Morphology and mineralogy of weathering crusts on highly porous oolitic limestones, a case study from Budapest. *Envir. Geol.* 46:333–349.

Torrent, J. 1995. *Genesis and Properties of the Soils of the Mediterranean Regions*. Napoli University Press, Napoli, Italy.

Torrent, J and A Cabedo. 1986. Sources of iron oxides in reddish brown profiles from Calcarenites in southern Spain. *Geoderma* 37:57–66.

Torrent, J and WD Nettleton. 1978. Feedback processes in soil genesis. *Geoderma* 20:281–287.

Torrent, J, Nettleton, WD and G Borst. 1980a. Genesis of a Typic Durixeralf of Southern California. *SSSAJ* 44:575–582.

Torrent, J, Schwertmann, U, Fechter, H and F Alferez. 1983. Quantitative relationships between soil color and hematite content. *SS* 136:354–358.

Torrent, J, Schwertmann, U and DG Schulze. 1980b. Iron oxide mineralogy of some soils of two river terrace sequences in Spain. *Geoderma* 23:191–208.

Toutain, F and JC Vedy. 1975. Influence de la végétation forestière sur l'humification et la pédogenèse en mileau acide et en climat tempéré. *Rev. Ecol. Biol. Sol* 12:375–382.

Tracy, CR, Streten-Joyce, C, Dalton, R, Nussear, KE, Gibb, KS and KA Christian. 2010. Microclimate and limits to photosynthesis in a diverse community of hypolithic cyanobacteria in northern Australia. *Environ. Microbiol.* 12:592–607.

Transeau, EN. 1903. On the geographic distribution and ecological relationships of the bog societies of northern North America. *Bot. Gaz.* 36:401–420.

Treadwell-Steitz, C and LD McFadden. 2000. Influence of parent material and grain size on carbonate coatings in gravelly soils, Palo Duro Wash, New Mexico. *Geoderma* 94:1–22.

Tremocoldi, WA, Steinhardt, GC and DP Franzmeier. 1994. Clay mineralogy and chemistry of argillic horizons, fragipans, and paleosol B horizons of soils on a loess-thinning transect in southwestern Indiana, USA. *Geoderma* 63:77–93.

Tricart, J. 1957. Observations sur le rôle ameublisseur des termites. *Rev. Geomorphol.* 8:170–172, 179.

Trifonova, TA. 1999. Formation of the soil mantle in mountains: The geosystem aspect. *Euras. Soil Sci.* 32:150–156.

Tripcevich, N, Eerkens, JW and TR Carpenter. 2012. Obsidian hydration at high elevation: Archaic quarrying at the Chivay source, southern Peru. *JAS* 39:1360–1367.

Troedsson, T and WH Lyford. 1973. Biological disturbance and small-scale spatial variations in a forested soil near Garpenburg, Sweden. *Stud. For. Suec.* 109. 23 pp.

Troughton, JH. 1972. *Carbon isotope fractionation by plants.* Proc. 8th Intl. Conf. Radiocarbon Dating (Wellington, NZ) E39–E57.

Trumbore, S. 2009. Radiocarbon and soil carbon dynamics. *Ann. Rev. Earth Planet. Sci.* 37:47–66.

Tsai, C-C and Z-S Chen. 2000. Lithologic discontinuities in Ultisols along a toposequence in Taiwan. *SS* 165:587–596.

Tsai, C-C, Chen, ZS, Kao, CI, Ottner, F, Kao, SJ and F Zehetner. 2010. Pedogenic development of volcanic ash soils along a climosequence in Northern Taiwan. *Geoderma* 156:48–59. Andisols

Tsai, H, Huang, W-S, Hseu, Z-Y and Z-S Chen. 2006. A river terrace soil chronosequence of the Pakua Tableland in central Taiwan. *SS* 171:167–179.

Tsikalas, SG and CJ Whitesides. 2013. Worm geomorphology: Lessons from Darwin. *PPG* 37:270–281.

Tsoar, H and K Pye. 1987. Dust transport and the question of desert loess formation. *Sedimentology* 34:139–153.

Tsukamoto, S, Rink, WJ and T Watanuki. 2003. OSL of tephric loess and volcanic quartz in Japan and an alternative procedure for estimation De from a fast component. *Rad. Meas.* 23:593–600.

Tugel, AJ, Herrick, JE, Brown, JR, Mausbach, MJ, Puckett, W and K Hipple. 2005. Soil change, soil survey, and natural resources decision making: A blueprint for action. *SSSAJ* 69:738–747

Tuhkanen, S. 1986. Delimitation of climatic-phytogeographical regions at the high-latitude area. *Nordia* 20:105–112.

Turcotte, DE and TH Butler Jr. 2006. Folists and Humods off the coast of eastern Maine. *SSH* 47:5–9.

Turk, JK, Chadwick, OA and RC Graham. 2011. Pedogenic processes. In: PM Huang, Y Li and ME Sumner (eds). *Handbook of Soil Sciences*, 2nd ed. CRC Press, New York. pp. 30–1 – 30–29.

Turk, JK, Goforth, BR, Graham, RC and KJ Kendrick. 2008. Soil morphology of a debris flow chronosequence in a coniferous forest, southern California, USA. *Geoderma* 146:157–165.

Turk, JK and RC Graham. 2009. Soil carbon and nitrogen accumulation in a forested debris flow chronosequence, California. *SSSAJ* 73:1504–1509.

2011. Distribution and properties of vesicular horizons in the western United States. *SSSAJ* 75:1449–1461.

Turunen, J, Tolonen, K, Tolvanen, S, Remes, M, Ronkainen, J and H Jungner. 1999. Carbon accumulation in the mineral subsoil of boreal mires. *Global Biogeochem. Cycles* 13:71–79.

Twidale, CR and AR Milnes. 1983. Aspects of the distribution and disintegration of siliceous duricrusts in arid Australia. *Geol. Mijn.* 62:373–382.

Twiss, PC, Suess, E and RM Smith. 1969. Morphological classification of grass phytoliths. *SSSAP* 33:109–115.

Ufnar, DF. 2007. Clay coatings from a modern soil chronosequence: A tool for estimating the relative age of well-drained paleosols. *Geoderma* 141:181–200.

Ugolini, FC. 1966. Soils of the Mesters Vig district, northeast Greenland. I. *The Arctic brown and related soils. Meddel. Grønland* 176:1–22.

1986a. Pedogenic zonation in the well-drained soils of the arctic regions. *QR* 26:100–120.

1986b. Processes and rates of weathering in cold and polar desert environments. In: SM Colman and DP Dethier (eds). *Rates of Chemical Weathering of Rocks and Minerals.* Academic Press, New York. pp. 193–235.

Ugolini, FC and C Bull. 1965. Soil development and glacial events in Antarctica. *Quaternaria* 7:251–269.

Ugolini, FC, Corti, G, Agnelli, A and F Piccardi. 1996. Mineralogical, physical, and chemical properties of rock fragments in soil. *SS* 161:521–542.

Ugolini, FC, Corti, G and G Certini. 2006. Pedogenesis in the sorted patterned ground of Devon plateau, Devon Island, Nunavut, Canada. *Geoderma* 136:87–106.

Ugolini, FC and R Dahlgren. 1987. The mechanism of podzolization as revealed by soil solution studies. In: D Righi and A Chauvel (eds). *Podzols et Podzolisation.* Association Française pour l'Etude du Sol and Institut National de la Recherche Agronom., Paris. pp. 195–203.

Ugolini, FC, Dahlgren, R, Shoji, S and T Ito. 1988. An example of andosolization and podzolization as revealed by soil solution studies, southern Hakkoda, northeastern Japan. *SS* 145:111–125.

Ugolini, FC, Dawson, H and J Zachara. 1977a. Direct evidence of particle migration in the soil solution of a Podzol. *Science* 198:603–605.

Ugolini, FC, Hillier, S, Certini, G and MJ Wilson. 2008. The contribution of aeolian material to an Aridisol from southern Jordan as revealed by mineralogical analysis. *J. Arid Env.* 72:1431–1447.

Ugolini, FC, Minden, R, Dawson, H and J Zachara. 1977b. An example of soil processes in the *Abies amabilis* zone of central Cascades, Washington. *SS* 124:291–302.

Ugolini, FC and AK Schlichte. 1973. The effect of Holocene environmental changes on selected western Washington soils. *SS* 116:218–227.

Ugolini, FC and RS Sletten. 1991. The role of proton donors in pedogenesis as revealed by soil solution studies. *SS* 151:59–75.

Ugolini, FC, Sletten, RS and DJ Marrett. 1990. Contemporary pedogenic processes in the Arctic: Brunification. *Sci. du Sol* 28:333–348.

Ugolini, FC, Stoner, MG and DJ Marrett. 1987. Arctic pedogenesis. I. Evidence for contemporary podzolization. *SS* 144:90–100.

Ujvari, G, Varga, A, Ramos, FC, Kovacs, J, Nemeth, T and T Stevens. 2012. Evaluating the use of clay mineralogy, Sr-Nd isotopes and zircon U-Pb ages in tracking dust provenance: An example from loess of the Carpathian Basin. *Chem. Geol.* 304:83–96.

Ulery, AL and RC Graham. 1993. Forest fire effects on soil color and texture. *SSSAJ* 57:135–140.

Ulmer, MG, Knuteson, JA and DD Patterson. 1994. Particle size analysis by hydrometer: A routine method for determining clay fraction. *SSH* 35:11–17.

Ulrich, HP. 1947. Morphology and genesis of the soils of Adak Island, Aleutian Islands. *SSSAP* 11:438–441.

Ulrich, R. 1950. Some chemical changes accompanying profile formation of nearly level soils developed from Peorian loess in southwestern Iowa. *SSSAP* 15:324–329.

Undul, O and A Tugrul. 2012. The influence of weathering on the engineering properties of dunites. *Rock Mech. Rock Engineer.* 45:225–239

Unger-Hamilton, RE. 1984. The formation of use-wear polish on flint: Beyond the "deposit versus abrasion" controversy. *JAS* 11:91–98.

USGS. 1997. Distribution of non-polar arid land. Available at http://pubs.usgs.gov/gip/deserts/what/world.html. US Geological Survey, Washington, DC.

Vacca, A, Adamo, P, Pigna, M and P Violante. 2003. Genesis of tephra-derived soils from the Roccamonfina Volcana, South Central Italy. *SSSAJ* 67:198–207.

Vadivelu, S and O Challa. 1985. Depth of slickenside occurrence in Vertisols. *J. Ind. Soc. Soil Sci.* 33:452–454.

Vahder, S and U Irmler. 2012. Effect of pure and multi-species beech (*Fagus sylvatica*) stands on soil characteristics and earthworms in two northern German forests. *Eur. J. Soil Biol.* 51:45–50.

Valentine, GA and CD Harrington. 2006. Clast size controls and longevity of Pleistocene desert pavements at Lathrop Wells and Red Code volcanoes, southern Nevada. *Geology* 34:533–536.

Valentine, KWG and JB Dalrymple. 1976. Quaternary buried paleosols: A critical review. *QR* 6:209–222.

Valette-Silver, JN, Brown, L, Pavich, M, Klein, J and R Middleton. 1986. Detection of erosion events using [10]Be profiles: Example of the impact of agriculture on soil erosion in the Chesapeake Bay area (USA). *EPSL* 80:82–90.

van Andel, TH. 1998. Paleosols, red sediments, and the Old Stone Age in Greece. *Geoarchaeol.* 13:361–390.

van Breeman, N. 1982. Genesis, morphology and classification of acid sulfate soils in coastal plains. In: JA Kittrick, DS Fanning, and LR Hossner (eds). *Acid Sulfate Weathering*. Soil Sci. Soc. Am., Madison, WI.

1987. Effects of redox processes on soil acidity. *Neth. J. Agric. Sci.* 35:271–279.

1995. How Sphagnum bogs down other plants. *Trends Ecol. Evol.* 10:270–275.

van Breeman, N and P Buurman. 1998. *Soil Formation*. Kluwer, London.

van Breeman, N, Lundström, US and AG Jongmans. 2000. Do plants drive podzolization via rock-eating mycorrhizal fungi? *Geoderma* 94:163–171.

van Breeman, N, Mulder, J and CT Driscoll. 1983. Acidification and alkalinization of soils. *Plant and Soil* 75:283–308.

van Breeman, N and R Protz. 1988. Rates of calcium carbonate removal from soils. *CJSS* 68:449–454.

Vance, E and NM Nadkarni. 1990. Microbial biomass and activity in canopy organic matter and the forest floor of a tropical cloud forest. *Soil Biol. Biochem.* 22:677–684.

Vance, GF, Boyd, SA and DL Mokma. 1985. Extraction of phenolic compounds from a Spodosol profile: An evaluation of three extractants. *SS* 140:412–420.

Vance, GF, Mokma, DL and SA Boyd. 1986. Phenolic compounds in soils of hydrosequences and developmental sequences of Spodosols. *SSSAJ* 50:992–996.

Vandenberghe, J. 1992. Cryoturbations: A sediment structural analysis. *Perm. Periglac. Proc.* 3:343–352.

2013. Grain size of fine-grained windblown sediment: A powerful proxy for process identification. *Earth-Sci. Revs.* 121:18–30.

Vandenberghe, J and P Van den Broek. 1982. Weichselian convolution phenomena and processes in fine sediments. *Boreas* 11:299–315.

VandenBygaart, AJ. 2011. Regosolic soils of Canada: Genesis, distribution, and classification. *CJSS* 91:881–887.

VandenBygaart, AJ and R Protz. 1995. Soil genesis on a chronosequence, Pinery Provincial Park, Ontario. *CJSS* 75:63–72.

van der Merwe, CR. 1949. A few notes with regard to misconceptions concerning soils of the tropics and sub-tropics. *Comm. Bureau Soil Sci. Tech. Pub.* 46:128–130.

van der Merwe, CR and HW Weber. 1963. The clay minerals of South African soils developed from granite under different climatic conditions. *S. Afr. J. Agric. Sci.* 6:411–454.

van Duin, RHA. 1963. The influence of soil management on the temperature wave near the surface. Wageningen Institute of Land and Water Management, Wageningen, the Netherlands. Tech. Bull. 29.

van Everdingen, R. (ed) 2005. Multi-language glossary of permafrost and related ground-ice terms. National Snow and Ice Data Center, Boulder, CO. http://nsidc.org/fgdc/glossary/

van Hees, PAW, Jones, DL, Finlay, R, Godbold, DL and US Lundström. 2005. The carbon we do not see – the impact of low molecular weight compounds on carbon dynamics and respiration in forest soils. *Soil Biol. Biochem.* 37:1–13.

van Hees, PAW, Lundström, US and R Giesler. 2000. Low molecular weight organic acids and their Al-complexes in soil solution: Composition, distribution and seasonal variation in three podzolized soils. *Geoderma* 94:173–200.

Van Herreweghe, S, Deckers, S, DeConinck, F, Merckx, R and F Gullentops. 2003. The paleosol in the Kerkom Sands near Pellenberg (Belgium) revisited. *Geol. Mijnbouw* 82:149–159.

Vaniman, DT, Chipera, SJ and DL Bish. 1994. Pedogenesis of siliceous calcretes at Yucca Mountain, Nevada. *Geoderma* 63:1–17.

Van Leeuwen, EP, Draaijers, GPJ and JW Erisman. 1996. Mapping wet deposition of acidifying components and base cations over Europe using measurements. *Atmos. Environ.* 30:2495–2511.

Van Nest, J. 2002. The good earthworm: How natural processes preserve upland Archaic archaeological sites of western Illinois, USA. *Geoarchaeol.* 17:53–90.

van Praag, HJ, De Smedt, F and TV Thanh. 2000. Simulation of calcium leaching and desorption in an acid forest soil. *EJSS* 51:245–255.

Van Ranst, E and F DeConinck. 2002. Evaluation of ferrolysis in soil formation. *EJSS* 53:513–519.

Van Ranst, E, Dumon, M, Tolossa, AR, Cornelis, JT, Stoops, G, Vandenberghe, RE and J Deckers. 2011. Revisiting ferrolysis processes in the formation of Planosols for rationalizing the soils with stagnic properties in WRB. *Geoderma* 163:265–274.

Van Reeuwijk, LP and JM de Villiers. 1985. The origin of textural lamellae in Quaternary coastal sands of Natal. *S. Afr. J. Plant Soil* 2:38–44.

van Rooyen, MW, Theron, GK, van Rooyen, N, Jankowitz, WJ and WS Matthews. 2004. Mysterious circles in the Namib Desert: Review of hypotheses on their origin. *J. Arid Environ.* 57:467–485.

van Ryswyk, AL and R Okazaki. 1979. Genesis and classification of modal subalpine and alpine soil pedons of south-central British Columbia, Canada. *AAR* 11:53–67.

Van Steijn, H, Coutard, JP and Filippo, HC. 1988. Simulation expérimentale de laves de ruissellement. *Bull. Assoc. Géog. Franç.* 1:33–40.

Van Vliet, B and R Langhor. 1981. Correlation between fragipans and permafrost with special reference to silty Weichselian deposits in Belgium and northern France. *Catena* 8:137–154.

Van Vliet-Lanoë, B. 1985. Frost effects in soils. In: J Boardman (ed). *Soils and Quaternary Landscape Evolution*. Wiley, Chichester, UK. pp. 117–158.

1988. The significance of cryoturbation phenomena in environmental reconstruction. *JQS* 3:85–96.

1991. Differential frost heave, load casting and convection: Converging mechanisms; A discussion of the origin of cryoturbations. *Perm. Periglac. Proc.* 2:123–139.

1998. Pattern ground, hummocks, and Holocene climate changes. *Euras. Soil Sci.* 31:507–513.

Van Wambeke, A. 1962. Criteria for classifying tropical soils by age. *JSS* 13:124–132.

1966. Soil bodies and soil classification. *Soils Fert.* 29:507–510.

1972. Mathematical expression of eluviation-illuviation processes and computation of effects of clay migration in homogeneous soil parent materials. *JSS* 23:325–332.

1992. *Soils of the Tropics: Properties and Appraisal.* McGraw-Hill, New York.

van Wesemael, B, Poesen, J and T de Figueiredo. 1995. Effects of rock fragments on physical degradation of cultivated soils by rainfall. *Soil Till. Res.* 33:229-250.

Van Zant, K. 1979. Late glacial and postglacial pollen and plant macrofossils from Lake West Okoboji, Northwestern Iowa. *QR* 12:358-380.

Varghese T. and G. Byju. 1993. Laterite soils. Their distribution, characteristics, classification. and management. Tech. Mon. 1, State Comm. on Science, Technology and Environment. Thiruvanathapuram, Sri Lanka.

Vasenev, II and VO Targul'yan. 1995. A model for the development of sod-podzolic soils by windthrow. *Euras. Soil Sci.* 27:1-16.

Vaughan, KL and PA McDaniel. 2009. Organic soils on basaltic lava flows in a cool, arid environment. *SSSAJ* 73:1510-1518.

Vaughn, DM. 1997. A major debris flow along theWasatch front in northern Utah, USA. *PG* 18:246-262.

Veenenbos, JS and AM Ghaith. 1964. Some characteristics of the desert soils of the UAR. *Trans. 8th Intl. Congr. Soil Sci.* 110:148-173.

Veenstra, JJ and CL Burras. 2012. Effects of agriculture on the classification of Black soils in the Midwestern United States. *CJSS* 92:403-411.

Velbel, MA. 1985. Geochemical mass balances and weathering rates in forested watersheds of the southern Blue Ridge. *AJS* 285:904-930.

1990. Mechanisms of saprolitization, isovolumetric weathering, and pseudomorphous replacement during rock weathering: A review. *Chem. Geol.* 84:17-18.

2011. Microdenticles on naturally weathered hornblende. *Appl. Geochem.* 26:1594-1596.

Velbel, MA and AI Losiak. 2010. Denticles on chain silicate grain surfaces and their utility as indicators of weathering conditions on Earth and Mars. *J. Sed. Res.* 80:771-780.

Vele, A, Frouz, J, Holusa, J and J Kalcik. 2010. Chemical properties of forest soils as affected by nests of Myrmica ruginodis (Formicidae). *Biologia* 65:122-127.

Veneklaas, EJ, Zagt, RJ, Van Leerdam, A, Van Ek, R, Broekhoven, AJ and M Van Genderen. 1990. Hydrological properties of the epiphyte mass of a montane tropical rain forest, Columbia. *Vegetatio* 89:183-192.

Veneman, PLM and SM Bodine. 1982. Chemical and morphological characteristics in a New England drainage-toposequence. *SSSAJ* 46:359-363.

Veneman, PLM, Jacke, PV and SM Bodine. 1984. Soil formation as affected by pit and mound microrelief in Massachusetts, USA. *Geoderma* 33:89-99.

Veneman, PLM, Vepraskas, MJ and J Bouma. 1976. The physical significance of soil mottling in a Wisconsin toposequence. *Geoderma* 15:103-118.

Vepraskas, MJ. 1999. *Redoximorphic Features for Identifying Aquic Conditions, N. Carolina Agric. Res. Serv. Tech. Bull. 301.* North Carolina State Univ., Raleigh.

Vepraskas, MJ and J Bouma. 1976. Model experiments on mottle formation simulating field conditions. *Geoderma* 15:217-230.

Vepraskas, MJ and SP Faulkner. 2001. Redox chemistry of hydric soils. In: JL Richardson and MJ Vepraskas (eds). *Wetland Soils.* CRC Press, Boca Raton, FL. pp. 85-105.

Vepraskas, MJ, He, X, Lindbo, DL and RW Skaggs. 2004. Calibrating hydric soil field indicators to long-term wetland hydrology. *SSSAJ* 68:1461-1469.

Vepraskas, MJ and LP Wilding. 1983a. Albic neoskeletans in argillic horizons as indices of seasonal saturation and iron reduction. *SSSAJ* 47:1202-1208.

1983b. Aquic moisture regimes in soils with and without low chroma colors. *SSSAJ* 47:280-285.

Verboom, WH and JS Pate. 2013. Exploring the biological dimension to pedogenesis with emphasis on the ecosystems, soils and landscapes of southwestern Australia. *Geoderma* 211/212:154-183.

Verheye, W. 1974. Soils and soil evolution on limestones in the Mediterranean environment. *Trans. 10th Intl. Congr. Soil Sci.* 6:387-393.

Vermeer, DE. 1966. Geophagy among the Tiv of Nigeria. *AAAG* 56:197-204.

Verosub, KL, Fine, P, Singer, ML and J TenPas. 1993. Pedogenesis and paleoclimate: Interpretation of the magnetic susceptibility record of Chinese loess-paleosol sequences. *Geology* 21:1011-1014.

Verrecchia, EP. 1987. Le contexte morpho-dynamique des croûtescalcaires: Apport des analysis séquentielles à l'escale microscopique. *ZG* 31:179-193.

1990. Litho-diagenetic implications of the calcium oxalate-carbonate biogeochemical cycle in semi-arid calcretes, Nazareth, Israel. *Geomicrobiol. J.* 8:87-99.

1994. L'origine biologique et superficielle des croûtes zonaires. *Bull. Soc. Géol. France* 165:583-592.

Verrecchia, EP, Freytet, P, Verrecchia, KE and J-L Dumont. 1995. Spherulites in calcrete laminar crusts: Biogenic CaCO$_3$ precipitation as a major contributor to crust formation. *J. Sed. Res.* A65:690-700.

Verrecchia, EP and MN LeCoustumer. 1996. Occurrence and genesis of palygorskite and associated clay minerals in a Pleistocene calcrete complex, SDE Boqer, Negev desert, Israel. *Clay Mins.* 31:183-202.

Vestin, JLK, Norström, SH, Bylund, D and US Lundström. 2008. Soil solution and stream water chemistry in a forested catchment II: Influence of organic matter. *Geoderma* 144:271-278.

Victoria, RL, Fernandes, F, Martinelli, LA, Piccolo, MDC, de Camargo, PB and S Trumbore. 1995. Past vegetation changes in the Brazilian Pantanal arboreal-grassy savanna ecotone by using carbon isotopes in the soil organic matter. *Global Change Biol.* 1:165-171.

Vidic, NJ. 1998. Soil-age relationships and correlations: Comparison of chronosequences in the Ljubljana Basin, Slovenia and USA. *Catena* 34:113-129.

Vidic, NJ and F Lobnik. 1997. Rates of soil development of the chronosequence in the Ljubljana Basin, Slovenia. *Geoderma* 76:35-64.

Vidic, NJ and KL Verosub. 1999. Magnetic properties of soils of the Ljubljana Basin chronosequence, Slovenia. *Chin. Sci. Bull.* 44:75-80.

Viereck, L, Dyrness, C and M Foote. 1993. An overview of the vegetation and soils of the floodplain ecosystems of the Tanana river, interior Alaska. *CJFR* 23:889-898.

Vilborg, L. 1955. The uplift of stones by frost. *Geog. Annal.* 37:164-169.

Viles, H. 1988. *Biogeomorphology.* Blackwell, New York.

1995. Ecological perspectives on rock surface weathering: Towards a conceptual model. *Geomorphology* 13:21-35.

Viles, H, Ehlmann, B, Wilson, CF, Cebula, T, Page, M and M Bourke. 2010. Simulating weathering of basalt on Mars and Earth by thermal cycling. *GRL* 37. doi:10.1029/2010GL043522

Viles, HA and AS Goudie. 2007. Rapid salt weathering in the coastal Namib desert: Implications for landscape development. *Geomorphology* 85:49-62.

Viles, HA, Goudie, AS, Grab, S and J Lalley. 2011. The use of the Schmidt Hammer and Equotip for rock hardness assessment in geomorphology and heritage science: A comparative analysis. *ESPL* 36:320-333.

Vincent, KR and OA Chadwick. 1994. Synthesizing bulk density for soils with abundant rock fragments. *SSSAJ* 58:455–464.

Vincent, KR, Bull, WB and OA Chadwick. 1994. Construction of a soil chronosequence using the thickness of pedogenic carbonate coatings. *J. Geol. Ed.* 42:316–324.

Visser, SA and M Caillier. 1988. Observations on the dispersion and aggregation of clays by humic substances. I. Dispersive effects of humic acids. *Geoderma* 42:281–295.

Vodyanitskii, YN, Shishov, LL, Vasil'ev, AA and EF Sataev. 2005. An analysis of the color of forest soils on the Russian Plain. *Euras. Soil Sci.* 38:11–22.

Vogel, DE. 1975. Precambrian weathering in acid metavolcanic rocks from the Superior Province, Vollebon Township, south central Quebec. *CJES* 12:2080–2085.

Vogel, JC. 1969. The radiocarbon time-scale. *S. Afr. Archaeol. Bull.* 24:83–87.

Vogt, T. 1927. Sulitjelmafeltets geologi og petrografi. *Norges Geol. Undersokelse* 121:1–560.

Voight, EE and MJ O'Brien. 1981. The use and misuse of soil-related data in mapping and modeling past environments: An example from the centrl Mississippi River valley. *Contract Abstracts CRM Archeol.* 2:22–35.

Voigt, GK. 1960. Distribution of rainfall under forest stands. *For. Sci.* 6:2–10.

Volk, OH and E Geyger. 1970. Schaumböden als Ursache der Vegetationslosigkeit in ariden Gebieten. *ZG* 14:79–95.

Volobuyev, VR. 1962. Use of a graphical method for studying the humus composition of the major soil groups of the USSR. *SSS* 1–3.

von Deimling, TS, Meinshausen, M, Levermann, A, Huber, V, Frieler, K, Lawrence, DM and V Brovkin. 2012. Estimating the near-surface permafrost-carbon feedback on global warming. *Biogeosci.* 9:649–665.

von Lützow, M, Kögel-Knabner, I, Ekschmitt, K, Matzner, E, Guggenberger, G, Marschner, B and H Flessa. 2006. Stabilization of organic matter in temperate soils: Mechanisms and their relevance under different soil conditions – A review. *EJSS* 57:426–445.

Vorenhout, M, van der Geest, HG, van Marum, D, Wattel, K and HJP Eijsackers. 2004. Automated and continuous redox potential measurements in soil. *J. Environ. Qual.* 33:1562–1567.

Vreeken, WJ. 1973. Soil variability in small loess watersheds: Clay and organic carbon content. *Catena* 1:181–196.

1975. Principal kinds of chronosequences and their significance in soil history. *JSS* 26:378–394.

1984a. Relative dating of soils and paleosols. In: WC Mahaney (ed). *Quaternary Dating Methods.* Elsevier, New York. pp. 269–281.

1984b. Soil landscape chronograms for pedochronological analysis. *Geoderma* 34:149–164.

Vucetich, CG. 1968. Soil-age relationships for New Zealand based on tephro-chronology. *Trans. 9th Intl. Congr. Soil Sci.* 4:121–130.

Wada, K. 1985. The distinctive properties of Andosols. *Adv. Soil Sci.* 2:173–229.

(ed). 1986. *Ando soils in Japan.* Kyushu Univ. Press, Fukuoka, Japan.

Wada, K and S Aomine. 1973. Soil development on volcanic materials during the Quaternary. *SS* 116:170–177.

Waegemans, G and S Henry. 1954. La couleur desLatosols en relation avec leurs des fer. *Trans. 5th Intl. Congr. Soil Sci.* 1/2:384–389.

Wagner, DP, Fanning, DS, Foss, JE, Patterson, MS and PA Snow. 1982. Morphological and mineralogical features related to sulfate oxidation under natural and disturbed land surfaces in Maryland. *SSSASP* 10:109–125.

Wagner, S, Costantini, EAC, Sauer, D and K Stahr. 2007. Soil genesis in a marine terrace sequence of Sicily, Italy. *Revista Mex. Ciencias Geol.* 24:247–260.

Wainwright, J, Parsons, AJ and AD Abrahams. 1999. Field and computer simulation experiments on the formation of desert pavement. *ESPL* 24:1025–1037.

Walbroeck, C. 1993. Climate-soil processes in the presence of permafrost: A systems modelling approach. *Ecol. Model.* 185–225.

Walch, KM, Rowley, JR and NJ Norton. 1970. Displacement of pollen grains by earthworms. *Pollen Spores* 12:39–44.

Walder, J and B Hallet. 1985. A theoretical model of the fracture of a rock during freezing. *GSA Bull.* 96:336–346.

Walker, MD, Everett, KR, Walker, DA and PW Birkeland. 1996. Soil development as an indicator of relative pingo age, northern Alaska, USA. *AAR* 28:352–362.

Walker, PH. 1962. Soil layers on hillslopes: A study at Nowra, New South Wales, Australia. *JSS* 13:167–177

1966. Postglacial environments in relation to landscape and soils in the Cary drift, Iowa. *Iowa. Agric. Exp. St. Res. Bull.* 549:838–875.

Walker, PH, Chartres, CJ and J. Hutka. 1988. The effect of aeolian accessions on soil development on granitic rocks in south-eastern Australia. I. Soil morphology and particle-size distributions. *AJSR* 26:1–16.

Walker, PH and P Green. 1976. Soil trends in two valley fill sequences. *AJSR* 14:291–303.

Walker, PH, Hall, GF and R Protz. 1968. Relation between landform parameters and soil properties. *SSSAP* 32:101–104.

Walker, PH and RV Ruhe. 1968. Hillslope models and soil formation. II. Closed systems. *Trans. 9th Intl. Congr. Soil Sci.* 58:561–568.

Walker, RB, Walker, HM and PR Ashworth. 1955. Calcium-magnesium nutrition with special reference to serpentine soils. *Plant Physiol.* 30:214–221.

Walkington, H. 2010. Soil science applications in archaeological contexts. A review of key challenges. *Earth-Sci. Revs.* 103:122–134.

Wallace, RW and RL Handy. 1961. Stone lines on Cary till. *Proc. Iowa Acad. Sci.* 68:372–379.

Wallinga, J. 2002. Optically stimulated luminescence dating of fluvial deposits: A review. *Boreas* 31:303–322.

Wallinga, J, Murray, AS, Duller, GAT and TE Törnqvist. 2001. Testing optically stimulated luminescence dating of sand-sized quartz and feldspar from fluvial deposits. *EPSL* 193:617–630.

Wallwork, JA. 1970. *Ecology of Soil Animals.* McGraw-Hill, London.

Waloff, N and RE Blackith. 1962. The growth and distribution of the mounds of *Lasius flavus* (F.) (*Hymenoptera: Formicidae*) in Silwood Park, Berkshire. *J. Anim. Ecol.* 31:421–437.

Walsh, PG and GS Humphreys. 2010. Inheritance and formation of smectite in a texture contrast soil in the Pilliga State Forests, New South Wales. *AJSR* 48:88–99.

Walters, JC. 1994. Ice-wedge casts and relict polygonal ground in northeast Iowa, USA. *Permafrost Periglacial Proc.* 5:269–282.

Walthal, PM, Day, WJ and WJ Autin. 1992. Ground-water as a source of sodium in the soils of the Macon Ridge, Louisiana. *SS* 154:95–104.

Walther, SC, Roering, JJ, Almond, PC and MW Hughes. 2009. Long-term biogenic soil mixing and transport in a hilly, loess-mantled landscape: Blue Mountains of southeastern Washington. *Catena* 79:170–178.

Waltman, WJ, Cunningham, RL and EJ Ciolkosz. 1990. Stratigraphy and parent material relationships of red substratum soils on the Allegheny Plateau. *SSSAJ* 54:1049–1057.

Wang, C and H Kodama. 1986. Pedogenic imogolite in sandy Brunisols of eastern Ontario. *CJSS* 66:135–142.

Wang, C, Beke, GJ and JA McKeague. 1978. Site characteristics, morphology and physical properties of selected ortstein soils from the Maritime Provinces. *CJSS* 58:405–420.

Wang, C, McKeague, JA and H Kodama. 1986a. Pedogenic imogolite and soil environments: Case study of Spodosols in Quebec, Canada. *SSSAJ* 50:711–718.

Wang, C, Nowland, JL and H Kodama. 1974. Properties of two fragipan soils in Nova Scotia including scanning electron micrographs. *CJSS* 54:159–170.

Wang, C, Ross, GJ and HW Rees. 1981. Characteristics of residual and colluvial soils developed on granite and of the associated pre-Wisconsin landforms in north-central New Brunswick. *CJES* 18:487–494.

Wang, C, Stea, RR, Ross, GJ and D Holmstrom. 1986b. Age estimation of the Shulie Lake and Eatonville tills in Nova Scotia by pedogenic development. *CJES* 23:115–119.

Wang, D, McSweeney, K, Lowery, B and JM Norman. 1995. Nest structure of ant *Lasius neoniger* Emery and its implications to soil modification. *Geoderma* 66:259–272.

Wang, H and LR Follmer. 1998. A polygenetic model for pedostratigraphic units in the Chinese Loess Plateau region. *QI* 51/52:52.

Wang, H, Lundstrom, CC, Zhang, ZF, Grimley, DA and WL Balsam. 2009. A Mid-Late Quaternary loess-paleosol record in Simmons Farm in southern Illinois, USA. *QSR* 28:93–106.

Wang, XS, Lovlie, R, Yang, ZY, Pei, JL, Zhao, ZZ and ZM Sun. 2005. Remagnetization of Quaternary eolian deposits: A case study from SE Chinese Loess Plateau. Geochem. *Geophys. Geosystems* 6. doi:10.1029/2004GC000901

Wang, Y, Amundson, R and S Trumbore. 1994. A model of 14CO2 and its implications for using 14C to date pedogenic carbonate. *Geochim. Cosmochim. Acta* 58:393–99.

Wang, Y, McDonald, E, Amundson, R, McFadden, L and O Chadwick. 1996a. Radiocarbon dating of soil organic matter. *QR* 45:282–288.

1996b. An isotopic study of soils in chronological sequences of alluvial deposits, Providence Mountains, California. *GSA Bull.* 108:379–391.

Ward, PA III and BJ Carter. 1998. Paleopedologic interpretations of soils buried by Tertiary and Pleistocene-age volcanic ashes: Southcentral Kansas, western Oklahoma, and northwestern Texas, USA. *QI* 51/52:213–221.

Ward, PA III, Carter, BJ and B Weaver. 1993. Volcanic ashes: Time markers in soil parent materials of the southern plains. *SSSAJ* 57:453–460.

Warke, PA, Smith, BJ and RW Magee. 1996. Thermal response characteristics of stone: Implications for weathering of soiled surfaces in urban environments. *ESPL* 21:295–306.

Warren, RE, McDonnell, CK and MJ O'Brien. 1981. Soils and settlement in the southern Prairie Peninsula. *Contract Abstracts CRM Archeol.* 2:36–49.

Warren-Rhodes, KA, Rhodes, KL, Boyle, LN, Pointing, SB, Chen, Y, Liu, S, Zhuo, P and CP McKay. 2007. Lithic cyanobacterial ecology across environmental gradients and spatial scales in China's hot and cold deserts. *FEMS Microbiol. Ecol.* 61:470–482.

Wascher, NE and ME Collins. 1988. Genesis of adjacent morphologically distinct soils in Northwest Florida. *SSSAJ* 52:191–196.

Wascher, HL, Humbert, RP and JG Cady. 1947. Loess in the southern Mississippi valley: Identification and distribution of the loess sheets. *SSSAP* 12:389–399.

Washburn, AL. 1956. Classification of patterned ground and review of suggested origins. *GSA Bull.* 67:823–866.

1980a. *Geocryology. A Survey of Periglacial Processes and Environments.* Wiley, New York.

1980b. Permafrost features as evidence of climatic change. *Earth-Sci. Revs.* 15:327–402.

Washer, NE and ME Collins. 1988. Genesis of adjacent morphologically distinct soils in Northwest Florida. *SSSAJ* 52:191–196.

Watchman, A. 2000. A review of the history of dating rock varnishes. *Earth-Sci. Revs.* 49:261–277.

Watchman, AL and CR Twidale. 2002. Relative and"absolute" dating of land surfaces. *Earth-Sci. Revs.* 58:1–49.

Waters, AC and CW. Flagler. 1929. Origin of small mounds on the Columbia River Plateau. *AJS* 18:209–224.

Waters, MR. 1992. *Principles of Geoarchaeology: A North American Perspective.* Univ. Arizona Press, Tuscon.

Watson, A. 1979. Gypsum crusts in deserts. *J. Arid Envs* 2:3–20.

1985. Structure, chemistry and origins of gypsum crusts in southern Tunisia and the central Namib Desert. *Sedimentology* 32:855–875.

1988. Desert gypsum crusts as palaeoenvironmental indicators: A micropetrographic study of crusts from southern Tunisia and the central Namib Desert. *J. Arid Envs* 15:19–42.

Watson, JAL. 1972. An old mound of the spinifex termite, *Nasutitermes triodiae* (Froggatt) (Isoptera: Termitidae). *Austral. J. Entomol. Soc.* 11:79–80.

Watson, JAL and FJ Gay. 1970. The role of grass eating termites in the degradation of a mulga ecosystem. *Search* 1:43.

Watson, JP. 1961. Some observations on soil horizons and insect activity in granite soils. *Proc. Fed. Sci. Congr. (Salisbury, S. Rhodesia)* 1:271–276.

1962. The soil below a termite mound. *JSS* 13:46–51.

1967. A termite mound in an iron-age burial ground in Rhodesia. *JE* 55:663–669.

1968. Some contributions of archaeology to pedology in central Africa. *Geoderma* 2:291–296.

1969. Water movement in two termite mounds inRhodesia. *JE* 57:441–451.

1977. The use of mounds of the termite *Macrotermes falciger* (Gerstacker) as a soil amendment. *JSS* 28:664–672.

Watts, FC and ME Collins. 2008. *Soils of Florida*. Soil Sci. Soc. Am., Book and Multimedia Publ. Comm., Madison, WI. 88 pp.

Watts, NL. 1980. Quaternary pedogenic calcretes from the Kalahari (southern Africa): Mineralogy, genesis and diagenesis. *Sedimentology* 27:661–686.

Watts, SH. 1977. Major element geochemistry of silcrete from a portion of inland Australia. *Geochim. Cosmochim. Acta* 41:1164–1167.

Watts, WA and HE Wright Jr. 1966. Late-Wisconsin pollen and seed analysis from the Nebraska Sandhills. *Ecology* 47:202–210.

Wayne, WJ. 1991. Ice-wedge casts of Wisconsinan age in eastern Nebraska. *Permafrost Periglacial Proc.* 2:211–223.

Webb, T III. 1974. A vegetational history from northern Wisconsin: Evidence from modern and fossil pollen. *AMN* 92:12–34.

Webber, PJ and JT Andrews. 1973. Lichenometry: A commentary. *AAR* 5:295–302.

Webster, R. 1965a. A catena of soils on the northern Rhodesia plateau. *JSS* 16:31–43.

1965b. A horizon of pea grit in gravel soils. *Nature* 206:696–697.

1968. Fundamental objections to the 7th approximation. *JSS* 19:354–365.

1994. The development of pedometrics. *Geoderma* 62:1–15.

2000. Is soil variation random? *Geoderma* 97:149–163.

Webster, R and M Lark. 2013. *Field Sampling for Environmental Science and Management*. Routledge, New York.

Webster, R and MA Oliver. 2007. *Geostatistics for Environmental Scientists*, 2nd ed. Wiley, New York.

Weerasinghe, HAS and MCM Iqbal. 2011. Plant diversity and soil characteristics of the Ussangoda serpentine site. *J. Natl. Sci. Foundation Sri Lanka* 39:355–363.

Wei, BG and LS Yang. 2010. A review of heavy metal contaminations in urban soils, urban road dusts and agricultural soils from China. *Microchem. J.* 94:99–107.

Weindorf, DC, Bakr, N, Zhu, Y, Haggard, B, Johnson, S and J Daigle. 2010. Characterization of placic horizons in ironstone soils of Louisiana, USA. *Pedosphere* 20:409–418.

Weisenborn, BN. 2001. A model for fragipan evolution in Michigan soils. MA thesis, Michigan State Univ., East Lansing.

Weisenborn, BN and RJ Schaetzl. 2005a. Range of fragipan expression in some Michigan soils. I. Morphological, micromorphological, and pedogenic characterization. *SSSAJ* 69:168–177.

2005b. Range of fragipan expression in some Michigan soils. II. A model for fragipan evolution. *SSSAJ* 69:178–187.

Weiss, T, Siegesmund, S, Kirchner, D and J Sippel. 2004. Insolation weathering and hygric dilatation: Two competitive factors in stone degradation. *Env. Geol.* 46:402–413.

Wells, SG, Dohrenwend, JC, McFadden, LD, Turrin, BD and KD Mahrer. 1985. Late Cenozoic landscape evolution on lava flow surfaces of the Cima volcanic field, Mojave Desert, California. *GSA Bull.* 96:1518–1529.

Wells, SG, McFadden, LD, Poths, J and CT Olinger. 1995. Cosmogenic 3He surface-exposure dating of stone pavements: Implications for landscape evolution in deserts. *Geology* 23:613–616.

Wenner, KA, Holowaychuk, N and GM Schafer. 1961. Changes in clay content, calcium carbonate equivalent, and calcium/magnesium ratio with depth in parent materials of soils derived from calcareous till of Wisconsin age. *SSSAP* 25:312–316.

Werchan, LE and JL Coker. 1983. *Soil Survey of Williamson County, Texas*. USDA-SCS, US Govt. Printing Off., Washington, DC.

Werts, SP and M Milligan. 2012. Controls on soil organic matter content and stable isotope signatures in a New Mexico lithosequence. *SS* 177:599–606.

Wescott, WA. 1993. Geomorphic thresholds and complex response of fluvial systems: Some implications for sequence stratigraphy. *Am. Ass. Petrol. Geol. Bull.* 77:1208–1218.

West, AJ, Bickle, MJ, Collins, R and J Brasington. 2002. Small-catchment perspective on Himalayan weathering fluxes. *Geology* 30:355–358.

West, KR. 2012. An assessment of a soil-based predictive model for locating buried Paleoindian-age cultural deposits in draws on the High Plains of eastern Colorado and western Kansas. MS thesis, Univ. of Kansas, Lawrence.

West, LT, Shaw, JN and FH Beinroth. 2011. Ultisols. In: PM Huang, Y Li and ME Sumner (eds). *Handbook of Soil Sciences*, 2nd ed. CRC Press, New York. pp. 33-167–33-177.

West, LT, Wilding, LP and CT Hallmark. 1988b. Calciustolls in central Texas. II. Genesis of calcic and petrocalcic horizons. *SSSAJ* 52:1731–1740.

West, LT, Wilding, LP, Strahnke, CR and CT Hallmark. 1988a. Calciustolls in central Texas. I. Parent material uniformity and hillslope effects on carbonate-enriched horizons. *SSSAJ* 52:1722–1731.

West, WF. 1970. The Bulawayo Symposium papers. II. Termite prospecting. *Chamb. Mines J.* 12:32–35.

Westergaard, B and HCB Hansen. 1997. Phosphorus in macropore walls of a Danish Glossudalf. *Acta Agric. Scandin. Sec. B* 47:193–200.

Westgate, JA and ND Briggs. 1980. Dating methods of Pleistocene deposits and their problems. V. Tephrochronology and fission-track dating. *Geosci. Can.* 7:3–10.

Westgate, JA, Easterbrook, DJ, Naeser, ND and RJ Carson. 1987. Lake Tapps tephra: An Early Pleistocene stratigraphic marker in the Puget Lowland, Washington. *QR* 28:340–355.

Westin, FC. 1953. Solonetz soils of eastern South Dakota: Their properties and genesis. *SSSAJ* 17:287–293.

Weyl, R. 1950. Schwermineralverwitterung und ihr Einfluss auf die mineralführung klastischer Sedimente. *Erdöl Kohle* 3:209–211.

White, AF, Blum, AE, Schulz, MS, Vivit, DV, Stonestrom, DA, Larsen, M, Murphy, SF and D Eberl. 1998. Chemical weathering in a tropical watershed, Luquillo mountains, Puerto Rico. I. Long-term versus short-term weathering fluxes. *Geochim. Cosmochim. Acta* 62:209–226.

White, AF, Blum, AE, Bullen, TD, Vivit, DV, Schulz, M and J Fitzpatrick. 1999. The effect of temperature on experimental and natural chemical weathering rates of granitoid rocks. *Geochim. Cosmochim. Acta* 63:3277–3291.

White, AF and SL Brantley (eds) 1995. Chemical weathering rates of silicate minerals. *Rev. Mineral.* 31:1–583.

White, AF, Vivit, DV, Schulz, MS, Bullen, TD, Evett, RR and J Aagarwal. 2012. Biogenic and pedogenic controls on Si distributions and cycling in grasslands of the Santa Cruz soil chronosequence, California. *Geochim. Cosmochim. Acta* 94:72–94.

White, EM. 1955. Brunizem-Gray Brown Podzolic soil biosequences. *SSSAP* 19:504–509.

1961. Calcium-solodi or planosol genesis from solodized-solonetz. *SS* 91:175–177.

1964. The morphological-chemical problem in solodized soils. *SS* 98:187–191.

1971. Grass cycling of calcium, magnesium, potassium, and sodium in solodization. *SSSAP* 35:309–311.

1978. Medium-textured soils derived from solodized soils. *Proc. S. Dakota Acad. Sci.* 57:81–91.

White, EM and RG Bonestell. 1960. Some gilgaied soils in South Dakota. *SSSAP* 24:305–309.

White, GN and JB Dixon. 2002. Kaolin – serpentine minerals. In: JB Dixon and DG Schulze (eds). *Soil Mineralogy with Environmental Applications*. SSSA Book Series no.7. Soil Sci. Soc. Am., Madison, WI. pp. 389–414.

White, EM and RI Papendick. 1961. Lithosolic Solodized-Solonetz soils in southwestern South Dakota. *SSSAP* 25:504–506.

White, EM and FF Riecken. 1955. Brunizem-Gray Brown Podzolic soil biosequences. *SSSAP* 19:504–509.

White, GN and LW Zelazny. 1988. Analysis and implications of the edge structure of dioctahedral phyllosilicates. *Clays Clay Mins.* 36:141–146.

White, K, Bryant, R and N Drake. 1998. Techniques for measuring rock weathering: Application to a dated fan segment sequence in southern Tunisia. *ESPL* 23:1031–1043.

White, RE. 1997. Soil science: Raising the profile. *AJSR* 35:961–977.

White, SE. 1976. Is frost action really only hydration shattering? A review. *AAR* 8:1–6.

Whitehead, DR. 1972. Development and environmental history of the Dismal Swamp. *Ecol. Monogr.* 42:301–315.

Whitehead, EM. 1925. *Science and the Modern World*. Macmillan, New York.

Whitehouse, IE, McSaveney, MJ, Knuepfer, PLK and TJH Chinn. 1986. Growth of weathering rinds on Torlesse sandstone, Southern Alps, New Zealand. In: SM Colman and DP Dethier (eds). *Rates of Chemical Weathering of Rocks and Minerals*. Academic Press, London. pp. 419–435.

Whitford, WG and FR Kay. 1999. Biopedturbation by mammals in deserts: A review. *J. Arid Envs.* 41:203–230.

Whitmeyer, SJ, Nicoletti, J and DG De Paor. 2010. The digital revolution in geologic mapping. *GSA Today* 20:4–10.

Whitney, MI and JF Splettstoesser. 1982. Ventifacts and their formation: Darwin Mountains, Antarctica. *Catena Suppl.* 1:175–194.

Whittaker, RH, Buol, SW, Niering, WA and YH Havens. 1968. A soil and vegetation pattern in the Santa Catalina Mountains, Arizona. *SS* 105:440–450.

Whittaker, RH and WA Niering. 1965. Vegetation of the Santa Catalina Mountains, Arizona: A gradient analysis of the south slope. *Ecology* 46:429–452.

Whittecar, GR. 1985. Stratigraphy and soil development in upland alluvium and colluvium north-central Virginia piedmont. *Southeast. Geol.* 26:117–129.

Whittig, LD and WR Allardice. 1986. X-ray diffraction techniques. In: A Klute (ed). *Methods of Soil Analysis. Part 1. Physical and Mineralogical Methods.* Agronomy Mon. 9, 2nd ed. Soil Sci. Soc. Am., Madison, WI. pp. 331–362.

Wieder, M, Givirtzman, G, Porat, N and M Dassa. 2008. Paleosols of the southern coastal plain of Israel. *J. Plant Nutr. Soil Sci.* 171:533–541.

Wieder, M and DH Yaalon. 1972. Micromorphology of terra rossa soils in northern Israel. *Israel J. Agric. Res.* 22:153–154.

Wieder, M, Yair, A and A Arzi. 1985. Catenary soil relationships on arid hillslopes. *Catena Suppl.* 6:41–57.

Wigley, TMA, Jones, PD and KR Briffa. 1987. Cross-dating methods in dendrochronology. *JAS* 14:51–64.

Wiken, EB, Broersma, K, Lavkulich, LM and L Farstad. 1976. Biosynthetic alteration in a British Columbia soil by ants (*Formica fusca* Linné). *SSSAJ* 40:422–426.

Wilde, SA. 1946a. *Forest Soils and Forest Growth.* Chronica Botanica Co., Waltham, MA.

1946b. Proposed additions to the terminology of forest soil genesis. *SSSAP* 10:416–418.

1950. Crypto-mull humus: Its properties and growth effects (a contribution to the classification of forest humus). *SSSAP* 15:360–362.

1966. A new systematic terminology of forest humus layers. *SS* 101:403–407.

Wilding, LP. 1967. Radiocarbon dating of biogenic opal. *Science* 156:66–67.

1994. Factors of soil formation: Contributions to pedology. *SSSASP* 33:15–30.

Wilding, LP, Jones, RB and GW Schafer. 1965. Variation in soil morphological properties within Miami, Celina, and Crosby mapping units in west-central Ohio. *SSSAP* 29:711–717.

Wilding, LP and H Lin. 2006. Advancing the frontiers of soil science towards a geoscience. *Geoderma* 131:257–274.

Wilding, LP, Odell, RT, Fehrenbacher, JB and AH Beavers. 1963. Source and distribution of sodium in Solonetzic soils in Illinois. *SSSAP* 27:432–438.

Wilding, LP and D Tessier. 1988. Genesis of Vertisols: Shrink-swell phenomena. In: LP Wilding and R Puentes (eds). Vertisols: Their Distribution, Properties, Classification and Management. Texas A&M UNiv. Printing Center, College Station, TX. pp. 55–81.

Wilding, LP, Williams, D, Miller, W, Cook, T and H Eswaran. 1990. Close interval spatial variability of Vertisols: A case study in Texas. In: JM Kimble (ed). *Characterization, Classification and Utilization of Cold Aridisols and Vertisols.* Proc. 6th Intl. Soil Correlation Mtg. USDA-SCS, Lincoln, NE. pp. 232–247.

Wildman, WE, Jackson, ML and LD Whittig. 1968. Iron-rich montmorillonite formation in soils derived from serpentinite. *SSSAP* 32:787–794.

Wilkinson, MT, Chappell, J, Humphreys, GS, Fifield, K, Smith, B and PP Hesse. 2005. Soil production in heath and forest, Blue Mountains, Australia: Influence of lithology and palaeoclimate. *ESPL* 30:923–934.

Wilkinson, MT and GS Humphreys. 2005. Exploring pedogenesis via nuclide-based soil production rates and OSL-based bioturbation rates. *Austr. J. Soil Res.* 43:767–779.

Wilkinson, MT, Richards, PJ and GS Humphreys. 2009. Breaking ground: Pedological, geological, and ecological implications of soil bioturbation. *Earth-Sci. Revs.* 97:257–272.

Willenbring, JK and F von Blanckenburg. 2010. Meteoric cosmogenic beryllium-10 adsorbed to river sediment and soil: Applications for Earth-surface dynamics. *Earth-Sci. Revs.* 98:105–122.

Williams, AJ, Buck, BJ and MA Beyene. 2012. Biological soil crusts in the Mojave Desert, USA: Micromorphology and pedogenesis. *SSSAJ* 76:1685–1695.

Williams, AJ, Buck, BJ, Soukup, DA and DJ Merkler. 2013. Geomorphic controls on biological soil crust distribution: A conceptual model from the Mojave Desert (USA). *Geomorphology* 195:99–109.

Williams, D, Cook, T, Lynn, W and H Eswaren. 1996. Evaluating the field morphology of Vertisols. *SSH* 37:123–131.

Williams, GE and HA Polach. 1969. The evaluation of 14C ages for soil carbonate from the arid zone. *EPSL* 7:240–242.

Williams, GP and HP Guy. 1973. Erosional and Depositional Aspects of Hurricane Camille in Virginia, 1969. USGS Prof. Paper 804.

Williams, MA. 1974. Surface rock creep on sandstone slopes in the northern territory of Australia. *Austral. Geog.* 12:419–424.

Williams, MAJ. 1968. Termites and soil development near Brocks Creek, Northern Territory. *Austral. J. Sci.* 31:153–154.

Williams, ME and ED Rudolph. 1974. The role of lichens and associated fungi in the chemical weathering of rock. *Mycologia* 66:648–660.

Williams, PJ and MW Smith. 1989. *The Frozen Earth: Fundamentals of Geocryology.* Cambridge Univ. Press, Cambridge.

Williams, SH and JR Zimbelman. 1994. Desert pavement evolution: An example of the role of the sheetflood. *J. Geol.* 102:243–248.

Williams, SK. 1994. Late Cenozoic tephrostratigraphy of deep sediment cores from the Bonneville basin, northwest Utah. *GSA Bull.* 105:1517–1530.

Williamson, TN, Gessler, PE, Shouse, PJ and RC Graham. 2006. Pedogenesis-terrain links in zero-order watersheds after chaparral to grass vegetation conversion. *SSSAJ* 70:2065–2074.

Willimott, SG and DWG Shirlaw. 1960. C horizon of the soil profile. *Nature* 187:966.

Willis, EH, Tauber, H and KO Munnich. 1960. Variations in the atmospheric radiocarbon concentration over the past 1300 years. *Radiocarbon* 2:1–4.

Willman, HB. 1979. Comments on the Sangamon soil. In: *Wisconsinan, Sangamonian, and Illinoian Stratigraphy in Central Illinois.* Illinois State Geological Survey Guidebook #13. Midwest Friends of the Pleistocene Field Conference Guidebook. pp. 92–94. Urbana-Champaign, IL.

Willman, HB and JC Frye. 1970. Pleistocene stratigraphy of Illinois. Illinois State Geol. Survey Bull. 94.

Willman, HB, Glass, HD and JC Frye. 1966. Mineralogy of glacial tills and their weathering profiles in Illinois. II. Weathering profiles. Illinois State Geological Survey Circular 400.

Wilshire, HG, Nakata, JK and B Hallet. 1981. Field observations of the December 1977 wind storm, San Joaquin Valley, California. *GSA Spec. Pap.* 186:233–251.

Wilson, K. 1960. The time factor in the development of dune soils at South Haven Peninsula, Dorset. *JE* 48:341–359.

Wilson, MA, Burt, R, Sobecki, TM, Engel, RJ and K Hipple. 1996. Soil properties and genesis of pans in till-derived Andisols, Olympic Peninsula, Washington. *SSSAJ* 60:206–218.

Wilson, MA, Indorante, SJ, Lee, BD, Follmer, L, Williams, DR, Fitch, BC, McCauley, WM, Bathgate, JD, Grimley, DA and K Kleinschmidt. 2010. Location and expression of fragic soil properties in a loess-covered landscape, Southern Illinois, USA. *Geoderma* 154:529–543.

Wilson, MJ. 1987. X-ray powder diffraction methods. In: MJ Wilson (ed). *A Handbook of Determinative Methods in Clay Mineralogy.* Blackie, London. pp. 26–98.

1999. The origin and formation of clay minerals in soils: Past, present and future perspectives. *Clay Mins.* 34:7–25.

2003. Clay mineralogical and related characteristics of geophagic materials. *J. Chem. Ecol.* 29:1525–1547.

Wilson, P. 2001. Rate and nature of podzolisation in aeolian sands in the Falkland Islands, South Atlantic. *Geoderma* 101:77–86.

Winchester, V. 1984. A proposal for a new approach to lichenometry. *Br. Geomorphol. Res. Group Tech. Bull.* 33:3–20.

Winkler, EM and EJ Wilhelm. 1970. Salt bursts by hydration pressures in architectural stone in urban atmosphere. *GSA Bull.* 81:567–572.

Winkler, S. 2005. The Schmidt hammer as a relative-age dating technique: Potential and limitations of its application on Holocene moraines in Mt Cook National Park, Southern Alps, New Zealand. *N. Zeal. J. Geol. Geophys.* 48:105–116.

Wintle, AG. 1973. Anomalous fading of thermoluminescence in mineral samples. *Nature* 245:143–144.

1993. Luminescence dating of aeolian sands: An overview. In: K Pye (ed). *The Dynamics and Environmental Context of Aeolian Sedimentary Systems.* Geological Society, London. pp. 49–58.

1996. Archaeologically relevant dating techniques for the next century. *JAS* 23:123–138.

2008. Luminescence dating: Where it has been and where it is going. *Boreas* 37:471–482.

Wintle, AG and JA Catt. 1985. Thermoluminescence dating of soils developed in Late Devensian loess at Pegwell Bay, Kent. *JSS* 36:293–298.

Wintle, AG, Shackleton, NJ and JP Lautridou. 1984. Thermoluminescence dating of periods of loess deposition and soil formation in Normandy. *Nature* 310:491–493.

Wittmer, MHOM, Auerswald, K, Bai, YF, Schaufele, R and H Schnyder. 2010. Changes in the abundance of C3/C4 species of Inner Mongolia grassland: Evidence from isotopic composition of soil and vegetation. *Global Change Biol.* 16:605–616.

Witty, JE and RW Arnold. 1970. Some Folists on Whiteface Mountain, New York. *SSSAP* 34:653–657.

Witty, JE and EG Knox. 1964. Grass opal in some chestnut and forested soils in north central Oregon. *SSSAP* 28:685–688.

Woese, CR. 1987. Bacterials evolution. *Microbiol. Rev.* 51:221–271.

Woese, CR, Kandler, O and ML Wheelis. 1990. Towards a natural system of organisms: Proposal for the domains Archaea, Bacteria, Eucary. *PNAS* 87:4576–4579.

Woida, K and ML Thompson. 1993. Polygenesis of a Pleistocene paleosol in southern Iowa. *GSA Bull.* 105:1445–1461.

Wolfe, SA, Muhs, DR, David, PP and JP McGeehin. 2000. Chronology and geochemistry of late Holocene eolian deposits in the Brandon Sand Hills, Manitoba, Canada. *QI* 67:61–74.

Wolff, EW. 2008. The past 800 ka viewed through Antarctic ice cores. *Episodes* 31:219–221.

Wolt, J. 1994. *Soil Solution Chemistry: Applications to Environmental Science and Agriculture.* Wiley, New York.

Wood, A. 1942. The development of hillside slopes. *Geol. Ass. Proc.* 53:128–138.

Wood, BW and HF Perkins. 1976. Plinthite characterization in selected southern Coastal Plain soils. *SSSAJ* 40:143–146.

Wood, TG and WA Sands. 1978. The role of termites in ecosystems. In: MV Brian (ed). *Production Ecology of Ants and Termites.* Cambridge Univ. Press, London. pp. 245–292.

Wood, TG, Johnson, AA and JM Anderson. 1983. Modification of soil in Nigerian savanna by soil-feeding *Cubitermes* (Isoptera: Termitidae). *Soil Biol. Biochem.* 15:575–579.

Wood, WR and DL Johnson. 1978. A survey of disturbance processes in archaeological site formation. In: MB Schiffer (ed). *Advances in Archaeological Method and Theory.* Vol. 1. Academic Press, New York. pp. 315–381.

Wood, YA, Graham, RC and SG Wells. 2005. Surface control of desert pavement pedologic process and landscape function, Cima Volcanic field, Mojave Desert, California. *Catena* 59:205–230.

Woods, WI. 1977. The quantitative analysis of soil phosphate. *Am. Antiq.* 42:248–252.

Woods, WI, Teixeira, WG, Lehmann, J, Steiner, C, WinklerPrins, AMGA and L Rebellato (eds). 2008. *Amazonian Dark Earths: Wim Sombroek's Vision.* Springer Publ., New York.

Woolf, D, Amonette, JE, Street-Perrott FE, Lehmann, J and S Joseph. 2010. Sustainable biochar to mitigate global climate change. *Nat. Commun.* 1, article 56. doi:10.1038/ncomms1053

Wright, RL. 1963. Deep weathering and erosion surfaces in the Daly River basin, Northern Territory. *J. Geol. Soc. Austral.* 10:151–163.

Wright, VP, Platt, NH, Marriot, SB and VH Beck. 1995. A classification of rhizogenic (root-formed) calcretes, with examples from the Upper Jurassic-Lower Cretaceous of Spain and Upper Cretaceous of southern France. *Sed. Geol.* 100:143–158.

Wright, VP and ME Tucker. 1991. *Calcretes. Intl. Assoc. Sedimentologists.* Blackwell Sci. Publ., Oxford.

Wright, WR and JE Foss. 1968. Movement of silt-sized particles in sand columns. *SSSAP* 32:446–448

Wu, F, Ho, SSH, Sun, QL and SHS Ip. 2011. Provenance of Chinese loess: Evidence from stable lead isotope. *Terrestrial Atmosph. Oceanic. Sci.* 22:305–314.

Wu, J, Nellis, MD, Ransom, MD, Price, KP and SL Egbert. 1997. Evaluating soil properties of CRP land using remote sensing and GIS in Finney County, Kansas. *J. Soil Water Cons.* 52:352–358.

Wu, SP and Z-S Chen. 2005. Characteristics and genesis of Inceptisols with placic horizons in the subalpine forest soils of Taiwan. *Geoderma* 125:331–341.

Wunsch, C. 2004. Quantitative estimate of the Milankovitch-forced contribution to observed Quaternary climate change. *QSR* 23:1001–1012.

Wurman, E, Whiteside, EP and MM Mortland. 1959. Properties and genesis of finer textured subsoil bands in some sandy Michigan soils. *SSSAP* 23:135–143.

Wynn, JG and MI Bird. 2007. C4-derived soil organic carbon decomposes faster than its C3 counterpart in mixed C3/C4 soils. *Global Change Biol.* 13:2206–2217.

Wysocki, DA, Schoeneberger, PJ, Hirmas, DR and HE LaGarry. 2012. Geomorphology of soil landscapes. In: PM Huang, Y Li and ME Sumner (eds). *Handbook of Soil Sciences,* 2nd ed. CRC Press, New York. pp. 29-1 – 29-26.

Wysocki, DA, Schoeneberger, PJ and HE LaGarry. 2000. Geomorphology of soil landscapes. In: ME Sumner (ed). *Handbook of Soil Science.* CRC Press, Boca Raton, FL. pp. E-5–E-39.

2005. Soil surveys: A window to the subsurface. *Geoderma* 126:167–180.

Xiao, JL, An, ZS, Liu, TS, Inouchi, Y, Kumai, H, Yoshikawa, S and Y Kondo. 1999. East Asian monsoon variation during the last 130,000 Years: Evidence from the Loess Plateau of central China and Lake Biwa of Japan. *QSR* 18:147–157.

Xu, B, Gu, ZY, Han, JT, Hao, QZ, Lu, YW, Wang, L, Wu, NQ and YP Peng. 2011. Radiocarbon age anomalies of land snail shells in the Chinese Loess Plateau. *QG* 6:383–389.

Yaalon, DH. 1969. Origin of desert loess. *8th Intl. Quaternary Assoc. Cong. (Paris)* 2:755.

1971. Soil-forming intervals in time and space. In: DH Yaalon (ed). *Paleopedology. Origin, Nature and Dating of Paleosols.* Israel Univ. Press, Jerusalem. pp. 29–39.

1975. Conceptual models in pedogenesis: Can soil-forming functions be solved? *Geoderma* 14:189–205.

1983. Climate, time and soil development. In: LP Wilding, NE Smeck, and GF Hall (eds). *Pedogenesis and Soil Taxonomy*. Elsevier, New York. pp. 233–251.

1987. Saharan dust and desert loess: Effect on surrounding soils. *J. Afr. Earth-Sci.* 6:569–571.

1992. On fortuitous results and compensating factors. *SS* 154:431–434.

1997a. Comments on the source, transport and deposition scenario of Saharan dust to southern Europe. *J. Arid Envs* 36:193–196.

1997b. Soils in the Mediterranean region: What makes them different? *Catena* 28:157–169.

2000. Down to Earth. Why soil – and soil science – matters. *Nature* 407:301.

Yaalon, DH and RW Arnold. 2000. Attitudes toward soils and their societal relevance: Then and now. *SS* 165:5–12.

Yaalon, DH and E Ganor. 1975. Rates of aeolian dust accretion in the Mediterranean and desert fringe environments of Israel. 9th Intl. Congr. *Sedimentol.* 169–174.

Yaalon, DH and D Kalmar. 1978. Dynamics of cracking and swelling clay soils: Displacement of skeletal grains, optimum depth of slickensides, and rate of intra-pedonic turbation. *ESPL* 3:31–42.

Yaalon, DH and J Lomas. 1970. Factors controlling the supply and the chemical composition of aerosols in a near-shore and coastal environment. *Agric. Meteorol.* 7:443–454.

Yaalon, DH and B Yaron. 1966. Framework for man-made soil changes: An outline of metapedogenesis. *SS* 102:272–277.

Yair, A. 1987. Environmental effects of loess penetration into the Negev Desert. *J. Arid Envs* 13:9–24.

1990. Runoff generation in a sandy area: The Nizzana sands, western Negev, Israel. *ESPL* 15:597–609.

1995. Short and long term effects of bioturbation on soil erosion, water resources and soil development in an arid environment. *Geomorphology* 13:87–99.

Yair, A and SM Berkowitz. 1989. Climatic and non-climatic controls of aridity: The case of the northern Negev of Israel. *Catena Suppl.* 14:145–158.

Yair, A and A Danin. 1980. Spatial variations in vegetation as related to the soil moisture regime over an arid limestone hillside, Northern Negev, Israel. *Oecologia* 47:83–88.

Yair, A and M Shachak. 1982. A case study of energy, water and soil flow chains in an arid ecosystem. *Oecologia* 54:389–397.

Yair, A, Yaalon, DH and S Singer. 1978. Thickness of calcrete (nari) on chalk in relation to relief factors, Shefela, Israel. *10th Intl. Congr. Sedimentol. (Jerusalem)* 2:754–755.

Yang, L, Jiao, Y, Fahmy, S, Zhu, AX, Hann, S, Burt, JE and F Qi. 2011. Updating conventional soil maps through digital soil mapping. *SSSAJ* 75:1044–1053.

Yang, TS, Hyodo, M, Yang, ZY, Ding, L, Fu, JL and T Mishima. 2007. Early and middle Matuyama geomagnetic excursions recorded in the Chinese loess-paleosol sediments. *Earth Planets Space* 59:825–840.

Yang, YH, Ji, CJ, Ma, WH, Wang, SF, Wang, SP, Han, WX, Mohammat, A, Robinson, D and P Smith. 2012. Significant soil acidification across northern China's grasslands during 1980s-2000s. *GCB* 18:2292–2300.

Yaro, DT, Kparmwang, T, Raji, BA and VO Chude. 2008. Extractable micronutrients status of soils in a plinthitic landscape at Zaria, Nigeria. *Comm. Soil Sci. Plant Anal.* 39:2484–2499.

Yaron, B, Dror, I and B Berkowitz. 2008. Contaminant-induced irreversible changes in properties of the soil-vadose-aquifer zone: An overview. *Chemosphere* 71:1409–1421.

Yassoglou, N, Kosmas, C and N Moustakas. 1997. The red soils, their origin, properties, use and management in Greece. *Catena* 28:261–278.

Yassoglou, NJ and EP Whiteside. 1960. Morphology and genesis of some soils containing fragipans in northern Michigan. *SSSAP* 24:396–407.

Yates, SR and AW Warrick. 2002. Geostatistics. In: JH Duane and GC Topp (eds). *Methods of Soil Analysis*. Part 4. *Physical Methods*. SSSA Book Series no. 5. Soil Sci. Soc. Am. Madison, WI. p. 81–118.

Yatsu, E. 1988. *The Nature of Weathering An Introduction*. Sozosha Publ., Tokyo.

Yavuz, H. 2011. Effect of freeze-thaw and thermal shock weathering on the physical and mechanical properties of an andesite stone. *Bull. Engineer. Geol. Environ.* 70:187–192.

Yemane, K, Kahr, G and K Kelts. 1996. Imprints of post-glacial climates and palaeogeography in the detrital clay mineral assemblages of an Upper Permian fluviolacustrine Gondwana deposit from northern Malawi. *PPP* 125:27–49.

Yerima, BPK, Calhoun, FG, Senkayi, AL and JB Dixon. 1985. Occurrence of interstratified kaolinite-smectite in El Salvador Vertisols. *SSSAJ* 49:462–466.

Yerima, BPK, Wilding, LP, Hallmark, CT and FG Calhoun. 1989. Statistical relationships among selected properties of Northern Cameroon Vertisols and associated Alfisols. *SSSAJ* 53:1758–1763.

Yoshida, H, Metcalfe, R, Nishimoto, S, Yamamoto, H and N Katsuta. 2011. Weathering rind formation in buried terrace cobbles during periods of up to 300ka. *Appl. Geochem.* 26:1706–1721.

Yoshinaga, N and S Aomine. 1962a. Allophane in some Ando soils. *JSSPN* 8:6–13.

1962b. Imogolite in some Ando soils. *JSSPN* 8:22–29.

You, C-F, Lee, T, Brown, L, Jiunsan Shen, J and J-C Chen. 1988. [10]Be study of rapid erosion in Taiwan. *Geochim. Cosmochim. Acta* 52:2687–2691.

Young, A. 1960. Soil movement by denudational processes on slopes. *Nature* 188:120–122.

1969. The accumulation zone on slopes. *ZG* 13:231–233.

Young, JA and RA Evans. 1986. Erosion and deposition of fine sediments from playas. *J. Arid Envs* 10:103–115.

Young, MH. 2002. The soil solution phase: Piezometry. In: JH Duane and GC Topp (eds). *Methods of Soil Analysis*. Part 4. *Physical Methods*. SSSA Book Series no. 5. Soil Sci. Soc. Am. Madison, WI. pp. 547–573.

Young, MH, McDonald, EV, Caldwell, TG, Benner, SG and DG Meadows. 2004. Hydraulic properties of a desert soil chronosequence in the Mojave Desert, USA. *Vadose Zone J.* 3:956–963.

Young, MH and JB Sisson. 2002. The soil solution phase: Tensiometry. In: JH Duane and GC Topp (eds). *Methods of Soil Analysis*. Part 4. *Physical Methods*. SSSA Book Series no. 5. Soil Sci. Soc. Am. Madison, WI. pp. 575–678

Yuretich, R, Knapp, E, Irvine, V, Batchelder, G, McManamon, A and SP Schantz. 1996. Influences upon the rates and mechanisms of chemical weathering and denudation as determined from watershed studies in Massachusetts. *GSA Bull.* 108:1314–1327.

Zaidelman, FR. 2007. Lessivage and its relation to the hydrological regime of soils. *EJSS* 40:115–125.

Zakharov, SA. 1927. *A Course of Soil Science*. Gosizdat, Moscow. [in Russian].

Zani, H, Rossetti, DF, Cohen, MLC, Pessenda, LCR and EH Cremon. 2012. Influence of landscape evolution on the distribution of floristic patterns in northern Amazonia revealed by delta 13C data. *JQS* 27:854–864.

Zarate, MA and A Tripaldi. 2011. The aeolian system of central Argentina. *AR* 3:401–417.

Zarate, MA. 2003. Loess of southern South America. *QSR* 22:1987–2006.

Zavalishin, AA. 1958. Dokuchayev's doctrine on factors in soil formation as a basis for the comparative geographical method of soil investigation. *SSS* 983–989.

Zavitsas, AA. 2005. Aqueous solutions of calcium ions: Hydration numbers and the effect of temperature. *J. Phys. Chem.* B109:20636–20640.

Zech, W, Hempfling, R, Haumaier, L, Schulten, H-R and K Haider. 1990. Humification in subalpine Rendzinas: Chemical analyses, IR and 13NMR spectroscopy and pyrolysis-field ionization mass spectrometry. *Geoderma* 47:123–138.

Zelazny, LW, Thomas, PJ and CL Lawrence. 2002. Pyrophyllite – talc minerals. In: JB Dixon and D Schulze (eds). *Soil Mineralogy with Environmental Applications.* Soil Sci. Soc. Am. Book Ser. 7. Soil Sci. Soc. Am., Madison, WI. pp. 415–430.

Zelazny, LW and GN White. 1989. The pyrophyllite-talc group. In: JB Dixon and SB Weed (eds). *Minerals in Soil Environments,* 2nd ed. Soil Sci. Soc. Am., Madison, WI. pp. 527–550.

Zerboni, A. 2008. Holocene rock varnish on the Messak plateau (Libyan Sahara): Chronology of weathering processes. *Geomorphology* 102:640–651.

Zhalnina, K, Dörr de Quadros, P, Camargo, FAO and EW Triplett. 2012. Drivers of archaeal ammonia-oxidizing communities in soil. *Frontiers Microbiol.* 3:1–9.

Zhang, H and S Schrader. 1993. Earthworm effects on selected physical and chemical properties of soil aggregates. *Biol. Fert. Soils* 15:229–234.

Zhang, H, Thompson, ML and JA Sandor. 1988. Compositional differences in organic matter among cultivated and uncultivated Argiudolls and Hapludalfs derived from loess. *SSSAJ* 52:216–222.

Zhang, M and AD Karathanasis. 1997. Characterization of iron-manganese concretions in Kentucky Alfisols with perched water tables. *Clays Clay Mins.* 45:428–439.

Zhang, X, An, Z, Chen, T, Zhang, G, Arimoto, R and BJ Ray. 1994. Late Quaternary records of the atmospheric input of eolian dust to the center of the Chinese Loess Plateau. *QR* 41:35–43.

Zhang, Y, Chen, WJ and J Cihlar. 2003. A process-based model for quantifying the impact of climate change on permafrost thermal regimes. *J. Geophys. Res. Atmosph.* 108. doi:10.1029/2002JD003354

Zhang, Y, Liu, T and S Li. 1990. Establishment of a cation-leaching curve of rock varnish and its application to the boundary region of Gansu and Xinjiang, western China. *Seismol. Geol. (Beijing)* 12:251–261.

Zhao, Z and DM Pearsall. 1998. Experiments for improving phytolith extraction from soils. *JAS* 25:587–598.

Zheng, DL, Hunt, ER and SW Running. 1996. Comparison of available soil water capacity estimated from topography and soil series information. *Landscape Ecol.* 11:3–14.

Zhongli, D, Rutter, N and L Tungsheng. 1993. Pedostratigraphy of Chinese loess deposits and climatic cycles in the last 2. 5 Myr. *Catena* 20:73–91.

Zhou, J and HS Chafetz. 2009. Biogenic caliches in Texas: The role of organisms and effect of climate. *Sed. Geol.* 222:207–225.

Zhou, J, Wu, YH, Jorg, P, Bing, HJ, Yu, D, Sun, SQ, Luo, J and HY Sun. 2013. Changes of soil phosphorus speciation along a 120-year soil chronosequence in the Hailuogou Glacier retreat area (Gongga Mountain, SW China). *Geoderma* 195:251–259.

Zhou, LP, Oldfield, F, Wintle, AG, Robinson, SG and JT Wang. 1990. Partly pedogenic origin of magnetic variations in Chinese loess. *Nature* 346:737–739.

Zhou, W, Donahue, D and AJT Jull. 1997. Radiocarbon AMS dating of pollen concentrated from eolian sediments: Implications for monsoon climate change since the late Quaternary. *Radiocarbon* 39:19–26.

Zhu, AX, Band, L, Vertessy, R and B Dutton. 1997. Derivation of soil properties using a soil land inference model (SoLIM). *SSSAJ* 61:523–533.

Zhu, AX, Hudson, B, Burt, J, Lubich, K and D Simonson. 2001. Soil mapping using GIS, expert knowledge and fuzzy logic. *SSSAJ* 65:1463–1472.

Zhu, AX, Yang, L, Li, BL, Qin, CZ, Pei, T and BY Liu. 2010. Construction of membership functions for predictive soil mapping under fuzzy logic. *Geoderma* 155:164–174.

Zhuang, QL, Melillo, JM, Sarofim, MC, Kicklighter, DW, McGuire, AD, Felzer, BS, Sokolov, A, Prinn, RG, Steudler, PA and SM Hu. 2006. CO_2 and CH_4 exchanges between land ecosystems and the atmosphere in northern high latitudes over the 21st century. *Geophys. Res. Lett.* 33. doi:10.1029/2006GL026972

Zielhofer, C, Espejo, JMR, Granados, MAN and D Faust. 2009. Durations of soil formation and soil development indices in a Holocene Mediterranean floodplain. *QI* 209:44–65.

Zimmerman, DW. 1971. Thermoluminescence dating using fine grains from pottery. *Archaeometry* 13:29–52.

Zi-Tong, G. 1983. Pedogenesis of paddy soil and its significance in soil classification. *SS* 135:5–10.

Zobeck, TM and A Ritchie Jr. 1984. Analysis of long-term water table records from a hydrosequence of soils in central Ohio. *SSSAJ* 48:119–125.

Zoltai, SC and C Tarnocai. 1975. Perennially frozen peatlands in the western Arctic and subarctic of Canada. *CJES* 12:28–43.

Zonn, SV. 1995. Use of elementary soil processes in genetic soil identification. *Euras. Soil Sci.* 27:12–20.

Zreda, MG and FM Phillips. 1995. Insights into alpine moraine development from cosmogenic [36]Cl buildup dating. *Geomorphology* 14:149–156.

Zreda, MG, Phillips, FM, Elmore, D, Kubik, PW, Sharma, P and RI Dorn. 1991. Cosmogenic [36]Cl production in terrestrial rocks. *EPSL* 105:94–109.

Zuzel, JR and JL Pikul Jr. 1987. Infiltration into a seasonally frozen agricultural soil. *J. Soil Water Conserv.* 42:447–450.

Glossary

Portions of this glossary have been adapted from Peterson (1981), Hole and Campbell (1985), Cady *et al.* (1986), Thomas (1996), and the Soil Science Society of America (1997).

α,α-dipyridyl A dye used to detect the presence of reduced iron (Fe^{2+}) in soils. A positive reaction (pink color) indicates the soil material in contact with the dye is reduced.

1938 soil classification system The soil classification/taxonomic system developed under the leadership of Curtis Marbut and used by the USDA in the United States from about 1938 to 1975. The system was genetically based and divided all soils into three orders: Zonal, Azonal, and Intrazonal.

A horizon The topmost mineral soil horizon, usually showing signs of organic matter accumulation and loss of other, mobile constituents such as clay, salts, and some cations.

ablation till Loose, relatively permeable material, either contained within or accumulated on the surface of a glacier deposited during the down wasting of nearly static glacial ice.

abrasion The physical weathering of a rock surface by running water, glaciers, or wind laden with fine particles.

absolute dating See *numerical dating.*

accelerator mass spectrometry dating See *AMS dating.*

accession See *pedogenic accession.*

accretion gley A type of dark, clay-rich, sticky (when wet) soil material, once termed gumbotil, that formed in swales on landscapes, by slopewash. Many are exposed today as clay-rich, gleyed, and overthickened paleosols.

acid A compound or molecule that can donate one or more H^+ cations (protons) to another compound or molecule. A strong acid completely dissociates in aqueous solutions, whereas a weak acid only partially dissociates.

acid dissociation constant (K_a) A chemical equilibrium constant that describes the degree to which an acid dissociates in water; *pKa* is the negative logarithm of the K_a.

acid sulfate soil A soil in which S derived from FeS has been oxidized to sulfate with the concomitant production of acid and precipitation of sulfate minerals.

acidic cations Cations with sufficient positive charge density that associated water molecules will donate protons to bulk solution water molecules, leaving OH^- to coordinate the metal ion and increasing acidity in the solution.

acid(ic) soil Soil with a pH value <7.0.

acid(ic) rock A usually light-colored, igneous rock that contains more than 60% silica and quartz.

acidification The formation of acidic conditions in a soil or sediment.

acidity The hydrogen ion activity in the soil solution expressed as a pH value.

actinomycetes A group of organisms intermediate between bacteria and true fungi, mainly resembling the latter. A form of filamentous, branching bacteria.

active factors Of the five state factors of Jenny (climate, organisms, relief, parent material, and time), the two that are not passive, i.e., climate and organisms.

active layer The top layer of soil subject to annual thawing and freezing in areas underlain by permafrost, i.e., frozen in the winter but thawed in summer. Sometimes called the *active zone.*

adsorbate Molecule or ion that can accumulate near a mineral or organic surface as a result of chemical or physical interactions with the solid.

adsorbent A mineral, e.g., a layer silicate or sesquioxide, or organic surface at which adsorbate molecules or ions accumulate.

adsorption The process of accumulation of a molecule or ion near a surface at a concentration greater than that in the bulk solution.

adsorption isotherm A regular series of adsorption experiments, conducted at a constant temperature, in which the equilibrium concentration of an adsorbate on a solid-phase adsorbent (such as soil) is plotted against its concentration in a fluid phase (such as a water solution).

aeolian See *eolian.*

aeration The process by which air enters the soil. In well-aerated soils, the soil air is very similar in composition to the atmosphere, whereas poorly aerated soils have higher CO_2 and lower O_2 contents than the atmosphere. The rate of aeration depends largely on the volume and continuity of air-filled pores within the soil, and the degree of waterlogging.

aerobic (i) Having molecular oxygen as a part of the environment. (ii) Growing only in the presence of molecular oxygen, such as aerobic organisms. (iii) Occurring only in the presence of molecular oxygen (said of chemical or biochemical processes such as aerobic decomposition).

aeroturbation The mixing of soil by gases, air, and wind.

age An interval of time measured back from the present.

aggradation The accumulation of sediment on the soil surface, thereby building up the surface by deposition.

aggregate A group of primary soil particles that cohere to each other more strongly than to other surrounding particles and are held together in a single mass or cluster. Discrete clusters of particles formed naturally are called peds. Types of aggregates include crumbs, granules, fecal pellets, and concretions. Aggregates produced by tillage or logging are called clods.

agric horizon A mineral soil horizon in which clay, silt, and humus derived from an overlying cultivated and fertilized layer have accumulated, occurring as horizontal lamellae or fibers, as coatings on ped surfaces or in wormholes. Wormholes and illuvial clay, silt, and humus occupy at least 5% of the horizon by volume.

ahumic soils Those with virtually no organic matter, originally defined for ultraxerous soils in Antarctica.

alban A cutan, light-colored because either the material composing it is dominated by quartz or the iron minerals within it have been reduced or translocated from it.

albic horizon A light-colored, eluvial horizon from which clay and free iron oxides have been removed, or in which the oxides have been segregated, to the extent that the color of the horizon is determined primarily by the color of the primary sand and silt particles, rather than by coatings on them.

albic neoskeletans Whitish coatings on ped faces that have been stripped of most of their clay and Fe-bearing minerals, usually by chemical reduction in an anoxic environment. Also called grainy gray ped coatings, gray ped and channel coatings, and albans.

alcrete A soil horizon cemented by aluminum-rich compounds.

Alfisols One of 12 major taxonomic orders in Soil Taxonomy. Alfisols have an umbric or ochric epipedon and an argillic horizon and are moist during at least 90 days during the growing season. Alfisols have a mean annual soil temperature of <8°C, or >35% base cation saturation in the lower part of the argillic horizon.

alkali soil (i) A soil with a pH of 8.5 or higher or with an exchangeable sodium ratio >0.15. (ii) A soil that contains sufficient sodium to interfere with the growth of most crops.

alkaline soil Soil with a pH value >7.0 (or >7.3 by some definitions).

alkalinity A measure of the capacity of a solution to accept protons and therefore to buffer the pH.

allochthonous A term that connotes that something, usually a sediment, is derived from someplace else, or is not indigenous to a site or area. See also *autochthonous*.

allophane An aluminosilicate with primarily short-range structural order. Occurs as exceedingly small spherical particles and is especially common in soils formed from volcanic ash.

alluvial Associated with transportation or deposition by running water.

alluvial fan Land counterpart of a delta. A constructional landform built of stratified alluvium, formed downslope from its source as streams carrying sediment, flowing rapidly along a steep gradient, enter a lowland with lower slopes and deposit sediment there.

alluvial soil A soil developing in, or developed from, fairly young alluvium, often exhibiting essentially no pedogenic development or modification of the recently deposited materials.

Alluvial soils (capitalized) A great soil group of the azonal order (1938 system of soil classification) consisting of soils with little or no modification of the recent alluvium in which they are forming. Roughly synonymous with Fluvents. Obsolete in the U.S. system of Soil Taxonomy Soil Taxonomy.

alluvial toeslope See *toeslope*.

alluvium Sediment deposited by running water of streams and rivers, typically occurring on terraces well above contemporary streams, on floodplains or deltas, or in alluvial fans.

Alpine Meadow soils A great soil group of the intrazonal order (1938 system of soil classification), composed of dark soils of grassy meadows at altitudes above the timberline. Obsolete in the U.S. system of Soil Taxonomy Soil Taxonomy.

alpine zone Areas in mountainous terrain that are devoid of trees, because of the cold temperatures of high altitudes. Areas above treeline.

aluminosilicates Minerals containing Al, Si, and O as their main constituents.

amorphous material Lacking crystalline structure. Noncrystalline soil constituents that either do not fit the definition of allophane or are unable to meet allophane criteria. Generally used with reference to Andisols and Spodosols.

amphibole A family of silicate minerals forming prism or needle-like crystals. Amphibole minerals generally contain Fe, Mg, Ca, and Al in varying amounts, along with water. Hornblende is a common dark green to black variety of amphibole.

AMS dating A form of radiocarbon dating, applicable to very small samples, that utilizes accelerator mass spectrometry (AMS) and involves the direct measurement of the ratio of different carbon isotopes.

anaerobic (i) Characterized by the absence of molecular oxygen. In soils, this is usually caused by excessive wetness. (ii) Growing in the absence of molecular oxygen (such as anaerobic bacteria). See also *anoxic*.

anaerobic respiration The metabolic process whereby electrons are transferred from a reduced, usually organic compound to an inorganic acceptor molecule other than oxygen. The most common acceptors are nitrate, Mn, Fe, and sulfate.

andesite A fine-grained volcanic rock of intermediate composition, consisting largely of plagioclase and mafic minerals.

andic Soil properties related to volcanic materials.

Andisols One of 12 major taxonomic orders in Soil Taxonomy Soil Taxonomy. Andisols are dominated by andic soil properties, inherited from ash, tephra, and other silica-rich volcanic parent materials.

anecic earthworms Earthworms that primarily burrow vertically, deep in the soil, and often inhabit one channel for long periods. Sometimes referred to as nightcrawlers.

angle of repose The maximum slope angle at which loose, cohesionless material will come to rest.

angular blocky A form of soil structure, common in clay-rich B horizons, with equant-shaped peds that have angular edges and sharp corners.

anion An ion having a negative charge, e.g., OH^-.

anion exchange capacity The sum of charges of exchangeable anions in a soil, usually expressed as centimoles of charge per kilogram of soil; it most commonly arises from adsorption of protons or water molecules at the surface of sesquioxides and at some N-containing functional groups in organic matter.

anisotropy A condition of a soil profile of being layered (horizonated) and in that sense, regularly heterogeneous, often contrasting with isotropic nature of the some original initial materials. In the context of place-to-place variability, anisotropic refers to the fact that some soil characteristics vary at different rates as direction varies.

anomalous fading In luminescence dating, the loss of electrons from the lattice traps of a mineral, on a timescale that is short compared to the lifetime predicted on the basis of trap size. It is primarily a problem with feldspar, leading to abnormally young age estimates.

anoxic See *anaerobic*.

anthric saturation Human-induced, aquic condition – a type of episaturation – that occurs in soils that are cultivated and irrigated, especially by flood irrigation, e.g., in rice paddies, cranberry bogs, and treatment wetlands.

anthrogenesis The effects that humans and human activity have on soils and pedogenesis, e.g., irrigation, cultivation, leveling, additions of nutrients, and drainage.

anthropic epipedon A surface horizon that has the same requirements as the mollic epipedon but is high in phosphorus or is dry >300 days (cumulative) during the period when not irrigated. Anthropic epipedons form under long-continued cultivation and fertilization.

anthroposequence A sequence of soils that vary in the degree to which they have been affected by human agency.

Anthrosols Soils that have been strongly impacted by humans, such as in cities or on mine spoil, or via long-term cultivation or irrigation. A taxonomic category in the WRB system of soil classification.

anthroturbation Soil and sediment mixing by human activities.

apedal Condition of a soil that has no structure, i.e., no peds, but rather is massive or composed of single grains.

apparent water table The longer-term regional water table that normally marks the top of a thick saturated zone in the subsurface. Contrast with *perched water table*.

aquaturbation Soil and sediment mixing by water, usually on a very small scale.

aquic conditions Continuous or periodic saturation and reduction, as indicated by redoximorphic features.

aquic soil moisture regime One of the five major soil moisture regimes in Soil Taxonomy Soil Taxonomy. Occurs when the soil is often in a reduced state and virtually free of dissolved oxygen because it is saturated by groundwater, or by water of the capillary fringe.

aquiclude A sediment body, rock layer, or soil horizon that does not transmit significant quantities of water under ordinary hydraulic gradients, i.e., it is nearly impermeable.

aquifer A saturated, permeable unit of sediment or rock that can transmit significant quantities of water under hydraulic gradients.

aquitard A slowly permeable body of rock or sediment that retards but does not prevent the flow of water through it.

arbuscular mycorrhizae A symbiotic, generally mutualistic, but occasionally weakly pathogenic association between a fungus and the interior parts of the roots of a vascular plant. See also *ectomycorrhizae*.

arenization Physical disintegration/breakup of rock induced by the chemical weathering of some of its weatherable minerals.

argillan A cutan composed predominantly of oriented phyllosilicate clay minerals.

argillic horizon A soil horizon that is characterized by the illuvial accumulation of phyllosilicate clays. Quantitatively, it has a minimum thickness (depending on the thickness of the solum) and a minimum quantity of clay (in comparison with an overlying eluvial horizon). It also usually has coatings of oriented clay on the surface of pores or peds, or clay bridges between sand grains.

argilliturbation Soil and sediment mixing by shrinking and swelling of clays, such as smectite, usually in a wet-dry climate.

aridic soil moisture regime One of five major soil moisture regimes in Soil Taxonomy Soil Taxonomy. Occurs where soils have no water available for plants for more than half the growing (warm) season, and during this same season there is no period as long as 90 consecutive days when there is water available for plants. Typical of arid regions (deserts). Same as *torric*.

Aridisols One of 12 major taxonomic orders in Soil Taxonomy Soil Taxonomy. Aridisols have an aridic (dry) moisture regime, an ochric epipedon, and other diagnostic horizons but lack an oxic horizon. Typical of deserts.

arkose A sedimentary rock dominated by sand-sized grains of feldspar and quartz.

arthropod An animal characterized by a hard, outer skeleton (exoskeleton) and jointed body parts (appendages), e.g., spiders, scorpions, crabs, crustaceans, millipedes, mites, centipedes, and insects such as termites and ants.

ash (volcanic) Unconsolidated, pyroclastic material < 2 mm in diameter.

aspect The direction toward which a slope faces with respect to the compass or rays of the sun.

association, soil See *soil association*.

autochthonous (i) Microorganisms and/or substances indigenous to a given site or ecosystem, i.e., the true inhabitants of an ecosystem. (ii) Sediment that is derived from that place, i.e., not from outside that place. See also *allochthonous*.

autotroph An organism that produces complex organic compounds, e.g., carbohydrates, fats, proteins, from simple substances present in its surroundings, generally using energy from light (photosynthesis) or inorganic chemical reactions (chemosynthesis).

available water That part of the soil water that can be taken up by plant roots.

available water capacity The amount of water released from soils, between in situ field capacity and the permanent wilting point (usually estimated by water content at a matric potential of –15 MPa). The weight percentage of water that a soil can store in a form available to plants. Commonly expressed as length units of water per length units of soil.

Azonal soils Soils without distinct genetic horizons, in the 1938 system of soil classification. Obsolete in the U.S. system of Soil Taxonomy Soil Taxonomy.

B horizon A major, subsoil mineral horizon with pedogenic, not rock, structure, that formed by illuviation of materials or by weathering in place.

backslope The hillslope position that forms the steepest, and generally linear, middle portion of the slope. In profile, backslopes are bounded by a convex shoulder (upslope) and a concave footslope (downslope).

backswamp A floodplain landform consisting of extensive marshy or swampy depressed areas between the higher natural levees (near the channel) and valley sides or terraces (farther from the channel).

basal till Unconsolidated material deposited and compacted beneath a glacier, usually with a relatively high bulk density. See also *till, ablation till, lodgement till*.

basalt A fine-grained, dark-colored igneous rock, formed from lava flows or minor volcanic intrusions. It is composed of calcic plagioclase, augite, and magnetite; olivine may be present. Extrusive equivalent of gabbro.

base A compound or molecule that is capable of accepting one or more hydrogen ions (protons).

base cation Alkali and alkaline earth cations, especially Ca^{2+}, Mg^{2+}, Na^+, and K^+, in the context of soil science.

base cycling The cycling of base cations between the soil and biosphere, as plants take them up and later release them back to the soil.

base cation saturation percentage The extent to which the exchange complex of a soil is occupied by exchangeable cations other than H and Al, i.e., with base cations. It is expressed as a percentage of the total cation exchange capacity.

bauxite A rock or sediment composed of aluminum hydroxides and impurities in the form of silica, clay, silt, and iron hydroxides. Usually a residual weathering product, exploited for aluminum ore.

bedding plane Surface separating layers of sedimentary rocks and deposits.

bedrock The solid rock that underlies the soil and regolith, or that is exposed at the surface.

beryllium-10 (^{10}Be) An unstable isotope of beryllium, whose content in soils and rocks is used for dating surface exposure. Both meteoric and in situ forms of ^{10}Be are used.

beta horizon, beta B horizon, ß horizon The second (lower) horizon of illuvial clay accumulation, e.g., a Bt horizon that is below and disjunct or separated from the main Bt horizon, higher in the solum.

biochemical processes Processes, impacts, and effects that are *chemically* mediated, by biota, in soils. Compare to *biomechanical processes*.

bioclast Stone or rock that has a partial biological origin, often referring to stones coughed up or passed through the guts of birds.

biocrust A soil crust formed at the surface, via the actions of microscopic soil organisms.

biocycling Translocation of minerals and elements from the soil to plants and back again.

biofabric A type of soil fabric that owes its existence and character to the biomechanical actions of soil fauna and/or flora.

biofunction The formula or quantitative expression of the differences that occur in soils as a function of changes in biota, assuming that the remaining soil-forming factors are held constant, or nearly so. See also *biosequence*.

biogeochemical cycling Like biocycling, but more broadly defined, including elemental transfers among not only the soil and biota, but also the atmosphere, hydrosphere, and rock below.

biomantle The differentiated zone in the upper part of soils that is produced largely by bioturbation and aided in its formation by subsidiary processes.

biomechanical soil processes Processes, impacts, and effects that are *physically* mediated, by biota, in soils. Compare to *biochemical processes*.

biopore A soil pore formed by biota, such as a worm burrow or a plant root.

biorelict Inherited biological feature (such as a mollusc shell or chitonic remnant of soil animal) in the mineral soil that is stable under the present soil conditions.

biosequence A group of related soils that differ from one another, primarily because of differences in kinds and numbers of plants and soil organisms as a soil-forming factor. When expressed as a mathematical equation, it is referred to as a *biofunction*.

biotite A common trioctahedral mica, ranging in color from dark brown to green in thin section, hence, its common name "black mica."

bioturbation The mixing of soils and sediments by organisms, both flora and fauna.

biscuit tops The name given to the rounded tops of columnar peds that are often coated with a residual, clay-poor material that is whitish colored. Commonly occur in soils that are high in Na.

bisequal soils Soils in which two sequa have formed, one above the other.

bleicherde A light-colored, leached E horizon in Podzolic soils.

blocky soil structure A type of soil structure where the peds take on a blocklike shape, i.e., many-sided with angular or rounded corners.

boehmite An Al-rich oxide clay found in humid, tropical soils, it is the most common crystalline form of alumina monohydrate, AlO(OH).

bog A peat-accumulating wetland that supports acidophilic mosses, particularly *Sphagnum*.

Bog soils A great soil group of the intrazonal order and hydromorphic suborder (1938 system of soil classification). Includes muck and peat soils. Roughly synonymous with Histosols. Obsolete in the U.S. system of Soil Taxonomy.

bombturbation The mixing of soils by explosive munitions, usually during warfare or military training activities.

bottomland The normal floodplain of a stream, subject to frequent flooding.

boulder A rock or mineral fragment >600 mm in diameter.

bouldery Containing appreciable quantities of boulders.

Bowen's reaction series A series of minerals formed during crystallization of a magma as it cools, in which the formation of minerals alters the composition of the remaining magma.

bowl In Vertisols, a cup- or trough-shaped subsurface feature, centered under and surrounding the microlow of a gilgai, commonly 3–5 m across and 1.5–3.0 m thick, containing numerous slickensides and bounded at its base by master slickensides. Substrate morphology is not preserved within bowls.

bowl and chimney topography See *gilgai*.

braided stream A channel or stream with multiple channels that interweave as a result of repeated bifurcation around interchannel bars, resembling (in plan view) the strands of a complex braid. Braiding is typical of broad, shallow streams of low sinuosity, high bedload, noncohesive bank material, and steep gradients.

braunification See *brunification*.

breccia A clastic rock composed mainly of angular, gravel-sized particles.

Brown Forest soils A great soil group of the intrazonal order and calcimorphic suborder (1938 system of soil classification), formed on calcium-rich parent materials under deciduous forest. These soils possess a high base status but lack a pronounced illuvial horizon. Roughly synonymous with Eutrepts. Obsolete in the U.S. system of Soil Taxonomy.

Brown Podzolic soils A zonal great soil group (1938 system of soil classification), similar to Podzols but lacking the distinct E horizon that is characteristic of the Podzol group. Acidic and formed under forest. Roughly synonymous with Eutrepts. Obsolete in the U.S. system of Soil Taxonomy.

Brown soils A great soil group (1938 system of soil classification) of soils found in temperate to cool, arid regions. They have a brown surface and a light-colored transitional subsurface horizon, over a horizon of $CaCO_3$ accumulation. Obsolete in the U.S. system of Soil Taxonomy.

brunification Pedogenic process bundle involving the release of iron from primary minerals, followed by the dispersion of particles of iron oxide in increasing amounts. The progressive oxidation or hydration of the soil produces mass brownish, reddish brown, and red colors. The process often forms cambic Bw horizons.

Brunisols A taxonomic order of soils that have undergone only slight development from the parent material, not unlike Inceptisols of the U.S. Soil Taxonomy. Brunisols are not an order in Soil Taxonomy but do occur in the classification system of Canada and several other countries. The main soil-formation process in Brunisolic soils is the leaching of soluble salts and carbonates, the formation of Fe and Al secondary minerals, and structure development.

Brunizems A synonym for Prairie soils (1938 system of soil classification), typical of tall and midgrass prairies. Obsolete in the U.S. system of Soil Taxonomy.

buffer A substance that prevents a rapid change in pH when acids or alkalis are added. In soils, common buffers include clay, humus, and carbonates.

bulk density Mass per unit volume (usually expressed as g cm^{-3} or Mg m^{-3}) of undisturbed soil, dried to constant weight at 105°C and calculated after removing the weight and volume of coarse fragments.

buried soil, buried paleosol Soil covered by a surface mantle of more recent depositional material, usually to a depth greater than 50 cm. Some consider a soil "buried" even if the burying deposit is thin, as long as it is identifiable. See also *geosol* or *paleosol*.

bypass flow See *macropore flow*.

^{13}C A stable isotope of carbon, whose fractionation in plants varies and, thus, can be used as an indicator of paleoenvironmental conditions, as the humus derived from plant residues accumulates in surface soils and those soils are later buried.

^{14}C See *radiocarbon*.

C horizon The soil horizon, minimally altered by pedogenesis, that lies immediately below the genetic soil profile, or *solum*. In many (though not all) soils, it represents the presumed parent material. Although many C horizons exhibit some alteration from their original state, the concept *sensu stricto* implies lack of alteration by surficial processes.

C3 pathway The most common pathway of carbon fixation in plants, typical of humid climate plants such as trees, most shrubs and herbs, and many grasses. Refers to the Calvin (C3) cycle plants use to fix CO_2.

C4 (Hatch-Slack) pathway An alternative carbon fixation pathway. The C4 pathway allows for more efficient carbon fixation in dry, warm environments and, thus, is mostly found among tropical, warm-season grasses and some sedges and herbs.

calcan A light-colored cutan composed of carbonates.

calcareous soil Soil containing sufficient free $CaCO_3$ and other carbonates to effervesce visibly or audibly when treated with weak HCl, usually suggesting at least 10 g kg^{-1} $CaCO_3$ equivalent.

calcic horizon A mineral soil horizon of secondary carbonate enrichment that is >15 cm thick, has a $CaCO_3$ equivalent >150 g kg^{-1}, and has at least 50 g kg^{-1} more calcium carbonate equivalent than the underlying C horizon.

calcification The pedogenic process of illuvial accumulation of calcium compounds in a soil horizon, such as in the calcic horizon of some Aridisols and Mollisols.

calcite Crystalline calcium carbonate, $CaCO_3$.

calcrete, caliche A layer, more or less cemented by illuvial, secondary carbonates of Ca or Mg that have been precipitated from the soil solution. It may occur as a soft thin soil horizon, as a hard thick bed, or as a surface layer exposed by erosion.

calibrated age An age, usually determined by using radiocarbon, that has been calibrated (adjusted) to reflect calendar years ago.

CAM (crassulacean acid metabolism) pathway The least common of the three carbon fixation pathways in plants. CAM plants, typical of desertlike conditions, include mainly plants in the Crassulaceae and Cactaceae families.

cambic horizon A nonsandy, mineral subsoil horizon that has soil structure rather than rock structure, contains some weatherable minerals, and is characterized by the alteration or removal of mineral material. Cambic horizons typically (i) have mottling or gray colors due to wetness, (ii) have stronger chromas or redder hues than in underlying horizons due to weathering, *or* (iii) have been decalcified by the removal of carbonates. They lack cementation or induration, and the evidence for illuviation is too weak to meet the requirements of an argillic or spodic horizon.

capillarity The process by which moisture moves in fine pore spaces and as films around particles in an unsaturated state as a result of the surface tension of water.

capillary water The water held in the "capillary" or small pores of a soil, usually with a tension >60 cm of water. It is held by adhesion and surface tension as films around particles and in the finer pore spaces. Surface tension is the adhesive force that holds capillary water in the soil.

carbon sequestration The transfer of atmospheric CO_2 into long-lived pools and its secure storage there, so that it is not immediately returned to the atmosphere.

carbonate, carbonate rock, carbonaceous rock A rock consisting primarily of a carbonate mineral such as calcite or dolomite, the chief minerals in limestone and dolostone, respectively.

carbonation A form of chemical weathering usually involving carbonic acid (H_2CO_3).

carpedolith, carpetolith, carpedolite See *stone line* or *lag concentrate*.

catena A sequence of soils along a slope, having predictable differences in characteristics mainly due to topography, variation in parent material, and drainage class, as well as the influence of slope processes on sediment transport. Milne (1935) originated the concept with his study of soils in East Africa. Bushnell (1942) suggested narrowing the concept to a topohydrosequence of soils developed in a single parent material, such as a glacial till.

cation An ion having a positive electrical charge. Common soil cations are Ca^{2+}, K^+, Mg^{2+}, Na^+, Al^{3+}, NH_4^+, Fe^{2+}, and H^+.

cation exchange The exchange among cations in solution and cations held on the negatively charged exchange sites of minerals and organic matter.

cation exchange capacity (CEC) The potential of soils for adsorbing cations, expressed in centimoles of charge per kilogram (cmol(-) kg^{-1}) of soil. Calculated as the sum of the charges of exchangeable base cations plus total soil acidity at a specific pH value, usually 7.0 or 8.2. In essence, soils with high CEC values have large amounts of negative charges per unit mass of soil. CEC in soils is influenced by the concentrations of organic matter and clay, and by the mineralogy of the clay fraction. See also *effective cation exchange capacity (ECEC)*.

cation ratio dating Used to date desert varnish on rock surfaces, it is the ratio of soluble (Ca and K) to insoluble Ti cations in the varnish. In theory, this ratio should decrease with time because the soluble cations are replaced or depleted relative to less mobile Ti.

cellulose Carbon-rich polymer of glucose that is a significant component of primary and secondary cell walls of plants; readily decomposed by many microorganisms.

cemented Having a hard, brittle consistency because the particles are held together by cementing substances such as humus, $CaCO_3$, silica, or the oxides of silicon, iron, and aluminum. The hardness and brittleness persist even when wet. See also *consistence*.

Cenozoic era The current geologic era, which began 66.4 million years ago and continues to the present.

chalk Soft white limestone composed of very pure $CaCO_3$, sometimes consisting largely of the remains of foraminifera, echinoderms, molluscs, and other marine organisms.

chamber In soil micromorphology, a relatively large circular or ovoid pore with smooth walls and an outlet through channels, fissures, or planar pores.

channel In soil micromorphology, a tubular-shaped pore or void.

channel neoferrans Coatings of Fe oxides on channel walls.

channer In Scotland and Ireland, gravel. In the United States, thin, flat rock fragments up to 150 mm long, typically shale or limestone.

channery Having large amounts of channers.

chelate-complex theory A theory of podzolization in which the mobility of oxidized Fe and Al cations within the soil, and particularly their transloaction to the B horizons, is ascribed to complexation by organic molecules (chelators), especially fulvic and low-molecular-weight organic acids. See also *protoimogolite theory*.

chelate A soluble complex composed of a central metal ion that is coordinated by two or more electron-rich ligands, all of which are part of one organic molecule or molecular ion, thus forming a ring structure. Soil organic matter can form chelate structures with some metals.

chelation The process of being chelated.

chemical weathering The chemical breakdown or decomposition of rocks and minerals into substances of different chemical composition or mineralogical structure. Often involves water and other soluble components in the soil solution, and/or changes in redox potential. For example, the transformation of orthoclase to kaolinite.

chemisorption Adsorption of a molecule or ion at a solid surface by means of a chemical bond.

Chernozems A soil of the Chernozemic order of the Canadian soil classification system. Also, a zonal great soil group in the 1938 system of soil classification, consisting of soils with a thick, nearly black or black, humus-rich A horizon high in exchangeable calcium, underlain by a lighter-colored transitional horizon above a zone of $CaCO_3$ accumulation. They occur in cool, subhumid climates under a vegetation of midgrass prairie. Many Chernozems are equivalent to Ustolls or Udic Ustolls. This term is not used in the U.S. system of Soil Taxonomy.

chert A cryptocrystalline, microscopically granular form of quartz. Occurs as nodules and as thin, continuous layers in limestone, dolostone, and mudstones.

Chestnut soils A zonal great soil group (1938 system of soil classification) consisting of soils with a moderately thick, dark brown A horizon over a lighter-colored horizon that is above a zone of $CaCO_3$ accumulation. Typical of mid- and short grass prairies. Many Chestnut soils are equivalent to Ustolls or Aridic Ustolls. Obsolete in the U.S. system of Soil Taxonomy.

chimney In Vertisols, a subsurface feature that forms a crude cone or wave-crest structure, centered under a microhigh and extending at least partially under adjacent microslope positions. Composed of substratum material that appears to upwell and reach close to the surface, chimneys are commonly bounded by master slickensides in the subsoil with maximum dip angles of 60°–75° under the microhigh.

chimney and bowl topography See *gilgai*.

chlorite A group of 2:1 layer silicate minerals that has the interlayer filled with a positively charged, metal-hydroxide octahedral sheet. Chlorites are both trioctahedral (e.g., M = Fe^{2+}, Mg^{2+}, Mn^{2+}, Ni^{2+}) and dioctahedral (M = Al^{3+}, Fe^{3+}, Cr^{3+}), and weather readily.

chroma The relative purity, strength, or saturation of a color, directly related to the dominance of the determining wavelength of the light and inversely related to grayness. One of the three variables of color.

chronofunction The formula or quantitative expression of the differences that occur in soils as a function of age or time of development, assuming that the remaining soil-forming factors are held constant, or nearly so. See also *chronosequence*.

chronosequence A group of related soils that differ from one another, primarily as a result of differences in time or soil age, as a soil-forming factor. When expressed as a mathematical equation, it is referred to as a *chronofunction*.

chronostratigraphic unit A sequence of rocks deposited during a particular interval of geological time.

clast A large fragment, such as a rock or pebble, that is significantly larger than the surrounding material.

clastic materials Rock or sediment composed mainly of fragments derived from preexisting rocks or minerals, i.e., not organic.

clay (i) A soil separate consisting of particles <2 μm in equivalent diameter. (ii) A soil textural class. (iii) In reference to clay mineralogy, a naturally occurring material composed primarily of fine-grained minerals, which is generally plastic at appropriate water contents and will harden when dried or fired. Although clay usually contains phyllosilicates, it may contain minerals with other structures.

clay coating See *argillan*.

clay domain A small stack of clay mineral platelets.

clay film See *argillan*.

clay flow An argillan that appears wavy or distorted, as if it had "flowed" along the ped face.

clay-free basis A mathematical treatment in which the concentration of one of the other particle size separates in a soil, e.g., sand, silt, is normalized not to the *entire* fine earth fraction (sand, silt, and clay) but to the clay-free fraction only (sand plus silt). Often used to detect lithologic discontinuities in soils.

clay-humus complex Complex formation between humic materials and clay, mechanically inseparable and chemically completely divisible only with great difficulty.

clay mineral Crystalline or amorphous mineral material <2 μm in diameter.

clay mineralogy (i) The mineral composition of clay-sized particles in a soil. (ii) The study of clay minerals.

clay pan A dense, compact, slowly permeable layer in the subsoil, with significantly more clay than the horizon above. It is usually very dense and has a sharp upper boundary. Clay pans generally impede drainage and are usually plastic and sticky when wet and hard when dry.

clay skin See *argillan*.

climofunction A quantitative expression of differences that occur in soils as a function of changes in climate, assuming that the remaining soil-forming factors are held constant, or nearly so. See also *climosequence*.

climosequence A group of related soils that differ one from another primarily as a result of differences in climate as a soil-forming factor. When expressed as a mathematical equation, it is referred to as a *climofunction*.

clod A compact, coherent mass of soil, usually produced by plowing, digging, or other processes, especially when these operations are performed on soils that are either too wet or too dry, clods are usually formed by compression or breaking off from a larger unit, as opposed to a building-up action, as in aggregation.

closed depression Generic name for any enclosed landform that has no surface drainage outlet and from which water escapes only by evaporation or subsurface drainage. See also *open depression.*

coalluvium Material that has some history of movement as alluvium and as colluvium.

coagulation In soils, the clumping due to an increase in multivalent cations in the soil solution.

coarse fraction, coarse fragments Unattached pieces of rock, >2 mm in diameter, that are strongly cemented or resistant to rupture. Includes gravels, cobbles, and boulders, among other forms of large fragments.

coarse-textured Soil texture term referring to the dominance of sand and coarse fragments in the soil.

coastal plain A plain of unconsolidated fluvial or marine sediment that had its margin on the shore of a large body of water, usually the sea, e.g., the coastal plain of the southeastern United States, extending from New Jersey to Texas.

cobbles Rounded or partially rounded rock or mineral fragments, 75–250 mm in diameter.

coefficient of linear extensibility (COLE) The volume change of a soil upon wetting and drying. Calculated as the percentage shrinkage in one dimension of a molded soil between two water contents, e.g., between its plastic limit and air-dry state. The ratio of the difference between the moist and dry lengths of a clod to its dry length, $(L_m - L_d)/L_d$, when L_m is the moist length (at a partial pressure of 1/3 bar) and L_d is the air-dry length.

collembola Small soil fauna more commonly referred to as springtails.

colloid (soil) Organic and inorganic material with very fine particle sizes (~0.1 to 0.001 μm) and therefore high surface areas.

colluvial Pertaining to material or processes associated with transportation and/or deposition by mass movement (direct gravitational action) and local, unconcentrated runoff on slopes.

colluviation The process of downslope, gravity-driven transport of sediment, resulting in a colluvial deposit downslope.

colluvium Unconsolidated, unsorted earth material transported under the influence of gravity, assisted by water, and deposited on lower slopes.

color hue The dominant spectral color. One of the three variables of color.

color value The degree of lightness or darkness of a color in relation to a neutral gray scale. On a neutral gray scale, value extends from pure black to pure white. One of the three variables of color.

columnar A shape of soil structure wherein the peds resemble long blocks, with their long dimension being the vertical one. Occurs in some subsoil horizons.

complex (map unit complex) A delineation on a soil map that contains the names of two or more series in its description, e.g., Spinks-Coloma sands, indicating that the map unit is a mix of at least two major soil components, along with other, unnamed minor components.

concretion Cemented, usually spherical or subspherical, concentration of a chemical compound, such as $CaCO_3$, gypsum, or iron oxide, that can be removed from the soil intact and that has crude internal symmetry. Concretions form by precipitation of mineral matter about a nucleus such as a leaf or a piece of shell or bone. In soil micromorphology, a concretion is a type of glaebule with a generally concentric fabric.

conductivity, electrical In soils, a measure of the soluble salt content and an index of ionic strength of the soil solution.

congeliturbation See *geliturbation.*

conglomerate A clastic sedimentary rock composed mainly of rounded boulders of various sizes, many larger than sand-sized.

conjugate base Portion of an acid that remains after a proton has been dissociated.

consistence The attributes of soil material as expressed in degree of cohesion and adhesion, or in resistance to deformation or rupture. *Soil strength.*

Examples:

cemented Hard; little affected by moistening.

brittle Ruptures, does not crumble, when pressed between thumb and finger.

loose Noncoherent when dry or moist and does not hold together in a mass.

friable When moist, crushes easily under gentle pressure between thumb and forefinger and can be pressed together into a lump.

firm When moist, crushes under moderate pressure between thumb and forefinger, but resistance is distinctly noticeable.

hard When dry, moderately resistant to pressure but can be broken with difficulty between thumb and forefinger.

soft When dry, breaks into powder or individual grains under only very slight pressure.

sticky When wet, adheres to other material and tends to stretch somewhat and pull apart rather than to pull free.

plastic When wet, readily deformed by moderate pressure but can be pressed into a lump; will form a "wire" when rolled between thumb and fore-finger.

consociation A soil map delineation dominated by a single soil taxon (or miscellaneous area) and similar soils. Generally, at least half of the pedons in each delineation of a soil consociation are of the same soil components that provide the name for the map unit. Most of the remainder of the delineation consists of soil components that are similar to the named soil.

constructional surface A surface owing its origin and form to depositional processes, with little or no subsequent modification by erosion.

continuous permafrost Permafrost occurring everywhere beneath the soils of a region.

contrasting soil A soil that does not share diagnostic criteria and does not behave or perform similarly to the soil being compared.

control section A portion of the soil on which some part of its classification is based. The location and thickness of control sections vary among different criteria and kinds of soil; for many soils, the mineralogical control section falls between depths of 25 and 100–200 cm.

convergent slope A slope in which the downslope flowlines of water converge at the base, e.g., a headslope.

coppice mound A small dune that forms around desert brush-and-clump vegetation that traps sand. Often partly due to human action that has disturbed nearby soils.

coprogenic material Remains of excreta and similar materials that occur in some organic soils.

corestone The generally rounded remnant of a rock, the remainder of which has weathered to saprolite or other by-products.

corrasion Physical erosion of rocks and sediment by frictional forces.

correlated age dating A type of dating that relies on correlating surfaces or sediments of unknown age to a sequence of others, for which numerical age data are known.

cosmogenic isotope An isotope that can be used to study the age and origin of the Earth.

cradle knoll The small mound of debris formed as soils falls from the root plate (ball) of an uprooted tree. Preexisting soil horizons are commonly obliterated or folded over, resulting in a heterogeneous mass of soil material. So named because lumberjacks used to place their babies to sleep inside the pits that adjoin the mounds, i.e., cradles.

cratermakers A group of soil fauna whose bioturbation activities are best expressed by the surface craters and depressions they make, via their digging, scratching, and furrowing activities.

creep Slow mass movement of soil and soil material down slopes, driven primarily by gravity but facilitated by saturation with water and by alternate freezing and thawing.

crest Synonym for summit or hilltop.

critical zone The near-surface environment in which complex interactions involving rock, soil, water, air, and living organisms regulate the natural habitat and determine the availability of life-sustaining resources. It includes the biosphere and extends through the pedosphere, through the unsaturated vadose zone, and into the saturated groundwater zone.

crotovina See *krotovina*.

crumb An antiquated term for a type of soil structure, typical of organic-rich A horizons. Crumb structure is today referred to as *granular*.

crust Surface layer in a soil that becomes harder than the underlying horizon.

cryic See *soil temperature regime*.

cryogenic soil Soil that has formed under the influence of cold soil temperatures.

Cryosols A soil in the Cryosolic order of the Canadian system of soil classification. A class of soils that are dominated by freeze-thaw activity and permafrost.

cryostatic pressures Expansion pressures created in soils and sediments due to the formation of ice.

cryoturbation Soil and sediment mixing by freeze-thaw activity, as ice crystals repeatedly form and melt.

crystalline rock A rock consisting of various minerals that have crystallized in place from magma, typical of igneous and metamorphic rocks.

crystalturbation Soil and sediment mixing by the growth and wastage of crystals, e.g., ice, salts.

cumulization The slow upward growth of the upper soil horizons (usually the A horizon) due to additions of sediment and their pedogenic incorporation. The additions, e.g., alluvium, loess, slopewash, must occur slowly enough so that pedogenesis can incorporate the sediment into the profile and thicken the A horizon.

cutan Coating on the surfaces of peds, stones, and so on. A common type of cutan is an argillan formed by translocation and deposition of clay particles on ped surfaces.

cyclosilicates A type of silicate mineral that has a ring structure; typically, six tetrahedra are arranged in a ring.

D horizon The unaltered soil parent material that lies below the C horizon; an informal term, D horizons are geogenic or non-pedogenic horizons of fresh sediment, excluding consolidated bedrock, characterized by original rock or sedimentary fabric, lack of tension joints, and lack of alteration features of biooxidation origin.

Darcy's law A law in soil physics that describes the flow of water through saturated porous media: $J_{sat} = -K_{sat} (\Delta H/\Delta z)$ where J_{sat} refers to volume transported per unit area per unit time (flux density, e.g., cm^3 of water per cm^2 of soil per day, which reduces to cm day^{-1}); K_{sat} is the saturated hydraulic conductivity of the soil, a coefficient whose value is dependent on the nature of the soil; Δz is the distance between the top of the saturated soil and a lower reference point where water freely drains and is defined to have zero hydraulic potential; and ΔH is the hydraulic head difference, which is the distance between the upper surface of the water table and the lower reference point in the soil

date A specific point in time, usually in the past. Contrast with *age*.

dealkalinization See *solodization*.

decalcification The removal of $CaCO_3$ or calcium ions from the soil, usually by leaching.

debris flux The movement of sediment downslope, usually referring to processes associated with a catena.

decarbonation See *decalcification*.

decay rate The rate, expressed in terms of half-life of the parent isotope, at which a population of radioactive atoms decays into stable daughter atoms.

decomposer Organism that breaks down dead or decaying organisms and in so doing performs the natural process of decomposition.

decomposition Breakdown of organic compounds by enzymatic activity; alternatively, mineral dissolution and transformation. See *chemical weathering*.

decomposition and synthesis The set of processes involving chemical weathering of primary and secondary minerals, into their constituent ions, followed by their rebuilding into clay minerals, sometimes using other clay minerals as a template.

deflation Erosion of fine particles by the action of wind.

deflocculation The inverse of flocculation. The state of soil particles when they act as dispersed, independent entities and not as clustered groups, or floccules. See also *dispersion*.

degradation Change of a soil or soil horizon to a more highly leached and weathered condition, usually accompanied by morphological changes such as development of an E horizon.

degraded B horizons Referring to the condition in which Bt or Bx horizons have lost clay and various cementing agents as a result of acidification and eluviation. Many degraded B horizons acquire a tonguelike (glossic) appearance as the E horizon engulfs them from above and widens along vertical cracks.

Degraded Chernozems A zonal great soil group (1938 system of soil classification) consisting of soils with a very dark brown or black A horizon underlain by a dark gray, weakly expressed E horizon and a brown B or Bt horizon. Formed in the forest-prairie transition of cool climates. Obsolete in the U.. system of Soil Taxonomy.

delineation An individual polygon on a soil map.

delta A fan-shaped body of alluvium deposited at or near the mouth of a river where it enters a body of relatively quiet water, usually a sea or lake.

dendrochronology The science of tree ring analysis.

dendrogeomorphology The study of geomorphic processes and surface exposure through the use of tree ring analysis.

denitrification The biochemical reduction of nitrogen oxides (usually nitrate and nitrite) to molecular (gaseous) nitrogen or nitrogen oxides, by bacterial activity (denitrification) or by chemical reactions involving nitrite (chemodenitrification). The biological reduction of nitrogen to N_2, NO, or N_2O, resulting in the loss of nitrogen into the atmosphere.

denudation Processes that result in the wearing away or the progressive lowering of the Earth's surface by weathering, erosion, mass wasting, and transportation of sediment.

depression A relatively low or sunken part of the Earth's surface, especially a low-lying area surrounded by higher ground. A closed depression has no natural outlet for surface drainage, whereas an open depression has a natural outlet.

depth function, depth plot A graphical way of presenting soil data, with the data shown on the *x*-axis and depth plotted (downward) on the *y*-axis.

desert crust A thin, hard surface layer, containing $CaCO_3$, gypsum, or other binding material, in a desert region.

desert pavement A natural concentration of wind-polished, closely packed (almost interlocking) pebbles, boulders, and other rock fragments, mantling the desert soil surface and protecting the underlying, finer-grained material.

desert varnish See *rock varnish*.

Desert Soils A zonal great soil group (1938 system of soil classification) consisting of soils with a very thin, light-colored surface horizon, which may be vesicular, ordinarily underlain by horizons enriched in carbonates. Formed in arid regions under sparse shrub vegetation. Roughly synonymous with Aridisols. Obsolete in the U.S. system of Soil Taxonomy.

desilication The loss of silica from a soil, primarily in solution and typical of hot, humid climates. Also known as *desilification*.

desorption The migration of adsorbed entities from adsorption sites. The reverse of adsorption.

deterministic uncertainty A perspective based on two fundamental axioms related to soil variability: (1) variability can be explained with more and better measurements of the soil system, and/or (2) variability may be an irresolvable outcome of complex system dynamics.

detrital grain In soil micromorphology, a mineral grain that was originally present in the parent material.

detrital sediments Sediments made of fragments or mineral grains weathered from preexisting rocks.

developmental upbuilding See *cumulization*.

diabase An intrusive igneous rock, intermediate between gabbro and basalt.

diachronous A term used to describe stratigraphic units whose bounding surfaces are not synchronous, i.e., uniform in age. See also *time transgressive*.

diagenesis All the physical, chemical, and biologic changes undergone by sediments from the time of their initial deposition, through their conversion to solid rock, exclusive of metamorphism.

diagnostic horizons Rigorously defined (in Soil Taxonomy) combinations of specific soil characteristics (in specific depth intervals) that are used to classify soils. Diagnostic horizons that occur at the soil surface are called epipedons, whereas those below the surface are called diagnostic subsurface horizons.

diapir A forced intrusion of material, usually malleable in nature, upward and into a surrounding matrix. Usually applied to rocks, but diapirlike materials are common in many Vertisols and some Gelisols.

diatoms Algae that possess a siliceous cell wall that remains preserved after the death of the organisms. They are abundant in both fresh and salt water and in a variety of soils.

diffuse double layer A conceptual and quantitative model that describes the distribution of ions near a charged solid surface, e.g., a clay or oxide surface. One layer of ions is immediately adjacent to the charged surface. The second layer consists of an ionic swarm in solution. Together they neutralize the surface charge.

diffusion The net movement of a substance (liquid or gas) from an area of higher concentration to one of lower concentration.

dilatation See *unloading*.

dilation The condition in which a soil or material has gained volume over time.

dioctahedral In a layer silicate, an octahedral sheet that has two-thirds of the octahedral sites filled by trivalent ions such as Al^{3+} or Fe^{3+}.

diorite Intrusive igneous rock made of plagioclase feldspar and amphibole and/or pyroxene. Like gabbro but less dark and containing less Fe and Mg.

discontinuity In a stack of sediments, any break or interruption in the sediment column, whatever its cause or length, usually a manifestation of nondeposition, a change in depositional systems, and/or accompanying erosion. See also *lithologic discontinuity*.

discontinuous permafrost Permafrost occurring beneath some areas, while other areas in the same region remain free of permafrost.

disintegration See *physical weathering*.

dispersion The state in which individual soil particles, e.g., clay and silt, act independently and do not adhere to one another in a suspension; in other words, they are neither coagulated nor flocculated, but remain suspended.

dissolution A chemical reaction in which a solid material is transformed to ions or molecules in a liquid, e.g., halite (NaCl) undergoes dissolution when placed in water.

dissolved load The mass of dissolved materials that water carries in solution.

dissolved organic carbon (DOC) Carbon in organic matter that is small enough to pass a 0.45 µm filter.

divergent slope A hillslope in which the downslope flowlines of water diverge at the base, e.g., nose slopes.

doline, dolina A closed depression in a karst region, often rounded or elliptical in shape, that forms by the solution and subsidence of the limestone near the surface, or by the collapse of an underground cave. Also known as a sinkhole.

dolomite, dolostone A carbonate sedimentary rock made up predominantly of the mineral dolomite, $CaMg(CO_3)_2$.

drainage class An assessment of the prevailing wetness conditions of a soil, related to the rate at which water is removed from a soil without artificial drainage. See Table 14.4 for definitions.

drain tile Porous, ceramic, plastic, or clay pipe or similar buried structure used to collect and conduct free water, i.e., that water below the water table, from a soil. Drain tiles are installed to lower the water tables in wet soils.

drift See *glacial drift*.

drumlin A low, smooth, elongated oval hill or ridge, usually composed of glacial till that may or may not have a core of bedrock or stratified drift. The long axis is parallel to the general direction of glacier flow. Drumlins, which usually occur in groups, are products of streamline (laminar) flow of glaciers, which molded the subglacial floor through a combination of erosion and deposition.

duff Litter that comprises the O horizon.

duff A forest humus/litter type, transitional between mull and mor, characterized by friable Oe horizons, reflecting the dominant zoogenous decomposers. Sometimes called *duff mull*, it is similar to mor in that it is generally an accumulation of partially to well-humified organic materials resting on the mineral soil.

It is similar to mull in that it is zoologically active. Also called moder (mostly in Europe).

dune A low mound, ridge, bank, or hill of loose, wind-blown sand, either bare or covered with vegetation.

duplex soil Term, commonly used in Australia, for a soil with clay-poor upper solum and a clay-rich B horizon. See also *texture-contrast soil*.

duricrust A duripan exposed at the surface by erosion.

durinode A soil nodule cemented or indurated with SiO_2. Durinodes break down in concentrated KOH after treatment with HCl to remove carbonates, but they do not break down on treatment with concentrated HCl alone.

duripan A subsurface soil horizon that is cemented by illuvial silica, usually as opal or other microcrystalline forms. Less than 50% of the volume of air-dry fragments will slake in water or HCl. When exposed at the surface by erosion it is commonly called a duricrust.

dust devil A small dust-bearing whirlwind, common in deserts.

dynamic denudation A general model of soil and landscape evolution and archaeological site formation. Developed by Donald Johnson, the model emphasizes the dynamic nature of soil and landscape evolution and integrates surface erosion-denudation, bioturbation, soil creep, throughflow, eluviation-illuviation, leaching, weathering, and saprolite production processes with several key theories of pedology and geomorphology, into a unified process framework.

dysic Soil situation characterized by a low level of bases, i.e., acidic.

E horizon A major soil horizon dominated by eluviation of clay, Fe, Al, and humus, among others, usually occurring above an illuvial B horizon. E horizons are usually light-colored because the coatings on the primary minerals have been stripped, revealing the natural, light color of quartz, which often dominates the mineralogy.

earthworm A tube-shaped, segmented animal found in many soils. Its digestive system runs straight through its body, allowing it to ingest and pass soil material through its gut. Earthworms function as important bioturbators.

eccentricity of the Earth's orbit A measure of the circularity of the Earth's orbit that varies in cycles of about 100,000 and 400,000 years.

ectomycorrhizae A symbiotic, generally mutualistic, but occasionally weakly pathogenic association between a fungus and the *exterior* parts of the roots of a vascular plant. See also *arbuscular mycorrhizae*.

edaphic Of or pertaining to the soil. Resulting from or influenced by factors inherent in the soil or other substrate, rather than by climatic factors.

edge effect Condition that occurs on the edges (usually shoulder slopes) of landforms; soils here are markedly different from those upslope or downslope. An edge can be either a *wet edge* or a *dry edge*.

effective cation exchange capacity (ECEC) The sum of cation charges exclusive of H^+, in a neutral salt extract (charges of Ca^{2+}, Mg^{2+}, Na^+, and K^+ as well as Al^{3+} in acid soils). It is usually expressed in centimoles of charge per kilogram of exchanger or (in older literature) millimoles of charge per 100 g of exchanger. Total CEC, in contrast, includes the exchangeable H^+ and is determined at a constant pH, such as 7.0 or 8.2.

electrical conductivity (EC) The conductivity of electricity through water or an aqueous extract of soil, which is commonly used to estimate soluble salt content and ionic strength.

electrolyte A molecule that separates into a cation and an anion when it is dissolved in a solvent, usually water, e.g., NaCl separates into Na^+ and Cl^- in water.

electron acceptor A compound that accepts electrons during biotic or abiotic chemical reactions and is thereby reduced.

electron donor A compound that donates or supplies electrons during metabolic or abiotic chemical reactions and is thereby oxidized.

elementary soil areal (ESA) A soil geography term, popularized in Russia by V. M. Fridland, for the simplest soil cover element. See also *polypedon*.

eluvial horizon, eluvial zone The horizons in the soil profile that are dominated by losses of constituents in solution and suspension, i.e., by eluviation. Typically refers to the A and B horizons, but the best developed part of the eluvial zone is the E.

eluviation The removal, in colloidal suspension or in solution, of mass from a soil horizon.

end moraine A ridgelike accumulation, chiefly composed of till, produced at the outer margin of an active glacier.

endogeic earthworms Earthworms that live at the soil surface and in the litter layer, ingesting organic matter and redistributing it among the upper horizons of the mineral soil.

endoliths Organisms that colonize pore spaces, cracks, and fissures in exposed bedrock.

endomycorrhiza A mycorrhizal association with intracellular penetration of the cortical cells of a host root by a fungus, as well as outward extension of the fungal hyphae into the surrounding soil

endosaturation The condition of saturation of a zone or soil horizon by groundwater (not perched water).

Entisols One of 12 major taxonomic orders in Soil Taxonomy. Entisols are mineral soils that lack distinct subsurface diagnostic horizons, i.e., they are weakly developed.

eolian Pertaining to material transported and deposited by the wind, including dune sand, sand sheets, loess, and parna. Also *aeolian*.

epifauna Animals that live, primarily, on and above the soil. Contrast to *infauna*.

epigeic earthworms Called "red wigglers" by fishermen, epigeic earthworms are small earthworms that live at the soil surface or in the litter layer.

epiliths Organisms that colonize the exposed surfaces of rock or mineral substrates.

epimorphic processes Subsurface processes associated with soil development, including weathering, leaching, and new mineral formation and inheritance.

epipedon In Soil Taxonomy, a strictly defined horizon at the soil surface, usually including the O or A genetic horizon. Comparable horizons below the surface are diagnostic *subsurface* horizons.

episaturation The condition in which the soil is saturated with water in one or more layers within 200 cm of the surface, which also has one or more unsaturated layers below, but above the true groundwater table. Episaturation is usually synonymous with perched water. See also *endosaturation*.

equifinality The situation in which a number of different processes all lead to essentially the same outcome. In geomorphology, the concept implies that different processes can produce similar landforms. In pedology, different pedogenic processes can lead to similar soil properties.

erosion The wearing away of the land surface by running water, wind, ice, or other agents that detach geologic materials and transport them elsewhere, usually downslope, including such processes as creep, slopewash, and erosion due to tillage.

erosion classes A grouping of soil properties based on the presumed degree of past erosion or on characteristic patterns of erosion. Applied to accelerated erosion due to human activity, not to normal geological erosion. Four erosion classes are recognized for water erosion and three for wind erosion.

erosion pavement A remnant layer of gravel or stones on the surface of the ground, after the removal of the fine particles by erosion.

erosion surface A geomorphic surface shaped by the erosive action of ice, wind, or (usually) running water.

erratic A stone or boulder glacially transported from place of origin and left in an area of different bedrock composition. Its presence confirms the prior glaciation of that site.

esker A long, narrow, usually sinuous steep-sided ridge composed of irregularly stratified sand and gravel that was deposited by a supraglacial stream flowing between ice walls, or in a subglacial ice tunnel, of a retreating but usually stagnant glacier. The esker materials are left behind when the ice melts.

etch pit Hole or hollow formed on a mineral surface by chemical weathering.

eubacteria Prokaryotes other than Archea.

euic Signifies a high level of base cations in a soil material, specified at family level of taxonomic classification.

eukaryote Cellular organism having a membrane-bound nucleus within which the genome of the cell is stored as chromosomes composed of DNA. Includes algae, fungi, protozoa, plants, and animals.

eustatic change in sea level A worldwide change in sea level caused by formation or melting of large continental glaciers.

eutrophic Lake or pond water with high concentrations of P and N. Growth of aquatic plants is promoted; sometimes excessive algal growth leads to depletion of dissolved O_2, threatening fish and other aquatic organisms.

evaporite Residue of salts (including gypsum and all more soluble species) precipitated by evaporation. Gypsum and halite are examples of evaporite minerals.

evapotranspiration The combined loss of water by evaporation from the soil surface and water bodies, as well as from transpiration by plants. The word is coined from the combined processes of evaporation and transpiration.

exchange complex The suite of negatively charged sites in a soil that can adsorb cations by exchange reactions.

exchangeable acidity Charge equivalents of H^+ and Al^{3+} that can be displaced by 1 M KCl solution, reported in cmol(+) kg^{-1} of soil.

exchangeable anion A negatively charged ion held on or near the surface of a solid particle by a positive surface charge that may be readily replaced by other negatively charged ions, e.g., Cl^-.

exchangeable base cations Cations at the surfaces of soil particles that can be readily replaced with a salt solution. In most soils, Ca^{2+}, Mg^{2+}, K^+, and Na^+ predominate. Ammonium is also a base cation, but it is usually present in small concentrations and may not be readily exchangeable. Historically, these cations were called *bases* because they are cations of strong bases. That usage is obsolete.

exchangeable cation A positively charged ion held on or near the negatively charged surface of a solid particle, which may be replaced by other positively charged ions in the soil solution. Usually expressed in centimoles of charge per kilogram.

exchangeable sodium percentage (ESP) The exchangeable Na concentration of a soil expressed as a percentage of the cation exchange capacity.

exchangeable sodium ratio (ESR) The ratio of exchangeable Na charges to the sum of charges of exchangeable Ca and Mg.

exfoliation A form of physical weathering in which thin-thick (5 mm–2 m thick) sheets of rock peel off the parent rock, sometimes resulting in broad, domelike features known as exfoliation domes.

exhumed paleosol A soil, geosol, or paleosol that formerly was buried and has been reexposed by erosion of the covering mantle.

extragrade A taxonomic term at the subgroup level of Soil Taxonomy. Soils in extragrade subgroups have properties that are not characteristic of any class in a higher category (any order, suborder, or great group) and that do not indicate a transition to any other known kind of soil. See also *intergrade*.

extrinsic Processes and inputs that are external (not internal) to a given system.

extrinsic threshold In pedogenesis, the point at which the limit of soil morphologic or developmental stability becomes exceeded; an extrinsic (external), subtle, but progressive change in one of the soil-forming factors, exclusive of time.

extrusive igneous rock Igneous rock that solidified from lava at the Earth's surface, characterized by very small mineral crystals, e.g., basalt. Compare with *intrusive* igneous rock.

F horizon Obsolete terminology for the Oe (fermentation) horizon.

fabric See *soil fabric*.

fabric contrast soil In Australia and other parts of the world, a soil that has a distinct change in soil fabric from the upper solum to the lower, usually in association with a Bt horizon.

factorial model See *state-factor model*.

facultative organism An organism whose metabolism can be both aerobic and anaerobic.

fault A fracture or fracture zone in the earth, with displacement along one side with respect to the other.

faunal passage In soil micromorphology, a small, tubular pore produced by soil fauna. See also *krotovina*.

faunalfunction The formula or quantitative expression of the differences that occur in soils as a function of changes in fauna, especially soil fauna, assuming that the remaining soil-forming factors are held constant, or nearly so. See also *faunalsequence*.

faunalsequence A group of related soils that differ from one another primarily in kinds and numbers of fauna, especially soil fauna, as a soil-forming factor. When expressed as a mathematical equation, it is referred to as a *faunalfunction*.

faunalturbation Soil and sediment mixing by the activities of animals, i.e., fauna.

Fe concentrations See *redox concentrations*.

Fe depletions See *redox depletions*.

fecal pellets Rounded and subrounded aggregates of fecal material produced by soil fauna.

Fe concentrations Small areas or splotches in a soil where Fe, generally Fe^{3+}, has been concentrated by precipitation as Fe oxide minerals. Commonly found in soils that undergo redox cycles. Also called *red mottles*.

Fe-Mn concretions Same as Fe concentrations but also including black spots and pellets of Mn.

feldspar Family of silicate minerals containing varying amounts of K, Na, and Ca, along with Al, Si, and O. K-feldspars contain considerable potassium, whereas plagioclase feldspars contain considerable Na and Ca.

felsic rock A light-colored, silica-rich igneous rock, also including some metamorphic crystalline rocks.

ferrallitization See *laterization* and *desilication*.

ferran A normally reddish or brownish colored cutan composed of iron oxides, hydroxides, or oxyhydroxides.

ferriargillan A cutan consisting of a mixture of silicate clay minerals and iron oxides, hydroxides, or oxyhydroxides.

ferricrete See *ironstone*.

ferrihydrite A dark reddish brown, poorly crystalline iron oxide mineral that forms in wet soils and commonly occurs as concretions. Various chemical formulas for ferrihydrite have been proposed, including $(Fe_2O_3) \cdot 5.9H_2O$ and $Fe_5O_7(OH) \cdot 4H_2O$.

ferrolysis A group of pedogenic processes involving the disintegration and dissolution of clay, largely due to the alternating periods of reduction and oxidation of Fe.

ferromagnesian Containing Fe and Mg, typically applied to mafic minerals, e.g., olivine.

ferrimagnet A magnetic mineral in soils, commonly magnetite or maghemite.

ferrous iron Iron in the reduced state (Fe^{2+}), which is usually more mobile than ferric iron (Fe^{3+}) in the soil solution.

ferrugination See *braunification*.

fersialitization A pedogenic process involving the inheritance and neoformation of smectitic clays and the immobilization of iron oxides due to alkaline soil conditions.

fibric material Organic soil material that, after rubbing between fingers, usually contains three-fourths or more recognizable fibers of undecomposed plant remains. It has a low bulk density and a very high water-holding capacity.

field capacity The content of water remaining in a soil, two or three days after it has been fully wetted with water and allowed to drain until free drainage is negligible, expressed as a percentage of oven-dry soil (weight or volume).

fifteen-atmosphere percentage The percentage of water contained in a soil that has been saturated, allowed to drain, and then allowed to enter into equilibrium with an applied pressure of 15 atm. Approximately the same as fifteen-bar percentage, it is commonly equated with *wilting point* water content.

fine-earth fraction Fraction of the soil that passes through a 2-mm sieve, i.e., the sand, silt, and clay fractions. Materials larger than 2 mm are referred to as the *coarse fraction*.

fine texture A broad group of soil textures consisting of or containing large quantities of silt and clay.

fine-textured soil A soil rich in clay and silt, and containing little sand or gravel. See *heavy soil*.

firm See *consistence*.

flaggy Containing appreciable quantities, usually >15% by volume, of flagstones.

flagstone A relatively thin, flat rock fragment, usually shale, limestone, slate, or sandstone, that is from 15 to 38 cm on the long axis.

flint A variety of chert, often dark in color.

flocculation The cohesion of colloidal soil particles in a suspension due to adsorbed polymers such as Fe or Al oxyhydroxides. Technically distinct, *flocculation* and *coagulation* are often used interchangeably. Opposite of *dispersion*. See *coagulation*.

floodplain The nearly level, comparatively broad alluvial plain that borders a stream and is subject to inundation under flooding, unless protected artificially. It is usually a constructional landform built of overbank sediment (alluvium) deposited during floods and of alluvium deposited on the banks of the stream during its lateral migration.

floralfunction The formula or quantitative expression of the differences that occur in soils as a function of changes in flora, or plant cover, assuming that the remaining soil-forming factors are held constant, or nearly so. See also *floralsequence*.

floralsequence A group of related soils that differ one from the other primarily because of differences in kinds and numbers of plants as a soil-forming factor. When expressed as a mathematical equation, it is referred to as a *floralfunction*.

floralturbation Soil and sediment mixing by the activities of plants, primarily due to uprooting, and root growth and expansion.

flow till A supraglacial till that is modified and transported by mass flows associated with glaciers.

footslope The colluvial, concave hillslope position at the base of a slope. In profile, footslopes are commonly slightly concave and are situated between the backslope (above) and toeslope (below).

forest floor All organic matter generated by forest vegetation, including litter and unincorporated humus, lying on the mineral soil surface.

fragipan A subsurface (Bx or Ex) soil horizon with very low organic matter content, high bulk densities, density, and/or high mechanical strength. Fragipans are root restricting and have hard or very hard consistence (seemingly cemented) when dry but are brittle when moist. Because they are slowly permeable, they often perch water and thus commonly have redoximorphic features. Many fragipans have coarse or very coarse polyhedronal or prismatic structure.

free face A slope that is nearly vertical, commonly formed in hard or cemented material like bedrock and associated with rockfalls.

free iron oxides Iron oxides that can be reduced and dissolved chemically by dithionite, generally including goethite, hematite, ferrihydrite, lepidocrocite, and maghemite, but not magnetite.

freezing front The bottom edge of a frozen soil layer, below which the soil temperature is assumed to be < 0°C.

Freundlich adsorption model A nonlinear, power-function model used to describe quantitatively the adsorption of an adsorbate to an adsorbent over a large range of concentrations.

friable See *consistence*.

frigid See *soil temperature regime*.

frost boil Upwellings of mud that occur through frost heave and cryoturbation in permafrost areas. They are typically 1–3 m in diameter with a bare soil surface and dominantly circular. Frost boils are also known as mud boils, frost scars, and mud circles.

frost heave Lifting or lateral movement of soil, caused by freezing processes and the formation of ice lenses or needles.

frost jacking See *upfreezing*.

frost shattering See *gelifraction*.

frost wedge V-shaped body of ground ice, usually less than 4 m in depth and 2 m in width, that typically forms from the surface downward, in areas of continuous permafrost.

frost wedge cast See *ice wedge cast*.

fulvate-complex theory See *chelate-complex theory*.

fulvic acid Organic material extracted from a soil with a strong base (pH ~11) that remains in solution after the extract is treated with a strong acid (pH ~1).

fulvic acid fraction Fraction of soil organic matter that is soluble in both alkaline and acidic solutions.

functional-factorial model A type of soil genesis model in which soil development is assumed to be a function of the interactions of a number of environmental factors. Also called a *state factor model*.

fungal hyphae Long, branching, usually tubular filamentous structures of a fungus.

fungi Simple organisms that lack chlorophyll and are composed of cellular filamentous growths known as hyphae.

gabbro A coarse-grained, intrusive igneous rock, chemically equivalent to a basalt.

galleries Interconnected tunnels made by termites, ants, and other soil fauna.

garden variety Colloquial term for the cosmogenic nuclide ^{10}Be that falls from the sky, collects on the soil surface, and accumulates there over time, as contrasted with in situ ^{10}Be that accumulates directly within quartz-rich rocks at or near the surface.

gas (soil) Nitrogen (N_2), oxygen (O_2), carbon dioxide (CO_2), methane (CH_4), and water vapor (H_2O) are the typically most abundant gases in soils.

gastroliths Clastic rock and gravel fragments ingested by an animal, usually a bird, in order to grind food in gastric digestion.

gelic See *soil temperature regime.*

gelifluction Form of mass movement in periglacial environments where a permafrost layer exists, characterized by the slow, wet downslope movement of soil material over the permafrost layer and the formation of lobe-shaped features. Also called *solifluction.*

Gelisols One of 12 major taxonomic orders in Soil Taxonomy. Gelisols form in cold regions where permafrost occurs at near the surface; as a result, they show evidence of mixing by cryoturbation.

gelifraction A form of physical weathering in which ice crystal growth shatters the rock.

geliturbation Mixing of soils and sediments by processes associated with ice and frost.

geoarchaeology The science that primarily encompasses the physical (geological, soils, etc.) aspects of archaeology.

geochronometric unit A division of time, based on the rock record. Types of geochronometric units include eons, eras, periods, epochs, and ages.

geographic information system/science (GIS) A method of overlaying spatial data of different kinds. The data are referenced to a set of geographical coordinates and encoded in a form suitable for handling in computer software programs.

geologic erosion Normal or natural erosion caused by natural weathering or other geological processes, as opposed to accelerated erosion due to human activities.

geomorphic surface A mappable area of the Earth's surface that has a common geologic history. The area is of similar age and is formed by a set of processes, be they erosional, constructional, or both, during an episode of landscape evolution. Geomorphic surfaces can be either *constructional* or *erosional.*

geomorphology The systematic examination of landforms and their interpretation as records of geologic history.

geophagy The deliberate ingesting of soil or sediment, often for religious or health reasons, as practiced by humans and some animals.

geosol Similar to a paleosol, but more rigorously defined, especially with respect to stratigraphic placement and nomenclature.

gibbsite $Al(OH)_3$. A mineral, typically clay-sized, with a platy habit that occurs in highly weathered soils and laterite. It may be prominent in residuum from feldspar-rich crystalline rocks.

gilgai A surface microrelief of small basins and knolls (or valleys and ridges) produced by argilliturbation – the expansion and contraction during wetting and drying (usually in regions with distinct, seasonal precipitation patterns) of clayey soils that contain large amounts of smectite.

gilgai A microfeature pattern typical of some Vertisols, composed of a succession of microbasins and microknolls on level areas, or of microtroughs and microridges parallel to the slope on sloping areas and produced by expansion/contraction and shear/thrust processes with changes in soil moisture. Various types of gilgai, based on the dominant shape of microhighs and microlows, are recognized.

glacial drift General term applied to all material transported and deposited by a glacier, whether it be by the ice or by meltwater from the ice.

glacial till Unsorted and unstratified material, deposited directly by glacial ice, which may consist of various mixtures of clay, silt, sand, gravel, stones, and boulders. Sometimes, till may be crudely sorted.

glaciofluvial deposits Material moved by glaciers and subsequently sorted and deposited by meltwater streams flowing from the ice. The deposits, typically coarse-textured, are stratified and may occur in landforms such as outwash plains, deltas, kames, eskers, and kame terraces.

glaciolacustrine deposits Material ranging from fine clay to sand derived from glaciers and deposited in glacial lakes, usually those that have one shoreline as the ice margin. Many are bedded or laminated with *varves.*

glaebule In soil micromorphology, a three-dimensional pedogenic feature within the soil matrix that is approximately prolate to equant in shape.

glassy A texture of extrusive igneous rocks that develops as the result of rapid cooling, so that crystallization is inhibited, giving the rock a shiny appearance, e.g., obsidian.

glauconite An Fe-rich dioctahedral mica. Mixtures containing an iron-rich mica as a major component can be called glauconitic.

gleization See *gleyzation.*

gleyed A soil condition, manifested by the presence of grayish, bluish, or greenish colors through the soil mass, or in mottles (spots or streaks, also known as redoximorphic features) among these same colors, all of which have resulted from prolonged soil saturation. Gleying occurs under reducing conditions, during which most of the free iron is reduced to Fe^{2+}.

gleyed soil Soil developed under conditions of poor drainage, resulting in reduction of iron and other elements and the formation of muted bluish or gray colors and other redoximorphic features.

Gleysols A taxonomic order of soils that are dominated by wetness and reduced conditions. Gleysols are not an order in Soil Taxonomy but do occur in the soil classification system of Canada.

gleyzation Processes involved in the gleying of soils, usually due to prolonged wetness. Associated with this process is the reduction of Fe and Mn.

glossic horizon An E horizon that protrudes in a tonguelike manner into a (usually) degrading Bt or Btx horizon.

gneiss A coarse-grained, foliated rock, commonly formed by the metamorphism of granite, in which bands of granular minerals (commonly quartz and feldspars) alternate with bands of flaky or elongate minerals, e.g., micas and pyroxenes. Generally less than half of the minerals are aligned in a parallel orientation.

goethite FeOOH. An iron oxide mineral that occurs in almost every soil type and climatic region and is responsible for the yellowish brown colors in many soils and weathered materials.

grain cutan Cutan associated with the surfaces of skeleton grains or other discrete units such as nodules and concretions.

grainy gray ped coatings See *albic neoskeletans*.

granite A light-colored, coarse-grained intrusive igneous rock characterized by the minerals orthoclase and quartz with lesser amounts of plagioclase feldspar and iron-magnesium minerals.

granular A shape of soil peds common to A horizons, with comparatively small, rounded, nearly spheroidal shapes; granular peds are typically high in organic matter.

gravel Clastic fragments larger than sand (>2 mm) but smaller than 25 mm in diameter.

gravelly Containing appreciable amounts of gravel.

gravimetric water content The amount of water in soil or sediment, as expressed on a mass basis, e.g., grams/gram. Contrast to *volumetric water content*.

gravitational water Water that freely moves into, through, or out of the soil under the influence of gravity, i.e., at low matric suction.

graviturbation Soil and sediment mixing by mass movements, which are driven mainly by gravity.

Gray-Brown Podzolic soils A zonal great soil group (1938 system of soil classification) consisting of soils with a thin, moderately dark A horizon and with a light-colored E horizon underlain by a Bt horizon. They occur on relatively young land surfaces, mostly base cation–rich glacial deposits, under deciduous forests in humid temperate regions. Roughly synonymous with Udalfs. Obsolete in the U.S. system of Soil Taxonomy.

Gray Desert soils A term used in Russia, and formerly in the United States, synonymously with Desert soil.

graywacke A variety of sandstone characterized by angularly shaped grains of quartz and feldspar, as well as small fragments of dark rock, all set in a matrix of finer particles.

great group One of the categories in the Soil Taxonomy system of soil classification; a category below the suborder, but above the subgroup and series levels. Great groups categorize soils according to soil moisture and temperature regimes, base cation saturation status, and expression of horizons.

great period In lichenometry, the initial period of rapid lichen growth that lasts about 10 to 100 years.

ground ice Term for any type of discrete ice body in a soil.

ground moraine A landscape mainly formed by, and owing its character to, subglacial processes.

Ground-water Laterite soils A great soil group of the intrazonal order and hydromorphic suborder (1938 system of soil classification), consisting of soils characterized by hardpans or concretional horizons rich in Fe and Al (and sometimes Mn) that have formed immediately above the water table. Obsolete in the U.S. system of Soil Taxonomy.

Ground-water Podzol soils A great soil group of the intrazonal order and hydromorphic suborder (1938 system of soil classification), consisting of soils with a thick O horizon over a very thin, acidic A horizon, which is underlain by a whitish gray, leached E horizon, which may be as much as 70–100 cm in thickness. The illuvial Bsm or Bhs horizon is brown or very dark-brown and commonly cemented. These coarse-textured soils form under forest vegetation in cool to tropical, humid climates under conditions of poor drainage. Obsolete in the U.S. system of Soil Taxonomy.

groundwater That portion of the water below the surface of the ground at a pressure equal to or greater than atmospheric pressure, i.e., in a saturated zone.

grus Weathered residuum from coarse-grained rocks.

grusification, grusivication Specifically, the formation of grus from hard granite. Generally, the formation of weathered rock from unweathered rock.

gumbotil Gray to dark-colored, leached, nonlaminated deoxidized clay, very sticky, and breaking with a starchlike fracture when wet, but very hard when dry. Obsolete term, replaced by "accretion gley." Gumbotils formed on old, stable landscapes and are often assumed to be buried paleosols.

gypcrete A soil horizon indurated or cemented by illuvial gypsum. Also known as a *petrogypsic horizon*.

gypsan A cutan composed of gypsum.

gypsic horizon A soil horizon of secondary $CaSO_4 \cdot 2H_2O$ enrichment that is >15 cm thick.

gypsification The process whereby a soil horizon becomes enriched in illuvial gypsum.

gypsum $CaSO_4 \cdot 2H_2O$. The mineral name for calcium sulfate.

gyttja A type of peat consisting of fecal material, strongly decomposed plant remains, shells of diatoms, phytoliths, and fine material particles, that usually forms in standing water.

H horizon Old terminology for the Oa (humus) horizon.

Half Bog soils A great soil group, of the intrazonal order and hydromorphic suborder (1938 system of soil classification), consisting of soils with dark-brown or black peaty material over gleyed and mottled mineral soil horizons. They form under conditions of poor drainage under forest, sedge, or grass vegetation in cool to tropical, humid climates. Obsolete in the U.S. system of Soil Taxonomy.

half-life The period required for half of an original population of atoms of a radioactive isotope to decay.

halloysite A member of the kaolin subgroup of clay minerals, similar to kaolinite in structure and composition except that hydrated varieties of halloysite have interlayer water molecules. Halloysite usually occurs as tubular or spheroidal particles and is most common in soils formed from volcanic ash.

Halomorphic soils (i) A suborder of the intrazonal soil order (1938 system of soil classification), consisting of saline and sodic soils formed on wet sites in arid regions, including the great soil groups Solonchak or Saline soils, Solonetz soils, and Soloth soils. (ii) In a general sense, a soil containing a significant proportion of soluble salts. Obsolete in the U.S. system of Soil Taxonomy.

halophyte A plant capable of growing in salty soil, i.e., a salt-tolerant plant.

haploidization Processes that lead to profile simplification. See also *horizonation*.

hard See *consistence*.

hardpan Colloquial term for a soil horizon with physical characteristics that limit root penetration and restrict water movement.

headslope A hillslope, as seen in plan view, i.e., from above, with concave boundaries above and below, situated in a hollow between interfluves and/or nose slopes.

heave In mass movement, the upward motion of material by expansion, e.g., the heaving of soils caused by freezing water.

heavy soil A colloquial term for a soil with a high content of the fine separates, particularly clay, so named because, when wet, these soils have a high drawbar pull and hence are difficult to cultivate.

hematite Fe_2O_3. A red iron oxide mineral that contributes to deep red colors in many soils; typical of soils in hot climates.

hemic material Organic soil material with an intermediate degree of decomposition that contains, after rubbing, one-sixth to three-quarters recognizable fibers of undecomposed plant remains.

Henderson-Hasselbalch equation Arrangement of the chemical expression that describes the dissociation of a weak acid to show that the degree of dissociation depends on the system pH and the pKa of the acid.

Henry's law constant A distribution coefficient relating the concentration of a molecule in the gas phase (measured in atmospheres) to its dissolved concentration in an adjacent liquid phase (measured in moles per liter) at a given temperature.

heterotroph An organism that cannot convert inorganic carbon to organic carbon and uses organic sources of carbon for its own growth.

histic epipedon An organic soil horizon at or near the surface that is saturated with water at some period of the year unless artificially drained. Typical of wet, very poorly drained soils in which accumulated organic materials do not completely decompose.

Histosols One of 12 major taxonomic orders in Soil Taxonomy, that has organic soil materials in more than half of the upper 80 cm, or that are of any thickness if overlying rock or fragmental materials that have interstices filled with organic soil materials. They are composed of mucks and peats with high concentrations of organic materials. Most Histosols occur on wet and/or cold sites.

Holocene epoch The current period of geologic time, extending from 10,000 radiocarbon yr BP (~11,700 cal years) to the present.

honeycomb frost Ice in porous soils in insufficient quantity to be continuous, thus giving the soil an open, porous structure and permitting the ready entrance of liquid water.

horizon A genetic soil layer approximately parallel to the land surface and differing from adjacent, genetically related layers in physical, chemical, and biological properties or characteristics, such as color, structure, texture, consistency, kinds and numbers of organisms present, and/or degree of acidity or alkalinity. It is assumed that these characteristics have been produced by soil-forming processes.

horizonation Processes that lead to profile complexity and/or horizonation. See *hapliodization*.

horizon-weighted A type of soil data in which the value for a soil horizon is multiplied by the horizon thickness; it is typically used to quantify pedogenesis.

hornblende A rock-forming ferromagnesian silicate mineral of the amphibole group.

hornblende etching The use of the etched or serrated edges that develop on hornblende, due to weathering, as a relative dating tool.

hue A measure of the chromatic composition, or wavelength, of light that reaches the eye. One of the three variables of color. In lay terms, the "color" of something.

humic acid The ill-defined dark brown to black organic material, composed of proteins and lignins, that can be extracted from soil by an alkaline solution (pH ~11) and that is subsequently precipitated by acidification to pH 1 to 2.

humic substances Relatively high-molecular-weight yellow-to-black colored organic substances formed by secondary synthesis reactions in soils; includes humic acid, fulvic acid, and humin.

Humic Gley soils Soil of the intrazonal order and hydromorphic suborder (1938 system of soil classification) that includes Wisenboden and related soils, such as Half-Bog soils, which have

a thin mucky or peaty O horizon above an A horizon. They form in wet meadows and forested swamps. Obsolete in the U.S. system of Soil Taxonomy.

humification The process whereby the carbon of organic residues is transformed and converted to humic substances through biochemical and abiotic processes, i.e., the decomposition of organic matter leading to the formation of humus. Also called *maturation*.

humin The fraction of the soil organic matter that cannot be extracted from soil with an alkaline solution (pH ~11).

humus Organic compounds in soil exclusive of undecayed plant and animal tissues, their "partial decomposition" products, and the soil biomass. The well-decomposed, relatively stable part of the organic matter found in soils, whose principal constituents are derivatives of lignin, protein, and cellulose. Humus, which is synonymous with "soil organic matter," has a high CEC.

hydration (i) The process whereby a substance takes up water. (ii) A form of chemical weathering involving the reaction of water molecules with a mineral, sometimes leading to structural instability and decomposition.

hydraulic conductivity A proportionality constant that relates to the flux of liquid water as it moves through a soil in response to a potential gradient.

hydraulic gradient (soil water) A vector (macroscopic) function that is equal to the decrease in the hydraulic head per unit distance through the soil, in the direction of the greatest rate of decrease. In isotropic soils, this will be in the direction of the water flux. In groundwater, this is the slope of the water table, measured by the difference in elevation between two points on the slope of the water table and the distance of flow between them.

hydraulic head, hydraulic potential, hydraulic pressure The sum of gravitational, hydrostatic, and matric water potential, usually expressed in units of length. The level to which groundwater in the zone of saturation will rise above a selected datum.

hydric soil A soil that is wet long enough to produce anaerobic conditions periodically, thereby influencing the growth of plants.

hydroconsolidation A model of fragipan formation that relies heavily on the formation of close packing in fragipans by self-weight collapse of preconditioned sediment, also known as physical ripening.

hydrogen bond An intermolecular chemical bond between a hydrogen atom of one molecule and a highly electronegative atom (e.g. O, N) of another molecule.

hydrolysis Literally, the dissociation of a proton from a water molecule; more generally, the breakdown of bonds in a mineral or compound by reaction with water.

Hydromorphic soils A suborder of intrazonal soils (1938 system of soil classification), consisting of seven great soil groups, all formed under conditions of poor drainage, including Bog, GroundWater Laterites, Ground Water Podzols, Half Bog, Humic Gley, Planosol, and Weisenboden soils. Obsolete in the U.S. system of Soil Taxonomy.

hydropedology The integration of classical pedology with soil physics and hydrology, for the purpose of studying the pathways, fluxes, storages, residence times, and spatiotemporal organization of water in the soil and subsoil.

hydrophilic Molecules or ions (or portions thereof) that have a strong polar character and thus attract water molecules; as a result they are normally soluble in water.

hydrophobic Molecules or ions (or portions thereof) that have little to no polar character and thus repel polar water molecules. As a result, they are normally not soluble or miscible in water.

hydrophobic soils Soils that are water repellent, often because of dense fungal mycelial mats or hydrophobic substances vaporized and reprecipitated during fire.

hydrosequence A sequence of related soils that differ from one another, primarily with regard to drainage class. One kind of *catena*.

hydrous mica See *illite*.

hydroxy-aluminum interlayers Polymers of Al hydroxide that are adsorbed at interlayer cation exchange sites of 2:1 clay minerals. Although not exchangeable by unbuffered salt solutions, they are responsible for a considerable portion of the titratable acidity (and pH-dependent charge) in some soils.

hydroxy-interlayered vermiculite (HIV) A vermiculite clay mineral with interlayer positions partially filled with an aluminum oxyhydroxide precipitate. Both the interlayer material and the octahedral sheet of the vermiculite layer are commonly dioctahedral. HIV is common in the coarse clay fraction of acid surface soil horizons. Obsolete terms are chlorite-vermiculite intergrade and vermiculite-chlorite intergrade.

hygroscopic water An obsolete term for the water adsorbed by a dry soil sample from an atmosphere of high relative humidity or for water remaining in a soil sample after "air drying."

hyperthermic See *soil temperature regime.*

hyphae Filamentous structures, common to fungi.

hypolith Organism that colonizes the undersides of translucent stones, usually in contact with the soil.

hysteresis effect The dependence of energy states on the path taken to reach those states. In the context of the relationship between soil water content and soil water matric potential, hysteresis means not only that the soil water content is a function of matric potential but also whether that potential has been reached by wetting the sample or by drying the sample.

ice segregation Ice formed by the migration of water, within pores, to the freezing plane, where it forms into discrete lenses, layers, seams, or masses.

ice wedge A wedge- or V-shaped mass of ice that develops from the soil surface downward, forcing the soil material beside it to become contorted and mixed.

ice wedge cast The morphological expression of an ice wedge after the ice has melted and sediment has filled in the wedge volume, partially preserving the wedge shape.

igneous rock Rock formed from the cooling and solidification of magma or lava, generally crystalline in nature.

illite A clay-size dioctahedral mica of detrital or authigenic origin.

illuvial horizon A subsurface soil horizon into which material from an overlying layer, i.e., an eluvial horizon, has been transported, either in solution (followed by precipitation) or in suspension (followed by coagulation).

illuvial materials Materials that have been transported into a horizon, usually in association with percolating water.

illuviation The translocation of material from one horizon and its deposition in another horizon of the same soil. Although the translocation usually takes place vertically, i.e., from an upper horizon to a lower horizon, movement may also occur laterally.

illuviation cutan Coating of illuvial material, often clay or humus or both, on the surfaces of peds and mineral grains, as well as lining pores.

imogolite An aluminosilicate mineral with an ideal composition $Al_2SiO_3(OH)_4$ that is mostly found in soils derived from volcanic ash and in weathered pumices. Imogolite is commonly found in association with allophane and is chemically similar to it. Imogolite consists of tubular threads, 10–30 μm in diameter.

imogolite-type material (ITM) Material that resembles imogolite or allophane, common in soils affected by podzolization,.

impacturbation Soil and sediment mixing that occurs as large objects, e.g., comets, asteroids, bombs, impact the surface and explode. Also see *bombturbation.*

Inceptisols A mineral soil order encompassing soils with weak pedogenic development that is typically expressed as a cambic Bw or Bg horizon, but no other diagnostic subsurface horizons that might be indicative of stronger development. Water is available to plants more than half of the year or for more than 90 consecutive days during the growing season.

inclusion Polypedon of soil that is different in some way from the dominant soil of a soil survey map unit. Inclusions are normally expected to occupy <25% of a mapping unit. Also see *minor component.*

index minerals In quantitative pedology or mineralogy studies, a mineral that is resistant to weathering and usually difficult to translocate within the soil.

indurated Very strongly cemented.

infauna See *soil infauna.*

infiltration The entry of water *into* a porous medium, namely, soil. Downward movement of water that has entered the soil is called *percolation.*

infiltration capacity The maximum rate at which water can infiltrate into a soil over a given period and under a given set of conditions.

inner layer In the context of the double layer of charges associated with the surfaces of suspended colloids, the *inner layer* refers to ions immediately adjacent to the surface of a charged colloid. Also see *outer layer.*

inner-sphere complex The chemical complex formed when a ligand penetrates the hydration sphere of a metal ion and coordinates directly with a charged adsorbate, with no intervening water molecule (contrast with outer-sphere adsorption).

inorganic In general terms, compounds that either do not contain carbon or are not derived from living organisms – but there are numerous exceptions.

inosilicates Single-chain silicate minerals.

insolation weathering A form of physical weathering driven by heat from the sun, coupled with cool-down cycles, that causes rocks to expand and contract differentially and ultimately to break into smaller pieces.

interfingering The condition, usually occurring in a glossic or degraded Bt or Btx horizon, in which thin, narrow tongues of eluvial material penetrate into the B horizon below. See also *tongueing.*

interflow Soil water that flows laterally through the upper soil horizons, usually because of a horizon or layer below that is less permeable. Also called *throughflow.*

interglacial The climatically warm period occurring between two glacial periods or advances. Longer than an *interstadial.* The most recent interglacial period is called the Sangamonian in North America and the Eemian in Europe.

intergrade A taxonomic term employed at the subgroup level of Soil Taxonomy. Intergrade classes have properties typical of the great group they belong to, but they also have properties that are transitional to other orders, suborders, or great groups. See also *extragrade*.

interlayer In phyllosilicate minerals, *interlayer* refers to the space between structural layers of the minerals, where hydrated cations, organic molecules, and oxyhydroxide precipitates may occur.

intermediate position In Vertisols, a subsurface feature that is transitional between the chimney and bowl, in which master slickensides have maximum angles of 20° to 40°. Intermediate positions are under the microslope surface feature and immediately surrounded by the bowl subsurface feature.

Intrazonal soils One of the three orders in the 1938 system of soil classification. Intrazonal soils have more or less well-developed soil characteristics, which reflect the dominating influence of some *local factor* of relief, parent material, or age, over the normal and more *regional* effects of climate and vegetation. Obsolete in the U.S. system of Soil Taxonomy.

intrinsic From within the system; internal.

intrinsic threshold In pedogenesis, the point at which the limit of soil morphologic or developmental stability becomes exceeded by some type of intrinsic (internal and developmental) change in the soil.

intrusive igneous rock An igneous rock that solidified from magma below the surface, characterized by large mineral crystals. Compare with *extrusive igneous rock*.

ionic radius The effective distance from the center of an ion to the edge of its electron cloud.

ionic strength An index of the concentration of all ion charges present in a solution.

ionic substitution The "replacement" of one or more ions in a crystal structure by others of similar size and electrical charge. Example: Fe^{2+} and Mg^{2+} occur in comparable structural positions in many minerals.

iron bacteria Bacteria that derive the energy they need to live and multiply by oxidizing dissolved ferrous iron.

iron concentrations See *redox concentrations*.

iron depletions See *redox depletions*.

iron oxides Group name for the oxides and hydroxides of iron, including the minerals goethite, hematite, lepidocrocite, ferrihydrite, maghemite, and magnetite.

iron pan A hardpan layer within a soil in which iron oxide is the principal cementing agent. See also *plinthite*.

ironstone Any in situ concentration of iron oxides that is at least weakly cemented.

isochronous A body of rock or a geomorphic surface that is all of the same age.

isofrigid See *soil temperature regime*.

isohyperthermic See *soil temperature regime*.

isolated paleosol A soil buried so deeply that it is essentially no longer influenced by surficial pedogenic processes.

isomesic See *soil temperature regime*.

isothermic See *soil temperature regime*.

isomorphous substitution The replacement of one atom by another of similar size in a crystal structure without disrupting or significantly changing the structure. When a substituting cation is of a smaller valence than the cation it is replacing, a net negative charge is realized in the mineral.

isotope Atom that differs from another in atomic mass number, but not in atomic number. For example, oxygen (atomic number 8) may have an atomic mass number of 16, 17, or 18, depending on whether it has 8, 9, or 10 neutrons. It therefore has three isotopes.

isotropic Homogeneous composition of a volume of material.

isovolumetric weathering Weathering in which the rock loses material but does not collapse, i.e., it retains its original volume. Isovolumetric weathering can occur when rock weathers to saprolite.

jarosite $KFe_3(OH)_6(SO_4)_2$. A pale yellow mineral found in acid sulfate soils.

joint The surface of a fracture in a rock, showing no displacement parallel to or along the fracture.

K-cycle Butler's theoretical construct that geomorphic surfaces undergo alternating periods of instability and stability, the former being periods of either aggradation or erosion.

kame A small hill composed of ice-contact stratified sand and gravel, deposited by glacial meltwater, but in contact with the ice, e.g. in a crevasse or as a fan into a proglacial lake.

kandic horizon Subsoil diagnostic horizon having more clay than some overlying horizons and dominated by low-activity clays, i.e., clay minerals with CECs $\leq$ 16 cmol(c) kg^{-1} clay, determined by a 1 M NH_4 acetate extract at pH 7. Common in Ultisols.

kaolinite A phyllosilicate clay mineral of the kaolin subgroup, with a 1:1 layer structure composed of shared sheets of Si-O tetrahedra and Al-(O,OH) octahedral, and with very little isomorphous substitution.

karst Topography or landscape with sinkholes, caves, and underground drainage that is formed by dissolution of limestone, gypsum, or other soluble rocks. Drainage in karst terrain is usually by underground streams.

K_d A coefficient that quantitatively describes a linear distribution of an adsorbate between solid and liquid phases, usually at low concentrations of the adsorbate.

kettle An irregularly shaped surface depression formed by the melting of a block of glacial ice that was once buried or partially buried by drift. If filled with water: *kettle lake*.

krotovina A filled faunal burrow or tunnel in a soil. Also spelled *crotovina*.

L horizon In the Canadian soil classification system, an organic horizon composed of litter residues. In Soil Taxonomy, the letter L is used to connote limnic soil materials. Old terminology for the Oi (litter) layer.

lacustrine deposit The clastic sediments and chemical precipitates deposited in a lake; usually thinly stratified and often fine-textured. If a glacial lake, then *glaciolacustrine sediment*.

lag concentrate (concentration) A layer or veneer of rocks and gravel that forms on a soil surface as finer materials are removed by erosion.

lamella (pl. **lamellae**) Thin, usually <3 mm in thickness, contorted and wavy illuvial clay bands or layers in sandy soils.

land capability class One of eight classes of land established by the U.S. Department of Agriculture's Natural Resources Conservation Service, distinguished according to their capability to produce common cultivated crops and pasture plants without deteriorating over a long period.

landform Any physical, recognizable form or feature on the Earth's surface, having a characteristic shape and produced by natural causes.

landscape position A particular location on a landscape, e.g., the summit or shoulder of a ridge, a ridge nose, a side slope, a backslope or toeslope, a cove or a drainageway.

lapilli Nonvesicular or slightly vesicular pyroclastics, 2 to 76 mm in at least one dimension, with an apparent specific gravity of ≥2.0.

laterite Red or mottled red/tan, Fe-rich soil material that has hardened and become cemented upon drying. Most laterite was a form of plinthite prior to its drying and hardening. Also called *peotroplinthite* or *ironstone*.

Lateritic soils A suborder of zonal soils (1938 system of soil classification) formed in warm-hot, humid regions and usually including the following great soil groups: Yellow Podzolic, Red Podzolic, Yellowish-Brown Lateritic, Reddish-Brown Lateritic, and Laterite. Obsolete in the U.S. system of Soil Taxonomy.

laterization A bundle of soil forming processes, common in humid tropical regions, whereby Fe is released by weathering from primary minerals and precipitates as secondary Fe oxides, leading to Fe concentrations at preferred sites and potentially to development of plinthite. Laterization places a premium on Fe mobility and is the suite of processes whereby soils develop into Latosols.

Latosols A suborder of zonal soils (1938 system of soil classification) including red soils formed under forested, tropical, humid conditions and characterized by low silica:sesquioxide ratios in the clay fractions, low base exchange capacity, low-activity clays, low contents of most primary minerals, and high degrees of aggregate stability. Roughly synonymous with Oxisols. Obsolete in the U.S. system of Soil Taxonomy.

latosolization The bundle of pedogenic processes that lead to the formation of oxic horizons, or soils formerly known as Latosols. In latosolization, residual sesquioxides accrue, as bases and silica are leached from the profile, under long-term weathering under a hot, humid climate. In-migration or translocation of Fe is minimal. Also see *laterization*.

layer In phyllosilicate mineral terminology, a combination of sheets in a 1:1 or 2:1 assemblage.

layer charge Magnitude of charge per formula unit of a mineral, which is balanced by ions of opposite charge external to the unit layer.

layer silicates See *phyllosilicates*.

leachate Liquid that has percolated through and out of a soil and therefore contains substances in solution and/or suspension.

leaching The removal of soluble materials from soil or other material by percolating water.

lee The side of a hill or feature that is sheltered from the prevailing winds; opposite of windward. Also called *leeward*.

lepidocrocite γ-FeOOH. An orange-colored iron oxide mineral that is usually found in mottles and concretions in wet soils.

lessivage The migration of clay particles suspended in soil water, from A and/or E horizons to B horizons, producing Bt horizons enriched in clay.

leucinization The lightening of a soil horizon, usually due to translocation of clay, Fe, and organic matter from that horizon, leaving behind mostly uncoated sand and silt grains, which are often predominantly light-colored.

lichen An organism formed by the symbiotic association of an alga and a fungus, functioning as a single organism.

lichenometry A relative dating technique that estimates the age of an exposed surface on the basis of the size of lichen thalli (bodies) that have developed on rocks that are exposed on that surface. Usually applied in alpine or cold regions.

ligand That portion of a solid surface, molecule, or ion that, because it is enriched in negative charge by ionization or by possessing one or more pairs of unshared electrons, coordinates with electron-deficient metal ions in a complex by a chemical bond.

light soil A colloquial term for a coarse-textured soil that has a low drawbar pull and hence is easy to cultivate. Contrast to *heavy soil*.

lignin The component of wood or other plant matter responsible for its rigidity.

lime, agricultural A soil amendment containing $CaCO_3$, $MgCO_3$, or other materials, used to neutralize soil acidity and furnish Ca and Mg for plant growth. Usually produced as finely ground limestone or dolostone.

limestone A sedimentary rock composed mostly of the mineral calcite, $CaCO_3$.

limnic material A common mineral component in some organic soils that was either deposited in water by precipitation or through the action of aquatic organisms or derived from underwater and floating aquatic plants and aquatic animals. A form of lacustrine sediment, but with a biogenic origin.

lipids Fats, oils, and related fatty compounds that are insoluble in polar liquids like water.

liquefaction The change of a solid substance into something that resembles a liquid, usually by the addition of large amounts of water.

liquid (soil) Usually water, but may include nonaqueous fluids such as gasoline in contaminated soils.

lithic contact The boundary between soil and continuous, coherent underlying material. The underlying material must be sufficiently coherent to make hand digging with a spade impractical.

lithification The process by which an unconsolidated deposit of sediments is converted into solid rock, involving compaction, cementation, and/or recrystallization.

lithobiont An organism that occupies a rock surface, e.g., a lichen or a moss.

lithofunction Quantitative expression of the differences that occur in soils as a function of changes in parent material, assuming that the remaining soil-forming factors are held constant, or nearly so. See also *lithosequence*.

lithologic discontinuity The contact, usually manifested as a horizontal or wavy surface within a soil, between two genetically dissimilar sediments. Soil horizon designations in the material below the discontinuity are preceded by the numeral 2 to indicate the second parent material in the vertical sequence. In the uppermost parent material, the numeral 1 is omitted.

lithorelict A micromorphological feature derived from the parent rock that can be recognized by its rock structure and fabric. It is usually surrounded by saprolite or soil materials.

lithosequence A group of related soils that differ from one another primarily as a result of differences in the parent material as a soil-forming factor. When expressed as a mathematical equation, it is referred to as a *lithofunction*.

Lithosols A great soil group of azonal soils (1938 system of soil classification) characterized by an incomplete solum or no clearly expressed soil morphology and consisting largely of freshly and weakly weathered rock fragments. Lithosols are often shallow to a lithic contact, i.e., bedrock. Obsolete in the U.S. system of Soil Taxonomy.

litter Organic materials at the surface of a soil that are not in an advanced stage of decomposition, usually consisting of freshly fallen leaves, needles, twigs, stems, bark, and fruit.

litter mat Fibric materials at the soil surface constituting an Oi horizon.

littering Addition of fresh litter and organic materials to the soil surface.

loamy A soil textural class modifier; also used as a group of textural classes at the family level of Soil Taxonomy. If a class modifier, loamy refers generally to textures that have at least some sand, silt, and clay, but are not dominated by any one of these, e.g., loam, sandy loam, silt loam, clay loam, sandy clay loam, and silty clay loam, among others. If a family particle-size class, loamy refers to soils with textures finer than very fine sandy loam but with <35% clay and <35% rock fragments.

loess Material transported and deposited by wind and consisting of predominantly silt-sized particles. It is usually yellowish brown and composed mainly of silicate minerals. It may also contain $CaCO_3$.

loose See *consistence*.

low-activity clay A clayey material, typical of oxic and kandic horizons and highly weathered soils in general, with an apparent CEC ≤16 cmol(-) kg^{-1} clay and ECEC ≤12 cmol(-) kg^{-1} clay.

luminescence dating A family of numerical dating techniques that is used to establish the point in time at which a substance was last exposed to light (and then buried) by measuring trapped electrons in crystal defects in certain minerals. Can be used to date loess, eolian sand, some types of alluvium and coastal sediments, fired clay, pottery, brick, and burned stones.

Luvisols A taxonomic order of soils that has strong evidence of lessivage, i.e., the translocation of clay, and the resulting presence of a clay-enriched Bt horizon. Luvisols are not an order in Soil Taxonomy but do occur in the classification system of Canada and several other countries. They are similar to Alfisols.

lysimeter A device for measuring percolation and leaching losses from a column of soil.

M horizon In some systems of tropical soil nomenclature, the M horizon is the uppermost mineral soil layer, which has usually been worked extensively by soil infauna such as termites. Roughly equivalent to the A horizon.

macrofauna Animals that fall within the arbitrary size range 2–20 mm body width. Larger than microfauna but smaller than megafauna.

macronutrient A plant nutrient found at relatively high concentrations (>500 mg kg^{-1}) in plants. Usually refers to N, P, and K but may include Ca, Mg, Fe, and S. Also called macroelement.

macropore A comparatively large, e.g., >75 μm diameter, pore in a soil. Contrast to *micropore*.

macropore flow The tendency for water that is rapidly applied to the soil surface to move into the profile mainly via saturated flow through macropores, thereby bypassing micropores and rapidly transporting any solutes to the lower soil horizons. Also called preferential flow and bypass flow.

mafic (i) Dark-colored igneous rocks with significant amounts of ferromagnesian minerals. (ii) A magma with significant amounts of Fe and Mg, but containing little quartz, feldspar, or muscovite mica.

maghemite γ-Fe_2O_3. A dark reddish brown magnetic iron oxide mineral chemically similar to hematite, but structurally similar to magnetite. Often found in well-drained, highly weathered soils of tropical regions.

magma Molten rock beneath the Earth's surface. Compare to *lava*.

magnetic declination Angle of divergence between true north and magnetic north, as measured in degrees east or west of true, or geographic, north.

magnetic pole The point on the Earth's surface to which compass needles point. Near, but not equivalent to, the geographic pole.

magnetic susceptibility The degree to which a sediment is affected by a magnetic field. Soils with higher amounts of magnetic susceptibility generally have more magnetic and ferromagnesian minerals.

magnetite Fe_3O_4. A black magnetic iron oxide mineral usually inherited from igneous rocks. Often found in soils as black magnetic sand grains.

magnetostratigraphy Use of the remnant magnetism in rocks to determine the general age of a rock body or sediment by correlation with the record of changes in the Earth's magnetic field during past geologic ages.

mangan A cutan composed of manganese oxide or hydroxide, usually identifiable because it will effervesce upon application of H_2O_2.

manganese concretions Small concentrations of MnO_2 or other Mn oxides in soils.

manganese oxides Oxides of manganese, e.g., birnessite and lithiophorite, typically black and frequently occurring in soils as nodules and coatings on ped faces, and usually in association with iron oxides.

map unit, mapping unit A polygon on a soil map, defined to characterize and delineate soil bodies. Limitations of map scale in relation to landscape complexity may mean that a map unit represents several kinds of soil, not all of which can be specified in the legend.

map unit complex See *complex*.

marble A metamorphic rock composed largely of calcite; the metamorphic equivalent of limestone.

marine isotope stage (MIS) Period in the Earth's past in which ^{18}O was either enriched or depleted from oceanic waters, as a result of cold-warm climate cycles and their effects on the formation of ice sheets. Also called *oxygen isotope stage*.

marl Soft and unconsolidated, $CaCO_3$-rich or lime-rich mud, usually mixed with varying amounts of clay or other impurities. Commonly forms as lacustrine sediment in alkaline, postglacial lake beds and is today found underlying some peat bogs.

mass balance analysis Procedure in which pedogenic changes in the masses of elements or minerals are calculated either on a horizon or on a profile basis.

massive A type of soil structure in which the soil breaks along no preferred planes, i.e., it has no pattern of repeating structural elements. Apedal.

mass movement The movement of material downslope under the influence of gravity, and not in flowing water, although water may lubricate and assist in the movement.

matric potential The potential energy imparted to soil water because of its attraction to solid surfaces.

matrix In soils, the fine material (generally <2 mm diameter), forming a continuous phase and enclosing coarser material and/or pores.

maturation See *humification*.

mature soil An obsolete term for a soil with well-developed soil horizons produced by processes of soil formation and essentially "in equilibrium" with its present climate and vegetation.

maximum-limiting date A numerical date on a surface or sediment that implies that the surface or sediment can be no older than that date.

mean residence time (MRT) The average period that carbon atoms spend in a reservoir such as soil organic matter. This term is usually estimated as the radiocarbon age determined for soil organic matter, which by definition is a composite of organic materials of various ages.

Mediterranean climate Climate typical of the west sides of continents at about 30° to 40° latitude, where summers are dry and hot, while winters are cool and rainy, and in which the soil moisture regime is usually xeric.

medium-textured Textural group consisting of very fine sandy loams, loams, silt loams, and silt textures.

melanization The darkening of a soil horizon, usually by additions of humus.

mesic *See soil temperature regime.*

metallophore Compound released by microorganisms or plants that forms strong, soluble complexes with essential micronutrient metals in the soil solution so that the metals can be taken up by the organisms.

metamorphic rock Rock derived from preexisting rocks that have been altered physically, chemically, and/or mineralogically as a result of natural geological processes, principally heat and pressure, deep within the Earth. During metamorphosis, rocks are altered but not completely melted.

metamorphism The processes of recrystallization and mineralogical change that take place in solid rock, under conditions of high temperature and pressure, usually well below the Earth's surface.

mica A group of layered 2:1 aluminosilicate minerals characterized by nonexpandability and high layer charge, which is usually satisfied by K. Common micas are muscovite and biotite.

microbial biomass The total mass of living microorganisms in a given volume or mass of soil.

microclimate Climate of a small area, e.g., a hillslope or a watershed, which is usually different from the regional climate (macroclimate), because it is modified by local differences in elevation, aspect, or other local phenomena.

microcrystalline In soil micromorphology, particles in which discrete crystals can only be seen with the aid of a microscope.

microdepression See *microlow.*

microfauna Protozoa, nematodes, arthropods, and other soil fauna of microscopic size.

microflora Organisms so small they can only be seen with a microscope, including bacteria, actinomycetes, fungi, algae, and viruses.

microhigh In Vertisols, a microrelief term applied to slightly elevated areas, with changes in relief from several centimeters to several meters. Microhighs have cross-sectional profiles that can be simple or complex and generally consist of gently rounded, convex tops with gently sloping sides.

microknoll Obsolete term; see *microhigh.*

microlow In Vertisols, a generic microrelief term applied to slightly lower areas relative to the adjacent ground surface, e.g., shallow depressions, with changes in relief from several centimeters to several meters. Microlows have cross-sectional profiles that can be simple or complex and generally consist of subdued, concave, open or closed depressions with gently sloping sides.

micronutrient A plant nutrient found in relatively small amounts (<100 mg kg^{-1}) in plants. Sometimes called trace elements or trace nutrients, including B, Cl, Cu, Fe, Mn, Mo, Ni, Co, and Zn. Also called *microelement* or *trace element.*

micropore Any pore smaller than about 30 μm in diameter. Contrast with *macropore.*

microrelief, microtopography The local, slight irregularities in surface form that are superimposed upon a larger landform, including features such as low mounds, swales, and shallow pits, e.g., gilgai, coppice dunes, and tree-tip mounds and pits. Generally, they are too small or intricate to delineate on a large-scale topographic or soil map.

microslope In Vertisols, a generic microfeature term that applies to a gently sloping area, including the lower part of a microhigh, that transitions between the slightly elevated microhigh and the outer edge on an adjacent microlow. Where present, it can make up a majority of the ground surface in areas of gilgai microtopography.

midden A pile of domestic refuse consisting of waste food, dung, and animal bones, i.e., anything that the local inhabitants wanted to dispose of.

Milankovitch cycles Earth's orbital cycles, first studied by Milutin Milankovitch, that have influenced global climate, especially the glacial cycles of the Pleistocene.

Mima mound Generally circular mounds, 1–4 m in diameter and named for the Mima Prairie in Washington State. They are common in the western United States. Although many theories have been postulated for the origins of these mounds, they are most likely due to point-centered bioturbation by rodents.

mineral A naturally occurring homogeneous solid with a definite chemical composition and an ordered atomic arrangement or crystalline structure. Minerals can be categorized as *primary* or *secondary.*

mineral soil A soil consisting predominantly of mineral materials of various sizes, e.g., clay, silt, sand, and gravel. Contrast with *organic soil.*

mineralization The conversion of a substance from an organic form to an inorganic state as a result of microbial activity.

mineralogically mature Having lost all or most of its weatherable minerals as a result of long-term weathering or corrasion.

minimum-limiting date A numerical date on a surface or sediment that implies that the surface or sediment can be no younger than that date.

minor component Term for a soil map "impurity," or another type of unnamed, different soil in an area delineated and labeled as a certain map unit. Also known as *inclusions.*

mites Very small arachnids that live in many O horizons and A horizons.

mixmasters A group of soil fauna whose bioturbation activities are best expressed by the subsurface mixing that they produce, usually with minimal surface expression, e.g., many species of earthworms.

moder See *duff mull.*

moisture characteristic function The mathematical relationship between soil water potential and water content in a soil sample.

moisture content The mass of water lost per unit of air-dry soil when it is dried at ~105°C for >8 h. When expressed as a percentage, moisture content is (water mass/dry soil mass) x 100.

mollic epipedon A dark-colored surface (A) horizon of mineral soil that is usually ≥25 cm thick, contains at least 6 g kg^{-1} organic carbon, is not massive, and hard or very hard when dry, and has a base cation saturation of ≥50% and <1,500 mg P_2O_5 kg^{-1}. Typical of soils formed under grassland vegetation.

Mollisols One of 12 major taxonomic orders in Soil Taxonomy. Mollisols have a mollic epipedon overlying a subsoils with a base saturation ≥50%. Typical of grassland environments.

monogenetic soil A soil that has formed under only one set of soil-forming factors, i.e., in its history there is little evidence of major changes in climate, vegetation, and so on. Contrast with *polygenetic soil*.

montmorillonite $Ca_{0.25}Si_4Al_{1.5}Mg_{0.5}O_{10}(OH)_2$. A 2:1 aluminosilicate mineral composed of two silica tetrahedral sheets and a shared Al and Mg octahedral sheet. Montmorillonite has a permanent negative charge that attracts interlayer cations that exist in various degrees of hydration. The interlayer cations such as Ca are readily exchangeable with other cations. Montmorillonite is one kind of *smectite*.

mor A type of acidic forest humus characterized by an accumulation or organic matter on the soil surface in matted Oe (F) horizons, reflecting the dominant mycogenous decomposers, usually with an abrupt lower boundary to the underlying mineral soil.

moraine A hummocky, ridgelike, glacial landform that forms when active ice moves glacial sediment forward and deposits it at the margin. Moraines that form beneath the ice are called ground moraines.

mottled zone A layer in a soil that is marked with mottles, i.e., reddish and grayish splotches, usually caused by redoximorphic (oxidation-reduction) processes, under alternating wet and dry conditions.

mottles Spots or blotches of different color or shades of color interspersed with the dominant (matrix) color.

moundmakers A group of soil fauna whose bioturbation activities are expressed on the surface by the mounds they make, e.g., ants and termites.

MRT See *mean residence time*.

muck Highly decomposed organic materials in soil, usually designated as an Oa horizon.

muck soil An organic soil (Histosol) in which the plant residues have been altered beyond recognition, i.e., they are predominantly sapric (Oa) materials. Contrast with *peat soil*.

mucky peat In organic soils and soil materials, an intergrade between muck and peat, but more like the latter.

mudstone A fine-grained detrital sedimentary rock made up of clay- and silt-sized particles, not unlike shale.

mull A forest humus type characterized by intimate incorporation of organic matter into the upper mineral soil, i.e., a well-developed A horizon, in contrast to accumulation on/above the surface. The crumbly nature of mull humus is often due to the actions of earthworms.

Munsell color system The primary system used to describe soil color. It specifies the relative degrees of the three simple variables of color: hue, value, and chroma. For example, 10 YR 6/4 is a color (of soil) with a hue = 10 YR, value = 6, and chroma = 4.

muscovite A clear, dioctahedral layer silicate of the mica group, sometimes called white mica, that has Al^{3+} in the octahedral sheet and Si and Al in a ratio of 3:1 in the tetrahedral sheet.

mutualist Organism that forms a symbiotic relationship with another organism – a relationship that is mutually beneficial, e.g., the association between cyanobacteria and fungi to create lichen.

mycelium (pl. **mycelia**) A mass of interwoven filamentous hyphae, such as that of the vegetative portion of the thallus of a fungus.

mycorrhiza (pl. **mycorrhizae**) Literally meaning "fungus root," mycorrhizae are an association, usually symbiotic, of specific fungi with the roots of higher plants.

natric horizon A mineral soil horizon that satisfies the requirements of an argillic horizon but that also has prismatic, columnar or blocky structure and at least one subhorizon

having >15% saturation of the cation exchange complex with exchangeable Na^+.

natural erosion Wearing away of the Earth's surface by water, ice, or other natural agents under natural environmental conditions of climate, vegetation, and so forth, undisturbed by humans. Also called *geologic erosion*.

natural soil drainage class The wetness class that is based on the frequency and duration of periods of saturation or partial saturation that existed during the development of the soils, as opposed to human-altered drainage (see Table 14.4).

nematode An elongated, cylindrical, unsegmented worm, common in soils and ranging in size from ~ 0.1 to > 20 mm in length.

nesosilicates Silicate minerals with crystals composed of isolated silica tetrahedra. Also called *orthosilicates*.

neutral soil A soil in which the pH of the surface layer is approximately between 6.5 and 7.3.

neutron A particle in the nucleus of an atom that lacks an electrical charge and has approximately the same mass as a proton.

nitrate reduction (biological) The process whereby nitrate is reduced by plants and microorganisms to ammonium for cell synthesis (nitrate assimilation, assimilatory nitrate reduction) or to nitrite by bacteria using nitrate as the terminal electron acceptor in anaerobic respiration (respiratory nitrate reduction, dissimilatory nitrate reduction).

nitrification Biological oxidation of ammonium to nitrite and nitrate, or a biologically induced increase in the oxidation state of nitrogen.

nitrogen fixation The transformation of elemental nitrogen to an organic form, usable by plants or microorganisms.

nodule (i) A small, irregularly shaped, cemented concentration of one or more minerals, e.g., $CaCO_3$ or Fe_2O_3, that can be removed from the soil intact, that has no orderly internal organization (like a concretion), and that differs in composition from the soil or sediment that surrounds it. (ii) In soil micromorphology, glaebules with an undifferentiated rock and/or soil fabric.

normal polarity A section of geologic time – such as the present – when the compass needle points to the magnetic north pole. See also *reversed polarity*.

normal soil See *zonal soil*.

north magnetic pole The point on the Earth where the north-seeking end of a magnetic needle points. It differs from the north geographic pole.

nose slope The projecting end of a ridgetop or interfluve, on which contour lines form convex curves. Flow of water across nose slopes is divergent.

not-soil A volume of material at the Earth's surface such as open water, glacial ice, flowing hot lava, large mass of salt, bedrock. Unconsolidated sediments qualify as not-soil, provided they have not been influenced by pedogenesis to the extent that they differ measurably from their state at $time_{zero}$.

numerical dating Establishing the actual or near-actual age of a surface or sediment, in calendar years, usually by radiometric means. Also known as chronometric or absolute dating. Compare to *relative dating*.

nutrient An element that is essential for organism growth and development.

O horizon A soil horizon composed predominantly of organic soil materials.

Oa horizon (H layer) An organic soil horizon consisting mainly of well-decomposed organic matter of unrecognizable origin, i.e., sapric material.

obliquity of the Earth's ecliptic Tilt of the Earth's rotational axis in relation to the plane in which the Earth circles the Sun. The obliquity cycles from about 21.5° through 24.5° and back to 21.5°, every 41,000 years.

obsidian hydration dating A relative dating technique that relates the thickness of a weathering (hydration) rind on obsidian (volcanic glass) to the time that the rock has been exposed at the Earth's surface.

ochric epipedon A weakly expressed surface horizon of a mineral soil. It is too light in color, too high in chroma, too low in organic carbon, or too thin to be a plaggen, mollic, umbric, anthropic, or histic epipedon. Most weakly expressed epipedons are ochric.

octahedral sheet Sheet of horizontally linked, octahedral-shaped units that are a basic structural component of phyllosilicate clay minerals. Each octahedral unit consists of a central, six-coordinated metal cation surrounded by six O ions or OH⁻ groups that, in turn, are linked to other neighboring metal cations. Also see *tetrahedral sheet*.

octahedron A basic structural unit of which many silicates are composed, consisting of an ion, e.g., Al, Fe, or Mg, surrounded symmetrically by six oxygen atoms.

Oe horizon (F layer) An organic soil horizon consisting mainly of partially decomposed organic matter with portions of plant structures still recognizable, i.e., hemic material. Formerly referred to as the fermentation layer.

Oi horizon (L layer (litter)) An organic soil horizon consisting mainly of organic matter that has undergone little or no decomposition, i.e., fibric material, e.g., freshly fallen leaves, needles, twigs, stems, bark, and fruits.

oligotrophic Environments in which the concentration of nutrients available for growth is limited, i.e., nutrient-poor habitats.

olivine Common silicate mineral found in rocks formed from mafic magma, varying in chemical composition between Mg_2SiO_4 and Fe_2SiO_4.

one-third-atmosphere percentage Obsolete term referring to the water content of a soil that has been saturated and then subjected to an applied pressure of one-third atmosphere. Approximately the same as one-third-bar percentage.

opal An amorphous variety of hydrous silica that occurs in various colors.

opal phytolith A microscopic body of noncrystalline silica (opal) that is secreted by a plant. Phytoliths sometimes accumulate in soils to such a degree that they can be isolated from paleosols to gather information about paleovegetational composition.

open depression Generic name for any enclosed or low area that has a surface drainage outlet whereby surface water can leave the enclosure. Compare to *closed depression*.

optically stimulated luminescence (OSL) dating A type of luminescence dating in which the equivalent dose is determined by exposing the sample to light

order The highest level taxonomic group in Soil Taxonomy. There are currently 12 soil orders.

organ, organan A cutan composed of organic matter or humus.

organic matter Portion of the soil that includes microflora and microfauna (living and dead) and residual decomposition products of plant and animal tissue. It generally consists primarily of *humus*.

organic soil A soil in which the thickness of organic soil materials is generally greater than the thicknesses of any mineral layers below, i.e., a soil that is composed predominantly of organic matter. In Soil Taxonomy, a Histosol.

organo-metallic complex A chemical association between an organic molecule and a metal, usually Fe or Al. Also called chelate-complex.

ornithogenic soils Soils dominated by bird feces, common in penguin rookeries.

ortstein A cemented spodic horizon, usually indicated as Bsm or Bhsm.

outer layer In the context of the double layer of charges associated with the surfaces of suspended colloids, the *outer layer* refers to diffuse zone of ions attracted to the surface by electrostatic forces but that are not retained at specific points on the colloid surface.

outer-sphere adsorption The electrostatic adsorption of a molecule or ion at a surface, such that at least one water molecule is present between the adsorbate and the adsorbent.

outwash Stratified and sorted detritus (chiefly sand and gravel) removed or "washed out" from a glacier by meltwater streams and deposited in front of the glacier, with the coarser material deposited nearer to the ice front. Also called *glaciofluvial material*.

outwash plain A constructional landform underlain by thick sequences of glacial outwash, sometimes pitted with kettles.

oven-dry soil Soil that has been dried at ~105°C until it reaches a constant mass.

overbank deposits Alluvial sediments deposited from floodwater, onto floodplains.

overburden Material, usually carried from an outside source, that overlies the topmost horizon of a soil.

overland flow Water that flows across the soil surface, but not in channels or laterally in subsurface throughflow.

oxbow lake A crescent-shaped, often ephemeral body of standing water on a floodplain, formed when the stream develops a neck cutoff and the ends of the original river channel silt in. In essence, a cut-off section of a river channel.

oxic horizon A mineral soil horizon characterized by the virtual absence of weatherable primary minerals or 2:1 layer silicate clays, by low CEC, and by small amounts of exchangeable base cations. They contain mainly 1:1 layer silicate clays and highly insoluble minerals such as quartz, or hydrated oxides of iron and aluminum, but little water-dispersible clay.

oxidation (i) Energy-releasing process involving the loss of one or more electrons by an ion or molecule. (ii) A chemical weathering process by which redox-sensitive metals such as iron or manganese in a rock or soil combine with oxygen to form residual oxide minerals. See also *reduction*.

oxidation-reduction reactions A class of chemical reactions in which one element loses and another element gains one or more electrons. The oxidation state increases or decreases, respectively. The element that is oxidized is called the *reductant*, and the element that is reduced is called the *oxidant*. Also called *redoximorphic processes*.

oxidation state The number of electrons to be added (or removed) from an ion to convert it to the elemental (zero valence) form.

oxides Minerals, often clay size, that are composed of cations, oxygen anions, and/or hydroxyl groups. The four main groups of oxides in soils are iron oxides, aluminum oxides, manganese oxides, and silicon oxides. The general term may also refer to true oxides, e.g., Fe_2O_3, hydroxides, e.g., $Al(OH)_3$, or oxyhydroxides, e.g., FeOOH.

Oxisols One of 12 major taxonomic orders in Soil Taxonomy. Oxisols have a highly weathered oxic horizon or continuous plinthite near the surface. Typical of old surfaces and old parent materials in hot, humid climates.

oxyaquic conditions Pertaining to soils that are saturated but not reduced and that do not contain redoximorphic features. Sometimes result from a perched water table.

oxygen isotope stage See *marine isotope stage*.

packing pore A kind of void or pore in soils that exists mainly between peds.

paha A loess-capped hill, usually elongate in plan view, that stands above an erosional surface. Paha are common erosional relicts on the Iowan erosion surface in Iowa, United States.

paleoclimate The climate of a period in the geologic past.

paleomagnetism The Earth's magnetism and magnetic record as it is recorded in rocks and sediments.

paleopedology The study of paleosols and the environments in which they formed.

paleosol (i) A soil that formed on a landscape in the past, often with distinctive morphological features resulting from a soil-forming environment that no longer exists at the site. The former pedogenic pathway was either altered because of external environmental changes, or was interrupted by burial. (ii) A buried soil. See also *relict paleosol* and *exhumed paleosol*. Roughly synonymous with *geosol*.

palimpsest A landscape feature or soil that preserves within itself some property indicative of its past and that has not yet been "written over" and obliterated by more recent processes.

paludification The expansion of a bog or thickening of a surface organic layer, caused by the gradual rising of the water table, as accumulations of peat impede drainage, or by continued accumulations of a thickening litter layer on a soil surface in a cold, nutrient-poor, and wet environment.

palygorskite $Si_8Mg_2Al_2O_{20}(OH)_2(OH_2) \cdot 4H_2O$. A fibrous clay mineral, common in soils of arid regions, composed of two silica tetrahedral sheets and one Al and Mg octahedral sheet that make up the 2:1 layer. Also referred to as *attapulgite*.

palynology The study of pollen as a key to past environments, by examining samples of various ages, usually from acidic, organic, sedimentation areas, such as bogs or lake beds.

pan A soil horizon that is strongly compacted, cemented, or has a high content of clay.

panspot See *slick spot*.

papules In soil micromorphology, glaebules composed dominantly of clay minerals with continuous and/or lamellar fabric and sharp external boundaries.

paralithic contact Similar to a lithic contact except that the underlying material is softer and can be dug with difficulty with a spade.

parent A radioactive element, e.g., ^{14}C, whose decay produces stable daughter isotopes.

parent material The relatively unaltered lower material in soils that is often similar to the material in which the horizons above have formed.

parna Silt and sand-sized aggregates of eolian clay.

partial bleaching A condition encountered in luminescence dating in which not all of the grain have been rezeroed by exposure to light, prior to burial, commonly resulting in an age estimate that is too old.

particle size The effective diameter of a particle.

particle size analysis Determination of the mass of the different soil separates in the fine earth (<2 mm diameter) fraction of a soil sample, usually by sedimentation, sieving, micrometry, laser diffraction, or combinations of these methods.

particle size distribution The fractions of the various soil separates in a soil sample, often expressed as mass percentages.

pathogen An organism that causes disease in its host.

pathway In pedogenesis, the direction that a soil is developing, as toward a particular end point or morphology. Pathways can be progressive (in which horizons are differentiated) or regressive (in which horizons are destroyed).

patterned ground A ground surface, characteristic of, but not confined to, permafrost regions or areas that have been subject to intense cryoturbation, exhibiting a discernibly regular pattern of microtopography, subsurface drainage, vegetation, or all of these.

peat Organic soil material, often formed in thick layers, in which most of the original plant parts are recognizable (fibric material) and only slightly decomposed, because of continuous waterlogging.

peat soil An organic soil (Histosol) that mainly contains fibric soil material in which the plant residues are not fully decomposed. Contrast with *muck soil*.

peaty muck In organic soils and soil materials, an intergrade between muck and peat, but more like the former.

pebbles Rounded or partially rounded rock or mineral fragments between 2 and 75 mm in diameter. Pebbles can be further refined as fine pebbles (2-5 mm), medium pebbles (5-20 mm), and coarse pebbles (20-75 mm).

ped The basic unit of soil structure, a ped is a natural soil aggregate, such as a block, column, granule, plate, or prism (in contrast with a clod, which is formed by human activity or artificially). Volumes of soil that are structureless (massive or single-grain) are called *apedal*.

pedalfer Obsolete term for a soil in which sesquioxides (of Fe and Al, hence the name) have increased relative to silica during soil formation. Typical of formation under humid climates.

pediment A gently sloping, erosional surface developed at the foot of a receding hill or mountain slope. The surface can be bare or thinly mantled with alluvium and colluvium in transport to the adjacent valley. The pediment backslope (the receding slope) is concave upward and rises from the pediment footslope to the upland.

pedimentation Formation of a pediment.

pedisediment A layer of sediment, eroded from the shoulder and backslope of an erosional slope, that lies on and is, or was, being transported across a pediment. Pedisediment is often coarser textured than the material being eroded, because the finer material gets transported farther downslope while the coarser material remains behind.

pedocal Obsolete term for soils in which calcium carbonate has accumulated during soil formation, as is typical in dry climates.

pedodiagenesis Combination of the words *pedogenesis* and *diagenesis* – any suite of processes that affect buried soils.

pedofeature A feature with a pedogenic origin, i.e., discrete fabric units present in soils and distinguishable from adjacent material by differences in grain size, organic matter, crystallinity, chemical composition, or internal fabric. Examples are cutans, fecal pellets, and glaebules (channel in-fillings).

pedogenesis All the natural processes involved in the formation of soils, including progressive and regressive processes.

pedogenic Related in some way to processes of soil genesis.

pedogenic accession A significant morphologic feature that is acquired by a soil through normal pedogenic processes.

pedogenic overprinting Emplacement of pedogenic properties upon a preexisting soil, similar to soil welding but the preexisting soil need not be a paleosol.

pedogenic pathway A set of pedogenic processes that can be either progressive or regressive and that normally lead to a given soil morphology if allowed to function for a sufficient period.

pedology The study of soils as naturally occurring phenomena, taking into account their their properties, origins, distribution, and occurrence on the landscape, as well as their evolution through time.

pedometrics The application of mathematical and statistical methods to the study of the distribution and genesis of soil properties.

pedon The smallest volume that can be called a "soil," a pedon is a three-dimensional body of soil with lateral dimensions large enough (1 to 10 m²) to permit the study of horizon shapes and relationships. The pedon is the primary sampling unit in pedology.

pedoplasmation A bundle of pedogenic processes associated with the transformation of saprolite to soil material with pedogenic fabric.

pedostratigraphic unit In stratigraphy, a body of rock or sediment defined by soil properties, typically associated with a named geosol. It usually consists of one (or more) pedogenic horizon(s) that is overlain by one or more formally defined stratigraphic units.

pedotubule A cemented, tubular structure in soil or sediment formed around roots and other organic or inorganic materials.

pedoturbation Soil mixing, accomplished by one of many vectors, e.g., plants, animals, gravity, impacts from meteoroids, shrink-swell clays, human activities, or frost action.

peraquic An aquic soil moisture regime in which the groundwater is always at or very close to the surface, i.e., with minimal fluctuation.

perched water table The upper surface of a saturated layer of soil material that is separated from an underlying saturated layer by an unsaturated layer. The upper surface is therefore said to be underlain by a *perched zone of saturation*. Contrast with *apparent water table*.

percolation The downward or lateral movement of water through (not INTO – that is *infiltration*) soil.

pergelic See *soil temperature regime*.

peridotite An ultramafic igneous rock, peridotite is the major constituent of the Earth's mantle.

periglacial Processes, conditions, areas, climates, and topographic features that occur near glaciers and ice sheets and are influenced by frost and cold temperatures created by the nearby ice.

permafrost Permanently frozen soil or unconsolidated surficial materials.

permafrost table The upper boundary of the permafrost, coincident with the lower limit of seasonal thaw.

permanent charge The net negative (or positive) charge of clay particles inherent in the crystal structure of the particle, not affected by changes in pH or by ion-exchange reactions.

permanent wilting point The largest water content of a soil at which indicator plants, growing in that soil, wilt and later are unable to recover even if placed in a humid chamber. Often estimated by the soil water content of -1.5 MPa soil matric potential.

permeability See *soil permeability*.

perudic A udic soil moisture regime in which precipitation exceeds evapotranspiration in all months, but the soil is not saturated for long periods.

pervection The pedogenic process whereby silt is translocated, usually as a result of freeze-thaw processes.

petrocalcic horizon A continuous, indurated soil horizon, impenetrable by roots, that is cemented by $CaCO_3$ and/or magnesium carbonate, that cannot be penetrated with a spade or auger when dry, and that consists of dry fragments do not slake in water. Usually designated Bkm or Bkkm, although in older terminology these horizons were designated as K.

petroferric contact A boundary between soil and a continuous layer of indurated soil that contains little or no organic matter and in which iron oxides are an important cement.

petrogypsic horizon A continuous, indurated, massive soil horizon, impenetrable by roots, that is cemented by calcium sulfate (gypsum). It can be chipped with a spade when dry, but the dry fragments do not slake in water.

petroplithite See *laterite*.

pH The negative logarithm of the chemical activity of H_3O^+ ions in solution. At low ionic strength, pH represents the negative logarithm of the *concentration* of H_3O^+ in moles per liter. It is a quantitative expression of the acidity or alkalinity of a solution, along a scale that ranges from 0 to 14: pH 7 is neutral, <7 is acid, and >7 is alkaline.

pH-dependent charge Charges on colloid surfaces that result from the adsorption or desorption of hydronium ions and therefore vary with pH of the solution.

phase See *soil phase*.

phenocryst A relatively large and usually conspicuous crystal in an extrusive igneous rock like granite; distinctly larger than the grains of the rock groundmass.

phyllite A metamorphosed mudstone, more coarse-grained than a slate and less coarse-grained than a schist, and with a silky sheen.

phyllosilicate A silicate mineral in which SiO_4 tetrahedra are linked together in sheets and are condensed with adjacent sheets of octahedral units containing Al, Mg, or other ions.

physical weathering. Break down of a rock into smaller fragments by frost action, salt-crystal growth, absorption of water, and/or other physical processes, but involving no chemical change. Also called *disintegration*.

physisorption The adsorption of molecules or ions at a solid-phase surface by van der Waals forces or by electrostatic forces.

phytolith An inorganic, usually opaline, body derived from replacement of plant cells with silica compounds. See *opal phytolith*.

phytotoxic Detrimental to plant growth.

phytoturbation See *floralturbation*.

piezometer A device used to measure the depth to the water table in an unconfined aquifer. It is usually a perforated tube placed in a soil to a depth below the highest the water table level, so that the water level in the piezometer tube reflects the water table depth in the soil.

pipe Subsurface tunnel or pipelike cavity formed by water moving through soil.

pit-and-mound topography A form of microrelief created by tree uprooting, usually formed over long periods of time in forested regions, including pits that mark the former position of the uprooted tree and adjacent mounds that formed as soil materials slumped off the root plate. Also called *cradle-knoll topography*.

placic horizon A black to dark-reddish mineral soil horizon, usually thin but may range from 1 to 25 mm in thickness, that is commonly cemented with iron and is slowly permeable or impenetrable to water and roots. Many placic horizons are wavy.

plaggen epipedon A man-made surface horizon >50 cm thick that is formed by long-continued manuring and mixing.

plan curvature Curvature of a slope as seen from above or as interpreted from contour lines on a topographic map.

Planosols A great soil group of the intrazonal order and hydromorphic suborder (1938 system of soil classification) consisting of soils with E horizons underlain by B horizons that are so strongly illuviated, cemented, or compacted that they perch water at certain times of the year. Obsolete in the U.S. system of Soil Taxonomy.

plasma Soil materials that are capable of being reorganized and/or concentrated by pedogenic processes, usually as driven by percolating water, including mineral or organic materials of colloidal size, as well as relatively soluble materials.

plasma concentration In soil micromorphology, concentration of any of the fractions of the plasma in various parts of the soil material.

plasma separation In soil micromorphology, features characterized by a significant change in the arrangement of the constituents rather than a change in concentration of some fraction of the plasma, e.g., aligning of plasma aggregates by stress at or near the surface of slickensides.

plastic See *consistence*.

platy A shape of soil structure in which the peds form crude plates that lie more or less horizontal to the soil surface.

playa An ephemerally flooded and usually barren area on a basin floor that is veneered with fine-textured sediment and acts as a temporary or final sink for drainage water. Playas commonly contain salts that accumulate as salty water evaporates from them.

Pleistocene epoch The geologic period following the Pliocene, extending from about 1.6 million years ago to the Holocene (10,000 years BP). Together, the Pleistocene and Holocene Epochs compose the Quaternary Period. During the Pleistocene, ice sheets repeatedly accumulated and melted back, across various parts of the Earth, in response to dramatic climate shifts.

pleochroism The changes in color when some transparent minerals are rotated in plane polarized light, usually under a petrographic microscope. Expressed in terms of the nature and intensity of the color change.

plinthite A weakly cemented iron-rich, humus-poor mixture of clay with other diluents that commonly occurs as dark red, redoximorphic concentrations that form platy, polygonal or reticulate patterns. Plinthite changes irreversibly to ironstone hardpans or to irregular, hard aggregates upon exposure to wetting and drying. Also called *laterite*.

Pliocene The last of the five geologic epochs of the Tertiary Period, extending from the end of the Miocene Epoch (ca. 5 million years ago) to the beginning of the Quaternary Period (and Pleistocene Epoch), about 1.6 million years ago.

plow layer The part of a soil profile that has been mixed by plowing and is designated as an Ap horizon.

plow pan A dense layer that forms immediately under the plow layer because of compression of the plow sole.

pluvial A climatically wet period, usually restricted to areas that are normally climatically dry, and generally thought of as occurring in the geologic past.

pluvial lake A lake formed during a pluvial (wet) climatic period.

pocosin A swamp, usually containing organic soils, and partly or completely enclosed by a sandy rim.

Podzols Soils in the Podzolic order of the Canadian soil classification system. Also, a great soil group of the zonal order (1938 system of soil classification) consisting of acidic soils formed in cool-temperate to temperate, humid climates, under coniferous or mixed forest, and characterized particularly by a highly leached, whitish E horizon and a B horizon enriched in illuvial humus, Al, and sometimes Fe. Roughly synonymous with Spodosols. This term is not used in the U.S. system of Soil Taxonomy.

podzolization Pedogenic processes resulting in the genesis of Spodosols, Podzols, and Podzolic soils, involving the translocation of Al and humus, and sometimes Fe, to a B horizon via percolating water. In older and Russian terminology, podzolization includes also the weathering of clays.

point of zero charge (PZC) The pH at which the concentration of adsorbed positive charges on the surface of a mineral or organic colloid equals that of negative charges. The point of zero proton charge (PZPC) is the PZC when H^+ or OH^- are the only ions reacting with the surface.

pollen The fine to coarse powder produced from the anthers (male gametes) of plants. Pollen accumulates in the bottoms of lakes and bogs and is fairly resistant to decay. Cores taken from these sites reveal the succession of plants that have inhabited the area. The study of pollen in this manner is called *palynology*.

polygenesis Formation of a soil under more than one pedogenic regime, brought about through extrinsic changes, e.g., climatic change, or through intrinsic changes within the soil, e.g., the development of a slowly permeable horizon. Although it is commonly assumed that polygenesis involves pedogenesis under at least two different climate or vegetation types, some scholars consider all soils to be polygenetic, depending on how one defines the term pedogenic regime.

polygenetic soil A soil that has been formed by two or more different and somewhat contrasting pedogenic regimes, so that all of its horizons or pedogenic properties are not genetically related to one another.

polypedon A theoretical construct designed to delimit a group of contiguous pedons that are sufficiently similar to allow the body to be classified as a single soil series. The boundaries of a polypedon are reached where there is no soil or where properties are those of a different taxonomic class. Polypedons can usually be classified to the soil series level, but a series normally has a wider range of properties than occur in a single polypedon.

pore A discrete volume of soil atmosphere completely surrounded by solid-phase soil material.

pore space The portion of soil bulk volume occupied by soil pores. Also called *pore volume*.

porosity The relative volume of a material that is occupied by pores, i.e., pore volume/bulk volume.

postdepositional modification (PDM) A time-dependent change in landform morphology involving physical and chemical changes in the rocks and soils on a geomorphic surface. PDM data can be used to assess relative age.

potassium fixation The process in which exchangeable or water-soluble potassium ions at colloid surfaces are moved to interlayer positions in clay minerals where they are retained by higher negative charge or where access to the bulk solution is blocked by hydroxy aluminum precipitates. Fixed K^+ may become slowly plant-available as redox and soil moisture conditions change during a growing season and the interlayer space expands.

Prairie soils A zonal great soil group (1938 system of soil classification) consisting of soils formed in temperate to cool-temperate, humid regions under tallgrass vegetation. Roughly synonymous with Udolls. Obsolete in the U.S. system of Soil Taxonomy.

precession of the equinox The wobble of the Earth as it spins, which changes the celestial direction in which its axis of rotation points. One "wobble" cycle takes about 23,000 years.

precipitation (chemical) A phase change in which ions in solution are sufficiently concentrated that they interact more strongly with one another than with the solution molecules and form a solid.

preferential flow See *macropore flow*.

primary mineral A mineral that has not been altered chemically since its crystallization from molten lava, i.e., as originally formed in an igneous rock. See also *secondary mineral*.

primary producer Organisms that produce biomass from inorganic compounds, i.e., autotrophs, or plants of various kinds.

principle of ascendancy and descendancy Stratigraphic principle that centers on geomorphic surfaces and how their location and relationship to each other can be used as a relative age dating tool, e.g., an erosion surface is younger than the youngest deposit or surface that it truncates.

principle of cross-cutting relationships Stratigraphic principle that states that a feature, e.g., a fault, igneous intrusion or a dike, that cuts across another body of rock must be younger than the host rock.

principle of original horizontality Stratigraphic principle that states that layers of sediment are usually, though not always, deposited in a horizontal position. As a result, rock layers that are parallel to the Earth's surface are interpreted to have not been disturbed and to have maintained their original horizontality.

principle of superposition The principle that in an undeformed sequence of sedimentary rocks or sediments, each bed is older than the one above and younger than the one below.

principle of uniformitarianism See *uniformitarianism*.

prismatic A shape of soil structure, common in B and upper C horizons, in which the peds are elongated in the vertical dimension (taller than they are wide); prisms typically have nearly flat tops and angular vertices on the sides.

proanisotropic pedoturbation A form of soil mixing involving processes that form or aid in forming/maintaining horizons, subhorizons, or genetic layers and/or cause an overall increase in profile order. See also *proisotropic pedoturbation*.

profile See *soil profile*.

profile-weighted A type of soil data in which the numerical values are weighted by horizon thickness and bulk density and then summed for the entire solum. This approach is typically used to quantify pedogenic development.

progressive pedogenesis Soil formation, development, and organization, including those processes and factors that singularly or collectively lead to organized and differentiated (more anisotropic) profiles. When progressive pedogenesis dominates, a soil develops more, thicker, and better-expressed genetic horizons.

proisotropic pedoturbation A form of soil mixing involving processes that disrupt, blend, or destroy horizons; subhorizons; or genetic layers and/or impede their formation and cause morphologically simplified profiles to evolve from more ordered ones. See also *proanisotropic pedoturbation*.

prokaryote Cell or organism lacking a membrane-bound, structurally discrete nucleus and other subcellular compartments, such as in bacteria. In prokaryotes, genetic material is in the form of loose strands of DNA found in the cytoplasm.

proto-imogolite theory In podzolization, one of two widely held theories of soil genesis in which the Fe, Si, and/or Al compounds (commonly referred to as imogolite-type materials or ITM) that have accumulated in the B horizon are assumed to have been translocated as positively charged, amorphous sols, under acidic conditions. Later, negatively charged organic matter compounds have moved into the B horizon and precipitated onto the ITM. See also *chelate-complex theory*.

proton A fundamental particle of matter, each of which holds a positive charge in the nucleus of an atom.

puff In Vertisols, a surface drape or exposure of up-welled substratum material forced to the surface and outcropping on a low mound or rim, i.e., the surface exposure of a chimney, or a type of diapir.

pumice A light, porous, volcanic rock that forms during explosive eruptions. It resembles a sponge because it consists of a network of gas bubbles frozen amid volcanic glass and minerals.

pyrite The mineral iron sulfide (FeS_2).

pyroclastic materials See *tephra*.

pyroxene A member of a group inosilicate (chain silicate) minerals that contains two metals, e.g., $CaMgSi_2O_6$, $CaFeSi_2O_6$.

quartz A light-colored or clear, resistant silicate mineral (SiO_2).

quartzite A metamorphic rock consisting largely of interlocking quartz grains; the metamorphic equivalent of sandstone or chert.

Quaternary Period The period of the Cenozoic Era of geologic time, extending from the end of the Tertiary Period (ca. 1.6 million years ago) to the present, and comprising the Pleistocene and Holocene Epochs.

radiocarbon The unstable radioisotope ^{14}C, which is derived from ^{14}N as cosmic ray bombardment adds a neutron to its nucleus and the nucleus emits a proton. Radiocarbon decays back to ^{14}N by beta decay, with a half-life of $5,730 \pm 30$ years.

radiocarbon dating Technique whereby the age of a sediment or geomorphic surface is estimated based on the radiocarbon (^{14}C) age of material within, on top of, or below it.

radioisotopic dating A family of numerical dating techniques in which various isotopes – both stable and some unstable – are used to track time (includes radiocarbon dating).

radionuclide A radioisotope; see *isotope*.

rainshadow A climatically dry area on the lee side of a topographic barrier, formed when moisture-bearing winds are blocked by this barrier, commonly a mountain range. Also called *rainshadow desert*.

reaction wood Wood that forms because of stress that is placed on the tree, often due to tilting. Reaction wood is often manifested as abnormally wide or narrow tree rings.

recarbonation The process whereby carbonate salts are added to an acidic soil profile, often via base cation cycling or additions of carbonate-rich dust.

recharge depressions Depressional landforms where water infiltrates into the soil and percolates to groundwater, thereby recharging it.

Red Desert soils Highly leached, red, clayey soils of the humid tropics (1938 system of soil classification), usually with very thick profiles that are low in silica and high in sesquioxides. Obsolete in the U.S. system of Soil Taxonomy.

Red-Yellow Podzolic soils A combination of the zonal great soil groups, Red Podzolic soils, and Yellow Podzolic soils (1938 system of soil classification), consisting of acidic, well-weathered soils formed under warm-temperate to tropical, humid climates, under forest vegetation, and usually, except for a few members of the Yellow Podzolic Group, under conditions of good drainage. These soils have a leached, light-colored E horizon and Bt horizon containing mainly 1:1 clays and oxides of aluminum and iron, varying in color from red to yellowish red to bright yellowish brown. Roughly synonymous with Ultisols. Obsolete in the U.S. system of Soil Taxonomy.

redox Adjectival abbreviation for *reduction-oxidation* (redoximorphic) processes, and the morphologies that have resulted from those processes.

redox concentrations Reddish or brownish zones of apparent accumulation of Fe and/or Mn oxides in soils. Also called red or brown mottles and *iron oxide concentrations*.

redox depletions Zones of low (≤ 2) chroma and high (≥ 4) value where Fe or Mn oxides have been dissolved by low redox potentials and lost, or where both Fe/Mn oxides and clay have been dispersed and lost by leaching. Also called *gray mottles* or *iron depletions*

redoximorphic feature A soil property associated with wetness that results from the reduction and oxidation of Fe and Mn compounds after saturation with water and desaturation, respectively. Informally called bright mottles and low-chroma (or gray) colors. Alternate spelling: redoxymorphic.

reduced matrix Condition in which the soil matrix has low-chroma colors, but undergoes a change in hue or chroma within 30 min after the soil material is exposed to air, implying that the color change is due to the oxidation of formerly reduced Fe.

reduction The gain of one or more electrons by an ion or molecule. See also *oxidation*.

regolith Unconsolidated materials and weathered rock that rest upon solid bedrock.

Regosols Soils of the Regosolic order of the Canadian system of soil classification. Also a soil of the azonal order (1938 system of soil classification) that lacks well-developed horizons and is developing in deep, unconsolidated, soft mineral deposits, e.g., sand, loess, or glacial drift. This term is not used in the U.S. system of Soil Taxonomy.

regressive pedogenesis Processes that reverse, stop, or slow soil development, including those processes and factors that singularly or collectively lead to simpler and less differentiated (more isotropic) profiles. When regressive pedogenesis dominates, soil horizons become thinner, blurred, eroded, and/or mixed, to the point of being no longer recognizable. Also called *soil regression*.

Regur soils An intrazonal group of dark, calcareous soils (1938 system of soil classification), high in clay, mainly smectite, and formed mainly from rocks low in quartz. The name comes from soils found extensively on the Deccan Plateau of India. Roughly synonymous with Vertisols. Obsolete in the U.S. system of Soil Taxonomy.

relative dating Dating of rocks, soils, stratigraphic layers, and geologic events by their relative position in some chronological order, without specific reference to the number of years before the present, or the absolute geochronometric age.

relict paleosol An old, usually highly polygenetic, soil that has never been buried and remains at the surface today, i.e.,

it has persisted on the land surface through multiple climatic changes.

relief The difference in elevation between the upland summits and the lowlands or valleys in a specified region or area.

remanent magnetism Magnetism acquired by a rock or sediment at some time in the past, as a result of its exposure to the Earth's magnetic field.

Rendolls A suborder of Mollisols known for brown-red colors, and often forming on limestone in Mediterranean climates.

Rendzinas A great soil group of the intrazonal and calcimorphic suborders (1938 system of soil classification) comprising shallow, stony soils that have brown-black, friable A horizons underlain by light gray to pale yellow, calcareous sediment that has weathered from highly calcareous parent materials, usually limestone, chalk or marl bedrock, under grass vegetation (or mixed grasses and forest) in humid and semiarid climates. Obsolete in the U.S. system of Soil Taxonomy.

reserve acidity Soil acidity (H^+ and Al^{3+}) that is not in the solution phase but that can be made soluble by cation exchange in a 1 M KCl solution. Reserve acidity can be neutralized by lime or a buffered salt solution, facilitating a rise in pH.

residual soil A soil formed from bedrock that has weathered in place, i.e., residuum, as evidenced by the fact that it is resting on consolidated rock of the same kind as that from which it was formed, and in the same location.

residual soil material, residuum Unconsolidated, weathered, or partly weathered material that has accumulated as consolidated rock has broken down in place.

resistant minerals Minerals that persist in soils after weathering, either because they were formed during the weathering process, or more likely, because they are resistant to the weathering processes. Compare to *weatherable minerals*.

retardant upbuilding Additions of material onto the soil surface that occur so rapidly that pedogenesis cannot readily incorporate them into the profile, essentially burying the soil.

reticulate mottling A network of red, gray, and brown mottles with no dominant color, most commonly found in deeper horizons of soils containing plinthite.

reversed polarity A period of geologic time when a magnetic needle would have pointed to the south magnetic pole.

rhizobia Bacteria under the collective common name for the genus *Rhizobium* that are able to live symbiotically in roots of leguminous plants, from which they receive energy and for which they fix molecular nitrogen.

rhizosphere The soil zone immediately adjacent to plant roots in which biological activity due to root exudates and microorganisms is higher than in the bulk soil.

rhyolite A fine-grained, silica-rich igneous rock; the extrusive equivalent of granite.

rill A small, intermittent watercourse with steep sides; usually only several centimeters deep.

riparian Land adjacent to a body of water or river that is at least periodically influenced by flooding.

rock A naturally occurring solid aggregate of one or more minerals in varying proportions.

rock flour Finely divided, mainly silt-size rock material ground by glacial action and found in streams issuing from melting glaciers. When dry, it may be suspended in wind as loess.

rock land A name on some soil maps and given to areas containing frequent rock outcrops and shallow soils in which rock outcrops occupy 25–90% of the surface.

rock outcrop In soil survey applications, a map unit that consists of exposures of bedrock protruding through the soil mantle.

rock varnish A thin, dark (brown-black), shiny film or coating of Fe and Mn oxides and silica, formed in arid regions on the surfaces of pebbles, boulders, rock fragments, and rock outcrops. Also called *desert varnish*.

root exudate Low-molecular-mass metabolites that enter the soil from plant roots.

root plate The soil + root mass that is torn up when a tree is uprooted.

root trace The path or conduit in a soil that remains after the root has decayed.

root zone The part of the soil that is, or can be, penetrated by plant roots. Also called rooting zone.

rough broken land Areas with very steep topography and numerous intermittent drainage channels, but usually covered with vegetation.

roundness The degree to which a particle's corners and edges are rounded. See also *sphericity*.

rubefaction, rubification The development of red color in soil, i.e., reddening.

runoff That portion of precipitation or applied irrigation water that does not infiltrate, but instead is discharged from the area, usually into stream channels, without entering the soil.

S horizon In some systems of tropical soil nomenclature, the S horizon refers to a subsurface stone line.

S-matrix In soil micromorphology, the plasma, skeleton grains, and voids that do not occur as pedological features, other than those expressed by specific extinction (orientation) patterns. Pedological features also have an internal S-matrix.

salic horizon A mineral soil horizon, ≥15 cm thick, enriched with at least 20 g kg^{-1} secondary salts that are more soluble than gypsum.

saline-alkali soil Obsolete term for a soil containing appreciable quantities of exchangeable Na and soluble salts, leading to high alkalinity conditions, such that the growth of most crop plants is reduced.

saline seep An area of saline water discharge at or near the soil surface under dry-land conditions, which reduces or eliminates crop growth.

saline soil A nonsodic soil containing sufficient soluble salts to adversely affect the growth of most crop plants.

salinization The pedogenic process whereby soluble salts accumulate in soils.

salt-affected soil A soil that has such high quantities of soluble salts, with or without high amounts of exchangeable Na, that the growth of most crop plants will be adversely affected.

saltation A type of eolian (wind) or fluvial (water) transport involving the rolling, bouncing, or jumping action of soil particles.

salt tolerance The ability of plants to resist the adverse, nonspecific effects of excessive soluble salts in soils.

sand Mineral particles, usually dominated by quartz, that in the USDA particle size system range in diameter from 0.05 to 2.0 mm.

sandstone A clastic sedimentary rock in which the particles are dominantly of sand size, from 0.05 to 2.0 mm in diameter.

sandy Texture classes of sand and loamy sand textures. A family particle-size class for soils with sand or loamy sand textures and <35% rock fragments in the upper subsoil.

Sangamon soil, Sangamon paleosol, Sangamon geosol A prominent paleosol or geosol in the United States, typically thick, well-developed, and usually buried within the Quaternary sediment column, and as such, serving as a prominent stratigraphic marker bed. It began forming during the Sangamonian interglacial period (marine isotope stage 5).

sapric material Organic soil (Oa) material that contains less than one-sixth recognizable fibers (after rubbing) of undecomposed plant remains. Sapric materials have very low bulk densities and very high water-holding capacities.

saprock Fractured bedrock with evidences of weathering restricted to fracture areas and margins. Gradational between saprolite and hard bedrock.

saprolite Soft, friable, isovolumetrically weathered bedrock that retains the fabric and structure of the parent rock, but exhibits extensive intercrystal and intracrystal weathering. Two types of saprolite are often described: massive and structured. *Massive saprolite* is a type of saprolite that has lost all rock structure, usually because of collapse and bioturbation.

saprolitization The formation of saprock and saprolite from unweathered rock.

saturated flow The movement of water in a soil whose pores are completely filled with water.

scarp Abbreviation for escarpment. A cliff or steep slope of considerable extent and any height.

schist A strongly foliated, coarsely crystalline, metamorphic rock, produced during regional metamorphism, that can readily be split into slabs or flakes because >50% of its mineral grains are parallel to each other.

secondary mineral A mineral, mainly phyllosilicate and oxide clays, resulting from the decomposition of a primary mineral or from the reprecipitation/recrystallization of the decomposition products of a primary mineral.

sedimentary facies An accumulation of deposits that grades laterally into other sedimentary accumulations that were formed at the same time but that exhibit different characteristics.

sedimentary rock Rock formed from the accumulation of sediment, which may consist of fragments and mineral grains of varying sizes from preexisting rocks, remains or products of animals and plants, the products of chemical action, or mixtures of these. Common sedimentary rocks are sandstones, shales, limestones, conglomerates, rock salt, and coal.

segregated ice Massive ice in a soil, which is relatively free of soil particles.

seismiturbation Soil and sediment mixing by earthquakes and the surficial settling that occurs after them.

self-mulching soil A soil in which the A horizon is so well aggregated that it does not crust and seal under heavy rain but instead serves as a surface mulch upon drying. Common in some Vertisols.

self-weight collapse The condition in which a soil or sediment, upon wetting and then drying, collapses, resulting in an increase in bulk density and a loss of pore space.

sequum (pl. **sequa**) The B horizon together with any overlying E horizons. It is essentially an eluvial and illuvial horizon couplet.

series See *soil series*

serpentine A group of trioctahedral, 1:1 layer silicate minerals with the general chemical formula $(Mg, Fe)_3Si_2O_5(OH)_4$. Polymorphs of serpentine include antigorite, chrysotile, and lizardite.

serpentinite A rock rich in serpentine.

sesquan A cutan composed of concentration of sesquioxides.

sesquioxide Oxides and hydroxides of Fe, Al, or Mn.

shale A sedimentary rock that is dominated by layer silicate minerals, resembles mudstone, and splits or fractures readily.

shear strength The resistance of a body to shear stress.

shear stress The stress on an object operating parallel to the slope on which it lies.

sheet In phyllosilicate mineral terminology, a flat latticework array composed of one level of linked coordination polyhedra. A sheet, as in a tetrahedral sheet or octahedral sheet, is thicker than a plane and thinner than a layer.

sheeting A form of physical weathering in which thin (< 5 mm thick) sheets of rock peel off the parent rock. Also called *spalling*.

sheet erosion, sheet flow The removal of a relatively uniform, thin layer of soil by rainfall and largely unchanneled surface runoff, i.e., sheet flow.

shoulder The hillslope element, dominantly convex in profile and erosional in origin, that forms the uppermost inclined surface near the top of a slope. A shoulder slope is the transition zone from backslope to summit.

shrink-swell The shrinking of soil when dry and the swelling when wet.

side slope The slope bounding a drainageway or valley, lying between the drainageway and the adjacent interfluve. Side slopes, as seen from above, have fairly straight boundaries above and below and are situated on the side of an interfluve.

Sierozems A zonal great soil group (1938 system of soil classification) found in temperate to cool, arid climates under desert vegetation and consisting of shallow soils with pale grayish A horizons grading quickly to calcareous material at depth. Obsolete in the U.S. system of Soil Taxonomy.

silan A cutan of silt-sized material. Also called *siltan*.

silcrete A silica-enriched hardpan, usually a duripan.

silica The basic chemical constituent common to all silicate minerals and silicic (felsic) magmas. As a pure, crystalline substance, silicon dioxide (SiO_2) comprises quartz and its microcrystalline forms, flint and chalcedony.

silica:alumina ratio The ratio of SiO_2 to Al_2O_3 in clay minerals and soils.

silica:sesquioxide ratio The ratio of SiO_2 to Al_2O_3 plus Fe_2O_3 in clay minerals and soils.

silicate A rock-forming mineral that contains silicon.

silica tetrahedron The basic structural unit of which all silicates are composed, consisting of a silicon atom surrounded by four oxygen atoms, i.e., a tetrahedron with an oxygen atom at each corner.

silicic A rock or material rich in silica.

silicification The process whereby a soil horizon becomes enriched in illuvial silica.

silt Mineral particles that range (in diameter) from 0.02 to 0.002 mm in the international system or 0.05 to 0.002 mm in the USDA system.

siltstone A sedimentary rock composed primarily of silt-sized clastic particles.

single-grain structure A soil structure classification, usually found only in extremely coarse-textured soils and sands, in which the soil particles occur almost completely as individual or primary particles with essentially no secondary particles or aggregates present. Also called *structureless*.

skeletan A cutan with a considerable amount of skeleton grains, usually silt-sized or larger, embedded within it.

skeleton Individual grains in soils that are relatively stable and not readily translocated, concentrated, or reorganized by soil forming processes, particularly by percolating water. They include mineral grains and resistant siliceous and organic bodies larger than about medium silt size (20–30 μm), or perhaps smaller. Often referred to as the *immobile fraction*.

slate A compact, fine-grained metamorphic rock that has slaty cleavage and forms by the low-grade metamorphism of shale.

slaty cleavage A style of foliation common in metamorphosed mudstones, characterized by nearly flat, sheetlike planes of breakage, similar in appearance to a deck of playing cards.

slick spot A small area of surface soil that is slick when wet because of alkalinity or high exchangeable Na. Slick spots have a puddled or crusted, very smooth, nearly impervious surface, while the underlying material is dense and massive. Also called *panspot*.

slickenside A slip face that is polished and striated, produced by one mass sliding past another, as when soil material expands in response to wetting. Slickensides are common below 50 cm in swelling clay soils that are subject to large changes in water content, i.e., wetting and drying.

slope (More correctly: **slope gradient**) The inclination of the land surface from the horizontal. Slope percentage is the vertical slope distance divided by horizontal distance, then multiplied by 100. Thus, a slope of 20 % has a rise of 20 m in 100 m of horizontal distance. Slope classes vary slightly among soil surveys, but this listing is typical:

Nearly level	0–3%
Gently sloping	1–8%
Strongly sloping	8–15%
Moderately steep	15–30%
Steep	30–50%
Very steep	50–95%

slope aspect See *aspect*.

slope element One of five parts of a slope: summit, shoulder, backslope, footslope, and toeslope (or alluvial toeslope).

slope gradient See *slope*.

slopewash Movement of material downslope by the combined impact of rainsplash and overland flow, possibly including small amounts of channelized flow.

slump Landform or mass of material that was produced by a mass movement process involving a shearing, sliding, and backward rotation along a curved, concave upward slip surface.

small-scale map A map having a scale smaller than 1:1,000,000.

smectite A 2:1 layer silicate with low layer charge but high CEC, including the dioctahedral members montmorillonite, beidellite, and nontronite, and the trioctahedral members saponite, hectorite, and sauconite.

sodic soil A nonsaline soil containing sufficient exchangeable Na to adversely affect crop production and soil structure, and with a sodium adsorption ratio ≥13.

sodicity The amount of exchangeable sodium in a soil.

sodium adsorption ratio (SAR) A quantitative relationship between Na and the divalent cations Ca^{2+} and Mg^{2+} in soil solution or in a saturated paste extract, where all concentrations are in moles per liter: $\dfrac{[Na^+]}{[(Ca^{2+} + Mg^{2+})^{0.5}]}$. The ratio is derived from the Gapon model for heterovalent exchange reactions.

soft See *consistence.*

soil Unconsolidated mineral and organic matter that is either presently or was at one time at the Earth's surface where it was influenced by the genetic and environmental factors of parent material, climate (including moisture and temperature variations), organisms, and topography, all acting over a period to produce a product with many physical, chemical, biological, and morphological properties that differ from those of the parent material. Most definitions of soil include the liquid and gas phases as well as the solid phase. Many definitions of soil also include the requirement that it exhibit genetic horizons and that it support vegetation.

soil aeration The exchange of soil air with air from the atmosphere. The air in a well-aerated soil is similar to that in the atmosphere; the air in a poorly aerated soil is considerably higher in CO_2 and lower in O_2.

soil amendment A substance added to soil to improve its chemical or physical characteristics, or as a means of treating a waste material.

soil association A soil map unit representing two or more soils that occur together in repeating and predictable patterns on the landscape.

soil body a collection of adjoining and genetically related pedons on the landscape, some or even most of which may fall within a single taxonomic unit.

soil boundary A line on a soil map, separating two contiguous mapping units, that represents points on the landscape that are transitional between those adjacent soil bodies. Along soil boundaries, gradients of mapping units change over a short distance. Map unit boundaries are an interpretation of soil mappers.

soil classification The systematic arrangement of soils into groups or categories on the basis of their measureable characteristics.

soil complex A soil map unit representing multiple soils that occur in a pattern too complex or poorly understood to be individually mapped at the scale of the survey.

soil cover The entirety of soils occurring in a region.

soil creep The slow downslope movement of soil and sediment due to surface heave as induced by freeze-thaw and wet-dry cycles.

soil endemism The idea that soils that are restricted to a particular geographic area because of a unique combination of soil-forming factors and processes that operate there.

soil erosion Removal of soil, usually by wind or water, faster than the soil-forming processes can replace it. Soil erosion is caused by natural, animal, and human activity (overgrazing, over-cultivation, forest clearing, mechanized farming, etc.). It may result in land infertility and lead to desertification or flooding. Compare to *geologic erosion.*

soil evolution The conceptual idea that soil development (pedogenesis) can ebb and flow, move forward toward more developed soils, or regress toward simpler soils, or remain static for short periods of time.

soil fabric In soil micromorphology, the spatial arrangement of solid particles and associated voids.

soil family In Soil Taxonomy, the classification category intermediate between the subgroup and the soil series. Families provide groupings of soils with common ranges in texture, mineralogy, temperature, and thickness.

soil fertility The ability of a soil to supply nutrients essential to plant growth.

soil formation factors See *state factors.*

soil-forming interval The period of time during which a soil is forming, usually from the initiation of soil genesis to the present (if the soil is still exposed at the surface) or until it becomes buried.

soil gas See *gas (soil).* Equivalent to soil air.

soil genesis See *pedogenesis.*

soil geography A subspecialization of physical geography concerned with describing, explaining, and mapping the areal distributions of soils.

soil geomorphology The science that studies the genetic relationships between soils and landforms.

soil horizon See *horizon.*

soil infauna Soil fauna that reside most of their lives within the soil proper.

soil landscape The soil portion of the landscape. *Soilscape* is an abbreviation.

soil landscape analysis The study of soil patterns that are related to landforms and pedogenesis.

soil liquid See *liquid (soil).* Equivalent to soil water.

soil micromorphology The study of undisturbed soil material by magnification, using light or electron microscopy (or other comparable instrumentation) to determine the arrangement and origins of soil components.

soil moisture regime In Soil Taxonomy, the long-term moisture status of a soil, roughly interpreted by the degree and annual period of saturation of a zone in the soil that corresponds to the rooting zone of most crops.

soil moisture tension Equal in magnitude but opposite in sign to the soil water pressure, soil moisture tension is equal to the pressure that must be applied to the soil water to bring it to a hydraulic equilibrium.

soil monolith A vertical section through a soil profile that has been preserved with resin and mounted for display.

soil morphology The physical entirety of a soil, including its horizonation, color, texture, structure, and other features that are describable in hand specimens.

soil order The highest level of soil classification in Soil Taxonomy. There are presently 12 soil orders: Entisols, Inceptisols, Spodosols, Ultisols, Alfisols, Vertisols, Oxisols, Histosols, Andisols, Aridisols, Mollisols, and Gelisols.

soil permeability The intrinsic capacity of the soil to allow movement of fluids like water or gas. Formally, permeability has units of length^2. In some older literature, the term was loosely equated with fluid flux density or with saturated hydraulic conductivity. Permeability classes of soils are considered obsolete and have been replaced by classes of saturated hydraulic conductivity.

soil phase A utilitarian grouping of soils of the same series, defined by soil or environmental features that are not class differentia in Soil Taxonomy, e.g., surface texture, surficial rock fragments, rock outcrops, substratum, special soil water conditions, salinity, physiographic position, erosion, thickness, and so on.

soil production The production of soil and regolith from solid rock, usually via weathering and biomechanical disintegration processes.

soil production function An equation that approximates the production of soil (residuum) from bedrock.

soil profile A vertical section of the soil through all its horizons and extending into the C horizon.

soil reaction Obsolete term for the acidity or alkalinity of a soil, usually expressed as a pH value. See also *pH.*

soil salinity The amount of soluble salts in a soil, as conventionally determined by the electrical conductivity of a saturation extract.

soilscape See *soil landscape*.

soil separates Mineral particles <2 mm in equivalent diameter and further subdivided between specified size limits. The names and sizes, in millimeters, of separates recognized by the USDA-NRCS are as follows:

Soil separate name	Size range (mm)
Very coarse sand	2.0–1.0
Coarse sand	1.0–0.5
Medium sand	0.5–0.25
Fine sand	0.25–0.10
Very fine sand	0.10–0.05
Silt	0.05–0.002
Clay	<0.002

soil series A basic unit of soil classification and the lowest category in Soil Taxonomy, a soil series is a subdivision of a family. Soil series are differentiated by soil properties important to soil use and management. A series is consistent in all major profile characteristics except texture of the A horizon.

soil solution The water surrounding soil particles and occurring in void spaces in soils.

soil structure The organization of the soil system as expressed by the degree of expression (grade), size, and shape of soil aggregates, i.e., peds. Secondarily, soil structure also considers the nature and distribution of pores and pore spaces.

soil survey The systematic examination and mapping of soils.

soil taxonomy A systematic arrangement of soils into groups or categories on the basis of their measureable characteristics.

Soil Taxonomy The USDA-NRCS system of soil classification used in the United States since 1975.

soil temperature regime In Soil Taxonomy, a category used to distinguish soils by the attribute of temperature. STRs are determined by the mean annual soil temperature (MAST) measured (or modeled) at 50 cm depth and its intraannual variability. See Table 8.13.

soil water content The water lost from the soil upon drying to constant mass at 105°C, expressed either as the mass of water per unit mass of dry soil, or as the volume of water per unit bulk volume of soil. Same as *soil moisture content*.

soil welding The processes that cause the horizons in a surface soil to deepen and connect, genetically, with a buried soil below, such that C horizon material no longer exists between the two.

solids (soil) Crystalline or poorly crystalline minerals and organic matter in soil.

solifluction Slow, viscous, downslope flow of water-saturated unconsolidated material, usually on top of a frozen substrate like permafrost, that is characteristic of, although not confined to, regions subjected to alternate periods of freezing and thawing.

Solods In the Canadian soil classification system, solonetzic soils that have been leached of most of their salts and Na, resembling Albolls or Aquolls.

solodization The process whereby Na is gradually removed from a profile and often replaced by Ca.

Solonchaks A great soil group of the intrazonal order and halomorphic suborder (1938 system of soil classification) consisting of soils with gray, thin, salty, surface crusts above a fine, granular A horizon, and at depth are grayish, friable, salty horizons. Solonchaks form under subhumid to arid climates, under conditions of poor drainage, and under a sparse growth of halophytic vegetation. Obsolete in the U.S. system of Soil Taxonomy.

Solonetz A great group in the Solonetzic order of the Canadian soil classification system. A great soil group of the intrazonal order and halomorphic suborder (1938 system of soil classification) consisting of soils with a very thin, friable, surface, which is underlain by a dark, hard, usually highly alkaline, B horizon. Solonetz soils form under subhumid to arid climates, and under a native vegetation of halophytic plants. They are slightly better drained than Solonchaks. This term is not used in the U.S. system of Soil Taxonomy.

Solonetzic soils A taxonomic order of soils in the Canadian soil classification system, that is characterized by the effects and activity of excess Na ions. Roughly equivalent to the informal term sodic soils.

solum (pl. **sola**) The genetic soil profile, usually referring to the A and B horizons.

sombric horizon A subsurface mineral horizon that is darker in color than the overlying horizon but that lacks the properties of a spodic horizon. Common in the cool, moist soils of the high altitude tropics.

sorosilicates A type of silicate mineral in which one corner O atom of each silicon tetrahedron is shared with an adjacent tetrahedron.

sorption The removal of an ion or molecule from solution by adsorption and absorption. This term is often used when the exact nature of the mechanism of removal is not known.

sorting A measure of the range of particle sizes in a sedimentary deposit, such that a deposit with a narrow range of particle sizes is termed well-sorted.

spalling A form of physical weathering in which thin (< 5 mm thick) sheets of rock peel off the parent rock. Also called *sheeting*.

sparmicritization Dissolution of large $CaCO_3$ crystals and reprecipitation of micritic crystals of $CaCO_3$ in the resulting voids.

spatial variability The variation in soil properties laterally and vertically across the landscape.

specific adsorption The strong adsorption of ions or molecules onto a surface, such that they are not readily removed by ion exchange.

specific heat capacity The heat energy required to raise the temperature of 1 g of a substance by 1°C.

specific surface area The solid-particle surface area (of a soil or porous medium) divided by the solid-particle mass.

sphericity The degree to which something conforms to a sphere, irrespective of the sharpness of its edges.

spodic horizon A mineral soil horizon that is characterized by the illuvial accumulation of amorphous materials composed of Al and organic carbon, with or without Fe. It is the diagnostic horizon for Spodosols.

Spodosols One of 12 major taxonomic orders in Soil Taxonomy. Spodosols are acidic soils that have a spodic horizon or a placic horizon that overlies a fragipan. They form in cool, humid climates, usually under forest or vegetation that produces abundant amounts of acidic litter.

springtails Very small insects of order Collembola that live near the soil surface and feed on organic matter.

stable isotope An isotope that does not decay radioactively over time.

stadial A part of geologic time in which a glacier makes a brief readvance.

state-factor model A model of soil development in which the soil (*S*) is envisioned as forming because of the complex interplay among five state factors. Also called *functional-factorial model*.

state factors In the state-factor model of soil genesis, $S = f$ (*cl*, *o*, *r*, *p*, *t*, ...), the five state factors are *cl* (climate), *o* (organisms), *r* (relief), *p* (parent material) and *t* (time). The dots represent other unspecified factors, such as fire, or inputs of dust.

static pedogenesis A condition in soils that occurs for only short periods of time, during which the energies of regressive and progressive pedogenic processes balance one another, such that the soil horizons neither develop nor are destroyed, and occur in a net *steady state*.

steady state A condition in soils and other systems in which the average condition of the system remains unchanged over a given period of time.

steppe Name given to the dry, cool, shortgrass grasslands of Russia and Ukraine.

sticky See *consistence*.

stone line A sheetlike concentration of coarse fragments in soils and sediments. In cross section, the line may be marked only by scattered fragments, or it may be a discrete layer of fragments. Also called *stone zone*, *stone layer*, and *carpetolith*.

stones Rock or mineral fragments between 250 and 600 mm in diameter if rounded, or 380 to 600 mm if flat.

stone zone See *stone line*.

stoniness Soil classes, used as a phase distinction in mapping soils, that are based on the relative proportion of stones at or near the soil surface.

stony A stoniness class in which there are enough stones at or near the soil surface to be a continuing nuisance during operations that mix the surface layer, but they do not make most such operations impractical.

strain The deformational changes that a soil undergoes through time, usually due to physical and chemical weathering, translocations, and leaching, and expressed as changes in horizon and profile thicknesses.

stratification The accumulation of material in distinct layers or beds.

stratified The characteristic of being in distinct layers or beds.

stratified drift Debris carried from a glacier by meltwater and deposited in well-defined layers or strata.

stratigraphy The succession and age relations of layered rocks and sediments.

stream terrace A relatively flat, former floodplain surface along a valley, with a steep bank separating it either from the floodplain or from a lower terrace. Terraces originally formed near the level of the stream, and represent the dissected remnants of an abandoned floodplain or valley floor.

structural charge The charge (usually negative) expressed at a mineral surface and resulting from isomorphous substitution within the mineral crystal.

subaerial A land surface that is not buried or covered with water but intersects the atmosphere. Compare to *subaqueous*.

subangular blocky A type of soil structure, common in B horizons, in which the peds resemble rounded cubes.

subaqueous A surface that is at present under water. Compare to *subaerial*.

Subarctic Brown Forest soils Soils similar to Brown Forest soils (1938 system of soil classification) except with thinner profiles and colder temperatures. Obsolete in the U.S. system of Soil Taxonomy.

subgroup In Soil Taxonomy, the fourth largest taxonomic grouping, a subset of great groups, but above the family and series taxonomic levels.

suborder In Soil Taxonomy, the second largest taxonomic grouping of soils, a subset of orders but above great groups and subgroups.

subsoil Technically, the B horizon; roughly, the part of the solum below the plow depth.

sulfidic material Waterlogged material or organic material that contains ≥ 7.5 g kg^{-1} of sulfide-sulfur.

sulfidization The accumulation of soil materials rich in sulfides, best exemplified in the anaerobic, humus-rich environments of tidal marshes, usually by the biomineralization of sulfate-bearing water.

sulfur cycle The sequence of transformations undergone by sulfur wherein it is used by living organisms, transformed upon death and decomposition of the organism.

sulfuric horizon A mineral or organic soil horizon composed of material that has both pH <3.5 and evidence of sulfuric acid in the form of jarosite mottles, $\geq 0.05\%$ soluble sulfate, or being overlain by sulfidic material.

sulfuricization The process whereby sulfuric acid is produced in soils or sediments as sulfide-bearing minerals (formed via sulfidization) are exposed to oxidizing conditions, become oxidized, and produce jarosite and/or sulfuric acid.

summit The highest point of any landform remnant, hill, or mountain.

supraglacial Carried upon, deposited from or pertaining to the top surface of a glacier or ice sheet.

surface area The area of the solid particles in a given quantity of soil or sediment.

surface exposure dating (SED) The technique used to determine the amount of time that a geomorphic surface has been stable and subaerially exposed, such that soils could have formed on it.

surface soil The uppermost part of the soil profile that is ordinarily cultivated during agricultural operations, or its equivalent in uncultivated soils. More correctly referred to as the surface horizon. Frequently designated as the plow layer, the surface layer, or the Ap horizon.

surface tension The elasticlike force in a body at its interface with another phase, especially notable between a liquid and the air, that tends to minimize or constrict the area of the surface.

swale A topographically low area.

talc $Si_4Mg_3O_{10}(OH)_2$. A trioctahedral magnesium silicate mineral with a 2:1 type layer structure, but without isomorphous substitution.

talf Geomorphic descriptor for a very flat area on an otherwise flat landscape.

talus Rock fragments, usually coarse and angular, derived from a cliff or very steep rock slope by falling, rolling or sliding, and currently piled up at the base of the cliff.

taxadjunct A soil that is correlated to an existing soil series for the purpose of expediency, because it is so like the soils of the defined series in morphology, composition, and behavior that little or nothing is gained by distinguishing it as a newseries, even though some of its properties would exclude it from the series.

taxon (pl. **taxa**) In the context of soil survey, a class at any categorical level in a system of taxonomy. Taxa are chosen because of their natural significance and their usefulness in discriminating among soils.

taxonomic generalization Simplification of a map pattern on a small-scale map by combining those map units on a more detailed, large-scale map that are pedologically similar.

taxonomic unit A subdivision within a taxonomic system. For example, Mollisols constitute a taxonomic unit. Contrast with *genetic unit* and *map unit.*

tectonic Rock structures produced by movements in the Earth's crust.

tectosilicate A type of silicate mineral that has a framework structure, i.e., all of the O ions in each tetrahedron are shared with adjacent tetrahedra.

tensiometer A device for measuring in situ soil-water matric potential. It has a porous, permeable ceramic cup connected through a water-filled tube to a manometer, vacuum gauge, pressure transducer, or other pressure-measuring device.

tephra Clastic volcanic materials that are ejected from a vent during an eruption and transported through the air, including ash (volcanic), blocks (volcanic), cinders, lapilli, scoria, and pumice.

tephrochronology Numerical dating technique that uses volcanic materials, usually tephra, of known age to date associated sediments, and to provide marker beds within the stratigraphic column.

termitarium (pl. **termitaria**) A mound made by a colony of termites.

terra rossa Residual, red, clayey materials and the soils formed from them, both of which usually overlie limestone bedrock at shallow depth. Terra rosa originates partly from the chemical weathering of limestone and partly from the influx of eolian sediments onto old, stable surfaces.

terrace (i) A steplike surface, bordering a stream or shoreline, that represents the former position of a floodplain, lake, or seashore. (ii) A raised, generally horizontal strip of earth and/or rock, constructed along a hill to make land suitable for tillage and to prevent accelerated erosion.

terrain The topography of the landscape, or "lay of the land."

terrane An area of a certain structure or rock type, such as granitic terrane or limestone terrane.

terrestrialization The closing up of a water body with organic and/or mineral sediment.

Tertiary A geologic period that occurred roughly 65 to 1.6 million years ago. The ensuing period is the Quaternary.

tessera (pl. **tesserae**) Like a pedon but including both soil and vegetation, a tessera is usually viewed as being smaller than a pedon, but larger than a hand specimen.

tetrahedral sheet Sheet of horizontally linked, tetrahedron-shaped units that are a basic structural component of phyllosilicate clay minerals.

tetrahedron See *silica tetrahedron.*

textural triangle A three-phase scale, i.e., a ternary diagram, with 100% contents of sand, silt, and clay at each corner, used to define soil textural classes.

texture, soil The relative proportions of sand, clay, and silt in a soil sample.

texture-contrast soil A soil that has a clay-impoverished upper solum and a finer-textured B horizon. Also called *duplex soils.*

thallus (pl. **thalli**) The vegetative part of simple plants, e.g., lichens, showing no differentiation into roots, stems, or leaves.

thermal expansion A form of physical weathering involving expansion of rocks due to heating or hear stresses.

thermic See *soil temperature regime.*

threshold A point (spatial or temporal) in a natural system at which a change occurs in the system. See also *extrinsic threshold* and *intrinsic threshold.*

throughflow See *interflow.*

tidal flats Areas of nearly flat, barren mud, normally containing an excess of soluble salt and periodically covered by tidal waters.

tile drain Concrete, ceramic, or plastic pipe, that is suitably porous and then placed at suitable depths and spacings in the soil to enhance and/or accelerate drainage of groundwater from the soil profile by collecting it and leading it to an outlet.

till See *glacial till.*

till plain An extensive, flat to undulating surface underlain by till. Also called *ground moraine.*

time transgressive As applied to geomorphic surfaces, the concept that parts of the same surface may be of different ages or that the age of the surface changes systematically across it. See also *diachronous.*

time $_{zero}$ The moment at which soil formation begins in fresh sediment. Theoretically each body of soil has one time$_{zero}$, although abrupt changes in environmental conditions may superimpose upon a preexisting soil body a succession of time$_{zeros}$, such that it is clearly *polygenetic.*

toeslope Hillslope position at the lowest part of the slope, below the footslope; the toeslope is typically a gently sloping, concave-upward, constructional surface that grades to a valley or closed depression. Also called *alluvial toeslope.*

tongueing The condition, usually occurring in a glossic or degraded Bt or Btx horizon, in which tongues of eluvial material penetrate into the B horizon below. Also called *interfingering.*

topofunction The formula or quantitative expression of the differences that occur in soils as a function of changes in topography, assuming that the remaining soil-forming factors are held constant, or nearly so. See also *toposequence.*

topography The relative positions and elevations of the natural or man-made features of an area that describe the configuration of its surface.

toposequence A sequence of related soils that differ from one another primarily because of topography as a soil-formation factor. One kind of *catena.* When expressed as a mathematical equation, it is referred to as a *topofunction.*

topsoil The layer of soil normally moved during cultivation and frequently designated as the A or Ap horizon.

torric A soil moisture regime, like aridic but used in a different hierarchical level of Soil Taxonomy.

tortuosity The nonstraight nature of soil pores.

total acidity Soil acidity that includes residual and exchangeable acidity, often calculated by subtracting the sum of charges due to exchangeable base cations from the cation exchange capacity.

trace elements See *micronutrient.*

translocation Migration of material in solution or suspension from one soil horizon to another.

tree-tip mound The small mound of debris sloughed from the root plate (ball) of an uprooted tree. Local soil horizons are commonly mixed, obliterated, or folded over, resulting in a heterogeneous mass of soil material. A mound is commonly adjacent to a pit that marks the former position of the tree. Also called *cradle-knoll.*

tree-tip pit The small pit or depression resulting from the area vacated by the root plate (ball) of an uprooted tree, commonly adjacent to small mounds composed of the displaced material. These pits are commonly sites of increased infiltration and enhanced soil development.

treeline In alpine areas, the ecotone between treeless alpine tundra and the forests in the warmer climate areas downslope.

trioctahedral An octahedral sheet, or a mineral containing such a sheet, that has all of the cation sites filled by divalent ions such as Mg^{2+} or Fe^{2+}.

trophic level The position that an organism occupies in a food chain.

truncated In soils, the condition of having lost, by erosion or excavation, all or part of the upper profile.

tubule In soil micromorphology, an infilled channel.

tuff A compacted deposit or rock that is ≥50% volcanic ash and dust, or a general term for all rocks formed from consolidated pyroclastic materials.

tundra A generally treeless habitat, found in high alpine and Arctic/Antarctic regions, that has an extremely cold climate, low biotic diversity, short growth and reproduction season and often overlies a layer of permafrost.

Tundra soils A zonal great soil group (1938 system of soil classification) consisting of soils with dark-brown peaty A and O horizons, over grayish, mottled horizons and having continually frozen substrata. Tundra soils form under cold, humid climates, with poor drainage, and with a native vegetation of lichens, moss, flowering plants, and shrubs. Obsolete in the U.S. system of Soil Taxonomy.

udic One of five major soil moisture regimes in Soil Taxonomy, indicative of humid climates and leaching conditions. Among other requirements, the soil moisture control section is not dry for as long as 90 cumulative days.

Ultisols One of 12 major taxonomic orders in Soil Taxonomy. Ultisols have an argillic horizon, a base cation saturation of <35% at a depth of 125 cm below the top of the argillic, and a thermic (or warmer) soil temperature regime. They are typical of older, weathered soils in warm, humid climates and are dominated by 1:1 clay minerals.

ultramafic Being very rich in Mg and Fe.

ultraxerous Extremely dry.

umbric epipedon A surface horizon of a mineral soil that is like a mollic epipedon but has a base cation saturation <50%.

unconformity A substantial break or gap in the geologic record where a rock unit is overlain by another that is not in direct stratigraphic succession, i.e., a buried erosion surface separates the two rock masses.

unconsolidated Sediments that are loose and not hardened.

uniformitarianism The geologic principle that assumes that the laws of nature are constant. It originally meant that the processes operating to change the Earth in the present also operated in the past. The meaning has evolved and today the principle of uniformitarianism acknowledges that past processes, even if the same as today, may have operated at different rates and with different intensities than those of the present.

unit structure In phyllosilicate mineral terminology, the total assembly of a layer plus the associated interlayer material.

unloading The release of confining pressure from rocks associated with the removal of overlying material, which may result in expansion of rock and the development of joints or sheets. Also known as *dilatation*.

unsaturated flow The movement of water in soils during which the pores are not filled to capacity with water.

upbuilding The upward growth of the soil surface by additions of sediment. Includes *developmental* and *retardant* types.

upfreezing The process whereby rocks and other particles are gradually moved upward by freeze-thaw activity.

urban land On soil maps, areas so altered or obstructed by urban works or structures that identification of individual soils is not feasible.

urbanthroturbation Human-initiated, nonagronomic soil mixing. Often refers to excavations such as mines and pits, road construction, urban land moving, or cut-and-fill operations.

ustic One of five major soil moisture regimes in Soil Taxonomy, intermediate between aridic and udic, that is common to temperate subhumid or semiarid regions, or in tropical and subtropical regions with a monsoon climate. Although a limited amount of water is available for plants, water is present at times when the soil temperature is optimum for plant growth.

vadose water Water in the soil or other unconsolidated material that occurs above the water table.

vadose zone See *zone of aeration*.

valency The number of bonds that can be formed by an atom of an element.

value See *color value*.

van der Waals forces The totality of intermolecular forces, i.e., the sum of the attractive or repulsive forces between molecules, other than ionic, covalent, or hydrogen bonds.

vapor flow The gaseous flow of water vapor in soils from a moist or warm zone of higher potential to a drier or colder zone of lower potential.

varnish See *rock varnish*.

variable charge A solid surface carrying a net electrical charge that may be positive, negative, or zero, depending on the activity of one or more species of potential-determining ions in the solution phase that is in contact with the solid.

varve A thin pair of glaciolacustrine sediment layers, each one seasonally deposited, usually by meltwater streams, in a glacial lake or other body of still water in front of a glacier. It usually consists of a lighter and a darker part, due to the change in sedimentation type and amount from the winter to the warm season.

ventifact A stone or pebble that has been shaped, worn, faceted, or polished by the abrasive action of wind-blown sand, usually under arid conditions, with the flat facets usually meeting at sharp angles.

vermiculite A highly charged, 2:1 layer silicate that forms from mica and may be di- or trioctahedral.

vermiform Wormlike; in soil micromorphology, referring to the fecal material of worms or fillings in faunal passages.

Vertisols One of 12 taxonomic orders in Soil Taxonomy, defined by extreme amounts of argilliturbation due to high amounts of smectite clays, in a wet-dry climate. Vertisols have 30% or more clay, deep and wide cracks when dry, and gilgai microrelief, intersecting slickensides, or wedge-shaped structural aggregates in the subsurface.

vesicle In geology, a cavity in a lava, formed by the entrapment of a gas bubble. In soil micromorphology, a circular, ovoid, or prolate soil pore with a smooth surface.

vesicular A horizon A type of A horizon containing innumerable, isolated, small-to-large, spherical pores (like foam), designated Av; found in desert soils or other soils where gases can be temporarily trapped as a soil dries or freezes.

viscosity The resistance to flow of a liquid.

void In soil micromorphology, a void is the same as a pore, which is the preferred term.

volcanic ash Fine, usually highly silicic, particles ejected during a volcanic eruption. They are dust-sized, sharp-edged, and glassy.

volcaniclastic Pertaining to the entire spectrum of fragmental materials of volcanic origin.

volumetric water content The amount of water in soil or sediment, as expressed on a volume basis, e.g., cm^3 of water / cm^3 of soil. Contrast to gravimetric water content.

vughs In soil micromorphology, relatively large pores, usually irregular and not normally interconnected with other pores of comparable size.

W horizon In some older systems of tropical soil nomenclature, the W horizon refers to a horizon, dominated by saprolite, i.e., a weathered rock horizon. This term is not used in Soil Taxonomy.

waterlogged Saturated (or nearly so) with water.

water suction See *matric potential*.

water table The upper surface of groundwater or that level in the ground where the water is at atmospheric pressure. The water table is the surface between the *zone of saturation* and the *zone of aeration*.

weatherable minerals Those minerals, usually primary, that are relatively easily decomposed or transformed in the soil/surficial environment. Compare to *resistant minerals*.

weathering The breakdown and changes in rocks and sediments at or near the Earth's surface produced by biological, chemical, and physical agents, or combinations of them.

weathering front, weathering discontinuity The lower edge of the weathering profile as it proceeds into soil parent material. Materials below the weathering front are either not weathered, or weathered to a much lesser degree, than are materials above.

weathering profile A vertical cut from the surface down through the weathered materials below, including the soil profile but not including unweathered parent materials or solid bedrock.

weathering rind The outer layer of a pebble, boulder, or rock fragment that has formed as a result of chemical weathering.

wedge-shaped soil aggregates Bicuneate, or wedge-shaped peds common to Vertisols.

welded soil A soil that is buried so shallowly and for a long enough period that pedogenesis has linked it to the surface soil above, i.e., there is no untransformed parent material between the solum of the surface soil and that of the buried soil.

wetland A transitional ecosystem between aquatic and terrestrial ecosystems that is inundated or saturated for long enough periods to produce hydric soils and support hydrophytic vegetation.

wetting front The boundary between the wetted region and the dry region of soil during the infiltration process.

whole regolith pedology A field of science that spans the interface between pedology and weathering science, by broadly characterizing and delineating regolith materials, so as to better manage them.

wilting point See *permanent wilting point*.

World Reference Base for Soil Resources (WRB) The international soil taxonomic system endorsed by the International Union of Soil Sciences (IUSS). It was developed by an international collaboration coordinated by the International Soil Reference and Information Centre (ISRIC) and sponsored by the IUSS and the FAO. The WRB borrows heavily from modern soil classification concepts, including Soil Taxonomy, the legend for the FAO Soil Map of the World 1988, the Référentiel Pédologique, and Russian concepts. It is based mainly on soil morphology as an expression of pedogenesis. Unlike Soil Taxonomy, soil climate is not directly a part of the taxonomic system.

xeric One of five major soil moisture regimes in Soil Taxonomy, common to Mediterranean climates that have moist cool winters and warm dry summers. A limited amount of water is present but does not occur at optimum periods for plant growth, i.e., in the cool season. In general usage, xeric refers to dry climatic conditions.

xerophyte Plant that grows in dry areas and does not have its roots set deeply in wet soil below the water table.

X-ray diffraction The method by which a beam of X-rays is diffracted by the three-dimensional array of atoms in a crystal structure and is used to identify and characterize minerals.

zircon $ZrSiO_4$. A hard, weathering-resistant mineral found as an accessory mineral in many kinds of rocks.

Zonal soils (i) Soils that are characteristic of a large region or zone. (ii) One of the three primary subdivisions (orders) in soil classification as used in the United States (1938 system of soil classification). The term is used today only in an historical context. Obsolete in the U.S. system of Soil Taxonomy.

zone of aeration Zone immediately below the ground surface but above the water table, within which pore spaces are partially filled with water and partially filled with air. Also called the *vadose zone*.

zone of saturation The zone below the *zone of aeration* and below the *water table*, in which all pore spaces are filled with water. Also called the *phreatic zone*.

Index